Advances in Intelligent Systems and Computing

Volume 1129

Series Editor

Janusz Kacprzyk, Systems Research Institute, Polish Academy of Sciences, Warsaw, Poland

Advisory Editors

Nikhil R. Pal, Indian Statistical Institute, Kolkata, India

Rafael Bello Perez, Faculty of Mathematics, Physics and Computing, Universidad Central de Las Villas, Santa Clara, Cuba

Emilio S. Corchado, University of Salamanca, Salamanca, Spain

Hani Hagras, School of Computer Science and Electronic Engineering, University of Essex, Colchester, UK

László T. Kóczy, Department of Automation, Széchenyi István University, Gyor, Hungary

Vladik Kreinovich, Department of Computer Science, University of Texas at El Paso, El Paso, TX, USA

Chin-Teng Lin, Department of Electrical Engineering, National Chiao Tung University, Hsinchu, Taiwan

Jie Lu, Faculty of Engineering and Information Technology, University of Technology Sydney, Sydney, NSW, Australia

Patricia Melin, Graduate Program of Computer Science, Tijuana Institute of Technology, Tijuana, Mexico

Nadia Nedjah, Department of Electronics Engineering, University of Rio de Janeiro, Rio de Janeiro, Brazil

Ngoc Thanh Nguyen, Faculty of Computer Science and Management, Wrocław University of Technology, Wrocław, Poland

Jun Wang, Department of Mechanical and Automation Engineering, The Chinese University of Hong Kong, Shatin, Hong Kong

The series "Advances in Intelligent Systems and Computing" contains publications on theory, applications, and design methods of Intelligent Systems and Intelligent Computing. Virtually all disciplines such as engineering, natural sciences, computer and information science, ICT, economics, business, e-commerce, environment, healthcare, life science are covered. The list of topics spans all the areas of modern intelligent systems and computing such as: computational intelligence, soft computing including neural networks, fuzzy systems, evolutionary computing and the fusion of these paradigms, social intelligence, ambient intelligence, computational neuroscience, artificial life, virtual worlds and society, cognitive science and systems, Perception and Vision, DNA and immune based systems, self-organizing and adaptive systems, e-Learning and teaching, human-centered and human-centric computing, recommender systems, intelligent control, robotics and mechatronics including human-machine teaming, knowledge-based paradigms, learning paradigms, machine ethics, intelligent data analysis, knowledge management, intelligent agents, intelligent decision making and support, intelligent network security, trust management, interactive entertainment, Web intelligence and multimedia.

The publications within "Advances in Intelligent Systems and Computing" are primarily proceedings of important conferences, symposia and congresses. They cover significant recent developments in the field, both of a foundational and applicable character. An important characteristic feature of the series is the short publication time and world-wide distribution. This permits a rapid and broad dissemination of research results.

** **Indexing: The books of this series are submitted to ISI Proceedings, EI-Compendex, DBLP, SCOPUS, Google Scholar and Springerlink** **

More information about this series at http://www.springer.com/series/11156

Kohei Arai · Supriya Kapoor ·
Rahul Bhatia
Editors

Advances in Information and Communication

Proceedings of the 2020 Future of Information and Communication Conference (FICC), Volume 1

 Springer

Editors
Kohei Arai
Faculty of Science and Engineering
Saga University
Saga, Japan

Supriya Kapoor
The Science and Information
(SAI) Organization
Bradford, West Yorkshire, UK

Rahul Bhatia
The Science and Information
(SAI) Organization
Bradford, West Yorkshire, UK

ISSN 2194-5357   ISSN 2194-5365 (electronic)
Advances in Intelligent Systems and Computing
ISBN 978-3-030-39444-8   ISBN 978-3-030-39445-5 (eBook)
https://doi.org/10.1007/978-3-030-39445-5

© Springer Nature Switzerland AG 2020
This work is subject to copyright. All rights are reserved by the Publisher, whether the whole or part of the material is concerned, specifically the rights of translation, reprinting, reuse of illustrations, recitation, broadcasting, reproduction on microfilms or in any other physical way, and transmission or information storage and retrieval, electronic adaptation, computer software, or by similar or dissimilar methodology now known or hereafter developed.
The use of general descriptive names, registered names, trademarks, service marks, etc. in this publication does not imply, even in the absence of a specific statement, that such names are exempt from the relevant protective laws and regulations and therefore free for general use.
The publisher, the authors and the editors are safe to assume that the advice and information in this book are believed to be true and accurate at the date of publication. Neither the publisher nor the authors or the editors give a warranty, expressed or implied, with respect to the material contained herein or for any errors or omissions that may have been made. The publisher remains neutral with regard to jurisdictional claims in published maps and institutional affiliations.

This Springer imprint is published by the registered company Springer Nature Switzerland AG
The registered company address is: Gewerbestrasse 11, 6330 Cham, Switzerland

Preface

We welcome all of you at the Future of Information and Communication Conference (FICC) 2020 which was held at The Park Central Hotel San Francisco on 5–6 March 2020.

The conference provides a platform to discuss information and communication technologies with participants from around the globe, both from academia and industry. The success of this conference is reflected in the papers received, with participants coming from several countries, allowing a real exchange of experiences and ideas. Renowned experts and scholars shared their excellent views and experiences through speeches. The main topics of the conference include Communication, Security and Privacy, Networking, Ambient Intelligence, Data Science, and Computing.

Each submission has been double-blind peer reviewed by two to four reviewers in the right area. More than 440 submissions have been received out of which only 135 (including 10 poster papers) were accepted for final presentation and will be published in the conference proceedings. This conference showcases paper presentations of new research, demos of new technologies, and poster presentations of late-breaking research results, along with inspiring keynote speakers and moderated challenge sessions for participants to explore and respond to big challenge questions about the role of technology in creating thriving, sustainable communities.

The conference success was due to the collective efforts of many people. Therefore, we would like to express our sincere gratitude to the technical program committee and the reviewers who helped ensure the quality of the papers as well as their invaluable input and advice. A special thanks goes to the keynote speakers and the conference organizers for their various support to FICC 2020.

We hope that the FICC 2020 experience was enjoyable and benificial for all the participants and the interested readers.

See you in our next SAI Conference, with the same amplitude, focus and determination.

Regards,
Kohei Arai
Saga University, Japan

Contents

The Control Method in Synchronous Frame for DVR to Mitigate
the Balanced and Unbalanced Voltage Sag/Swells Phenomenon
in Power Network 1
Vu Thai Hung, Shu Hong Chun, and Le Ngoc Giang

Performance Evaluation of IPv6 and IPv4 for Future Technologies ... 15
Augustus E. Ibhaze, Obinna Okoyeigbo, Uyi A. Samson, Paschal Obba,
and Ignatius K. Okakwu

Investigate the Performance of 240 GHz Millimeter:
Wave Frequency over Fiber with 10 and 20 Gbps 23
Fawziya Al Wahaibi and Hamed Al Raweshidy

Low Power BCH Decoder Using Verification Algorithm
and Two-Step Parallel Chien Search Architecture 31
Noha K. Shebl, Saleh M. Eisa, Hanady H. Issa, and Khaled A. Shehata

A Videogame Driven by the Mind: Are Motor Acts
Necessary to Play? 40
Luigi Bianchi

Evaluation of RSSI as a Non-visual Target Tracking Technique
for Drone Applications 51
Christopher Lee and Sudhanshu Kumar Semwal

Power Efficient Multi-relay Cooperative Diversity in Wireless
Network Using Hybrid Relaying Protocol 60
Shital Joshi and Malaykumar Shitalkumar Bhakta

Exploring the Evolutionary Bispectrum 79
Abdullah I. Al-Shoshan

A Novel Approach to Blockchain-Based Digital Identity System 93
Md Abdullah Al Mamun, S. M. Maksudul Alam, Md. Shohrab Hossain,
and M. Samiruzzaman

Towards Blockchain-Based GDPR-Compliant Online Social Networks: Challenges, Opportunities and Way Forward 113
Javed Ahmed, Sule Yildirim, Mariusz Nowostawski,
Mohamed Abomhara, Raghavendra Ramachandra, and Ogerta Elezaj

Systematization of Knowledge on Scalability Aspect of Blockchain Systems 130
Parth Anand Shukla and Saeed Samet

Smart Dam: Upstream Sensing, Hydro-Blockchain, and Flood Feature Extractions for Dam Inflow Prediction 139
Takato Yasuno, Akira Ishii, Masazumi Amakata, and Junichiro Fujii

Using Blockchain in IoT: Is It a Smooth Road Ahead for Real? 159
Sonali Chandel, Song Zhang, and Hanwen Wu

Developing a Blockchain-Enabled Collaborative Intrusion Detection System: An Exploratory Study 172
Daniel Laufenberg, Lei Li, Hossain Shahriar, and Meng Han

An Adaptive Context Modeling Approach Using Genetic Algorithm in IoTs Environments 184
Ahmed A. A. Gad-Elrab, Shereen A. El-aal, Neveen I. Ghali, and Afaf A. S. Zaghrout

Online Sports Activities Travel Guide (SATG) 205
Ola Hegazy

A Heterogeneous Scalable-Orchestration Architecture for Home Automation 218
Jeferson Apaza-Condori and Eveling Castro-Gutierrez

Ambient Intelligence Applications in Architecture: Factors Affecting Adoption Decisions 235
Maryam Abhari and Kaveh Abhari

Photovoltaic Mobile System Design for Non-interconnected Zones of Meta's Department 251
Obeth Romero

An IoT Based e-Health Platform Using Raspberry Pi 257
El-Hadi Khoumeri, Rabea Cheggou, and Kamila Ferhah

Li-Fi Prospect in Internet of Things Network 272
Augustus E. Ibhaze, Patience E. Orukpe, and Frederick O. Edeko

A Multidimensional Control Architecture for Combined Fog-to-Cloud Systems 281
Xavi Masip-Bruin, Vitor Barbosa Souza, Eva Marín-Tordera, Guang-Jie Ren, Admela Jukan, and Jordi Garcia

Study of Polynomial Backoff for IEEE 802.11 DCF 300
Bader A. Aldawsari

**Quality of Service Provision Within IEEE 802.11
CSMA/CA Protocol** 313
Kamil Samara, Hossein Hosseini, Zaid Altahat, Joseph Stewart,
David Ehley, and Miguel Estrada

**Realistic Cluster-Based Energy-Efficient and Fault-Tolerant
(RCEEFT) Routing Protocol for Wireless Sensor Networks (WSNs)** ... 320
Emmanuel Effah and Ousmane Thiare

**Efficient Cache Architecture for Packet Processing
in Internet Routers** 338
Hayato Yamaki

**Optimization of Cultural Heritage Virtual Environments
for Gaming Applications** 353
Laura Inzerillo, Francesco Di Paola, Yuri Alogna,
and Ronald Anthony Roberts

**Performance Evaluation of the Update Messages of Locator
Identifier Split Protocols Using an IP Paging Mechanism
at the End Networks** 372
Aadarsh Bussooa and Avinash Mungur

**Business and Environmental Perspectives of Submarine Cables
in Global Market** 392
N. Aishwarya

**Conditional Convolutional Generative Adversarial Networks Based
Interactive Procedural Game Map Generation** 400
Kuang Ping and Luo Dingli

**Example of the Use of Artificial Neural Network
in the Educational Process** 420
Suleimenov Ibragim, Bakirov Akhat, Matrassulova Dinara,
Grishina Anastasiya, Kostsova Mariya, and Mun Grigoriy

**Hybrid Recommendation System for Young Football Athletes
Customized Training** 431
Paulo Matos, João Rocha, Ramiro Gonçalves, Filipe Santos,
David Abreu, Hugo Soares, and Constantino Martins

AALADIN: Ambient Assisted Living Assistive Device for Internet 443
Priscila Cedillo, Wilson Valdez, and Andrés Córdova

**Target Localization with Visible Light Communication in High
Ambient Light Environments** 456
Kofi Nyarko and Emmanuel Shedu

A Minimal Social Weight Wearable Device for Thermal Regulation ... 467
Javier Benedicto Serrano and Sudhsnahu Kumar Semwal

Virtual Teleportation of a Theatre Audience Onto the Stage:
VR as an Assistive Technology................................. 477
Saikrishna Srinivasan and Gareth Schott

Eliciting Evolving Topics, Trends and Foresight about Self-driving
Cars Using Dynamic Topic Modeling........................... 488
Workneh Y. Ayele and Gustaf Juell-Skielse

Quality Assurance/Quality Control Engine for Power
Outage Mitigation 510
S. Chan

Smart Policing for a Smart World Opportunities, Challenges
and Way Forward 532
Muhammad Mudassar Yamin, Andrii Shalaginov, and Basel Katt

Midnight in Tokyo: Mobility Service for Bar-Hopping 550
Chihiro Sato and Naohito Okude

Urban Sensibilities, Sharing, and Interactive Public Spaces:
In Search of a Good Correlation for Information
and Communication in Smart Cities 563
H. Patricia McKenna

Visual Odometry from Omnidirectional Images
for Intelligent Transportation................................. 576
Marco Marcon, Marco Brando Mario Paracchini, and Stefano Tubaro

Drivers and Barriers for Open Government Data Adoption:
An Isomorphic Neo-Institutional Perspective 589
Henry N. Roa, Edison Loza-Aguirre, and Pamela Flores

"Seeking Privacy Makes Me Feel Bad?": An Exploratory
Study Examining Emotional Impact on Use
of Privacy-Enhancing Features................................ 600
Hsiao-Ying Huang and Masooda Bashir

An Adaptive Security Architecture for Detecting Ransomware
Attack Using Open Source Software 618
Prya Booshan Caliaberah, Sandhya Armoogum, and Xiaoming Li

Shilling Attack Detection Scheme in Collaborative Filtering
Recommendation System Based on Recurrent Neural Network 634
Jianling Gao, Lingtao Qi, Haiping Huang, and Chao Sha

Runtime API Signature for Fileless Malware Detection 645
Radah Tarek, Saadi Chaimae, and Chaoui Habiba

**Efficient Implementation and Computational Analysis
of Privacy-Preserving Auction Protocols** 655
Ramiro Alvarez and Mehrdad Nojoumian

**Dynamic Programming Approach in Conflict Resolution Algorithm
of Access Control Module in Medical Information Systems** 672
Hiva Samadian, Desmond Tuiyot, and Juan Valera

**Two-Factor Authentication Using Mobile OTP
and Multi-dimensional Infinite Hash Chains** 682
Uttam K. Roy and Divyans Mahansaria

**IT Security for Measuring Instruments: Confidential Checking
of Software Functionality** 701
Daniel Peters, Artem Yurchenko, Wilson Melo, Katsuhiro Shirono,
Takashi Usuda, Jean-Pierre Seifert, and Florian Thiel

**Hardware Transactional Memory as Anti-analysis Technique
for Software Protectors** 721
Federico Palmaro and Luisa Franchina

**Development of the Technique for the Identification, Assessment
and Neutralization of Risks in Information Systems** 733
Askar Boranbayev, Seilkhan Boranbayev, and Askar Nurbekov

**Estimation of the Degree of Reliability and Safety
of Software Systems** 743
Askar Boranbayev, Seilkhan Boranbayev, and Askar Nurbekov

**A Novel Image Steganography Using Multiple LSB Substitution
and Pixel Randomization Using Stern-Brocot Sequence** 756
Md. Abdullah Al Mamun, S. M. Maksudul Alam, Md. Shohrab Hossain,
and M. Samiruzzaman

Detecting PE-Infection Based Malware 774
Chia-Mei Chen, Gu-Hsin Lai, Tzu-Ching Chang, and Boyi Lee

Network Security Monitoring in Automotive Domain 782
Daniel Grimm, Felix Pistorius, and Eric Sax

Android Malware Detection in Large Dataset: Smart Approach 800
Qudrat E. Alahy, Md. Naseef-Ur-Rahman Chowdhury, Hamdy Soliman,
Moshrefa Sultana Chaity, and Ahshanul Haque

**Design and Implementation of an e-Voting System Based
on Paillier Encryption** 815
Miaomiao Zhang and Steven Romero

AI-Enabled Digital Forensic Evidence Examination 832
Jim Q. Chen

Stained Visible Watermarking: A Securely Tunable Way of Joint Image Copyright and Privacy Protection 842
Xiaoming Yao and Hao Wang

sD&D: Design and Implementation of Cybersecurity Educational Game with Highly Extensible Functionality 857
Yoshiyuki Kido, Nelson Pinto Tou, Naoto Yanai, and Shinji Shimojo

Mobile System for Determining Geographical Coordinates for Needs of Air Support in Cases of GPS Signals Loss 874
Karol Jędrasiak, Aleksander Nawrat, Przemysław Recha, and Dawid Sobel

Cyber Manhunt 883
Sonali Chandel, Yanjun Chen, Jiale Dai, and Jianyan Huang

Smart Agriculture with Advanced IoT Communication and Sensing Unit 903
David Krcmarik, Reza Moezzi, Michal Petru, and Jan Koci

Author Index 913

The Control Method in Synchronous Frame for DVR to Mitigate the Balanced and Unbalanced Voltage Sag/Swells Phenomenon in Power Network

Vu Thai Hung[1,2(✉)], Shu Hong Chun[1], and Le Ngoc Giang[2]

[1] School of Electrical Engineering, Kunming University of Science and Technology, Kunming, China
vuthaihungqs@yahoo.com
[2] AD-AF Academy of Viet Nam, Son Tay, Ha Noi, Viet Nam

Abstract. Nowadays great attention is paid to the problem of power quality. With the modern equipment, their operation is very sensitive to the quality of the power supply. When the power quality is not guaranteed, such as a nonstandard voltage, current or frequency, which will damage the equipment or the equipment will operate improperly. One of the basic problems here is Sag/Swells voltage. To solve this problem, the dynamic voltage Restorer (DVR) is put into use in the power distribution network. DVR will reduce or add the voltage in series with supply voltage through injection transformer to correct the amplitude, voltage phase and harmonic components into line. This paper proposes a control method for the DVR based on space vector control, with two feedback loops according to voltage and current. The authors use independent control methods, separation of components of direct and reverse in d-q Synchronous Reference Frame for DVR. This method has effectively controlled for the voltage sags and swells phenomenon in balanced and unbalanced. The research results are simulated and verified on Matlab - Simulink software.

Keywords: Dynamic Voltage Restorer · Voltage sags · Voltage swells

1 Introduction

Power quality has received increasingly more attention by researchers due to its impacts on utilities as well as both industrial and commercial electrical consumers [1, 3]. Power quality problem is defined as "Any power problem manifested in voltage, current, or frequency deviation that results in the failure or misoperation of customer equipment" [2, 3]. Among various power quality problems, voltage sag is the most frequent one. Voltage sag is a decrease in root mean square (rms) voltage magnitude in the range of 0.1 to 0.9 per unit (p.u) at fundamental frequency, with the duration of half cycle to one minute [1, 3]. It is often caused by balanced or unbalanced faults in the distribution systems or by large induction motors startup [2, 3]. Voltage swell, on the other hand, is an increase in rms voltage magnitude, typically between 1.1 and 1.8 per unit, at fundamental frequency and duration of half cycle to one minute [2, 3]. Although effects of a voltage swell is often more destructive than sag, it is not as important as voltage dip. This is because voltage swells are less common in comparison with sags in distribution systems.

Custom power devices are based on power electronic technology which is much faster than conventional electromechanical based protection devices. These devices mitigate voltage sags/swells originated from supply side and improve power quality of customers, especially critical customers, at distribution level [2]. There are different types of custom power devices such as Dynamic Voltage Restorer (DVR), Distribution Static Synchronous Compensators (DSTATCOM), Static Var Compensator (SVC) and Uninterruptible Power Supplies (UPS). Each of them has its own advantages and disadvantages. Among all of them, DVR is considered to be the most efficient and effective one for voltage sag/swell mitigation. DVR is a series compensator which injects voltage to the point of common coupling (PPC) to maintain the voltage of sensitive load at nominal voltage. DVR is a power electronic converter based series compensator that can protect critical loads from almost all supply side disturbances other than voltage interruptions.

When a voltage Sag/Swells occurs, it usually results in changes in both amplitude and phase angle. For an unbalanced Sag/Swells voltage, to ensure the DVR can compensate the voltage with amplitude, phase angle and frequency accurately to improve the amplitude of stable operating voltage on the load, it requires the DVR controller to correctly control both the positive sequence, negative sequence and zero sequence. In addition, the control signals used for DVR include voltage and current signals, in which the current signal in the control process has a large instantaneous fluctuation which, if not with the good control, becomes interference source which greatly affects the control quality.

In [6] uses the voltage-based loop in combination with Hysterisis control, its advantage is very good kinetic response, very high sustainability, simple and reliable control. However, its disadvantage is that the output current is always undulating due to the characteristics of the controller that the relay has delay and switching frequency constantly changes, depending on the variation of the input signal, which can be dangerous for the semiconductor valves.

In [9–11] use voltage-based loop in combination with PI control with feedforward control, or resonance control with feedforward control to improve the dynamics of the controller, however these works have not mentioned the current controller, which greatly affects the control accuracy.

In [8] improves the control method by the implemented in the synchronous frame with two loops, an inner loop which controls the current and an out loop which controls the voltage, however with the use of proportional controller for the current loop circuit, in order for the control error decrease, the proportional coefficient K_c must be large, the correction of K_c coefficient is only allowed within a certain range because when the larger K_c is, leading to the reduction of the phase reserve and the reserve amplitude of the system, to a certain threshold value, which will cause the system to become unstable, therefore, the system always has control errors.

In [5, 7] refer to the DVR controller using the vector control method on the rotation frame dq0 using the PI controller, PR controller and the repeating controller with two voltage and current control loops, however, both of these articles stop at controlling each of the positive sequence components and mainly fit only with the balance Sag/Swells voltage.

The said works do not have any solution to compensate for the reverse sequence component or zero sequence, leading to inaccurate and deformed or asynchronous voltage because the controller calculates to insert into the grid resulting in the changes on the grid during compensation, adversely affecting the load. Therefore, it is necessary to develop the structure and load voltage control algorithm for the DVR that can handle the sequence components that appear in all balanced and unbalanced Sag/Swells voltage types.

In order to overcome the above unsatisfactory points, in this article, a complete compensation strategy is used, with the vector control method using two adjustable loop circuits connected by the voltage and current, separate control separation according to the positive sequence component and the negative sequence component on the dq0-synchronous frame.

2 DVR Basic Configuration and Components

The main components of the DVR are shown in Fig. 1 and are summarized hereafter [3, 4, 8, 10, 12]:

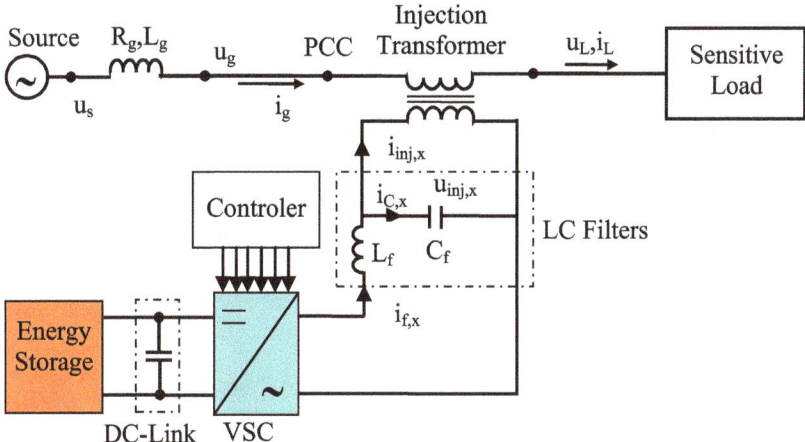

Fig. 1. DVR basic configuration

Where: u_S is the grid source voltage described by a voltage source u_g in series with the resistance and resistance of the grid R_g, L_g.

u_L, i_L: Three-phase voltage and current of the load
R_f, L_f, C_f: Resistance, inductance and capacitance of the filter
u_{inj}, i_{inj}: Three-phase voltage and current inserted by DVR
i_f, i_C: Three-phase current flows through the inductor and capacitor of the LC filter.

Series Injection Transformers: The three single-phase injection transformers are used to inject missing voltage to the system. The electrical parameters of series injection transformer should be selected correctly to ensure the maximum reliability and effectiveness.

Filter Unit: The nonlinear characteristics of semiconductor devices cause distorted waveforms associated with harmonics at the inverter output. To overcome this problem and provide high quality energy supply, filter unit is used. The inverter side and line side filtering are basic types of filtering scheme.

The inverter side filter is closer to the harmonic source and low voltage side thus it prevents the harmonic currents to penetrate into the series injection transformers. This can cause voltage drop and phase shift in the fundamental component of the inverter output. The line side is closer to high voltage side so higher rating on transformer is needed.

Voltage Source Converter: VSC is a power electronic device consists of storage device and switching devices used to generate the compensating sinusoidal voltage of required magnitude, duration, in phase as that of system and instantaneously. In DVR voltage source converter provides the missing voltage during voltage sag.

Storage Device: It is basically used to supply the necessary energy to VSC to generate the compensating voltage.

Control Circuit: Control circuit continuously monitors the supply voltage. The function of control system is to detect the disturbance in the supply voltage, compare it with the set reference value and then generate the switching pulses to the VSC to generate the DVR output voltages which will compensate the voltage sag/swell.

3 Control Model and Algorithm

It is assumed that the serial transformer in Fig. 1 is ideal; there is no magnetization and leakage inductance, at a ratio of 1:1. In addition, the transformer, the grid voltage and the load are replaced by an equivalent current source, the inserted voltage of the DVR u_{inj} is viewed primarily by the voltage u_c on the capacitor of the filter L_f C_f.

The simple model of VSC and LC filter connected to grid as mentioned in Fig. 2, [5, 12]: VSC converter is represented as a voltage source u_{inv}, i_f is the current running through the filter inductor, u_{dc} is DC-Link voltage.

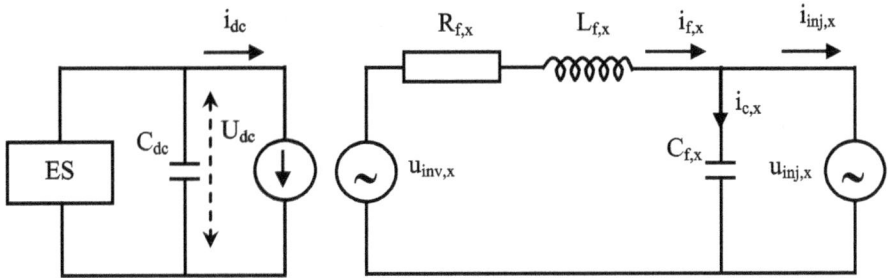

Fig. 2. Model of VSC and LC filter connected to grid [5]

From the structure diagram of Fig. 2, Kirchhoff's law applies to three-phase voltage and current, we have the following equations:

$$i_{fx}(t) = i_{Cfx}(t) + i_{injx}(t) = C_f \frac{d}{dt} u_{injx}(t) + i_{injx}(t) \tag{1}$$

$$u_{invx}(t) - u_{injx}(t) - R_f i_{fx}(t) - L_f \frac{d}{dt} i_{fx}(t) = 0 \tag{2}$$

Apply Clacrke conversion on static frame $\alpha\beta$ as follows:

$$\frac{d}{dt} i_f^{(\alpha\beta)}(t) = \frac{1}{L_f} u_{inv}^{(\alpha\beta)}(t) - \frac{1}{L_f} u_{inj}^{(\alpha\beta)}(t) - \frac{1}{L_f} R_f i_f^{(\alpha\beta)}(t) \tag{3}$$

$$\frac{d}{dt} u_{inj}^{(\alpha\beta)}(t) = \frac{1}{C_f} i_f^{(\alpha\beta)}(t) - \frac{1}{C_f} i_{inj}^{(\alpha\beta)}(t) \tag{4}$$

Apply Park conversion from static frame $\alpha\beta$ to rotating frame dq with a Phase Locked Loop (PLL) in sync with grid voltage vector [5, 12].

$$\frac{d}{dt} u_{inj}^{(dq)}(t) = \frac{1}{C_f} i_f^{(dq)}(t) - \frac{1}{C_f} i_{inj}^{(dq)}(t) \mp j\omega u_{inj}^{(dq)}(t) \tag{5}$$

$$\frac{d}{dt} i_f^{(dq)}(t) = \frac{1}{L_f} u_{inv}^{(dq)}(t) - \frac{1}{L_f} u_{inj}^{(dq)}(t) - \frac{1}{L_f} R_f i_f^{(dq)}(t) \mp j\omega L_f i_f^{(dq)}(t) \tag{6}$$

Conduct interruption:

$$\frac{1}{T_S} (u_{inj}^{(dq)}(k+1) - u_{inj}^{(dq)}(k)) = \frac{1}{C_f} i_f^{(dq)}(k) - \frac{1}{C_f} i_{inj}^{(dq)}(k) \pm j\omega u_{inj}^{(dq)}(k) \tag{7}$$

$$\frac{1}{T_S}(i_f^{(dq)}(k+1) - i_f^{(dq)}(k)) = \frac{1}{L_f}u_{inv}^{(dq)}(k) - \frac{1}{L_f}u_{inj}^{(dq)}(k) - \frac{1}{L_f}R_f i_f^{(dq)}(k) \mp j\omega i_f^{(dq)}(k) \quad (8)$$

In which T_S is the sample period.

The load voltage control algorithm is described in the form of mathematical equations and the following diagrams:

Current Regulator

From the Eq. (8), we develop the equation describing the current regulator when positive and negative sequence components are controlled separately, using PI:

+ The equation describing of the positive sequence current regulator:

$$u_{inv}^{dq+}(k) = u_{inj}^{dq+}(k) + R_f i_f^{dq+}(k) \mp j\omega L_f i_f^{dq-}(k) + G_I^P \frac{L_f}{T_S}(i_f^{dq*+}(k+1) - i_{inj}^{dq+}(k)) \quad (9)$$

+ The equation describing of the negative sequence current regulator:

$$u_{inv}^{dq-}(k) = u_{inj}^{dq-}(k) + R_f i_f^{dq-}(k) \pm j\omega L_f i_f^{dq+}(k) + G_I^N \frac{L_f}{T_S}(i_f^{dq*-}(k+1) - i_{inj}^{dq-}(k)) \quad (10)$$

With: G_I^P; G_I^N is the transfer function of the negative and positive sequence current controller.

Voltage Regulator

From the Eq. (7), the equation describes the voltage regulator when adding PI for the positive sequence and negative sequence on the dq frame as follows:

+ The equation describing of the positive sequence voltage regulator:

$$i_f^{dq+}(k+1) = i_{inj}^{dq+}(k) + G_u^P \frac{C_f}{T_S}(u_{inj}^{dq*+}(k) - u_{inj}^{dq+}(k)) + j\omega C_f u_{inj}^{dq-}(k) \quad (11)$$

+ The equation describing of the negative sequence voltage regulator:

$$i_f^{dq-}(k+1) = i_{inj}^{dq-}(k) + G_u^N \frac{C_f}{T_S}(u_{inj}^{dq*-}(k) - u_{inj}^{dq-}(k)) + j\omega C_f u_{inj}^{dq+}(k) \quad (12)$$

G_u^P; G_u^N: The transfer function of the negative and positive sequence voltage controller.

The positive and negative sequence regulator structures are implemented from the Eqs. (9), (10), (11), (12) as shown in Figs. 3 and 4.

The synthesized controller structure is based on two corresponding regulators of positive and negative sequence components as shown in Fig. 5.

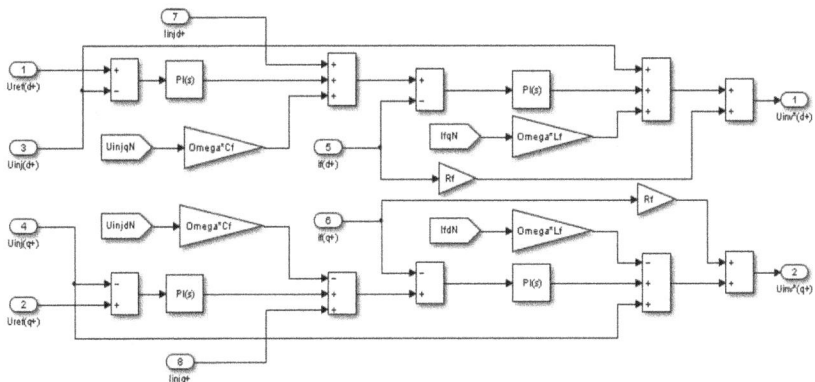

Fig. 3. Structure of positive sequence regulator of DVR

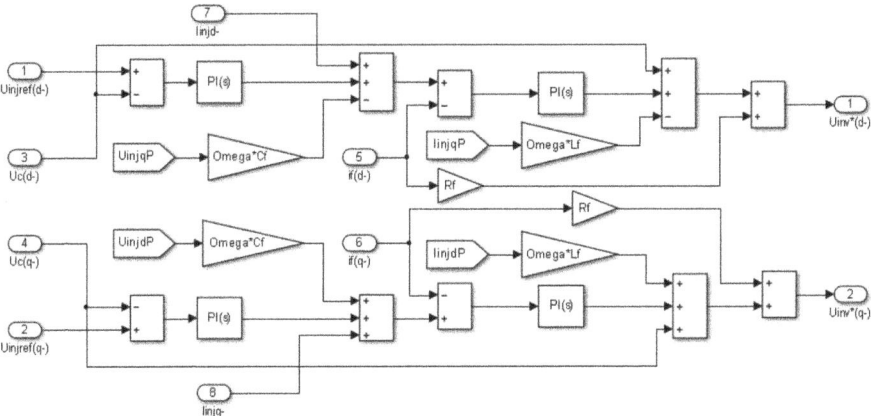

Fig. 4. Structure of negative sequence regulator of DVR

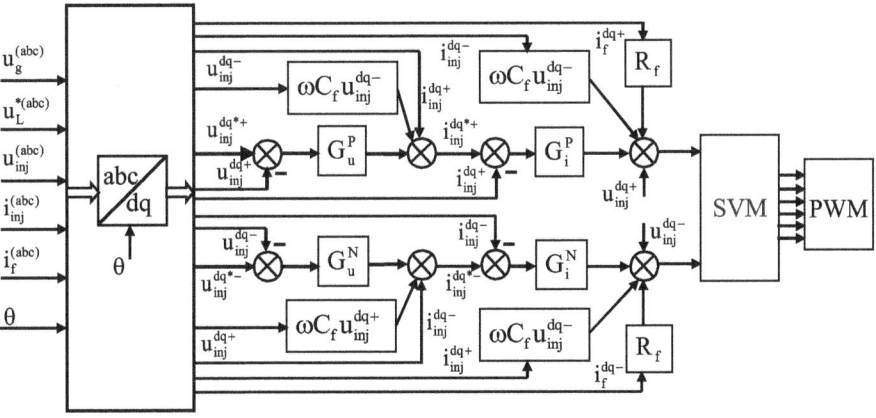

Fig. 5. Controller structure diagram of DVR

4 Simulation Results

To illustrate a typical response of DVR with the proposed control strategy, a simple 50 Hz power distribution system with a sensitive load as shown in Fig. 1 is considered. In order to see the performance of DVR with the proposed control strategy, the simulation results are compared with a double closed-loop control strategy of two voltage and current-adjusted circuits, which does not separate the sequence component [7], and the voltage loop control strategy [2, 12] to see the advantages of this control strategy:

– Compare the proposed control strategy with the control strategy of two voltage and current – based circuits that do not separate the sequence components, at the same conditions on the grid, load parameters and controller parameters.

+ Case 1: During t = 0.02 − 0.06 s, the 3-phase voltage source is unbalanced, phase A voltage is 0.5 pu, phase B voltage is 0.705 pu, phase C voltage is 0.865 pu.

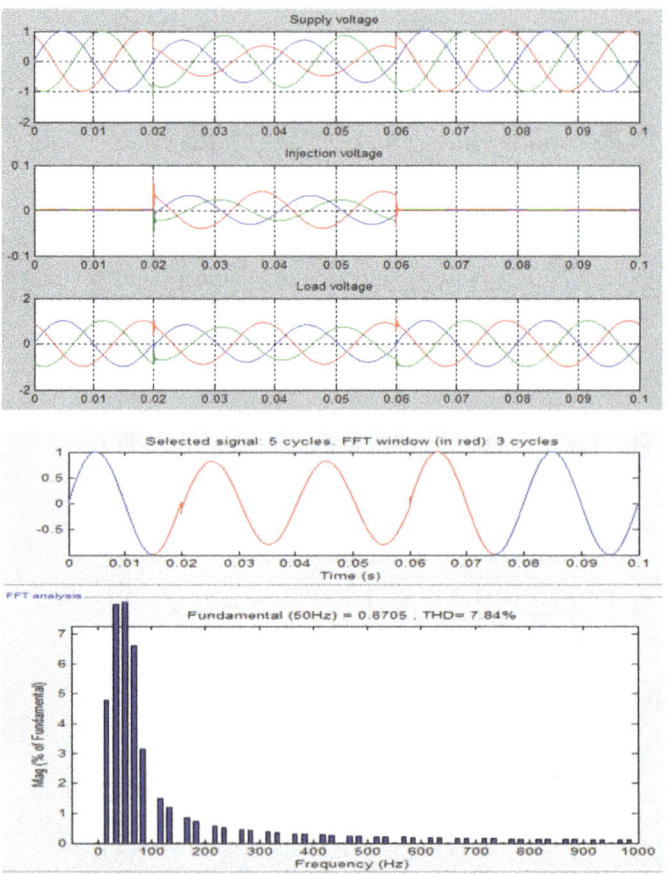

Fig. 6. DVR response to the control strategy of two conventional loops

From the simulation results shown in Figs. 6 and 7, we can see, if the DVR has a control strategy of 2 conventional loop circuits, the back-up compensation voltage is restored to 0.805 pu and the harmonic distortion coefficient of the THD source voltage is 7.84%; If the DVR with the control strategy is proposed, then the load compensation voltage is restored to 1 pu and the harmonic distortion coefficient of THD source voltage is 0.4%.

+ Case 2: A balanced 3-phase grid with noise disturbance. During t = 0.02 − 0.06 s, 3-phase voltage is down 0.5 pu; During t = 0.06 − 0.1 s, 3-phase voltage is 1.2 pu high; During t = 0.04 − 0.08 s, there is an interference; THD of the source voltage is 20.83%.

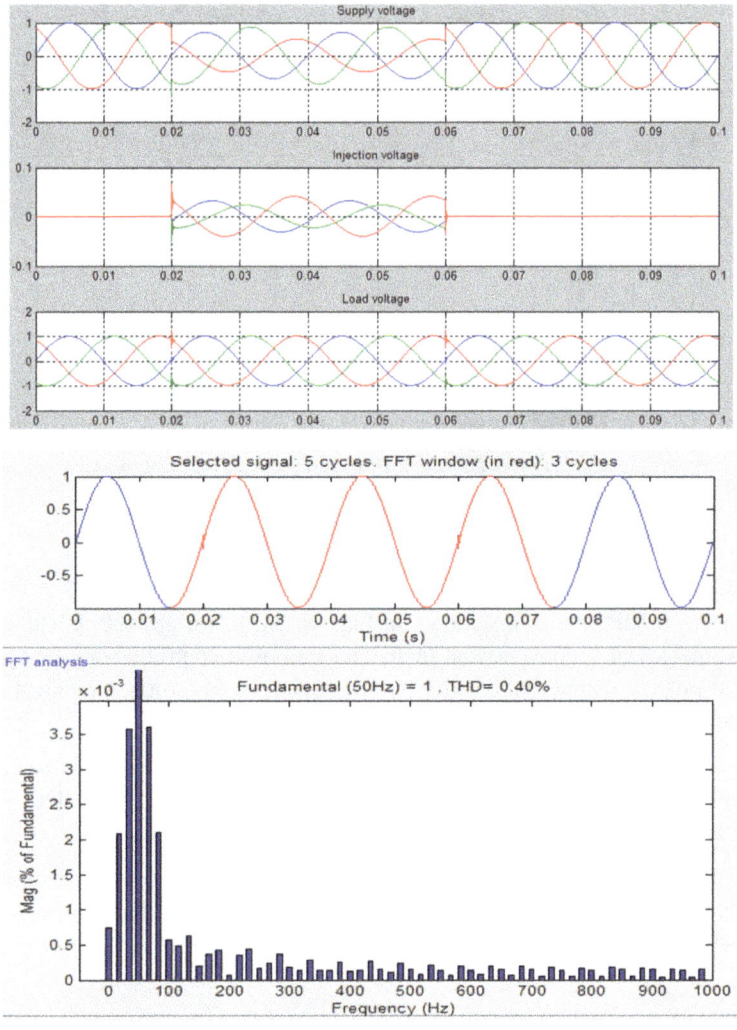

Fig. 7. DVR response to the proposed control strategy

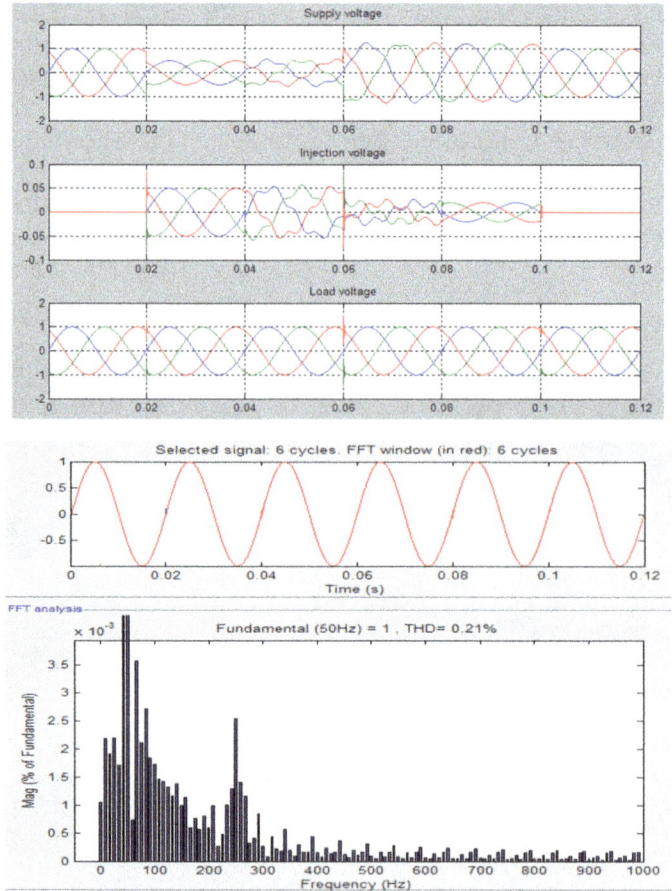

Fig. 8. DVR response to the proposed control strategy

From the simulation results shown in Figs. 8 and 9, we can see, DVR with both control strategies of 2 conventional loops and proposed control strategy, after recovering load voltage compensation to 1 pu and harmonic distortion coefficient of THD source voltage is 0.21%.

- Compare the proposed control method with the loop control method according to voltage using PI stage, not separating the positive and negative sequence components.

For a balanced 3-phase voltage source, in which t = 0.02 − 0.06 s, 3-phase voltage is 0.5 pu; in which t = 0.06 − 0.1 s, 3-phase voltage is 1.2 pu; During t = 0.04 − 0.08 s, there is an interference, THD of the source voltage is 24.28%.

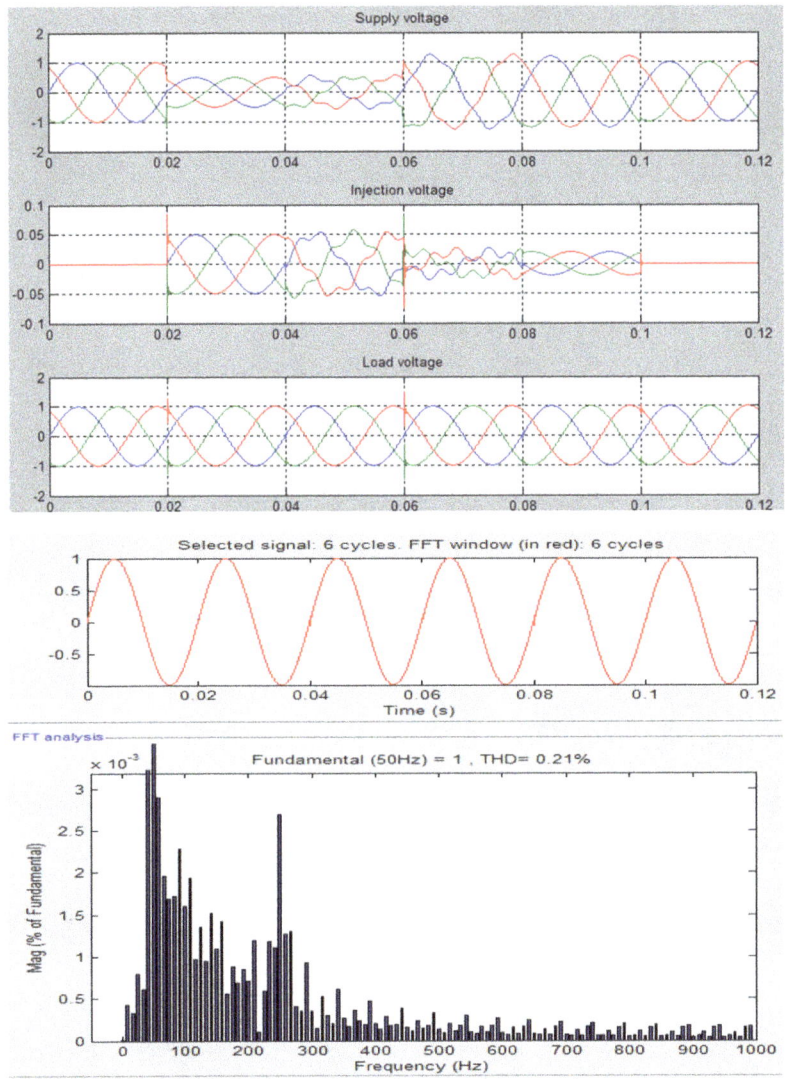

Fig. 9. DVR response to the control strategy of two conventional loops

From the simulation results shown in Figs. 10 and 11, we can see, if the DVR with loop control strategy according to voltage, the load compensation voltage is restored to 0.9816 pu and THD is 1.41%. If the DVR with the proposed control strategy, the load compensation voltage is restored to 1 pu and THD is 0.21%.

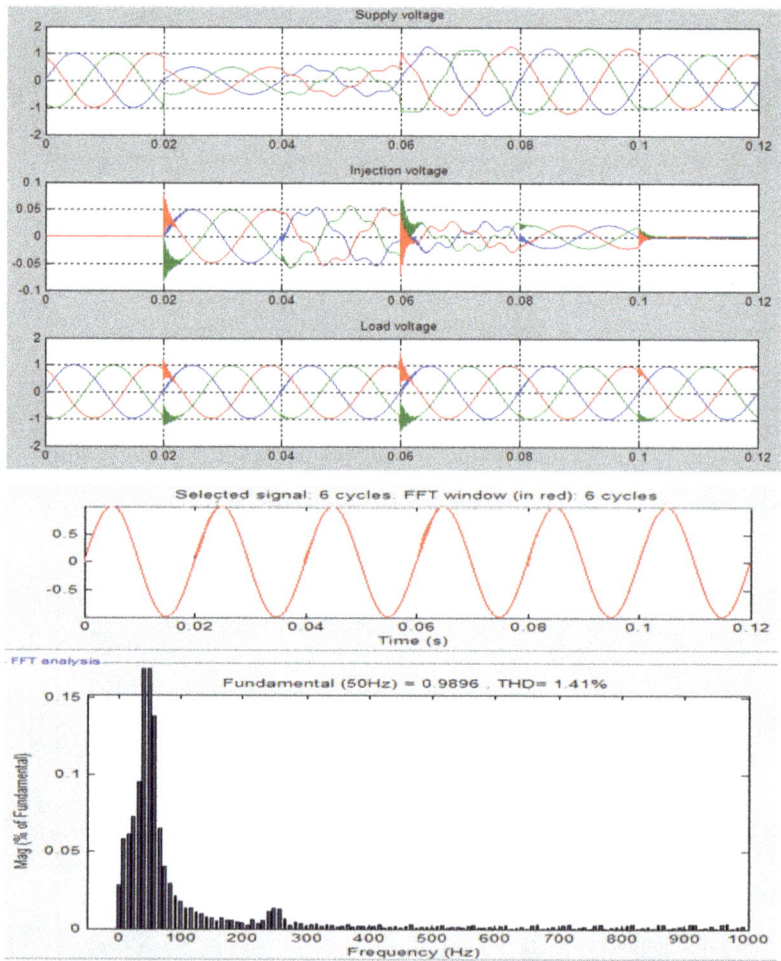

Fig. 10. DVR response with loop control strategy according to voltage

Through the simulation results, with the control method of 2 conventional loops, when the grid source appears Sag/Swells unbalanced voltage, the compensation load voltage is not restored to the working voltage (load voltage is still unbalanced and unrecoverable on 1 pu, with large THD index) while the proposed control method has been overcome very well due to measures to control each sequence component. With balanced Sag/Swells voltages, both methods provide equally good compensation efficiency for fast control response, distortion of small load compensation voltage (low THD index), high control accuracy (load voltage is restored 1 pu).

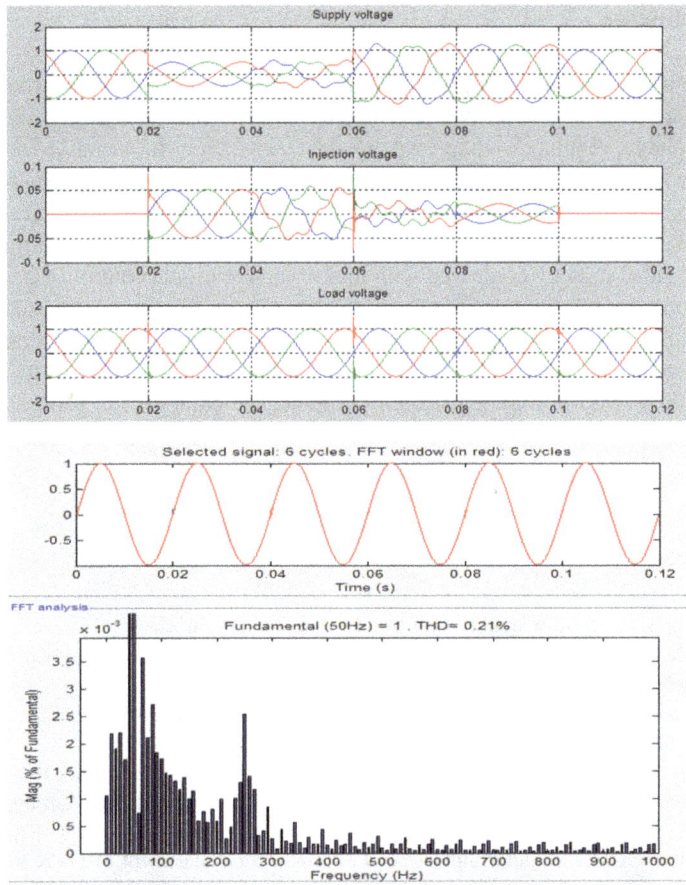

Fig. 11. DVR response to the proposed control strategy

For the voltage loop control strategy, due to the large variability of the current during the control process, the selection of the filter parameter and PI control stage parameters becomes difficult. The resistance value of the filter circuit coil requires very small to suppress the fluctuation of the current; control errors always exist (the load recovery voltage is always smaller than 1 pu) and the control characteristic is always slower. With the proposed control strategy, the current control loop acts as a filter to suppress the electrical current interference on the system, reducing the effect of interference to the output insert voltage of the DVR; therefore, the control characteristics are fast with high control accuracy.

5 Conclusion

The article has proposed an improved control technique in the dq synchronous frame, based on the control circuit of two adjustment loops, inner current adjustment loop and external voltage regulation loop, independent separation of negative and positive sequence component. Through the simulation results, the proposed control strategy gives the good control performance for both Sag/Swells voltage of balance as well as unbalanced one, with or without harmonic interference. The control performance of the proposed control strategy compared with the two conventional control strategies results in better control with different Sag/Swells voltage cases. However, we can clearly see the disadvantages of this approach, that is: Must use many transformation operation between coordinate systems, when the grid is unbalance, it is required to use double controllers and cross decoupling, leading to increased computation and complexity in digital signal processing, which adversely affects to the dynamic response of the system.

References

1. Polycarpou, A.: Power quality and voltage sag indices in electrical power systems. In: Electrical Generation and Distribution Systems and Power Quality Disturbances, pp. 139–159. Frederick University (2011). ISBN 978-953-307-329-3
2. Bollen, M.H.J.: Understanding Power Quality Problems, 672 p. Wiley-IEEE (1999). ISBN 978-0-7803-4713-7
3. Baggini, A.: Handbook of Power Quality. Wiley, University of Bergamo, Italy (2008)
4. Nielsen, J.G.: Design and Control of a Dynamic Voltage Restorer. Aalborg University (2004)
5. Bhumkittipich, K., Mithulananthan, N.: Performance enhancement of DVR for mitigating voltage sag/swell using vector control strategy. Energy Procedia **9**, 366–379 (2011)
6. Hadi, E., Abdolreza, S.: Simulation of Dynamic Voltage Restorer Using Hysteresis Voltage Control. Babol Noshirvani University of Technology (2009)
7. Reshmi, V., Mabel, E., Jayasree, M.S.: Mitigation of Voltage Sag, Harmonics and Voltage Unbalances Using Dynamic Voltage Restorer. ResearchGate (2013)
8. Ping, P.X.: The research on detection and control method of dynamic voltage restorer. Doctoral thesis (2014)
9. Li, Z., Wu, Z., Xia, L., Zhou, W.: Compound resonant control for dynamic voltage restorers under arbitrary load conditions. Proc. CSEE **33**(25), 130–138 (2013)
10. Ming, P.X.: Theory and simulation research on dynamic voltage restorer, Doctoral thesis (2005)
11. Hui, X.: Study on dynamic voltage restorer applied to distribution system, Master thesis (2004)
12. Trinh, T.D.: Control of dynamic voltage restorer for protection of industrial sensitive load from voltage sags, Doctoral thesis (2014)

Performance Evaluation of IPv6 and IPv4 for Future Technologies

Augustus E. Ibhaze[1(✉)], Obinna Okoyeigbo[2], Uyi A. Samson[2], Paschal Obba[2], and Ignatius K. Okakwu[3]

[1] Department of Electrical and Electronics Engineering, University of Lagos, Akoka, Nigeria
eibhaze@unilag.edu.ng
[2] Department of Electrical and Information Engineering, Covenant University, Ota, Nigeria
{obinna.okoyeigbo, aiyudubie.uyi}@covenantuniversity.edu.ng,
obbap@ymail.com
[3] Department of Electrical and Electronics Engineering, University of Benin, Benin City, Nigeria
igokakwu@yahoo.com

Abstract. Due to the ever increasing number of devices connected to the Internet, several methods, protocols and strategies have been developed to accommodate the increasing traffic. One of these methods is the supposed transition between Internet Protocol version 4 (IPv4) and Internet Protocol version 6 (IPv6) thereby extending the addressing capabilities of the protocols. In this study, the two addressing schemes are compared. Simulation results showed that IPv6 has better performance with respect to selected protocols such as Hyper Text Transfer Protocol, File Transfer Protocol and Jitter in Voice over Internet Protocol (VoIP) calls. Both IPv4 and IPv6 had a mean object score of 3.2, a metric for measuring the quality of VoIP calls.

Keywords: Internet of Things · Jitter · Quality of Service · Traffic · IPv6 · IPv4

1 Introduction

Research has shown that if not for several schemes such as subnetting, route summarization and most importantly Network Address Translation (NAT) Overload or Port Address Translation (PAT), that have been introduced, IPv4 addresses would have been exhausted over a decade ago. This as a direct consequence of the immense growth of the Internet is far beyond what the original designers had anticipated. With the evolution of IoT (Internet of Things), experts have predicted that over 30 billion devices would be connected to the Internet by the year 2020. Obviously, IPv4 addressing scheme would be insufficient for the over 30 billion device connections, hence the need for a solution.

There has been lots of development in IoT technologies in recent years; very cardinal of these is the transport of IP over low power radio technologies which is ride mostly on IPv6 [1, 2]. Majority of these works, contributions and developments see the IPv6 as an alternative option for IPv4 considering the vast unique address space acclaimed globally and the less complex nature of the network configuration of the IPv6.

IPv6 is a 128-bit addressing scheme and can contain approximately 340 undecillion addresses which is extremely large compared to IPv4 that is a 32-bit addressing scheme with approximately 4.3 billion addresses [3]. Hence, the IPv6 is a potential solution.

Therefore, IPv4 and IPv6 addressing schemes are compared in order to see the benefits and flaws of both addressing schemes [4], and determine the feasibility of IPv6 as a potential solution.

IPv4 and IPv6 are compared based on the following applications that are dominant on the World Wide Web (WWW):

- File Transfer Protocol (FTP)
- Hyper Text Transfer Protocol (HTTP)
- Jitter in VoIP calls
- Mean Object Score (MOS)

IPv4 addresses pool which is available for allocation is decreasing rapidly, from the last/8 pool of available IPv4 addresses, report shows that four out of five Internet Registries are already allocated. A transition from IPv4 to IPv6 is the most probable solution to the problem of IPv4 address shortage. The need for this transition from IPv4 to IPv6 is indeed inevitable, hence a number of IPv6 transition technologies are proposed in current research, like the dual-stack, tunnel and translation [5–7]. The feasibility of the coexistence and interoperability of IPv4 and IPv6 was investigated in [8] using tunnel-based and dual-stack technique, while stating address translation possibilities. According to [9], translation improves efficiency by providing interconnection between IPv4 and IPv6 network thereby making them interoperable.

Research efforts into reducing the tunneling overhead, an issue in IPv6, have also been made. In [10], FlowLAN an IPv6-based non-tunneling distributed virtual network called was proposed. It reduces the tunneling overhead by rewriting the headers of outgoing packets and recovering the packets when receiving them.

There has also been recent research on the process of adoption of IPv6 in terms of characterizing and measuring, in terms of performance, routing, traffic, and topology [11]. During the World IPv6 Day, investigation and analysis of tunneled traffic, application mix and traffic was carried out [12]. The investigation result showed that native IPv6 traffic produced a near double with respect to traffic volume. The result showed also similarity between the IPv6 and IPv4 application mix. Dhamdhere et al. in [13], investigated whether IPv6 and IPv4 converged in terms of performance, routing, traffic, and topology, and showed that it routing scheme presents more pathological behavior, compared to the IPv4. This is also confirmed in [11].

IPv4 being the fourth version of the Internet Protocol (IP) that is use for network node identification. It is one of the core standards-based internetworking protocols and also the first version deployed for production of ARPANET in 1983 [14]. IPv4 addresses are represented by 32-bit integer that is written in the dot-decimal notation, consisting of four octets separated by dots. IPv4 operates on best effort delivery model, which does not guarantee delivery, avoidance of duplicate delivery and proper sequencing.

IPv6 often called IPng (Internet Protocol next generation) developed by the Internet Engineering Task Force (IETF) with the purpose of overcoming the IPv4 address exhaustion problem. An IPv6 address is a 128-bit numerical address, which consists of a 64-bit subnet prefix and a 64-bit IID (Interface Identifier) portion, used for address configuration. The length of the IID and the subnet prefix may vary based on the link type. IPv6's 128-bit addresses can be represented by eight groups of four hexadecimal digits, with each group separated by colons, example: 3011:0db7:0000:0034:0000: 8a4e:0360:7344, although, there are methods to abbreviate this full notation. IPv6 address scheme not only solves the problem of address scarcity of the IPv4, but also achieves an improvement in performance, security, and QoS, among others [15]. In the nearest future, it is certain that IPv6 will take over IPv4, as the core network layer protocol for next generation networks.

In recent past, it has been proposed that everything on the universe will be assigned IP address via IPv6 address. Several issues came up as a result of this proposition, these issues includes data sovereignty, security concerns and data privacy [16]. The devices in IoT interact with one another via the wireless network technologies using Zwave, Bluetooth and Zigbee [17] or through cloud or gateway using TLS, HTTP, TCP, UDP, CoAP, IP [18]. With the high relevance of the internet in our day to day activities, several devices around the globe are becoming connected via the network. It is now a generally accepted notion that IPv4 addressing space is really limited to cater for the demand which necessitates an upgrade from the 32-bit address space to 128-bit address space having higher capacity and better capability.

Based on the classification of 5G network done by Next Generation Mobile Networks (NGMN) group in which 5G network was classified into eight [19] families namely; Massive internet of thing, Broadband access everywhere; High user mobility; Broadband access in dense areas; Lifeline communication; Ultra-reliable communication; Extreme real-time communication; and Broadcast-like services. The implication of this classification simply means high data rate and increase device connectivity within the Internetwork and higher data usage, which IPv4 can obviously not handle [20].

The remainder of this paper is organized as follows, Sect. 2 presents the research method used, model simulation and protocol performance validation. Section 3 presents detailed comparison of IPv6 and IPv4 performance evaluation outcome, while conclusions are drawn in the last section.

2 Research Method

The Network shown in Fig. 1 is a star topology network, used to simulate the quality of protocols on IPv4 and IPv6. In this research work, this network topology is tagged as Modelled Star Topology Network (MSTN) represented accordingly in the legends of Figs. 2, 3, 4 and 5. The simulation was conducted using OPNET (Optimized Network Engineering Tool) simulator based on its reliability for network simulation and modelling for both industry and academic purposes. The simulation ran for an hour to ensure performance validation in relation to actual productive networks.

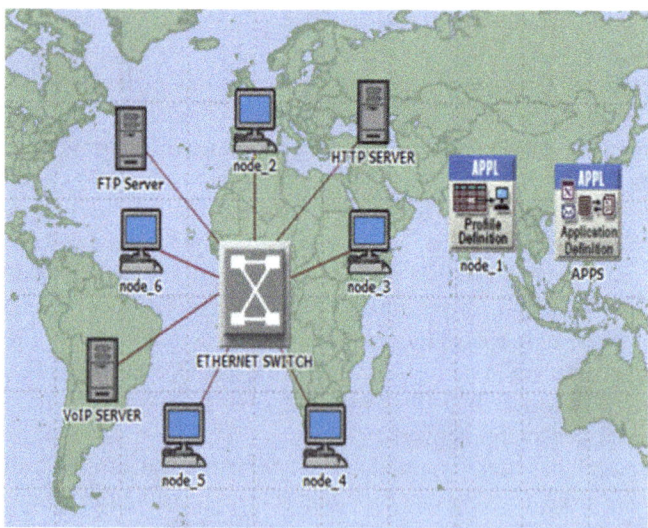

Fig. 1. Modelled star topology network (MSTN)

2.1 File Transfer Protocol (FTP)

The upload response time for the IPv6 and IPv4 were compared, and the result shown in Fig. 2. The result shows that IPv6 has a faster response time than its IPv4 counterpart.

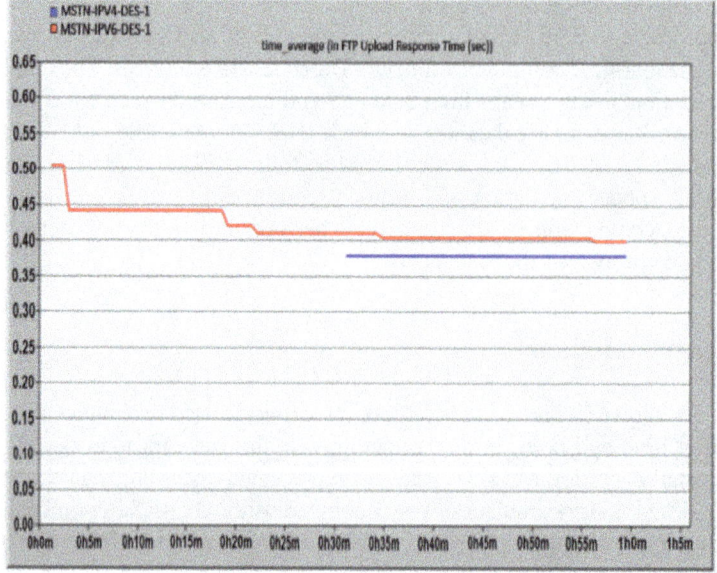

Fig. 2. IPv6 vs IPv4 in terms of upload response time for FTP

Performance Evaluation of IPv6 and IPv4 for Future Technologies   19

2.2 Hyper Text Transfer Protocol (HTTP)

The page response time for web pages which use the Hyper Text Transfer Protocol was simulated, and the result is shown in Fig. 3. The HTTP had a faster response with IPv6 than with the IPv4.

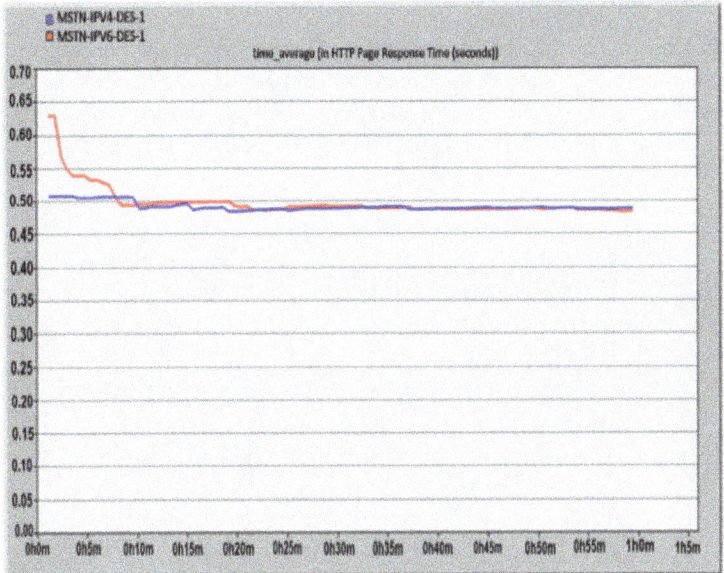

Fig. 3. IPv6 vs IPv4 in terms of page response time for HTTP

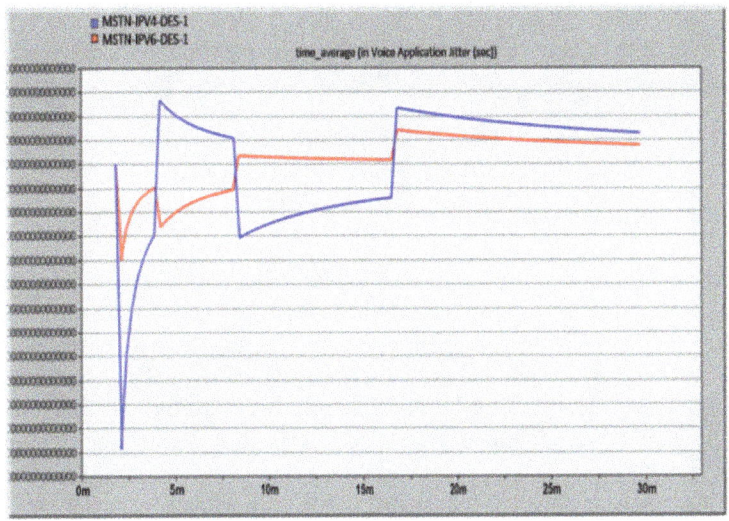

Fig. 4. IPv6 vs IPv4 in terms of jitter

2.3 Jitter in VoIP Calls

Jitter is simply the variation of delay of each packet. It is measured by the variance of the time latency in a network. Jitter affects the QoS in VoIP calls. Simulation results show that IPv4 has a higher Jitter rate than IPv6 as shown in Fig. 4.

2.4 Mean Object Score (MOS)

The Mean Objective score (MOS) is used to measure the quality of VoIP calls. On a scale of one to five, the MOS standard ranges from 3.5 to 4.2 according to ITU [21]. In this simulation, using GSM quality of speech and all schemes of voice encoders available, IPv4 and IPv6 had the same MOS value of 3.2 shown in Fig. 5.

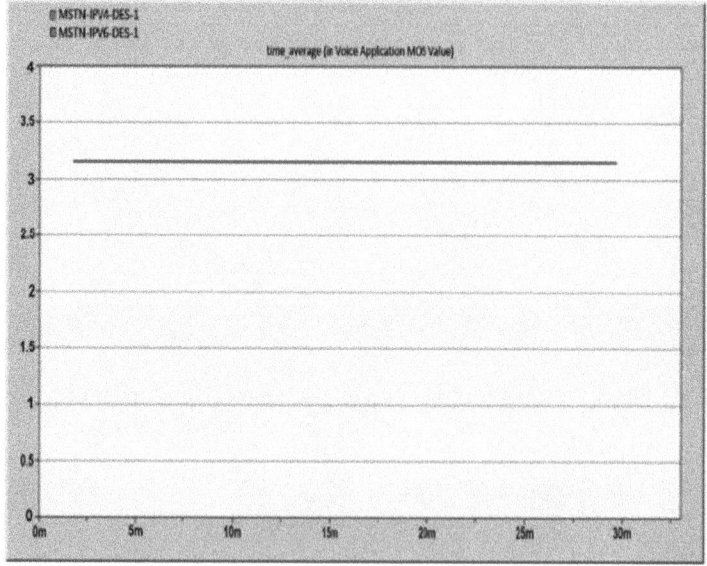

Fig. 5. IPv6 vs IPv4 in terms of MOS value

3 Results and Discussion

The results are gotten from statistics about certain factors that determines the performance of IPv4 and IPv6. FTP upload response time was faster on IPv6 than on IPv4. This means that FTP server and client operators like FileZilla will respond faster on IPv6. HTTP web pages responded faster on IPv6 than on IPv4. This could be due to so many reasons and statistics used on the Network. If the ISP doesn't possess a robust network, IPv6 could respond at a slower pace. Depending on the kind of applications, operating systems, network equipment and when the ISP relies on 6 to 4 tunnels. All things being equal and with a robust ISP, IPv6 may actually provide better performance. In terms of VoIP calls, IPv6 and IPv4 had the same MOS value based on the network Jitter is a major issue when it comes to VoIP calls and IPv6 responded better with a lower jitter rate than IPv4, therefore the IPv6 would give better QoS. Tables 1 and 2 summarizes the findings.

Table 1. Comparison of VoIP parameters

	MOS	Jitter
IPv4	3.2	Increasing at a higher rate with time
IPv6	3.2	Increasing at a slower rate when compared to IPv4

Table 2. Comparison using different kinds of traffic

	FTP upload response time	HTTP page response time
IPv4	Slower response time	Slower response time
IPv6	Faster response time	Faster response time

4 Conclusion

Migration from IPv4 to IPv6 certainly presents significant and relevant improvements in the evolution of the technological advances hosted by the network. The unprecedentedly large traffic envisaged in the emerging chaotic Internet of Things Network will invariably require a large pool of addressing scheme available in the IPv6 scheme. Despite this enormous advantage inherited from the IPv6 framework, the similarities of performance relative to IPv6 can as well be harnessed through IP tunneling schemes. This study was done using VoIP, FTP and HTTP traffic in order to check their efficiency when simulated on the two IP addressing schemes. The tunneling of IPv4 and IPv6 among other innovations cannot be over rated as vast majority of protocols will run efficiently.

Acknowledgment. The authors gratefully acknowledge the sponsorship and funding of this project by the University of Lagos.

References

1. Idachaba, F.E., Olowoleni, J.O., Ibhaze, A.E., Oni, O.O.: IoT enabled real-time fishpond management system. In: Proceedings of the World Congress on Engineering and Computer Science, San Francisco, USA (2017)
2. Ibhaze, A.E., Idachaba, F.E., Eleanor, E.C.: E-health monitoring system for the aged. In: IEEE International Conference on Emerging Technologies and Innovative Business Practices for the Transformation of Societies (EmergiTech), Mauritius (2016)
3. Edelman, B.: Running out of numbers: scarcity of IP addresses and what to do about it. In: First International ICST Conference, AMMA, pp. 95–106 (2009)

4. Ahmed, M., Litchfield, A.T., Ahmed, S., Mahmood, A., Meazi, M.E.H.: VoIP performance analysis over IPv4 and IPv6. Int. J. Comput. Netw. Inf. Secur. **6**(11), 43–48 (2014)
5. Maojia, S., Congxiao, B., Xing, L.: A SDN for multi-tenant data center based on IPv6 transition method. In: IEEE Information Technology, Networking, Electronic and Automation Control Conference (2016)
6. Bao, C., Huitema, C., Bagnulo, M., Boucadair, M., Li, X.: IPv6 addressing of IPv4/IPv6 translators. IETF RFC 6052 (2015)
7. Zhai, Y., Bao, C., Li, X.: Transition from Ipv4 to Ipv6: a translation approach. In: 2011 6th IEEE International Conference on Networking, Architecture and Storage (NAS). IEEE (2011)
8. Govindarajan, K., Setapa, S., Meng, K.C., Ong, H.: Interoperability issue between IPv4 and IPv6 in OpenFlow enabled network: IPv4 and IPv6 transaction flow traffic. In: 2014 International Conference on Computer, Control, Informatics and Its Applications (IC3INA), pp. 58–63. IEEE (2014)
9. Han, G., Sheng, M., Bao, C., Li, X.: Stateless communication mechanism between an IPv4 network and the IPv6 internet. J. Comput. Appl. **8**(35), 2113–2117 (2015)
10. Yi, B., Congxiao, B., Xing, L.: FlowLAN: a non-tunneling distributed virtual network based on IPv6, pp. 229–234. IEEE (2016)
11. Livadariu, I., Elmokashfi, A., Dhamdhere, A.: Characterizing IPv6 control and data plane stability. In: IEEE INFOCOM 2016 - The 35th Annual IEEE International Conference on Computer Communications (2016)
12. Sarrar, N., Maier, G., Ager, B., Sommer, R., Uhlig, S.: Investigating IPv6 traffic: what happened at the world IPv6 day? In: PAM, March 2012
13. Dhamdhere, A., Luckie, M., Huffaker, B., Claffy, K.C., Elmokashfi, A., Aben, E.: Measuring the deployment of IPv6: topology, routing and performance. In: ACM SIGCOMM IMC, November 2012
14. John, N.S., Anoprienko, A.A., Ndujiuba, C.U.: TCP/IP protocol-based model for increasing the efficiency of data transfein computer networks. IGI Global, USA (2013)
15. Gallaher, M.P., Rowe, B.R.: The costs and benefits of transferring technology infrastructures underlying complex standards: the case of IPv6. J. Technol. Transfer **5**(31), 519–544 (2006)
16. Ibhaze, A.E., Akpabio, M.U., John, S.N.: A review on smart grid network security issues over 6LoWPAN. In: Proceedings of the Second International Conference on Internet of Things, Data and Cloud Computing, Cambridge, United Kingdom (2017)
17. Ibhaze, A.E., Akpabio, M.U., Atayero, A.A.A.: Smart meter solution for developing and emerging economies. Int. J. Autom. Smart Technol. **8**(3), 119–126 (2018)
18. Al-Fuqaha, A., Guizani, M., Mohammadi, M., Aledhari, M., Ayyash, M.: Internet of things: a survey on enabling technologies, protocols, and applications. IEEE Commun. Surv. Tutor. **17**(4), 2347–2376 (2015)
19. Boccardi, F., Heath Jr., R.W., Lozano, A., Marzetta, T.L., Popovski, P.: Five disruptive technology directions for 5G. IEEE Commun. Mag. **52**(2), 74–80 (2014)
20. Popoola, S.I., Faruk, N., Atayero, A.A., Oshin, A.A., Bello, O.W., Adigun, M.: 5G radio access network technologies: research advances. In: World Congress on Engineering and Computer Science, vol. 1 (2017)
21. Arjona, A., Westphal, C., Ylä-Jääski, A., Kristensson, M.: Towards high quality VoIP in 3G networks - an empirical approach. In: IEEE Fourth Advanced International Conference on Telecommunications, pp. 143–150 (2008)

Investigate the Performance of 240 GHz Millimeter: Wave Frequency over Fiber with 10 and 20 Gbps

Fawziya Al Wahaibi[(✉)] and Hamed Al Raweshidy

Brunel University London, London, UK
{Fawziya.al-wahaibi,hamed.al-raweshidy}@brunel.ac.uk

Abstract. In this work, a new approach is introduced to generate 240 GHz millimeter-wave signals using dual Parallel Mach-Zehnder modulators only. 216 GHz, 240 Ghz, 264 GHz and 288 GHz can be generated by tuning the input frequency of local oscillator to 18 Ghz, 20 GHz, 22 GHz and 24 GHz individually. By properly setting of the MZM biasing point, RF LO frequencies and phases shift, eighths order optical sidebands are generated with OSSR of 36.7 dB. At the photo detector 12 tupled frequencies are obtained with RFSSR of 30.01 dB. Further, the performance of generated 240 GHz is investigated by modulated the generated signals with baseband signal of 10 Gbps and 20 Gbps by using electro Absorption modulator. The max Q factor of different transmission distance is measured and analyzed.

Keywords: Upper band of millimeter-wave · Dual parallel Mach-Zehnder modulators (DP-MZM) · Optical sideband suppression ratio · Frequency 12-tupling · Millimeter wave (mm-wave) · Single mode fiber (SMF)

1 Introduction

The demand for high-speed wireless communications continues among the fast development of various users applications such as; big data analysis, artificial intelligence and Internet of Everything. Therefore, huge amount of data traffic is a case that is needed to be solved. Based on Ericsson predication data traffic is predicted to be increased to 50 Petabytes in 2021. Current mobile networks are growing so rapidly to deliver "anywhere and anytime communication". Extensive researches are required in developing current mobile system to handle vast amounts of data with high throughput per users' device which could be reach up to several Tera-bps (Tbps) for beyond the fifth generation (B5G). Millimeter waves (mm-waves) generally refer to the frequency range of 30 to 300 GHz (wavelengths of the range 1–10 mm) [1]. The upper band of mm-wave frequencies 0.20–0.300 THz in wireless communications is an attractive candidate for providing capacity improvements for future communication systems. These have garnered attention, because of the huge bandwidth that can be delivered in wireless communication and this band has not been allocated for any application yet in wireless communication systems [2]. There has been a lot of research that has shown the potential of THz communications, which will deliver ultra-fast future wireless communication systems accomplishing rates of 10 Gbps and 20 Gbps [3–6]. Because of optical generation of a millimeter (mm-wave) signal is the key stages in implementing mm-wave signals over fiber system technology. There are various approaches have been reported lately for generating such waves including: direct modulation, optical heterodyning, and external modulation [7]. However; external modulation technique is most probably accurate, with such schemes having revealed the most potentiality for high-frequency signal generation owing to the large tunable range, excellent stability, high spectral purity and lower system cost [8]. In addition to this, external modulation technique utilizing MZM have been deployed in generating mm-waves, for instances; single sideband (SSB), double sideband (DSB) and optical carrier suppression (OCS), with each having different capabilities in resistance for the fiber dispersion. Based on research [9], it is found that OCS is preferred because of its lead of processing the highest receiver sensitivity. A lot of research considered mm-waves generation methods by using external modulation which evolved carrier suppression with a different frequency multiplication conception. In [10], the researcher implemented two parallel MZMs based on a multiplication factor of four of local oscillator frequency to generate the 36 GHz mm-wave frequency. Specifically, a 36 GHz mm-wave was generated with an OSSR of 30 dB and RFSSR of 24.6 dB. In [11], the researcher used two Parallel MZMs based on a multiplication factor of four to generate a photonic microwave frequency with an OSSR of 38.2 dB. A multiplication factor of six was demonstrated to generate mm-waves at 60 GHz for ROF with an OSSR of 23 dB in [12]. A tunable system was implemented in [13] to generate mm-waves at 20 GHz to 100 GHz by employing the uniform fiber Bragg grating based acousto-optic tunable filter (UFBG-AOTF). Moreover, a 240 GHz mm-wave signal was generated by using SiGe:C BiCMOS technology in [14], where 30 GHz was used as the input frequency in order to generate 216–256 GHz, which means the generated frequency was the 8th multiple of the input frequency. In [15], 120 GHz was generated by using a 12-tupling mm-wave generation system with two cascaded dual-parallel Mach–Zehnder modulators without an optical filter, which included four MZMs. This was a tunable system up to 216 GHz with an OSSR of 25.1 dB and RFSSR of 19.1 dB. An optical mm-wave signal with 26.4 dB OSSR and 20.3 dB RSSR was achieved by using a 12-tupling mm-wave generation system based on two cascaded dual-parallel Mach-Zehnder modulators in [16]. Recently, in [17], a filterless approach for generating a 160 GHz mm-wave with a multiplication factor of 16 being applied to the input local oscillator, based on only two MZMs and two optical attenuators (OATT), with an OSSR of 30.34 dB and RSSR of 23.79 dB, has been achieved. In this study, a scheme of filterless frequency-tupling mm-waves with Dual Parallel MZMs (DP-MZMs) was used in order to generate 12-tupling mm-waves of the higher band of mm-wave frequencies. By correctly setting the MZM bias point, RF LO voltages and extinction ratio, only eighth order optical side-bands were generated, which resulted in a 12-tupled mm-wave at the photodetector. Simulations were completed by using optisystem software, with 216, 240, 264, 288 and 300 GHz mm-waves being achieved by tuning the input local oscillator to 18, 20, 22, 24 and 25 GHz, respectively.

2 Principle of the Proposed Design

The principle of proposed design in generating upper band of mm-wave signals is summarized in Fig. 1.

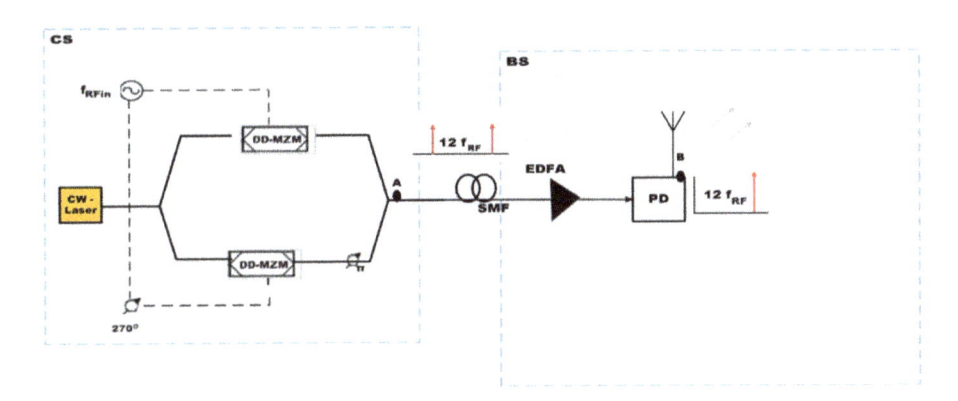

Fig. 1. Block diagram of the mm-wave frequency with 12-tupling techniques. CW, continuous wave laser; DD-MZM, dual-drive Mach–Zehnder modulator; EDFA, erbium-doped optical fiber amplifier; SMF, single mode fiber; PD, photodiode.

The optical field is emitted to Dual Parallel - MZMs (DP-MZM) s using a continuous wave laser (CW) laser, which is denoted by [18]:

$$E_n(t) = E_0 cos(\omega_0 t) \qquad (1)$$

where, E_0 represents the amplitude of optical field and ω_0 indicates the angular frequency of the optical carrier. Both DP-MZMs are based at peak transmission point. Then, electrical driving signals are sent to the upper MZM and lower MZM, which are denoted by:

$$V_0 = V_m \cos \omega t \text{ and } V_m \cos(\omega t + 3\frac{\pi}{2}), \text{ respectively.} \qquad (2)$$

Then, the phase shift is applied at the output of the lower MZM. Thus, the output of the Parallel DD-MZMs is:

$$E_{out}(t) = E_0 \{cos(\omega_0 t) cos[mcos(\omega t)] - cos(\omega_0 t) cos[mcos(\omega t + 3\frac{\pi}{2})]\} \qquad (3)$$

Where, m indicates that the phase modulation index m of DP-MZM is $\frac{\pi V_m}{2V_\pi}$. Expanding $cos mcos(\omega t)$ and $cos(m\omega t + 3\frac{\pi}{2})$ using the Bessel function, the output optical field can be written as:

$$E_{out}(t) = -E_0 \sum_{n=1}^{\infty} J_{4n-2}(m)(\cos\{[\omega_0 + (4n-2)\omega]t\} + \cos\{[\omega_0 - (4n-2)\omega]t\}) \quad (3)$$

where, J_n indicates the Bessel function of the first kind of order n and m equals 1.635 π which is corresponding values of J_2. The output optical signal can be simplified to:

$$E_{out}(t) = -E_0 J_6(m)(\cos\{[\omega_0 + 6\omega]t\} + \cos\{[\omega_0 - 6\omega]t\}) \quad (4)$$

After square law detection using photodiode (PD) with responsivity, then:

$$i_{12} = R|E_0 J_6(m)|^2 \cos(12\,\omega t) \quad (5)$$

3 Simulation and Discussion

The design for generation 240 GHz mm-wave is implemented using optisystem16 software. With sane implemented design, the performance of 240 GHz over SMF in downlink data transmission is investigated.

3.1 Generating 240 GHz

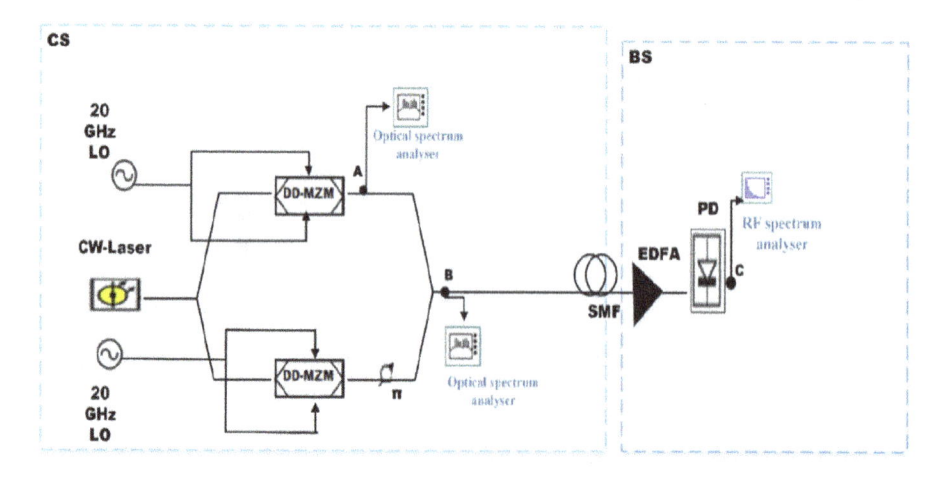

Fig. 2. Simulation parameter set up for generating 240 GHz mm-wave signal

The design for generating 240 GHz mm-waves is illustrated (see Fig. 2). 193.1 THz optical signal is injected into the MZMs via a CW laser source with spectral width of 10 MHz. An input optical power is applied to the optical signal is 0 dBm. Both of DD-MZM is driven by input electrical signal of 20 GHz. The phase shift of the upper

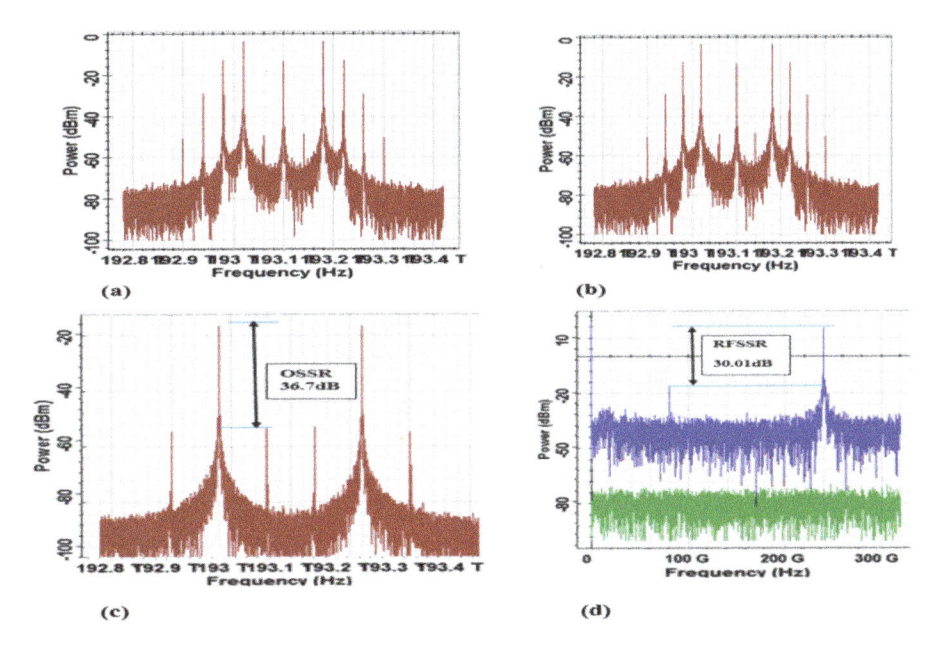

Fig. 3. (a) Optical mm-wave related to 193.1 THz at upper MZM. (b) Optical mm-wave related to 193.1 THz at lower MZM. (c) Obtained optical spectrum signal at the optical adder. (d) Achieved 240 GHZ mm-wave.

DD-MZM is 0, whilst a phase shift of 270° is applied to the lower MZM GHz. The extinction ratio is set to 60 dB in order to achieve best OSSR between the generated harmonics that is associated with desired signal. The switching bias voltage of the upper and lower MZMs is 3 V. The setting of both DD-MZMs is at the peak operation point to obtain output optical spectrum which includes carrier and even order sidebands. The obtained optical spectrum at the output of optical coupler (is seen in Fig. 3(a)) with OSSR of 36.7 dB (see Fig. 3(c)). Then, the achieved optical mm-wave frequencies are sent over 20 km of SMF-28. At the receiver, EDFA is implemented to compensate the power loss of the generated optical signal. At the PD, 240 GHz mm-wave signal in electrical domain of a 12 times that of the input RF LO is obtained. The 4th order optical sideband powers are 30 dB below the 8th order optical sideband and the undesired harmonic electrical signals are effectively suppressed, (see Fig. 3(d)).

Thus, it has been proven that, 240 GHz can be generated by using the proposed system without the requirement of a filter to remove an unwanted harmonics signal. Additionally, the proposed system is tunable, which means that frequencies at the upper bands of mm-waves of 216 GHz, 240 GHz, 276 GHz, 288 GHz and 300 GHz, can be generated by tuning the input LO signals with frequencies of 18 GHz, 20 GHz, 22 GHz, 24 GHz and 25 GHz, respectively.

3.2 Downlink Data Transmission at 240 GHz

Moreover, performance of achieved higher band of mm-wave frequencies over SMF is investigated using optisystem software with same parameters set up that have been used in Sect. 3.1. The optical adder output shows ±8th sidebands at 193.22 THz and 192.02 THz. The obtained optical sidebands are separated by 240 GHz (0.24 THz). Then, the obtained signal is modulated with 10 Gbps baseband signal using an electro absorption modulator (EAM). The data are modulated on both the optical sidebands (see Fig. 4). Modulated optical mm-waves are then sent over different lengths of SMF, with 16.75 ps/nm/km dispersion and the fiber attenuation is 0.2 dB/km which is specified by common industry limit to stimulate the actual environment as possible. EDFA is used to compensate the losses over the transmission distance. PD at the receiver is used to detect the modulated 240 GHz mm-wave signal. The extracted signal contains a 240 GHz mm-wave signal and the baseband signal. Then, an electrical band pass filter, with a bandwidth of 1.5 times the bitrate, is used to distinct the baseband signal from 240 GHz mm-wave. An eye diagram analyser is used to assess the performance of the received mm-wave by measuring the max Q factor for the different transmission distances (see Fig. 5). It shows the max Q factor of the obtained 240 GHz signal when it is modulated with two different data rates of baseband signal 10 Gbps and 20 Gbps. The max Q factor at 240 GHz with a 10 Gbps baseband signal is 27.11 and with 20 Gbps, it is 11.26 over a 5 km transmission distance, as shown in Fig. 5(b). The max Q factor over 10 km for 10 Gbps is 16.7, while for 20 Gbps it is 2.1.

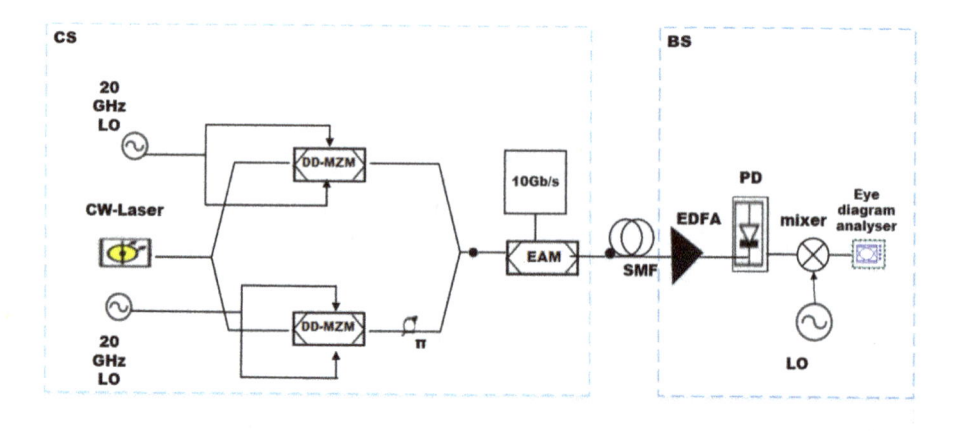

Fig. 4. Simulation scheme of modulating the optical mm-wave with 10 Gbps using an electro absorption modulator (EAM)

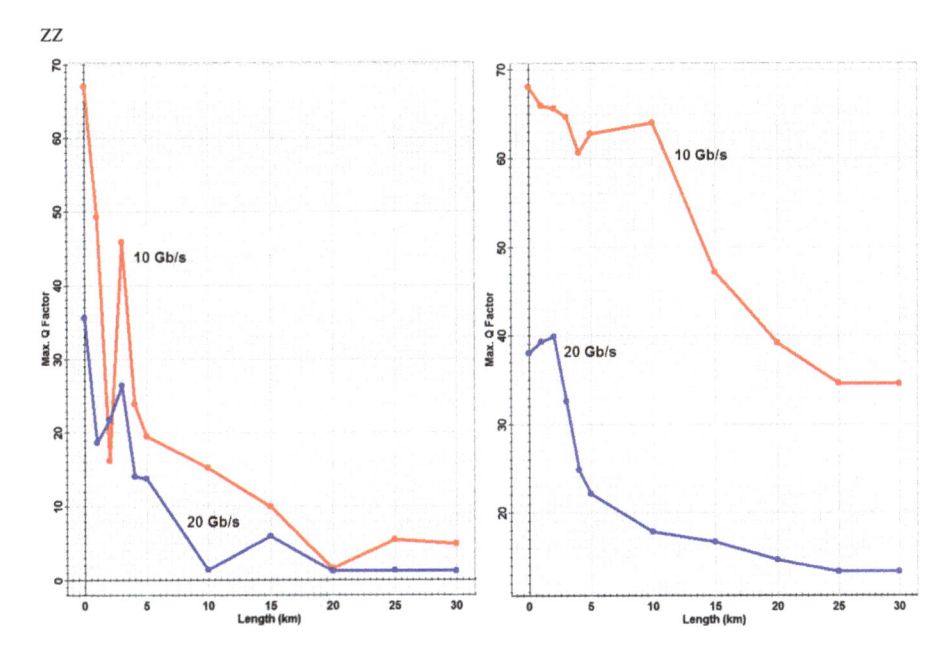

Fig. 5. Max Q factor of 240 GHz for different SMF lengths: (a) without a dispersion effect; (b) with a dispersion effect.

The decrease in the value of max Q factor when the fiber is length extended is because of a dispersion effect and attenuation in the fiber itself. As seen in Fig. 5(a), the max Q factor for 10 Gbps is 39 at 10 km fiber length, while when there is a dispersion effect it is 16.7. Because of the dispersion effect of the fiber, the time slot and profile of an optical pulse are changed in the course of propagation which make the signal spread, causing inter symbol interference which it causes bit errors at the receiver.

4 Conclusions

In this proposed system, 240 GHz is generated using 12-tupling techniques. This frequency can be achieved by using a low input frequency, which minimizes the cost of the design for generating mm-wave over fiber. High quality upper band mm-wave frequencies have been obtained with an OSSR of 36.7 dB and RFSSR of 30.01 dB. Further, the performance of generated frequencies 240 GHz over SMF-28 was investigated. Based on the max Q factor that has been obtained, the signal performance is high quality. 216 GHz, 240 Ghz, 264 GHz and 288 GHz and 300 GHz are generated by tuning the input frequency of local oscillator to 18 Ghz, 20 GHz, 22 GHz, 24 GHz and 25 GHz individually. Thus, the upper band of mm-wave frequencies can handle a high data rate over a long distance, having achieved a satisfactory max Q factor at these desired frequencies.

References

1. Thomas, V.A., El-Hajjar, M., Hanzo, L.: Millimeter-wave radio over fiber optical upconversion techniques relying on link nonlinearity. IEEE Commun. Surv. Tutor. **18**(1), 29–53 (2016)
2. Song, H.J., Nagatsuma, T.: Present and future of terahertz communications. IEEE Trans. Terahertz Sci. Technol. **1**(1), 256–263 (2011)
3. Martin, K.: Teraherz Communications: A 2020 Vision, vol. 35, pp. 325–338. Springer, Heidelberg (2007)
4. Seeds, A.J., Shams, H., Fice, M.J., Renaud, C.C.: TeraHertz photonics for wireless communications. J. Light. Technol. **33**(3), 579–587 (2015)
5. Boulogeorgos, A.A.A., Papasotiriou, E.N., Kokkoniemi, J., Lehtomaeki, J., Alexiou, A., Juntti, M.: Performance evaluation of THz wireless systems operating in 275–400 GHz band. In: IEEE Vehicular Technology Conference, vol. 2018-June, pp. 1–5 (2018)
6. Federici, J., Moeller, L.: Review of terahertz and subterahertz wireless communications. J. Appl. Phys. **107**(11), 111101 (2010)
7. Beas, J., Castanon, G., Aldaya, I., Aragon-Zavala, A., Campuzano, G.: Millimeter-wave frequency radio over fiber systems: a survey. IEEE Commun. Surv. Tutor. **15**(4), 1593–1619 (2013)
8. Giotas, T.: Optical Generation and Distribution of a Wide-Band Tunable Millimeter-Wave Signal With an Optical External Modulation Technique_and Mach Zehnder_Proof_Pages_27, vol. 53, no. 10, pp. 3090–3097 (2018)
9. Yu, J., et al.: Optical Millimeter-Wave Generation or Up-Conversion Using, vol. 18, no. 1, pp. 2005–2007 (2006)
10. Yang, A., Gu, W., Yu, S., Wang, C., Jiang, T.: A frequency quadrupling optical mm-wave generation for hybrid fiber-wireless systems. IEEE J. Sel. Areas Commun. **31**(12), 797–803 (2014)
11. Lin, C.T., Shih, P.T., Chen, J., Xue, W.Q., Peng, P.C., Chi, S.: Optical millimeter-wave signal generation using frequency quadrupling technique and no optical filtering. IEEE Photonics Technol. Lett. **20**(12), 1027–1029 (2008)
12. Mohamed, M., Zhang, X., Hraimel, B., Wu, K.: Frequency sixupler for millimeter-wave over fiber systems. Opt. Express **16**(14), 10141 (2008)
13. Wang, Y., Pei, L., Li, J., Li, Y.: Millimeter-wave signal generation with tunable frequency multiplication factor by employing UFBG-based acousto-optic tunable filter. IEEE Photonics J. **9**(1), 1–10 (2017)
14. Eissa, M.H., Malignaggi, A., Ko, M., Schmalz, K., Borngräber, J., Ulusoy, A.C., Kissinger, D.: A 216–256 GHz fully differential frequency multiplier-by-8 chain with 0 dBm output power. Int. J. Microwave Wirel. Technol. **10**(5–6), 562–569 (2018)
15. Zhu, Z., Zhao, S., Zheng, W., Wang, W., Lin, B.: Filterless frequency 12-tupling optical millimeter-wave generation using two cascaded dual-parallel Mach-Zehnder modulators. Appl. Opt. **54**(32), 9432 (2015)
16. Chen, X., Xia, L., Huang, D.: Optical generation of 12-tupling millimeter-wave signal without optical filtering. J. Opt. Commun. **37**(3), 295–299 (2016)
17. Wang, D., Tang, X., Xi, L., Zhang, X., Fan, Y.: A filterless scheme of generating frequency 16-tupling millimeter-wave based on only two MZMs. Opt. Laser Technol. **116**, 7–12 (2019)
18. Chen, H., Ning, T., Li, J., Pei, L., Zhang, C., Yuan, J.: Study on filterless frequency-tupling millimeter-wave generator with tunable optical carrier to sideband ratio. Opt. Commun. **350**, 128–134 (2015)

Low Power BCH Decoder Using Verification Algorithm and Two-Step Parallel Chien Search Architecture

Noha K. Shebl[✉], Saleh M. Eisa, Hanady H. Issa, and Khaled A. Shehata[✉]

Department of Electronics and Communication, Arab Academy for Science and Technology, Cairo, Egypt
nohakamal90@gmail.com, k_shehata@aast.edu

Abstract. Bose-Chaudhuri-Hocquenghem (BCH) code is normally utilized in communication systems in order to enhance its reliability. The main computational complexity and high power consumption stages in BCH come from its two main stages, the key equation solving (KES) and Chien search (CS). This paper presents two different algorithms to reduce the computational process and hence reduce the total power consumption. These algorithms are the verification algorithm and the two-step parallel CS architecture which are utilized in KES and CS stages, respectively. The whole proposed system is implemented using LABVIEW tools from NI. The results show that the proposed BCH algorithm provides an enhanced performance when compared to the conventional one.

Keywords: Bose-Chaudhuri-Hocquenghem (BCH) codes · Key equation solving (KES) · Chien search (CS) · BM · LABVIEW

1 Introduction

Bose-Chaudhuri-Hocquenghem (BCH) code is vastly employed in wireless communication because of its relatively simple decoding complexity and powerful error correction capability [1, 2]. Also, (BCH) code is a standout amongst the most famous error correction codes (ECC). However, the computational power of BCH decoders is considerably high. Therefore, latest research work focuses on reducing the computational power of the BCH decoder. Chien search (CS) consumes almost half of the power of the BCH, so it became a major contributor to the overall power consumption.

Many different techniques were used to reduce power consumption such as Early termination technique detailed in [3, 4] and Polynomial order reduction (POR) technique developed in [5]. All these techniques have a problem that the power saving relies upon the error positions. The two-step CS discussed in [6] divides the parallel CS into two steps, the first step is accessed every cycle and the second step is accessed only if the first step is successful which leads to the reduction of the number of access and as a result reduces the computational power. A more suitable verification algorithm presented in [7] is utilized also for further improvement. This algorithm depends on finding the invalid polynomials without the need to spend any effort to calculate their roots which reduces the computational power [8].

In this paper a new approach that consolidates both algorithms used in KES and CS stages so as to reduce the BCH computational power consumption is presented. The first algorithm is the verification algorithm and the second one is the two-step parallel CS architecture. The proposed system is introduced in Sect. 2. The LABVIEW implementation of the proposed system is shown in Sect. 3. In Sect. 4, the system validation is presented and the simulation results are shown in Sect. 5. Finally, this work conclusion is summarized in Sect. 6.

2 Proposed System

A binary BCH (n, k, t) code over the Goalie Field GF (2^m) can correct t bits for n code length for k message length. The BCH decoder mainly comprises of three stages: syndrome calculation (SC), key equation solving (KES) and Chien search (CS) as shown in Fig. 1. The SC module starts to calculate the 2t syndromes once it receives the r(x) code. The role of KES stage is to find the error locator polynomials using Berlekamp-Massey (BM) algorithm. The final stage CS is used to get the roots of the error locator polynomial in order to find the location of errors and then correct the errors to get the corrected code c(x) [8]. Each operation stage in BCH works simultaneously and autonomously [9]. The proposed system modifies the main high power consumption modules in BCH decoders. In the proposed KES module, a verification algorithm is utilized to efficiently check the eligibility of each error locator polynomial generated by the BM algorithm. Accordingly, only the eligible polynomials are processed. In addition, a two-step parallel CS architecture is applied to the CS module. This architecture is based on dividing the CS algorithm into two steps in order to reduce its power consumption, where the second step is accessed only when the first step is successful which saves power. The following subsections explain these two modifications.

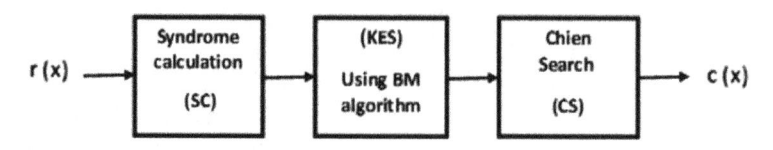

Fig. 1. Basic BCH decoder structure

2.1 KES Using Verification Algorithm

The verification algorithm is used to check if the error located polynomial calculated from the BM algorithm is capable to generate a correct error pattern for a valid corrected code word. All polynomials Λ (x) that do not have any non-zero roots are considered as error locator polynomials [8, 10]. This algorithm defines an auxiliary polynomial d(x) which can also save the computational power. The polynomial d(x) is defined by Eq. (1).

$$d(x) = gcd\left((x^{2^m} - x) mod \wedge (x), \wedge(x)\right) \quad (1)$$

where the gcd (a, b) is the greatest common divisor operator, and $x^{2^m} - x$ includes all components on GF(2^m) as d(x) roots.

Equation (2) is the main condition which should be verified to grantee that $\Lambda(x)$ has different non-zero roots GF (2^m).

$$x^{2^m} mod \wedge (x) = x \quad (2)$$

The modulo operation in Eq. (2) is calculated by using the polynomial inversion algorithm (PIA) [8].

2.2 CS Using the Two-Step Parallel CS Architecture

Conventional p-parallel CS architecture is implemented in order to reach high throughput, where the parallel factor p is the number of α^i substitutions performed at the same time. This number α^i is substituted iteratively for $1 \leq i \leq n$, where α is a primitive element of Galois field GF (2^m) [6, 10].

Equation (3) describes the error locator polynomial equation in the conventional p-parallel CS architecture.

$$\Lambda(x) = \sum_{j=1}^{t} \lambda_j x^j + 1 = Y(x) + 1 \quad (3)$$

The p-parallel CS equation is rearranged in the Two-step architecture, where i ranges from 1 to p. When Y (α^{wp+i}) is 1, an error existence is detected by the CS. Therefore, α^{wp+i} is a root of the error locator polynomial. Finally, the proposed two-step architecture is defined in Eq. (4) to detect the no errors case early [6, 8].

$$Y(\alpha^{wp+i}) = concat \left\{ \Omega(w) \begin{bmatrix} A_{i1,(m-1:m-l)} \\ A_{i2,(m-1:m-l)} \\ \vdots \\ A_{it,(m-1:m-l)} \end{bmatrix}, \Omega(w+1) \begin{bmatrix} A_{(2^m-1)+(i-p)1,(m-l-1:0)} \\ A_{(2^m-1)+(i-p)2,(m-l-1:0)} \\ \vdots \\ A_{(2^m-1)+(i-p)t,(m-l-1:0)} \end{bmatrix} \right\}$$
(4)

3 LABVIEW Implementation of the Proposed BCH Decoder

The main block diagram of the proposed BCH decoder is implemented in LABVIEW as indicated in Fig. 2. The three stages of the BCH decoder shown in the block diagram will be discussed in the following sub-sections.

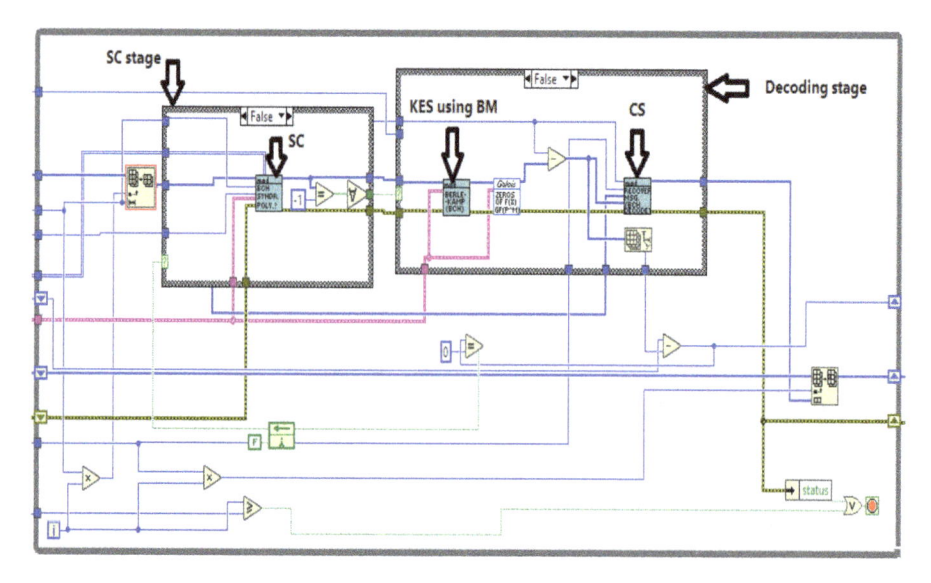

Fig. 2. The proposed BCH decoder block diagram in LABVIEW

3.1 Syndrome Calculation (SC)

The SC stage shown in Fig. 3 can be acknowledged by a linear feedback shift register (LFSR) detailed in [1, 10], which is a simple structure. Therefore, this stage consumes a less computational power.

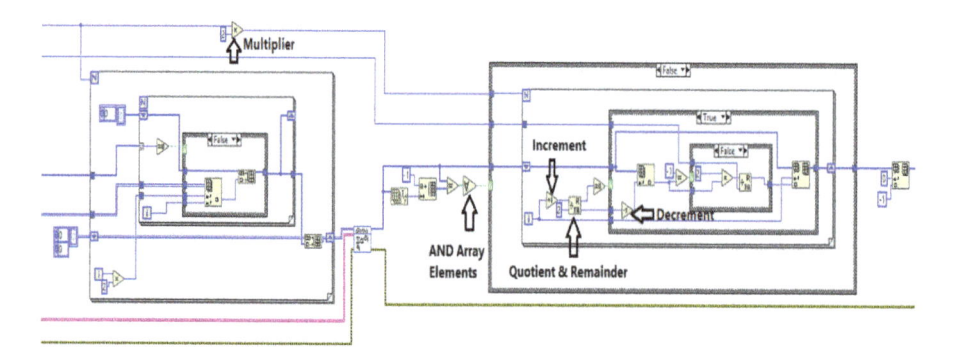

Fig. 3. SC block diagram in LABVIEW

3.2 Key Equation Solving (KES) Module

The most intricate stage in the BCH decoding is the KES stage shown in Fig. 4, consume a large power. Usually it can be acknowledged by the BM algorithm which requires large number of multipliers and dividers. One pass Chase decoding was applied in order to save the computational power where the BM algorithm is performed only once to get a starting error locator polynomial. Then, all the required error locator polynomials are driven from the first one by using polynomial update algorithm [11]. Finally, the eligibility verification algorithm [10] is utilized to get a better employment of the chase decoding algorithm as it saves computational power by finding the invalid polynomial without calculating its roots [7].

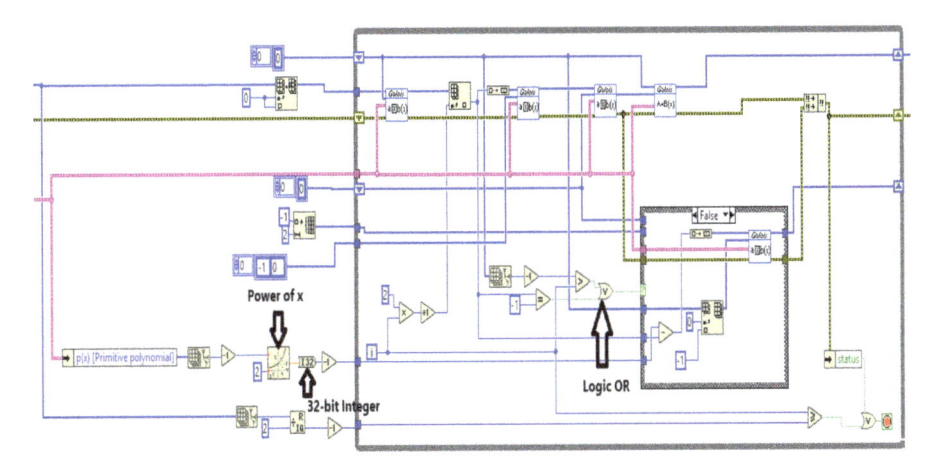

Fig. 4. KES block diagram in LABVIEW

3.3 Chien Search (CS) Module

The CS shown in Fig. 5 is last stage in the BCH decoding. The highly parallel CS performs thorough calculations which leads to spending a huge power consumption in this stage. The two-step parallel CS architecture described in [6, 8] is used in order to overcome this problem. In this architecture, the procedures of finding the error locator polynomial roots is decomposed into two steps based on the binary matrix representation. The second step is not accessed unless the first step is successful which leads to a significant power saving.

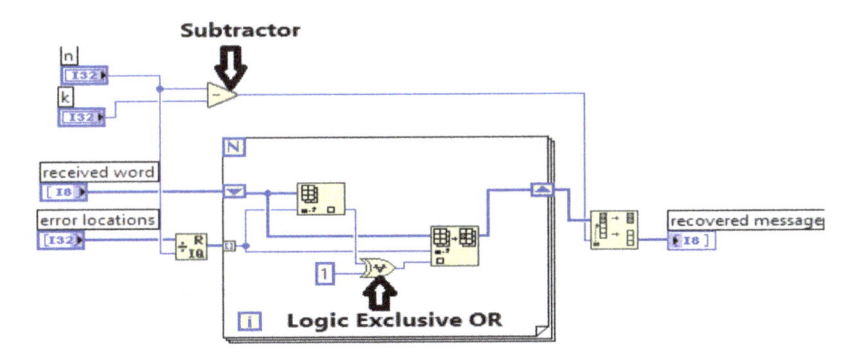

Fig. 5. CS block diagram in LABVIEW

4 System Validation

For simplification, the proposed BCH decoder is performed for n = 15 (code length), k = 7 (message length) and t = 3 (maximum corrected bits), where these lengths can be adjusted easily for different and larger BCH decoder sizes. The system validation results are shown in Table 1. The message bit stream with length k is set manually to the BCH decoder, where errors location are known previously in order to check if the errors are detected and corrected after decoding.

When the BCH decoder received the message with no errors, the output of the decoder was the original massage as no errors were detected by the decoder. By applying 1 or 2 or 3 bit error to the message, the BCH decoder was able to detect and correct all errors even when applying the errors to different locations in each case. However, by applying 4 bit error to the message, the decoder was able to correct only the first three errors in the message bit steam leaving the last error without detection or

Table 1. System validation results

No. of errors (t)	Message	Decoded Message
No errors (t=0)	1100101	1100101
1 bit error (t=1)	1000101	1100101
	1101101	1100101
	1100100	1100101
2 bit error (t=2)	1001101	1100101
	0100001	1100101
	1100110	1100101
3 bit error (t=3)	0110001	1100101
	1001111	1100101
	1110110	1100101
4 bit error (t=4)	0111111	1100111
	1001110	1100100
	1111011	1100111

correction, so the output decoded message was not completely corrected. The main reason for this is that the BCH decoder was designed for maximum number of corrected bits (t = 3), so it can correct only the first three detected errors regardless of the total number of errors. Therefore, any massage with more than three errors will not be corrected.

5 Results and Discussion

The simulation of the proposed BCH decoder is performed on LABVIEW and compared with our previous work detailed in [8]. The conventional BCH (15, 7, 3) decoder is performed with adaptive white Gaussian noise (AWGN) and QPSK modulation technique. The decoder size was chosen for simplification and our LABVIEW simulation can be adjusted easily for different and larger BCH decoder sizes (different n, k and t). Figure 6 shows that the SNR/bit (Eb/No) is 3 dB when Bit Error Rate (BER) is equal to 10^{-2}. When the SNR/bit becomes 7 dB, the BER reaches 10^{-3} [8].

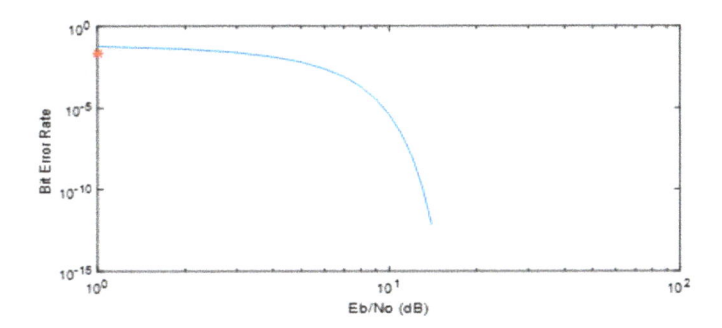

Fig. 6. BER performance of conventional BCH decoder

Figure 7 shows the BER curve of the proposed BCH decoder compared with that of the BCH decoder with the two-step parallel CS architecture only discussed in [8]. It can be noticed that in the case of applying the two-step parallel CS architecture, the SNR/bit are equal to 15 dB and 20 dB for BER of 10^{-2} and 10^{-3}, respectively. When applying both techniques, the verification algorithm and the two-step parallel CS architecture, the SNR/bit gets better. The SNR/bit of the proposed BCH decoder are equal to 8 dB and 12 dB for BER equal to 10^{-2} and 10^{-3}, respectively.

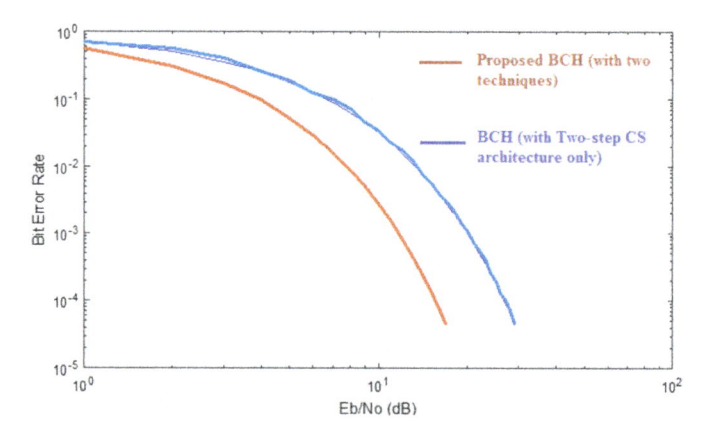

Fig. 7. BER performance of proposed BCH decoder

Therefore, the proposed algorithm gives better BER performance than the two-step parallel CS architecture for the same SNR/bit [8] as shown in Fig. 7. However, the BER performance of the conventional BCH decoder shown in Fig. 6 is better than that of the proposed one shown in Fig. 7 as we had to sacrifice better BER performance in order to reduce the computational power in the proposed BCH decoder compared with the conventional one [6–8].

6 Conclusion

A new approach is presented in this work which combines two different techniques together to reduce the power consumption in the conventional BCH decoder by reducing the number of computations. The verification algorithm is the first technique utilized in the BM algorithm to check the obtained error locator polynomial eligibility by counting the number of its roots. An iterative polynomial inversion algorithm is also utilized in order to decrease the required commutations. The two-step parallel CS architecture is the second technique utilized in the CS in order to reduce its power consumption. It decomposes the two-step parallel CS architecture, where the second step is not accessed unless the first step is successful leading to a significant power saving. The results show that the SNR/bit of the proposed algorithm is better than that of the two-step parallel CS architecture by 20% for 10^{-2} BER and by 60% for 10^{-3} BER. Therefore, the proposed BCH decoder gives better BER performance than the two-step parallel CS architecture and also reduces the computational power compared with the conventional BCH. For future work, the proposed BCH decoder will be designed and implemented to be suitable for Digital Video Broadcasting-Satellite Second Generation (DVB-S2). A hardware implementation for the proposed BCH decoder will be designed and tested.

References

1. Moreira, J.C., Farrell, P.G.: Essentials of Error-Control Coding. Wiley, Hoboken (2006)
2. Sklar, B.: Digital Communications, Fundamentals and Applications, 2nd edn. Prentice-Hall, Upper Saddle River (1988)
3. Wu, Y.: Low power decoding of BCH codes. In: Proceedings of the IEEE ISCAS, pp. II-369–II-372, May 2004
4. Lee, K., Lim, S., Kim, J.: Low-cost, low-power and high throughput BCH decoder for NAND flash memory. In: Proceedings of the IEEE ISCAS, pp. 413–415, May 2012
5. Wong, S., Chen, C., Wu, Q.M.: Low power Chien search for BCH decoder using RT-level power management. IEEE Trans. Very Large Scale Integr. Syst. **19**(2), 338–341 (2011)
6. Yoo, H., Lee, Y., Park, I.: Low-power parallel Chien search architecture using a two-step approach. IEEE Trans. Circ. Syst. II: Express Briefs **63**(3), 269–273 (2016)
7. Zheng, N., Mazumder, P.: An efficient eligible error locator polynomial searching algorithm and hardware architecture for one-pass chase bch codes decoding. IEEE Trans. Circ. Syst. II: Express Briefs (2016). https://doi.org/10.1109/TCSII.2016.2581587
8. Shebl, N.K., Eisa, S.M., Isaa, H.H., Shehata, K.A.: A low power BCH based on error locator polynomial searching algorithm and two step parallel Chien search algorithm. In: 3rd International Conference on Advanced Technology & Applied Sciences (ICaTAS), Malaysia-Japan International Institute of technology, Malaysia (2018)
9. Lin, S., Costello, D.J.: Error Control Coding, Fundamentals and Applications. Prentice-Hall, Inc., Englewood Cliffs (1983)
10. Wicker, S.B.: Error Control Systems for Digital Communication and Storage. Prentice-Hall, Englewood Cliffs (1994)
11. Lee, Y., Yoo, H., Park, I.-C.: High-throughput and low-complexity BCH decoding architecture for solid-state drives. IEEE Trans. Very Large Scale Integr. Syst. **22**(5), 1183–1187 (2014)

A Videogame Driven by the Mind: Are Motor Acts Necessary to Play?

Luigi Bianchi[✉]

Department of Civil Engineering and Computer Science, "Tor Vergata"
University of Rome, Via del Politecnico 1, 00133 Rome, Italy
luigi.bianchi@uniroma2.it

Abstract. In this manuscript, the architecture of a PC based videogame driven by just electroencephalographic (EEG) brain signals is described. It bypasses the natural pathways of nerves and muscles, thus theoretically allowing also people affected by severe motor disorders to play with it. It is built on top of a framework designed for implementing neuro-feedback and Brain-Computer Interface (BCI) systems: spontaneous self-induced modifications of the EEG signals are detected and converted into an analog-like value, which is then used to control the speed of a puppet that competes in a virtual race against others whose speed is controlled by the PC. Preliminary results from five healthy volunteers and comparing three different regression rules show that is possible, after a short calibration phase, to take the control of the puppet by performing simple and repetitive motor imagery mental tasks. Actually, it is under testing in clinical contexts, to rehabilitate children affected by ADHD syndrome and autism but it can be also used as an inclusive game, to allow motor disabled people to play with the same rules with their acquaintances, relatives, and friends.

Keywords: Neuro-feedback · Regression · Videogame · Inclusion

1 Introduction

In the last decade, neurofeedback (NFB) has gained the interest of several researchers because it has been proved that it could support not only medical doctors in making diagnoses but also therapists and caregivers in the rehabilitation process of people affected by a wide range of neurological disorders: there are several examples in which NFB has been successfully adopted in the treatment of Attention Deficit Hyperactivity Disorder (ADHD) [1, 2], epilepsy [2, 3], schizophrenia [4], cerebral palsy [5], stroke [6–8] and traumatic brain injuries [9]. Moreover, in recent years, even non-clinical applications of NFB have been proposed, such as those aiming at improving some motor skills in athletes [10].

In principle, NFB is based on the fact that people can be trained to self-regulate their brain signals processed in quasi-real-time through a system that provides visual or acoustic feedback modulated by the ongoing neural activity. This can also be achieved by performing different mental tasks. Then, brain signals are converted in some form of feedback according to a well-defined rule that relates the neural signals with the feedback and that is specific and characterizes every NFB system: there are various methods to perform this, and several different brain signals can be used. Among them, noninvasive systems are preferred such as electroencephalography (EEG) and functional Near Infra-Red Spectroscopy (fNIRS) that are the most widely used, mostly because they provide a good balance between costs and users safety. However, magnetoencephalography (MEG) and fMRI have been also used as a brain signal recording methods, but they are used only in a limited number of research laboratories because they represent a too expensive solution.

The aim of this work is to show that it is possible to build a videogame in which users do not use a console to play it but just their brain activity. This allows severely disabled people, such as locked-in or tetraplegic patients to play either against the computer or against their friends or relatives. This last will occur allowing all competitors to play with the same rules, thus breaking down the physical barriers of a mouse, keyboard or console and promoting social inclusion.

The proposed videogame was implemented to provide a rehabilitation tool for children affected by ADHD. It was designed to train them to keep their attention level high while performing a freely chosen mental task. In practice, they were asked to stay relaxed for a few seconds (relax epoch) and then to perform a mental task (performance epoch). It can be seen that spectral signals computed from the ongoing EEG activity differ between the two epochs. Figures 1 and 2 illustrate two spectra computed by averaging several epochs relative to the C3 and C4 electrode positions and the relax (red line) and performance epochs (green line).

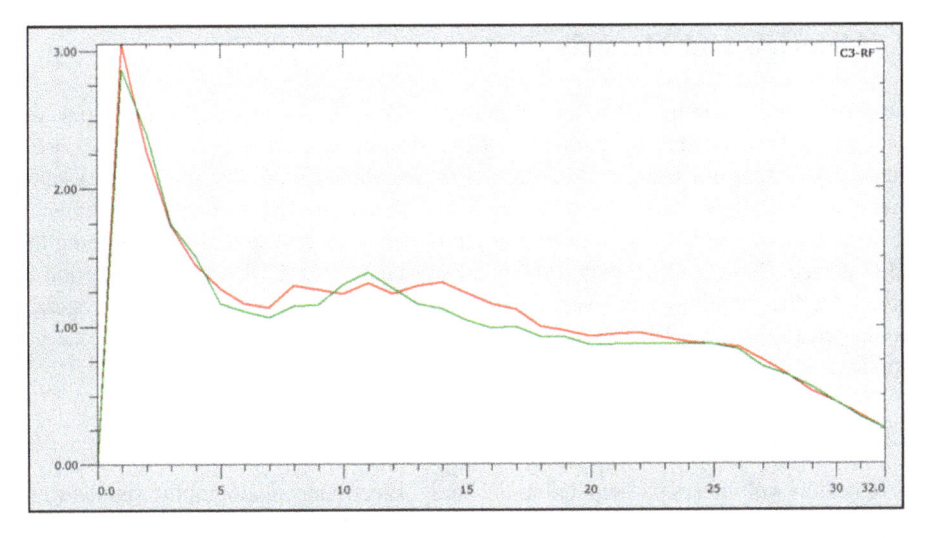

Fig. 1. Power spectrum of the C3 EEG electrode acquired during the relax (red line) and performance (motor imagery) mental tasks epochs.

The chosen mental task was motor imagery, and the spectra are consistent to it in both location (C3 and C4 collect signals from the motor cortex) and band (a reduction of the beta activity, in the 12–30 Hz frequency band, is observed as expected). The idea of this application is to train subjects to execute the mental tasks in periods whose length increases across successive recording sessions.

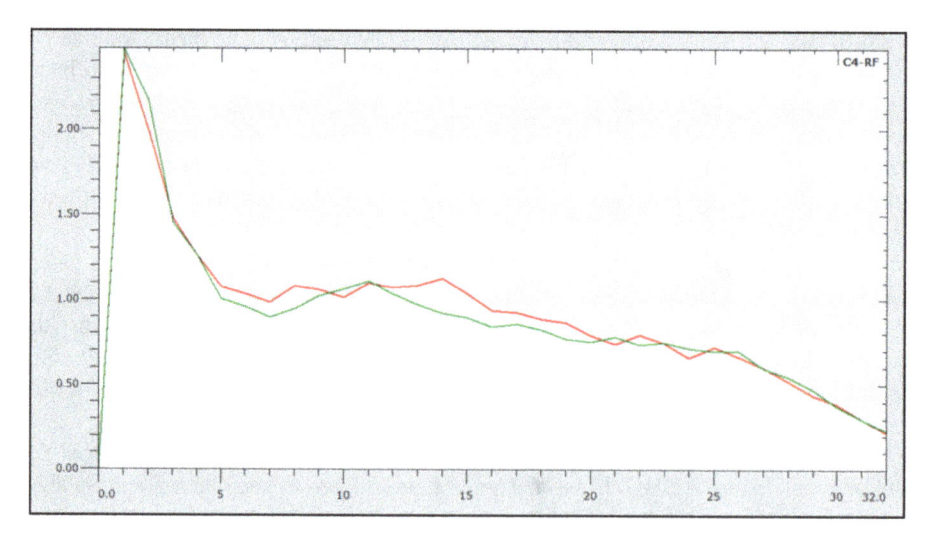

Fig. 2. Power spectrum of the C3 EEG electrode acquired during the relax (red line) and the performance (motor imagery) mental tasks epochs.

In the following paragraph, the whole experiment will be described in details.

2 Materials and Methods

Two computers running Microsoft Windows were used for this experiment: one for data acquisition, hosting a commercial EEG system, and another one for the video game. The reason for using two computers is motivated by the fact that EEG systems manufacturers allow sharing the acquired data with external applications, but they want to be sure that these lasts do not cause any problem to their system that could hang the EEG device. Usually, then, they provide a classical TCP or UDP socket connection to broadcast the acquired EEG signals to external applications hosted by a separate computer. In this way, EEG systems vendor are in some way assured that the external application could not harm their acquisition system.

2.1 Data Acquisition

Five healthy volunteers participated in the study. Electroencephalographic signals were acquired with a Mizar 40 channels EEG system by EBNeuro (Florence, Italy). Twenty-one electrodes were positioned according to the 10–20 International System, with the ground and reference electrodes positioned at the left and right mastoids respectively. The sampling rate was 1024 Hz for each sensor. Data packets were then transmitted through a plug-in mechanism, which handled a bidirectional TCP/IP connection to the videogame PC. The approximated rate of this transmission was 32 Hz, which corresponds to send a signals packet every 31.25 ms: due to the intrinsic non-real-time behavior of the TCP-IP protocol, some packet transmission could be further delayed by few more ms. This, however, only affects the visualization, and not the processing, of the feedback which can present a jitter of few ms, which is considered negligible for this kind of protocol: data, as it will be described later, are analyzed by FFTs whose length may vary from 500 to 2000 ms and while the subject is performing the same task for several seconds, so with an EEG devoid of transitions.

Finally, all data were stored in either the EBNeuro file format (on the EEG computer) or in NPX file format [13] on the videogame computer for further processing [14] and completed with additional information such as the position of the six racers, the selected, features, all timing events, etc.

2.2 The Videogame

The videogame [11] was implemented using a software framework (the BF++) aimed at implementing BCI and BioFeedback systems [12]. Written in the C++ programming language, it provides several facilities for data processing, such as filtering either in the time or in the space domain, as well as several classifiers and regressors, to name a few, are also available. Microsoft Visual Studio 2017 was used as a programming platform and all the software was written in C++ [15]. Finally, the adoption of the TCP protocol is motivated by the fact that it is minimally platform dependent so that the videogame can be easily recompiled for supporting other operating systems, thus overcoming the problem of the lack of standards in BCI [16].

In the videogame, six puppets run in a virtual race from the left to the right of the screen, on six virtual horizontal lines (Fig. 3). The speed of five of them is controlled by the computer, while the speed on the sixth one is controlled by the player and driven by the neurofeedback mechanism: the more he can perform a specific mental task, the faster its puppet.

Fig. 3. A screenshot of the videogame: 6 puppets run from left to right. Five of them are controlled by the computer while one (the forth from top) is driven by the neurofeedback.

2.3 The Protocol

Three different operating modalities were implemented: Exploring, Calibrating and Racing. The completion of each of them represents a run, and it is formed by a collection of trials. Finally, each trial is composed of three epochs:

(a) Inter Trial Interval (ITI), during which no data is processed and is provided to allow users to stretch, blink, speak, etc.;
(b) Relax, during which the user has to stay relaxed. This epoch is devoted to the collection and storage of some reference data;
(c) Performance, during which the user has to perform a mental task and whose data are then compared to those of the relax epoch.

In Fig. 4, the relationship between run, trials and epochs is illustrated, whereas in Table 1, the number of trials necessary to complete a run and the epoch's duration for each operating modality is reported.

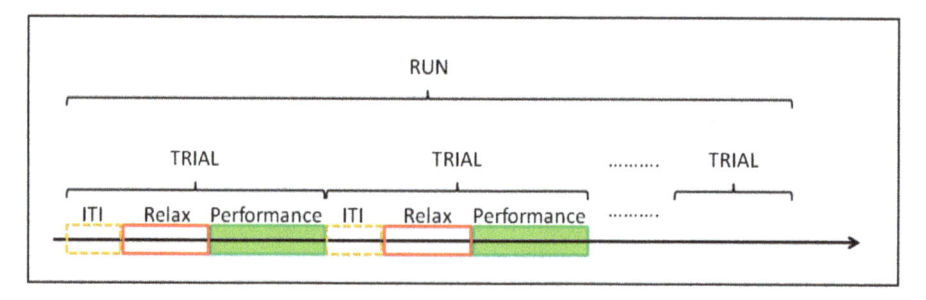

Fig. 4. A Run is formed by a collection of trials and a trial is formed by a sequence of epochs: Inter-trial Interval (ITI), Relax and Performance. During the Performance epoch, the subject performs the mental task.

Table 1. Number of trials to complete a run and epochs durations for each of the three operating modalities

	Exploring	Calibrating	Racing
Trials count	6	16	20
ITI [ms]	6000	6000	6000
Relax [ms]	5000	8000	8000
Performance [ms]	5000	8000	8000

The meaning and purposes of the three operating modalities are now illustrated:

(a) Exploring (Fig. 5): a short modality used to allow users to familiarize with the video game and to find a mental task that can be used to play the game. Users can perform several different tasks to find the one that is the most comfortable for them. Data from six relax and performance epochs (single epoch duration 5 s) are compared: a paired t-test on their respective spectral data calculated after averaging 30 FTTs computed on 1 s of signal for each channel is performed. Then, for each spectral sample and each sensor, it is determined if the chosen mental task can be distinguished from the relax on some electrodes and at certain frequencies. This was determined whenever the p-value was less than 0.05. No feedback is provided to the user, but at the end of the run, it is notified which spectral features, if any, were identified. If no features are found, users can try a different mental task until an appropriate one is recognized.

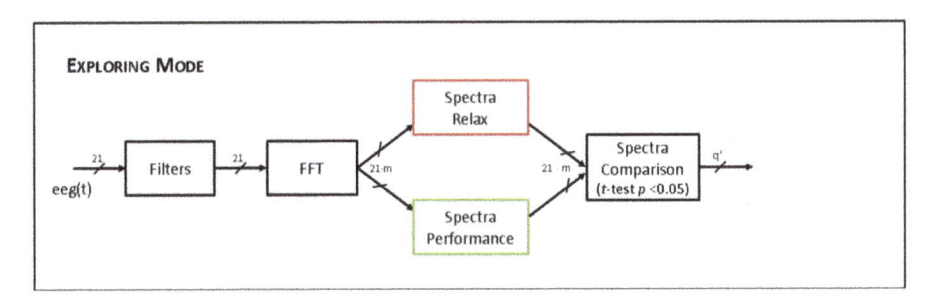

Fig. 5. Processing pipleline in the exploring modality. 21 channels are acquired and after the FFT computation m spectral features are obtained for each channel. Then, after a paired t-test ($p < 0.05$), only q' features are selected.

(b) Calibrating (Fig. 6): the mental task identified during the Exploring mode run is selected. Then, similarly to the exploring modality, spectral data are compared, but after a larger number of trials (16) and with longer epoch lengths (8 s). The selected features (after a paired t-test, p threshold value = 0.05) were used to train a regressor to establish a relationship between the spectral data collected during the relax and the performance epochs (FFT computed on 1 s window length). To reduce the probability to select noisy channels, only the spectral features in the [8–30 Hz] band were considered: eye artifacts are usually low-frequency components (<8 Hz), while EMG can be observed at higher frequencies. A multiple linear regressor is trained by assigning a value of 0 for the relax epoch and a value of 1 for the performance epoch. In this way, if future data such as those of the racing mode are similar to those of the performance epoch then a value close to 1 will be predicted, whereas if the spectral data are similar to those of the relax epoch, then a value close to 0 will be expected. In practice, the regressor will determine the weights that will be used on the selected features to predict the mental task executed by the user. Again, no feedback is provided.

46   L. Bianchi

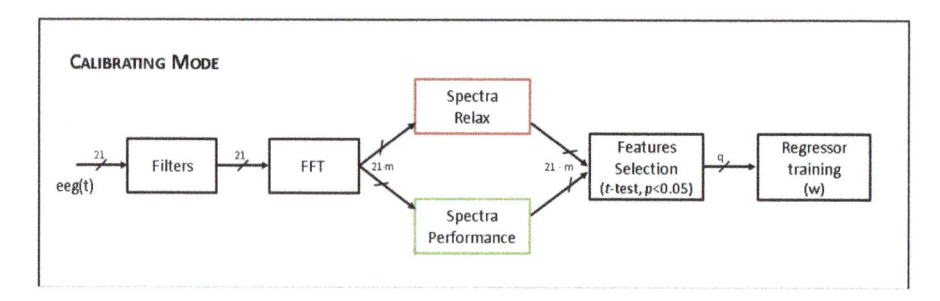

Fig. 6. Processing pipeline in the calibrating modality. 21 channels are acquired and after the FFT computation m spectral features are obtained for each channel. Only features in the alpha [8–12 Hz] and beta bands [12–30 Hz] are considered. Then, after a paired *t*-test (p < 0.05), only q features are selected to train a regressor and generate the weights vector w.

(c) Racing (Fig. 7): the regression rule computed during the calibration phase (features selection, weight matrix w) is applied to the puppet controlled by the neurofeedback mechanism: the more a user can replicate the mental task executed during the performance epoch in the calibration phase, the faster the racer.

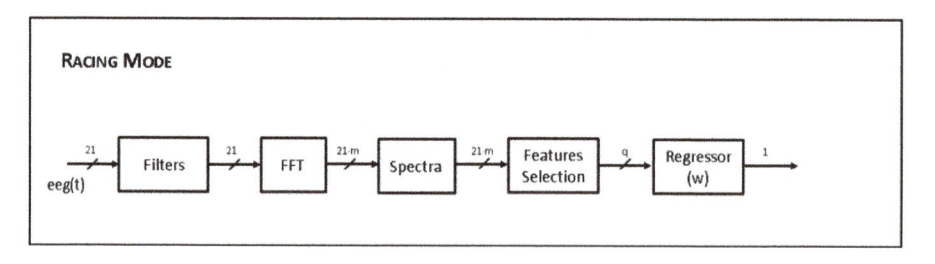

Fig. 7. Processing pipeline in the racing modality. 21 channels are acquired and after the FFT computation m spectral features are obtained for each channel. Features selected after the calibration procedure will then be used and multiplied by the w weights vector.

Finally, while the videogame was designed to process data online, to allow a reliable interaction on the player with the game, data were also processed offline, to try to investigate what are the best regression laws to use in future sessions. In this last case, then, different regression algorithms were compared.

2.4 On-Line Processing

In the online version of the game, calibration was performed by comparing spectra computed by FFT on 1 s length temporal windows. As a regression rule, a multiple linear regression on features selected after the computation of the *t*-test and belonging to the [8–30 Hz] band was performed, thus determining the weights vector w. Its size was then equal to the number of the selected features. In the racing modality, then, the spectral features were continuously computed, whenever a new data packet is received, and multiplied by the weights vector. This determines the motion of the puppet controlled by the neurofeedback.

In Fig. 8, a screenshot of a single race (a trial in the racing modality) is shown. On the right part of the screen, it is possible to (optionally) see the ongoing spectral activity (green line) compared with the mean spectra relative to the relax epoch (red curve) of four selected sensors: (C3, C4, Oz and Pz). The 11 small red dots on the red curves indicate the features selected during the calibration phase. All subjects but one asked to hide the panel because they found it distracting.

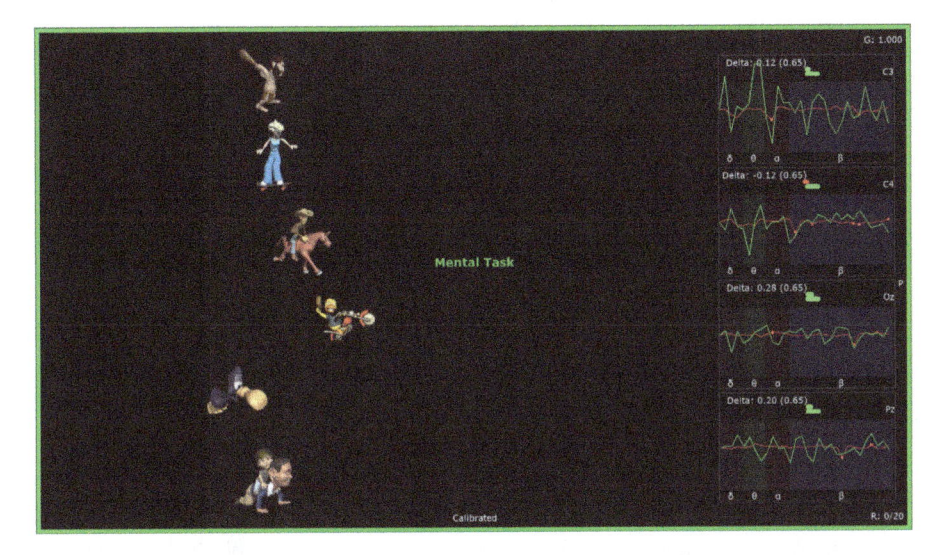

Fig. 8. Screenshot of a single race with additional information on the right pane on the ongoing spectral activity relative to 4 EEG sensors.

2.5 Off-Line Processing

In this case, in addition to the multiple linear regression, also an Artificial Neural Network (ANN) and a StepWise Linear Discriminant Analysis (SWLDA) were used. Candidate features were the same selected in the online version so that only the weights vectors were different.

3 Results

In Fig. 9, it is shown a file re-opened with the freely available NPXLab software tool [14] for reviewing the acquired data. Twenty seconds of data are shown during the Racing operating modality. In addition to the 21 acquired sensors (blue traces), it is possible to see in the last six (black) curves the variations of the positions of the racers: they start to move approximately one second after the beginning of the Performance epoch, and stop at its end. They are stored as virtual channels to keep track of the exact position of the feedback provided to the users.

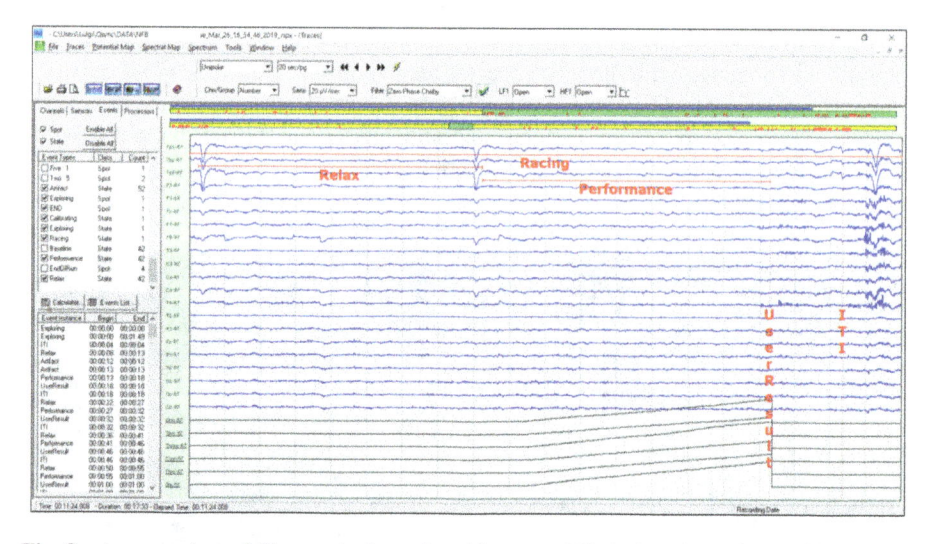

Fig. 9. A screenshot of file acquired on the videogame PC during the racing modality: it is possible to observe the presence of the position of the racers in the last six (black) traces that were stored as virtual channels in addition to the 21 (blue) EEG traces. Markers and events are also stored for offline analysis and visualized in red text.

3.1 On-Line Processing

Results relative to the online version of the game are shown in Table 2. They represent the difference, expressed in percentage, between the distance covered by NFB controlled puppet and the mean distance covered by the five puppets controlled by the PC. Note that chance level should provide a value of −50%, whereas all values are far above that threshold demonstrating that all of them were able to play the game.

Table 2. Difference among the distance covered by the NFB controlled puppet and the mean distance covered by the five puppets controlled by the PC by using the Multiple Linear Regression (MLR)

Regressor	MLR
FFT size (sec)	1.0
S1	−10.2%
S2	+12.2%
S3	+24.5%
S4	−5.3%
S5	−4.8%

3.2 Off-Line Processing

Results relative to the offline analysis are shown in Table 3. They represent the difference, expressed in percentage, between the distance covered by NFB controlled puppet and the mean distance covered by the five puppets controlled by the PC. The MLR column relative to the online performances is reported for comparing the results among the three different regressors.

Table 3. Difference among the distance covered by the NFB controlled puppet and the mean distance covered by the five puppets controlled by the PC. Two different regressors are compared with the online used Multiple Linear Regression (MLR): Artificial Neural Networks (ANN) and Step-Wise Linear Discriminant Analysis (SWLDA).

Regressor	MLR	ANN	SWLDA
S1	−10.2%	6.2%	−20.2%
S2	12.2%	7.1%	−6.8%
S3	24.5%	6.8%	45.6%
S4	−5.3%	5.9%	−39.1%
S5	−4.8%	6.5%	−25.9%
Mean	3.2 ± 14.6%	6.5 ± 0.5%	−9.8 ± 32.8%

It can be seen that in all cases results are above the chance level (−50%). ANN provides the best mean value across subjects, whereas MLR and SWLDA are characterized by a higher variability. SWLDA provides also the worst mean value. In all cases, however, the control of the videogame can be achieved in all five subjects in a limited amount of time.

4 Discussion and Conclusions

In the present work, it has been demonstrated that it is possible to learn to play a videogame through brain activity, without using the natural pathways of nerves and muscles in a few minutes.

The social impact that the proposed approach can have on the community of disabled people is also outstanding even if a lot of work remains to be done. For example trying different regressors, features selection methods, pre-processors, in the online condition of the game will be of fundamental importance as well as extending the testing to different population of players. This, however, necessitates a lot of time because the amount of different combination of methods to be tested can increase exponentially.

Also, adding multi-player support to the videogame will represent an exceptional inclusive way for making disabled people playing with the same rules with friends, relatives and acquaintances, bringing "fair-play" meaning to a new dimension.

References

1. Johnstone, S.J., Roodenrys, S.J., Johnson, K., Bonfield, R., Bennett, S.J.: Game-based combined cognitive and neurofeedback training using Focus Pocus reduces symptom severity in children with diagnosed AD/HD and subclinical AD/HD. Int. J. Psychophysiol. **116**, 32–44 (2017)
2. Bakhtadze, S., Beridze, M., Geladze, N., Khachapuridze, N., Bornstein, N.: Effect of EEG biofeedback on cognitive flexibility in children with attention deficit hyperactivity disorder with and without epilepsy. Appl. Psychophysiol. Biofeedback **41**(1), 71–79 (2016)
3. Marzbani, H., Marateb, H.R., Mansourian, M.: Neurofeedback: a comprehensive review on system design, methodology and clinical applications. Basic Clin. Neurosci. **7**(2), 143–158 (2016)
4. Nan, W., Wan, F., Chang, L., Pun, S.H., Vai, M.I., Rosa, A.: An exploratory study of intensive neurofeedback training for schizophrenia. Behav. Neurol. **2017**(2), 1–6 (2017)
5. Alves-Pinto, A., Turova, V., Blumenstein, T., Hantuschke, C., Lampe, R.: Implicit learning of a finger motor sequence by patients with cerebral palsy after neurofeedback. Appl. Psychophysiol. Biofeedback **42**(1), 27–37 (2017)
6. Reichert, J.L., Kober, S.E., Schweiger, D., Grieshofer, P., Neuper, C., Wood, G.: Shutting down sensorimotor interferences after stroke: a proof-of-principle SMR neurofeedback study. Front. Hum. Neurosci. **15**(10), 348 (2016)
7. Renton, T., Tibbles, A., Topolovec-Vranic, J.: Neurofeedback as a form of cognitive rehabilitation therapy following stroke: a systematic review. PLoS One **12**(5), e0177290 (2017)
8. Kober, S.E., Schweiger, D., Witte, M., Reichert, J.L., Grieshofer, P., Neuper, C., Wood, G.: Specific effects of EEG based neurofeedback training on memory functions in post-stroke victims. J. Neuroeng. Rehabil. **12**, 107 (2015)
9. Bennett, C.N., Gupta, R.K., Prabhakar, P., Christopher, R., Sampath, S., Thennarasu, K., Rajeswaran, J.: Clinical and biochemical outcomes following EEG neurofeedback training in traumatic brain injury in the context of spontaneous recovery. Clin. EEG Neurosci. **49**(6), 433–440 (2018)
10. Jeunet, C., Glize, B., McGonigal, A., Batail, J.M., Micoulaud-Franchi, J.A.: Using EEG-based brain computer interface and neurofeedback targeting sensorimotor rhythms to improve motor skills: theoretical background, applications and prospects. Neurophysiol. Clin. pii: S0987-7053(18)30259-4 (2018)
11. Bianchi, L.: Speedy'O'Brain: a neuro-feedback videogame driven by electroencephalographic signals. Int. J. Biol. Biomed. Eng. **12**, 229–234 (2018)
12. Bianchi, L., Babiloni, F., Cincotti, F., Salinari, S., Marciani, M.G.: Introducing BF++: a C++ framework for cognitive bio-feedback systems design. Methods Inf. Med. **42**, 104–110 (2003). ISSN 0026-1270
13. Bianchi, L., Quitadamo, L.R., Marciani, M.G., Maraviglia, B., Abbafati, M., Garreffa, G.: How the NPX data format handles EEG data acquired simultaneously with fMRI. Magn. Reson. Imaging **25**(6), 1011–1014 (2007). ISSN 0730-725X
14. Bianchi, L.: The NPXLab suite 2018: a free features rich set of tools for the analysis of neuro-electric signals. WSEAS Trans. Syst. Control **13**, 145–152 (2018). Art. #18. ISSN/E-ISSN 1991-8763/2224-2856
15. Bianchi, L.: The software model of the 'Speedy'O'Brain' neuro-feedback videogame. Int. J. Circ. Syst. Signal Process. **13**, 543–549 (2019). ISSN 1998-4464
16. Bianchi, L.: Brain-computer interface systems: why a standard model is essential on BCI standards. In: 2018 IEEE Life Sciences Conference, Montreal (2018)

Evaluation of RSSI as a Non-visual Target Tracking Technique for Drone Applications

Christopher Lee and Sudhanshu Kumar Semwal[(✉)]

Department of Computer Science, University of Colorado,
Colorado Springs, CO, USA
{clee7,ssemwal}@uccs.edu

Abstract. Visual tracking algorithms for drone applications is a common technique for tracking and following targets. However, partial or full occlusion can cause visual trackers to lose its target or even begin following the wrong target. To combat this challenge, non-visual tracking algorithms can be used in parallel to provide more robust tracking in environments where line-of-sight cannot be guaranteed. In this study, a non-visual tracking technique using RSSI (Received Signal Strength Indicator) is evaluated for its suitability in drone applications. The study results show a RMS error of 1.17 ft. for the RSSI technique within a 10 foot radius in outdoors environments, which is an improvement over the current reported accuracy of GPS. However, interference from multipath fading in indoors settings remains a significant challenge for the RSSI technique, and modifications to the RSSI technique to mitigate multipath fading are proposed as future work.

Keywords: Unreal · Action-adventure · Game design

1 Introduction

As drone technology becomes more advanced and economically feasible for general consumers, one potential application is personal security. A robotic "guardian" that can follow a target while recording video of its surroundings could act as a strong deterrent for robbers and kidnappers, particularly those targeting children and the elderly. In the event of a crime, the recorded video can be used to aid law enforcement in tracking the criminals. A significant amount of work has been done on visual tracking of arbitrary objects using machine learning and neural networks. However, continued tracking through partial or full occlusion, which can occur frequently in crowded or indoors settings, remains a challenge. Tracking a person through a crowd where people can have similar appearances and clothing poses further difficulty, particularly if there is malicious intent to confuse the algorithm. In our application, however, we have the advantage of cooperation between the drone and the target, meaning that the tracking algorithm does not necessarily need to work for arbitrary targets. For example, the target can wear a beacon device that periodically broadcasts a signal to the drone. This can lift the line-of-sight requirement between the drone and the target and make for a robust tracking algorithm in a wide variety of settings. Indoor positioning frameworks that can locate objects using various techniques exist [2, 4, 7], but our application would need to work without any specialized infrastructure in the building. Historically, GPS has been a successful method of personal tracking, and its ubiquity makes it appealing for this type of application. However, GPS tends to have poor performance indoors due to signal attenuation. Therefore, we need a method that can work in arbitrary environments. One potential technique is RSSI (Received Signal Strength Indicator), which uses the signal strength from three or more receivers to calculate the location of the signal source. Since the receivers can be mobile (i.e. drones), it could circumvent the problem of signal attenuation due to building walls. In the event it is not possible to rely on either visual tracking or non-visual tracking alone, a combination of the two may yield satisfactory results. For example, if occlusion causes the visual tracker to fail, the approximate location of the target can be calculated from the target's beacon signal to help restore visual tracking. Even when the visual tracking algorithm is working, the beacon signal can be used to verify that it has not misidentified the target.

2 Objectives

The goal of this study is to evaluate the feasibility of RSSI as a non-visual target tracking algorithm for use in drone applications. Since the application requires operation in arbitrary environments, the primary metric of this evaluation will be accuracy as a function of environmental parameters. As GPS is currently reported to have an average of 0.715 m (2.3 ft.) by the US government as of 2017 [3], this will be the standard by which RSSI will be evaluated in this study. The accuracy of RSSI will be tested in a variety of environments, such as empty outdoor open fields, indoor open rooms, and indoor hallways. Since occasional occlusion is inevitable, the signal attenuation from various common materials such as glass and fabric will also be tested.

3 RSSI Techniques and Results

A. Background

RSSI (Received Signal Strength Indicator) is simply the intensity of a received signal from some signal source. the signal intensity at a receiver is proportional to the inverse square of the distance to the source, RSSI can be used to measure this distance given the output power of the source. However, since the signal is often subject to variation due to a multitude of environmental factors such as multipath interference and signal attenuation [6, 8], distance-ranging techniques based on RSSI is often complemented by some a priori empirical data regarding the environment, especially in indoors settings [10]. Since RSSI can only be used to calculate distance to the source and not orientation, a triangulation method needs to be used to determine the relative location of the transmitter. Specifically, for relative location in n dimension, $n + 1$ receivers are needed.

B. Experimental Setup: Signal Transmitter

Bluetooth 4.0, also known as Bluetooth Low Energy (BLE), is the signal protocol used in this study as it is well-suited and thus commonly supported by modern low-energy consumption devices. BLE beacons are commercially available and inexpensive. For this study, the BC037-iBeacon sold by Blue Charm Beacons was chosen due to its relative stability in output signal strength. This beacon can transmit up to +4 dBm (2.5119 mW).

C. Experimental Setup: Signal Receiver

The Raspberry Pi 3B+ was used as the signal receiver due to its low cost and flexibility, and has been shown to have adequate processing power for RSSI data collection and processing [5]. While the Raspberry Pi 3B+ features built-in Bluetooth 4.0 hardware support and can therefore detect and communicate with BLE devices out of the box, its location on the board and close proximity to the other electrical components may cause unwanted measurement error. Furthermore, the Raspberry Pi requires some form of housing to prevent damage to exposed circuitry, especially in outdoors settings. While this would be unlikely to cause interference with normal usage, low measurement error is needed for distance-ranging RSSI techniques. Therefore, an external USB Bluetooth 4.0 adapter was also used, and RSSI noise was compared between the onboard adapter and the external USB adapter. For this study, the Panda Bluetooth 4.0 USB Nano Adapter sold by Panda Wireless was used.

4 Results and Analysis

By paying attention to the genres of games listed above, we took inspiration for the development of our game by combining play styles from the survival and the action-adventure subcategory. The initial goal was to create a game that randomly spawns enemies at specific intervals in waves. The complexity of enemy behavior increases with time. As soon as an enemy is spawned in the environment, it starts looking for the main player character and uses a set of actions to try and take down the player.

5 Methodology

Onboard adapter vs Extenal Adapto: to compare the contribution to overall system noise between the Raspberry Pi's onboard BLE adapter and an external USB adapter, RSSI measurements were taken as a function of distance in an empty outdoors grass field using the RadBeacon Dot as the signal transmitter. The beacon was set to transmit once per second at 0 dBm power. Measurements were taken for 90 s at each distance, and the mean and standard deviation of each data set are shown in Fig. 1. The external USB adapter exhibits much lower error at all measured distances. This suggests that the onboard adapter is subject to significant interference from signal reflection within the

Raspberry Pi's plastic housing, which is also supported by the overall higher measured RSSI values. There may also be interference from other nearby components on the circuit board. As the USB adapter is external to the case and separated from the other circuitry, it is not subject to such interference. For drone applications, this suggests that the signal receiver should be a separate component located on the outside of the drone hull, unobscured by any drone housing. This also means the signal receiver will need to be weatherproofed separately unless a suitable, non-interfering housing material can be used. At distances greater than 10 ft., the system begins to suffer significant data loss, so 10 ft. is considered the effective range of this hardware configuration.

A. **Outdoors vs Indoors**

Based on the results in Fig. 1, the error appears to be increase with distance, with the exception of the data at 10 ft. - this may be because −90 dBm is considered to be the noise floor for wireless signals and data points under this threshold are likely to be lost, artificially capping the lower range to −90 dBm. We can see that within the beacon's range, the maximum standard deviation at any distance is 1.87 dBm, whereas the minimum difference in RSSI between 1-foot intervals is 1.6 dBm. This gives us an RMS error of 1.17 ft., which is lower than the reported RMS error of GPS (2.3 ft.). Therefore, the RSSI-positioning technique appears to be a viable competitor to existing outdoors tracking techniques.

To characterize the RSSI-positioning technique in indoors environments, the same experiment was performed in both a conference room and a hallway. In both cases, the area was unoccupied by people and the environment remained constant. This data is shown in Fig. 2. We can see that the overall system error is much larger than in the outdoors data and the curves are much farther removed from the expected inverse square shape, implying a much higher noise floor due to multipath fading. Furthermore, the data sets in two different indoors environments differ significantly, likely due to a combination of high overall system error and different multipath characteristics. Therefore, outside very short ranges (less than 1 foot), the basic RSSI-positioning technique does not appear to be viable indoors. Mitigating of multipath fading is a field that has seen significant research, and there exist techniques that may allow RSSI-positioning to be more viable indoors. While their evaluation and implementation are not a part of this study, some of these techniques will be briefly discussed in the Future Work section.

6 Occlusion

While resistance to occlusion is one of the perceived strengths of a non-visual tracking method such as RSSI, there is still some expected signal attenuation if there are physical obstacles blocking the signal path, particularly from materials that strongly reflect or absorb within the Bluetooth frequency range. To better understand how occlusion might affect signal accuracy, the beacon was placed inside containers of various materials to test their effect on the mean RSSI. A baseline measurement was taken in open air using the Panda Bluetooth 4.0 USB Nano Adapter and the BC037-iBeacon at a distance of 2 ft. The beacon was then placed in enclosed cylindrical containers with 0.125 in. thick walls of the following materials: glass, plastic (polyethylene), and metal (stainless steel). The beacon was also wrapped in 0.5 in. of fabric (cotton) and grasped in a human hand. The resulting attenuation (difference in mean RSSI over 90 s) is shown in Fig. 3.

Unsurprisingly, the highest attenuation is from metal, which is well-known to be highly reflecting of wireless signals. Glass, on the other hand, shows a slight amplification instead of attenuation - there may be a slight focusing effect from the container shape combined with weak attenuation of the signal. It should be noted that differently manufactured glass, such as thermal windows, may exhibit different characteristics due to differences in composition and processing. Humans and fabric, which can be expected to make up a significant portion of occlusion for our applications (crowds of people), show a moderate amount of attenuation. If combined with a visual tracking algorithm, a sudden drop in RSSI may be able to confirm that occlusion has occurred. Since RSSI also recovers quickly from occlusion, this can prevent the visual tracker from mistakenly tracking another target as well as assist in recovering the original target. Additionally, the signal attenuation from occlusion can be beneficial for maneuvering purposes - if an obstacle that is transparent to light but not RF is present between the drone and the target, such as a clear plastic window, the discrepancy between the reported distances from the visual and non-visual tracking algorithms can help determine the presence of the obstacle, which may be difficult with a visual tracking algorithm alone.

The signal attenuation from occlusion can also potentially be used to our advantage in another situation. Typically, triangulation to determine the relative location of the transmitter is implemented with distantly spaced receivers to improve accuracy. For drone applications, however, this would not only require multiple drones per target, increasing cost and clutter, each drone would need to track all other drones' positions, increasing processing time and overall system error. Therefore, all receivers required for triangulation would ideally be located on a single drone. The close proximity of the receivers decreases accuracy as the expected RSSI difference between receivers is small, causing noise to have a larger effect. However, if the receivers are distributed evenly around the drone body, it would be expected that for any given beacon position, at least one receiver would experience signal attenuation due to occlusion with the drone body itself. Since this would cause a noticeable difference in RSSI, larger than would be expected from a difference in distance alone, this can be used to determine orientation while using RSSI values from the other receivers to determine distance. Experiments with multiple Raspberry Pi receivers operating simultaneously 1 foot apart showed negligible difference in RSSI to isolated receivers, suggesting that placing multiple receivers on a single drone would not cause inter-sensor interference.

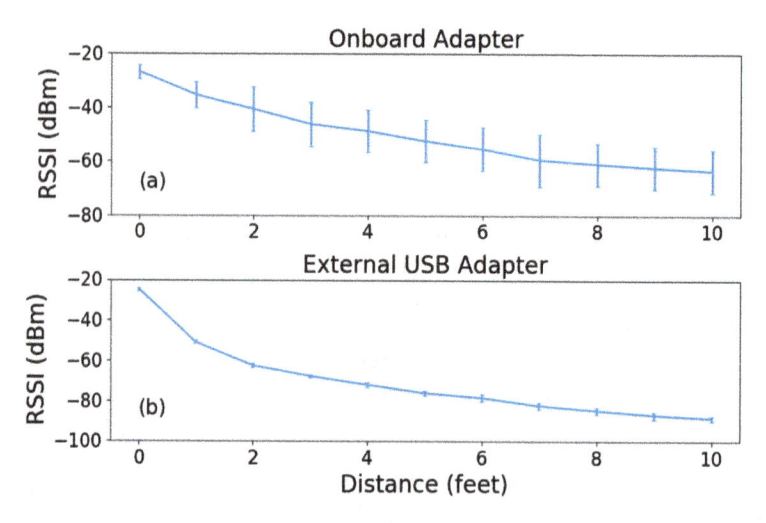

Fig. 1. Mean measured RSSI values from a RedBeacon Dot beaconover 90 s as a function of distance in an empty outdoors grass field. Error bars are given as standard deviation. (a) Raspberry Pi 3B+ onboard Bluetooth 4.0 adapter. (b) Panda Bluetooth 4.0 USB Nano adapter.

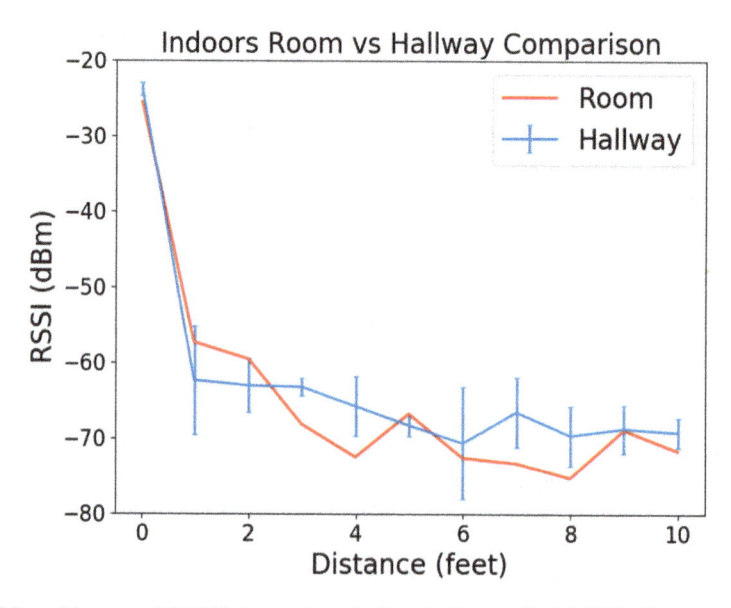

Fig. 2. Mean Measures RSSI Values using the Panda Bluetooth 4.0 USB Nano Adapter and BC037-iBeacon over 90 s as a function in indoor environments. Errors bars are shown on only one data set for clarity.

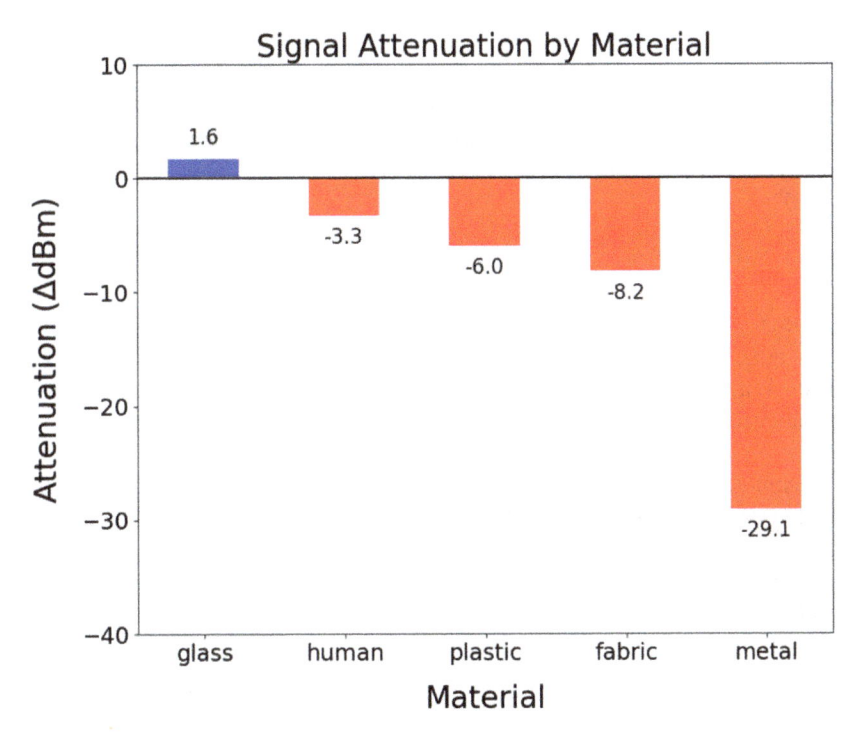

Fig. 3. Measures signal attenuation (difference in RSSI from an open baseline measurement) by obstacles of various materials.

7 Future Work

The primary shortcoming of the RSSI technique is, as expected, multipath fading in indoors settings. Mitigation of multipath fading has seen significant research due to its application for general wireless networks and not just indoors positioning systems, and techniques involving filtering [1] or signal modulation [9] have been proposed. A relatively simple technique that was proposed in 2008 by Pu et al. specifically for RSSI applications [6] may show promise for our particular application. Since the multipath fading should be the same regardless of signal power, the authors propose alternating between high and low power output signals. Since the difference between the two signals should be constant, the amount of noise in a measurement can be estimated by comparing the actual difference from the expected difference. Performing filtering based on this criteria showed noticeably improved stability in RSSI readings, and the computation is light enough to be suitable for resource-limited applications such as ours. Usage of this technique would require a more complex beacon capable of programmatically altering its output signal characteristics in real-time. This may require the design and fabrication of a custom beacon. While this would represent an investment of time and cost, it would also allow for greater control in output signal error, which was found to vary significantly between commercial products. Since this value is not often reported, a custom beacon with known properties would improve system performance and reliability. Finally, combining the RSSI technique with a visual tracking technique can have several benefits in handling occlusion, as described in the Occlusion section. Therefore, a complete implementation would likely involve some coupling of these techniques, which would require control algorithms that use the combined data to make decisions.

8 Conclusion

In this study, RSSI with a Bluetooth signal was evaluated as a potential non-visual tracking technique for drone applications in arbitrary environments. Results showed that RSSI is a viable technique for outdoors applications, demonstrating comparable or better performance to existing techniques such as GPS. The basic technique is not well-suited for indoors applications due to high noise from multipath fading, but a modified RSSI technique using more sophisticated filtering may show improved performance. For best system performance, an externally located receiver and a transmitter with low signal variation are required. Testing of signal attenuation due to occlusion showed that such signal variation may actually be advantageous in some situations, and can be used for applications such as determining orientation to the target and occlusion detection for light-transparent barriers.

References

1. Ali, Z., Landolsi, M.A., Deriche, M.A.: Comparison of derivative based and derivative free Kalman filters for multipath channel estimation in CDMA networks. In: 2009 5th International Conference on Wireless Communications, Networking and Mobile Computing, pp. 1–4 (2009)
2. Chen, W., Kao, W., Chang, Y., Chang, C.: An RSSI-based distributed real-tile indoor positioning framework. In: 2019 IEEE International Conference on Applied System Invention (ICASI), pp. 1288–1291 (2018)
3. GPS.gov.2017.GPS Accuracy (2017)
4. Herrera, J.C.A., Ploger, P.G., Hinkenjann, A., Maiero, J., Flores, M., Ramos, A.: Pedestrian indoor positioning using smartphone multi-sensing, radio beacons, user positions probability map and IndoorOSM floor plan reprsenation. In: 2014 International Conference on Indoor Positioning and Indoor Navigation (IPIN), pp. 636–645 (2014)
5. Jais, M.I., Ehkan, P., Ahmad, R.B., Ismail, I., Sabapathy, R., Jusoh, M., Rahim, H.A., Malek, M.S.: Hardware comparison capturing received signal strength indication (RSSI) for wiresless sensors network (WSN). In: 2015 IEEE Student Conference on Research and Develoment (SORReD), pp. 278–282 (2015)
6. Pu, C., Chung, W.: Mitigation of multipath fading effects to improve indoor RSSI performance. IEEE Sensor J. 8(11), 1884–1886 (2008)
7. Song, J., Jeong, H., Jiang, Y., Zhang, L., Park, Y.: Improved indoor position estimation algorithm based on geo-magnetism intensity. In: 2014 International Conference on Indoor Positioning and Indoor Navigation (IPIN), pp. 741–744 (2014)
8. Xu, L., Fang, Y., Jiang, Y., Zhang, L., Feng, C., Bao, N.: Variation of received signal strength in wireless sensor network. In: 2011 3rd International Conference on Advanced Computer Control, pp. 151–154 (2011)

9. Yu, H., Cho, H., Kang, C., Hong, D.: A new multipath interference mitigation techniques for high-speed packet transmission in WCDMA downlink. IEEE Signal Process. Lett. **12**(9), 601–604 (2005)
10. Zhang, K., Zhang, Y., Wan, S.: Research of RSSI indoor ranging algorithm based on Gaussian – Kalman linear filtering. In: 2016 IEEE Advanced Information Management, Communicates, Electronic and Automation Control Conference (IMCEC), pp. 1628–1632 (2016)

Power Efficient Multi-relay Cooperative Diversity in Wireless Network Using Hybrid Relaying Protocol

Shital Joshi[✉] and Malaykumar Shitalkumar Bhakta

Oklahoma State University, Stillwater, OK 74078-1053, USA
Shital.Joshi@okstate.edu, mbhakta@ostatemail.okstate.edu

Abstract. Cooperative diversity technique using relays can improve the reliability of the network as well as improve the performance in the event of fading. For systems employing such cooperative diversity technique, outage probability and transmission power are very important performance measure. This paper uses multiple relays employing both Amplify-and-forward (AF) and decode-and-forward (DF) as a relaying protocol to compute the outage probability of the network and formulates an optimization problem to minimize the overall transmit power of the network. Since the cooperation can be costly for the nodes which act as the relay node, it is very important that the transmission power of the relay nodes are also within the bound. The results show that using both AF and DF are advantageous over using only AF or DF as a relay protocol in a multi-relay cooperative network. The overall transmit power of the HDAF network is 17% and 4% less as compared to that with only AF or DF protocols, respectively. Similarly, the transmit power of the source node decreases by 14% and 6.4% as compared to that with only AF or DF protocols, respectively.

Keywords: Cooperative communication · Power allocation · Outage probability · Multi-relays cooperation · Hybrid decode-amplify-forward relaying

1 Introduction

The number of users in mobile radio have been increasing significantly over the past decade and so is the demand for higher speed and larger bandwidth. These requirements have placed a stringent requirement on the frequency usage and the transmission power. With the advent of 5G services, it is possible to provide a maximum peak download speed of up to 20 Gbps ($\sim$ 20 times faster than 4G network), 10 times lower latency than 4G network and 100% connectivity: all with subsequent reduction in the energy consumption ($\sim$ 90% less energy) as compared to 4G network [1]. However, multipath fading can be one of the major limiting factors. Often multipath fading is considered as a negative phenomenon in wireless communication which makes wireless links highly unreliable by hindering proper radio communication. Due to the motion in the vicinity of the wireless nodes or the presence of obstacle in the line-of-sight (LOS) path, the transmitted signal from the radio transmitter divides and takes multiple paths to reach to the receiver node. So, when the receiver receives multiple copies of the transmitted signal, these signals are often out-of-phase with each other. If these received signals are out-of phase then it can cause problems with phase distortion and inter-symbol interference, which results in overall signal deterioration. It is hence very important to account for multipath fading, when the ultimate goal is to provide seamless connectivity of wireless devices especially for a 5G network where the reliability of the higher speed communication is vital.

Multiple input multiple output (MIMO) [26] and Massive MIMO [4] are capable of providing high performance gain in the presence of multipath fading, but due to the small form factor and limited power supply of the terminal nodes, their implementation is not practically feasible. Several techniques have been developed under time, frequency and spatial diversity which can be applied at either the transmitter side or the receiver side. With little to no correlation between the diversity branches, the network can achieve a significant performance gain. Recently cooperative communication have gained a lot of attention due to its ability to mitigate fading in wireless networks. In cooperative communication, a number of nodes are selected which act as a relay station. These nodes help the transmitting node by forwarding the data they receive to its destination, thereby forming a virtual antenna array. It is particularly beneficial in small terminal devices like in IoT where it is difficult to mount multiple antennas to mitigate fading. This technique effectively combats the adverse effect of multipath fading, path loss, shadowing to provide large coverage, throughput, system capacity and power [22].

Power consumption of an individual node as well as the overall network have become one of the most important design aspect and the design constraints for the future communication systems. In 2007, Information and Communication Technologies (ICT) contributed to around 1.3% of the global greenhouse gas emissions (GHGE) with global electricity consumption of 3.9% [24]. The footprint of ICT is continuously growing and with the advent of 5G networks, it will expand even more. As the number of ICT devices increases, if unchecked, the global greenhouse gas emissions (GHGE) from ICT could exceed 14% of 2016-level by 2040 [5]. However with the smart application, it is estimated in [23] that ICT could infact reduce the global GHGE by around 1–4% by 2030. Apart from the environmental aspect, power consumption is one of the most important design aspect for cooperative network. In the traditional centralized communication system, all the communications route through the centralized base station, which is not constrained with limited power supply. However, in the distributed networks like cooperative communication system, one or more user terminals can act as relay stations for the source node. The nodes in these system can be battery-constrained, and it may not be convenient, feasible or even practical to charge or replace the battery. Hence, relaying can be quite expensive for nodes in terms of its lifetime and can significantly decrease the overall network service time in the applications like Internet-of-Things (IoT) [2]. One of the potential solution for that can be to harvest radio frequency energy so that these devices can be made self-powered and eliminate any battery dependencies [11,36]. This paper, however, considers that each node is battery powered and hence are subjected to a limited power source.

In the cooperative communication network, the transmitter (i.e. the source node, S) communicates with the receiver (i.e. the destination node, D) in two basic phases. In the first phase, also called the broadcast phase, the source node broadcasts the signal, which can be heard by all the neighboring nodes. Out of these several neighbor nodes, some are selected as a relay nodes R. In the second phase, also called the relaying phase, these relay nodes first process the received signal depending upon the strategy they employ then forward the processed signal to the destination node. Relay operation technique can be categorized into two modes: half duplex (HD) and full duplex (FD) relaying mode. The relays receive and forward the received signal in orthogonal time slots (i.e. orthogonal channel), if HD mode is employed. If FD mode is used then the relays can receive and forward the signal at the same time. HD mode utilizes twice the channel resources as compared to direct transmission (DT) mode and hence is more spectral inefficient as compared to FD mode. However, the spectral efficiency in FD comes at the cost of increased self-interference due to self-induced signal leakage at the input and output of the relay. This degrades the system performance. The advantage of cooperative communication becomes evident at the destination when it retrieves the data. When the source-destination link experience noise or extreme fading, then the destination may not be able to decode the data correctly. With the cooperative setting, the destination can make use of the additional information from the relay and helps in improved decoding.

Two of the most common cooperation strategies employed during the relaying phase are decode-and-forward (DF) and amplify-and-forward (AF). In DF, the relay node decodes, re-modulates and re-transmits the received signal. This implies that each relay nodes need to have a full processing capability, just like a base station (BS) and hence can be expensive in terms of receiver complexity as well as the processing power requirement. In the event when the relays fail to correctly decode the source information, the decoding error may propagate to the destination [19]. In the case of AF, the relay node amplifies the received signal and re-transmit forward without decoding. So, AF is computationally not as expensive as DF. However, when AF amplifies the received signal, it amplifies the noise as well and decreases the system performance [6]. The advantage of AF is the soft information representation while that of the DF is the coding gain [3].

This paper considers the multi-relay cooperative system employing dual cooperation strategy. One relay node employs DF while another employs AF such that the network can obtain the advantage of both AF as well as DF. Since it is equally important to efficiently combine the diversity signals at the receiver to take advantage of the diversity gain obtained from the diversity scheme, maximal-ratio combining (MRC) is used at the receiver to combine the signals from diversity paths. The unique contributions of this paper are:

1. Multi-relay cooperative system employing dual cooperation strategy to forward the received source signal.
2. Computation of outage probability analysis for a dual cooperation strategy.
3. Formulation of an optimization problem to minimize the total power consumption of the overall network with a minimum outage constraint and per relay power limit.

To the best of authors' knowledge, this is the first time network performance analysis for such a system is being analyzed.

The remaining of the paper is organized as follows: Sect. 2 covers a brief review of related work. System model is described in Sect. 3, followed by the outage probability analysis in Sect. 4. Section 5 formulates the power optimization problem considering the outage probability of the network. Section 6 discussed the Simulation results. Conclusions and direction for future work are given in Sect. 7.

2 Related Work

Relay network forms the basis of cooperative diversity and this section covers some of the relevant literature on this area. There are several cooperative diversity techniques proposed and analyzed for wireless networks. In [17], four low-complexity cooperative diversity protocols: amplify-and-forward, decode-and-forward, selection relaying scheme and incremental relaying scheme were analyzed for delay constrained scenario where the terminals were restricted to half-duplex transmission and channel state information (CSI) considered available only at the receiver. It has been shown that AF achieves full diversity of order two with one relay node based on the analysis from outage probability. In [32,33], closed-form symbol error rate (SER) formulation have been derived for single-relay DF and AF cooperative techniques respectively. SER for multi-node DF scheme has been considered in [28]. Even though the SER improves with the increase in the number of relays, [28] showed that it reduces the bandwidth (BW) efficiency as the degree of freedom decreases to transmit the information.

Recently hybrid cooperation protocol has gained much interest due to its optimum performance. In [25], hybrid decode-amplify-forward (HDAF) relaying protocol have been investigated for multiple parallel relays and closed-form bounds for outage probability and BER have been derived. In [21], the hybrid relay selection (HRS) scheme have been analyzed for the frame error rate (FER), where the relay node is chosen between the two strategies (AF and DF) to forward the source information. In [9], a hybrid decode-amplify-forward incremental cooperative diversity protocol have been proposed using signal-to-noise power ration (SNR)-based relay selection and the expressions for outage probability and average channel capacity have been derived. In [10], the performance for hybrid decode-amplify-forward relaying protocol for underlay cognitive radio [14] have been studied.

Power and resource allocation for multinode DF scenario was considered in [18]. In [30] outage probability for multinode AF relay network was considered. In [15], multinode AF cooperative network is considered to perform power allocation for minimization of outage probability. Relay nodes are feedback with the channel information for optimum power allocation. In [34], power and bandwidth are allocated to sources and relays for cooperative AF and DF network by a centrally controlled access point or the base station. In [38], the joint optimization of relay selection and power allocation for orthogonal multiuser systems using AF relay is studied. [35] proposed the joint optimization of relay selection and power allocation for DF uplink. Author in [37] considered a dual-hop multi-relay cooperative system for joint power allocation and relay position optimization to minimize system outage. Considering the node power spent during relaying action, simultaneous wireless information and power transfer (SWIPT) technique was proposed in [36] where a transmitted signal carries the source information as well as used to power the IoT devices. The result showed an increase in the spectral efficiency of the network.

One of the critical decision in cooperative communication is the selection of node for relay. Several protocols have been proposed in the literature. In [12], multiple relay selection scheme have been proposed based on the signal-to-noise power ratio (SNR). Outage probability for dual-hop [13] cognitive AF relay network has been derived in [8] which concluded that the diversity order is defined by the fading severity between the two hops in the secondary network and is independent of the primary network.

3 System Model

A basic relay communication network as shown in Fig. 1 (a) is considered, which consists of a source node (S), a destination node (D) and two relay nodes (R_1 and R_2). The network topology is considered to be static, deterministic where source and destination are considered to be one hop away i.e. the information from source can directly reach to the destination and via relays. Two relays are used to forward the received signals to the destination where one relay uses DF protocol and another uses AF protocol. Both relays operate in half duplex mode i.e. the source transmits signal to both relays and destination during the first time slot and during the second time slot, the relays forward the information to the destination. The source node do not transmit new information during the second time slot. A perfect channel state information (CSI) is assumed to be available at the destination terminal. It is thus possible to perform the channel estimation required for MRC to combine the signals from the diversity branches and is shown in Fig. 1(b).

It is assumed that the wireless channel is subjected to Rayleigh frequency-flat fading and the coherence time is considered to be long as compared to the time required by the source to complete one block of information transfer. During the

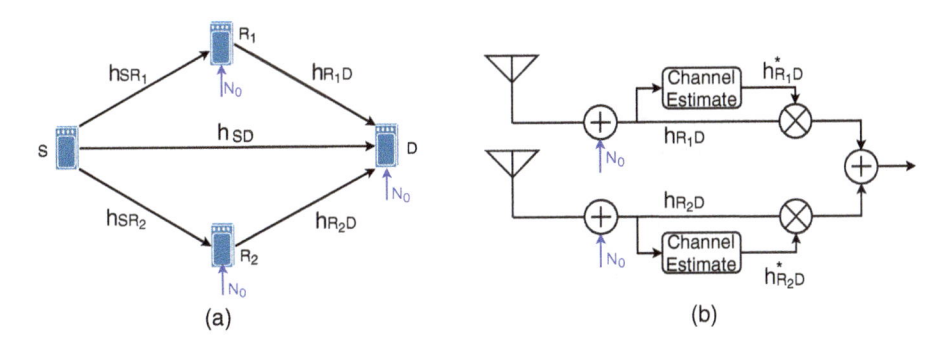

Fig. 1. (a) Two relays cooperative communication model (b) MRC at the destination [16]

broadcast phase of the transmission, the signal received at the destination from the source is [20]:

$$y_{SD} = \sqrt{P_S} h_{SD} x + \eta_{SD} \quad (1)$$

The instantaneous SNR at the destination is given by:

$$\gamma_{SD} = \frac{P_S |h_{SD}|^2}{N_0} \quad (2)$$

while the signal received at the relay R_i is:

$$y_{SR_i} = \sqrt{P_S} h_{SR_i} x + \eta_{R_i} \quad \forall i \in \{1, 2\} \quad (3)$$

where P_S is the transmit power of the source, x is the signal transmitted by the source, the channel gain h_{ij} ($\sim \mathcal{CN}(0, \delta_{ij})$) is a complex channel gain with mean zero and variance δ_{ij} from the link $i \to j$. η_{R_i} is the additive white Gaussian noise at node i with mean zero and variance N_0. It is assumed that the channel gains and noise are independent.

During the relaying phase of transmission, the signal received at the destination from each relay is given by:

$$y_{R_i D} = \sqrt{P_{R_i}} h_{R_i D} \hat{x}_i + \eta_{R_i D} \quad \forall i \in \{1, 2\} \quad (4)$$

where P_{R_i} is the transmit power at the ith relay node, $\hat{x}_i$ is the processed data at the ith relay based upon the selected cooperative relaying scheme. Similarly, $\eta_{R_i D}$ is the additive white Gaussian noise (AWGN) with zero mean and variance of N_0 at $R_i \to D$ link.

In the following subsection, various cooperation relaying scheme will be discussed. In addition to DF and AF, Hybrid Decode-Amplify-Forward (HDAF), which combines the benefits of both DF and AF is discussed. For the remainder of the paper, AF network refers to a network where both relays employ AF protocol for relaying the source information and DF network refers to when both relays employ DF protocol.

3.1 Decode and Forward (DF)

When both the nodes employ DF relaying scheme, the relays first decode the received signal, encode it and then forward it to their destination. The relays can correctly decode the received signal only if the received signal-to-noise ratio is above a certain threshold value in which case $\hat{x} = x$. Thus, the final signal received at the destination node through relay R_i is:

$$y_{DF_i} = \sqrt{P_{R_i}} h_{R_i D} x + \eta_{R_i D} \tag{5}$$

The instantaneous SNR at the destination using DF due to relay R_i is given by:

$$\gamma_{DF_i} = \frac{P_{R_i} |h_{R_i D}|^2}{N_0} \tag{6}$$

3.2 Amplify and Forward (AF)

When both the nodes employ AF relaying scheme, the relays first amplifies the received signal and then forward it to their destination. In the process of amplifying the received signal, it also amplifies the noise signal. The final signal received at the destination node through the relay R_i is:

$$y_{AF_i} = \sqrt{P_{R_i}} h_{R_i D} \beta_i + \eta_{R_i D} \tag{7}$$

where β_i is the amplification gain due to AF at i^{th} relay node and is given by [27]:

$$\beta_i \leq \sqrt{\frac{1}{P_S |h_{SR_i}|^2 + N_0}}, \quad i \in \{1, 2\} \tag{8}$$

The instantaneous SNR at the destination using AF dueto relay R_i is given by [31]:

$$\gamma_{AF_i} = f\left(\frac{P_S |h_{SR_i}|^2}{N_0}, \frac{P_{R_i} |h_{R_i D}|^2}{N_0}\right) \tag{9}$$

where, $f(x, y) = \frac{xy}{1+x+y}$.

3.3 Hybrid Decode-Amplify-Forward (HDAF)

In hybrid decode-amplify-forward (HDAF) scheme, the system can take the advantage of both AF and DF scheme. Like AF scheme, it can have the merit of soft information representation as well as the coding gain of DF [22]. In this paper, the two independent relay nodes adopt the HDAF cooperation, where one relay node adopts AF cooperation while another relay node adopts DF cooperation mode. If the relay node cannot decode the source data correctly then it is converted to direct transmission mode. Under this strategy, the signal received at the destination is given by:

$$y_{HDAF_1} = \sqrt{P_{R_1}} h_{R_1 D} \beta_1 + \eta_{R_1 D} \tag{10}$$

$$y_{HDAF_2} = \sqrt{P_{R_2}} h_{R_2 D} \hat{x} + \eta_{R_2 D} \tag{11}$$

The SNR at the output of MRC is the sum of SNRs on each branch.

Since the MRC output signal (r) is a sufficient statistics for the original transmitted source signal (x), the mutual information between the received signal (y) and x is equal to the mutual information between the x and r [7] i.e.,

$$I(x;r) = I(x;y) \tag{12}$$

The average mutual information then satisfies:

$$I_{DF} \leq I(x;r) \tag{13}$$

4 Outage Probability Analysis

This section provides the analysis of outage probability for HDAF protocol employing different cooperation strategy. An outage occurs in a network when the maximum mutual information of the source and the destination falls below the spectrum utilization, R.

4.1 Decode and Forward (DF)

The probability that the relay nodes correctly decode the source information for a given R is given by [29]:

$$\begin{aligned} P_c &= Pr\left\{\tfrac{1}{3} I_{SR_i} > R\right\} \qquad \forall i \in \{1,2\} \\ &= Pr\left\{\tfrac{P_S |h_{SR_i}|^2}{N_0} > 2^{3R} - 1\right\} \qquad \forall i \in \{1,2\} \end{aligned} \tag{14}$$

Let the random variable, $\omega_i = \frac{P_S |h_{SR_i}|^2}{N_0}$, then it is an exponential random variable with rate, $\lambda_i = \frac{N_0}{P_S \delta_{SR_i}^2}$. Then, the Eq. (14) can be obtained as:

$$P_c = exp\left(-\frac{N_0 (2^{3R} - 1)}{P_S \delta_{SR_i}^2}\right) \qquad \forall i \in \{1,2\} \tag{15}$$

When the relays adopt DF as a relay protocol, then there can be four cases of network interruption:

1. both relay nodes correctly decode the source information (i.e. $\hat{x} = x$) but the destination node fails to decode the relay as well as the source information,
2. out of two relay nodes, the relay node R_1 fails to decode the source information while R_2 correctly decodes the source information. However, the destination node fails to decode the R_2 information,

3. out of two relay nodes, the relay node R_2 fails to decode the source information while R_1 correctly decodes the source information. However, the destination node fails to decode the R_1 information,
4. both the relay nodes fail to decode the source information and the destination node fails to decode the source information.

The probability of the network interrupt can be expressed as:

$$P_{DF}^{outage} = (1 - P_{d_1})(1 - P_{d_2}) Pr\left(\frac{1}{3}log_2\left(1 + 3\gamma_{SD}\right) < R\right)$$

$$+ (1 - P_{d_1}) P_{d_2} Pr\left(\frac{1}{3}log_2\left(1 + 2\gamma_{SD} + \gamma_{DF_2}\right) < R\right)$$

$$+ (1 - P_{d_2}) P_{d_1} Pr\left(\frac{1}{3}log_2\left(1 + 2\gamma_{SD} + \gamma_{DF_1}\right) < R\right)$$

$$+ P_{d_1} P_{d_2} Pr\left(\frac{1}{3}log_2\left(1 + \gamma_{SD} + \gamma_{DF_1} + \gamma_{DF_2}\right) < R\right) \quad (16)$$

At high SNR, it is easy to show:

$$1 - exp\left(-\frac{(2^{3R} - 1)N_0}{P_S \delta_{SR_i}}\right) \approx \left(\frac{(2^{3R} - 1)N_0}{P_S \delta_{SR_i}}\right) \quad (17)$$

$$exp\left(\frac{(2^{3R} - 1)N_0}{P_S \delta_{SR_i}}\right) \approx 1 \quad (18)$$

On solving, Eq. (16) can be re-written as:

$$P_{DF}^{outage} = \frac{g(R)}{P_S \delta_{SD}}\left[\frac{1}{3 P_S^2 \delta_{SR_1} \delta_{SR_2}} + \frac{1}{4 P_S P_2 \delta_{SR_1} \delta_{R_2 D}} + \frac{1}{4 P_S P_{R_1} \delta_{SR_1} \delta_{R_1 D}} + \frac{1}{6 P_{R_1} P_{R_2} \delta_{R_1 D} \delta_{R_2 D}}\right] \quad (19)$$

where, $g(R) = [(2^{3R} - 1)N_0]^3$

4.2 Amplify and Forward (AF)

The maximum mutual information between the source and the destination when both the relays use AF protocol is given by:

$$I_{AF} = \frac{1}{3}log_2(1 + \gamma_{SD} + \gamma_{AF_1} + \gamma_{AF_2}) \quad (20)$$

where γ_{AF_1} and γ_{AF_2} represent SNR of the two relay links in the cooperative AF protocol.

The system's outage probability in this case is given by:

$$P_{AF}^{outage} = Pr(I_{AF} < R)$$

$$= Pr\left(\frac{P_S|h_{SD}|^2}{N_0} + f\left(\frac{P_S|h_{SR_i}|^2}{N_0}, \frac{P_{R_1}|h_{R_1D}|^2}{N_0}\right)\right.$$

$$\left. + f\left(\frac{P_S|h_{SR_2}|^2}{N_0}, \frac{P_{R_2}|h_{R_2D}|^2}{N_0}\right) < 2^{3R} - 1\right)$$

$$= \frac{g(R)}{3P_S\delta_{SR_1}}\left(\frac{1}{P_S\delta_{SR_1}} + \frac{1}{P_{R_1}\delta_{R_1D}}\right)\left(\frac{1}{P_S\delta_{SR_2}} + \frac{1}{P_{R_2}\delta_{R_2D}}\right) \quad (21)$$

4.3 Hybrid Decode-Amplify-Forward (HDAF)

In dual-relay HDFA cooperative communication, one node i.e. relay node R_1 adopts AF protocol while another node R_2 adopts DF protocol to forward the received source signal. If the relay node R_2 fail to decode the source information correctly then the source signal re-transmits the signal to the relays R_2. Thus the network's outage probability is given by:

$$P_{AF}^{outage} = P_{c_2}Pr\left(\frac{1}{3}log_2(1 + \gamma_{SD} + \gamma_{AF_1} + \gamma_{DF_2}) < R\right)$$

$$(1 - P_{C_2})Pr\left(\frac{1}{3}log_2(1 + 2\gamma_{SD} + \gamma_{AF_1}) < R\right)$$

$$= \frac{g(R)}{P_S\delta_{SD}}\left(\frac{1}{P_S\delta_{SR_1}} + \frac{1}{P_{R_1}\delta_{R_1D}}\right)\left(\frac{1}{6P_S\delta_{SR_2}} + \frac{1}{4P_{R_2}\delta_{R_2D}}\right) \quad (22)$$

5 Power Optimization

System performance can be improved in collaborative communication through resource sharing. Since the wireless links between the $S \to D$, $S \to R_1$, $S \to R_2$, $R_1 \to D$ and $R_2 \to D$ are time-varying in nature and are subjected to different channel conditions, not all the nodes use same transmission power. Therefore, it is necessary to allocate optimum power to all the transmit nodes so as to reduce the total power consumption of the network and the total interference. This section formulates an optimization problem to reduce the total transmission power at all the nodes subjected that the minimum network outage as well as the power limit at each node are maintained. Thus, the optimization problem is given as:

$$[P_S^*, P_{R_1}^*, P_{R_2}^*] = \min_{[P_S, P_{R_1}, P_{R_2}]}(P_S + P_{R_1} + P_{R_2})$$

$$\text{s.t.} \quad P_{HDAF}^{outage} \leq \zeta$$

$$0 \leq P_k \leq P_{max} \quad \forall k \in \{S, R_1, R_2\} \quad (23)$$

where ζ is the upper limit of the network outage probability that the network can afford, P_{DF}^{outage} is the outage probability of the cooperative network and P_{max} is the maximum transmit power limit at each transmitter, which must be non-negative. The optimization problem in Eq. (23) is a convex function where each of the constraint forms a convex set. Thus, the Eq. (23) has an optimal solution which can be most effectively solved using Lagrange multiplication. The optimization problem can then be expressed as:

$$L(P_S, P_{R_i}, \lambda) = P_S + P_{R_1} + P_{R_2} + \lambda(P_{DF}^{Outage} - \zeta) \tag{24}$$

where $\lambda(\geq 0)$ is the Lagrange operator. Then the KKT conditions are:

$$\frac{\partial L(P_S, P_{R_i}, \lambda)}{\partial P_S} = \frac{\partial L(P_S, P_{R_i}, \lambda)}{\partial P_{R_1}} = \frac{\partial L(P_S, P_{R_i}, \lambda)}{\partial P_{R_2}} = \frac{\partial L(P_S, P_{R_i}, \lambda)}{\partial \lambda} = 0 \tag{25}$$

Considering the stationary condition, Eq. (25) gives,

$$\lambda(P_{DF}^{outage} - \eta) = 0 \tag{26}$$

6 Simulation Results

This section presents the simulation results of the multi-relay cooperative relay network where one relay uses DF protocol while another uses AF protocol. All simulation results have been obtained using MATLAB 2018a on Dell PowerEdge c6320 server with four 2.4 GHz Intel Xeon E5-2680 v4 fourteen-core processors on Linux distro. All figures were obtained after averaging the instances of data obtained from 100 iterations. MATLAB optimization toolbox is used to compute the power optimization. The value for the parameters used in the simulation are given in Table 1.

The simulation results considers three cases:

1. Case 1: Both relay nodes use AF to forward the source information.
2. Case 2: Both relay nodes use DF to forward the source information.
3. Case 3: One node (i.e. R_1) uses AF node while another node (i.e. R_2) uses DF to forward the source information.

Figure 2 compares the source node transmit power as a function of outage probability for the three cases. As the network outage constraint is relaxed, the source node transmit power decreases in all the three cases. When the network does not permit any outage (i.e. $P_{outage} = 0$), then the source node uses maximum available transmit power (= 25 dBm) in all the three cases. As the network allows outage, the source node transmit power decreases from 25 dBm to 7.33 dBm, 6.72 dBm and 6.29 dBm for AF, DF and HDAF protocol respectively.

Table 1. Typical values for parameters [4]

Parameters	Values
Transmission rate, R	1 bps/Hz
Maximum transmit power at each node	25 dBm
Noise variance, N_0	1
Channel coefficients: $\delta_{SD}, \delta_{SR_1}, \delta_{SR_2}, \delta_{R_1D}, \delta_{R_2D}$	1–10
Outage probability, η	0–0.05

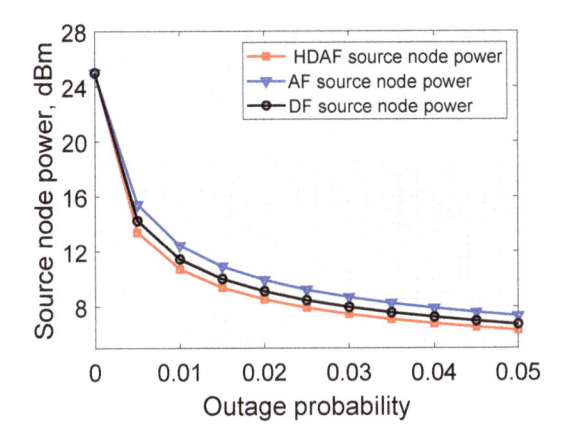

Fig. 2. Source node power with different outage probability

With HDAF, the improvement of 14% can be achieved with respect to AF and 6.4% with respect to DF.

Figure 3 compares the total transmit power of the source as well as the relay nodes as a function of outage probability for the three cases. Under no outage scenario, all of the three scheme consume a maximum power of 75 dBm. With the outage probability of 0.05, the total transmit power of the network reduces to 13.37 dBm, 11.55 dBm and 11.1 dBm for AF, DF and HDAF protocol respectively. With HDAF, the improvement in the total network transmit power of 16.98% can be achieved with respect to AF and 3.89% with respect to DF.

Figure 4 compares the individual relay node transmit power as a function of outage probability for the three cases. Both DF and HDAF outperform AF when the network allows outage. Under HDAF, the transmit power for relay node which employs DF protocol (i.e. relay R_2, given by dotted red line) is less than the case when both relay nodes use DF protocol. However, the transmit power for relay node which employ AF protocol (i.e. relay R_1, given by solid red line) is insignificantly more than the case when both relay nodes use DF protocol but significantly less than when both the relays use AF protocol.

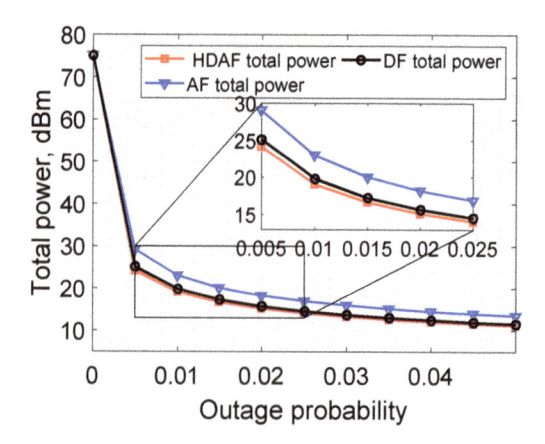

Fig. 3. Total network transmit power with different outage probability

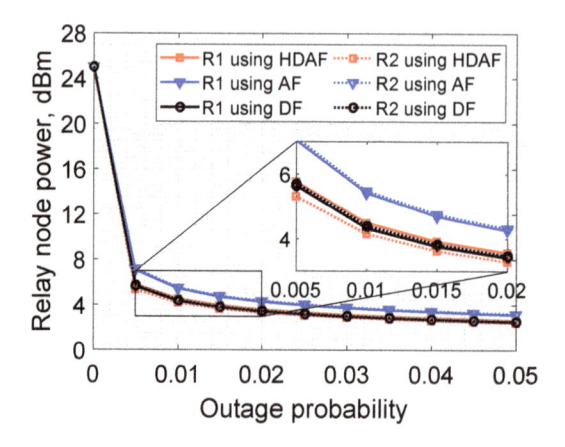

Fig. 4. Relay node transmit power with different outage probability

Figures 5, 6, 7, 8 and 9 consider the network's performance under different channel condition between $S \rightarrow R_i$ and $R_i \rightarrow D$ while keeping same channel condition between $S \rightarrow D$. The different channel conditions are:

1. Channel condition 1: $\delta_{R_2D} > \delta_{R_1D}$; $\delta_{R_1D}, \delta_{R_2D} > \delta_{SR_i}$ and $\delta_{SR_1} = \delta_{SR_2} = 4$
2. Channel condition 2: $\delta_{R_2D} < \delta_{R_1D}$; $\delta_{R_1D}, \delta_{R_2D} > \delta_{SR_i}$ and $\delta_{SR_1} = \delta_{SR_2} = 4$
3. Channel condition 3: $\delta_{SR_2} > \delta_{SR_1}$; $\delta_{SR_1}, \delta_{SR_2} > \delta_{SD}$ and $\delta_{R_1D} = \delta_{R_2D} = 4$
4. Channel condition 4: $\delta_{SR_2} > \delta_{SR_1}$; $\delta_{SR_1}, \delta_{SR_2} > \delta_{R_iD}$ and $\delta_{R_1D} = \delta_{R_2D} = 4$

In all these four cases, $\delta_{SD} = 1$. The range of all these channel coefficients are considered between 1 and 10.

In all Figs. 5, 6, 7 and 8, the performance of HDAF and DF network outperform the AF network in terms of relay node transmit power. When the channel coefficient between $R_2 \rightarrow D$ is larger than $R_1 \rightarrow D$, then the relay R_2 employing

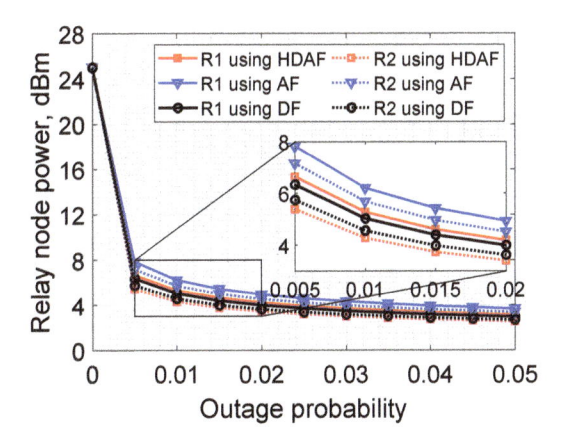

Fig. 5. Individual relay node transmit power for channel condition 1

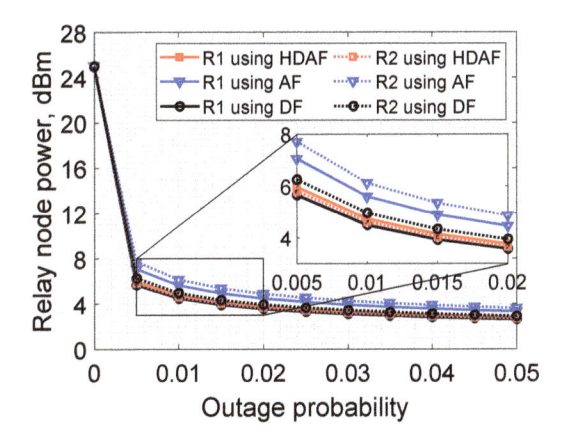

Fig. 6. Individual relay node transmit power for channel condition 2

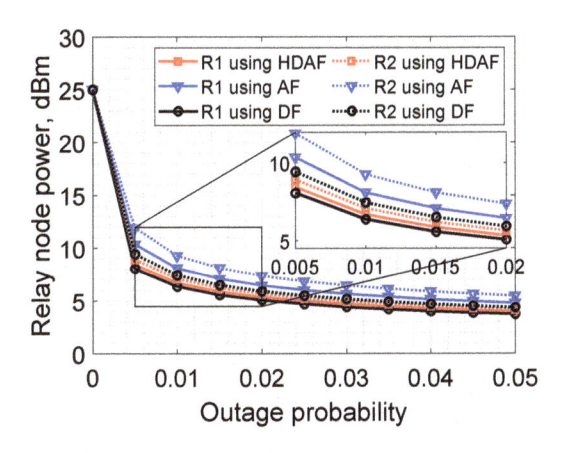

Fig. 7. Individual relay node transmit power for channel condition 3

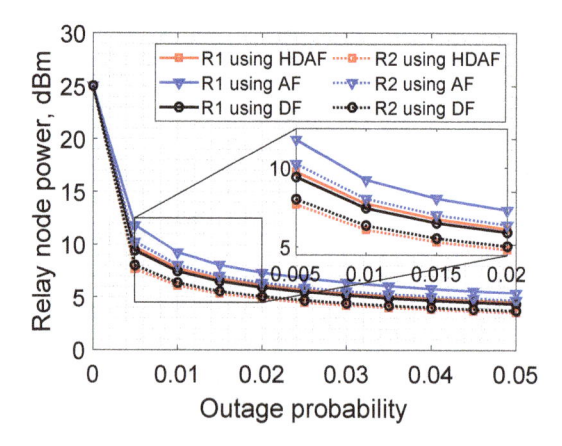

Fig. 8. Individual relay node transmit power for channel condition 4

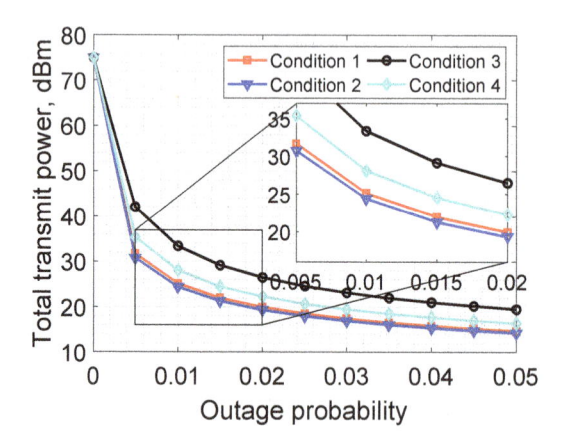

Fig. 9. Total transmit power with different outage probability for different channel condition

DF in HDAF network uses minimum transmit power as compared to the relays in the DF network as evident in Figs. 5 and 8. The relay R_1 in HDAF network uses more transmit power than the relays in the DF network but less than the relays in the AF network. In Figs. 6 and 7, the individual relay node transmit power in HDAF lies between the relay nodes transmit power of DF network which illustrates the marginal advantage of using AF protocol in one relay over DF protocol being used at both the relays.

Figure 9 compares the performance of HDAF with respect to outage probability for various channel conditions specified earlier. When the channel variance between $R_i \to D$ is large compared to the variance between $S \to R_i$, the advantage of employing HDAF protocol is clearly evident in the total transmit power of the network. It is shown from the result that when the channel coefficient between $R_2 \rightarrow D$ is large, it is more advantageous to employ DF protocol in the corresponding relay.

7 Conclusion and Future Work

The results from this paper show that when the network does not permit any outage then it does not make any difference whether the relay uses AF, DF or HDAF protocol to forward the source information in terms of the source node transmit power, individual relay node transmit power and the total network transmit power. All the node uses maximum available transmit power to satisfy the zero outage constraint of the network. The advantage of choosing among the protocols becomes evident when the network allows some outage. With respect to source node transmit power, individual relay node transmit power and the total transmit power of the network, the performance with AF protocol is worse as compared to DF and HDAF protocol. HDAF outperforms the DF protocols in terms of total network transmit power and the source node transmit power. However for the individual relay node transmit power, the performance of DF and HDAF are very close. The result also showed that the performance improvement from HDAF is clearly evident when there is a large channel variance between the relays and the destination as compared to the case between the source and the relays.

As a future work, the similar analysis will be performed for N relay network with multiple antennas scenario at the source and the relays . A distributed algorithm will be developed such that the nodes can coordinate among themselves to choose DF and AF protocol accordingly.

References

1. Agiwal, M., Roy, A., Saxena, N.: Next generation 5G wireless networks: a comprehensive survey. IEEE Commun. Surv. Tutorials **18**(3), 1617–1655 (2016). https://doi.org/10.1109/COMST.2016.2532458
2. Albalawi, U., Joshi, S.: Secure and trusted telemedicine in Internet of Things IoT. In: 2018 IEEE 4th World Forum on Internet of Things (WF-IoT), pp. 30–34, February 2018. https://doi.org/10.1109/WF-IoT.2018.8355206
3. Bao, X., Li, J.: Efficient message relaying for wireless user cooperation: decode-amplify-forward (DAF) and hybrid DAF and coded-cooperation. IEEE Trans. Wireless Commun. **6**(11), 3975–3984 (2007). https://doi.org/10.1109/TWC.2007.06117
4. Barnes, S.D., Joshi, S., Maharaj, B.T., Alfa, A.S.: Massive MIMO and femto cells for energy efficient cognitive radio networks, pp. 511–522. Springer, Cham (2015). https://doi.org/10.1007/978-3-319-24540-9_42
5. Belkhir, L., Elmeligi, A.: Assessing ICT global emissions footprint: trends to 2040 & recommendations. J. Cleaner Prod. **177**, 448–463 (2018). https://doi.org/10.1016/j.jclepro.2017.12.239. http://www.sciencedirect.com/science/article/pii/S095965261733233X

6. Chen, Y., Shi, R., Long, M.: Performance analysis of amplify-and-forward relaying with correlated links. IEEE Trans. Veh. Technol. **62**(5), 2344–2349 (2013). https://doi.org/10.1109/TVT.2013.2241093
7. Cover, T.M., Thomas, J.A.: Elements of Information Theory. Wiley, New York (1991)
8. Duong, T.Q., da Costa, D.B., Elkashlan, M., Bao, V.N.Q.: Cognitive amplify-and-forward relay networks over nakagami-m fading. IEEE Trans. Veh. Technol. **61**(5), 2368–2374 (2012). https://doi.org/10.1109/TVT.2012.2192509
9. Duy, T.T., Kong, H.: Performance analysis of hybrid decode-amplify-forward incremental relaying cooperative diversity protocol using SNR-based relay selection. J. Commun. Netw. **14**(6), 703–709 (2012). https://doi.org/10.1109/JCN.2012.00036
10. Duy, T.T., Kong, H.Y.: On performance evaluation of hybrid decode-amplify-forward relaying protocol with partial relay selection in underlay cognitive networks. J. Commun. Netw. **16**(5), 502–511 (2014). https://doi.org/10.1109/JCN.2014.000089
11. Jayakumar, H., Lee, K., Lee, W.S., Raha, A., Kim, Y., Raghunathan, V.: Powering the Internet of Things. In: 2014 IEEE/ACM International Symposium on Low Power Electronics and Design (ISLPED), pp. 375–380, August 2014. https://doi.org/10.1145/2627369.2631644
12. Jing, Y., Jafarkhani, H.: Single and multiple relay selection schemes and their achievable diversity orders. IEEE Trans. Wireless Commun. **8**(3), 1414–1423 (2009). https://doi.org/10.1109/TWC.2008.080109
13. Joshi, S., Albalawi, U.: Energy efficient routing considering link robustness in wireless sensor networks. In: 2017 IEEE International Conference on Power, Control, Signals and Instrumentation Engineering (ICPCSI), pp. 3116–3120, September 2017. https://doi.org/10.1109/ICPCSI.2017.8392299
14. Joshi, S., Albalawi, U.: Energy efficient cognitive radio network using high altitude platform station. In: Arai, K., Kapoor, S., Bhatia, R. (eds.) Advances in Information and Communication Networks, pp. 190–202. Springer, Cham (2019)
15. Khabbazi, A., Nader-Esfahani, S.: Power allocation in an amplify-and-forward cooperative network for outage probability minimization. In: 2008 International Symposium on Telecommunications, pp. 241–245, August 2008. https://doi.org/10.1109/ISTEL.2008.4651307
16. Kim, S.W., Wang, Z.: Maximum ratio diversity combining receiver using single radio frequency chain and single matched filter. In: IEEE GLOBECOM 2007 - IEEE Global Telecommunications Conference, pp. 4081–4085 (2007)
17. Laneman, J.N., Thitae, D.N.C., Wornell, G.W.: Cooperative diversity in wireless networks: efficient protocols and outage behavior. IEEE Trans. Inf. Theor. **50**(12), 3062–3080 (2004). https://doi.org/10.1109/TIT.2004.838089
18. Larsson, E.G., Cao, Y.: Collaborative transmit diversity with adaptive radio resource and power allocation. IEEE Commun. Lett. **9**(6), 511–513 (2005). https://doi.org/10.1109/LCOMM.2005.1437354
19. Lee, I., Lee, H., Choi, H.: Exact outage probability of relay selection in decode-and-forward based cooperative multicast systems. IEEE Commun. Lett. **17**(3), 483–486 (2013). https://doi.org/10.1109/LCOMM.2013.020413.122335
20. Li, J., Petropulu, A.P., Weber, S.: On cooperative relaying schemes for wireless physical layer security. IEEE Trans. Signal Process. **59**(10), 4985–4997 (2011). https://doi.org/10.1109/TSP.2011.2159598
21. Liu, T., Song, L., Li, Y., Huo, Q., Jiao, B.: Performance analysis of hybrid relay selection in cooperative wireless systems. IEEE Trans. Commun. **60**(3), 779–788 (2012). https://doi.org/10.1109/TCOMM.2012.011312.110015

22. Liu, Y., Pan, G., Zhang, H., Song, M.: Hybrid decode-forward amplify-forward relaying with non-orthogonal multiple access. IEEE Access **4**, 4912–4921 (2016). https://doi.org/10.1109/ACCESS.2016.2604341
23. Malmodin, J., Bergmark, P.: Exploring the effect of ICT solutions on GHG emissions in 2030. In: EnviroInfo/ICT4S (2015)
24. Malmodin, J., Moberg, S., Lundén, D., Finnveden, G., Lövehagen, N.: Greenhouse gas emissions and operational electricity use in the ICT and entertainment & media sectors. J. Ind. Ecol. **14**(5), 770–790 (2010). https://doi.org/10.1111/j.1530-9290.2010.00278.x. https://onlinelibrary.wiley.com/doi/abs/10.1111/j.1530-9290.2010.00278.x
25. Olfat, E., Olfat, A.: Performance of hybrid decode-amplify-forward protocol for multiple relay networks over independent and non-identical flat fading channels. IET Commun. **5**(14), 2018–2027 (2011)
26. Pachori, K., Mishra, A.: Performance analysis of MIMO systems under multipath fading channels using linear equalization techniques. In: 2015 International Conference on Advances in Computing, Communications and Informatics (ICACCI), pp. 190–193, August 2015. https://doi.org/10.1109/ICACCI.2015.7275607
27. Nguyen, Q.T., Nguyen, D.T., Sinh, C.L.: Outage probability analysis of amplify-and-forward cooperative diversity relay networks. CoRR abs/1604.06164 (2016). http://arxiv.org/abs/1604.06164
28. Sadek, A.K., Su, W., Liu, K.J.R.: Multinode cooperative communications in wireless networks. IEEE Trans. Signal Process. **55**(1), 341–355 (2007). https://doi.org/10.1109/TSP.2006.885773
29. Seddik, K.G., Sadek, A.K., Su, W., Liu, K.J.R.: Outage analysis of multi-node amplify-and-forward relay networks. In: IEEE Wireless Communications and Networking Conference, 2006, WCNC 2006. vol. 2, pp. 1184–1188, April 2006. https://doi.org/10.1109/WCNC.2006.1683638
30. Seddik, K.G., Sadek, A.K., Su, W., Liu, K.J.R.: Outage analysis and optimal power allocation for multinode relay networks. IEEE Signal Process. Lett. **14**(6), 377–380 (2007). https://doi.org/10.1109/LSP.2006.888424
31. Song, X., Zhang, M., Liu, W., Liu, F.: Threshold-based hybrid relay selection scheme. In: 2016 12th World Congress on Intelligent Control and Automation (WCICA), pp. 222–227, June 2016. https://doi.org/10.1109/WCICA.2016.7578619
32. Su, W., Sadek, A.K., Liu, K.J.R.: SER performance analysis and optimum power allocation for decode-and-forward cooperation protocol in wireless networks. In: IEEE Wireless Communications and Networking Conference, 2005, vol. 2, pp. 984–989, March 2005. https://doi.org/10.1109/WCNC.2005.1424642
33. Su, W., Sadek, A.K., Ray Liu, K.J.: Cooperative communication protocols in wireless networks: performance analysis and optimum power allocation. Wireless Pers. Commun. **44**(2), 181–217 (2008). https://doi.org/10.1007/s11277-007-9359-z
34. Upadhyay, M.A., Kothari, D.K.: Optimal resource allocation techniques for cooperative AF and DF wireless networks. In: 2014 International Conference on Signal Processing and Integrated Networks (SPIN), pp. 382–387, February 2014. https://doi.org/10.1109/SPIN.2014.6776983
35. Vardhe, K., Reynolds, D., Woerner, B.D.: Joint power allocation and relay selection for multiuser cooperative communication. IEEE Trans. Wireless Commun. **9**(4), 1255–1260 (2010). https://doi.org/10.1109/TWC.2010.04.080175
36. Wang, D., Zhang, R., Cheng, X., Yang, L.: Relay selection in power splitting based energy-harvesting half-duplex relay networks. In: 2017 IEEE 85th Vehicular Technology Conference (VTC Spring), pp. 1–5, June 2017. https://doi.org/10.1109/VTCSpring.2017.8108276

37. Zhang, X.J., Gong, Y.: Joint power allocation and relay positioning in multi-relay cooperative systems. IET Commun. **3**(10), 1683–1692 (2009). https://doi.org/10.1049/iet-com.2008.0738
38. Zheng, G., Zhang, Y., Ji, C., Wong, K.: Optimizing relay selection and power allocation for orthogonal multiuser downlink systems. In: 2009 International Conference on Wireless Communications Signal Processing, pp. 1–5, November 2009. https://doi.org/10.1109/WCSP.2009.5371668

Exploring the Evolutionary Bispectrum

Abdullah I. Al-Shoshan[(✉)]

Computer Engineering Department, College of Computer, Qassim University,
P.O. Box 6688, Qassim 51452, Saudi Arabia
drshoshan@gmail.com

Abstract. Since most of the natural phenomena have some characteristics varying with time, these types of phenomena or signals are classified as time-varying, or non-stationary signals. Therefore, dealing with these types of phenomena, like modeling or estimation, needs time-varying algorithms. For non-stationary processes, the conventional power spectrum does not reflect the time variation of the process characteristics. With the introduction of bispectrum in digital signal processing, a new approach to the solution of non-minimum phase system (NMP) identification problem is devised. This approach exploits the fact that the bispectrum contains information regarding both the phase and the magnitude of the system. Although bispectrum has been applied in identification of non-minimum phase LTI systems, it requires the assumption of stationarity and restricts the process to have non-symmetric probability density. When the input/output of the system are non-stationary, neither the power spectrum nor the bispectrum can handle this problem because they do not reflect the time variation of the process characteristics. In this paper, some algorithms based on the evolutionary spectrum to solve these problems are proposed. Our methodology to solve this problem is to use the theory of the evolutionary spectral and to introduce the evolutionary bispectral analysis. In this paper, a definition of the evolutionary bispectrum (EB) with its properties, especially the ability of removing non-stationary Gaussian noise, is discussed, and several algorithms for reconstructing the signal from the evolutionary bispectrum are also discussed.

1 Introduction

In our real life, we have many phenomena and signals that have some or all of their characteristics vary with time. Therefore, these variations need to be catched up and processed. In recent years, the estimation of parameters of a non-stationary process has been considered by various authors [1, 2 and 3]. According to the Wold-Cramer decomposition [4], we can represent a discrete-time non-stationary process, $x(n)$, as the output of a causal, linear, and time-varying (LTV) system, with impulse response $h(n,m)$, to a discrete-time stationary zero-mean, unit-variance white noise process, $e(n)$, as shown in Fig. 1. A stationary signal may be viewed as a sum of sinusoids with random amplitudes and phases. Thus, a stationary process $x(n)$ can be expressed as [3].

80   A. I. Al-Shoshan

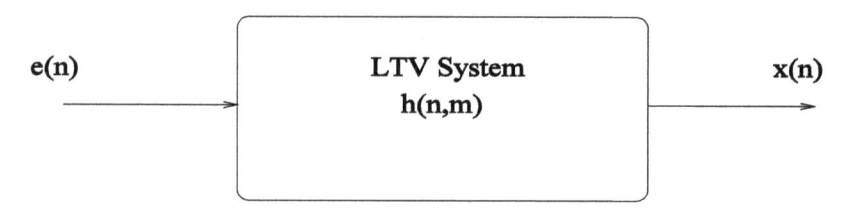

Fig. 1. The Wold-Cramer representation of a non-stationary signal

$$x(n) = \int_{-\pi}^{\pi} e^{jwn} dZ(\omega) \qquad (1)$$

where $Z(\omega)$ is a process with orthogonal increments such that

$$E\{dZ^*(\omega_1)dZ(\omega_2)\} = \begin{cases} \frac{S(\omega)d\omega}{2\pi}, & \omega_1 = \omega_2 = \omega \\ 0, & \omega_1 \neq \omega_2 \end{cases} \qquad (2)$$

where $S(\omega)$ is the power spectral density function of $x(n)$. Hence the power of $x(n)$ is given by

$$E\{|x(n)|^2\} = \frac{1}{2\pi}\int_{-\pi}^{\pi} S(\omega)d\omega = R(0) \qquad (3)$$

If the process is non-stationary the family of sinusoids only is no longer valid. Given that the spectral characteristics of a non-stationary process change with time, the above representation does not reflect the instantaneous changes in the energy of the signal. Consequently, we need a more general form to represent it. When the additive noise $\eta(n)$ shown in Fig. 2 is non-stationary, the traditional stationary techniques will not be able to remove this type of noise. However, it will simply reduce the effect of non-stationary noise with symmetric probability density like Gaussian, Laplace, uniform, and Bernouli-Gaussian. To remove these types of noise, we propose using the evolutionary bispectrum (EB) recently introduced by Priestley [3].

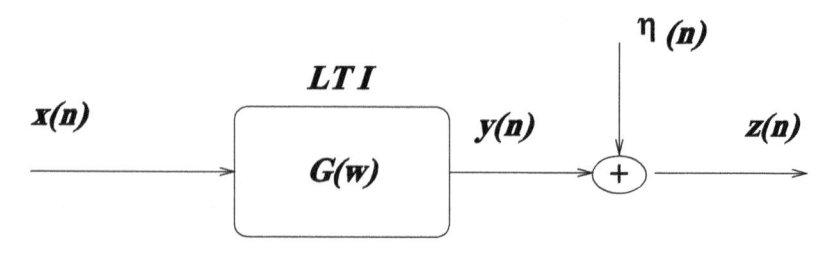

Fig. 2. A noisy output of an LTI system

2 Evolutionary Bispectrum

The EB is defined as follow. Let $x(n)$ be a zero-mean non-stationary process represented as in Fig. 1 where $e(n)$ is a stationary white noise. The third-order moment of $x(n)$ is given by

$$R(n, m_1, m_2) = E\{x(n)x(n+m_1)x(n+m_2)\} \quad (4)$$

which, by using the Wold-Cramer representation, becomes

$$R(n, m_1, m_2) = \int_{-\pi}^{\pi}\int_{-\pi}^{\pi}\int_{-\pi}^{\pi} H_x(n, \omega_1)H_x(n+m_1, \omega_2)H_x(n+m_2, \omega_3) \\ e^{j(\omega_2 m_1 + \omega_3 m_2)} e^{jn(\omega_1 + \omega_2 + \omega_3)} E\{dZ(\omega_1)dZ(\omega_2)dZ(\omega_3)\} \quad (5)$$

where

$$E\{dZ(\omega_1)dZ(\omega_2)dZ(\omega_3)\} = \begin{cases} S_e(\omega_1, \omega_2)d\omega_1 d\omega_2, & \omega_1 + \omega_2 + \omega_3 = 0 \\ 0, & \omega_1 + \omega_2 + \omega_3 \neq 0 \end{cases} \quad (6)$$

i.e., it vanishes except along the plane $\omega_1 + \omega_2 + \omega_3 = 0$, then we have that

$$R(n, m_1, m_2) = \int_{-\pi}^{\pi}\int_{-\pi}^{\pi} H_x(n, \omega_1)H_x(n+m_1, \omega_2)H_x(n+m_2, -\omega_1-\omega_2) \\ S_e(\omega_1, \omega_2)e^{j(\omega_2 m_1 - (\omega_1 + \omega_2)m_2)} d\omega_1 d\omega_2 \quad (7)$$

and setting $m_1 = m_2 = 0$, we get

$$R(n, 0, 0) = E\{x(n)^3\} = \int_{-\pi}^{\pi}\int_{-\pi}^{\pi} S_x(n, \omega_1, \omega_2)\, d\omega_1 d\omega_2 \quad (8)$$

therefore, the EB of $x(n)$ is defined as

$$S_x(n, \omega_1, \omega_2) = H_x(n, \omega_1)H_x(n, \omega_2)H_x(n, -\omega_1-\omega_2)S_e(\omega_1, \omega_2) \quad (9)$$

Assuming that $e(n)$ is a zero-mean non-Gaussian white noise with $E\{e(n)e(n+m)e(n+k)\} = \delta(m,k)$ then $S_e(\omega_1,\omega_2) = 1$ and therefore Eq. (9) becomes

$$S_x(n, \omega_1, \omega_2) = H_x(n, \omega_1)H_x(n, \omega_2)H_x(n, -\omega_1-\omega_2) \quad (10)$$

When $e(n)$ is zero-mean Gaussian, then $S_e(\omega_1,\omega_2) = 0$ and accordingly $S_x(n,\omega_1, \omega_2) = 1$. Therefore, the EB of a zero-mean non-stationary Gaussian process is identically zero. If the system $h_x(n,m)$ is not time varying, then $H_x(n,\omega)$ will also be time-invariant, therefore, the EB in this case will reduce to

$$S_x(\omega_1, \omega_2) = H_x(\omega_1)H_x(\omega_2)H_x(-\omega_1 - \omega_2) \tag{11}$$

which has a similar interpretation to that of the conventional bispectrum of a stationary process, namely that it is the triple product of the $H_x(n,\omega)$ at frequencies ω_1, ω_2, and $(-\omega_1-\omega_2)$. Let $y(n)$ be a zero-mean, non-Gaussian process corrupted by an i.i.d. zero-mean, Gaussian noise, $\eta(n)$, such that

$$z(n) = y(n) + \eta(n) \tag{12}$$

where $y(n)$ and $\eta(n)$ are independent, then the third order moment of $z(n)$ is

$$\begin{aligned}E\{z(n)z(n+m_1)z(n+m_2)\} &= E\{y(n)y(n+m_1)y(n+m_2)\} \\ &+ E\{\eta(n)\eta(n+m_1)\eta(n+m_2)\}\end{aligned} \tag{13}$$

Since $y(n)$ has a zero-mean value, its first, second, and third order time-varying cumulants (TVCs) reduce to its first, second, and third order time-varying moments, respectively, so

$$c^y(n, m_1, m_2) = E\{y(n)y(n+m_1)y(n+m_2)\}$$

therefore, the third-order TVC of $z(n)$ becomes

$$c^z(n, m_1, m_2) = c^y(n, m_1, m_2) + c^\eta(n, m_1, m_2) \tag{14}$$

and

$$E\{z(n)^3\} = E\{y(n)^3\} + E\{\eta(n)^3\} \tag{15}$$

but since $\eta(n)$ is a zero-mean, Gaussian noise, then $E\{\eta(n)^3\} = 0$ and $S_\eta(n,\omega_1,\omega_2) = 0$, so

$$E\{z(n)^3\} = E\{y(n)^3\} \tag{16}$$

and hence

$$S_z(n, \omega_1, \omega_2) = S_y(n, \omega_1, \omega_2) \tag{17}$$

which indicates that the additive Gaussian noise does not affect the third-order TVC or the EB of $y(n)$. We will show later an example that the additive i.i.d. zero-mean non-stationary Gaussian noise can be removed using the EB. The estimation of the EB can be done as follows. We estimate the time-dependent function $H(n,\omega)$ using any time-frequency distribution estimator to get $\hat{H}(n, \omega)$, then from the triple product of $\hat{H}(n, \omega)$ at frequencies ω_1, ω_2, and $(-\omega_1-\omega_2)$, considering that the EB of $y(n)$ is equivalent to the EB of $z(n)$, we get

$$\hat{S}_z(n,\omega_1,\omega_2) = \hat{S}_y(n,\omega_1,\omega_2) = \hat{H}_y(n,\omega_1)\hat{H}_y(n,\omega_2)\hat{H}_y(n,-\omega_1-\omega_2) \quad (18)$$

where $\hat{H}_z(n,\omega)$ is an estimate of the time-varying amplitude. In the case when $y(n)$ is stationary, i.e., $w_n(m) = 1/N$, the evolutionary bispectrum reduces to

$$\hat{S}_y(\omega_1,\omega_2) = \frac{1}{N^3} Y(\omega_1)Y(\omega_2)Y(-\omega_1-\omega_2) \quad (19)$$

which is the averaged triple product estimator of the bispectrum.

3 EB Properties

From the definition of the EB in Eq. (18), similar to the conventional bispectrum [8], the EB has the following properties:

Property 1: $S(n,\omega_1,\omega_2)$ is generally complex, i.e.,

$$S(n,\omega_1,\omega_2) = |S(n,\omega_1,\omega_2)|e^{j\psi(n,\omega_1,\omega_2)} \quad (20)$$

Property 2: $S(n,\omega_1,\omega_2)$ is doubly periodic with period 2π

$$S(n,\omega_1+2m\pi,\omega_2+2k\pi) = S(n,\omega_1,\omega_2) \quad (21)$$

where m and k are integers.

Property 3: Symmetry of $S(n,\omega_1,\omega_2)$:

$$\begin{aligned}S(n,\omega_1,\omega_2) &= S^*(n,-\omega_1,-\omega_2) = S(n,\omega_2,\omega_1) = S^*(n,-\omega_2,-\omega_1) \\ &= S(n,\omega_1,-\omega_1-\omega_2) = S(n,-\omega_1-\omega_2,\omega_1) \\ &= S(n,-\omega_1-\omega_2,\omega_2) = S(n,\omega_2,-\omega_1-\omega_2)\end{aligned} \quad (22)$$

Therefore, the knowledge of the evolutionary bispectrum in the triangular region $\omega_2 \geq 0$, $\omega_1 \geq \omega_2$, $\omega_1+\omega_2 \leq \pi$ for each n shown in Fig. 3 is enough for a complete description of the EB. This advantage will reduce the computation of the EB which done over one sector of the $\omega_1 \times \omega_2$ domain and use the above symmetry properties to generate the whole EB. In the stationary case, Eq. (22) reduces to

$$\begin{aligned}S(\omega_1,\omega_2) &= S^*(-\omega_1,-\omega_2) = S(\omega_2,\omega_1) = S^*(-\omega_2,-\omega_1) \\ &= S(\omega_1,-\omega_1-\omega_2) = S(-\omega_1-\omega_2,\omega_1) \\ &= S(-\omega_1-\omega_2,\omega_2) = S(\omega_2,-\omega_1-\omega_2)\end{aligned} \quad (23)$$

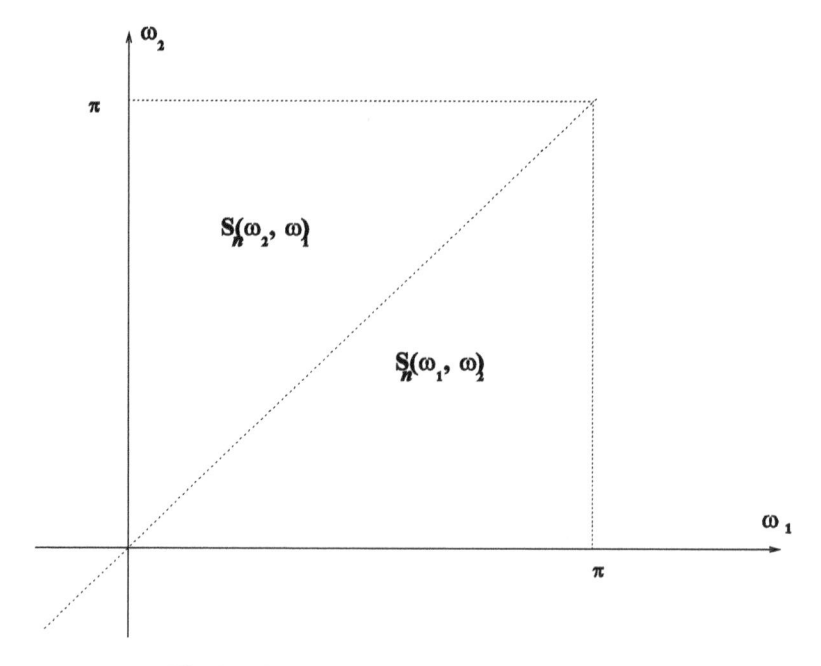

Fig. 3. The symmetry regions of the EB at each n

Property 4: Gaussian Process. If $\{x(n)\}$ is a non-stationary zero-mean Gaussian process, its evolutionary bispectrum is identically zero, i.e., $S_x(n,\omega_1,\omega_2) = 0$.

Property 5: Linear Phase Shift. Given $\{x(n)\}$ with evolutionary bispectrum $S_x(n,\omega_1,\omega_2)$, the process $y(n) = x(n-D)$, where D is a constant integer, has a relation with $x(n)$ as follows. We have

$$x(n) = \int_{-\pi}^{\pi} H_x(n,\omega) e^{j\omega n} dZ(\omega) \tag{24}$$

therefore

$$y(n) = \int_{-\pi}^{\pi} H_y(n,\omega) e^{j\omega n} dZ(\omega) = \int_{-\pi}^{\pi} H_x(n-D,\omega) e^{j\omega(n-D)} dZ(\omega) \tag{25}$$

which yields

$$H_y(n,\omega) = H_x(n-D,\omega) e^{-j\omega D} \tag{26}$$

therefore, the evolutionary bispectrum of $y(n)$ becomes

$$S_y(n, \omega_1, \omega_2) = H_y(n, \omega_1)H_y(n, \omega_2)H_y^*(n, \omega_1 + \omega_2)$$
$$= H_x(n - D, \omega_1)H_x(n - D, \omega_2)H_x^*(n - D, \omega_1 + \omega_2) \quad (27)$$
$$= S_x(n - D, \omega_1, \omega_2)$$

which suppresses $e^{-j\omega D}$ but preserve the shift in the EB as a shift in time, therefore, the EB preserve the time shift in the signal. If $x(n)$ is stationary, then the above equation reduces to

$$S_y(\omega_1, \omega_2) = S_x(\omega_1, \omega_2) = \frac{1}{N^3} X(\omega_1)X(\omega_2)X(-\omega_1 - \omega_2) \quad (28)$$

which shows that the time shift has no effect on the classical bispectrum.

Property 6: Evolutionary Bicoherence Index. The evolutionary bicoherence function (or the normalized evolutionary bispectrum) combines two completely different entities; the evolutionary spectrum $S(n,\omega)$ and the evolutionary bispectrum $S_y(n,\omega_1,\omega_2)$ and is defined as

$$S_y^B(n, \omega_1, \omega_2) = \frac{S_y(n, \omega_1, \omega_2)}{\sqrt{S_y(n, \omega_1)S_y(n, \omega_2)S_y(n, \omega_1 + \omega_2)}} \quad (29)$$

This function may be very useful in the detection and characterization of nonlinearities in time series and in discriminating linear processes from nonlinear ones. A signal is said to be a linear non-Gaussian process of order 3 if the magnitude of the third order coherency, $|B(n,\omega_1,\omega_2)|$, is constant over all frequencies: otherwise, the signal is said to be a nonlinear process. If $y(n)$ is stationary, Eq. (29) reduces to

$$S_y^B(\omega_1, \omega_2) = \frac{S_y(\omega_1, \omega_2)}{\sqrt{S_y(\omega_1)S_y(n, \omega_2)S_y(\omega_1 + \omega_2)}} \quad (30)$$

For an impulse,

$$x(n) = \delta(n - n_0),$$
$$H_x(n, \omega) = \delta(n - n_0) \quad (31)$$

we have

$$S_x(n, \omega_1, \omega_2) = \begin{cases} 1, & n = n_0 \\ 0, & \text{otherwise} \end{cases} \quad (32)$$

From the above properties, the evolutionary bispectrum is shown to preserve the essential properties of the stationary bispectrum.

4 Signal Reconstruction

Another important matter when dealing with the EB is the reconstruction of the signal $y(n)$ from $\hat{S}_y(n,\omega_1,\omega_2)$ (or equivalently $\hat{S}_z(n,\omega_1,\omega_2)$). In this section, we will show how to reconstruct $\hat{H}_y(n,\omega)$ from $\hat{S}_y(n,\omega_1,\omega_2)$, and how will the time-varying amplitude $H(n,k)$ of a process be recovered from its EB where to recover $H(n,k)$, we need to recover both of its magnitude and phase. In this section, we propose three algorithms for magnitude recovery and one algorithm for phase recovery.

4.1 Magnitude Reconstruction

For magnitude recovery we propose three algorithms, one of them is recursive. Suppose we are given samples $S(n,\frac{2\pi k}{N},\frac{2\pi l}{N}) = S(n,k,l); k,l = 0,1,\ldots,N-1$ of the EB $S(n,\omega_1,\omega_2)$ of sequence $x(n)$. Therefore Eq. (27) can be rewritten as

$$S(n,k,l) = H(n,k)H(n,l)H^*(n,k+l) \qquad (33)$$

From the above equation, we propose the following algorithms for reconstructing $H(n,k)$ from $S(n,k,l)$.

4.1.1 Magnitude Recovery Algorithm 1

This algorithm is non-recursive in which the magnitude of $H(n,k)$ can be reconstructed by using only one slice of the evolutionary bispectrum. Let $l = 0$ then from Eq. (33)

$$S(n,k,0) = H(n,0)|H(n,k)|^2 \qquad (34)$$

Therefore, the magnitude of $H(n,k)$ can be obtained as follows:

$$|H(n,k)| = \sqrt{\frac{S(n,k,0)}{H(n,0)}} \qquad (35)$$

for all values of n and k. To be able to reconstruct the magnitude of $H(n,k)$, $H(n,0)$ must differ from zero, and is found from

$$H(n,0) = \sqrt[3]{|S(n,0,0)|} \qquad (36)$$

Therefore, we first estimate $S(n,k,0)$, then from Eqs. (35) and (36), we can recover the magnitude of $H(n,k)$.

4.1.2 Magnitude Recovery Algorithm 2
A similar recovery algorithm of $H(n,k)$ can be obtained recursively as follows.

Exploring the Evolutionary Bispectrum   87

$$|S(n,k,l)| = |H(n,k)||H(n,l)||H(n,k+l)| \qquad (37)$$

then

$$|H(n,k+l)| = \frac{|S(n,k,l)|}{|H(n,k)||H(n,l)|} \qquad (38)$$

and letting $l = 1$, we have

$$|H(n,k+1)| = \frac{|S(n,k,1)|}{|H(n,k)||H(n,1)|} \qquad (39)$$

where

$$|H(n,1)| = \sqrt[6]{\frac{|S(n,1,1)|^3 |S(n,1,3)|}{|S(n,1,2)||S(n,2,2)|}} \qquad (40)$$

Equation (40) can be obtained as follows: From Eq. (33), we have

$$|S(n,1,1)| = |H(n,1)|^2 |H(n,2)| \qquad (41)$$

and multiplying both sides of Eq. (41) by $|H(n,1)||H(n,3)|$ we get

$$|S(n,1,1)||H(n,1)||H(n,3)| = |H(n,1)|^2 |H(n,2)||H(n,1)||H(n,3)| \qquad (42)$$

or

$$|S(n,1,1)||H(n,1)||H(n,3)| = |H(n,1)|^2 |S(n,1,2)| \qquad (43)$$

and multiplying both sides of Eq. (43) by $|H(n,4)|$ we get

$$|S(n,1,1)||H(n,1)||H(n,3)||H(n,4)| = |H(n,1)|^2 |S(n,1,2)||H(n,4)| \qquad (44)$$

or

$$|S(n,1,1)||S(n,1,3)| = |H(n,1)|^2 |S(n,1,2)||H(n,4)| \qquad (45)$$

Finally, multiplying both sides of Eq. (45) by $|H(n,2)|^2$ we get

$$|S(n,1,1)||S(n,1,3)||H(n,2)|^2 = |H(n,1)|^2 |S(n,1,2)||H(n,4)||H(n,2)|^2 \qquad (46)$$

or

$$|S(n,1,1)||S(n,1,3)||H(n,2)|^2 = |H(n,1)|^2 |S(n,1,2)||S(n,2,2)| \qquad (47)$$

Multiplying both sides of Eq. (47) by $|H(n,1)|^4$ we get

$$|S(n,1,1)||S(n,1,3)| \quad | \quad H(n,1)|^2|H(n,2)|H(n,1)|^2|H(n,2)|$$
$$= |H(n,1)|^6|S(n,1,2)||H(n,2)|^2|H(n,4)| \tag{48}$$

which can be written as

$$|S(n,1,1)|^3|S(n,1,3)| = |H(n,1)|^6|S(n,1,2)||S(n,2,2)| \tag{49}$$

therefore,

$$|H(n,1)| = \sqrt[6]{\frac{|S(n,1,1)|^3|S(n,1,3)|}{|S(n,1,2)||S(n,2,2)|}} \tag{50}$$

From Eqs. (39) and (50), we observe that to recover $|H(n,k+1)|$ we need $|H(n,k)|$, $|H(n,1)|$, $|S(n,1,2)|$, and $|S(n,2,2)|$ to be different from zero. Therefore, to start this recursive algorithm, we estimate $S(n,k,1)$, $S(n,1,3)$, $S(n,1,2)$, and $S(n,2,2)$, then we use Eq. (50) to get $|H(n,1)|$, then substituting in Eq. (39) for $k = 1, 2, \ldots, N-1$ to recover $|H(n, k+1)|$.

4.1.3 Magnitude Recovery Algorithm 3

This algorithm is non-recursive. In Eq. (37) let $k = 1$, then we have

$$|S(n,k,k)| = |H(n,k)|^2|H(n,2k)| \tag{51}$$

and going in a similar way as in the previous recovery algorithm, we get

$$|H(n,k)| = \sqrt[6]{\frac{|S(n,k,k)|^3|S(n,k,3k)|}{|S(n,k,2k)||S(n,2k,2k)|}} \tag{52}$$

From Eq. (52), we observe that to recover $|H(n,k)|$ using this algorithm we need $|S(n,k,2k)|$ and $|S(n,2k,2k)|$ to be different from zero. Therefore, we first estimate $S(n,k,k)$, $S(n,k,3k)$, $S(n,k,2k)$ and $S(n,2k,2k)$, then we use Eq. (50).

4.2 Phase Reconstruction

To obtain an estimate of $H_y(n,k)$ we need also to estimate its phase. Let $\psi_n(\omega_1, \omega_2)$ be the unwrapped phase of $S(n,\omega_1,\omega_2)$ and $\phi_n(\omega)$ be the unwrapped phase of $H(n,\omega)$ for each n. From Eq. (27), we have the following relation:

$$\psi_n(\omega_1, \omega_2) = \phi_n(\omega_1) + \phi_n(\omega_2) - \phi_n(\omega_1 + \omega_2) \tag{53}$$

Discretizing the frequencies we get the discretized versions $\psi_n(i\Delta\omega_1, j\Delta\omega_2)$ and $\phi_n(i\Delta\omega)$, where $i, j = 0,\ldots, N-1$, $\Delta\omega = \frac{2\pi}{N}$, and N is the total number of frequency points. For simplicity, we will use the notations i, j instead of $i\Delta\omega$ and $j\Delta\omega$, respectively, to get

$$\psi_n(i,j) = \phi_n(i) + \phi_n(j) - \phi_n(i+j) \tag{54}$$

Letting $j = 1$ in Eq. (54), we have

$$\psi_n(i,1) = \phi_n(i) + \phi_n(1) - \phi_n(i+1) \tag{55}$$

and for $i = 1,\ldots, N - 1$ we get

$$\psi_n(1,1) = 2\phi_n(1) - \phi_n(2)$$
$$\psi_n(2,1) = \phi_n(1) + \phi_n(2) - \phi_n(3)$$
$$\psi_n(3,1) = \phi_n(1) + \phi_n(3) - \phi_n(4)$$

$$\cdot$$

$$\cdot$$

$$\psi_n(N-1,1) = \phi_n(1) + \phi_n(N-1) - \phi_n(0)$$

where in the last entry of the above equation we used the periodicity of the phase so that $\phi_n(N) = \phi_n(0)$. As in those algorithms, the value of $\phi_n(0)$ can be either zero or $\pm\pi$. The value of $\pm\pi$ indicates a sign reversal, and thus we let $\phi_n(0) = 0$. The above equations can be written in a matrix form as

$$\Psi_n(1) = A_N \Phi_n \tag{56}$$

where $\Psi_n(1) = [\psi_n(1,1)\ldots \psi_n(N-1,1)]^T$, $\Phi_n = [\phi_n(1)\phi_n(2)\ldots -\phi_n(N-1) + \phi_n(0)]^T$, and

$$A_N = \begin{pmatrix} 2 & -1 & 0 & \ldots\ldots & 0 \\ 1 & 1 & -1 & 0\ldots & 0 \\ 1 & 0 & 1 & -1\ldots 0 & \vdots \\ \vdots & \vdots & \ddots & \ddots & -1 \\ 1 & 0 & \ldots & 0 & 1 \end{pmatrix} \tag{57}$$

is an $(N-1) \times (N-1)$ matrix. It can be shown that A_N is an invertible matrix. Indeed, using the structure of A_N we have that post-multiplying it by the following matrix

$$J_N = \begin{pmatrix} 1 & 1 & 1 & \ldots\ldots & 1 \\ 0 & 2 & 2 & \ldots & 2 \\ \vdots & \ddots & \ddots & \ddots & \vdots \\ \vdots & \vdots & \ddots & N-2 & N-2 \\ 0 & 0 & \ldots & 0 & N-1 \end{pmatrix} \tag{58}$$

we get

$$P_N = A_N J_N = \begin{pmatrix} 2 & 0 & 0 & \cdots & 0 \\ 1 & 3 & 0 & \ddots & 0 \\ 1 & 1 & 4 & 0 \cdots & \vdots \\ \vdots & \vdots & \ddots & \ddots & \vdots \\ 1 & 1 & \cdots & 1 & N \end{pmatrix} \quad (59)$$

so that $det(A_N) = N$. It is even furthermore true that the matrices $\{A_k, k \geq 3\}$ are also invertible. Assuming that

$$A_k J_k = P_k \quad (60)$$

$$A_{k+1} J_{k+1} = \begin{pmatrix} A_k & & \theta_{k-2}^T \\ & & -1 \\ \cdots & \cdots & \cdots \\ 1 & \theta_{k-2} & \vdots & 1 \end{pmatrix} \begin{pmatrix} J_k & \vdots & b_{k-1} \\ \cdots & \cdots & \cdots \\ \theta_{k-1} & \vdots & k \end{pmatrix} \quad (61)$$

where $b_{k-1}^T = [1\ 2\ \ldots\ k-1]$, θ_k is a $k \times 1$ *null* vector. The above is so since $A_k\, b_{k-1} = (0\ 0\ \ldots\ 0)$. In order for any of the phase estimation procedures to work, the phase $\psi_n(\omega_1,\omega_2)$ needs to be unwrapped. We thus calculate $\Phi_n = A_N^{-1} \Psi_n(1)$ for every n, which is the phase of the time-varying amplitude $H(n,l)$.

5 Simulation

To show the power of the EB in removing the non-stationary additive Gaussian noise, consider this example. According to Fig. 2, we have $z(n) = y(n) + \eta(n)$, and let $y(n)$ be represented as $y(n) = A(n)e^{j\omega_0 n}$, where $A(n)$ is decomposed as

$$A(n) = \sum_{i=0}^{M-1} a_i \beta_i(n)$$

with $M = 3$, $a_i = 1/i$, $\omega_0 = 0.25\pi$ and $\{\beta_i(n)\}$ is a set of orthonormal polynomials (the Fourier orthonormal polynomials are considered). Also let $\eta(n)$ be a zero-mean, non-stationary Gaussian noise such that $\eta(n) = d(n)w(n)$, where $d(n)$ is a deterministic signal and $w(n)$ is a zero-mean, unit variance stationary Gaussian noise and independent of $y(n)$. Obviously, $E\{\eta(n)\} = 0$ and $\sigma_\eta^2(n) = d^2(n)\sigma_w^2 = d^2(n)$. The real part of the non-stationary signal $y(n)$ is shown in Fig. 4(a), its EP is shown in Fig. 4(b). The real part of the noisy signal $z(n)$ is shown in Fig. 5(a). Figure 5(b) displays the EP of the noisy signal with SNR = 0 dB. Using 60 Monte-Carlo runs in such a way that $z^{(i)}(n) = y(n) + \eta^{(i)}(n)$, $i = 1, 2,\ldots, 60$, Fig. 6(a) shows the real part of the reconstructed signal $\hat{y}(n)$ versus the original one, $y(n)$, and Fig. 6(b) displays the EP of $\hat{y}(n)$ after utilizing the EB to remove the noise.

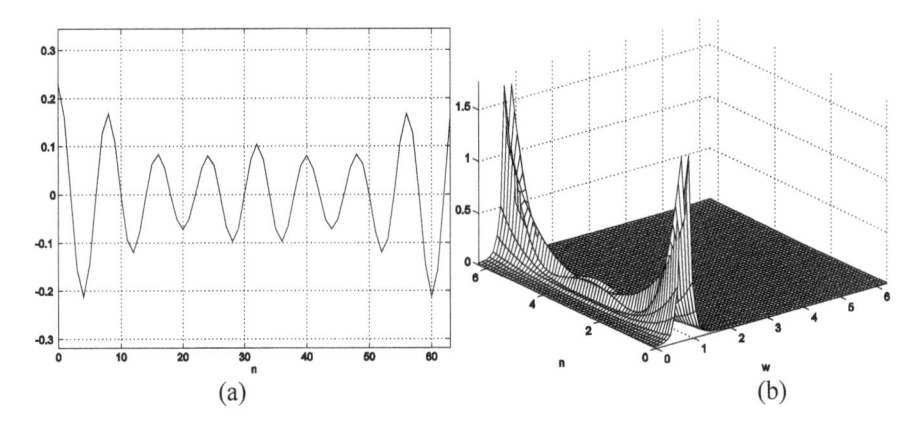

Fig. 4. (a) The real part of $y(n)$, (b) The EP of $y(n)$

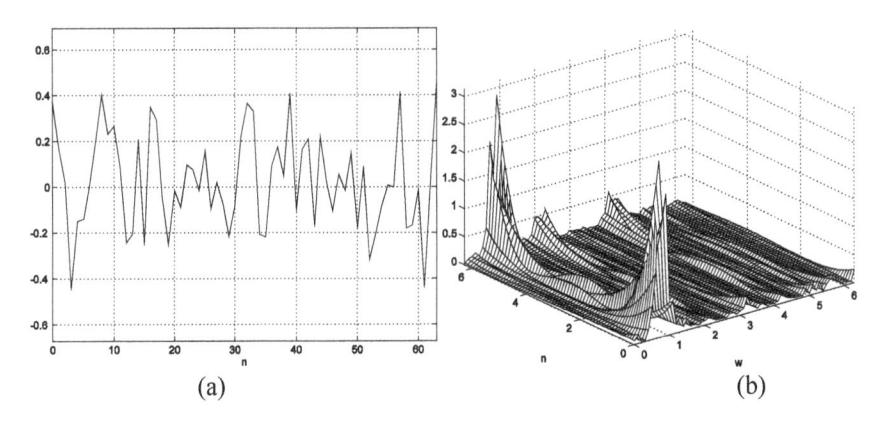

Fig. 5. (a) The real part of $z(n)$ with SNR = 0 dB, (b) The EP of $z(n)$

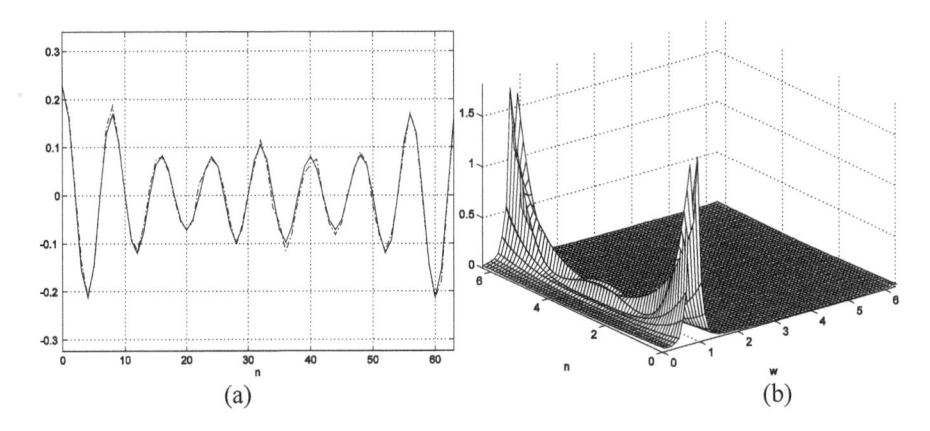

Fig. 6. (a) The real part of $\hat{y}(n)$ vs. $y(n)$ (smooth), (b) The EP of the signal $\hat{y}(n)$ processed by the EB

6 Conclusions

In this paper, the evolutionary bispectrum was explored and some of its properties were discussed, especially the ability of removing non-stationary Gaussian noise. Several algorithms for reconstructing the signal from the evolutionary bispectrum were given. Although many of its properties were introduced and proved, not all of them have been used yet, such as the evolutionary bicoherance function, which has been used for testing the linearity of the system in the stationary case. Moreover, since the evolutionary bispectrum of a Gaussian process is identically zero, it can be used as a test of Gaussianity of a non-stationary process. In addition, from the linear phase shift property of the evolutionary bispectrum, one may try to apply the evolutionary bispectrum for finding the direction of arrival of a signal.

References

1. Grenier, Y.: Time-dependent ARMA modeling of nonstationary signals. IEEE Trans. Acoust. Speech Signal Process. **31**(4), 899–911 (1983)
2. Rao, S.T.: The fitting of nonstationary time series models with time-dependent parameters. J. Stat. Soc. Ser. B **32**, 312–322 (1970)
3. Priestley, M.B.: Non-linear and Non-stationary Time Series Analysis. Academic Press, New York (1988)
4. Kenny, O.P., Boashash, B.: Time-frequency analysis of backscattered signals from diffuse radar targets. In: IEE Proceedings F Radar and Signal Processing, vol. 140, no. 3, pp. 198–208, June 1993
5. Wood, J.C., Buda, A.J., Barry, D.T.: Time-frequency transform: a new approach to first heart sound frequency dynamics. IEEE Trans. Biomed. Eng. **39**(7), 730–740 (1992)
6. Haykin, S.: Advances in Spectrum Analysis and Array Processing. Prentice-Hall, Englewood Cliffs (1991)
7. Giannakis, G.B., Mendel, J.M.: Identification of nonminimum phase system using higher order statistics. IEEE Trans. ASSP **37**(3), 360–377 (1989)
8. Nikias, C.L., Petropulu, A.P.: Higher-Order Spectra Analysis. Prentice-Hall, Englewood Cliffs (1993)
9. Priestley, M.B.: Spectral Analysis and Time Series. Academic Press Inc., Cambridge (1981)
10. Kayhan, A.S., El-Jaroudi, A., Chaparro, L.F.: Evolutionary periodogram for non-stationary signals. IEEE Trans. Signal Process. **42**, 1527–1536 (1994)
11. Pincinbono, B. (ed.): Time and Frequency Representation of Signals and Systems. Springer, New York (1989)
12. Ye, Y., Tugnait, J.K.: Noisy input/output system identification using integrated polyspectrum. In: ICASSP, vol. IV, pp. 452–455 (1993)
13. Nikias, C.L., Mendel, J.M.: Signal processing with higher-order spectra. IEEE Signal Process. Mag. **10**, 10–37 (1993)
14. Cramer, H.: On some classes of nonstationary stochastic processes. In: Proceedings of Statistics and Probability, pp. 57–78 (1961)
15. Liporace, L.A.: Linear estimation of nonstationary signals. J. Acoust. Soc. Am. **58**, 1288–1295 (1975)
16. Sundaramoorthy, G., Raghuveer, M., Dianat, S.: Bispectral reconstruction of signals in noise: amplitude reconstruction issues. IEEE Trans. ASSP **38**(7), 1297–1306 (1990)
17. Matsuoka, T., Ulrych, T.J.: Phase estimation using the bispectrum. Proc. IEEE **72**(10), 1403–1411 (1984)

A Novel Approach to Blockchain-Based Digital Identity System

Md Abdullah Al Mamun[1], S. M. Maksudul Alam[1], Md. Shohrab Hossain[1(✉)], and M. Samiruzzaman[2]

[1] Department of CSE, Bangladesh University of Engineering and Technology, Dhaka, Bangladesh
{1405077.muam,1405087.smma}@ugrad.cse.buet.ac.bd,
mshohrabhossain@cse.buet.ac.bd
[2] Research scientist, BioNanoTech, London, UK
samiruzzaman@gmail.com

Abstract. In today's life, the demand for privacy and transparency is everywhere. Blockchain is going to lead the modern world with its revolutionary technology because of its security, transparency altogether. Identity is an important aspect in our daily life as a citizen and digital identity will play an important role to make our life easy and hassle free. The use of blockchain in digital identity is important to keep it secure and undoubtedly prevents the violation of privacy. However, there is lack of works that includes the most vital bio-metric information of an individual for creating a blockchain-based digital identity. In this paper, we have proposed a system on blockchain-based digital identity for individuals using bio-information. We have implemented our proposed digital identity system using ethereum smart contract. The system has fulfilled all the criteria of an identity system. Our results show that an intruder cannot access the personal data of a citizen; any unauthorized access attempt is denied instantly, thereby ensuring the privacy of private citizen data. All of the functionalities and the security issues have been tested. Our proposed blockchain-based digital identity system can be useful for the government of a country to provide its citizen a highly secured digital identity. It will make the life of all citizens hassle free without requiring them to carry any paper document.

Keywords: Blockchain · Proof of work · Smart contract · SHA256 · Hashing · Address · Smart contract deployment · Testnet

1 Introduction

Blockchain is a new technology to decentralize [1] the centralized systems. It was first proposed by a group of researchers in 1991. Simply blockchain means the chain of *blocks*. Each block contains information that can be of different types. Tampering blocks becomes difficult because to tamper a block, it is needed to recalculate the *proof of work* of the following blocks. That is why the blockchain technology is more secured and trusted. Another reason of security is, as it is a distributed system, creating a new block should be validated by all of the nodes in the network.

Most of us spend a very considerable amount of our lives online. We try to use online services in order to avoid wasting time. So, We badly need an digital identity which will uniquely identify ourselves. The current world population is 7.7 billion as of February 2019 according to the most recent United Nations estimates elaborated by Worldometers [2]. It is a major concern that we have to deal with huge number of population of the world for giving each of them a digital identity. Now, we need a system that will store all the necessary information of a citizen of a country. Then every citizen will be responsible individually for his course of actions. On the other hand we also need to provide privacy to a citizen. We want no tampering of personal data stored in database and also want assurance that no one will ever be able to pretend to be someone else. This throws us a challenge of big data cause huge data of population needs to be handled very carefully. We need such system that can handle big data undoubtedly. In [3] Elena Karafiloski et al. have already explained about the solution for big data challenges in blockchain technology. Considering these circumstances blockchain will be best for implementing digital identity because the application of blockchain is increasing [4] day by day for its high security protection. Most trusted digital identification is like one of the main challenges facing the internet ever since it was invented, because none of the traditional, offline means of verifying that someone is who they say they are applying [5]. Besides, a digital identity may generate logical questions about central points of failure and surveillance states if these identities are created, stored, and managed by a central authority. So we need to move to a decentralized system for ensuring transparency.

There have been few research works on the creation of digital identity using blockchain technology. However, those works do not include bio-information. Takemiya et al. [6] proposed a method of creating digital identity using blockchain, in which, a private key and AES-256 algorithm are used for storing an individual information in hyperledger Iroha. Mudliar et al. [7] introduced blockchain-based identity system for Indian national identity, Adhaar. Yasin et al. [8] explains about online identity and smart contract management system. [9] presents the contribution of digital identity on blockchain to a smart city environment.

To the best our knowledge, there has been no earlier work on blockchain based digital identity using bio-metric information of an individual. Takemiya et al. [6] is the only the previous work that proposes a method to create a digital identity using private key. Other works on blockchain-based digital identity system in [6–9] did not use bio-metric information of an individual. Using only the private key of an individual is not enough for creating identity. If anyone knows the private key, he or she will manage to pretend to be someone else whose key has been lost. That is why individual's privacy is not protected. Therefore, extra protection is needed beyond the private key.

The *limitations* of the existing works are as follows. It is less secure since private Key may get lost or stolen and No bio-metric information have been used. Only Private Key has to be maintained. It is difficult for a person to always maintain the private key.

We, therefore, propose a system to use both bio-metric information, i.e. finger and iris pattern which are taken from each individual and a private key, defined by the corresponding citizen, for creating blockchain-based digital identity.

The *aim* of our work is to propose a digital identity system using the bio-metric information of an individual and a private key, defined by the corresponding citizen to identify the specific person uniquely.

Our *objective* of this work is to build a system which will be highly secured. As the information of every citizen can be stored altogether in a decentralized way in blockchain technology without tampering of the data, that is why Without one's fingerprint, iris pattern and private key, no one can try to be someone else.

The main *contributions* of our works can be summarized as follows:

- We have proposed blockchain-based digital identity system using individual's bio-information.
- We have implemented the digital identity system using Ethereum smart contract.
- We have presented the robustness of our implemented system.

Our proposed system ensures the highest security by using three step verification: finger pattern, iris pattern and a private key of an individual for creating digital identity using blockchain platform so that no intruder can impersonate others.

Our proposed system performs better than previous works because we have embedded three step verification in our system. We do not require the use of high CPU-intensive encryption-decryption computation using public or private key, rather we have only used SHA256 hashing algorithm that is used by blockchain technology itself. Our first step verification is to match the finger pattern; the second step is matching of the iris pattern and the final step is matching the private key of an individual.

The main advantages of our blockchain-based digital identity system are (i) every citizen is uniquely identifiable, (ii) decentralized, (iii) secured platform, and (iv) feasible to implement.

The rest of the paper is organized as follows. Section 2 describes about some basic blockchain terminologies. In Sect. 3, we explain about major research works which have been done so far on digital identity based on blockchain technology, followed by the proposed digital identity model using blockchain technology in Sect. 4. Section 5 contains the implementation of our works using Ethereum smart contract on online Remix IDE. In Sect. 6, we explain the deployment of our Ethereum smart contract. Finally, Sect. 8 has the concluding remarks.

2 Blockchain Terminologies

As Blockchain is a new technology compared to other technology. There are some core concepts about blockchain [10] which will be discussed below in details.

2.1 Blockchain

Blockchain is a collection of *blocks* which contain information. A blockchain is maintained and managed by a peer to peer network. A large number of computers/devices are connected to the network. A computer/device connected to the network is called a node. The nodes basically store the blockchains and ensure the data protection. All the nodes over the network have the same copy of the blockchains. As a result just not like other centralized database system, blockchain database is distributed to every nodes in the network. Data security and validity are given the most priority in a blockchain network. This is why the whole network follows some rules and protocols.

Blocks: A block in blockchain is directly connected to the previous block by holding the hash of previous block. In Fig. 1, We can see three consecutive blocks. The first block is called *genesis block* which is created by the blockchain network. If anyone tries to tamper any block in the network, the hash of the tampered block will change automatically. As a result, there will be a mismatch between the hash of current block and hash of previous block located in next block as consecutive blocks are cryptographically linked together. If mismatch occurs then all the blocks, following the tampered block, will be invalid. Therefore, all those invalid blocks will be discarded. Highest security is ensured in this way.

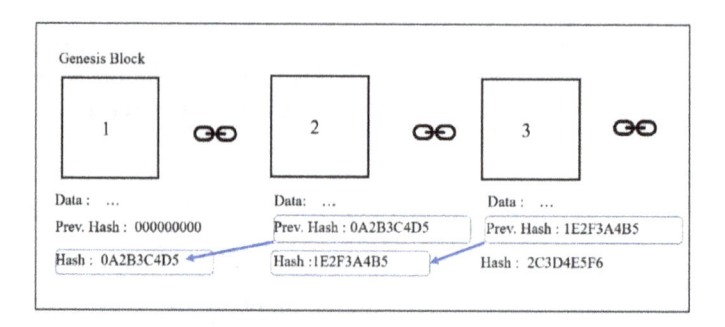

Fig. 1. Blocks are cryptographically linked together

But now a days, computers are very efficient in calculation and it can easily validate the following blocks by recalculating their hash. To get out of this problem, *proof of work* is maintained to give higher security to the network. *Proof of work* is the mechanism that slows down the creation of the new blocks in the network. There is another segment in a block which is called nonce. Basically a nonce is a number. We can only change the nonce to change the hash of whole block for mining the block. Mining block means legally establishing the block to the ledger which will be distributed over the whole network.

SHA-256 Hashing: Blockchain uses SHA-256 for its hashing algorithm to provide the highest security of its data. Hash is generated by solving highly computational mathematical problem in SHA-256 hashing algorithm. Any change in the block leads to generate different hash for the same block. Every hashing algorithm requires some features that provide highest security for private data. This SHA-256 hashing algorithm possesses some characteristics. They are given below:

- Original input in hashing function can't be retrieved from the hashed output.
- Output is always same for same input in the hashing algorithm.
- Algorithm provides output quickly.
- A small change in input causes dramatically change in output which is known as avalanche effect.

2.2 Smart Contract

Smart contract has been developed on the blockchain technology. Smart-contract is a blockchain based program that performs a specific task. Smart-contract runs into blocks in a blockchain. User requests to perform the particular task to a smart-contract for which it has been designed. As it is unbiased and logically fair, everyone can trust a smart-contract. A smart-contract plays role just like a human individual in online world. Even most of the cases, it is much more fast, efficient, correct and honest than a person. It can be used as a substitute of a person. Smart contract prevents an intruder from monopolizing system data and to keep securing blockchains [11].

3 Previous Works

There are four recent works that are closely related to our blockchain-based digital identity systems. They are given below:

- Private key method.
- Integration of existing identity with blockchain system.
- Online identity and smart contract management system.
- Contribution of blockchain on digital identity.

Takemiya et al. [6] presented a way of creating digital identity using blockchain. In their work they have used a private key for encryption and stored that encrypted data on central server. They have encrypted the data using AES-256 algorithm and stored in hyperledger Iroha. Then the hash of the submitted data is checked with the hash of previously stored hyperledger data. Kumaresan Mudliar et al. [7] has proposed a model about integration of existing national identity with blockchain system. They have integrated Adhaar(Indian national identity) with blockchain and showed different applications of blockchain based identity. Affan Yasin et al. [8] have explained about online identity and smart contract management system. Their idea is to analyze a person's online activities to have a rough idea about the person using the peer to peer network to detect appropriate person to be engaged in a task. Rogelio Rivera et al. [9] introduced the contribution of blockchain on digital identity in a smart city environment.

4 Proposed Blockchain-Based Identity System

In this modern age of science, our demand is that we will make every movement of our life independently without leaking our privacy and also we expect transparency everywhere. Our blockchain-based digital identity system will provide the following advantages:

- No way to temper the identity of a person.
- Create a fair database of a whole nation in a country.
- No need of a person to have any carriable thing with him for the proof of his identity and flexible enough in changing private key.

In this section, we have proposed a model for creating a digital identity for the citizen of a country using blockchain technology which will be implemented in Sect. 5.

4.1 Interaction Diagram of Digital Identity System

In order to be a citizen of a country, A sequence of procedures for achieving citizenship must be followed. In Fig. 2, The interaction diagram has been illustrated properly.

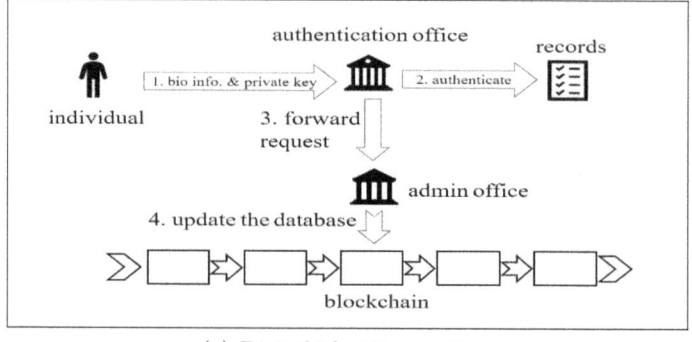

(a) Digital identity creation

Fig. 2. Interaction diagram of Digital identity creation

Firstly, Each person will provide his personal details, finger Prints, iris pattern and a private key to the local govt. office. **Secondly,** The local govt. office will verify the personal details of an individual. **Then,** Local govt. office will approve the request if the personal details are valid then forward the individual's request to admin office. **After that** Admin will approve the local govt. office request and generate digital identity of the corresponding person using his bio-information and private key. At last admin updates the database (ledger) with appropriate information. **Finally,** After successfully completing all of these procedures, A person will be able to be a citizen of a particular country.

4.2 Assumptions of Our Digital Identity System

In order to simplify the process of taking fingerprint and iris pattern inputs of an individual, we have assumed that after scanning the fingerprint on a trusted suitable device (Finger print Scanner), it returns a string as output, denoted by α_F in this paper. The iris pattern scanner device also returns a string based on a person's iris pattern, denoted by α_I in this paper. Naturally, the output string given by both the finger print detector and iris pattern detector is unique for each person since finger pattern and iris pattern is unique for each individual person all over the world.

4.3 Proposed Algorithm

We use bio-metric information to generate a person's identity (hash of finger print, iris pattern and a private key). The recorded data is used to check the validity of a person's identity. In this section, we explain the detail steps of algorithm for creating digital identity.

Step 1:
taking personal details, finger prints, iris pattern and a private key from the person.
Step 2:
verify the personal details by local govt. office.
Step 3:
approved the person's request by local govt. office based on person's valid information.
Step 4:
forward the request to admin office.
Step 5:
admin generates final string from finger prints, iris pattern and the private key provided by the person.
Step 6:
apply SHA256(Hashing) to the final string and assign a unique address by the system.
Step 7:
approve the request and store all activities to ledger with this identity by the admin.

While completing the above steps if any error or discrepancy occurs then the process will not be completed and database will not be effected with the actions completed so far. The above steps are described elaborately in the followings:

Taking Personal Details: Each person will provide his all necessary personal information e.g. name, father's name, mother's name, birth certificate, religion, permanent and current address to the local govt. office. The local govt. office will verify the personal details status by enquiring the information provided by the person.

Taking Finger Prints and Iris Pattern: We need a fingerprint scanner for scanning the finger pattern of a person and also an iris pattern scanner for scanning the iris pattern of the eyes of a person. After scanning the finger pattern, the fingerprint scanner will return a string as output which has been introduced before as α_F and then after scanning the iris pattern of a person, the iris pattern scanner will return a string as output which was previously known as α_I.

Fix a Private Key: Each person will select a private key which can be any number or an array of characters. This private key is denoted by α_P. This key along with α_F and α_I are used to create digital identity. This private key can be changed further if the person thinks that the key might have been stolen or lost. The person will direct request to admin office with his own bio information for changing the key again in future.

Change a Private Key: When a person is already a valid citizen of a country then he can change his private key if he thinks that private might have been stolen or lost. The citizen will provide his/her bio-metric information to the admin office. Admin office will verify whether its a valid citizen's bio-information or not. If the information is valid then using the smart contract, admin changes the previous private key to new private key provided by the citizen. The newly generated hash is not even known by admin. Its all maintained by smart contract. In this case admin only verifies the citizen but nothing else, thereby ensuring the highest level of privacy.

Verify the Personal Details and Approve the Request by Local Govt. Office: The local govt. office will approve the request of the person for being citizen by validating the information provided by the person. After proper verification from the appropriate authority, e.g, local chairman, police stations then local govt. office will authenticate the person's request for the approval of being a citizen of the country and forward the request to admin office for finalizing the process. If the information provided by the person is found invalid then the request is discarded by the local govt. office.

Generate Final String: For ensuring the highest security for the digital identity system, Admin appends the string α_F with α_I and α_P. So, our final string is the combination of three identifiers of a person.

$$\text{Final String}, \Psi_{ID} = \text{Append}(\alpha_F, \alpha_I, \alpha_P)$$

SHA-256 Hashing and Assign a Unique Address by the System:
Finally, after getting the Final String (Ψ_{ID}), Admin performs the SHA-256 hashing algorithm on 'Ψ_{ID}'. The hashing function returns 64 characters where each character is 4 bits. Each character is in range from 0 to F in hexadecimal. This array of 64 characters are combinedly unique for each individual as 'Ψ_{ID}' is different for each person. Thus, the system uses the bio-metric information of a person to generate his digital identity. The system then stores the SHA-256 hashed value of 'Ψ_{ID}', i.e, hash of finger Print, iris pattern and a private key of the corresponding person in the ledger.

$$\alpha_{ID} \leftarrow SHA-256(\Psi_{ID})$$

So this 'α_{ID}' for a person stored in ledger must be unique person to person for sure. No matter the private key differs or not. Then system generates a unique address for 'α_{ID}' and Points 'α_{ID}' to a unique address. A citizen can only recover or change his key with the help of his bio-information because only the hash of his bio-information is pointed to a permanent unique address which is assigned to the citizen initially by smart contract. Once an address is used that will not be used by another citizen. Address varies citizen to citizen and address generation is automatic and it is generated by smart contract only. So, the generated address must be distinct.

Approves the Request of the Person of Being a Citizen by the Admin:
Finally admin approves the request of the person for being a citizen of a country and stores all private data of a citizen to ledger with this identity. Each person's ID (α_{ID}) is saved on the ledger on the Blockchain. When a person performs any online activities or transactions, he/she will give his finger pattern, iris pattern and private key. Based on the inputs, the recorded data will be used to check the validity of that person's identity by the smart contract. The given ID will be considered to be invalid if it does not match any of the stored ID in the ledger. If the ID is invalid, then all actions of the person will be aborted. Moreover, there is no way to impersonate since there is three step verification (finger pattern, iris pattern and private key). If the citizen forgets the private key or doubts that the key might have been stolen then he can apply to admin office directly for changing his private key in future. The admin office has only the authority to change the private key of a citizen. As the system can detect each person individually, every person is responsible for his own online activities. There is no way to temper the private data of others in the ledger of the blockchain as we described earlier in Sect. 2. After approving the digital identity of a person, all activities performed by the person will be stored and it is not public to others for privacy reasons.

5 Implementation of Digital Identity System

As blockchain is very new compared to other technology that is why the syntax of programming language and other utilities are changing day by day. It is extending its features more and more. The platform and utilities we have used are given below:

Ethereum Smart contract:
Ethereum [12,13] is an open blockchain platform that lets anyone build and use decentralized applications that run on blockchain technology.
Solidity:
Solidity [14] is an object-oriented, high-level programming language for implementing smart contracts.
Remix:
Remix [15] is a latest online integrated development environment (IDE) for solidity language.

We have implemented our code on Ethereum Smart contract for creating digital identity. Smart contract has been developed on the blockchain technology. We have written the functionality in the specified programming language (solidity), compiled the code and deployed it in the blockchain platform.

5.1 Function Implementation of Digital Identity Creation

All screenshots of implementation of the function that are called in our system while creating a digital identity for a citizen in remix IDE are given below:

We have made a smart-contract named "digitalIdentity". In the contract, an individual is treated as a person.

```
struct person
{
    string dateOfBirth;
    string privateKey;
    address[] parents;
}

address admin;
mapping(address=>person) citizens;
mapping(string=>address) hashToAddress;
mapping(string=>address) recoveryAddress;
mapping(string=>string) keys;
mapping (address => bool) validCitizen;
```

Fig. 3. Function implementation of digital identity creation (1)

In Fig. 3, an individual is represented by the "person" structure. The "citizens" is the list of all valid citizens of a country. Each citizen will have a unique bio-metric hashed value and a unique address. Each unique bio-metric hashed value(string type) of a citizen is pointed to a unique address of that citizen in "hashToAddress" list and this unique address is pointed to that particular person structure in "citizens" list. "RecoveryAddress" list is used to recover one's address if any citizen loses his key. It can only be done by admin. All private keys of citizens of a country is stored in "Keys" list that cannot be seen by anyone cause it is only accessed by smart contract. "ValidCitizen" is used to identify whether a person is a citizen or not. We have used modifier [16] "onlyAdmin" in our smart contract. The modifier is used to put the restriction to any function that the function can be used by the admin only. It ensures that only admin can call the modified function and access the database (ledger). There is only one admin. The address of Admin is "admin" here. A modifier is used so that only admin can make the changes to the ledger. The function "getAdmin" is used to get the address of admin. By calling this function We can see who the admin is.

```
function setHashToAddress(string memory hashWithkey, string memory hashWithoutkey, address _address)
onlyAdmin public
{
    hashToAddress[hashWithkey] = _address;
    recoveryAddress[hashWithoutkey]=_address;
}

function changePrivatekey( string memory hashWithoutkey, string memory newPrivateKey)
onlyAdmin public
{
    address _citizenAddress = recoveryAddress[hashWithoutkey];
    string memory oldPrivatekey = keys[hashWithoutkey];
    string memory oldhashWithkey = string(abi.encodePacked(hashWithoutkey, oldPrivatekey));
    delete keys[hashWithoutkey];
    delete hashToAddress[oldhashWithkey];
    citizens[_citizenAddress].privateKey = newPrivateKey;
    keys[hashWithoutkey] = newPrivateKey;
    string memory newhashWithkey = string(abi.encodePacked(hashWithoutkey, newPrivateKey));
    hashToAddress[newhashWithkey]=_citizenAddress;
}
```

Fig. 4. Function implementation of digital identity creation (2)

In Fig. 4, we can see two functions of the contract. Both are accessible by admin only. The function "setHashToAddress" takes the hashWithKey(combined hash of iris pattern, finger print and private key) and hashWithoutKey (combined hash of iris pattern, finger print), and a system generated unique address as parameters. The function maps the hashWithKey and hashWithoutKey to the same unique address. "hashWithoutKey" is used in order to change the private key of a citizen if the private key is lost or stolen. The function "changePrivateKey" is used to change a person's private key.

In Fig. 5, the function "setCitizenInfo" is used to set the personal information of a citizen. Only admin can access to this function. The function takes the personal information of a citizen, i.e., bio-metric hashed value with private key (hashWithkey), bio-metric hashed value without private key (hashWithoutkey), private Key, father's address, mother's address, date of birth and the unique address assigned to the citizen generated by smart contract as input. This function assigns all the information taken as input to the citizen's digital identity. It also validate the person's identity as valid citizen after assigning the information.

In Fig. 6, the function "getCitizenInfo" is used to get the information of a citizen. But for this, we have to pass the bio-metric hashed value with private key (hashWithKey) of the person as the parameter to the function and it will return the information of the person that was stored in ledger. The function

```
function setCitizenInfo
(
    string memory  hashWithkey,
    string memory  hashWithoutkey,
    string memory  _dateOfBirth,
    string memory  _privateKey,
    address _fatherAddress,
    address _motherAddress,
    address _citizenAddress
)onlyAdmin public {
        setHashToAddress(hashWithkey, hashWithoutkey, _citizenAddress);
        keys[hashWithoutkey] = _privateKey;

        citizens[_citizenAddress].dateOfBirth = _dateOfBirth;
        citizens[_citizenAddress].privateKey = _privateKey;
        citizens[_citizenAddress].parents.push(_fatherAddress);
        citizens[_citizenAddress].parents.push(_motherAddress);
        validCitizen[_citizenAddress] = true;

}
```

Fig. 5. Function implementation of digital identity creation (3)

```
function getCitizenInfo(string memory  hashWithkey)
onlyAdmin view public
returns (string memory, address, address)
{
    address _citizenAddress = hashToAddress[hashWithkey];
    person memory p = citizens[_citizenAddress];
    return (p.dateOfBirth, p.parents[0], p.parents[1]);

}

function getHashToAddress(string memory hashWithkey)
onlyAdmin public view
returns (address)
{
    return hashToAddress[hashWithkey];
}

function isCitizen(string memory  hashWithkey) view public
returns (bool)
{
    return validCitizen[hashToAddress[hashWithkey]];
}
```

Fig. 6. Function implementation of digital identity creation (4)

"getHashToAddress" function is used to get which hashWithKey is mapped to which address. It takes hashWithKey as input at return the corresponding address. The function "isCitizen" is used to verify if a person is the valid citizen or not. The function takes the hashWithKey of a person and returns *true* if the person is a valid citizen, else returns *false*.

6 Results

We have implemented our proposed algorithm in Sect. 4 using solidity language. We have done the experiment on Remix IDE.

We have deployed our smart contract into a test net from admin address: E and the smart contract address: G.

Our smart contract has been successfully deployed. We have called all the functions for setting and getting the information from blockchain ledger through our smart contract. We have also tried to intrude our system from unauthorized address to check the vulnerability of our system. Our system has prevented the unauthorized access successfully.

Here, we have used 9 addresses and 4 hashes for our testing purposes: The addresses are described below:

A: Random Citizen's address
B: Random Citizen's father's address
C: Random Citizen's mother's address
P1: Random Citizen's initial private key
P2: Random Citizen's new private key
P3: Intruder provided private key
E: Admin's address
F: Intruder's address
G: Contract address

The hashes are described below:

$\alpha 1$: Random Citizen's hashed value of bio-metric information with private key
$\alpha 2$: Random Citizen's hashed value of bio-metric information without private key
$\alpha 3$: Fake hashed value of bio-metric information with private key provided by intruder
$\alpha 4$: Fake hashed value of bio-metric information without private key provided by intruder

6.1 Deployment of Our Smart Contract

In order to test the functionality of our smart-contract, we have deployed it into a blockchain network.

In Fig. 7, it shows that the smart contract "digitalIdentity" has been deployed into the test net successfully. The contract was deployed from the admin address 'E'. The contract is deployed at the address 'G'.

Figure 8 shows the proof that all of our functions of the deployed smart contract are working properly. Here we can see the log of the remix IDE. The log shows that the function have been called for testing.

6.2 Assign a Unique Address for Each Citizen by the System

In Fig. 9, here the admin is mapping the hashWithKey ($\alpha 1$) and hashWithoutKey ($\alpha 2$) to the address "A" by calling the function "setHashToAddress". The admin address is "E". The Address "A" is generated uniquely by the system and it is assigned uniquely to each citizen.

Fig. 7. Smart contract has been deployed successfully

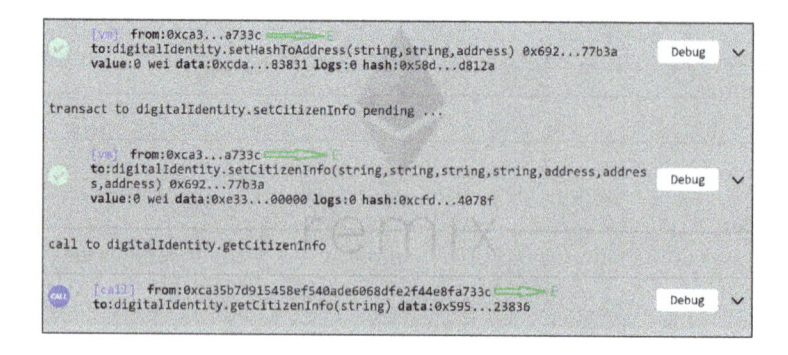

Fig. 8. Functions have been executed successfully

Blockchain-Based Digital Identity   107

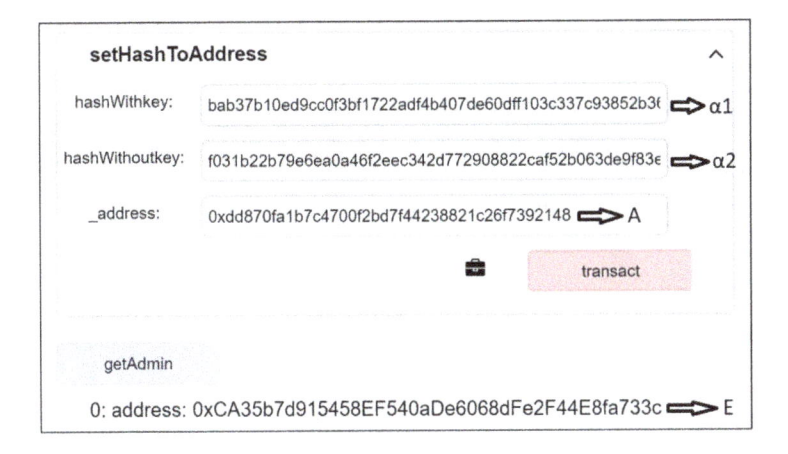

[Figure showing setHashToAddress form with hashWithkey: bab37b10ed9cc0f3bf1722adf4b407de60dff103c337c93852b3(→α1, hashWithoutkey: f031b22b79e6ea0a46f2eec342d772908822caf52b063de9f83e→α2, _address: 0xdd870fa1b7c4700f2bd7f44238821c26f7392148→A, transact button, and getAdmin returning 0: address: 0xCA35b7d915458EF540aDe6068dFe2F44E8fa733c→E]

Fig. 9. Mapping 'α1' and 'α2' to address 'A' by Admin 'E'

6.3 Entering Verified Citizen Entry by the Admin

In Fig. 10, The admin has set the information, i.e. hashWithKey (α1), hashWithoutKey (α2), person's provided private key "P1", father's address "B", mother's address "C" and person's unique address "A" generated by smart contract to create the digital identity of a citizen.

[Figure showing setCitizenInfo form with hashWithkey: bab37b10ed9cc0f3bf1722adf4b407de60dff103c337c93852b368d74850a286→α1, hashWithoutkey: f031b22b79e6ea0a46f2eec342d772908822caf52b063de9f83e300a8a55b881→α2, _dateOfBirth: "01Jan1900", _privateKey: "testingPrivateKey"→P1, _fatherAddress: 0x583031d1113ad414f02576bd6afabfb302140225→B, _motherAddress: 0x4b0897b0513fdc7c541b6d9d7e929c4e5364d2db→C, _citizenAddress: 0xdd870fa1b7c4700f2bd7f44238821c26f7392148→A, transact button]

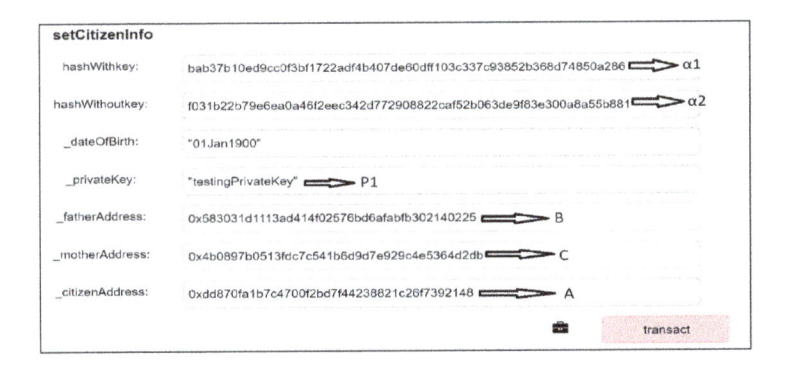

Fig. 10. Admin sets details of the citizen having address 'A'

6.4 Checking the Validity of a Citizen

In Fig. 11, The function "isCitizen" verifies the validity of a person as a citizen. It returns *true* if the person is a valid citizen, else returns *false*. The function is called with the hashWithKey (α1) and it returns whether the person with hashWithKey (α1) is a valid citizen or not.

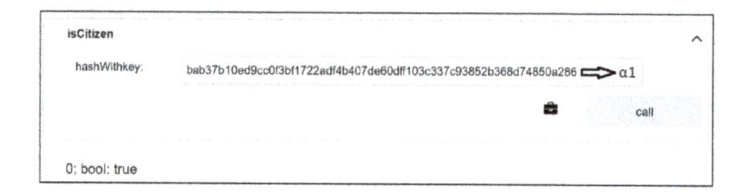

Fig. 11. Checking the citizenship status of a person having hash '$\alpha 1$'

6.5 Views a Citizen Information

In Fig. 12, Admin has called the function "getCitizenInfo" to get the information of the particular citizen with hashWithKey ($\alpha 1$). The function returned his information, i.e. date of birth, father's address, mother's address. The address of the particular person can be gained by the admin call of function "getHashToAddress" with the hashWithKey. We can see the address for the particular citizen is "A".

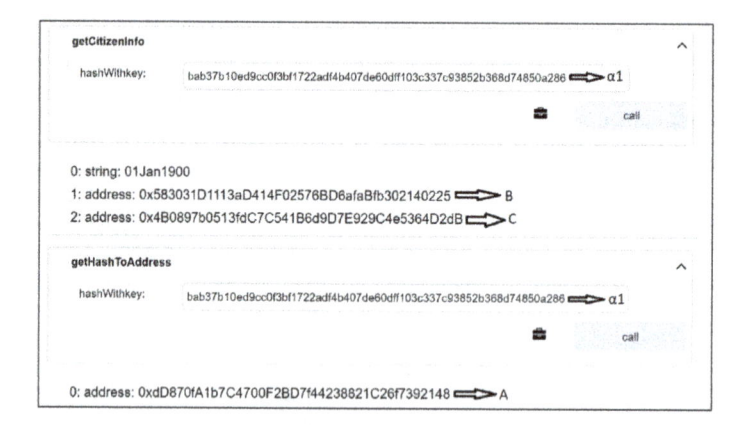

Fig. 12. View the information and address of a citizen having hash c by Admin

Fig. 13. Change the private key of a citizen having hash without Private key '$\alpha 2$'

6.6 Changing Private Key of a Citizen

A citizen can change his private key in future if he wants to. In Fig. 13, we can see that a the function "changePrivateKey" is being called to change the private key of the person whose hashWithoutKey is "$\alpha 2$". If everything is okay(admin calls the function), the person's new private key will be "P2".

6.7 Attack by an Intruder

In Fig. 14, we can see that an intruder (any random person except the Admin) whose address is "F" (intruder's Address) has tried to create a fake digital identity with the provided information, i.e., hashWithKey "$\alpha 3$", hashWithoutKey "$\alpha 4$", intruder's private key "P3", father's address "B", mother's address "C" and citizen's address "A".

6.8 Fraud Detection

In Fig. 15, we can see the privacy protection of our system. The system does not allow any person except admin to modify the private data of a citizen. An intruder with the address "F" tried to call the function "setCitizenInfo" to create a fake digital identity. But our system denied the request. Thus unauthorized access is aborted. If an unauthorized person tries to access or modify any information in blockchain ledger then he will not be allowed to do that. As we can see that an intruder with address "F" who tried to create a fake citizen entry but then the unauthorized access was denied. So our smart-contract did not let him/her to do the unauthorized task An error message has been shown in Fig. 15. Thus the highest security of personal data is ensured cause no intruder can access any confidential information of others except the proper authority. Here admin is the authority.

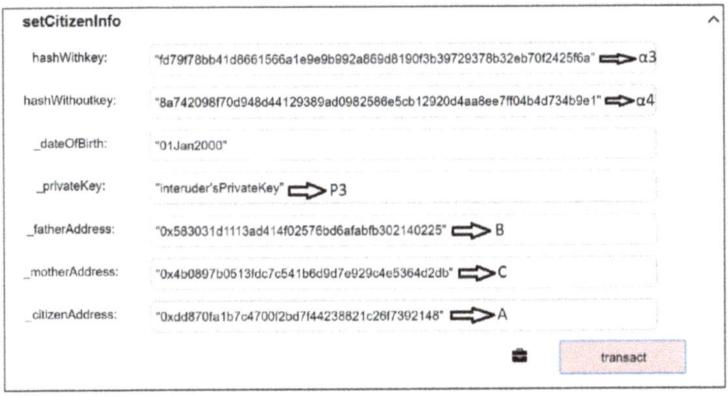

Fig. 14. Intruder is trying to create fake digital identity

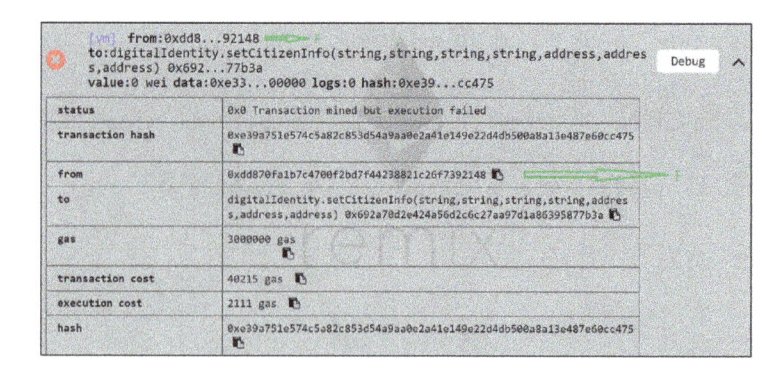

Fig. 15. System denied the intruder's request

6.9 Summary of Results

Smart contract "digitalIdentity" was successfully deployed into the test net. Digital identity was created using bio-information and private key of a person. Personal data was saved in ledger with this identity. Valid access to the information was ensured. Unauthorized access and modification of any data was impossible. System denied the unauthorized request.

So finally, we have ensured some important features of our blockchain-based digital identity described as follows:

- One person cannot pretend to be someone else as it is impossible to match two different bio-information.
- Government can keep track all activities and information of a citizen using the identity.
- We have not used too much complex model. So a person will not face much difficulty while using it.
- a person can be easily be identified if he is a valid citizen or not.
- There is no way of tempering private data because blockchain provides highest security protection.

The following Table 1 shows the proof of our work:

Table 1. Findings of our digital identity creation

Features	Result	Comments
Uniquely identification	Yes	shown in Fig. 9
False citizen entry	No	shown in Fig. 10
Validation of a Citizen	Yes	shown in Fig. 11
Confidentiality and flexibility ensured	Yes	shown in Fig. 13
Successfully attack by Intruder	No	shown in Fig. 14
Privacy protection	Yes	shown in Fig. 15

7 Future Works

We have provided three step verification layer, i.e. finger pattern, iris pattern and private key. But if a person is physically disable then we may face difficulties to take finger pattern or iris pattern. On the other hand if a citizen becomes physically disable in future then he may not be able to provide his finger pattern or iris pattern. In that case private Key is only the protection layer for them in our algorithm. We can also apply this digital identity system on other systems for making our life more easy and secure.

8 Conclusion

Blockchain is really undeniably a great invention of modern technology. The main mechanisms of blockchain are no more in vague to us. The core concepts of blockchain technology such as ledger, blocks, hashes, blockchain protocol, mining methods and so on have been introduced here. Our aim was to create a secured, feasible and meaningful digital identity. We have deployed our smart contract into a test net and Our smart contract was successfully deployed. We have also tried to intrude our system from unauthorized address to check the vulnerability of our system. Our system has handled the unauthorized access successfully. The ethereum platform of blockchain technology has been used in our work. All our process have been performed by programs(smart contract) in the blockchain. All of our information have also been stored in blockchain. Our model has included the bio-metric information, i.e. iris pattern, finger print of a person to create his identity. To add an extra layer of protection we have included private key. Still our system is much better than any of existing manual processes. So far, our work is mainly focusing on creating a secured and feasible blockchain-based digital national identity. Our algorithm is not so complex but much more applicable in real life. Considering knowledge level of majority of the people, we tried to keep our system simple. This can be the initiative of our march to the digitization of our lives.

Acknowledgment. This research was supported by SciTech Consulting and Solutions, Dhaka, Bangladesh.

References

1. Buterin, V.: Blockchain: The Meaning of Decentralization (2017). https://medium.com/@VitalikButerin/the-meaning-of-decentralization-a0c92b76a274?fbclid=IwAR0e2zSjubCjzM25PMsb3Y-G5g490Vpsv7U9GBnHPu5Xion-4dWzMMhIQYI
2. Current World Population: World Population Clock-Worldometers (2019). http://www.worldometers.info/world-population/
3. Karafiloski, E., Mishev, A.: Blockchain solutions for big data challenges: a literature review. In: 17th International Conference on Smart Technologies (EUROCON), 6–8 July, Ohrid, Macedonia, pp. 763–768. IEEE (2017)

4. Dai, F., Shi, Y., Meng, N., Wei, L., Ye, Z.: From bitcoin to cybersecurity: a comparative study of blockchain application and security issues. In: 4th International Conference on Systems and Informatics (ICSAI), 11–13 Nov, Hangzhou, China, pp. 975–979. IEEE (2017)
5. knowing me, knowing you: Self-Sovereign Digital identity and the future for charities. https://www.cafonline.org/about-us/blog-home/giving-thought/the-future-of-doing-good/self-sovereign-digital-identity-and-the-future-of-charity
6. Takemiya, M., Vanieiev, B.: Sora identity: secure, digital identity on the blockchain. In: Annual Computer Software and Applications Conference (COMPSAC), 23–27 July, Tokyo, Japan, pp. 582–587. IEEE (2018)
7. Mudliar, K., Parekh, H., Bhavathankar, P.: A comprehensive integration of national identity with blockchain technology. In: International Conference on Communication information and Computing Technology (ICCICT), 2–3 February, Mumbai, India, pp. 1–6. IEEE (2018)
8. Yasin, A., Liu, L.: An online identity and smart contract management system. In: 40th Annual Computer Software and Applications Conference (COMPSAC), 10–14 June, Atlanta, GA, USA, vol. 2, pp. 192–198. IEEE (2016)
9. Rivera, R., Robledo, J., Larios, V., Avalos, J.: How digital identity on blockchain can contribute in a smart city environment. In: International Smart Cities Conference (ISC2), 14–17 Sept, Wuxi, China, pp. 1–4. IEEE (2017)
10. Zheng, Z., Xie, S., Dai, H.-N., Chen, X., Wang, H.: Blockchain challenges and opportunities: a survey. Int. J. Web Grid Serv. **14**(4), 352–375 (2018)
11. Watanabe, H., Fujimura, S., Nakadaira, A., Miyazaki, Y., Akutsu, A., Kishigami, J.: Blockchain contract: Securing a blockchain applied to smart contracts. In: IEEE International Conference on Consumer Electronics, 7–11 Jan, Las Vegas, NV, USA, pp. 467–468 (2016)
12. ETHEREUM, a next generation of Blockcahin and working mechanism of ethereum as a virtual machine (2018). http://www.ethdocs.org/en/latest/introduction/what-is-ethereum.html
13. ETHEREUM development Tutorial (2018). https://github.com/ethereum/wiki/wiki/Ethereum-Development-Tutorial
14. SOLIDITY-solidity 0.5.6 Documentation (2018). https://solidity.readthedocs.io/en/v0.5.6/
15. REMIX- solidity IDE (2018). https://remix.ethereum.org/
16. Solidity Modifier Tutorial-Control Functions with Modifiers. https://coursetro.com/posts/code/101/Solidity-Modifier-Tutorial---Control-Functions-with-Modifiers

Towards Blockchain-Based GDPR-Compliant Online Social Networks: Challenges, Opportunities and Way Forward

Javed Ahmed[1,2](✉), Sule Yildirim[1], Mariusz Nowostawski[1], Mohamed Abomhara[1], Raghavendra Ramachandra[1], and Ogerta Elezaj[1]

[1] Norwegian University of Science and Technology, Gjovik, Norway
javed.ahmed@ntnu.no
[2] Sukkur IBA University, Sukkur, Pakistan

Abstract. Online Social Networks (OSNs) are very popular and widely adopted by the vast majority of Internet users across the globe. Recent scandals on the abuse of users' personal information via these platforms have raised serious concerns about the trustworthiness of OSN service providers. The unprecedented collection of personal data by OSN service providers poses one of the greatest threats to users' privacy and their right to be left alone. The recent approval of the GDPR (General Data Protection Regulation) presents OSN service providers with great compliance challenges. A set of new data protection requirements are imposed on data controllers (OSN service providers) by GDPR that offer greater control to data subjects (OSN users) over their personal data. This position paper investigates the link between GDPR provisions and the use of blockchain technology for solving the consent management problem in online social networks. We also describe challenges and opportunities in designing a GDPR-compliant consent management mechanism for online social networks. Key characteristics of blockchain technology that facilitate regulatory compliance were identified. The legal and technological state of play of the blockchain-GDPR relationship is reviewed and possible ways to reconcile blockchain technology with the GDPR requirements are demonstrated. This paper opens up new research directions on the use of the disruptive innovation of blockchain to achieve regulatory compliance in the application domain of online social networks.

Keywords: Privacy · Data protection · GDPR · Blockchain · Online social network · Transparency · Subject rights

1 Introduction

The recent Cambridge Analytica scandal on the abuse of personal data by social media for leveraged political influence has raised serious concerns regarding technical, commercial, political and ethical aspects of personal data collection. The scandal brought up the fact that service providers, such as Facebook and Google, in particular, exhibit enormous social influence that can shake or derail the democratic foundation of western societies. Since its inception, Facebook has collected 300 petabytes of personal data which is increasing at the speed of 4 new petabytes of data per day [27]. In the current Big Data era, data is an immensely valuable asset in an economy. It is commonly considered as the oil of the 21st century, which is not only fueling the success of the tech giants (i.e. Facebook, Google, Apple, Amazon) but also driving innovation and economic growth. The current situation is that the benefits of a data-driven society are reaped by a few multinational organizations that make the majority of their profit through services offered to users who pay for them with their personal data. Users have little or no control over how their personal data is stored and used. In recent years, mainstream media has been recurrently addressing controversial incidents related to privacy breaches [18].

The GDPR [28,30] came into force across Europe on 25th May 2018. It aims to give back control over personal data to the data subjects by imposing new data protection requirements on data controllers and processors. GDPR presents various challenges for Online Social Networks (OSNs), not only from a legal perspective but also from a technical view, mainly in the areas of data management and automation. OSNs are not fully prepared to comply. And when they attempt to comply, there are often major gaps [4]. GDPR recognizes data subjects' consent as a legitimate ground for data processing. The main aim of promoting the notion of consent is to provide data subjects control over their personal data. At present, consent management mechanisms in OSNs are either non-existent or not GDPR compliant [11]. It is not an easy task for OSNs to attain consent compliance. A consent compliance mechanism must have certain characteristics that are acceptable to both data subjects and data controllers. These issues uncover new research directions and pose interesting research challenges.

In this paper, we identify challenges imposed by the new GDPR data protection requirements for OSN service providers. These requirements aim to offer OSN users more control over their personal data, while at the same time enabling transparency in data processing and sharing activities carried out by the service providers and third parties. We also identify some of the opportunities offered by the disruptive innovation of blockchain that facilitates regulatory compliance by maintaining tamper-evident audit logs for information accountability. We also explore whether it is possible to reconcile blockchain with new requirements of GDPR. This research paves the way for designing a block-based GDPR compliant consent management model for personal data processing and sharing in online social networks. The key characteristics of blockchain technology including transparency and decentralization add value to the consent management model for regulatory compliance.

An explanation for the use of blockchain technology for GDPR compliance is still necessary, as the current research literature suggests that two initiatives (GDPR and Blockchain) are at odds [20]. They appear to be at odds with each other until the underlying principles of GDPR and Blockchain are observed. Both share common principles of data privacy and give data subjects more control over their digital private data. Both GDPR and blockchain aspire to increase integrity, trust, and transparency in a generally unsafe environment. GDPR does so by imposing responsibilities upon data controllers and processors. GDPR assumes to the extent that data controllers and processors are centralized actors with control over the system. GDPR compliance approaches based on a centralized architecture result in limited transparency and a lack of trust. On the other hand, blockchain ensures trust and transparency by utilizing the computational power of the masses and by sharing the register with all peers in the P2P network. The unprecedented transparency provided by blockchain technology sits uneasily with GDPR obligations related to privacy and information confidentiality. The dilemma with adopting blockchain for consent management is in finding the trade-off between transparency and information confidentiality. One of the solutions is to use private blockchain that allows only permitted parties to have access to all transactions. However, private blockchain thus loses the primary advantage of decentralization. Moreover, a dishonest central authority is capable of tampering the transaction history for personal gain. Wang et al. [32] proposed a framework that preserves information confidentiality without compromising transparency using zero-knowledge proof (ZKP). We conclude that prominent features of the blockchain technology can be effectively utilized to manage personal data full compliance with the GDPR legislation.

The rest of the paper is organized as follows. In Sect. 2, we provide the theoretical background of GDPR, Blockchain and Online social networks and how they interplay to achieve regulatory compliance. This section also presents a comparative analysis of existing literature in the domain. In Sect. 3, we identify challenges imposed by the new regulatory requirements of GDPR for OSN service providers. To better understand legal obligations, we also identify various stake-holders involved in data processing and sharing activities carried by social web systems. In Sect. 4, we discuss some of the opportunities offered by inherent blockchain technology features for designing GDPR compliant online social networks. Finally, we conclude the paper with future research directions and open research questions in Sect. 5.

2 Theoretical Foundation

This section describes the basic building blocks of online social networks, blockchain technology and general data protection regulation (GDPR). We also discuss their interplay that facilitate regulatory compliance.

2.1 Online Social Networks

Online social networks have undergone exponential growth in the last decade. Topmost visited sites by internet surfer are online social networks[1]. Surfing social

---

[1] Alexa http://www.alexa.com/topsites.

media is the fourth most popular activity on the internet nowadays[2]. Socializing with friends and family across the globe via online social networks is a cost-effective mechanism for the masses. A large proportion of the success of these platforms can be attributed to the fact that they allow their users to create their own space and a great way to connect with like-minded people, learn and share knowledge. Online social networks promote the vision of a human-centric web that is a key breakthrough attributed to these platforms. The primary source of information in the human-centric web is users, their network and interests that reside entirely in social networking services [2]. A widely used definition given by Boyd et al. [8] captures all the key elements of OSNs. The authors define OSNs as follows:

> An online social network is a web-based service that allows individuals to construct a public or semi-public profile within the service; articulate a list of other users with whom they share a connection, and view and traverse their list of connections and those made by others within the system.

According to this definition, every OSN user can create his/her own profile. An OSN user maintains his digital persona using profile that contains a number of attributes related to user such as demographics information, interests, preferences and various types of user generated content. Connections is another important feature offered by online social networks. Connections refer to existing social relationships of the OSN users. Many online social networks label connections as a friend; however, it is problematic due to the reason that all connections in the network are treated equally. Whereas, connections established between many users whose relationship may be better described by a different label such as family, friends, colleagues, acquaintances, etc. Labeling connections as friends precludes differentiation for selective information sharing. According to the aforementioned definition, the third feature of OSNs is traversing connection. This feature allows OSN users to find each other and construct a networked community within which they can share information.

The most popular fora for self-representation and user interactions are online social networks these days. Internet surfers join social networks to present themselves and share huge amounts of personally identifiable information. This behavior of OSN users causes a serious privacy threat to them. Online social network usage gives rise to several privacy threats including privacy threats related to OSN users [15], third party applications [3] and OSN service providers [16]. Online social networks provide a multitude of privacy tools to mitigate privacy threats concerning to OSN users. Despite the array of privacy controls, current online social networks fail to provide an effective mechanism to manage access to uploaded user content. The main reason for this failure is the shortcoming of online social networks to represent diverse social relationships. This problem has been addressed extensively in existing literature [1]. In this paper, we are addressing the issue of privacy threats related to service providers and third parties.

---

[2] Nielsen http://www.nielsen.com/.

Privacy threat related to a service provider involves the relationship between user and service provider that is based on trust. The service provider has full access to any user data because the OSN's underlying system is designed and configured by the service provider. The important privacy threats concerning service providers are data retention, data selling, and targeted marketing. Recent scandals like Cambridge Analytica have shaken the trust that users have put in their service providers. Moving towards a decentralized architecture is one of the most straightforward solutions to mitigate service provider related privacy threats. Decentralization is among the distinct blockchain features that play a vital role in challenging the monopoly of financial institutions by introducing cryptocurrencies. The blockchain-based solution may provide decentralization to address service provider related privacy threats.

With the emergence of Web 2.0, online social networks are offering open platforms to enable external developers to build applications that provide seamless integration of profile data with third-party applications. Facebook[3] and Google's Open Social[4] are leading this effort. These platforms have opened doors for external developers to launch their applications for social networking sites. These third-party applications pose serious privacy risks for online social network users because installed applications receive privileges equal to those of profile owners and can access users' profile data. Online social network users are unaware of the amount of data being exposed to external developers because such information flow is hidden from, or not clear to users. The mainstream media also recognized this data breach[5]. Enabling transparent data flow between service providers and application developers can mitigate privacy threats related to third parties. Transparency is a distinct feature of blockchain that may have a vital role in consent-based privacy compliance.

2.2 Blockchain

Due to the hype and success of cryptocurrencies, blockchain attracted significant interest from governments, businesses, capital markets, and the research community. It is foreseen as the core backbone of many future technologies such as the internet of things, smart cities, etc. Originally, the term blockchain was coined by a person using the name Satoshi Nakamoto in 2008 [22]. Blockchain technology is still in its infancy stage of development. The first phase of its development is termed as Blockchain 1.0. This embryo phase of blockchain development deals with cryptocurrencies such as bitcoin, ether, etc. Blockchain 2.0 refers to the second phase of its development which introduces the concept of smart contracts. Blockchain technology employs smart contracts to deals with issues of mutual trust and identity among participants. The next generation of blockchain technology will become a powerful tool for Industry 4.0 [24]. Blockchain technology

---

[3] Facebook for Developers, https://developers.facebook.com/.
[4] Open Social, https://www.getopensocial.com/.
[5] Millions of Facebook user records exposed in data breach, https://www.telegraph.co.uk/technology/2019/04/03/millions-facebook-user-records-exposed-data-breach/.

can be understood as an emerging distributed ledger technology (DLT) that enables applications to operate in a fully decentralized fashion without the need for any trusted central authority. Blockchain is secure and transparent by design and relies on well-known cryptographic tools and distributed consensus mechanisms to provide key characteristics such as anonymity, immutability, auditability, transparency and trust [34]. To better understand the core concepts of the blockchain, this section presents an overview of the technology.

Blockchain is a sequence of blocks and each block is cryptographically linked to the previous block after validation through a consensus decision. The genesis block of the blockchain has no previous block. All the blocks are linked together in the chronological order i.e. from the genesis block to the latest block. A block consists of a block body and block header. The block body is composed of a transaction counter and transactions. The block header includes various headers such as nonce, timestamp, parent block hash, Merkle tree root hash, etc. The concept of consensus mechanism is central to blockchain technology. What blocks are added to the ledger will be decided by an agreed-upon consensus mechanism. There exist a different kind of consensus mechanism such as proof of work, proof of stake, practical byzantine fault tolerance, etc. Instead of trusting a central authority, trust is placed in the algorithms underlying the consensus mechanism. This is the basis for the characterization of blockchain as a trust-less system. A digital signature scheme based on asymmetric cryptography is used in an untrustworthy environment to validate the authentication of transactions. Consensus mechanism and asymmetric cryptography are implemented to achieve ledger consistency and security.

Current blockchain systems are divided into three categories: public blockchain, private blockchain, and consortium blockchain [34]. Public blockchain is also known as permissionless due to its open nature. Any participant node can read, write and engage in the consensus process. Public blockchain is completely transparent and decentralized in nature. The typical example of a public blockchain is Bitcoin [31]. Private blockchain is also known as permissioned due to its closed nature. The private blockchain is limited to a specific organization. In private blockchain write permission is kept centralized and read permission may be public or restricted to specific nodes. Private blockchain is closed and centralized in nature. Consortium blockchain is also known as a hybrid. It is a partially decentralized blockchain. Pre-selected nodes will engage in the consensus process. In consortium blockchain read permission may be public or restricted to specific nodes. The typical example of consortium blockchain is Hyperledger.

With the emergence of blockchain technology, many decentralized services and applications are proposed that build on or use blockchain to achieve independence from centralized service providers' monopoly. Blockchain-based online social networking services can be designed to address the aforementioned privacy threats from third party and service providers. Blockchain technology enhances transparency. Each user has complete transparency over what data is being collected about him/her and how it is accessed. Blockchain asserts data ownership and user privacy by enabling transparency. At the same time, blockchain provides anonymity to its users by allowing them to create pseudo-anonymous transactions without the need for revealing personally identifiable information about them. Blockchain-based solutions not only provide greater transparency to OSN users regarding their personal data but also offer advantages towards regulatory compliance. In the context of newly enforced GDPR, the consent of the user is the legitimate ground of data processing. At present, most of the OSNs lack GDPR compliant consent mechanism which means users' lack of control over their personal data. Before introducing a blockchain-based consent management for OSNs, we present a brief overview of the legal obligations of GDPR on service providers and third parties.

2.3 General Data Protection Regulation

Since digital technology has profoundly changed the way how personal data is collected, accessed and used, on 25th May 2018, the European Commission (EC) implemented a new legislative framework called the General Data Protection Regulation (GDPR) (EU) 2016/679 [28,30]. The aim of this Regulation was to remedy the shortcomings of Directive 95/46/EC and to further harmonize the data protection rules within the EU as well as raise the privacy and data protection levels of affected individuals. GDPR has reshaped the way organizations approach data protection and data privacy. Organizations dealing with personal data of EU citizens must ensure they are compliant with the new GDPR requirements. The GDPR regulations are described in 99 articles that cover all aspects of personal data processing by organizations. GDPR extends the responsibility and accountability requirements of organizations involved in processing personal data of the EU citizens.

In the context of GDPR, three main roles are identified: data subject, data controller and data processor. A data subject is the owner of personal data. Personal data means any information pertaining to an identified or identifiable natural person. Data controller determines the purposes and means of processing personal data. The data controller is the point of accountability in GDPR. A data processor is responsible for processing personal data on behalf of a controller. GDPR sets out six core data processing principles that facilitate the protection of personal data processing. The first principle deals with the lawful, fair and transparent processing of personal data. The purpose limitation principle ensures the collection of data for specific purposes and prohibits processing for incompatible purposes. The data minimization principle discourages excessive collection of personal data and only adequate data collection is advised. The collection of accurate and up to date data is handled with the principle of accuracy. As per the storage limitation principle, data should not be stored for longer than necessary. Finally, the integrity and confidentiality principle deals with the secure processing of personal data. In addition to these data processing principles, GDPR recognizes the consent of a user as a legitimate ground for data processing. Data processing principles are used to derive a set of rights for data subjects. The most important rights of data subjects under GDPR are:

right of rectification, right of access, right to erasure and rights pertaining to automate processing.

GDPR is applicable to all organizations that process personal data of EU citizens whether based within the jurisdictional boundaries of the European Union or any third country. Compliance with the regulations is enforced by public authorities. Apparently, the supervisory authority in each EU member state is responsible for monitoring GDPR compliance. Data processing organizations are required to demonstrate their compliance with GDPR only in cases of data subjects lodging complaints with the supervisory authority about the misuse of their personal data. The supervisory authority has also extensive rights to access personal data processing activities in cases of suspected violations by organizations. Failure to comply with GDPR results in huge fines up to 20 million euros or 4% of the total annual profit. Due to the irregular nature of GDPR compliance verification by a supervisory authority, each organization has to prove and document that it has been continuously abiding by GDPR requirements. Moreover, due to the lack of transparency, it is beyond the capability of data subjects to recognize whether the data controller fully complies with GDPR and effectively protects his/her personal data. Therefore, a prospective GDPR compliant mechanism must inherit features of transparency and auditability to enable data subjects to oversee what data is collected and how it is processed by the data controller or processor.

3   Challenges of Blockchain-Based OSNs Under GDPR

With the introduction of blockchain technology, a plethora of decentralized services have been proposed that achieve independence from centralized entities. Blockchain can also be used in online social networks. Decentralization, transparency and distributed consensus give blockchain the potential to address most of the prevalent privacy concerns in OSNs. One of the initial efforts in this direction is Steem[6]. Steem is a blockchain-based social media platform that supports community building and social interaction with cryptocurrency rewards. Such OSN can be further made self-healing by a blockchain-based reputation system. One such system was proposed by Qin et al. [25]. The authors presented a blockchain-based academic social network and proposed a new consensus algorithm named proof of reputation (PoRe). Chen et al. [10] proposed a blockchain-based trusted social network that ensures privacy by doing peer-to-peer information exchange. The proposed model uses blockchain for limiting large-scale rumors spreading via online social networks. Blockchain technology has a set of very attractive features for applications in this domain. However, such applications are required to comply with the GDPR. Blockchain-based applications pose several challenges to regulatory compliance in the light of new changes imposed by GDPR. The following subsections present a detailed description of such challenges.

---

[6] Steem, https://steem.com/SteemWhitePaper.pdf.

3.1 Challenge of Informed Consent

The concept of consent originated in the field of medicine. Consent is aimed at providing the data controllers legitimate grounds for personal data processing. Consent has various forms, such as informed, explicit, unambiguous, broad, etc. Each of these forms is quite diverse in nature. GDPR promotes the notion of consent to provide data subjects full control of their personal data. The absence of consent means a lack of control for data subjects over their personal data. Consent ensures that the data subject has ownership of their personal data and they choose how to navigate in the digital data world. In this section, the aim is to present a set of guidelines to manage consent in online social networks taking into account the provisions of the GDPR.

Online social networks like Facebook have a long history of using the personal data of their users without obtaining explicit or direct consent. One such example is the US midterm election in 2010, where the company experimented with 61 million users and in attempts to influence their voting behavior [7]. GDPR came into force in May 2018 and is applicable to all companies (data controllers) that process personal data of EU citizens, whether operating in Europe or any third country. GDPR compliance is the only way forward for data controllers handling personally identifiable information of EU data subjects. Whereas, non-compliant data controllers are subject to hefty fines and may suffer loss of reputation among the increasingly privacy-conscious users. The current state of the art reveals that consent management mechanisms in online social networks are either non-existent or not GDPR compliant. Scientific literature also reveals the case of consent misuse in OSNs and one such example is the contagion study conducted by Facebook [19]. Facebook did not require any explicit user consent for the study on the grounds that users have already given broad consent when they signed up to use the social network. The study provoked extended criticism and Facebook publicly acknowledged and apologized for its fault. Being a leading OSN service provider, Facebook has been questioned by regulators over the years about its privacy practices[7]. It is yet to be seen how Facebook and other OSN service providers manage to comply with new changes imposed by GDPR.

A consent management mechanism constitutes a major step towards becoming compliant with GDPR. It is challenging to achieve consent compliance in current online social networks. A valid consent under GDPR must be freely given, specific, informed and unambiguous. Current state of the art reveals that consent given by users to online social networks lacks granularity because OSN service providers seek consent for several purposes bundled together and users do not have the freedom to give or deny consent for each purpose. One such example is Facebook's facial recognition feature that requires users to accept all purposes even if the user finds only one of them acceptable. If a user consents in such a setting, then it is not freely given consent. Another important feature of valid consent under GDPR is specificity that promotes transparency. As per the

---

[7] Mark Zuckerberg Testimony: Senators Question Facebook's Commitment to Privacy, https://www.nytimes.com/2018/04/10/us/politics/mark-zuckerberg-testimony.html.

aforementioned example of Facebook's facial recognition feature, single consent is sought for multiple purposes and the OSN service provider does not allow users to give specific consent for each purpose. Informed consent also promotes transparency by enabling data subjects to understand the nature of processing and the data collected and used in an intelligible format. However, the current terms and conditions of service providers do not represent a minimal set of information in plain and clear language. Data subjects should take clear affirmative action to give consent. Inactivity, silence or pre-ticked boxes do not make valid consent under GDPR.

Recently, several tools that focus on GDPR compliance have been released.[8],[9] These tools are designed to target self-assessment after completing standard questionnaire. The research community should focus on designing a consent management model that automatically checks if existing data processing and sharing activities are GDPR-compliant. Transparency with respect to the collection, processing, and sharing of personal data is a key enabler for OSN service providers to achieve GDPR compliance. Blockchain can play a vital role in order to provide said transparency with respect to personal data processing and sharing.

3.2 Challenge of Data Erasure and Amendment

GDPR introduces the concept of the right to be forgotten which empowers data subjects to request the data controller for the erasure of their personal data completely. This right is introduced in article 17 of the GDPR, but, it is not an absolute right and only applies in certain circumstances. Data subjects may request their data erasure only when one of the six legislative grounds outlined in the article applies. Moreover, the data controller can retain personal data for archiving purposes in the public interest under the legal protection of the GDPR. However, a data controller must be able to prove when one of the exceptions to the right to erasure applies. Therefore, the data controller must be able to erase the data subject's personal data from their records. The blockchain-based solution for OSNs brings many benefits, but it introduces a feature of immutability which guarantees that stored data is tamper-proof. This feature of immutability prohibits the straightforward application of the right to be forgotten. However, the regulation only applies to personal data stored in the blockchain. If the data is rendered completely anonymous, it falls out of the scope of the GDPR legal framework. Nevertheless, encryption is considered a pseudonymization technique and pseudonymized data continues to qualify as personal data and falls under the scope EU data protection regime. This section aims at setting guidelines on how to mitigate the challenge of data erasure and amendment taking into account the provisions of GDPR.

Recently, several solutions are being developed to achieve the goal of designing GDPR-compliant blockchain use cases. The straight forward solution may be

---

[8] GDPR Compliance Toolkit, https://info.nymity.com/gdpr-compliance-toolkit.
[9] Microsoft GDPR Detailed Assessment, https://aka.ms/gdrpdetailedassessment.

storing personal data off-chain where the blockchain merely holds proof that the data is valid [21]. Off-chain storage facilitates the right to erasure and amendment. Personal data could be restricted to a private permissioned blockchain instead of a public permissionless blockchain. Private permissioned blockchain will find it easier to apply the letter of the GDPR than public permissionless blockchain. Current state of art suggests many data obfuscation, encryption and aggregation techniques that can be used to turn personal data into digital signatures that are cryptographically linked to original data without actually revealing that data [13]. When applying such techniques to process personal data should take into consideration reversal and linkability risks. Farshid et al. [12] purposed design for a forgetting blockchain. The authors implemented a proof concept prototype that maintains most of the key features of blockchain technology but facilitates data erasure. The technique is applied to permissioned blockchains and it is evaluated with the help of domain experts. Bayle et al. [6] proposed a modular architecture that ensures GDPR compliance by providing the means to enforce the right to be forgotten. The mechanism to implement the right to be forgotten relies on the centralized infrastructure of the data controller. Geelkerken et al. [29] identified two ways in which the blockchain technology could be utilized to store personal data in compliance with the requirements of the GDPR. According to the authors the opaque blockchain instead of the transparent blockchain ideal for erasing and altering personal data, but it would require a trusted third party to comply with legal requirements.

From the GDPR perspective, public keys also constitute personal data. The pertinent question is why public keys cannot qualify as anonymous data. To qualify as anonymous data, the public key must irreversibly prevent the identification of a specific data subject. Blockchain history reveals that despite asymmetric encryption identification remains possible by connecting public keys with additional information that means public keys are pseudonymous data and fall under the scope of GDPR. It is not straight forward to design GDPR compliant solutions for public keys. The keys are an essential component of blockchain technology and constitute part of the transaction's metadata that is required for validation. Therefore, public keys cannot be moved to off-chain like transactional data. Despite the difficult nature of the problem, the research community suggested various techniques which include mixing services, ring signature, and zero-knowledge proof. A comprehensive overview of these techniques is presented in the research survey by Feng et al. [13]. These techniques are capable of anonymizing public keys for GDPR compliance that is evident from existing solutions based on these techniques such as CryptoNote, Monero, Zerocoin, and Zerocash.

3.3 Challenge of Identifying the Data Controller

The main roles identified in the GDPR context are the data subject, data controller and data processor. The data controller has a key role in determining the purposes and means of processing personal data in accordance with the constraints imposed by GDPR. The data controller is the point of accountability in the new legal framework. Therefore, it must be possible to identify a data controller, which is not always easy in the context of blockchain technology. From a technical perspective, a blockchain is a network consisting of nodes that send and receive messages in decentralized fashion without a central point of control, which makes it difficult to assign roles. These roles are designed to fit the traditional centralized client-server scenario whereby a single entity offers some services to data subjects pertaining to the collection and processing of personal data. In this section, the aim is to evaluate who qualifies as a data controller in different variants of blockchain technology.

To identifying a data controller public permissionless blockchain is an object of debate and less straight-forward. Public permissionless blockchains are distributed and decentralized peer-to-peer networks, where each node can read, write and engage in the consensus process. The basic idea is to replace the traditional client-server model with one based on the collective processing of data via shared protocols. In such a setting, either no node qualifies as the data controller or every node qualifies as a data controller. Another interesting question related to data controller identification is whether all participating nodes can qualify as potential joint controllers. In principle, to qualify as a joint controller under the GDPR legal framework, the nodes jointly determine the purposes and means of processing. Whereas, permissionless blockchain is shaped by the nodes' behavior. They do not determine the modalities for data processing of other nodes. Therefore, they do not qualify as joint data controllers. The data subject adds personal data to the blockchain and controls its data using asymmetric cryptography. This prompts the question of whether a data subject himself/herself can qualify as a data controller. As a matter of fact, a data subject may be able to qualify as a data controller in some situations where he/she is adding personal data to a blockchain [14]. In the case of online social networks, the data subject (OSN user) is a content producer and manager rather than a content consumer [2], thus qualifying as a data controller. Fortunately, the situation is quite clear in private permissioned blockchains and there are two possible scenarios. In the first scenario, the community decides together with the validation rules that the blockchain implements, in which case all nodes fall under the definition of joint data controllers and share the responsibility of compliance. In the second scenario, the blockchain accepts the contribution of validators that do not participate in defining the validation rules, and these nodes fall under the definition of data processors [17].

4 Opportunities of Blockchain-Based OSN Under GDPR

Despite the multitude of privacy controls, the current online social networks fail to provide an effective mechanism to manage access to the uploaded content of users. The issue of privacy has received significant attention in both the research literature and the mainstream media. OSNs users are also becoming conscious of their online presence and expect to have more control, traceability, accountability, and ownership of their data. The emergence of blockchain technology has led the computing domain towards a decentralization, transparency, and autonomy. Most of the prevalent privacy concerns in OSNs can be addressed by the disruptive innovation of blockchain technology using the aforementioned features. Blockchain-based solutions for OSNs enable users to control, trace and claim ownership of every piece of content they share. The following subsections describe in detail some of the opportunities offered by blockchain-based privacy protection mechanisms for online social networks.

4.1 Decentralized Protection of Personal Data

The most widely used online social networks are centralized services that are controlled and managed by single-large corporation. This allows a single entity to collect and control an unprecedented massive amount of personal and sensitive data of users from across the globe. Such an unprecedented collection of personal data constitutes a major threat to users' privacy and to their rights to be left alone. Moreover, OSNs users have lost control of what happens with their data afterward and they cannot withdraw permission in the current privacy setting. To address the issue of privacy, the decentralized online social network architectures were proposed [5,35]. Decentralization has been considered as the panacea to privacy issues, especially in the realms of online social networks [5]. Major privacy concerns related to the centralized model are addressed by properly achieving decentralization. However, research reveals that decentralization architecture brings new technical challenges such as control and coordination, reliability and authenticity, etc. With the advent of blockchain technology, most of these issues can be addressed with inherent features of the blockchain such as immutability, transparency, and peer-2-peer (P2P) consensus.

Blockchain is designed to operate without the need for a central authority. The validation of transactions is done through peer-based consensus which is suitable for authentication of ownership rights as the history of all transactions is validated. Some of the advantages of using blockchain-based solution for OSNs are control over the ledger is distributed across many mining nodes, therefore, no one can monopolize the network and reduces the dependence on centralized service providers. P2P consensus mechanism ensures data integrity and manipulation of any information in the distributed ledger is rendered practically impossible. Monitoring and surveillance are much harder to achieve in blockchain-based solutions for OSNs because social communication is peer to peer without involving a third party controlling process. Blockchain-based online social networks intend to promote individual privacy and data sovereignty if properly implemented. Using blockchain-based solutions users remain in full control of their personal data. Blockchain offers the potential to provide a truly open and free service architecture for social communication services where users are not locked into any distributed service maintained in a centralized fashion. Adaption of decentralized OSNs was slow due to their federated social communication architecture which locked the users into their service platform that hampered a truly open and heterogeneous ecosystem for social communication.

Some of the earliest blockchain-based solutions for decentralized online social networks are Ushare and Tawki. Ushare [9] is a blockchain-based solution for user-controlled social media. It is a user-centric blockchain supported social media network that enables users to control, trace and claim ownership of every piece of content they share. The proposed solution consists of: a hash table with encrypted content shared by a user, a system for controlling the maximum number of shares performed by the user's circle members, a local personal certificate authority (PCA) that manages the user's circles and the Blockchain. However, this study presents only conceptual design and technical details on how to develop the platform are missing. Tawki [33] is a decentralized service architecture for social communication proposed by Westerkamp et al. Tawki that allows users full control of their personal data which is stored and managed by personal data storage. Tawki data storage is implemented using common REST-based API which facilitates users to send and request data to and from other users' personal data storage. Tawki uses the Ethereum blockchain to manage user identities.

4.2 Enabling Privacy Through Transparency

Online social networks collect a huge amount of personally identifiable information of the users. There is implicit trust by the users that OSNs will not misuse their sensitive personal information. The process of information usage by these web systems is fairly complex and users are not completely aware of what is happening with their data. We have previously described a recent scandal about the misuse of users' data from the social web. This raises the serious need for information accountability where appropriate use of the personally identifiable information can be determined after reviewing the usage pattern from audit logs. Furthermore, these audit logs can be used to check compliance with user's usage restrictions that assert no unauthorized data usage has taken place and also enable OSN service providers to be transparent with regards to data usage. Thus, we stress on implementing transparency to achieve information accountability with provenance mechanisms. Blockchain technology is regarded as a tamper-evident database that comprises a log of all transactions with transparency as an inherent feature. The transparency of blockchain provides a greater degree of control to end-users, who no longer need to trust OSN service providers with opaque data processing and usage mechanisms. According to [33], the blockchain-based solution provides data sovereignty by enhancing user control over personal data. Enabling transparency in the social web systems is a necessity to assert data ownership and privacy of users. Nissenbaum [23] introduces the notion of contextual integrity as a new benchmark for privacy. Transparency plays a pivotal role to achieve better contextual integrity. Seneviratne [26] also argues that transparency is a key component in achieving privacy and compliance.

4.3 Tamper Evident Ledger Based Regulatory Compliance

With the enforcement of GDPR, OSN providers are obliged to comply with this new regulation. At present, they are not ready to comply or even had major gaps in GDPR-compliance. Non-compliance with this regulation imposes hefty fines apart from other legal issues while operating in the EU or processing the personal data of EU citizens. Thus, GPDR compliance is the only way forward to operate in the EU or process the data of EU citizens. GDPR imposes great challenges for the OSN providers concerning their current business model. Such an obligation is the recognition of users' consent as a legitimate ground for their personal data processing. Current OSN providers lack the GDPR compliant consent management model which means OSN users have a lack of control over their personal data. GDPR promotes the notion of consent to provide users more control over their personal data. Achieving consent compliance is not an easy task in online social networks. However, blockchain technology offers features that can add value to the consent management model for the processing of personal data in the context of OSNs. It aims to facilitate users to assert their rights and get bigger control over their personal data.

5 Conclusion and Future Work

Most of the online social networks lack an effective consent management mechanism that is compliant with the newly enforced GDPR. In the absence of such a mechanism, data subjects lose control over their personal data that poses a serious privacy problem. GDPR compliant consent management is an important step towards protecting user privacy in online social networks. In this paper, we presented some of the opportunities for using blockchain technology to address this issue. Blockchain technology provides certain features that offer OSNs users fine-grained control over their personal data. We also took into consideration the challenges imposed by new EU regulations (GDPR) for OSN providers and its apparent conflicts with blockchain technology. The paper opens up new research directions to be explored. In the future, we intend to develop a proof of concept prototype for blockchain-based GDPR compliant consent management model for online social networks.

Acknowledgment. This work was carried out at the department of information security and communication technology, Norwegian University of Science and Technology, Gjovik, Norway during the tenure of an ERCIM 'Alain Bensoussan' Fellowship Programme.

References

1. Ahmed, J.: Privacy in online social networks: an ontological model for self-presentation. In: International Conference on Knowledge Engineering and the Semantic Web, pp. 56–70. Springer (2016)

2. Ahmed, J., Governatori, G., van der Torre, L.W.N., Villata, S.: Social interaction based audience segregation for online social networks. In: ECSI, pp. 186–197 (2014)
3. Ahmed, J., Shaikh, Z.A.: Privacy issues in social networking platforms: comparative study of facebook developers platform and opensocial. In: International Conference on Computer Networks and Information Technology, pp. 179–183. IEEE (2011)
4. Alizadeh, F., Jakobi, T., Boldt, J., Stevens, G.: GDPR-Reality check on the right to access data: claiming and investigating personally identifiable data from companies. In: Proceedings of Mensch und Computer 2019, pp. 811–814. ACM (2019)
5. Bahri, L., Carminati, B., Ferrari, E.: Decentralized privacy preserving services for online social networks. Online Soc. Netw. Media **6**, 18–25 (2018)
6. Bayle, A., Koscina, M., Manset, D., Perez-Kempner, O.: When blockchain meets the right to be forgotten: technology versus law in the healthcare industry. In: 2018 IEEE/WIC/ACM International Conference on Web Intelligence (WI), pp. 788–792. IEEE (2018)
7. Bond, R.M., Fariss, C.J., Jones, J.J., Kramer, A.D.I., Marlow, C., Settle, J.E., Fowler, J.H.: A 61-million-person experiment in social influence and political mobilization. Nature **489**(7415), 295 (2012)
8. Boyd, D.M., Ellison, N.B.: Social network sites: definition, history, and scholarship. J. Comput. Mediat. Commun. **13**(1), 210–230 (2007)
9. Chakravorty, A., Rong, C.: Ushare: user controlled social media based on blockchain. In: Proceedings of the 11th International Conference on Ubiquitous Information Management and Communication, pp. 99. ACM (2017)
10. Chen, Y., Li, Q., Wang, H.: Towards trusted social networks with blockchain technology. arXiv preprint arXiv:1801.02796 (2018)
11. De, S.J., Imine, A.: On consent in online social networks: privacy impacts and research directions (short paper). In: International Conference on Risks and Security of Internet and Systems, pp. 128–135. Springer (2018)
12. Farshid, S., Reitz, A., Roßbach, P.: Design of a forgetting blockchain: a possible way to accomplish GDPR compatibility. In: Proceedings of the 52nd Hawaii International Conference on System Sciences (2019)
13. Feng, Q., He, D., Zeadally, S., Khan, M.K., Kumar, N.: A survey on privacy protection in blockchain system. J. Netw. Comput. Appl. **126**, 45–58 (2018)
14. Finck, M.: Blockchains and data protection in the european union. Eur. Data Prot. L. Rev. **4**, 17 (2018)
15. Gross, R., Acquisti, A.: Information revelation and privacy in online social networks. In: Proceedings of the 2005 ACM workshop on Privacy in the electronic society, pp. 71–80. ACM (2005)
16. Ho, A., Maiga, A., Aïmeur, E.: Privacy protection issues in social networking sites. In: 2009 IEEE/ACS International Conference on Computer Systems and Applications, pp. 271–278. IEEE (2009)
17. Ibáñez, L.-D., O'Hara, K., Simperl, E: On blockchains and the general data protection regulation (2018)
18. Isaac, M.: Facebook security breach exposes accounts of 50 million users (2018)
19. Kramer, A.D., Guillory, J.E., Hancock, J.T.: Experimental evidence of massive-scale emotional contagion through social networks. Proc. Nat. Acad. Sci. **111**(24), 8788–8790 (2014)
20. Millard, C.: Blockchain and law: incompatible codes? Comput. Law Secur. Rev. **34**(4), 843–846 (2018)

21. Molina-Jimenez, C., Sfyrakis, I., Solaiman, E., Ng, I., Wong, M.W., Chun, A., Crowcroft, J.: Implementation of smart contracts using hybrid architectures with on and off–blockchain components. In: 2018 IEEE 8th International Symposium on Cloud and Service Computing (SC2), pp. 83–90. IEEE (2018)
22. Nakamoto, S.: Bitcoin: a peer-to-peer electronic cash system (white paper) (2008). https://bitcoin.org/bitcoin.pdf. Accessed 28 May 2019
23. Nissenbaum, H.: Privacy as contextual integrity. Wash. L. Rev. **79**, 119 (2004)
24. Onik, M.M.H., Kim, C.-S., Yang, J.: Personal data privacy challenges of the fourth industrial revolution. In: 2019 21st International Conference on Advanced Communication Technology (ICACT), pp. 635–638. IEEE (2019)
25. Qin, D., Wang, C., Jiang, Y.: RPchain: a blockchain-based academic social networking service for credible reputation building. In: International Conference on Blockchain, pp. 183–198. Springer (2018)
26. Seneviratne, O., Kagal, L.: Enabling privacy through transparency. In: 2014 Twelfth Annual International Conference on Privacy, Security and Trust, pp. 121–128. IEEE (2014)
27. Smith, K.: 53 incredible facebook statistics and facts (2019)
28. Tikkinen-Piri, C., Rohunen, A., Markkula, J.: EU general data protection regulation: changes and implications for personal data collecting companies. Comput. Law Secur. Rev. **34**(1), 134–153 (2018)
29. van Geelkerken, F.W.J., Konings, K.: Using blockchain to strengthen the rights granted through the GDPR. In: Litteris et Artibus, pp. 458–461 (2017)
30. Voigt, P., Von dem Bussche, A.: The EU general data protection regulation (GDPR). In: A Practical Guide, 1st edn. Springer, Cham (2017)
31. Vujicic, D., Jagodic, D., Randic, S.: Blockchain technology, bitcoin, and Ethereum: a brief overview. In: 2018 17th International Symposium INFOTEH-JAHORINA (INFOTEH), pp. 1–6. IEEE (2018)
32. Wang, Y., Kogan, A.: Designing confidentiality-preserving blockchain-based transaction processing systems. Int. J. Account. Inf. Syst. **30**, 1–18 (2018)
33. Westerkamp, M., Göndör, S., Küpper, A.: Tawki: towards self-sovereign social communication (2019)
34. Yaga, D., Mell, P., Roby, N., Scarfone, K.: Blockchain technology overview. arXiv preprint arXiv:1906.11078 (2019)
35. Yeung, C.A., Liccardi, I., Lu, K., Seneviratne, O., Berners-Lee, T.: Decentralization: the future of online social networking. In: W3C Workshop on the Future of Social Networking Position Papers, vol. 2, pp. 2–7 (2009)

Systematization of Knowledge on Scalability Aspect of Blockchain Systems

Parth Anand Shukla[✉] and Saeed Samet

School of Computer Science, University of Windsor, Windsor, ON, Canada
{shukl11g,Saeed.Samet}@uwindsor.ca

Abstract. Blockchains has redefined the way software industry's core mechanisms operate. Advent of blockchains have intrigued the industry and research community by the properties like immutability, reliability and availability it adheres. Since then, community has observed extensive research to make this technology viable and replace the existing computing paradigms. Blockchain technology promises global, immutable, self-governed system of records with no intermediaries. Technology with such properties have strong use cases where a secure audit trail is quintessential. But the performance of blockchains is not at par with the existing industry standards making industry reluctant to surge towards the blockchains. This paper presents a comprehensive analysis on the recent approaches used to enhance the performance of blockchain systems.

Keywords: Blockchain · Byzantine fault · Consensus algorithms · Cryptography · Decentralized systems · Distributed systems

1 Introduction

In the present years, Blockchain technology has gained strong attention in the tech community and other industries where it seems to be a viable refinement, an addition or a replacement of conventional computing paradigm in the technology stack. The substantial reason behind this charisma is the three quintessential properties that it adheres. First, an *immutable ledger* formed by hashing the transaction blocks with each other forming a chain which is also responsible for the second property i.e., *transparent audit trail* which can help organizations to trace the operations and state changes inside the blockchain and third, *high availability of data* which is attained by replicating the data across every peer in the network making system tolerant to the single point of failure.

Blockchain is a completely decentralized system which is governed by set of instructions that makes sure that all the nodes in the system maintain a common state. These instructions are preached in form of consensus algorithms. A decentralized system without a common consensus will disintegrate, regardless of the participants in the system trust each other particularly or do not trust at all. A solid governance is the key to a working decentralized system which is fulfilled by consensus algorithms. There is no specific standard for consensus algorithms. It essentially depends on the use-case scenario and requirements of the system.

Blockchain technology has seen three major iterations till now. **First generation** was originated with *Bitcoin* [18] which essentially laid foundations of a fully functional decentralized & trust-less technology. Bitcoin was an implementation of blockchain technology which was based on concept of cryptocurreny. First generation brought realizations regarding extending the technology to a higher spectrum of use-cases, which lead to the **second generation** of blockchains which involved *Smart Contracts & Tokenization* which was brought in by Ethereum [5]. Second generation proposed a view to Tokenize any physical and digital assets without enforcing platform ownership. Finally, **third generation** is essentially industry ready design [1] of blockchains which are having performance at par with present systems and can be work as a valuable addition to technology stack of the organizations. Third Generation is the future and present industry stands on the transition phase, where the research is surging ahead making blockchain viable for industry. Currently blockchains have not been widely accepted despite wide spectrum of use-cases is because of the major problem that this technology faces, i.e., performance scaling which is resultant of various aspects starting from the computational complexity of the consensus algorithms, fault tolerance of system in terms of Byzantine faults and expensive storage needs as the continuously growing ledger with respective data entities needs to be replicated over entire network nodes.

This paper systematizes and presents the research done to increase the scaling ability of blockchains. Section 2 discusses the major improvements done in the field of optimization of consensus algorithms, inclined towards byzantine consensus algorithms. Section 3 presents the current research encompassing the idea of *off-chaining*, i.e. outsourcing expensive computations and storage expenses from blockchain network without compromising the root properties of blockchain systems. Section 4 briefly describes the research scope with regards to industry ready blockchain designs and finally, Sect. 5 concludes the paper.

2 Approach Towards Scaling Byzantine Consensus

Consortium blockchains have been using Practical Byzantine Fault Tolerance [4] (PBFT) as state of the art consensus algorithms, which was ideal choice given the system setup. PBFT managed to tolerate f byzantine nodes by using total $3f + 1$ nodes in the network. Nodes internally performed internal atomic multicast messages to ensure the state of the system and validity if messages. The problem with PBFT is the resource foot-print and communication complexity of the system. The communication complexity of the system is of $O(n^4)$ where n is the number of nodes in the system. Hence this approach cannot be extended to a large network of nodes.

Scaling Byzantine consensus is a modular approach where different aspects of the consensus mechanism are taken into consideration and optimized based on the need. A recent survey [2] shows that while architecting a blockchain, Quorum Election, Communication Topology, Cryptography techniques and Trusted

Execution Environments (Specialized Hardware Technologies) are prime factors leading to scaling of Byzantine consensus when applied in various phases of consensus. The current research uses amalgamation of these techniques to formulate new solutions.

2.1 Quorum Formation

Scalability gets hindered in state of the art Byzantine consensus algorithms where the entire network participates in processing and validating a single transaction. Making the whole network to verify the transactions not only increases the complexity in achieving the consensus but also leads to a high amount of communication overhead. So the notion behind the quorum formation is to use a subset of nodes to validate the transaction and rest of the nodes in the network learn from them. This can be formally represented as follows.

Let n be the complete set of nodes in the blockchain network. n' number of nodes out of n forms the consensus committee which is responsible for running consensus among them and remaining nodes, the learner nodes learn from the consensus committee. Here the learner nodes have no interaction with the consensus committee which makes learner nodes a part of passive replication schemes as they accept the results computed by n'. This essentially reduces the resource foot-print and communication complexity of the system as only subset of nodes have to communicate with each other.

This approach was first used in CheapBFT [12] and ReBFT [6] where a secure Field Programmable Array (FPGA) based subsystem of nodes were responsible for validating transactions and rest of the nodes were part of passive replication. Instead of $3f+1$, these approaches used $2f+1$ active replicas and remaining f are passive, but there are assumptions. System will only rely on $2f+1$ replicas during normal state (no byzantine faults or arbitrary failures). In case of failures, fall back protocols are activated which again uses $3f+1$ nodes.

2.2 Communication Topology

Communication topology governs the way nodes interact with each other in the system. It can be static or dynamic depending on the implementation of the system. As discussed before, the prime reason of PBFT not scaling to the mark is the communication design.

Recent approaches use balanced trees as part of the communication design where the leader of the quorum is root of the node and rest of the nodes essentially forms a balanced tree which can optimize the communication. ByzCoin uses balanced tree for entirety of the system and FastBFT [16] uses it in the message aggregation phase. Use of tree as communication topology reduces the communication complexity but it compromises with the liveliness of the system. In case of arbitrary node failure in a tree, the entire subsystem connected with that node will be unavailable. Bigger the height of the node more adverse the failure becomes.

LinBFT [21] presents a paradigm to cut short $O(n^2)$ communication complexity of PBFT to $O(n)$ by aggregating all the results from the network in every phase. Instead of nodes interacting with each other, node sends cryptographically signed messages to leader and leader validates the messages. This approach bottlenecks as leader has to aggregate, verify and disseminate the messages to rest of the network. Similar but more optimistic rather speculative approach is used in Zyzzyva [13] where the once the leader submits the transaction to the nodes in the network, all the nodes speculatively validate and execute the transaction and returns the result back to the leader. Leader on finding inconsistencies in execution among the nodes corrects nodes converging to a common state of system, reducing replication cost to theoretically null (in absence of byzantine nodes).

Gosig [14] on other hand uses probabilistic gossip paradigm to disseminate information in the system, this essentially form a random graph topology at the core as node randomly disseminates the transaction in the network leading to $O(n.log(n))$ overhead for each node in the various rounds of the system.

2.3 Cryptographic Techniques

Cryptographic primitives have played an essential role in reducing computational and communication overheads in achieving the consensus. Cryptography and reduced computation might sound oxymoronic, but it has substantially helped in many different ways. Algorand [9] in its cryptographic sortition makes use of Verifiable Random Functions [17] (VRFs), where every node in the network runs the function to know his role in the consensus. Not only does this eliminate the communication overhead between the nodes with regards to selection of committee to initiate consensus but also makes leader and committee election free from any malign activities adaptive adversaries can possibly do to disrupt the process.

Advanced signature schemes have also played an essential role in reducing authentication payload by trimming signatures of multiple nodes form of a committee to a single signature. One of the prominent use-cases is in FastBFT where Collective Signing [20] (CoSi) method is used to construct Schnorr multisignature [19]. CoSi is used in FastBFT's message aggregation phase where the quorum forms a balanced tree having leader node at the root and messages starts aggregated from leaf nodes and at the end root has a single message signed by all the nodes of the quorum. A similar approach is used in ByzCoin for aggregation of messages.

2.4 Trusted Hardware Components and Trusted Execution Environments

Recent research shows use of trusted hardware components and secure computing schemes to reduce the computational resources used in validating the operations within the peers. One of the earlier use-case was observed in the

ReBFT and CheapBFT. In these protocols, a smaller subsystem of nodes essentially decided the state of the system as rest of the nodes simply replicated the results formulated by the consensus committee, to make sure the subsystem is not malicious, FPGA based trusted system is incorporated to authenticate and verify the protocol messages.

A much more enhanced approach is observed in FastBFT which uses Intel SGX service provisioned by CPU for creating a secure Trusted Execution Environment (TEE) which is immune to sybil attacks and tamper resistant from any other system calls happening inside the node. FastBFT leverages a optimized light weight secret sharing scheme over TEE with efficient message aggregation protocol making it presently one of the most scaled byzantine consensus protocol in the literature. However using specialized hardware schemes in the Byzantine consensus is also a constraint in terms of acceptance and incorporation in existing system as a ready to go functionality.

3 Off-Chain Paradigm

Off-chaining has been a new area of research observed in blockchains which has potential to enhance scalability and privacy of the system. Off-chaining focuses on outsourcing the data storage (not the ledger) and computations to a third party trusted subsystem without sacrificing the fundamental essence of blockchains i.e. **immutability** and **availability** of data. This research is especially relevant to pubic blockchains as every transaction processed in the system is evaluated and executed on every node in the network, leading to less throughput. And as far as user specific data is concerned, keep data available on all the nodes of the system does not guarantee privacy of the data. Off-chaining has capabilities to extirpate the limitations expressed above.

3.1 Off-Chain Storage

It has been pointed out that with increasing use-cases of the blockchain technology, storing data on chain becomes expensive. The data includes essentially assets and tokens representing the data directly related with users. To validate the transactions, not only the ledger but also the this user specific data needs to replicated over entire network. In public blockchains this leads towards reduced privacy over the personal data and chance of data leaks are high.

The way to address this issue is to export the user specific data to a highly available secure off-chain storage system capable of handling failures. Whenever blockchain has to execute a transaction that involves state change in the data, first the blockchain network retrieves data from the highly available storage ensuring the integrity of the data. After retrieval, transaction is processed and state update on data is performed and finally third party storage writes changes on the data and updates the state. Here system relies on the third party to update the data on the storage system. To make sure that changes are made properly and there is no malice in the third party system, a content addressable storage system [8] can be taken into consideration. This can by following steps shown below while accessing the data:

1. Each data-file (d) is stored in the content-addressable storage and mapped with its hash-value ($hx(d)$).
2. User stores reference address of d i.e. $ref(d) = hx(d)$, and keeps $ref(d)$ privately mapped in it's smart-contract.
3. While querying the data, user can use $ref(d)$ to access d. On retrieving d it can calculate the hash again and verify the integrity of data.

This approach solves secure out-sourcing data to a third party storage system but the problem prevailing with liveliness of system still prevails. In case of data loss, blockchain cannot do anything to retrieve the data back. Hence, for now off-chaining approach for exporting data seems an active research area needing substantial and strong improvements encompassing liveliness of blockchain system.

3.2 Off-Chain Computing and Verifiable Computation Schemes

Off-chain computing is a paradigm in which the execution part of blockchain gets outsourced to an off-chain network and verification part is performed on-chain [7]. Suppose time taken to perform an on-chain computation be t_{on} and off-chain be t_{off}. And let time taken to verify the computation be t_{ver}. Off-chain computation guarantees that $t_{off} <<< t_{on}$. But the complete notion is useful when, $t_{off} + t_{ver} < t_{on}$.

Off-chaining introduces trust issues because consensus computation is dependent on the untrusted third-party. The crucial part is the verification phase because the resultant computation of off-chain network determines the next state of the blockchain. So if the off-chain network successfully supplies. false verification proofs, chain state gets corrupted. This approach reduces the computational overhead but maximizes trust issues.

To solve this quest, verifiable computation schemes can be used which use cryptography in its base which can improve the trust in the verification phase with high magnitudes. Verification algorithm used in this approach should be cheap making t_{ver} less and algorithm should be non-interactive, i.e. the prover should prove its computation in a single message or with very few message exchange leading to reduced communication overhead. Zcash [11] uses verifiable computations to validate the transaction in their blockchain. They use Zero-Knowledge Succinct Non-Interactive Argument of Knowledge (ZkSNARKs) [3], which is a special kind of zero-knowledge proof [10] which lets a prover prove its computation within a fraction of seconds, and the proof size is also minimal. However, the initial setup phase, which is a one-time setup is computationally expensive than the actual execution of the computation.

4 Research Focus Towards Third Generation

The work that is discussed in the above sections represents the transition phase of laying a strong foundations for the Industry ready blockchain designs, but there is more to it. There are many other areas where technology needs tweaks and improvements. Industry ready designs imply inter-operable blockchain solutions, where organizations running their specific blockchains setups should share data a blockchain that is completely different from it's own. A recent work shows value transfer ledger [15] system where blockchains running their own consensus algorithms share data as they need by the policies enforced on the global network of blockchains. Also, there are other aspects that are lacking in the blockchains system that prevail in existing systems, like inter-organizational data management, information-silo reduction, shared data validation, and so on. This are the possible technical aspects of third generation blockchain technology which has not been vividly addressed in current research and are directly connected to the scaling property of the system.

5 Conclusion and Future Work

This paper highlights the fundamental research aspects which are observed in the blockchain system which try to optimize the efficiency and scaling property of the blockchain starting from fundamental consensus algorithms to the system design. At the end, this paper also briefly explains the research focus towards industry ready blockchain designs. Based on the areas touched by this paper describing problems that blockchain faces such as byzantine adversaries, fault tolerance, and information dissemination make communication paradigm in blockchains more likely to have tendencies of a P2P network. Hence, approaches used in scalability for P2P systems if tuned concerning blockchains can be a beneficiary measure for a scalable blockchain design. Additionally, there is a wide and mostly unexplored scope of research in making off-chaining approach steady and reliable by incorporating highly available storage systems and algorithms to compute computations over the external system by using zero-knowledge proofs and lightweight verifiable computing schemes. As per the requirement specification, scaling blockchains have different terminal definitions. However, with a proper amalgamation of refinements mentioned in this survey, when used for the desired use-case can form a scalable consensus mechanism can increase the overall throughput of the system.

Blockchain technology has hyped the technology community. Perhaps this inundation of hype is due to how easy it is to think of a high level use-cases of the applications of blockchain technology, but there is a lot to it. This paper tries to provide a succinct view of the research observed in this area and will essentially help reduce the steepness of the learning curve for a neophyte in this field.

References

1. Androulaki, E., Barger, A., Bortnikov, V., Cachin, C., Christidis, K., De Caro, A., Enyeart, D., Ferris, C., Laventman, G., Manevich, Y., Muralidharan, S., Murthy, C., Nguyen, B., Sethi, M., Singh, G., Smith, K., Sorniotti, A., Stathakopoulou, C., Vukolić, M., Cocco, S.W., Yellick, J.: Hyperledger fabric: a distributed operating system for permissioned blockchains. In: Proceedings of the Thirteenth EuroSys Conference, pp. 30:1–30:15. ACM, New York (2018)
2. Berger, C., Reiser, H.P.: Scaling Byzantine consensus: a broad analysis. In: Proceedings of the 2nd Workshop on Scalable and Resilient Infrastructures for Distributed Ledgers, pp. 13–18. ACM, New York (2018)
3. Bowe, S., Gabizon, A., Green, M.D.: A multi-party protocol for constructing the public parameters of the Pinocchio zk-SNARK. In: Zohar, A., et al. (eds.) Financial Cryptography and Data Security. FC 2018. LNCS, vol. 10958. Springer, Heidelberg (2019)
4. Castro, M., Liskov, B.: Practical Byzantine fault tolerance and proactive recovery. ACM Trans. Comput. Syst. **20**, 398–461 (2002). https://doi.org/10.1145/571637.571640
5. Dannen, C.: Introducing Ethereum and Solidity: Foundations of Cryptocurrency and Blockchain Programming for Beginners. Apress, Berkeley (2017)
6. Distler, T., Cachin, C., Kapitza, R.: Resource-efficient Byzantine fault tolerance. IEEE Trans. Comput. **65**, 2807–2819 (2016). https://doi.org/10.1109/TC.2015.2495213
7. Eberhardt, J., Heiss, J.: Off-chaining models and approaches to off-chain computations. In: Proceedings of the 2nd Workshop on Scalable and Resilient Infrastructures for Distributed Ledgers, pp. 7–12. ACM, New York (2018)
8. Eberhardt, J., Tai, S.: On or Off the blockchain? Insights on off-chaining computation and data. In: De Paoli, F., Schulte, S., Broch Johnsen, E. (eds.) Service-Oriented and Cloud Computing, pp. 3–15. Springer (2017)
9. Gilad, Y., Hemo, R., Micali, S., Vlachos, G., Zeldovich, N.: Algorand: scaling byzantine agreements for cryptocurrencies. In: Proceedings of the 26th Symposium on Operating Systems Principles, pp. 51–68. ACM, New York (2017)
10. Goldreich, O., Micali, S., Wigderson, A.: Proofs That yield nothing but their validity or all languages in NP have zero-knowledge proof systems. J. ACM **38**, 690–728 (1991). https://doi.org/10.1145/116825.116852
11. Hopwood, D., Bowe, S., Hornby, T., Wilcox, N.: Zcash protocol specification. Technical report 2016-1.10. Zerocoin Electric Coin Company (2016)
12. Kapitza, R., Behl, J., Cachin, C., Distler, T., Kuhnle, S., Mohammadi, S.V., Schröder-Preikschat, W., Stengel, K.: CheapBFT: resource-efficient byzantine fault tolerance. In: Proceedings of the 7th ACM European Conference on Computer Systems, pp. 295–308. ACM, New York (2012)
13. Kotla, R., Alvisi, L., Dahlin, M., Clement, A., Wong, E.: Zyzzyva: speculative Byzantine fault tolerance. In: Proceedings of Twenty-first ACM SIGOPS Symposium on Operating Systems Principles, pp. 45–58. ACM, New York (2007)
14. Li, P., Wang, G., Chen, X., Xu, W.: Gosig: scalable byzantine consensus on adversarial wide area network for blockchains. arXiv:1802.01315 (2018)
15. Li, W., Sforzin, A., Fedorov, S., Karame, G.O.: Towards scalable and private industrial blockchains. In: Proceedings of the ACM Workshop on Blockchain, Cryptocurrencies and Contracts, pp. 9–14. ACM, New York (2017)

16. Liu, J., Li, W., Karame, G.O., Asokan, N.: Scalable Byzantine consensus via hardware-assisted secret sharing. IEEE Trans. Comput. **68**, 139–151 (2019). https://doi.org/10.1109/TC.2018.2860009
17. Micali, S., Rabin, M., Vadhan, S.: Verifiable random functions. In: 40th Annual Symposium on Foundations of Computer Science (Cat. No. 99CB37039), pp. 120–130. IEEE Comput. Soc., New York (1999)
18. Nakamoto, S.: Bitcoin A Peer-to-Peer Electronic Cash System (2008)
19. Schnorr, C.P.: Efficient signature generation by smart cards. J. Cryptol. **4**, 161–174 (1991). https://doi.org/10.1007/BF00196725
20. Syta, E., Tamas, I., Visher, D., Wolinsky, D.I., Jovanovic, P., Gasser, L., Gailly, N., Khoffi, I., Ford, B.: Keeping authorities "honest or bust" with decentralized witness cosigning. In: 2016 IEEE Symposium on Security and Privacy (SP), pp. 526–545 (2016)
21. Yang, Y.: LinBFT: linear-communication byzantine fault tolerance for public blockchains (2018)

Smart Dam: Upstream Sensing, Hydro-Blockchain, and Flood Feature Extractions for Dam Inflow Prediction

Takato Yasuno[✉], Akira Ishii, Masazumi Amakata,
and Junichiro Fujii

Research Institute for Infra. Paradigm Shift,
5-20-8 Asakusabashi Taito-ku, Tokyo, Japan
{tk-yasuno,akri-ishii,amakata,
jn-fujii}@yachiyo-eng.co.jp

Abstract. Heavy rain occurs frequently during extreme weather, and the associated resultant flood damage represents a social problem. The present study aims to redefine the dam watershed as a smart dam and attempts to systematize the technology for flood prediction by integrating upstream sensing, dam inflow prediction, and a hydro-blockchain. In order to detect high water levels, we devise an upstream sensing method to observe the water level at the uppermost stream of a dam watershed, and summarize potential implementation hurdles. We also propose a hydro-blockchain configuration that provides a basis for the fair transaction of water rights. We implement a field study in the Kanto region, Japan, to observe the upstream water level using the devised water level sensor. We analyze the relationship between the measured water levels and the dam inflow and also analyze the response of the time difference. Furthermore, we propose a flood-feature extraction using a 20 year hydro-dataset of rainfall and water levels, and propose dam inflow prediction models using various time series machine learning algorithms. We demonstrate the application of our model results and discuss their usefulness.

Keywords: River sensing · Hydrologic blockchain · Flood feature extraction · Time series machine learning

1 Introduction

1.1 Background and Related Work

Floods in a Changing Climate and Hydrological Modeling. In recent years, extreme flooding has caused massive damage worldwide [1, 2]. In Japan, sudden rainfall and extreme river flooding have occurred in regions such as Joso City, Ibaraki Prefecture [3]. Owing to such extreme and sudden rainfall events, it is difficult to predict the inflow to a dam and to set the operational dam outflow. In flood hydrology, empirical models such as regression models and artificial neural networks (ANNs) have gained popularity [4, 5]. There are several approaches for water resource time series prediction modeling that use statistical and machine learning methods. For example, ANNs have been applied to monthly reservoir inflow time series [6, 7] using data that included 500 monthly streamflow measurement over a period of 40 years as well as climate and land cover data, whereby rainfall-runoff modeling simulates the streamflow of watersheds.

There are currently more than 70 hydrologic models that have varying degrees of data requirements and can be used for numerous applications, for example, estimation of flood runoff and assessment of inundation. An exhaustive review of hydrological models was published in 2002 [8]. The temporal and spatial derivatives of hydrological processes that occur in nature are important in modeling, and models are classified as either distributed (spatial) or lumped (non-spatial). Commonly used temporal scales used in hydrological models are daily or monthly scales [9–11]. In Japanese watersheds, the distance between upstream and downstream locations is usually relatively short; thus, the time scale should be on the scale of hours or even minutes for flood forecasting in Japan. Hydrological models are physically based and can perform simulations for the estimation of water quantity. However, model accuracy is insufficient for recent flood forecasting due to the extreme nature of phenomena related to climate change.

Water Smart City and Smart Dam. Today, smart urbanization is part of thousands of urban projects around the world. A decade ago, the promise of improving and optimizing urban services through the application of information and communication technology (ICT) was largely a techno-utopian fantasy [12]. Many examples of empirical studies of smart cities from 2013 to 2019 are detailed elsewhere [12]. Rob et al. [13] outlined the typology of a smart city for a region of Dublin, Ireland, which included entrepreneurship innovation, green energy, sustainability, resilience, e-government, open data, transparency, accountability, evidence-informed decision making, better service delivery, quality of life, safety, security, risk management, intelligent transport, inclusiveness, and empowerment.

Other studies have focused on water resources, for example, Karwot et al. reported the problems of water distribution and usage as well as problems relating to the management of the technical infrastructure in Poland [14]. The authors provided some ideas and concepts associated with the use of modern tools for supporting water management and the creation of end-user participation for improving the water distribution system [15]. More recent concepts and projects include those in Australia (water-supply city, water-cycle city and water-sensitive city), Korea (water-efficient infrastructure), and in Denmark and the Netherlands (water-smart cities).

However, methods for implementing and realizing water-smart cities remain unclear. Dams have key roles in water-smart cities because they create reservoirs, and enable the optimization of reservoir operation to prevent floods and also generate hydropower. Despite these roles, there is a need for practical methods for the development of a smarter dam for a sustainable basin and the efficient use of water resources (i.e., for agriculture, industry, everyday life, and hydropower).

1.2 Study Objectives

The present study aims to redefine the dam watershed as a smart dam and attempts to organize the technology for flood forecasting through the development of key requirements, for example, upstream sensing, dam inflow prediction, and hydro-blockchain. Specifically, in order to detect flood signs, we propose an upstream sensing method to observe the river water level in the uppermost stream of a dam watershed and implement this in a watershed in the Kanto region, Japan. We summarize five main hurdles for implementation the upstream sensor based on lessons learned in our study installation. In addition, we propose a hydro-blockchain structure that is critical for the fair, conflict-free transaction of water rights among stakeholders. We report on a field study in the selected region whereby the upstream water level was measured by the devised sensor. We aim to analyze the co-relations between the midstream water level and the dam inflow, as well as the response of the time difference between upstream and downstream locations. Furthermore, we propose (i) a flood-feature extraction using a 20 year hydro-dataset of rainfall and water levels from the study area, and (ii) dam inflow prediction models via time series machine learning and deep learning algorithms with the aim of applying the models to the study area to demonstrate their usefulness.

2 Upstream River Sensing and Hydro-Blockchain

2.1 Advantage of Upstream River Sensing

The flood control of a dam is based on operational rules and is carried out according to the amount of flood inflow into a reservoir. Dams have a role as the most upstream point of flood countermeasure because they allow the lowering of the downstream river water level before shifting to disaster prevention operations during a dam flood. Dams can therefore secure sufficient time for the safe evacuation of residents. However, in recent years, localized and concentrated heavy rains in Japan, which are understood to be related to climate change, have in some cases caused a larger than anticipated inflow into the dam reservoir. Therefore, it is necessary to prepare countermeasures against flooding on the downstream side of the case study dam in the event that the usual flood control operation cannot be handled because of localized and concentrated heavy rains, or in the event that the dam itself is included in flood protection plans.

To achieve such countermeasures, it is important to accurately predict the inflow to the case study dam. However, the upstream area of the dam is mountainous, and the spatial arrangement of the river sensors is comparable to an urban area. Existing sensing systems are insufficient for ensuring the accuracy of predicting the inflow to the dam. Thus, if the water level change in the upstream area of the dam can be detected by a large number of sensors, data would be available for flood forecasting that can provide an indication of the imminence of a flood. Such a system and the generated data would be useful for dam operations and flood countermeasures in the study region and potential also in other regions.

2.2 Hurdles for Upstream River Sensing

Remote sensing using artificial satellites is one method for acquiring information for the upstream area of a dam; however, it is not suitable for predicting inflow to a dam when observation frequency and real-time performance is required. Upstream river sensing is therefore the only effective way to overcome this limitation. In order to install river sensors in the upstream area of a dam, the following five main hurdles may need to be addressed.

Sensing at an Effective Point for Flood Forecasting. In steep mountainous areas such as in the present study, the path of the river flow is likely to change due to river bed evolution and sedimentation that occurs during a flood. Since it is impractical and ineffective to monitor all outflows (e.g., from small valleys and swamps), it is preferable to select points with abundant flow rates (i.e., main rivers) that are located downstream of tributaries. However, because it can be difficult to access rivers in mountainous areas, and the water level needs to be measured vertically from directly above, the location for installing water level sensors can be limited (e.g., to bridges) and not necessarily optimal. In some situations, an angle sensor, which can measurement the water level from an oblique angle without restricting the position of the water level sensor, may also be required, and is being development in other research not reported here.

Construction of a Communication Environment. A communication environment is necessary for data because flood forecasting requires real-time performance. However, mobile communication radio waves are not generally reliable in mountainous areas. Use of satellite communications, installation of mobile base stations, and networking of low-power wide-area networks (LPWAN) are also unrealistic because of the cost involved. Since digital convenience radios (DCR) depend on geographical features and communication distance their effectiveness can be unknown. In order to reliably transmit water level data, it can therefore be necessary to install a communication line to the location where a water level sensor is installed.

Securing Power. The power requirements of a water level sensor and the communication equipment means that solar power generation and batteries are insufficient; thus, it can be necessary to install electrical wires to the sensor location to ensure reliable and continuous measurement throughout the year.

Installations in Limited Space. The prop and the box for storing communication equipment must be installed in a limited flat area in steep mountainous areas. Therefore, a quick and simple foundation such as the ground screw method is required.

Application for Installation Permission. In the present study, the water level sensors were installed on public land, and occupancy permission based on Japanese road and river laws was required for each installation location. If the location of a water level sensor is in a designated national park or conservation area, an application based on relevant laws (i.e., natural park law, forest law) may also be required.

2.3 Field Case Study

Figure 1 shows the sensor arrangement in the upstream area of the dam area in the present study. The catchment area of the in Kanto region is ~ 100 km^2. Previous to the present study, there was only one water level sensor in this basin, but on July 10, 2019, an up-stream water level sensor point was installed. On July 27, 2019, there was a flood due to Typhoon No. 6 that was measured by the sensors between 00:00 to 12:00 (Fig. 2). Figure 2 shows that the peak difference between rainfall observation point 3 and the upstream water level sensor point was 50 min, and that the peak difference between the upstream water level sensor point and the midstream water level sensor point was 40 min. Consequently, we were able to capture a flood signal 40 min compared to before by installing the upstream water level sensor, which highlights the effectiveness of the sensor method as a high water countermeasure. With respect to the peak inflow to the dam, we presumed that inflow from the basin where rainfall observation point one is located was dominant since this experienced higher rainfall (Fig. 2).

The confirmation of flood propagation (i.e., the flood arrival time) from the uppermost stream to the middle stream suggests that flood forecast signal information can provide an indication of the imminence of a flood. Continued monitoring by the sensors could therefore provide useful data for the future that might be used for studying the relationship between the upstream and midstream water levels in order to set flood thresholds at these locations. To fully understand local flood phenomena, in the future we intend to optimize the balance, position, and number of water level sensors. When forecasting flood events ahead of the propagation time, we note that it is necessary to consider the weather forecast.

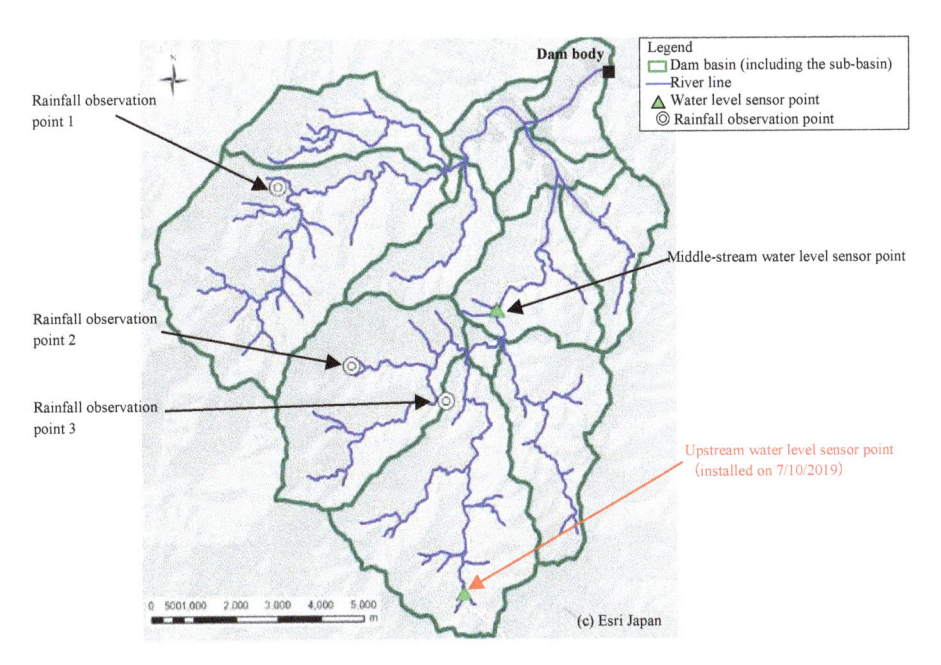

Fig. 1. Catchment map of the study area, Kanto region, Japan, showing the locations of water level sensors, rainfall observation points, and the dam.

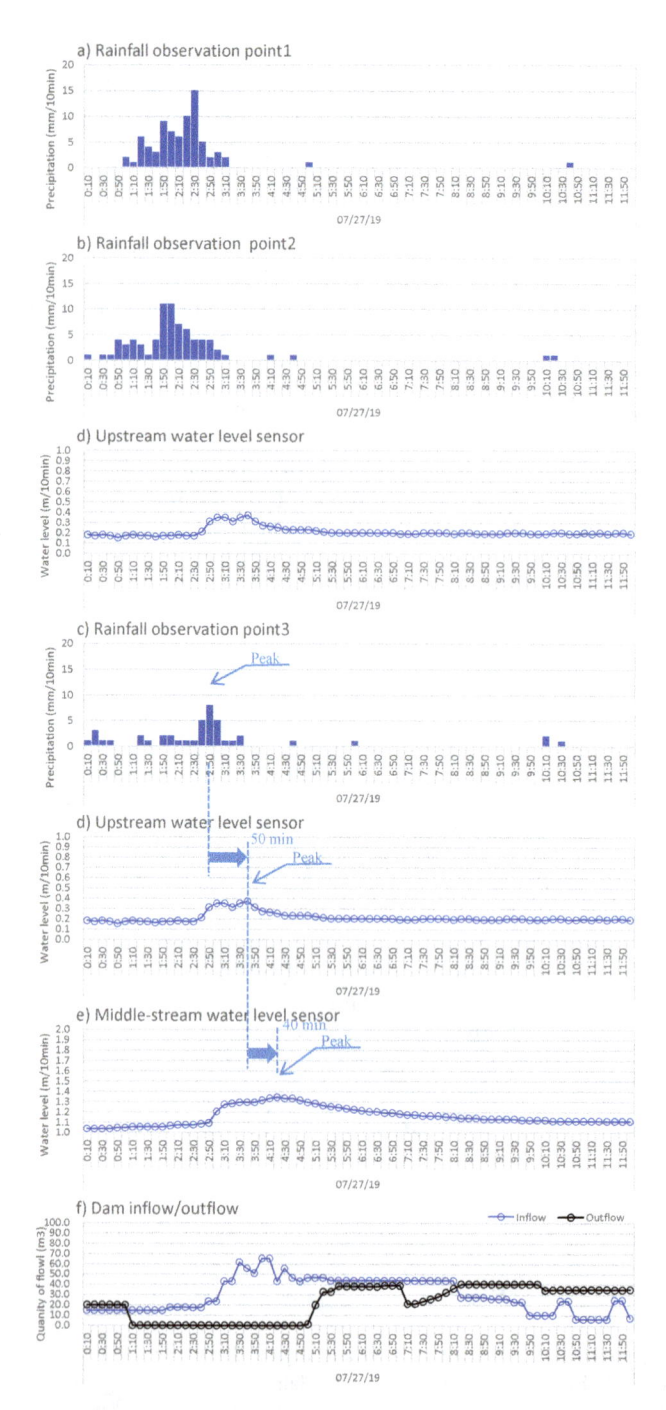

Fig. 2. Rainfall at observation points (a) one, (b) two, and (c) three in the study area, Kanto region, Japan. River water levels at the (d) upstream, and (e) middle stream sensors. (f) Dam inflow/outflow data (Typhoon No. 6, July 27, 2019).

2.4 Hydro-Blockchain for Flood Forecasting

In order to improve the accuracy of dam inflow predictions, it is necessary to use various data such as rainfall, water level, artificial water inflow, and reservoir operation. However, various players and stakeholders are involved in a dam, and data is generally distributed and managed by each player's own system. Data is not necessarily shared, and may incur a cost due to restrictions on the use of private and sensitive data. In order to change this situation, it is necessary to build a system that uses data fairly and securely.

Here, we propose a future system based on blockchain technology [16, 17] termed "hydro-blockchain for flood forecasting" (Fig. 3), which can be developed to aggregate the registration and use of large spatiotemporal hydrological datasets. This system enables data management, exchange, and sharing under the robust security of blockchain technology.

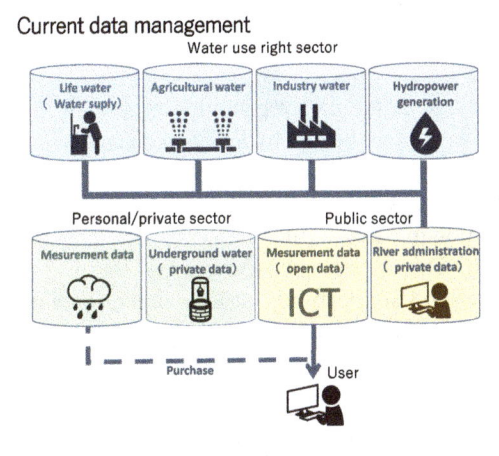

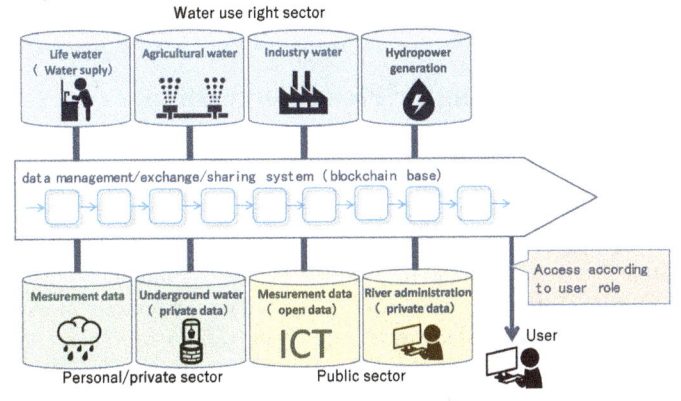

Fig. 3. Schematic of the hydro-blockchain data management/exchange/sharing system for flood forecasting (Top: Current data management, Bottom: Hydro-blockchain data management).

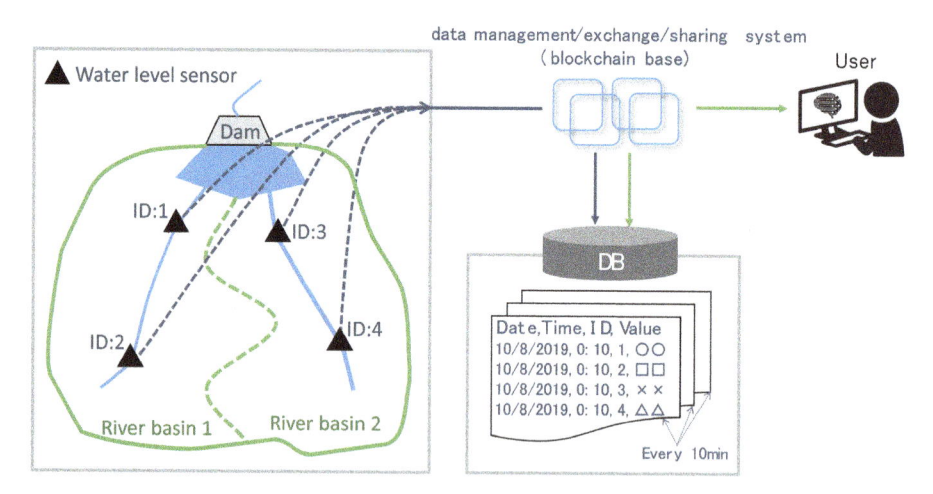

Fig. 4. Schematic of the data registration process from the water level sensors to the database (DB) and data usage via the hydro-blockchain.

Figure 4 shows a schematic of the data registration process from the water level sensors to the database and data usage via the hydro-blockchain. Data is identified by a sensor's ID and recorded in the database via the blockchain every 10 min. The data user is given data usage authority according to their role. As data is exchanged through encrypted transactions, it is possible to use data seamlessly and fairly while maintaining privacy and data protection. This system helps to prevent the dispersion and disappearance of data that are useful for learning various flood characteristic, and also improves the accuracy of dam inflow predictions. In addition, by using the smart contract function, information sharing, including dam inflow prediction results, enables (i) a quick and smart coordination between conflicting stake holders, and (ii) the smooth and timely dissemination of disaster prevention information. Importantly, the system can not only operate in just one basin but also can be integrated to operate in multiple basins.

3 Time Series Prediction for Flood Forecasting

3.1 Related Works on Machine Learning Weather Data

Hydrologic Machine and Deep Learning Modeling. During the mid-2010s, many machine learning algorithms were applied to water level forecasting problems, for example, Gaussian generalized additive models (GAMs), multivariate additive regression splines (MARSs), random forest (RF), support vector regression (SVR), genetic programming (GEP), and Gaussian process regression (GPR). Several models have been compared by others, including comparisons between Gaussian linear regression, GAMs, MARSs, ANN models, random forests (RFs), and regression tree models [18].

This comparison suggested that GAMs and RF can effectively capture some non-linear relationships. A study undertaken by Li et al. [19] compared several water level forecasting models (i.e., RF, support vector regression, ANNs, and a linear model) using daily lake water levels from five lake gauges over 50 years. The results suggested that the RF provided a more reliable and accurate prediction of lake water levels [19].

A variable selection approach can improve the forecasting efficiency of daily reservoir discharge by machine learning methods. Yang et al. [20] used 2854 daily reservoir levels covering a period of eight years to test five methods: ANN, instance-based classifier, k-nearest-neighbor classifier, RF, and the random tree. Key variables that influenced the reservoir water level were selected and improved models were developed. The experimental results indicated that the RF forecasting model, when applied to a variable selection with full variables, performed better than the other models [20]. Vamsi proposed the use of GPR for forecasting weather conditions [21], but used a narrow dataset of just two years. The parameters that were considered for the weather prediction included precipitation, wind speed, and temperature. The main advantage of GPR is that it is easily understood. Adiya used a hybrid approach by combining discriminatively trained predictive model GPR with a deep neural network that modeled the joint statistics of a set of weather-related variables [22]. However, the authors hybrid modeling only used five years of data, which consisted of balloon observations at 60 locations. Unfortunately, the applicability of GPR is unclear for flood forecasting to predict the inflow to a dam based on a long-term dataset.

In the late 2010s, deep learning approaches such as long short-term memory (LSTM) were applied to flood forecast datasets. Zhang et al. [23] compared the simulation performance of a recurrent neural network, LSTM, and a gated recurrent unit. The authors showed that the three models could accurately predict reservoir inflow; however, they only collected data every four or six hours over four years. Xuan-Hien et al. [24] suggested a LSTM model for flood forecasting that used the daily discharge and rainfall as input data. The authors used datasets from 1961 to 1984 (24 years) for the Da River basin, Vietnam, to forecast flow rates over one, two, and three days. However, the historical dataset was daily and many under-prediction error (UPE) points remained in their forecast, which would lead to a delay in reservoir operations in the event of a flood hazard. Unfortunately, the usefulness of LSTM is unclear for recent flood forecasting for the prediction of inflow to a dam. Hence, many researchers have undertaken studies on pre-processing, combined modeling, and evaluation index in order to improve modeling accuracy. However, knowledge on time series machine learning modeling for flood forecasting remains incomplete.

In the following Sects. 3.2, 3.3, and 3.4, we propose three methods, respectively: (1) flood feature extractions as pre-processing, (2) time series regression modeling using machine learning algorithms towards a hydro-dataset, and (3) computing the sum of an under-prediction index for evaluating flood hazard. We also demonstrate several applied results for the study area.

3.2 Flood Feature Extraction

Trend Feature Extractions Using Weighted Least Squares. Existing time series analysis of hydrological data relies on various filtering procedures for pre-processing, for example, standardized, auto-regressive, moving average, and weighted moving average. However, these procedures do not focus on the trend and peak of a hydrograph, which could represent key features of a flood's scale and continuous high water. Here, we highlight how to extract features associated to the hydrograph peak and trend change, which can in turn be used for supporting decision making regarding announcing a flood alert.

Firstly, we propose the trend estimates of coefficients from the historical dam inflow dataset for the study area. For example, the past 3 h before the current time step has a batch dataset containing 12 inflow time steps and 15 min units. This trend coefficients can be estimated every time step using the weighted least squares (abbreviated as WLS) algorithm. The older an observation is, the smaller the WLS becomes, which infers a weak memory. The WLS trend feature can also be used to represent the shift-up change among past time steps and can predict the kurtosis change around the peak of a flood hydrograph.

Volume Extractions Using Log-Trapezoids Integration. Secondly, we propose the trapezoids integration as the hydro-stock volume from the past time steps. If the trend features are extracted in a limited way it may result in the hydro-volume features being mistaken as the flood scale and the continuous high water. Hydro-volume features can be computed at every time step using the trapezoids integration from the historical dam inflow values, for example, for the past three hours with 12-time steps and 15-min units. These values of trapezoids integration are of a higher numerical order than of those in the hydro-dataset (e.g., rainfall, river level, and historical dam inflow). Thus, we propose the logarithm of the trapezoids integration and transform them into standardized variables whose original values are subtracted by the mean and divided by the standard deviation based on the integration series from the past time steps. In Sect. 3.3, this hydro-volume feature extraction is termed the log of trapezoids integration (abbreviated as LTI).

3.3 Dam Inflow Prediction Method for Flood Forecasting

Regression Tree and Ensemble Algorithms. We employed a hydrological time series as a baseline model using a regression tree [25]. Although a regression tree model can provide a fast computation, it offers a limited representation of flood features. We also simulated the hydrological dataset using the ensemble models of bagging and least square-base boosting (abbreviated as LSBoost) [26–28], which involve optimizing hyper parameters based on the cross-validation function.

GPR Algorithm. GPR is a nonparametric Bayesian learning model that can be placed directly over the space or nonlinear function instead of specifying a parametric family of nonlinear functions [29, 30]. The covariance matrix is a kernel function (e.g., linear, exponential, squared exponential, Gaussian, Uhlenbech, and quadratic functions). An exponential kernel function is more accurate than other kernels for dam inflow time series. The hyper parameter sigma is optimized using the cross-validation function.

LSTM Regression Algorithm. In the case of a simple recurrent neural network, long-term dependencies are impossible to learn because the network of vanishing gradient problem has deep layers. The theoretical reasons for this effect were studied by Hochreiter and Schmidhuber in the 1990s [31]. The LSTM is a deep learning model for representing a long-term dependence between inflow time steps, and is designed to solve the vanishing gradient problem [31, 32]. One block of the LSTM framework provides a good performance for a lumped dam region. Before the final regression layer, it is possible to insert a dropout layer at a scale of 0.3 in order to avoid a limited pattern of training data every mini-batch.

3.4 Flood Risk Index UPE

Under-Prediction Index for Evaluating Flood Hazard. Table 1 shows two types of dam inflow prediction error. Minimization of the UPE (shown in the upper right-hand side of Table 1) is important for minimizing the risk of flood damage [33–35].

Sum of UPE for the Flood Risk Index. The UPE at time t is indicated by Eq. (1):

$$UPE_t = \widehat{y}_t - y_t \quad (1)$$

where $\widehat{y}_t$ is the dam inflow prediction, and y_t is the actual inflow value. If UPE is negative then $\widehat{y}_t < y_t$, which appears to be an optimistic forecast at time t. The sum of under-prediction errors (sUPE) among a term $t \in \{1, \ldots, T\}$ is formulated using Eq. (2):

$$sUPE = \sum_{t=1}^{T} \min\left(\widehat{y}_t - y_t, 0\right) \quad (2)$$

A negative value can be computed as the predicted case when there are only inflows that are lower than the actual inflow. If under-predictions occur frequently, i.e. if the absolute value of UPE_t is due largely to these negative values, then the mitigation policy may be insufficient for the risk of an over-flow scenario. These errors would correspond to the risk of flood damages.

Table 1. Two types of dam inflow prediction errors.

		True, actual value	
		Low water	High water
Prediction	Low water	True low water	**Under-prediction error (Flood damage risk)**
	High water	**Over-prediction error (Hydropower opportunity loss)**	True high water

3.5 Applied Results

Hydro-Dataset for Flood Forecasting. The case study dam is a 32 m high concrete gravity dam (PG) with the purpose of hydropower generation (electrical capacity of 279 MWh/year). The reservoir capacity is 1.5 million m^3, and the dam catchment area is 112 km^2. There are three river height sensors are installed. It takes 90 min for river water to flow from sensor one to the dam, 60 min from sensor two to the dam, and 20 min from sensor three to the dam.

Table 2 lists the hydro-dataset names and variable profiles. The model is set to a 3-h forward forecast. Each of the window sizes is set to 12 (i.e. the historical data of the past three hours), and the unit time intervals are 15 min.

Table 2. Hydro-dataset names and variable profiles.

Data name	Data role and feature profile
Dam inflow time series	Target variable. The time series of the dam inflow volume from the upstream region with a measurement interval of 15 min
Rain gauge	The quantity of rainfall is measured at nine points in the upstream region
River height sensor	Main stream and river features. The water height is sensed at three points and includes an upstream sensor

Applied Result One: Use of the Hydro-Dataset. Table 3 shows root-mean-squared error (RMSE) and runtime results for fives models: regression tree, ensemble, GPR, SVR, and LSTM that were developed using a standard hydro-dataset. This is the benchmark scenario for comparison to scenarios with additions. Figure 5 shows dam inflow predictions results using a hydro-dataset, where these are applied by methods such as regression tree, ensemble Gaussian process regression, support vector regression, and long short-term memory. Figure 6 shows the prediction error plot of predicted values minus the actual dam inflow using a standard dataset. There are a lot of underpredictions less than the actual dam inflow, in other words this means that these errors influences much flood damage risk.

Table 3. Test evaluations and Bayes optimization based on the cross-validation function applied to a standard hydro-dataset. RMSE is the root-mean-squared error.

Model	RMSE	Runtime (minutes)
Regression tree	17.49	1
Ensemble model (bagging, LSBoost)	14.01	26
Gaussian process regression (GPR)	12.05	55
Support vector regression (SVR)	12.58	43
Long short-term memory (LSTM)	11.09	70

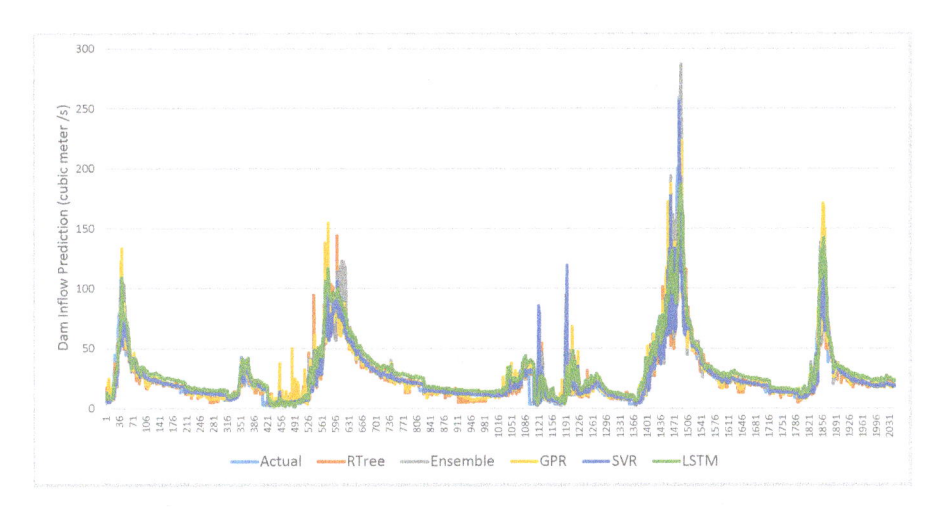

Fig. 5. Predictions (regression tree, RTree; ensemble; Gaussian process regression, GPR; support vector regression, SVR; long short-term memory LSTM) and actual dam inflow series using a hydro-dataset.

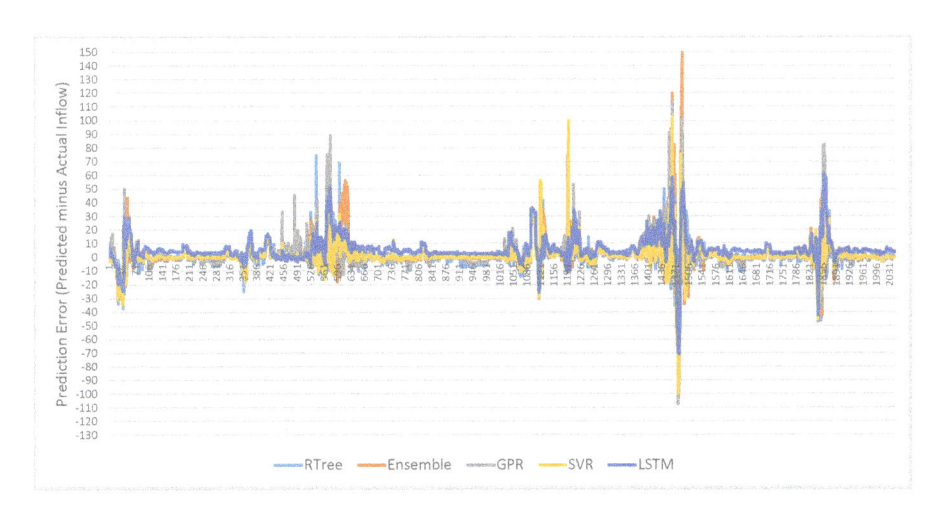

Fig. 6. Prediction error plot of predicted values minus the actual dam inflow using a standard dataset. Showing regression tree (RTree), ensemble, Gaussian process regression (GPR), support vector regression (SVR), and long short-term memory (LSTM) models.

Applied Result Two: Flood-Feature Extraction Added. Table 4 shows the RMSE and runtime results for the five models that were developed using a hydro-dataset with WLS trend and LTI volume flood-feature extractions added. The RMSE in Table 4 are lower than those obtained using the standard hydro-dataset (Table 3). The flood feature extractions are effective for improving the accuracy of each model, especially the LSTM model. Figure 7 shows dam inflow predictions results using a hydro-dataset also added with hydro-features extractions, where these are applied by the above noted five methods. Figure 8 shows the prediction error plot of predicted values minus the actual dam inflow using a standard dataset added with hydro-features extractions. Compared with the previous applied results one, under-predictions less than peak inflow are reduced, so this indicates that the better predictions improve the less flood damage risk, although under-prediction errors are not completely eliminated.

Table 4. Test evaluations and Bayes optimization based on the cross-validation function applied to a hydro-dataset with weighted least squares (WLS) trend and log of trapezoids integration (LTI) volume flood-feature extractions added. RMSE is the root-mean-squared error.

Model	RMSE	Runtime (minutes)
Regression tree	5.39	2
Ensemble model (bagging, LSBoost)	2.72	34
Gaussian Process regression (GPR)	5.54	56
Support vector regression (SVR)	4.99	48
Long short-term memory (LSTM)	1.80	71

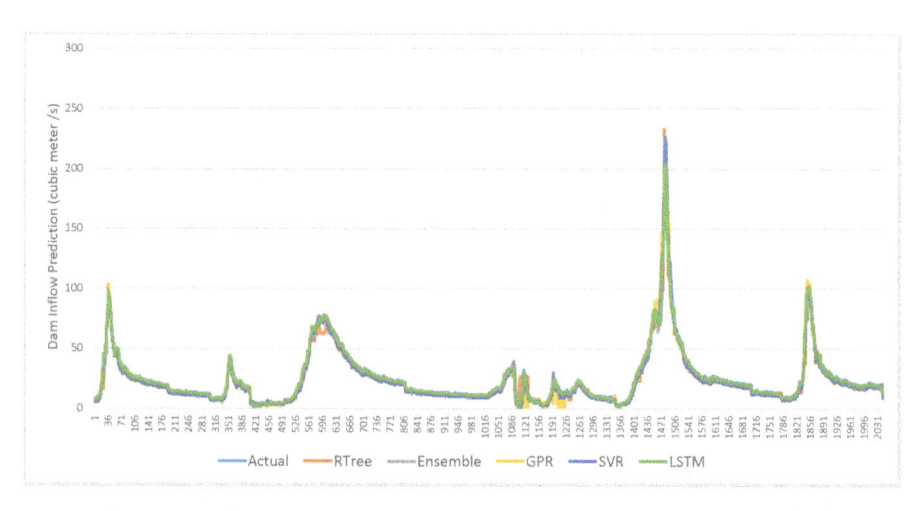

Fig. 7. Predictions (regression tree, RTree; ensemble; Gaussian process regression, GPR; support vector regression, SVR; long short-term memory LSTM) and actual dam inflow series added with hydro-features extractions.

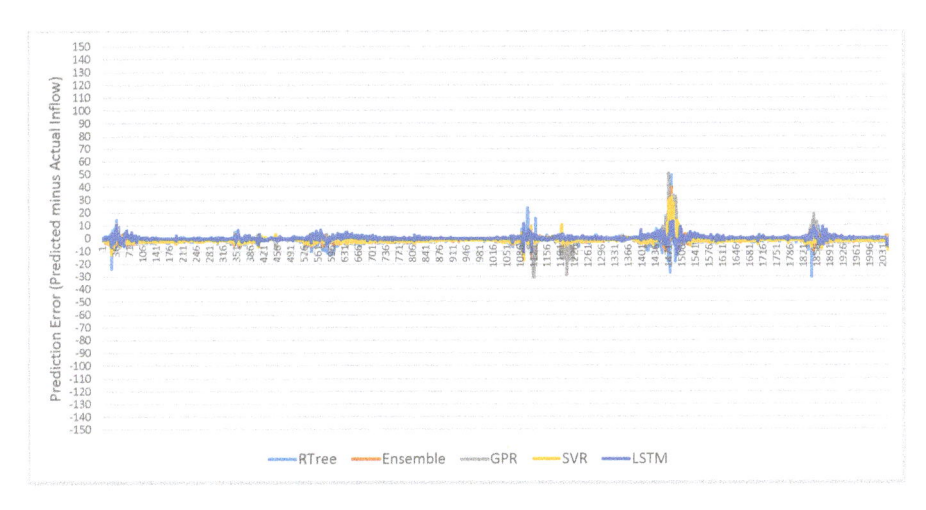

Fig. 8. Prediction error plot of predicted values minus actual dam inflow added to hydro-features extractions. Showing regression tree (RTree), ensemble, Gaussian process regression (GPR), support vector regression (SVR), and long short-term memory (LSTM) models.

Applied Result Three: Upstream Sensor Variable Added. Table 5 shows the RMSE and runtime results for the five models when developed using the hydro-dataset with WLS trend and LTI volume flood-feature extractions and an upstream sensor variable added. The RMSE in Table 5 show a slight improvement in comparison to those in Table 4. Thus, the upstream sensor variable is somewhat effective for improving the accuracy of each model, with the LSTM model being the most accurate. Figure 9 shows dam inflow predictions results using a hydro-dataset also added with hydro-features extractions and an upstream sensor variable, where these are applied by the previous noted five models. Figure 10 shows the prediction error plot of predicted values minus the actual dam inflow using a standard dataset added with hydro-features extractions and an upstream sensor variable. Therefore, under-predictions less than extreme inflow is strongly decreased, so this suggests that the higher accuracy improves the smaller flood damage risk among three applied results.

Table 5. Test evaluations and Bayes optimization based on the cross-validation function applied to a hydro-dataset with weighted least squares (WLS) trend and log of trapezoids integration (LTI) volume flood-feature extractions and an upstream sensor variable added. RMSE is the root-mean-squared error.

Model	RMSE	Runtime (minutes)
Regression tree	5.30	2
Ensemble model (bagging, LSBoost)	2.56	35
Gaussian process regression (GPR)	5.47	57
Support vector regression (SVR)	4.55	49
Long short-term memory (LSTM)	1.61	71

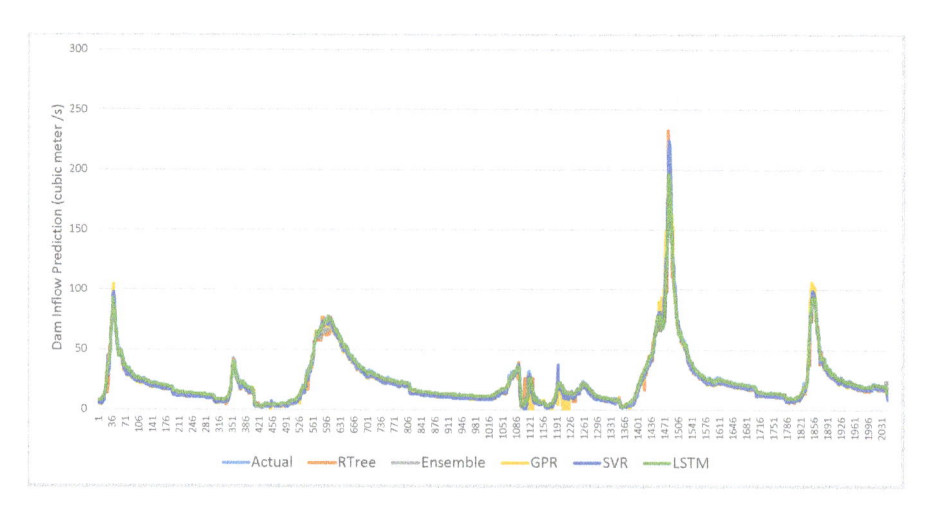

Fig. 9. Predictions (regression tree, RTree; ensemble; Gaussian process regression, GPR; support vector regression, SVR; long short-term memory LSTM) and actual dam inflow series added to flood-feature extractions such as weighted least squares (WLS) trend, log of trapezoids integration (LTI) volume, and an upstream sensor variable.

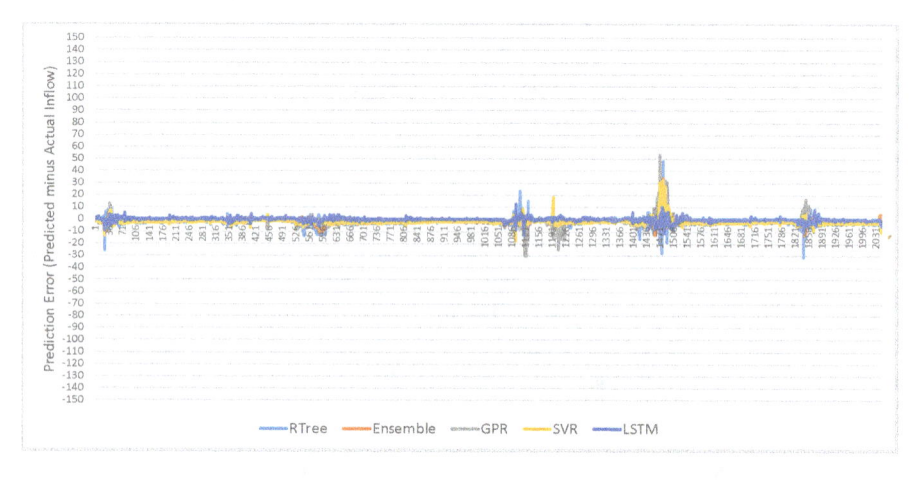

Fig. 10. Prediction error plot of predicted values minus actual dam inflow added to flood-feature extractions such as weighted least squares (WLS) trend, log of trapezoids integration (LTI) volume, and an upstream sensor variable. Showing regression tree (RTree), ensemble, Gaussian process regression (GPR), support vector regression (SVR), and long short-term memory (LSTM) models.

Evaluation sUPE on the Best Model. The LSTM model was the most accurate of the five machine and deep learning algorithms assessed here. Comparison of the LSTM sUPE for each of the three modeling scenarios is shown in Fig. 11. Comparison of the hydro-dataset standard with the hydro-dataset standard with hydro-feature extraction added shows that our proposed flood feature extractions improved the sUPE and hence allows for a reduced risk of flood damage. Furthermore, the addition of the upstream sensing variable is effective for minimizing the sUPE to avoid potential delays that may result in the event that an inflow prediction is lower than the actual dam inflow.

sUPE (sum of Under Prediction Error)

Hydro dataset standard	Hydro + Flood features	Hydro + Feature + Z Upstream
-1498.2	-1434.2	-901.7

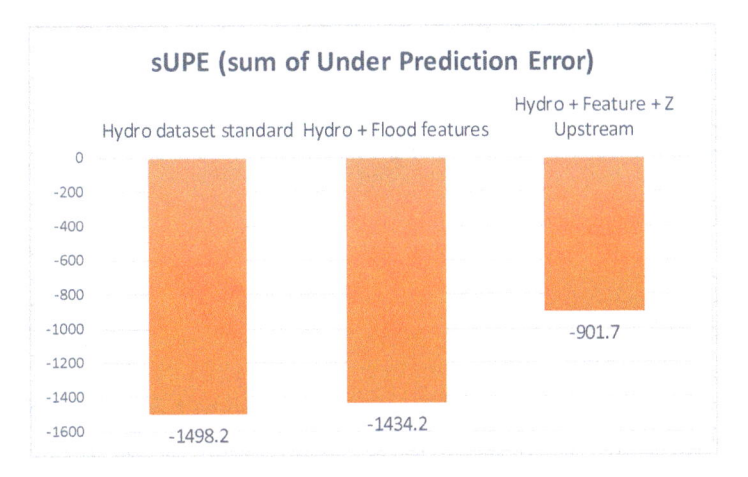

Fig. 11. Comparison of sum of under-prediction errors (sUPE) for the long short-term memory (LSTM) model: (a) hydro-dataset standard, (b) hydro-dataset standard with hydro-feature extraction added, and (c) hydro-dataset standard with hydro-feature extraction and an upstream sensor added.

4 Conclusion

4.1 Concluding Remarks

We applied a smart dam technique for flood forecasting, which involves upstream sensing, flood feature extracted time series regression modeling, and a hydro-blockchain. In addition to the existing 20 year hydro-dataset for our case study, we found that it was possible to add upstream features by installing a water level sensor at the previously unmonitored uppermost stream. We summarized hurdles that may need to be overcome for smart dams, and detail the installation requirements for a water level sensor. Using the water level data measured by a sensor installed in the most upper stream of the dam in the in Kanto study region, we analyze the relationship between the upstream, middle, and downstream water levels during a typhoon in July 2019.

Our study highlights the problems of predicting dam inflow over a forecast period of 3 h using data from the past 3 h. We used WLS and LTI flood feature extractions methods to define flood trend and flood volume features, respectively. These methods vastly improved the prediction accuracy of the 3 h forecast in comparison to that using the standard hydro-dataset only. We also found that the addition of a sensor resulted in a further, slight improvement in the RMSE. Hence, we conclude that upstream sensing and flood feature extractions are useful for improving the accuracy of flood prediction.

4.2 Future Works and Opportunities

In the future, we aim to address low water forecasting during periods outside of the flood season. In order to forecast more than 12 h ahead, a deeper stream network will be constructed by incorporating the neighboring watershed to the study area.

In a situation of multiple dams, water level changes and the volume of reservoir water can become more complex, and can make it difficult for dam managers to make decisions about the discharge operation. Although upstream sensing for dam watersheds is an undeveloped field, it can increase data availability and offers the possibility of improving the accuracy of flood peak and volume predictions that are helpful for flood forecasting. Increasing the density of observations would provide data that could increase understanding of flood features and improve prediction accuracy.

Rainfall, river water levels, and other natural phenomena are complex and watershed-specific, and case studies on floods are site specific. Unsupervised generative learning may allow for the discovery of flood scenarios previously unexperienced. Our study highlights the opportunity that exists for improving the accuracy of flood forecasting by setting up synthetic flood scenarios that use observational data.

Furthermore, we have opportunities to create cyber-physical architectures [36], i.e. digital twin control for dam outflow towards various flood scenarios. We continue to be more robust the above proposed techniques to expand upstream sensing, to extract more densely flood features on hydro-blockchain, and to optimize the dam outflow task installing a hybrid learning technique between a relaxed physical generator and a faster and more accurate inflow predictor.

Acknowledgments. We thank Mr. Shinichi Kuramoto and Mr. Takuji Fukumoto (MathWorks Japan) for providing stable resources for machine and deep learning.

References

1. UN News Centre: EM-DAT International Disaster Database. https://reliefweb.int/disasters. Accessed 10 Aug 2019
2. Insurance Information Institute: World Natural Catastrophes. https://www.iii.org/fact-statistic/facts-statistics-global-catastrophes. Accessed 10 Aug 2019
3. Japan floods: City of Joso hit by 'unprecedented' rain. BBC, 10 September 2016. http://www.bbc.com/news/world-asia-34205879. Accessed 10 Aug 2019
4. ASCE: task committee on application of artificial neural networks in hydrology, II: hydrological applications. J. Hydrol. Eng. **5**, 124–137 (2000)
5. Mujumdar, P.P., et al.: Floods in a Changing Climate Hydrologic Modeling, International Hydrologic Series. Cambridge University Press, Cambridge (2012)
6. Othman, F., et al.: Reservoir inflow forecasting using artificial neural network. Int. J. Phys. Sci. **6**(3), 434–440 (2011)
7. Attygalle, D., et al.: Ensemble Forecast for Monthly Reservoir Inflow: A Dynamic Neural Network Approach (2016). https://doi.org/10.5176/2251-1938_ors16.22
8. Singh, V., et al.: Mathematical modeling of watershed hydrology. J. Hydrol. Eng. **7**(4), 270–292 (2002)
9. Kouwen, N., et al.: Grouped response units for distributed hydrologic modeling. J. Water Resour. Plann. Manage. **119**(3), 289–305 (1993)

10. Ewen, J., et al.: SHETRAN: distributed river basin flow and transport modeling system. J. Hydrol. Eng. **5**(3), 250–258 (2000)
11. Fortin, J.P., et al.: A distributed watershed model compatible with remote sensing and GIS data, I: description of model. J. Hydrol. Eng. **6**(2), 91–99 (2001)
12. Andrew, K., et al.: Inside Smart Cities: Places, Politics and Urban Innovation. Routledge, London (2019)
13. Rob, K., Claudio, C., et al.: Actually existing smart dublin: exploring smart city development in history and context. In: Andrew, K. (ed.) Inside Smart Cities (2019)
14. Karwot, J., Kazmierczak, J., et al.: Smart water in smart city, hydrology and water resources. In: 16th International Multidisciplinary Scientific GeoConference SGEM (2016)
15. Smart Cities of Tomorrow, Water (2014). http://www.smartcitiesoftomorrow.com/water/. Accessed 10 Aug 2019
16. Drescher, D.: Blockchain Basics: A Non-Technical Introduction in 25 Steps, 1st edn. Apress (2017)
17. Information System Authority: Data Exchange Layer X-tee. https://www.ria.ee/en/state-information-system/x-tee.html. Accessed 10 Aug 2019
18. Shortridge, J.E., et al.: Machine learning methods for empirical streamflow simulation: a comparison of model accuracy, interpretability and uncertainty in seasonal watersheds. Hydrol. Earth Syst. Sci. **20**, 2611–2628 (2016)
19. Li, B., et al.: Comparison of random forests and other statistical methods for the prediction of lake water level: a case study of the Poyang Lake in China. Hydrol. Res. **47**(S1) (2016)
20. Yang, J.-H., et al.: A time-series water level forecasting model based on imputation and variable selection method. Comput. Intell. Neurosci. **2017**, 1–11 (2017)
21. Vamsi, K.G.: Modeling of weather data using gaussian linear regression. Int. J. Comput. Sci. Inf. Technol. **6**(4), 3693–3696 (2015)
22. Adiya, G., et al.: A deep hybrid model for weather forecasting. In: ACM Conference on Knowledge Discovery in Databases, KDD 2015, Sydney, Australia (2015)
23. Zhan, D., et al.: Simulating reservoir operation using recurrent neural network algorithm. Water **11**, 865 (2019)
24. Xuan-Hien, L., et al.: Application of long short-term memory (LSTM) neural network for flood forecasting. Water **11**, 1387 (2019)
25. Brieman, L., Friedman, J., et al.: Classification and Regression Trees. Wadsworth (1984)
26. Friedman, J., Hastie, T., Tibshirani, R.: Additive logistic regression: a statistical view of boosting (with discussion). Ann. Stat. **28**, 307–337 (2000)
27. Friedman, J.: Greedy function approximation: a gradient boosting machine. Ann. Stat. **29**(5), 1189–1232 (2001)
28. Scholkopf, B., Freund, Y.: Boosting: Foundations and Algorithms. MIT Press, Cambridge (2012)
29. Rasmussen, C.E., Williams, C.K.I.: Gaussian Processes for Machine Learning. MIT Press, Cambridge (2006)
30. Sergios, T.: Machine Learning: A Bayesian and Optimization Perspective. Elsevier (2015)
31. Hochreiter, S., Schmidhuber, J.: Long short-term memory. Neural Comput. **9**(8), 1735–1780 (1997)
32. Francois, C.: Deep Learning with Python, Manning (2018)
33. Domingos, P.: MetaCost: a general framework for making classifiers cost-sensitive. In: ACM Conference on Knowledge Discovery in Databases, vol. 165–174 (1999)
34. Ling, C.X., et al.: Cost-sensitive learning and the class imbalance problem. In: Sammut, C. (ed.) Encyclopedia of Machine Learning. Springer, Heidelberg (2008)

35. Yasuno, T: Dam inflow time series regression models for minimizing loss of hydro power opportunities. In: Workshop of Data Mining for Energy Modeling and Optimization, PAKDD2018, Melbourne (2018)
36. Zandi, K., Ransom, E.H., Topac, T., Chen, R., Beniwal, S., Blomfors, M., Shu, J., Chang, F-K.: A framework for digital twin of civil infrastructure – challenges & opportunities. In: 12th International Workshop on Structural Health Monitoring, pp. 1659–1665, DEStech (2019)

Using Blockchain in IoT: Is It a Smooth Road Ahead for Real?

Sonali Chandel[✉], Song Zhang, and Hanwen Wu

College of Engineering and Computing Sciences,
New York Institute of Technology, Nanjing, China
{schandel,szhang33,hwu26}@nyit.edu

Abstract. The rise of our smart life, along with the popularity of smart devices, has made the Internet of Things (IoT) a prevalent research topic in recent years. Even though it increases the intelligence level of society to a very great extent, it still has many flaws, especially when it comes to the security domain. The birth of blockchain has brought a new evolution for IoT. The decentralized and traceable feature of blockchain is entirely in sync with the need for the decentralized IoT devices in the future. Currently, there are many applications of blockchain that include but not limited to, smart homes, smart cities, banking, insurance, and financial services. However, there are still many challenges regarding the use of blockchain in IoT, as many things require to be proved and realized. Therefore, the paper is going to research the challenges of blockchain in IoT to find out about its causes and the cures. The changes in the identity of IoT during its lifetime is also covered and how to supervise the governance of data and privacy in a smart device. Data protection law is also taken into consideration when it comes to legal and compliance issues. To solve these difficulties, several perspectives are worth considering, like how to optimize the data authentication, ways to secure communication, methods to improve the smart contract, etc. The paper also aims to help the people from the industry to have better clarity and understanding of the scope of using blockchain in IoT.

Keywords: Blockchain · Internet of Things · Security issues · Decentralized · Data authentication · Data privacy · Data protection · Data governance

1 Introduction

With the rapid growth and evolution in the network, electronics, and wireless communication technologies, the Internet of Things (IoT) has attracted the attention of many researchers as well as countries. It is a paradigm based on the Internet that comprises of many interconnected technologies like RFID (Radio Frequency Identification) and WSAN (Wireless Sensor and Actor-Network) to exchange information [1]. IoT devices are controlled remotely, and the information is shared over the IoT network. IoT plays a very significant role in turning our society into a smarter one. However, currently, it is just at the beginning stage. While IoT has created a promising future for us where devices can communicate and interact with each other, there also exists plenty of challenges in a centralized IoT network that needs proper attention from the industry and the researchers to get it solved. For example, one of the most significant issues in dealing with IoT is its centralized feature.

The development of society has witnessed tremendous benefits from information transmission. With the growth of IoT, it cannot be denied that centralized networks encounter trouble in handling the heavy network traffic, expensive devices, and human labor costs. It is challenging to patch IoT devices and update them with vulnerability and security updates [2]. According to the Gartner report, worldwide spending on the IoT security market has reached $1.5 billion in 2018 and will reach up to $3.1 billion in 2021. This already proves that security can be a significant issue for IoT [43].

Blockchain lays down a clear path for the future of IoT. Blockchain is a type of distributed ledger technology, which aims at how to organize different entities to communicate directly with each other. It is immutable, as the transaction cannot be changed once the record has been confirmed valid. In a decentralized digital age, the blend of blockchain and IoT technology is relatively appealing. During the first digital trend, bitcoin transaction was implemented using blockchain technology. IBM has summed up three essential advantages in blockchain, which is, building trust, reducing cost, and accelerating transactions [4].

On the contrary, blockchain also has its limitations. Enormous electronic energy costs, low transaction speed, small storage capacity, and some privacy issues are the most common. Despite blockchain's limitations, it can still be regarded as a considerable and promising choice for IoT because it is decentralized and its security features suit the need of IoT devices perfectly. By the use of smart contracts, blockchain can help IoT devices to build their communication without needing any central management [4].

One advantage of having a decentralized IoT network is that it can primarily improve the speed of data transmission and connectivity among devices. Earlier people used to connect their electronic devices and transmit information mostly via the Internet. However, a decentralized IoT network allows nearby devices to send and receive data, thus optimizing speed and connectivity among devices [5]. IoT can benefit from blockchain technology in many fields. For example, an irrigation system can use blockchain technology to control the water flow rate due to sensors sending information about the conditions of the crops.

Similarly, oil platforms leverage this technology to interact among devices and make some adjustments based on weather conditions [5]. These examples present a promising future of this integration between IoT and blockchain. Although this integration sounds very promising for the future, the reason for choosing this topic is to let people know that the reality is not as bright as it looks. The integration between these two latest technologies is not able to fix all the problems of IoT. Besides, it also generates other challenges during the integration. Researchers are now coming out with some soluti~ to strengthen this powerful integration of the two technologies.

organized as follows. Section 2 presents the related works mainly on this integration. Section 3 discusses the principle, applications, and mplementing blockchain in IoT. Section 4 analyzes the challenges in different aspects. In Sect. 5, we introduce some solutions for the ased on others' research and our considerations before concluding

2 Related Work

Since the birth of blockchain, its promising future with IoT has been explored by many researchers. However, the road for this integration is not that smooth, and several papers have already highlighted the challenges in this area. The author in [38] points out the current security issues in IoT systems and the drawbacks in the existing approaches to solve these issues. At the same time, it compares three architectural security designs, analyzing its advantages and disadvantages. The author in [28] presents the perspective of the legal problems to address the vulnerabilities in the cybersecurity environment in the IoT field. Those issues mainly relate to Europe's new cybersecurity law GDPR, which makes the privacy data harder to spread through the network. Because the IoT devices grasp the data from the real world and contain sensitive data, it will be one of the main targets of those data protection laws. The author in [26] gives us a broad overview of challenges and potential benefits when integrating IoT with blockchain. The author in [30] summarizes the major security issues and problems in different categories of the IoT field and proposes the solution of using blockchain to solve those issues. For example, the paper covers the challenges like Sybil, spoofing attacks, and insecure physical interface. Those challenges will be discussed in detail in Sect. 4.

The author in [39] has surveyed the security of IoT and has put forward eight IoT frameworks to ensure safety. However, none of those frameworks involves the solution of utilizing blockchain technology. In [25], the author concludes that the combination of blockchain with IoT paves a new way of exploring a market filled with connected devices. However, it ignores the challenges and security problems brought by blockchain. The author in [40] attempts analyzing the possibility of whether blockchain can be applied in the IoT area. Also, they claim that four cases of blockchain characteristics match well with IoT. The author in [41] describes the privacy and security issues in IoT and how blockchain can be applied in the IoT field. One of the threats in privacy issues is hacking, and it will happen when IoT device is collecting data and processing them. The author in [42] focuses on how blockchain satisfies IoT needs.

Concluding from the above research works, we found that there are two main obstacles in the integration of IoT and blockchain. The first is that the data collected from the IoT network is not compatible with the typical size of the blockchain. Then, it causes the speed and storage issue. Another one is associated with the data, IoT devices have collected. Detrimental data will cause security issues, and private data contain hidden risks of privacy issues. However, in most of the research works above, they have failed to go into the details or introduced just a specific field of challenge. In this paper, we have tried to sum up all the problems and analyze them according to different issue types. Due to the limitation of available information, we could not explain much about legal and compliance issues. The author in [26] talks about the challenges in IoT and blockchain in detail, but it does not mention the concrete solutions for those challenges. However, we have tried to find specific solutions for those challenges as well. This is another point the above works fail to take care of as well. Therefore, following the different issue types, we have summarized and analyzed the solution technologies mentioned in [34–37].

3 Implementing Blockchain in IoT

3.1 The Architecture of IoT Blockchain

Due to a vast application area of IoT, we have picked just one specific area to introduce the architecture of implementing blockchain in IoT. The diagram of IoT using blockchain in the business area is shown in Fig. 1 [6]. The function of the underlying technical layer is to manage the portfolios (it is a grouping of financial assets such as stocks, bonds, and commodities) by using the goods' classification mechanism. The data from the real world of the credit system and classification mechanism both can be stored on the endpoint side. The former system can be completed by ordinary statistics, but the latter may involve a data analysis algorithm, which may cause some computational problems in practice. The IoT and smart contract platforms are located in the infrastructure layer. This layer gives the system an overall stable testing platform instead of building the system from scratch. Different platforms have different specific targets. Therefore, it is better to choose a blockchain platform that focuses on the financial area for this system. In the content layer, entities serve as the consumers, and commodity serves as goods. The primary function of this layer is to make abstract objects represent roles in the real world. The difference between human beings and DACs (Distributed Autonomous Corporations) is that DACs can buy things from each other without human interference. The exchange layer is the core of the whole model, which contains the primary system and currency for deals. This layer ensures the transaction is P2P (peer-to-peer) without needing the interface of a third-party, and the value of this transaction is measured by cryptocurrency. Although there may exist a little difference in the architecture between every IoT blockchain industry, the core of this architecture for everyone is the smart contract and P2P protocol, which makes the communication between machines possible.

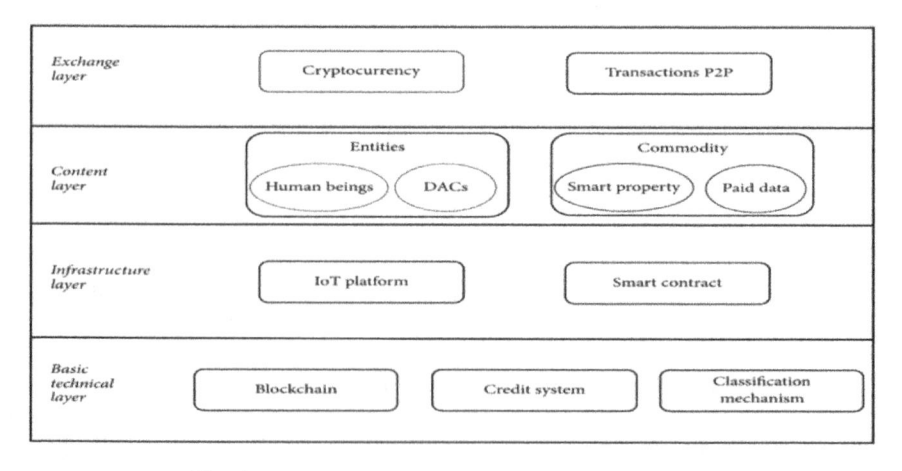

Fig. 1. IoT blockchain architecture for business model

3.2 Current IoT Blockchain Application

IoT blockchain can be used in different areas. Some of the applications shown in Table 1 are based on different focus areas. These applications can be divided as follows:

- For applications focused on the improvement of the fundamental part of the IoT blockchain system [8, 14–16], the main focus for improvement is security, speed, and coordination.
- For energy-focused applications, the first focus is on how to save energy [7, 23], and the other is on creating a marketplace for selling energy [22].
- For the supply chain [11–13, 17], the functions of those applications differ from each other, but they all serve to improve the strength or the variety of this chain.
- For the data sharing area, some focus on the business ecosystem [18], but others may focus on the world's real-time data [19].
- For the hardware area [9, 20], people can manage and deal with their data more conveniently without hiring a third party.
- For the location area, FOAM [21] is mainly used for location service, which may replace GPS in the future.
- For the insurance area, Aigang [10] targets an automated insurance application without the interface of center management.

3.3 Overall Impact of Blockchain Implementation in IoT

The implementation of blockchain in IoT not only brings significant benefits to IoT itself, but it also influences the whole society in a very positive way. Figure 2 shows some significant impacts of using blockchain in IoT. Inside the oval, five typical aspects that blockchain has brought great benefits to IoT are listed. The smart contract can make devices in the IoT network communicate with each other. This can make them more stable with the help of blockchain as the data can be shared among devices in this network. Besides, the P2P (peer-to-peer) protocol in blockchain can be transformed into M2M (machine-to-machine) protocol in the IoT network and provide the decentralized service. Without the additional need for a third party to handle the data from different IoT devices, money and labor force can be saved. Moreover, the IoT devices in a blockchain network are in consensus because they need to communicate and share data, and it makes the whole network more sensitive to the threats or errors.

Outside the oval is the impact that the integration of blockchain and IoT will have on the society. As the actual impact involves plenty of different areas and industries, some of them are put into discussion in detail. The hottest trend at present is everything becoming smart, and that includes smart home, smart city, and every other option that can be replaced by smart devices. Although a centralized IoT network seems to work well in many areas with the number of endpoints growing, blockchain will show its strength, especially in the smart home field. Another promising area for this integration is healthcare. For example, at present, patients' health data will be collected through IoT devices and uploaded into the central servers. However, this may cause privacy issues, and data breach may happen in case the servers are attacked (which is quite common). With the help of blockchain, first, the security will be strengthened. Besides, the patients can use the apps on their phones to collect their personal health information and send it to the doctor directly without the need for a central server [24].

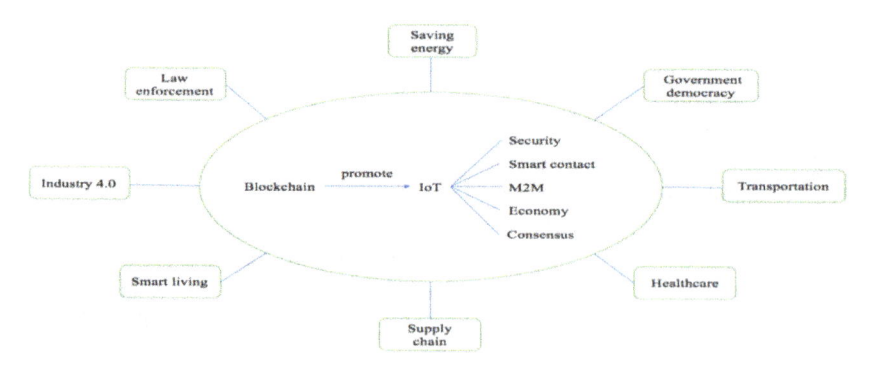

Fig. 2. The interactions of IoT and blockchain

Table 1. Different IoT blockchain applications and features

Applications	Features
LO3 Energy [7]	Manage the energy at a local level, especially for smart grids and buildings
IBM ADEPT [8]	Offer greater scalability and security for the IoT, current test in smart homes
Slock.it [9]	Empower anyone to easy rent, share or sell anything that can be locked
Aigang [10]	Fully automated insurance, especially for insurance innovation
AERO Token [11]	Drone delivery and flight mapping
Chronicled [12]	Power end-to-end smart supply chain
Modum [13]	Provide data integrity and authenticity for the supply chain, especially for industries
Blockchain of Things [14]	Secure the IoT connections
Atonomi [15]	Provide security protocol
IoTeX [16]	Faster speed, nested blockchain, and autonomous coordination
Ambrosus [17]	Supply chain especially for food and pharmaceutical enterprises
Waltonchain [18]	Completely data sharing and absolute information transparency business ecosystem
Streamr [19]	A platform for the free and fair exchange of the world's real-time data
Helium [20]	A green mining machine using radio waves
FOAM [21]	Secure location service, no more GPS
Power Ledger [22]	Marketplace for renewable energy
Grid+ [23]	Dramatic savings in-home energy cost

4 Challenges of Using Blockchain in IoT

4.1 Challenges in General

This section summarizes the general challenges when integrating blockchain technology with IoT. Challenging issues like security domain and data reliability become a significant barrier when these two technologies meet each other [26]. Despite the advantages of integrating blockchain with IoT, their combination also leads to many drawbacks. Some blockchain and IoT based challenging issues are presented below.

- **Speed:** Although on one side, IoT is fast enough to transmit data in the Gigabyte unit, on the other hand, transactions in blockchain still cause some latency time because of blockchain's decentralized architecture. For example, if a piece of transaction information is confirmed by a specific part of the blockchain network earlier but has not been confirmed by the whole network, then the overall performance will still be affected [26]. Besides, the feature of limited bandwidth makes the real-time transaction more difficult and time-consuming to be handled for numerous networks of nodes [3].
- **Storage capacity:** Each transaction's information generated by network users who want to exchange money with each other is stored in the blockchain. As a result, its length will keep growing day by day. In such a case, more storage space needs to be allocated, posing a challenge of expanding the storage dramatically. Take the cost of time by validating transactions in blockchain as an example. With the increasing number of transactions in the blockchain, the amount of time that transaction record needs to be confirmed increases as well [26]. Storage is also a huge hurdle for smart devices. This is because, as the size of ledger increases, devices such as sensors will have limited capabilities due to their low storage capacity [5]. At the same time, when data is transmitted among IoT devices, redundant information will increase the burden for storage under the blockchain technology. Therefore, blockchain may not be a suitable vessel to store huge amounts of data collected from IoT devices.
- **Security issues:** With the introduction of distributed technology, current challenges not only come from data protection, smart contracts, devices, and networks, but it also includes security issues such as loss of privacy, authentication of identity, and spoofing [3]. Security issues can be divided into three categories: low-level security issues, intermediate-level security issues, and high-level security issues [31]. For low-level security issues, one may encounter Sybil and spoofing attacks, insecure physical interface, etc. Sybil attacks can also happen in intermediate-level security issues. High-level security issues cover vulnerable software and cloud violation based on IoT.

IoT is a proliferation of online-connected devices with different architectures to store data, while blockchain can guarantee that data generated by IoT is immutable. It seems that the integration of blockchain and IoT makes the best use of each other's advantages. However, a risk may arise when the data has already been corrupted in IoT before it is transmitted to the blockchain. In such a case, the blockchain must preserve the wrong data from the beginning. Also, data corruption caused by IoT not only arise security issues at the initial stage, but the failure of sensors and improper working devices can also damage the reliability of data made by IoT [26].

Furthermore, as Wireless Sensor Networks (WSN) become the primary technology applied to IoT, security issues bring much concern to the public because of constrained available resources caused by WSN [27]. For example, how to secure wireless communication links against eavesdropping or traffic analysis in IoT is a massive concern for many people and organizations. Another severe problem in security issues is privacy. It may pose a significant threat to IoT during different stages. These issues can be summarized as data collection, data procession, and data feedback. In the stage of data collection, raw data is gathered by sensors and actuators, processed, and uploaded to the Internet. However, due to some misinterpretation of Internet protocols, such data process flow can be intercepted by hackers, which can lead to data leakage [41]. For the last two stages, data privacy issues can be caused by the lack of standard configurations for interfacing a large number of IoT devices. In addition, for legal and right issues, a lack of comprehensive laws' regulation on the IoT data makes it possible for hackers' manipulation of flowing data.

- **Smart contract:** For one thing, a smart contract is called a computerized protocol, which allows reliable transactions to be carried out among users who want to exchange money without third parties. For another, they are called "scripts" stored in the blockchain [25]. However, the challenges we present here should be paid attention to. First, data will be saved in a smart contract safely only if the code to write a smart contract is precise and perfect [6]. However, human clerical errors can quickly bring mistakes into the system making it less safe. Second, even if the smart contract can take custody over transactions, large amounts of data to be accessed can make the smart contract overburdened, which may slow down the transaction speed.

4.2 Legal and Compliance Issues

The European Commission has devoted itself to protecting Network and Information Security (NIS) since 2001. Besides, they have set up a committee on Critical Information Infrastructure Protection in 2009, which aims at keeping Europe from online attacks and raising cybersecurity levels to the maximum [28]. So far, the draft of the NIS directive proposed by the European council has been published and already come into force. Three goals are covered in the law to ensure the protection of users' data, and privacy: (1) Improve national cybersecurity capabilities, (2) strengthen EU-level cooperation, (3) ensure security and incident notification requirements.

Scholar Marie-Helen Maras stated that the US Federal Trade Commission took action against "TRENDnet" (all kinds of potential information leakage incidents caused by online-connected IoT devices) among IoT devices including surveillance cameras in 2013 [29], the intention of which was to protect consumers' privacy and security from being hacked. The law also suggests that IoT infrastructure should be equipped with an access control system so that only authenticated users can get permission to access the data on sensitive devices.

5 Solutions for Existing Challenges of Blockchain in IoT

Although the integration of IoT and blockchain brings many challenges, researchers have already come out with many optimization strategies to overcome these challenges. Most of these challenges are caused by the limitation of blockchain itself instead of having some brand-new problems. So, the key point to solve these challenges is to advance blockchain technology. This section is going to introduce some solutions based on our research from already published projects or studies. There are three typical solutions that can be applied to the problems in the integration of IoT and blockchain. These are explained in the following sub-sections.

5.1 Transaction Speed and Resource Utilization Improvement

The low transaction speed of traditional blockchain largely hinders the communication between the IoT devices, which decreases the bandwidth of the IoT network. However, the technology behind the IOTA, a cryptocurrency for the IoT industry, has the solution for this low-speed issue, which is called the tangle [31]. Unlike traditional blockchain, which uses the single direction chain to store each transaction, the tangle is a directed acyclic graph for storing transactions. Its structure is shown in Fig. 3 [31].

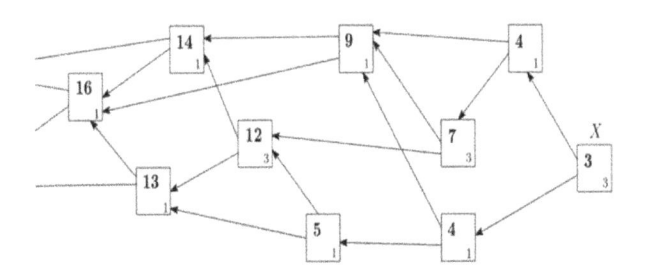

Fig. 3. The structure of the tangle

In Fig. 3, each square represents the transaction. The number in the bottom right side in each square denotes its weight, and the one on the top left denotes the cumulative weight concerning its incoming paths/sources. For example, the box that has weight 9 comes from its incoming past weights: 3, 3, 1, 1. By adding the weight of itself, 9 becomes its cumulative weight. Therefore, if a user issues a transaction to the network, it will immediately change the cumulative weights in each transaction. So, the nodes can check whether the transaction conflicts with the tangle history or not to make the verification decision. This is because, if the transaction is being manually changed, it will cause a mismatch between the tangle history and the current situation.

While the tangle structure improves the transaction speed up to 1000 tx/s (transaction per second) [32], it also makes the system vulnerable to "large weight" attacks [31]. In this attack, the attacker will perform two fraudulent transactions. One is a typical transaction, that will be accepted by the network. Another one is a double-spending transaction (money can be spent twice), which is not allowed in the network.

However, attackers can use their immense computational power to create many small transactions to increase the weight of double-spending transaction. Finally, this transaction will outpace other transactions and get accepted.

Aside from changing traditional blockchain structure, HDAC (a blockchain technology company) uses a hybrid blockchain platform (the integration of public and private blockchains) to speed up transactions [33]. Public blockchain exists among the public network whose consensus mechanism is open to all while private blockchain pervades among certain authorized entities in a closed network. The comparison of transaction speed between different blockchains is shown in Table 2 [33]. The data in Table 2 indicates that the HDAC blockchain has better transaction speed performance than the other two blockchains. Meanwhile, HDAC's ePoW (Electronic Proof of Work) consensus algorithm contributes to the efficient use of energy by reducing the number of nodes required to achieve consensus instead of wasting plenty of electricity like PoW (Proof of Work) algorithm.

Table 2. Comparison of transaction speed among different blockchain platforms

Features	Transaction speed
Bitcoin blockchain	7 tx/s
HDAC blockchain	160 tx/s (public) 500 tx/s (private) 1000 tx/s (target)
Ethereum blockchain	25 tx/s

5.2 Secure Communication and Authentication Reinforcement

Although blockchain can help IoT devices mitigate some security issues, it is still not immune to some external attacks. Smart devices use built-in protocols such as Transport Layer Security (TLS) or Secure Sockets Layer (SSL), which have various security systems against control signals [33]. However, these security protocols are complex for an IoT device and need centralized management. On the one hand, a runtime upgrading strategy can enable dynamic application and network-level upgrades. On another hand, by introducing the run-time reconfiguration mechanism into the WSN, which is used for the communication of IoT devices, it would make the whole system more robust.

To deal with the authentication issues in IoT devices, the out-of-band two-factor authentication scheme for IoT blockchain can be a functional trial for managing the security problems in a smart home. The scheme uses a secondary authentication factor to separate a home IoT device from an external malicious device, which enhances the authentication and authorization process to a great extent [34, 35, 37]. Although centralized authentication technologies are still prevailing currently, those improvements would make the decentralized IoT network more robust and popular in the near future.

5.3 Smart Contract Perfection

When it comes to IoT's blockchain's smart contract perfection, things become much more complicated. The system has to create different contracts for different situations and manage them at the same time. These management problems should be solved by a smart system. One approach to solve this problem is to import a cognitive system to mitigate management challenges [36]. The whole structure of this system and each layer's function is shown in Fig. 4.

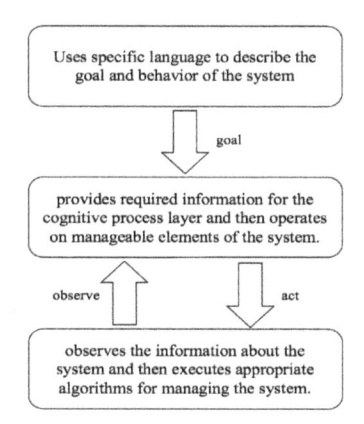

Fig. 4. The structure of the cognitive IoT system for IoT blockchain

The first layer, also called the requirement layer, is like a programming level. It is like using a specific programming language to abstract the objects in the real world into the system. The second layer, the cognitive process layer, is to observe the status of things and trigger the smart contract. It is the core part of mitigating the challenges in the management. The last layer, things management layer, focuses on selecting optimal algorithms in smart contracts for different tasks required by the second layer.

6 Conclusion and Future Work

Blockchain's decentralized feature may perfectly suit IoT, but the process of implementation of these two technologies together is not a smooth path. Therefore, many things should be taken care of during this integration. This paper mainly researches the current significant challenges that IoT and blockchain must address for them to work successfully. There are other issues that also need to be considered, but these issues are less pressing than these major challenges, so they are not covered in this paper.

In the challenges field of combining blockchain with IoT, the current issues such as speed, storage capacity, security and privacy issues, and smart contract are restricting the further development of the IoT industry. Meanwhile, the guarantee of data protection law is the key to push forward the IoT field successfully. Since there are so many challenges waiting to be solved, it is still too early to announce that blockchain is the best fit for IoT. Nevertheless, we still foresee a promising future integrating IoT with blockchain, if the above issues and solutions can be handled appropriately.

The current solutions for challenges in IoT blockchain are summed up for future work. The key solution for improving transaction speed is around blockchain's inherent mining algorithm, such as PoW and PoS. Similarly, the main point to secure IoT and blockchain network is to secure IoT devices themselves instead of a blockchain. Besides, the management of smart contact in IoT blockchain should apply a smart method, which refers to the AI domain. With the elimination of those challenges and the assistance of the advanced technology in other fields, a reliable, efficient, and scalable IoT blockchain will dominate the IT industry soon.

References

1. Díaz, M., Martín, C., Rubio, B.: State-of-the-art, challenges, and open issues in the integration of the Internet of things and cloud computing. J. Netw. Comput. Appl. **67**, 99–117 (2016)
2. IoT security: smart business requires smarter Internet of Things security (2018). https://www.i-scoop.eu/iot-security-smarter-internet-of-things-security/
3. Groopman, J.: Six challenges facing blockchain and IoT convergence (2018). https://internetofthingsagenda.techtarget.com/blog/IoT-Agenda/Six-challenges-facing-blockchain-and-IoT-convergence
4. Blockchain and the IoT: the IoT blockchain opportunity and challenge (2017). https://www.i-scoop.eu/blockchain-distributed-ledger-technology/blockchain-iot/#Blockchain_as_the_answer_to_IoT_challenges
5. Dickson, B.: The benefits and challenges of using blockchain in IoT development (2016). https://bdtechtalks.com/2016/06/09/the-benefits-and-challenges-of-using-blockchain-in-iot-development/
6. Jesus, E.F., et al.: A survey of how to use blockchain to secure internet of things and the stalker attack. Secur. Commun. Netw. **2018**, 9675050:1–9675050:27 (2018)
7. LO3 Energy (2017). https://lo3energy.com
8. Higgins, S.: IBM reveals proof of concept for blockchain-powered IoT (2015). https://www.coindesk.com/ibm-reveals-proof-concept-blockchain-powered-internet-things/
9. Prisco, G.: Slock.it to introduce to smart locks to smart Ethereum contracts, decentralize the sharing economy (2015). https://bitcoinmagazine.com/articles/slock-it-to-introduce-smart-locks-linked-to-smart-ethereum-contracts-decentralize-the-sharing-economy-1446746719/
10. Aigang (2017). https://aigang.network
11. AeroTaken (2017). https://medium.com/@aerotoken
12. Chronicled (2017). https://chronicled.com
13. Modum (2017). https://modum.io
14. Blockchain of Things (2017). https://blockchainofthings.com
15. Atonomi (2017). https://atonomi.io
16. IoTex (2017). https://iotex.io
17. Ambrosus (2017). https://ambrosus.com/#features
18. Waltonchain (2017). https://www.waltonchain.org
19. Streamr (2017). https://www.streamr.com
20. Helium (2017). https://www.helium.com
21. FOAM (2017). https://foam.space

22. Power Ledger (2017). https://powerledger.io
23. Grid + (2017). https://gridplus.io
24. Esposito, C., et al.: Blockchain: a panacea for healthcare cloud-based data security and privacy? IEEE Cloud Comput. **5**(1), 31–37 (2018)
25. Christidis, K., Devetsikiotis, M.: Blockchains and smart contracts for the internet of things. IEEE Access **4**, 2292–2303 (2016)
26. Reyna, A., et al.: On blockchain and its integration with IoT. Challenges and opportunities. Future Gener. Comput. Syst. **88**, 173–190 (2018)
27. Perrig, A., Stankovic, J., Wagner, D.: Security in wireless sensor networks. Commun. ACM **47**(6), 53–57 (2004)
28. Weber, R.H., Studer, E.: Cybersecurity in the Internet of Things: Legal aspects. Comput. Law Secur. Rev. **32**(5), 715–728 (2016)
29. Maras, M.-H.: Internet of things: security and privacy implications. Int. Data Privacy Law **5**(2), 99 (2015)
30. Khan, M.A., Salah, K.: IoT security: review, blockchain solutions, and open challenges. Future Gener. Comput. Syst. **82**, 395–411 (2018)
31. Popov, S.: The tangle. cit. on, p. 131 (2016)
32. Price, C.: On blockchain and the Internet of Things (2018). https://blockchainatberkeley.blog/on-blockchain-and-the-internet-of-things-f6b0b2deb528
33. HDAC (2017). https://hdac.io/
34. Ruckebusch, P., et al.: Gitar: generic extension for internet-of-things architectures enabling dynamic updates of network and application modules. Ad Hoc Netw. **36**, 127–151 (2016)
35. Taherkordi, A., et al.: Optimizing sensor network reprogramming via in situ reconfigurable components. ACM Trans. Sens. Netw. (TOSN) **9**(2), 14 (2013)
36. Saghiri, A.M., et al.: A framework for cognitive Internet of Things based on blockchain. In: 2018 4th International Conference on Web Research (ICWR). IEEE (2018)
37. Wu, L., et al.: An out-of-band authentication scheme for internet of things using blockchain technology. In: 2018 International Conference on Computing, Networking and Communications (ICNC). IEEE (2018)
38. Sha, K., Wei, W., Yang, T.A., Wang, Z., Shi, W.: On security challenges and open issues in Internet of Things. Future Gener. Comput. Syst. **83**, 326–337 (2018). https://doi.org/10.1016/j.future.2018.01.059
39. Ammar, M., Russello, G., Crispo, B.: Internet of Things: a survey on the security of IoT frameworks. J. Inf. Secur. Appl. **38**, 8–27 (2018). https://doi.org/10.1016/j.jisa.2017.11.002
40. Conoscenti, M., Torino, D., Vetr, A., Torino, D., Martin, J.C. De., (e.d.): Blockchain for the Internet of Things: A Systematic Literature Review
41. Manoj, N., Kumar, P.: ScienceDirect blockchain technology for security issues and challenges in IoT. Procedia Comput. Sci. **132**, 1815–1823 (2018). https://doi.org/10.1016/j.procs.2018.05.140
42. Fernández-caramés, T.M., Member, S.: A review on the use of blockchain for the internet of things **3536**(c), 1–23 (2018). https://doi.org/10.1109/ACCESS.2018.2842685
43. IoT security spending 2018–2021: regulatory compliance becomes key (2019). https://www.i-scoop.eu/internet-of-things-guide/iot-security-spending-2018-2021/

Developing a Blockchain-Enabled Collaborative Intrusion Detection System: An Exploratory Study

Daniel Laufenberg, Lei Li[✉], Hossain Shahriar[✉], and Meng Han[✉]

Kennesaw State University, Marietta, GA 30060, USA
dklaufe@gmail.com,
{lli13,hshahria,mhan9}@kennesaw.edu

Abstract. A Collaborative Intrusion Detection System (CIDS) is a system which a set of IDS work together to defend the computer networks against increasingly sophisticated cyber-attacks. Despite more than decade of research on CIDS, trust management and consensus building among IDS hosts remain as challenging problems. In this paper, we conducted an exploratory study to tackle those two challenges by leveraging the inherent immutability and consensus building capability of blockchain technology. We proposed an architecture for a blockchain-enabled CIDs and implemented a preliminary prototype system using open-source projects such as Hyperledger and Snort. Our initial evaluation on a benchmark testing showed the proposed architecture offers a feasible solution by addressing the issues of trust management, data sharing and consensus building, as well as insider attacks in the network environment of CIDSs.

Keywords: Blockchain · Collaborative Intrusion Detection Systems · HyperLedger · Snort

1 Introduction

Interconnected computer networks have been the engine for economic growth and innovation for the past few decades. It has become increasingly important to protect the digital infrastructure of our society against cyber-attacks. The Intrusion Detection System (IDS), has been widely adopted by individuals, business and organizations to protect their computer networks.

Working with existing firewalls and anti-virus systems, an IDS is a device or software application which monitors network traffic, identifies attacks by building normal network profiles (anomaly-based IDS) or matching the patterns of malicious behavior or violations (signature-based IDS) that protects computer networks against attacks [34]. An IDS can offer real-time, cross-platform, and pre-host protection and is a viable solution to mitigate some malicious attacks [13, 29]. Anomaly-based IDSs are prone to having many false positives [32]. Signature-based IDSs are generally better with the precision rate but can often miss attacks if the signature database is outdated or incomplete [28, 33].

As cyber-attacks are becoming more sophisticated and being launched at a larger scale and across platforms [23], an intrusion detection system would be more effective if it works with other IDSs. For example, IDS hosts can exchange information such as network traffic data and alerts and share signature databases. [7, 10, 23, 30, 31]. Such a system is referred as a collaborative intrusion detection system (CIDS). Despite promising benefits of CIDS, the underlying trust behind sharing of resources remains a major concern. An attacker host may join in a collaborative IDS system network and provide inaccurate or malicious signatures. Moreover, a host environment may be tampered with to alter the data files that actually store signature s (Snort IDS saves the rules in plain text files, which can be easily altered).

Recently, there has been a spike of interest in the blockchain technology where distributed data structure is shared and replicated among the participants in a peer-to-peer network [1–5, 8, 12]. The built-in immutability and consensus building make the blockchain technology a viable solution to develop collaborative IDS and overcome trust management and consensus building among IDS [26]. Alexopoulos et al. [23] proposed a general framework for block-based collaborative IDS which is focused on using blockchain for alert sharing and consensus building.

Inspired by the effort of Alexopoulos et al.'s work [23], we introduce a blockchain-enabled architecture for a signature-based IDS. In addition to alert exchange, we also propose to use the blockchain technology for signature management such as signature sharing, creation and verification among hosts in a CIDS. We also present the implementation strategies of the architecture. Based on our knowledge, the proposed architecture is the first kind for CIDSs.

The remainder of the paper is organized as follows: Sect. 2 introduces related work on Blockchain and intrusion detection systems, Sect. 3 discusses an architecture for a collaborative Signature-based IDS based on the blockchain technology. Section 4 presents the implementation consideration of the proposed architecture. Finally, Sect. 5 concludes the paper.

2 Background and Related Work

2.1 Overview of Blockchain

Blockchain can loosely be defined as a data structure, database, or a growing list of records, called blocks, which are linked using cryptography [27]. There are three types of blockchains: public, private and consortium [6, 27].

A public blockchain such as BitCoin is an open system [26] where anyone can join and participate in the system. The two advantages of public blockchain are its characteristics of Permission-less and immutability. Having a public blockchain removes the necessity for a access control protocol. Applications can be added to the network without approval, and blockchain becomes the transport layer of these applications [26]. A public blockchain is stored typically on a peer-to-peer network. This allows for the data to be nigh unchangeable due to many computers storing the data and agreeing on what is legitimate data and what could possibly be illegitimate.

A private blockchain is a closed system in which the use of the blockchain is controlled. There are limited applications of private blockchain as the central control work against the decentralization aspect which is key to the blockchain concept.

The consortium blockchain is a mixture of both public and private blockchain. Typically, a consortium is public but the number of nodes who can change the data in the blocks is limited. Consortium's are sometimes invite-only for this purpose.

Blockchain tends to fall short when it comes to scalability, depending on the consensus algorithm used. Speed is a big concern as well for any application of a blockchain system.

2.1.1 Consensus Algorithm

As a distributed structure, consensus building is very important for blockchain where nodes of the blockchain construct and support the decision that works best for the rest of them [40]. It's a form of resolution on how to add blocks, data, or do anything to the blockchain. There are many consensus algorithms, however, this paper will only cover those pertinent to the paper: Proof of Work, Proof of Stake, Delegated Proof of Stake, Proof of Authority, Byzantine Fault Tolerance, Proof of Elapsed Time.

Proof of Work (PoW). In a proof of work system, the new blocks in the chain are created by those that have the computational power to solve complex mathematical problems. PoW has limitations with power consumption and inefficiency. This system is used in BitCoin and would not be ideal for the proposed IDS.

Proof of Stake. In a proof of stake system, the new blocks are created in a distributed consensus. The next block is chosen by combination of random selection and wealth range. Ethereum has a proof of stake currently in development called Casper.

Delegated Proof of Stake. In a delegated proof of stake, those that have "stake" in the blockchain can vote for others to have control of the chain. It is not all about who owns the most cryptocurrency or most stake in the blocks. It is about having democratic votes to mitigate the risks of the original proof of stake.

Proof of Authority (PoA). PoA consensus is built further off of Proof of Stake. Instead of voting or allowing someone who was an early adopter of a blockchain to have "stake" in it, the proof of authority puts the onus on those with the reputation to be in control of the chain. These are trusted individuals within the community or network that are well-respected.

Byzantine Fault Tolerance (BFT). Named after an old adage of Byzantine General problems, this algorithm has been around for some time. The idea is that two generals were attempting to communicate between enemy lines and can never be 100% sure that their messages are received. BFT at its simplest form is a way to avoid nodes or blocks in the chain doing something that they were not supposed to do. BFT can be found in many popular consensus algorithms in blockchain.

Proof of Elapsed Time (PoET). The PoET consensus algorithm is designed to be a production-grade protocol capable of supporting large network populations. PoET algorithm relies on secure consensus without the power consumption drawbacks of the

Proof of Work algorithm. Each person in the network waits a random amount of time, whoever finishes waiting first becomes the leader of the new block in the chain [25].

2.1.2 Blockchain Applications

There are many different applications being conceived by researchers in the field of blockchain, such as consensus algorithm research Proof of Majority [2], ProductChain, a scalable blockchain framework for supply chains [16], AutoBotCatcher a system proposed to protect the infrastructure of IoT devices [4], TickEth, a proposed system for using blockchain to buy sporting event tickets [22], a package delivery system [14], and a networking trading system [17]. There is also work showing how blockchain can be used with machine learning in [21, 24] where the ledger self-adapts to transaction demands. As we can see there is a body of work done on how blockchain can be used for things other than cryptocurrencies.

2.2 Intrusion Detection Systems (IDS)

There is a wealth of studies on Intrusion Detection System (IDS) because of its impact on cybersecurity. IDSs can be categorized as host-based IDS and network-based IDS. There are pros and cons for each type of IDS. In a host-based system the IDS runs on a single host, this allows for the IDS to directly monitor that host and which resources were attacked. Host-based can make it difficult to analyze the intrusion attempts on multiple computers and will be difficult to work in a large network environment. In a network-based IDS a network sensor is installed on the network interface card and allows for an entire network to be monitored. All of these packets are analyzed; however, this can take a lot of time and resources, and can miss packets going to a specific host.

IDSs can also be divided by the detection methods: signature-based intrusion detection systems and anomaly-based intrusion detection systems. A signature-based IDS relies on patterns of malicious behavior or violations to recognize the attacks. Signature-based IDS could ideally identify 100% of the attacks with no false alarms as long as signatures are specified ahead of time. However, each signature, even if it leads to the same attack, has the potential to be unique from any other signatures. This is the most commonly implemented IDS [11, 18, 34–36].

The other common type of IDS is an anomaly detection system. This type of IDS focuses on the system's normal behaviors instead of focusing on attack behaviors, as seen with signature-based intrusion detection systems. To implement this type of IDS, the approach is to use two phases. The first phase is the training phase where the systems behavior is observed in the absence of any type of attack. Normal behavior for the system is identified into a profile. After this, the second phase or detection phase, begins. In this phase, the stored profile is compared to the way the system is currently behaving and deviations from the profiles are considered potential attacks on the system. This can lead to several false positives [15, 37–39].

There has been a growing trend of research towards CIDS due to the speed and efficiency of peer-to-peer networks [31, 32]. As the Internet becomes faster the shareability of an application becomes more likely. We are no longer bound to slower speeds, or forced to store data locally, we can store and share data seamlessly among many networks. CIDS is a part of this evolution and is a major reason why we chose this in our research. There are many other new ideas being conceived, such as the research done on securing Internet of Things devices with a blockchain-based collaborative IDS [5]. There is promise for CIDS to aid in securing intelligent electronic devices (IED) with intrusion detection [10]. Kademlia (a peer-to-peer overlay) also shows proof of concept with a CIDs-like system in [19].

2.3 Intersection of Blockchain and IDS

Given the built-in immutability and consensus building of blockchain technology, researchers [23, 26] have started to apply blockchain technology to tackle issues in Collaborative IDSs.

Meng *et al.* [26] conducted comprehensive survey on applicability of blockchain technology in intrusion detection and identified several open challenges in the field such as latency, complexity, security, privacy, and limited signature coverage. Our work addresses these challenges by establishing a peer-to-peer network of signatures, implementing security and trust policies via consensus building in an blockchain-enabled environment. Alexopoulous et al. [23] proposed a system that uses blockchain technology for trust building and alarm data exchange in CIDSs and discussed some design considerations.

3 The Blockchain-Enabled CIDS Architecture

Building on Alexopoulus et al. [23] 's work, we introduce a comprehensive architecture specifically for a signature-based CIDS. As signature/rule exchange and protection are a critical part of a CIDS, blockchain technology can be used to facilitate the rule exchange and secure the ruleset of each host IDS (hence we label each IDS as blockchain-IDS).

As shown in Fig. 1, the proposed architecture applies consortium blockchain infrastructure to build trust among participating IDS hosts and enable secure storage and exchange of rule sets. Similarly, to a traditional IDS, there are three components inside of a host in the proposed NIDS architecture: (1) a sniffer that reads and breaks down network traffic and sends them to the detection engine; (2) a detection engine that compares the packets received from the sniffer with rules/signature; (3) a rule manage that handles the maintenance of the rules in a host and rule exchange with other hosts.

Each host IDS creates block to store its rule set and alarm data while in a traditional IDS these rules are stored in ASCII format .txt files. Figure 2 shows how an IDS store rules in the blocks.

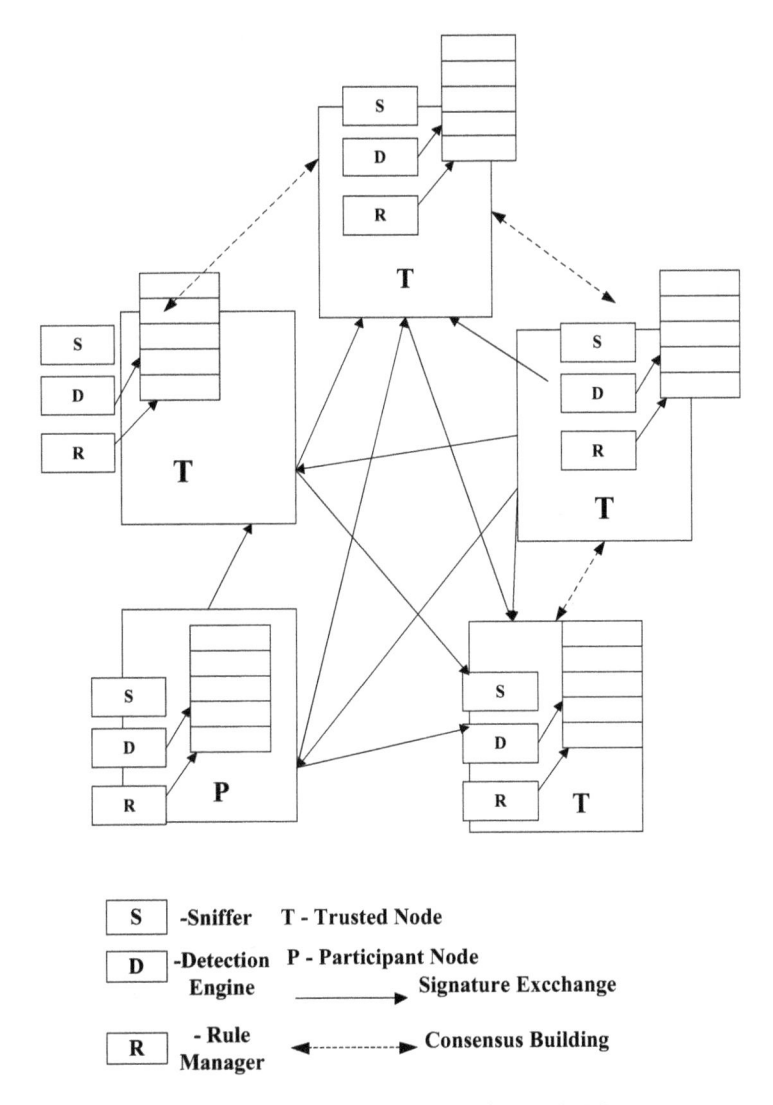

Fig. 1. Architecture of blockchain-based IDS

In the proposed architecture, there are two types of host IDS: trusted nodes (T) and participating nodes (P). All nodes can make a request to change the rule set such as adding new rules, modifying or deleting existing rules. However, only trusted nodes are involved in consensus building process which approve or reject the change request. Delegated proof of stake is used as the consensus algorithm. The rule change approval process is illustrated in Fig. 2.

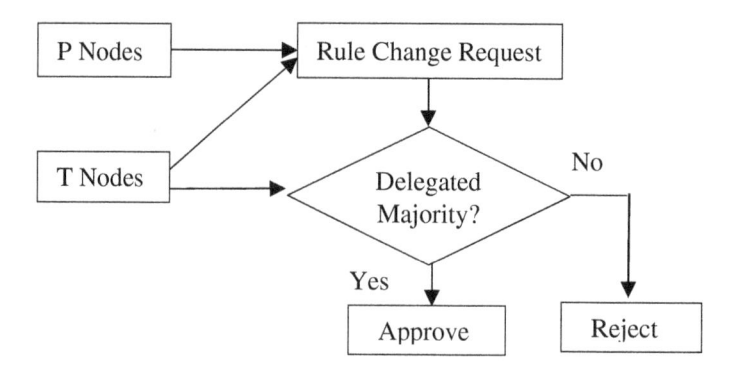

Fig. 2. Consensus building process

Below is the general flow for adding a new rule to the system. The process for updating or removing a rule is similar to rule addition.

(1) T or P node makes an add-rule request.
(2) Notifications are sent to all T nodes on the network that a new request needs to be voted on.
(3) T nodes analyze the pending request and vote within a predesignated time frame.
(4) Votes are tallied automatically by the system and the request either is approved or rejected.
(5) Multiple requests can be voted on at once due to the blocks being immutable. The approved requests are implemented in batch sequence.

4 Implementation

4.1 Prototype Development

We built a preliminary prototype for the proposed architecture. The rule set is adapted from Snort, an open-source, free and lightweight network intrusion detection system (NIDS). There are three popular options for blockchain implementation: Ethereum Virtual Machine, Truffle Suite, and HyperLedger.

Ethereum is best suited to cryptocurrency and would cost money to use the network. Truffle Suite is only capable of development on the Ethereum network which would be counterproductive to a prototype for this research paper [20]. HyperLedger is an umbrella project of open source blockchains and related tools, started in December 2015 by the Linux Foundation and supported by big industry players like IBM, Intel and SAP to support the collaborative development of blockchain-based distributed ledgers [9]. There are many tools and frameworks available via HyperLedger. We want to have to an open and consortium blockchain, hence the HyperLedger is a better choice to build our prototype.

The HyperLedger Sawtooth or Fabric framework could work well for the purposes of creating a blockchain-based IDS. The chaincode can initialize a ledger IDS rules using blockchain. We can then implement with javascript the following classes which are not an exhaustive list and new or existing classes can be added/remove as needed.

CreateRule- Creates a rule in the blockchain.
RemoveRule – Remove a rule from the blockchain.
QueryAllRules – This query will return all rules currently in the blockchain.
QueryRuleProperties – Will return the properties of a ruleID # such as Port/Protocol/Owner/Etc.
UpdateRuleProperties – Allow update to the properties of a rule if something changes.
UpdateRuleOwner – Update the owner of a rule.

A snippet of the CreateRule function written in JavsScript for the HyperLedger framework is listed as below.

```
async createRule(ctx, ruleNumber, RuleAction, protocol, sourceIP, sourcePort, Direction, destIP, destPort, msg
    sid, Revision, ClassType, Reference, RuleOwner) {
        console.info('============ START : Create Rule ============');
        const rule = {
            RuleAction,
            docType: 'rule',
            protocol,
            sourceIP,
            sourcePort,
            Direction,
            destIP,
            destPort,
            msg,
            sid,
            Revision,
            ClassType,
            Reference,
            RuleOwner,
        };

        Await. ctx.stub.putState(ruleNumber, Buffer.from(JSON.stringify(rule)));
        console.info('============ END : Create Rule ============');
    }
```

These classes can be used to manipulate the blockchain from the backend. The frontend instantiates the consensus algorithm and allow for the consortium to take place on a larger scale.

Our prototype is built on a machine with following configurations: Intel Core i7-3630QM 2.4 GHz with 6 MB L3 Cache, 8 GB DDR3 Memory, Dual NVIDIA GeForce GT 650M SLI. Running Ubuntu 16.04. The source code for our prototype can be accessed from the Github repository, https://github.com/delerak/bbids.

4.2 Benchmark Testing

We conducted benchmark testing to evaluate the performance of our research prototype. The test was done on Hyperledge Caliper platform which allows users to test different blockchain solutions with predefined use cases and get a set of performance test results [41].

The Caliper benchmark first loads a blockchain configuration file predefined by the user; then creates context in the preparing stage, generate transactions in the testing stage and return statistics in the reporting stage [41]. Once the benchmark test is run Caliper begins sending transactions to the blockchain network. These transactions are simply communication packets being sent between the network nodes and ensuring that the blockchain can function under network stress. None of the blockchain data is altered during these tests.

The configuration used in our benchmark was a "simple config" included with the Caliper framework. These config files define variables which are used during the benchmark process. Some examples of the variables are txNumber, txDuration and rateControl, these variables were left at default values for the tests that were run.

The summary in Table 1 shows several outputs. The name of the tests is open/query; these are simply labels that are used to differentiate testing variables that can be defined in the configuration file. We defined open and query with differing txNumbers, txDuration and rateControl variables to test for various different transaction numbers. The transaction send rate is how many transactions are being sent per second, the latency is based on the time it takes for a transaction or query from the submission by the client until it is processed and written on the ledger. Throughput is the number of transactions or queries per second (TPS) that was processed by the blockchain network itself.

Table 1. Benchmark tests of preliminary prototype

Test	Type	Send rate	Latency Max	Latency Min	Latency Avg.	Throughput
1	Open	50.3	78.16	1.26	42.43	10.3
2	Open	100.5	71.13	1.22	36.81	12.3
3	Open	149.5	74.63	1.08	38.10	12.3
4	Query	100.2	0.10	0.01	0.01	1002
5	Query	199.8	0.02	0.01	0.01	199.4

5 Conclusions

Collaborative IDS (CIDS) has been proposed to detect increasingly complex cyber-attacks. Overcoming the issue of trust and sharing of rulesets has remained to a challenge. In this paper, we proposed a CIDS architecture leveraging blockchain technology's record immutability and tamper proof properties of stored data in the distributed ledger. We provided details on various workflows for our CIDS including how to add or update rulesets. We implemented our prototype using HyperLedger frameworks and evaluated using an available benchmark. The initial results look promising, including running over 1000 transactions on the blockchain, latency numbers were low and acceptable range for a small peer-to-peer network.

Our future work plan includes applying our prototype within a real network, testing with simulated attack network traffics, and evaluating the performance with a large number of IDS nodes in place.

References

1. Ranganthan, V.P., Dantu, R., Paul, A., Mears, P., Morozov, K.: A decentralized marketplace application on the ethereum blockchain. In: 2018 IEEE 4th International Conference on Collaboration and Internet Computing (CIC), Philadelphia, PA, pp. 90–97 (2018)
2. Kim, J.-T., Jin, J., Kim, K.: A study on an energy-effective and secure consensus algorithm for private blockchain systems (PoM: Proof of Majority). In: 2018 International Conference on Information and Communication Technology Convergence (ICTC), Jeju, pp. 932–935 (2018)
3. Xu, J.J.: Are blockchains immune to all malicious attacks? Financ. Innovation **2**, 1 (2016). https://doi.org/10.1186/s40854-016-0046-5
4. Sagirlar, G., Carminati, B., Ferrari, E.: AutoBotCatcher: blockchain-based P2P botnet detection for the internet of things. In: 2018 IEEE 4th International Conference on Collaboration and Internet Computing (CIC), Philadelphia, PA, pp. 1–8 (2018)
5. Singla, A., Bertino, E.: Blockchain-based PKI solutions for IoT. In: 2018 IEEE 4th International Conference on Collaboration and Internet Computing (CIC), Philadelphia, PA, October 2018, pp. 9–15 (2018)
6. Dannen, C.: Bridging the blockchain knowledge gap. In: Dannen, C. (ed.) Introducing Ethereum and Solidity, pp. 1–20. Apress, New York (2017)
7. Golomb, T., Mirsky, Y., Elovici, Y.: CIoTA: collaborative anomaly detection via blockchain. In: Proceedings 2018 Workshop on Decentralized IoT Security and Standards, San Diego, CA (2018)
8. Pop, C.: Decentralizing the stock exchange using blockchain an ethereum-based implementation of the Bucharest Stock Exchange. In: 2018 IEEE 14th International Conference on Intelligent Computer Communication and Processing (ICCP), Cluj-Napoca, pp. 459–466 (2018)
9. Hyperledger - Open Source Blockchain Technologies, Hyperledger. https://www.hyperledger.org/. Accessed 20 Feb 2019
10. Hong, J., Liu, C.-C.: Intelligent electronic devices with collaborative intrusion detection systems. IEEE Trans. Smart Grid **10**(1), 271–281 (2019)
11. Al-Utaibi, K.A., El-Alfy, E.-S.M.: Intrusion detection taxonomy and data preprocessing mechanisms. J. Intell. Fuzzy Syst. **34**(3), 1369–1383 (2018)

12. Xin, W., Zhang, T., Hu, C., Tang, C., Liu, C., Chen, Z.: On scaling and accelerating decentralized private blockchains. In: 2017 IEEE 3rd International Conference on Big Data Security on Cloud (BigDataSecurity), IEEE International Conference on High Performance and Smart Computing, (HPSC) and IEEE International Conference on Intelligent Data and Security (IDS), Beijing, China, pp. 267–271 (2017)
13. Czirkos, Z., Hosszú, G.: P2P based intrusion detection. In: Encyclopedia of Information Communication Technology (2019)
14. Ngamsuriyaroj, S.: Package delivery system based on blockchain infrastructure. In: 2018 Seventh ICT International Student Project Conference (ICT-ISPC), Nakhonpathom, July 2018, pp. 1–6 (2018)
15. Junjoewong, L., Sangnapachai, S., Sunetnanta, T.: ProCircle: a promotion platform using crowdsourcing and web data scraping technique. In: 2018 Seventh ICT International Student Project Conference (ICT-ISPC), pp. 1–5 (2018)
16. Malik, S., Kanhere, S.S., Jurdak, R.: ProductChain: scalable blockchain framework to support provenance in supply chains. In: 2018 IEEE 17th International Symposium on Network Computing and Applications (NCA), Cambridge, MA, pp. 1–10 (2018)
17. Wanjun, Y., Yuan, W.: Research on network trading system using blockchain technology. In: 2018 International Conference on Intelligent Informatics and Biomedical Sciences (ICIIBMS), Bangkok, October 2018, pp. 93–97 (2018)
18. Marteau, P.-F.: Sequence covering for efficient host-based intrusion detection. IEEE Trans. Inf. Forensics Secur. **14**(4), 994–1006 (2019)
19. Czirkos, Z., Hosszú, G.: Solution for the broadcasting in the Kademlia peer-to-peer overlay. Comput. Netw. **57**(8), 1853–1862 (2013). https://doi.org/10.1016/j.comnet.2013.02.021
20. State of the DApps A list of 2,551 blockchain˘ apps for Ethereum, Steem, EOS, and more. https://www.stateofthedapps.com/. Accessed 20 Feb 2019
21. Anceaume, E., Guellier, A., Ludinard, R., Sericola, B.: Sycomore: a permissionless distributed ledger that self-adapts to transactions demand. In: 2018 IEEE 17th International Symposium on Network Computing and Applications (NCA), pp. 1–8 (2018)
22. Corsi, P., Giovanni, L., Marina, R.: TickEth, a ticketing system built on ethereum. In: SAC, April 2019
23. Alexopoulos, N., Vasilomanolakis, E., Ivánkó, N.R., Mühlhäuser, M.: Towards blockchain-based collaborative intrusion detection systems. In: Critical Information Infrastructures Security, pp. 107–118 (2018)
24. Carmen, H.: Understanding blockchain opportunities and challenges. eLearn. Softw. Educ. **4**, 275–283 (2018). 9p
25. Rilee, K.: Understanding Hyperledger Sawtooth — Proof of Elapsed Time. Medium (2018)
26. Meng, W., Tischhauser, E.W., Wang, Q., Wang, Y., Han, J.: When intrusion detection meets blockchain technology: a review. IEEE Access **6**, 10179–10188 (2018)
27. Yli-Huumo, J.: Where is current research on blockchain technology?—A systematic review. PLoS ONE **11**(10), e0163477 (2016). https://doi.org/10.1371/journal.pone.0163477
28. Warzynski, A., Kolaczek, G.: Intrusion detection systems vulnerability on adversarial examples. In: 2018 Innovations in Intelligent Systems and Applications (INISTA), Thessaloniki, pp. 1–4 (2018)
29. Intrusion Detection Systems - Techotopia. https://www.techotopia.com/index.php/Intrusion DetectionSystems. Accessed 04 Mar 2019
30. Vasilomanolakis, E., Stahn, M., Cordero, C.G., Muhlhauser, M.: On probe-response attacks in collaborative intrusion detection systems. In: 2016 IEEE Conference on Communications and Network Security (CNS), Philadelphia, PA, pp. 279–286 (2016)
31. Jin, R., He, X., Dai, H.: Collaborative IDS configuration: a two-layer game-theoretic approach. IEEE Trans. Cogn. Commun. Netw. **4**(4), 803–815 (2018)

32. Ficke, E., Schweitzer, K.M., Bateman, R.M., Xu, S.: Characterizing the effectiveness of network-based intrusion detection systems. In: MILCOM 2018 - 2018 IEEE Military Communications Conference (MILCOM), Los Angeles, CA, pp. 76–81 (2018)
33. Massicotte, F., Labiche, Y.: On the verification and validation of signature-based, network intrusion detection systems. In: 2012 IEEE 23rd International Symposium on Software Reliability Engineering, Dallas, TX, USA, pp. 61–70 (2012)
34. Vigna, G., Robertson, W., Balzarotti, D.: Testing network-based intrusion detection signatures using mutant exploits. In: Proceedings of the 11th ACM Conference on Computer and Communications Security - CCS 2004, Washington DC, USA, p. 21 (2004)
35. Accorsi, R., Stocker, T., Müller, G.: On the exploitation of process mining for security audits: the process discovery case. In: ACM Symposium of Applied Computing (SAC), Coimbra, Portugal, pp. 1462–1468 (2013)
36. King, J., Williams, L.: Log your CRUD: design principles for software logging mechanisms. In: Proceedings of the 2014 Symposium and Bootcamp on the Science of Security - HotSoS 2014, Raleigh, North Carolina, pp. 1–10 (2014)
37. Sekar, R.: Specification-based anomaly detection: a new approach for detecting network intrusions. In: Proceedings of the 9th ACM Conference on Computer and Communications Security - CCS 2002, Washington, DC, USA, p. 265 (2002)
38. Mashima D., Ahamad, M.: Using identity credential usage logs to detect anomalous service accesses. In: Proceedings of the 5th ACM Workshop on Digital Identity Management (DIM), Chicago, Illinois, USA, pp. 73–79 (2009)
39. Liu, Y., Zhang, L., Guan, Y.: A distributed data streaming algorithm for network-wide traffic anomaly detection. ACM SIGMETRICS Perform. Eval. Rev. **37**(2), 81–82 (2009)
40. de Vries, A.: Bitcoin's growing energy problem. Joule **2**(5), 801–805 (2018)
41. Hyperledger Caliper: Architecture (2019). https://hyperledger.github.io/caliper/docs/2_Architecture.html. Accessed 16 June 2019

An Adaptive Context Modeling Approach Using Genetic Algorithm in IoTs Environments

Ahmed A. A. Gad-Elrab[1,2]($\boxtimes$), Shereen A. El-aal[2], Neveen I. Ghali[3], and Afaf A. S. Zaghrout[2]

[1] King Abdul-Aziz University, Jeddah, Saudi Arabia
[2] Faculty of Science, Al-Azhar University, Cairo, Egypt
{Asaadgad,shereen.a.elaal}@azhar.edu.eg, afaf211@yahoo.com
[3] Faculty of Computers and Information Technology,
Future University in Egypt, Cairo, Egypt
neveen.ghali@fue.edu.eg

Abstract. Internet of Things (IoTs) is the future of ubiquitous and personalized intelligent service delivery. It depends on installing intelligent sensors to sense and control physical environment to generate enormous amount of data with various data types. Context aware computing is employed for transforming these sensor data into knowledge through three stages: collection, modeling and reasoning. In context modeling, raw data represents in according meaningful manner statically. Furthermore, with growth of IoTs live applications, static modeling is not convenient because of changing context data structure overtime. The work in this paper is dedicated to propose a new dynamic approach for context modeling based on genetic algorithm and satisfaction factor. In addition, flexibility indicator property and context based are defined to measure the performance of the proposed approach.

Keywords: Internet of Things (IoTs) · Context modeling · Genetic algorithm · Satisfaction degree · Flexibility indicator

1 Introduction

Recently, the concept of IoTs [1] has widely expanded to create a world in which all smart objects are connect and communicate with each other with a minimum human intervention. To meet this challenge, a massive number of sensors are needed to collect row sensor data then turn it to context information. Subsequently, context aware system [2] is required to transform these context information to service for end user according to specific task. The main function of context aware system is transforming raw data to knowledge by collecting, modeling and reasoning. It is used in various framework such as applications and middlewares to provide services to end-users. Many researchers have addressed IoTs data types according to different perspectives. For example, authors in [4] categorized context data into eight areas: RFID, address/unique identifiers, descriptive data, positional and environmental data, sensor data, historical data, physics models, and command data. While authors in [3] classified it as two areas, primary and secondary.

Context modeling is an essential step for context awareness. It is accountable of analysis, validate and design of contextual information. Many studies proposed various context modeling techniques and they discussed merits, demerits and applicability [3] of each technique. Furthermore, developers choose context modeling technique for a specific context through context aware applications statically. The main problem in IoTs is that the structure of context data changes overtime, therefore static modeling can not adaptable for modeling these changes. In this paper, a new dynamic approach for context modeling based on genetic algorithm and satisfaction factor is proposed. The main goal of the proposed approach is to make the application or middleware system point out the adaptive model for different context types based on optimal selection computing algorithm. Also, to capture the implicit and explicit context information and to implement the scalability demands of context model for different users, systems, and applications.

2 Related Work

Context aware systems use context to provide adequate information and services to end users depends on stakeholders tasks to control physical environments [5]. The big challenge of IoTs applications is managing and analyzing massive amount of data provided by IoTs heterogeneous devices efficiently in order to integrate them as vital information. In literature, several context modeling approaches are introduced. Perera et al. [3] surveyed the six most popular context modeling techniques and compared them based on some of ubiquitous computing such as interoperability, partial validation and applicability.

On the other hand, many research demonstrated various context aware pervasive computing system in smart environments [7] and embedded interactions [8]. For example, smart living room [9] in which the authors built the system based on ontology model. Also, healthcare environment is presented in [10] to provide healthcare service in which the authors modeled the context information using ontology model. Modeling approaches under area of ubiquitous computing are divided [3] according to the using of data structure. In spite of complexity of ontology in context representation and information retrieval, ontology mechanism is preferred in managing and representing context especially in real human situation systems.

The most context modeling techniques are surveyed in [3]. The authors displayed a comparison of context modeling and representation techniques. This research is performed based on the three most popular context modeling techniques used: ontology based, logic based and object based modeling. Ontology based model uses semantic technologies to regulate contexts into ontologies.

It model the context based on the relationships defined by the ontology and Data can be stored in convenient database while context structure is provided by ontology. Based on the previous surveys, ontology model is preferred in context representation for context aware system. It provides an easier representation of context and sharing model. In addition, validation of ontology is strongly provided due to standardizations availability. Representation and retrieve information are obstacles of the ontology model.

Logic based model represents context information based facts, prediction and rules. Logic based uses these rules and facts to express policies and constrains to drive new facts. Therefore, it extracts new high level context information based on low level context. It supports logical reasoning and any user can add logic to system in the run time. Drawbacks of logic based model are partial validation difficult to maintain, no standard and strongly coupled with applications.

Object based model provides hierarchies and relationship modeling. It supports encapsulation and reusability as it integrated well into context aware systems using high level programming language. However, it is hard to use in retrieving information, it supports data transformation over network as it can be used in run time context modeling and storage mechanism. One of its flaws has no standards so, it is difficult to use in validation.

Modeling approaches demand set of requirements such as Distributed composition, Partial validation, Incompleteness and ambiguity, Level of formality and Applicability to existing environments [11]. In addition, Bettini et al. [12] studied the relevant requirements of context modeling and reasoning techniques and they presented techniques to deal with two issues of any framework for context modeling and reasoning: situation abstractions and uncertainty of context information. Furthermore, one of the main issues of context modeling is there is no standard for handling large number of different information data types for representation and there is no standard in choosing model for every context in which every model is tailored for a particular application. The main focus of this paper is mapping between different context types and set of models to accomplish tasks in short time.

3 Context Modeling Problem (CMP)

In context representation phase for IoTs, the problem is how to determine for each context the best model which will be used to represent it to satisfy its requirements and improve the IoTs services and processes. This problem is called Context Modeling Problem (CMP). In this section, the assumptions and models are introduced then the CMP problem will be formulated.

3.1 System Model and Assumptions

IoTs becomes a new challenge in many of the recent applications. Main drawback of current context modeling for ambient intelligence in IoTs is object dependent model which means most of the current models depend on context object and they cannot fit to represent another context object. Moreover, context modeling is a principle phase in any context aware middleware architecture. Current context models works in static mode in which the system cannot change the representation model for a certain context at the run time. Every model is chosen for a certain context in a specific environment by system programmer. One of the greatest challenges for context modeling is how to find the optimal model for representing certain context in IoTs areas. The main goal of this representation is satisfying all satisfaction requirements of reasoning processes that will be done in this context to designate the optimal model for a certain context based on optimal selection algorithm.

Here, the system model consists of a set of contexts, $C = \{c_1, c_2, .., c_i, ..., c_n\}$, where each c_i represents a certain context in IoTs as parking data, image data, audio data, traffic data, or weather data respectively. A set of models, $M = \{m_1, m_2, ..., m_j, ..., m_k\}$, where each m_j represents a certain model as key value, markup schema, graphical, object oriented, logic, or ontology which can be used to model some contexts in IoTs. Assume that each context c_i needs number of requirements to be satisfied in context modeling phase and is denoted as $Req(c_i)$ and is defined as follows:

$$Req(c_i) = \{r_{i1}, r_{i2}, r_{i3}, ...r_{il}...r_{iy}\}, \quad (1)$$

where r_{il} represent a requirement and y is the number of requirements. Here, the satisfaction requirements for a context c_i by a model m_j is denoted as $SReq(c_i, m_j)$. This satisfaction requirements value evaluates context modeling efficacy. $SReq(c_i, m_j)$ is defined as follows:

$$SReq(c_i, m_j) = \{sr_{i1}, sr_{i2}, sr_{i3}, ...sr_{ih}...sr_{ix}\}, \quad (2)$$

where r_{ih} represent a satisfied requirement and x is the total number of satisfied requirements. Not that

$$SReq(c_i, m_j) \subseteq Req(c_i), \quad (3)$$

Based on the set of requirements, $Req(c_i)$, and the satisfaction requirements, $SReq(c_i, m_j)$, the utility function for context modeling which is used to measure the efficiency of modeling the provided contexts and defined as follows.

$$u(c_i, m_j) = w_1 * Sat(c_i, m_j) + w_2 * \frac{1}{cost(c_i, m_j)} \quad (4)$$

where $Sat(c_i, m_j)$ represents the satisfaction ratio of a context c_i by a model m_j and is defined as follows:

$$Sat(c_i, m_j) = \frac{|SReq(c_i, m_j)|}{Req(c_i)} \quad (5)$$

and $cost(c_i, m_j)$ represents the cost of modeling the context c_i by the model m_j and this cost can be calculated in different ways as delay time, storage size, or by other means. Finally, w_1 and w_2 are the weights for satisfaction ratio and cost value, respectively. These weights represent the importance of satisfaction ratio and cost value for a user or a developer. In addition, the values of w_1 and w_2 must satisfy the following condition:

$$w_1 + w_2 = 1 \tag{6}$$

3.2 Problem Formulation

The system model aimed to maximize the utility function provided that each context is modeled by only one model approach. Based on system models and assumptions, the CMP problem can be formulated as follows.

$$Maximize\ U(C, M) = \sum_{i=1}^{n} \sum_{j=1}^{k} u(c_i, m_j) * x_{ij} \tag{7}$$

such that,

$$x_{ij} \in \{0, 1\}, \tag{8}$$

$$\sum_{j:(i,j)} x_{ij} = 1, \tag{9}$$

$$\sum_{i:(i,j)} x_{ij} \leq |c| \tag{10}$$

Constraint (8) represents the decision variable x_{ij}, where if x_{ij} is equal to 0, this means that a context c_i is not modeled by a model m_j while if x_{ij} is equal to 1, this means that a context c_i is modeled by a model m_j. Constraint (9) means that each context c_i is modeled by only one model m_j. Constraint (10) means that the number of contexts that are modeled by different models is less then or equals the number of contexts.

Based on this formulation, CMP is an optimization problem and the value of decision variable x_{ij} must be determined to solve this problem. In the next section, the proposed approach will be introduced to solve CMP.

4 Proposed Adaptive Context Modeling Approach

In this section, to solve the CMP problem that being formulated in the previous section, a new approach called *Dynamic Genetic-Based Context Modeling Approach* (DGBCMA) is proposed.

4.1 Basic Idea

DGBCMA is a heuristic approach to solve the optimization CMP problem for maximizing the context modeling satisfaction and minimizing the modeling cost. To satisfy these goals, the basic idea of DGBCMA is based on 4 issues: (1) determining the set of requirements properties of each context type, (2) determining the set attributes of model, (3) calculating the satisfaction degree of each context type based on its requirements properties, and (4) selecting the most appropriate model for each context by using genetic algorithm which will maximize the satisfaction degree of the context and minimize its modeling cost.

4.2 Proposed Approach

Based on the basic idea of DGBCMA, the proposed approach consists of four phases: (1) Determination phase, (2) Calculating phase, (3) Selection phase, and (4) Matching phase. These phases are described as follows.

(a) **Determination phase:** In this phase, DGBCMA determines for each context type c_i all its related requirements which are represented as a set of required attributes $RC_i = \{rc_{i1}, rc_{i2}, ...rc_{in}\}$. In addition, the DGBCMA determines for each context model m_j all of its related properties which are represented as a set of attributes $AM = \{ar_{j1}, ar_{j2}, ...ar_{jk}\}$. This set of attributes represents the properties that can be satisfied by model m_j.

(b) **Calculating phase:** In this phase, DGBCMA calculates the satisfaction degree of each context c_i with respect to each model m_j by using Eq. (5). Also, DGBCMA calculates the modeling cost of each context c_i to be modeled by a model m_j. Finally DGBCMA calculates the utility function $u(c_i, m_j)$ by using Eq. (4).

(c) **Selection phase:** In this phase, to select the most appropriate model for each context, DGBCMA uses a genetic algorithm to find the value of a decision variable x_{ij}. Here, a genetic algorithm creates a selection scale based on an evaluation criterion for each pair of context type and context model (c_i, m_j) by using the calculated utility function in the previous step.

(d) **Matching phase:** In this phase, fitness degree is used from the previous phase based on specified requirements to match each context type with a compatible model for representation.

Algorithm 1. DGBCMA algorithm to represent context with a compatible model works as follows:

Input:
1: $C = \{c_1, c_2, ...c_n\}$ is a set of all available contexts.
2: $M = \{m_1, m_2, ...m_k\}$ is a set of all available models.
3: $RC_i = \{rc_{i1}, rc_{i2}, ...rc_{in}\}$ is a set of required context attributes.
4: $AM_j = \{ar_{j1}, ar_{j2}, ..., ar_{jk}\}$.

Process:
1: Calculate the $Sat(c_i, m_j)$ by Equ:(5).
2: Calculate $u(c_i, m_j)$ by equ.(4).
3: Calculate x_{ij} by genetic algorithm.
4: Calculate $U(C, M)$ by equ.(7).
5: Match each context with a compatible model by $Max\ U(C_i, m_j)$.

5 Simulation and Results

This section evaluates the performance of the proposed schema DCMT which maps every context data type to only one model of modeling techniques.

5.1 Simulation Setting

To show the performance of DGBCMA for different contexts, we conducted a model system based on twenty contexts for different environment with six context models. The set of simulated contexts are parking, audio, images, weather, road traffic audio and mobile data. Subsequently, satisfaction requirements for contexts and models are extracted according to parametric evaluation matrix [6] where context requirement for every model is subset of context requirements. The utility function $u(c_i, m_j)$ for every context and model is calculated by using Eq. (4). Here, the utility function $u(c_i, m_j)$ is considered as a fitness function of genetic algorithm. The objective of genetic algorithm is to select the optimal model for every context based on maximizing fitness function such that each context is represented by only one model. All simulation experiments are conducted by using MATLAB.

5.2 Performance Matrices

To measure modeling approach system effectiveness, two flexibility indicators are defined based on the number of context properties and the number of contexts. These flexibility indicators are called *Property-based Flexible Stability* and *Context-based Flexible Stability* indicators. The set of symbols that are used to define the two proposed flexibility indicators and their meaning are shown in Table 1.

Table 1. The set of symbols and their meaning for the two proposed flexibility indicators

Terminologies (parameters and its description)	
n_j	The number of total contexts at step j
m_j	The number of assigned contexts for certain models at step j
x_{ij}	The number of properties of context i at step j
avX_j	The average number of properties for all contexts at step j
avY_j	The average number of properties for all assigned contexts for certain models at step j
FS_{PB}	The Property-based Flexible Stability
FI_{CB}	The Context-based Flexible Stability

The two proposed flexibility indicators are defined as follows:

1. *Property-based Flexible Stability Indicator, FS_{PB}*
   FS_{PB} measures the effect of the changing number of properties on the modeling approach flexibility. FS_{PB} is calculated as follows.

- Assume that the average number of properties for all contexts at step j is denoted as avX_j and is given by the following equation:

$$avX_j = \frac{\sum_{i=1}^{n_j} x_{ij}}{n_j} \quad (11)$$

- Assume that the average number of properties for all assigned contexts for certain models at step j is denoted as avY_j and is given by the following equation:

$$avY_j = \frac{\sum_{i=1}^{m_j} x_{ij}}{m_j} \quad (12)$$

Here, the property-based flexible stability FS_{PB} is defined as the ratio between the value of changing in average number of properties for all assigned contexts at cascading steps, j and $j+1$ to the value of changing in average number of properties for all contexts at cascading steps, j and $j+1$. So, FS_{PB} is defined as follows.

$$FS_{PB} = \frac{avY_{j+1} - avY_j}{avX_{j+1} - avX_j} \quad (13)$$

Based on the value of FS_{PB}, there are two cases:
- Case 1: $FS_{PB} \geq 0$, this means that the molding approach is property-based flexible.
- Case 2: $FS_{PB} < 0$, this means that the molding approach is not property-based flexible.

2. *Context-based Flexible Stability Indicator FS_{CB}*

FS_{CB} measures the effect of the changing number of contexts on the modeling approach flexibility. Here, FS_{CB} is defined as the ratio between the value of changing in average number of assigned contexts at cascading steps, j and $j+1$ to the value of changing in average number of all contexts at cascading steps, j and $j+1$. FS_{CB} is defined as follows.

$$FS_{CB} = \frac{m_{j+1} - m_j}{n_{j+1} - n_j} \quad (14)$$

Based on the value of FS_{CB}, there are two cases:
- Case 1: $FS_{CB} \geq 0$, this means that the modeling approach is context-based flexible.
- Case 2: $FS_{CB} < 0$, this means that the modeling approach is not context-based flexible.

In addition to these flexibility indicators, a satisfaction degree and modeling cost, which are defined in Sect. 3.1, are used in the evaluation.

5.3 Results and Discussion

Here, the results of conducted simulations of the proposed approach DGBCMA will be introduced and discussed. The conducted simulations are based on three changing parameters: (1) changing the number of context properties, (2) changing the number of contexts and (3) changing the number of models. In every change the data is tested based on satisfaction degree, context modeling cost, FS_{PB}, and FS_{CB}. In addition, in every change, there are two cases of cost will be considered: (a) Context-based cost, which considers only the cost of number of contexts and (b) Property-Context-based cost, which considers the cost of number of contexts and the cost of number of properties of each context. The performance of DGBCMA is compared to ontology, object based and logic modeling approaches. Every experiment is run for five times then the average result is taken for analysis.

Changing Number of Context Properties. In this section number of properties is ranged from 2 to 80 with constant number of models which is 5 and constant number of contexts which is 20. The simulation results will be presented for Context-Based Cost and Property-Context-Based Cost cases as follows:

(a) *Context-Based Cost*

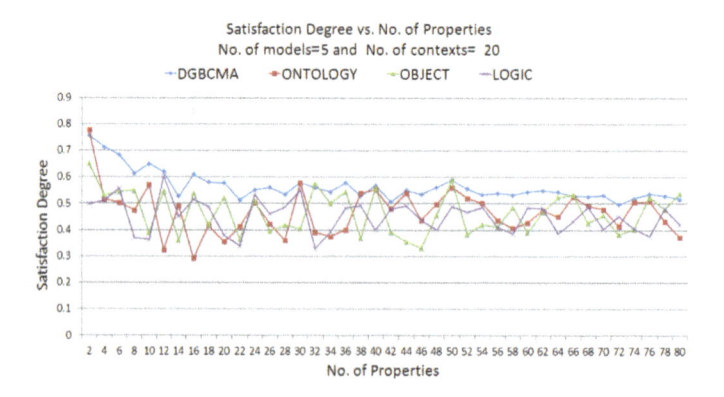

Fig. 1. Satisfaction degree based on context cost with changing number of properties.

Figure 1 shows the satisfaction degree against different number of properties. As shown in Fig. 1, the satisfaction degree of DGBCMA is larger than ontology, object and logic models. This is because, DGBCMA can dynamically adapt its selection based on the requirements of each context while other model do not.

Figure 2 shows the modeling cost against different number of properties. As shown in Fig. 2, the modeling cost increases as number of context properties increases. This is because when the number of properties increases, an

An Adaptive Context Modeling Approach Using Genetic Algorithm   193

additional cost is needed to model these new properties. It is clear that the modeling cost is almost started with 150 and it climbed gradually with increasing the number of properties while ontology model achieved lower cost comparing to the other three models. The modeling cost of DGBCMA is slightly larger than ontology, object and logic models.

Figure 3 shows the property-based flexible stability indicator, FS_{PB}, against different number of properties based on context cost. As shown in Fig. 3, the FS_{PB} value of DGBCMA was started with 0 when the number of properties was equal 2 then it varied between 0 and 1 with different number of properties along the experiment while the FS_{PB} values for ontology, object and logic models were unstable and they ranged between −10 and 5. As a result, DGBCMA has a higher flexibility with respect to different number of properties. This means that DGBCMA can adapt its matching dynamically in efficient way better than other modeling approaches.

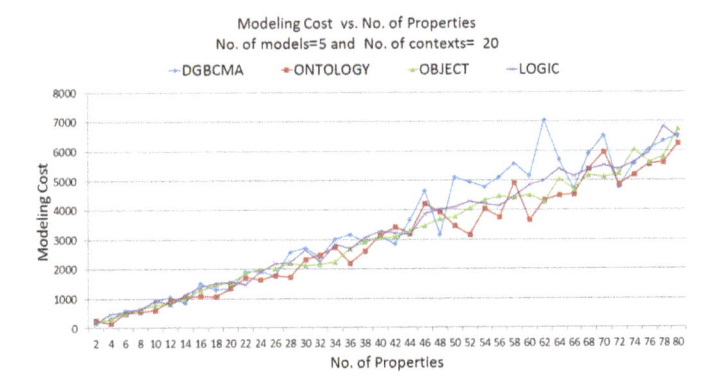

Fig. 2. Modeling cost based on context cost with changing number of properties.

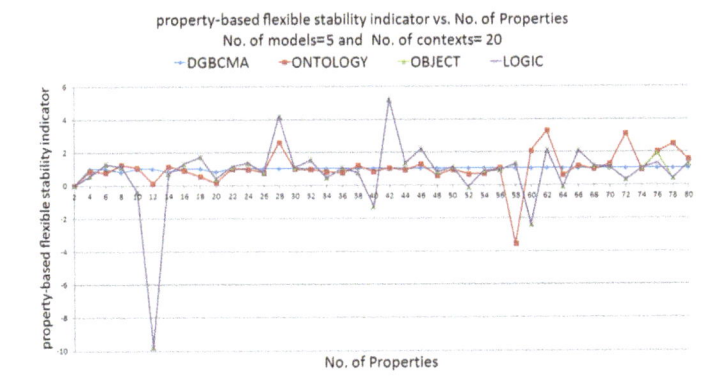

Fig. 3. Flexible stability for property based on context cost with changing number of properties.

(b) *Property-Context-Based Cost*

Figure 4 shows the satisfaction degree against different number of properties. As shown in Fig. 4, the satisfaction degree of DGBCMA is larger than ontology, object and logic models. This is because, DGBCMA can dynamically adapt its selection based on the requirements of each context while other model do not.

Figure 5 shows the modeling cost against different number of properties. As shown in Fig. 5, the modeling cost increases as number of context properties increases. As mentioned in Fig. 2 description, more properties require additional cost, this means, modeling cost is proportional to the number of properties. It is clear that the modeling costs are almost started with 150 and they climbed gradually with increasing the number of properties while logic model achieved higher cost comparing to the other three models. The modeling costs of object, ontology and DGBCMA are convergent however object model cost is slightly lower.

Figure 6 demonstrates the property-based flexible stability indicator, FS_{PB}, against different number of properties based on context and property costs. As shown in Fig. 6, the FS_{PB} value of DGBCMA was started with 0 when the number of properties was equal 2 then its rate was almost steady by 1 with different number of properties along the experiment. In addition, the FS_{PB} values for ontology, object and logic models were unstable and their results were increased and slumped dramatically with changing number of properties where they ranged between −4 and 5 along the experiment. As a result, DGBCMA has a higher flexibility with respect to different number of properties. This means that DGBCMA can adapt its matching dynamically in efficient way better than other modeling approaches.

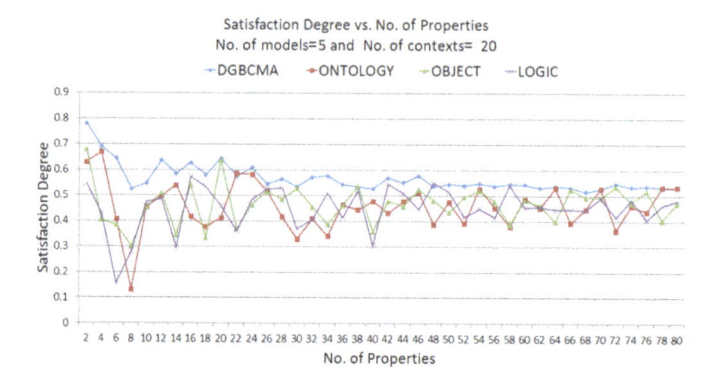

Fig. 4. Satisfaction degree based on property and context costs with changing number of properties.

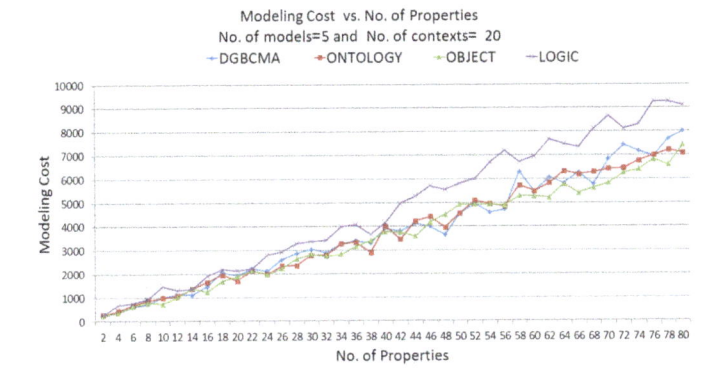

An Adaptive Context Modeling Approach Using Genetic Algorithm 195

Fig. 5. Modeling cost based property and context costs with changing number of properties.

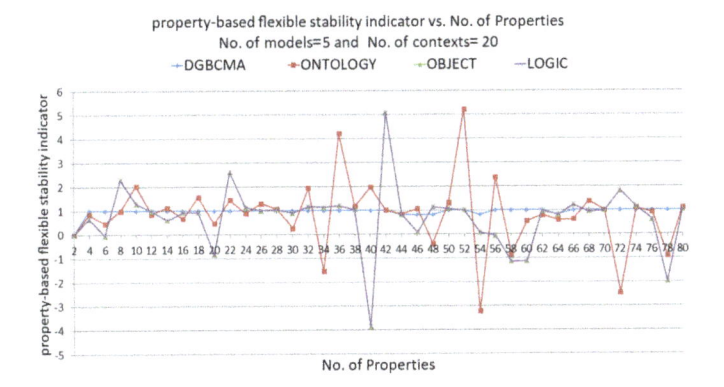

Fig. 6. Flexible stability for property based on context and property costs with changing number of properties.

Changing Number of Contexts. In this section number of contexts is ranged from 3 to 41 with constant number of properties which is 20 and constant number of models which is 5. The simulation results will be presented for Context-based Cost and Property-Context-based Cost cases as follows:

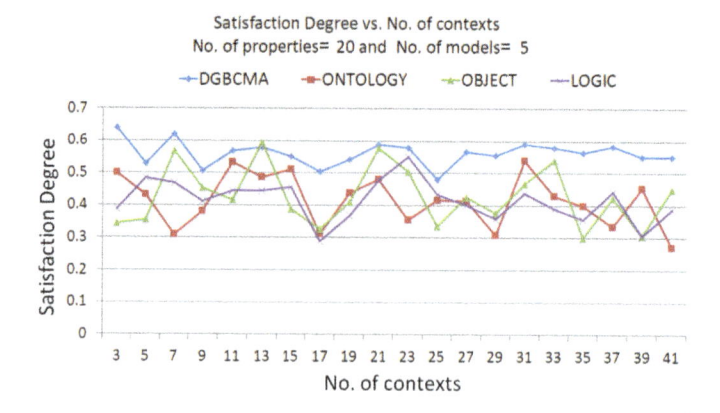

Fig. 7. Satisfaction degree based context cost with changing number of contexts.

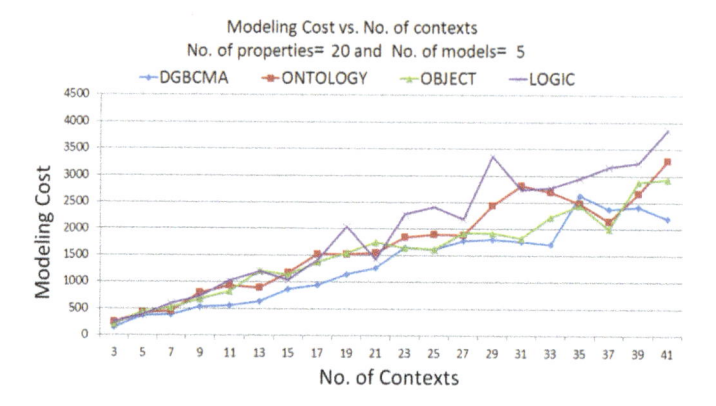

Fig. 8. Modeling cost for context based context cost with changing number of contexts.

(a) *Context-Based Cost*

Figure 7 shows the satisfaction degree against different number of contexts. As shown in Fig. 7, the satisfaction degree of DGBCMA is ranged between 0.3 and 0.65 approximately and it achieved higher degree than ontology, object and logic models. This is because, DGBCMA can dynamically adapt its selection based on the requirements of each context while other model do not.

Figure 8 shows the modeling cost against different number of contexts. As is presented in Fig. 8, the modeling cost increases as number of contexts increases. This is because when the number of contexts increases, an additional cost is needed to model these new contexts. It is clear that the modeling cost is almost started with 150 and it climbed gradually with increasing the number of contexts while DGBCMA approach achieved lower cost comparing to the other three models. The modeling cost of logic model is slightly higher than ontology, object models and DGBCMA.

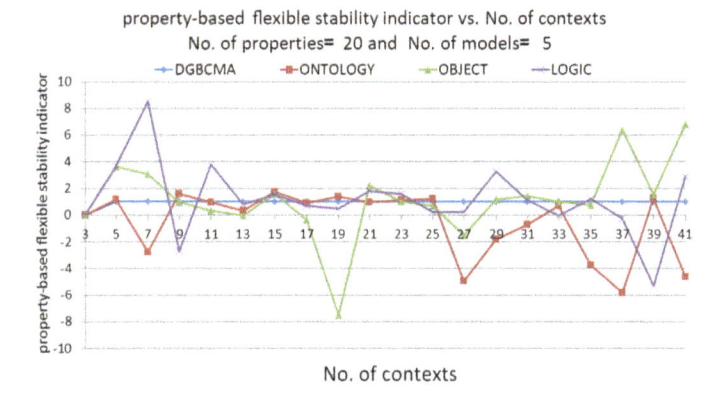

An Adaptive Context Modeling Approach Using Genetic Algorithm 197

Fig. 9. Flexible stability for property based on context cost with changing number of contexts.

Figure 9 shows the property-based flexible stability indicator, FS_{PB}, against different number of contexts based on context cost. As shown in Fig. 9, the FS_{PB} value of DGBCMA was started by 0 when the number of contexts was equal 3 and became stabilized at 1 with increasing number of contexts along the experiment while the FS_{PB} values for ontology, object and logic models were unstable and they ranged between -8 and 8, approximately. As can be seen, ontology, object based and logic models are flexible but unstable where the FS_{PB} for these three models climbed and dropped sharply with increasing number of contexts while the FS_{PB} for the proposed approach DGBCMA achieved higher flexibility and stability. This means that DGBCMA can adapt its matching dynamically in efficient way better than other modeling approaches.

Figure 10 shows the context-based flexible stability indicator, FS_{CB}, against different number of contexts. As shown in Fig. 10, the FS_{CB} value of

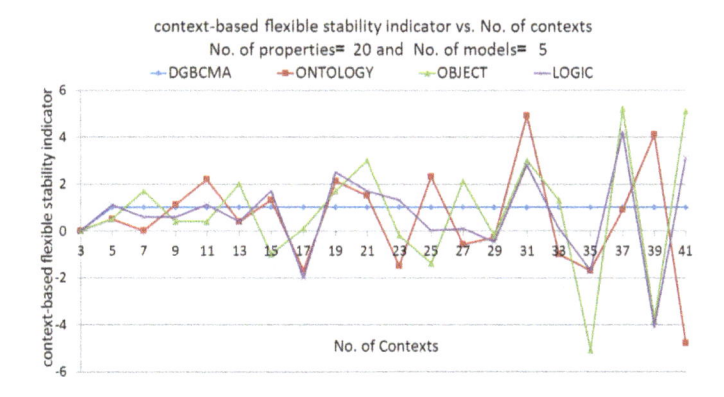

Fig. 10. Flexible stability for context based on context cost with changing number of contexts.

DGBCMA was started with 0 when the number of contexts was equal to 2 then it became stabilized at 1 with different number of contexts along the experiment while the FS_{CB} values for ontology, object and logic models were unstable and they ranged between −5 and 5, approximately, and they were climbed and dropped dramatically with increasing number of contexts. As a result, DGBCMA has a higher flexibility with respect to different number of contexts. This means that DGBCMA achieves the optimum matching dynamically better than other modeling approaches.

(b) *Property-Context-Based Cost*

Figure 11 shows the satisfaction degree against different number of contexts. As shown in Fig. 11, the satisfaction degree of DGBCMA achieved the optimum degree with changing number of contexts where it is larger than ontology, object and logic models. This is because, DGBCMA can dynamically adapt its selection based on the requirements of each context while other model do not.

Figure 12 demonstrates the modeling cost against different number of contexts. As shown in Fig. 12, modeling cost is proportional to the number of contexts. It is clear that the modeling cost is almost started with 150 and it climbed gradually with increasing the number of contexts while object model achieved lower cost comparing to the other three models. The modeling cost of DGBCMA is slightly larger than object and logic models while ontology achieved the largest cost.

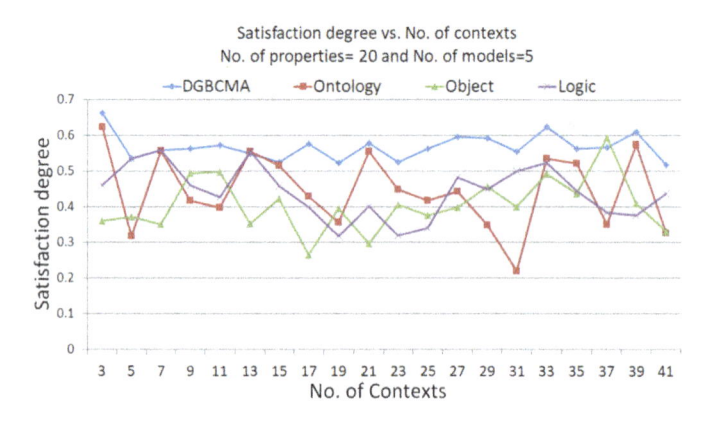

Fig. 11. Satisfaction degree based context and property costs with changing number of contexts.

Figure 13 shows the property-based flexible stability indicator, FS_{PB}, against different number of contexts based on property and context costs. As shown in Fig. 13, the FS_{PB} value of DGBCMA was started with 0 when the number of contexts was equal 3 then it rose to 1 and became stable with increasing number of contexts along the experiment. On the other hand, the FS_{PB} values for ontology, object and logic models were changed up

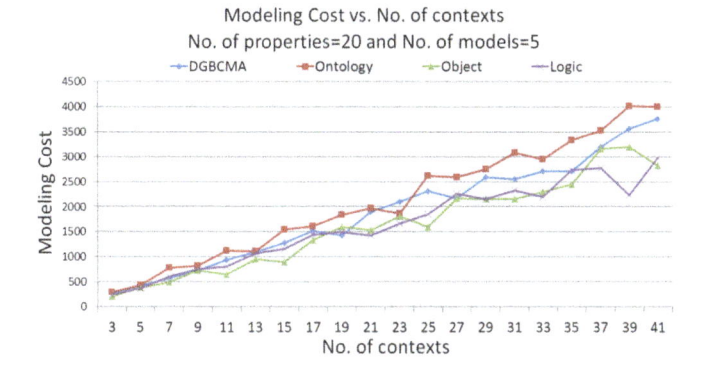

An Adaptive Context Modeling Approach Using Genetic Algorithm   199

Fig. 12. Modeling cost for context based on context and property costs with changing number of contexts.

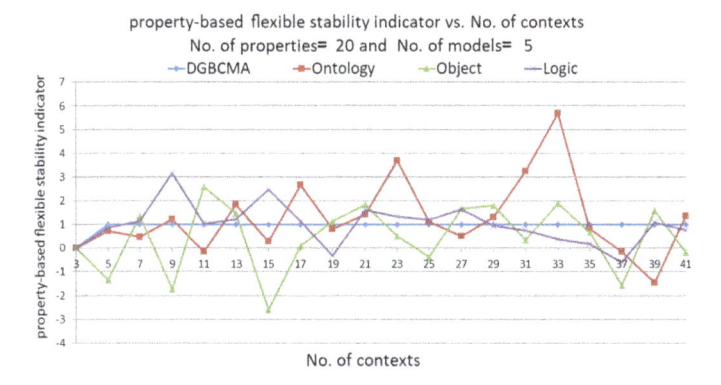

Fig. 13. Flexible stability for property based on context & property costs with changing number of contexts.

and down with increasing number of contexts and they ranged between −3 and 6, approximately. As a result, DGBCMA has a higher flexibility with respect to different number of contexts. This means that DGBCMA can adapt its matching dynamically in efficient way better than other modeling approaches.

Figure 14 shows the context-based flexible stability indicator, FS_{CB}, against different number of contexts. As shown in Fig. 14, the FS_{CB} value of DGBCMA was started with 0 when the number of contexts was equal 3 then it varied to 1 with different number of context along the experiment while the FS_{CB} values for ontology, object and logic models were unstable and they ranged between −4 and 8. As a result, DGBCMA has a higher flexibility with respect to different number of contexts. This means that DGBCMA can adapt its matching dynamically in efficient way better than other modeling approaches.

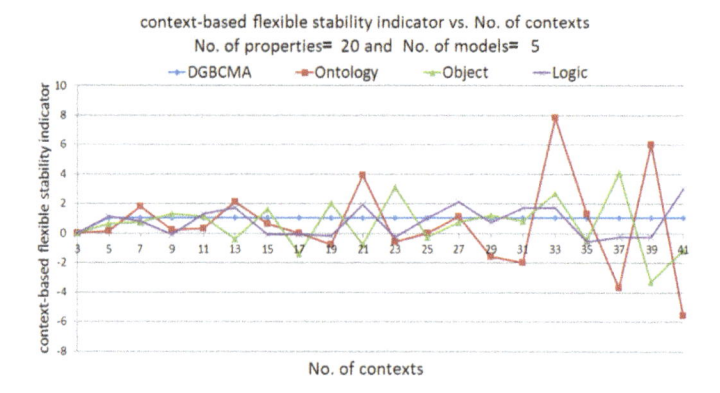

[Chart: context-based flexible stability indicator vs. No. of contexts, No. of properties= 20 and No. of models= 5, with series DGBCMA, Ontology, Object, Logic]

Fig. 14. Flexible stability for context based on context and property costs with changing number of contexts.

Changing Number of Models. In this section number of models is ranged from 3 to 21 with constant number of properties which is 20 and constant number of contexts which is 20. The simulation results will be presented for Context-Based Cost and Property-Context-Based Cost cases as follows:

(a) *Context-Based Cost*

Figure 15 shows the satisfaction degree against different number of models. As presented in Fig. 15, the satisfaction degree of DGBCMA is larger than ontology, object and logic models. This is because, DGBCMA can dynamically adapt its selection based on the requirements of each context while other model do not.

Figure 16 shows the modeling cost against different number of models. As shown in Fig. 16, the costs of four models were unstable as the model numbers increased. They approximately started with the highest cost and then

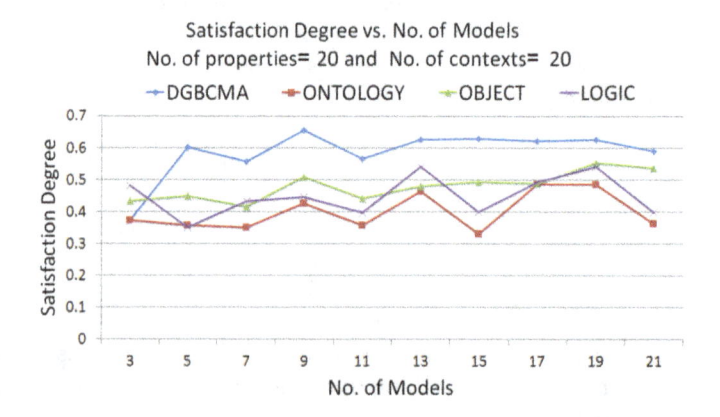

[Chart: Satisfaction Degree vs. No. of Models, No. of properties= 20 and No. of contexts= 20, with series DGBCMA, ONTOLOGY, OBJECT, LOGIC]

Fig. 15. Satisfaction degree for context based on context cost with changing number of models.

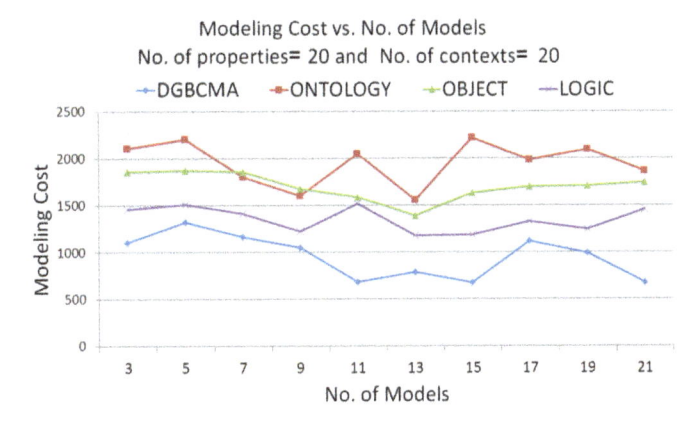

Fig. 16. Modeling cost for context based context cost with changing number of models.

decreased slightly; thereafter, they climbed again slightly. It means, the modeling cost is unsteady. It is clear that the modeling cost of ontology model achieved the largest cost while the modeling cost for DGBCMA is much lower comparing to the other three models.

Figure 17 demonstrates the property-based flexible stability indicator, FS_{PB}, against different number of models based on context cost. As shown in Fig. 17, the FS_{PB} value of DGBCMA was started with 0 when the number of models was equal 3 then it varied between 0 and 1 with different number of models along the experiment while the FS_{PB} values for ontology, object and logic models were unstable and they ranged between −3 and 7. As a result, DGBCMA has a higher flexibility with respect to different number of models. This means that DGBCMA can adapt its matching dynamically in efficient way better than other modeling approaches.

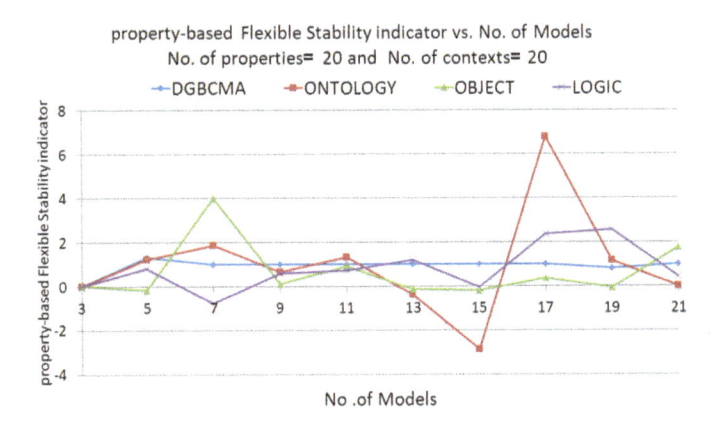

Fig. 17. Flexible stability for property based on context cost with changing number of models.

(b) *Property-Context-Based Cost*

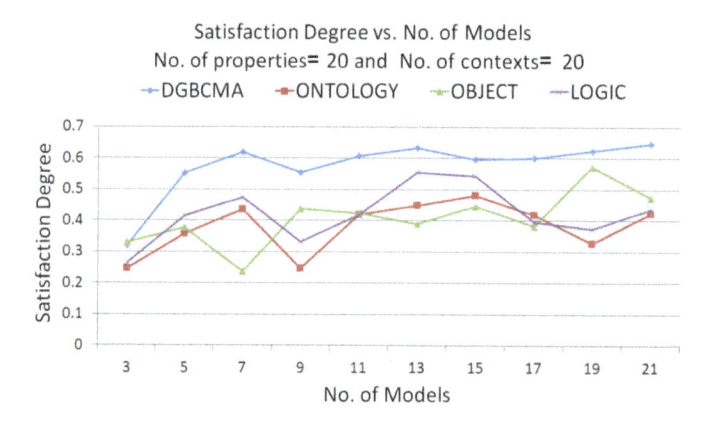

Fig. 18. Satisfaction degree based on context and property costs with changing number of models.

Figure 18 shows the satisfaction degree against different number of models. As presented in Fig. 18, the satisfaction degree of DGBCMA is much higher than ontology, object and logic models. This is because, DGBCMA can dynamically adapt its selection based on the requirements of each context while other model do not.

Figure 19 shows the modeling cost against different number of models. As shown in Fig. 19, more models require additional cost. It is clear that the modeling cost is climbed gradually with increasing the number of models while DGBCMA approach achieved optimal cost comparing to the other three models. The modeling cost of object model is the largest with increasing models number.

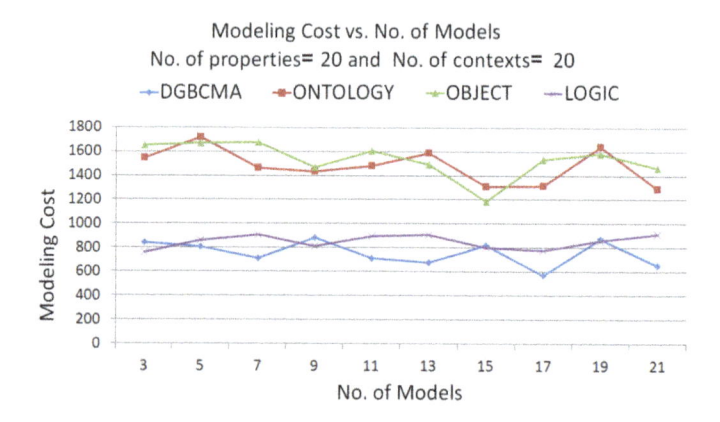

Fig. 19. Modeling cost based on context and property costs with changing number of models.

property-based flexible Stability indicator vs. No. of Models
No. of properties= 20 and No. of contexts= 20

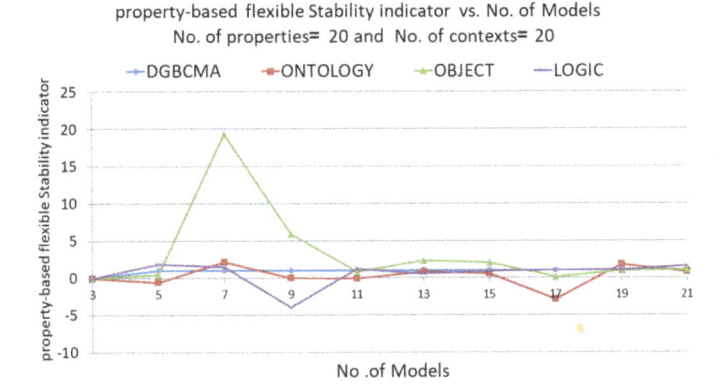

Fig. 20. Flexible stability for property based on context and property costs with changing number of models.

Figure 20 demonstrates the property-based flexible stability indicator, FS_{PB}, against different number of models based on property and context costs. As shown in Fig. 20, the FS_{PB} value of DGBCMA was started with 0 when the number of models was equal 3 then it varied to 1 with increasing number of models along the experiment while the FS_{PB} values for ontology, object and logic models were unstable and they ranged between -4 and 20. As a result, DGBCMA has a higher flexibility with respect to different number of models. This means that DGBCMA can adapt its matching dynamically in efficient way better than other modeling approaches.

6 Conclusion

In this paper, we study the context modeling problem for IoTs and propose Dynamic Genetic-Based Context Modeling Approach (DGBCMA) to solve it. The main challenge of the proposed approach is to match between set of contexts and available context models dynamically based on maximizing satisfaction degree and minimizing the modeling cost. In addition, flexibility indicator based property and based context are utilized to measure the proposed approach efficiency. Moreover, different simulation are proceeded based on three changes: different number of context properties, different number of contexts, and different number of models. The experiments results show that DGBCMA is effective, outperforms and more flexible than other comparing approaches. In the future work, studying and evaluating the proposed approach in real scenarios will be considered. Also, the effect of new context parameters will be studied.

References

1. Ashton, K.: That "internet of things" thing in the real world, things matter more than ideas. RFID J. (2009). http://www.rfidjournal.com/article/print/4986
2. Schilit, B., Theimer, M.: Disseminating active map information to mobile hosts. IEEE Netw. **8**(5), 22–32 (1994). http://dx.doi.org/10.1109/65.313011
3. Perera, C., Zaslavsky, A., Christen, P., Georgakopoulos, D.: Context aware computing for the internet of things: a survey. IEEE Commun. Surv. Tutorials **16**(1), 414–454 (2014)
4. Cooper, J., James, A.: Challenges for database management in the internet of things. IETE Tech. Rev. **26**(5), 320–329 (2009)
5. Sain, M., Lee, H., Chung, W.Y.: Designing context awareness middleware architecture for personal healthcare information system. In: Proceedings of the 2010 12th International Conference on Advanced Communication Technology (ICACT), Phoenix Park, Korea, pp. 1650–1654, February 2010. ISBN: 978-1-4244-5427-3
6. Moore, P., Hu, B., Zhu, X., Campbell, W., Ratcliffe, M.: A Survey of context modeling for pervasive cooperative learning. In: IEEE Information Technologies and Applications in Education (2007)
7. Ali, A., Shirehjini, N., Semsar, A.: Human interaction with IoT-based smart environments. Multimedia Tools Appl. **76**(11), 13343–13365 (2017)
8. Kranz, M., Holleis, P., Schmidt, A.: Embedded interaction: interacting with the Internet of Things. IEEE Internet Comput. **14**(2), 46–53 (2010)
9. Miraoui, M., El-etriby, S., et al.: Ontology-based context modeling for a smart living room. In: Proceedings of World Congress on Engineering and Computer Science, vol. I (2015)
10. Kim, J., Chung, K.: Ontology-based healthcare context information model to implement ubiquitous environment. Multimedia Tools Appl. **71**(2), 873–888 (2014)
11. Sagaya, K.S., Kalpan, Y.: A review on context modelling techniques in context aware computing. Int. J. Eng. Technol. (IJET) **8**(1), 429–433 (2016)
12. Bettini, C., Brdiczka, O., Henricksen, K., et al.: A survey of context modelling and reasoning techniques. Pervasive Mobile Comput. **6**, 161–180 (2010)

Online Sports Activities Travel Guide (SATG)

Ola Hegazy[✉]

Imam Abdulrahman Bin Faisal University, Dammam, Kingdom of Saudi Arabia
ohegazy@iau.edu.sa

Abstract. Nowadays, the E-tourism became a very popular way for most tourists to plan their vacation travels, and most of those applications or websites that help people in this are only for reservations; places in hotels, tickets for airplane, etc. Here in this paper I will represent a graduation project of some students under my supervision. This project provides a trending tourism application website that demonstrates the most famous tourism sport activities' places in different countries. Our website will help the tourist who is interested in a certain activity to find the places, countries where this activity is practice and choose the preferable place or country; he/she wants to go to practice this activity. Also, this application allows the tourists to reserve in their interested activities through the professional companies in the countries they choose to plan and design all the trip for them. This new idea makes our website different than other traditional tourism websites or applications, and also, we allow the customer to contact these companies that support the sport activities' trips 24/7 to help them with their reservation.

Keywords: Sport Activities Travel Guide (SATG) · Sport activities tourism · Tourism industry

1 Introduction

As tourism is one of the fastest growing industries today, thus the sport activities that making a new tourism industry is getting more and more important. People have become more interested in sports activities of all kinds (like; hiking, rock climbing, mountain climbing, skiing, diving, etc.), and they wiling to travel far away to participate in sports activities that they find them interesting. Sports activities can offer various economic and social benefits for hosts; therefore hosts responsible can and should employ events effectively in this tourism. It has become widely accepted that every community and destination need to adopt a long-term strategic approach to sports activities tourism, by planning and developing in order to realize the full tourism potential of sports activities. This could be accomplished by listing the famous activities tourism places in different countries and facilitate the way for the tourists to contact any tourism company in the chosen country for the chosen sport activity. From here we got our idea to make online Sports Activities Travel Guide (SATG) website that demonstrates the most famous tourism places offering sports activities in different countries to help the interested tourists find and reserve in their wanted activities. And we made our website different from other tourism websites by allowing the customers to contact those companies 24/7 to help them with their reservations. In addition, we

© Springer Nature Switzerland AG 2020
K. Arai et al. (Eds.): FICC 2020, AISC 1129, pp. 205–217, 2020.
https://doi.org/10.1007/978-3-030-39445-5_16

will try to provide simple and organized searching way that helps the tourist to find what he/she wants.

Using our online SATG website, we can issue the most important travel companies in every country that offer any kind of sports activities in this country to provide many options for tourists to choose which making search easier and more efficient for them. Also, on the other hand, our website will help in increasing the marketing for those companies, and in turn increase the economy for the whole country.

2 Background and Related Works

2.1 Development of e-Tourism

At the last decades, tourism in general had influenced too much with the development of the "Information Communication Technologies" (ICTs). The determination of the (CRSs) "Computer Reservation Systems" in 1970s, the (GDSs) "Global Distribution Systems" in the end of 1980s, and the in the late 1990s "Internet" had converted dramatically the operational and strategical practices in tourism. According to the statistics published by "World Tourism Organization" (WTO) in the near future, countries that don't inhabit the ICT in their tourism infrastructure would not be able cope up with the leap of tourism evolution of those countries that apply a substantial role in ICT infrastructure [1].

During the 1990s, the wide implementation of the Internet formed new avenues of spreading and approaches of reaching tourists straight, causing the Internet to start reshaping the structure of tourism sectors. In the early 21st century, mobile technologies along with social media intensely altered the ways that customers access information and handling transactions, thus changing the influence structure and equilibrium between customers and businesses [2].

In the early 2010s, the view of the application to tourism of incredible amounts of online data has the prospect of converting "big data" into knowledge, intensely enhancing the tourism experience and giving a deeper understanding of behavioral patterns and the construction of this industry.

From the above studies, the status now is obviously showed that the tourism became the world's biggest industry and its development shows a continuous year to year increase. The Internet has been the main source of tourist destination information for travelers. About 95% of Web users use the Internet to collect travel related data, and about 93% stat that they check tourism websites for their scheduling for holidays.

2.2 Survey and Related Work

Throughout our search for any related works, and surveying about same or related ideas, we found that most of the companies are making websites and applications just for reservations, any kind of reservations (rooms in hotels, tickets for airplanes, etc.) And most of these reservations and trips that offered by these companies are only for the main tourism branches of site scenes tours, or tours for visiting the most famous touristic monuments, historical or holy places in different countries.

So, for examples, we found:

Frommer's. Frommer's is a travel guidebook series founded by Arthur Frommer (see Fig. 1). Frommer's has extended to contain more than 350 guidebooks throughout 14 series, alike other media it contains the radio show Frommer's Travel Show and the website Frommers.com [3]; founded in New Jersey, United States of America.

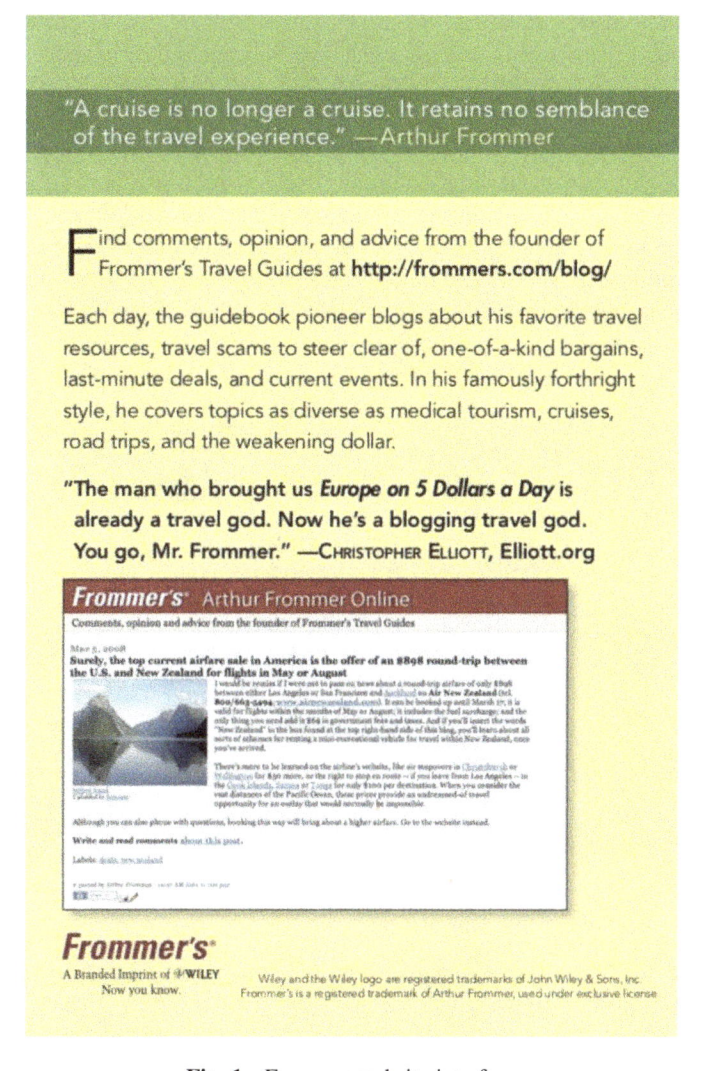

Fig. 1. Frommer website interface

The features of this travel guidebook and website are:

1. If you can't purchase any of their travel guides, then you can use their website to know any destination on the map.

2. There are large maps that you can navigate on the site.
3. They provide travel and trip ideas.
4. The articles on every location cover anything that you would need to know.
5. They have forums that you can engage in a discussion with travelers.

But the main problems with this that they don't reserve you the trips or contacting you with any travel agent that could do so, also, they didn't support any sports activities places, only the touristic places or historical ones.

Trip Advisor. From its name it's clear that Trip Advisor is clearly about evaluations, advices and reviews on places like resorts, hotels, and also traveling flights, holidays rentals, travel guides, vacation packages, etc. (see Fig. 2). On the website of each destination you can find out data about things to do, top rated hotels, vacation rentals, restaurants… etc. with supporting videos and images. As it is a popular website, you can subscribe to the free newsletter to send you the latest information about your preferable places and reviews on each aspect of your planned travel Traipse over to assistance from real traveler's section for precious tips. You can also download a free PDF city guide for the destinations [4]; founded in Massachusetts, United States of America.

Fig. 2. Trip advisor website interface

The main features of this website are:

1. Trip Advisor is considered as an exclusive, timely, general resource with both reviews and pictures posted by travelers.
2. Trip Advisor enables a variety of thoughts to be stated and shared.
3. Functionality has been added now to help users to book hotels, vacation rentals, flights, and even restaurants directly from their site.
4. Hotel managers have the option to respond to reviews, good and bad.

But the defects of it are as follows:

1. Difficult to understand the use of the site.
2. Indicates only places and of course not activities.
3. Multi path that may make the tourist confused.
4. There is no sequence in the search process.

Rather than these, there are a lot more and most of them concentrate on the places of famous sight scenes but not the sports activities that could be practiced in those places, if any.

Only one exception we found which is Tourdust.

Tourdust. Tourdust is a specialist tour operator. They focus on creating magical tailor-made journeys that bring travel & places alive and do not scrimp on the smiles. Born in a nomad's camp on the Mongolian Steppe and homed in the Atlas Mountains of Morocco, they have been creating private tailor-made journeys for adventurers and families for the past 6 years [5]; founded in Tring, United Kingdom.

It provides examples of tourists' attractions in several countries including mountain climbing, camping, safari, games city visits and family adventures (see Fig. 3).

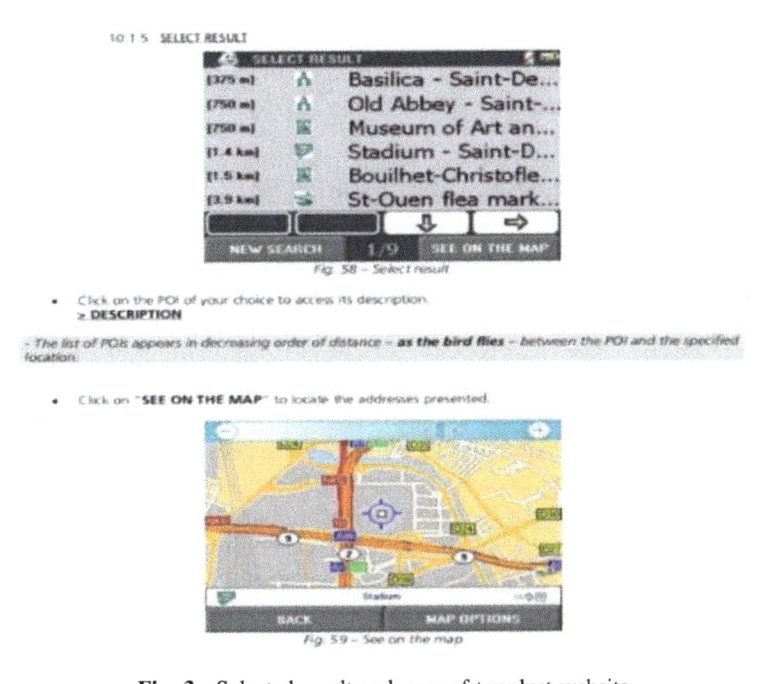

Fig. 3. Selected result and map of tourdust website

Fig. 4. Tourdust website interface

The features of this tour operator are:

1. Luxurious safaris.
2. Privately guided adventure with superb local guides.
3. Tailor—made journeys with a spirit of adventure.
4. Family holidays.

So, this is the only website that mentions something about the activities, but its defect is they don't classify the activities provided by the site based on the state/city in which it is located and the company that supports this activity (see Fig. 4).

3 Motivation and Problem Statement

Due to all above reasons, we got our idea to activate the sports activities tourism as a new trend in the tourism industry. Throughout the thinking to enrich the knowledge about sports activities tourism places, and from our desire to encourage the trend of traveling to practice new sport activity, we got the idea of listing the famous sports activities places in different countries in our website, as an application that facilitates the way for the tourist to contact any tourism company that supports his/her chosen sport activity in a chosen country, and get more information about how to practice this activity in that country. Also, he/she can reserve his/her trip through this company if they like the supplied offer. By applying these measurements, our website will demonstrate the most important sports activities travel companies in different countries worldwide and make the search easier and more efficient.

This will provide many options for tourists on one side and increasing the marketing for these kind of companies on other side, where this is satisfied by making them more popular in that field, which e with era of working on the tourism as a high source of economics for a lot of countries.

4 Approach and Methodology

Here we will define the approach and methodology we used to design and implement our website application.

4.1 Approach

First, our approach was to analyze the system with all its requirements and components, and then define the functions that our application could perform to support the user with the information he/she needs to know about for their chosen activity. This information will be briefed to the countries and the companies inside each country that practice and support this activity. This will be applied by the sequence demonstrated in the following section of our methodology.

4.2 Methodology

Our methodology to apply the above approach started by defining the functional requirements (as follow), and after that we will display samples of the drawings of the Use Cases for these functions, and the Sequence Diagrams that shows how these functions will be applied in sequence and the result of each of them.

Functional Requirements: Functional requirements show the operations and activities the system must be able to perform. So, our application functional requirements are:

- *Activity*: It's a button that has a drop-down menu with all activities that have been added to the system by the administrator of the application website. Its function is allowing the user (tourist) to search and choose throughout this menu which sport activity he/she wants to practice. Then, the visitor should make the next choice.
- *Country*: It's a button that has a drop-down menu with all countries that have been added, again, by the administrator of the application. Its function is allowing the user (tourist) to search and choose between available countries through this menu, the country he/she prefers to practice the chosen activity in it. Then, when he/she chooses a specific country from this list, the page will be refreshed and demonstrate all the companies inside this country that offer trips for the selected sport activity.
- *Company*: This is a link that appears on the page of search after choosing the first two choices. When the web page refreshed it gives the result of search as a screen of the selected sport activity with the countries that practicing this activity and the link of companies that offer trips for this activity in each country.
- *Reserve*: This function is offered through the link of the company that visitor find in the previous page, when he/she click on this link the page will be transferred to the private page of the company where he/she can find a form of three textboxes that he/she should fill in; Full name, email, and a message box. The message box he/she can contact the company through writing their preferable dates for the journey. And then click a Reserve button.

- *Rating the trip*: This is a function that permits the tourist to evaluate his/her journey after finishing it, mainly evaluating the company that supported his/her trip.
- *Rating the App*: Here is another function that allows the tourist again, to evaluate our website through his experience of travelling using our application in searching and reserving his/her journey.

Use Case: After defining the different functions that our application could support, we started to design a Use Case for each of them. For the limited space here, we chose some of these diagrams to display.

The following figure is the general Use Case diagram, that demonstrate the functions which are supported by the administrator of the website, and the functions that offered to the user (tourist) to use by our application, to get his/her request (see Fig. 5).

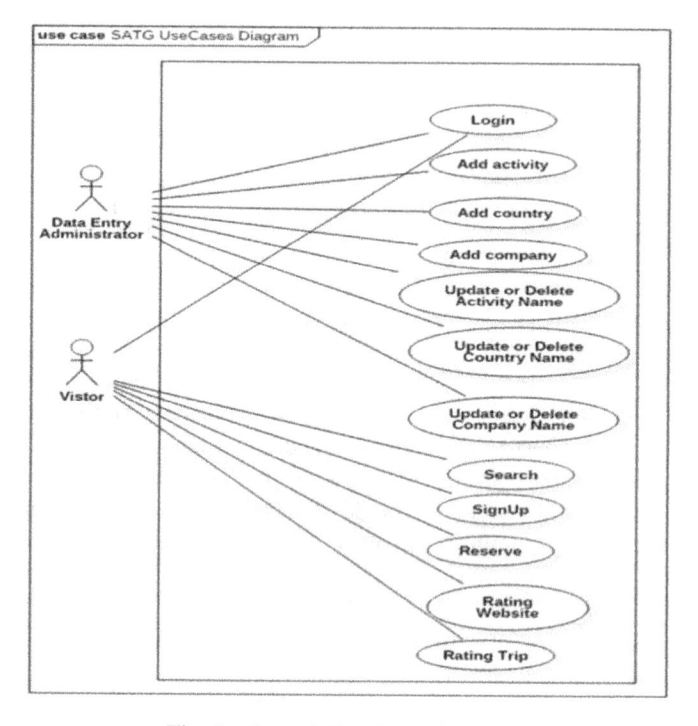

Fig. 5. General Use Case of our App

First; Functions of the Administrator: Here are the functions of the admin who is maintaining and updating website all the time.

- *Login*: This is the first function that our website supports, it allows the admin to log inside the application. Whereas the functions of the admin are different from the functions of the user (visitor/tourist), then, in this function the application check the username and the password for the person to validate the rights and authorities for

him/her. For example, the admin can add a new activity to the application or delete one already found. This of course is not allowed to the other users (tourists).
- *Add Activity:* This function allows the admin to add a new activity to the list of the activities that our application offers to the tourists to choose of them.
- *Add Country:* This function allows the admin to add a new country to the list of countries that our application offers for each activity we support.
- *Add Company:* This function allows the admin to add a new company to the companies list that found in each country and offer trips for the activities that this country supports.
- *Update or Delete Activity:* This function allows the admin to delete a certain activity that found in the activities' list that the application support. It also allows him/her to update or change some information about any activity, like the name or the photo, etc.
- *Update or Delete Country:* This function allows the admin to delete a country from the countries' list that support a certain activity. It also allows him/her to update the information about this country like; name, companies inside this country that support a certain activity, etc.
- *Update or Delete Company:* This function allows the admin to delete a company from the companies' list that found in each country and support a certain activity or activities. And it also allows the admin to update or modify the about a certain company, like; name, address, email, etc.

Second; Functions of the Visitor: Here are the functions of the admin who is maintaining and updating website all the time.

- *Sing up*: This function allows the visitor to use the application for the first time by creating an account to start, otherwise, he/she can't use the application.
- *Login*: After registering by creating an account, every time the visitor need to get into the application again, he/she just need to login from the home page by using the same username and password he/she created for his/her account in the beginning. Everytime, the application check the username and the password for any person to validate the rights and authorities for him/her as a user or an admin.
- *Search*: This function allows the visitor to search in the of activities that our website support to choose what is preferable for him/her. Then after choosing the activity, he/she will open the menu for the country to see which ones supporting this activity and choose one of them. Then, the page will be autorefresh to display the this specific country and the companies inside it that could help the user in reserving his trip.

- *Reserve*: When the visitor click on the link of one the company he choose, the page will change to the company reservation page to give him/her a dialogue box where he/she has to fill one more simple details that are full name & email with a box for a message to right all requested information for example; date and time preferred for his/her trip, number of persons, level of activity, etc.
- *Rating the trip*: After finishing the trip, our website gives the right to the visitor to rate his trip with the chosen company, which will affect in the future for continue supporting this company in our application or not, also give an imprison for other people about this sport activity and help in marketing this kind of tourism and those different available places for practicing certain sport activities that may not be well known before.
- *Rating our website*: This function gives the availability to the visitor to rate our website to help us to continue improving and updating our services to the tourists, and make our quality transparent to all users who not yet try our services.

Sequence Diagram: In this section we will demonstrate some sequence diagrams that shows how some important functions of the above are working in sequence.

The first diagram is for the login process that represents the steps and the procedure of checking the authentication and authorities of the person who ask to the login (see Fig. 6).

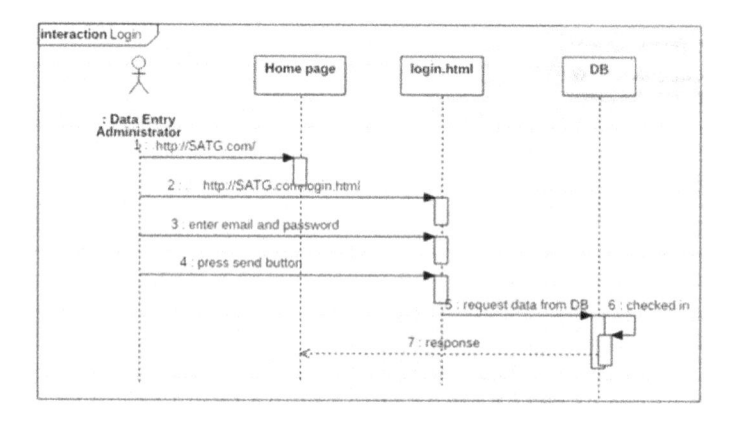

Fig. 6. Sequence diagram of login function

Second diagram is for the admin functions to add a new activity in the list of activities that supported by our application. This example is repeated in the same manner by the admin in all of his/her functions of adding, deleting or updating all other elements of the application (activity, country or company) (see Fig. 7).

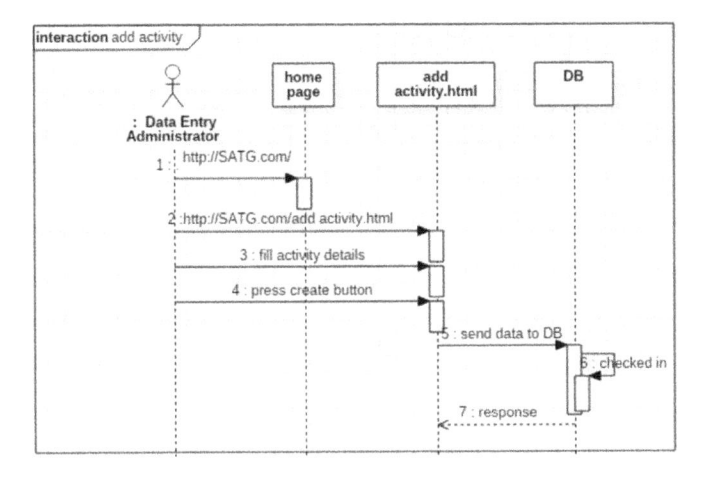

Fig. 7. Sequence diagram of "Add Activity" function

Third diagram is for the steps of the visitor has to follow to search for the preferable activity his/her wants to practice it, along with the steps to choose the of it place (country) and the enterprise to support his/her trip to complete the search process (see Fig. 8).

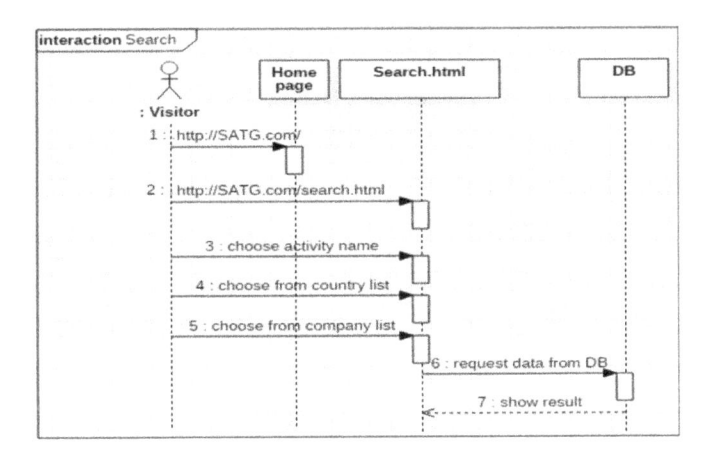

Fig. 8. Sequence diagram of "Add Activity" function

5 Interfaces of Our Website

Now we will demonstrate some interfaces of our website that shows the home page first then the following pages that support some functions of that mentioned previously.

This page is the main page of the website which appears automatically when the user opens our website for the first time (see Fig. 9).

Fig. 9. SATG home page

This is the sign-up page for creating an account to start use the application (see Fig. 10).

Fig. 10. SATG sign up page

This page is the searching page for the user, where he/she can search his/her preferable activity after he/she is registering in the website by creating an account (see Fig. 11).

Fig. 11. SATG searching page

This is the interface for the admin though which he/she can add a new activity with its data about the country that support it and the companies that are offering trips for this activity inside that country (see Fig. 12).

Fig. 12. SATG interface page for the admin

6 Conclusions and Future Work

By this website application we could satisfy our main objective of this project, which is listing some existing sport activities that found worldwide in different countries and facilitate the way to the sport activities' lovers to reach their aim in a most simple, easy and interesting way. Also, we could satisfy our idea of creating and encouraging a trending tourism branch of sport activity tourism instead of only sight seen tourism or just visiting famous and historical places worldwide.

In future, we aim to update and add more technical services for these sport activities as requested on our website, more details and more attractive and elegant interfaces, with a big orientation to apply this website in mobile application to make it easier and increase its availability.

References

1. Alcántara-Pilar, J.M., del Barrio-García, S., Crespo-Almendros, E., Porcu, L.: Toward an understanding of online information processing in e-tourism: does national culture matter? J. Travel Tour. Mark. **34**(8), 1128–1142 (2017)
2. Pan, B.: E-Tourism Encyclopedia, December 2014 (2015)
3. Frommer website. https://www.frommers.com/
4. Tripsavvy website
5. Tourdust website

A Heterogeneous Scalable-Orchestration Architecture for Home Automation

Jeferson Apaza-Condori[1,2] and Eveling Castro-Gutierrez[1,2(✉)]

[1] Universidad Nacional de San Agustin de Arequipa, Arequipa, Peru
{japazacondori,ecastro}@unsa.edu.pe
[2] CiTeSoft-UNSA, Arequipa, Peru

Abstract. Internet of Things is represented by the large number of smart devices connected to the internet and the number of devices is constantly growing, according to Gartner this amount would reach 20.4 billion devices by 2020, there are many industries that make use of this technology, among them is the home automation, where we can find several architectural proposals that try to solve the implementation of devices connected to the Internet in an environment, however most of these proposals only give solutions for specific requirements using a certain technology, avoiding problems such as network management, security, heterogeneous devices scalability, etc. In this paper we present a novel Architecture for Home Automation based on the guidelines proposed by ISO/IEC 30141:2018 (Internet of Things), as well as use cases proposed by the OneM2M. The implementation of this architecture will guarantee us an easy integration, orchestration of IoT devices adapting to the environment context as well as managing control in communications.

Keywords: Internet of Things (IoT) · Home automation architecture · OneM2M · ISO/IEC 30141:2018 · Quality of Service (QoS) · Ambient intelligence

1 Introduction

In nowadays there is an increase in the production electronic devices connected to the internet [1], these devices must be integrated into an environment to adapt to new needs for the end user. There are many alternatives to solve the problem of the heterogeneous device such as proposed by the authors [2,3]. A possible solution is through the implementation of a model based on SOA (Architecture Oriented Services) [17–19], where device functionalities are exposed as services. The devices inclusion will create a high consumption in the network for a Home Automation System, for this reason it's very important the Quality of Services (QoS) management specifically for those devices that send data in real-time for the users (p.e. Security Cameras) so that our devices always respond without generate information loss. Also an intelligent environment is a system centered on the user and it's effectiveness is to understand preferences, needs and abilities of the users in such a way that doesn't realize about their conscious perception [4,28]. The aim of this paper is designed an architecture for home automation based on qualities described above, so that it's possible to add different devices IoT, communication management and context environment, the proposed architecture is based on the guidelines by ISO/IEC 30141:2018 (Internet of Things) and moreover of the use cases from OneM2M, in order to take referential models and a standard conceptualization for the develop architecture.

The rest of this work is structured as follows: Sect. 2 presents an overview of the state-of-the-art and related works as well as the trend of emerging technologies in IoT projects; Sect. 3 presents our architectural proposal; Sect. 4 presents a use case to validate this Architecture; Sect. 5 has the conclusions and finally Acknowledgment section details our recommendations and future works.

2 State-of-the-Art

In this section, we will focus in a review of the current research on topics related to the Internet of Things related to the objective of this work, the related works are classified into five categories (2.1) Review of Proposed IoT Architectures, (2.2) SOA applied to IoT projects, (2.3) Importance of QoS in IoT, (2.4) Environmental Intelligence.

2.1 Review of Proposed IoT Architectures

The design of Architectures for projects in IoT, is taking a great importance in recent years solving specific problems in different industries (health, commerce, homes, etc.). Miladinovic et al. [3] proposed an Architecture model based on NVF (Virtualization Network Function), its proposal consists of 2 layers where the first layer is a simple Gateway that doesn't store any logic, its function is to set the IoT devices, the Gateway implements the ZigBee protocol due to its low consumption, and the second layer is the Data-Center it's function is creating virtual machines for the applications and managing their life cycle, the communication of the 2 layers is through REST due to its simplicity. Author in [7] implement an Architecture for Intelligent Health Systems, the input data are physiological parameters of the patient also sensors were installed in the environment (temperature, humidity, etc.), the authors consider a hybrid layer that's formed by RFID technologies and WSN to store patient data, the application management layer takes the hybrid layer as input and takes the data periodically from the patient sensors. In order to give a semantic knowledge, finally they consider a layer where information is stored temporarily to the sensors. In Wang et al. [8] develop an architecture based on ontologies, their main contribution is to facilitate the spontaneous interoperability of devices being able to communicate without the need to standardize their Information models, use the OWL language to represent ontologies, its architecture presents a central layer called SesGat whose objective is to provide knowledge about the ontology comparison process. In Wang et al. [9] implement an architecture based on FPGA programmable doors, the architecture take the ISO standard IEEE 1451.2, in order to separate the design with the communications, considers a master module that provides functions for network communication, moreover of local storage and interfaces to access high-speed devices such as digital camera, Implement a system vehicle monitoring to validate its architecture. Desai et al. [10] propose an Architecture based on a Semantic Gateway that provides interoperability between systems, this architecture adapts to privacy problems by allowing users to control the sensor data in the gateway, the authors cites that an scalable architecture must be independent of messaging protocol standards, at the same time it must provide integration and translation between various protocols. Khan et al. [11] propose a generic architecture for IoT projects in addition to distinctive features and possible future applications, the architecture is made up of 5 layers: Perception Layer, Network Layer, Middleware Layer, Application Layer and Business Layer. Coelho et al. [12] propose a smart home architecture to support the care of people with special needs (physical and cognitive disabilities), this proposal has 2 layers the first layer manages the devices locally in real time while the second layer uses an engine of analysis in the cloud, the engine uses tracking algorithms and take as input a set of observations of sensors, the algorithm has a graphical model of the set of possible routes between sensors. Finally, Krco et al. [13] analyze the architectural proposals for IoT projects carried out by the European Commission (especially the FP7 Program), and analyze diversity of Architectures, emphasizing that the current approach only analyzes solutions based on specific requirements for applications, in addition to the different terminology that is used to define its architectures this causes to a limited interoperability between systems, also mentions the IoT-A architecture as an effort to provide a reference model together with a set of best practices to help IoT project developers.

2.2 SOA Applied on IoT Projects

Hamzei et al. [14] propose efficient techniques on composition of Services oriented to the use of Internet of Things, the approach divides the services into 4 types: SOA, RESTFUL, Heuristic and Model-Based, the author concludes that this approach is not very researched but it's very necessary due to the great variety of technologies and new applications that will unite physical objects. Avila et al. [15] make an analysis of functionalities for healthcare applications, the author implements remote monitoring systems to the patient, where use an SOA model with the objective of creating federated, distributed and scalable architectures. Ruta et al. [16] present a framework inspired by the interaction of social networks to increase the self-configuration of devices and self-orchestration in homes, the devices become autonomous social agents, where they interact and coordinate automatically on the environment, the framework is based on an SOA model for the encapsulation, exchange and composition in a semantic services description.

2.3 The Importance of QoS in IoT

Rahman et al. [20] propose a communication model for IoHt (Internet of Health Things), based on the cloud and supported by Fog-Computing, the novelty of this model its ability to handle data routing separately for different data types coming from conventional data and in real time. The results show an improvement in the QoS parameters obtaining a low energy consumption and a reduced latency. In Zhang et al. [21] propose a Q-Learning model to design a data transmission scheme in vehicular cognitive networks to minimize the transmission cost while using different communication modes and resources, the model considers the maintenance of the QoS specifically based on "Delay" characteristic, its analytical results indicate a reduction in transmission costs by helping to deliver data under "Delay" restriction. In Yao et al. [22] analyze the interconnection increasing problem of internet-connected devices generating large data amounts as a consequence of increasing traffic on the network where it proposes a model called "Cloud Radio Access Network (C-RAN)" considering the future for network architecture helping to counteract traffic and performance in IoT services, the model is also based on Edge Computing, relieving QoS indicators, in addition to avoiding frequent sensors activation as a result of reducing energy consumption. Gu et al. [23] designed a random access method set and transmission data protocols in "Massive Machine Type Communication (mMTC)" to guarantee the quality of service (QoS) in different types of network traffic, the model is based on a priority queue with class constraint (ACB) priority queue not only alleviates network traffic but also guarantees optimal QoS.

2.4 Ambiental Intelligence

Bylykbashi et al. [24] implement an intelligent environment to improve human sleep conditions, the dataset test is implemented by a sensor node, a receiver node and an actor node, the sensor node emits microwaves in the direction of the human subject, these results are taken to the actor node that's used to cool and warm the user's bed, its results conclude that factors such as lighting, humidity and temperature of the room have proportional effects to the quality of the user's sleep. Moseley et al. [25] implement an environmental intelligence network using a fused reality system and a process based on recognition/perception patterns, the objective is to be able to trace an agent (who could be human or not) and who isn't connected to the system, in addition to predicting behavior and in a certain sense reacting appropriately, its results show the behaviors learned from the agent according to a given environment. Echanobe et al. [26] propose a multiprocessor system based on FPGA that take environmental characteristics that are taken to a PWM-ANFIS system based on neuro-fuzzy that has the ability to learn, reason and adapt to changes, to validate their proposal they implemented tests in a bedroom where they demonstrated the capacity of learning and adaptation by the system. Kanagarajan et al. [27] design ontologies for intelligent environments, the ontologies are based on user behavior, as a test case it places sensors in an environment that capture user information where they are stored and then refined and form a temporal sequence histogram with in order to obtain new improved sequences, managing to create the ontology of user behavior.

3 Proposed Architecture

For the design and development of this novelty architecture we adopt the guidelines offered by ISO 30141: 2018 (Internet of Things), these guidelines define a set of generic features and parts that can be used as a starting point to create specific architectures in applications IoT. In addition we take into consideration the effort that OneM2M made in developing a technical report of use cases for several M2M industries, these use cases are focused on a sequence of interactions with the users and actors that intervene according to the field of application specifically we'll focus on the use cases for home automation [6].

As shown in Fig. 1, this architecture is divided into four layers: Service Layer, Communication Layer, Application Layer and Business Layer, that are described in detail.

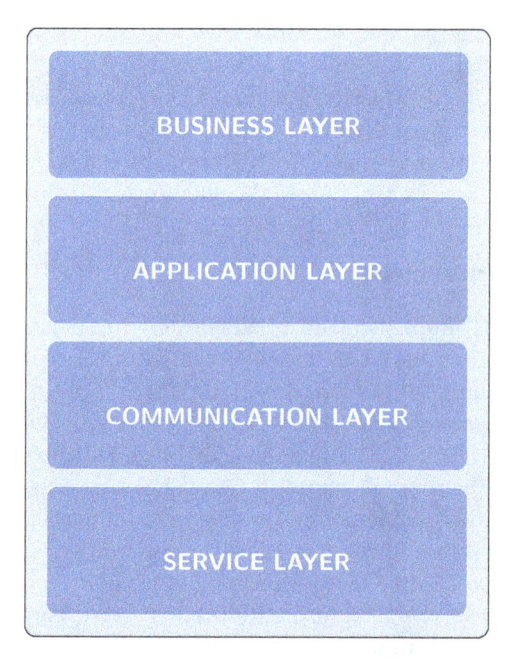

Fig. 1. Proposed architecture.

3.1 Service Layer

As shown in Fig. 2, Service Layer registers the IoT devices installed in a home environment with the aim integrating with the rest architecture, this layer consists of a Gateway that set registry and gateway functions, the devices are anchored to this by a Generic Router according to the communication protocol used (MQTT, HTTP, CoAP, etc.). As suggested by OneM2M [6] The Gateway also add additional functions to the IoT devices (if required), so that any modification or adjustment in the devices should only be developed in this layer without setting device (e.g. changing the measurement scale for the temperature sensor from Celsius to Fahrenheit), additional development functionalities are performed in the component "Service Provider" that use an SOA model inside, an advantage of this model is independence of platforms for its implementation.

In an environment, it's important handling a control about predefined events for our devices, in such a way that users are notified or alerted when a condition has been met for any event described, for this case we implement a component called "Management Event" that is communicated directly to the Gateway because it has the device information, an example case would be sent to user average readings of energy consumption produced during the month.

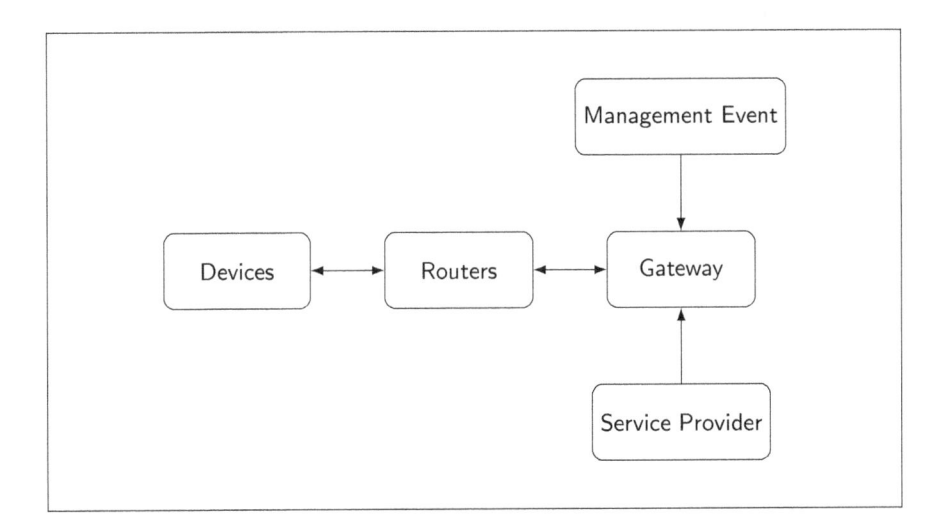

Fig. 2. Service Layer components

3.2 Communication Layer

The aims of this layer is to manage communications in a way that allows the safe and optimal message delivery from the application layer to the service layer, for the development of this layer we take as architecture MQTT reference, remember that aims of this architectural proposal is not restricted to a specific protocol, but rather the real aim is to create a base model that would be used for the development home automation without focus on a specific technology or tool. The philosophy MQTT considers a broker as a central communication point and is responsible for dispatching all messages to recipients who have subscribed to a specific publication. OneM2M suggests existence of an authorization component as well as to a session component for each request, in such a way that guarantee us security between the applications and the IoT devices. As shown in Fig. 3, we have a broker to which is connected to 3 components. If we want that message arrives at its destination (users), first we must evaluate the time connection for each request, that work will do "Sessions" component, QoS (Quality of Services), establishes priorities for message delivery according to the IoT device in this way we avoid loss of information and overload in the network, finally the Security component establish unsafe encryption between users and devices.

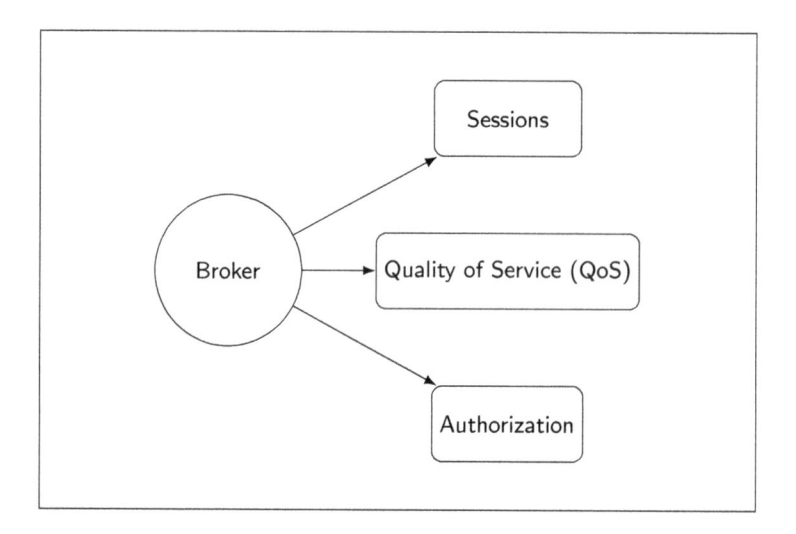

Fig. 3. Communication Layer components

3.3 Application Layer

The aims of this layer is to implement a set of applications for IoT-user that manipulate these devices registered in an environment, ISO/IEC 30141:2018 defines a IoT-user as an entity that could be of two types (Human User and Digital User) where the digital user represents a specialization of an IoT-user and interacts with one or more services offered by the IoT System, as shown in Fig. 4. The implementation of applications can be very varied [29–31] (Cloud Services, Mobile APP, Voice Assistants and even other IoT-devices), and the developer's freedom to implement them, a main component of this layer is the "Middleware" component, which is responsible for making interoperability possible at the application level between two or more devices.

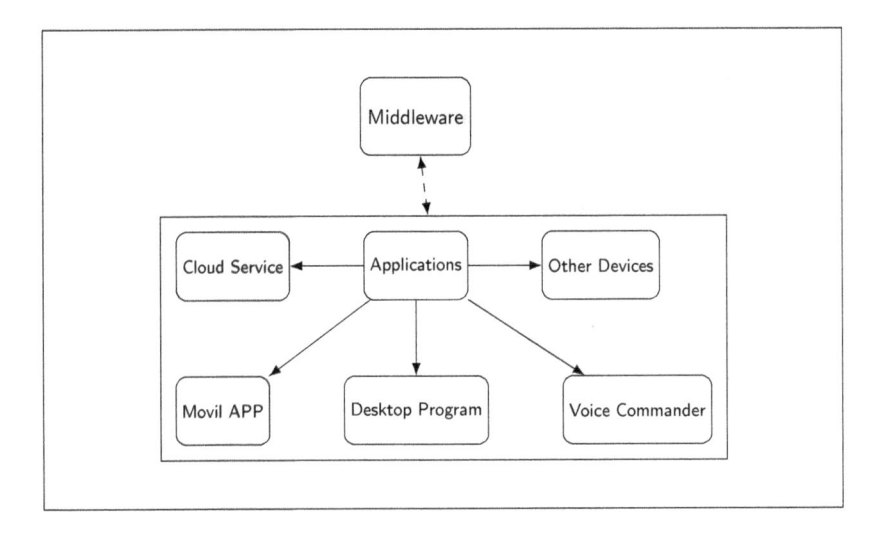

Fig. 4. Application Layer components

3.4 Business Layer

The aims of this layer is to implement intelligent autonomous models for the devices, OneM2M [6] mentions that an M2M system must support a semantic data model that can be represented by ontologies and that has discovery properties, however ISO/IEC:30141 represents an autonomous model by interoperability levels (syntactic and semantic), semantic interoperability being responsible for establishing the meaning between messages, considering this in our layer we represent a component called "Ontologies", which take as a starting point characteristics of the IoT devices previously registered in the Service Layer, the ontologies implementation can be very varied, however most of these proposals are often project-specific or discontinued, to overcome these problems W3C-SSN introduced a generic and independent model domain called semantic sensor network ontology (SSN) and is based on the OWL (Web Ontology Language) language. Another important component is the business models whose function is to implement applications based on ontologies (according to the context), finally we consider a component whose function is to monitor all the activities that are done with the devices in order to enrich our business models, the components distribution for this layer can be seen in Fig. 5.

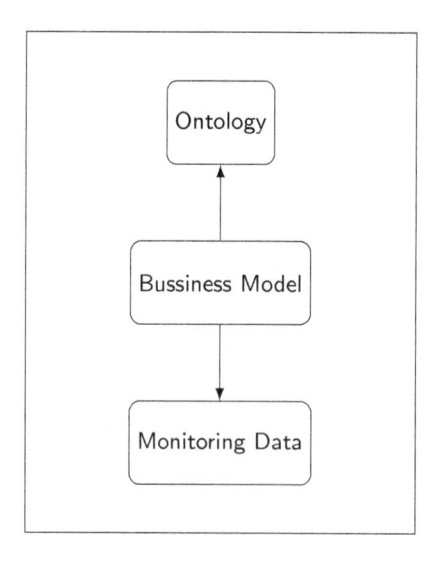

Fig. 5. Business Layer components

4 Use Case

In this section, we will describe a use case to validate this architecture. A scenario was created with 8 devices installed in different parts of home (rooms, garden and living room), where the devices were programmed and configured with different processing and communication technologies, the selected technology will depend mainly on the quantity of information to process, in addition to the energy type source that we use (stacks, batteries or solar cells), The devices are registered in the Service Layer through a visual interface where the main characteristics device are detailed, as well as we want to add functionality to any of devices or execute scheduled tasks we would do in this same layer. The implementation of Service-Layer was performed with a Raspberry-PI Model 3B, where services were created for each communication protocols used by the devices, the sensors information is pre-processed and then go through another device that manages communications delivering optimal data from the sensors to their respective clients or applications to make use of this information, Finally in the Business-Layer layer we developed examples of ontologies for some sensors that needed a use according to the environment.

4.1 Implemented Devices

We installed sensors in the bedroom that obtain the temperature and light of environment, these use the Web-Socket communication protocol this protocol selected for its latency reduction f allowing the teams to attend more simultaneous connections [35]. In the garden a water consumption measurement sensor and a taps actuator were installed, the first one uses a solar energy source (5.5 V) and both use the MQTT protocol because it considers a low power consumption and a low coupling in data transfer which is ideal for devices where the power source is limited [36]. On the home façade we installed a motion sensor that uses the CoAP protocol, and an RFID actuator system that uses the HTTP protocol due to its ease of implementation and which is waiting for the RFID-Card reading to enable entry into the home. Finally in the living room we installed an electrical energy measurement sensor, the measurement was the Watt and the communication protocol used is the CoAP. The summary of devices together with the technologies used is found in Fig. 6.

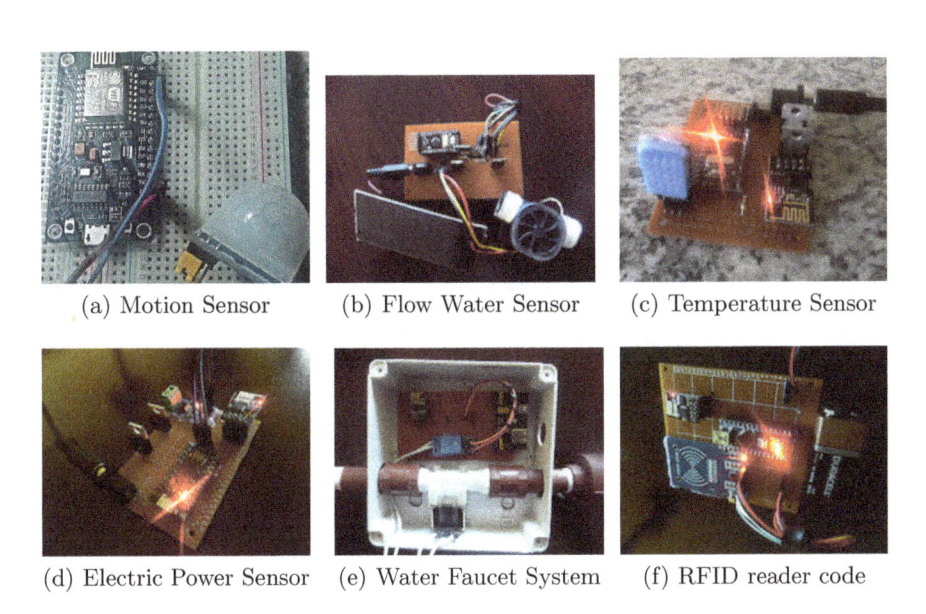

(a) Motion Sensor (b) Flow Water Sensor (c) Temperature Sensor

(d) Electric Power Sensor (e) Water Faucet System (f) RFID reader code

Fig. 6. Devices IoT implemented for use case.

4.2 Service Layer

To implement this layer, a Raspberry Pi computer was acquired for its versatility and economic value on IoT Projects [32–34], where the Raspbian Operating System was installed, also a system was developed in Java, where the Protocols necessary for the interaction with the devices were configured (MQTT, CoAP, HTTP, Web-Socket). The messages arrival can go through a pre-processing filter, and will depend exclusively on the additional device requirements, for this use case we obtained the readings from the consumption of water liters, and created an additional event so that the end the day calculate the average amount of liters consumed, the Management-Event component is responsible for sending the message to the user (going through the communications layer) only when the end of day condition is fulfilled, the information flow for this layer can be seen in Fig. 7.

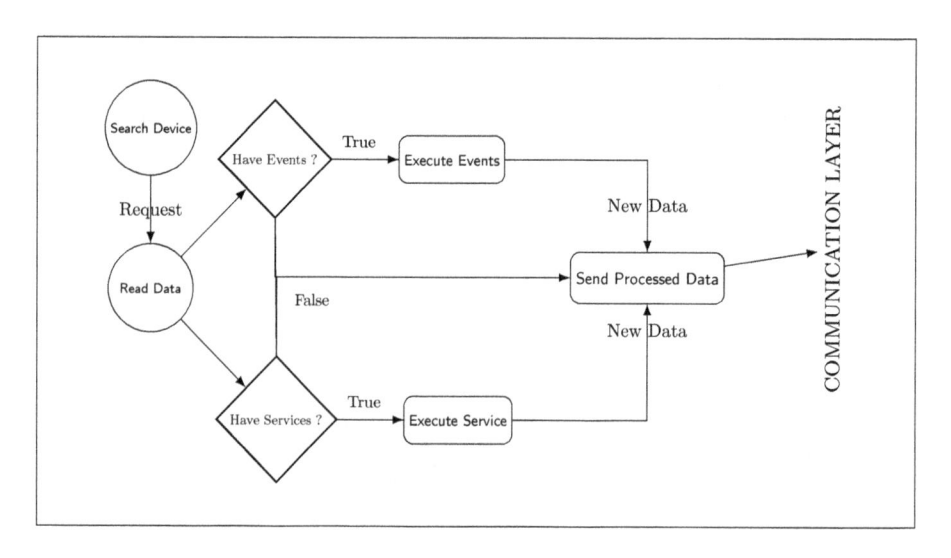

Fig. 7. Flow diagram for Service Layer

4.3 Communication Layer

According to ISO/IEC 30141: 2018 (Internet of Things), it states that IoT systems require network management and that the form will depend on the network and communication type that is carried out through the network. Same, it states that importance of creation profiles, management QoS (Quality of Services), Handling Faults and the security control [5].

For the development of this layer we use on the RabbitMQ tool which helped us to manage communications, this tool supports different protocols such as HTTP, MQTT, CoAP, among others. The main functionality of RabbitMQ is the messages queuing acting as middleware between senders and recipients, providing benefits such as the guarantee of delivery and messages order, as well as elimination of redundancy, decoupling this due to queue balancing [37].

In any communication type, a producer and a consumer are required; in addition to a address to where the information is going to be sent, the producer needs that address that in RabbitMQ is called "exchanges", we create an exchange for each device-IoT, the consumer obtains these messages by subscribing to queues, the queues are linked to the "exchanges" through a process called "Binding" if not doing so the message is lost in the "Exchange". The modeling distribution of "exchanges" and Queues for our use case can be seen in Fig. 8.

To guarantee security this tool presents two authentication types (user-password and X.509 certificates), it is also advisable to establish authorization levels to virtual hosts as well as roles for each user.

In the case that several consumers share the same queue, it is advisable to specify how many messages can be sent to each consumer or customer at a time, for this RabbitMQ manages the QoS (Quality of Service) where you can specify the message priority for the channel current or for all connections channels.

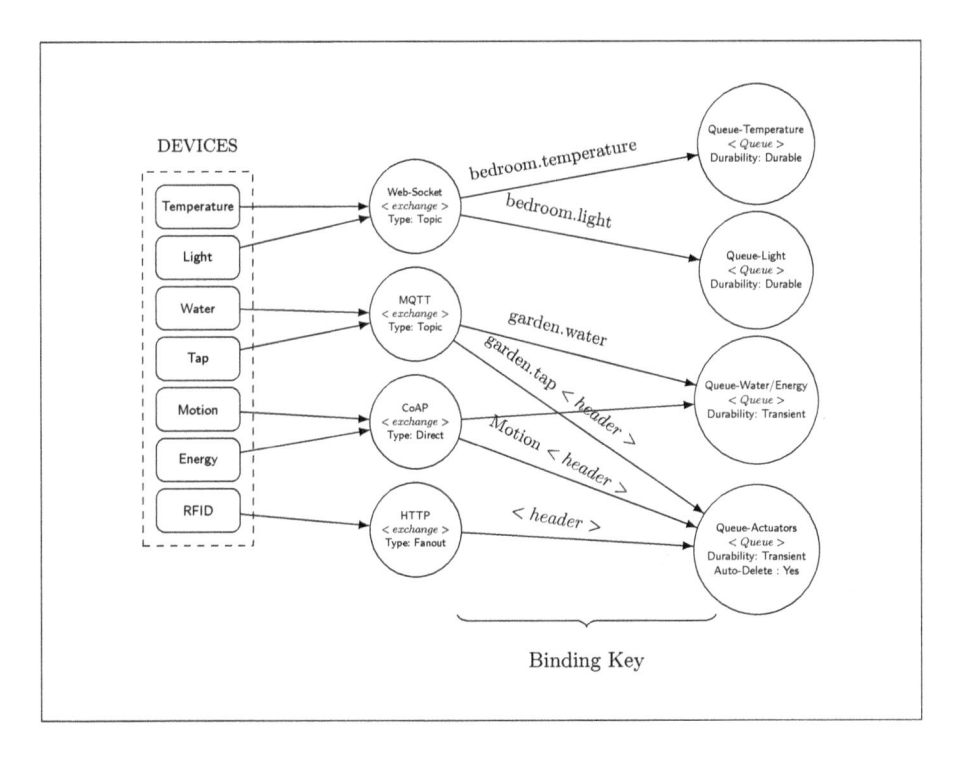

Fig. 8. Broker component implemented on RabbitMQ

4.4 Application Layer

For this layer we implement a middleware that consists of a service that consumes data from Communication Layer and sends it to different applications that subscribe to the middleware. In this way we can implement different applications independent of platform or tool that use. As shown in Fig. 9, we managed to implement applications for several platforms such as (a) Cloud Services to send our sensors to ThinkSpeak, (b) Desktop applications that monitor Water-Sensor values, (c) Mobile Applications that enable opening of faucets or (d) even IoT devices to control access to a house.

(a) Cloud Service by ThingSpeak

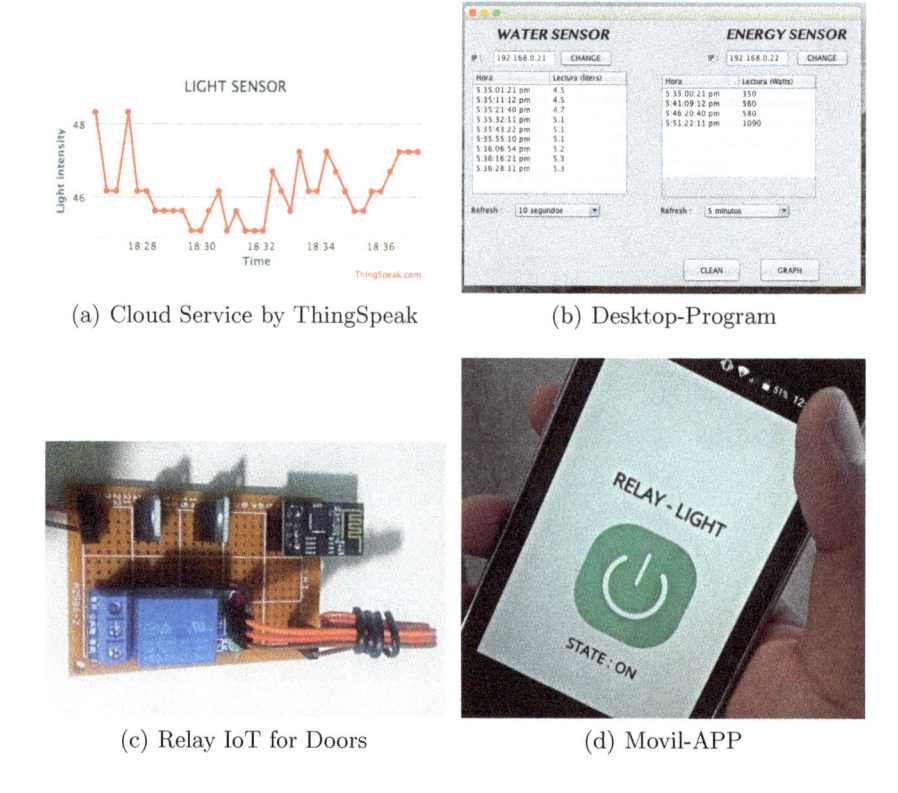

(b) Desktop-Program

(c) Relay IoT for Doors

(d) Movil-APP

Fig. 9. Applications implemented for use case

4.5 Business Layer

The main components of this layer are ontologies development in such a way that they are adapted to the context. The characteristics devices were obtained from the Service layer, ontologies were defined for: devices, physical spaces, home environments and users as shown in Fig. 10.

We use an OWL API to be able to use these ontologies where also has a reasoner that allows finding inconsistencies, the ontologies go through a rules engine that helps us discover inferences and create new rules if it's required.

The Business Model component has the function of creating Applications for these ontologies according to the context we wish to interpret, as shown in Fig. 10 for this use case, we developed location models, user search, and the preference of devices by environment and climate.

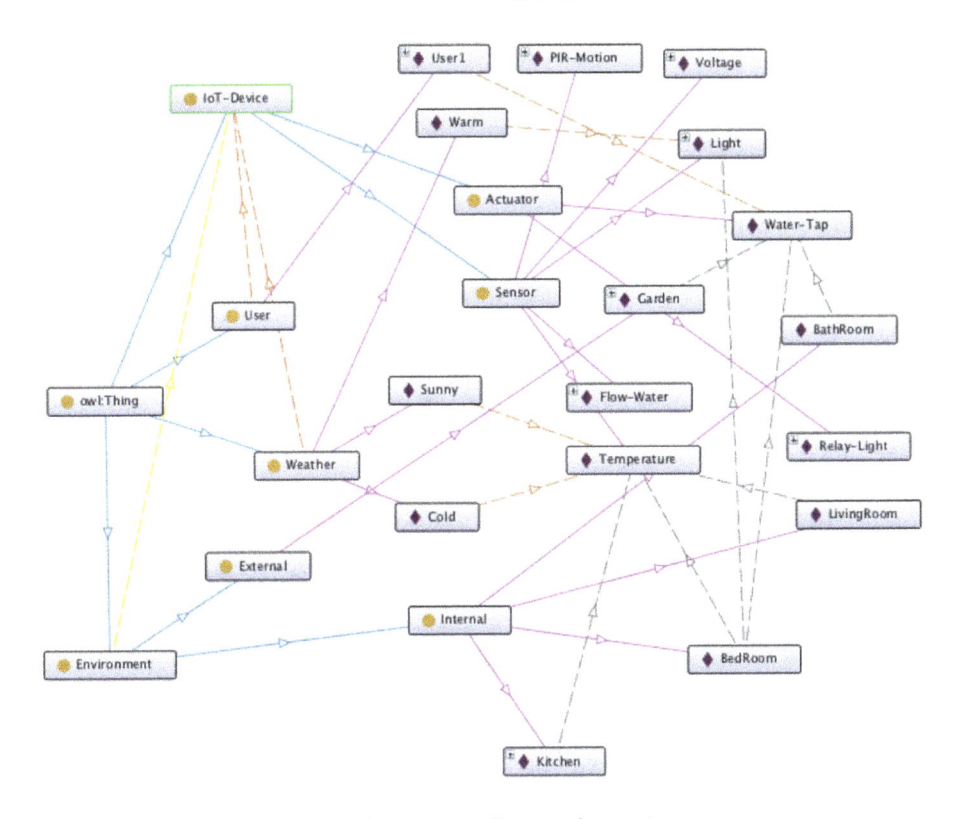

Fig. 10. Ontologies design according to the environment context

5 Conclusions and Recommendations

This work presents an Architecture for IoT Devices Installed in a Home. The use of concepts and good practices defined by ISO/IEC 30141:2018 OneM2M, as well as the analysis of Architectures proposes by other authors helped us to define a stable, scalable and appropriate architectural model to the context in order to guarantee integration and orchestration facilities for devices. It was important to consider the role of communication protocols that exist in IoT and see their most relevant qualities according to application field. It is important to mention that in a real scenario, the performance of each layer may be more relevant than another, which means that it will depend exclusively on the complexity and application that we develop, in this work our devices used all the layers component only for explanatory purposes.

Acknowledgment. We want to thank Universidad Nacional de San Agustin de Arequipa for financing this project through the contract No. TP14-2018, as well as the recognition to the CiTeSoft Research Center (EC-0003-2017-UNSA), for providing us, the environment and the equipment necessary to develop this project.

References

1. Gartner Inc.: Gartner Says 20.4 Billion Connected "Things" Will Be in Use in 2020, Up 30 Percent From 2015. https://www.gartner.com/en/newsroom/press-releases/2017-02-07-gartner-says-8-billion-connected-things-will-be-in-use-in-2017-up-31-percent-from-2016
2. Maciel, F., Thangaraj, C.: A scalable modular heterogeneus system for home and office automation. In: IEEE MIT Undergraduate Research Technology Conference (URTC), November 2016. https://doi.org/10.1109/URTC.2016.8284079
3. Miladinovic, I., Schefer, S.: NFV enabled IoT architecture for an operation room enviroment. In: 2018 IEEE 4th World Forum on Internet of Things, May 2018. https://doi.org/10.1109/WF-IoT.2018.8355128
4. Augusto, J.: Reflections on ambiente intelligence systems handling of user preferences and needs. In: 2014 International Conference on Intelligent Enviroments, June 2014. https://doi.org/10.1109/IE.2014.70
5. ISO/IEC 30141:2018(IoT). https://www.iso.org/standard/65695.html
6. OneM2M - Technical report TR-0001-V 2.4.1. http://www.onem2m.org
7. Catarinucci, L., Donno, D., Mainetti, L., Palano, L., Patrono, L., Stefanizzi, M., Tarricone, L.: An IoT-aware architecture for smart helathcare systems. IEEE Internet Things J. (2015). https://doi.org/10.1109/JIOT.2015.2417684
8. Juarez, J., Rodriguez, J., Garcia, R.: An ontology-driven communication architecture for spontaneous interoperability in home automation systems. In: IEEE Emerging Technology and Factory Automation, September 2014. https://doi.org/10.1109/ETFA.2014.7005270
9. Wang, S., Hou, Y., Gao, F., Ji, X.: A novel IoT access architecture for vehicle monitoring system. In: 2016 IEEE 3rd World Forum on Internet of Things (ES-IoT), December 2016. https://doi.org/10.1109/WF-IoT.2016.7845396
10. Desai, P., Sheth, A., Anantharam, P.: Semantic gateway as a service architecture for IoT interoperability. In: 2015 IEEE International Conference on Mobile Services, June 2015. https://doi.org/10.1109/MobServ.2015.51
11. Khan, R., Ullah, S., Zaheer, R., Khan, S.: Future internet: the internet of things architecture, possible applications and key challenges, December 2012. https://doi.org/10.1109/FIT.2012.53
12. Coelho, C., Coelho, D., Wolf, M.: An IoT smart home architecture for long-term care of people with special needs. In: IEEE 2nd World Forum on Internet of Things (WS-IoT), December 2015. https://doi.org/10.1109/WF-IoT.2015.7389126
13. Krco, S., Pokric, B., Carrez, F.: Designing IoT architecture(s): a European perspective. In: 2014 IEEE World Forum on Internet of Things (WF-IoT), April 2014
14. Hamzei, M., Jafari, N.: Towards efficient service composition techniques in the internet of things. IEEE Internet Things J. (2018). https://doi.org/10.1109/JIOT.2018.2861742
15. Avila, K., Sanmartin, P., Jabba, D., Jimeno, M.: Applications based on Service-Oriented Architecture (SOA) in the field of home healthcare. Sensors (Basel) (2017). https://doi.org/10.3390/s17081703
16. Ruta, M., Scioscia, F., Loseto, G., Sciascio, E.: A semantic-enabled social network of devices for building automation. IEEE Trans. Industr. Inf. (2017). https://doi.org/10.1109/TII.2017.2697907
17. Soon, J., Kumar, S., Shanmugavel, M.: OSGi-based, embedded, distributed, telematics framework for flexible service provisioning in cyber-physical production systems. In: 2016 IEEE International Conference on Computational Intelligence and Computing Research (ICCIC), May 2017. https://doi.org/10.1109/ICCIC.2016.7919566

18. Malarvizhi, R., Kalyani, S.: SOA based open data model for information integration in smart grid. In: 2013 Fifth International Conference on Advanced Computing (ICoAC), December 2013. https://doi.org/10.1109/ICoAC.2013.6921941
19. Ochoa, A.: Gestionando datos heterogeneos provenientes de sensores para medir la calidad del aire en Bogota. Universidad Nacional de Colombia - Departamento de Ingenieria de Sistemas y Computacion, February 2018. https://doi.org/10.13140/RG.2.2.30507.34086
20. Rahman, A., Afsana, F., Mahmud, M., Shamim, M., Ahmed, M., Kaiwartya, O., Taylor, A.: Towards a heterogeneous mist, fog, and cloud based framework for the internet of healthcare things. IEEE Internet Things J. (2018). https://doi.org/10.1109/JIOT.2018.2876088
21. Zhang, K., Leng, S., Peng, X., Pan, L., Maharjan, S., Zhang, Y.: Artificial intelligence inspired transmission scheduling in cognitive vehicular communication and networks. IEEE Internet Things J. (2018). https://doi.org/10.1109/JIOT.2018.2872013
22. Yao, J., Ansari, N.: Joint content placement and storage allocation in C-RANs for IoT sensing service. IEEE Internet Things J. **6**(1) (2018). https://doi.org/10.1109/JIOT.2018.2866947
23. Gu, Y., Cui, Q., Ye, Q., Zhuang, W.: Game-theoretic optimization for machine-type communications under QoS guarantee. IEEE Internet Things J. **6**(1) (2018). https://doi.org/10.1109/JIOT.2018.2856898
24. Bylykbashi, K., Spaho, E., Obukata, R., Ozera, K., Liu, Y., Barolli, L.: Implementation and evaluation of an ambient intelligence tested: improvement of human sleeping conditions by a fuzzy-based system. Int. J. Web Inf. Syst. **14**(1), 123–135 (2017). https://doi.org/10.1108/IJWIS-12-2017-0082
25. Moseley, R.: Creating an ambient intelligence network using insight and merged reality technologies. In: 2017 Computing Conference, January 2018. https://doi.org/10.1109/SAI.2017.8252139
26. Echanobe, J., Campo, I., Basterretxea, K., Martinez, M.: An FPGA-based multiprocessor-architecture for intelligent environments. Microprocess. Microsyst. (2014). https://doi.org/10.1016/j.micpro.2014.07.005
27. Kanagarajan, S., Ramakrishnan, S.: Development of ontologies for modeling user behaviour in ambient intelligence environment. In: IEEE International Conference on Computational Intelligence and Computing Research (ICCIC), March 2016. https://doi.org/10.1109/ICCIC.2015.7435736
28. Mikulecky, P.: Formal models for ambient intelligence. In: 2010 Sixth International Conference on Intelligent Environments, July 2010. https://doi.org/10.1109/IE.2010.76
29. Singh, A., Mehta, H., Nawal, A., Gnana, O.: Arduino based home automation control powered by photovoltaic cells. In: 2018 Second International Conference on Computing Methodologies and Communication (ICCMC), February 2018. https://doi.org/10.1109/ICCMC.2018.8488144
30. Kaur, B., Kumar, P., Kumar, M.: An illustration of making a home automation system using Raspberry Pi and PIR sensor. In: 2018 International Conference on Intelligent Circuits and Systems (ICICS), April 2018
31. Abbas, N., Mansoori, I.: Smart door system for home security using Raspberry pi3. In: 2017 International Conference on Computer and Applications (ICCA), October 2017. https://doi.org/10.1109/COMAPP.2017.8079785
32. Vujovic, V., Maksimovic, M.: Raspberry Pi as a Sensor Web node for home automation. Comput. Electr. Eng. (2015). https://doi.org/10.1016/j.compeleceng.2015.01.019

33. Patchava, V., Babu, H., Ravi, P.: A smart home automation technique with Raspberry Pi using IoT. In: 2015 International Conference on Smart Sensors and System (ICSSS), December 2015. https://doi.org/10.1109/SMARTSENS.2015.7873584
34. Rostyslav, K., Tkatchenko, S., Golovatsyy, R.: Features home automation system development based Raspberry Pi using Java ME SDK. In: The Experience of Designing and Application of CAD System in Microelectronics, February 2015. https://doi.org/10.1109/CADSM.2015.7230912
35. Vujovic, M., Savic, M., Stefanovic, D., Pap, I.: Use of NGINX and WebSocket in IoT. In: 2015 23rd Telecommunications Forum Telfor (TELFOR), January 2016. https://doi.org/10.1109/TELFOR.2015.7377467
36. Kishore, R., Soratkal, S., Stefanovic, D., Pap, I.: MQTT based home automation system using ESP8266. In: 2016 IEEE Region 10 Humanitarian Technology Conference (R10-HTC), April 2017. https://doi.org/10.1109/R10-HTC.2016.7906845
37. Ionescu, V.: The analysis of the performance of RabbitMQ and ActiveMQ. In: 2015 14th RoEduNet International Conference - Networking in Education and Research (RoEduNet NER), October 2015. https://doi.org/10.1109/RoEduNet.2015.7311982

Ambient Intelligence Applications in Architecture: Factors Affecting Adoption Decisions

Maryam Abhari[✉] and Kaveh Abhari

Department of Management Information Systems, Fowler College of Business,
San Diego State University, San Diego, CA 92182, USA
mabhari@sdsu.edu

Abstract. Ambient Intelligent systems enabled by recent advancements in Artificial Intelligence (AI) and Internet of Things (IoT) technologies, offer new mechanisms for data-driven decision making in architectural design. Understanding why architects are willing to adopt these systems and what factors affect their decisions is a critical first step towards assessing the viability and informing the design of these systems. We develop and contextualize a theoretical model that examines the impact of architects' perceptions of value and risk associated with ambient intelligence (AmI) on their intention to adopt these systems during the design phase. The model also examines the role that commitment to learning about AmI and commitment to collaboration with IT professionals play in mediating such impacts. We validate the model via a field survey of architects across the North America. Our findings indicate that both perceived value and perceived risk play varying roles in adopting AmI and that commitment to learning and commitment to collaboration each acts as a mediator between architects' perceptions and their behavioral intentions. This study presents an attempt to examine the slow adoption rate of AmI in architectural design and thus offers a path to future theoretical developments and practical insights into AmI application design in architecture.

Keywords: Ambient intelligence · Architecture · Adoption · Perceived values · Perceived risks · Commitment to learning · Commitment to collaboration

1 Introduction

The efficiency of design process is one of the key priorities for architecture firms and it refers to delivering quality service to clients in a timely manner and at a reasonable cost. To achieve this efficiency, architects need to tackle various challenges to bridge between project priorities, regulatory requirements and end-users needs. New technologies such as Ambient Intelligent systems can be instrumental in responding to these challenges since they can assist in identifying and prioritizing design requirements and developing design recommendations. These technologies challenge prior design practices and provide new opportunities to experiment with new design methods. While technological innovations in this domain have the potential to facilitate the design process and improve design outcomes, there are a limited number of applications reported in theory and practice due to various technical and organizational barriers [1, 2]. It is essential to understand and overcome these barriers to encourage the use of ambient intelligence (AmI) in design process.

Increased use of intelligent systems in various contexts is leading architects to recognize their potential in the design process. While AmI systems have been around for a while, architecture firms have not fully understood and systematically explored their applications in planning and design. Although literature in other disciplines (e.g., healthcare, energy, urban planning) has reported on the applications of AmI, limited studies have been conducted in the architecture context. Hence, this study aimed at clarifying why architectural firms are slow in adopting AmI systems. We investigated how architects' perceptions of value and risk associated with AmI affect their intention to adopt AmI systems as a design tool in their practice, especially in the initial phase of design processes. Considering AmI as a design tool requires training and establishing a new form of collaboration between architects and IT professionals. Therefore, we also identified the role of commitments to learning and collaboration in driving AmI applications.

2 Theoretical Background

Over the past few decades, the architecture industry has faced significant technological transformations. Introduction of each technological advancement has opened new opportunities as well as challenges for architects and architectural firms. The architecture, engineering, and construction industry started adopting new technical concepts and implementing new methods of information sharing such as building information modeling (BIM), virtual reality (VR), 3D printing, smart materials, and internet of things (IoT) to advance and support the practice. An international study conducted by Chaos Group in 2017 reveals how new technologies are transforming the architecture industry worldwide. This study showed that the major changes in the industry correspond with the increased dependence on technology. However, utilizing new technologies are limited with projects' tight deadlines, limited budgets, and lack of resources including access to IT professionals [3]. The importance of training to effectively implement new technologies is also frequently cited in the literature [4].

Introducing new technologies to the design process has led to a paradigm shift in architecture practice. The conventional approach where architects, engineers, and contractors were largely operating independently of one another called for a new integrated approach to improve the efficiency and reliability of design processes. This need has triggered the rise of technologies such as BIM to connect architect and other consultants from the very early stages of design projects. The new approach assists projects' stakeholders to make informed decisions which in turn, positively affect buildings' performance and users' experiences, especially in large-scale projects. However, BIM and similar technologies have been mainly utilized as tools to support project planning, design documentation and construction management rather than systems to recommend data-driven design solutions.

2.1 Ambient Intelligence

AmI refers to intelligent systems that sense environmental signals such as people's presence, movements, temperature, noise, and light, and then respond accordingly. These systems are collections of hardware and software that enable environmental or spatial data collection. With the advancements in AI and IoT technologies, AmI systems are now capable of collecting more complex signals in real time, running more advanced algorithms, and generating more multifaceted responses. Newly developed AmI systems can communicate and analyze environmental signals over the Internet and autonomously modify or trigger responses. AmI systems can consist of connected nodes within a number of autonomous or automated sensors. These systems are capable of informing design processes by collecting and processing data about the physical environment, running different experiments in built environments, and verifying design assumptions in similar settings. Data collected provide a valuable source of insight for design professionals to solve complex design problems in similar projects' settings. Design problems can range from space planning (e.g. circulation. layout and utilization) to energy and water management, to regulatory requirements (e.g. accessibility, safety and security).

2.2 Current Applications of AmI in Architecture

AmI systems have been increasingly employed to minimize the operational costs related to buildings' energy consumption and facility occupancy. For example, learning thermostats, smart lights and shades, motion sensors, energy tracking switches and smart cameras are examples of AmI systems found in energy efficient buildings. Smart sensors connected to the Internet collect, transmit and process ambient data to either directly maintain the users' comfort or indirectly inform decision-makers to make inform decision about resource utilization and facility management [5–9]. These devices help improve not only overall buildings' performances, such as energy and water efficiency, but also occupants' well-being in general [4]. Despite the increasing applications of smart sensors, design professionals rarely think of using AmI proactively as an integral part of their design processes.

AmI systems have great potentials to assist architects from the very beginning of the design process in three proactive ways. The first way is to employ AmI to optimize design solutions for new buildings by collecting data from the existing buildings with similar scale and scope. For example, data collected from existing public buildings such as hospitals, airports, stadiums, or concert halls, provide architects with ample evidence on how these spaces are being used, how the traffic flows, and how users experience different areas. The collected information can be used to estimate needs for different spaces, traffic, and interactions and therefore assist designers to better formulate their design solutions. Collecting data from the buildings that are going for remodeling or renovation is the second possible way. The data collected from the current usage and condition of a building have a great potential to reveal both positive and negative aspects of the existing design and therefore to better guide the future design. The third way is to utilize AmI as an experimental tool to explore new ideas in programming and early design phase and to verify existing design assumptions in multi-phase projects. For example, architects can use AmI to experiment with different materials, design concept, and space planning. They can also collect objective data from users before the next phase of project rather than relying on their assumptions and intuition.

An example of an AmI application is IBM IoT for built environment. This system uses sensors to collect data and then uses IBM Watson to provide recommendations based on the data captured related to the usage of different spaces and other properties such as temperature. Integrated space management systems such as Tririga can process the data generated by Watson to improve the operational and environmental performance of the spaces. Tririga is traditionally used to combine, analyze, and visualize environmental data to inform space configuration, energy preservation, and maintenance optimization [10]. This is a successful example for integration of sensor technologies, energy management systems and AI that helps with the efficient management of diverse and difficult-to-service buildings [11]. A growing number of firms have adopted these technologies to not only improve their buildings' performance and lower operational cost, but also to enhance the users' experience and well-being [12, 13].

With the rise of AmI applications in buildings, the demand for these technologies will continue to grow. Although architects acknowledge the impact of utilizing AmI systems on building management and occupants' well-being, they are still cautious in proactively using the technology as a design tool. Research on AmI's slow adoption rate in architectural design is also limited. The existing research mainly focuses on opportunities and challenges [2, 8] rather than explaining why architects are hesitant to adopt AmI as an integral part of their design. Even though, there exists several studies on challenges of the IoT [14–17], very few reflect the opinions and behaviors of the architects and design professionals as the key players in design process. Therefore, the main goal of this study was to narrow this gap by examining the challenges that architects are facing to adopt AmI systems in design process.

3 Theoretical Model

The literature and our preliminary interviews with 29 architects and design professionals helped identify four constructs that may influence the intention to adopt AmI: perceived risk, perceived value, commitment to learning, and commitment to collaboration (see, [16] for details). This model is supported by the Technology-Organization-Environment literature [17–21] and provides a better understanding of why architecture firms are slow in adopting new technologies such as AmI. The hypotheses are presented in Fig. 1.

We identified the (perceived) value of AmI related to (a) data-driven or evidence-based design, (b) systematic design processes, and (c) design reputation or portfolios [16]. Architects can offer more realistic design solutions if they have access to factual sensor-generated evidence (e.g. traffic, occupancy, light, temperature). Access to this kind of data helps architects to not only solve functional problems such as layout but also effectively justify and communicate the design solutions. This also facilitates the collaboration between the architects, consultants and contractors. AmI can also improve the efficiency of design processes during site and requirements analysis since architects can refer to the data collected from similar spaces. Furthermore, architects can gain reputation by offering data-driven design solutions. This reputation can improve architects' design portfolios and attract more clients who are interested in (or require) data-driven design solutions. For example, an architect can show how her data-driven design can minimize energy consumption and thus save money for her client. This saving in turn can justify why the architect is charging a higher fee for her design. Considering these values, we posit:

> H_1: *Higher perception of AmI's application value among architects is associated with higher levels of AmI adoption in architectural design.*

Recent studies reported time and cost are the main considerations in adopting a new technology in building design [3]. By using new technologies, architects may lose their control over these important factors because of the resources that they need to spend on learning and incorporating new technologies. Addressing the new requirements imposed by new data can be both time-consuming and expensive. Utilizing AmI also imposes additional constraints on design outcomes. For example, the system-generated recommendations might contradict with the architects' choices and preferences during the schematic design phase. This limits architects in following the conventional design process and creatively formulate their design solutions. Creativity in design can be also limited by various regulatory obligations beyond building codes that are better observed by AmI than design professionals (e.g. sustainability and accessibility).

Perceived risks are also associated with design complexity. While AmI systems help with design simplification in some cases, the technology is not mature enough to handle complex design problems. For example, AmI may fall short in addressing unforeseen design issues or special design cases since the existing technologies have been programmed for standard applications such as monitoring than design.

The time needed to educate architects on how to use the technology and coordinate with technologists may increase the risk of project delay. Therefore, some architects may show little willingness in using AmI whilst they are under pressure for on-time project delivery. Likewise, project cost is as important as project time, if not more. Using AmI inevitably increase the cost of project because of the cost associated with training, equipment setups, data collection and data analysis. This investment does not appear to have immediate return (unless its related to energy consumption). Some factors influencing organizations' expenses include the cost of digital tools and setting up equipment such as sensors, the budget for team training, the cost of equipment maintenance, R&D budgets, and labor cost. Architects are not able to charge a higher service fee in many cases unless the client asks for a data-driven design solution. Therefore, charging a higher fee would negatively affect firm competitiveness and client retention rate. All these risks considered, we expected that the perceived risk of using AmI in the design process would affect its adoption rate. Therefore,

> H_2: *Higher perception of AmI's application risk among architects is associated with lower levels of AmI adoption in architectural design.*

Incorporating AmI is associated with upfront training costs for architects. Lack of training hinders the use of AmI. The organizational learning literature suggests that commitment to learning is one of the key drivers of new technology adoption [22].

Similarly, it is difficult to adopt AmI if there is no or limited willingness to learn about its applications even when the necessary training is provided for the design team. Therefore,

H_3: *Higher perception of AmI's application value among architects is associated with higher levels of commitment to learning about AmI.*
H_4: *Higher perception of AmI's application risk among architects is associated with lower levels of commitment to learning about AmI.*
H_5: *Higher commitment to learning (about AmI) among architects is associated with higher levels of AmI adoption in architectural design.*

Communicating the value and addressing the risks associated with AmI are not instrumentally favorable if they do not lead to a higher interest among the architects to learn about AmI. The positive perception of its value without this commitment is insufficient to drive adoption. In other words, commitment to learning can be an underlying mechanism justifying the relationship between perceived value and intention to adopt. The opposite scenario can be correct as well. The high perception of risk seems one of the reasons behind the low adoption rate, partially because it discourages actors from learning about AmI. Therefore,

H_{6a}: *Commitment to learning mediates the relationship between perceived value and intention to adopt.*
H_{6b}: *Commitment to learning mediates the relationship between perceived risks and intention to adopt.*

Collaboration between architects and other AmI stakeholders a play positive role in driving the adoption of this technology and achieving project goals [23]. This collaboration facilitates project coordination and above that innovation in architectural practice [24, 25]. Collaboration presents more impediments for small architectural firms who have less technical resources at their disposal [26]. In such organizations, it is uncommon to include IT professionals as design team members. However, when architects learn about the value of AmI in the design process, they may show a higher willingness to collaborate with IT professionals to actualize the values associated with AmI. On the contrary, the willingness to collaborate with IT professionals may decrease if the architects perceive no or limited value, or high risks, in employing AmI in design. It is unreasonable to expect that architects would effectively adopt AmI without collaboration with IT professionals. Commitment to collaboration is also crucial to establishing trust between architects and IT professionals. Therefore, we expect the perception of AmI value and risk to influence architects' commitment to collaboration with IT professionals, and in return, we expect that commitment to collaboration affects intention to adopt. Therefore,

H_7: *Higher perception of the value of AmI's application among architects is associated with higher levels of commitment to collaboration with IT professionals.*
H_8: *Higher perception of AmI's application risk among architects is associated with lower levels of commitment to collaboration with IT professionals.*
H_9: *Higher commitment to collaboration with IT professionals among architects is associated with higher levels of AmI adoption in architectural design.*

To benefit from AmI while mitigating its risks, architects may find considerable benefit in partnering with IT professionals. This partnership may help them to take a more active role in defining the applications of AmI, for example, in collecting and processing data or verifying the design assumptions. Architects may have a greater intention to adopt AmI partly because of the likelihood that their collaboration with IT professionals can improve the design process and enhance the end users' experience. Likewise, architects may find no value in adopting AmI when their perception of risk prevents them from collaboration with IT professionals. Therefore,

H_{10a}: *Commitment to collaboration with IT professionals mediates the relationship between perceived value and intention to adopt.*
H_{10b}: *Commitment to collaboration with IT professionals mediates the relationship between perceived risks and intention to adopt.*

Perceived Value
- Evidence-based Design
- Design Process (convenience)
- Design Portfolio

Perceived Risk
- Design Complexity
- Design Concept
- Design Cost
- Design Time

Mediation effects:
H_{6a}: PVL → CLR → INT
H_{6b}: PRS → CLR → INT
H_{10a}: PVL → CCL → INT
H_{10b}: PRS → CCL → INT

Control variables: Trust and Knowledge

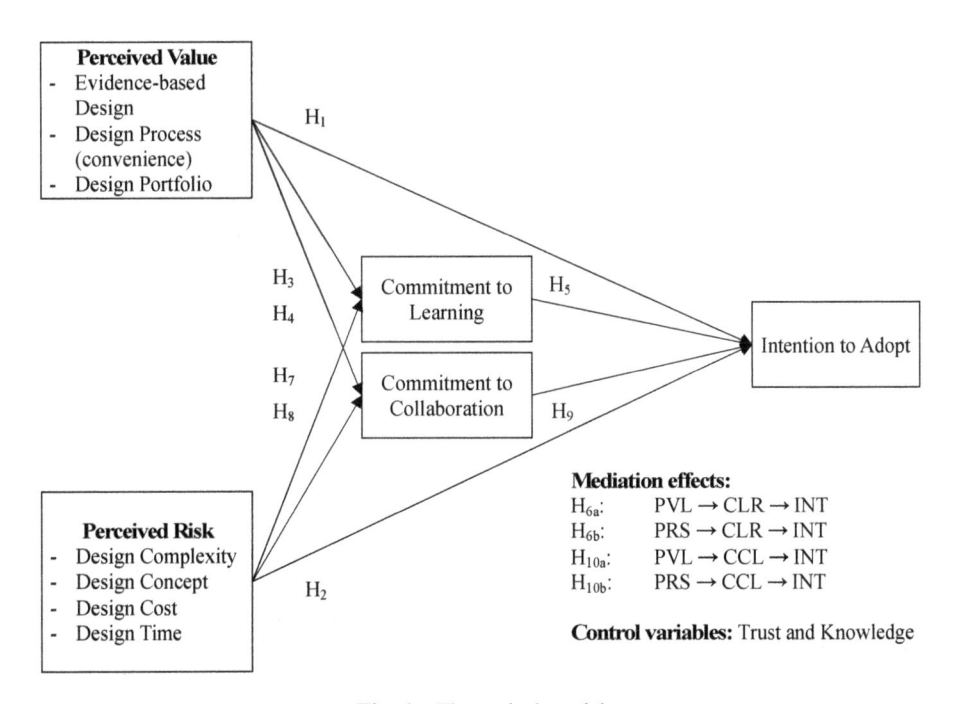

Fig. 1. Theoretical model

Lastly, we identified knowledge and trust as two important factors that may affect architects' perception of AmI and thus we included them as control variables in testing the proposed hypotheses. Trust is associated with architects' confidence in the capability of this technology to support decision making and design process. Knowledge is related to architects' understanding of this technology and its domain concept and applications.

4 Method

We used a field survey to gather data on the current adoption of AmI in architecture and validate our research model. The Partial Least Squares (PLS) modeling technique was used to evaluate our model [27]. SmartPLS [28] was employed to first validate our measurement model and then test the hypotheses [29, 30]. PLS modeling is chosen over other analytical methods because of its capability in the modeling of formative constructs [31–33].

Our literature review and preliminary interviews provided insight into how to contextually operationalize the constructs of interests [16]. The reflective measurement items for 'intention to adopt' were adapted from previous studies [32–34]. The measurements for the 'commitment to learning' and 'commitment to collaboration' constructs were respectively adapted from studies on firm innovation capability [22] and organizational partnership [35, 36]. The construct measurements for 'perceived value' and 'perceived risk' were newly developed as formative constructs based on our preliminary interviews to fully capture the context of this study. Perceived value was measured formatively by three items: evidence-based design, streamlined design process, rich design portfolio (while the fourth item, better user experience, was dropped after the pilot study). Four items were used to measure perceived risk as a formative construct: increasing design complexity, design time, and design cost as well as limiting creativity.

We first pre-tested the measurement model and then pilot tested the survey instrument before the field survey. The pre-test was designed to reveal respondent concerns and questionnaire issues. The survey questionnaire was circulated among 11 professionals familiar with the concepts of this study to solicit feedback on the wording and presentation of the questions. A pilot study in the form of an online survey was then conducted to evaluate the proper functioning of the instrument. The pilot study collected data from 70 professionals and helped to establish the initial instrument quality including the required reliability and validity for all the constructs.

The data for the field survey were collected from the members of three online communities where the members were invited via direct message to participate in our online survey. We used 236 responses after screening out incomplete data or data from unqualified respondents. To identify qualified respondents, participants were first asked to indicate their own experience with AmI technologies. After that, respondents were presented a set of Likert-type questions related to their perception of value and risk, their commitment to learning and collaboration, and their intention to adopt AmI systems. Finally, the respondents were asked to rate their own and their organization's trust and knowledge in AmI technologies as control variables. Demographic information (e.g., gender, age, education, position, employment) was lastly collected.

5 Results

5.1 Measurement Model

The evaluation of reflective constructs involved the test of reliability and internal consistency, factorability, convergent validity and discrimination validity). As shown in Table 1, the test of factor loading indicates acceptable item reliability [27]. Cronbach's alpha and the composite reliability of all the constructs are higher than 0.7, that suggested adequate internal consistency among the items [27]. The results also revealed adequate convergent and discriminant validity because (a) all Average Variance Extracted (AVE) values are higher than 0.50 [27]; (b) the square root of the AVEs were larger than the correlations of corresponding variables with the other variables [37]; and (c) the inter-construct correlations were well below the 0.90 threshold [27]. The test of common method bias using a full collinearity assessment also showed that the pathological VIFs were lower than the 3.3 threshold [38]. This results suggested the absence of common method bias [39].

Table 1. Psychometric properties of reflective constructs.

Reflective constructs	Loadings	α	CR	AVE	CCL	CLR	INT
Commitment to collaboration	CCL1: 0.825 CCL2: 0.841 CCL3: 0.861 CCL4: 0.808	0.854	0.901	0.696	**0.834**		
Commitment to learning	CLN1: 0.906 CLN2: 0.929 CLN3: 0.815 CLN4: 0.847	0.899	0.929	0.766	0.604	**0.875**	
Intention to adopt	INT1: 0.897 INT2: 0.871 INT3: 0.741 INT4: 0.855	0.863	0.907	0.711	0.627	0.588	**0.843**

We used PLS method with a bootstrapping to evaluate the indicators' validity and multicollinearity of formative measurements—measurements of perceived value and perceived risk. As Table 2 shows the weights and loadings of all the indicators are significant which suggest the indicator validity [27]. The indicators' weights represent the partialized effect of the subscales on the affordance construct, controlling for the effect of all other indicators [40]. The significances of weights indicated the importance of all items in measuring the perceived values and risks of AmI application in architecture. Multicollinearity was then calculated for both formative constructs by estimating the Variance Inflation Factor (VIF) of each indicator. All computed VIF values are well below the conservative threshold of 5.0 (tolerance of 0.20), suggesting that multicollinearity is not a threat to the validity of the study's findings [27].

Table 2. Loadings and weights of the formative items

Construct	Indicator	VIF	Loadings*	Weights*
Perceived values	PVL1	2.132	0.913	0.489
	PVL2	1.949	0.839	0.318
	PVL3	1.642	0.815	0.352
Perceived risks	PRS1	1.191	0.629	0.332
	PRS2	1.369	0.839	0.554
	PRS3	1.374	0.686	0.259
	PRS4	1.239	0.606	0.244

* $p < 0.001$ level.

5.2 Structural Model

We tested the hypotheses with controlled effects of knowledge of AmI and trust in AmI. To test the model, we first examined the significance of the direct effect of perceived value, perceived risk, commitment to learning, and commitment to collaboration on intention to adopt. The results revealed that the perceived values construct is positively associated with intention to adopt ($\beta = 0.266$, $p < 0.01$), commitment to learning ($\beta = 0.233$, $p < 0.05$), and commitment to collaborate ($\beta = 0.346$, $p < 0.001$). The findings thus supported hypotheses H_1, H_3, and H_7 (Table 3). The perceived risk construct also exerted significant negative impact on commitment to learning ($\beta = -0.406$, $p < 0.001$), commitment to collaboration ($\beta = -0.254$, $p < 0.01$) and intention to adopt ($\beta = -0.184$, $p < 0.05$). Therefore, the results supported H_2, H_4 and H_6 (Table 3). The findings also supported the significant positive effect of commitment to learning ($\beta = 0.170$, $p < 0.01$) and commitment to collaboration ($\beta = 0.321$, $p < 0.01$) on intention to adopt. Thus, the results supported hypotheses H_5 and H_9 (Table 3). These finding suggest that perceived value and perceived risks along with commitment to learning and commitment to collaborate all have a significant impact on intention to adopt ($R^2 = 0.57$).

Table 3. Results of the structural model assessment.

Dependent variable	Hypothesis	Support	β	R^2	Q^2
Intention to adopt	H_1: PVL → INT	Supported	0.266**	0.57	0.37
	H_2: PRS → INT	Supported	−0.184*		
	H_5: CLR → INT	Supported	0.170**		
	H_9: CCL → INT	Supported	0.321**		
Commitment to learning	H_3: PVL → CLR	Supported	0.233*	0.34	0.24
	H_4: PRS → CLR	Supported	−0.406***		
Commitment to collaborate	H_7: PVL → CCL	Supported	0.346***	0.30	0.19
	H_8: PRS → CCL	Supported	−0.254**		

* $p < 0.05$; ** $p < 0.01$; *** $p < 0.001$; ns = no significance β = path coefficients; R^2 = determination coefficient; Q^2 = predictive relevance (calculated by blindfolding).

The test of mediation effects was then carried out to examine the role commitment to learning and commitment to collaboration play in mediating the impact of perceived value and risk on intention to adopt. Following recent practices for testing indirect influence [41–43], we adopted the bootstrapping method as the more rigorous and powerful approach to assess the mediating role [41]. In this study, the 95% confidence interval of the indirect effects is obtained with 5,000 bootstrap resamples. The results of our analysis reveal that both commitment to learning and commitment to collaboration significantly carry the influence of perceived value and perceived risk on the intention to adopt. Specifically, the results confirmed the role of commitment to learning in mediating the relationship between perceived value and intention to adopt (H_{6a}: $\beta_{Indirect} = 0.040$, $CI = 0.005$ to 0.084) and the relationship between perceived risk and intention to adopt (H_{6b}: $\beta_{Indirect}$ = −0.069, CI = −0.142 to −0.013), as none of the bias-corrected 95% confidence intervals contains zero (Table 4). Similarly, the results confirmed the role of commitment to collaboration in mediating the relationship between perceived value and intention to adopt (H_{10a}: $\beta_{Indirect}$ = 0.111, CI = 0.031 to 0.207) and the relationship between perceived risk and intention to adopt (H_{10b}: $\beta_{Indirect}$ = −0.082, CI = −0.194 to −0.019). The total indirect effect of perceived value and perceived risk were respectively 0.151 ($p < 0.01$) and 0.151 ($p < 0.05$). Therefore, our data support H_{6a-b} and H_{10a-b} (Table 4). Lastly, the calculation of the Predictive Relevance (Q^2) in the presence of the control variables confirmed the importance of the proposed constructs in predicting intention to adopt (Q^2 = 0.307), commitment to learning (Q^2 = 0.239), and commitment to collaboration (Q^2 = 0.190).

Table 4. Results of the mediation assessment.

Mediator	Hypothesis	Indirect effect	Confidence interval	Supported
Commitment to learning	H_{6a}: PVL → CLR → INT	0.040	0.005 to 0.084	Supported
	H_{6b}: PRS → CLR → INT	−0.069	−0.142 to −0.013	Supported
Commitment to collaboration	H_{10a}: PVL → CCL → INT	0.111	0.031 to 0.207	Supported
	H_{10b}: PRS → CCL → INT	−0.082	−0.194 to −0.019	Supported

6 Discussion

The objectives of this study were to explain the adoption of AmI in architectural practice by accounting for architects' perceptions of value and risk associated with AmI systems. We also introduced and tested the importance of commitment to learning and commitment to collaboration as the mediators between perceived value and risk and intention to adopt. We then examined the relationships between perceived value and risks, commitment to learning and collaboration, and intention to adopt via a survey of architects. The findings of this study indicate that perceived value improves architects' commitment to learning as well as their commitment to collaboration. However, the higher the perceived risks, the lower the commitment to learning and commitment to collaborate. The comparison of effect sizes interestingly suggests that perceived value is a better predictor of commitment to collaboration while perceived risk is a better predictor of commitment to learning. This result highlights the importance of understanding both perceived value and perceived risk before planning to adopt AmI.

As hypothesized, perceived value and perceived risk are the predictors of intention to adopt. However, the comparison of effect size suggested that perceived value is a better predictor of intention to adopt than perceived risk. Moreover, commitment to learning and commitment to collaboration both positively influence the intention to adopt among architects. Interestingly, the effect size of commitment to collaboration was higher than the effect size of commitment to learning. That finding supports the importance of willingness to collaborate with IT professionals in this domain.

The results substantiated the mediating effects of both commitment to learning and commitment to collaboration. This result confirms the role of learning as a mechanism through which architects' perceptions of AmI systems partially influences their adoption behavior. This suggests that architects' perceptions play a part not only in driving their adoption decisions but also in shaping their learning and collaboration behavior. Our findings also suggest a more nuanced understanding of organizational learning in the context of new technology adoption and use that goes beyond traditional adoption drivers such as ease-of-use and utility. Therefore, a better understanding of action-oriented attitudes, such as commitment, can advance technology use research as well as technology design practice. In the next section, we discuss how this study contributes to practice as it pertains to AmI.

7 Implications

This study offers six implications for AmI systems adoption in architectural practice.

First, the starting point for architecture firms is to acknowledge the fact that the application of AmI in their project is a blend of conventional design and new technological approaches. Hence, firms require to offer new forms of training and commitment to change in order to better address this dual responsibility. AmI-related Projects' leaders including principal architects and project managers must be among those who have the experience to effectively collaborate with non-traditional consultants such as IoT specialists and sensor network engineers. Prior to proposing or initiating any project involving AmI systems, firms should invest in training their project team for effective coordination with IT professionals who may not necessarily be familiar with architectural design processes.

Second, it is essential for architecture firms to conduct a cost-benefit analysis during the pre-planning phase of each project to decide on the feasibility of AmI adoption. If the architecture firms allocate this analysis time up-front, they can better address the risks associated with project cost, time, and quality. For example, AmI adoption for some projects may not seem feasible when it requires significant financial investments and the return on these investments is not guaranteed in the short term. Likewise, AmI adoption may add unnecessary complexity to a simple project. Hence, conducting a detailed cost-benefit analysis is necessary for architects to learn about the potential benefits and therefore encourage them to adopt AmI and collaborate with IT professionals. Therefore, this approach seems viable either when conventional methods have failed to offer an optimal solution or when sufficient data are available from projects of similar scope. Allowing time for pre-planning would help architects to better estimate the value and risk of AmI before application.

Third, to explore potential avenues for AmI systems adoption, it is essential for architects to work in a team with adequate knowledge about data-driven design processes. To deliver a successful data-driven design solution, the entire process must be clearly defined and well-executed. This can be achieved through a shared framework and language among the project teams and consolidation of their various and sometimes potentially contradictory perspectives. Developing appropriate and successful AmI adoption requires more focused teams with enough knowledge of new technologies and available resources. For example, dividing up human resources into smaller and more focused teams is more effective to develop a successful AmI-related project. This approach could minimize unnecessary added complexity to a project and provide leading architects with a greater degree of control over the process.

Fourth, architects need the help of IT professionals to select the most appropriate AmI technologies for a project. Different systems require different expertise and implementation techniques and they provide different set of data. The type of technology to be adopted may have significant impact on a project's outcome as it may change the direction of the project entirely (e.g. multiple sensor combinations, data collection methods, and communication protocols). Hence, it is important to effectively compare and contrast different systems in terms of utility, reliability, and viability and to clearly communicate these analyses with architects and other decision-makers.

Fifth, establishing trust in AmI systems is significant not only within the project team but also among the firm's top management who are key players in supporting this new approach. One method to establishing trust is to highlight and review the best practices. By identifying successful AmI-supported projects, architects can provide evidence about whether and how the introduction of AmI systems would make economic sense. When trust is established, architects can benefit from both management and client support to secure the required resources to effectively carry out the project.

Lastly, flexibility is a key to effectively incorporate AmI technologies in architectural practice. Architects should be flexible throughout the duration of a project in terms of project planning, design options, design process, and design outcome. They should understand the trade-off between a data-driven design and the conventional design process. Like adopting any new technology, this degree of flexibility encourages architects to embrace new opportunities to better address their clients' needs and expectations.

8 Conclusion

Architects have been constantly faced technological innovations. While each technological transformation creates new challenges and offers new opportunities, some have more profound impact on the practice. AmI is among technologies that their prevalence is not widespread, and its potential remains untapped in architectural practice. This study was an attempt to narrow this gap by examining the reasons behind the low adoption of AmI in architecture.

This study contributes to both theory and practice pertaining to the technology adoption literature by shedding light on the risk-return trade-off in this field. The findings offer a path to future theoretical developments to encourage effective applications of AmI by better understanding of its values and risks as well as architects' commitment to learning about this new technology and commitment to collaboration with a new group of consultants. From the practical standpoint, our findings provide a guiding framework to educate architects on AmI applications.

Future studies can expand this investigation to lead the development of AmI and its applications. For example, the analysis of best-practices can help with the identification of success factors in AmI-supported projects. Studying specific use cases will be also instrumental to provide a more practical guide to AmI adoption. Another possible area of research is the protection of privacy. AmI cannot deliver value added services without collecting data about individuals' behavior, preferences and activities. This can lead to data security and privacy challenges that may discourage architects from fully adopting the technology. Further research is thus needed to identify the risks associated with data management and define necessary procedures to protect privacy and access control.

References

1. Johnson, R.E., Laepple, E.S.: Digital innovation and organizational change in design practice (2003)
2. Intrachooto, S.: Technological innovation in architecture: effective practices for energy efficient implementation. Massachusetts Institute of Technology, USA (2005)
3. Chaos Group: 2017 Architectural Visualization Technology Report (2017)
4. Khaddaj, M., Srour, I.: Using BIM to retrofit existing buildings. Procedia Eng. **145**, 1526–1533 (2016). Pre-Project Planning for sustainable urban development View project Dispute Resolution View project
5. Minoli, D., Sohraby, K., Occhiogrosso, B.: IoT considerations, requirements, and architectures for smart buildings – energy optimization and next generation building management systems. IEEE Internet Things J. **4**, 269–283 (2017)
6. Zhou, B., et al.: Smart home energy management systems: concept, configurations, and scheduling strategies. Renew. Sustain. Energy Rev. **61**, 30–40 (2016)
7. Neuhofer, B., Buhalis, D., Ladkin, A.: Smart technologies for personalized experiences: a case study in the hospitality domain. Electron. Mark. **25**(3), 243–254 (2015)
8. McIntyre, L.: Smart technologies in buildings: exploring the conflicts and opportunities for health and wellbeing, September 2018
9. Hui, T.K.L., Sherratt, R.S., Sánchez, D.D.: Major requirements for building Smart Homes in Smart Cities based on Internet of Things technologies. Future Gener. Comput. Syst. **76**, 358–369 (2017)
10. IBM: TRIRIGA. https://www.ibm.com/products/tririga
11. Verdanix: Siemens and IBM Collaborate to Enhance Smart Buildings with IoT (2016)
12. Chen, Y.-K.: Challenges and opportunities of Internet of Things. In: 17th Asia South Pacific Design Automation Conference, pp. 383–388 (2012)
13. Spencer, B.F., Ruiz-Sandoval, M.E., Kurata, N.: Smart sensing technology: opportunities and challenges. Struct. Control Health Monit. **11**(4), 349–368 (2004)
14. Chong, C., Kumar, S.P.: Sensor networks: evolution, opportunities, and challenges. Proc. IEEE **91**(8), 1247–1256 (2003)
15. Mukhopadhyay, S.C., Suryadevara, N.K.: Internet of Things: challenges and opportunities. In: Mukhopadhyay, S.C. (ed.) Internet of Things. Smart Sensors, Measurement and Instrumentation, pp. 1–17. Springer, Cham (2014)
16. Abhari, M., Abhari, K.: Design with perfect sense: the adoption of smart sensor technologies in architectural practice. In: Hawaii International Conference on System Sciences (2019)
17. Leung, D., Lo, A., Fong, L.H.N., Law, R.: Applying the Technology-Organization-Environment framework to explore ICT initial and continued adoption: an exploratory study of an independent hotel in Hong Kong. Tour. Recreat. Res. **40**(3), 391–406 (2015)
18. Racherla, P., Hu, C.: eCRM system adoption by hospitality organizations: a Technology-Organization-Environment (TOE) framework. J. Hosp. Leis. Mark. **17**(1–2), 30–58 (2008)

19. Chau, P.Y.K., Tam, K.Y., Tam, K.Y.: Factors affecting the adoption of open systems: an exploratory study. MIS Q. **21**(1), 1–24 (1997)
20. Oliveira, T., Martins, M.: Literature review of information technology adoption models at firm level. Electron. J. Inf. Syst. Eval. **14**(1), 110–121 (2011)
21. Jia, Q., Guo, Y., Barnes, S.J.: Enterprise 2.0 post-adoption: extending the information system continuance model based on the technology-Organization-environment framework. Comput. Hum. Behav. **67**, 95–105 (2017)
22. Calantone, R.J., Tamer, C.S., Yushan, Z.: Learning orientation, firm innovation capability, and firm performance. Ind. Mark. Manag. **31**, 515–524 (2002)
23. Vermesan, O., Friess, P.: Internet of Things: Converging Technologies for Smart Environments and Integrated Ecosystems. River Publishers, Gistrup (2013)
24. Guy, S., Farmer, G.: Reinterpreting sustainable architecture: the place of technology. J. Archit. Educ. **54**(3), 140–148 (2001)
25. Chiu, M.-L.: An organizational view of design communication in design collaboration. Des. Stud. **23**(2), 187–210 (2002)
26. Ramilo, R., Bin Embi, M.R.: Critical analysis of key determinants and barriers to digital innovation adoption among architectural organizations. Front. Archit. Res. **3**(4), 431–451 (2014)
27. Hair, J.F., Hult, G.T.M., Ringle, C.M., Sarstedt, M.: A Primer on Partial Least Squares Structural Equation Modeling (PLS-SEM). SAGE Publications, Thousand Oaks (2013)
28. Ringle, C., Wende, S., Becker, J.: SmartPLS 3. SmartPLS GmbH, Boenningstedt (2015)
29. Chin, W.W.: The partial least squares approach to structural equation modeling. Mod. Methods Bus. Res. **295**, 295–336 (1998)
30. Hair, J.F., Ringle, C.M., Sarstedt, M.: PLS-SEM: indeed a silver bullet. J. Mark. Theory Pract. **19**(2), 139–152 (2011)
31. Wetzels, M., Odekerken-Schröder, G., van Oppen, C.: Using PLS path modeling for assessing hierarchical construct models: guidelines and empirical illustration. MIS Q. **33**, 177–195 (2009)
32. Chen, I.Y.L.: The factors influencing members' continuance intentions in professional virtual communities a longitudinal study. J. Inf. Sci. **33**(4), 451–467 (2007)
33. Bhattacherjee, A., Premkumar, G.: Understanding information systems continuance: an expectation-confirmation model. MIS Q. **25**(3), 351–370 (2001)
34. Zhang, Y., Fang, Y., Wei, K.-K., Chen, H.: Exploring the role of psychological safety in promoting the intention to continue sharing knowledge in virtual communities. Int. J. Inf. Manag. **30**(5), 425–436 (2010)
35. Wu, F., Cavusgil, S.T.: Organizational learning, commitment, and joint value creation in interfirm relationships. J. Bus. Res. **59**(1), 81–89 (2006)
36. Salam, M.A.: The mediating role of supply chain collaboration on the relationship between technology, trust and operational performance: an empirical investigation. Benchmarking **24**(2), 298–317 (2017)
37. Fornell, C., Larcker, D.F.: Structural equation models with unobservable variables and measurement error: algebra and statistics. J. Mark. Res. **18**(3), 382–388 (2006)
38. Kock, N.: Common method bias in PLS-SEM: a full collinearity assessment approach. Int. J. e-Collab. **11**(4), 1–10 (2015)
39. Kock, N., Lynn, G.S.: Lateral collinearity and misleading results in variance-based SEM: an illustration and recommendations. J. Assoc. Inf. Syst. **13**(7), 546–580 (2012)
40. Cenfetelli, R.T., Bassellier, G.: Interpretation of formative measurement in information systems research. MIS Q. **33**(4), 689–707 (2009)

41. Preacher, K.J., Hayes, A.F.: Asymptotic and resampling strategies for assessing and comparing indirect effects in multiple mediator models. Behav. Res. Methods **40**(3), 879–891 (2008)
42. Rucker, D.D., Preacher, K.J., Tormala, Z.L., Petty, R.E.: Mediation analysis in social psychology: current practices and new recommendations. Soc. Pers. Psychol. Compass **5**(6), 359–371 (2011)
43. Hayes, A.F.: Introduction to Mediation, Moderation, and Conditional Process Analysis: A Regression-Based Approach. Guilford Press, New York (2013)

Photovoltaic Mobile System Design for Non-interconnected Zones of Meta's Department

Obeth Romero[✉]

Corporación Universitaria del Meta, Cra 32 # 34-76, Villavicencio, Colombia
r_obeth@hotmail.com

Abstract. The latest generation technologies and information systems are currently considered as a fundamental resource for the economic and social growth of any nation; aware of the present events and transformations on the planet, as a result of adverse changes in the environment, and in the quest of contributing to the mitigation of climate change through the design and development of more complex machines through electronic, mechanical and electronic means, that generate lower impact on the environment focused on the design of an automated mobile photovoltaic system to supply of energy to populations located in geographical areas or places of difficult access to bring electricity, and which are not connected to any electrical network. Taking advantage of alternative sources of energy based on the use of surrounding natural resources in each region, favors the universal access to electricity and promotes the use of solar energy as a source of electricity, a fundamental element of photovoltaic energy that can be automated, systematized with monitoring and control tools.

Keywords: Arduino · Internet of Things · Energy · Photovoltaic · Programming

1 Introduction

Renewable energies have become technologies for the development of countries, many nations have adopted the use of them as state policy. Agreements and protocols arranged to mitigate the negative environmental impact on the environment associated with energy consumption from fossil energy. Where the concepts of centralization are currently unattached from our day by day, this referring to the large electric generating plants and in the same way to the transmission in the electric networks, taking into account that in the transmission there is loss of energy, due to the great distances to cover carrying the electric service, therefore what is being looked for is to reduce the gaps in the non-interconnected areas that some Latin American and Caribbean countries have. On the other hand, distributed systems become energy solutions for remote sites where electrical networks cannot arrive due to the different given situations that may arise, whether due to the topology or another factor [1].

The natural resources of each country become an important and profitable factor for the production of electric energy, starting by identifying the natural resources of the region (sun, water, air, etc.), in such a way that the beneficial characteristics of the regions become an alternative sources of energy that meet the energy demand of each country (photovoltaic, wind, thermal and hydraulic). Investing in access to energy also means more lit schools, functioning health clinics, waterpumps for and sanitation, cleaner indoor air, faster food processing and more income-generating opportunities, according to the UNDP action4energy initiative (United Nations Development Program, 2018) [1].

The generation and transmission of electrical energy is not far from systematization processes, especially in the development of computer technologies, automation and digital processes, that is, they are tools that facilitate work in energy processes.

2 Materials and Methods

Materials

12 V 50RP 47 kg high torque Electrical Gear Train
100 W polycrystalline solar panel
100 Ah battery - 12 V
Drive L298N for DC motors
600 W inverter
USB Malecable
Male-Female USB cable extension
Power cables
Arduino Mega Card

Methods

The design of a mobile photovoltaic system for non-interconnected areas in the Meta department is developed in three fundamental phases.

A first phase determines the environmental radiation conditions - UV rays.

A second phase is to consult the meteorological databases available specifically for radiation values and analyze their behavior in recent years.

In this second stage an electro-mechanical mobile system for the photovoltaic structure is designed.

Outline a mobile pilot system with electrical and electronic devices that control the movement of solar panels.

In the last stage an algorithm is developed in Arduino sketches for the movement of the panels.

The electronic devices are programmed to interact with the system, and in this way the information provided executes the movement of the panel.

3 Results

The radiation data is obtained from the solar atlas of the Institute of Hydrology and Environmental Studies - IDEAM. Data acquired from the website which were consulted for their annual analysis of daily average horizontal global irradiation, and classified in the graph by their monthly values, unit of measurement in kwh/m^2 [2].

Photovoltaic Mobile System Design for Non-interconnected Zones 253

The data in Fig. 1 is interpreted and it is determined that the Meta department has an approximate solar radiation range of 4.5 out of 5.0 [2].

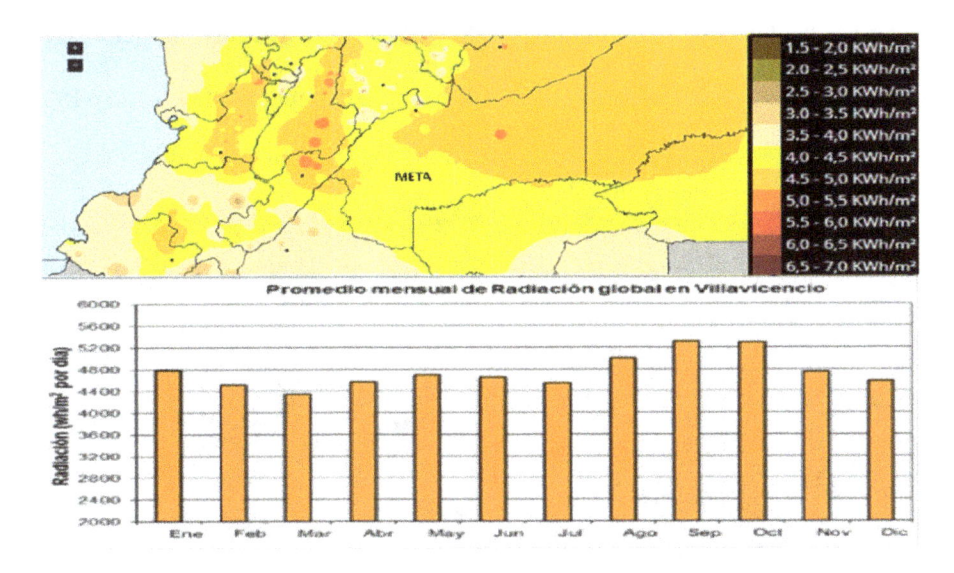

Fig. 1. Goal radiation [2]

Design Mobile Structure
What we are looking for in this section is to provide a practical mobile system for the displacement of solar panels, with a mobile structural design [8].

In Fig. 2, the structural prototype is designed from the measurements of the other devices that interact with the mobile structure, specifically measures of width, length and weight of the solar panels as batteries and also each of the components that are part of the photovoltaic system. The transport trailer measures 120 cm long by 60 cm wide and 30 cm height and with 14″ wheels: 36 cm high, 18.5 cm width [4].

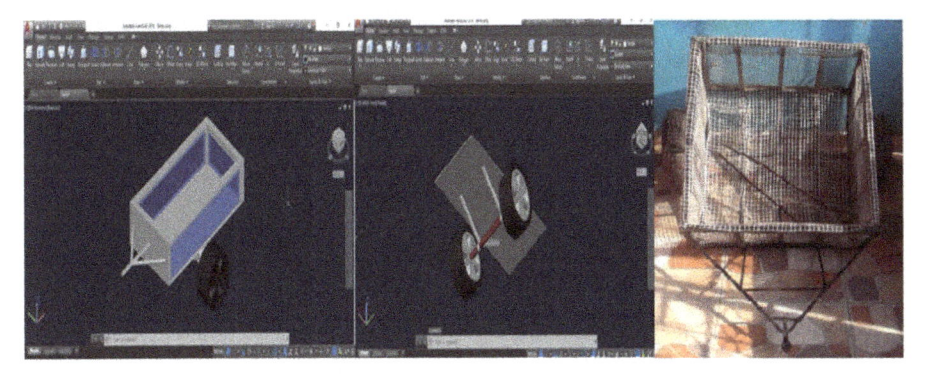

Fig. 2. Mobile structure design.

4 Design Mobile Photovoltaic Structure

The system is designed to lift a panel 30–40 cm high, as shown in Fig. 3; the gearmotor has the function of moving the panel vertically and is articulated to the trailer structure, controlling the high by the Arduino mega card panel. Photovoltaic devices such as batteries and inverters are located in the trailer as a photovoltaic system [6].

Fig. 3. Mobile photovoltaic system - with automated vertical movement.

5 System Architecture

The automated photovoltaic system is based on devices connected through a Megamicroprocessor card communicating different hardware. A high-level system architecture, consisting of hardware and software, is presented. The presented system was conceived using hardware capable of supporting different devices and flexible enough to adapt adaptive software [3].

Hardware

As illustrated in Figs. 2, 3 and 4, the hardware devices adopted for the implementation of the mobile photovoltaic system consist of an MEGA 2560 Arduino card, a portable computer with processing and memory capacity, which supports programmable programs, input-output ports, and uses standard peripheral gateways, as the DC motor allows vertical movement [7].

Fig. 4. System architecture

Software

Important software for the proposed architecture is presented. They will be mentioned below.

Language C - Arduino Sketches

This programming language controls the motor lifting movements of the panel; and it allows communication and configuration of the hardware devices involved, such as the generation of the code to control the mobile photovoltaic system [9].

In Fig. 5, the programming code where the motor is parameterized is displayed, the code is designed so that the system responds to the vertical up and down mobility when a computer key is activated [5].

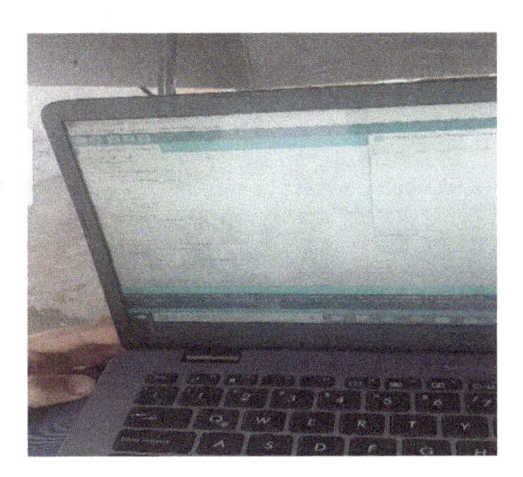

Fig. 5. Arduino sketches code

6 Conclusions

The developed system so far, meets the characteristics and purpose of the proposal, providing the user with displacement of ubiquity and obtaining electricity in places where there is no electric service. However, the automation of the system can be improved or considered as a pilot model. In addition, additional functionalities for the system are considered, such as the possibility of creating rule sets directly from programming, and a tool for the 4.0 industry as additional artificial intelligence.

The system can be scalable over time as it can be adjusted to the number of sensors or devices added and desired, as a matter for continuous improvement, such as converting it into a web information system, storing and controlling the information obtained from the different devices or sensors.

References

1. UN program for development, 24 de julio de 2018. http://www.undp.org/content/undp/es/home/presscenter/articles/2015/02/24/el-pnud-el-bid-y-la-cepal-se-unen-para-impulsar-el-acceso-a-la-energ-a-sostenible-en-am-rica-latina-y-el-caribe.html
2. Instituto de Hidrología y Estudios Ambientales – IDEAM. http://atlas.ideam.gov.co/visorAtlasRadiacion.html
3. Le Vinh, T., Bouzefrane, S., Farinone, J.-M., Attar, A., Kennedy, B.P.: Middleware to integrate mobile devices, sensors and cloud computing. Procedia Comput. Sci. **52**, 234–243 (2015)
4. Pease, S.G., Trueman, R., Davies, C., Grosberg, J., Yau, K.H., Kaur, N., Conway, P., West, A.: An intelligent real-time cyber-physical toolset for energy and process prediction and optimisation in the future industrial Internet of Things. Future Gener. Comput. Syst. **79**, 815–829 (2018)
5. Suarez, J., Quevedo, J., Vidal, I., Corujo, D., Garcia-Reinoso, J., Aguiar, R.L.: A secure IoT management architecture based on information-centric networking. J. Netw. Comput. Appl. **63**, 190–204 (2016)
6. Mahbubur Rahman, M., Selvaraj, J., Rahim, N.A., Hasanuzzaman, M.: Global modern monitoring systems for PV based power generation: a review. Renew. Sustain. Energy Rev. **82**, 4142–4158 (2018)
7. Gómez, A., Cuiñas, D., Catalá, P., Xin, L., Li, W., Conway, S., Lack, D.: Use of single board computers as smart sensors in the manufacturing industry. Procedia Eng. **132**, 153–159 (2015)
8. Sanchez, M.A.: Energia Solar Fotovoltaica. Editorial Limusa, España (2010)
9. Luca, J.I.: Programación de Arduino Introducción a Sketches. ediciones Anaya Multimedia, España (2018)

An IoT Based e-Health Platform Using Raspberry Pi

El-Hadi Khoumeri[✉], Rabea Cheggou, and Kamila Ferhah

Ecole Nationale Supérieure de Technologie, Algiers, Algeria
{elhadi.khoumeri,rabea.cheggou,kamila.ferhah}@enst.dz

Abstract. In our modern world, millions of people die each year because of a lack of information about their health. In addition, according to the WHO (World Health Organization) in 2020 most of these deaths in the world will be due to diseases chronic diseases such as diabetes, hypertension or cardiovascular diseases [1]. The evolution of these pathologies requires many and expensive care. Rising costs in health systems could be reduced, if this gave more attention to disease prevention through regular assessment of the health status of patients and their treatment in the early stages, with the use of e-health. e-Health is about distance care practice. It relies on telecommunication infrastructures or the Internet of Things to allow the exchange or collection of medical information from patients to remote health centers. This allows for example the intervention on isolated patients or cannot easily access health centers. More specifically, we talk about telemedicine when a doctor performs a remote consultation. Medical remote assistance refers to the case where a distant colleague for a diagnosis helps a doctor or the possibility of performing a surgical operation at distance, still in the experimental state. She made spectacular achievements, especially college islands they involve infrastructure complex. The goal of this project is to create an e-health platform for the public through the Internet of Things, which allows people and objects to be connected anywhere, anytime, anywhere. Anyone and anything, ideally using any network.

Keywords: e-Health · Raspberry Pi · Electrodiagram · Electromyogram · Spo2 · Airflow

1 Internet of Things in the Medical Field

1.1 Introduction

The Internet of Things (IoT) is at the center of consumer and business attention. Moreover, for good reason, the promise of a world populated by connected objects offers countless opportunities through the possibilities offered, both as a user and as a service provider [2]. The applications of the Internet of Things result in many "new or improved" practical uses that significantly affect the lives of individuals, businesses and communities. The expected potential benefits facilitate its adoption by this diversity of users [3].

Today, hospitals such as clinical experiencing increases without continually sick, who regularly consult with specialized medicines, patients travel far to make cost-intensive, time-consuming consultations between round trips. The need to keep elderly patients at home and very interesting from a budgetary point of view in patient health. For this, a low-cost embedded solution is necessary. We try to detail our solution in what follows.

1.2 Proposed Solution

Improving human health and well-being is the ultimate goal of any economic, technological and social development.

The solution lies with the Internet of Things, no need to move, no need to have another doctor, no more infernal waiting in hospitals, and all done in anonymity.

Devices connected to the Internet, presented to patients in different forms, make it possible to follow information on the state of health of the patients. This creates an opening that allows smart devices to provide accurate data, reducing the need for direct interaction, between the patient and their doctor.

With better understanding, providers can improve patient care, chronic disease management, hospital administration and supply chain efficiencies to provide medical services to the most people at reduced costs [4].

IoT is emerging technology breakthrough will provide promising solutions for healthcare, creating a more revolutionary archetype for the healthcare industry built on a privacy/security model [5].

In general, as shown Fig. 1, e-health solution powered by the IoT has the following features: Surveillance and follow-up (patient supervision, personal care in case of chronic illness, elderly supervision or well-being and preventive care).

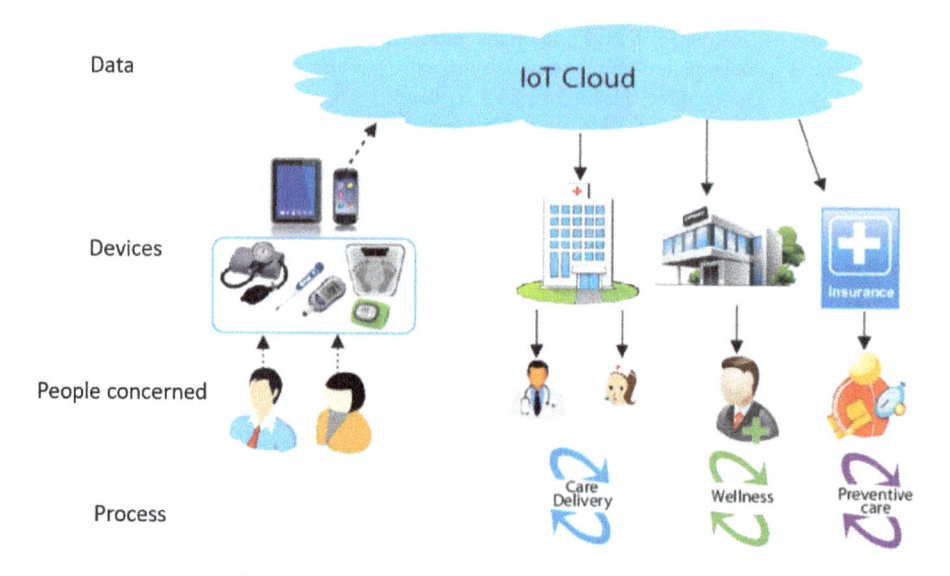

Fig. 1. e-Health through the Internet of Things [6]

- Remote service.
- Information management.
- Inter-organizational integration.

2 Related Works

In recent years, a renewed interest seems to have emerged in the field of telemedicine with the development of several projects more or less advanced around this theme [7]:

- SCAD v2 project (remote cardiological follow-up), deployed in Basse Normandy in France, supported by the University Hospital of Caen and Professor Grollier; an evolution of SCAD was reserved only to insufficient cardiac. Today this device allows for interactive educational monitoring and uses remote monitoring for diabetes, respiratory or renal insufficiency, severe obesity, etc. thanks to new connected equipment [8].
- OSICAT project (Optimization of outpatient surveillance for cardiac insufficiency by tele-cardiology), based on 12 local investigative centers, and coordinated by the University Hospital of Toulouse and Prs Galinier and Pathak. This project consists in strengthening the management of patients suffering from chronic diseases combining comprehensive care, therapeutic education, daily collection of physiological data and telephone follow-up by doctors or specialized nurses [9].
- In Sweden, Yan Hu and Sara Eriksén have proposed a hybrid cloud conceptual model for chronic home care, discussing possible future opportunities and challenges for applying cloud models with the huge population in order to provide care for patients with chronic home diseases [10].
- The e-care project, conceived as part of the French government's national program in 2013, is set up by the University Hospital of Strasbourg. The main objective of the project is to optimize the follow-up of the patients, by detecting the signs precursors of cardiac decompensations, by a system of telemedicine, associated with tools of motivation and education, which will make it possible to decrease the number of re-hospitalizations [11].

2.1 Telemedicine in Algeria

The project "Telemedicine for remote regions in Algeria", operational since December 2007, led and led by the Center for Advanced Technology Development (CDTA) of Baba-Hassen (Algiers), in collaboration with the Ministry of Health, population and hospital reform, Algeria Telecom Satellite and funded by the Research Center for International Development (IDRC) [12].

The project involves the implementation of a pilot network that interconnects two selected hospitals and a health center in different regions of the country, with the help of new information and communication technologies and satellite systems, a Telemedicine platform between the hospitals of Birtraria and Ouargla Hospital and the hospital center of Hadjira were born (see Fig. 2).

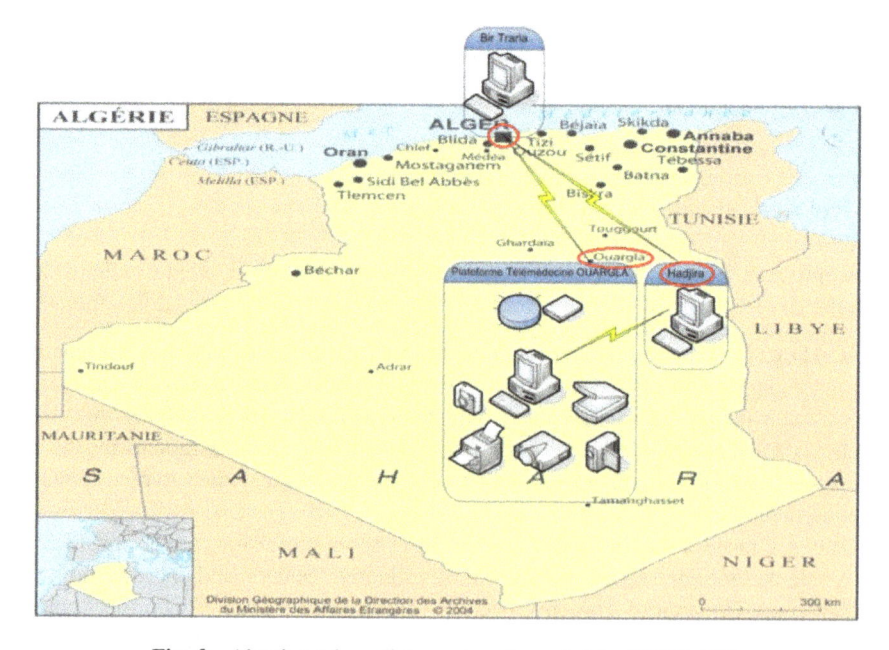

Fig. 2. Algerian telemedicine network created by CDTA [15]

Telemedicine has allowed patients in Ouargla, to follow up for several weeks until their cures. Better yet, surgeries took place online in Ouargla, while being followed and supervised from Algiers by professors from Birtraria hospital, which made it possible to avoid any complications and especially allowed the patients of Ouargla not to aggravate their cases by moving north [13].

So far, no application where remote monitoring platform has been developed in Algeria, we find only applications developed exclusively for health professionals, such as remote training and telediagnosis performed by the Advanced Technology Development Center (CDTA).

These applications still require the assistance of doctors or nurses (cooperation, meeting a distance between doctors), not a direct relationship between Doctors/patients, which should often plan and organize specific dates and times to make conferences, adds to this the postponement of these conferences due to power failure or internet connection problem which generates another appointment for the patient.

The application we propose is a medical telemonitoring platform that does not require a physical presence with patients, but rather the use of sensors connected to the patient who transmit data to doctors to diagnose the patient without even talking to him. In the next chapter, we detail a little more our platform.

3 Hardware

Before the realization of our platform, we will present in this chapter the different components of our application: Raspberry Pi 2 card, bridge, Arduino Shields and the sensors used.

3.1 The Raspberry Pi Card

The Raspberry Pi is a small computer the size of a bankcard. It was designed by a not-for-profit educational foundation to help you discover the world of computing from a different angle [14].

The heart of this computer is an FPGA (Broadcom 2835) incorporating an ARM11 processor clocked at 700 MHz and many peripherals.

They can be, directly connected to a classic HMI, mouse/keyboard/HDMI display or composite video, however like any Linux computer, Raspberry Pi can integrate its own development tools and a human-machine interface based on controllable SSH from another computer via Ethernet or WIFI [15].

The expansion connector supports parallel I/O as well as most communication buses. A particularly economical and powerful support can be easily, implemented in small systems requiring access to the physical world by sensors/actuators having digital interfaces (see Fig. 3).

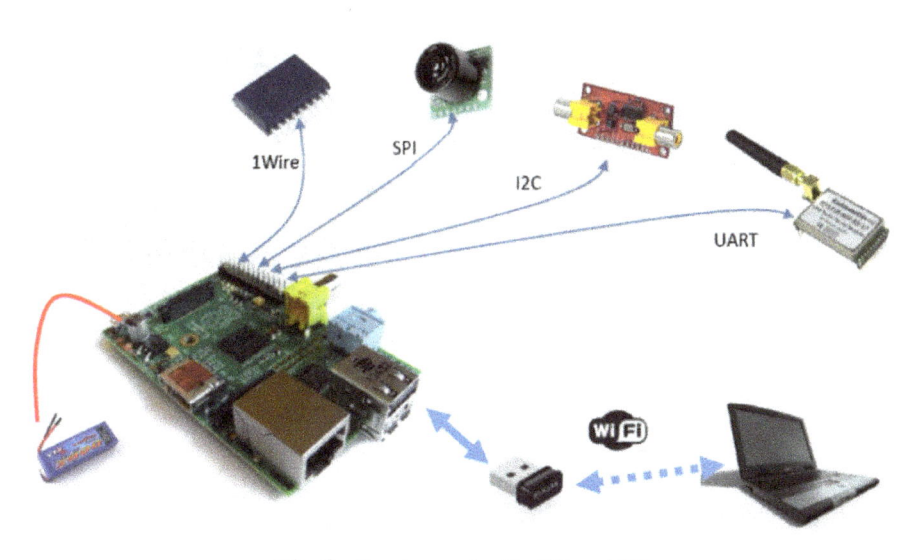

Fig. 3. Raspberry pi with different I/O

Uses of the Raspberry Pi Card
The potential applications of this tool are without end. A little tour of the internet leaves no doubt about the possibilities of offered developments. It allows use in office with open office, internet access with a web browser, learning programming with Scratch,

Python and Minecraft Pi or Rubyet Sonic Pi. Further, there is the completely world of free utilities and free under GNU/Linux. To give some interesting quick examples:

- You can create your media center with OSMC.
- We can do - retrogaming - with Recalbox.
- We can design his robot.
- You can take pictures of the Stratosphere.
- You can build your own weather station.
- It is possible to monitor from a distance his connected Henhouse.

Finally, we can even send his scientific experiments in Space! In the spirit of the recent week of a - Code Hour - in the AstroPi project, two Raspberry Pi, have been sent to the International Space Station with sensors and code created during a contest by children. Primary and secondary education in Great Britain [16]. What a dream!

3.2 The Raspberry Pi/Arduino Bridge

In our project, we will use the product of "Cooking-Hacks", an invention of the Libelium wireless communication company. This matching circuit we will simplify the task, as it allows connecting Shields designed for Arduino directly on Raspberry Pi (see Fig. 4).

Fig. 4. The Raspberry Pi/Arduino Bridge

A dedicated library allows us to use and control the Analog, Digital, UART, SPI and I2C I/O that are connected to the GPIO pins of the Raspberry Pi.

3.3 The E-Health Shield Platform

Libelium the Spanish company has developed the platform "e-Health Shield Sensors" for the educational community and developers of embedded systems and Internet of Things. This platform is dedicated to biomedical applications using different sensors needed for monitoring the physiological signs of the human body [17]. Among the information that we offer these sensors (see Fig. 5), are:

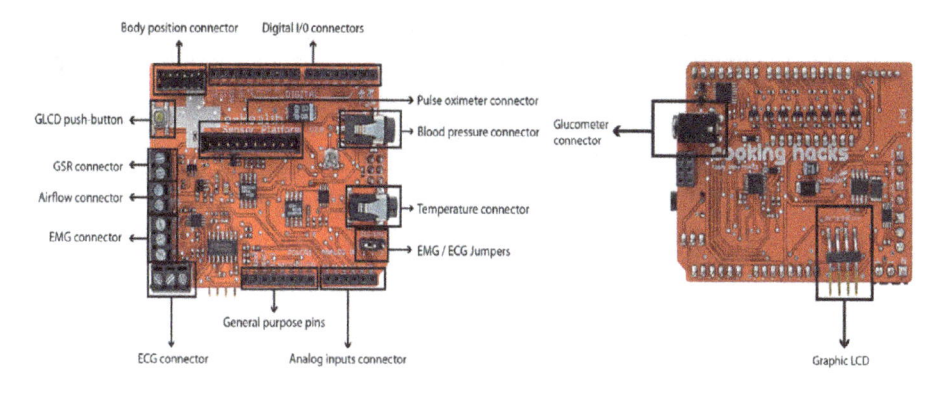

An IoT Based e-Health Platform Using Raspberry Pi 263

Fig. 5. Ports of different sensors on the E-Health Shield [17]

- The electrocardiogram (ECG), glucometer,
- The galvanic response of the skin (GSR - perspiration),
- Pulse, oxygen in the blood (SpO2),
- The air flow (breathing),
- Body temperature,
- The blood pressure (sphygmomanometer),
- The position of the patient (accelerometer) and the muscle/electromyography sensor (EMG).

The data provided by E-Health Shield can be sent to the cloud in order to make a permanent storage or use in real time to monitor or diagnose the state of a patient by sending the data directly to a laptop or smartphone the attending physician.

3.4 The ECG, EMG, SpO 2 and AirFlow Sensors

The electrocardiogram/electromyogram (ECG/EMG) sensor shown in Fig. 6, measures the electrical potential generated by the contraction of the heart or muscle.

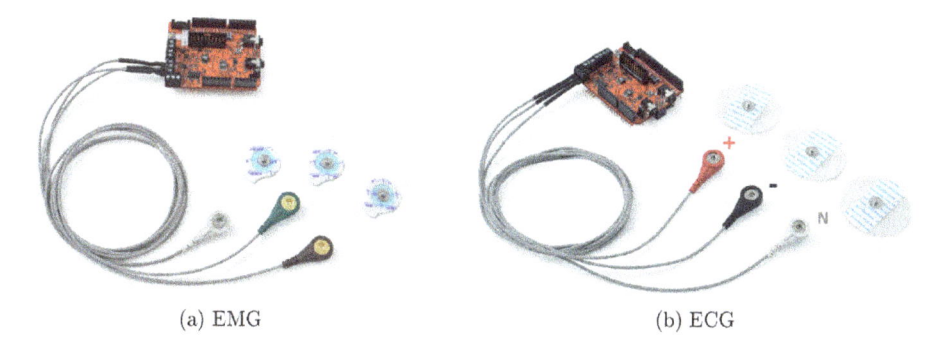

(a) EMG (b) ECG

Fig. 6. Electromyogram/electrocardiogram (EMG/ECG) sensor

The function of the Pulsed Oxygen Saturation (SpO2) sensor is to represent the percentage of oxygenated hemoglobin in relation to the total amount of hemoglobin in the blood.

The sensor AirFlow or the sensor flow nasal air is a device used to monitor a patient's airflow requiring respiratory assistance (see Fig. 7).

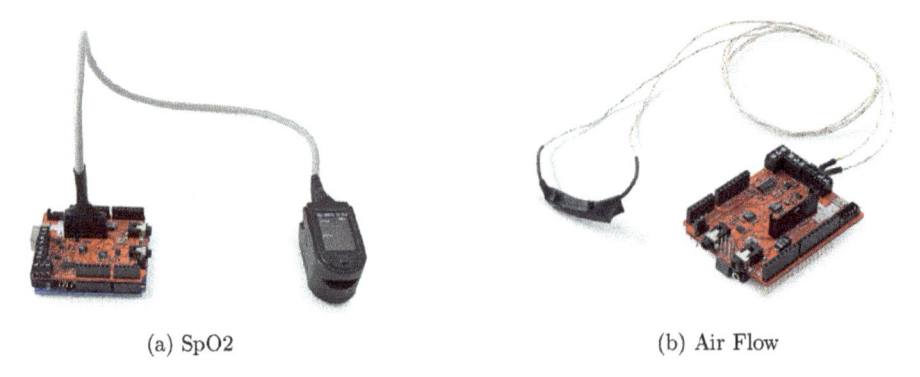

(a) SpO2   (b) Air Flow

Fig. 7. SpO2 and AirFlow sensor

4 System Description

The main function of our platform is telemedicine, to make medical diagnostics for distant patients (foreign/home/workplaces) outside clinics or hospitals.

This function puts into interaction between three modules:

- The patient module: this part is installed in the patient; it is composed of the different sensors mentioned above. They provide information constantly about the patient's condition.
- The server module: receives the data from all the patient modules and manages the authentication and the security of the communications, as well as the coherence of the data. These are stored in a way that preserves the privacy of patients.
- The module (supervisor): the doctor receives the patient's records; it can remotely access data collected from the server from a computer interface.

The platform includes a Raspberry Pi 2 card, which is the computing and processing unit, an e- Health shield card that converts sensor signals - ECG -, - EMG -, - SpO2 - and - Air Flow -. Thus, the retrieved data is transmitted from the Raspberry Pi to a server (database), and available on an interface via our server (see Fig. 8).

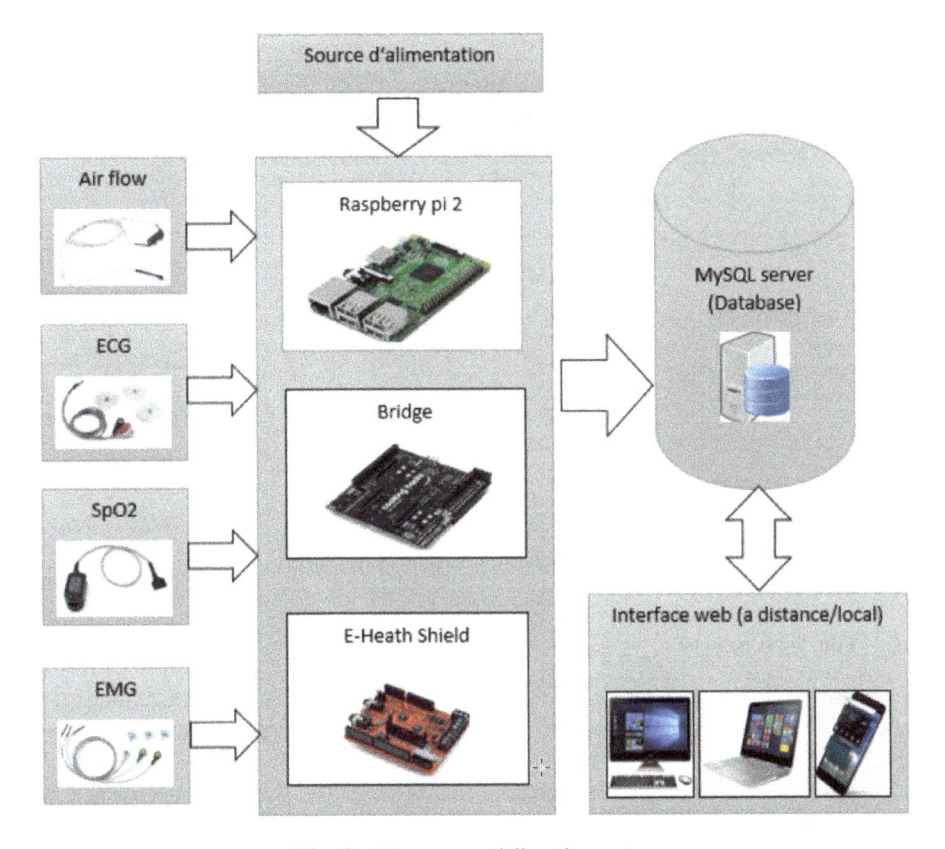

Fig. 8. Diagram modeling the system

5 Software Architecture

At the time of taking measurements, the Raspberry Pi sends an HTTP request to the server via the GET or POST method. The Common Gateway Interface (CGI) scripts run on the server process the request and read the data sent by Raspberry Pi. The data will be transferred and stored in the MySQL database. The server returns the response message to the Raspberry Pi card to indicate whether the data is received (see Fig. 9).

When the user requests the data, the user machine (i.e., the web browser) sends an HTTP request to the server. The web server receives the request and searches for the requested data in the database. The scripting engine running on the Web server processes the instructions to create an HTML page. The HTML feed is then returned to the web browser. The browser then displays the page requested by the user.

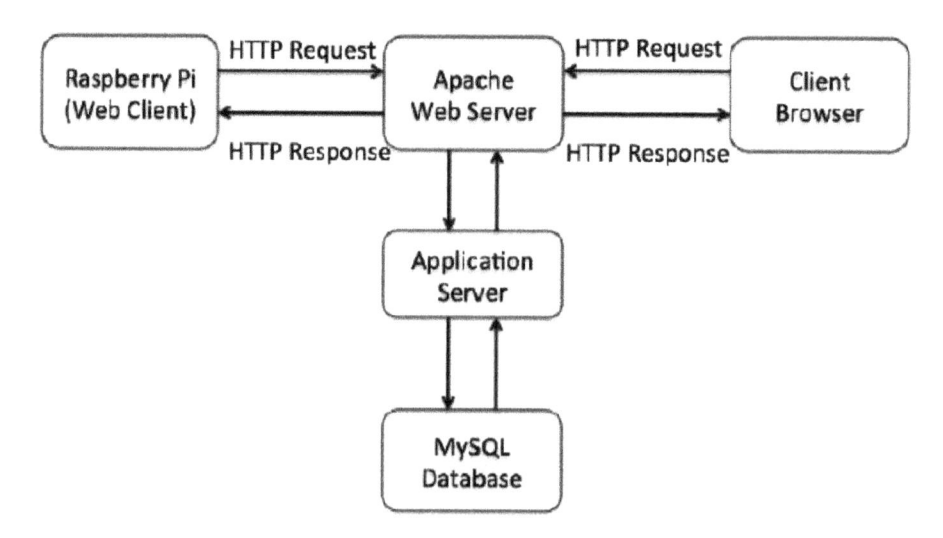

Fig. 9. Software architecture of the system

5.1 The Web Server

A web server is a computer program that processes a server-side application, establishes two-way and/or unidirectional and synchronous or asynchronous connections with the client.

The web server provides static content to a browser for example downloads a file and sends it over the network to a user's browser. This exchange is supported by the browser and the server; it is done using the HTTP protocol. The server is a term used to refer to the computer running the program.

5.2 Database

A database is a structure in which one can store and organize the sharing of data in a structured way [18].

The main page: This page allows the doctor or other persons authorized to consult the patient information. Figure 10 shows the information about the patient on the page home just after the authentication of the doctor.

This page contains tabs that are different data collected by the sensors (see Fig. 11) and their visualization as graphs and histogram, which will enable healthcare professionals to make their medical reports.

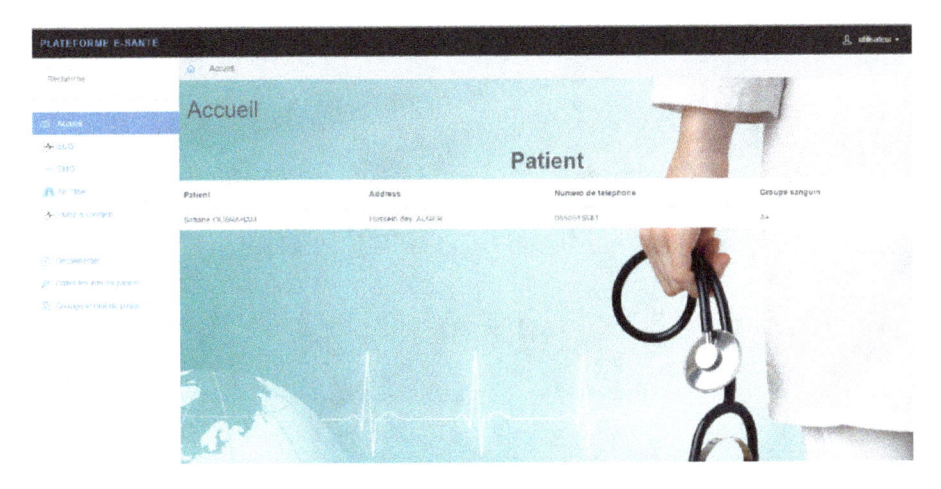

Fig. 10. The main page of the platform

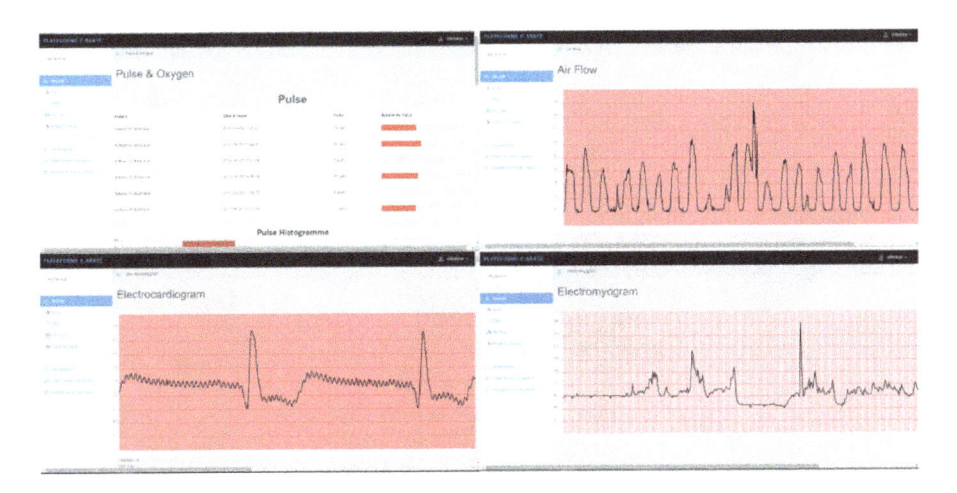

Fig. 11. Presentations of different information collected

6 Test and Results

In this part, we discuss the results and tests performed. This is to evaluate and validate the proper functioning and stability of our system.

To evaluate the performance of our system, we first put the sensors in place as in Fig. 12 for the ECG and EMG and Fig. 13 for the SpO2 and AirFlow sensors.

(a) EMG (b) ECG

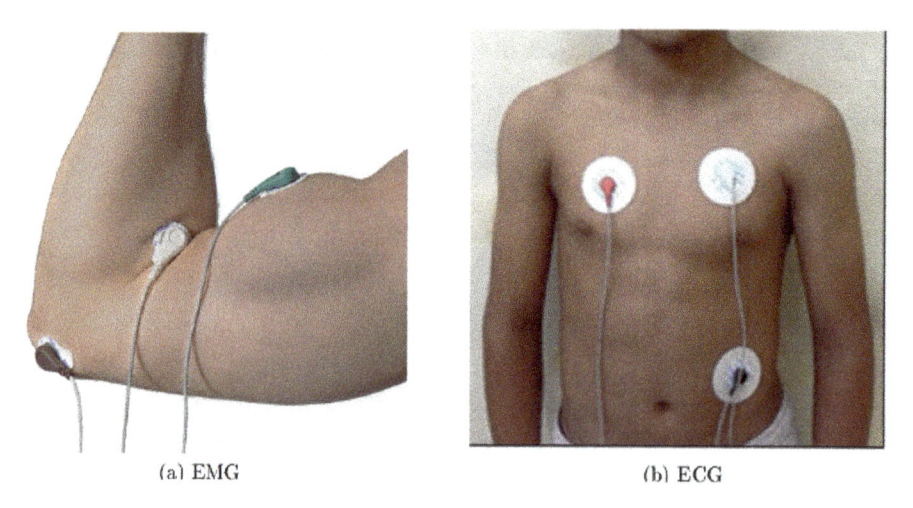

Fig. 12. The method of using ECG and EMG sensors

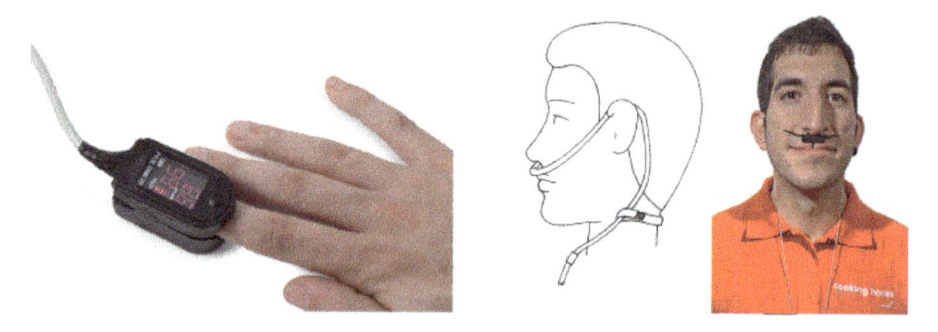

Fig. 13. How to use Spo2 and AirFlow sensors

6.1 Real-Time Display and Interpretation of Data

As soon as the sensor data sent to the web server, the latter will process it and automatically save it to the database with date and time of receipt. The latter will directly process the data and automatically save it on the data base with the date and time of its receipt.

The doctor consults the graphs displayed on the web page in real time. To do this, the browser sends a request to the server for the data to be retrieved and display on the screen on the browser as shown in Figs. 14, 15 and 16.

Whenever the sensors collect the vital data of a patient, they will be sent back to the web server. The database will archive these data so that the physician will have a history of the data and follow the patient's progress (see Fig. 17), to make it easier for the doctor to search by date in the history of the data collected., we added a filter for the display of the data over a given period (see Fig. 18).

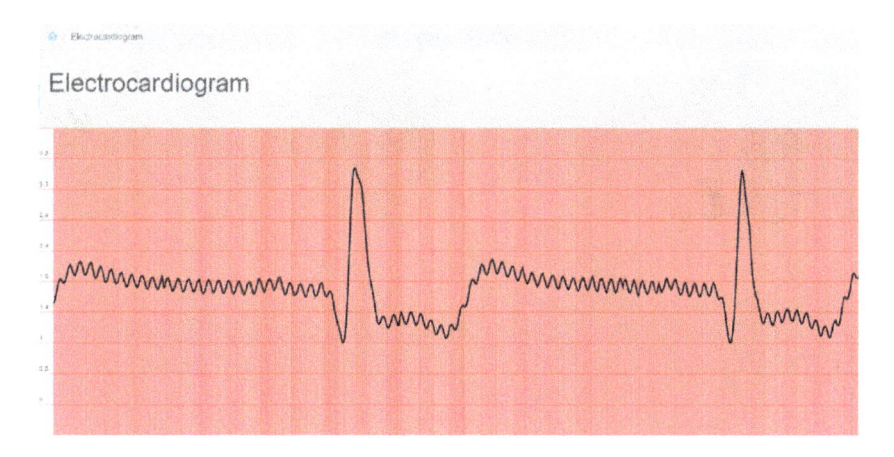

Fig. 14. Interpretation of ECG graph data

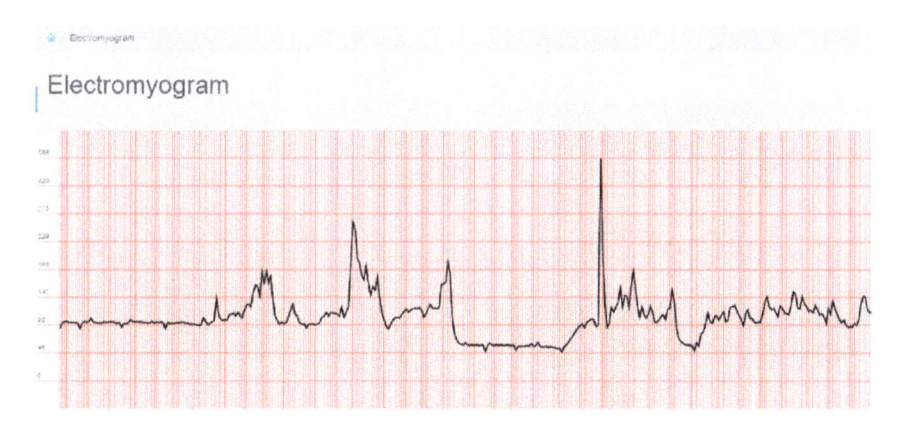

Fig. 15. Interpretation of data in EMG graph

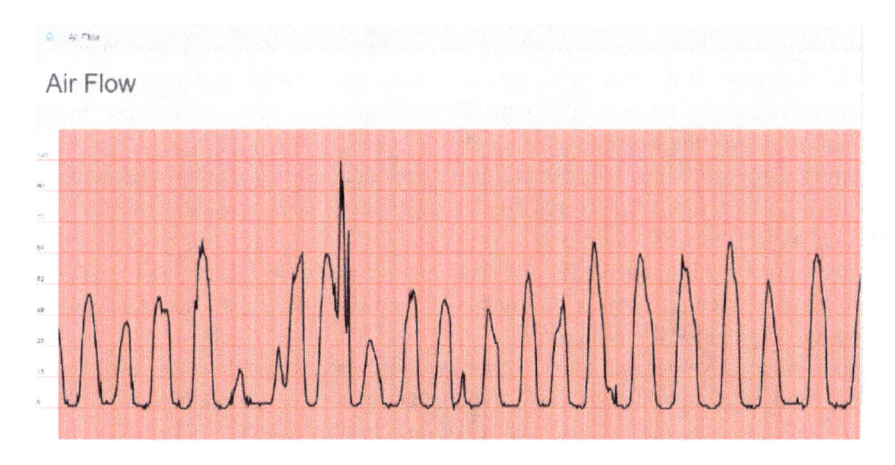

Fig. 16. Interpreting Air Flow graph data

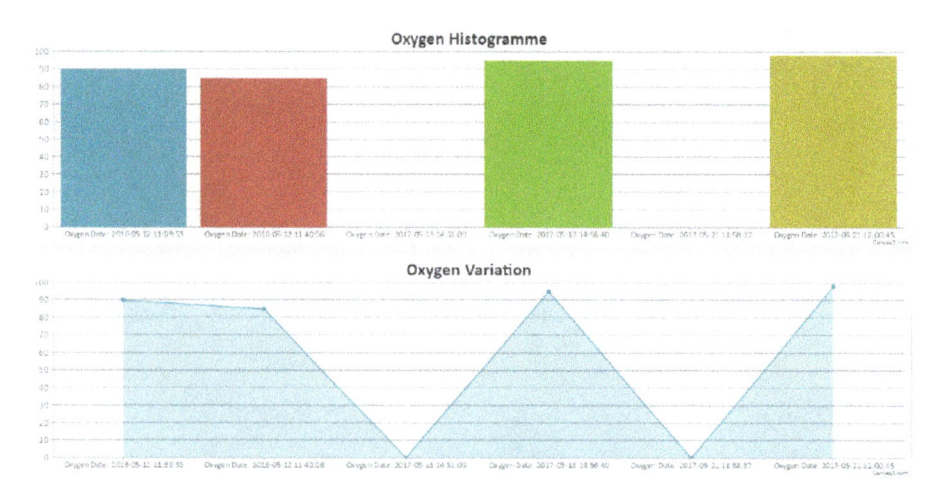

Fig. 17. Data history for Spo2 sensor

Fig. 18. Filtering data over a period

7 Conclusion

The objective of this project was the design and implementation of an e-health platform, which aims at monitoring and medical monitoring of patients at a distance, and for this, we used the Internet of Things. The solution proposed, in this work, makes it possible to ensure the objective effectively.

Our e-health platform proposes to collect medical data thanks to biomedical sensors, able to measure certain physiological parameters such as heart rate, respiratory rate and oxygen level in the blood.

The Raspberry Pi transmits in real time the data collected by sensors to a server, in order to store them. Allowing the doctor to retrieve these data, to ensure medical follow-up of patients at any time.

As a perspective, several works could be considered to continue this project. First, we can think of integrating our platform on an ambulance as a black box, which will be used by the paramedic to make primary diagnoses to the patient, thus informing the doctors of the patient's condition and preparing a treatment or intervention medical.

In the case where our solution will be generalized in the field of health, such as patient monitoring remotely or even inside health facilities in Algeria.

The nurses and doctors may be busy with other things, which means reducing the workload and concentrate for patients with advanced disease.

References

1. WHO. https://www.who.int/fr/news-room/fact-sheets/detail/cardiovascular-diseases-(cvds). Accessed 21 Aug 2019
2. Samuel Ropert: Internet of Things - A key pillar of digital transformation, Idate research, October 2015
3. ARCEP: Livre blanc IoT - préparer la révolution de l'internet des objets (2016). ISSN 2258-3106
4. Saviance Technologies: Will the Internet of Things Analytics Revolutionize the Healthcare Industry? A Saviance Technologies Whitepaper (2014)
5. Kalarthi, Z.M.: A review paper on smart health care system using internet of things. IJRET Int. J. Res. Eng. Technol. **5**, 80–84 (2016)
6. Pang, Z.: Technologies and architectures of the Internet-of-Things (IoT) for health and well-being. Ph.D. thesis in Electronic and Computer Systems, KTH-Royal Institute of Technology, Stockholm, Sweden, January 2013
7. http://www.lafnim.com/fnim-rendez-vous.asp?id=60. Accessed 17 Mar 2017
8. http://www.telesante-basse-normandie.fr/services-de-telemedecine/scad-2/scad-2,1923,1948.html. Accessed 27 Mar 2017
9. https://www.osicat.fr/index.php. Accessed 27 Nov 2017
10. Hu, Y., Eriksén, S., Lundberg, J.: Future directions of applying healthcare cloud for home-based chronic disease care. In: eTELEMED: The Ninth International Conference on eHealth, Telemedicine, and Social Medicine, Nice, France, March 2017
11. Andrès, E., Talha, S., Hajjam, M., Hajjam, J., Ervé, S., Hajjam, A.: Telemedicine to monitor elderly patients with chronic diseases, with a special focus on patients with chronic heart failure. J. Gerontol. Geriatr. Res. **5**, 311 (2016)
12. http://www.eldjazaircom.dz/index.php?id_rubrique=215&id_article=1101. Accessed 10 Aug 2019
13. Sekkai, L., Abbassene, A.: MPTIC: Plateforme Télémédecine du CDTA Expérience, points forts, points faibles et perspectives. In: Workshop International sur la Télémédecine: Usage et Défis, Algiers, Algeria, October 2008
14. https://ecolebranchee.com/2016/10/18/raspberry-pi-ordinateur-apprentissageprogrammation. Accessed 12 Apr 2016
15. https://eduscol.education.fr/sti/sites/eduscol.education.fr.sti/files/ressources/pedagogiques/4346/4346-1-rpi-presentation.pdf. Accessed 10 Aug 2019
16. http://binaire.blog.lemonde.fr/2015/12/28/raspberry-pi-la-petite-histoire-dunegrande-idee/. Accessed 27 Nov 2017
17. https://www.cooking-hacks.com/documentation/tutorials/raspberry-pi-to-arduinoshields-connection-bridge. Accessed 27 Nov 2017
18. Cornuéjols, A.: Bases de données concepts et programmation. AgroParis-Tech, Spécialité Informatique, France (2009)

Li-Fi Prospect in Internet of Things Network

Augustus E. Ibhaze[1(✉)], Patience E. Orukpe[2],
and Frederick O. Edeko[2]

[1] Department of Electrical and Electronics Engineering, University of Lagos,
Akoka, Yaba, Lagos, Nigeria
`eibhaze@unilag.edu.ng`
[2] Department of Electrical and Electronics Engineering, University of Benin,
Benin City, Nigeria
`patience.orukpe@uniben.edu, frededeko@yahoo.co.uk`

Abstract. With the innovation and maturity of Light Fidelity (Li-Fi) networks, there will be a paradigm shift on how people communicate through the Internet. While Internet of Things (IoT) is seen as the interconnection between embedded intelligence, Things/Everything and the Internet, the bandwidth requirement for its deployment cannot be underestimated. Despite the frequency reuse requirement of the congested radio frequency (RF) spectrum, the utility requirements of certain critical devices prohibit the use of radio frequency signal within its environment although requiring high speed/bandwidth efficient techniques. By considering the vast majority of things that will be connected via the Internet, Li-Fi remains the most promising technique for IoT deployment in high speed/large bandwidth requirement and RF interference prohibited areas. This paper therefore provides an overview on the deployment of the Visible Light Communication network, highlighting its challenges, potentials and the future of Internet of Things.

Keywords: Internet of Things · Light-Fidelity · Bandwidth · Capacity · VLC

1 Introduction

The Internet of Things is a platform that enables many devices such as sensors, wearables, appliances, utilities, etc. connect to a network in order to communicate with data centers, controllers and other devices. IoT is a hyper-connection of virtually everything with the capacity to extend broadband Internet access beyond devices like computers, tablets and smartphones to daily used objects/devices to communicate and interact with the external environment [1] Some of these devices are cars, vending machines, electronic appliances etc.

The term Li-Fi was coined by Professor Harald Haas of the University of Edinburgh during his TED talk in 2011 where he envisioned light bulbs as wireless routers [2] Li-Fi is a form of wireless communication that makes use of visible light to transmit high speed data. Speeds up to 1 Gb/s were recorded in Estonia, which is about 100 times faster than current home Wi-Fi speeds [3].

Although Li-Fi and Wi-Fi share similarities in their electromagnetically transmitting data, Wi-Fi makes use of radio waves while Li-Fi operates within the visible light spectrum. It uses Light Emitting Diodes (LEDs) as the visible light source in VLC systems. Li-Fi is promising because of its fast switching characteristics, low energy consumption, large bandwidth requirement and is readily available.

As IoT deployment continues to grow, and sensors are added to more and more things, a faster and sufficient data transmission source will be required. To be able to achieve this, a communication infrastructure that aims at controlling the devices' behavior and provide communication reliability should be ubiquitous and cost effective. This is where the choice of Visible Light Communication (VLC) for IoT becomes relevant.

VLC offers some benefits such as higher bandwidth, ability to reuse already existing light infrastructure, free from electromagnetic interference, improved privacy and security as light cannot penetrate barriers and many more. The VLC system makes use of Light Emitting Diodes (LEDs) which allow the construction of low-cost communication systems [4] Besides providing wireless connectivity in homes, the combination of the illumination and transmission function can be utilized in scenarios where safe and non-interfered communication is necessary such as underground mine, hospital, airplane, underwater, in petrochemical industries as well as radio frequency restricted areas.

2 Li-Fi/Wi-Fi for Use in Internet of Things

Certainly, Wi-Fi is a tremendous technology that has vast applications that has impacted our lives both personally and professionally. But Radio Frequency (RF) signals on which Wi-Fi is based on has its pros and cons. On one hand it provides connectivity among 5 billion cellular phones globally and 70,000 TB of data transfer annually [5]. And on the other hand, the use of RF signals is restricted at hospitals and aircrafts because of its hazardous effects due to signal interference which could impair normal functioning of classified equipment within such domains thereby imposing undue risk on well-meaning personnel. While the Wi-Fi network may not be completely replaced by Li-Fi networks owing to its shorter range among other limiting variables of Li-Fi technology, both networks can serve as complementary networks in a larger heterogeneous architecture for improved capacity. Other issues with RF technology in relation to Li-Fi are highlighted as follows.

2.1 Capacity

Presently, wireless data is transmitted via radio waves which are scarce, expensive and limited in its bandwidth requirement. With the evolution from First Generation and now towards 5G technology, there is an increasing demand on wireless communication infrastructure while the radio spectrum is overcrowded and rapidly decreasing. The visible light spectrum ranging from 430 THz to 790 THz possess the capacity for larger bandwidth compared to its radio frequency counterpart occupying 3 kHz to 300 GHz bandwidth [6]. The available large bandwidth associated with the visible light spectrum will provide an enabling platform for the deployment of Internet of Things initiatives as large number of "Things" and "Everything" interconnect/communicate with themselves and the Internet.

2.2 Security

Radio waves can penetrate through walls and buildings thereby enabling a third party intercept the travelling signal. But light cannot penetrate through walls and objects thereby making this mode of data transmission very secured and unbreachable by distant network intruders. Li-Fi application is especially an advantage for use by the military who from time to time need to convey information securely. An external person cannot intercept the information being transmitted without being within visible range due to light's inability to penetrate opaque objects.

2.3 Speed

Li-Fi offers greater speed of wireless data with over 1 Gb/s data rate recorded which is about 100 times faster than the current Wi-Fi speeds and is expected to reach Terabyte per second. With this fast connectivity, IoT with Li-Fi is very promising as the anticipated billion-node connection will pose no worries regarding network congestions.

2.4 Electromagnetic Interference

Light waves do not cause interference. This makes Li-Fi useable in RF restricted areas like airplanes where RF interference may tamper with other RF sources/equipment within such domains which may cause damage.

3 Application Areas of Li-Fi Network

Visible Light Communication (VLC) network regarded as Light-Fidelity (Li-Fi) network could be deployed both indoors and outdoors to complement other communication techniques.

3.1 Indoor Applications

Multimedia data in the form of picture, audio or video can be transmitted via Li-Fi network at a much faster rate than the use of RF. VLC can also be used to provide broadband Internet access to users at home, workplace and so on. This network can be used to enable smart indoor lighting required in smart networks. Smart lighting based on VLC system provides the integrated system of illumination, communication and control which could reduce the hard wiring and energy consumption within a facility. Also, in the development of intelligent homes which involves the use of many sensors that will require a large use of Internet of Things concept can be run effectively on the communication platform provided by VLC.

In hospitals and healthcare centers, patient's critical state and other important medical details must be monitored. The use of devices that make use of the radio frequency is highly prohibited with respect to some critical health care applications such as operations in theatres, MRI scanners. Since light does not impose undue interference, Li-Fi network becomes a promising solution for data exchange in such zones.

VLC can be applied in indoor positioning (Indoor Global Positioning System GPS) as the data source. Smart phones and devices manufactured in recent years contain navigation features through the use of GPS satellite signals. But these signals bounce around in buildings thereby making indoor GPS positioning ineffective. With the installation of VLC enabled LED, general lighting in buildings, indoor positioning can be achieved.

In security conscious applications, VLC offers a solution to data security. Since light cannot pass through objects, information cannot leave the confinement of a room. This renders eavesdropping and hacking impossible outside the room. Military applications that require high level of security may especially take advantage of this confinement property.

3.2 Outdoor Applications

In the development of intelligent transport systems and smart street lighting, Li-Fi networks can be deployed. Traffic management and safety can be achieved by replacing the car headlights and tail lights by LEDs to enable vehicle to vehicle communication in which vehicles can download information and have real time perspectives about the situation on the road, for example like identifying the best route to take to avoid traffic. And also, in the long run, public lighting which includes street lights can serve as broadband Internet access point.

VLC can be used in EMI sensitive environments like aircrafts where infrared, Bluetooth, Wi-Fi are disallowed to avoid creating electromagnetic inference with sensitive radio equipment. On aircrafts, also where the use of cell phone is prohibited due to potential interference with the aircraft navigation and communication devices from an RF emitting device, VLC provides an alternative by providing data transmission without electromagnetic interference.

In explosion prone areas like in the petroleum and chemical plants, where transmission of other frequencies like radio frequency, microwaves, etc. can be hazardous because they deal with highly flammable products and these mobile devices due to heat and other emissions can cause sparks in these environments. But the use of VLC technology in these hazardous environments can eliminate all that.

Unlike radio waves that attenuate rapidly under water, acoustic waves had been previously deployed for communication. Since the available bandwidth for acoustic wave is low, VLC provides a larger bandwidth which results in faster data communication [7]. Light can penetrate for relatively large distances enabling communication from diver to diver, diver to drilling rig, etc. and light could also be used as a source of illumination simultaneously underwater.

4 Challenges in the Implementation of Li-Fi Networks

For Li-Fi network deployment, some of the challenges to be surmounted are highlighted as follows.

4.1 Uplink and Radio Frequency Augmentation

Most research currently in VLC focus more on downlink operations i.e. from the transmitter (LED) to the receiver (photodiode), as against the uplink. In practice, constantly turning on the LED consumes significant amount of energy and could cause visual disturbance for the users if implemented outside regulations [8]. The uplink operation also requires the user-device maintain directional contact with the receiver. In the case of mobile devices which are continuously moving and rotating this can result in the reduction of throughput. To address these issues, it was proposed to use other types of communication (RF and infrared) to transmit uplink data [9].

4.2 Line of Sight (LOS)

Visible light signals cannot penetrate through objects. This brings about a disadvantage in signal coverage for multiple users. Though light can be reflected but it does not penetrate and this characteristic is a disadvantage preventing the signal from spreading among multiple rooms. In multiple rooms there would be barriers like walls disrupting the line of sight of the signals thereby causing a total obstruction to the signal transmission. In practice, the unpredictable nature of users' actions when using a VLC-based mobile device could cause misalignment of the receiver with the transmitter and this leads to a drop or total loss of data transmission. Design and development of methods to provide LOS alignment is required.

4.3 Ambient Light Noise/Interference

During the day, sunlight luminance is hundreds of thousands of lux much higher than that of a typical LED used in indoor-VLC system which is lower than 1000lux [10]. It would be hard for the photodiode to detect the LED light signal under direct sunlight due to the saturation of the photodiode. Photodiode saturation occurs when the photodiode is not able to convert excess incident optical power into a photocurrent output and all the excess optical power received will be absorbed as heat and this leads to the damage of the photodiode. Also, apart from sunlight, the skylight alone has a luminance that is 10 times higher than that of the LEDs. This necessitates the design of efficient detectors for Li-Fi networks.

4.4 Mobility and Coverage

User mobility in VLC poses a new issue that significantly differs from those of RF. In the place of RF, there exists cellular technology that provides connectivity within a cell and from cell to cell. For ubiquitous VLC mobile technology, infrastructure has to be put in place if the cellular technology in RF has to be replicated and implemented in

VLC. Also bearing in mind that VLC is used as a complementary technology to RF networks, user mobility requires protocols for vertical (RF-VLC) and horizontal (VLC-VLC) handoff [11].

5 Potentials of Visible Light Communication

Using VLC can address problems that could not be solved by the RF waves mainly due to electromagnetic interference and health challenges developed due to excessive exposure to the waves. There is no known side effect of visible light exposure on the human body. Instead the use of VLC could serve two purposes as an illumination and communication source. VLC is an eco-friendly technology.

Visible light spectrum occupies the frequency range from 430 THz to 790 THz which is 1000 times greater than the radio waves portion of the spectrum which ranges from 3 kHz to 300 GHz. This at least solves the shortage of the radio frequency bandwidth issue. VLC system has been modified to achieve very high data rates [12]. Intensive research and experimentation are in progress to attain higher data rates. The visible light frequency band is unlicensed and serves as an entirely free infrastructure for the development of a complex communication network.

6 Investigation into Li-Fi Technologies

With the abundance of license free spectrum at terahertz (visible light) for optical wireless communication, its inherent large bandwidth translates into higher carrier frequency applications. Unlike its radio frequency counterpart situated for lower carrier frequencies, the terahertz frequency band possesses the potentials for high data rate deployments. The flip in technology modelling and implementation for optical wireless communication system lies with the nature of the processed signal. Due consideration is therefore needed in the transceiver design requirements as the transmitted signal is real valued and unipolar. The bipolar transmitted signal will have to be appropriately processed and converted to a unipolar signal for application in a Li-Fi network environment.

Depending on the data rate requirement, signal processing (modulation and demodulation) could be achieved through either intensity modulated/direct detection (IM/DD) [13] or external modulation [14]. Direct modulation technique employs laser diodes or light emitting diodes (LED) which make them suitable for data rates up to 2 Gb/s while external modulation employs Mach-Zehnder interferometric modulator suitable for data rates beyond 2.5 Gb/s.

To achieve higher data rates, there exist a compromise between single-carrier and multi-carrier techniques in both academia and industry. Although single-carrier techniques are marked for lower data rate applications, its implementation with suitable equalizer can achieve data rates equivalent of multi-carrier techniques – with a tradeoff in equalizer design complexity. Some of the investigated single-carrier techniques ranges from on-off shift keying (OOK) to diverse pulse modulation schemes [5] with the downside of non-linear signal distortion in the transmitter front end caused by frequency selectivity in the dispersive channel as data rate demand increases. To mitigate this effect, multi-carrier scheme such as orthogonal frequency division multiplexing (OFDM) [15] is deployed where parallel sub-bands are orthogonally modulated for signal transmission.

7 Future of Internet of Things

The impact of IoT in the coming years will change the way we live and work. Being that there would be digital identity for every human, creature, objects and things. Everybody and everything would be a part of the Internet making them traceable and trackable. There would be an overall ecosystem of devices which will be connected to one thing – the Internet, although inheriting the security challenges posed by it. Since certain intermediaries will be eliminated due to the eventual deployment of Internet of things (IoT) and Internet of Everything (IoE), everything will be done in real time be it information transmission, actions and decision making.

In healthcare, IoT will give healthcare professionals the ability to remotely monitor their patients' health [16] and proffer remediation. Also, these connected devices can monitor and alert medical professionals and call for emergency services based on the patient's heartbeat, temperature and other parameters.

In power systems, IoT has become an enabler for smart grid. The idea is that smart meters, smart thermostats and appliances receive information on pricing and the total demands on the system so that they can draw the power they need from the grid at off peak times when it is cheapest. This makes power networks less prone to black-outs. Also, power lines and pipelines would be embedded with sensors to be able to collect data, detect and analyze maintenance issues.

IoT will bring about intelligent transportation systems in which traffic lights can be integrated with sensors that can adjust their luminance according to the position of the cars and the period of the day. This reduces congestion and fuel consumed in urban areas. In airplanes, sensors put in their engines will be able to detect and isolate developing problems and communicate with the pilot and the ground crew while the plane is still in flight. In vehicles, IoT encourages vehicle-to-vehicle communication in which vehicles and roadside units such as street lights, traffic lights provide each other with information. Information provided could be safety warnings, traffic information and so on [17].

One key factor missing from the IoT space is the standardization around security and privacy. Once information is on the Internet, it is almost impossible to remove it completely. According to [18], IoT allows easy access to personal and corporate data, which could lead to life threatening situations if hackers gain access to your vehicles, hack into medical systems like the life support system and so on. The growth of IoT brings about innovation and at the same time vulnerability. There need to be some sets of strict standards around how this information is being used and how to protect it. The deployment of IoT will therefore bring about an eventual innovative research stride in security and privacy of interconnected "Things" and "Everything".

8 Conclusion

With all that has been achieved so far with IoT, we are just scratching the surface because there is still a lot more to come. We are still in the early days of IoT and just to think back on how the Internet was in 1990s so is where we are at in the IoT platform. With the incorporation of Li-Fi in IoT, the possibilities will be endless and with the predicted growth rates of IoT deployment, Li-Fi could be a viable technology to accommodate the huge bandwidth requirement for the deployment of IoT framework.

Acknowledgment. The authors gratefully acknowledge the sponsorship and funding of this project by the University of Lagos and Petroleum Technology Development Fund with award No. P4567720076521527

References

1. Aswin, K.M., Aswin, R.G., Lokesh, V., Sugacini, M.: IOT enabled by Li-Fi technology. In: National Conference on Communication and Informatics, Sriperumbudur (2016)
2. Haas, H.: Wireless data from every light bulb. In: TED Global, Edinburgh (2011)
3. Ovenden, J.: LiFi Will Make The Internet of Things Possible. Innovation Enterprise (2015)
4. Schmid, S., Mangold, S., Corbellini, G., Gross, T.R.: LED-to-LED visible light communication networks. In: Proceedings of the Fourteenth ACM International Symposium on Mobile Ad Hoc Networking and Computing, pp. 1–10 (2013)
5. Haas, H., Yin, L., Wang, Y., Chen, C.: What is LiFi? J. Lightwave Technol. **34**(6), 1533–1544 (2016)
6. Karunatilaka, D., Zafar, F., Kalavally, V., Parthiban, R.: LED based indoor visible light communications: state of the art. IEEE Commun. Surv. Tutor. **17**, 1649–1678 (2015)
7. Uema, H., Matsumura, T., Saito, S., Murata, Y.: Research and development on underwater visible light communication systems. Electr. Commun. Jpn. **98**(3), 9–13 (2015)
8. BS EN: 62471:2008, Photobiological Safety of Lamps and Lamp Systems. BSI British Standards, September 2008
9. Shao, S., Khreishah, A., Rahaim, M.B., Elgala, H., Ayyash, M., Little, T.D.C., Wu, J.: An indoor hybrid Wi-Fi-VLC internet access system. In: IEEE 11th International Conference on Mobile Ad Hoc and Sensor Systems, pp. 569–574 (2014)
10. Do, T.-H., Yoo, M.: Potentialities and challenges of VLC based outdoor positioning. In: IEEE International Conference on Information Networking (ICOIN), pp. 474–477 (2015)
11. Pathak, P.H., Feng, X., Hu, P., Mohapatra, P.: Visible light communication, networking and sensing: a survey, potential and challenges. IEEE Commun. Surv. Tutor. **17**(4), 1–2 (2015)
12. Dursun, I., Shen, C., Parida, M.R., Pan, J., Sarmah, S.P., Priante, D., Alyami, N., Liu, J., Saidaminov, M.I., Alias, M.S., Abdelhady, A.L., Ng, T.K., Mohammed, O.F., Ooi, B.S., Bakr, O.M.: Perovskite Nanocrystals as a color converter for visible light communication. ACS Photonics **3**(7), 1150–1156 (2016)
13. Cox III, C., Ackerman, E., Helkey, R., Betts, G.E.: Techniques and performance of intensity-modulation direct-detection analog optical links. IEEE Trans. Microw. Theory Tech. **45**(8), 1375–1383 (1997)
14. Cox, C.H., Ackerman, E.I., Betts, G.E.: Relationship between gain and noise figure of an optical analog link. In: IEEE MTT-S International Microwave Symposium Digest, vol. 3, pp. 1551–1554 (1996)

15. Komine, T., Haruyama, S., Nakagawa, M.: Performance evaluation of narrowband OFDM on integrated system of power line communication and visible light wireless communication. In: IEEE 1st International Symposium on Wireless Pervasive Computing, p. 6 (2006)
16. Paessler, D.: The future of IoT is unwritten. IoT Agenda, September 2016
17. Tarannum, S.: Data transmission through smart illumination via visible light communication technology. Int. J. Tech. Res. Appl. **IV**(2), 136–137 (2016)
18. Mark, F.: Three things you probably didn't know about the future of IoT. Forbes, pp. 1–2, March 2015

A Multidimensional Control Architecture for Combined Fog-to-Cloud Systems

Xavi Masip-Bruin[1(✉)], Vitor Barbosa Souza[2], Eva Marín-Tordera[1], Guang-Jie Ren[3], Admela Jukan[4], and Jordi Garcia[1]

[1] CRAAX Lab, Universitat Politecnica de Catalunya,
08800 Vilanova i la Geltrú, Spain
{xmasip,eva,jordig}@ac.upc.edu
[2] Universidade Federal de Viçosa, Viçosa, Minas Gerais 36570, Brazil
vitorbs@dpi.ufv.br
[3] IBM Almaden Research Center, San Jose 94002, USA
gren@us.ibm.com
[4] Technischen Universität Braunschweig, 38106 Brunswick, Germany
a.jukan@tu-bs.de

Abstract. The fog/edge computing concept has set the foundations for the deployment of new services leveraging resources deployed at the edge paving the way for an innovative collaborative model, where end-users may collaborate with service providers by sharing idle resources at the edge of the network. Combined Fog-to-Cloud (F2C) systems have been recently proposed as a control strategy for managing fog and cloud resources in a coordinated way, aimed at optimally allocating resources within the fog-to-cloud resources stack for an optimal service execution. In this work, we discuss the unfeasibility of the deployment of a single control topology able to optimally manage a plethora of edge devices in future networks, respecting established SLAs according to distinct service requirements and end-user profiles. Instead, a multidimensional architecture, where distinct control plane instances coexist, is then introduced. By means of distinct scenarios, we describe the benefits of the proposed architecture including how users may collaborate with the deployment of novel services by selectively sharing resources according to their profile, as well as how distinct service providers may benefit from shared resources reducing deployment costs. The novel architecture proposed in this paper opens several opportunities for research, which are presented and discussed at the final section.

Keywords: Fog computing · Combined F2C systems · Virtual control architecture

1 Introduction and Motivation

The unstoppable growth of devices at the edge of the network including wearables, smartphones, vehicles, sensors, actuators or appliances, along with the development of distinct network technologies enabling wireless sensor networks, machine-to-machine (M2M) communication or pervasive computing, just to name a few, paved the way for the so-called Internet of Things (IoT) [1]. Simultaneously, core technologies have substantially evolved including enhanced distributed and high performance computing, data center networks or self-manageable resources, among others. Leveraging such technological evolution along with wide network ubiquity, higher network availability and, finally, the severe needs imposed by innovative (foreseen but also unforeseen) services, cloud computing was developed to enable remote requests for service execution, anywhere and anytime with seamless integration of distinct end-user devices [2]. When put together, Cloud and IoT pave the way for deploying new highly demanding services, benefitting from both real-time data collection from devices at the edge and processing power and long-term storage both brought up by cloud providers at the core of the network. Novel scenarios, such as smart cities, smart home, smart transportation or smart agriculture, are remarkably evolving by adopting the smart capacities brought by such a technological deployment. However, the increasing demand for real-time IoT services, such as dependable services in e-Health, traffic control in smart transportation and optimized tracking in Industry 4.0, whose requirements include not just real-time data collection, but also real-time data processing, has put in check cloud computing as the appropriate solution for provisioning real-time sensitive IoT applications. Indeed, the large distance between cloud data centers and end-user devices undoubtedly introduces a considerable latency for remote service execution.

In order to cope with the delay added by the employment of cloud premises, fog computing [3] has been recently proposed aiming at diminishing the network delay, by bringing computing resources close the end-user, through highly virtualized micro data centers (see also [4] for a recent survey on the topic). In fact, with the unstoppable growth of edge devices' capabilities in distinct aspects –processing power, storage, autonomy and connectivity–, the set of services that may be offered by idle edge resources increases significantly, rather than being only considered for data provisioning. Nevertheless, it is worth noticing that the set of services benefitting from fog computing may be limited by the capacities brought by very constrained edge resources as well as by their dynamicity and volatility. Thus, the execution of services demanding high processing capacity, for instance, should not dismiss reliable cloud resources. Therefore, it seems obvious that the role of fog is not to compete with cloud, but to set a global collaborative scenario where services execution, regardless their demands and requirements, may benefit from both, cloud and fog, allocating those resources best suiting specific service demands, be it either at cloud, fog or a combination of both.

Achieving an optimal (fog and/or cloud) resource allocation empowering QoS-aware service execution, low network load, green computing, and scalability– requires a novel control mechanism intended to considering the characteristics of each resource type, the set of provided services, as well as the end-user service requirements. Recent efforts are being devoted to design a general architecture dealing with fog and cloud resources control, turning into two main directions. An active work is led by the OpenFog Consortium through the recently published OpenFog Reference Architecture [5]. Another approach, referred to as Fog-to-Cloud (F2C), was proposed in [6], aiming at designing a solution for coordinated management of fog and cloud resources through a layered topology, hierarchically organized, enabling parallel service execution in any layer of the envisioned resources topology. More recent references related to the F2C approach propose solutions for F2C layers distribution [7, 8] according to distinct devices characteristics, such as processing capacity, energy availability, mobility profile, and communication technologies, and also evaluate the benefits brought by its deployment [9]. This paper puts the focus on the roots of the F2C architecture and specifically analyzes to what extent a unique, single management architecture may handle any control demand (whatever it be, and whatever it demands) to come from services, scenarios, devices and users. Thus, the question is, "may a unique control architecture support any potential demand?". The main rationale for such a question is twofold, devices mobility and system heterogeneity. Figure 1 intended to illustrate the problem, shows a 3-layered F2C resources topology considering both mobility and proximity to the end-user as the main attributes for the F2C layers definition. Two areas are graphically included, preliminary splitting resources according to the geographical distribution, although additional policies must be defined (still an ongoing work). Indeed, the lower layer (fog layer 1 or simply Fog-1) is composed mainly of resources on the move, such as vehicles and smartphones, providing low network latency at the cost of a higher disruption probability. The upper fog layer (fog layer 2 or Fog-2) is composed of permanently or temporarily static resources in a smart scenario whose network latency may be higher than the one perceived in the Fog-1 resources, albeit it is still considerably lower than the cloud communication delay. This layer may embrace both fixed resources in a building and resources provided by cars in a parking lot, for instance. Finally, the upper layer (cloud layer) is composed of reliable resources provided by cloud data centers.

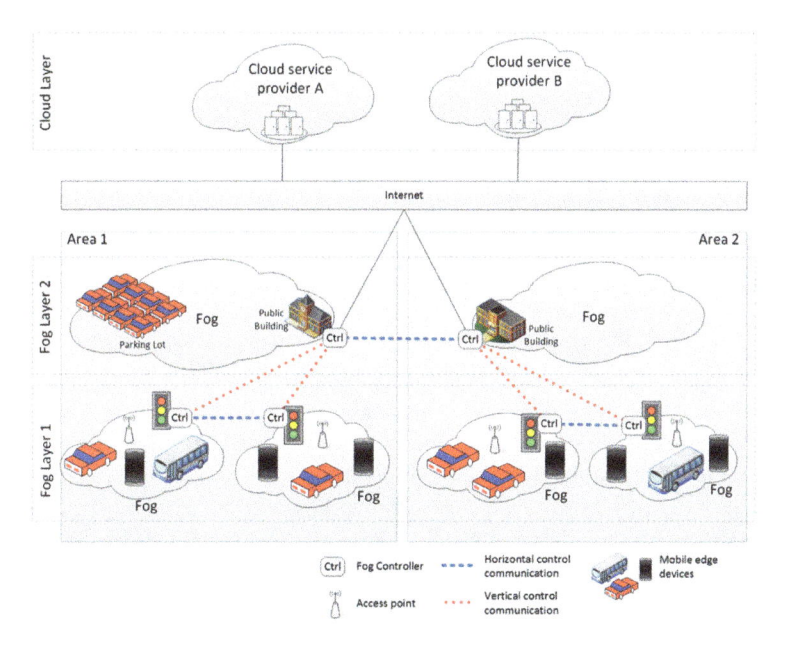

Fig. 1. F2C layered architecture.

Albeit the general F2C topology may follow the characteristics presented in Fig. 1, there is no way to guarantee that such management topology is the most appropriate to meet the entire set of requirements for any potential service. In fact, looking at the expected high mobility, large heterogeneity, collaborative models based on sharing policies and also assuming that the current world of innovative smart services is only in its infancy, the management topology for a proper global service execution might rely on distinct management configurations to best meet specific service requirements and the existing resources capacities. It is our believe that one logically centralized management and control topology, handling any service to be executed and any real-time resource topology, may not be able to guarantee near-optimal resources selection and services orchestration. Therefore, we envision distinct control topologies able to optimize the management of heterogeneous and dynamic resources, tailored to manage service demands optimally, thus setting a sort of a multidimensional control architecture, set by different views of the real control devices into virtual control instances. This paper is intended to introduce this multidimensional approach for the F2C architecture, mainly emphasizing on:

- Positioning the need for a multidimensional architecture, leveraging virtual control instances and discussing its concept and main benefits.
- Providing the main characteristics of the multidimensional approach regarding control and data plane.
- Presenting research opportunities as well as open challenges for the successful deployment of the proposed multidimensional architecture.

The rest of this paper is organized as follows. Section 2 presents previous works in the literature regarding the decoupling of control and data planes in scenarios with high dynamicity. In Sect. 3, the multidimensional architecture being proposed in this paper is presented, and some preliminary results are drafted in Sect. 4. The main research opportunities in this area are presented in Sect. 5, whilst Sect. 6 concludes the paper.

2 The Roots Endorsing the Multidimensional Approach

This section aims at presenting the distinct inputs that have contributed as a background for the positioning of this multidimensional approach for F2C systems –it could be applied to any strategy designed to managing the full stack of resources from the edge up to the cloud though. One of the main concepts behind the multidimensional architecture is related to the decoupling of control plane from the data (or forwarding) plane, as proposed by Software Defined Networking (SDN). Albeit the big enhancement introduced by the advent of SDN on existing network topologies has been initially limited to scenarios presenting high stability, such as data center networks, several works have foreseen the benefits of SDN adoption in geographically distributed environments presenting wireless communications and high dynamicity, such as VANET or IoT [10]. Authors in [11] propose an SDN architecture for VANETs where control communication may follow distinct approaches including centralized, distributed and hybrid. Whilst centralized and distributed concepts are similar to conventional SDN and VANETs respectively, in the hybrid operation mode, a centralized

SDN controller delegates control functions to local agents, such as policy rules and routing protocols' parameters dissemination. The main aim of the proposed architecture is to provide resilience in SDN-based routing for VANETs. Fog and SDN are combined in [12] in order to deploy a logically centralized SDN controller for orchestrating IoT services in distinct fogs. To that end, an updated view of the underlying resources is required by the SDN controller enabling real-time detection of policies violation, resource reservation, and flow rules dissemination, among others. Therefore, local SDN agents are deployed at each fog node in order to control intra-fog communication through policies obtained by the SDN controller.

The high dynamicity and heterogeneity of the existing resources in terms of computing performance, storage capacity, energy consumption, mobility, reliability and volatility, among others, may demand the specialization of local controllers in order to support the diverse and heterogeneous nature of the service demands. For instance, controllers deployed for the management of services to be executed within vehicles on the move, such as infotainment, traffic control, urban resilience, etc.–, may store the city topology map as well as vehicles position and speed in order to perform accurate handover predictions [12]. On the other hand, controllers handling services demanding communication with environment monitoring resources, such as those deployed through WSN, may keep information about the energy profile of edge resources, hence maximizing service lifespan.

Leveraging SDN, Service-Oriented Architectures (SOA) and fog computing concepts, the Control as a Service (CaaS) concept introduced by [8], aims at using idle resources at the edge of the network, e.g., processing, storage or network resources, to store and keep an updated view of underlying resources, map service requests into the most suitable resources, and enable efficient inter-controller communications, among others. The proposed strategy aims at enabling control decisions to be taken closer to end-users. Therefore, near-zero delay demanded by real-time services can be successfully achieved from the first steps of the service allocation process, when the edge resources selection and reservation are performed by controllers at the edge of the network rather than achieving reduced delay only on the service execution and the data transmission. The assessment of the F2C architecture control topology presented in that work has shown the tradeoffs between controllers' capacity, number of controllers, and number of control layers. Although preliminary results have shown the feasibility of the proposed paradigm, several issues are still unsolved, mainly dealing with the dynamic selection of controllers, control topology, and resilience strategies, just to name a few, all fostering new research avenues and thus seeking for future work.

Moving the control decisions to the edge has also been assessed by [13], where authors present a framework whose resource management is sent to the edge rather than letting it to the service providers. The management or resources, which are organized as Mini datacenters (MDCs), is assigned to the Edge Computing Infrastructure Provider (ECIP), which establish contracts with service providers enabling an auction-based edge resource sharing. In that work, the relationship between ECIP and MDC is 1-to-N, i.e., one MDC can be operated exclusively by one single ECIP, what does not meet the conditions of the multidimensional control architecture proposed in this paper.

3 Multidimensional Architecture

In this section we introduce the multidimensional control concept, as envisioned for the F2C architecture, designed to represent distinct control plane instances through overlapped planes, bringing in a key difference with the traditional utilization of a single control plane. For the sake of comprehension, we initially go over the multidimensional representation of the data plane. Later in this section, we dig into the multidimensional control architecture through the introduction and analysis of distinct scenarios. It is worth mentioning that even though the proposed multidimensional control solution is designed to be applied on a F2C system, the rationale behind the concept supports its deployment on any control strategy mixing highly demanding services, mobility and resources heterogeneity.

3.1 Data Plane

Nowadays, Service-Oriented Architectures (SOAs) leverage the diversity of available resources, be it either distributed or centralized and offering distinct capabilities to support effective services execution. To that end, resources may be allocated to either a whole service or a simple task (part of a more complex service), the latter requiring an orchestration strategy to enable the successful execution of the whole service. This task aggregation process may consider distinct execution strategies, either sequential, parallel, or a combination of both. Therefore, assuming the fact that services may be composed of distinct service chains, two or more services may be represented by distinct logical planes, each one with the respective data path topology, as illustrated by *Service A* and *Service B* in Fig. 2. On one hand, *Service A*, which has its data path presented in foreground, may be a general representation of services responsible for obtaining and preprocessing data collected from devices at the edge of the network. On the other hand, *Service B*, which has its data path presented in background, may represent services requiring resources with high processing capability in order to generate a detailed and low-delay analysis of data stored in distinct datasets.

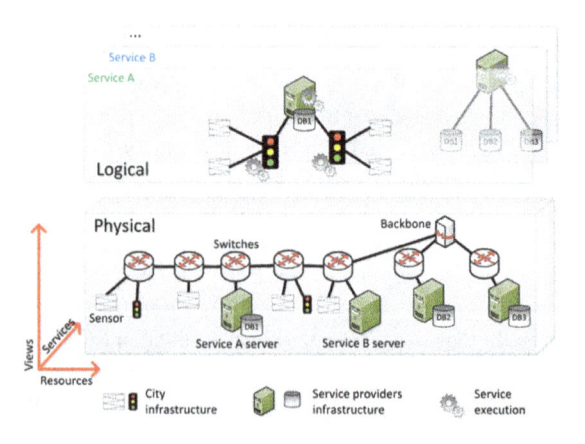

Fig. 2. Multidimensional data plane view for the execution of distinct services.

For instance, let's consider *Service A* collects data through several sensors deployed in a smart city obtaining air pollution data, which is later aggregated by distributed computing resources deployed in city traffic lights, generating the input for the creation of a pollution map for a neighborhood in real-time. The pollution map may be spread to subscribed users or third-party services according to a certain policy (e.g., when reaching a threshold predetermined according to their profiles), so that subscribers may take the expected actions. Service subscribers may include, for instance, applications related to e-Health, Smart Transportation, Smart Environment and Smart Industry, each one requiring specific update intervals and map coverage area. On the other hand, *Service B* aims at selecting the best route for vehicles according to several real-time information and user preferences, including reduction of route time, traffic jams, pollution, tolls, number of traffic lights (reduction of overall number of stops), and number of public services vehicles in the selected route (such as garbage collector trucks) among others.

3.2 Control Plane

Albeit scalability concerns in conventional SDN networks have been largely discussed in the recent literature (see for example [14] and [15]), control mechanisms demanded by novel F2C systems envision the coordinated management of the whole set of resources in order to optimally map available resources into the services requirements. However, the conventional SDN control plane is responsible for the communication among distinct networks, that is, the specific information regarding end-points of each network is usually not considered when setting a path between two networks and, consequently, resources information kept by controllers is limited to conventional network information, such as network addresses, interfaces, and costs, among others. This condition may substantially impact on scenarios requiring edge devices information, such as fog computing and F2C systems. Indeed, when putting together fog computing and SOAs, novel IoT services may be deployed, leveraging not only network aspects (network communication technologies, available network interfaces, bandwidth, etc.), but also the set of resources available at the edge of the network, including for example processing and storage resources. In fact, considering such additional set of resources enables the execution of the desired network services, but also new highly interesting features, such as the offloading of services deployed on end-user devices.

Hence, in IoT scenarios leveraging edge approaches (fog, edge, F2C), additional control information must be also contemplated to both optimally map services into resources and enable new performance functionalities. Thus, novel control architectures must be designed, considering the whole set of information needed to efficiently support services demands while optimizing resources availability. The required additional information must represent characteristics inherent to each resource type, such as:

- Processor: architecture, clock rate, number of shared cores, cache size.
- Storage: type, capacity, read/write velocity.

- Sensor/actuator: type of data/action produced by the device as well as the inherent characteristics of each type, such as range, resolution, sensitivity, just to name a few.

Distinct control topologies have been studied aiming at the deployment of SDN networks, including centralized, decentralized and hierarchical topologies with distinct number of layers [15]. Each topology may present benefits and drawbacks regarding deployment simplicity, scalability, cost, response time, or manageability, just to name a few. We argue that the control scenario envisioned by F2C systems does not fit into a single control topology approach. Moreover, the issue is not only what the control topology will be (centralized, decentralized, hierarchical) but also whether a unique one may cover the whole set of control needs and specifications. Indeed, recognized the diversity in services demands, the constraints brought by adopting resources mobility and the increasing heterogeneity of resources, we definitely argue that using one single and unique control topology for the management of the diversity of service categories enabled by smart scenarios in an IoT world turns to be unfeasible –the mapping of services requirements into the most suitable resources must be met taking into consideration several resource characteristics, such as the aforementioned ones. Therefore, the envisioned scenario enforces the adoption of new control approaches, such as the creation of specific resources lookup tables, containing resources able to support distinct service requirements.

Aligned to this trend, we push for the creation of distinct control plane instances (kind of control virtualization), each one managing specific resources according to a distinct set of characteristics. For instance, one control instance may manage resources able to provide high performance computing whilst another one may manage resources able to provide green computing. Several approaches must be considered aiming at the optimization of control planes deployment. It is worth mentioning that the relationship between control and resources is not exclusive, i.e., one resource characteristic may match more than one control instance. Each control plane instance may be composed by a distinct set of controllers, thus presenting distinct control topologies intended to optimally manage the underlying resources, thus setting different control views or dimensions (i.e., multidimensional control). In this paper, we follow the work done by the EU mF2C project [16] intended to design a control architecture for F2C systems, and thus we adopt the naming defined within the project. Accordingly, the communication among controllers and underlying resources is done by means of novel elements, referred to as Agents [17], whose deployment on real devices may be performed either as an application download – executed on each resource, or as an ad-hoc light functionality built on devices with higher simplicity, such as some sensors. The Agent is the element responsible for running all required control functionalities on all the devices participating in the F2C system and will be deployed as different "suites" according to the target device's hardware (it looks obvious that a deployment on a laptop will not be the same than in a Raspberry Pi controlling a sensor in a city). The Agents are organized into a hierarchy, according to the hierarchical view of the F2C architecture (see Fig. 1). Each fog belonging to a Fog Layer will select one Agent (i.e., device) to act as the Leader, taking over the main control responsibilities for its area of coverage and included elements (see the traffic lights in Fig. 1 for Fog Layer 1).

The policies designed to handle the Leader selection process are out of the scope of this paper and its design is actually an ongoing work for the authors.

In Fig. 1, we show a single control topology, where Leaders are deployed at the traffic lights, and the fogs (two per area) are set meeting a specific policy or a set of deployed rules, yet to be defined. We pose the fact that such a static control topology might not suit all potential demands coming from all services to be executed and available resources. In fact, this is the main rationale to suggest the multidimensional control view, where the control topology may vary to optimally suit resources capacities and services demands. In short, a device may play as Leader for a set of services and as a normal Agent for another one. This means that the Agent software must support such a multidimensional view, defining policies and strategies to guarantee the best control topology for any service.

In order to illustrate the expected behavior for the envisioned multidimensional control approach, we introduce three distinct scenarios, namely different companies, different SLAs and finally the Control as a Service. For the sake of understanding and to perfectly define the different control roles devices may play, we will keep using the terminology and naming used in the mF2C project, as defined previously in this section.

Distinct Companies Employing Shared Resource. This scenario poses a smart city putting together different infrastructure components consisting of both the own city infrastructure and the one offered by the envisioned sharing model. First, we consider a smart city whose IoT infrastructure is already deployed and consequently is made available for service providers. Second, beyond the resources deployed by the city, the set of available resources might also include idle resources from users' devices, in a sharing model, where users willing to contribute to the whole set of resources may participate in such a collaborative framework. Therefore, we may assume that the infrastructure to be deployed by a service provider to enable services execution at the available edge resources would simply consist in the resources where the Leader would be deployed at, for a particular set of services and this would only happen when the service provider wants to have a strict and total control on the services, i.e. a sort of "proprietary" control. It is also worth noting that the usage of shared resources by distinct companies opens opportunities for new business models both between distinct companies and between companies and users. In this scenario, contracts established among parties must define which companies can make use of edge devices, resources provided, priorities, prices, policies, etc. This scenario can be certainly extended for different service providers turning into the distinct control instances depicted in Fig. 3. We may also see in Fig. 3 that whilst several physical resources, e.g., sensors, are shared by distinct providers, some of them are only available for one of the service providers, e.g., traffic lights, according to contracts previously established by parties. On the other hand, resources initially available for both providers, such as the yellow cab, must implement some sort of resources monitoring function, since for example the complete utilization of its shared resources by Company A (see Fig. 3) would make it not available for Company B. Hence, distinct control strategies must be sought considering scalability, required table accuracy, business models, overhead, etc.

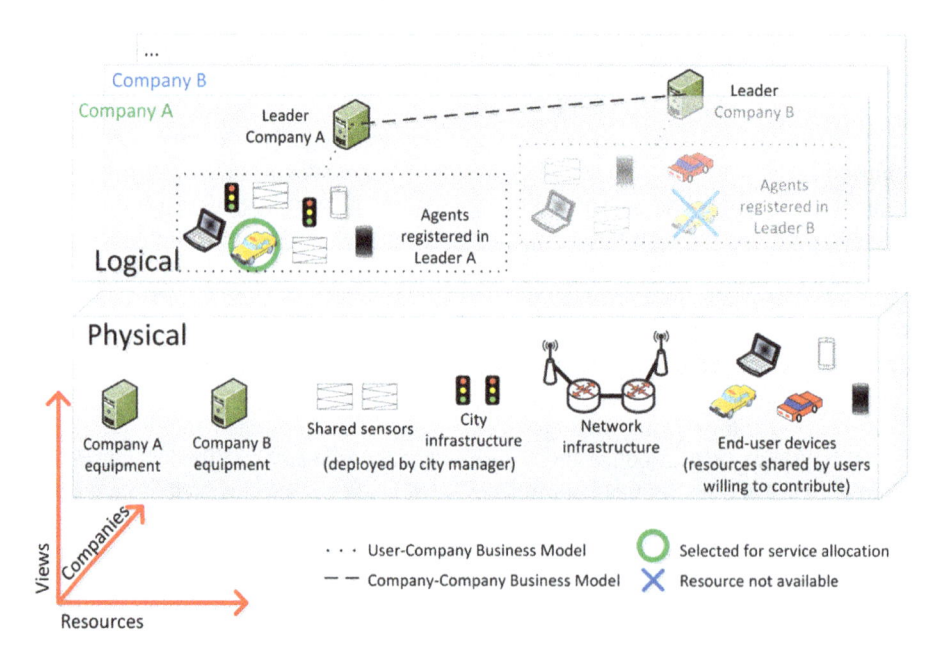

Fig. 3. Multidimensional control plane for management of shared resources by distinct companies.

Distinct SLA Provisioning. Different from the first scenario, this scenario shows the benefits obtained by a single service provider (company) when deploying multiple control instances intended to meet different SLAs in the resources provisioning process. Since the mapping of suitable resources for each service type may require distinct control data, such as resources characteristics specific for each individual service demands, the deployment of Leaders able to manage a large set of resources producing huge volume of control data in a specific area turns to be non-realistic. This means that different Leaders should have to be deployed in a specific area, what undoubtedly would require Leaders communication to keep a synchronized view of the underlying resources, driving a non-negligible communications overhead. However, the deployment of distinct control instances managing resources and providing distinct QoE meeting the required SLAs, may drastically reduce the number of resources that a Leader needs to manage. For instance, distinct control instances may manage resources demanding distinct requirements, such as green computing, high security communication, high performance computing, and free usage, as illustrated side by side in Fig. 4 for the sake of simplicity. It is also worth noting that distinct topologies for Leaders may be assumed for each SLA, since several factors may be considered in the topology definition, such as amount of underlying resources, their categories, capacities, etc. Moreover, albeit Agents may be controlled by more than one Leader, resources lookup table in the Leader may be considerably reduced, hence, easing the synchronization among

Leaders. Finally, we must also remark that in this second scenario, this synchronization also brings benefits when considering the fact that Leaders are not competing unlike the first scenario, where competitors may restrict smart agents' inter-communication.

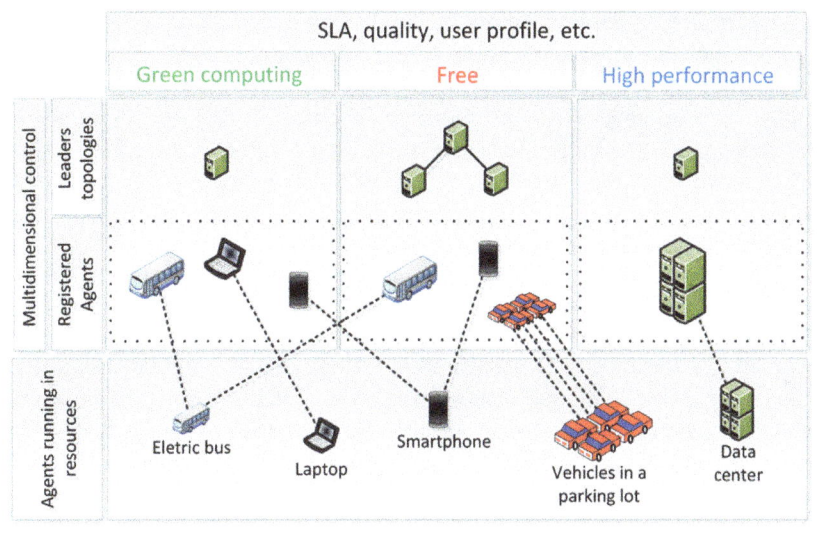

Fig. 4. Multidimensional control plane for provisioning of resources according to SLAs.

Control as a Service. As described in the previous section, the CaaS concept, introduced in [8], aims at using devices at the edge of the network as controllers – leveraging fog computing concepts in order to bring control decisions closer to the end-user, thus enabling real-time resource selection for sensitive IoT services execution. Unfortunately, the limited capacity inherent to edge devices makes usually unrealistic such devices to simultaneously run distinct services and, more important from a control perspective, to play different control roles, i.e., Agent or Leader for different services. Therefore, the categorization of resources, followed by the creation of distinct resource databases compliant to distinct service requirements, will clearly show the capacity to use constrained resources in terms of both processing and storage, to play as Leaders for specific services. This databases categorization will make resource selection and provisioning to be performed with a reduced latency, due to the reduced resources database size, constituted exclusively by resources able to provide the services managed by that Leader.

In such scenario, for each control instance, Agents deployed at edge devices must use strategies for the selection of the most suitable device to play the Leader role according to service characteristics and broadly speaking, according to a policy or strategy to be defined to that end. It must be remarked that, as the control plane may be formed by multiple control instances, the device playing as the Leader in one instance, may or may not be used as Leader for other control instances. Therefore, an edge device selected as Leader for a service may use part of the shared resources to run the Leader and further use idle shared resources for the execution of other services. This scenario is shown in Fig. 5, where edge devices highlighted with green and blue circles are selected as Leaders for service A and B respectively. Analyzing the logical view of the distinct control topologies, two key aspects deserve special attention. First, each service makes use of resources provided by distinct devices, according to their suitability to execute that service, regardless where the resources are. Second, as previously introduced, devices serving as Leader for a particular service may play a different role or simply share resources for other services, for example, the traffic light serving as Leader for service A can further share idle resources for service B provisioning. Finally, it is worth noting that the Agent (i.e., a functionality embedded in the Agent software) should be responsible for handling the sharing of local resources for service execution as well as their utilization as Leader.

4 Preliminary Results

In order to present a preliminary assessment of the concepts presented in this paper, two distinct experiments are carried out in an in-lab testbed deployed at the lab. The two proposed experiments, referred to as distinct SLA provisioning and CaaS, are inferred from the set of scenarios introduced in Sect. 3.2. The first experiment is based on the scenario where distinct SLA are provided by a company through the deployment of distinct control instances, each leveraging the most suitable resources.

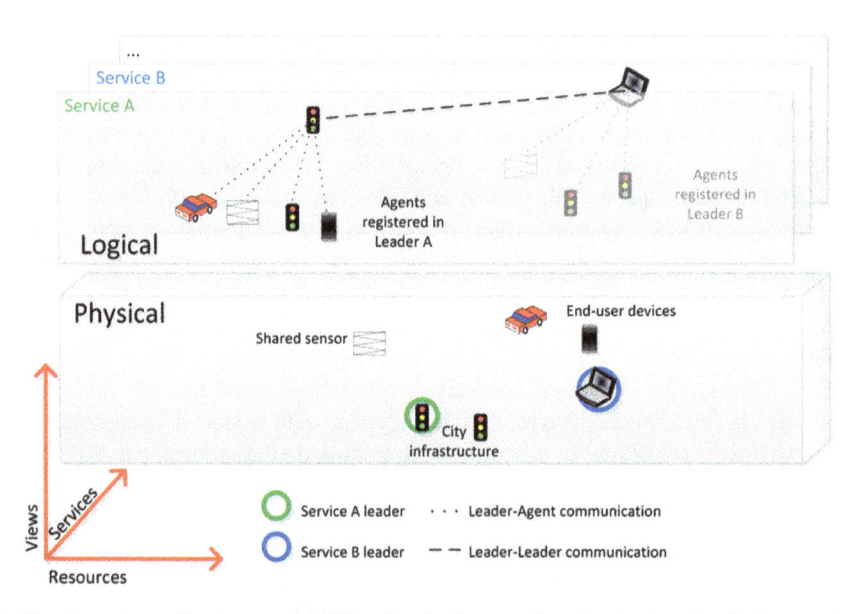

Fig. 5. Control as a Service provided by edge devices employed as leaders for distinct services.

As previously described, the resources are shared among distinct leaders. Thus, an efficient allocation demands an updated view of the underlying resources by all leaders. In this experiment, we assume two static leaders (deployed by one Service Provider) presenting wired connectivity. Both leaders can communicate to synchronize the resources allocation, thus enabling each leader to have an updated view of the shared resources usage. In this approach, we consider that, upon receiving a service request, a leader relies on the current view of underling resources and selects a set of the most suitable ones for the service provisioning, ordered according to their suitability which relies on both the service and resources categorization and the previously determined mapping policies. In the next step, the leader establishes wireless communication with the first resource, i.e., the most suitable one of the ordered set, asking for the required resources reservation and keeps waiting for a reply. In case of failure or negative reply, the procedure is repeated for the next resource of the ordered set, and so on. In case of positive reply, the second leader is informed about the allocation of the shared resource which may include additional information, such as the estimated allocation time, according to the accepted service categorization.

It is worth noting that, as described by the second scenario included in this paper, each leader in this experiment constitutes a distinct control dimension, where underlying resources may or may not be shared between dimensions. Figure 6 compares the presented approach and a single dimensional control topology for the first scenario in terms of the processing delay, showing a considerably lower request processing latency for the multidimensional approach proposed in this paper.

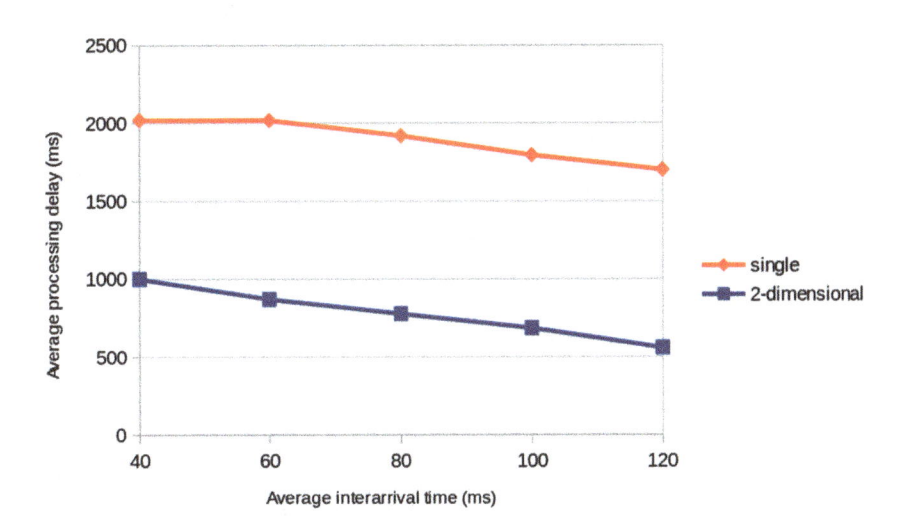

Fig. 6. Comparison between a single and 2-dimensional control for the SLA scenario.

In the second experiment, the CaaS scenario (third use case in Sect. 3.2) is considered. Assuming that the policy to select the Leaders is out of the scope of this paper, two aspects are considered affecting the Leaders selection process. First, in this scenario shared resources at the edge are utilized for the deployment of Leaders. Second, we assume 2 devices are deployed to play the Leader role with the aim of improving services execution (i.e., two control dimensions). Hence, we assume the first Leader is used for processing the requests from the early beginning of the experiment execution, whilst the second Leader is activated by the first one, according to 2 distinct policies, highlighting the differences brought in when including the multidimensional concept, defined as follows:

- **Forward**: Policy used by the first Leader when it cannot process a received request due to its limited capacity. Hence, the received request is forwarded to the second Leader, which may either accept or deny the request for a service execution according to its current load (additional policies may be added here to enrich this decision). Notice that we consider only 2 Leaders in the proposed experiment, thus impeding further additional request forwarding. This policy does not implement the multidimensional concept, since all Leaders must be ready to run any service and the decision of acceptance is based exclusively on the load of the Leader in terms of amount of concurrent requests. In other words, each Leader should be aware of all the underlying resources offering the services required by the requests received by the first Leader.
- **Split**: This policy considers a multidimensional view where distinct services may leverage distinct control topologies. In this approach, a Leader may decide to accept a particular service or a specific set of services. Therefore, if a Leader cannot handle the management of requests from a new service type, a new Leader is selected among the Agents through an election mechanism (yet to be defined). The new Leader, then, defines a new control topology, taking into account the service requirements and the set of available resources, responsible for the service provisioning. To that end, different approaches may be applied (yet an ongoing work), such as, for example, broadcasting welcome messages containing relevant information about the service and waiting for compliant Agents willing to share resources for that particular service and thus willing to join the new control instance, further enabling service clients to know the leader responsible for the management of that service type.

In addition to these two approaches and for the sake of comparison, a single topology approach is further deployed, in order to analyze the obtained results when no extra Leaders are deployed. In the single approach, no second Leader is considered, hence when the first Leader cannot handle a new request (due the limited capacity of the Leader), the request is denied. Therefore, if a service request does not go through, the service client employs an exponential-back off-based strategy for retrying the service request. It is worth noting that Forward and Split strategies do not guarantee the successful reservation of edge resources, hence, the exponential-back off retransmission scheme may be employed by clients each time a service request is not accomplished, regardless the policy employed by Leaders.

The capacity of each Leader is defined regardless the utilized approach. Therefore, the maximum capacity of a Leader is set as the maximum amount of service requests it is willing to process concurrently. Indeed, we consider, for simplicity, that all service types have the same complexity for resource selection. Therefore, as shown in Fig. 7,

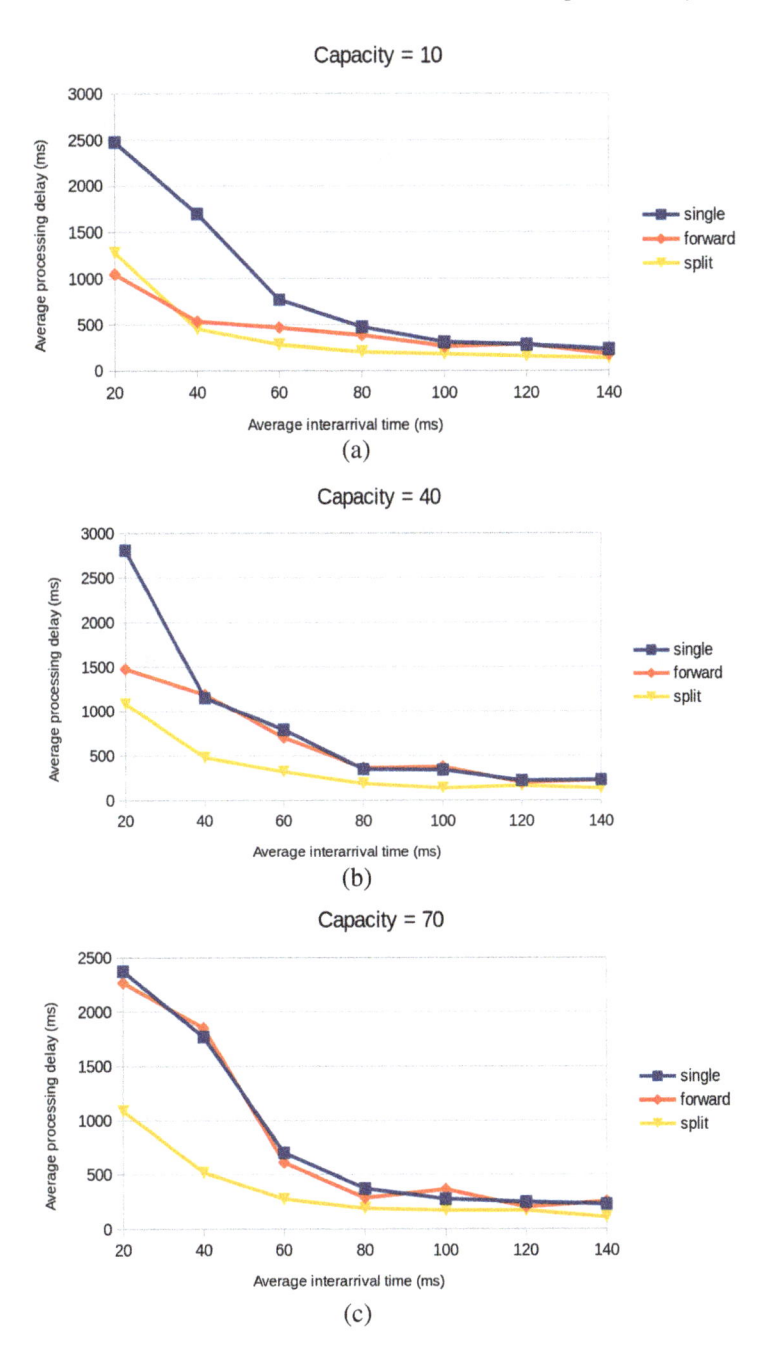

Fig. 7. Comparison of distinct strategies for processing service requests for distinct leader capacities: (a) 10, (b) 40, and (c) 70 concurrent requests.

for each comparison the capacity of all leaders used is the same. The analysis of the plots in Fig. 7 shows that the best results in terms of delay for resource reservation (average request processing delay) into distinct scenarios set by playing with both the amount of concurrent requests and the requests interarrival time, are obtained when deploying a two dimensional control (split). Furthermore, the increment of capacity of the Leaders (in terms of amount of service requests) does not result in a reduced average request processing delay. This may be justified by the fact that the constrained processing capacity of the Leaders, along with the large number of requests, leads to a high competition and overload of resources used for processing the received requests. Moreover, the increment of the capacity on the Leaders turns into a higher delay when forwarding requests. This is justified by the fact that, with higher capacity, the split strategy reduces the requests forwarding rate, which makes the Leader's behavior to tend to the one presented by the single approach.

5 Opportunities and Challenges

The deployment of the multidimensional control architecture introduced in this paper raises several challenges for its successful deployment. In this section, we assemble the described challenges in order to provide distinct opportunities for future research in this topic, described as following.

In a multidimensional architecture, where shared resources are managed by Agents in distinct dimensions, the scalability assessment is crucial. Indeed, once an Agent acting as Leader selects a resource for service execution, and since the latter can be initially available for more than one Leader, distinct strategies to keep an updated view of resources in distinct Leaders must be assessed.

Since distinct control topologies, such as centralized, distributed and hierarchical, present advantages and disadvantages, strategies for the topology definition for distinct dimensions according to service needs and available resources are required, enabling the optimal deployment of Leaders, minimizing signaling and latency whilst enabling control decisions to be taken closer to end-user making use of updated resources information.

In scenarios where Leaders are dynamically assigned, strategies for runtime selection may be assessed. It is worth mentioning that distinct strategies may be employed for Leader selection considering the layer they are located in control topology, service to be provisioned, amount of Agents to be controlled and their characteristics, among others. In addition, Leaders coordination is an added challenge when considering both intra and inter-control instances coordination.

The deployment of distinct Leaders for the provisioning of distinct services yields new challenges regarding the knowledge of available Leaders by clients. Alternatives must be defined in order to enable each client to discover which Leader is managing resources able to provide the required service. Whilst solutions such as the deployment of brokers may be effective, the added delay must be assessed, especially for highly sensitive services.

The deployment of such a F2C collaborative model fuels the establishment of novel business models not only among service providers, but also between service provider and clients. For the latter, SLA between each pair client-provider may comprise expected QoS, user preferences regarding shared resources, schedules, privacy, and other preferences that shall be available in user profile. For the former, besides directives for sharing private resources, an SLA may include rules for data sharing while respecting clients' preferences, such as privacy. The definition of novel business models is the basis for the successful deployment of such collaborative model.

As Agents are responsible for managing resources available at edge devices, its implementation shall require the definition of policies to enable proper resource allocation according to preferences defined at user profile.

In traditional host-oriented networking, an SDN controller does not need to have knowledge about the edge devices. Rather, the controller keeps only information regarding forwarding devices topology and, among others, it can define the switches (forwarding devices) that should be used in order to establish communication between two distinct networks. On the other hand, next generation service-oriented IoT applications will require the edge resource selection according to the services offered by them. Moreover, the amount of information regarding the edge resources whose controllers should keep will increase according to the amount of offered services.

Several security concerns must be considered in such a collaborative model. That includes but is not limited to privacy, authentication, access control, identity management, integrity, and availability.

Finally, besides the challenges arisen from the deployment of the proposed multidimensional architecture, several challenges are still not completely solved by considering F2C systems. Therefore, resources discovery, resources monitoring, devices tracking, service allocation, efficient services orchestration, or optimal service-resources mapping, are just some of the research topics that still deserve efforts in order to enable the successful deployment of F2C systems.

6 Conclusion

In the upcoming years, new business models shall arise based on collaborative models where mobile and non-mobile end-users will share idle resources whilst distinct service providers will benefit not only from end-users resources but also from resources deployed on the ground (cities, transport systems, etc.). In this paper, we raise concerns about deploying one single control topology able to provide optimal management of the tremendous amount of envisioned shared resources, while providing distinct QoS requirements according to distinct SLAs for the large set of potential services. Aligned to this concern, this paper positions a multidimensional control architecture for novel combined Fog-to-Cloud (F2C) systems, as a potential solution for a service-tailored management of the devices deployed at the edge of the network. This multidimensional architecture envisions the coexistence of distinct control plane instances enabling optimal management of available resources while fulfilling QoS requirements regarding deployed services and end-user profiles. Through three distinct scenarios, the multidimensional control concept is presented in a comprehensive manner, and is later evaluated in two of them, intended to highlight its potential benefits. Nevertheless, we reinforce the fact that, due to the novelty of the proposed concept, the multidimensional concept opens many different research avenues, opportunities and challenges, most of them listed in the last section of this paper to provide the reader with a complete picture of the overall scenario that must be addressed for a successful deployment of the proposed concept.

Further work will go to many directions, including security provisioning (designing an architecture responsible for providing security to the whole set of F2C systems, considering their specific characteristics, i.e., mobility, heterogeneity, etc.), deployment in specific verticals (including health and vehicular systems) and the design of clustering policies using ML strategies (to identify the proper solution to optimize resources consumption and services performance).

Acknowledgment. This work was supported by the H2020 EU mF2C project, ref. 730929 and for UPC authors, also by the Spanish Ministry of Economy and Competitiveness and the European Regional Development Fund under contract RTI2018-094532-B-I00.

References

1. Ahmed, E., Yaqoob, I., Gani, A., Imran, M., Guizani, M.: Internet-of-things-based smart environments: state of the art, taxonomy, and open research challenges. IEEE Wirel. Commun. **23**(5), 10–16 (2016). https://doi.org/10.1109/MWC.2016.7721736
2. Gubbi, J., Buyya, R., Marusic, S., Palaniswami, M.: Internet of Things (IoT): a vision, architectural elements, and future directions. Future Gener. Comput. Syst. **29**, 1645–1660 (2013). https://doi.org/10.1016/j.future.2013.01.010
3. Bonomi, F., Milito, R., Zhu, J., Addepalli, S.: Fog computing and its role in the internet of things. In: MCC 2012 Proceedings of the First Edition of the MCC Workshop on Mobile Cloud Computing, pp. 13–16. ACM (2012). https://doi.org/10.1145/2342509.2342513
4. Hu, P., Dhelim, S., Ning, H., Qiu, T.: Survey on fog computing: architecture, key technologies, applications and open issues. J. Netw. Comput. Appl. **98**, 27–42 (2017)
5. OpenFog Consortium: OpenFog Reference Architecture for Fog Computing, February 2017. https://www.openfogconsortium.org/wp-content/uploads/OpenFog_Reference_Architecture_2_09_17-FINAL.pdf
6. Masip-Bruin, X., Marín-Tordera, E., Tashakor, G., Jukan, A., Ren, G.J.: Foggy clouds and cloudy fogs: a real need for coordinated management of fog-to-cloud computing systems. IEEE Wirel. Commun. **23**(5), 120–128 (2016). https://doi.org/10.1109/MWC.2016.7721750
7. Masip-Bruin, X., Marín-Tordera, E., Jukan, A., Ren, G.J.: Managing resources continuity from the edge to the cloud: architecture and performance. Future Gener. Comput. Syst. **79, part 3**, 777–785 (2018)
8. Souza, V.B., Gómez, A., Masip-Bruin, X., Marin-Tordera, E., Garcia, J.: Towards a fog-to-cloud control topology for QoS-aware end-to-end communication. In: 2017 IEEE 25th International Symposium of Quality of Service (IWQoS), Vilanova i la Geltrú, Barcelona (2017)
9. Ramirez, W., Masip-Bruin, X., Marín-Tordera, E., Souza, V.B.C., Jukan, A., Ren, G.J., González de Dios, O.: Evaluating the benefits of combined and continuous fog-to-cloud architectures. Comput. Commun. **113**, 43–52 (2017)
10. Sood, K., Yu, S., Xiang, Y.: Software-defined wireless networking opportunities and challenges for Internet-of-Things: a review. IEEE Internet Things J. **3**(4), 453–463 (2016). https://doi.org/10.1109/JIOT.2015.2480421

11. Ku, I., Lu, Y., Gerla, M., Gomes, R.L., Ongaro, F., Cerqueira, E.: Towards software-defined VANET: architecture and services. In: 2014 13th Annual Mediterranean Ad Hoc Networking Workshop (MED-HOC-NET), Piran, pp. 103–110 (2014). https://doi.org/10.1109/MedHocNet.2014.6849111
12. Tomovic, S., Yoshigoe, K., Maljevic, I., Radusinovic, I.: Software-defined fog network architecture for IoT. Wireless Pers. Commun. **92**(1), 181–196 (2017). https://doi.org/10.1007/s11277-016-3845-0
13. Xu, J., Palanisamy, B., Ludwig, B.H., Wang, Q.: Zenith: utility-aware resource allocation for edge computing. In: IEEE Edge 2017, Honolulu, Hawaii, 25–30 June 2017 (2017)
14. Yeganeh, S.H., Tootoonchian, A., Ganjali, Y.: On scalability of software-defined networking. IEEE Commun. Mag. **51**(2), 136–141 (2013). https://doi.org/10.1109/MCOM.2013.6461198
15. Hu, J., Lin, C., Li, X., Huang, J.: Scalability of control planes for Software defined networks: modeling and evaluation. In: 2014 IEEE 22nd International Symposium of Quality of Service (IWQoS), Hong Kong, pp. 147–152 (2014). https://doi.org/10.1109/IWQoS.2014.6914314
16. mF2C project. http://www.mf2c-project.eu. Accessed May 2019
17. Deliverable 2.6, mF2C EU project. http://www.mf2c-project.eu/wp-content/uploads/2017/06/mF2C-D2.6-mF2C-Architecture-IT-1.pdf. Accessed May 2019

Study of Polynomial Backoff for IEEE 802.11 DCF

Bader A. Aldawsari[✉]

Department of Computer Science and Engineering, University of Colorado Denver,
Denver, CO, USA
bader.aldawsari@ucdenver.edu

Abstract. This paper proposes and studies performance of the IEEE 802.11 distributed coordination function (DCF) with new backoff functions, which are analyzed under saturation conditions. A unified Markov chain model is presented, which is used to obtain theoretical results of saturation throughput for all backoff functions which includes the newly proposed cubic backoff (CB) function. The theoretical results are also verified through simulation experiments. Based on the results, performance of all backoff functions can be improved significantly by setting the minimum window size to typically large values in Basic access mechanism, while a typically small windows size is optimal for RTS/CTS access mechanism. The results also show that CB has significantly higher saturation throughput compared to the standard binary exponential backoff in Basic access mechanism, while it performs better in RTS/CTS access mechanism in cases where number of stations is typically high. One key finding from this study is that performance of the IEEE 802.11 DCF can be improved by optimizing the minimum window size parameters and modifying the window backoff function.

Keywords: IEEE 802.11 DCF · Saturation throughput · Markov chain

1 Introduction

Wireless local area networks (WLAN) are one of the most widely used types of network and are based on the IEEE 802.11 standard. As a result, numerous researchers have focused on analyzing the performance of the IEEE 802.11 distributed coordination function (DCF) protocol, which governs access to the channel for all connected stations in the network. The objective of these researches is to find ways to improve performance of the network in various performance measures such as developing new communication algorithms or through optimization of current algorithm parameters.

The binary exponential backoff (BEB) function is the standard backoff function used in IEEE 802.11 for both Ethernet and WLANs. When a station has packets to transmit, it first chooses a random wait-time interval based on the current window size. The backoff window size is then multiplied by 2 for every failed transmission attempt. The goal of increasing the window size is to increase the probability of avoiding collisions for future re-transmissions.

In this paper, the saturation throughput for IEEE 802.11 DCF is studied and analyzed with four backoff functions that have varying window growth rate. The goal is to investigate the relation between the window size and saturation throughput by comparing several backoff functions with different window sizes, and find optimal window size settings for different access mechanisms. Saturation conditions are assumed in which a station always has a packet to transmit at all times. This implies that, when a transmission of a packet is successful in round t for a station, the same station immediately has another packet that is ready for transmission at round $t + 1$. Saturation conditions help in analyzing a contention resolution protocol in worst-case conditions where the throughput becomes dependent on the number of stations contending for access to the same shared channel.

This paper presents a comparison of saturation throughput for IEEE 802.11 DCF using four different backoff functions along with the validation of the results through simulations. The results show that both quadratic backoff (QB) and cubic backoff (CB) functions have significantly better performance compared to the BEB function for the Basic access mechanism. Further analysis of the results shows that in general a large window size improves saturation throughput in Basic access mechanism. Thus, there is a potential to improve the BEB throughput by choosing a typically larger minimum window size that is different than the IEEE 802.11 standard value of 32.

This paper is organized as follows. Section 2 presents a review of previous and related work. Section 3 presents an overview of studied backoff functions. Sections 4 and 5 presents an analytical model used for obtaining saturation throughput. The performance of all backoff functions are studied and analyzed in Sect. 6. Finally, a conclusion of this study is presented in Sect. 7.

2 Previous and Related Work

In early analyses of backoff functions, the performance was analyzed under Poisson arrival of packets, such as the work by Hastad et al. [4]. They studied the stability of both the polynomial backoff and BEB functions regarding the packet arrival rate and found their stability results.

Bianchi [1] introduced one of the most widely used models to analyze the DCF, where the IEEE 802.11 DCF is analyzed under saturation conditions. In saturation conditions, a station always has a packet to transmit in its queue. The analytical model found in [1] uses a Markov chain model and was proved through simulations to be highly effective in capturing the real saturation throughput. It has been widely adopted by numerous researchers [2,9].

The quadratic backoff (QB) function has been compared with the BEB function with both the bounded value k and unbounded k, where k is the maximum backoff stage [10]. The results showed that the QB function can achieve saturation throughput comparable to the BEB function.

Xu et al. [11] studied the DCF of IEEE 802.11e, which was designed for a WLAN, in which a packet is dropped when the maximum retry limit is reached. They proposed an analytical model of the enhanced distributed channel access (EDCA) mechanism of IEEE 802.11e with a single access class. The polynomial backoff (PB) and linear backoff (LB) functions have been analyzed and compared using performance measures, such as the saturation throughput, channel access delay, and pack drop rate, and it was found that different backoff functions can lead to better performance compared to the BEB function with a tradeoff between the packet delay and packet drop performance.

Several researchers have used similar models to analyze the IEEE 802.11 DCF, such as the work in [6] analyzing the BEB function with both finite and infinite numbers of stages. They focused on analyzing the general exponential backoff function with a base of r and found the optimal value to be $\frac{1}{1-e^{-1}}$. Other works have focused on analyzing the BEB function under different channel conditions and or with different configurations of the DCF [3,5,7].

3 Backoff Functions Overview

A backoff function is defined as part of the IEEE 802.11 DCF, where the window size is changed based on feedback from the channel. A station uniformly chooses a random backoff time slot from the backoff window before transmitting a packet. The window size is initially set to a predefined size and is denoted as W_0. In this section, the four backoff functions included in the analysis are defined. For the BEB function, the window size W_i is defined as follows:

$$W_i = W_0 \cdot 2^i, \qquad for \qquad i = 0, ..., M$$

For the LB function, the window size is defined as follows:

$$W_i = W_0 \cdot (i+1), \qquad for \qquad i = 0, ..., M$$

For the QB function, the window size is defined as follows:

$$W_i = W_0 \cdot (i+1)^2, \qquad for \qquad i = 0, ..., M$$

For the CB function, the window size is defined as follows:

$$W_i = W_0 \cdot (i+1)^3, \qquad for \qquad i = 0, ..., m$$

For the general PB function, the window size is defined as follows:

$$W_i = W_0 \cdot (i+1)^c, \qquad for \qquad i = 0, ..., m \qquad c = 1, 2, 3$$

4 Analytical Model

This section describes the analytical model based on Bianchi's model used for the saturation throughput analysis of the IEEE 802.11 DCF protocol with the

BEB function [1]. It is assumed that the system consists of a fixed number of stations denoted n that are contending for access to the channel with ideal conditions. A single station is first studied independently of other stations using a Markov model, which makes it possible to derive the stationary probability of transmitting a packet in a randomly chosen time slot τ. The stationary probability τ is then used to compute the throughput for the two access methods, Basic and RTS/CTS, that are considered in this paper.

The time interval of time slot t is assumed to be a constant denoted as σ. To analyze the performance under saturation conditions, the system is modeled as a Markov chain model consisting of several states. A state is represented by both backoff counter b and backoff stage s, where the backoff stage is bounded by M. A time slot t is represented by integer values, where t and $t+1$ are two consecutive time slots. The backoff counter of a selected station is represented by the stochastic process $b(t)$, whereas the backoff stage is represented by the stochastic process $s(t)$, where $0 \leq s < M$.

Initially, the window size is set to a predefined size W_0, which is then modified based on the feedback from the channel. The value i denotes the backoff stage, so by having $i = M$, which is the maximum backoff stage, it defines the maximum window, which is denoted by CW_{max}.

Next, several variables are defined in this analytical model, such as p, which is the probability of a collision for any transmission attempt of a packet regardless of the number of transmission attempts. State transition probabilities in the Markov chain model are as follows:

$$P(x, y | x\backslash, y\backslash) = z$$

The pair $(x\backslash, y\backslash)$ represents the source state where the two values in the pair are the backoff stage and backoff counter, respectively, at time slot t. The pair (x, y) represents the new values in the target state at time slot $t + 1$. The probability of transition between the two states is z.

Next, the Markov chain model used for the analysis of all backoff functions is defined, which is based on the following transitional probabilities

$$\begin{cases} P\{i, k | i, k+1\} = 1 & k \in (0, W_i - 2), i \in [0, M] \quad (1) \\ P\{0, k | i, 0\} = (1-p)/W_0 & k \in [0, W_0 - 1], i \in [0, M] \quad (2) \\ P\{i, k | i-1, 0\} = p/W_i & k \in [0, W_i - 1], i \in [1, M] \quad (3) \\ P\{M, k | M, 0\} = p/W_M & k \in [0, W_M - 1] \quad (4) \end{cases}$$

A state transition that is caused by decrementing the backoff counter with probability 1 is represented in (1). The second Eq. (2) represents a state transition of successful packet transmission where stage i is reset to stage 0 with a probability of $(1-p)/W_0$, and a new backoff counter is randomly chosen from the range $(0, W_0 - 1)$. The third Eq. (3) is for a transition from backoff stage $i-1$ to i when the transmission was unsuccessful. The final Eq. (4) is for unsuccessful transmission when the backoff stage is m, such that the backoff stage remains the same and a new random backoff counter is selected randomly. The stationary distribution is defined as follows:

$$b_{i,k} = lim_{t\to\infty}\{Ps(t) = i, b(t) = j\}, \quad i \in [0, M], j \in [0, W_i] \tag{5}$$

Based on the Markov chain of a station's state, the following equations can be derived:

$$b_{i,0} \cdot p = b_{i+1,0}$$

$$b_{i,0} = p^i b_{0,0} \qquad i > 0, \ i < M$$

This shows that for a station to be in state $b_{i,0}$, it must go through retransmission of the same packet exactly i times, where each transmission is unsuccessful. In the following expressions, three transitional equations are defined based on the backoff stage value:

$$b_{i,k} \begin{cases} \dfrac{W_i - k}{W_i} \cdot (1-p) \sum_{j=0}^{m} b_{j,0} & i = 0 \tag{6} \\[6pt] \dfrac{W_i - k}{W_i} \cdot p \cdot b_{i-1,0} & 0 < i < M \tag{7} \\[6pt] \dfrac{W_i - k}{W_i} \cdot p \cdot (b_{M-1,0} + b_{M,0}) & i = M \tag{8} \end{cases}$$

Next, the following equation is derived based on Eq. (6):

$$\sum_{i=0}^{m} b_{i,k} = \frac{b_{0,0}}{1-p} \tag{9}$$

From (9), the three Eqs. (6, 7, and 8) are rewritten as follows:

$$b_{i,k} = \frac{W_i - k}{W_i} b_{i,0} \tag{10}$$

Finally, the sum of all state probabilities is equal to 1, which can be used to obtain the individual probabilities $b_{i,0}$ at every backoff stage. In total, this represents the probability of packet transmission denoted as α in an arbitrary time slot.

$$\sum_{i=0}^{M} \sum_{k=0}^{W_i - 1} b_{i,k} = 1 = \sum_{i=0}^{M} b_{i,0} \sum_{k=0}^{W_i - 1} \frac{W_i - k}{W_i} \tag{11}$$

$$= \sum_{i=0}^{M} b_{i,0} \frac{W_i + 1}{2} \tag{12}$$

$$= \frac{b_{0,0}}{2} (W(\sum_{i=0}^{M-1} (ip^i)^c + \frac{(Mp^M)^c}{1-p}) + \frac{1}{1-p}) \tag{13}$$

The variable c denotes the polynomial exponent, which can be linear, quadratic, or cubic. Based on the above equation, the following equation is derived:

$$b_{0,0} = \frac{2(1-p)}{(W((1-p)\sum_{i=0}^{M-1}(i+1)^c(1+M)^c p^M) + 1)} \quad (14)$$

Next, we can derive the probability τ, which is the probability that a station transmits in a randomly selected time slot, because a station transmits its packet when $b = 0$, regardless of the backoff stage. The general equation for calculating τ for the PB function is calculated as follows:

$$\tau = \sum_{i=0}^{M} b_{i,0} = \frac{b_{0,0}}{1-p} = \frac{2}{(W((1-p)\sum_{i=0}^{M-1}(i+1)^c(1+M)^c p^M) + 1)} \quad (15)$$

From the previous equation, τ depends on the probability p that a transmitted packet experiences a collision. Assuming that the system reaches a steady state where every transmission experiences the system in the same state, the probability p can be computed based on the fact that, in a random time slot, at least one of the remaining stations transmits a packet. Thus, the probability p is computed using the following equation:

$$p = 1 - (1-\tau)^{n-1} \quad (16)$$

Finally, the two Eqs. (13 and 14) represent a nonlinear system with two unknown variables, τ and p, which can be solved using mathematical techniques.

5 Saturation Throughput

Before obtaining the saturation throughput S, the value P_{tr} is defined, which is the probability that at least one station transmits a packet in a time slot. Because there are n stations contending on the channel, where each station transmits with probability τ, the value P_{tr} is computed as follows:

$$P_{tr} = 1 - (1-\tau)^n \quad (17)$$

Next, the probability that a transmission is successful P_s is defined, which is based on the probability that exactly one single station transmits on the channel and that at least one station transmits which is computed as follows.

$$P_s = \frac{\tau n(1-\tau)^{n-1}}{P_{tr}} = \frac{\tau n(1-\tau)^{n-1}}{1-(1-\tau)^n} \quad (18)$$

Finally, the saturation throughput S is defined as the percentage of time where the channel is used to transmit packet bits. It is computed as the ratio of the successfully transmitted number of bits over the time slot duration:

$$S = \frac{E[\text{size of successfully transmitted payload}]}{E[\text{time slot duration}]} \quad (19)$$

The expected size of the transmitted payload bits $E[P]$ is calculated as the average of the number of transmitted payload bits for both successful transmissions and collision cases. Thus, the expected number of successfully transmitted payload bits is calculated as $P_{tr}P_s E[P]$. On the other hand, the expected time-slot duration is calculated based on the three observed types of feedback from the channel: successful transmission, collision, and silence. The probability of silent feedback from the channel is $(1 - P_{tr})$, where σ is the duration of a silent time slot. For any time slot, the probability of having a collision is $P_{tr}(1 - P_s)$, and the number of transmitted payload bits is T_c, whereas the probability of a successful transmission is $P_{tr}P_s$ and the number of transmitted payload bits is T_s. Thus, the Eq. (17) can be expressed as follows:

$$S = \frac{P_{tr}P_s E[P]}{(1 - P_{tr})\sigma + P_{tr}(1 - P_s)T_c + P_{tr}P_s T_s} \quad (20)$$

In the Eq. (20), the values T_s, T_c, σ and $E[P]$ are represented in the same unit (bits or μ). The two values T_s and T_c can vary based on the channel access mechanism. Let's consider that the total header of a packet is $Header^* = PHY_{header} + MAC_{header}$ and that the propagation delay is denoted γ. For the Basic access mechanism, the two values T_s and T_c are computed as follows:

$$T_s^{basic} = H + E[P] + SIFS + \gamma + ACK + DIFS + \gamma$$
$$T_c^{basic} = H + E[P^*] + DIFS + \gamma$$

When using the RTS/CTS access mechanism, collision can be experienced only in RTS frames. Thus, T_s and T_c are computed as follows:

$$T_s^{rts/cts} = RTS + SIFS + \gamma + CTS + SIFS + \gamma$$
$$\qquad + H + E[P] + SIFS + \gamma + ACK + DIFS + \gamma$$
$$T_c^{rts/cts} = RTS + DIFS + \gamma$$

The average size of the longest packet payload involved in a collision is denoted as $E[P^*]$. The saturation throughput S is now computed using Eq. (20) based on the IEEE 802.11b parameters found in Table 1.

6 Results and Analysis

The saturation throughput of different backoff growth functions has been obtained using the analytical model found in [1] for both Basic access and RTS/CTS mechanisms. The accuracy of the model has been verified by comparing the results with simulation experiments with all backoff functions using the network simulator tool [8].

In all figures (Figs. 1, 2, 3, 4, 5, 6 and 7), the simulation results are plotted with a dotted line for each backoff function. The theoretical results almost

Table 1. System parameters based on the IEEE 802.11b standard which are used to obtain saturation throughput of all backoff functions.

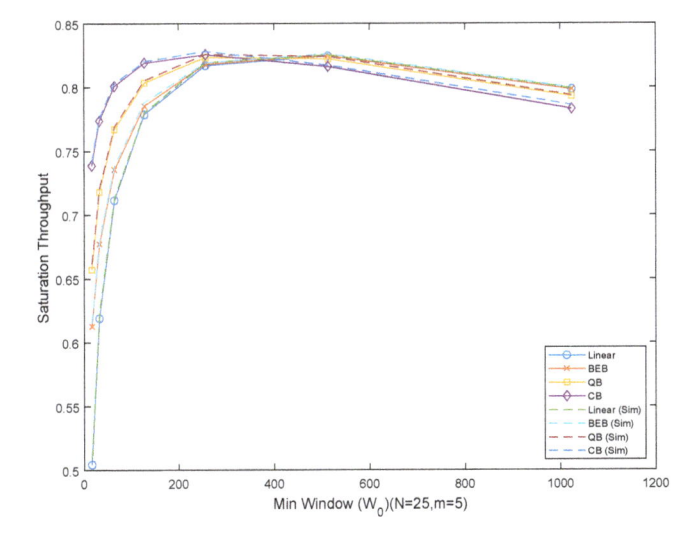

Slot time (σ)	50 µs
Packet payload	8184 bits
PHY header	128 bits
MAC header	272 bits
Header*	400 bits
ACK	240 bits
RTS	288 bits
CTS	240 bits
Propagation delay (γ)	1 µs
SIFS	28 µs
DIFS	128 µs

Fig. 1. Saturation throughput for different minimum window sizes W_0 and Basic access mechanism ($n = 50, m = 5$)

matches the simulation results, which shows how accurate the model is in computing the saturation throughput.

In this study, the effect of the window growth function on the saturation throughput has been analyzed for three different PB functions along with the BEB function. Figure 3 shows a comparison of the saturation throughput for all backoff functions with $W_0 = 32$ and using Basic access mechanism. It shows a clear difference in performance with CB having the highest throughput while the LB has the worst performance. Comparing the window size growth as seen in

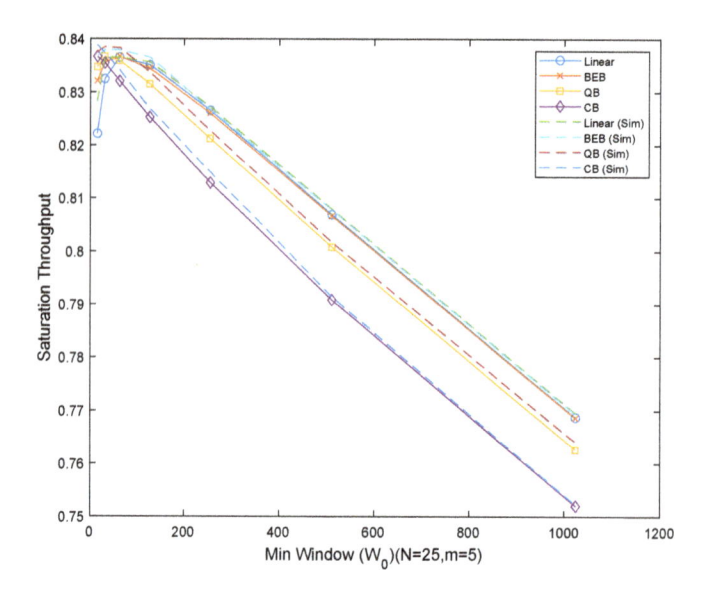

Fig. 2. Saturation throughput for different minimum window sizes W_0 using RTS/CTS access mechanism ($n = 50, m = 5$)

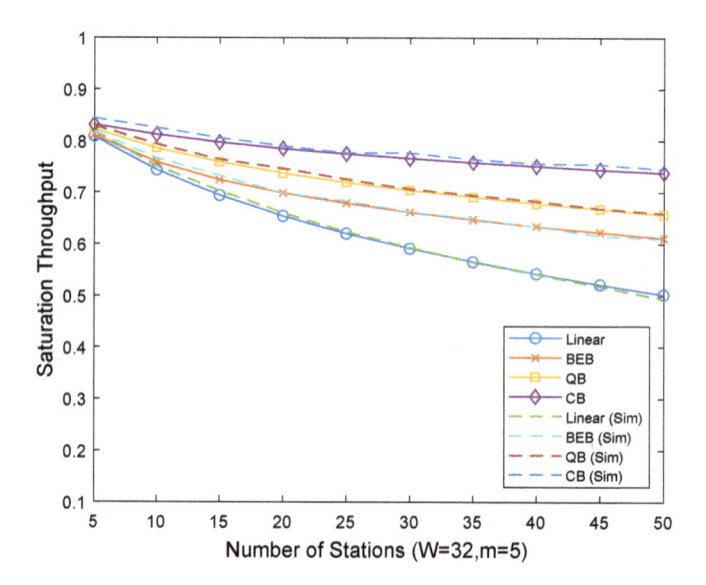

Fig. 3. Saturation throughput for different number of stations ($n = 5, .., 50$) for Basic access mechanism and $W_0 = 32$.

Fig. 7, the LB function has the worst performance among all backoff functions because its window size grows significantly slower compared to all other backoff functions, while CB grows significantly faster. This shows that large window size

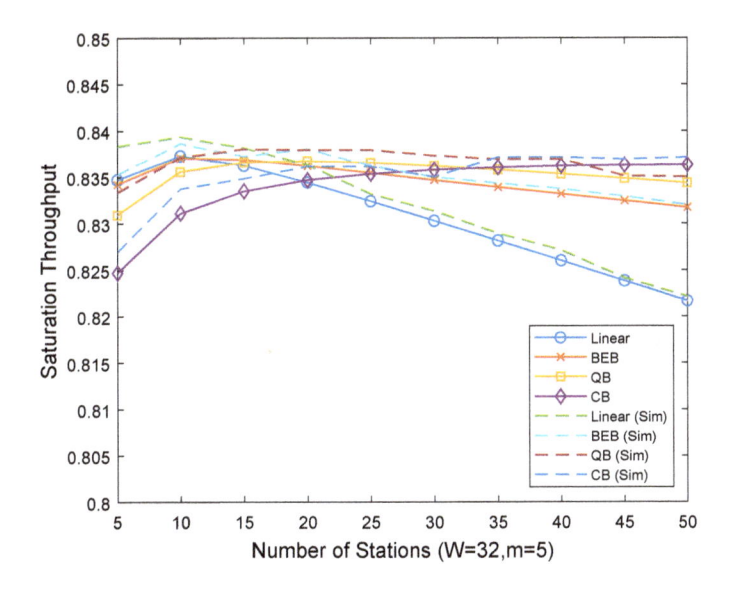

Fig. 4. Saturation throughput for different number of stations ($n = 5,..,50$) for RTS/CTS access mechanism.

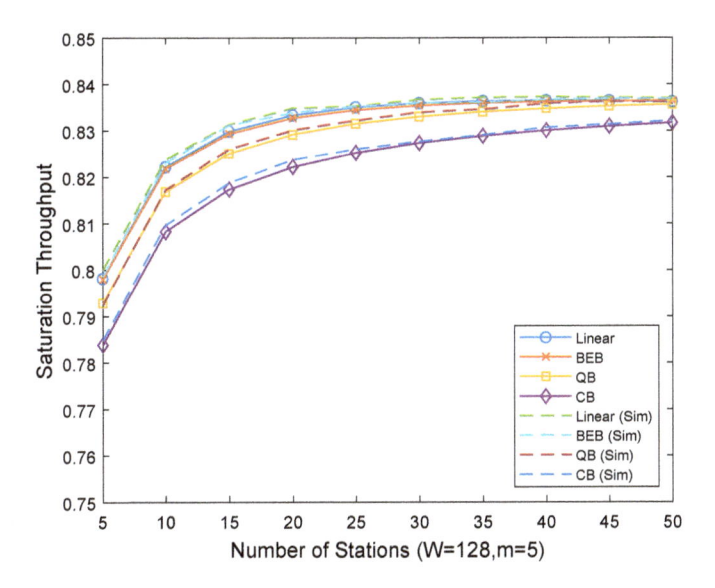

Fig. 5. Saturation throughput for different number of stations ($n = 5,..,50$) for RTS/CTS access mechanism and $W_0 = 128$.

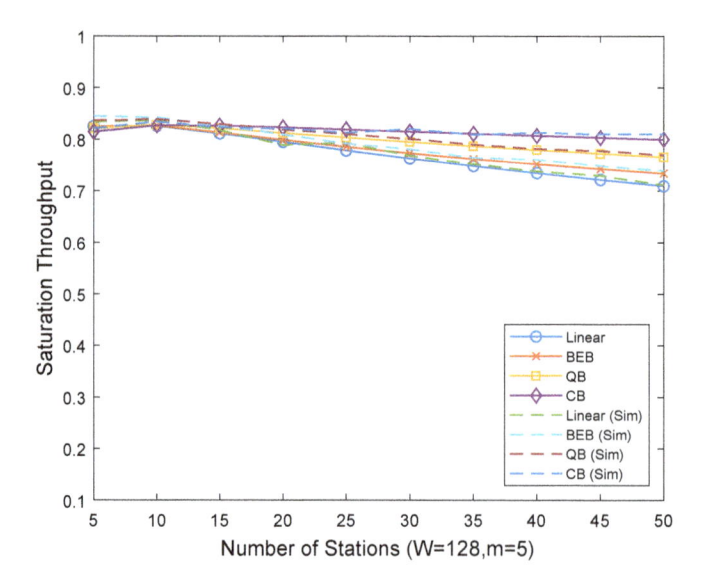

![Figure 6]

Fig. 6. Saturation throughput for different number of stations (n = 5, .., 50) for Basic access mechanism and $W_0 = 128$.

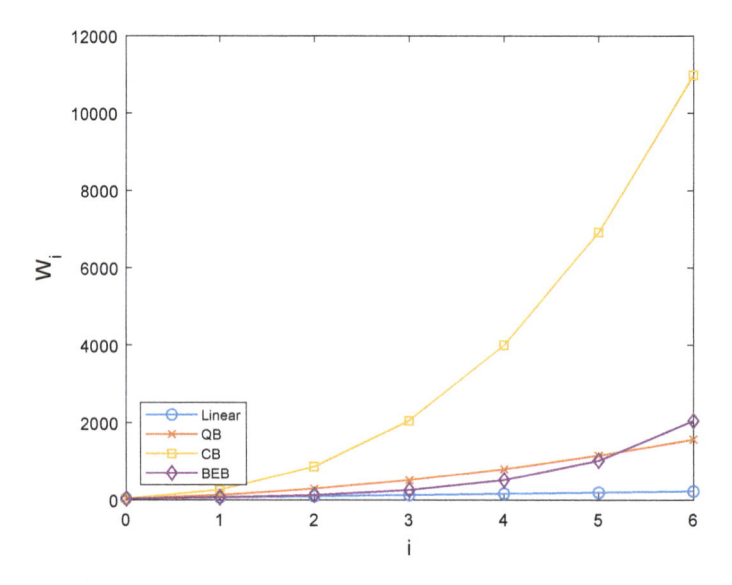

![Figure 7]

Fig. 7. Comparing window size of all backoff functions. i is the backoff stage number.

in early backoff stages brings significant performance improvement as CB has the highest window size in early stages.

Saturation throughput is also compared in RTS/CTS access mechanism as seen in Fig. 4. It shows that the throughput is slightly affected when the number of stations is increased compared to the Basic access mechanism. In addition, CB only performs better compared to other backoff functions when $n \geq 25$.

Figure 1 compares saturation throughput in Basic access for the different backoff functions with different starting window sizes in the range $(16, ..., 1024)$ and $n = 25$. It can be seen that CB has the highest throughput only when $W_0 < 512$, and then its saturation throughput becomes the lowest of all backoff functions. This can be seen as an effect of its minimum windows size becoming larger compared to other backoff functions. It can be said that the optimal W_0 value for Basic access should be chosen from the range (256–512). Thus, it was found that a typically small value for W_0 yield better performance in RTS/CTS access mechanism, while in Basic access a typically large value for W_0 improves the throughput.

Comparing saturation throughput for different minimum window sizes in RTS/CTS access mechanism yield contradicting outcome as seen in Fig. 2. Saturation throughput starts to decrease with $W_0 = 32$ for CB while all other backoff functions begin to decrease starting from $W_0 = 64$. The optimal W_0 for RTS/CTS access should be typically in the low range of 16–64.

As described previously, in RTS/CTS access mechanism for any $W_0 > 32$, the saturation throughput is decreasing which is why both CB and QB have the lowest performance in Fig. 5.

To summarize, the result of the experiments shows that it is preferable to have a backoff function with high growth curve for Basic access, while it is the opposite for the RTS/CTS access method. The reason for this variance is that in Basic access, the time overhead for a collision is significantly higher than the RTS/CTS access mechanism. Therefore, the saturation throughput of basic access is improved significantly due to collision avoidance, whereas in RTS/CTS access method the throughput is improved by having a station reattempts transmission of a packet after a short amount of time.

7 Conclusions

In this paper, saturation throughput of three backoff functions were analyzed and compared to the BEB function. A Markov chain model was used to obtain the theoretical results, which were validated through simulations. Even though all backoff functions have been compared and analyzed under small number of network settings, some interesting results have been found. The results showed that using Basic access mechanism both the QB and CB functions have a higher saturation throughput compared to the BEB function with any number of stations. Analysis of the results shows that saturation throughput can be improved significantly by choosing a larger minimum windows size compared to the current standard value of the IEEE 802.11. The findings in this paper shows that the performance of the IEEE 802.11 DCF can be improved through using a modified backoff and/or optimizing window size parameters. Furthermore, a typically large window size improves the throughput performance, which explains the gain in the saturation throughput for both the QB and CB functions, as both functions have larger window sizes in the earlier stages compared to the BEB function. It would be interesting to evaluate the performance of the new proposed backoff function based on realistic network traffic instead of saturation condition and compare it with the standard BEB.

References

1. Bianchi, G.: Performance analysis of the IEEE 802.11 distributed coordination function. IEEE J. Sel. Areas Commun. **18**(3), 535–547 (2000)
2. Carvalho, M.M., Garcia-Luna-Aceves, J.J.: Delay analysis of IEEE 802.11 in single-hop networks. In: 11th IEEE International Conference on Network Protocols (ICNP 2003), Atlanta, GA, USA, 4–7 November 2003, pp. 146–155 (2003)
3. Daneshgaran, F., Laddomada, M., Mesiti, F., Mondin, M., Zanolo, M.: Saturation throughput analysis of IEEE 802.11 in the presence of non ideal transmission channel and capture effects. IEEE Trans. Commun. **56**(7), 1178–1188 (2008)
4. Håstad, J., Leighton, F.T., Rogoff, B.: Analysis of backoff protocols for multiple access channels. SIAM J. Comput. **25**(4), 740–774 (1996)
5. Kumar, P., Krishnan, A.: Saturation throughput analysis of IEEE 802.11b wireless local area networks under high interference considering capture effects. CoRR abs/1002.1689 (2010)
6. Kwak, B., Song, N., Miller, L.E.: Performance analysis of exponential backoff. IEEE/ACM Trans. Netw. **13**(2), 343–355 (2005)
7. Ni, Q., Li, T., Turletti, T., Xiao, Y.: Saturation throughput analysis of error-prone 802.11 wireless networks. Wirel. Commun. Mob. Comput. **5**, 945–956 (2005)
8. Open Source: Network simulator. https://www.nsnam.org
9. Sakurai, T., Vu, H.L.: MAC access delay of IEEE 802.11 DCF. IEEE Trans. Wirel. Commun. **6**(5), 1702–1710 (2007)
10. Sun, X., Dai, L.: A comparative study of quadratic backoff and binary exponential backoff in IEEE 802.11 DCF networks. In: 45st Annual Conference on Information Sciences and Systems, CISS 2011, The John Hopkins University, Baltimore, MD, USA, 23–25 March 2011, pp. 1–6 (2011)
11. Xu, D., Sakurai, T., Vu, H.L.: An analysis of different backoff functions for an IEEE 802.11 WLAN. In: Proceedings of the 68th IEEE Vehicular Technology Conference, VTC Fall 2008, Calgary, Alberta, Canada, 21–24 September 2008, pp. 1–5 (2008)

Quality of Service Provision Within IEEE 802.11 CSMA/CA Protocol

Kamil Samara[1(✉)], Hossein Hosseini[2], Zaid Altahat[1], Joseph Stewart[1], David Ehley[1], and Miguel Estrada[1]

[1] University of Wisconsin-Parkside, Kenosha, USA
samara@uwp.edu
[2] University of Wisconsin – Milwaukee, Milwaukee, USA

Abstract. The goal of this work is to propose and test a Quality of Service (QoS) provision within the IEEE 802.11 CSMA/CA protocol. The proposed QoS provision divides the network traffic into three distinct classes: normal, urgent, and critical. From the three traffic classes, critical data is transmitted fastest with the highest priority, and then followed by urgent data transmission with the next priority level, and subsequently normal data with the lowest priority level. The prioritization of different data is achieved by assigning different delay times for each class. This research utilizes the network simulation framework NS3 to simulate traffic in order to test performance of the protocol. Our simulation results indicate improvement in transmission of urgent and critical data.

Keywords: QoS · MAC layer · Differentiated services · Networking

1 Introduction

The IEEE 802.11 (WiFi) was designed as a simple and effective protocol to achieve wireless transmission of data. IEEE 802.11 Medium Access Control (MAC) protocol is composed of two protocols: Distributed Coordination Function (DCF) built on Carrier Sense Multiple Access/Collision Avoidance (CSMA/CA) protocol and Point Coordination Function (PCF) which is built on the top of DCF protocol stack as shown in Fig. 1 [1].

DCF is a distributed protocol which does not provide QoS and treat all traffic the same while PCF is a centralized protocol that tries to provide conflict free access for users or some form of QoS.

The DCF layer is basically a simple CSMA (carrier sense multiple access) design. If a node has a frame to send, it listens to the channel. If the channel is idle, the node can start transmitting; if not the node must wait until the current transmission is completed before transmitting [1].

To minimize collisions and achieve fairness, DCF uses a set of delays. In fact, there are three different delays based on an IFS (Interframe space) value. These delays are:

1. SIFS (short IFS): The shortest IFS, used for all instantaneous response activities
2. PIFS (point coordination function IFS): A medium-length IFS, used by PCF to issue polls

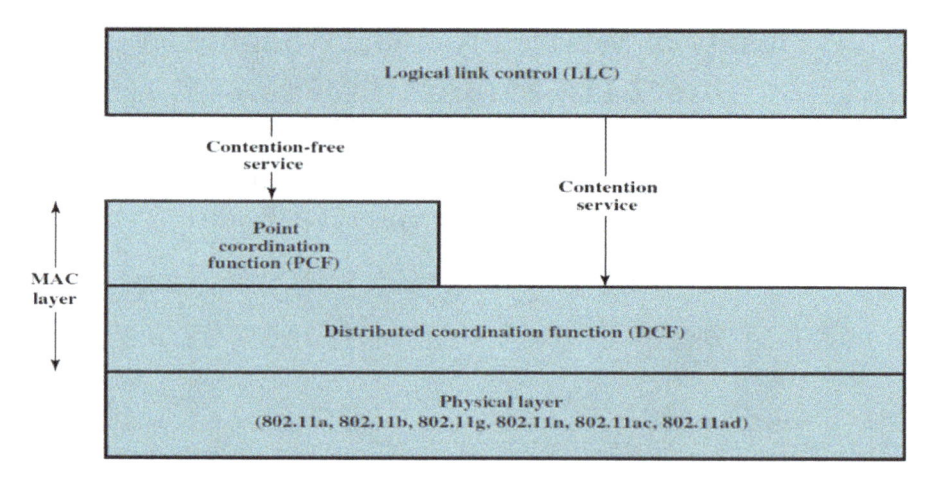

Fig. 1. IEEE 802.11 protocol architecture [1]

3. DIFS (distributed coordination function IFS): The longest IFS, used as a minimum delay for data frames competing for access

Using IFS different delays, as shown in Fig. 2, DCF imposes its transmission logic. Our focus will be on the DIFS delay. Any node trying to send a data frame will have to wait a DIFS regardless of data type being send. This will result in treating all data equally [1].

In this work, first, we propose an amended version of DCF protocol that provides QoS, and then we introduce simulation results to evaluate the performance of the new algorithm. The QoS is achieved by using different DIFSs according to data class being sent. This will result in giving higher priority for data with shorter DIFS.

2 Related Work

QoS approaches for IEEE 802.11 [1] can be categorized into two main groups:

- Differentiated services
- Integrated Services.

Here, we will be concentrating on differentiated services (DS) because it is strongly associated to our work. QoS is DS in accomplished by categorizing traffic into distinct groups where upper priority groups are provided advantage over lower ones by favoring them when competing for wireless access. The main differentiated services methods could be summarized as follows:

In [2] the authors present a Persistent Factor DCF protocol, in this protocol each traffic group is assigned a persistent factor P where low-priority groups have a greater P and high-priority groups have smaller P. In a back-off stage, an evenly distributed random number r is generated in every slot time. Transmission reassumes only if (r > P) in the current slot time.

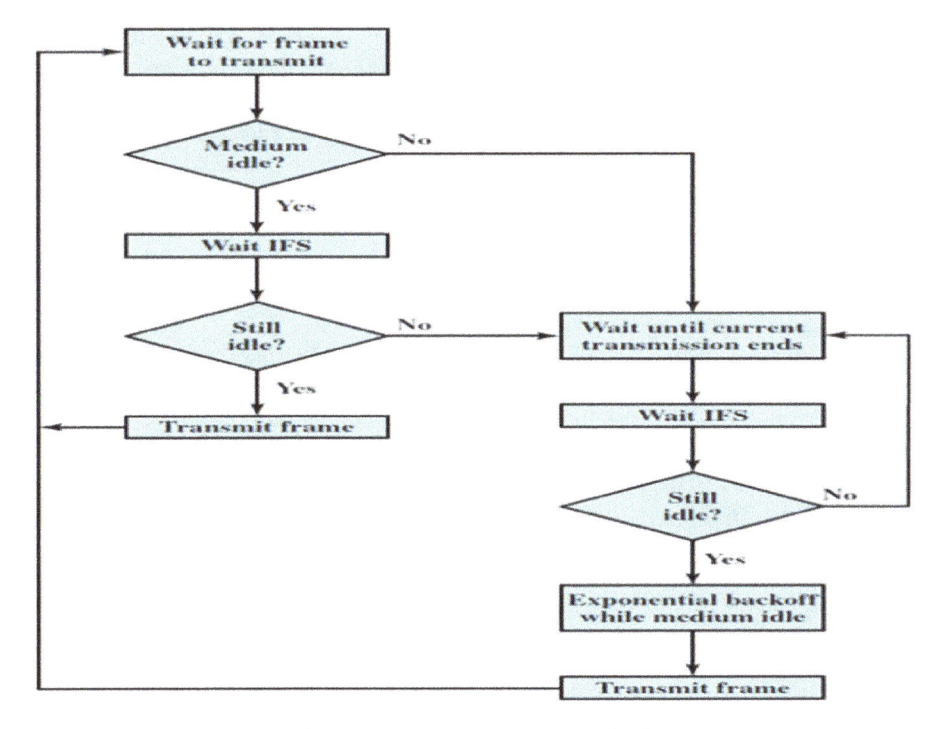

Fig. 2. IEEE 802.11 channel access logic [1]

In [3] a Distributed Weighted Fair Queue (DWFQ) protocol is introduced. In DWFQ, the back-off window size of any traffic category is increased to decrease the category's priority and decrease the window size to increase the category's priority.

Distributed Fair Scheduling (DFS): The main idea of DFS is to allocate high-priority category a smaller back-off interval which will result is traffic flow with lower back-off interval to transmit first [4].

Another research team from Shanghai Jiao Tong University that was able to improve the 802.11 network stack by integrating TCP acknowledgements into the MAC layer. While their research wasn't concerned with establishing a quality of service, they were able to refine interlayer network communications by making the MAC self-aware of Network/Internet Layer information. This was done by incorporating ACK information within the CTS packet at the MAC layer [5].

3 DCF Protocol with QoS Provision

Distributed Coordination Function (DCF) is based on Carrier Sense Multiple Access/Collision Avoidance (CSMA/CA) protocol. However, DCF protocol treats all data the same as ordinary data and does not provide QoS provision either. This is a direct result of using the same DIFS delay for all types of data [1].

AS was mentioned earlier, DCF protocol does not provide QoS and treat all the data the same as ordinary. In the following we propose a modified version of DCF protocol that provides QoS. The proposed QoS divides the network traffic into three different classes with different priority levels: normal, urgent, and critical. Normal data is the same as ordinary data in original DCF protocol. Urgent data is a type of data, like real time data, that has to be transmitted right away. Critical data is a type of data that has to be transmitted with high reliability without any corruption, like medical records. Normal data or ordinary data is regular data with no extra demand on the speed of delivery or reliability. Our protocol sends the critical data with highest priority, then urgent data, and finally normal data.

In our implementation of this paradigm, we are dividing the network traffic into three different groups: normal, urgent, and critical. This division is designer and application specific. We chose this classification for testing purposes.

We found that each Wi-Fi version (a, b, g, n, ac) had a different established value for these time delays. For instance, 802.11a has a slot time of 9 μs, while 802.11b has a slot time 20 μs [6], for our simulation purposes we used the 802.11a values.

DIFS = SIFS + (SlotTime * 2), so by modifying the slot time we were able to configure three different DIFS delays, described as follows:

$$\text{Normal traffic} = (\text{Slot Time} - 0\%) = \text{No Difference} \quad (1)$$

$$\text{Urgent Traffic} = (\text{Slot Time} - 15\%) \quad (2)$$

$$\text{Critical Traffic} = (\text{Slot Time} - 30\%) \quad (3)$$

Since all nodes must wait a DIFS delay before transmitting a data packet, packets which wait less will be transmitted first resulting in higher priority for those types of packets.

4 Simulation

4.1 Simulator

To evaluate the performance of our proposed provision we used the NS3 network simulator. The performance is measured by testing throughput, and delivery time.

The network simulator NS3 was chosen because of its extensive documentation library and broad range of use. NS3 also supports virtualization, making it easier to use within a virtual environment. Also, NS3 is loaded with features [7].

In order to setup and use NS3 it is essential to have a good working knowledge of Linux/Unix operating system, and expertise with C++ and Python. For this research, Ubuntu 16.04 LTS was used as the primary operating system.

4.2 Simulation Results

Two simulation experiments were performed. The first simulation experiment emphasis was to compare the throughput between the three different data classes; normal, urgent, and critical.

The second simulation experiment emphasis was to compare the average delivery time between the three different data classes; normal, urgent, and critical.

Experiment One: Throughput. In this experiment, we are measuring the average throughput (Mbits/s) over a period of 15 min for the three different traffics; normal, urgent, and critical.

As seen in Fig. 3, altering the Slot Time does actuality change throughput. The throughput increases as the DIFS delay decrease. Since higher priority classes have lower delay, priority 3 (critical data) has the highest throughput followed by priority 2 (urgent) and lastly priority 1 (normal data).

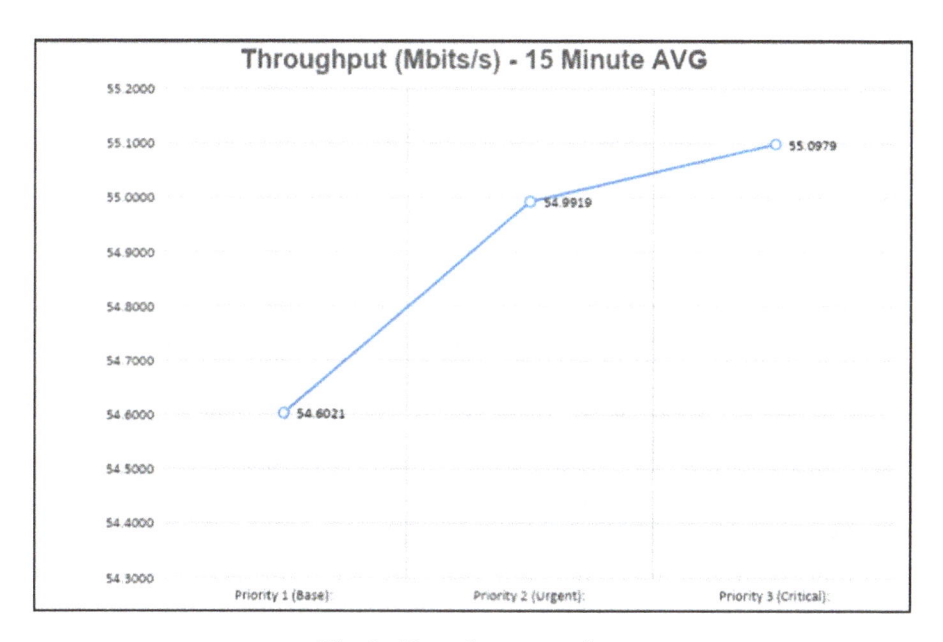

Fig. 3. Throughput comparison

Experiment Two: Delivery Time. In this experiment, we are measuring the average delivery time in seconds needed to transmit 2048 packets, all of size 4096 bytes for the three different traffics; normal, urgent, and critical.

Again, decreasing the Slot Time improved performance for higher priority classes. As shown in Fig. 4, in each higher priority from priority 1 (normal) to priority 3 (critical), a drop in average delivery time is observed.

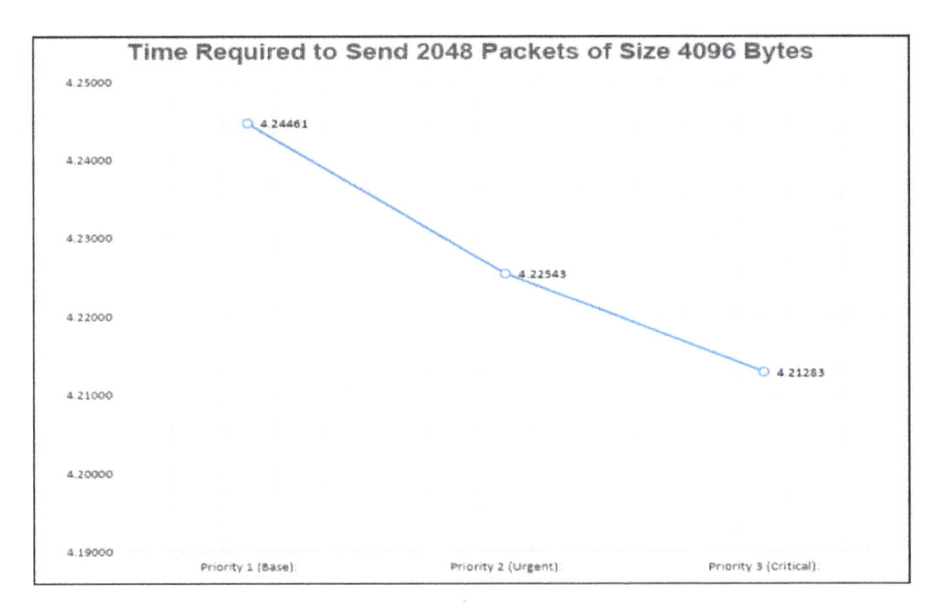

Fig. 4. Delivery time comparison

5 Conclusion

Without QoS, all traffic is treated the same and assigned the same resources and when congestion occurs, any packet can be dropped without regard as to what the packet contains. Implementing QoS at MAC layer will help preserve important data and can result in improved performance on the upper layers.

We believe the IEEE 802.11 could be improved by provisioning differentiated services according to traffic type. In this paper, we proposed a differentiated services scenario that divides the network traffic into three different traffic classes with different priority levels: normal, urgent, and critical. The priorities were established by changing the DIFS delay time, assigning higher priority classes a shorter waiting delay.

The performance of the proposed algorithm was tested using NS3 simulator. Two experiments were performed. The first experiment compared average throughput among the three different classes while the second experiment compared average delivery time. In both experiments, higher levels classes performed better by achieving higher throughput ang lower delay.

References

1. Stallings, F.: Data and Computer Communications, 10th edn. Pearson, London (2014)
2. Ge, Y., Hou, J.: An analytical model for service differentiation in IEEE 802.11. In: IEEE ICC 2003, vol. 2, pp. 1157–1162. IEEE (2003)
3. Banchs, A., Pérez, X.: Distributed weighted fair queuing in 802.11 wireless LAN. In: IEEE ICC 2002, vol. 5, pp. 3121–3127. IEEE (2002)

4. Vaidya, H., Bahl, P., Gupta, S.: Distributed fair scheduling in a wireless LAN. In: ACM MOBICOM 2000, pp. 167–178. ACM (2009)
5. Ding, L., Zhang, W., Yu, H., Wang, X., Xu, Y.: Incorporating TCP acknowledgements in MAC layer in IEEE 802.11 multihop ad hoc networks. In: GLOBECOM 2009 - 2009 IEEE Global Telecommunications Conference, pp. 1–5. IEEE (2009)
6. NS3 Class Reference. https://www.nsnam.org/doxygen/classns3_1_1_wifi_mac.html. Accessed 27 Jan 2019
7. Nsnamorg. https://www.nsnam.org/docs/tutorial/html/index.html. Accessed 27 Jan 2019

Realistic Cluster-Based Energy-Efficient and Fault-Tolerant (RCEEFT) Routing Protocol for Wireless Sensor Networks (WSNs)

Emmanuel Effah[✉] and Ousmane Thiare

Gaston Berger University, Saint Louis, Senegal
emmanueleffah08@gmail.com, ousmane.thiare@ugb.edu.sn
https://www.ugb.sn/

Abstract. Despite the numerous research advances in multichannel event reporting (ER) protocols in Wireless Sensor Networks (WSNs), the core issues such as effective energy consumption management, balanced network-wide energy depletion rates, fault-avoidance-based fault tolerance (FT) for network reliability and elimination of needless data redundancies in ER remain challenges that have not received adequate and holistic research considerations commensurable with the recent technological advancements and the skyrocketing demands and permeance of WSN's applications in all fields of life. RCEEFT is an adaptive cluster-based data acquisition and multichannel routing protocol for WSNs which integrates unique measures to ensure peerless improvement in WSNs' energy consumption, balanced network-wide energy depletion rates (improved sensor field coverage), fault-avoidance-motivated FT and elimination of needless data redundancies in ER (enhanced accuracy and precision in ER) to extend the WSNs' lifespan. The experimental results obtained through simulations established RCEEFT as a more realistic protocol and also performs better than EESAA, DEEC, E-DECC, SEP, M-GEAR, D-DEEC, T-DEEC and LEACH in terms of energy efficiency and WSN lifetime, wider coverage sensitivity, balanced network-wide energy depletion rates, FT and elimination of needless data redundancies in ER.

Keywords: Spatial correlation model (SCM) · Mass measurement pattern and accuracy concept (MMP&AC) · Event reporting (ER)

1 Introduction

WSN comprises a group of specialized transducers called sensor nodes (SN) fitted with communication and processing infrastructures intended for data acquisition, condition monitoring or surveillance services [1]. The industrial and civilian application areas of WSNs include battlefield surveillance, home automation, industrial process monitoring and control, machine health monitoring, healthcare applications, smart space and traffic control, among others. WSN technology has received high adoption and pervasiveness for countless reasons including ease of implementation (no long cable runs are needed), ability to operate in harsh environments, easy troubleshooting and repair and high levels of performance [2]. However, WSNs are highly prone to faults and network failures due to these reasons:

- The SNs are in limited available power which restrict them to low-power processors, limited memory storage, low-power radio transceivers leading to low data transmission rates and limited coverage range [3].
- Lack of post-deployment maintenance services of the SNs to replace exhausted units such as drained batteries [4].
- Lack of efficient energy optimization techniques and realistic auto-fault detection and fault-avoidance-based FT mechanisms that resolve the root causes of faults and not their effects.
- Lack of energy-efficient routing protocols equipped with realistic and simplistic techniques to ensure balanced network-wide energy consumption and manage energy depletion rates of SNs, and elimination needless ER redundancies.

In many WSN applications, a SN's failure due to the stated reasons could result in calamities of an unimaginable magnitude. However, these challenges are best addressed based on the fault detection and fault tolerance (FD&FT) mechanisms embedded in the associated routing protocol. The present study therefore examines the state of the art and proposes a realistic cluster-based, energy-efficient and fault-tolerant (RCEEFT) solution to address the stipulated challenges in a unique way. In addition, our solution distinctively achieves higher ER accuracies and precisions, FT and network reliability without extra complexity cost by deploying novel approaches that control needless redundancies using alternating node-pairing and correlation control techniques that are founded on spatial correlation of event information and mass measurement pattern and accuracy concept (MMP&AC).

It has been well documented that cluster-based routing paradigms have higher potential to enhancing network lifespan, prolonged event coverage period, improve FT and nodes energy savings by minimizing the regular distant-communication cost [2] in WSNs. In clustering as illustrated in Fig. 1, the whole network is divided into fixed or variable sized clusters containing SNs in it and each cluster is denoted by a cluster head (CH) that aggregates the cluster data and send it to the BS directly or via other relay CHs (RCHs). member nodes(MN) of the cluster communicate only with their respective CHs in their proximity hence minimizing energy consumption due to the reduced distances of data transmission. This implies that forming the optimal number of clusters is beneficial in minimizing the energy wastages in WSNs.

The rest of this paper is organized as follows: Sect. 2 presents a systematic review of related works on cluster-based MCPs in WSNs to uncover their main challenges for our consideration while Sect. 3 presents and cements our proposed theoretical background of RCEEFT routing protocols. The proposed routing protocol (RCEEFT) is discussed and explained in Sect. 4; Sect. 5 presents results from simulation, discussions and comparisons of RCEEFT with similar protocols while Sect. 6 concludes the paper with future scopes.

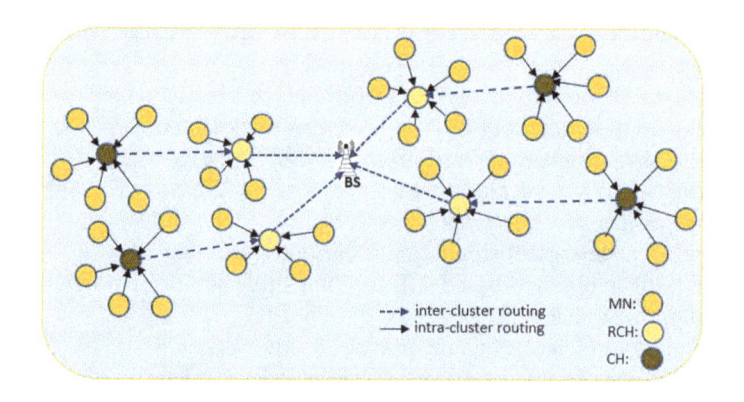

Fig. 1. Clustering architecture in WSN

2 Related Literature

Cluster-based multichannel routing protocols in WSNs can be classified under homogeneous WSNs if SNs have the same initial energies and heterogeneous WSNs if nodes have different initial energy levels [5,6]. For instance in homogeneous WSNs such as Energy Efficient Sleep Awake Aware (EESAA) [7], Low Energy Adaptive Clustering Hierarchy (LEACH) [7,8], Hybrid Energy-Efficient Distributed (HEED) [9], Power Efficient Gathering in Sensor Information Systems (PEGASIS) [10], SNs have equal initial energies whereas SNs in heterogeneous WSNs such as Stable Election Protocol (SEP) [11], Distributed Energy Efficient Clustering (DEEC) [12], Developed DEEC (DDEEC) [13], Enhanced DEEC (EDEEC) [14], Threshold DEEC(TDEEC) [15] have different initial energies [5,6].

SEP [11] is designed to deal with heterogeneous networks by introducing the concept of advance and normal nodes for CH election. The advanced SNs have more energy than normal ones and so are mostly selected as CHs. Though SEP extends the network's stability duration, it is unfit for WSNs with more than two types of energies.

To curb the possibility of non-distributive CH selection in WSNs, the authors in [16,17] proposed a special kind of rechargeable node, called gateway node, at the center of the WSN that collects and aggregates data from CHs and nodes near gateway before sending it to the BS. Though this approach improved network lifespan, it is limited in realistic applications.

DEEC [12,18] is a cluster-based algorithm in which a CH is selected on the basis of probability of ratio of residual energy and average energy of the network.

In this algorithm, a node having more energy has more chances to becoming a CH so that the lifetime of the WSN is prolonged in an energy balanced manner. However, DEEC over-burdens the advanced SNs with the CH tasks which hinders effective network-wide energy distribution.

DDEEC assumed similar method used by DEEC to estimate the average energy of the network and select CH. It was affirmed that SNs with more residual energy at a given round r were more likely to become CH, so, advanced SNs performed the CH task more frequently than the normal SNs. At a point where the residual energies of the advanced SNs are in equilibrium with those of the normal SNs, DEEC continues to punish the advanced SNs with the CH roles which is not an optimal way for network-wide energy distribution. To avoid the unbalanced power depletion problem in DDEEC, Threshold DEEC(TDEEC) [15] introduces threshold residual energy approach that ensured that the CH task is equally shared among all SNs beyond the equilibrium level. TDEEC [15] adopted the same CH selection and average energy estimation methods in DEEC. Per round, SNs decide whether to become a CH or not by choosing a random number (r_n) between 0 and 1. If this number is below a given threshold (Th in Eq. 1) which varies per round, then that SN becomes CH for that round.

$$Th = \begin{cases} \frac{p_d}{1-p_d*((firstround) mod \frac{1}{p_d})} & if\ n \in ActiveSNs \\ 0, if otherwise \end{cases} \quad (1)$$

where: p_d Desired percentage of CHs per round.

EDEEC [14] also inherits DEEC and introduces a new group called supernodes to the existing normal and advanced nodes to increase the network heterogeneity to the third level. EDEEC outperformed DEEC and DDEEC when evaluated by [18] using their simulation results.

PEGASIS [10] is a distributed algorithm whereby each SN in the WSN sends its data to one of its neighbors such that data is gathered in a chain manner before a dedicated SN finally sends all to the BS. The main challenge of PEGASIS is that every single node must obtain the global topology map of the network. CHs in LEACH [7,8] protocol are selected periodically and energy drains uniformly by role rotation. In PEGASIS energy load is distributed by forming a chain itself or being organized by BS. Since global knowledge about the network is essential in the chain formation, the results of PEGASIS depicted wastage of resources.

EESAA also a distributed clustering method and introduces a SNs pairing concept whereby all highly-correlated SNs with the same application and at a minimum distance between them are paired in one active-sleep mode schedule plan and vice versa for data sensing and communication. EESAA protocol used improved CHs selection technique that is based on the remaining energy of SNs but, the pairing concept of EESAA creates coverage loopholes or gaps tendencies in randomly deployed event fields of uneven SNs densities.

Results from the performance assessment of these cluster-based algorithms (e.g. [7] LEACH, PEGASIS, DEEC, SEP, EESAA, DDEEC, EDEEC, TDEEC) using the basic WSN evaluation metrics such as stability period, clustering process and throughput yielded substantial improvements in energy savings in their respective homogenous and heterogeneous WSNs. However, the recent advances in WSN technology and the associated applications are gradually rendering these algorithms obsolete and craving the need for more improved versions to match the existing developments. It can clearly be unveiled from the reviewed algorithms that they are highly constrained in applications. Thus, the homogeneous-based algorithms perform best only in homogeneous WSNs and vice versa. However, it is impractical for a WSN to maintain its ideal homogeneity throughout its lifetime or for a heterogeneous WSN to maintain its ideal number of energy levels. Hybrid WSN algorithms are therefore of higher demand for research considerations and also in the future scope for realistic advancement in WSNs.

Notably, fault avoidance is expected to be the most important aspect of all FT mechanisms in routing protocols in WSNs [19,20], and since the commonest root-cause of most faults in WSNs emanate from mismanaged power depletion rates [2] of the network, the best FT designs in WSNs are those that maximize energy efficiency in addition to FT task. This is because, the reliability of the entire WSN, as shown in Fig. 2, is compromised beyond the network stability period even under efficient fault-tolerant solutions that integrate redundancy-based FT methods, clustering-based FT measures and deployment-based FT mechanisms [19,21]. The FT mechanism embedded in RCEEFT does not only focus on the network's ability to operate properly in the presence of faults or absorbs the topological dynamics but also inculcates adequate fault avoidance mechanisms to ensure better energy efficiency so that the reliability goal and the business function of WSNs are not hindered.

In RCEEFT, different clusters are formed at each round which result in high network scalability/reconfigurability, auto-fault isolation and tolerance and energy savings. Aside the fault-tolerant phase, CHs also perform the needless redundancy checks during ER at the data transmission phase of RCEEFT to optimize the network's total energy consumption by eliminating the communication cost of transmitting these unwanted data.

2.1 Performance Evaluation Metrics of RCEEFT

Performance evaluation parameters used to assess the competence of cluster-based routing protocols in WSNs are network lifetime, network reliability or stability period, network instability duration, energy distribution and energy depletion patterns of the network, cluster optimality and precise packet delivery rates [7] as elaborated below. These are deduced from the characteristic plots of the number of dead/alive nodes, the number of data packets sent to BS during rounds, WSN's remaining energy or energy depletion rates during rounds and the number of CHs and cluster optimality per round.

1. WSN lifetime: it is the duration from start till the round of network failure that is determined by the total number of rounds.
2. WSN's reliability or stability period: It is duration of network operation from start until first node dies. This represent the most reliable duration of WSNs as illustrated in Fig. 2.

3. Instability period: It is duration from the first SN's death count till the entire network fails and depicts the period of varying network reliability in Fig. 2.
4. Number of Alive SNs: It is a parameter that describes number of alive nodes during each round.
5. Packets Delivery Rates: It is the number of packets received by BS successfully.
6. SN failure patterns: It is the outline of SNs' death pattern and its effects on network's coverage loss.
7. Remaining energy (RE) of WSN: This defines the energy depletion rate of the entire WSN from start to the point of network failure. For a well-balanced energy depletion rate, RE is expected to reduce at a least constant rate.
8. Number of CHs and cluster optimality ensures that the total number of clusters or CHs per round is limited to the domain of optimal number of clusters as proposed by [22] so that the communication cost can be kept at the most minimum.

3 Theoretical Background of RCEEFT

The present study is founded on two concepts: 1. the spatial correlation model (SCM) concept which states that the sensory measurements or event information of SNs due to an event source(s)-ES are spatially correlated and 2. mass measurement pattern and accuracy concept (MMP&AC) which states that the sensory data recorded of a group of healthy or normal SNs under the coverage of an ES(s) must follow a given regular model or pattern and their accuracy levels be determined by their relative distances from the ES(s). Thus, under the same environmental conditions, a pair of normal and equidistant SNs from a given

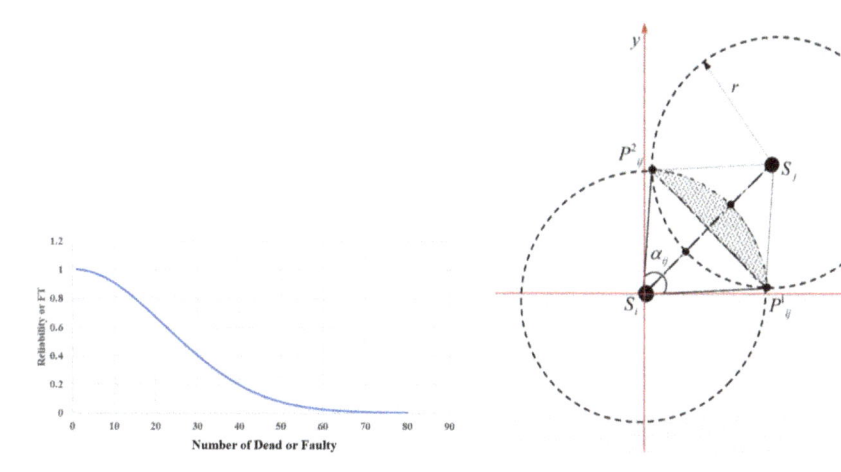

Fig. 2. WSNs reliability versus Dead/Faulty/Failed SNs plot

Fig. 3. CM diagram for two SNs

event source (ES) in a sensor field will record approximately similar event information regardless of whether or not their coverage ranges overlap. MMP&AC is a novel concept proposed by the present study.

3.1 Spatial Correlation Model (SCM)

Many correlation models have been are proposed in literature, however, this study adopts SCM [23] for its suitability and superiority in efficiency over the rests. SCM truly deals with the physical phenomena of interest, the features of the sensed events and the actual conditions of the SNs such as sensing range, locations of SNS and inter-SNs distances ($d_{(i,j)}$). For SNs with omnidirectional radio transceivers with disc-like coverage range r, it can be assumed that the nodes have static or defined transmission ranges (T_x) s.t. $T_x >> r$ and also the nodes detect nothing outside r. The terms retain their definitions as in [23]. If all reported sensory data to the BS from the SNs are jointly Gaussian, then the covariance between two readings from nodes n_i and n_j at locations s_i and s_j respectively can be expressed as:

$$Cov s_i, s_i = \sigma_S^2 K_\vartheta(||s_i - s_j||) \tag{2}$$

where: σ_S^2 is the samples variance observed for SNs which is assumed to be the same for all SNs and $Cov(.)$ denotes the mathematical covariance. $||.||$ is the Euclidean distance ($d_{i,j}$) between n_i and n_j and $K_\vartheta(.)$ represents the correlation function with $\vartheta = \theta_1, \theta_2, \theta_3, \ldots$ control parameters.

If coverages ranges describing circles of radii r and areas A, and $A(n_i, n_j)$ overlap as shown in Fig. 3 then the relationship between the ratio of the overlapped coverage area to a coverage area of either n_1 or n_2 defines the correlation function shown in Eq. 3 if n_i and n_j are separated by $d_{i,j}$.

$$\rho(i,j) = K_\vartheta(d) = \frac{A_i^j + A_j^i}{A} \tag{3}$$

where: $K_\vartheta(d)$ increases to 1 if $d_{i,j} = 0$ and decreases to 0 if $d_{i,j} \geq 2r$. Given that

$$A_i^j = A_j^i = r^2 cos^{-1}(\frac{d_{(i,j)}}{2r}) - \frac{L_{ij} d_{(i,j)}}{4} \tag{4}$$

If the chord length of the intersected coverages is also given by $L_{ij} = 2.\sqrt{(r^2 - \frac{d_{(i,j)}^2}{4})}$, then the $\rho(i,j)$ can be restated as:

$$\rho(i,j) = K_\vartheta(d) = \frac{2cos^{-1}(\frac{d_{(i,j)}}{2r})}{\pi} - \frac{d_{(i,j)}}{\pi r^2} \cdot \sqrt{(r^2 - \frac{d_{(i,j)}^2}{4})} \tag{5}$$

Assuming a correlation control parameter $\vartheta = 2r$ for regulating the degree of correlation among nodes.

$$\rho(i,j) = K_\vartheta(d) = \frac{2cos^{-1}(\frac{d_{(i,j)}}{\vartheta})}{\pi} - \frac{2d_{(i,j)}}{\pi \vartheta^2} \cdot \sqrt{(\vartheta^2 - d_{(i,j)}^2)} \tag{6}$$

The correlation model can be generalized as shown in Eq. 7

$$\rho(i,j) = \begin{cases} \frac{2\cos^{-1}(\frac{d(i,j)}{\vartheta})}{\pi} - \frac{2d_{(i,j)}}{\pi\vartheta^2} \cdot \sqrt{(\vartheta^2 - d_{(i,j)}^2)} & if\ 0 \leq d_{(i,j)} < \vartheta \\ 0\ if\ d_{(i,j)} \geq \vartheta \end{cases} \quad (7)$$

So if a correlation threshold ξ is defined as $0 < \xi \leq 1$, then n_i and n_j will be strongly correlated if $\rho(i,j) = K_\vartheta(d) \geq \xi$ and weakly correlated if $\rho(i,j) = K_\vartheta(d) = K_\vartheta(||s_i - s_j||) < \xi$. The average number of nodes per cluster will depend on SNs density at the event area as well as the sensing radius ($\frac{\theta}{2}$).

The correlation function approaches zero if $d_{(i,j)} \geq 2r$ which implies that n_i and n_j are not correlated implying non-overlapping sensing coverages (Eq. 7) and different measurement values, however, MMP&AC still holds provided these SNs are spatially equidistant from ES(s). This because, there are possibilities of having SNs n_k that are adjacent or at the opposite side of an ES without intercepting As yet, well-correlated measurement values if $d_{(i,ES)} = d_{(j,ES)} = d_{(i,ES)}$. Such situations are uncovered by SCM. MMP&AC provides a broader perspective to accommodate such flaws in SCM. Secondly, it can also be affirmed that if the correlation function is equal to 1 it implies that the nodes are perfectly correlated and must record approximately equal sensory data. This is also not absolutely true because the sensory data is dependent on $d_{i,j}$.

3.2 The Mass Measurement Pattern and Accuracy Concept (MMP&AC) in WSNs

The event data which serves as the input to most FD&FT algorithms in WSNs represent the measure of the change in a physical quantity about the phenomenon of interest. If this event information is assumed to be spatially correlated, then MMP&AC that under the same environmental conditions, a pair of normal and equidistant SNs from a given event source (ES) in a sensor field will record approximately similar and specifically patterned event information regardless of whether or not their coverage ranges overlap.

Thus, if $\Delta\theta_{if} = \theta_f - \theta_i$ represents the average change in physical quantity due to $ES(i)$ of SNs n_i and n_j within the coverage range and at distances $d_{(i,ES)}$ and $d_{(j,ES)}$ respectively away from ES, then the magnitude of the sensory readings $X(i)$ and $X(j)$ of nodes i and j are inversely proportional to $d_{(i,ES)}$ and $d_{(j,ES)}$ respectively. Therefore, if i and j are non-defective SNs and $d_{(i,ES)} < d_{(j,ES)}$, then their respective readings must obey Eq. 8, otherwise, i and j are defective or faulty.

$$\forall_{i,j}, X(i) \geq X(j) \quad (8)$$

This firmly establishes the fact that non-faulty SNs in close proximity to an ES under the MMP&AC, regardless of whether their coverage ranges overlap or not, must record approximately similar or patterned sensory data if $d_{(i,ES)} = d_{(j,ES)}$ or if $d_{(i,ES)} \neq d_{(j,ES)}$ than the nodes which are farther away from the coverage zone of ES, because the signal strength due to ES generally decays with distance and obeys the isotropic power model illustrated in Eq. 9.

$$P_i = \frac{P_0}{1 + bd_{ij}^a} = P_0 \times \psi_{ij} \qquad (9)$$

Where: P_i = the strength of signal received at node i, P_0 = the strength of signal received at node j which is the same as the signal strength of ES, assuming that j and ES are located at the same point and so the difference in signal strengths is negligible. $d_{(i,ES)}$ = Euclidean distance between nodes i and ES. $a(a \in \Re^+)$ denotes the signal attenuation exponent and b = constant.

RCEEFT relied on MMP&AC and SCM concepts to realize the fault-avoidance-motivated FT algorithm, select the optimal proxy ER data for an enhanced global ER accuracy and precision of the network, eliminate unwanted redundancies in ER data, implement our new pairing and duty-cycling concept of alternately-correlated SNs to optimize the network stability and lifespan. In RCEEFT, the nodes that are in close proximity to the ES are better and accurate candidates for ER than the distant ones.

With the aid of the MMP&AC, if D_{min} defines the minimum threshold measurement $\Delta\theta_{if}$ of i nodes, then the measurement accuracy test (MAT) can be performed with the aid of Eq. 10 to classify $X(i)$ as faultless (if Eq. 10 is true) or faulty (if Eq. 10 is false). The degree of correlation is also estimated using Eqs. 7 and 9 where ξ signifies the minimum correlation threshold used to implement our active-sleep mode pairing concept, FD&FT algorithm and the technique used to eliminate unwanted ER data as shown in Fig. 5.

$$X(i) \geq D_{min}, s.t \ X(i) \in \{0,1\} \qquad (10)$$

4 RCEEFT: The Proposed Routing Protocol

RCEEFT protocol is designed for hybrid (single-level and multilevel) WSNs, and inherits the advanced CH selection and clustering techniques of EESAA as its underlying mechanisms and proposes unique enhancements solutions that:

1. Eliminate needless data redundancies in ER to the BS which minimize the WSN's extra communication cost for transmitting these redundant data in order to improve the network stability and prolong the lifespan of the network. As well-established in [2], unoptimized power depletion rates of SNs is the principal cause of most faults in WSNs. Therefore, the matchless network stability and reliability (Figs. 2 and 7) of RCEEFT reveals its unique fault avoidance capabilities which is extremely crucial but least addressed in most FT algorithms.
2. Detect and tolerate SN failures and errors in sensory data because aside a SN's failure, any other fault in SNs is manifested as a sensory/processor's data output.

3. Exploit spatial correlation theory and MMP&AC to achieve alternate-pairing and duty-cycling of SNs for ER in such a manner to address the intermittent coverage loss challenge in EESAA [7] due to erratic SNs death pattern that divides the WSN into disjointed segments in large-scale WSNs.
4. Enhance energy efficiency and sensor field coverage of the RCEEFT WSN by ensuring equal and uniformly distributed minimal energy depletion rates of all SNs at no extra computational cost. This is because if some nodes deplete their energies faster than others and die, the network might be partitioned into disjointed segments in terms of communication and fail abruptly [7]. Therefore, to ensure that SNs survive as long as possible, we ensure that the protocol is source initiated, uses event-driven reporting and superb computational simplicity.
5. Ensures that RCEEFT is scalable, simple and runs efficiently in a unique and realistic way in both homogeneous and heterogeneous WSNs without any significant variance in operational efficiencies.

No previous protocol including LEACH [7,8], SEP, EESAA, WideMAC [24], WiseMAC [25], DEEC and post-DEEC algorithms is addressing these issues to the extent of RCEEFT.

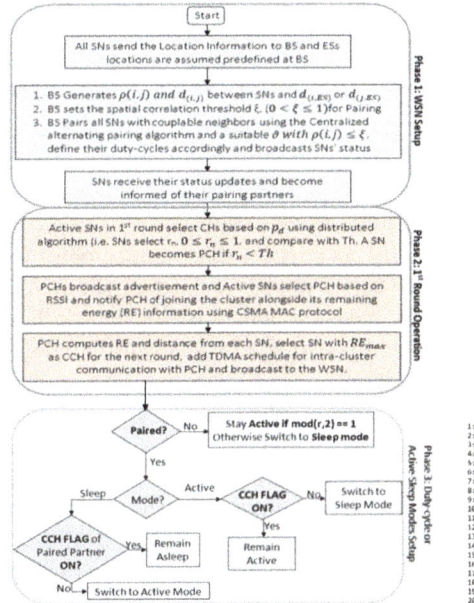

Fig. 4. Flowchart of RCEEFT

Fig. 5. Algorithm 1: FD&FT and elimination of unwanted ER data redundancies

4.1 The Algorithmic Phases of RCEEFT

1. As illustrated in Fig. 4, the setup phase of RCEEFT protocol estimates the network operation parameters, design pairing, and duty-cycle plan and broadcast to the SNs in the WSN. Unique from the concept of pairing and scheduling highly correlated SNs which creates loopholes or coverage gaps in the sensor field during data sensing and communication as proposed by [7], our concept instead schedules highly correlated nodes alternately based on the MMP&AC and SCM if $\rho(i,j) \leq \xi$ (i.e spatial correlation threshold) and also schedule isolated nodes using a suitable correlation control parameter in order to minimize data redundancies in both sensing and communication, maximize sensor field coverage loss and reduce the extra overhead at the sink. The isolated SNs are scheduled to have longer duty-cycles in order to ensure very negligible coverage loss. During the 1st round operation phase, active SNs self-select themselves as CHs using the predefined Pd (desired percentage of CHs value) and the distributed algorithm whereby each active SN selects a random number (r_n) from the range $(0 \leq r_n \leq 1)$ and compare with a threshold Th in Eq. 1. A SN becomes a parent CH (PCH) if $r_n < Th$. As shown in Fig. 4, the PCH selects the child CH (CCH) for the next round.

    Phase 3 presents how the duty-cycle plan of the SNs in the WSN is implemented. As Illustrated in Fig. 6, the optimal number of clusters and CHs per round is determined and predefined as simulation parameter using the cluster quality evaluation indexes in our earlier article [22] as 10 and 12 for the homogeneous and the heterogeneous WSNs respectively. An enhanced CH selection technique based on the remaining energy of nodes and relative distance of SNs is deployed where the CH of the current round will predict the CH for the next active round.

2. During the data assessment, aggregation and transmission phase presented in Fig. 5, a node within a cluster first of all checks if its sensed data is up to the predefined measurement threshold of the network D_{min} using the MAT in order to detect and tolerate erroneous measurements. If no, it switches its transceiver into sleep-mode. If yes, it will proceed and transmit the data to its PCH (Fig. 5) during its TDMA- schedule while all active SNs in the non-transmission mode switch their transceivers to sleep-mode. During data aggregation, the CH performs the ER data redundancy checks depicted in Fig. 5. In next round, nodes in Active-mode with exception of the selected CCHs switch into sleep-mode and those in the sleep-mode switch into active-mode so that the algorithm repeats. Energy consumption is minimized due to the deployed duty-cycling technique; ER data redundancy is eliminated using the redundancy check technique, and erroneous measurements and node failures are identified and tolerated using the MAT and the recursive clustering algorithm respectively.

3. Scalability: RCEEFT adopted the flat network structure in order to ensure that our approach scales better than the classical hierarchical clustering structures predominantly used in WSNs.

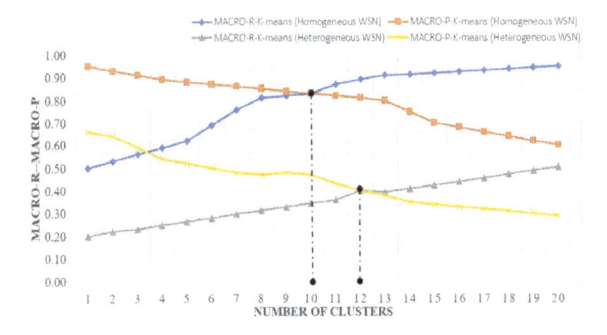

Fig. 6. *Optimal number of clusters using K-Means:* $R_M - P_M$ indexes against No. of clusters

5 Simulation Results and Discussion

Table 1. Simulation parameters

Parameter	Value
Network Field	(100, 100)
Number of SNs (Normal)	100
Number of SNs (Densed)	150
Network Field (Densed)	(120, 120)
Number of Defective SNs	4, 5, 6, 7, 8, 10, 12
Initial Energy of Normal SNs	0.5 J
Optimal No. of Clusters per Round & CHs	10(*homog.*), 12(*Heterog.*)
Correlation Threshold for pairing	$\xi < 0.50$
Desired percentage of CHs per round (Pd)	0.1(*Homog.*), 0.12(*Heterog.*)
Data Aggregation Energy cost	50 pJ/bit J
Packet size (Normal SNs)	4000 bit
Packet size (Defective SNs)	*BitsFluctuateWithRE*
Packet transmission Energy (Tx)	50 nJ/bit
Packet Reception Energy (Rx)	50 nJ/bit
Transmit amplifier (Eamp)	100 pJ/bit/m^2

The vital simulation parameters for RCEEFT are shown in Table 1.

In order to confirm the reliability and consistency of RCEEFT routing protocol, RCEEFT was first simulated under different operational conditions and monitored the possible variations in its performance evaluation metrics. As illustrated in Fig. 7, RCEEFT protocol demonstrates peerless performance stability and reliability in terms network lifetime, network stability period, packets delivery rates, superb pattern in CHs selection under the cluster optimality indexes and high FT in dense, up-scaled and normal conditions. According to Fig. 7, scaling up the number of SNs by 50% and network size by 20% as shown in the dense deployment resulted in a commensurable increase in optimal number of clusters and CHs to 15, the number of delivered packets from 45000 to about 79000 (75% *Approx.*) and the network stability duration by 31%. This positive responses of RCEEFT justify its ability to ensure balanced or evenly distributed energy depletion rates, scalability and its capacity to operate efficiently in both homogeneous and heterogeneous WSNs (hybrid networks). Results in Fig. 7 further affirms that RCEEFT protocol is practically realistic because it adapts efficiently to the network dynamics without any dependencies on nodes' capabilities or network layout.

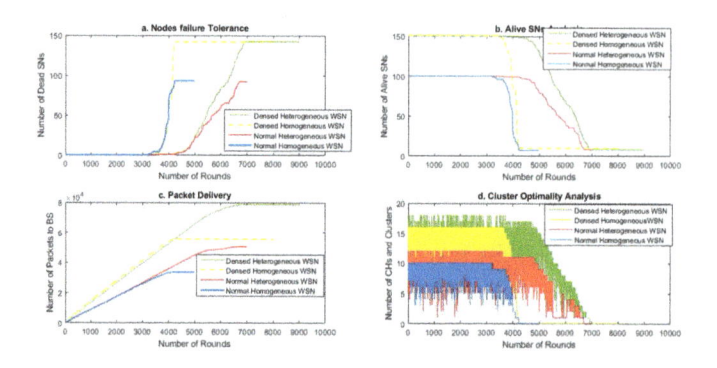

Fig. 7. Performance evaluation plot of RCEEFT under WSNs conditions

To measure and analyze the performance of the proposed protocol-RCEEFT, comparative simulations are performed with EESAA, LEACH, SEP, DEEC and the post-DEEC protocols and compared using the assessment metrics such as network stability period, lifetime, instability period, Optimal number of clusters and stability in CHs selection patterns, packet delivery rates to BS, remaining energy of network and SNs failure patterns.

Variations in the number of alive SNs under the various WSNs' evolution are examined, and as Fig. 8 illustrates, RCEEFT has a longer stability period as compared to the other protocols. The first node of RCEEFT dies around the 4278th round which is more than the entire network lifetime of the other protocols (Fig. 8). Stability period of RCEEFT is almost 280% more than EESAA and over 300% greater than the other protocols (SEP, LEACH, DEEC and the post-DEECs as Figs. 8a and 8b present. It is because of the sleep-awake property of nodes, alternate-pairing, and scheduling concepts and effective cluster optimality and CH selection procedure deployed. The optimization of the duty-cycles of SNs also helped to improve energy consumption and improve network stability period and lifespan Fig. 9. Per Figs. 8 and 10 we notice that the first node dies around

4278th and the network fails after 6657th round which clearly shows that the unstable section of RCEEFT starts very late as compare to the other protocols. In addition, Fig. 11 further shows that there is no abrupt coverage loss of the monitored field in RCEEFT as compared to the other protocols. This due to the balanced and more stable network-wide energy depletion rate of RCEEFT depicted in Fig. 9 which is superior to other related protocols.

Figure 14 also shows that SNs in RCEEFT die at a constant rate whilst those of SEP, LEACH, EESAA, DEEC and the post-DEEC protocols occur at erratic rates. This observation affirms that energy dissipation is properly distributed among all the nodes in the RCEEFT-based network than the others in comparison which resulted in much-prolonged network lifetime.

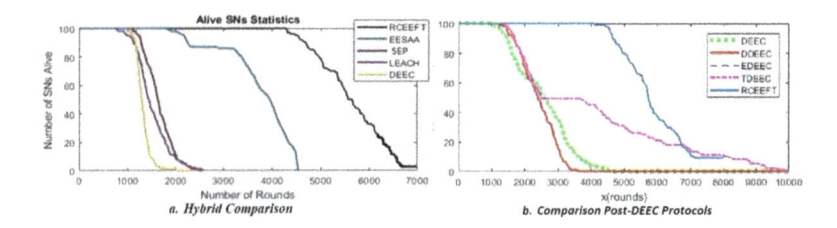

Fig. 8. Comparison of network stability periods using the statistics on alive SNs variations

Using the $R_M - P_M$ cluster quality evaluation Indexes in Fig. 6, we analyze the optimal number of clusters/CHs per round for all the routing protocols. According to Fig. 15, RCEEFT demonstrated matchless certainty in both clustering and CHs selection when compared with the other protocols. Though this CH selection process is random, RCEEFT portrays some patterns and control in the CHs selection which explains the constancy and efficiency in data deliveries rates to BS (Fig. 13).

Fig. 9. Comparison of network's energy depletion rates

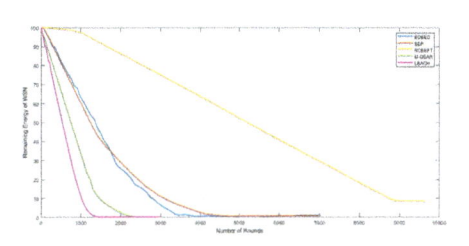

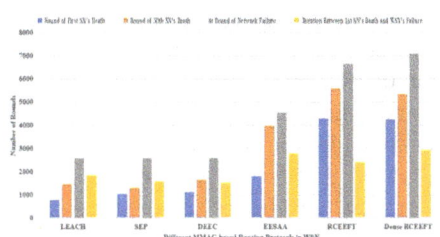

Fig. 10. Comparison analysis of network's operational patterns RCEEFT and state of the art in WSNs

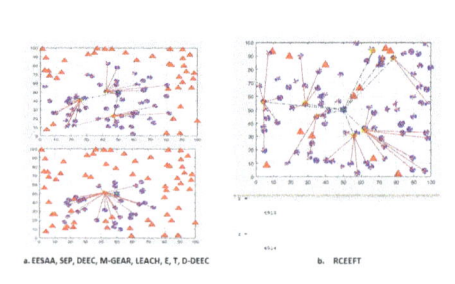

Fig. 11. Comparison of SNs failure (death) pattern: implications on balanced network-wide energy depletion rates

Fig. 12. FT analysis plot of RCEEFT

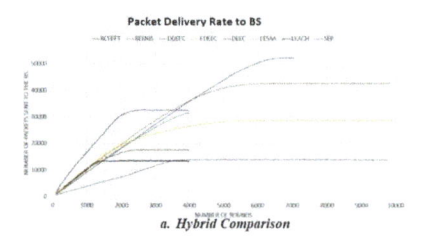

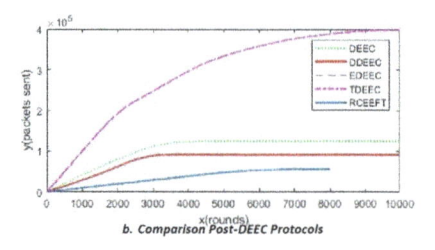

Fig. 13. Data delivery and needless redundancy control of RCEEFT and other protocols

The congestion in data transmission and the quantum of data transmitted to the BS is well-controlled and down-scaled by the unwanted redundancy control algorithm. This explains why the quantum of data of RCEEFT in Fig. 13b is way below those of post-DEEC protocols. Comparing the amount of data packets delivered to BS in the post-DEEC protocols vis-a-vis the network lifetime, we can also deduce that there are a lot junk data transmissions in the post-DEECs Fig. 13b. The duty-cycling approach, redundancy checking algorithm, and faulty sensory data detection and tolerance algorithms ensured that less data (just the accurate and precise ER data) is transmitted to the BS. Despite the elimination of junk data transmission, the successful data delivery to BS by RCEEFT is much better than SEP, DEEC, EESAA and LEACH due to the prolonged network lifetime of RCEEFT. To assess the FT capabilities of RCEEFT, it was operated at different p_d values and its associated number of clusters and CHs and allowed 4% defective SNs. As shown in Fig. 12, RCEEFT tolerated all defective SNs (100%) at the optimal number of clusters and depicted reduced FT when operated outside the cluster optimality zones.

The results and description firmly establish that the RCEEFT protocol shows much simplicity in computation (because SNs do not do any complex computations) and implementation (because RCEEFT is an easy and comprehensible protocol even during operation).

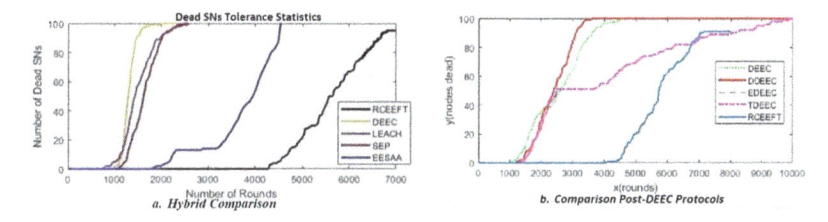

Fig. 14. Comparison of network instability using the statistics on dead SNs variations

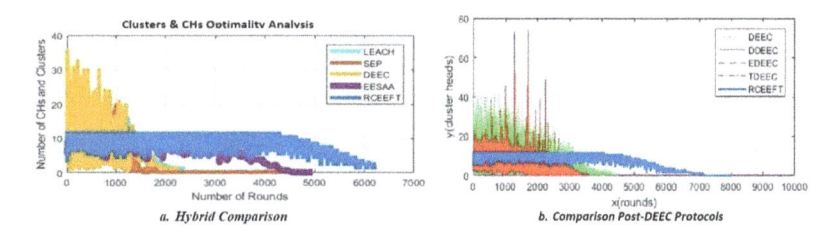

Fig. 15. Clusters and CHs selection analysis

6 Conclusion

A more optimized, cluster-based data acquisition and multichannel routing protocol for WSNs, RCEEFT, that is founded on MMP&AC and integrates unique measures to ensure peerless improvement in network energy efficiency and prolonged lifetime, balanced network-wide energy depletion rates with much prolonged network stability period, fault-avoidance-motivated FT and elimination of needless ER data redundancies and maintain the network coverage, accuracy and precision throughout the network lifetime has been proposed. With the aid of spatial correlation of event information and MMP&AC (proposed), the alternately correlated SNs pairing and scheduling concept, fault-avoidance-motivated FT and needless ER data redundancy check approaches were implemented without compromising simplicity and scalability of the network. Per the simulation results, RCEEFT significantly outperformed some existing protocols such as LEACH, SEP, DEEC, M-GEAR, EDEEC, DDEEC, TDEEC, and EESAA using the stipulated evaluation metrics in both normal, dense or up-scaled hybrid energy-leveled WSNs. We recommend that future researches should consider more improved measures to reduce message complexities and packet collisions, minimize idle times of the radio transceivers and the optimize the volumes of data transmitted by SNs without disregarding accuracies and precisions in ER data using testbed experiments. The RCEEFT protocol depicted high FT because the right mix of FT mechanisms was incorporated. In the future scope, we are working on a multi-hop version of RCEEFT to address possible distant-hop communication challenges large-scale WSNs.

References

1. Yu, Q., Li, G., Hang, X., Fu, K.: An energy efficient mac protocol for wireless passive sensor networks. Future Internet **9**(2), 1–12 (2017)
2. Effah, E., Thiare, O.: Survey: faults, fault detection and fault tolerance techniques in wireless sensor networks. Int. J. Comput. Sci. Inf. Secur. (IJCSIS) **16**(10), 1–14 (2018). ISSN 1947-5500
3. Darwish, I.M., Elqafas, S.M.: Enhanced algorithms for fault nodes recovery in wireless sensors network. Int. J. Sens. Networks Data Commun. **6**(1), 1–9 (2016). https://doi.org/10.4172/2090-4886.1000150
4. Titouna, C., Ari, A.A.A., Moumen, H.: FDRA: fault detection and recovery algorithm for wireless sensor networks. In: International Conference on Mobile Web and Intelligent Information Systems, pp. 72–85. Springer, Cham (2018)
5. Souissi, I., Ben Azzouna, N., Ben Said, L.: A multi-level study of information trust models in WSN-assisted IoT. Comput. Netw. **151**, 1–12 (2019)
6. Xu, G.C.R., O'Hare, L.: A smart and balanced energy-efficient multihop clustering algorithm (smart-BEEM) for MIMO IoT systems in future networks. Sensors **17**, 1574 (2017)
7. Shah, T., Javaid, N., Qureshi, T.: Energy efficient sleep awake aware (EESAA) intelligent sensor network routing protocol. In: International Multitopic Conference (INMIC), pp. 317–322 (2012). https://doi.org/10.1109/INMIC.2012.6511504
8. Mehmood, A., Mauri, J.L., Noman, M., Song, H.: Improvement of the wireless sensor network lifetime using leach with vice-cluster head. Ad Hoc Sens. Wirel. Netw. **28**, 1–17 (2015)
9. Younis, O., Fahmy, S.: HEED: a hybrid, energy-efficient, distributed clustering approach for ad hoc sensor networks. IEEE Trans. Mob. Comput. **3**(4), 660–669 (2004)
10. Lindsey, S., Raghavendra, C.S.: PEGASIS: power-efficient gathering in sensor information systems, vol. 3, p. 3 (2002). https://doi.org/10.1109/AERO.2002.1035242
11. Smaragdakis, G., Matta, I., Bestavros, A.: SEP: a stable election protocol for clustered heterogeneous wireless sensor networks, pp. 1–11 (2004)
12. Qing, L., Zhu, Q., Wang, M.: Design of a distributed energy-efficient clustering (DEEC) algorithm for heterogeneous wireless sensor network. Comput. Commun. **29**, 2230–2237 (2006)
13. Elbhiri, B., Saadane, R., El Fkihi, S., Aboutajdine, D.: Developed distributed energy-efficient clustering (DDEEC) for heterogeneous wireless sensor networks. In: 5th International Symposium on I/V Communications and Mobile Network (ISVC), pp. 1–4. IEEE (2010)
14. Saini, P., Sharma, A.K.: E-DEEC- enhanced distributed energy efficient clustering scheme for heterogeneous WSN. In: 2010 1st International Conference on Parallel, Distributed and Grid Computing (PDGC - 2010), pp. 205–210 (2010). https://doi.org/10.1109/PDGC.2010.5679898
15. Pankaj, K., Barwar, C.: Performance and comparative analysis of distributed energy efficient clustering protocols in wireless sensor networks. Int. J. Adv. Electron. Comput. Sci. **3**(7), 17–22 (2016). ISSN: 2393-2835
16. Nadeem, Q., Rasheed, M., Javaid, N., Khan, Z., Maqsood, Y., Din, A.: M-GEAR: gateway-based energy-aware multi-hop routing protocol for WSNs. In: Eighth International Conference on Broadband and Wireless Communication and Application, pp. 164–169, July 2013. https://doi.org/10.1109/BWCCA.2013.35

17. Gurpreet, K., Sukhpreet, K.: Enhanced M-GEAR protocol for lifetime enhancement in wireless clustering system. Int. J. Comput. Appl. **147**(14), 30–34 (2016)
18. Rajeev, K., Rajdeep, K.: Evaluating the performance of DEEC variants. Int. J. Comput. Appl. **97**(7), 9–16 (2014)
19. Chouikhi, S., Elkorbi, I., Ghamri-Doudane, Y., Saidane, L.A.: A survey on fault tolerance in small- and large-scale wireless sensor networks. Comput. Commun. **69**, 22–37 (2015)
20. Titouna, C., Aliouat, M., Gueroui, M.: FDS: fault detection scheme for wireless sensor networks. Wirel. Pers. Commun. **86**(2), 549–562 (2016)
21. Kakamanshadi, G., Gupta, S., Singh, S.: A survey on fault tolerance techniques in wireless sensor networks. In: 2015 International Conference on Green Computing and Internet of Things (ICGCIoT), Noida, pp. 168–173 (2015). https://doi.org/10.1109/ICGCIoT.2015.7380451
22. Effah, E., Thiare, O.: Estimation of optimal number of clusters: a new approach to minimizing intra-cluster communication cost in WSNs. Int. J. Innovative Technol. Exploring Eng. **8**(7S2), 521–524 (2019)
23. Shakya, R.K.: Spatial correlation-based efficient communication protocols for wireless sensor networks. Ph.D. thesis, Department of Electrical Engineering, Indian Institute of Technology, Kanpur, India, pp. 1–137 (2014)
24. Pak, W.: Ultra-low-power media access control protocol based on clock drift characteristics in wireless sensor networks. Int. J. Distrib. Sens. Netw. **13**(7), 1–5 (2017)
25. Kumar, A.: WiseMAC protocol for wireless sensor network-an energy-efficient protocol (2014)

Efficient Cache Architecture for Packet Processing in Internet Routers

Hayato Yamaki[✉]

The University of Electro-Communications, 1-5-1 Chofugaoka, Chofu, Tokyo, Japan
yamaki@uec.ac.jp

Abstract. Recent internet traffic growth becomes a serious problem for internet routers from the aspects of both packet processing throughput and power consumption. Packet Processing Cache (PPC) is one of the promising approaches to solve this problem. PPC caches table lookup results in packet processing and enables to reduce the number of TCAM (Ternary Content Addressable Memory) accesses, which are known as the major cause of degrading the throughput and power efficiency of packet processing. Although PPC can process a packet at fast with significant low energy if the packet hits in the cache, the cache miss rate of PPC is still high due to the small cache capacity. Furthermore, because conventional PPC has no tolerance to attacks, the cache miss rate of PPC is drastically increased when attacking. For these problems, we propose two different cache architecture, called Port-aware Cache and Victim IP Cache. Our simulation showed that the combination of them can achieve 1.81x higher throughput with 45% smaller energy per packet when compared to conventional PPC.

Keywords: Internet router · Packet processing · Cache architecture

1 Introduction

Recently, an internet traffic amount increases explosively due to applications such as video streaming, IoT (Internet of Things), and cloud services. As a result, throughput requirement of internet routers is getting higher and higher. To make matters worse, power consumption of internet routers is also non-negligible [1,2]. According to the report [3,4], the power consumption of network devices will reach several percentages of power generated in the world. Internet routers must achieve both high throughput and low power consumption.

In routers, table lookups are known as the major cause of degrading the throughput and energy efficiency of packet processing [5–7]. When a packet is processed, routers need to retrieve tables such as a routing table, an ARP (Address Resolution Protocol) table, an ACL (Access Control List), and a QoS (Quality of Service) table, to determine how to process the packet (e.g., where to forward or whether to filter). Recent routers store these tables into TCAM

(Ternary Content Addressable Memory) to realize fast lookups. TCAM is a specialized memory for fast data search and enables to search data with one cycle by comparing all bits in TCAM simultaneously. For this operation, TCAM consumes significant large energy compared to same-sized SRAM [8]. It is reported that the power consumption of TCAM accounts for 40% of all the power consumed in a router [9,10]. To make matters worse, search performance of TCAM is not sufficient to satisfy the requirement of future line speed (i.e., 400 Gbps and more) due to the low operation frequency. To realize high-throughput and low-energy table lookups, existing TCAM-based approach is not suitable.

Packet Processing Cache (PPC) is one of the promising approaches to solve this problem [11,12]. PPC stores TCAM lookup results into a small cache memory and reuses them to process subsequent packets. When many packets can be processed by using the cache, PPC can achieve high throughput with high energy efficiency because of the fast and low-energy cache memory.

The throughput and energy consumption of the table lookups with PPC depend on the cache miss rate. Thus, reducing the number of cache misses is an important issue for PPC. However, from the aspect of it, PPC has two problems: the high average cache miss rate and the drastic increase in the cache misses by attacks. In this study, we propose two different cache architecture, called Port-aware Cache and Victim IP Cache, to solve these problems.

The rest of this paper is organized as follows. We first explain more details and problems of PPC in Sect. 2 and introduce the relative studies of PPC in Sect. 3. After that, we propose Port-aware Cache and Victim IP Cache in Sect. 4 and Sect. 5, respectively. The evaluation of the proposed architecture is shown in Sect. 6. Finally, we conclude this paper in Sect. 7.

2 Packet Processing Cache

2.1 Outline

Packet Processing Cache (PPC) enables to reduce the number of TCAM accesses by caching the TCAM lookup results into a cache memory. When a router retrieves tables, one or several fields in a packet header are used as a key for table lookups. Conventionally, five tuples (i.e. source/destination IP addresses, source/destination port numbers, and protocol number) are used in most tables. This nature results that packets which have the same five tuples get the same table lookup results. From this fact, PPC defines packets which have the same five tuples as a flow and stores the table lookup results per flow into the cache. By processing the first packet of a flow and caching the table lookup results, the second and following packets can be processed without TCAM accesses.

Figure 1 shows the outline of packet processing flow with PPC. The cache of PPC consists of 13 bytes flow information (i.e., five tuples) and 15 bytes table lookup results. The table lookup results typically include 1 byte output port information as a result of the routing table, 12 bytes MAC addresses as a result of ARP table, 1 byte filtering information as a result of the ACL, and 1 byte QoS information as a result of the QoS table. Conventionally, PPC is constructed by

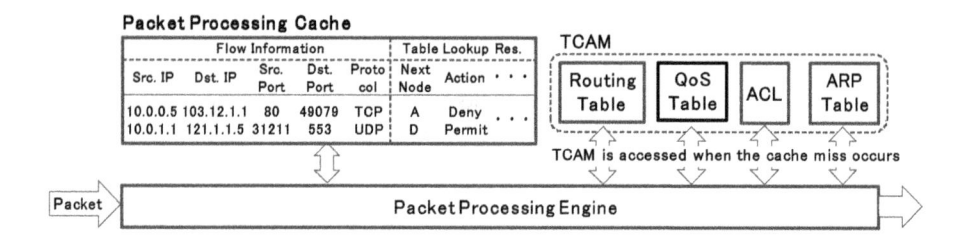

Fig. 1. Outline of packet processing flow with PPC.

32 KB small SRAM from the aspect of the access latency. This is because the latency gap between the cache and TCAM is small, unlike processor caches. For example, the latency of an L2 cache in CPU (almost 256 KB) is around 5 ns, and it is the same as that of TCAM.

2.2 Problems

The throughput and energy consumption of the table lookups with PPC depend on the cache miss rate of PPC because the latency and access energy of the cache are almost negligible when compared to TCAM. Thus, the cache miss rate is an important factor for PPC. From this aspect, PPC has two problems:

Tolerance to Attacks. When attacks such as port-scan attacks or attacks on vulnerabilities in an application pass through a router, a large number of PPC entries are evicted by the enormous attack flows. This situation causes drastic increase in the number of cache misses. In most attacks, attack flows are composed of a small number of packets and have little potential to hit in PPC. In addition, attack flows are scattered enormously in a short period. As the result, the cache miss rate is drastically increased in a short period, and the table lookup performance is greatly decreased.

A Large Number of TCAM Accesses. PPC can reduce the number of TCAM accesses by reusing the cached table lookup results. However, state-of-the-art PPC still shows the average cache miss rate of approximately 30% [13,14], and it means that 30% of all packets are still required to access TCAM. Make matters worse, several TCAM accesses are required to process one packet while PPC can process a packet with one cache access. From this reason, further reduction in the number of TCAM accesses is important, especially from the aspect of power consumption. One of the reasonable approach is to improve the cache miss rate of PPC; however, it is difficult due to the small cache capacity such like an L1 cache in microprocessors. Moreover, the number of cache entries in PPC is restricted due to the large entry size (28 bytes per entry).

3 Related Works

In this section, we introduce several studies of improving the cache miss rate in PPC. Although there are no studies which focuses on attack tolerance of PPC, various approach has been proposed to improve the cache misses in PPC.

There are studies of increasing the number of cache entries by compressing the cache tag. Chang et al. proposed digest cache, which used not the 5-tuple but the hash value of the 5-tuple as the cache tag [15]. Ata et al. compressed the cache tag by using only the 3-tuple as the cache tag (i.e. the source IP address, the destination IP address, and the smaller port number) [16]. Although these studies can improve the cache miss rate, there are many cache conflicts due to the lack of the flow information. As a result, the additional hardware is required for preventing the cache conflicts.

Liao et al. pointed out that CRC hash function, conventionally used in PPC as the cache index, is not suitable for PPC and leads the cache conflict. To solve this problem, they proposed effective cache indexing scheme [17]. The proposed indexing scheme divided the cache space into two areas and assigned different cache indexes to each area. They showed two universal hashes [18], which selects hash functions at random, are suitable for PPC to achieve the low cache miss rate. However, the implementation cost of the universal hashes is high because of the large input data size (i.e., 13-byte flow information).

4 Port-aware Cache

We first propose Port-aware Cache, which focuses on protocol-level flow behavior. It can prevent an increase in the cache misses when attacking by isolating the influence of each protocol.

4.1 Motivations

For PPC, the characteristics of internet protocols are important because the constitution and behavior of flows are mainly determined by them. We show it about HTTP (HyperText Transfer Protocol) and DNS (Domain Name System), which are major protocols in internet applications, as examples. HTTP flows are composed of not a few packets because they require three-way handshake and carry web contents, such as web pages, pictures, and videos. On the other hand, DNS flows are composed of one packets in many cases because of the simple request-reply model.

Figure 2 shows the comparison of the number of flows and packets with DNS, HTTP, and other protocols. In this measurement and following analyses, we used our simulator and WIDE traffic trace, which is a real core-network traffic. Details of them are explained in Sect. 6. This graph indicates that the amount of DNS packets is not large on the aspect of the number of packets, but the amount of DNS flows is largest in the network. The amount of HTTP packets is largest in networks, but the amount of HTTP flows is small. These results mean that

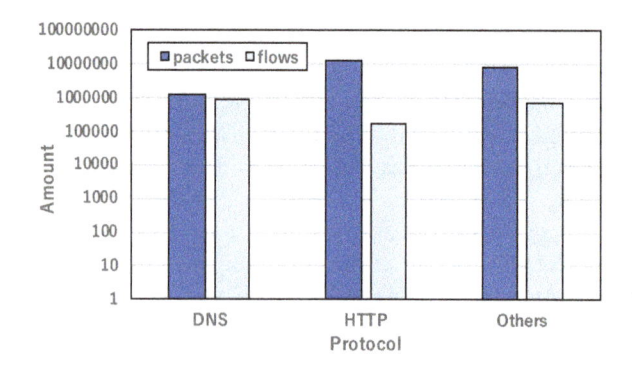

Fig. 2. The number of packets and flows of DNS, HTTP, and other protocols.

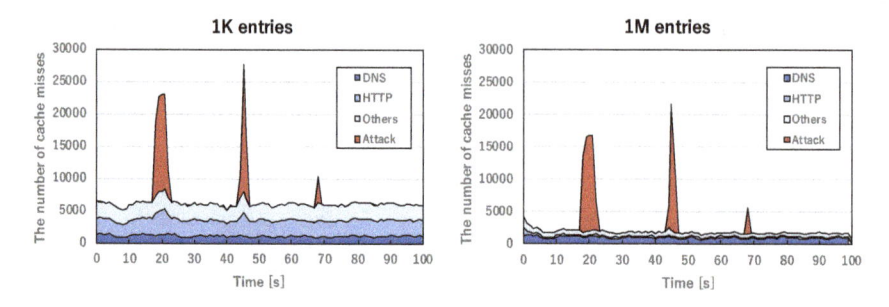

Fig. 3. The number of cache misses in PPC and the breakdown (comparison between 1K and 1M entries PPCs).

HTTP and DNS show not only remarkable for the number of packets or flows but also distinctive constitution.

Besides major internet protocols, Attack flows also have distinctive constitution and behavior. As mentioned in Sect. 2.2, attack flows are composed of a small number of packets and scattered drastically. As the result, the cache entries are polluted. We analyzed the impact of attacks on the number of cache misses of PPC in Fig. 3. The left graph shows the number of cache misses and the breakdown in 1K entries conventional PPC, and the right graph shows that in 1M entries PPC. Figure 3 indicates that attack flows cause not only drastic increase in the cache misses of attack flows but also increase in the cache misses of HTTP and Other protocol flows if PPC has a small number of entries. This is because useful entries are evicted by attacks. Note that cache misses of DNS flows are not affected by attacks and independent of the number of cache entries because DNS flows hardly hit in PPC anyway. On the other hand, HTTP and other protocol flows have potential to hit for the most part by assigning a large number of entries.

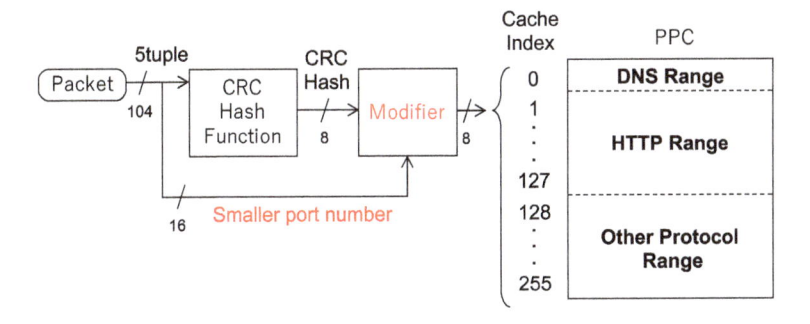

Fig. 4. Block diagram of Port-aware Cache.

4.2 Architecture of Port-aware Cache

Based on the analyses in Sect. 4.1, we propose Port-aware Cache, which divides PPC entries to several protocol ranges. In this paper, we try to divide PPC into three protocol ranges: DNS, HTTP, and other protocol (including attacks) ranges. This is because these groups are remarkable for an influence on PPC. Figure 4 shows the architecture of Port-aware Cache. To realize the access to each protocol range, the CRC-hash value of a five tuple is modified based on the smaller port number. Packets access to the corresponding cache range by using the modified CRC-hash values.

Port-aware Cache has two advantages. First, appropriate number of cache entries can be assigned to each protocol range. As mentioned in Sect. 4.1, the cache misses of DNS flows can not be reduced by increasing the number of entries because DNS flows are composed of almost one packet. Thus, it is better to assign a small number of entries as the DNS range. Unlike DNS, HTTP flows are composed of a large number of packets, and thus it is better to assign a large number of entries as the HTTP range. The second advantage is that Port-aware Cache isolates an influence of each protocol flows on PPC. As shown in Fig. 3, attacks impede cache hits of HTTP flows by evicting useful HTTP entries. Port-aware cache enables to prevent this problem by restricting entry insertion from a range to other ranges.

5 Victim IP Cache

In this paper, we also propose victim IP cache to reduce the number of TCAM accesses. The victim IP cache is placed after PPC, and packets which miss in PPC are accessed.

5.1 Motivations

As mentioned in Sect. 2, a large number of TCAM accesses are still required of PPC due to the high average cache miss rate and the several TCAM accesses

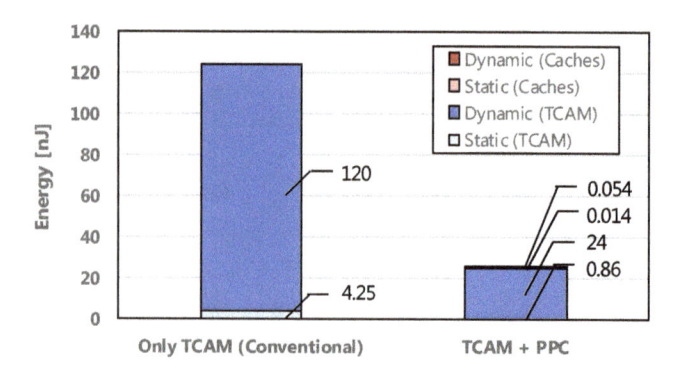

Fig. 5. Breakdown of power consumed by the table lookups.

for processing a packet. We analyzed the energy consumption of TCAM-based architecture and conventional PPC and showed the breakdown in Fig. 5. The graph shows the breakdown of power consumed by the table lookups when processing a packet. The energy consumption was calculated from our simulation results and the PPC energy model explained in Sect. 6. Figure 5 shows that PPC can reduce the energy of table lookup processing significantly by reducing the dynamic energy of TCAM. However, the dynamic energy of TCAM is still dominant in table lookups, while that of the cache is negligible. Thus, to reduce the number of TCAM accesses is an important issue for PPC. Cache miss reduction is one of the reasonable approaches to meet this requirement: however, it is difficult for PPC as mentioned in Sect. 2.

As another study of caching packets, IP caching has been proposed. Unlike PPC, the IP cache stores only routing table lookup (and ARP table lookup) results and reuses them by using destination IP addresses of packets as the cache tag. Past studies show that the IP cache can achieve the cache hit rate of more than 90% [6]. In this paper, We try to combine the IP cache with PPC and reduce the whole number of TCAM accesses.

5.2 Architecture of Victim IP Cache

We propose Victim IP Cache, which assists PPC by caching packets which miss in PPC. Unlike PPC, Victim IP Cache stores lookup results of two tables, i.e., the routing table and the ARP table, and thus only the destination IP address is required to store as the cache tag. Victim IP Cache is accessed when a packet misses in PPC. If the packet hits in Victim IP Cache, the packet can be processed without retrieving the routing table and the ARP table. As mentioned in Sect. 5.1, because IP caches have higher locality than flow-based caches such like PPC, it is expected that many packets which miss in PPC hit in Victim IP Cache.

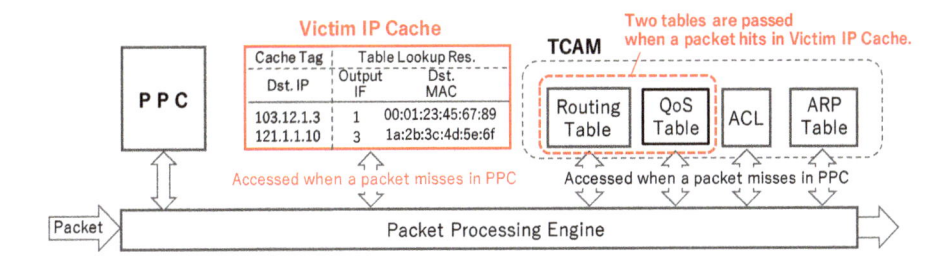

Fig. 6. Block diagram of Victim IP Cache with PPC.

Figure 6 shows the block diagram of Victim IP Cache with PPC. An entry of Victim IP Cache consists of 11 byte: 4 bytes destination IP address as the cache tag, 1 byte output interface as the routing table lookup result, and 6 bytes destination MAC address as the ARP table lookup result. Because the entry size of Victim IP Cache is smaller than that of PPC, Victim IP Cache can store larger entries than PPC. Thus, it is expected that using Victim IP Cache can achieve higher cache hit rate than using a flow-based victim cache. Unlike typical victim caches, the entries in Victim IP Cache are not moved to PPC if the entries hit. This is because a packet does not always hit in PPC even if the packet hits in Victim IP Cache due to the difference of the locality between the flow and the destination IP address.

6 Evaluations

In this section, we show the evaluations of Port-aware Cache and Victim IP Cache from our simulation of packet processing with PPC.

6.1 Experimental Setup

For the simulation, we used our in-house PPC simulator written in C++ and 11 types of real-network traffics. The clock frequency, the PPC configuration, and the network-trace name were set in the simulator as the configuration parameters. Each value is shown in Tables 1 and 2. In this paper, PPC was configured as a 4-way 1K-entry cache, which was conventionally used. The latency of the cache was estimated using CACTI 6.5 [19], which was widely-used memory simulation tool. The network traces used in the simulation were acquired from various universities or laboratories networks (obtained from RIPE Network Coordination Centre [20]) and a Japanese core-network (obtained from WIDE MAWI WorkingGroup [21]).

The simulator runs based on the clock frequency and simulates cycle-accurate operations. First, the flow information (i.e., the five tuple) is extracted from an arrived packet based on the timestamp of the packet. Next, the cache access is simulated. If the flow exists in the cache (i.e., a cache hit), the processing of the packet is finished. If a cache miss occurs, the packet is sent to TCAM modules.

Table 1. Simulator parameters.

Parameter		Value
Cache	Entry	1 K
	Way	4
	Replacement	LRU
	Latency	0.5 ns
TCAM latency		5 ns
Clock frequency		1 GHz

Table 2. Details of network traces.

Trace name	Packets/sec
IPLS [20]	99,264
UFL [20]	51,319
PUR [20]	42,300
MRA [20]	41,372
FRG [20]	32,722
CNIC [20]	31,023
WIDE [21]	24,657
APN [20]	20,793
TXG [20]	11,610
COS [20]	7,972
BUF [20]	7,827

After the TCAM access latency, the processing of the packet is finished, and the cache entry is updated. The simulator measures the cache miss rate every seconds (Tables 1 and 2).

6.2 Evaluation of Port-aware Cache

Best Configuration of Each Protocol Range. First, we investigated the appropriate ratio of each protocol range in Port-aware Cache. Figure 7 shows the average cache miss rates when the number of cache entries in HTTP range was set from 4 to 1,004 at an interval of 40. Note that the number of cache entries in DNS range was always set to 8 because it showed the best performance. Our simulation results showed that the combination of 8 entries DNS range, 800 entries HTTP range, and 216 entries other protocol range is the best configuration of Port-aware Cache. It means that HTTP is the most influential in PPC, and it is the best to assign as many entries as possible in the HTTP range. The best configured Port-aware Cache can reduce the average cache miss ratio by 9.41% compared to conventional PPC.

Attack Tolerance. Next, we evaluated the attack tolerance of Port-aware Cache. Figure 8 shows the comparison of the cache miss rates between conventional PPC and Port-aware Cache. In WIDE trace, eight attacks were observed at 19 s, 46 s, 69 s, 286 s, 465 s, 723 s, 737 s, and 783 s. When compared to conventional PPC, Port-aware Cache improved the cache miss rate by 3.77 points (8.64%) on average against attacks.

Furthermore, Fig. 9 shows the number of cache misses and the breakdown in WIDE trace from 0 s to 100 s. When compared to Fig. 3, Port-aware Cache can not only reduce the average number of HTTP flow misses but also prevent the increase in HTTP flow misses by attacks. However, other-protocol flow misses

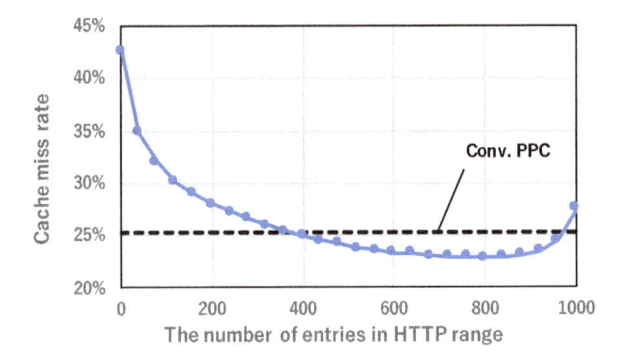

Fig. 7. Impact of the number of entries in HTTP range on cache performance.

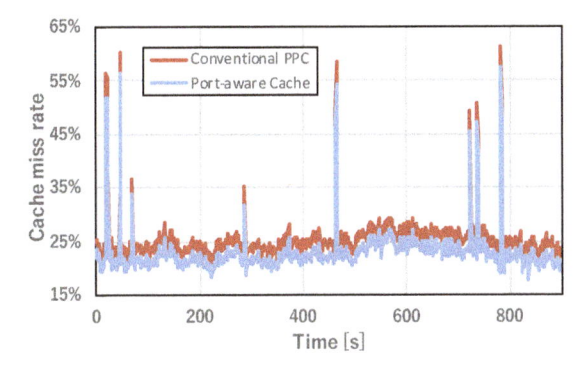

Fig. 8. Comparison of cache miss rate between conv. PPC and Port-aware Cache in WIDE trace.

were increased by attacks because attack flows were included in the other protocol range. More fine-grained division of the other-protocol range is one of the reasonable solutions. This is our future work.

6.3 Evaluation of Victim IP Cache

Improvement in Average Cache Miss Rate. To show the effectiveness of Victim IP Cache clearly, we implemented a typical victim cache in our simulator and compared the cache miss rates. Figure 10 shows the average cache miss rates for various networks. The dotted bars in the graph show the cache miss rates of the upper level cache (i.e., conventional PPC). The graph indicates that Victim IP Cache can improve the cache miss rate of PPC remarkably. Victim IP Cache reduces the number of cache misses by 85.5% on average, while the typical victim cache reduces them by 26.0%. Thus, it is better for PPC to use Victim IP Cache rather than the typical victim cache.

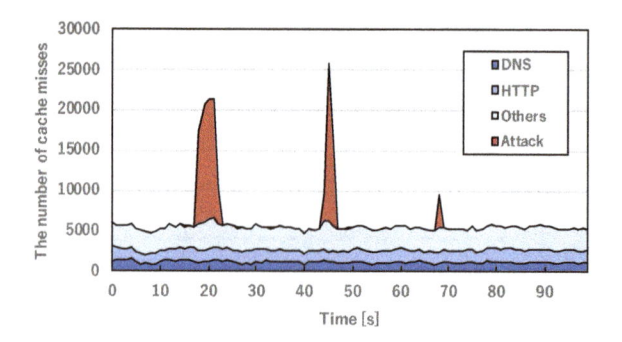

Fig. 9. The number of cache misses and the breakdown with Port-aware Cache.

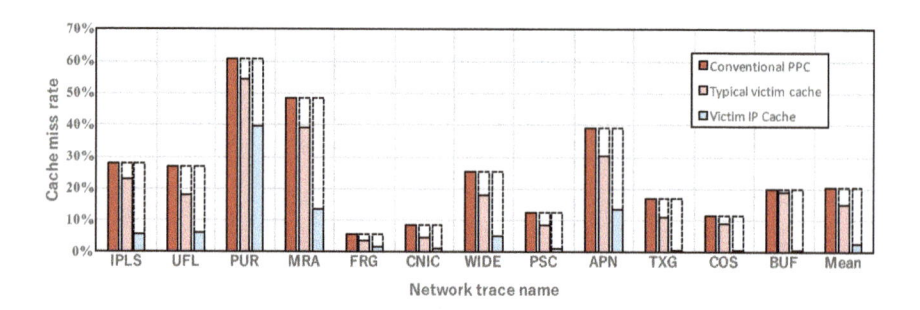

Fig. 10. Cache miss rates for various networks.

Impact on Throughput and Energy. As shown in Fig. 10, Victim IP Cache can take care of most packets which miss in PPC. However, only two tables are bypassed when packets hit in Victim IP Cache, and several TCAM accesses are still required per packet (e.g., the ACL and the QoS table). Thus, the cache miss rates shown in Fig. 10 do not directly represent the improvement in the throughput and energy consumption of the table lookup processing. In this subsection, we defined throughput and energy models of PPC with Victim IP Cache based on the past study [14] and estimated the throughput and energy consumption of the table lookup processing.

First, we extended the throughput model. The throughput of the table lookup processing with PPC and Victim IP, represented as T, is calculated from Eq. 1.

$$T = \frac{l}{t_{Avg}} = \min\left(\frac{l}{t_{PPC}}, \frac{l}{t_{Victim} m_{PPC}}, \frac{l}{t_{TCAM}(2 m_{Diff}/n + m_{Victim})}\right) \quad (1)$$

Here, t_{PPC}, t_{Victim}, t_{TCAM}, and t_{Avg} represent the latency of PPC, Victim IP Cache, TCAM, and the average, respectively. m_{PPC} and m_{Victim} represent the cache miss rates in PPC and Victim IP Cache, respectively. In addition, m_{Diff} represents the difference between m_{PPC} and m_{Victim}. n and 2 in the equation mean the number of table lookups required to process a packet when occurring the cache miss in

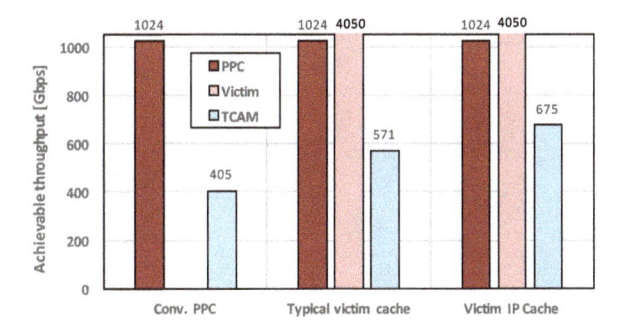

Fig. 11. Breakdown of achievable throughput of PPC, Victim cache, and TCAM. The throughput of whole architecture is determined by the minimum throughput.

PPC and Victim IP Cache. l means the packet length and is set to 64 bytes as the shortest length of a packet in this paper. Equation 1 indicates that the throughput of whole architecture is restricted by minimum achievable throughput of PPC, Victim IP Cache, and TCAM. When compared to the conventional throughput model, Victim IP Cache can increase the achievable throughput of TCAM by reducing the number of TCAM accesses.

Next, we extended the energy model. The energy consumption of the table lookup processing with PPC and Victim IP Cache, represented as E_{All}, is calculated from Eq. 2.

$$E_{All} = E_{PPC}^{Dynamic} + E_{Victim}^{Dynamic} m_{PPC} + E_{TCAM}^{Dynamic}(2m_{Diff} + nm_{Victim}) + P^{Static} t_{Avg}. \qquad (2)$$

Here, $E_{PPC}^{Dynamic}$, $E_{Victim}^{Dynamic}$, and $E_{TCAM}^{Dynamic}$ represent the dynamic energy of each memory per access, while P^{static} represents the sum of the static power of all memories. Equation 2 indicates that introducing Victim IP Cache can reduce the dynamic energy of TCAM while it requires additional dynamic and static energy of the victim cache memory.

Based on Eqs. 1 and 2, we estimated the throughput and energy consumption of table lookup processing. The latency and energy of each memory was estimated from CACTI. Figures 11 and 12 show the breakdown of the achievable throughput and the energy consumption. From the Fig. 11, Victim IP Cache can achieve 1.67x higher throughput than conventional PPC. However, the throughput of whole architecture is still restricted by the achievable throughput of TCAM. Furthermore, Fig. 12 showed that Victim IP Cache can also reduce the energy consumption by 40.1% while the typical victim cache reduced by 29.0%. Although introducing a victim cache causes the additional increase in the energy due to the victim cache memory, the dynamic energy of TCAM is still dominant.

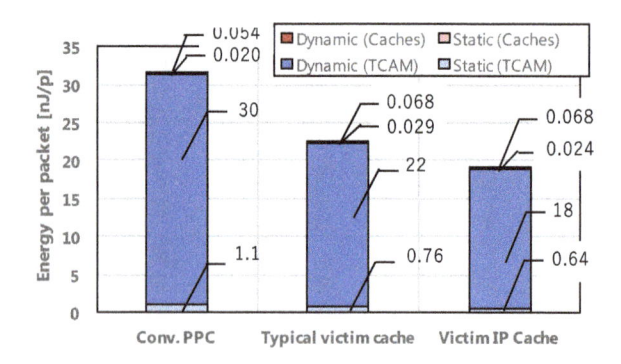

Fig. 12. Breakdown of energy consumption in conventional PPC, PPC with typical victim cache, and PPC with Victim IP Cache.

6.4 Combination of Proposed Architecture

Finally, we evaluated the combination of Port-aware Cache and Victim IP Cache. We summarized the estimates of the average cache miss rate, throughput, and energy consumption in Table 3. The combination of Port-aware Cache and Victim IP Cache can achieve 1.81x table-lookup throughput with 45.0% smaller energy per packet. Table 3 showed that the throughput and energy consumption were improved by combining Port-aware Cache and Victim IP Cache though Victim IP Cache mainly performed when attacking.

Table 3. Estimates of avg. cache miss rate, throughput, and energy consumption.

		Conv. PPC	Port-aware Cache	Victim IP Cache	Combination
Avg. miss rate	PPC	25.3%	22.9%	25.3%	22.9%
	Victim	n/a	n/a	5.02%	5.00%
Throughput		405 Gbps	447 Gbps	675 Gbps	734 Gbps
Energy consumption		31.5 nJ	28.5 nJ	18.9 nJ	17.4 nJ

7 Conclusion

Packet Processing Cache (PPC) is promising approach to realize both high-throughput and low-power packet processing in routers. However, PPC has two problems: a large number of TCAM accesses and attack tolerance. To tackle these problems, we proposed two different cache architecture for PPC.

Port-aware Cache can isolate the influence of each internet application and assign appropriate number of entries to applications. Our simulation result showed that the best configured Port-aware Cache (800 HTTP entries, 8 DNS entries, and 216 other protocol entries) can reduce the average cache miss rate by 9.41% and the number of cache misses in attacks by 8.64%. For further improvement, more fine-grained division is effective, and it is a future work.

Victim IP Cache adds an extra IP Cache after PPC and enables to save packets missed in PPC. Our simulation result showed that 85.5% of all the packets missed in PPC can hit in Victim IP Cache. As a result, the energy consumption of the table lookup processing per packet is reduced by 40.1% compared to conventional PPC.

Finally, we considered the combination of Port-aware Cache and Victim IP Cache. Our simulation showed that the combined architecture performed the best and can achieve 1.81x higher table-lookup throughput with 45.0% smaller energy per packet when compared to conventional PPC.

Acknowledgements. This work was supported by JSPS KAKENHI Grant Number JP18K18022.

References

1. The Ministry: Tabulation and Estimation of Internet Traffic in Japan (2016). http://www.soumu.go.jp/main_content/000462459.pdf
2. METI: Green IT Initiative in Japan. http://www.meti.go.jp/english/policy/Green ITInitiativeInJapan.pdf
3. Fan, J., Hu, C., He, K., Jiang, J., Liuy, B.: Reducing power of traffic manager in routers via dynamic on/off-chip scheduling. In: 2012 Proceedings of the IEEE INFOCOM, Orlando, FL, pp. 1925-1933 (2012)
4. Zheng, X., Wang, X.: Comparative study of power consumption of a NetFPGA-based forwarding node in publish-subscribe Internet routing. Comput. Commun. **44**, 36–43 (2014)
5. Gamage, S., Pasqual, A.: High performance parallel packet classification architecture with popular rule caching. In: 18th IEEE International Conference on Networks (ICON), Singapore, pp. 52–57 (2012)
6. Talbot, B., Sherwood, T., Lin, B.: IP caching for terabit speed routers. In: IEEE Global Telecommunications Conference (GLOBECOM 1999), Brazil, vol. 2, pp. 1565–1569 (1999)
7. Guinde, N.B., Rojas-Cessa, R., Ziavras, S.G.: Packet classification using rule caching. In: 2013 Fourth International Conference on Information, Intelligence, Systems and Applications (IISA 2013), Piraeus, pp. 1–6 (2013)
8. Agrawal, B., Sherwood, T.: Ternary CAM power and delay model: extensions and uses. IEEE Trans. Very Large Scale Integr. (VLSI) Syst. **16**(5), 554–564 (2008)
9. Nawa, M., et al.: Energy-efficient high-speed search engine using a multi-dimensional TCAM architecture with parallel pipelined subdivided structure. In: 2016 13th IEEE Annual Consumer Communications & Networking Conference (CCNC), Las Vegas, NV, pp. 309–314 (2016)
10. Hewlett-Packard Development Company.: Energy Efficient Networking - Business white paper. http://h17007.www1.hp.com/docs/mark/4AA3-3866ENW.pdf
11. Okuno, M., Nishi, H.: Network-processor acceleration-architecture using header-learning cache and cache-miss handler. In: The 8th World Multi-Conference on Systemics, Cybernetics and Informatics (SCI 2004), pp. 108–113 (2004)
12. Yamaki, H., Nishi, H.: An improved cache mechanism for a cache-based network processor. In: Proceedings of the International Conference on Parallel and Distributed Processing Techniques and Applications (PDPTA 2012), Las Vegas, NV, pp. 1–7 (2012)

13. Yamaki, H., Nishi, H.: Line replacement algorithm for L1-scale packet processing cache. In: Adjunct Proceedings of the 13th International Conference on Mobile and Ubiquitous Systems: Computing Networking and Services (MOBIQUITOUS 2016), Hiroshima, Japan, pp. 12–17 (2016)
14. Yamaki, H., Nishi, H., Miwa, S., Honda, H.: Data prediction for response flows in packet processing cache. In: 2018 55th ACM/ESDA/IEEE Design Automation Conference (DAC), San Francisco, CA, pp. 1–6 (2018)
15. Chang, F., Feng, W.C., Li, K.: Efficient packet classification with digest caches. In: Proceedings of Third Workshop Network Processors and Applications (NP-3) (2005)
16. Ata, S., Murata, M., Miyahara, H.: Efficient cache structures of IP routers to provide policy-based services. In: IEEE International Conference on Communications (ICC 2001), Helsinki, vol. 5, pp. 1561–1565 (2001)
17. Liao, G., Yu, H., Bhuyan, L.: A new IP lookup cache for high performance IP routers. In: Proceedings of the 47th Design Automation Conference (DAC), Anaheim, CA, pp. 338–343 (2010)
18. Carter, J.L., Wegman, M.N.: Universal classes of hash functions (extended abstract). In: Proceedings of the Ninth Annual ACM Symposium on Theory of Computing (STOC 1977), New York, NY, pp. 106–112 (1977)
19. Muralimanohar, N., et al.: Optimizing NUCA organizations and wiring alternatives for large caches with CACTI 6.0. In: Proceedings of the 40th Annual IEEE/ACM International Symposium on Microarchitecture (MICRO 40), Chicago, USA, pp. 3–14 (2007)
20. RIPE Network Coordination Centre: Réseaux IP Européens Network Coordination Centre RIPE NCC. http://www.ripe.net/
21. WIDE MAWI WorkingGroup: MAWI Working Group Traffic Archive. http://mawi.wide.ad.jp/mawi/

Optimization of Cultural Heritage Virtual Environments for Gaming Applications

Laura Inzerillo[✉], Francesco Di Paola, Yuri Alogna, and Ronald Anthony Roberts

University of Palermo, Palermo 90128, Italy
{laura.inzerillo,francesco.dipaola,
ronaldanthony.roberts}@unipa.it,
yuri.alogna@community.unipa.it

Abstract. Serious games are games with a purpose beyond entertainment and are widely acknowledged as fruitful tools for learning and developing skills across multiple domains, including educational enhancement. In the last few years, the world of serious games has widely increased. The use of these types of games can aid in classrooms to not only help the students learn concepts but also to improve their motivation to do so. However, designing games necessitates very specialized personnel and the process can often be costly and slow. The adaptions of the design to the implantation phase are also difficult and the process needs more focus. The challenge of this study was to create a game within the confines of cultural heritage and preservation that was an adventure game as well, available on any platform with a friendly user interface and free to download on the web. To achieve this great goal, it was necessary to overcome the limits and problems due to the virtual environment starting from 3D acquisition and finishing with image stability. In this paper, we show the flowchart and methodology for the optimization of a cultural heritage virtual environment for gaming applications.

Keywords: Video games · Cultural heritage · Virtual reality · 3D survey · 3D modeling

1 Introduction: The Value of Serious Video Games

Nowadays there are many popular serious games. Amongst them, some have changed the world fighting for humanity, politics, environmental causes and more [1]. Serious games (SG) are consistently featured in the news. Computer games are an effective way to retaining interest of players (learners) by attracting their attention for much more time than traditional approaches [2]. Videogames creating both tangible and intangible cultural artefacts in highly interactive and dynamic ways should provide a powerful and increasing appeal and engagement for all users by possessing an integrated form of fun and play. Nevertheless, sometimes, in games this doesn't happen. The games addressed to the training/educational have more an application/app aspect and feel than an adventure game making them less appealing to users. This means that these games are not widely used among younger people and this represents a challenge for the developer.

2 Aim of the Paper

In this research project, the goal was to design a serious videogame that is an adventure/educational game as well, within the scope of cultural heritage applications. The prerequisite was to advance the gamer's knowledge base. For this reason, more sophisticate techniques about the photorealistic virtual models within games were considered. Whilst developing a game there are numerous graphical elements that must be deliberated. These include Presentation, Perspective, Color, Dimensionality, and Realism.

However, the primary difficulty of constructing a video game, where cultural assets are not a scenario but are part of the game itself, is in photorealism which has become crucial to games. The design of a photorealistic adventure game needs the use of optimization techniques and strategies. This paper shows how to achieve this.

Amongst a human's senses, vision can be considered as the most controlling. For a human to understand information, diverse visual cues are heavily relied on. These are critical for alignment and perception. Given the breakthroughs in hardware, most games utilize 3D based gaming engines which allow for projections in a perspective manner with high quality optics. One landmark game which visualized a fully textured environment was Quake. From this point onwards it is typical for games to use these techniques [3].

Realism is typically defined by the photo-realistic look and feel of the game. To obtain this there are particular factors that must be considered. These include sound, character animation, authentic object behavior and game individuals. The way in which this is presented can define the gaming environment and how the player is visualized on the screen. Considering a higher though level, the linking of the game's storyline also plays a significant role in the player's opinion of the game's realism.

This paper addresses the photorealism of the graphics for a cultural heritage game and optimizes the pipeline used to ensure the final product is appealing to users. The general software pipeline for the generation of these games is depicted in the flowchart in Fig. 1.

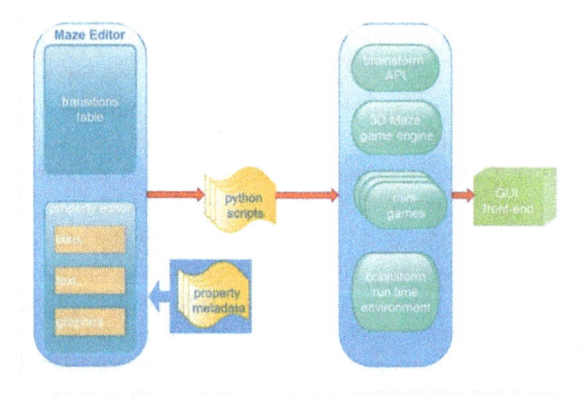

Fig. 1. Generic Software pipeline for the generation of games

3 Considerations for Creating Cultural Heritage Games

Serious video games for cultural heritage use interactive simulations of realistic virtual heritage scenarios using virtual and augmented reality and artificial intelligence for content creation and adaption. Interactive virtual museums use gaming technology for both entertaining and educating visitors usually by incorporating some exploration and reassembling tasks and quizzes. The most common problems of serious games are connected to the higher development cost, lower attractiveness compared to entertainment games, transition between instructional design and actual game design implementation. This is depicted in the approaches needed for Cultural heritage (CH) applications shown in Fig. 2.

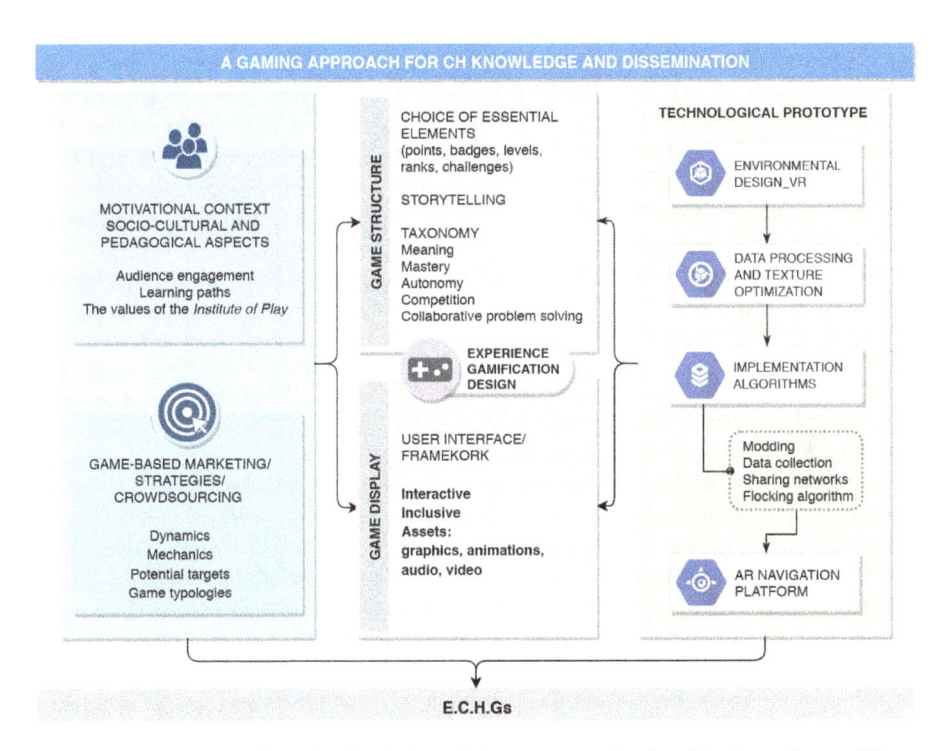

Fig. 2. Pipeline utilized for the design of the structure of cultural heritage Games [4]

Solutions to problems of SG are the rise of cheap, ubiquitous hardware, the use of robust networks that allow for connectivity without the administrative constraints of the past and the adoption of artificial intelligence tools and software.

Video games can form part of portable cultural heritage. The games can preserve a snapshot of technological and socio-economic achievements and provide evidence of progress in the game industry evolution and the landscape of cultural heritage sites and monuments. There are existing video games museums which also have active online archives which can be utilized as a platform for futures works with examples such as the Videogame History Museum: http://www.vghmuseum.org/; Computerspiele museum: http://www.computerspielemuseum.de/; Museo Games (Paris); and also the game adaption, Assassin's Creed, which highlights a tour of Ancient Egypt: https://store.steampowered.com/app/775430/Discovery_Tour_by_Assassins_Creed_Ancient_Egypt/.

4 Workflow and Tools Employed for Optimizing Techniques for Cultural Heritage Gaming

In the wide context of Digital Heritage and in this gaming scenario, the visual communication of a 4.0 cultural experience implies a process of digitization and integration of all the tools to promote, attract and to involve the user to create effective multidisciplinary analysis and study platforms.

By combining a spectrum of computer vision and graphics interfaces and techniques and integrating them onto high-performance devices, one can generate high resolution environments. Within these environments, the user can move around and explore virtual but photorealistic models and objects in real-time.

The goal of this study is to examine game structure design approaches in handling 3D data and creating a platform that would allow this data to be handled on gaming platforms for the advancement of the cultural heritage landscape [5]. The pipelines introduced in this study identifies the issues in the acquisition of 3D data relative to complex objects, ornaments, works of art and museum pieces and provides solutions on how this data can be easily acquired and analyzed.

Within this environment, objects require specific features with regards to their texture quality, topology and polygonal weight. These features have to adaptable with respect to the type of model that is being produced. As a result of this, it is clear to see the importance of developing appropriate design and deployment systems for the 3D content. This will ensure that when the elements are reproduced they visualize realistic features and can adequately compete with the typical gaming development environment in non-educational gaming [6, 7].

In many instances, the texture generated onto 3D models is worse than those seen in 2D imagery. This is due to the fact that the 3D texture is coming from several different sources. The sources can have differing internal camera parameters and can be affected by a multitude of natural factors such as lighting and exposure [8, 9].

These naturally occurring factors can be an impediment to the high resolution texture generation and cause correspondences that are not accurate or realistic when compared to the real-life object [10, 11], and [12].

This study was generated to overcome these types of problems to ensure the final product contains objects that are realistic for the virtual environment within games.

A prototype of a virtual reality museum within a cultural heritage game was created to evaluate the results of the techniques used in the study. Within this environment the definite environmental design and motion choices were designed. The entire process for this to be done was broken down into three distinct phases.

The first of these was the 3D modeling of the actual environment. This was done utilizing 2D documentation developed through surveys and actual design drawings.

Once this was achieved, the second part dealt directly with the digital acquisition. This was done utilizing typical photogrammetric techniques such as Structure-from-Motion (SfM) and structured light techniques, when needed. These techniques were utilized for all the necessary museum articles and collections. Additionally, at this stage some of the optimization techniques were employed such as the reduction of the meshes and texture enhancement using sculpting methodologies [13, 14], and [15].

The final stage was coding the structure of the game utilizing UnReal Engine 4. At this stage there was also the development of specific add-ons and parametric-procedural algorithms in Python to allow for more productivity within the workflow.

With respect to the first stage of the process, some of the important concepts that were utilized in constructing the workflow choice need to be examined. One such concept is that of the Level of Details (LOD). The LOD is a concept which dictates the effectiveness of the virtual environment in creating a game which is realistic, playable and interactive. Once LODs are used appropriately they allow for easy and interactive game navigation, manipulating both the available software and hardware resources for efficient performance. With reference to the hardware, the Video card is a key component as its task to create the environment around which the character needs to navigate. As a result, LODs need to be different for every created object. Multi-resolution techniques are generally used to control this and to ensure that the GPU's load is reduced. The process involves loading the meshes, of varying polygon numbers, into the game engine.

For high polygonal load models, it is not a requirement for the same number of polygons to be attained for the meshes. Instead, it is more important for the data to go through optimization and corrections for the next stages of processing. This will be examined further in the study.

The LOD methodology was not applied on all the objects in the created prototype and instead the loads were lowered utilizing procedural texture techniques which are also described in the study. The use of the procedural techniques still ensured that the final products could be visualized even in scenarios where there are hardware availability restrictions.

Furthermore, the entire structure of the architecture and environment was done within the Blender (stable v2.79 and beta v 2.8). This software is open-source and is commonly distributed under the GPL license. The software contains modifiers that allow for effects to be applied to the meshes under analysis. Some common modifiers used in this type of 3D scene created are subdividing the surface, smoothing the mesh and the mirror and bevel modifiers (Fig. 3).

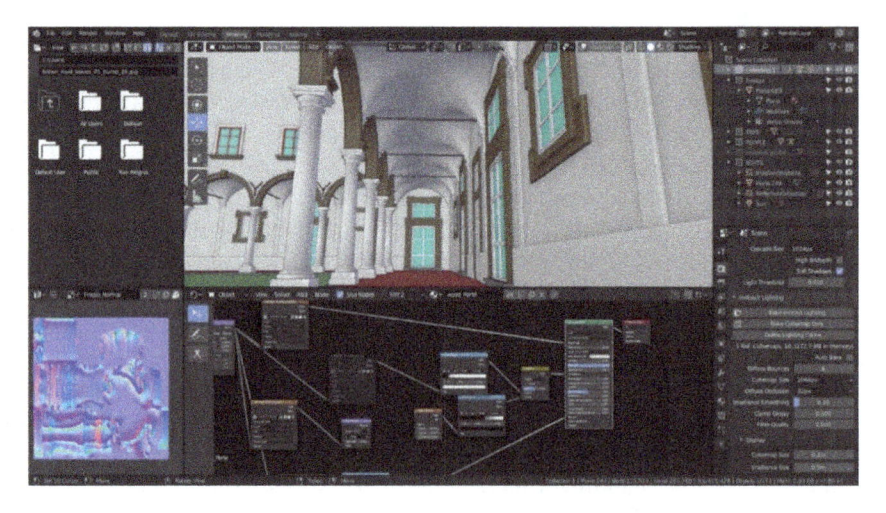

Fig. 3. Visualization of gaming design utilizing tools from Blender 3D

In addition, using proprietary 3D sculpture tools, the architectural elements (arches ashlars, moldings and friezes) have been modelled, adopting some methodologies that allowed obtaining a graphic rendering close to reality, simultaneously preserving the computational load of the processor (Fig. 4).

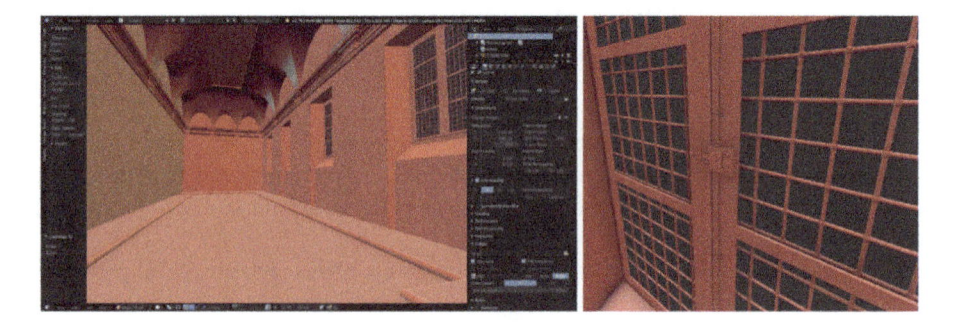

Fig. 4. The game environment in the Matcap display

To allow for these actions, Retopology and Remeshing processes were adapted using MeshLab and Instant Meshes. The processes were done on the polygonal models. During this, the meshes are replaced with updated versions that will now reflect stable and precise elements. The new meshes have reduced polygonal faces (low-poly). They must also maintain original shape of the previous mesh (the high poly version) and they must also have a similar topology in line with the project aims. It must be noted that in instances where there is organic geometry, several details are lost at the retopology stage. However, they can be displayed utilizing procedural textures at the Baking stage. These types of textures are developed to collect and keep information in the 2-

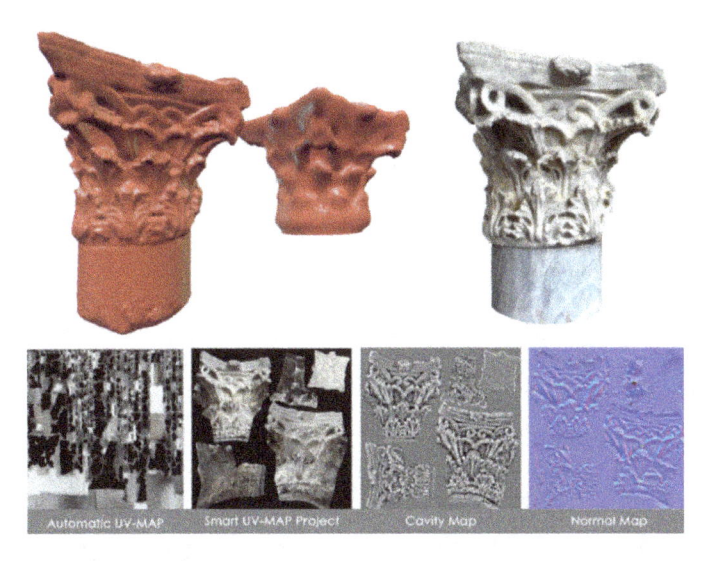

Fig. 5. Retopology and Baking phases on gaming environment elements

Dimensional maps without the need to add to the geometry. Examples of this are the Displacement map, the Cavity map and MeshLab and Instant Meshes (Fig. 5).

With regards to the setting of the parameters of the Lightmap, a series of tests were carried out in order to establish which the best resolution values were. This was based on a computational viewpoint, for the export from Blender to the Unreal Engine platform. The result of the tests showed that, with the same number of polygons, two meshes with Lightmap with a resolution of 512 pixels are more efficient than a single Lightmap of 1024 pixels (Fig. 6).

Fig. 6. The addition of procedural textures allowing for the information storage in the 2D maps

Another optimization strategy utilized in order to obtain a real-time photorealistic visualization was to represent the areas of the scene not accessible to the player with visual devices. As an example, the windows placed on floors higher than 10 m or the areas not accessible by character, were geometrically built through a flat rectangular surface. This was then calculated using the technique of baking the mesh containing the details shaped with image-based methods (Fig. 7).

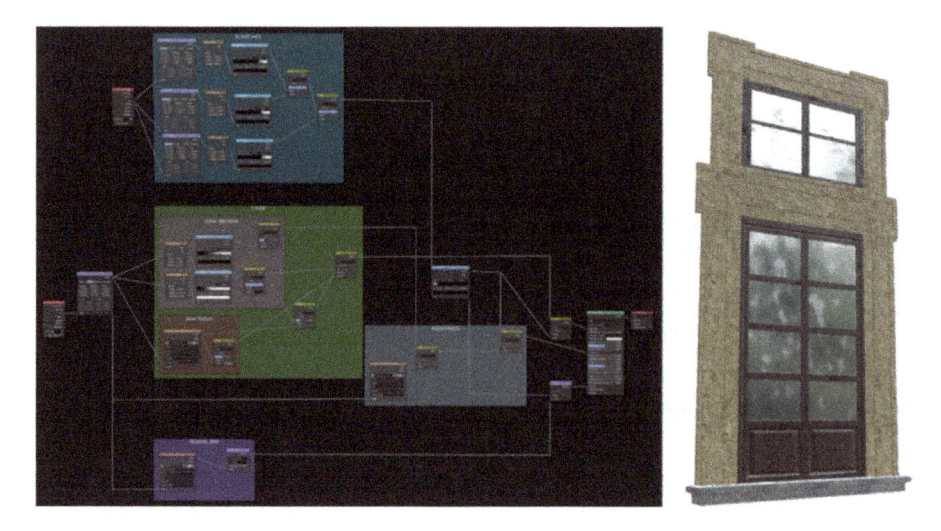

Fig. 7. A visualization of the creating of particular elements in the Blender environment

Using the procedural mode, the necessary textures were developed. This includes the different materials required for the layers within a particular game scene. Node editor was employed to carry out this process. Utilizing mixing tools, the procedural maps were collectively pooled. For this to be done there were different types of sockets deployed within the nodes and these are given in Table 1 below.

Table 1. Constitution of the types of sockets involved in the analysis

Socket	Color	Descriptions
Color node	Yellow	Transfer of color information using RGB values but not considering the alpha channel
Numerical	Grey	Displays numerical information, using floating variables from 0 to 1. Conversion to greyscale is done here
Vector	Blue	Displays coordinate information using general vector types
Shader node	Green	Internal operation, recognition and conversion information is stored inside the rendered cycles of earlier nodes (Fig. 8)

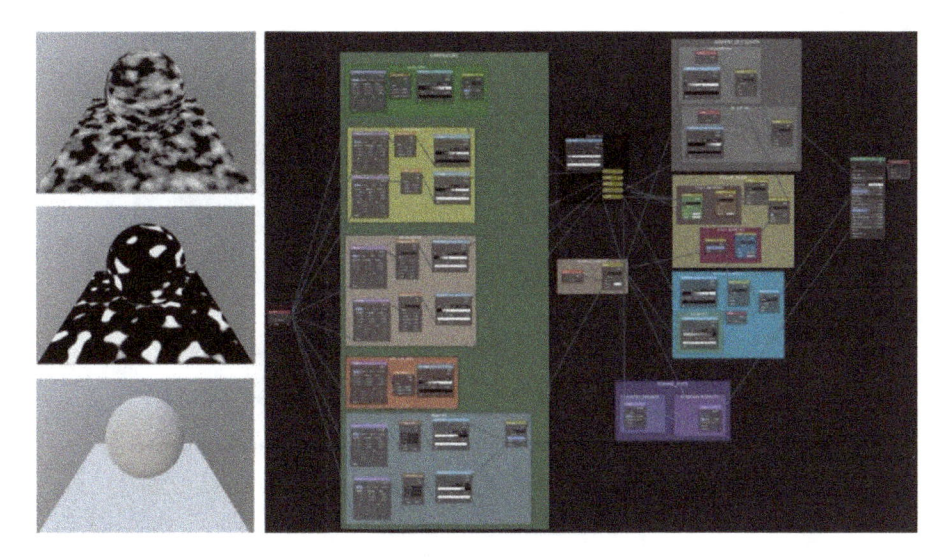

Fig. 8. The reproduction of the layers of particular gaming elements within Blender

5 Sculpting and Mesh Reduction

After the 3D acquisition of the artefacts, which were done using the photogrammetric techniques and a structured light scanner, the optimization of the geometry and texture of the objects was the next step to be deployed [16, 17].

From literature and from experience it is known that within the texture mapping, the most common issues are: the low texture in the resolution, missing mesh portion colors, photo mesh boundary inconsistencies; the topological errors due to formal geometric complexity and finally the integration with different acquisition technologies (Fig. 9) [7, 18].

The tested process aimed to overcome these critical issues. The texture optimization workflow includes multiple steps: the collision detection of the mesh, mesh corrections and remeshing; Polygonal mesh parameterization; Model quality for final user; Mesh segmentation unwrap; development of island and the optimization of the UV Map's layout. The phase involving the UV Mapping can be associated with the "flattening" phase of a solid that is represented in a two-dimensional mode according to logical and mathematical criteria.

Before continuing the project to the next step of mesh segmentation, considered the "unwrap", topological errors must be identified. This is to ensure that the mesh structure is optimal. Indeed, often, by exporting a model acquired with Reverse Modeling techniques, the resulting mesh may have vertices that are overlapped, or not connected to each other. It is possible that there are visualization problems during the reduction of the weight.

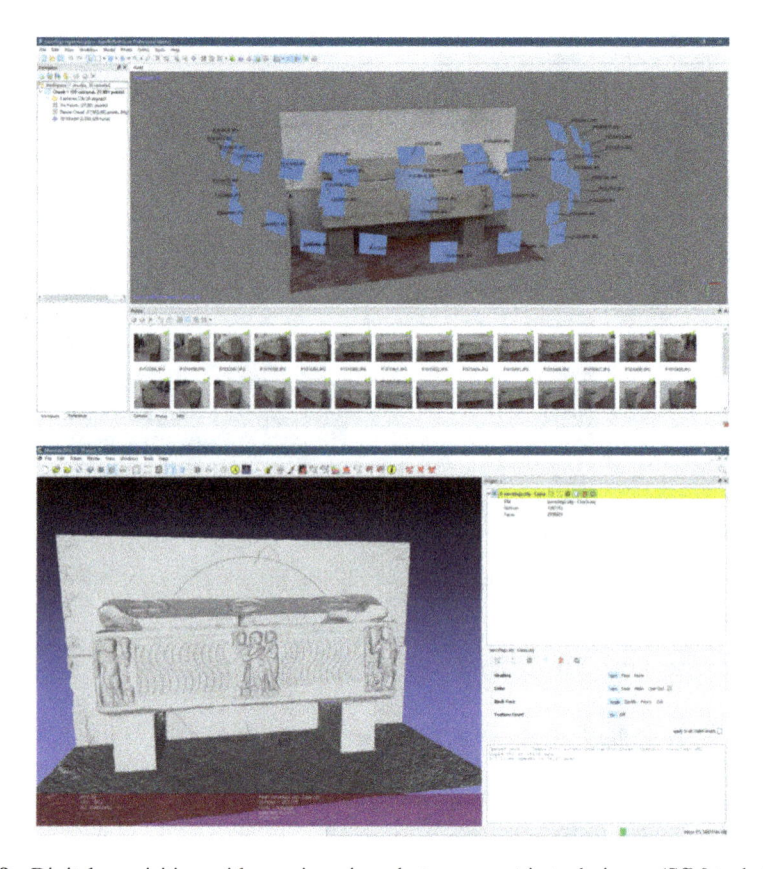

Fig. 9. Digital acquisition with non-invasive photogrammetric techniques (SfM technique)

Therefore, in order to correct any errors in the structure of the mesh vertices are eliminated through a specific tool for detecting redundant vertices. This allows for a lighter polygonal load which aids in mesh visualization without the need to alter geometry (Fig. 10) [19, 20].

6 Mesh and Applied Texture Optimization

Once the real-time visualization and gamification is completed, it is necessary to establish which path to follow for the next stage. This depends on the use of the visualized objects. There are essentially two approaches at this stage: to progress using a model with a full load, to engage in further multidisciplinary research, or aim for a lighter one for a real-time display and use in gaming.

Within the context of this study, the second approach is necessary and as a result the next stage involved the use of "mesh partitioning". This process allocates a mesh weight, to allow for the creation of a UV Map. Within the included blender libraries

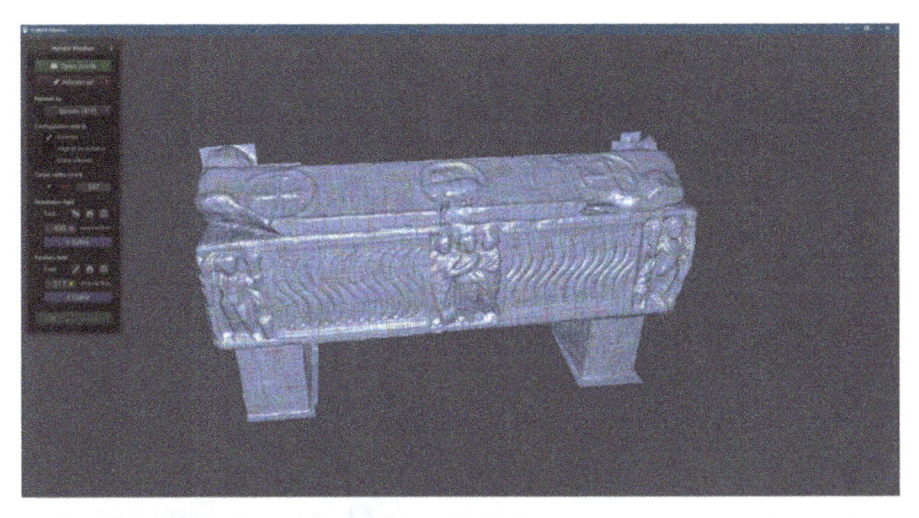

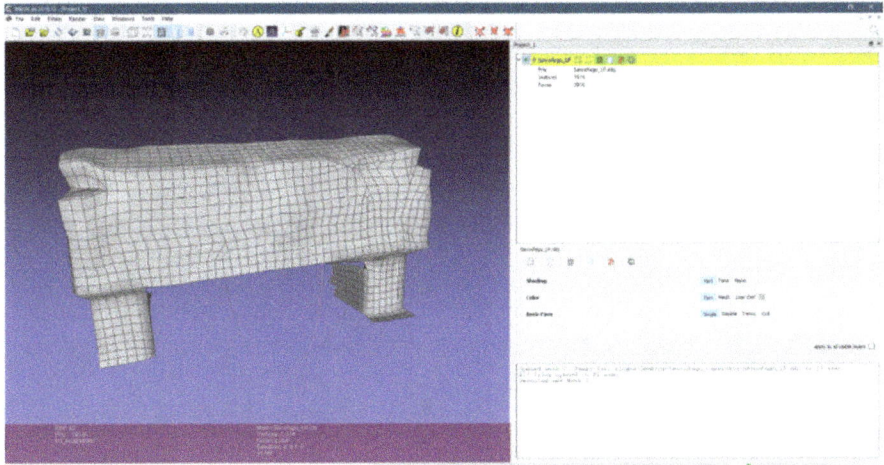

Fig. 10. Mesh collision detection, Re-meshing and polygonal weight reduction.

there is an automatic procedure for creating UV Maps and island segmentation; the tool is called Smart UV Project.

With the utilization of this tool, disjointed maps displaying triangular patches can be created. Tools are utilized within the software which are committed to creating vertex maps which are defined as UV vertex maps. The function of this is to allocate texture with the numerically developed models which have a multifaceted geometrical shape. The UV vertex maps create a definite relationship between the 3D mesh's vertices and the image's pixels. This is due to the texture being that of a planar 2D figure. The calculation of the vertices on the 2D domain is based on the approximation of the minimum squares according to mathematical formulas that minimize the distortion of the triangles along the perimeters of the patches. In Blender, it is possible for the "Angle Limit" to be adjusted. This is done within the operator menu. The aim of

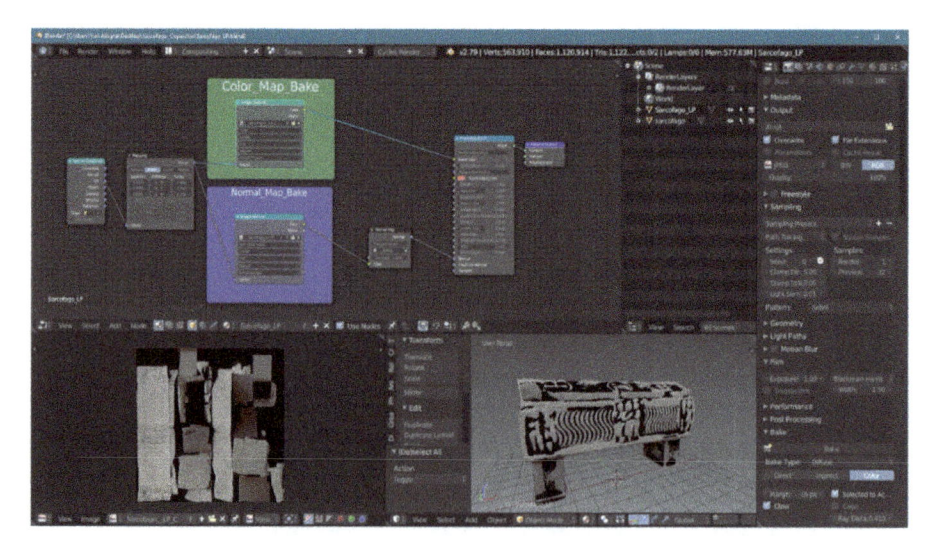

Fig. 11. Baking node, UV-remapping and texture optimization. Mesh with final baked UV-Map disposition

this is to allow for minimal distortion. However, this process in many instance is not able to handle the object's geometrical features or the obtained unwrap does not optimize the 2D domain at all. This leaves significant empty portions on the object (Fig. 11).

Therefore, there is a need to divide the produced meshes into sections and then to allow for the unwrap on a planer surface. This is done with respect to the semantic criteria which is dictated by the user and considers information on the topology, color, display and object shape.

An additional goal of this workflow is to develop manageable solutions for the "UV vertex project". These solutions must allow for the complete UV parameter bitmap map to be encompassed in the environment with the model having a high image resolution. Following the phases of this method for each of the acquired 3D models the following steps were developed: Removing Doubles Vertices from the acquired Mesh and closing seam, and edges connections review; Mesh partitioning process for better UV-Mapping Working process; Mesh marking seams process for a manual unwrapping mode. These steps allow optimizing the polygonal weight of the object compared to the initially acquired mesh, dividing the geometry in a custom manner (Fig. 12) [21].

Baking is a tool that is wildly utilized to solve Computer Graphics problems. It allows users to manage high density and high polygonal load models. Within Baking the UV vortex map is defined and then the model's data is acquired in the high resolution state. After this it is 'cooked' in a model that with low polygonal load using morphological data (heavy simulation calculations concerning light to matter ratio, illumination and radiosity); normal RGB and displacement maps [22].

In the final phase of texture optimization, after baking, some areas of interest have been selected using the Texture Painting mode tool and emphasized by creating masks whilst editing some parameters such as saturation, contrast and brightness. Following

Fig. 12. The optimization process from the "original" UV-Map to the final one

this process of editing and optimizing the geometry and texture applied. Each model is then assigned: the initial mesh with original UV-Map Arrangement; the Matcap Visualization with final UV-Map Arrangement and the final Mesh, with the final baked UV-Map disposition.

7 Design and Implementation of Specific Add-ons Using Python

The aforementioned mesh partitioning technique can be surmised as a technique that fundamentally allots a specific weighting to each mesh under analysis. The tool used for this is the "Weight Paint Mode". In the Blender environment, there are two methods to use this tool: in a manual mode and in an automatic one. In the automatic mode, a color variation is used. The goal of this is to produce a selection of the mesh that is both automatic and semantic in nature. This study presents an implementation of this tool using python [23]. The specific add-on is described in detail below. The weight is translated into shades of color using red, yellow, green and blue (Fig. 13).

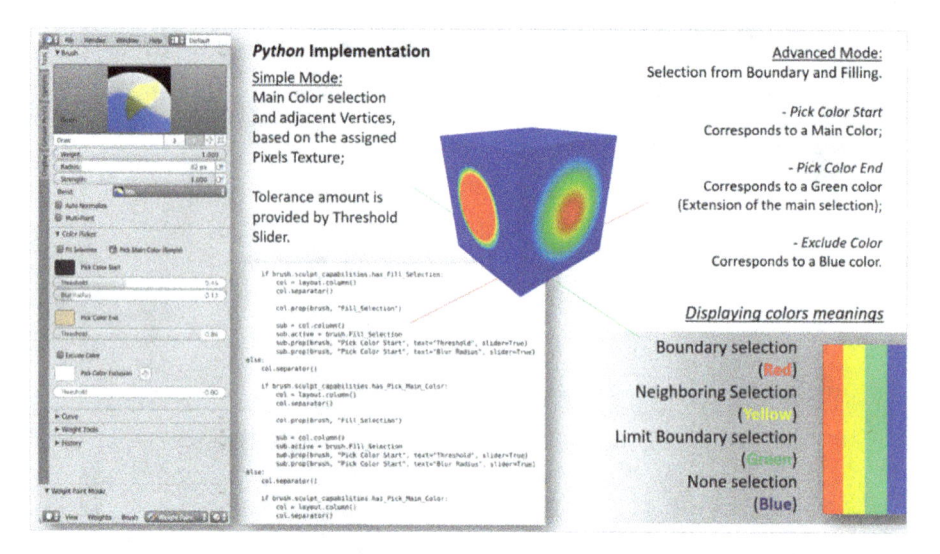

Fig. 13. "Weight Paint" Mode, automatic selection, with the implemented Color Picker tool. Python implementation Add-on

Each color has a specific meaning:

- The Red color is decoded by the tool as a user-defined selection and represents the boundary selection.
- The Yellow color is decoded by the tool as an extension or expansion selection. This occurs after selecting and choosing inside the box color "pick color start" the selection's color tonality.
- The Green color identifies the end of the selection that is chosen by the box color "pick color end". The slider threshold manages the visibility as well as the extension.
- The Blue color indicates that no selection (or exclusion) of pixels and, consequently, of mesh vertices has been made.

The algorithm has been defined according to two modes of automatic interaction and this is displayed given the rendered item shown in Fig. 14.

The first is the simple mode made using "pick main color selection". This mode selects the main color and also automatically the adjacent vertices corresponding to the pixels of the texture assigned to the object. The chosen color is displayed inside the "Pick Color Start" Color Box. The parameter that manages the Pick function and the selection tolerance is the "Threshold" slider, while the "Blur Radius" function manages the selection blur. Modifying the "Threshold" and "Blur Radius" parameters the extension changes on the adjacent similar pixels, and therefore parametrically the vertices below.

The second mode is called the "Advanced Mode", which makes selection by the boundary "Weight Paint". After having selected the boundary through "Weight paint mode", the command will fill the mesh area within this selection. There are two Color picker modes: "Pick Color Start" Function (picks the vertex with the primary color

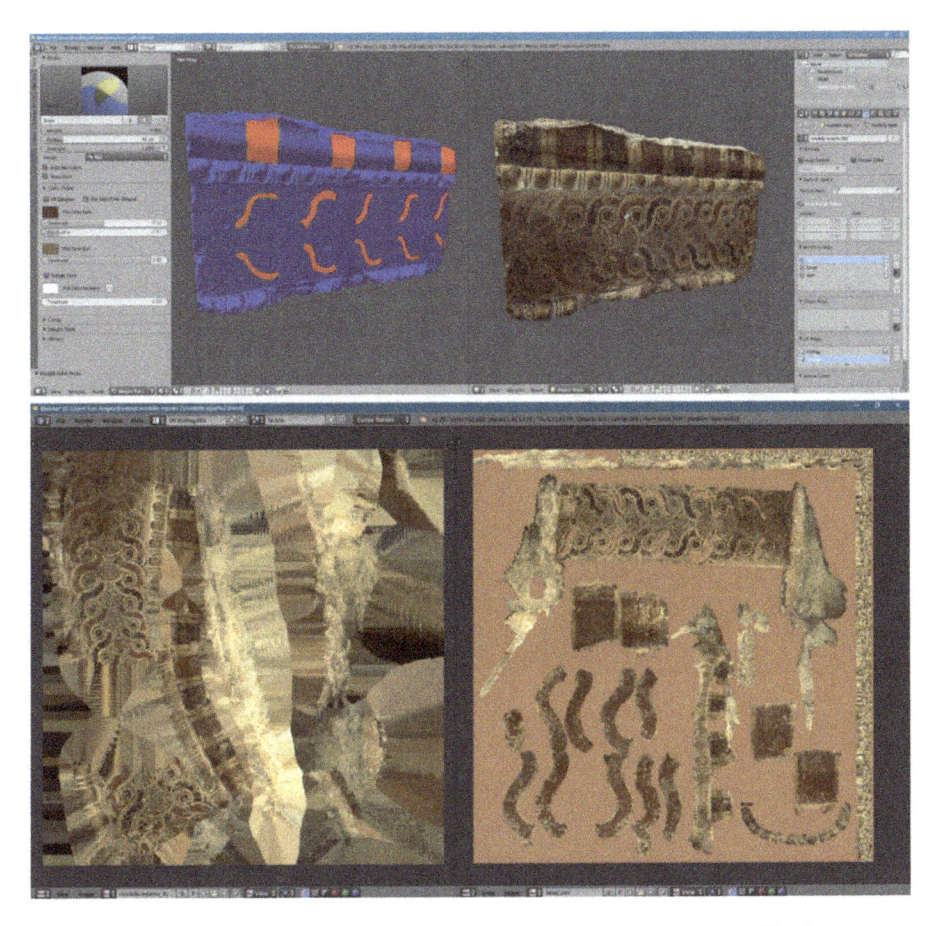

Fig. 14. Automatic boundary selection by "Weight Paint" Advanced Mode, Blender Environment (above); comparison between the automatic and custom UV maps (bottom image)

assigned to it) and "Pick Color End" function (pick the vertex with the primary color assigned at the end of the selection).

Finally, the "Exclude Color" and "Pick Color Exclusion" functions have the role to select the vertexes which are assigned as "external" colors to the selection criterion. This occurs if the vertices with different colors are automatically selected within the selection area (Fig. 15).

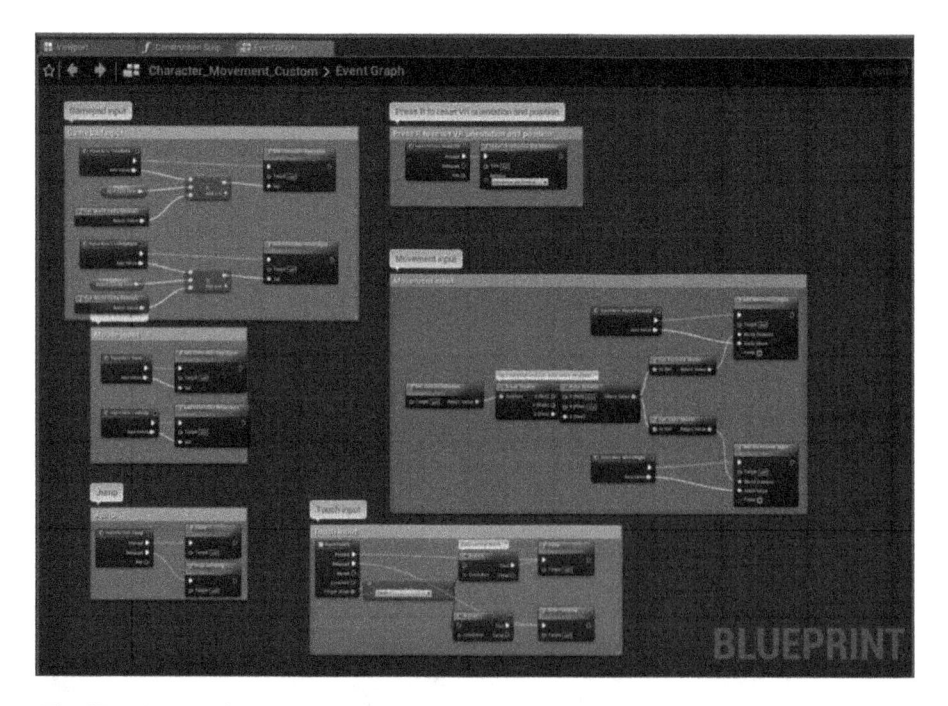

Fig. 15. The node interface showing the arrangement of the scripts within Unreal Engine 4

8 Conclusions

There are tremendous opportunities for adapting and evolving cultural heritage games by with the introduction of enhanced realism and photorealistic elements to them. Applied games for cultural heritage with non-linear and interactive storytelling will be used more and more for education and interactive cultural displays. This research helps to bridge some of the gaps in promoting these types of games.

The novel elements of the study lie in the optimization strategies developed to overcome the issues of producing lightweight but photo-realistic models to be used in the gaming environment. The techniques developed in Blender and using the python add-on allow for this to be possible and are effective at easily generating 3D visuals that are both accurate and realistic enough to be appealing to users. This is important to capture details of the artefacts, not just for preservation but also to capture the attention of the users by creating realistic replications in the gaming environment.

Within the study and using Blender, texture reconstruction was carried out utilizing advanced segmentation methods. Additionally, an algorithm developed and written in python was established as an alternative method for the similar goal. This method functions by creating the mesh unwrap which is more functional with the geometry of the intricate environment and the required defined semantic benchmarks (Figs. 15 and 16). Using this pipeline, it can be seen that there is high potential for the production of models with varying resolutions and realistic textures for the final product that satisfy the needs of the gaming developers and the final end users.

Optimization of Cultural Heritage Virtual Environments for Gaming Applications 369

In Fig. 16 below, screenshots from the developed gaming prototype are shown. This depicts the user interface and the high resolution of the elements within the environment. Given the origins of the objects and the pipeline used this represents a further step in the advancement of cultural heritage games and making them accessible and user friendly.

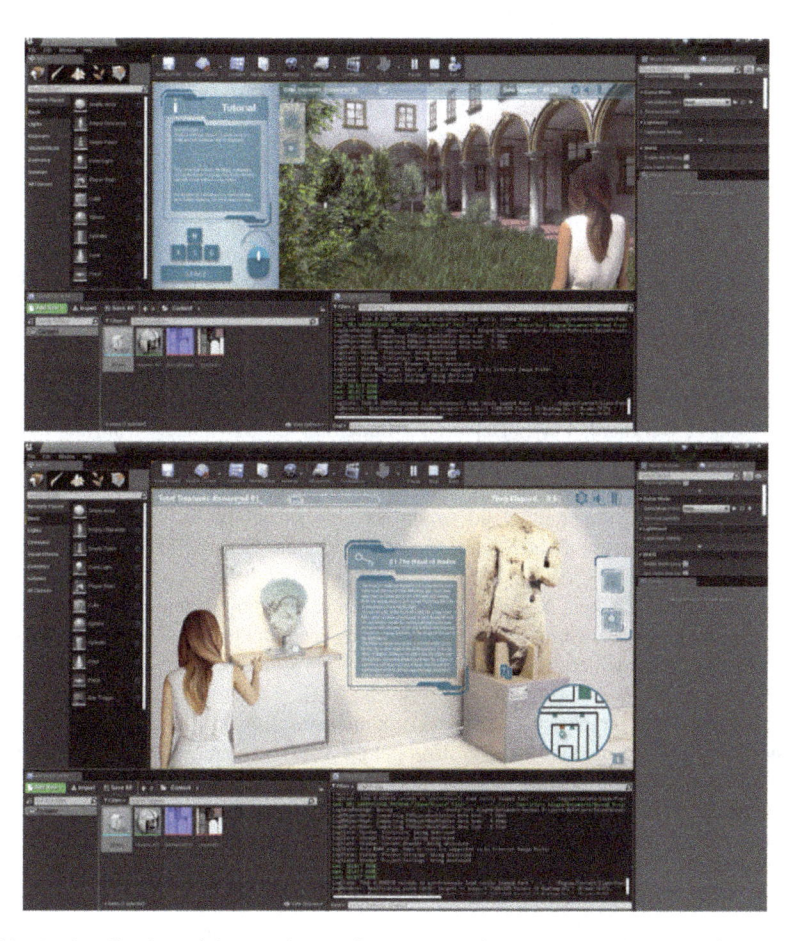

Fig. 16. A visualization of the gaming environment on the UnReal Engine 4 platform showing the movement of a user within the game.

Acknowledgments. The research presented in this paper was carried out as part of the H2020-MSCA-ETN-2016. This project has received funding from the European Union's H2020 Programme for research, technological development and demonstration under grant agreement number 721493.

References

1. Janarthanan, V.: Serious video games: games for education and health. In: Proceedings of the 9th International Conference on Information Technology, ITNG, pp. 875–878 (2012)
2. Laamarti, F., Mohamad, E., El Saddik, A.: An overview of serious games. Int. J. Comput. Games Technol. **2014**, 15 (2014). Article no. 358152
3. Barbosa, A.F.S., Pereira, P.N.M., Dias, J.A.F.F., Silva, F.G.M.: A new methodology of design and development of serious games. Int. J. Comput. Games Technol. **2014**, 8 (2014)
4. Di Paola, F., Inzerillo, L., Alognaa, Y.: A gaming approach for cultural heritage knowledge and dissemination. In: ISPRS - International Archives of the Photogrammetry, Remote Sensing and Spatial Information Sciences, vol. XLII-2/W15, pp. 421–428, September 2019
5. Di Paola, F., Milazzo, G., Spatafora, F.: Computer aided restoration tools to assist the conservation of an ancient sculpture. The colossal Statue of Zeus enthroned. International Archives of the Photogrammetry, Remote Sensing and Spatial Information Sciences - ISPRS Archives, CIPA, XLII, vol. XLII 2/W5, pp. 177–184. Copernicus Publications (2017)
6. Parrinello, S., Bercigli, M., Bursich, D.: From survey to 3D model and from 3D model to "videogame". The virtual reconstruction of a Roman Camp in Masada, Israel. DISEGNARECON Archaeol. Draw. **10**(19), 1–19 (2017)
7. Valanis, A., Fournaros, S., Georgopoulos, A.: Photogrammetric texture mapping of complex objects. In: Euromed Conference, Limassol, Cyprus (2010)
8. Minh, H.N., Burkhard, C.W., Delmas P., Lutteroth, C.: 3D models from the black box: investigating the current state of image-based modeling. In: Communication Proceedings, WSCG, Pilsen, Czech Republic, pp. 249–258 (2012)
9. Remondino, F.: From point cloud to surface, the modeling and visualization problem. International Archives of the Photogrammetry, Remote Sensing and Spatial Information Sciences, vol. XXXIV-5/W10 (2003)
10. Chen, Z., Zhou, J., Chen, Y., Wang, G.: 3D texture mapping in multi-view reconstruction. In: Advances in Visual Computing. Lecture Notes in Computer Science. vol. 7431, pp. 359–371. Springer, Heidelberg (2012)
11. Finlay, J., Wilson, A., Milburn, K., Nguyen, M.H., Lutteroth, C., Wünsche, B.C.: An investigation into graph cut parameter optimisation for image-fusion applications. In: Proceedings of Image and Vision Computing New Zealand, IVCNZ, Dunedin, New Zealand, pp. 480–485 (2012)
12. Goldluecke, B., Cremers, D.: Super resolution texture maps for multiview reconstruction. In: Proceedings of the 12th International Conference on Computer Vision, ICCV, pp. 1677–1684 (2009)
13. Minh, H.N., Burkhard, C.W., Delmas, P., Lutteroth, C.: Modelling of 3D objects using unconstrained and uncalibrated images taken with a handheld camera. In: Computer Vision, Imaging and Computer Graphics - Theory and Applications, pp. 1–16. Springer, Heidelberg (2012)
14. Nguyen, H.M., Wünsche, B., Delmas, P.: High-definition texture reconstruction for 3D image-based modeling. In: 21st International Conference on Computer Graphics, Visualization and Computer Vision (2013)
15. Phothong, W., Wu, T.C., Lai, J.Y., Yu, C.Y., Wang, D.W., Liao, C.Y.: Quality improvement of 3D models reconstructed from silhouettes of multiple images. Comput. Aided Des. Appl. J. **15** (2018)
16. Inzerillo, L.: Smart SfM: salinas archaeological museum. International Archives of the Photogrammetry, Remote Sensing and Spatial Information Sciences - ISPRS Archives, vol. 42 2W5, pp. 369–373 (2017)

17. Schonberger, J.L., Frahm, J.M.: Structure-from-motion revisited. In: Proceeding CVPR (2016)
18. Stathopoulou, E.K., Valanisa, A., Lermab, J.L., Georgopoulos, A.: High- and low-resolution textured models of complex architectural surfaces. International Archives of the Photogrammetry, Remote Sensing and Spatial Information Sciences, ISPRS, vol. XXXVIII-5/W16 (2011)
19. Valkenburg, R., Alwesh, N.: Seamless texture map generation from multiple images. In: Proceedings of the 27th Conference on Image and Vision Computing New Zealand, IVCNZ, NY, USA, ACM, pp. 7–12 (2012)
20. Xu, L., Jianguo, E.L., Chen Y., Zhang, Y.: A general texture mapping framework for image-based 3D modeling. In: Proceeding of the 17th IEEE International Conference on Image Processing, ICIP, pp. 2713–2716 (2010)
21. Colburn, A., Agarwala, A., Hertzmann, B., Curless, M., Cohen, F.: Image based remodeling. IEEE Trans. Vis. Comput. Graph. **19**(1), 56–66 (2013)
22. Lai, J.Y., Wu, T.C., Phothong, W., Wang, D.W., Liao, C.Y., Lee, J.Y.: A high-resolution texture mapping technique for 3D textured model. Appl. Sci. **8**(11), 2228 (2018)
23. Minh, H.N., Burkhard, C.W., Delmas, P., Lutteroth, C.: A hybrid image-based modelling algorithm. In: Proceeding of the 36th Australasian Computer Science Conference, ACSC, Adelaide, Australia, pp. 115–123 (2013)

Performance Evaluation of the Update Messages of Locator Identifier Split Protocols Using an IP Paging Mechanism at the End Networks

Aadarsh Bussooa[(✉)] and Avinash Mungur

Department of Information and Communication Technologies,
Faculty of Information, Communication and Digital Technologies,
University of Mauritius, Reduit 80837, Mauritius
aadarsh.bussooa@umail.uom.ac.mu, a.mungur@uom.ac.mu

Abstract. This paper surveys two popular protocols, namely, LISP and HIP, which are used in decoupling the IP namespace into distinct locator and identifier namespaces, with the aim of addressing the challenges associated with the anticipated exponential growth of Internet-connected devices. One such challenge pertains to how the design of the Internet infrastructure can be re-engineered to support fast host mobility across distinct networks while maintaining session connectivity and minimising packet loss. The paper lays emphasis on the update mechanisms employed by these protocols, which are used to inform administrative servers of location changes of mobile nodes. The update mechanisms of these protocols are applied to an IP paging architecture and their performances are evaluated. The use of an IP paging mechanism shifts the focus of these protocols onto the optimisation of the end networks and how well they cope with update messages. Indeed, such an architecture lays emphasis on fast inter-paging area mobility, which is occurring in a restricted hierarchical IP paging network topology. Therefore, this paper evaluates the performance of update messages on different levels of network traffic and IPsec cryptography at the end networks.

Keywords: Locator identifier split protocols · IP paging mechanism · Locator Identifier Separation Protocol · Host Identity Protocol · Location update · IPsec · Map-Register message · HIP Update packet

1 Introduction

In the past, the Internet consisted, mainly, of computers connected by cables to the Internet backbone. Over the years, wireless communication technologies were introduced and computers became mobile nodes (MNs) in a network. As the smartphone industry boomed, the number of wireless MNs scaled up and the IPv4 address space entered a state of progressive depletion. While today's Internet still depends almost entirely on IPv4 addresses, a much larger address space, known as IPv6, is gradually being adopted. Looking forward, much more devices would be joining the Internet with the advent of the Internet of Things concept. While the IPv6 communication protocol seems like a logical step towards the issue of depleting IPv4 addresses, the protocol suffers from the coupled locator and identifier properties of the IP address. Indeed, when a MN moves from one area to another, its IP address changes and all established communications to peers are broken. This dual role of the IP address hinders smooth host mobility.

Therefore, there is a need to make changes to the core of the Internet architecture, such that higher level protocols inherit from these changes. Multiple solutions have been sought, including network-based approaches and host-based approaches. These solutions will be referred to as Locator Identifier Split (LIS) protocols in this paper. The LIS protocols, which are under active development are summarised and critically reviewed in this paper. The performance of LIS protocols with respect to their update mechanism is also investigated.

This paper is organised as follows: Sect. 2 provides an in-depth literature review of two popular LIS protocols, namely LISP and HIP. Section 3 analyses the requirements of the experiment. Section 4 states how the experiment is designed. Section 5 shows the experiment results and evaluates the performance of LIS update messages. Finally, Sect. 6 concludes this paper.

2 Literature Review

There are various LIS protocols, some under active development while others are inactive and archived. Examples of LIS protocols include the Locator Identifier Separation Protocol (LISP) [1, 2], Internet Vastly Improved Plumbing (Ivip) [3, 4], Host Identity Protocol (HIP) [5], Six/one router [6], Identifier Locator Network Protocol (ILNP) [7, 8], GLI-Split [9] and Mobile IP [10, 11] amongst others. A comprehensive review of LIS protocols is provided in [12]. In this section, the emphasis is laid on the location update mechanism which LISP and HIP employ. These two protocols were chosen since the former is a network-based approach while the latter is a host-based approach to the IP namespace decoupling. Moreover, their respective location update mechanism plays a vital role in the operation of the protocol.

2.1 Locator Identifier Separation Protocol (LISP)

The Locator Identifier Separation Protocol (LISP) uses a network-based approach to host mobility. The main aspect of the LISP protocol is its use of tunneling as its mode of operation [1]. The LISP protocol introduces the concept of Endpoint Identifier (EID) and Routing Locator (RLOC). EIDs are assigned to MNs and RLOCs are assigned to the routers in the network, achieving the desired decoupling of the IP addressing scheme.

The EID-to-RLOC mappings are important to enable packets to be routed through the tunnel to the correct destination. To store and propagate these mappings, the existence of a distributed database is assumed [13]. The LISP mapping service consists of components as described in Table 1.

Table 1. Summary of LISP Map-Server Interface Components

Component	Description
Map-Resolver	It accepts the Map-Requests from the Ingress Tunnel Router (ITR) and attempts to find a match in the EID-to-RLOC mapping cache. The operation of the Map-Resolver shares similarities with DNS caching servers
Map-Server	It stores the authoritative EID-to-RLOC mappings which are learnt from Egress Tunnel Routers (ETRs) by an update mechanism. The operation of the Map-Server shares similarities with authoritative DNS servers
Map-Register Message	It is a message sent by an ETR to update the definite EID-to-RLOC mappings of the Map-Server, which controls the EID prefix used by the ETR
Map-Notify Message	It is an acknowledgement sent in response to the successful receipt of the Map-Register message
Map-Request Message	It is a message sent by an ITR to query the EID-to-RLOC mappings of the Map-Server which contains these definite records
Map-Reply Message	It is a message sent in response to a Map-Request containing the RLOCs of the requested ETR

The implementation of the LISP protocol depends on the EID-to-RLOC mapping service, without which the network traffic to a LISP site cannot happen. It is, therefore, crucial that the mapping service is up and running, together with knowledge of the most recent EID-to-RLOC records. LISP employs an update mechanism at the Egress Tunnel Router (ETR) to advertise the EIDs under its control. Map-Register messages are used to update the EID-to-RLOC records of an ETR at the Map-Server.

Upon initial contact of an ETR with a Map-Server or after a long period of absence of communication, the ETR registers itself by sending Map-Register messages at a rate of 3 messages per minute for a maximum duration of 5 min. Afterwards, the ETR sends Map-Register messages every minute to the Map-Server to maintain and update its EID-to-RLOC records. If a Map-Register message from an ETR takes longer than 3 min to reach the Map-Server, that ETR's registration at the Map-Server is removed. The RFC documentations of LISP [13] acknowledges that the above mentioned rules, regarding Map-Register messages and the registration process at the Map-Server, does not favour fast host mobility. However, these RFC documentations discuss the core architecture of LISP and, most probably, assumes that the ETR of a LISP site remains stationary most of the time.

While the other LISP messages are equally important in running a LISP architecture, the Map-Register message is of prime importance in this paper's study. Indeed, supporting host mobility is tightly coupled with how fast an ETR can update its Map-Server using Map-Register messages. The Map-Register message is simply a UDP packet with the LISP message being the data it carries. The Map-Register packet structure complies with the standard structure of any IPv4/IPv6 UDP packet. The minimum size of the Map-Register message was calculated to be approximately 120 bytes. Knowledge of the packet size of a LISP message is useful in performing network simulations, as was the case in [14] where the packet size was determined and the location management cost of the LISP protocol was calculated. This paper would subsequently make use of the Map-Register message packet size in the network experiments that would be carried out.

The LISP Mobile Node architecture is built on top of LISP and the LISP Map-Server Interface [15]. It enables MNs to roam and to be discoverable while travelling through distinct networks. In a normal LISP architecture, the ITR and ETR would connect to the end nodes. However, in the LISP MN architecture, the MN itself acts as a LISP site with both an ITR and an ETR. Such implementation is made possible by creating a sub-layer between the IP layer and the Link layer of the protocol stack.

As shown in Fig. 1, when a MN roams to a new network, the ETR immediately updates the Map-Server about a change in its RLOC address using Map-Register messages. Moreover, since the previous EID-to-RLOC record is cached in multiple ITRs, the MN informs these ITRs of this newly invalid entry through different mechanisms. The MN's ETR continuously sends Map-Register messages based on a timer, which is made to run as a background process. It is claimed to provide robustness on the network with which the MN is connected [15]. The LISP Mobile Node RFC documentation does not specify the time interval for sending these Map-Register messages. In the core architecture of LISP, ETRs usually update the Map-Server every minute. If such a time value is applied to a LISP network supporting host mobility, the LISP control-plane functionalities would become a bottleneck to the packet flow in the network. In contrast with the separate entities of the ETR and ITR in LISP's topology, LISP MN is implemented in the MN's protocol stack itself. There is the issue of the MN having varying computing resources and different connection speeds, which would have a negative impact on the MN's performance and the network's performance. Therefore, this paper would investigate the performance of the LIS update messages on varying levels of network traffic.

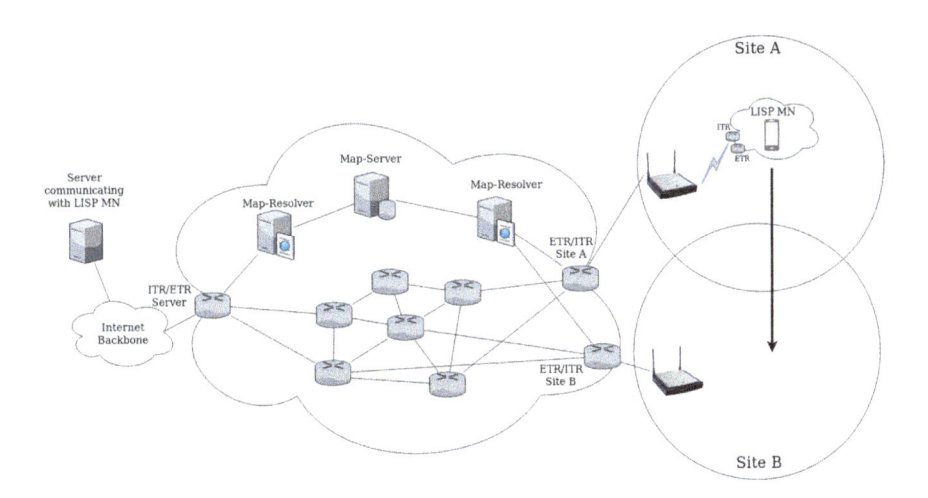

Fig. 1. LISP MN moving from Site A to Site B with bidirectional network traffic flow

From a security standpoint, the Map-Register message is protected using an authentication mechanism. However, it is still susceptible to attacks that could undermine the security and performance of a LISP network. The Map-Register message is of high importance in MNs with fast mobility. Compromising the Map-Register message could prevent the delivery of packets to MNs or flood the network with wrong information on the status of ETRs since records are propagated and cached in the LISP network. To assess how the Map-Register message can be compromised, one should understand how the control-plane authentication mechanism of LISP works. LISP makes use of the Message Authentication Code (MAC) to protect the integrity of the Map-Register message. The following steps details how the MAC algorithm works.

1. At the sender's end, the LISP message is used as input to a MAC algorithm together with a shared key to produce a hash of that message known as the MAC data tag (t1).
2. The LISP message and the MAC data tag (t1) are sent to the destination.
3. At the receiver's end, the LISP message undergoes the same process as in step 1 with the same key to produce the MAC data tag (t2).
4. The computed tag t2 is compared with the received tag t1.
5. If the tags are similar, then the LISP message was not modified during transmission.
6. Else, there is an issue with the LISP message. It could have been compromised as in a man-in-the-middle attack or simply received with errors.

Without compromising the ETR or without knowledge of the shared key, the extent of an attack on the LISP network is limited. If the contents of a LISP encapsulated packet is altered, a receiving LISP node would compute the MAC data tag and invalidate the received packet. However, such an invalid packet could be sent in numerous amount causing a denial-of-service attack by overloading the computing resources of LISP nodes across the network.

If an ETR is compromised, then the LISP message could be modified before it is hashed by the MAC algorithm at the sender's end. An attacker could de-aggregate the assigned EID prefix of an ETR and break it into multiple EID prefixes and registering each of them with the Map-Server [16]. Moreover, a compromised ETR could over-claim its EID prefixes to include those not legitimately under its control. The de-aggregation or over-claiming of the EID prefixes causes the Map-Server to propagate wrong mappings to ETRs, which would be stored in the ETRs' EID-to-RLOC caches, enabling the attacker to perform a multitude of attacks. The de-aggregation or over-claiming of EID prefixes also works the other way round, where the Map-Server is compromised and accepts wrong mappings from an attacker.

The different entities in the LISP topology, an ETR amongst others, trust each other [17]. It is made possible by the use of pre-configured shared keys to establish these trust relationships. The exact mechanism of how these keys are distributed is not discussed in the RFC documentations of LISP Control-Plane. Other security features of Map-Register messages is out of scope of the documentation as well. However, one can understand how some aspects of LISP-SEC could be applied in the context of Map-Register messages.

LISP-SEC builds itself on top of these trust relationships. One-Time Keys (OTKs) are encrypted using the shared keys that were used to establish these trust relationships

[18]. These keys are updated periodically. The OTKs provide for anti-replay attack protection while its encryption provides for confidentiality. Since integrity is provided for by the MAC data tag, the protocol fulfils the requirements of the CIA triad. The exchange of Map-Request and Map-Reply messages, where the Map-Server is not acting as proxy and OTKs are used, is summarised as follows:

1. An ITR, willing to transmit a Map-Request message (M1), generates an OTK (ITR-OTK).
2. The ITR-OTK is encrypted, using the pre-shared key that was used for establishing trust relationships between the ITR and the Map-Resolver, and sent to the Map-Resolver.
3. The Map-Resolver decapsulates the Encapsulated Control Message (ECM) and forwards the Map-Request as a new ECM message (M2) together with the ITR-OTK.
4. Upon receipt of the Map-Request message M2 by the Map-Server, it retrieves the longest EID-prefix for the destination EID. This EID-prefix is added to the ECM of the Map-Request (M3), together with an HMAC computed using the ITR-OTK for integrity checks (EID-prefix authorisation data).
5. A new OTK (MS-OTK) is derived using a Key Derivation Function (KDF) applied on the ITR-OTK. The MS-OTK is included in the ECM message M3 and forwarded to the ETR over their established trust relationship.
6. Upon receipt of the Map-Request message M3 by the ETR, the MS-OTK is decrypted. The relevant EID-to-RLOC mapping is retrieved.
7. The ETR replies to the Map-Request with a Map-Reply to the requesting ITR, together with a computed HMAC using the MS-OTK for integrity checks. The ECM of the received Map-Request is also included in the Map-Reply.
8. Upon receipt of the Map-Reply by the ITR, the integrity of the EID-prefix authorisation data is verified using the ITR-OTK. The MS-OTK is computed by the ITR using the same KDF used by the Map-Server and the integrity of the Map-Reply is verified. The Map-Reply is discarded if the integrity checks fail.

It can be concluded that LISP-SEC secures the exchange of LISP control-plane packets, through the use of pre-shared keys that establish trust relationships between LISP entities. The pre-shared keys are changed periodically, although the mechanism by which it is done is not discussed. Like Map-Request messages, Map-Register messages could be secured in a similar way.

2.2 Critical Review of LISP

LISP is primarily motivated with improving the scalability of the global routing table, rather than enhancing the mobility of nodes [19]. Indeed, through the exponential growth of multi-homed sites, the issue of the ever increasing size of routing tables is causing concern. LISP addresses this issue by decoupling the identifier namespace and the locator namespace of the IP address, in such a way that a multi-homed site would have multiple locator values but a single identifier value.

A major benefit of LISP lies in the fact that it can be implemented without making changes to end nodes, except in the case of LISP MN. The mechanism for sending a packet remains unchanged and only the core of the Internet architecture requires modifications. A LISP network can, therefore, work alongside the Internet's infrastructure [20].

The Map-Register message plays an important role in LISP's architecture. In the case of stationary nodes in a LISP network, periodically sending a Map-Register message every minute is claimed to increase the robustness of the network, with the knowledge about which nodes are active and which ones are not. Knowing about the robustness of intermediate nodes is useful in LISP. When mobility is present, the host sends a Map-Register message, whenever it roams into a network with a different RLOC. These MNs also continuously sends Map-Register messages at a regular interval of time, as is the case for a stationary node.

However, when we speak of mobility, especially fast host mobility, the robustness of a specific communication link is no longer a major factor for maximising the network's performance, since the link to the MN would be constantly changing. Indeed, if a MN is moving fast, it will only have time to send a few Map-Register message before it associates itself to another network. Therefore, it is more efficient for the MN to only send one Map-Register message triggered by the detection of a new network rather than sending messages continuously.

A LISP MN is made to act as a LISP site, implementing both ITR and ETR functionality in the sub-layers of its protocol stack. Some MNs have limited resources, given that there is a myriad of Internet-connected devices with low computing power and low communication speeds. Sending too many messages at one time could overload both the network and the MN. Therefore, this paper would investigate the performance of LIS update messages across a wireless link to a device with low computing resources.

2.3 Host Identity Protocol (HIP)

The Host Identity Protocol (HIP) is a protocol, used to establish keys and to negotiate parameters. HIP employs a host-based approach to host mobility. In contrast to LISP's use of tunneling, HIP uses address rewriting as its main method of operation. However, HIP does share some similarities with LISP. Indeed, the desire to decouple the IP namespace into distinct locator and identifier namespaces, because of the overloaded semantics of the Internet Protocol, is still prevalent.

In HIP, the IP addressing scheme is used as is, without bringing any changes to its characteristics. It is viewed as being the locator namespace of HIP, a valid characteristic of IP addresses. Therefore, the core architecture of the Internet remains intact. The identifier namespace is implemented by introducing a sub-layer in the host's TCP/IP protocol's stack. This HIP sub-layer sits in between the Transport layer and the Network layer [21].

The IP address binds to the Network layer while the Host Identity (HI) binds to the HIP layer. The HIP layer has the ability to maintain connectivity of sessions, despite changes in IP addresses (locator). This is done by taking advantage of the assumption of the Socket API [22] that IP addresses are static. Therefore, to maintain socket session connectivity while the IP address is a variable, the layer directly above the transport layer should have a static address. In the case of HIP, the HIP layer provides for this static address, in the form of the Host Identity.

The Host Identity (HI) is the public key from the asymmetric public/private key pair of a signature algorithm that uniquely identifies a host. The private key is kept by the host to sign outgoing packets. These packets are verified, on receipt by a peer, using the HI (public key) of the sender. A cryptographically secure key pair would entail the use of long key lengths. Therefore, the public key, which would be used as HI, might not be able to fit a standard IPv4/IPv6 packet or might become an overhead in the transmission of packets, taking much space compared to the actual packet contents.

To enable the public key to be used in standard Internet packets, the Host Identity is hashed using the Host Identity Hash (HIH) algorithm to produce the Host Identity Tag (HIT), whose length is 128 bits and equivalent to the length of an IPv6 address. The HIH also takes into consideration the IPv6 prefix allocated to a particular network in generating the HIT. Although each host independently generate their HIT, the probability of collision, that is two hosts in a particular network generating the same HIT is very low.

The HIP Base Exchange (BEX) forms part of the control-plane key negotiation protocol of HIP. The BEX is used for key exchange between two peers. The two main functions of the BEX include the generation of shared symmetric keys for secure control-plane parameter negotiation and the generation of keys to create Security Associations (SA) between peers for data-plane packet transmission.

As shown in Fig. 2, the BEX is a four packet exchange between the Initiator and the Responder, which primarily establishes the shared symmetric keys to be used for communication [5]. On completion of such an exchange, a HIP Association is created. During the four packet exchange, Diffie-Hellman keys are generated and Security Associations are created.

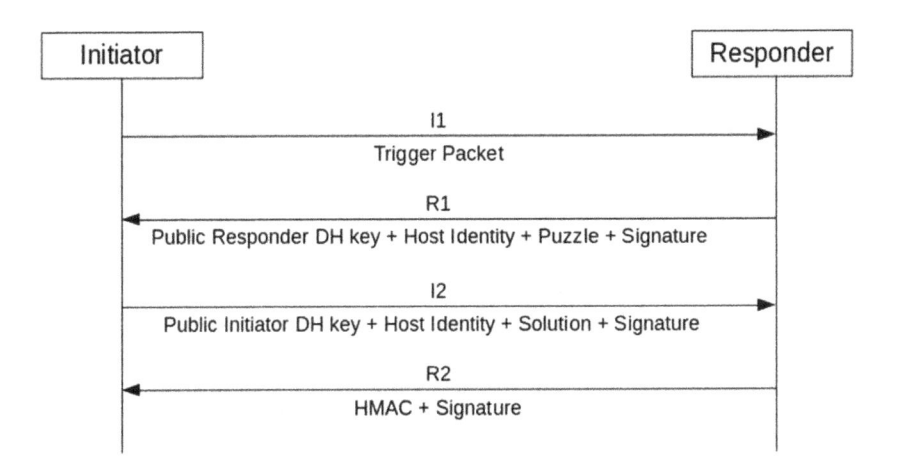

Fig. 2. HIP Base Exchange with a four packet exchange

It is worth noting that the standard Diffie-Hellman exchange can be done in just three packet exchanges. The reason behind HIP's use of four packets is part of its security strategy. Indeed, such a design increases the cost required to establish communication at the initiator's side. Moreover, in the R1 packet exchange, a cryptographic puzzle is sent to the initiator. The puzzle requires that the initiator uses up some CPU cycles in solving it, consolidating the status of the initiator as a genuine peer. Since solving the puzzle requires time and the right solution, a Denial-of-Service attack is less likely to take place.

The Host Identity (public key) of peers is published in the Resource Record (RR) of a DNS server. The initiator performs a normal DNS lookup, a name-to-IP-address lookup, to obtain the IP address (locator) of the destination. It also performs a name-to-HI lookup as an additional step. The initiator constructs an HI-to-IP-address mapping in its HIP layer, facilitating the creation of the trigger packet I1 [23].

The subsequent packet exchanges (R1, I2 and R2) contain the standard Diffie-Hellman exchange and the creation of keys for the SA. Each of the packet is signed by the sender's private key and is verified by the receiver using the sender's public key (public HI), which is published in the DNS RR. Therefore, one can verify that incoming packets was genuinely sent by a particular peer.

The symmetric keys generated, on completion of the DH key exchange, are used to protect the control-plane messages using a Hash-based Message Authentication Code (HMAC) algorithm, similar to the one used in LISP. After a successful Base Exchange, packets are transmitted in Encapsulating Security Payload (ESP) format in HIP's data-plane.

The packet flow sequence for key negotiation and data-plane communication with the HIP protocol, together with the use of ESP, is as follows.

1. The IP address of the destination is looked up using DNS, as part of HIP BEX.
2. The HI of the destination is looked up using DNS, as an additional step in HIP BEX.
3. An HI-to-IP-address mapping is created as a result of the DNS lookups.
4. HIP performs a Base Exchange to establish keys for control-plane and data-plane communications.
5. Packets are created with IP addresses before being handled by HIP's adaptation of the IPsec ESP implementation.
6. The checksums created in upper layers are recalculated using HITs as addresses.
7. The source and destination addresses are replaced with IP addresses.
8. The packet is encrypted using ESP and sent over the network.
9. On receipt, the destination peer replaces the source and destination IP addresses with HITs.
10. The checksums are verified and the packet is handed over to upper layers.

The HIP protocol supports host mobility by updating peers of locator (IP address) changes when the host roams to other networks. The HI binding in the HIP layer maintains transport-layer connectivity. However, connectivity in the lower layers is temporarily interrupted while the host sends a HIP Update packet to notify its peers, modifies its existing local bindings and generates new keys for control-plane and data-plane transmission [24].

The HIP Update packet informs peers of locator value changes. In the worst case, two peers may be changing their IP addresses at the same time, preventing each of the two Update packets from reaching their destination peers. Moreover, if DNS is used as the method for a peer to update its IP address, the changes may take time to propagate in the DNS infrastructure. HIP proposes the use of a Rendezvous mechanism alongside the use of DNS as a solution to these contentions.

Since the IP address of a highly mobile host would be changing constantly and a DNS service may not be able to keep up, both the DNS and the host agree on a static location, where they would communicate through. That static location is where the Rendezvous Server (RVS) is. It acts as an intermediary node between the DNS and the host.

As shown in Fig. 3, the host publishes its HI, along with the IP address of the Rendezvous Server (RVS) it communicates to, in the DNS RR. The host updates the Rendezvous Server, using a HIP Update packet, whenever its IP address changes (1). Therefore, the RVS has knowledge of the most up-to-date locator value of the host. When a peer (initiator) wishes to communicate with a mobile host (responder), it queries the DNS for the HI and the RVS of the host (2). The initiator then sends the trigger packet I1 to the RVS (3), instead of being sent directly to the mobile host [25].

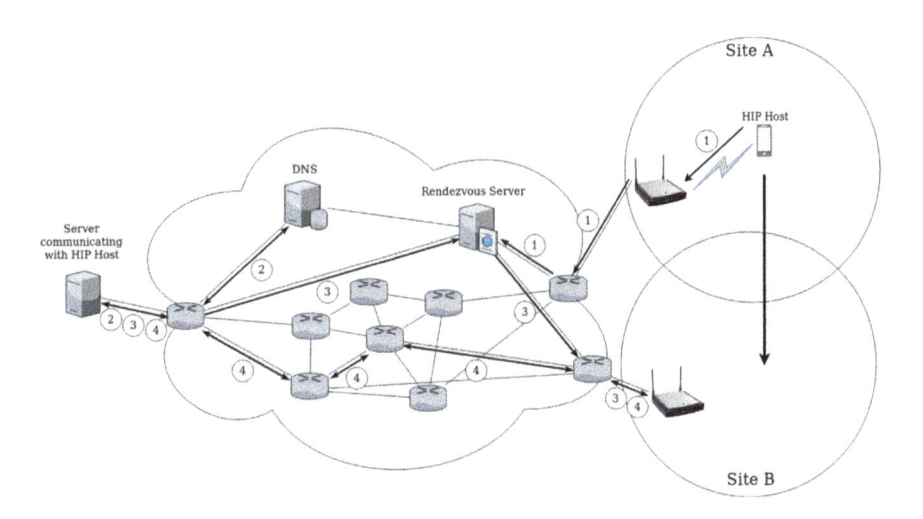

Fig. 3. HIP Host Mobility using DNS and Rendezvous Server

Upon receipt of the I1 packet, the RVS determines if the packet is destined to itself or a host under its administrative control. A mismatch between the packet's destination HIT and the RVS's HIT indicates that the received packet needs to be forwarded.

The RVS checks its registrations for a match, rewrites the destination IP address as the one the intended host registered and forwards it. The responder receives the I1 packet and performs the standard HIP Base Exchange (4), via a direct communication link between the initiator and the responder.

The HIP Update packet is critical to the mobility of a host. Similar to LISP's use of Map-Register messages to update the Map-Server of a LISP MN, a HIP host registers itself and maintains the validity of its registration with the Rendezvous Server using HIP Update packets [26]. A HIP packet is essentially an IPv6 packet with extension headers. The minimum size of the HIP Update packet was found to be approximately 144 bytes, larger than the size of LISP Map-Register message. This paper would, subsequently, evaluate the performance of the update messages of LIS protocols by creating UDP packets with a definite size of 144 bytes.

2.4 Critical Review of HIP

HIP takes a different approach from LISP in decoupling the IP namespace into distinct locator and identifier namespaces. Implemented as a sub-layer in the protocol stack of each participating host, HIP spares changes to the current architecture of the Internet. It is designed to work alongside the DNS infrastructure of the Internet. A host-based approach also implies that such a protocol could be implemented with a system upgrade of the hosts' operating system.

The HIP Base Exchange is the initial step in communicating with the HIP protocol. The HIP BEX performs the Diffie-Hellman key exchange, the building block towards the use of more advanced cryptographic tools. The content of control-plane packets is protected from changes using the HMAC algorithm. Moreover, the use of a signature algorithm and public key cryptography is the standard approach in determining the origin of a packet, similar to how session keys are generated higher in the protocol stack.

Increasing the initial cost at the initiator's side for starting a HIP association is a good deterrent, decreasing the likelihood of Denial-of-Service attacks. The I1 packet, which triggers the responder to start the three-way HIP association, adds up to the cost of initiating communication. The cryptographic puzzle, that the initiator is required to solve by using up CPU cycles, consolidates its status as a genuine peer.

The current DNS infrastructure supports host mobility, together with the use of a Rendezvous server. The RVS acts as a static location, through which a peer can initiate a HIP association with a mobile host. The RVS is updated of IP address changes of the mobile host using HIP Update packets. The delay involved with the HIP Update packet to reach the RVS is critical for a smooth handover with minimal interruption between distinct networks. The amount of traffic to the RVS affects this delay. If the delay is too long and a highly mobile host switches networks within this delay, then the host might never get the opportunity to communicate any data through the network.

2.5 Summary of LISP and HIP

Table 2 summaries the main differences between LISP and HIP approach.

Table 2. Summary of LISP and HIP

	LISP	HIP
Approach	Network-based	Host-based
Operation	Encapsulation (Tunneling)	Address Rewriting
Addressing Scheme	RLOC (locator)	IP (locator)
	EID/IP (identifier)	HIT (identifier)
Host Implementation	Sub-layer between Network and Data Link (L2.5)	Sub-layer between Transport and Network (L3.5)
Update Mechanism	Map-Request Messages	HIP Update Packets
Update Database	Map-Server	DNS + Rendezvous Server
Control-Plane Security	Trust Relationships + HMAC + One-Time Keys	Trust Relationships (DNS) + HMAC + Public Key Cryptography
Data-Plane Security	IPSec ESP	IPSec ESP

3 Analysis

3.1 IP Paging Mechanism

A MN is normally battery-operated and optimised for energy efficiency. Indeed, a long battery life is a desirable property for MNs. While much effort has been put into minimising the energy consumption of MNs through processor optimisation and electronic component upgrades, the communication process of MNs still bears a cost to their battery life.

MNs typically communicate through wireless links such as Wi-Fi and 4G. Most MNs do not communicate all the time and there is a need to prolong their battery life by switching the device into a sleep or stand-by mode. Sleep mode means in no way that the MN will stop communicating entirely. Instead, the MN communicates at a reduced frequency such that there is a balance between mobility and battery life.

A MN informs its network attachment point that it is switching into sleep mode, such that it can be woken up again when data packets destined for it arrives. The time at which the wireless Access Point (AP) will trigger the MN to possibly wake up is precisely known by both the AP and the MN. However, when there is a sleeping MN with fast mobility, roaming to other networks rapidly, then the MN might never get the opportunity to properly enter sleep mode or may be disrupted by continuous handover processes and it would negatively impact on the battery life.

The issue of sleeping MNs worsens when LIS protocols are introduced. Indeed, each LIS protocol has its own way of updating its location database. The ETRs, in a LISP MN, update its Map-Server when it moves into a new network. In HIP, the host itself updates the Rendezvous Server when it changes location. The fact that update messages of LIS protocols could be sent periodically implies that a sleeping MN would be waking up at these times only to transmit a packet, which is not efficient.

If IP paging mechanisms are implemented in LIS protocols, their use would allow a sleeping MN's location to be known while it roams across networks. An IP paging mechanism divides groups of access routers (ARs) into paging areas. If a MN becomes inactive, the paging area is searched for the MN. In a way, when the precise location of a MN is unknown, a set of locations would be defined as the location of the MN. Therefore, only an approximate location of the MN would be known immediately when IP paging is used [27, 28]. The paging area is then searched to find the exact location of the MN using an IP paging algorithm, such that buffered packets can be delivered.

The use of an IP paging mechanism spares the MN from having to send update messages too frequently, preserving the battery life of the MN. Moreover, since less update messages travel through the network, more bandwidth becomes available for other useful purposes. An IP paging mechanism is especially useful in large networks, where an update message from a MN would require much time to propagate to other routers. The use of an IP paging mechanism also shifts the focus of LIS protocols onto the optimisation of the end networks and how well they cope with update messages.

The network topology to be implemented would consist of the Wireless LAN Paging Agent (WLPA) and the Core Paging Agent (CPA) as shown in Fig. 4. The purpose of the WLPA is to register the location changes of the MN. In contrast to the update messages sent to a, possibly distant, administrative server in LISP or HIP, the update messages in an IP paging architectures travel a much shorter distance to the WLPA, thus taking less time overall for the host to successfully send an update message. This property is particularly desirable for fast host mobility. Indeed, bringing some functionality of the Map-Server or Rendezvous Server closer to the MN has the effect of increasing the reachability of a MN when it is given a very small amount of time to update its administrative server, in environments where handovers occur at a high frequency. In such cases, the update message might not get enough time to reach or get acknowledged by the administrative server.

The purpose of the CPA is to act as the authoritative record of mappings within the IP paging network. In the event of fast host mobility, where the location of the MN is precisely known by the WLPA and not by the CPA, the CPA could initiate a search process using an IP paging algorithm to find the MN. In the context of LIS protocols, the CPA could act on behalf of MNs to update their respective administrative servers, the Map-Server or Rendezvous Server.

The combined functions of the WLPA and the CPA would help in effectively redirecting in-flight packets to the correct paging area when a MN is travelling very fast across distinct networks within the IP paging mechanism. The delivery of in-flight packets with minimal packet loss and the reduction in the cost of updating the location database are the main benefits behind the use of an IP paging mechanism.

In the context of the use of IP paging, the LIS update message will be referred to as a Page Beacon Message (PBM). The term "Page Beacon Message" was proposed in a performance evaluation paper [29].

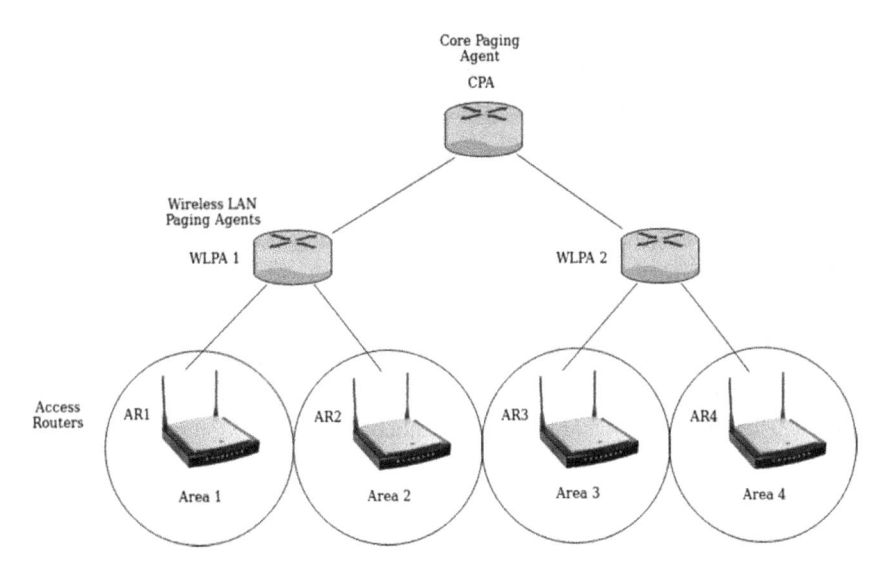

Fig. 4. Implemented IP paging mechanism in a hierarchical network topology

3.2 Time Calculation

The performance evaluation in this paper would require a method for calculating the time taken for packets to travel through a given communication link. The end-to-end delay is the time taken for a packet to be transmitted from source to destination. In contrast, the round-trip time is the time taken for a packet to be transmitted from source to destination and for the acknowledgement of the receipt of the packet to be received by the source from the destination.

The round-trip time is a meaningful measure of the bidirectional network traffic. However, for the unidirectional measurement of network traffic, the end-to-end delay is more suitable. The end-to-end delay can be approximated as half the value of the round-trip time. However, it makes the bold assumption that the bandwidth, congestion and quality of service are the same in both directions. Moreover, it assumes that the clocks at the source and destination are synchronised.

Given that components in a network would not have the same clock time, determining the end-to-end delay for a packet from a source to a destination is a challenge. Indeed, no two clocks on different machines can ever be exactly synchronised. However, they can be synchronised, albeit using costly equipment, at a certain acceptable level and their clock drifts corrected continuously. Therefore, round-trip time (RTT) would be used as a time metric in determining the performance of LIS update messages.

3.3 Transport Protocol

LIS protocols may use a transport protocol of their choice to transmit update messages. Some LIS protocols may employ the User Datagram Protocol (UDP) while others may use the Internet Control Message Protocol (ICMP). The possibility of using the

Transmission Control Protocol (TCP) against the UDP protocol is debated below. The main argument in favour of TCP is the fact that a transmitted packet is always acknowledged, sparing the development of an acknowledgement mechanism when implementing the update messages of LIS protocols.

However, TCP is a connection-oriented protocol which requires connection establishment between two hosts for its operation. Since update messages are sent whenever a host moves to another network, any established connection using TCP would be broken. The host would have to re-establish a TCP connection before being able to send update messages again. Moreover, if a TCP packet is not acknowledged, it is buffered and resent, while maintaining in-sequence delivery.

When TCP is used, packets are guaranteed to be delivered and at a later time, if necessary. An effective location update mechanism requires real-time requests and responses. Therefore TCP is not a suitable protocol to send location updates. The UDP protocol would be used instead.

3.4 Cryptography

LIS protocols divides the administrative functions of their protocol into the control-plane functionalities while the functions related to data transmission is categorised as the data-plane functionalities. The data-plane is secured by IPsec [30, 31], which is a recommendation of IPv6 security [32]. The control-plane of these protocols is secured mainly by trust relationships. However, in this paper, the possibility of using IPsec for control-plane functionalities, along with the associated overheads, is investigated. The suitability of IPsec for this paper's study is mainly attributed to the inter-paging area mobility in a restricted hierarchical network topology and the fact that a particular WLPA would always be reachable from the paging area under its control.

3.5 Summary

This paper would, subsequently, make use of an IP paging mechanism, through which LIS update messages would be sent between the CPA, WLPA and MN. The IP paging mechanism emphasises the importance of design considerations of LIS protocols at the end networks, where computing resources are limited. The LIS update message would be referred to as a PBM in the context of IP paging. To measure the performance of PBMs, round-trip time would be used as a time metric. The PBMs are essentially UDP packets of size 144 bytes. The PBMs are secured using IPsec from the MN to the WLPA. The PBMs are evaluated with varying levels of network traffic and IPsec cryptography.

4 Experiment Design

The experiment consists of sending PBMs through a network topology supporting an IP paging architecture and calculating the round-trip time with different levels of network traffic and IPsec cryptography. The experiment is split into two parts. In the first part of the experiment, PBMs are sent through the wireless communication link between the MN and the WLPA. In the second part of the experiment, PBMs are sent through the wired communication link between the WLPA and the CPA.

The Wi-Fi access router supports connection speeds of up to 300 Mbps on the 2.4 GHz band. Cat5e Ethernet cables, which support connection speeds of up to 1000 Mbps, are used throughout this experiment. However, it should be noted that not all the network interface cards in the topology support such a speed. In some machines, the connection speed is limited to 100 Mbps. Therefore, the experiment would be made to generate network traffic based on the maximum capacity of any given communication link.

The network components are set up in such a way that the lower one goes down the topology, the lesser the amount of network traffic the component would be able to handle. The length of Ethernet cables are kept to a minimum, such that the least possible transmission delay through the communication medium is recorded irrespective of the distance covered. Indeed, the farther the distance covered, the greater the transmission delay and the greater the round-trip time of update messages.

Network traffic is generated using iPerf based on the maximum supported connection speed of any given communication link. Network traffic is generated unidirectionally from the fast sender to the slow receiver. Therefore, network traffic is generated from the WLPA to the MN, in the first part, and from the CPA to the WLPA, in the second part.

The effects of IPsec between the MN and the WLPA are investigated. However, from the WLPA to the CPA, IPsec is not used since the architecture leverages the trust relationships between the WLPA and the CPA, similar to trust relationships in LIS protocols. Therefore, in the second part of the experiment, the PBMs are only tested under different levels of network traffic.

Since the size of the acknowledgement packet received would be the same as that of the corresponding PBM that was sent, the RTT values would give a precise measure of the bidirectional traffic and an approximate measure of the unidirectional traffic (by halving). It should be noted that although the size of the packet stays the same in both directions, there is a difference in the size of the packet when IPsec cryptography is used. Indeed, the size of the packet, initially at 144 bytes of data, nearly doubles to 248 bytes. The size of the packet increases owing to the encryption and encapsulation of the packet.

Although the data in the PBMs remains almost the same, it contains a counter which changes part of the data, which in turn changes the checksum of the packet. Moreover, the encryption process (through IPsec), based on two important concepts of confusion and diffusion (avalanche effect), changes the whole packet's content when even the slightest bit is modified. Therefore, no two packets sent in this experiment are the same. A total of 300 packets were transmitted for a duration of five minutes at a rate of one packet per second in each experiment.

5 Performance Evaluation

5.1 Average RTT of UDP PBM from MN to WLPA

As shown in Fig. 5, when network traffic is at 0%, PBMs take, on average, 5.53 ms to travel from the WLPA to the MN and back, with IPsec disabled. There is a slight increase on the average RTT to 5.79 ms when IPsec is enabled. This increase of 4.70% can be interpreted as negligible. Indeed, at 0% network traffic, much bandwidth is available to address the overhead involved with IPsec.

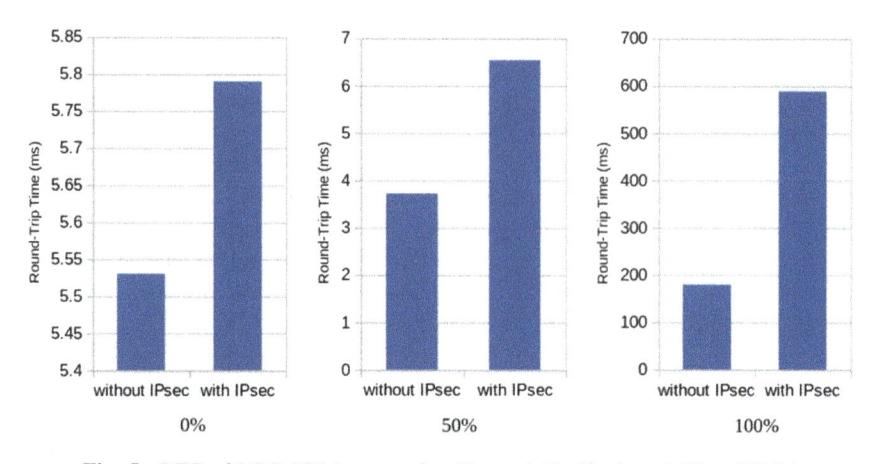

Fig. 5. RTT of UDP PBM on varying Network Traffic from MN to WLPA

When network traffic is at 50%, PBMs take, on average, 3.72 ms without IPsec for the round-trip transmission, which is lower than the RTT of 5.53 ms at 0% network traffic. This could be explained by the overhead involved with the initial cost of starting communication when network traffic is at 0%. When IPsec is enabled, the average RTT significantly increases, by the order of 75.8%, to 6.54 ms. Therefore, the use of IPsec impacts on the round-trip time.

When network traffic is at 100%, PBMs take, on average, 179 ms without the use of IPsec to perform the round-trip distance. A heavily loaded communication link drastically delays the unencrypted PBMs by approximately 30 times compared to an idle communication link. When IPsec is established, the delay significantly increases to 589 ms. It represents an increase of 228%. Therefore, the impact of IPsec usage on heavily loaded networks is noticeable.

Since the RTT values take into consideration the processing of packets, that is, the time it takes to create a packet and to process it at the different network layers, the 228% increase includes the overhead in generating the IPsec encrypted packets. Moreover, the MN, being a slow sender and receiver with low computing power, could significantly increase the packet processing time.

5.2 Average RTT of UDP PBM from WLPA to CPA

The RTT values recorded between the WLPA and the CPA are consistent with the trend that higher levels of network traffic implies longer delays with the PBMs as illustrated in Fig. 6. At 0% network traffic, the PBM takes, on average, 0.47 ms for round-trip transmission. When the network traffic is loaded to 50%, the average RTT nearly doubles to 0.73 ms. A full traffic load of 100%, the time taken for round-trip transmission is 2.81 ms, equivalent to an increase of 498% compared to an idle communication link.

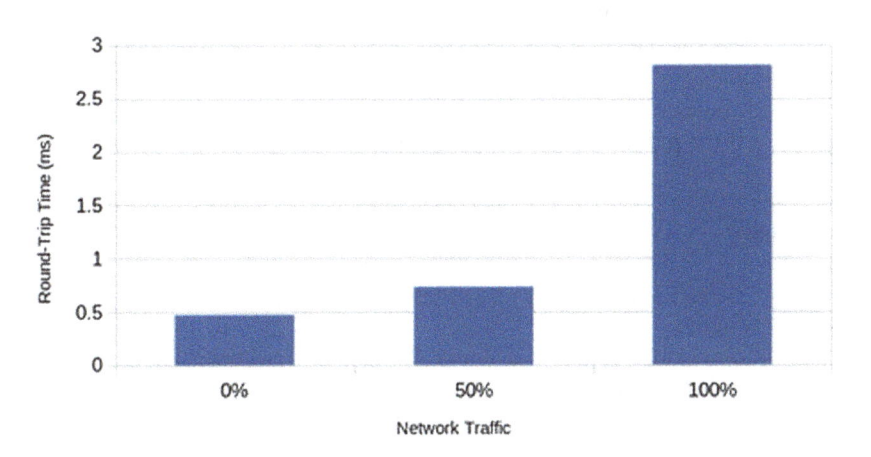

Fig. 6. RTT of UDP PBM on different levels of Network Traffic from WLPA to CPA

6 Conclusion

This paper has surveyed the approaches of different LIS protocols in updating the location changes of connected MNs. It has investigated the use of LIS update messages in a network topology containing an IP paging architecture. The implications of the use of cryptography to the delays in delivering the update message were also investigated through experiments, involving variations in the level of network traffic.

The type of communication medium at the end networks has an effect on the performance of LIS update messages. The RTT values recorded suggest that the wireless communication link between the MN and the WLPA becomes highly congested and delays the unencrypted or encrypted LIS update messages more than when the wired communication link between the WLPA and the CPA is stress-tested.

The experiments that were performed shows that the use of IPsec cryptography, a recommendation for IPv6 implementation, affects the performance of LIS update messages the higher the level of network traffic. In this scenario of congested network traffic, MNs with low computing resources could become overwhelmed with sending or receiving LIS update messages, hindering host mobility and negatively impacting on the network performance.

The paper has assessed the use of an IP paging mechanism together with LIS protocols and hints at the possible implementation of IP paging in LIS protocols, under active development, for improved mobility. The importance of update messages in LIS protocols were highlighted and the security mechanisms designed to protect these messages were discussed. Although the update messages of LISP and HIP are well documented, there is still room for improvement through the adoption of emerging network technologies and continued research.

The distinct approaches to the IP namespace decoupling of LISP and HIP have their own benefits. While HIP mainly requires changes on the host device, LISP needs changes in the Internet architecture and the use of upgraded or new networking equipment. The Internet is a vast network that supports the operation of different protocols seamlessly. There is absolutely no reason to choose one LIS protocol over the other, simply because they can effortlessly co-exist with each other. The choice of using any of these protocols would eventually lie in the hands of the user.

References

1. Farinacci, D., Fuller, V., Meyer, D., Lewis, D.: The Locator/ID Separation Protocol (LISP), RFC 6830 (2013)
2. Fuller, V., Farinacci, D., Meyer, D., Lewis, D.: Locator/ID Separation Protocol Alternative Logical Topology (LISP+ALT), RFC 6836 (2013)
3. Whittle, R.: Ivip (Internet Vastly Improved Plumbing) Architecture, draft-whittle-ivip-arch-04 (archived) (2010)
4. Mungur, A.: Analysis of locator identity split protocols in providing end-host mobility. Int. J. Innov. Eng. Technol. (IJIET) 7(2), 365–375 (2016)
5. Moskowitz, E.R., Heer, T., Jokela, P., Henderson, T.: Host Identity Protocol Version 2 (HIPv2), RFC 7401 (2015)
6. Vogt, C.: Six/one router: a scalable and backwards compatible solution for provider-independent addressing. In: Proceedings of the 3rd International Workshop on Mobility in the Evolving Internet Architecture - MobiArch 2008, p. 13 (2008)
7. Atkinson, R.: ILNP Concept of Operations, draft-rja-ilnp-intro-11 (archived) (2011)
8. Atkinson, R., Bhatti, S., Hailes, S.: A proposal for unifying mobility with multi-homing, NAT, & security, p. 74 (2007)
9. Menth, M., Hartmann, M., Klein, D.: Global locator, local locator, and identifier split (GLI-Split). Future Internet 5(1), 67–94 (2013)
10. Perkins, E.C.: IP Mobility Support for IPv4, Revised, RFC 5944 (2010)
11. Perkins, E.C., Johnson, D., Arkko, J.: Mobility Support in IPv6, RFC 6275 (2011)
12. Komu, M., Sethi, M., Beijar, N.: A survey of identifier-locator split addressing architectures. Comput. Sci. Rev. 17, 25–42 (2015)
13. Fuller, V., Farinacci, D.: Locator/ID Separation Protocol (LISP) Map-Server Interface, RFC 6833 (2013)
14. Seo, E., Sarang Wi, S., Zalyubovskiy, V., Chung, T.M.: The scalable LISP-Deployed software-defined wireless network (LISP-SDWN) for a next generation wireless network. IEEE Access 6, 66305–66321 (2018)
15. Farinacci, D., Lewis, D., Meyer, D., White, C.: LISP Mobile Node, draft-ietf-lisp-mn-04 (2018, work in progress)

16. Saucez, D., Iannone, L., Bonaventure, O.: Locator/ID Separation Protocol (LISP) Threat Analysis, RFC 7835 (2016)
17. Fuller, V., Farinacci, D., Cabellos, A.: Locator/ID Separation Protocol (LISP) Control-Plane, draft-ietf-lisp-rfc6833bis-22 (2018, work in progress)
18. Maino, F., Ermagan, V., Cabellos, A., Saucez, D.: LISP-Security (LISP-SEC), draft-ietf-lisp-sec-17 (2018, work in progress)
19. Yang, J., et al.: IER: ID-ELoc-RLoc based architecture for next generation internet. J. Electron. 31(6), 519–536 (2014)
20. Saucez, D., Iannone, L., Bonaventure, O., Farinacci, D.: Designing a deployable internet: the locator/identifier separation protocol. IEEE Internet Comput. 16(6), 14–21 (2012)
21. Moskowitz, E.R., Komu, M.: Host Identity Protocol Architecture, draft-ietf-hip-rfc4423-bis-19 (2018, work in progress)
22. Gilligan, R., Thomson, S., Bound, J., McCann, J., Stevens, W.: Basic Socket Interface Extensions for IPv6, RFC 3493 (2003)
23. Laganier, J.: Host Identity Protocol (HIP) Domain Name System (DNS) Extension, RFC 8005 (2016)
24. Henderson, E.T., Vogt, C., Arkko, J.: Host Mobility with the Host Identity Protocol, RFC 8046 (2017)
25. Laganier, J., Eggert, L.: Host Identity Protocol (HIP) Rendezvous Extension, RFC 8004 (2016)
26. Laganier, J., Eggert, L.: Host Identity Protocol (HIP) Registration Extension, RFC 8003 (2016)
27. Ramjee, R., Li, L., La Porta, T., Kasera, S.: IP paging service for mobile hosts. Wireless Netw. 8(5), 427–441 (2002)
28. Mungur, A., Edwards, C.: Performance of a tiered architecture to support end-host mobility in a locator identity split environment. In: Proceedings of Conference on Local Computer Networks, LCN, vol. 26–29–Octo, no. Imi, pp. 446–449 (2015)
29. Mungur, A., Tuhaloo, M., Jawarun, M.: Performance evaluation of a hybrid paging mechanism to support locator identity split end-host mobility. In: 2016 8th IFIP International Conference on New Technologies Mobility and Security, NTMS 2016 (2016)
30. Kent, S., Seo, K.: Security Architecture for the Internet Protocol, RFC 4301 (2005)
31. Frankel, S., Krishnan, S.: IP Security (IPsec) and Internet Key Exchange (IKE) Document Roadmap, RFC 6071 (2011)
32. Deering, S., Hinden, R.: Internet Protocol, Version 6 (IPv6) Specification, RFC 8200 (2017)

Business and Environmental Perspectives of Submarine Cables in Global Market

N. Aishwarya

CHRIST (Deemed-to-be University), Bangalore, India
infoaishu@gmail.com

Abstract. If an individual uses any of the social media networking sites, such as Facebook, Instagram, YouTube, Twitter and the like, a subsea cable is involved there. Submarine cables are considered as critical global communications infrastructure. These cables are used by various telecom providers and content provider companies such as Google, Facebook, and Microsoft to provide seamless transmission of data for their services. Growing internet users and increasing internet traffic for various social media sites is the major reason for the growth of this market. Submarine cables enable data services such as the email, internet banking, social media networking, search engines and all other aspects related to internet that are taken for granted in daily life. These submarine cables scales up the ubiquity of cloud computing and builds digitization of activities. Undersea cable network is the new economic trade route and acts a commodity in Information age. This paper reviews the business and environmental impacts of submarine cables in the global market.

Keywords: Submarine cable · Satellite · Undersea cable · Social media · Subsea cables

1 Introduction

Change is the only entity which is ever constant. We live in a world where new products are launched every day, new technologies are introduced every month and new companies are hurled every year. Being in the internet era, companies are forced to embrace web-enabled enterprise models where they closely monitor and engage the customer. Year on year, we experience newer gadgets from these companies which ease the lives of millions of people, bringing on innovative customer experiences and constantly focusing to improve customer satisfaction and profits. With the advent of technologies like Augmented Reality, Virtual Reality, Cloud computing, Data mining and Automation, companies are leveraging their resources for the benefit of their growth as well as the economy they thrive in.

Internet is the backbone for today's business. It was developed in the year 1990's by Tim Berners Lee. Internet has changed the way how business activities are done. In contrast to the early years of its development, internet is no longer just a privilege for the inhabitants of the developed economies, but instead is a worldwide phenomenon [1]. However, during the last decade, the enormous users of internet have emerged from the developing economies. According to Suboptic.org, the total number of internet users has increased from 44% to 62% (2006 to 2011) in developing nations [2]. Almost 25% of the world's total internet usage comes from China and they represent 37% of the developing economies' internet users. Youngsters below the age group of 25, are ruling the internet world with 45% usage, a whole new generation of so-called "Digi-Natives" are created [3]. This Digi-natives use internet for everything and also attach it to simple physical devices in the form of IoT, which has led the global data traffic to upsurge tremendously.

There is high competition prevailing in this industry among the various Internet service providers (ISP) to provide data at high speed, less latency and at reduced cost. Globally, in last decade, most of the Internet data transmission is handled by underground fiber optic communication. Apart from them, satellite and submarine cables are used. If one can take into account the number of cables used for internet, it merely looks like spaghetti work of really long wires. This paper presents views on the rise of submarine cables and the major players in the industry along with the business and environmental impacts of submarine cables on data transmission. This study becomes more relevant on the current trend as major players in the Industry 4.0 rely on data transmission for their business efficiency. Submarine cables are considered as the global highways of information and given the attracting features of them, technology companies are moving more towards deploying them in oceans than satellite technology, hence it becomes more compelling to analyse their effectiveness and impacts in the industry.

2 Rise of Submarine Cables

The improvements in satellite and submarine technology during the past two decades is astounding. It is believed that the submarine cable technology have surpassed those in satellite technology in the innovation and development arena [4]. However, today, submarine fiber-optic cables provide more than that 95% of the world's international telecommunications [5]. The overall submarine cable system market was valued at USD 10.83 billion in 2017 and is expected to reach USD 20.93 billion by 2023, at a CAGR of 12.25% from 2018 to 2023 [5]. Submarine cables are were launched first in 1858 by Cyrus west Field, which worked only for three weeks, further attempts in 1865 and 1866 made them to be successful. The capability to transfer both structured and unstructured data rapidly over large terrestrial scales has overpowering impacts on the political, social, economic, technological, financial and legal dimensions. International law, in particular, the law of the sea, has recognized the freedom to lay submarine cables and perform associated operations and has placed certain obligations on states related to the protection of submarine cables [6]. TeleGeography lists nearly 350 cables as of 2019 [7] - Some are transatlantic and transpacific cables traveling across different oceans and coasts along regions. The whole network of submarine cables spans more than 55,000 miles, with some being buried as far underwater as Mount Everest towers above ground [8]. Submarine cables can be used as power cables used to transmit electricity and as communications cables to transmit data communications traffic. Both are designed for underwater use and are usually laid on or buried under the seabed [9].

The laying of submarine cables in seabed is an interesting phenomenon. They are laid by connecting them to "landing stations" along the seaboard. Companies have developed laying boats and hi-tech cable installing technology are used for cable installation under various laying circumstances [10]. Massive boats are used in the process of burying the sea-cables and their routes are strategized to move along flat seabed as much as possible, avoiding coral reefs and ship wrecks as well the deep trenches or undersea mountains. The inside look of an undersea cable shows a crux made of fiber levels and wires shielded in protective armoring to prevent damage from water, shipwrecks and shark bites. Protection for the cables may be smeared in form of a rock-mattress cover, cast iron shells, cable anchoring, ducting or rock dumping. Other protection measures are the use of special backfill materials for cable burial or to cover cables with reinforced concrete slabs or steel plates [10]. The thickness and thinness of the fiber in an undersea cable depends on their depth in ocean. At the lowest levels of ocean, the fiber is thinner and at the near shore they are thicker. This is because of the threats they have in shallow waters.

A comparison made between submarine cables and landlocked cable systems shows that in data transmission both systems equally capable but the landlocked systems face difficulties for international connections, especially routing through numerous sovereign territories [3]. Some of the factors which make submarine cables the superior medium for internet are speed of the signal, bandwidth limitation, data transmission price and reliability [11]. When compared to satellite transmission, submarine calls have less delay and promotes functions such as "high-frequency trading". On a similar note, the average per unit cost per Mb\s is way lesser for a submarine cable when compared to satellite transmissions. Also the operating life of satellite is said to be ten years, whereas the submarine cables are expected to work twice their lifetime. Once the satellite is launched, there is no cent percent guarantee that they will work to their efficiency. If the launch fails or if there is signal loss, then there is typically no substitute available to ensure the constant data transmission.

Submarine cables can facilitate most of the services which are earlier provided by the landlocked systems. Some of them include Internet, Messaging, social media, or any service which can be availed using internet. Apart from telecommunications, fiber-optic cables have also been used by the military for intelligence gathering through acoustic sensing and for marine scientific research [12]. Also, submarine cables has a huge economic and developmental impact such as economic growth, customer benefit, increased bandwidth, increased government revenue and regional stability among others [13].

Historically, subsea cables were funded and used for by telecom associates. But as a modern trend, all the big digital companies like Google, Microsoft, Facebook and Amazon have started getting in on the game, putting huge money behind the infrastructure that's made the shift to an always connected world possible. Figure 1 shows the submarine cable map across the world.

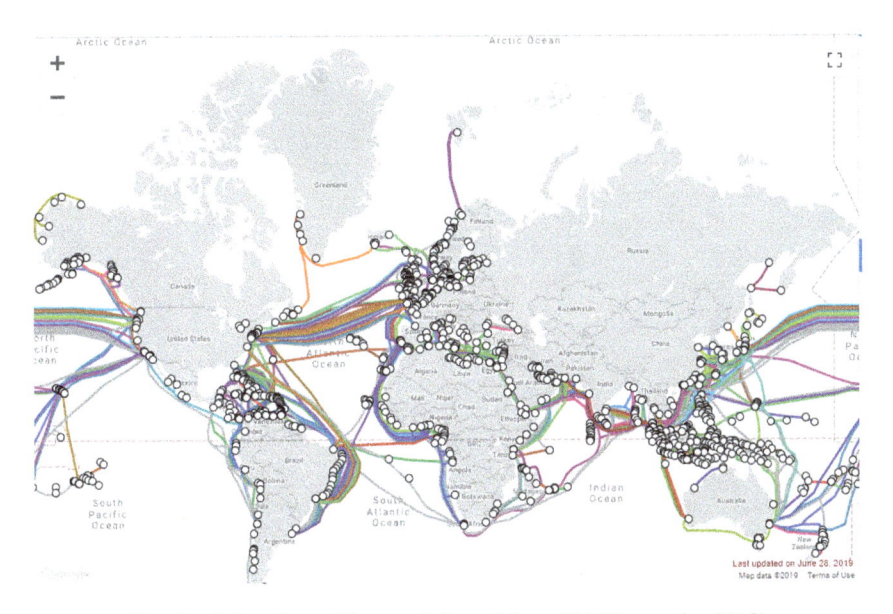

Fig. 1. Submarine cable map (adopted from TeleGeography 2019)

3 Major Players in Submarine Cable Business

The prime driver of the new system demand in undersea cable has happened after the dot-com bubble burst and the growth rate is highest in Southeast Asia, Australia, Africa and Latin America [13]. Trans-Atlantic, Trans-Pacific, Americas, Europe-Asia, and EMEA are considered to be the major submarine cable market [14]. Content providers such as Google, Facebook, Microsoft, and Amazon are the major investors in submarine communication cables [14]. These companies are investing hugely to have incredible bandwidth demand for their data center traffic and route prioritization for new cables. Most industries such as Banking, Shipping, Airlines, supply chain, manufacturing businesses and entertainment will snap to halt when the communication networks goes down, there is no plan B available for most of the industries as well as most of the countries [15]. World major powers such as U.S, Australia, Japan and Singapore heavily rely on submarine cables for their international communications.

Amazon owns three cables namely JUPITER, BtoBE and Hawaiki, in which JUPITER will connect the route from Asia to US in 2020 at 60 terabits of data per second. Facebook part-owns nine submarines cables, also they own part-own JUPITER along with Amazon. Google owns three cables as sole-owner and part-owns eleven cables. Microsoft owns four major cables named GTT Express, MAREA, AEConnect-1 and NCP cable system. Among all the major players, Google holds the major shares of 63,605 miles of submarine cables. Most of these content providers focus on sole-owned cables to support their cloud services. Figure 2 provides the various content providers and their investments.

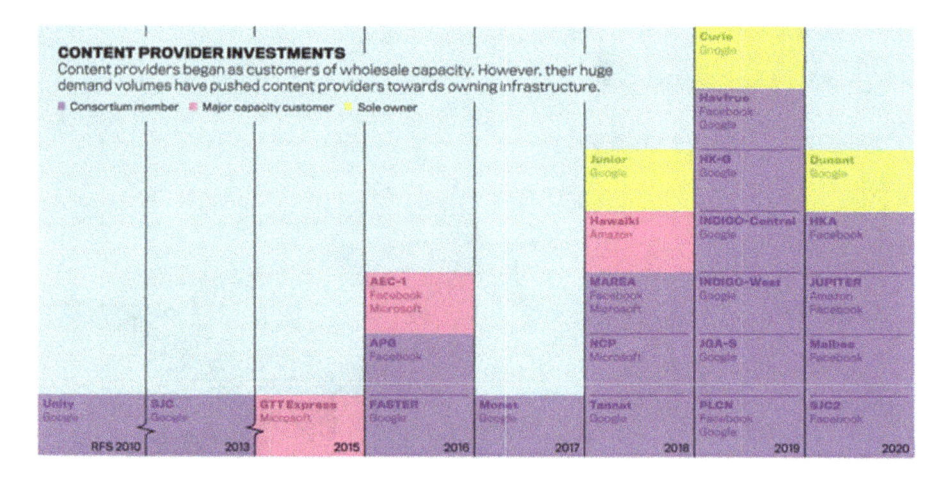

Fig. 2. Content Provider Investments (adopted from TeleGeography 2019)

Currently there are fifteen subsea cables across five cities in India namely Mumbai, Chennai, Cochin, Trivandrum and Tuticorin. Out of which, five are owned by Tata Communications, Global Cloud eXchange owns two, Reliance JIO owns one and Bharti Airtel owns three cable landing stations.

4 Business Impact of Submarine Cables

Majority of this shift is due to the businesses interest in digital transformation and cloud-based approach. Content generators purely focus on maximizing their cloud computing services to quickly deliver the audio, image, news and video to their demanding customers. Meanwhile cloud provider companies such as Amazon web services, Microsoft Azure should be able to provide these services and hence investing in the submarine cables makes a lot more sense for these companies. These companies also enjoy lot more autonomy on laying, managing and maintaining these cables instead of renting or paying lease to another vendor.

These submarine cables scales up the ubiquity of cloud computing and builds digitization of activities. This paves way for the global companies in digital economy to associate with their stake-holders right away, irrespective of the distance. Also, submarine cables handles the user's expectations for high-performance data connectivity anytime and anywhere. These cables bring companies to the digital edge, where the cables are laid directly to their data centers enabling the systems to provide constant data transmission and ensure global connectivity. The demand for interconnection may not be as old as the submarine cable industry [16], but they are linked now, both in their accelerated growth rates today and the ways they will empower businesses tomorrow. Figure 3 shows the approximate global submarine cable cost involved in construction. Chinese players have rapidly expanded their business around submarine cables, both as vendor as well as the buyer. They also participate in financing the project either through consortium model or by private financing. Huawei Marine network, which has in total

90 submarine cable projects are majorly present in Africa region. However, amid pressure from US government to blacklist the company over spying accusation, the company is poised to sell its majority stake in the submarine business [18].

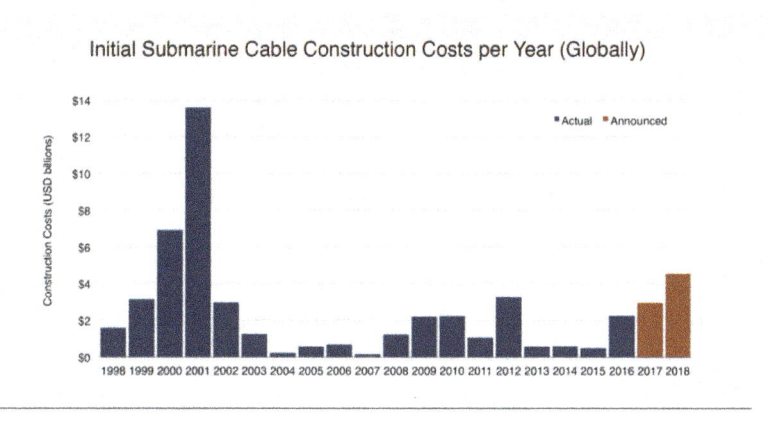

Fig. 3. Approximate global submarine cable construction cost (Stronge 2017) [17]

Cloud computing services, OTT video content, online administrative services, videoconferences, social media networks and other connected services; all of this is possible with the help of submarine cables.

5 Environmental Impact

By definition, submarine cables are not "pollution of the marine environment", nor can cables realistically cause such pollution [19]. They do not produce any "significant harmful changes" to the marine environment. However, it remains a legal responsibility for countries to ensure that it is in control and protect the marine environment. Submarine cables, with their strategic position at the bottom of the oceans, would have many new uses in the near future. According to International Telecommunication Union, an initiative has been setup to build sensors over the submarine cables to measure and analyse the real-time aspects of climate change across the oceans. Also, researchers believe that it could also be used for detecting Tsunami across the world.

Noise impacts of installing and removing a submarine cable is always questioned among researchers. However, there are only few reports which delve on this issue. This may be because of the fact that noise is not regarded a key environmental issue in association with subsea cables by most environmentalists [20].

Submarine cables are considered to have a long positive track record in the marine environment with their small carbon footprint. Also they give positive contribution in reducing greenhouse gases, stand uniquely apart from high impact uses that are of concern to the area beyond the limits of national jurisdiction such as shipping, deep seabed mining, fishing, pipeline and energy [19].

6 Major Challenges Associated with Submarine Cables

The entire concept of submarine cables providing long distance internet transmission at relatively low cost and high bandwidth looks all appealing to the companies. However, there are many challenges in successfully implementing this system and maintaining it. One of the major task, organizations face in laying an undersea cable is that it is a slow process, and a tedious work. The internet transmission is still vulnerable underwater as it is in the ground. Submarine cables face large threats from sharks bites, ship anchors, ship wrecks, boats and cruise ships. Also, maintaining a submarine cable when it is under repair is a wearying task for the organizations. Apart from these, there are various theories stating that countries tap the submarine cables of other countries for information. Also, cyber warfare can be executed as easily as a deep sea diver cutting the submarine cables with a simple wire cutter. On the same note, there is a possibility of internet outage due to submarine cable disruptions.

7 Future Scope

At present, eight trans-atlantic fiber optics cable systems like AMENCAS-I, COLUMBUS-II, CARAC, BAHAMAS and the like are connecting Europe and North America are either in operation or under construction [21]. The Japanese have laid extensive underwater fiber-optic cables to connect their various islands, besides other forms of communication with neighbouring nations. In future, Submarine optical communication can be utilized in the area of diver-to-diver communication links, diver-to-submarine links, submarine-to-UAV (Unmanned Aerial Vehicles) links, Submarine-to-submarine links, UAV-to-UAV links and Submarine-to-satellite links and the like [22]. Numerous strategic partnerships are happening across countries such as Airtel – Telecom Egypt, Airtel – Gulf Bridge international to diversify their global network to serve the massive growth in demand for data services, particularly in emerging markets across South Asia, Africa and Middle East [23]. The need to provide broadband wireless communications for underwater applications will be increasing in the forthcoming years [24]. Also, the needs of the content providers and cloud companies is growing strong, particularly in trans-ocean projects, hence to boost local geographical market, they may require further connectivity. Creation of new trans-ocean routes will be implemented where content and cloud providers are not addressing at present also markets which are under-served will be developed to increase the availability of inter-data center connectivity. The next chapter in undersea cable competition would be on the wide availability of new processing and sensor technology to detect climate changes, Tsunami prevention, managing unmanned undersea vehicles and seabed warfare [25]. Also, the increase in undersea cable usage will lead to growth in marine renewable energy sector [26].

References

1. APEC: Economic Impact of Submarine cable disruptions: Asia-Pacific Economic cooperation (2013)
2. Suboptic Homepage: SubOptic. https://suboptic.org/. Accessed 5 Apr 2019
3. ibid. See Ref. 1
4. Agarwal, N., Vyas, A.K.: Submarine optical communication. Int. J. Electron. Comput. Sci. **1**, 370–374 (2012)
5. UNEP/ICPC: Submarine Cable System Market by Application (Communication, Power), Component (Dry Plant, Wet Plant (Cables, Branching Units, Repeaters)), Offering (Installation & Commissioning, Maintenance, Upgrades), and Region - Global Forecast to 2023 (2018)
6. Davenport, T.: Submarine communications cables and law of the sea: problems in law and practice. Ocean Dev. Int. Law **43**, 201–242 (2012)
7. TeleGeography: (n.d.). https://blog.telegeography.com/this-is-what-our-2019-submarine-cable-map-shows-us-about-content-provider-cables. Accessed 4 May 2018
8. Peterson, A.: Everything you need to know about the undersea cables that power your internet – and why they're at risk of breaking. Independent, October 2015
9. ibid. See Ref. 7
10. Karin: Impacts of submarine cables on the marine environment—a literature review. Federal Agency of Nature conservation (2006)
11. ibid. See Ref. 13
12. Chave: The NEPTUNE Scientific Submarine Cable System. National Science Foundation's Ocean Observatories Initiative (2015)
13. Ruddy, M.: Economic Perspectives on the Global Submarine Cable Market (2016)
14. ibid. See Ref. 5
15. Burnett, D., Carter, L.: International Submarine Cables and Biodiversity of Areas Beyond National Jurisdiction: The Cloud Beneath the Sea. International Submarine Cables and Biodiversity of Areas Beyond National Jurisdiction: The Cloud Beneath the sea, pp. 1–72
16. Poole, J.: Submarine cable boom fueled by new tech, soaring demand. Network World (2017). https://www.networkworld.com/article/3260784/submarine-cable-boom-fueled-by-new-tech-soaring-demand.html
17. Stronge, T.: Submarine Cables: Are We in a New Bubble? TeleGeography (2017)
18. Bernal, N.: China's Huawei to sell undersea cable business. The Telegraph. https://www.telegraph.co.uk/technology/2019/06/03/chinas-huawei-sell-undersea-cable-business/. Accessed 3 June 2019
19. ibid. See Ref. 20
20. ibid. See Ref. 11
21. Rao: Optical Communication. Universities Press (2015)
22. ibid. See Ref. 4
23. The Economic Times: Airtel partners with Telecom Egypt for global submarine cable systems. The Economic Times, p. 56, 06 August 2018
24. Gkoura, L.K., et al.: Underwater optical wireless communication systems: a concise review. Konstantin Volkov, IntechOpen (2017)
25. Taormina, B., et al.: A review of potential impacts of submarine power cables on the marine environment: Knowledge gaps, recommendations and future directions. Renew. Sustain. Energy Rev. **96**, 380–391 (2018)
26. Smarter Society: Hello Future. https://hellofuture.orange.com/en/submarine-cables-global-highways-information/. Accessed 16 Sept 2019

Conditional Convolutional Generative Adversarial Networks Based Interactive Procedural Game Map Generation

Kuang Ping and Luo Dingli[✉]

University of Electronic Science and Technology of China,
Chenghua 610051, Sichuan, People's Republic of China
kuangping@uestc.edu.cn, godofpen05@gmail.com

Abstract. There is a strong need for a procedural map design system which can generate complex detail game maps but with simple user control. We propose an interactive real-time design system made with Conditional Generative Adversarial Network and Convolutional Neural Network. This system takes user-defined game-play area map as input, and generate a complex game map with the same design pattern as training samples automatically. It can output an abstract label map which can be used in other procedural generator called theme renderer. The impacts of our obtained results show the potential of deep learning methods used in procedural game map generation.

Keywords: Generative adversarial networks · Procedural generation · Convolutional neural networks

1 Introduction

It is necessary to design a great game map when you develop a great game. As the size of the game world is getting bigger and bigger, the dimension of game maps becomes larger and larger. Considering the huge cost of manually design the big game world, developers are more interested in programming game maps with the help of algorithms. However, the design of game map needs human ideas, which means it cannot be replaced easily by simple algorithms. It needs to consider the game-play design, the sense of vision and the style of the whole game. These are not easy to express manually by rules or codes. Therefore, a more reasonable approach is to use neural networks to learn these rules from a set of examples and to control the basement shape through Designers' interactive tools. Finally, the network generates detailed results depending on both the map style from examples and the shape from designers.

There are many methods for procedural generation, which can be divided into three kinds: rule-based generation, noise-based generation and example-based

---

This work supported by Sichuan Science and Technology Program 2019ZDZX0009, 2019ZDZX0005, 2019GFW1116 and 2018GZ0008.

generation. The first kind of methods usually use a set of manually designed generation rules like 'generate a **door** at a **wall** between two **rooms**'. Take dungeon generation as an example, the dungeon consists of a different set of different usage rooms, each room has regular and straightforward shapes. This makes rules can describe the dungeons. Rules-based approaches have been proved to be useful in dungeon generation. Noise-based methods are more popular for generating outdoor worlds, such as giant planets with complex terrain features. These methods can generate non-redundant scenery for the result, which is more suitable for terrain generation than rule-based method. The example-based approach has been used over the years, and its basic idea is to learn rules from a series of examples and produce new results. With the improvement of machine learning and deep learning, many developers use neural network as a powerful tool based on example method. Generating antagonistic network (GAN) is a network that can generate results based on a set of given examples without giving any rules manually, which is a powerful tool to accomplish this task.

The combination of these three approaches is a more useful solution for industrial use. We introduce a solution including cGAN (extension of GAN), convolutional neural network (CNN) and traditional procedural methods as postprocessing. In order to learn the design style of a given example, the neural network is trained by a large number of designed game maps. Also a conditional mask drawn by the designer will be used as the input of the cGAN to control the generation. As a result, the solution takes into account the randomization of the generated results, the play-ability area style of the game and the control of the designer.

In order to achieve our solution, we need a designer's basic mask. cGAN will use masks as input to generate a generated label map, each label is represented as a map element. Label maps will be processed by other layers of processors, such as CNN, which is used to connect each game area processor, or CNN, which is based on traditional processors used to generate terrain heights. Finally, the theme rendering program will be used to generate the final 2D game map or even 3D game map. All the above steps can be completed in 1–2 s. This solution enables designers to change results interactively by editing a part of the input mask map.

The main contributions are as follows:

1. We propose an interactive procedural game map generation solution based on Deep Learning. This solution can automatically divide the rooms and areas, and generate terrain and decorators based on a mask map.
2. We propose a group of neural networks contains cGAN and CNN in order to generate a label based game map.
3. We show a set of different testing in order to prove our method can be used as many different kinds of game map generation.

2 Related Works

Procedural content generation has been discussed in many papers. Article [6] contains an overview about different kind of procedural generation methods. We just make a summary of these methods by dividing them into three kinds: rule-based generation, noise-based generation and example-based generation. This part also introduces the related information of deep learning and neural network.

2.1 Rule Based Generation

Rule-based generation is based on the idea that the generation schedule can be described by a set of relevant rules, most of which are described as grammar trees. Each rule contains some operation and a set of sub-rules. By executing the rules from root to leaf, it can generate objects or modifies the generated results. This method can be used to generate game maps, dungeons, etc. Article [8] and [24] are examples of projects generated for role-playing games. For industry purposes, many games are powered by these methods such as Diablo and Darkest Dungeon. These methods usually require designers to build a lot of small parts of the game map and combine them according to manually defined rules. Also, Diablo also uses this method to generate names for inventory. These names are divided into prefixes, types and other parts, and then, use rules to generate names and the ultimate ownership of inventory, so that the inventory volume corresponds to its name. In another area, the L-system is a powerful tool for generating trees and plants.

2.2 Noise Based Generation

In order to generate randomized results, noise is a good choice. Perlin noise; for example, has been widely used in different procedural generation methods. Author in [25] shows a method for procedurally generating clouds with Perlin noise. Researchers have introduced many terrain generation solutions based on different kinds of noise combinations; for example, [28] are using noise functions to generate terrain with complex features. These methods tend to use noise functions as following steps: using a noise function as an original height map, which defines the height at each position of terrain, and then using many noise functions to modify the result in order to simulate the erosion. By computing on GPU, terrain can be generated in real time [23]. However, it is difficult to control the generation results directly, even if we can easily control the mixture method or mixture weight of different noise function. Therefore, even though the noise-based generation can generate good random results for terrain and cloud, it is still difficult to directly control the result generation based on noise.

2.3 Example Based Generation

Because designers want better control over the results, and they do not want to spend too much time on details, a reasonable approach is asking the user to make sketches to show the basic idea of their design and use methods to adding details from existing examples. This method is also widely used in terrain generation; for example, [30] considers this problem to be a patch matching problem. By matching the height field from real terrain into user-defined sketch, a new terrain contains complex detail has been produced. Methods allowing users to design the result interactively are called interactive terrain editing. Papers like [9,10] introduce good solutions in this kind of methods.

Traditional patch fitting methods may produce correct results in some cases, yet deep learning approaches can provide better results in example-based generation area; for example, [13] provides an interactive terrain editing system based on the conditional generative adversarial network. The network in this paper is trained based on a set of real terrain information and predicts the results in real time according to the designer's sketch.

2.4 Deep Learning in Procedural Generation

By the improvement of machine learning, deep learning approaches are widely used for generation. A really powerful network is Generative Adversarial Network [12]. By training two adversarial networks: generator and discriminator, the generator will get the ability for generating same style results from given examples. Since the discriminator is training during this process, it provides a gradient for optimizing generator. However, original GAN only takes a random input vector Z as the generator's input, which means the user can not control the results. Conditional generative adversarial network [22], an improvement of GAN, allows a user control by concatenating the user control input vector Y with original input Z as a new input vector. To use an image as input, a popular solution is using a U-Net [27] network architecture. This network is based on Convolutional Neural Network [19], which take one or more images as input, and processed by many convolution layers, then provide a final result. A combination of U-Net network and conditional GAN is Pix2Pix [16]. It can be used to take one image as input and generate another image based on the input. This feature makes this network suitable for procedural terrain generation, like in [13].

3 Overview

As shown in Fig. 1, our pipeline contains three different steps. The user's input is a mask map, where the white area represents the shape of the game area. This map will be given to conditional GAN(cGAN) as input. Then, cGAN will generate an image with many channels(usually more than one channel), which we call *LabelMap*. Each channel represents a kind of game-play label. Gameplay labels are an abstract representation of game-play element like the ground, mountain, tree and so on. We call cGAN generated game-play label map as *InitialLabelMap*. Then one or more convolution neural network(CNN) will use some label channels to process additional a label channel or modify a label channel. These CNN networks are called post-process CNN. Finally, a program called *ThemeRenderer* is used to convert abstract game-play label into real game elements in the game. Like we can calculate the height of mountains based on mountain label channel. Alternatively, we can produce random size trees on the tree label position. This allows us to generate 3D scenes based on the results although the networks learns from examples in 2D games.

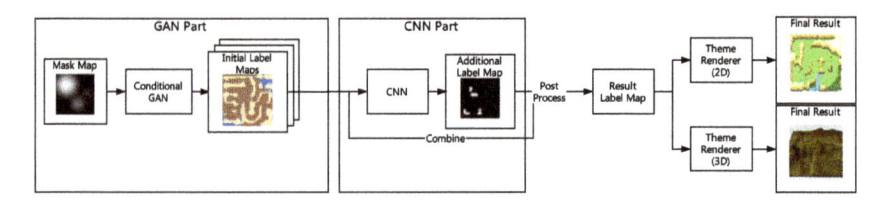

Fig. 1. Our system consists of three main steps. Designer draws a mask as input to conditional GAN, and then it generates an original label map with multi-channels. Each channel represents a label, such as water, ground or mountains. Then, the original label map will be processed by one or more CNN networks to modify the map or add additional maps. Finally, the label map will be rendered by the theme renderer, and different theme renderer will produce different results, such as 2D maps and 3D terrain.

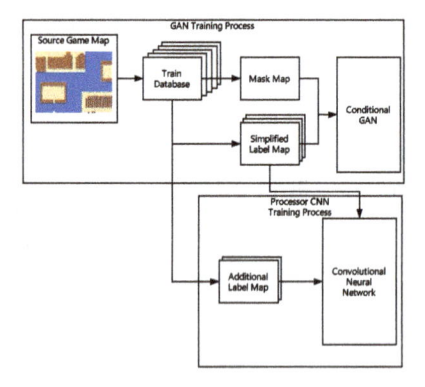

Fig. 2. The source game map is converted into a training database map, each channel containing a label data. Then it is simplified to a lower channel number map and a mask map for training conditional GAN(cGAN). Then, another label channel map is used as input to train a processor CNN network with a simplified label map.

Before use to produce results, we need to train these neural networks. The training steps are shown in Fig. 2. We use game maps from games on Nintendo Entertainment System (NES). NES game maps are usually constructed by tiles, each tile usually represents one kind of game element, which makes it easy to convert into a label map. We simplify the converted label map and then calculate the game-play area mask from labels. For example, take Ground Label Channel used as the game-play area. Then we use the game-play area mask as input and the corresponding label map as output to train the cGAN. For training CNN, it depends on the usage of this network, and details of the training process will be explained in the following sections.

4 Neural Network Architecture

We introduce two types of neural network architecture to generate label maps. In this section, we will discuss the structure of the network and the reasons for our selection.

4.1 Conditional Generative Adversarial Network

We use a network based on conditional GAN [22], which is based on the original Generative Adversarial Network [12] and the U-Net encoder-decoder framework [16,27]. Like we shows in Fig. 3, our cGAN contains two deep networks: U-Net encoder-decoder network generator G and Convolution Neural Network discriminator D. For each training, there will be a pair of images obtained from training database: Y is a mask image convert from the gamep-lay area, X is a simplified label image. The goal is to train G with real image pair (Y, X), and then it can use the input mask image Y to generate the output label map image $G(Y)$. The discriminator network D is trained to identify the given image pair (Y, X) and $(Y, G(Y))$ and give the result about which it is. This process can be described as optimizing both two network parameters to play this min-max game:

$$IoU = \frac{\sum_{ijk}\left[I\left(y'_{ijk} > p\right) * I(y_{ijk})\right]}{\sum_{ijk}\left[I\left(I\left(y'_{ijk} > p\right) + I(y_{ijk})\right)\right]} \quad (1)$$

In fact, in the original conditional GAN paper, it not only needs to input Y but also Z, which makes the formula contains conditional parameters. However, according to the suggestion from [13], we do not use the noise vector z and use dropout to provide randomize.

Author in [16] introduce a method to improve the training: slice the image pair (Y, X) and $(Y, G(Y))$ into a set of sub-patches with smaller size like 32×32, then use discriminator to discriminate each patch and combine them by this formula:

$$\sum_{(Y_n, X_n) \in (Y, X)} D(Y_n, X_n) + \sum_{(Y_n) \in Y} D(Y_n, G(Y_n)) \quad (2)$$

We also follow this advise and use a small number of patches since the size of our label map is small but channels are more than regular images.

We use a combination of mean squared error and categorical cross-entropy as loss function to train generator, and use binary cross-entropy as loss function to train discriminator. Since MSE measures the difference between the original map and generated results, it makes the generator provide images that close to a given mask. Our masks have the same shapes because they are converted from original label maps. Also, cross entropy measures the style difference which is

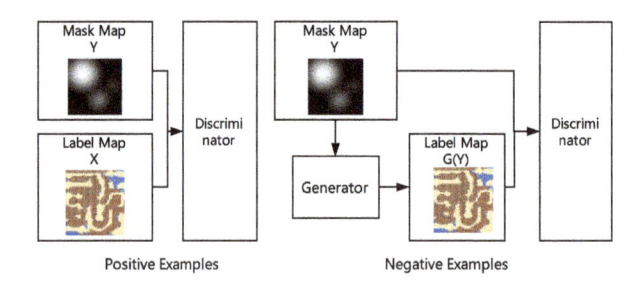

Fig. 3. Conditional Training overview: generator G takes Mask Map Y to produce $G(Y)$, discriminator D takes a pair of (Y, X) or $(Y, G(Y))$ and try to discriminate which it is.

discriminated by discriminator D. And We use RMSProp as optimize function. During our test, it provides better results than Adam, but the difference between them is not statistically significant.

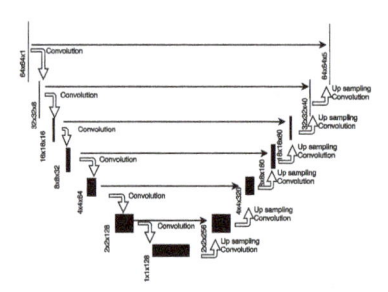

Fig. 4. The basic shape of our U-Net convolutional neural network. Compared with the original Pix2Pix paper, our layer is smaller.

4.2 U-Net Based Convolutional Neural Network

Since the generator network G and the processor CNN network shares the basic U-Net network architecture. The basic structure is in Fig. 4. we introduce them both in this part.

Basically, this network is an encoder-decoder network. However, as described in [22] and [16], U-Net concatenate the correspond size layer with lower layer's output as input for up convolution layer. This approach keeps the detail information of higher positions to avoid losing position information during encode and decode operations.

In our case, we adjusted the network from [16], but we did not use a large and deep U-Net structure because our label map's size was small (64 × 64), although it contained more channels. Pix2Pix architecture outputs a pixel-based image, which means that small differences do not really affect the results. However, our approach is to category labels, which means that small differences can have an better impact. For example, changing the red channel of a pixel by 0.1 does not really change the final result, but it may lead to a larger label channel value, which will provide different elements, such as changes from tree to water. Therefore, we intend to use a simpler network architecture to make the results more stable and accurate.

Figure 7 shows an example of CNN called *Connection CNN*. It shows the basic training and usage of this network. Details are explained in the next section.

4.3 Other Networks

We also build and test an another different network as a generator called Deep Convolutional Generative Adversarial Network (DCGAN) [26]. In some cases, people do not or can not control the generate result manually, but want to generate result from a random vector Z directly. For example, in some games, there are dungeons or areas generated each time when player comes in.

We use a network adjusted from DCGAN, and followed advice from Wasserstein GAN [7], and provide more stable training progress and get usable results successfully. In this network, the generator is a decoder with a randomize input vector Z and provide an output label map the same as generator G in cGAN. We limit the weight of discriminator in $[-0.01, 0.01]$ after each training, and then train D 5 times and G 1 time in each epoch. Finally, we use softmax function as the activation function, categorical cross-entropy as the loss function, and RMSprop as an optimizer.

This network can be used to replace the generator G in our pipeline, then it can be used as a random map generator. There is no need to change the rest of our pipeline.

5 Training

Normally our network's training database contains a set of image pairs (Y, X). Y is mask image and X is a label map. However, it is not easy to draw label maps directly manually, and it is not affordable for drawing a large number of maps by professional artists. So we choose to use game maps from NES games, such as Legend of Zelda. These old game maps consist of small tiles. Each tile looks like a small image representing game elements, or in our case, a label. But there are still many kinds of tiles, we need to simplify the original label database to produce more stable training. In this section, we use maps from Legends of Zelda to show the basic training progress of our networks.

5.1 Training Database Preparation

As shown in Fig. 5, the construction of training database is completed before start training the network. Training Database is built in the following steps:

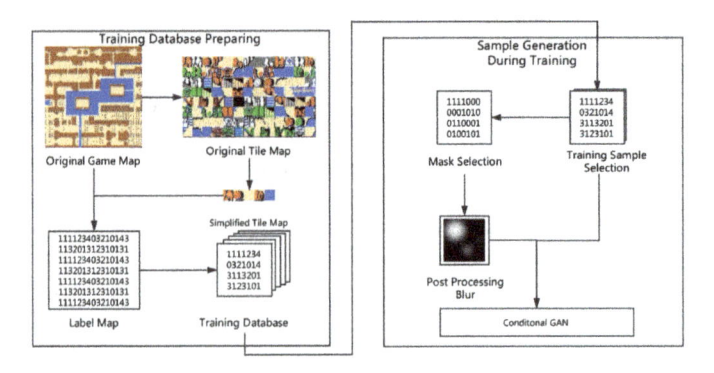

Fig. 5. This image shows the basic sample processing for training database construction and conditional GAN training. Main database processing is divided into two parts. The main database processing is divided into two parts. The first part is from the original game map to a set of label maps as training database, which is completed before training. The second part is the training input and goal from the selected samples to the conditional GAN in each epoch, which is done in each epoch.

- **Labeling.** Building a directory map containing each kind of 16 × 16 tiles. Each 16 × 16-pixel region is then marked with a directory index, which converts the result into a single-channel integer map.
- **Simplify.** Usually we have a large number of types for tiles, and we want to simplify them. In our case, we only use five kinds of tile types: Ground, Tree, Water, Mountain, Rock.
- **Slicing.** After conversion, we will have a basic label map. We want to slice the result into small parts so that they can be added to the training database. We converted the map into 64 × 64 patches, 32 × 32 patches, 16 × 16 patches. Patches smaller than 64 × 64 will be scaled into 64 × 64.
- **Rotate and Flip.** In order to add more training samples, we rotate and flip each patch.
- **Sample filtering.** We use a filter to separate some unusable patches, which do not contain any game-play areas.

Finally, we obtain a single-channel training database, which will be extended to a five-channel training database at run time.

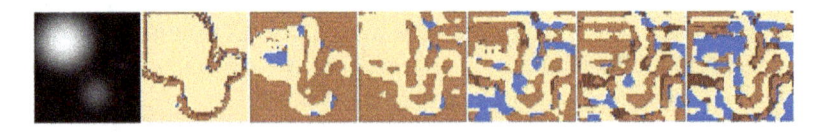

Fig. 6. This image shows the same mask (first image) as input and the results in training progress. Epoch number from left to right is 10, 100, 190, 290, 390, 550. It shows that the network can generate small rooms if the mask area is large.

5.2 Conditional GAN Training

For training the conditional GAN, we choose to generate the training samples on every epoch instead of before the training begins. We have tested preparing all of the training samples at the beginning. It takes up huge memory. So we decided to prepare this single channel 64×64 image as a sample database for middle results. In each training epoch, we only convert the images selected by this epoch into the two image pair lists (Y, X).

Mask Image (Y). In order to generate the mask image, we must first define the meaning of this mask. As we expand before, the designer draws this mask to define *Game-Play Area*. In our case, we define *Game-Play Area* as an area containing space for placing enemies, items, events and so on. It does not mean a room, because after the designer draws a large game area, we want to make the neural network automatically divide the area into small rooms. In conclusion, the game-play area defines a high-intensity empty room space.

After we select the channel of empty space (e.g. the ground layer), we must do some image processing operations to reduce the wall between the rooms and decorative objects (e.g. rocks). In practice, we use one level close operation and three level open operation to reduce paths, small holes and decorative objects. Then we use a sigma = 5 Gaussian filtering to blur the results. For validation, we use a manual mask image to test the network during the training progress. Figure 6 shows an example.

Label Vector Image (X). Each label is treated as categorical vectors rather than an integer. Therefore, in each epoch, the selected one-channel image is converted to a categorical vector image. Theme rendering will uses the maximum value channel to be the category. That is to say, the network can even produce negative output vector, and the theme renderer still produces results.

Training Details. We trained each epoch on 10000 selected label image from the training database, and the batch size is 250. We trained 1000 epoch for results. The discriminator's learning rate is 0.0001 and the generator's learning rate is 0.0005.

5.3 Connection CNN Training

We use the example of connection CNN to explain the idea of training CNN. There can be different usage of CNN to post-process the output from conditional GAN. One of the important CNNs is *Connection CNN*, which is used to connecting areas as much as possible. In conditional GAN, we removed small paths and blur empty spaces for the mask, which make the mask lacks small details. GAN can generate different areas like mountains, lands, waters based on the mask but it is hard to generate paths and connections. So we used another CNN to generate them. Like we show in Fig. 7, it takes the output from GAN

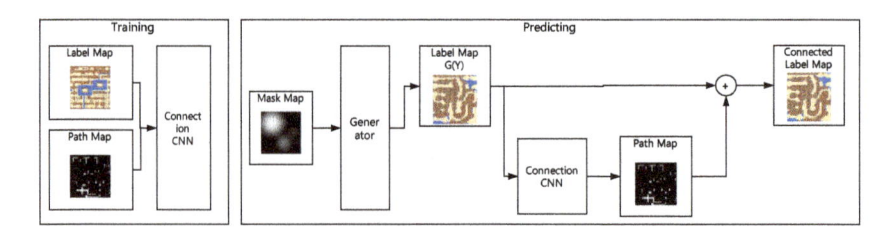

Fig. 7. *Connection CNN*, an example of post-processing CNN network. It takes a label map and a manually marked path map to train. And it takes conditional GAN's output as input to generate a path map. The path map is added to the generated result in order to connect areas.

and produces a one-channel image, which is called *Path Map*. These areas can be connected by adding a path map with manual control weights to the output ground channel.

We are not using the Pix2Pix framework since it is not a style transfer problem. The output of this CNN is not the same as the input. So we using an Encoder-Decoder U-Net CNN and use a binary cross-entropy as loss function to train this network directly. We manually marked the path in the game map to create training samples. Because the path map database contains smaller information and most area is black, we do not use image patches smaller than 64×64. We suggest limiting the value range of the path map to $[0.01, 1.0]$, rather than $[0.0, 1.0]$, which will provide more stable and faster training.

6 Authoring

We will talk about the authoring and theme rendering progress in this section. The authoring progress is shown in Fig. 1. This progress contains these steps:

- **Mask Capturing.** A web-based sketcher captures the designer's input and convert it into 0-1 mask image. It will be transferred to the back end to do the next steps.
- **Pre-Processing.** Pre-processing is applied to the mask contains blur and some other operation.
- **Network Processing.** Each network processes the previous output. The first network is always conditional GAN. And then it is post-processed by other networks like connection CNN.
- **Post Processing.** The results will be post-processed by traditional methods to avoid some strange results. Also, it combines outputs of all networks. This step will output a multi-channel label map.
- **Theme Rendering.** This process will select a theme renderer to generate images or 3D scenes based on the output label map.

So usually the designer's working progress is like this:

1. Draw a mask
2. Watch the theme renderer's result, which is sent to the website in 1–2 s.
3. Change some part of the mask.
4. Back to step 2.

6.1 Network Processing and Combine

After getting the mask image from designer, we convert that into floating-point format and apply a blur kernel on that. Then the blurred image is sent to conditional GAN to obtain an initial label map output, which will be sliced into different channels and used by other networks.

There are two different kinds of label maps, one is raw data from conditional GAN, whose value range much larger than $[0.0, 1.0]$, and the other is a clamped map. Which is these will be selected depends on the type of post-processing CNN. In our example, the connection CNN will use the range of values in $[0.0, 1.0]$. This value can be scaled the by divide the maximum number of all these values, or can be directly clamp in $[0.0, 1.0]$, which is our choice. However, the first method may show more information about the prediction info from conditional GAN because the ratio remains unchanged. Figure 8 compares the results before and after the connection CNN. The connection CNN take 'Ground' channel from output, and use that to generate a path map. We usually multiply the path map with a controlling weight and add it directly into 'Ground' output channel. The output from connection CNN sometimes contains a large number of writing areas, which is definitely not path. So we used an operation to remove this parts:

(Matrix B is 3×3 identity matrix.)

$$\begin{aligned} C &\leftarrow A \ominus B \\ A &\leftarrow Clamp(A - C, 0.0, 1.0) \\ A &\leftarrow Blur(A \oplus B) \end{aligned} \quad (3)$$

A is the output from path CNN, and B is the operation kernel.

The controlling weight is used to control the probability of paths. We used 100 and clamp them back to $[0.0, 1.0]$, which provided a very clear boundary for the path.

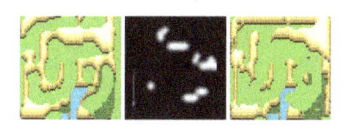

Fig. 8. This image shows one example of original output from conditional GAN and the connection CNN processed result. The first image is the original output. The second image is the path image produced from connection CNN, the third image is the connected areas. We can see the first area is not connected, and the final result is connected. The first and the final image is rendered by 2D height field theme renderer.

6.2 Theme Renderer

A theme renderer takes the final output label map as input, and generates an image or a 3D scene as output. At present, we have implemented a 2D theme renderer similar to Zelda style, a 2D height field theme renderer and a 3D terrain renderer. The 2D Zelda style theme renderer is used as a debugger to visualize each label. This theme renderer generates label map images. The 2D height field theme renderer uses a more detailed tile to draw and calculates the height of the mountain. Finally, we use World Machine to generate a complex terrain based on the height field to show it can be used to generate 3D game scenes.

Given a 2D height map from output without height information, we can calculate the height result by the Algorithm 1.

Algorithm 1. Calculate height result
---
set $H \leftarrow 0.0$
set $A \leftarrow Map_{mountain\ channel}$
while A has non-zero value **do**
    $H \leftarrow H + A$
    $A \leftarrow Binary\_Erosion(A)$
end while

A 3D theme renderer can take one or more channels to build a 3D scene. For example, it could take an output image from 2D height map renderer and build a 3D terrain by applying more operations like modify and erosion, as shown in Fig. 9.

7 Results and Discussion

The training database building was implemented in Python. The network has been implemented in Keras [4] based on TensorFlow backend. The training was performed on an Nvidia GeForce GTX 1080Ti with 11 Gb memory. We used a web-based designer front end based on HTML5 canvas and Ajax. However, since we can send JSON post from any application, so it can be replaced into others. The interactive designing system runs on a standard desktop computer with an Intel Core i5 CPU clocked at 3.00 GHz and with an Nvidia GeForce GTX 1050Ti. We use world machine 2 to build 3D terrain in order to show the 3D scene.

In this section, we will introduce the usage of this system, show the results of cGAN, WGAN, CNN, and the performance of our system. We will also talk about the limitation and the future works. Also we show the results in Figs. 13 and Fig. 14.

Fig. 9. A 3D scene theme renderer can use height map output from 2D height field renderer and then build a 3D scene. First image shows the 2D height field theme renderer's output, the middle image shows the height map dumped from the output. We build an algorithm in World Machine to produce the right image which shows a detailed terrain based on the height map. We do not use the path map to affect the height map in this case since we can consider the path map as a cave.

7.1 Usage

According to our goal, designers can use the system as a quick sketch game map designer. According to the masked image of the game area of the designer, the system can be used to generate complex game maps with multiple types of areas. After iteration, when the designer thinks it is good enough, the system can export different label channels to images and use them in other procedural generation systems to produce the final complex game scenarios. Because the steps from the label channel image to the final complex game scenario take time, our system allows designers to iterate quickly before making the final step.

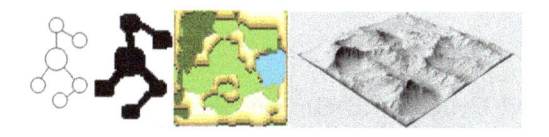

Fig. 10. Designers can use our system to fast iterate the map design. They think considering the map as a connected graph of the game-play areas, then use a sketch to draw this graph, and the network produces a 2D game map. If they decided, another procedural system produces a detailed 3D game scene, which may take more time. In this case, it took 1 min 8.8 s to generate.

We choose to use *Game-play Area* instead of terrain sketch like [13], because we think for game designer especially level designer, are more likely to view the entire game map structure as a connection map. They want to define game-play areas and connect them, and then they want to view game maps based on this structure. We show this progress in Fig. 10. However, for game artists like level artists, [13] is definitely a good solution to build complex terrain.

7.2 Design Pattern Learned by Network

We hope that our neural network can learn the design pattern from the given examples. In our case, maps of Legends of Zelda actually are rooms connected with small path and holes. In each room, there are enemies, NPCs and items. This means the map is actually not a realistic style map from the real world but a designed map based on the game-play need. Usually, designers want to design maps which encourage players to explore, fight or rest. This requires that the map divided into individual rooms. Each room has its own functions. For example, a fighting area contains an enemy, or a safety room contains recovery items and saving points. Then the rooms are connected with other rooms by small holes or narrow paths. Enemies usually not pass these connections. It reduces the complexity of enemy AI and gives the player a chance to recover. The network has successfully learned this pattern and produce the same feature. We can see the rooms in the generated map like we show in Fig. 11. Sometimes, the network may even produce a "secret area" of a narrow door.

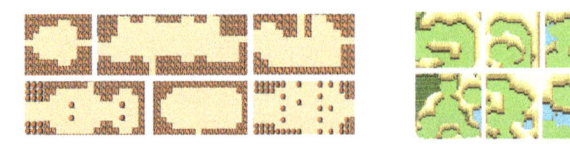

Fig. 11. Images in the first two raw are from maps of Legend of Zelda. It shows the room based design pattern. Rooms are individual areas connected by narrow paths. Images in the 3 and 4 row are the results of our network, which shows it successfully learned the pattern. We can recognize the rooms in these maps and they are connected by generated paths from conditional GAN and connection CNN.

7.3 Label Based Procedural Generation

We choose to generate label maps instead of results directly. Here are the reasons:

Training Stable. In our test, according to our label-based solution, it is really difficult to get good results for images generated directly from conditional GAN at 1024 × 1024 resolution. This is because the network needs to go deep enough to understand the meaning of each tile first, and then the relationship between the tiles. In fact, the design patterns which we want the network to learn is the relationship. Reducing the size and number of blocks will make the training remarkably stable and achieve better results.

Interact with Traditional Procedural Solution. There are many kinds of image-based procedural solution for generating game assets. By exporting specific kind of labels as an images, more complex scene can be generated. For example, we can use world machine 2 to generate terrain based on mountain label image. Or we can generate foliage based on tree label image. The label map is a guide for other procedural methods. This avoids networks to generate a huge map with complex details because it is tough.

Instead of generating the result label map in one network, we choose to use multi-networks. The label is a public data sharing by each network. Some networks can even produce additional labels. In our test, the more label kinds, the more unstable the network is, so it is better to reduce the label kinds and generate more labels step by step.

7.4 Comparisons

In this section, we will compare our methods with other methods in different ways, like network architecture and other traditional or neural network procedural solutions. We also test the differences caused by the number of label kinds.

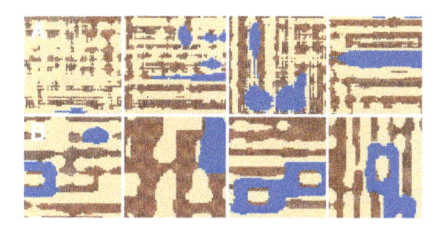

Fig. 12. Comparison between WGAN and CGAN. Upper raw contains results from WGAN output. Lower raw contains results from CGAN output.

Network Architecture. We tested the network with WGAN solution and conditional U-Net GAN solution. In fact, in the same number of label kinds, WGAN can also produce receptive results. We show the comparison in Fig. 12. The results look repeat without human control input. Also, we tested the encoder-decoder framework for DCGAN without U-Net. The control of input image looks little effect on the final results.

Other Traditional Methods. There are many other procedural ways to generate game maps or scenes. Methods without a neural network can be seen in [5]. Two most popular ways are as follows:

- **Grid Dungeon Generation** Generate a set of box rooms, and them move them until there is no collision. Then using paths to connect these rooms.
- **Cellular Automata** Using cellular automata to generate rooms with more randomize and complex dungeons.

Images for these methods can be seen on [2] and [3]. Our methods can generate rooms contains hard edges like the first method or rough edges like the second method. And almost both of these methods need manually define generation rules, but the difference lies in the type of rules. Directly controlling the generation results is not easy in these methods.

There are also industry software. One of them is Dungeon Architecture [1], which is a hybrid of grid dungeon generation and marker-based generation. It first generates dungeons based on the first method and then generates markers by human-designed rules. This software can produce a good result for dungeons. However, if build complex outside world, it needs a highly skilled designer to code for complex generation rules. This inspires the idea of label-based generation.

Neural Network-Based Methods. There are also many neural network-based methods. One of that is [13], which inspires us for using conditional GAN and U-Net framework. This method mostly focuses on generating terrain based on user's input image. This image defines the rough sketch of terrain. Then neural networks produce the large and detailed terrain shape. Our approach does not directly focus on terrain generation but can generate more than that. And also our input is not the rough sketch but the game-play area. We consider the usage is for game map designer to iterate the map fast, and other methods are for level artist to make terrain. And finally their output is a 3D terrain model but ours is an abstract label map for using in other processes.

7.5 Performance

Database and Training. We build the database convert from Legend of Zelda 1 on NES. The database contains 23040 64×64 samples, 3920 32×32 samples and 980 16×16 samples. All of these samples are scale into the same 64×64 size. The database is prepared in 30 min 20 s.

Our training label map is 64×64 resolution containing 5 channels. We use the integer format to store the label value and expand dynamically during each training epoch. The training is in 16 h with 1000 epoch in GTX 1080 Ti.

Run-Time Performance. In each update request, our system's performances are as follows:

- **Request Prepare Time: 0.004 s.** Contains the time getting from the designer front end and translate into the input image with blur filter.
- **conditional GAN Time: 1.468 s.** From conditional GAN which takes input and produces an output label map in 64×64 contains 5 channels.
- **connection CNN Time: 0.144 s.** From taking the output from conditional GAN and produce a path map in 64×64 contains one channel.

7.6 Limitations and Future Work

Our system still has some limitations, like the follows:

1. Areas inside the small rooms does not contain many details. It lacks elements. We think this can be fixed by another decorator CNN to produce decorators like rocks and trees inside the empty rooms.

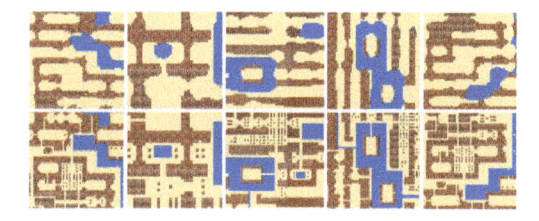

Fig. 13. This image shows the conditional GAN output and the ground truth. For each column, the upper one takes the lower one's game-play area, and generate an output, the lower one is original map clips from train database.

Fig. 14. This is more generated results based on the randomly chosen input. It shows even regular input can generate complex results with different area types. We can also recognize a room based map design pattern in these results.

2. 3D theme renderer cannot be produced in real time. The most time cost thing is the traditional erosion operation to produce complex details for terrain map. This can be fixed by integrating a system like [13].
3. Side running games like Super Mario do not fit by our cGAN system. It cannot be described easily with game-play areas. However, our WGAN system can still support this type. How to integrate human control into this type of game is still a question in our system.
4. By using some real-time methods like [11,17,18,20,21], the system's performance can be improved. Also, many new network architectures can be used to in this system like [14,15,29].

7.7 Conclusion

This paper introduces a framework for procedurally generating game maps based on human control and given example samples. The framework can learn map design pattern from given examples, take human input control into account, and generate an abstract label map with the same pattern type. This process can be done in real time.

Designers treat the game map design as draw game-play areas and connect them, and then will get images contains ground, mountains, trees and water areas. Finally, we introduce how to use Theme Renderer to convert the label map into 2D maps and 3D scenes, which shows how this system interacts with transitional procedural generation systems to produce high detailed results. We have explained the details about the neural network used in this system. And we hope to show the potential of using Deep Learning methods in procedural content generation.

References

1. Content detail
2. Create a procedurally generated dungeon cave system
3. Generate random cave levels using cellular automata
4. Keras: The python deep learning library
5. Procedural content generation wiki
6. Procedural generation, December 2017
7. Arjovsky, M., Chintala, S., Bottou, L.: Wasserstein gan. arXiv preprint arXiv:1701.07875 (2017)
8. Ashlock, D., Mcguinness, C.: Automatic generation of fantasy role-playing modules. In: 2014 IEEE Conference on Computational Intelligence and Games (2014)
9. Cordonnier, G., Galin, E., Gain, J., Benes, B., Guerin, E., Peytavie, A., Canie, M.-P.: Authoring landscapes by combining ecosystem and terrain erosion simulation. ACM Trans. Graph. **36**(4), 134 (2017)
10. Gain, J., Marais, P., Straßer, W.: Terrain sketching. In: Proceedings of the 2009 Symposium on Interactive 3D Graphics and Games, pp. 31–38. ACM (2009)
11. Gentile, C., Li, S., Kar, P., Karatzoglou, A., Zappella, G., Etrue, E.: On context-dependent clustering of bandits. In: Proceedings of the 34th International Conference on Machine Learning, vol. 70, pp. 1253–1262. JMLR. org (2017)
12. Goodfellow, I., Pouget-Abadie, J., Mirza, M., Xu, B., Warde-Farley, D., Ozair, S., Courville, A., Bengio, Y.: Generative adversarial nets. In: Advances in Neural Information Processing Systems, pp. 2672–2680 (2014)
13. Guérin, É.: Interactive example-based terrain authoring with conditional generative adversarial networks. ACM Trans. Graph. (TOG) **36**(6), 228 (2017)
14. Henderson, P., Islam, R., Bachman, P., Pineau, J., Precup, D., Meger, D.: Deep reinforcement learning that matters. In: Thirty-Second AAAI Conference on Artificial Intelligence (2018)
15. Imani, M., Ghoreishi, S.F., Allaire, D., Braga-Neto, U.M.: MFBO-SSM: multi-fidelity Bayesian optimization for fast inference in state-space models. In: Proceedings of the AAAI Conference on Artificial Intelligence, vol. 33, pp. 7858–7865 (2019)
16. Isola, P., Zhu, J.-Y., Zhou, T., Efros, A.A.: Image-to-image translation with conditional adversarial networks. arXiv preprint arXiv:1611.07004 (2016)
17. Kar, P., Li, S., Narasimhan, H., Chawla, S., Sebastiani, F.: Online optimization methods for the quantification problem. In: Proceedings of the 22nd ACM SIGKDD International Conference on Knowledge Discovery and Data Mining, pp. 1625–1634. ACM (2016)
18. Korda, N., Szörényi, B., Shuai, L.: Distributed clustering of linear bandits in peer to peer networks. In: Journal of Machine Learning Research Workshop and Conference Proceedings, vol. 48, pp. 1301–1309. International Machine Learning Society (2016)
19. Krizhevsky, A., Sutskever, I., Hinton, G.E.: Imagenet classification with deep convolutional neural networks. In: Advances in Neural Information Processing Systems, pp. 1097–1105 (2012)

20. Li, S.: The art of clustering bandits. Ph.D. thesis, Università degli Studi dell' Insubria (2016)
21. Li, S., Karatzoglou, A., Gentile, C.: Collaborative filtering bandits. In: Proceedings of the 39th International ACM SIGIR Conference on Research and Development in Information Retrieval, pp. 539–548. ACM (2016)
22. Mirza, M., Osindero, S.: Conditional generative adversarial nets. arXiv preprint arXiv:1411.1784 (2014)
23. Olsen, J.: Realtime procedural terrain generation (2004)
24. On, C.K., Foong, N.W., Teo, J., Ibrahim, A.A.A., Guan, T.T.: Rule-based procedural generation of item in role-playing game. Int. J. Adv. Sci. Eng. Inf. Technol. **7**(5), 1735 (2017)
25. Qi, Y., Shen, X.-K., Duan, M.-Y., Cheng, H.-L.: A method of rendering clouds with perlin noise [j]. Acta Simulata Systematica Sinica **9**, 023 (2002)
26. Radford, A., Metz, L., Chintala, S.: Unsupervised representation learning with deep convolutional generative adversarial networks. arXiv preprint arXiv:1511.06434 (2015)
27. Ronneberger, O., Fischer, P., Brox, T.: U-Net: convolutional networks for biomedical image segmentation. In: International Conference on Medical Image Computing and Computer-Assisted Intervention, pp. 234–241. Springer (2015)
28. Schaal, J.: Procedural terrain generation. A case study from the game industry. In: Game Dynamics, pp. 133–150 (2017)
29. Wang, C., Bernstein, A., Le Boudec, J.-Y., Paolone, M.: Explicit conditions on existence and uniqueness of load-flow solutions in distribution networks. IEEE Trans. Smart Grid **9**(2), 953–962 (2016)
30. Zhou, H., Sun, J., Turk, G., Rehg, J.M.: Terrain synthesis from digital elevation models. IEEE Trans. Vis. Comput. Graph. **13**(4), 834–848 (2007)

Example of the Use of Artificial Neural Network in the Educational Process

Suleimenov Ibragim[1,2], Bakirov Akhat[1(✉)], Matrassulova Dinara[1], Grishina Anastasiya[3], Kostsova Mariya[4], and Mun Grigoriy[5]

[1] Almaty University of Power Engineering and Telecommunications, Almaty, Kazakhstan
esenych@yandex.ru, axatmr@mail.ru
[2] Institute of Information and Computational Technologies, Almaty, Kazakhstan
[3] V.I. Vernadsky Crimean Federal University Sevastopol Institute of Economics and Humanities (Branch), Sevastopol, Russia
[4] Sevastopol State University, Sevastopol, Russia
[5] Al-Farabi Kazakh National University, Almaty, Kazakhstan

Abstract. An example of an artificial neural network intended for use in the educational process (in such disciplines as "The socio-political importance of artificial intelligence systems", "History and philosophy of science", etc.) is presented. The neural network provides automatic processing of critical reviews written by students for pseudoscientific works, presented in abundance in the current periodical press. This makes it possible to transfer such an innovative form of study as the writing of critical reviews by students to the distance learning mode. An additional function of this neural network is testing of students in order to identify individuals with a psychological type that is appropriate to the scientist in the true meaning of the word.

Keywords: Hirsch index · Pseudoscience · Critical thinking · Passionarity · Neural network · Profanation of science

1 Introduction

Currently in Kazakhstan there is a clear oversupply of higher education institutions [1]. At the beginning of the 2017/2018 academic year, 127 higher educational institutions operated in the republic with a population of about 18 million (or 1 university for a little over 140 thousand people). For comparison, in countries that are recognized leaders in the field of higher education, there are 650–700 thousand people per university [2]. For example, in the UK there are 89 universities for 60.4 million people, in Finland for 5.2 million - 20 universities. The excess of higher education institutions in Kazakhstan is partly due to objective reasons, since the mass consciousness continues to see higher education as a social elevator.

The massive nature of higher education (university students are more than 50% of those of the corresponding age group [3]) leads to a sharp drop in the quality of educational services, and to a significant drop in the level of university science. Attempts by the Ministry of Education and Science of Kazakhstan to rectify the situation, unfortunately, often cause the opposite effect, at least, if we speak about the effectiveness of research activities in Kazakhstan's universities. In particular, the introduction into widespread use of such a scientometric indicator as the Hirsch index led to a sharp increase in the number of pseudoscientific works published by university professors for the sake of formal reporting, for the sake of membership in dissertation councils, the ability to qualify for budget funding, etc. [4–6].

Improving the quality of scientific works performed by university professors in the current conditions is a very urgent task, since the primary focus on formal indicators not only leads to an increase in the number of pseudoscientific publications [6], but also the use of relevant materials as the basis for graduation, master's and doctoral dissertations that de facto corrupts students and serves as a breeding ground for corruption.

However, numerous pseudoscientific works presented in the current literature can be considered, including, as a well-defined resource for developing critical thinking skills [4, 5]. Namely, the writing of critical reviews is a promising form of academic work, especially if we take into account that under current conditions, writing essays, which traditionally remain an important part of the educational process, has already become a profanation [4].

The purpose of this work is to develop an artificial neural network (ANN) that provides a parallel solution to the following tasks: the development of critical thinking among students, countering the growth of pseudoscientific publications and dissertations (both master's and doctoral), as well as identifying among students with pronounced passionarity and other personal qualities necessary for the implementation of fruitful scientific activities in modern conditions. It is assumed that the developed ANN will be the basis for the artificial intelligence system of the same purpose.

2 Program Development and Results

Recall that by Gumilev [7] refers to passionarity as a characterological dominant, an unbreakable inner desire (conscious or, more often, unconscious) for activities aimed at the realization of a goal (often illusory).

As experience with young scientists shows, in modern conditions passionarity is often much more serious than formal competences (for example, there are plenty of examples where a budding young scientist threw science at the occurrence of some or other obstacles) the factor of passionarity for achieving success in science is demonstrated by its entire history [8].

Thus, Tycho de Brahe, who spent all his personal fortune on astronomical exploration, as well as many other Enlightenment figures who are ready for many personal sacrifices to attain Truth, should be referred to the clearly expressed passionaries.

The developed ANN is focused on analyzing the content of critical reviews written by undergraduates on pseudoscientific works. In developing assignments, emphasis was placed on criticizing pseudoscientific works, however, neural networks based on this principle can also be used to analyze critical reviews of publications of any type (i.e., including truly scientific, but containing some miscalculations or elements conscientious delusions).

Critical reviews written by undergraduates of the Almaty University of Power Engineering and Telecommunications (AUPET) were used as the primary material for the development of the ANN as part of the practical tasks of the innovation discipline "Artificial Intelligence as a Driver of the Fourth Technological Revolution". Performance of just such tasks is provided for by the work program for this discipline, developed in parallel with the textbook [9].

As an experiment, the task was performed remotely, which allowed to fully cover the entire flow of the Institute of Space Engineering and Telecommunications AUPET.

The analysis of the texts of the reviews allowed us to choose groups of words, phrases and speech turns, expressing the degree of critical thinking on the text of the reviewed publication. Review analysis also showed that undergraduates are clearly divided into several groups according to the degree of readiness to critically interpret publications in the open press, according to the degree of readiness to defend their point of view, etc. Direct analysis of the texts also showed that there is a well-defined correspondence between the nature of the words used, phrases and speech patterns, and the categories to which students can be attributed.

It should be noted that the question of assessing psychological qualities in terms of identifying the propensity for fruitful scientific work is not for the first time put. Thus, an experimenter who is an adherent of a scientific hypothesis often unknowingly introduces a distortion in the interpretation of the data, providing confirmation of this hypothesis. This effect is called the "Pygmalion Effect" (this is a special phenomenon in psychological science, which consists in the fact that a person's expectations regarding another object or situation largely determine the specifics of his own actions, interpretation of the behavior of others, which allows these expectations to be realized).

One of the professional competencies of a researcher is the ability to create an image of a future research, the so-called ability to anticipate, predict. This contributes to the systematization and structuring of scientific ideas, as well as the possibility of its implementation. It is also appropriate to emphasize that an important tool of scientific knowledge, along with analysis, synthesis, classification, induction, deduction, is also reflection. A.V. Karpov defined reflection as meta-ability: "Reflection ... manifests itself as the representation in the human mind of the mechanisms and forms of voluntary control over the processes of information generation, its development and functioning. ... The ability to reflect can be understood as the ability to reconstruct and analyze a plan understood in a broad sense to build your own or someone else's thoughts; as the ability to distinguish in this regard its composition and structure, and then to objectify them, to work out according to the goals set" [10].

You can talk about the so-called professional reflection as the process of implementation of the internal mental activity of the subject - the reflective activity in relation to scientific activity, the subject of scientific knowledge, which is based on professional activity as an ergatic system, which is represented by a variety of subject-object and self-subject relations.

The subject of analysis of professional reflection are: "The image of the object of professional activity" and "The image of the subject of professional activity." The psychological criterion for the success of scientific research is the formation of an attitude towards oneself as the subject of a chosen activity: the professional and emotional orientation of an individual determine each other.

The researcher realizes emotional intentions, which, as a rule, are unsaturated. According to B.I. Dodonov, it can be a whole range of emotional states:

- altruistic: the scientist wants to help humanity;
- communicative: research for the implementation of scientific dialogue, the search for interdisciplinary connections;
- gloric: studies for recognition;
- praxical: dedication to this business;
- frightening: overcoming various obstacles that lie in the way of the research;
- romantic: the pursuit of science to all that is unknown;
- gnostic: the desire to find something new in the familiar, to create your innovative product;
- aesthetic: a sense of the beauty of the truth of the study;
- hedonistic: science creates mental comfort;
- acquisitic: the accumulation of theories, attitudes, principles, etc.

The listed personal aspects of the researcher are determinants of true science, the absence of these personal indicators leads to pseudoscience.

The classification according to the degree of critical attitude towards pseudoscientific texts is presented in Table 1, examples of indicative words and phrases are also presented there.

Table 1. Classification of indicative words and phrases according to the degree of critical attitude of students to the material of pseudoscientific work.

Classification/description of the feature	Characteristics of used expressions/examples of indicative phrases
Understanding the content of the work while rejecting the very idea of criticism/The rejection of the question of the need for criticism from the "younger"	The desire of the author to focus on the positive aspects of the work *The material is presented logically, the results were reviewed, studies were conducted, the results show, these options allow, the conclusions reflect, the conclusions are justified, there is no doubt about the relevance, etc.*
Departure from answer/expressed desire to get away from a sharp discussion of the shortcomings; understanding of the importance of criticism of flaws in the desire not to carry out this criticism personally	Statements with positive semantic color dominate over negative ones (positive ones have a tinge of doubt, negative ones reflect a desire to criticize insignificant flaws) *- It would be, in my opinion, it would be possible, the average scientific value, but the authors did, the article is consistent, the topic is relevant, the relevance of the article is beyond doubt, meets all the requirements, problems in the article are solved, the article is structured, framed in accordance with the requirements, material it is stated logically,*

(*continued*)

Table 1. (*continued*)

Classification/description of the feature	Characteristics of used expressions/examples of indicative phrases
	the article is written in an understandable and accessible language, can be used as theoretical material, etc. *- the topic was not disclosed, not enough informative spelling errors, grammatical errors, goals were not achieved, the article needs some work, the output is not meaningful, is not fully disclosed, there is no novelty, as a remark, no references are indicated, it requires serious work, not recommended for publication, translation inaccuracies, distant relation to the text, no conclusions, negative aspects, not fully disclosed, etc.*
Soft criticism/Ready for critical assessments of flaws in soft and smooth form.	Negative expressions prevail, focusing on the criticism of minor flaws. Positive ratings are also present, but less common *Not fully disclosed, no novelty, as a remark, no references are indicated, no literature is indicated, difficult to use the article, completely inconsistent with the content, the language of the article is not clear, the topic is not disclosed, the article needs to be improved, the conclusion is inconsistent, the goals are not achieved, there is no novelty, the author could not, do not disclose the whole scientific nature, the calculations are not justified, requires serious improvement, not recommended for publication, distant relation to the text, no conclusions, negative sides, not fully disclosed, etc.*
Tough criticism/readiness for harsh criticism of shortcomings, readiness to defend one's opinion	Reviews contain ONLY negative expressions: *Not recommended, incomprehensible, unclear, problem not disclosed, inconsistent, illogical, does not reveal the essence, does not satisfy the requirements, I do not recommend publishing, easy reading, strange manner of presentation, a number of comments, no logical conclusion, the main article is missing, the article requires self-analysis of the article requires complete substantive processing, etc.*
Misunderstanding of the essence of the reviewed work	Direct retelling of the article's test lack of indicative words and phrases

The revealed correspondence allows using well-known methods for the construction and training of ANNs [9, 11]. For the training of the neural network, reviews obtained in the course of the experiment described above were used. In parallel, for the purpose of control, students were surveyed, providing additional confirmation of their ability to develop critical assessments.

An example of one of the questionnaires is presented in Table 2. In the preamble of the questionnaire the respondents (undergraduates) were provided with the following information, which was further clarified during practical exercises.

The article "Academic unworthiness of 2018" was published on the website of the Internet resource Mirror of the Week (Ukraine). It reflects the results of the "anti-competition" conducted by the Ukrainian colleagues. This "anti-competition" is based on the idea of an outstanding physicist S. Sharapov, who proposed to distinguish academically dishonest scientists, giving them a kind of "black mark". Thus, the scientific community acquires the ability to control "its own territory". The idea was brought to life by Ukrainian activists of the Dissergate initiative; The first such "award" took place in 2016.

Respondents were asked to select only one of the presented answers, each of which corresponds to the "black marks" awarded by Ukrainian colleagues in the nominations listed below.

Table 2. The list of questions in the questionnaire proposed to undergraduates during the survey and the absolute number of undergraduates N who chose this answer.

№	Possible answer	N
1	"Plagiarist of the Year"- a black mark is sent to colleagues and officials who most shamelessly used other people's results	9
2	"Pseudo-scientist of the year" - nomination for those who published and defended the most malicious nonsense	5
3	"Murzilka year" - the magazine is awarded, which most eagerly published pseudoscientific works (most often for cash)	7
4	"Scandal of the Year" - for the loudest scandal in the scientific community	1
5	"Posipaka year" - awarded to organizations (for example, academic councils) for "significant contribution" to the production of plagiarists, falsifiers and fabricators	3
6	Kazakhstan is not Ukraine, we definitely don't need such "anti-competition"	8
7	I do not consider myself entitled to pass judgment on this issue, undergraduates should not even think about criticizing elders	5

The distribution of responses received in percent is presented in Fig. 1.

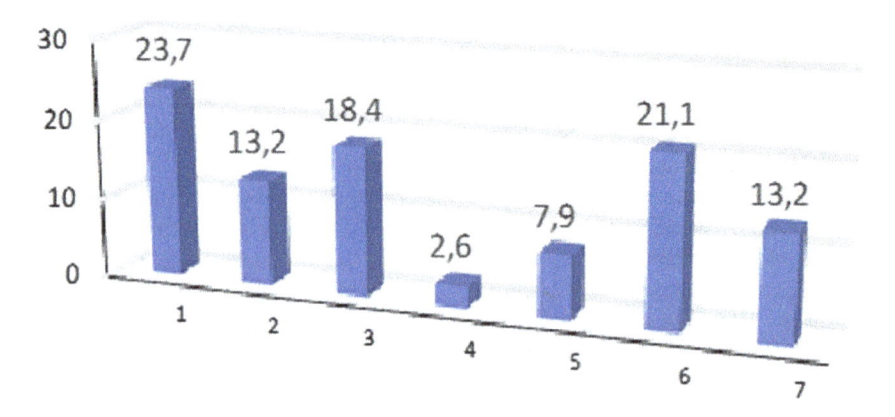

Fig. 1. The percentage distribution of answers on the options.

A trained neural network allows classifying students according to the signs that determine their readiness for scientific work. We emphasize that in modern conditions one of the most essential features is the willingness to overcome obstacles (created by both the conservatism of a significant part of the teaching staff and the bureaucratic routine) and other character traits, which together form a passionate psycho type [12].

To implement the artificial neural network of the proposed type, the programming language Python was used, with the open neural network library for deep learning Keras. It is a superstructure over the Deeplearning4j, TensorFlow and Theano frameworks. This library is aimed at efficient work with the networks of in-depth training, at the same time designed to be compact, modular and expandable. It was created as part of the ONEIROS research effort (the Open-ended Neuro-Electronic Intelligent Robot Operating System) [13]. TensorFlow was used as a backend, which is an open software library for machine learning developed by Google for solving the problems of building and training a neural network with the goal of automatically finding and classifying images, achieving the quality of human perception. It is used both for research and for the development of Google's own products [14].

The developed system is focused on working with textual samples (review texts), therefore, a preliminary processing of the training sample for the created deep artificial neural network in a convenient form was carried out. The available sample of reviews was divided into 5 groups according to the degree of understanding of the material under review in accordance with the classification criteria reflected in Table 1.

The training sample reviews were placed in one csv file. The structure of a csv file is a string consisting of two columns separated by an "@" sign. The first column contains the review itself in one line, the second indicates the group to which this review belongs. Later, the Keras library was used to preprocess text: using the Tokenizer function (the tokenize module is a lexical scanner for Python source code implemented in Python). The scanner of this module returns comments as tokens; the words in the sample were replaced with digits that could be used by artificial neural network.

The implemented network is a sequential recurrent artificial neural network consisting of three layers. Schematically such a neural network is shown in Fig. 2.

The first input layer is a fully connected layer consisting of 256 neurons, with the type of activation ReLu (Fig. 3). This activation function has the name "rectifier" (rectifier, by analogy with a one-half-period rectifier in electrical engineering). Neurons with this activation function are called ReLU (rectified linear unit). The ReLU has the following formula $f(x) = \max(0, x)$ and implements a simple threshold transition at zero.

Consider the positive and negative sides of ReLU.

Positive sides:

- The calculation of sigmoids and hyperbolic tangent requires demanding operations, such as exponentiation, while ReLU can be implemented using a simple threshold transformation of the activation matrix at zero. In addition, the ReLU is not saturated.
- The use of ReLU significantly increases the rate of convergence of stochastic gradient descent (in some cases up to 6 times) compared with sigmoid and hyperbolic tangent. It is believed that this is due to the linear nature and the lack of saturation of this function [15].

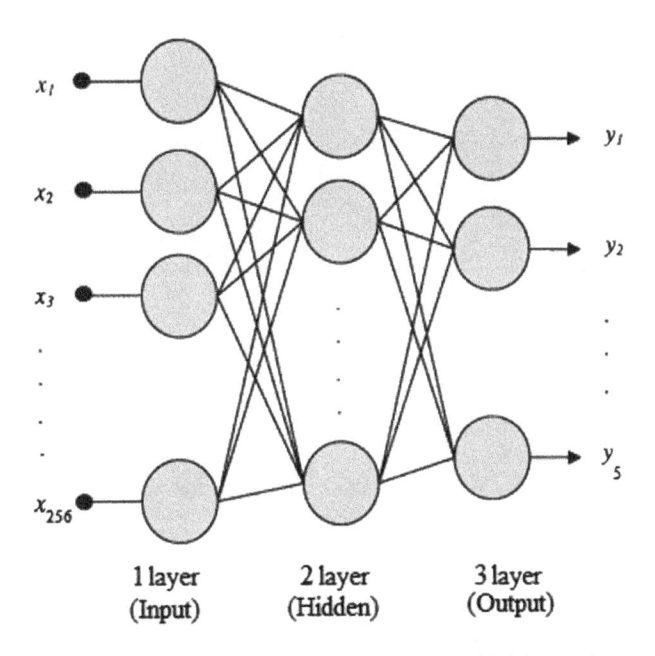

Fig. 2. The scheme of the primary version of the artificial neural network.

Negative sides:

Unfortunately, ReLUs are not always reliable enough and can fail ("die") in the learning process. For example, a large gradient passing through a ReLU can lead to such an update of the balance that the neuron is never activated again. If this happens, then, from now on, the gradient passing through this neuron will always be zero. Accordingly, this neuron will be irreversibly disabled. For example, if the learning rate is too high, it may turn out that up to 40% of the ReLUs are "dead" (that is, never activated). This problem is solved by choosing the appropriate learning rate, which in this paper was chosen empirically.

The second layer of the constructed network is a layer of regulation with the probability of neuron shutdown of 50%. The output layer consists of 5 neurons corresponding to the number of defined categories. The activation type for the output layer is Softmax, the total sum of the outputs of all 5 neurons of the output layer is 1, which corresponds to the classification procedure.

The SoftMax function (soft maximum function) is often used in neural networks as an activation function when solving a classification problem. Softmax is defined by the following formula: $\sigma(z)_i = \frac{e^{z_i}}{\sum_{k=1}^{N} e^{z_k}}$, where z_i – the value at the output of the i-th neuron before activation, and N - the total number of neurons in the layer.

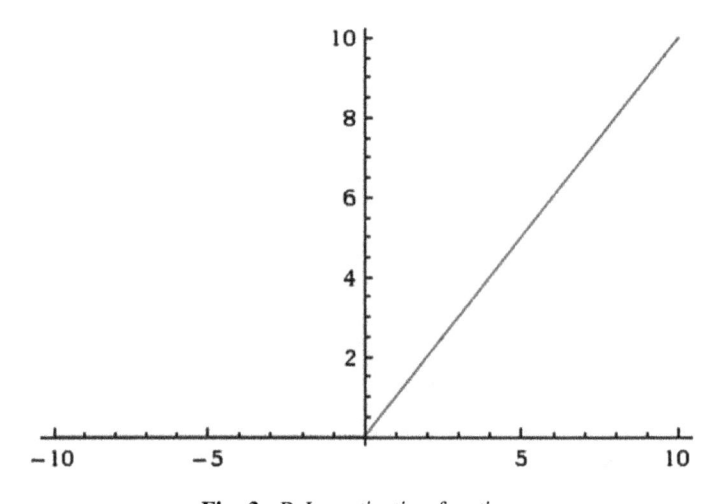

Fig. 3. ReLu activation function.

As a function of the error when compiling the model of an artificial neural network, categorical cross-entropy[1] was chosen, which is well suited for the case when at the output of an artificial neural network there are probabilities of the appearance of a class

---

[1] Cross-entropy is one of many possible loss functions (another is the loss of the SVM loop). These loss functions are usually written as J (theta) and can be used in gradient descent, which is an iterative basis for moving parameters (or coefficients) to optimal values.

[16]. The model is optimized using the method of adaptive moment estimation (Adam), the accuracy is chosen as the model metric.

There were 40 reviews in the training set. The size of the mini-sample for each epoch of learning was 18 reviews, the number of learning epochs was 15. The option of automatic separation of the sample into training and verification was also chosen. The size of the test sample was 10% of the main.

According to the results of 15 epochs of learning, the accuracy of determining the class of reviews on test data was 94.4%, in the training sample 50% (Fig. 4). The relatively low value of the accuracy of determining the class on the training sample is caused by the small size of the original sample, however, the results show that the developed tool is quite efficient and can be further brought to the level of real practical use by increasing the training sample.

```
Train on 18 samples, validate on 2 samples
Epoch 1/15
18/18 [==============================] - 2s 96ms/step - loss: 2.0089 - acc: 0.1111 - val_loss: 1.5148 - val_acc: 0.0000e+00
Epoch 2/15
18/18 [==============================] - 0s 445us/step - loss: 1.4680 - acc: 0.3889 - val_loss: 1.8863 - val_acc: 0.0000e+00
Epoch 3/15
18/18 [==============================] - 0s 389us/step - loss: 1.1337 - acc: 0.4444 - val_loss: 2.4155 - val_acc: 0.0000e+00
Epoch 4/15
18/18 [==============================] - 0s 389us/step - loss: 0.9287 - acc: 0.6667 - val_loss: 2.7458 - val_acc: 0.0000e+00
Epoch 5/15
18/18 [==============================] - 0s 445us/step - loss: 0.6838 - acc: 0.8889 - val_loss: 3.0558 - val_acc: 0.5000
Epoch 6/15
18/18 [==============================] - 0s 445us/step - loss: 0.6847 - acc: 0.8889 - val_loss: 3.3143 - val_acc: 0.5000
Epoch 7/15
18/18 [==============================] - 0s 445us/step - loss: 0.5302 - acc: 0.8889 - val_loss: 3.5213 - val_acc: 0.5000
Epoch 8/15
18/18 [==============================] - 0s 445us/step - loss: 0.4458 - acc: 0.9444 - val_loss: 3.6991 - val_acc: 0.5000
Epoch 9/15
18/18 [==============================] - 0s 445us/step - loss: 0.3468 - acc: 0.9444 - val_loss: 3.8344 - val_acc: 0.5000
Epoch 10/15
18/18 [==============================] - 0s 445us/step - loss: 0.3257 - acc: 0.9444 - val_loss: 3.9366 - val_acc: 0.5000
Epoch 11/15
18/18 [==============================] - 0s 389us/step - loss: 0.2584 - acc: 1.0000 - val_loss: 4.0200 - val_acc: 0.5000
Epoch 12/15
18/18 [==============================] - 0s 445us/step - loss: 0.2476 - acc: 1.0000 - val_loss: 4.0929 - val_acc: 0.5000
Epoch 13/15
18/18 [==============================] - 0s 389us/step - loss: 0.2384 - acc: 1.0000 - val_loss: 4.1628 - val_acc: 0.5000
Epoch 14/15
18/18 [==============================] - 0s 389us/step - loss: 0.2074 - acc: 1.0000 - val_loss: 4.2180 - val_acc: 0.5000
Epoch 15/15
18/18 [==============================] - 0s 389us/step - loss: 0.2080 - acc: 0.9444 - val_loss: 4.2882 - val_acc: 0.5000
```

Fig. 4. Learning outcomes of an artificial neural network.

3 Conclusion

Thus, there is a real opportunity to transfer a significant part of practical classes in such disciplines as "History and Philosophy of Science" [8] into a remote mode, in parallel testing students for aptitude for independent scientific work with the aim of recruiting staff for the implementation of scientific and technical projects and programs including interdisciplinary nature.

A completely similar tool can also be used to stimulate interdisciplinary cooperation focused on specialties related to electronics and information and communication technologies. In this case, students are invited to write critical reviews on the nature of the use of measurement methods (in terms of electronic support, for example), as well as in the use of infocommunication methods in related (and not only) fields of knowledge.

Criticism of techniques actually used in works in chemistry, nanotechnology, ecology, etc. allow students to see the points of application of their efforts, which in turn is a guarantee for fruitful project activities within the framework of the concept of the "triangle of knowledge".

References

1. Shyryn, M.K.: About some problems of higher education in Kazakhstan. In: Proceedings on Global Challenges and Modern Trends in the Development of Higher Education 2013 (2013)
2. Estimated number of universities worldwide as of July 2018, by country. https://www.statista.com/statistics/918403/number-of-universities-worldwide-by-country/. Accessed 21 May 2019
3. National report on the state and development of the education system of the Republic of Kazakhstan (following the results of 2017). http://iac.kz/sites/default/files/nacionalnyy_doklad_za_2017_god_s_oblozhkami_dlya_sayta.pdf. Accessed 01 June 2019
4. Mun, G.A.: Ecology and alternative energy - the battlefield of information war. In: Koryo Ilbo, 49, pp. 12–13, 7 December 2018
5. Suleimenov, I., Gabrielyan, O., Egemberdyeva, Z., Kopyshev, E., Tasbolatova, Z.: Implementation of educational information technology to develop critical thinking skills. In: News of the Scientific and Technical Society "Kakhak", vol. 1, no. 64, pp. 63–71 (2019)
6. Mun, G.A., Tasbolatova, Z.S., Suleimenov, I.E.: Pseudoscience as a resource: non-standard approaches in educational information technologies. In: News of the Scientific and Technical Society "Kakhak", vol. 1, no. 64, p. 43 (2019)
7. Gumilev, L.N.: Ethnosphere: the history of people and the history of nature; Ethnogenesis and the biosphere of the Earth (2012)
8. Suleimenov, I.E., Gabrielyan, O.A., Sedlakova, Z.Z., Mun, G.A.: History and philosophy of science (2018)
9. Kalimoldayev, M.N., Mun, G.A., Pak, I.T., Bakirov, A.S., Baipakbayeva, S.T., Suleimenov, I.E.: Artificial intelligence as a driver of the fourth technological revolution. A manual for undergraduates (2018)
10. Karpov, A.V.: Reflexivity as a mental property and a method for its diagnosis. Psychol. J. **24** (5), 45–57 (2003)
11. Khaikin, S.: Neural Networks: A Full Course, 2nd edn. (2008)
12. Savelyeva, O.V., Maslova, M.V.: Passionality as a measure of human activity and society. Philos. Educ. **3**, 163–167 (2008)
13. Keras: The Python Deep Learning library. https://www.keras.io/. Accessed 23 May 2019
14. Abadi, M., et al.: Tensorflow: a system for large-scale machine learning. In: 12th (USENIX) Symposium on Operating Systems Design and Implementation (OSDI 16), pp. 265–283 (2016)
15. Li, Y., Yuan, Y.: Convergence analysis of two-layer neural networks with ReLU activation. In: Advances in Neural Information Processing Systems, pp. 597–607 (2017)
16. Zhang, Z., Sabuncu, M.: Generalized cross entropy loss for training deep neural networks with noisy labels. In: Advances in Neural Information Processing Systems, pp. 8778–8788 (2018)

Hybrid Recommendation System for Young Football Athletes Customized Training

Paulo Matos[1,3(✉)], João Rocha[1], Ramiro Gonçalves[2,4], Filipe Santos[1,3], David Abreu[1], Hugo Soares[1], and Constantino Martins[1,3]

[1] Computer Science Department, Engineering Institute - Polytechnic of Porto, Porto, Portugal
{psm,jsr,jpe,1140272,1140445,acm}@isep.ipp.pt
[2] Trás-os-Montes e Alto Douro University, Vila Real, Portugal
ramiro@utad.pt
[3] GECAD Research Group on Intelligent Engineering and Computing for Advanced Innovation and Development, Porto, Portugal
[4] INESC TEC - Institute for Systems and Computer Engineering, Technology and Science, Porto, Portugal

Abstract. Information and communication technologies (ICTs) are been increasingly used in sports over the past decades, especially in professional football, with the goal of enhancing preparation and improve the athletes' performance. Training programs, however, are not accessible for young and amateur athletes. Most of the systems available, don't have learning skills to adjust, develop and find new suggestions for training, specifically designed for each athlete. In this paper we present the *Smart Coach* architecture and user adaptation model and describe our hybrid recommendation system to help the development of young athletes. It simplifies the relationship between the team's technical staff leaders and their young athletes, enhancing the counselling of the young person and their development as an athlete. The system allows performance evaluation for young athletes utilizing various measurements. The match information is captured intuitively and adaptively by acquaintances, relatives and staff, using a comfortable smartphone interface.

Keywords: Recommender systems · User modelling · Personalized coaching · Reasoning

1 Introduction

In the world of sports, especially in football, information and communication technologies (ICTs) are increasingly being used to enhance athlete training methods, improve team outcomes and support sport decisions and refereeing.

For the most part, though, training systems in young and (semi)amateur athletes, do not consider their performance and results in training or competition. The young athlete physical and mental condition, match performance, technique and tactics aren't considered either, throughout the training selection and recommendation process.

On the other hand, in order to adapt, evolve and find new training suggestions for each young person, such programs need to have learning capacity. The lack of learning capabilities, render the effects of these systems, unadapted and not centered on the specific characteristics of the players.

In this way, the *Smart Coach* Recommendation System aims to innovate and benefit from the use of artificial intelligence technology and techniques, to help coaches and technical staff, allowing them to further assess their youth athletes, and improve their skills, training and development.

In this article we discuss *Smart Coach* user adaptation model focusing on young athletes' evolution. In essence, *Smart Coach* will allow the technical, tactical, physical, and emotional characteristics of young athletes to be represented, adjusting a Dynamic Training Model to define a training schedule, optimized to improve the performance and skills of a young athlete.

Section 2 provides a brief description of User Modelling and Recommendation Systems, describing in a concise manner some applications specifically developed for football coaching.

In Sect. 3, the investigation method is concisely described and the adaptation to the current project is presented.

In Sect. 4, we explain the *Smart Coach* proposed architecture in detail and in Sect. 5, we draw some conclusions and discuss future research.

2 State-of-the-Art

In this section we resume the current state-of-art of user modelling and recommendation systems. There is also a review of five coaching software programs, focusing mainly on football.

2.1 User Profiling and Recommendation Systems

User profiling is usually carried out using two groups of techniques, the behavioural and the knowledge-based [9]. Knowledge-based adaptation typically results from the information gathered through forms, enquiries and other user research, with the purpose of producing a collection of heuristics. Behavioural adjustment is associated with user analysis during their daily tasks and behaviours [16].

In a historical approach, Allen, Choen, Perrault and Elaine Rich carried out in the 70s, one of the first user profiling research. During this state-of-art, it became evident that two of the most important research scientists in this area, were Rich [15] and more recently Kobsa [8]. Many different systems have been developed over the past decades to store different kinds of user knowledge. In works by Morik, Kobsa, Wahlster and McTear in 2001, some of these programs were analysed and reviewed [10].

Different user modelling strategies and methods have been used to classify knowledge, some of these are oriented towards data representation, others are oriented towards data inference [17].

User profiling approaches (decision trees, linear methods, text mining, neural networks, data mining and Bayesian networks) are all examples of predictive statistical models, as they are used in fields containing several thousand articles (clients, goods, behaviours, etc.) and will also benefit of the latest developments in machine learning [24]. Therefore, due to their specific characteristics, not all of them could be applied in some areas.

Linear models are probably the most common technique, and it may even be said that nearly all systems use them. There are even systems, that are based entirely on linear models. These systems are easy to construct and comprehend; they are efficient and assume probabilistic information to have reliable results, a concept that has so far been successfully implemented [24]. They usually use weighted sums or means, of commonly accessed items to determine user preferences. In the previously described product applications they can predict new unknown objects' likelihood [17].

You may describe a Recommendation System as one set of various methods, used by several programs, to organize or filter objects and to choose the optimal or the most appropriate item, according to the user preferences [14]. While the most likely situation would be when the program will have to select the ideal thing within a specific category, that would otherwise be chosen arbitrarily (without filtering), there are other more important situations where certain objects or categories of objects cannot be presented to the consumer at a given time, for instance, due to the position of the player in the field. Therefore, a complete recommender system should be prepared to deal with both types of situations. The operational mode usually employed by recommendation systems is to utilize a knowledge base (the user profile) as the foundation for a set of formulas to decide which will be, among all the items available, those that will suit best the user, according to a wide range of theories or strategies [16]. The ideal approach to satisfy the athletes in this work, is to recommend trainings to develop the athlete skills and reduce their deficiencies in game-play.

Recommending something to someone, bears an inherent responsibility, as it is essential to ensure reliability and consistency in the quality of the recommendation, in order to obtain the users trust. Essentially, these systems are based on three paradigms (content, collaborative or knowledge-based) and all their possible combinations [3,6,7,18]. Content-based filtering attempts to capture knowledge from the content of unstructured or unorganized items, such as textual or descriptive attributes, and generally includes potent text mining algorithms, obtained from the information recovery area. Collaborative filtering (also known as social-filtering) is one of the most frequently employed techniques and has been heavily influenced by the development of Web 2.0 ("digital web"). It relies on the information of other users, to recommend items to the current user [3]. Knowledge-based filtering is almost unavoidable, because it involves the use of any type of domain knowledge in a recommendation system [17].

In some systems, the referenced techniques are merged, taking advantage of the procedures best features and also minimize their weaknesses. These systems are classified as hybrid approaches.

Those two areas, recommendation systems and user profiling, were the basis used to build this research. The recommendation system, is complemented by an application, used to acquire important athlete's performance information during matches.

2.2 Analysis of Similar Software

The five coaching software programs discussed in this section, were designed to assist football coaches in the organization of their teams, and shortly describe the applications key functionalities. No artificial intelligence mechanism, particularly recommendation systems, is included in any of these applications. Table 1 shows some of the training related functionalities, implemented in *Smart Coach* and the applications presented in this section. After an extensive research, no recommendation systems linked to sport were found.

Table 1. Coaching software functionalities

Functionalities	Tactical soccer	My coach football	Dossier do treinador	SportEasy	Soccer coach - team sports manager	*Smart coach*
Training recommendation	–	–	–	–	–	•
Training management	•	•	•	•	•	•
Exercise editor	•	•	•	–	–	•
Fixed exercises	–	–	–	–	•	–

• Implemented   – Not implemented

Tactical Soccer. Is a commercial mobile/desktop application focused on football, that has the ability to handle individual players and teams. Can be used to build workouts, plan sessions of training, and establish line-ups, match formations and tactics. The program was designed with the primary goal of enhancing individual and collective train quality and productivity [21].

MyCoach Football. Is a free mobile/desktop application for professional and amateur football coaches. The software will maintain data from teams and matches, allowing the coaches to design and manage training sessions. One advantage of this program is the ability to share workouts and details about training sessions with a coach community that also utilizes the software. A smartphone application is also available, that can be used in real time to gather match statistics [12].

Dossier Do Treinador. Is a football and futsal centered commercial web application. The program includes a full workout editor, automated statistics, player profiles, schedule and appearance tracking, and multiple reports production (e.g. match updates, schedules, player profiles) [5].

SportEasy. Is a web and mobile free/commercial software used by amateur teams to manage their squad. The software can be used in any sport and all club members may access it. The program promises to be "the best way to manage the sports team" but has no workout or training managing functionalities [20].

Soccer Coach: Team Sports Manager. It's a commercial iOS application for football coaches (is available a free limited version), providing the ability to manage squads and athletes, establish training sessions, and select exercises from a list of available workouts (depending on the version used) [19].

As presented in Table 1, *Smart Coach* is the only system with an intelligent system to recommend players what skills to improve and suggests suitable exercises to improve those skills.

3 Methodology

After a bibliographical review of some research methodologies, the one that's believed to be the best fit for the presented project was identified and is known as Design Science. This methodological approach is appropriate for problem solving, leading to innovation creation [4].

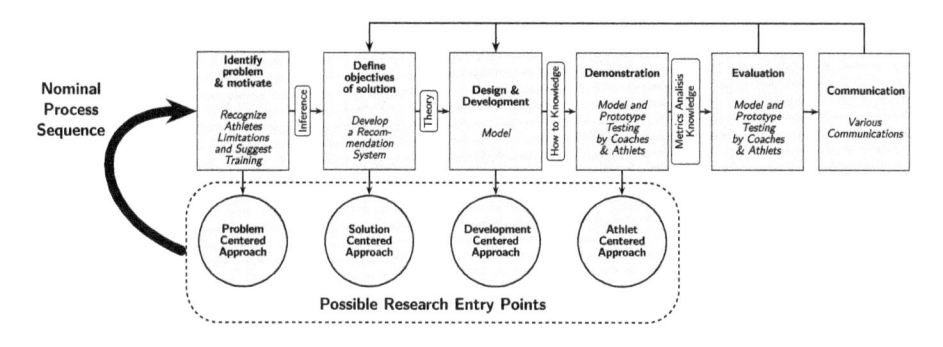

Fig. 1. Adaptation of peffers DSR methodology to *Smart Coach* [13]

There are different reasons for choosing this methodology, in the course of this research and development work, namely:

- The Design Science methodology favours, a more practical approach to problem solving, using experimental processes that supports knowledge extraction from young sport athletes [22].

- The participation of different categories of persons involved, such as coaches, technical staff, young players, with their experiences and questions can add diversity to the collected data [2].
- This methodology allows a continuous improvement of the solution, since the design is progressive, allowing to include developments from evidences identified during the several experimental phases.

The research project development, is presented in Fig. 1, derived from Peffers Design Science Research methodology and is divided in six stages [13]. In the initial stage, this research was directed to recommendation systems analysis and the possibility of application in football training. The required needs of athletes match performance information, was also defined in this stage. Stage two comprised the solution objectives definition. In stage three, the software solution was designed and the first prototypes for data acquisition and recommender systems were implemented. The project is now entering stages four and five, with the initial "in field" data collection and the first training recommendations and their evaluation will follow. This paper is one of the stage six outcomes.

4 *Smart Coach* Proposal

This paper proposes the development of a training recommendation system based on the athlete's performance during matches, to improve the young football player's skills. The performance information is gathered by family and friends, using a web responsive software and saved in a database for the recommendation system later usage. The system architecture (Fig. 2) consists of three levels. The first level, clusters the information in three groups, and associates it to models used by the recommendation system. The second level, the recommendation and learning module, uses data from the knowledge and interface levels and recommends the skills the player should improve, and the exercises suited to improve those skills. The Interface level allows the input/output of data to the system, namely, the player statistic collection and feedback.

Considering the players field position, physical attributes (*e.g.* height, weight, pace, jumping height, etc.) and match statistics, user profiles will be produced and fine-tuned. Table 2 displays the statistics considered the most important, that are obtained during matches for each young player. Some of these statistics, depending on the position the athlete occupies during matches, are more or less significant accordingly.

The player modelling module, discussed in Sect. 4.1, generates a player profile based on all previously collected data, and the appropriate training for that specific athlete is chosen by the training recommendation module. At the end of Sect. 4.1, this recommendation process is discussed and is represented in Fig. 3.

4.1 Smart Coach User Profiling

This section presents the characteristics to allow concepts modelling. These concepts are then applied to define the young athletes' profiles and recognize their

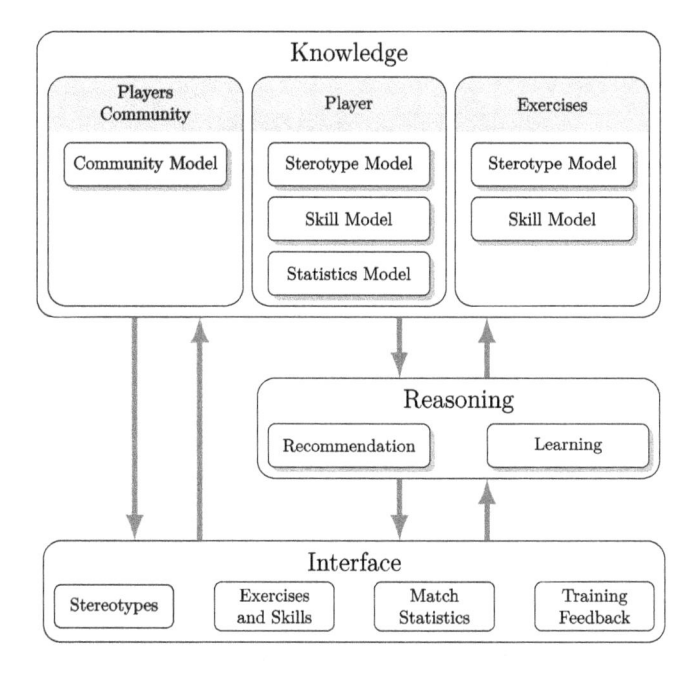

Fig. 2. Diagram of proposed architecture

different characteristics. Finally, the athlete profile is used, to pinpoint the most suitable exercises to meet the needs of the young athletes.

The solution implemented for the *Smart Coach* young athletes User Profiling, initially involved the definition of stereotypes for the athletes. The stereotypes have been established through interviews with some football coaches and the application of the K-Means clustering algorithm [1] to statistics gathered during matches within two football academies in Portugal. These statistics were collected according to the attributes specified in Table 2. Each cluster has been classified with a set of attributes, with different weights and mapped according to their training relevance and influence in the performance of a football team. The outcomes of this clustering were the profiles of the young football players, *e.g.* goalkeeper, defender midfielder, attacker. Each cluster has a type of user that is associated with certain sets of attributes/tasks, which usually need to be executed by him, throughout a football match.

In the proposed solution, some domain-independent data was not considered in player modelling, but is stored, since it may have future importance and can be used to produce better reports. Thus, domain-independent data stored in *Smart Coach* defines characteristics that are common to most user profiles and are generally referred to as generic profiles (see Table 3).

The key attributes that describe the particularities of each young athlete and is origins, affect the actual performance during a football match and can determine his game-play style (*i.e.* a more or less defensive attitude, a more

Table 2. Performance attributes collected during matches and their importance by player position [11]

Attribute	Goalkeeper	Central defender	Full-backs	Defensive midfielders	Centre midfielders	Attacking midfielders	Wingers	Strikers
(Un)Successful saves	•	–	–	–	–	–	–	–
Passing precision	•	•	•	•	•	•	•	○
Clearances	○	•	•	•	•	○	○	○
(In)Complete interception	○	•	•	•	○	○	○	○
Ball recuperation	–	○	○	•	•	•	○	○
(Un)Successful challenges	–	•	•	•	•	•	○	○
Fouls committed	•	•	•	•	•	•	•	•
Fouls suffered	○	○	○	○	•	•	•	•
(Un)Successful dribbling	–	○	○	○	○	•	•	•
Gained/Lost Duels	–	•	•	•	•	•	•	•
(Un)Successful crossings	–	○	•	○	○	•	•	○
Shots (on/off goal)	–	○	○	○	•	•	•	•
Offside	–	○	•	○	•	•	•	•
Assists	–	•	•	•	•	•	•	•
Goals	–	•	•	•	•	•	•	•

• Relevant indicator ○ Applicable indicator – Inapplicable

Table 3. Domain-independent data collected by *Smart Coach*

Characteristic	Description
Personal details	Name, address, e-mail, etc.
Demographic details	Age, gender, etc.
Education	Academic degree, Technological *vs* Social studies
Life experience	Jobs (current and past), hobbies (sports or others), etc.
Disabilities	Hearing, visual, other

skilled or athletic player, etc.). Such goals and objectives are derived directly from the domain model and the user domain-dependent information is redefined. All of these stereotypes correlate with a set of objectives, tasks or functions.

The goals and objectives for young athletes are successfully achieved, if they perform a set of actions required for a certain training to be accomplished. To all the trainings (despite of granularity) is assigned a percentage of success in the user model, that allows the verification of the successful execution of the workouts. The player also gives feedback on trainings and exercises, allowing the system to adjust the recommendations. As a consequence, the user experience is enhanced with recommendations of different workouts, based on his personal performance/training statistics (Fig. 3).

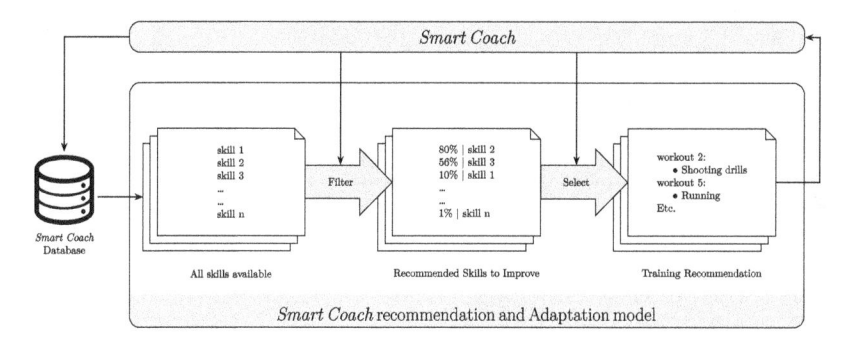

Fig. 3. *Smart Coach* recommendation and Adaptation model (evolution from [11])

4.2 Training Determination Process

The training selection mechanism, attempts to determine the collection of workouts, that the participant needs to perform, depending on the young athlete condition, observed at a certain point in time (Fig. 3).

It is an unpredictable and imprecise process to decide which activity is being carried out. The athlete's actions observations might not have an accurate perception, and as such, may contribute to incorrect conclusions.

In the context of the *Smart Coach* system, the relationships between training exercises, will have as parameters elements associated with defined stereotypes in the young athlete profile. Therefore, a hierarchical structure is established for each stereotype, associating all the workouts within its skill, and describing the relationships between them.

Considering a list of all possible skills the athlete can improve, the adaptation model will, in the first step, use information about the user's stereotypes and match statistics, considered relevant by the staff, and apply our filtering algorithm. This will order the list of skills that can be trained, giving a greater percentage to the ones the athlete needs to improve. Therefore, an ordering is made, using data such as the expected course of training and athletes' personal performance history, according to the likelihood of each skill contribute for the athlete improvement. In a next step, using this ordered list, and the young athletes' workout preferences and feedback information, a list of workouts is chosen, that will optimize the improvement of the skills that have higher percentages in the list. The training plan the young athlete should perform, with the suggested workouts and respective information is then presented to him (Fig. 3).

The recommendation algorithm is presented in Algorithm 1. When a player or staff member requests a training/exercise recommendation, the list of skills is filled with all the skills the system can recommend being improved. Then the individual recommendation algorithms are called to collect the weight for each skill regarding the three approaches (stereotypes, statistical analysis and player similarity). The following step calculates the averaged skill weights and subsequently the list of skills is ordered. The final step is to classify the exercises using the skill weights and adjust the score with previous received player feedback. The list of exercises is then ordered, and a training recommendation is presented to the player/staff member.

Algorithm 1. Training recommendation algorithm

```
// Apply recommendation algorithms to skills
skills ← applyRecomendation(ra1) ;            // Sterotypes
skills ← applyRecomendation(ra2) ;            // Statistical Analysis
skills ← applyRecomendation(ra3) ;            // Player Similarity
// Calculate the global skill weights
foreach skill ∈ skills do
    foreach tech ∈ techniques do
        skill_weight ← ∑(weight(tecValue(tech)));
    end
end
order(skills, playerProfile);
// Rank exercises using recomended skills to improve and players
    feedback
foreach exer ∈ exercises do
    weightExercise(exer, skills);
    adjustWeightFromPreviousFeedback(exer, playerFeedback);
end
order(exercises, playerProfile);
```

---

5 Evaluation and Conclusion

The framework was already tested and evaluated in two Under-19 football teams:

team A: from a academy school (doesn't have any senior teams);
team B: from a semi-professional club (with senior team).

The duration of this case study has run over the course of two months and the age of the athletes is within 15 and 18 years old.

The prototype evaluation process has been split into four phases. In the first phase, each team was randomly divided into two groups: experimental group and control group. The random process obeyed some criteria to distribute the athletes by the different groups. These criteria were as follows:

- The number of athletes in each group is approximately the same;
- The field positions of the young players, Forward, Defender, Midfielder, etc. were distributed in a similar proportion across the groups.

The second phase consisted in a set of questionnaires, conducted to gather missing data needed to instantiate the young football player model variables of *Smart Coach*. The integrity of each survey was determined using the Cronbach

Alpha coefficient of reliability [23]. The third phase consists in the use of *Smart Coach* by the two experimental groups, over the course of two months.

Finally, the fourth and last phase will consist in the sports results evaluation and the assessment of athletes' improvement (both experimental and control groups). For this purpose, staff members will be interviewed, matches statistical data of each athlete will continue to be collected throughout two months, and a survey will be carried out to evaluate the suitability, usability and acceptance of *Smart Coach* functionalities.

The fourth phase is expected to start soon. It was already possible to observe with the data obtained, that the innovative solution developed to implement the young athlete profile, seems to be valid. The represented young athlete characteristics definition and the hybrid solution using stereotypes for the representation of the player's knowledge and to suggest what training/activity should be performed at some point, appear to be getting positive results. This validates our conviction that the present work allows the definition of a new model and process for young athletes profiling to be used in recommendation systems to support training in football. This model enables young athletes to enhance their skills in order to accelerate their evolution as a football player.

Acknowledgments. This work was supported by National Funds through the FCT—Fundação para a Ciência e a Tecnologia (Portuguese Foundation for Science and Technology) Project UID/EEA/00760/2019 and by FLAD—Fundação Luso-Americana para o Desenvolvimento (Luso-American Development Foundation) Project 52/2020.

References

1. Anderberg, M.R.: Cluster analysis for applications. Technical report, Office of the Assistant for Study Support Kirtland AFB N MEX (1973)
2. Barab, S., Squire, K.: Design-based research: putting a stake in the ground. J. Learn. Sci. **13**(1), 1–14 (2004)
3. Berka, T., Plößnig, M.: Designing recommender systems for tourism. In: Proceedings of ENTER 2004, pp. 26–28 (2004)
4. Cheong, C., Cheong, F., Filippou, J.: Using design science research to incorporate gamification into learning activities. In: PACIS, p. 156 (2013)
5. Dossier do treinador de football (2018). http://www.dossierdotreinador.com/
6. Felfernig, A., Gordea, S., Jannach, D., Teppan, E., Zanker, M.: A short survey of recommendation technologies in travel and tourism. OEGAI J. **25**(7), 17–22 (2007)
7. Kabassi, K.: Personalizing recommendations for tourists. Telematics Inform. **27**(1), 51–66 (2010)
8. Kobsa, A.: User modeling: recent work, prospects and hazards. Hum. Factors Inf. Technol. **10**, 111–111 (1993)
9. Kobsa, A.: Generic user modeling systems. User Model. User-Adap. Inter. **11**(1–2), 49–63 (2001)
10. Martins, C., Faria, L., Carvalho, C.V.D., Carrapatoso, E.: User modeling in adaptive hypermedia educational systems. Educ. Technol. Soc. **11**(1), 194–207 (2008)

11. Matos, P., Rocha, J., Gonçalves, R., Almeida, A., Santos, F., Abreu, D., Martins, C.: Smart coach–a recommendation system for young football athletes. In: Novais, P., Lloret, J., Chamoso, P., Carneiro, D., Navarro, E., Omatu, S. (eds.) Ambient Intelligence - Software and Applications,10th International Symposium on Ambient Intelligence. Advances in Intelligent Systems and Computing, vol. 1006, pp. 171–178. Springer, Cham (2019)
12. My coach football - the digital assistant for educators (2018). https://www.mycoachfootball.com/en/
13. Peffers, K., Tuunanen, T., Rothenberger, M.A., Chatterjee, S.: A design science research methodology for information systems research. J. Manage. Inf. Syst. **24**(3), 45–77 (2007)
14. Porter, J.: Watch and learn: how recommendation systems are redefining the web, May 2006. https://articles.uie.com/recommendation_systems/. Accessed 01 Feb 2019
15. Rich, E.: User modeling via stereotypes. Cogn. Sci. **3**(4), 329–354 (1979)
16. Santos, F., Almeida, A., Martins, C., Gonçalves, R., Martins, J.: Using POI functionality and accessibility levels for delivering personalized tourism recommendations. Comput. Environ. Urban Syst. **77**, 101173 (2019)
17. Santos, F., Almeida, A., Martins, C., Oliveira, P., Gonçalves, R.: Tourism recommendation system based in user functionality and points-of-interest accessibility levels. In: Mejia, J., Mirna, M., Rocha, Á., San Feliu, T., Peña, A. (eds.) Trends and Applications in Software Engineering, pp. 275–284. Springer, Cham (2017)
18. Schafer, J.B., Konstan, J., Riedl, J.: Recommender systems in E-commerce. In: Proceedings of the 1st ACM Conference on Electronic Commerce, pp. 158–166. ACM (1999)
19. Soccer coach — the definitive coaching app (2018). http://www.teamsportsmanager.com
20. Online sports team management software - sporteasy (2018). https://www.sporteasy.net/en/home/
21. Tactical soccer - complete suite for soccer coaches (2018). https://tacticalsoccer.co.uk/
22. Van den Akker, J., Gravemeijer, K., McKenney, S., Nieveen, N.: Educational Design Research. Routledge, London (2006)
23. Woodward, C.A., Chambers, L.W.: Guide to Questionnaire Construction and Question Writing. Canadian Public Health Association, Ottawa (1983)
24. Zukerman, I., Albrecht, D.W.: Predictive statistical models for user modeling. User Model. User-Adap. Inter. **11**(1–2), 5–18 (2001)

AALADIN: Ambient Assisted Living Assistive Device for Internet

Priscila Cedillo[✉], Wilson Valdez, and Andrés Córdova

Computer Science Department, University of Cuenca, Cuenca, Ecuador
{priscila.cedillo,wilson.valdezs,
andresf.cordovac}@ucuenca.edu.ec

Abstract. Accidents during elderly are one of the greatest geriatric syndromes due to the frequency with which they appear between patients. There are incidents that mark a milestone for acute and chronic deterioration in the health during old age. Among the instruments explored over time by the medical area in order to know the factors that lead to falls in elderly people, the Tinetti test is the most complete and validated measurement instrument. Tinetti is an observational scale that allows the evaluation by means of two sub-scales (i) the march (9 items) and (ii) the balance (13 items) in static position and when the patient is moving. Generally, this test is performed by specialized medical personnel (doctors or physiotherapists) and the acquisition of data used in each one of the items in this test is recorded manually. This paper presents the design and construction of an Assistive Technology Device, which supports medical personnel during the application of the balance sub-scale in the Tinetti test. This solution will consist of a carpet equipped with sensors that allows to know the position of the patient through heat maps and the body weight that is distributed in each area during the performance of the Tinetti test. In addition, it will allow the acquisition of data in a digital form, and since it is an object of the Internet of Things, it will allow the monitoring of the information coming from the test locally and remotely, as well as registering the information and making it available on the cloud.

Keywords: Tinetti test · Ambient intelligent · Assistive technology · Ambient assisted living

1 Introduction

Ambient Intelligence (AmI) represents a new technological paradigm, which is becoming a multidisciplinary area related to the ubiquitous computing that has lead changes in the design of communication protocols, devices, systems, etc., leading on new environments where people are surrounded with networks of the embedded intelligent devices that can sense their state, anticipate, and perhaps adapt to their needs [1, 2]. AmI is rapidly positioning as an area where various topics are covered (e.g., medicine, home and environment automation, healthcare, instrumentation, networks) to help society, including several priority groups (i.e., pregnant, people with catastrophic diseases, special needs, pregnant and development disabilities, kids and elderly) [3] through technology [4].

Assistive Technology (AT) is defined as any device, system or service related to delivery of assistive products that allows patients to live healthy, productive and independent life. In this way, letting them perform a task with increasing ease and safety, which otherwise would have been difficult for them. [5–7]. AT takes part in Ambient Intelligence because of its importance in improving people's lifestyle. Besides AT has considerable potential as partial substitutive for several social and health care interventions [5]. Beyond being a substitute, AT can also be considered a complement to facilitate many habitual tasks, if done by themselves or with another person's help (e.g., medical practices) [8].

The intelligent and suitable environments can foster the participation and inclusion of disabled individuals in social, economic, political and cultural life [7]. In this context, the Internet of Things (IoT), defined by the International Telecommunication Union (ITU) as "a global network infrastructure, linking physical and virtual objects through the exploitation of data capture and communications capabilities" [9], creates enabling environments by offering people with disabilities assistance in building access, transportation, information, communication and health care [10]. IoT plays an important role in the digitalization and management of the data [9]. In a concrete way, in the Assistive Technology Devices, the connectivity allows to obtain information constantly from the devices hosted on the cloud and letting it be available anytime and anywhere.

Falls in elderly people are one of the most important geriatric syndromes because of the frequency which it appears between patients and in half of such cases, the falls are recurrent [11]. These are frequent problems in this population and set a milestone for acute and chronic deterioration in the health of elderly people [12–14]. Because falls can be a consequence of the combination of several factors, there are many studies in the medical area that address the main factors which lead to falls in these vulnerable sector [12–17]. The Tinetti test presented by Dr. Mary Tinetti in 1986 [17] is an observational scale which allows to evaluate patients through two sub-scales: (i) march (9 items), and (ii) balance by considering static and moving positions (13 items) [17]. The screening characteristics, shown by the Tinetti test support its inclusion in periodic health examination of the health for elderly people [18]. This test is commonly performed by specialized medical personnel such as doctors or physiotherapists. It suffers the typical limitations of any qualitative evaluation scale as: (i) it is subject to a roof effect of the scale that makes it impossible to identify individuals, with very few balance problems; (ii) its results depends on intuitive judgments, so that cannot provide a correct appraisal [18]. In order to combat these limitations, considering instrumental measurement may guarantee an objective measurement, and it will provide a scientific basis for communication between professionals, documentation of treatment efficacy, and scientific credibility within the medical community. In effect, instrumental measurements may ensure fine resolution as well as avoid floor and ceiling effects [19].

Based on those considerations, in this paper, it is presented an Ambient Assisted Living Assistive Device for INternet (AALADIN), which represents an Assistive Technology prototype equipped with internet connection for supporting the Tinetti Test during the evaluation sub-scale of balance in order to make more exhaustive appraisals. It consists of a carpet conformed by infrared sensors distributed along the surface which allows the measurement of the foot posture, positioning and it's exactly weight distribution in each area of this carpet, through heat maps. Also, the internet connection allows to storage the information from the test on a cloud platform by user profile and test made, in order to collect information for supporting health personnel in taking preventive actions that avoid future accidents related to falls of elderly people.

The structure of this paper is as follows: Section two presents the related work, Section three contribution of the device, Section four the design of the prototype, Section five presents a functional evaluation of the prototype, Section six presents future steps to be followed, and finally, in Section seven the conclusions.

2 Related Work

The Tinetti test [17] was developed by Dr. Mary Tinetti in 1986; it is oriented to study the functional mobility capacity of elderly people. Its study emerges from the need to make a timely practical assessment of patient mobility [17]. The test is developed in two sub-scales: (i) balance (static and dynamic), which addresses 13 items, and (ii) March, consisting of 9 items. The scope of this test is to detect subjects at risk of falls within a year, based solely on the risk factors such as problems in balance and March.

There are several implementations that have been carried out based on the Tinetti test, for example, Ballesteros et al., [20] present the development of a robotic walker that allows prediction of the value of the Tinetti mobility test starting from spatiotemporal parameters of the march obtained from the walker while the patients walk. The prediction generated from the device provides continuous information about the balance condition of the user and this parameter is used to modify the amount of help that is given to the patient through the walker. The same authors in [21] propose an endowment of sensors, from which it is possible to obtain walkers devices parameters of March and the force exerted on the handles, and therefore, apply to the Tinetti test.

Moreover, in Van Vooren et al., [22] a study is presented with 46 patients between 81 and 87 years old, divided into three groups, each one developing different tasks: (i) playing physiotherapy and video games, (ii) doing physiotherapy and dual task training, (iii) doing standard physiotherapy, and another physiotherapy session with the same duration of dual task training and video games. Those tests were carried out before and after the completion of the Tinetti test, standing and walking time, and standing in a chair support. Among the conclusions of the study, it was highlighted that, as a consequence of the previous training by physiotherapy and video games, patients show improvements in the Tinetti test, however, it does not process and store data in a database for future data analysis.

In another study presented by Giansanti et al., [23] the construction of a prosthesis for tele-monitoring and tele-rehabilitation is indicated. It presents a system for step counting and sensing of the force exerted in the foot plantar area. The tests of this prosthesis were carried out in three subjects who underwent 100 steps (quickly and slowly) after brain rehabilitation of one of the balance tests of the Tinetti test.

Moreover, Jaume et al., [24] present a new experimental system which focuses on improving the control of posture and balance in elderly people with cerebral palsy. It consists on the development of a game for balance rehabilitation therapies. This game was developed in order to obtain feedback, adaptability, motivational components and monitoring. In addition, the interaction technology used is an artificial vision system, as it is not invasive and prevents patients from wearing devices. This game was evaluated during 24 days in sessions of 20 min based on Tinetti test scores, obtaining positive results of the therapy.

Among the studies reviewed, it is highlighted that several of them are based only on walking, most are devices for collecting information on therapies, and evaluations to patients are made pre and post therapies. However, among the solutions found, there are no devices that provide support during the test. Therefore, the prototype that is presented in this paper is to fill that gap and provide a technological tool at the time of the evaluation of the Tinetti Test, allowing the collection of information in a digital format, giving a better treatment to it and getting more detailed results of the Tinetti Test.

3 AALADIN Contribution

Tinetti test is widely applied in clinical areas as medical or specialist assessment and functional rehabilitations [25, 26]. The evaluations are based on the execution tasks through direct, organized and structured observation which leads on more objective information that can be obtained related to the physical function of elderly people in order to make an initial assessment. Also, it establishes intervention objectives in geriatric centers or institutions as part of the comprehensive income assessments and defines intervention and fall prevention programs [27, 28].

This proposal looks to improve the perception, and strengthen the evaluation results based on the sub-scale of balance by applying the Tinetti test. The information obtained is collected from the sensors located in the developed device. The design of the AALADIN device represents a complement during the execution of the Tinetti test which involves reliability and usability such as (correct data reading, safety for elderly patients, and a plug and play device) with a non-invasive device for patient.

In Fig. 1 is shown a block diagram which represents the contribution of AALADIN device. Firstly, it is necessary to follow all the Tinetti test steps (1) by inviting the elderly patient (2) using the AALADIN device (3) then, the sensors inside the carpet collect data; after that, (4) analog data is sent and processed in the microcontroller, (5) the processed data is presented through heat maps and weight distributed values in a graphical user interface (6) the information is available for the evaluator; (7) thus, the evaluator approves the item or decides if the item needs to be repeated. Finally, the information of the test will be stored on the cloud in order to access it remotely.

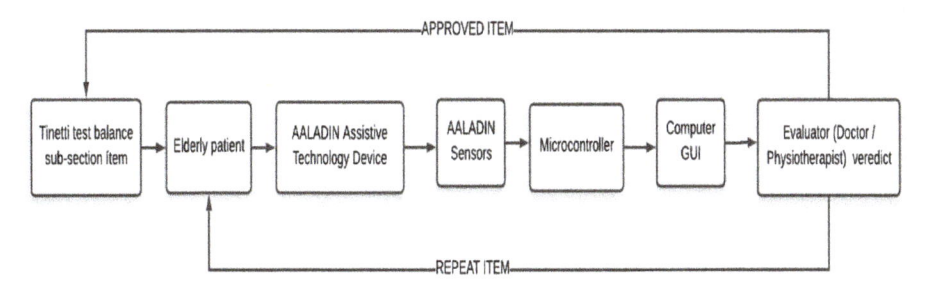

Fig. 1. AALADIN block diagram.

4 AALADIN Design

The device consists on a carpet which has 64 force sensors placed homogeneously in quadrangular sections of ten centimeters each side (see Fig. 2). With these 64 sensors the carpet has a resolution of 8x8 pressure points. The measurements of the carpet are shown in Fig. 3. Agglomerated sponge was the principal material in the construction of the carpet, this is represented with yellow in the diagrams shown in Fig. 2b. About the position of each sensor, they are located in the center of a cylindrical hole drilled in the center of each square. To avoid the false readings, the sensors were placed on the sponge, avoiding so the base of the carpet.

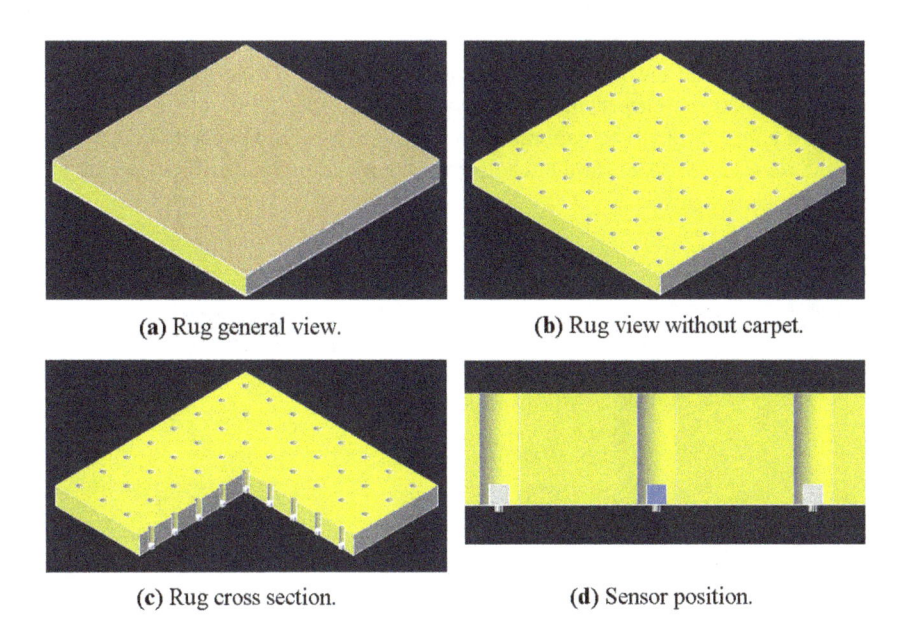

(a) Rug general view.

(b) Rug view without carpet.

(c) Rug cross section.

(d) Sensor position.

Fig. 2. Carpet general scheme.

4.1 Prototype Operation

With 64 sensors, the carpet senses the force exerted in each one of the 64 generated sections. Each sensor measures distance (infrared sensor) by relating it with the compression in the carpet. The sponge acts like a spring and with the infrared sensor the sponge "elongation" can be obtained. To obtain the force with the elongation a linear regression model is necessary. Each infrared sensor is connected to an ADC port of a microcontroller; in this case the possible values of the sensors are between 0 y 1024, considering the greater value as the highest compression on the carpet surface.

Each microcontroller has eight analogical ports; therefore, the system needs eight of this circuit. Considering this limitation, a serial communication protocol is necessary. The purpose of this protocol is to transmit eight numerical values from the eighth microcontroller to the previous one, until the first one sends the data to a computer. The transmission process is accumulative, for example, the seventh microcontroller receives data coming from the corresponding eight sensors which are in the carpet; this data is attached to the data coming from to the sixth microcontroller; this process repeats until the complete package of data is received by the first microcontroller. The information that needs to be transmitted contains the value of each sensor and a code of the position (number of analogical port and the respective microcontroller). With this information a heat map is generated; here, the first position corresponds to the first sensor in the first microcontroller (see Fig. 6).

Also, it is necessary to calibrate the 64 sensors that operate in the carpet. They provide an estimate of the weight that the carpet supports. Also, a regression model is necessary to know which is the force applied to each sensor. The process to obtain the model implies the use of different known weights to generate a reference set of points, with this information the model can be easily calculated. Because the used weights were smaller than the carpet, the calibration of the carpet needs to be executed one section at a time. Through the calibration process previously explained, it is possible to obtain the regression model for the whole carpet, depending on the force exerted at any point in the carpet surface.

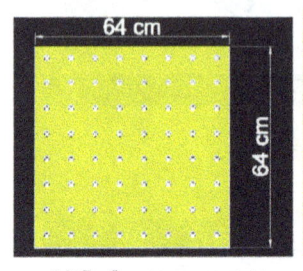

(a) Surface measurements 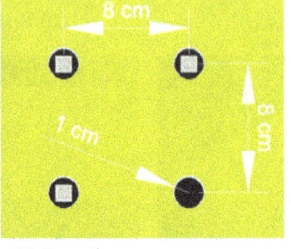 (b) Space between sensors, hole to place sensors radious (c) Rug height, without cover.

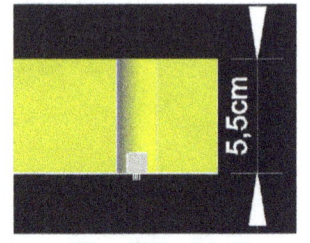

Fig. 3. Carpet measurements.

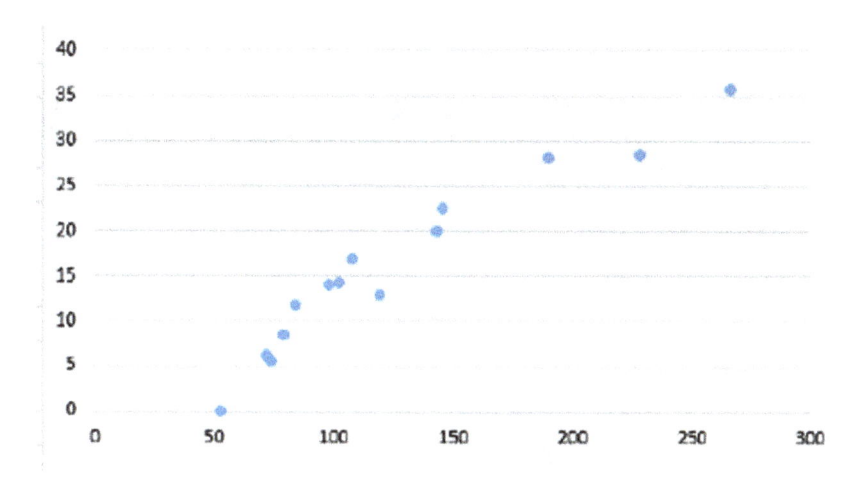

Fig. 4. Set of points obtained with one sensor

The regression model was obtained by measuring one sensor each time, this measuring was performed on sensor in the section where the weight was applied. One example of the set of points obtained in this process is shown in Fig. 4. With this set of points, the model show in Fig. 5 was calculated, in this case it has been used a logarithmic model in order to adjust the obtained values in an appropriative way.

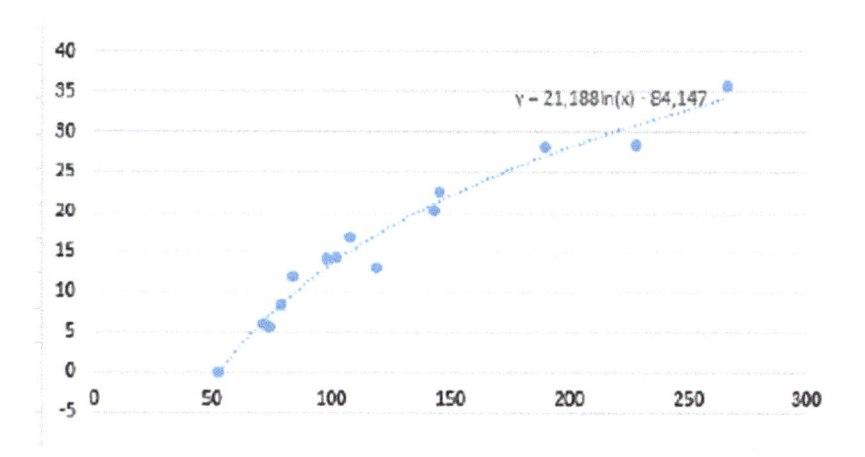

Fig. 5. Regression model for sensor data

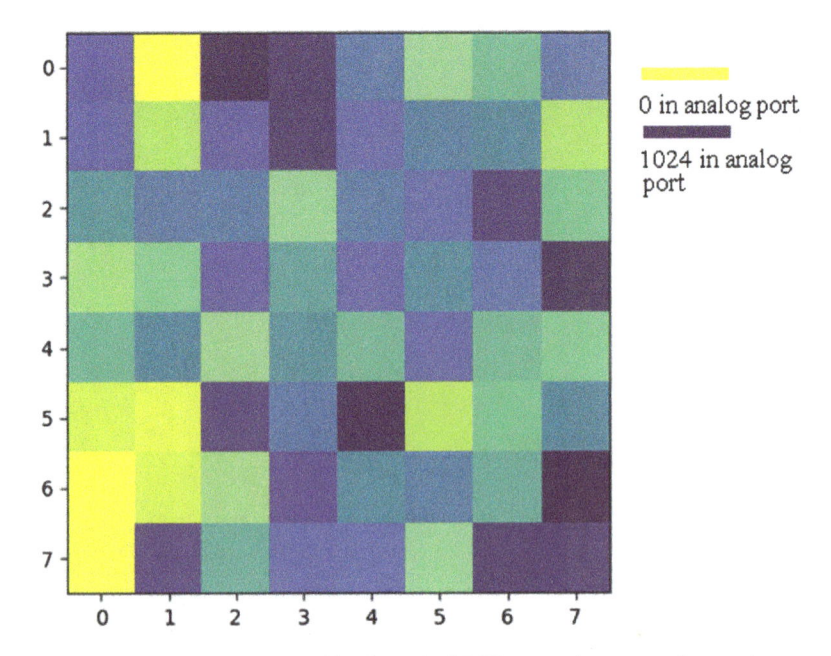

Fig. 6. Heat map generated in the AALADIN carpet in an experimentation

The information obtained from the 64 sensors which were distributed on the carpet is presented through a heat map (Fig. 6). Here it is possible to observe the force applied to each sensor, and it is represented by a color scale. This scale has the lowest value (0) in the analog port represented in purple, while on the other hand, the highest value on the analog port (1024) is represented in yellow. Based on this system, is possible to observe the behavior of the patient on the carpet, in terms of balance.

5 Functional Evaluation

To evaluate the functionality of the device, the methodology presented by Paunović et al. [29] was applied. It proposes the Product Test Plan (PTP) testing which supports the evaluation of a device in production process. It has four phases: (i) Hardware testing, (ii) Functionality testing, (iii) Stress testing, (iv) Robustness testing. The phases should be executed sequentially and the transition to the next stage is possible only with positive report from previous stage. The advantage of this testing approach is that it has a sequential testing approach, along with a good test plan, provides structured finding and fixing errors, and gives exact results [29].

A detailed block flow diagram for testing AALADIN is shown on Fig. 7, said testing is performed by considering the elements which conforms the device and that should be evaluated.

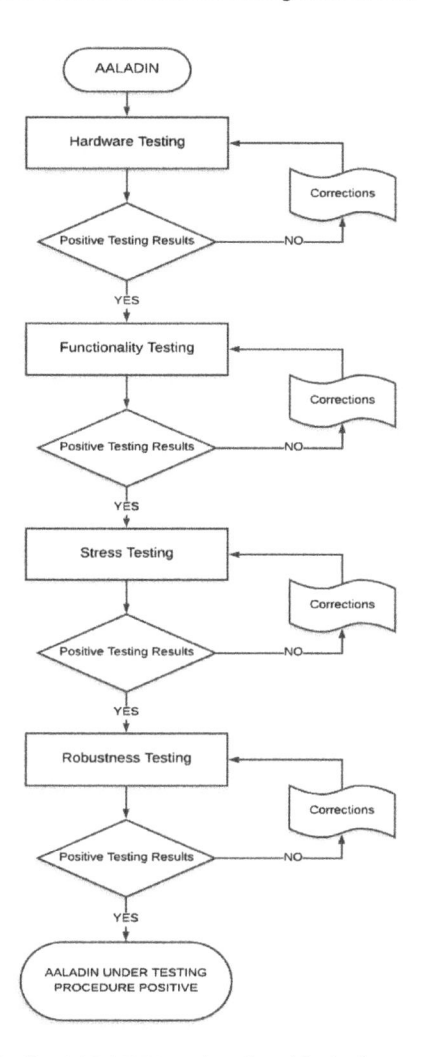

Fig. 7. AALADIN testing flow block diagram.

5.1 Testing Hardware

The first step consists on examining the electronic devices that conforms AALADINO. It means to verify the state of infrared sensors and microcontroller´s Printed Circuit Board (PCB) looking for possible manufacturing errors [29]. In the case of sensors, it was tested for possible short circuits; also, the PCBs have been tested for continuity and resistance (0 Ω of resistance at points with equal potential) in link connections on conductor layers and finally assembling as recommends Paunović et al., [29]. The evaluations were made with a professional multimeter.

5.2 Testing Functionality

The second step consists on the verification of the functionality and correct performance of the components of AALADIN, which are associated to the characteristics and features of the device. It involves the simulation of a realistic working environment of the sub systems of the device (i.e., ADC data acquisition, communication between microcontrollers, communication between computer and microcontroller) [29]. In order to test the correct performance on the microcontrollers, demo programs for each sub system was installed.

Paunović et al. [29], suggests the development of a summary table where performance of the components is presented. In Table 1, the result in the functionality testing of the components is shown.

Table 1. Components functionality test report.

	Lecture Notes		
Item	Test Case	Date & Time	Result
1	Microcontroller ADC data acquisition	22/04/2019	PASS
2	Communication between microcontrollers	25/04/2019	PASS
3	Communication between computer and microcontroller	10/05/2019	PASS
4	Microcontroller - internet connection and data sending	24/05/2019	PASS

Tested cases	4	
Passed	4	100%
Failed	0	0%
Inconclusive	0	0%

5.3 Testing Stress

This phase aims to verify the stability of the system in environment conditions different from the nominal, generally defined by device specification datasheet. It is necessary to evaluate the device performance beyond normal operation [29]. In this case the microcontrollers sending data operation and latency have been tested in a warm environment.

5.4 Testing Robustness

This phase consists on examine the performance of the device in conditions beyond the nominal, by improper use of the device [29]. In this instance, the weight was applied to the whole surface of the carpet, which is an abnormal operation. The microcontroller response was satisfactory.

6 Discussion

The development of AALADIN represents the first step to obtain an efficient supporting device for evaluators during the balance sub-scale assessment in the Tinetti test. However, being a first approach, it presents promissory results although there are certain aspects to be improved.

The first challenge of this device was to obtain a correct physical and functional design of the prototype. In order to reach those goals, the adoption of a carpet was considered because it allows to have a non-invasive and comfortable device for the final user. About its functionality, several options were considered in order to better obtain the information. Firstly, the ideal material used for the carpet was agglomerated sponge since it allows for an even spring effect throughout the surface, Also, it allows to analyze each part of the carpet independently, depending of the contraction in each zone. For it, strength sensors were used, where the 64 sensors allow for the device to have an 8×8 point resolution. Then, the force applied in each sensor permits to create a heat map according to the data obtained from sensors and represented by a color scale. However, it is possible to improve the resolution of the device with more sensors and therefore to obtain detailed information related to foot posture and weight distribution over the carpet to improve the AALADIN sensibility. Also, for the evaluation process it is necessary to assess the solution with more stakeholders to reach people with different physical and social conditions.

7 Conclusions and Further Work

This paper presents a first approach of an assistive technology device for ambient assisted living environments oriented to health personnel. This non-invasive device conforms a complement to the evaluators of the Tinetti test in the balance sub-scale, applied to elderly people. The representation in heat maps of the distributed weights on the carpet surface will provide a complement and a strengthening of information for the health personnel at the moment of evaluate each item of the test. The capability of the device to send the obtained data to the internet represents an important opportunity for it to become an IoT device and also access to the data remotely.

Besides, considering the relevance and potential on the field of action of this development it is necessary to extend this study, mainly taking into account the importance of the Tinetti test for supporting medical evaluations with elderly patients.

As further work, also it is interesting to evaluate user perceptions of the device. Then, it will be necessary to state controlled experiments related to measure the ease of use, usefulness, and intention to use in the future of the device. The measurements must be oriented to elderly patients and evaluators (doctors, physiotherapists). According to the obtained results work on quality of use, considering the accessibility and ergonimics in order to model the device to the perceived needs from the users. Even though, the developed device is focused on supporting the balance scale of the Tinetti test, it is also necessary to develop a complementary device oriented to measure the March scale of elderly people. The development of the future device will be aligned to the technologies used in AALADIN prototype. On the other hand, AALADIN takes part of a research project which works in the development of Fog Computing technology applied to AAL environments, therefore to assemble it as part of the project is another further work. The data obtained from the carpet will be part of a data mining analysis which will find repetitive patterns or rules in order to predict behavior of the patients.

As mentioned on the introduction of this paper, Assistive Technologies has considerable potential as partial substitutive or complement for several social and health care interventions [5, 8]. In this paper it has been shown that the development of AALADIN represents a complementary solution which contributes to obtain a better analysis for the elderly people when been evaluated in the Tinetti test.

Acknowledgment. This development is part of the research projects called (i) Design of architectures and interaction models for assisted living environments aimed at older adults. Case study: playful and social environments, winner of the XVIII DIUC Call for Research Projects; and, (ii) Fog Computing applied to monitoring devices used in assisted living environments; Case study: platform for the elderly, winner of the XVII DIUC Call for Research Projects. Therefore, thanks to the sponsor "DIUC, Direccion de Investigación de la Universidad de Cuenca" because of the support during this work. Besides, very special thanks to the Education Coordination Zonal 6 and educational authorities of the participating educational institutions that enabled the execution of this research.

References

1. Vasilakos, A., Pedrycz, W.: Ambient Intelligence, Wireless Networking, And Ubiquitous Computing. Artech House Inc., Norwood (2006)
2. Tapia, D.I., De Salamanca, U.: An ambient intelligence based multi-agent system for Alzheimer health care. Int. J. Ambient. Comput. Intell. (IJACI) **1**(1), 15–26 (2009)
3. de Salud, M.: Manual del Modelo de Atención Integral del Sistema Nacional de Salud Familiar Comunitario e Intercultural (MAIS - FCI). Minist. Salud publica del Ecuador, pp. 64–72 (2012)
4. Cook, D.J., Augusto, J.C., Jakkula, V.R.: Ambient intelligence: technologies, applications, and opportunities. Pervasive Mob. Comput. **5**(4), 277–298 (2009)
5. McCreadie, C., Tinker, A.: The acceptability of assistive technology to older people. Ageing Soc. **25**(1), 91–110 (2005)
6. Cowan, D.D., Turner-smith, D.A., Engineering, C.O.R.: The role of assistive technology in alternative models of care for older people. Res. HMSO **2**, 325–346 (1999)
7. World Health Organization: Assistive Technology. http://www.who.int/mediacentre/factsheets/assistivetechnology/%0Aen/
8. Roelands, M., Van Oost, P., Buysse, A., Depoorter, A.: Awareness among community-dwelling elderly of assistive devices for mobility and self-care and attitudes towards their use. Soc. Sci. Med. **54**(9), 1441–1451 (2002)
9. ITU: The Internet of Things. Itu Internet Rep., p. 212 (2005)
10. Domingo, M.C.: An overview of the Internet of Things for people with disabilities. J. Netw. Comput. Appl. **35**(2), 584–596 (2012)
11. Tinetti, M.E.: Preventing falls in elderly persons. N. Engl. J. Med. **348**, 42–49 (2003)
12. Nevitt, M.C., Cummings, S.R., Hudes, E.S.: Risk factors for injurious falls: a prospective study. J. Gerontol. **46**(5), M164–M170 (1991)

13. Sattin, R.W.: Falls among older persons: a public health perspective. Annu. Rev. Public Health **79**, 489–508 (1992)
14. Persons, O.: Risk factors for serious injury during falls by older persons in the community. J. Am. Geriatr. Soc. **43**, 1214–1221 (1995)
15. Tinetti, M.E., Speechley, M., Ginter, S.F.: Risk factors for falls among elderly persons living in the community. N. Engl. J. Med. **319**(26), 1701–1707 (1988)
16. Nevitt, M.C., Cummings, S.R., Kidd, S., Black, D.: Risk factors for recurrent nonsyncopal falls: a prospective study. JAMA **261**(18), 2663–2668 (1989)
17. Tinetti, M.E.: Performance-oriented assessment of mobility problems in elderly patients (1986)
18. Raîche, M., Hébert, R., Prince, F., Corriveau, H.: Screening older adults at risk of falling with the Tinetti balance scale. Lancet **356**(9234), 1001–1002 (2000)
19. Panella, L., Tinelli, C., Buizza, A., Lombardi, R., Gandolfi, R.: Towards objective evaluation of balance in the elderly : validity and reliability of a measurement instrument applied to the Tinetti test. Int. J. Rehabil. Res. **31**, 65–72 (2008)
20. Ballesteros, J., Urdiales, C., Martinez, A.B., Tirado, M.: Online estimation of rollator user condition using spatiotemporal gait parameters, pp. 3180–3185 (2016)
21. Ballesteros, J., Urdiales, C., B. Martinez, A., Tirado, M.: Automatic assessment of a rollator-user's condition during rehabilitation using the i-Walker platform **4320**(c) (2017)
22. Bruno, B., Christophe, B., Véronique, F., Bart, J.: A preliminary study of the integration of specially developed serious games in the treatment of hospitalized elderly patients (2017)
23. Giansanti, D., Member, I., Tiberi, Y., Maccioni, G., Member, I.: The Codivilla-Spring for daily activity monitoring, pp. 4720–4723 (2008)
24. Jaume-i-capó, A., Martínez-bueso, P., Moyà-alcover, B., Varona, J.: Interactive rehabilitation system for improvement of balance therapies in people with cerebral palsy. IEEE Trans. Neural Syst. Rehabil. Eng. **22**(2), 419–427 (2014)
25. Pina, J.A.L.: Análisis psicométrico de la escala de marcha y equilibrio de Tinetti con el modelo de Rasch. Fisioterapia **31**(5), 192–202 (2009)
26. Conroy, S., et al.: A multicentre randomised controlled trial of day hospital-based falls prevention programme for a screened population of community-dwelling older people at high risk of falls. Age Ageing **39**(6), 704–710 (2010)
27. Blank, W.A., et al.: An interdisciplinary intervention to prevent falls in community-dwelling elderly persons: protocol of a cluster-randomized trial [PreFalls]. BMC Geriatr. **11**, 7 (2011)
28. de Vries, O.J., et al.: Multifactorial intervention to reduce falls in older people at high risk of recurrent falls: a randomized controlled trial reducing falls in older people at high risk. JAMA Intern. Med. **170**(13), 1110–1117 (2010)
29. Paunovic, N., Kova, J., Rešetar, I.: A methodology for testing complex professional electronic systems *. Serb. J. Electr. Eng. **9**(1), 71–80 (2012)

Target Localization with Visible Light Communication in High Ambient Light Environments

Kofi Nyarko[1(✉)] and Emmanuel Shedu[2]

[1] Morgan State University, 1700 East Cold Spring Ln.,
Baltimore, MD 21251, USA
kofi.nyarko@morgan.edu
[2] Johns Hopkins University, 3400 N. Charles Street,
Baltimore, MD 21218, USA
eshedu1@jhu.edu

Abstract. This paper builds on prior research that demonstrated how inexpensive commercial off-the-shelf lighting components and microcontrollers could be used to construct a solution for occupant and asset localization and tracking through visible light communication (VLC). A significant challenge encountered in the prior work was mitigating the effect of ambient light optical interference. This paper describes the implementation of several techniques to negate the effect of ambient light interference under varying conditions. The techniques involved modifications to the modulation scheme employed, redundant transmissions over multiple wavelengths, and an adaptive digital filtering technique. Furthermore, this paper discusses the challenges involved with implementation the approach on a severely resource constrained microcontroller and the optimization strategies employed. The overall effectiveness of the system is measured and discussed.

Keywords: Visible light communication · Indoor localization · Indoor position estimation · Ambient light compensation

1 Introduction

A recent study shows that about 17% of total electricity consumed in the residential and commercial sector is used for lighting [1]. The primary method of reducing this energy consumption is by replacing traditional incandescent and compact fluorescent light fixtures with low-energy consuming Light Emitting Diode (LED) fixtures. A few years ago, this method proved to be costly, however due to the economies of scale, the costs have been greatly reduced and are finally competitive with traditional lighting methods, such as incandescent and compact fluorescent lighting. In addition, the development of white LEDs (WLEDs) have helped spur their adoption. Due to their very nature, LEDs provide efficient yet highly controllable light that is well suited for visible light communication (VLC) and its potential as a method for occupant and asset localization.

In previous work [2], we presented an indoor localization technique that overcame the limitations of radio frequency (RF) based location sensing methods, which typically suffer from positioning error from multipath fading, and infrared (IR) methods, which are typically limited by non-line of sight (LOS) conditions in indoor environments [3]. We proposed an indoor positioning system that uses an infrastructure of LED lighting units where positioning is accomplished through trilateration of sensed units with approximated distance vectors from the receiver. Since ceiling lights can be seen from all illuminated areas within a tracking volume, the system performs under LOS conditions where multipath fading is not a limiting factor. The use of infrastructure lighting presents a cost-effective solution for wide scale deployment suitable in RF limited areas, such as hospitals.

However, under additional testing, a drawback from this approach became apparent. In the presence of a significant amount of ambient light, error is induced at the receiving tags. Essentially, the presence of high intensities of daylight, or other non-coded light sources, degrading the signal to interference ratio (SIR) and effectively drowns out the incident coded light at the receiver. In order to mitigate this effect, several techniques were implemented and tested to improve SIR under hash ambient lighting conditions. The following sections detail the outcomes of these implementations.

2 Background

VLC is data communication using safe electromagnetic radiation visible to the human eye. Unlick RF wireless communication, VLC does not interfere with radio frequencies, there is no regulation on bandwidth allocation, eavesdropping is limited since communication is confined by opaque walls, and radiation is generally considered safe [4]. Like other communication systems, VLC requires a transmitter and a receiver. The transmitter transmits an encoded signal through the modulation of visible light intensity, which is sensed by the receiver and processed accordingly.

One issue unique to VLC systems, is that there may be other optical visible light sources that contribute interference at the receiver. A common technique used to reduce optical noise interference is adaptive filtering [5]. This technique needs to estimate the channel characteristics and the interference signals. It then uses a linear prediction coefficient technique to equalize the received signal. This approach requires continuous adaptive monitoring and feedback.

Recent research [6] showed that it may be possible to accomplish reduction of noise interference without adaptive monitoring through non-return-to-zero inverted (NRZI) modulation. A similar effect may also be possible through Manchester coding and Hadamard error correcting code [7].

Another approach to mitigating the effect of optical signal degradation is through the implementation wavelength filtering [8]. This approach requires the use of LEDs capable of illumination across 2 or more wavelengths. Specifically, this approach would require the use of Red-Green-Blue (RGB) LEDs.

3 Target Localization Platform

The previously developed system [2] consists of a set of lighting modules and receiving tags that are comprised of inexpensive off-the-shelf components, such as low-cost microcontrollers capable digital I/O operations, high power LED capable of producing 230 lumens of illumination at 3 V, field effect transistor (FET) based driver circuits, IR transmitters/receivers, and standard phototransistors. The lighting modules are arranged in a grid, where each unit is interconnected with two adjacent neighbors at a fixed distance from each neighbor. Each unit has four pairs of IR LED diodes and photo detectors on each face, which enable each module to communicate with two adjacent units as shown in Fig. 1.

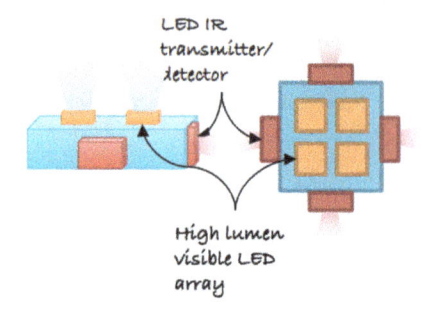

Fig. 1. Visible light communication units

Lighting units are grouped into chains, where a virtual token (a special communication packet) is initialized to the group number and ID of the next module in the chain and broadcasted by the "Head of Chain" using all four connected IR diodes. Upon receipt of the token, the module whose ID matches the ID indicated in the token, sends out its unique ID over the VLC array, modifies the token to reflect the next module ID in the chain and rebroadcasts the token. This process is repeated until the "Tail of Chain" module receives the ID. This module modifies the token by setting the next ID to the ID of the previous module in the chain and toggles the direction of transmission to indicate that the token will be traveling in the reverse direction. This sequence is repeated when the token arrives at the "Head of Chain" module so as to cause the token to continuously travel up and down the chain. Through this method, the VLC communication of each module is time division multiplexed based on the location of the token, which illuminates packet collision on a single chain.

The receiver is a photo detector equipped microcontroller that detects the unique IDs of the closest n number of units within visible range and calculates its relative position through trilateration. Once each signal is received, its power at the receiver is used to determine the approximate distance between transmitter and receiver.

In order to assess target localization effectiveness in real-world conditions, three sites were identified on the campus of Morgan State University for testing. The first site (1) was in a 320 sq. ft. research lab with three exterior east facing windows with artificial illumination provided solely by the VLC modules. The second site (2) was a

500 sq. ft. classroom with no exterior windows and artificial illumination from preexisting fluorescent lamps as well as the VLC modules. The third site (3) was a 1,500 sq. ft. stock room with artificial illumination provided by incandescent lighting as well as the VLC modules. As discussed in the results section, the prior VLC decoding scheme proved effective in situations where the only illumination provided was through the VLC modules, as is the case in site 2. Hence further research was needed to implement an ambient light filter to improve performance in situations akin to sites 1 and 3.

4 Implementation

The target location system employs a variation of Pulse Position Modulation (PPM) that has proven to be efficient and less complex. In DPPM, data encoding is done regardless of a clock. Each pulse takes reference from the rising or falling edge of the previous pulse. This simplifies synchronization between the devices. In this work, DPPM is implemented using inverted pulses since the LED must be on for most of the time for illumination and to prevent flickering.

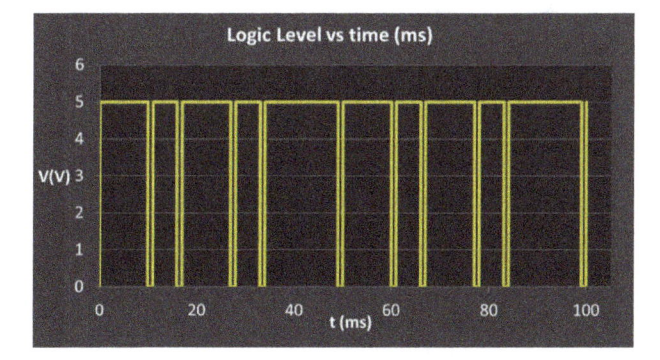

Fig. 2. Encoded (10101) visible light reception with DPPM

As illustrated in Fig. 2, when the bit is 0, the pulse is delayed for some time t_0 while when the bit is 1, the pulse is delayed for some time t_1 which is greater than t_0. According to Rufo et al., other modulation techniques require complex synchronization system to ensure correct detection [9]. Even though DPPM was chosen as the preferred encoding scheme for the system based on the computational limitations of the hardware, attempts were made to implement other modulation and coding schemes in order to determine their effectiveness in ambient lighting conditions indicative of commercial operating environments. The general specifications of the microcontrollers used are given in Table 1.

One of the first coding schemes implemented in place of DPPM was Manchester coding, which encodes both clock and data into a single signal and transmits serially. The algorithm used to implement this encoding is as follows: the data are represented

Table 1. Low cost microcontroller specifications

Operating voltage	5 V
Input voltage	7–12 V
DC current per I/O pin	40 mA
Flash memory	32 KB
EEPROM	1 KB
Clock speed	16 MHz

with line transitions, where a logic 0 is represented by a transition from HIGH to LOW, and a logic 1 is represented by a transition from LOW to HIGH. Manchester encoding is considered a non-return-to-zero (NRZ) encoding that is exclusively-ORed with the clock. NRZ has a level the matches the logical signal. In this case the clock would latch the value being transmitted at the start of each clock cycle and it would be sampled at the receiving end in the middle of the clock cycle.

Since other research [6] suggests that noise reduction is also possible with NRZI modulation, this scheme was also implemented for comparison. The logic level transitions for Manchester and NRZI are displayed in Fig. 3. Normally, the NRZI can be decoded by using XOR gate at the receiver with two consecutive bits sampled at synchronized timing. As recommended by [6], waveform processing was used instead of the XOR gate to optimize the equalization and decoding process.

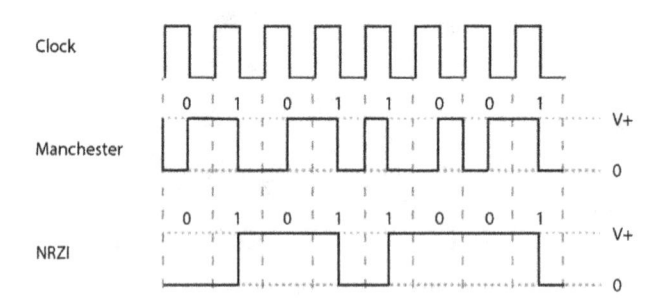

Fig. 3. Logic level transitions for manchester and NRZI encoding

As later discussed in the Results section, the implementation of both of these encoding schemes did not result in any significant resiliency of the system in the presence of ambient illumination levels in test site 1 and 3. Hence another attempt at ambient light rejection as made using the Hadamard code with the aforementioned encoding schemes.

The Hadamard code is an error-correcting code commonly used for error detection and correction of transmitted data over unreliable and noisy channels. The Hadamard code is an example of a linear code over a binary alphabet that maps messages of length k to code words of length 2^k. Normally, Hadamard codes are based on Sylvester's construction of Hadamard matrices, but the term—Hadamard code is also used to refer to codes constructed from arbitrary Hadamard matrices. The Hadamard code is a locally decodable code, which provides a way to recover parts of the original message with high probability [7]. Hadamard code uses Hadamard matrices for encoding and decoding error-correcting codes. Figure 4 shows sample Hadamard matrix where each row is a possible code.

Fig. 4. Sample hadamard matrices for encoding/decoding [7]

While this method did prove that single bit errors could be easily corrected, it did not significantly impact the resiliency of the tracking system in the presence of ambient light interference. In addition, this method proved to be too computationally intensive to maintain the update rate of 44 Hz with the low-cost microcontrollers that were used in the VLC system. Specifically, the ATMega328 microcontroller was implemented in the both the transmitter lighting modules and the receiver tags.

Another approach that was explored was based on research that suggests that transmitting at different wavelengths may reduce optical interference for VLC systems [8]. In order to proceed with this implementation, the white LEDs which were used for the luminaire modules had to be replaced with RGB LEDs. The specifications of the white and RGB LEDs are given in Table 2.

Table 2. White and RGB LED specifications

LED color	Neutral white	Red (645 nm)	Green (540 nm)	Blue (485 nm)
Brightness	230 lm	96 lm	190 lm	70 lm
Current draw	700 mA	700 mA	700 mA	700 mA
Beam angle	120°	125°	125°	125°
Maximum forward voltage	3.25 Vf	2 Vf	3.5 Vf	3.5 Vf

The photodiode of the receiving tag was replaced with an RGB color sensor that consisted of a 3-channel Si photodiode with the following sensitivities: 460 nm (blue), 540 nm (green), and 620 nm (red). The photosensitive area of the sensor is 1 mm. The DPPM, Manchester, and NRZI encoding schemes were utilized in the lighting module with the RGB LEDs. For robustness, the same data stream was modulated across all three wavelengths in the manner that the optical interference from an ambient lighting source would have to significantly degrade SIR across all three wavelengths to affect the receiver's ability to decode the VLC transmission. Initial testing shows that this method may prove to be viable option as discussed in the results section, however, the drawback of this approach requires upgrading all white LEDs in the lighting modules to RGB LEDs which carry a slightly higher cost.

The final approach attempted was one in which no system hardware changes were required. Rather, a continuously adaptive digital filter was applied to the received signal. The receiver models ambient light through the use of a digital running average filter. There are several implementation methods possible, such as a mean filter [10], and a median filter [11], however, these implementations require large memory capacity, which is a resource that is severely lacking in the microcontrollers used in this project. Other implementations, such as Eigen-background [12] and Mixture of Gaussian [13, 14] have additional computational complexity, which is also prohibitive on small microcontrollers. Due to its high memory compactness and low computational complexity, the running average method, also called the cumulative moving average (CMA) method, was used.

With CMA, light intensity, I, is sampled in an ordered stream and the average is calculated across the last n intensity points.

$$CMA_n = \frac{I_1 + \ldots + I_n}{n} \quad (1)$$

This brute-force approach requires the storage of n intensity samples in a buffer of fixed length n, such as an array, where all the values are continuously shifted as new data points arrive. However, this method is memory and data intensive for the microcontroller. Rather the cumulative average can be updated with each new sample point by simply updating the CMA as follows:

$$CMA_{n+1} = \frac{I_{n+1} + n.CMA_n}{n+1} \quad (2)$$

With an appropriately chosen value for n, CMA can effectively characterize the ambient light. Filtering out the ambient light simply becomes a matter of subtracting the CMA from the current intensity value. The receiver analyzes the resulting signal for the last k clock cycles of the DPPM encoded signal to establish a new, possibly non-zero, threshold value.

$$\hat{I} = I_{n+1} - CMA_n \quad (3)$$

$$I_{th} = \langle \min(\hat{I}_{1\ldots k}) + \max(\hat{I}_{1\ldots k}) \rangle \quad (4)$$

Since DPPM guarantees logic level transitions every clock cycle, the mean minimum and maximum values of the signal $(\hat{I}_{1...k})$ can be easily determined, where the transition threshold value (I_{th}) is just the mean of the sum of the minimum and maximum values. As discussed in the results section, this method proved the most effective in the 3 test cases while retaining the baseline position update frequency of 44 Hz.

5 Results and Discussion

As mentioned in the previous section, multiple modulation schemes were implemented to ascertain their effect on optical interference from ambient light. The modulation schemes tested were DPPM, Manchester and NRZI. The performance of each modulation scheme was assessed through the percentage of frames that were either not received at all or were not successfully decoded. A cyclic redundancy check (CRC) was performed on all frames to detect errors and correct if possible. The system was run for a period of 5 min during which, the receiver tag was moved in a raster scan pattern (where possible) within the environment under test, as shown in Fig. 5. The blue line represents the path of receiver motion in one of the 3 test cases, where the room was either illuminated by lighting from the VLC modules (red squares) or a combination of VLC modules and ambient lighting from windows or another lighting source.

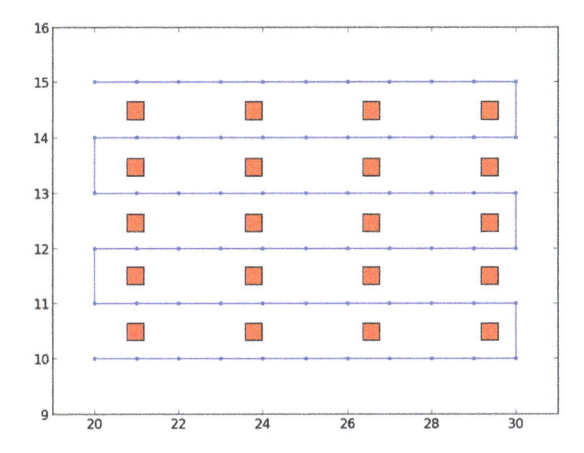

Fig. 5. Raster pattern of receiver motion (blue line) across test environment illuminated by VLC tracking modules (red squares)

Based on the expected update rate of the lighting chain used, the receiver calculates the number frames that should have been received and successfully decoded over the test period. Table 3 shows the results obtained with each modulation scheme. The ambient light illumination range for the 320 sq. ft. research lab with exterior windows (case 1) was 50–830 lx. The range for the 500 sq. ft. classroom with no exterior windows (case 2) was 0 lx (only lighting is from the VLC modules). The range for the 1500 sq. ft. stock room with incandescent lighting (case 3) was 200–275 lx. The NRZI and Manchester modulation schemes do not appear to provide any additional resiliency against optical interference from ambient light. However, the result shows that NRZI did perform better in the case of incandescent ambient lighting. This may be due to the fact that the interference frequency of this type of ambient light is effectively below 200 kHz. Prior research suggested that NRZI can effectively reduce the optical interference in low-frequency band from DC to over 200 kHz. This would imply that this type of ambient light interference has low-pass characteristics and hence can be effectively rejected by using NRZI modulation.

Table 3. VLC performance with DPPM, manchester and NRZI modulation

Modulation scheme	Percent frame loss	Test case	Ambient illumination (lux)
DPPM	46.2	1	50–830
Manchester	49.1	1	50–830
NRZI	48.9	1	50–830
DPPM	1.8	2	0
Manchester	1.9	2	0
NRZI	1.6	2	0
DPPM	32.6	3	200–275
Manchester	35.2	3	200–275
NRZI	29.1	3	200–275

A similar test was conducted using RGB LEDs and RGB color sensors. Duplicate frames were sent across all three wavelengths for redundancy. The receiver was modified to independently decode the VLC transmission from each color channel and simply select the first successfully decoded frames across the three wavelengths. This process was performed across all three aforementioned modulations (see Table 4).

Table 4. VLC performance using RGB LEDs and color sensors

Modulation scheme	Percent frame loss	Test case	Ambient illumination (lux)
DPPM	35.8	1	50–830
Manchester	36.4	1	50–830
NRZI	38.7	1	50–830
DPPM	1.9	2	0
Manchester	1.7	2	0
NRZI	1.9	2	0
DPPM	23.9	3	200–275
Manchester	26.3	3	200–275
NRZI	28.5	3	200–275

On average, redundant VLC transmission across multiple wavelengths resulted in an approximate 11% improvement in frames dropped. It is possible that these results could be further improved by adaptive digital filtering on each channel to remove the ambient light contribution on the corresponding wavelength, and constructively combining the resulting signal.

The final approach tested across the same 3 test cases was the cumulative moving average adaptive digital filter. The approach of updating the cumulative average with each new sample point rather than storing all samples within a fixed buffer was used. From Eq. 2, $n = 128$ resulted in the least dropped packets across all test cases (see Table 5).

Table 5. VLC performance using the CMA digital filter with n = 128

Modulation scheme	Percent frame loss	Test case	Ambient illumination (lux)
DPPM	26.2	1	50–830
Manchester	28.1	1	50–830
NRZI	29.8	1	50–830
DPPM	1.8	2	0
Manchester	1.9	2	0
NRZI	1.9	2	0
DPPM	8.2	3	200–275
Manchester	10.1	3	200–275
NRZI	9.3	3	200–275

The cumulative moving average digital filter managed to reduce the number of dropped frames by approximately 20% in test cases 1 and 3, which contained ambient light interference. This was accomplished with no further changes to hardware and with little impact on the update rate of the VLC system, due to the fact that the algorithm employed wasn't particularly memory or computationally intensive. Further work needs to be conducted to evaluate the effect of combining the CMA digital filter with multi-wavelength VLC transmissions.

6 Conclusion

This paper has evaluated several methods for reducing the optical interference from ambient light for indoor localization based on visible light communication. The methods that proved the most effective were broadcasting redundant information across multiple wavelengths and using a cumulative moving average adaptive digital filter to dynamical set the threshold for detecting transition changes in the VLC signal. The multi-wavelength approach reduced optical interference by approximately 11% and the CMA digital filter reduced optical interference by approximately 20%. There could be a possibility for greater gains if both approaches are combined.

References

1. How Much Energy is Used for Lighting. EIA. http://www.eia.gov/tools/faqs/faq.cfm?id=99&t=3
2. Nyarko, K., Emiyah, C., Mbugua, S.: Building occupant and asset localization and tracking using visible light communication. In: Proceedings of the SPIE, Automatic Target Recognition XXVI, vol. 9844, pp. 98440B, 12 May 2016. https://doi.org/10.1117/12.2224089
3. Bahl, P., Padmanabhan, V.N.: RADAR: an in-building RF-based user location and tracking system. In: Proceedings of the IEEE INFOCOM 2000, March, vol. 2, pp. 775–784 (2000)
4. Yang, Y., Chen, X., Lin, Z., Liu, B., Chen, H.D.: Design of indoor wireless communication system using LEDs. In: Communications and Photonics Conference and Exhibition (ACP), Asia, vol. 2009, pp. 1–2, 2–6 November 2009
5. Yánez, V.G., Torres, J.R., Alonso, J.B., Borges, J.A.R., Sánchez, C.Q., González, C.T., Jiménez, R.P., Rajo, F.D.: Illumination interference reduction system for VLC Communications. In: Proceedings of the WSEAS International Conference on Mathematical Methods, Computational Techniques and Intelligent Systems, pp. 252–257 (2009)
6. Liu, Y.F., Yeh, C.H., Wang, Y.C., Chow, C.W.: Employing NRZI code for reducing background noise in LED visible light communication. In: 18th OptoElectronics and Communications Conference Held Jointly with International Conference on Photonics in Switching (2013)
7. Gour, S., Murarka, S., Kumar, S.: Review on reduction of optical background noise in light-emitting diode (LED) optical wireless communication systems using Hadamard error correcting code. Intern. J. Emerg. Technol. Adv. Eng. 4(9), 233–235 (2014). ISSN 2250-2459, ISO 9001:2008 Certified Journal
8. Yang, S.-H., Kim, H.-S., Son, Y.-H., Han, S.-K.: Reduction of optical interference by wavelength filtering in RGB-LED based indoor VLC system. In: Proceedings of the 16th Opto-Electronics and Communication Conference, OECC, pp. 551–552, July 2011
9. Rufo, J., Quintana, C., Delgado, F., Rabadan, J., Perez-Jimenez, R.: Considerations on modulations and protocols suitable for visible light communications (VLC) channels: low and medium baud rate indoor visible light communications links. In: 2011 IEEE Consumer Communications and Networking Conference (CCNC), pp. 362–364, 9–12 January 2011
10. Lo, B.P.L., Velastin, S.A.: Automatic congestion detection system for underground platforms. In: International Symposium on Intelligent Multimedia, Video and Speech Processing, pp. 158–161 (2000)
11. Cucchiara, R., Grana, C., Piccardi, M., Prati, A.: Detecting moving objects, ghosts and shadows in video streams. IEEE Trans. Pattern Anal. Mach. Intell. 25(10), 1337–1342 (2003)
12. Oliver, N.M., Rosario, B., Pentland, A.P.: A Bayesian computer vision system for modeling human interactions. IEEE Trans. Pattern Anal. Mach. Intell. 22(8), 831–843 (2000)
13. Stauffer, C., Grimson, W.E.L.: Adaptive background mixture modelsfor real-time tracking. In: IEEE International Conference on Computer Vision and Pattern Recognition (CVPR), pp. 246–252 (1999)
14. Dong, W.T., Soh, Y.S.: Image-based fraud detection in automatic teller machine. Int. J. Comput. Sci. Netw. Secur. (IJCSNS) 6(11), 13–18 (2006)

A Minimal Social Weight Wearable Device for Thermal Regulation

Javier Benedicto Serrano and Sudhsnahu Kumar Semwal[✉]

Department of Computer Science, University of Colorado, Colorado Springs, USA
`{jbenedic,ssemwal}@uccs.edu`

Abstract. Temperature changes can be severe during mountain-trekking based on the sun location and trail path. The task of keeping the body warm and comfortable to avoid hypothermia can be mitigated by a wearable device which records real-time data and analyzes it. In this paper, we propose a wearable device for body temperature regulation called warmUp which has minimal social weight. This device can predict, trained on the previously collected users' data, whether or not it is cold for that person. The wearable device is compact and can be hidden inside a vest. A simple binary classifier indicated that the user is getting cold by using present temporal temperature variations. We train the neural network by first collecting the data using the input sensors hidden inside the vest and asking the user to classify the data. Future applications of our device are: automatically warming the vest which the user is wearing; fire fighting applications, and wilderness survival.

Keywords: Wearable computing · Outdoor applications · HCI

1 Introduction

Technological advances are improving computing power, reducing hardware sizes, and decreasing the price of wearable and intelligent devices. Our goal is to assemble and program a wearable computer capable of predicting if a person is getting cold or not, and provide notification. Hence the term warmUp for our system. The applications of this project are wide and have the potential of impacting several real life areas. Our first idea was to conceive a device meant to be worn by hikers and people who love outdoors. If during an activity the users were in risk of suffering health issue due to cold, our device would help by automatically and notify the person to make provisions. We notify the users when cold predictions are made by sending them alert messages directly to their smartphones. During extreme conditions, this may also provide a way to connect to emergency service if the need arise.

There are at least two other specific areas in which our device could be helpful: Aged and Elderly could be provided another level of defense against suffering hypothermia or other issues related due long time cold exposure. Another interesting area of usage would be the smart housing market, as we could monitor in

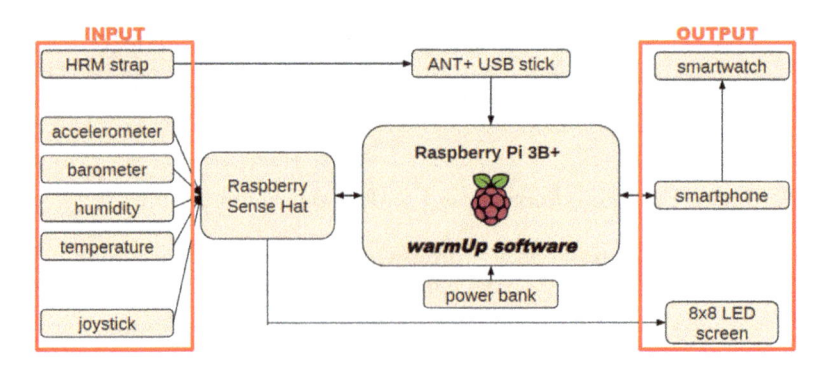

Fig. 1. WarmUp input/output diagram.

real time a person's thermal sensation [13]. This intelligent ambient temperature regulation application would lead to sustainable houses capable of using energy in a more efficient way. Third application where our research could be useful is for fire-fighters as once.

2 Related Work and Motivation

With new devices becoming smaller, sleeker, and fashionable wearable-devices are becoming popular in our culture, specially forming social computing. Smartphones are only a few example of devices exploding in the social scene in just a last few years. Wearable computing surrounds an individual with technology, and so most importantly must have minimal social weight [4]. Wearable computing has also found applications in fashion [4], perspective maps [1], annotate meetings [2], contextual [3], be part of kindergarten activities [5] and tangible interfaces such as beads [6], and crossover applications [7], among others. Wearable-environments allow us an opportunity to express, be creative, and quickly perform tasks which were otherwise impossible. At the minimum these systems are expected to understand verbal commands, provide feedback through a variety of output displays such as sound or displays mounted on the hand itself. RFID Tags and Tangible interfaces can communicate with the technology around it and aids with contextual information. PhysioHMD [10] used a comfortable HMD to collect physiological data. A multi-material mono-filament is proposed towards creating advance smart garments with minimal social weight [9]. A special recurrent neural network is proposed for human activity recognition [8]. Another wearable device called Embr [11] is a watch capable of raising or reducing the user's thermal sensation by 5 °C upon user's explicit feedback. Our goal is different as we want to be able to predict whether a person is getting cold or not. Recently new devices measure temperature using temperature sensors near the nostrils [12].

3 System Overview

System overview of our system is described in Fig. 1. The basic constraints when looking for hardware were the size, the price, the ability to connect with the central processing unit and the usefulness of the data provided by the sensors. The hardware diagram of the system can be seen at Fig. 1.

The main component of the project consists of Raspberry Pi 3B+, running Raspbian OS (Linux distribution) [18]. In addition, other components, which are part of our implemented wearable system are as follows:

- *Raspberry Sense HAT*: This device is connected on top of the Raspberry Pi using its 40 GPIO pins and provide us multiple ambient sensors, a joystick to control the program interface and a 8 × 8 LED display helpful to provide a graphical interface to the end user. Sensors used by our software were the accelerometer unit, the barometer, the humidity and the temperature sensor [14].
- *Garmin HRM and ANT+ USB stick*: The Garmin Heart Rate Monitor works as an input sensor providing real-time data of the user's heart beat. Since this sensor uses a proprietary wireless system, an ANT+ USB stick attached to the Raspberry Pi was needed. An external library *ANT+* was needed to communicate and extract beats per minute (bpm) data from the HRM strap. Some "random" instability results due to bugs produced by the aforementioned library in our implementation.
- *Smartphone and Smartwatch*: In order to receive notifications, a smartphone was needed. When pairing a smartwatch with it, we are able to receive cold predictions on our wrist without the necessity of looking at a phone. The mobile device works as a hot-spot itself providing Wi-Fi to the Raspberry, enabling wireless communication. This way, the system can send messages to the user through Internet.
- *Power bank*: this device was simply used to enable portability to our Raspberry based system.

Excluding the smartphone and smartwatch from the equation, the approximate price of the whole system was $250. The warmUp software was based on Python version 2.7. The choice of Python was due to its simplicity and extended usage over the scientific community.

4 Operation Modes

Three indispensable operation modes where provided within our user interface. Those are, the recording mode, the training mode and the prediction mode. A simplified scheme on how those modes work is showcased in Fig. 2.

4.1 Recording

The recording method builds a training dataset. Each of them are normalized to be between 0 and 1 in order to train the model properly.

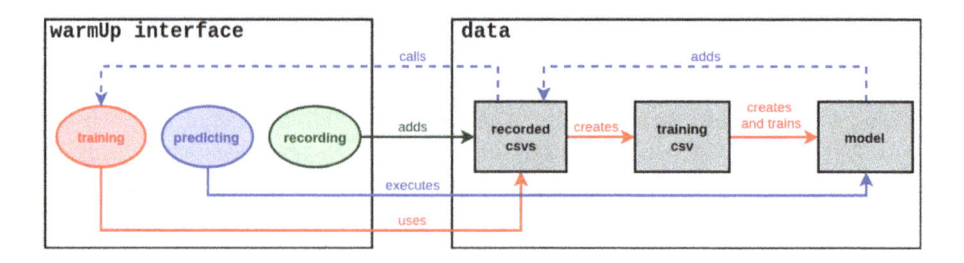

Fig. 2. Operation modes available within the warmUp interface.

Dynamic Data. Five types of dynamic data are considered in our system:

- *Temperature*: In Celsius degrees, this attribute is one of the most important inputs of our system. Our temperature sensor can read values from −40 °C to 160 °C.
- *Humidity*: Captures the relative humidity expressed as percentage. The values of humidity can range from 0% to 100%.
- *Pressure*: Captures the pressure of the ambient measured in hectopascals. For normalization considerations, we know that our sensor can capture ranges of pressure between 260 hPa and 1260 hPa.
- *Movement*: This metric denotes whether the user is doing some kind activity or is idle. Since the accelerometer readings can be quite inaccurate, we decided to compute the squared difference on the 3 acceleration axis between the last time and actual time step of readings. If the difference appears to be higher than a certain threshold, we consider the attribute as 1 (there is movement), otherwise we save it as a 0. Our assumption about this metric is that users doing a non-static activity should be slightly less prune to feel cold. Due to arbitrary spikes sometimes in our accelerometer sensor, averaging helps to keep things in perspective.
- *Heart Rate*: This metric captures the user's heart rate in beats per minute. It denotes the level of activity carried out by the user. We assume that the higher the heart rate is, the more activity is done by the user meaning that the cold tolerance should increase. This metric is a custom touch to our system and we think it might make a difference in the cold prediction task. To normalize this input data we consider the formula $max_{HR} = 207 - 0.7 \cdot age$.

Static Data. It is a known fact that each person can have different cold perceptions due to multitude of factors. So WarmUp collects the gender, age, height, weight values of the user. Our assumption is that people with higher body fat (high weight and short height) might have more tolerance to cold. We consider a maximum height of 250 kg for normalization purposes but that can be changed as needed.

The last parameter recorded by the system is the perception of cold. This is explicitly inputed by the user at each time step. This is not an input attribute but the ground truth needed to label each our training samples for eventual use

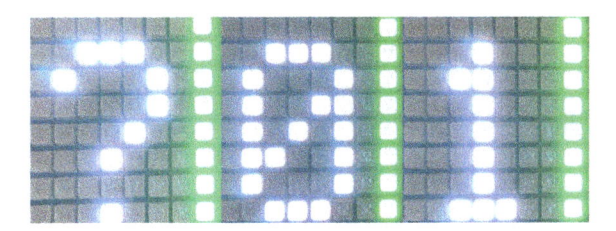

Fig. 3. Ground truth values displayed on screen during the recording mode. The "?" denotes that the system is waiting for the user's initial ground truth input.

for our classified. We consider 0 as the user being fine and 1 as the user having cold thermal sensation. As seen in Fig. 3, the user can use the joystick to change the ground truth to 1 (up) or to 0 (down). At the initial stage, the interface shows the symbol "?" until the user generates a joystick event and we will consider the label of this special event as 0. Every 15 s the recording program reads all the input values from sensors and generates a training sample.

Before starting to record any training sample, the program needs 1 min to calibrate the sensors. This is important since some sensors return a value of 0 during initial readings. Also, the HRM requires some time to stabilize as to produce accurate heart rate readings. After the user terminates the recording session (upon using the left joystick), the program generates a JSON file and CSV file with all the recorded training samples. Those files are named with the name of the user who recorded the data plus the timestamp value of the time the recording mode initiated. While the CSV file is used to train the model, the JSON file is just useful to check the values of the samples since we store unnormalized values in it.

4.2 Training

We expect users to run several recording sessions before proceeding to train the actual model. The training mode first merges all the CSV files created over multiple executions of the previous mode into a single one. This merged CSV will have 10 columns (1 label and 9 attributes) and as many rows as data samples. Then, the file is fed to our model for training purposes. This model is created using the Keras library, a high level library built on top of Tensorflow [15,16]. Training module defines our model architecture, initializes it with random weights, trains it with the merged train CSV by performing 150 epochs with batch size of 10 and finally saving the model structure (as a JSON file) as well as the trained weights (as a H5 file) within two separate files.

The Model: In between the input and the output of our network, we have two hidden layers with 5 neurons each. One of those neurons is a bias neuron. The model then, is defined as a feed-forward network (no loops or backward connections) and as fully connected as all the neurons have weighted connections with all the nodes of its next layer (bias neurons are excluded from that definition). At the end, we end up with an ANN with 64 trainable parameters, also known as internal parameters or weights. The activation functions used for the hidden layers are Rectifier Linear Units (ReLu) and the Sigmoid function for the output layer. Instead of the common Stochastically Gradient Descent optimizer, our model uses an Adam optimizer minimizing a binary cross entropy based in the accuracy metric of our predictions.

4.3 Predicting

The warmUp wearable device predicts the cold thermal sensation of users. Its work flow is similar to the recording mode in the sense that the system has to be calibrated and that at each time step, 9 attributes are recorded in order to make a guess.

As to make predictions, once the system has been calibrated, we need to load both the model and the trained weights first initialized during the training stage. We feed the 9 recorded attributes into our binary classifier network. Since our network returns a single output ranging between 0 and 1, we classify the prediction as cold when the value is higher than 0.5 or as not cold when the results is lower than 0.5. We wait for four consecutive cold readings before showing to the user the prediction swap (either from not cold to cold or the opposite way). In this manner, the user will perceive smoother system predictions. This four consecutive readings technique worked well and created better accuracy. Also such averaging somewhat eliminates isolated prediction/classification errors, and predictions are more robust. Also such averaging creates smoother and gradual transitions from cold to not cold, and vis-a-vis. This also represent the idea that humans do not feel suddenly cold all of a sudden or suddenly warm. When four straight cold predictions indicate change of state, the system executes a method to notify the user about the issue.

In order to show feedback to the user, the warmUp display shows the two digits ambient temperature readings on screen [14–16, 18]). Then, to show the predicting value we use a color coding (see Fig. 4). This means that when the temperature values are displayed using white light, the system is predicting a not cold thermal sensation. Otherwise, the system system displays blue temperature

Fig. 4. Prediction feedback on the sense HAT screen during the prediction mode.

Fig. 5. Prediction feedback marked as wrong during the prediction mode.

values when cold predictions are made. We also provide a way to manually correct a prediction. For example, a user can let the system know about a prediction error. To do so, if the user uses the up joystick (see Fig. 5), the system will show a red bar above the temperature readings (down joystick can be used to go back to accepted predictions). Upon doing so, the system saves the time-step recorded attributes in conjunction with the correct prediction label, meaning that if the prediction showed was not cold and the user explicitly said it was wrong, all the input attributes are labeled as cold instead. See the broken blue arrows in Fig. 2): We can run the train mode again on the updated set of training samples. The binary classifier model will be initialized and trained again from scratch.

5 Evaluation of Results

To address the accuracy tests, a training dataset of roughly 3000 labeled samples was used. This dataset was split into 70% test and 30% training. Our initial results had an accuracy results performing a 5-fold cross validation were at 65.7%; the best accuracy obtained by a single test was 69.3%. Usually 85%–90% of accuracy should be considered achievable. In future, we plan to look at: (a) overfitting due to the relative small size of our training dataset. Working on building a larger samples dataset would contribute to reduce the overfitting problem of our system; (b) The samples are too clustered within the dataset space even after adding the mentioned Gaussian noise to training samples. More dataset samples and variety could improve this situation. (c) Equal number of samples of cold and not feeling cold will eliminate the bias as our samples were mostly classified as not feeling cold. Over the training dataset used, approximately 70% of the samples were labeled as no cold in contrast with a 30% labeled as cold. We need roughly 50% of each label within the training samples dataset. (d) The network architecture could be improved.

6 Reacting

The ultimate goal of the warmUp application is to let the user know the cold prediction. We implement this feature by using a Telegram account. Telegram is a free message application and allows the use of Telegram bot. Since Telegram has a well documented and extended API support, the process of creating a bot and sending messages was fairly simple [17,18].

The only special parameters we need are a bot token (obtained when the bot is created) and a chat ID. The chat ID can be obtained by the user after starting a conversation with the bot. This is compulsory since this way Telegram prevents bot spamming. At practice, this means that a real user has to start a conversation first with a bot in order to receive messages from it. After this is done, the two parameters are inputed into the warmUp system to properly configure the bot notifications. With the creation of an API request, the final notification can be send to the user's profile immediately upon cold transitions.

With the purpose of making the warmUp notifications more useful, some ambient and user's metrics at the moment of the cold prediction are also sent in conjunction with the cold warning (see Fig. 6). Those metrics are the ambient temperature, humidity, pressure and user's heart rate. It is important to mention that this messaging communication is done via Internet connection, so a hotspot to share Wi-Fi with the Raspberry Pi device would be needed. We could have a smartwatch paired with our smartphone and receive the cold alerts on our wrist. This is really handy since we do not need to look at our phone, making our overall usage experience more enjoyable.

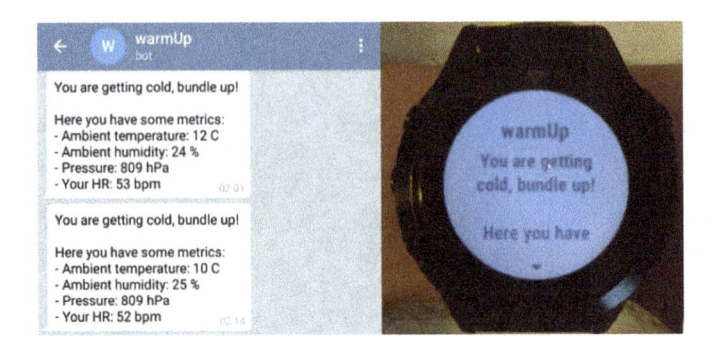

Fig. 6. Telegram bot notifications displayed on the Telegram phone application (left) and on a smartwatch (right).

7 Resulting Vest Has Minimal Social Weight

Our wearable computer has minimal social weight as it can be safely hidden inside a vest as shown in (see Fig. 7). The warmUp system can be carried easily with almost no social weight in the context of outdoors activities. In fact, the only piece of technology that can be spotted at clear sight is the cable connecting the Raspberry Pi module with the power bank. Our vest achieves a goal for creating a first prototype with almost no social weight, which was our first priority.

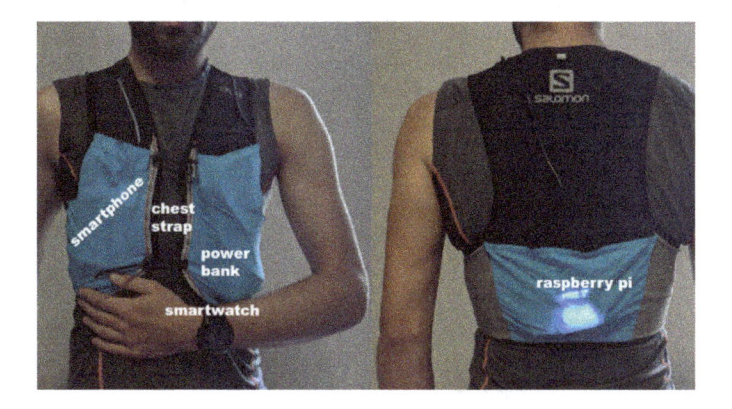

Fig. 7. Proposed presentation of the warmUp wearable computer.

8 Conclusions and Future Work

We have come up with a wearable device with almost no social weight and which is capable of predicting whether a person is getting cold or not. Although our predictions were modest (65%), we think that classification techniques can be improved by collecting more data and even sampling. Also, being able to control the application through chat communication with our Telegram bot would be extremely valuable.

In future, we expect the size and price of the device to decrease with better hardware. We also imagine our vest design to be integrated into smart watches and predictions to be more meaningful when combined with heart predictions and barometer readings. Our final goal is to automatically heat users' bodies without the need of explicit command, such as suits for fire-fighters to keep bodies comfortable.

References

1. Lehikoinen, J., Suomela, R.: Perspective maps. In: Sixth International Symposium on Wearable Computing, Seattlle, WA, October 2002, pp. 171–178 (2002)
2. Bern, N., Schiele, B., Junker, H., Lukowicz, P., Troster, G.: Wearable sensing to annotate meeting recordings, International Symposium on Wearable Computing, Seattlle, WA, October 2002, pp. 186–193 (2002)
3. Bristow, H.W., Baker, C., Cross, J., Woolley, S.: Evaluating contextual information for wearable computing, Sixth International Symposium on Wearable Computing, Seattlle, WA, October 2002, pp. 179–185 (2002)

4. Toney, A., Mulley, B., Thomas, B.H., Piekarski, W.: Minimal social weight user interactions for wearable computers in business suits. In: ISWC 2002 Proceedings, pp. 57–64 (2002)
5. Park, S., Locher, I., Savvides, A., Srivastave, M.B., Chen, A., Muntz, R., Yuen, S.: Design of wearable sensor badge for smart kindergarten, In: IEEE ISWC 2002 Proceedings, pp. 231–238 (2002)
6. Barry, B.: Story Beads: a wearable for distributed and mobile computing. MS thesis, MIT Media Labs (2000)
7. Brian, W., Jhonson, K., Metzgar, T., Semwal, S.K., Snyder, B., Yu, K., Neafus, D.: Crossover applications. In: IEEE VR Research Demo 2009, Lousiana, USA (2009)
8. Zeng, M., Gao, H., Yu, T., Mengshoel, O.J.: Understanding and improving recurrent networks for human activity recognition by continuous attention. In: ACM International Symposium on Wearable Computers, pp. 56–63, Singapore (2018)
9. Ozbek, S., Molla, M.T.I., Crompton, C., Holschuh, B.: Novel manufacturing of advanced smart garments: knitting with spatially-varying multi-material monofilament. In: ACM International Symposium on Wearable Computers, pp. 120–129, Singapore (2018)
10. Bernal, G., Yang, T., Jain, A., Maes, P.: PhysioHMD: a comfortable modular toolkit for collecting physiological data from head mounted displays. In: ACM International Symposium on Wearable Computers, pp. 160–167, Singapore (2018)
11. Smith, M.J., Warren, K., David, C.-T., Shames, S., Sprehn, K.A., Schwartz, J.L., Zhang, H., Arens, E.A.: Augmenting smart buildings and autonomous vehicles with wearable thermal technology. In: HCI (2017)
12. Kodama, R., Terada, T., Tuskamoto, M.: A context recognition method using temperature sensors in the Nostrils. In: ACM International Symposium on Wearable Computers, pp. 220–221, Singapore (2018)
13. Serrano, J.B.: warmUp: an intelligent wearable computer for automatic body temperature regulation, wearable computing and complex systems (CS5790) Instructor. SK Semwal, University of Colorado at Colorado Springs, pp. 1–8 (2019)
14. Getting started with the Sense HAT. https://projects.raspberrypi.org/en/projects/getting-started-with-the-sense-hat. Accessed 05 Nov 2019
15. Building Neural Network using Keras for Classification. https://medium.com/datadriveninvestor/building-neural-network-using-keras-for-classification-3a3656c726c1. Accessed 05 Nov 2019
16. Binary Classification Tutorial with the Keras Deep Learning Library. https://machinelearningmastery.com/binary-classification-tutorial-with-the-keras-deep-learning-library/. Accessed 05 Nov 2019
17. How to create a Telegram bot, and send messages with Python. https://medium.com/@ManHay_Hong/how-to-create-a-telegram-bot-and-send-messages-with-python-4cf314d9fa3e/. Accessed 05 Nov 2019
18. Display two digits numbers on the Raspberry PI Sense HAT. http://yaab-arduino.blogspot.com/2016/08/display-two-digits-numbers-on-raspberry.html. Accessed 05 Nov 2019

Virtual Teleportation of a Theatre Audience Onto the Stage: VR as an Assistive Technology

Saikrishna Srinivasan[✉] and Gareth Schott

University of Waikato, Hamilton, WA 3216, New Zealand
sks6l@students.waikato.ac.nz

Abstract. For more than a decade, virtual reality (VR) has been employed to enrich and heighten media experiences. Despite the recognized potential and promise of VR, and ample investment, it has yet to fully transform or replace existing screen-based experiences (e.g. film or gaming). This research forms a part of a larger project to shift VR applications beyond otherwise apparent areas of screen-based media, in order to enhance audience access and propinquity to a live performance. This study is being conducted in the field of theatre, a dramatic medium in which audience are traditionally static and where an individual's seating position determines their perceptual experience (distance, angle, lights, obstructed vs clear view). This paper introduces the broader project and its experimentation with VR as an assistive technology. The project seeks to utilize VR as a means of converting an otherwise static experience to provide collective moments of visual teleportation, onto stage, into props and on actors. VR offers a non-invasive means of introducing variance in viewer proximity or position relative to performance. This paper reports on the early development and use of a three-dimensional theatre prototyping in order to explore the technical requirements for application to a VR theatre experience.

Keywords: Virtual reality · 360-degree camera · Assistive technology · Emancipation · 3D-prototype · Perspective · Theatre · Invisible

1 Introduction

1.1 Project Scope and Aims

The scope of our first set of experiments in a broader VR-theatre project that aims to incorporate virtual reality (VR) technology into a theatre experience, is to establish functional and impactful camera position/placement that will ultimately transform, augment and enhance a static experience of a dynamic performance. The aim of employing VR is to alter the nature of audience engagement and alter their sense of presence in a positive way. Initial experiments constitute a vital first step as they provide a finite number of practical test-case results that will later be implemented in live testing in conjunction with theatre practitioners. Our ultimate aim is to have VR stimulate new approaches to theatre practice and performance by offering and empowering theatre practitioners with a means to experiment with audiences' proximity and presence. In order to achieve this aim, key parameters like camera positioning in or around the stage, relative to the performers and the number of performers on the stage will play an important role. The first set of experiments sought to ascertain potential heightened and concentrated perspectives through which a two-person on-stage interaction could be experienced via VR. It is posited that through on-stage presence an audience may gain greater sense of intimacy, typically achieved in film via close up and extreme close ups, or engender new perspectives in theatre such as first-person perspective or providing perspective from an inanimate object found on stage.

1.2 Experiment

This experiment was focused on the production of a 3-D prototype of a theatre setting in order to identify possible areas for camera (and in doing so, audience) placement on stage. Charlton and Moar [3] conducted a comparable study in which they too introduced theatre audience to the experience of being inside the stage and amidst performance. They employed VR with the aim of inducing a state of deeper engagement from bodily presence. The play 'Fellow Creature', consisted of only two actors plus a camera (GoPro rig) that was placed in a central position on stage. The camera viewpoint, viewed through a Head Mounted Display (HMD), permitted audience members to experience the play amid the actors. This experiment focused on the effect of the perspective change on audience, sense of voyeurism and its similarities in theatre and VR. This experiment helped to bring audience 'onto the stage' and transform their status and role from non-participatory witnesses outside the play to co-presence. However, it simply aimed to place the audience at the center of the stage. The series of experiments supporting the current project pursue a similar line of thinking to Charlton and Moar, but with greater focus given to the technical installation and application of VR on what is the best position/angle for the camera on stage, determining the appropriate vertical height for the camera view and assessing the effect of camera movement and motion blur on audience viewing. The project also seeks to extend the use of augmented objects (AO), visible through the HMD, to enrich the setting of the performance. Finally, the current project will also examine the application of VR in two distinct theatre conditions - a more traditional theatre experience in which a story is enacted or played out uninterrupted on the stage and a devised theatre or a collective creation which emerges from collaboration and improvisation. A comparative analysis will be conducted of the relative merits of VR to both theatrical contexts.

The primary aim of the experiment reported here is to begin the process of narrowing down the number of viable camera positions that can be made available to theatre practitioners to work around, and for an audience to teleport into from a seated position. A simple stage-bound performance comprised of two actors in conversation can be captured in a finite number of angles, typically conveying 25–40° angles when a standard camera is used [5]. When introducing VR into theatre performance, the possibilities for wider perceptual field increases exponentially due to its 360-degree field of view. This brings an additional set of considerations concerning the visibility of otherwise unrevealed offstage areas such as the wings, crossovers or voms, lighting rigs and also the audience. The aim of this first experiment was to use the 3D prototype of a simple theatre-stage context to filter out a finite number of inciting camera positions, that will act as a centroid for all adjacent and approximate camera positions. Successive experiments will aim to evaluate, test and validate the impact, appeal and advantage associated with different camera positions as well as its impact for the actors/performers. Ultimately the project seeks to explore how the addition of a dynamic perspective changes the audience experience of a theatre experience and its future evolution. The intention is to follow prototype experiments with a physically staged experiment that consists of real performers (who will also serve as research participants). The acquired camera positions will eventually be tested and evaluated by an in-house audience (who will also serve as research participants).

The prototype is designed with only two static characters placed on the stage, to mimic a real-time conversation between two actors. Since this was a pilot study the complexity of a real-world theatre performance is purposely reduced and simplified to a situation comparable to black box theatre. This experiment aimed to identify possible at pivotal or centroidal positions for camera placement to provide an inciting stage view available to all audience members seated from downstage to upstage. From initial discussions with theatre production companies based in Hamilton (New Zealand), it has been noted that employing subtle facial expressions to convey emotion or feeling is difficult in a theatre context as it is difficult to guarantee that audience will perceive them or if they will even be attentive to the particular and relevant zone of information conveying a potentially crucial aspect of the play. Also if we consider the experience of a stage show, a sports event or live concert in person compared to viewing a framed and edited version on screen, while a live experience allows the audience to experience the energy and absorb the atmosphere of an event, there is a real likelihood that smaller gestures, looks or moments on stage or on field will be lost due to audience distance and where attention is being directed. For this reason, large digital display screens have now become standard in sports or concert stadiums (even when not televised), to some extent acknowledging what is lost or missed when viewing solely from the crowd. Similarly, the application of VR to theatre provides a window of opportunity for an otherwise quiet and static audience to gain even greater intimacy via proximity and presence. Indeed, Wendell Cole writes: "since the final quarter of the nineteenth century it has been apparent that major efforts of architects and designers have been directed toward bringing audience and actors into more intimate contact with one another" [18].

In the broad context of the study, prototyping will serve as a key tool and methodology over the course of the research project – as it provides a means to evaluate and test camera use, placement and viewing prior to testing and application with real actors and viewers within a theatrical rehearsal or performance. Focus groups, workshops (testing experiment with actors), interviews and questionnaires will be used to evaluate limitations and potential of VR, at a later stage of the research. Once prototyping is completed, it will be followed by focus groups with selected theatre production company participants/artists to determine the perceived affordances of the medium and identify any significant challenges that needs to be resolved prior to adoption and application by practitioner for use with audience. On completion of focus groups and analysis of the data collected, experimental workshops will be conducted in which VR will be assimilated into theatre practice, guided by the results of prototyping and consultation with practitioners concerning the perceived need, application and challenges. Experimental workshop will aim to reach rehearsal in front of a test audience in order to assess the performance of the VR technology, actors' interaction with or performance in the presence of VR technology on stage, and audience engagement with VR alongside traditional viewing. Once actor and audience testing are complete, participants will evaluate their experience via a questionnaire and short feedback interviews in order to provide insights into the success of the experience aesthetically and technically, as well as the level of absorption and immersion provided by VR.

1.3 Prototype

As stated above prototyping aims to help in the evaluation, and determination of the number of camera positions that can provide a 360-degree view of the performance from the stage. Evaluation of the prototype is based on theoretical analysis of the stage and the performers placement. The findings will be further evaluated through data collection with practitioners and mock-audience during rehearsals before commencing successive experiments.

The aim is to assist the audience to achieve absorption and immersion in their engagement with theatre performance. As theatre has continued to evolve, more immersive modes of theatre have been explored. Nicola Shaughnessy suggests immersive theatre provides an idea of engaging the audience/spectator in a common shared space with performers, in doing so, submerging them in an experimental space which neglects real and fictional boundaries [14]. An example of immersive (and also promenade) theatre can be found in British theatre company Punchdrunk's adaptation of Shakespeare's Macbeth in 2011 entitled *Sleep No More*. For the New York performance of the play, warehouses in Manhattan's Chelsea neighborhood were transformed into hotel-like performance spaces allowing audience to move freely through five floors and settings, allowing them to interact with props, or observe the actors at their leisure [16]. However, like a seated audience the *Sleep No More* audience were asked to remain silent (and also masked) at all times, refrain from using phones or cameras and maintain a respectful distance from the performance. Nevertheless, a performance accessible in this way catered to senses like touch, smell, sound and physical proximity to the actors (presence). This example, experimented with the idea of group immersion or 'productive participation' drawing on audience's capacity and inherent productivity to build up the story of their own [1].

In 1974, philosopher Robert Nozick [12] invited us to imagine an 'experience machine' capable of artificially stimulating the brain to induce our desired experiences. As De Brigard outlines Nozick "seems to suggest that most people's intuitive reaction after considering the thought-experiment would be just like his: they would feel very little inclination to plug in" permanently [4]. Nozick posits people care too much about "living in contact with reality" [12]. The application of VR proposed in this project instead offers multiple perspectives as a witness to a dramaturgical performance taking place in front of the audience. Not confined to a seat or single position the audience, via VR, is able to employ a type of locomotion that allows them to access the stage without requiring any physical transfer (teleportation). Like Alston's notion [1] of collaborative theatre immersion, VR immersive experiences hold the potential to offer an escapist experience from within the real world, occupying the audience in a constructed immersive experience and 'tricked out spaces' that allows the audience escape their seats and penetrate the stage as an imperceptible spectator and voyeur in a performance.

2 Technical Evaluation Criteria

A prototype was developed in 3D modelling and animation software Maya. The 3D characters and the models employed are royalty-free and do not require copyright. The 3D characters and set are retopologized and change in textures were applied as per the experiment's requirements, such as various lighting setups (spot light, area light). The benefits of a three-dimensional setup, is an infinite possible camera positions were broken down into potential camera positions that can enhance audience viewing. Considering two basic orthographic views - top view and front view, both orthographic views play a significant role in determining and drawing the three-dimensional view of any object [7]. A two-dimensional top view and front view for the whole setup and an imaginary circle around the two characters was developed, as shown in Fig. 1. The imaginary circle is formed by the camera when it is revolved around, with the performance as its center. This provided two variables 1. The center of the circle which will be the focal point and 2. the radius of the circle which will be the focal length of the camera. Still, there are infinite possible camera positions along the circumference of the circle. The circle is divided into four equal quadrants with respect to the straight line between the two characters. Since all four quadrants are identical, I have taken the first quadrant (0–90°) as my sample. This reduces the number of camera position to a finite number, unlike the previous scenario of all four quadrants.

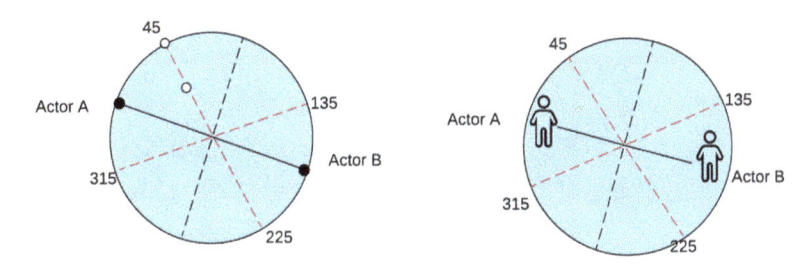

Fig. 1. Top view and front view of the performance space to breakdown the camera placement position

By evaluating the single quadrant, the best camera position inside the quadrant can be determined. Using the same principle, the camera positions in the other three quadrants can also be determined. There are two possible ways to determine the best camera position. Since it is considered as a defined shape quadrant, the centroid of the quadrant would serve as one of the best possible positions for the camera.

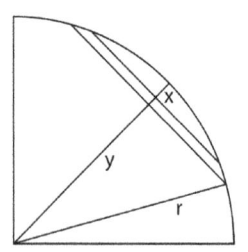

Fig. 2. Quadrant 1 - 0 to 90°

A single quadrant from the entire circle is considered, as shown in Fig. 2. Considering an infinitely small horizontal strip of thickness 'dy', at a distance y from the base.

The length of the strip will be 2x.

The moment of all such strips of the semicircle about the base divided by the area of the semicircle would give us the distance of the centroid from the base.

$$\bar{y} = \frac{4}{\pi r^2} \int_0^r 2xy\, dy$$

By Pythagoras theorem, $x = \sqrt{r^2 - y^2}$

$$\bar{y} = \frac{4}{\pi r^2} \int_0^r 2y\sqrt{r^2 - y^2}\, dy$$

$$= -\frac{4}{\pi r^2} \left[\frac{2}{3}(r^2 - y^2)^{\frac{3}{2}} \right]$$

$$= -\frac{4}{\pi r^2} \left[-\frac{2r^3}{3} \right]$$

$$= \frac{8r}{3\pi}$$

So the centroid would be (0, 8r/3 π). The other possible position would be at an angle of 45° at the edge of the circle [5]. The best positions for the other quadrants can also be determined the same way. Using this method of calculation, there would be eight possible 'optimal' positions from the top view.

Similarly, with the same calculation, the best position from the front view can be determined, and it will be four positions, unlike top view since the other four positions in quadrant 3 and 4 will fall beneath the ground. When the findings of the top view and front view positions are correlated that results in eight possible positions, which will be evaluated further in the successive experiment using real-time stage, actors and theatre practitioners. Some of the camera positions that have been tested using the prototype are tabulated in Table 1.

Table 1. Various iterations of camera position from the prototype design

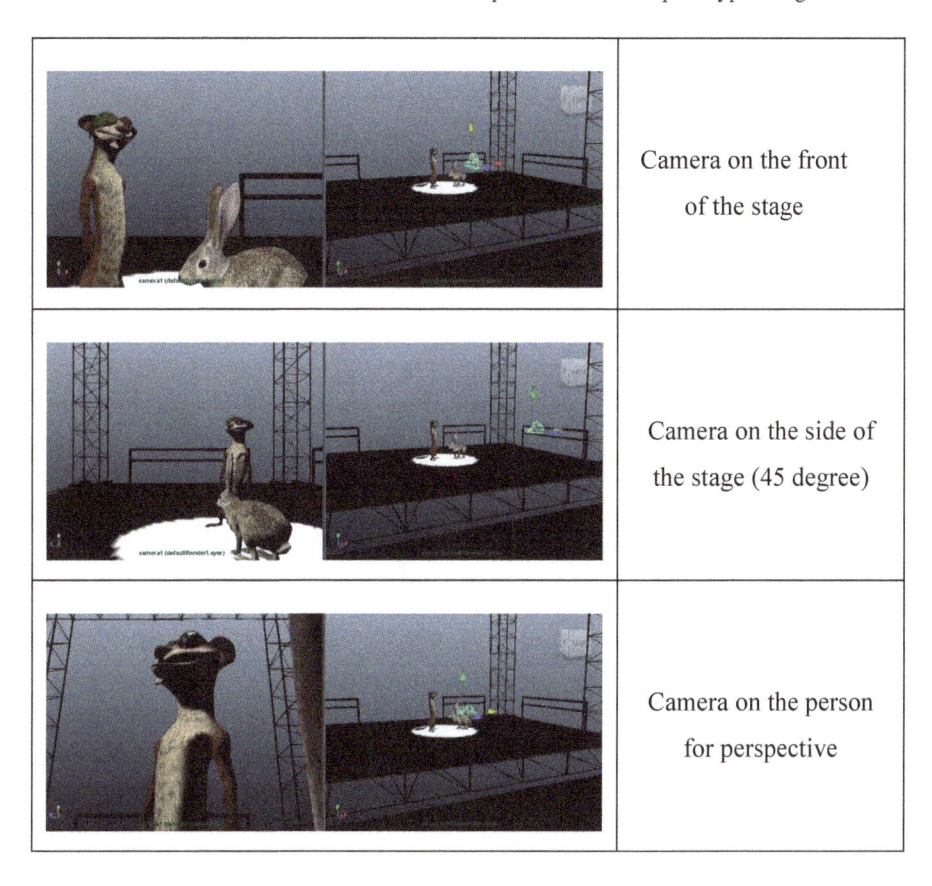

	Camera on the front of the stage
	Camera on the side of the stage (45 degree)
	Camera on the person for perspective

Evaluation of filmed theatre or live performance or music concert provides useful information regarding traditional camera positioning and camera angles that audience may be familiar with. In order to add a layer of filter with the concluded results of the camera placement, a video analysis of the 'Hamlet - Rehearsing the sword fight' directed by Robin Lough was done [10]. The reason for the choice of this particular piece of artefact is because it stands in the similar interest of this research and similar to the type of experiments which are proposed further in the research. The sword fight takes place between two characters within a defined circle in the stage, restricting many complex movements. The two characters, Kobna Holdbrook-Smith and Benedict Cumberbatch are represented as character A and character B respectively. The analysis result of the *Hamlet - Rehearsing the sword fight*' is tabulated in Table 2.

Table 2. Findings from the analysis of '*Hamlet - Rehearsing the sword fight*'

Camera Placement	Camera Height	Action/ performance	Representation
Shot 1- Left of Character B	Eye level	Character A is attacking and advance, character B retreating	
Shot 2- Right of Character A	Eye level	Character B is attacking and advance, character A retreating	
Shot 3- Left of Character A	Mid level (to focus the hand movement and sword) (Character B is right handed)	Character B is attacking and showing some sword skills	
Shot 4- Left of Character B	Eye level (to show the facial expression)	Character A advancing aggressively	

When comparing the findings from this analysis of '*Hamlet: Rehearsing sword Fight Analysis*' against findings of the prototype experiment, it coincides with most of the valid camera placement. However there are a few more new interesting camera positions and placements found in the analysis and in the prototype. All of these camera placements will be put to test in the successive physical experiments to derive a better understanding and usage of VR 360° camera in theatre performance.

Audience experience of fictional events or a performance can be altered radically when it is conveyed from a first or second-person perspective, point of view (POV). A subjective experience is widely accepted as spontaneous and distinct [17]. Indeed, the subjective or first-person POV is often used to emphasize reactions and response to a protagonist, instill immediacy and connection, and garner empathy from an audience [9]. The idea of permitting an audience to experience a close-up camera position or a first-person POV opens up the possibilities of theatre benefitting from cinematic insights.

As cinematographer Benoit Delhomme explains, when discussing his approach to filming *At Eternity's Gate*, a biographical re-imaging of Vincent Van Gogh: The camera "can be like a microscope. The way I was using close-ups in this film was to capture van Gogh's soul ... The face becomes a landscape. What is more interesting than the face? So much to see in the face." The introduction of 360-degree view from a character's perspective will only enhance the audience's experience and involvement [6]. Theatre is often conceived as a 'third-person' art form in which 'on stage' characters do not see, they are typically seen. POV through VR is a means for the audience to teleport into the actor's space and adopt a similar perspective thus also gaining a psychological reading. Since individual audience typically consume performances from a single position, this project helps to amplify the idea of liberation from a single site and consider when it is most effective for freedom to be given [13]. Individual immersion via VR intends to heighten spatial presence and give prominence to the feeling of 'being there' or the illusion that what is happening is occurring to the spectator as much as the performer [15]. Even though this idea of audience spectatorship would be considered unorthodox, particularly when contrasted with traditional spectatorship, it has been explored as a topic of discussion for some time, seeking to open the door for experimentation in theatre performances [6].

Camera placement could become highly valuable as a means to allow the audience to interrogate, study and relish the drama and intensity of any performance [8]. Studies show that multiple camera placement is vital for Image MAGnification (IMAG) in any stage performance. The purpose of Image MAGnification is to magnify the person's "image" so people farther from the stage can more easily see them [19]. This is one of the reasons, various camera positions and composition rules have been followed for recording stage performance. Some of the basic camera positions followed are close in on a subject, closeup headshots - to capture facial expressions, moderate close-ups - to capture body language and movement, wide angle shots in combination with a low angle to create a more dynamic feel of the entire stage, triangle or split group formations - to capture group performance [8]. These findings from the existing literature coincides with majority of the results from this experiment. All of these camera positions will be examined and evaluated with a physically staged experiment that will consist of real performers (who will also serve as research participants) under various evaluation criteria of the number of performers, lighting setup, nature of performers (height, age, sex) in the successive experiments and the effect of application of VR in theatre performance will be evaluated.

3 Conclusion and Findings

The fundamental purpose of this experiment was to begin examination and evaluation of the infinite number of (stage 1) camera positions and filter them to a finite number, which has been achieved. Since the acquired camera positions for an inciting viewing are finite, we can proceed to further check the actual viability of the camera positions in the subsequent experiments. Charlton and Moar [3] have already conducted a similar experiment to examine the effect of having a static 360-degree camera at the center of the stage. However, in this scenario there was no defined or specific reason for the selection of the camera placement, as they were aiming to examine the invisible, passive, voyeuristic viewer 'in the room' whilst a dramatic scene unfolds. The present work stands in contrast to Charlton and Moar's work as it focused more on determining a viable or centroidal camera position and inciting a view from inside the stage. This research will not only prompt the audience to explore the performance and construct their own impressions of performances but also help to evaluate how the virtual presence of an audience can induce a sensation of being inconspicuous yet close at hand [11] and quite removed from the audience. From this experiment eight positions from an orthographic view, camera positions from the actor's perspective, were identified to offer a change of perspective to an audience, emancipating the story from a single-perspective experience.

Future audience that experience VR immersion during a live dramaturgical performance will be introduced to a different modality by which a performance is conveyed. Moving forward, we aim to examine how such dynamic perspective changes in a theatre performance, will effect and affect both performers and audience and whether it is to the benefit or detriment to a theatre experience.

4 Reflections

From a media technician perspective, it is very intriguing to investigate the affordances that the VR medium possesses, and how that can be applied to a more traditional and historical storytelling medium that conventionally has its audience wedded to a single space. There is some significant technical testing and research being conducted on VR by the British Royal National Theatre who are exploring the use of VR studio to develop dramatic models for viewing in the 360° medium. [2]. However, very few theatre practitioners and performers to-date have been given the opportunity to expand their practice by moving the audience to view a performance from different vantage points throughout a performance and the opportunity experiment with this technical opening. It is our hope, that if the risk of investing in the application of a new technology is broken down, testing is completed on small units of activity/presentation with convincing findings, and presented linearly, then adoption of VR might happen sooner. By exploring and evaluating VR, and promoting prototyping as a pre-production practice prior to testing within theatre practice, it is hoped that VR might assume a role in the process of theatre practice.

References

1. Alston, A.: Beyond Immersive Theatre Aesthetics, Politics and Productive Participation. Palgrave Macmillan, London (2016)
2. Brown, M.: National Theatre Creates Virtual Reality Studio for New Projects (2016). https://www.theguardian.com/stage/2016/aug/02/national-theatre-creates-virtual-reality-studio-for-new-projects. Accessed 18 Jul 2019
3. Charlton, J.M., Moar, M.: VR and the dramatic theatre: are they fellow creatures? Int. J. Perform. Arts Digital Media **14**(2), 187–198 (2018)

4. De Brigard, F.: If you like it, does it matter if it's real? Philos. Psychol. **23**(1), 43–57 (2010)
5. Freer, I.: Film studies 101: the 30 camera shots every film fan needs to know (2018). https://www.empireonline.com/movies/features/film-studies-101-camera-shots-styles/. Accessed 26 Jun 2019
6. Gardner, L.: The Roman Tragedies. The Guardian, London (2009)
7. Hu, Y., Xuliang, G., Baoquan, Z., Xiaonan, L., Shujin, L.: Lecture Notes in Computer Science (including Subseries Lecture Notes in Artificial Intelligence and Lecture Notes in Bioinformatics), vol. 8509, pp. 612–619 (2014)
8. Minoia, A.: Capture the drama – 7 theatre and stage photography tips. https://expertphotography.com/theatre-stage-photography-tips/. Accessed 01 Sept 2019
9. Muller, V.: Film as film: using movies to help students visualize literary theory. Engl. J. **95**(3), 32–38 (2006)
10. National Theatre.: National Theatre Live|Hamlet - Rehearsing the sword fight (2019). https://www.youtube.com/watch?v=k1sThbRvwF0. Accessed 01 Sept 2019
11. Newton, K., Soukup, K.: The storyteller's guide to the virtual reality audience (2016). https://medium.com/stanford-d-school/the-storyteller-s-guide-tothe-virtual-reality-audience-19e92da57497. Accessed 25 Jun 2019
12. Nozick, R.: Anarchy, State, and Utopia. Blackwell, Oxford (1974)
13. Ranciere, J.: The Emancipated Spectator. Verso, London (2009)
14. Shaughnessy, N.: Applying Performance: Live Art, Socially Engaged Theatre and Affective Practice. Palgrave Macmillan, Basingstoke (2012)
15. Slater, M.: Place illusion and plausibility can lead to realistic behaviour in immersive virtual environments. Philos. Trans. R. Soc. B **364**(1535), 3549–3557 (2009)
16. Soloski, A.: Sleep no more: from avant garde theatre to commercial blockbuster (2015). https://www.theguardian.com/stage/2015/mar/31/sleep-no-more-avant-garde-theatre-new-york
17. Thomas, E.A.: Camera grammar: first-person point of view and the divided "I" in Rouben Mamoulian's 1931 Dr. Jekyll and Mr. Hyde. Q. Rev. Film Video **32**(7), 660–666 (2015)
18. Wendell, C.: Some contemporary trends in theatre architecture. Educ. Theatre J. **7**(1), 16 (1955)
19. Wright, P.: What Is IMAG and when should you use it? (2018). https://www.ignitermedia.com/blog/2018/10/01/what-is-imag-2/. Accessed 09 Aug 2019

Eliciting Evolving Topics, Trends and Foresight about Self-driving Cars Using Dynamic Topic Modeling

Workneh Y. Ayele[✉] and Gustaf Juell-Skielse

Department of Computer and Systems Sciences, Stockholm University,
Stockholm, Sweden
workneh@dsv.su.se

Abstract. Self-driving technology is part of smart city ecosystems, and it touches a broader research domain. There are advantages associated with using this technology, such as improved quality of life, reduced pollution, and reduced fuel cost to name a few. However, there are emerging concerns, such as the impact of this technology on transportation systems, safety, trust, affordability, control, etc. Furthermore, self-driving cars depend on highly complex algorithms. The purpose of this research is to identify research agendas and innovative ideas using unsupervised machine learning, dynamic topic modeling, and to identify the evolution of topics and emerging trends. The identified trends can be used to guide academia, innovation intermediaries, R&D centers, and the auto industry in eliciting and evaluating ideas. The research agendas and innovative ideas identified are related to intelligent transportation, computer vision, control and safety, sensor design and use, machine learning and algorithms, navigation, and human-driver interaction. The result of this study shows that trending terms are safety, trust, transportation system (traffic, modeling traffic, parking, roads, power utilization, the buzzword smart, shared resources), design for the disabled, steering and control, requirement handling, machine learning, LIDAR (Light Detection And Ranging) sensor, real-time 3D image processing, navigation, and others.

Keywords: Dynamic Topic Modeling · Topic modeling · NLP · Self-driving cars · Topic evolution · Topic trends · Forecasting in topics

1 Introduction

Vehicle manufacturers are currently building their capability to manufacture state-of-the-art vehicles by including self-driving features and artificial intelligence (AI). For example, Volvo Group is teaming up with US chipmaker Nvidia to develop AI for its self-driving trucks[1]. Similarly, Ford and Volkswagen disclosed that they are teaming up to produce self-driving and electric cars together in an effort to reduce costs for new

---

[1] https://www.bbc.com/news/topics/c90ymkd8lglt/driverless-cars.

technologies[2]. Furthermore, the use of self-driving cars has a positive impact on the quality of life, safety, fuel cost reductions, travel optimization, accident reductions, and efficient parking which in turn contributes to urban planning activities [1]. Also, connected cars are part of the connected world [2], a phenomenon that is gaining attention of the industry and academia. One example is the discourse on smart cities [3]. Smart cities can be viewed as ecosystems of interacting entities where self-driving cars are becoming important types of entities [4].

However, there are emerging concerns regarding the use of self-driving technology such as the impact on transportation systems, safety, affordability, control and liabilities [1], ethical issues [5], and the impact on urban planning [6]. Therefore, it is essential to undertake research and innovation regarding self-driving technologies. Innovation in the automotive industry relies on innovation and R&D done by suppliers [7]. Also, a case study by IBM shows that for a breakthrough in innovation, cross-industry alliances with non-suppliers are vital [7]. Besides, innovation can be stimulated through innovation contests [8–10]. Also, technology transfer agents, incubators, and collaborative research centers could contribute to idea generation by acting as intermediaries between academia and the industry [11]. The quality of ideas can be evaluated using standards related to originality and quality aspects [12].

Despite increasingly growing scholarly literature as a result of research activities, it is becoming harder to generate ideas [13]. To unlock the potential of this scholarly literature, trend analysis can be used to generate innovative ideas. Furthermore, eliciting and analyzing evolving trends can be helpful for decision-makers and stakeholders in academia and the industry [14], and elicitation of topics about emerging trends is vital for making decisions [15]. Trend analysis can be used for forecasting trends in technology [16], and forecasting during idea generation improves idea evaluation [12]. In this research, data is collected from a scholarly literature, and the generation of trends is conducted.

To explore the growing collection of scholarly literature tools are needed [17]. Blei and Lafferty proposed Dynamic Topic Modeling (DTM) based on probabilistic time-series models for analyzing the evolution of topics [18]. Also, DTM enables technology agents to identify innovative ideas. Machine learning can be used to generate ideas along with AI and analytics by predicting relevance and providing valuable insights [19]. Similarly, it is possible to elicit trends and temporal patterns using scientometric, visual analytics, and machine learning [20].

Therefore, in this study, DTM is used to generate evolving topics and elicit ideas about self-driving cars, and through trend analysis, to provide a method for technology scouting and idea evaluation and to offer insights to the industry; by doing so we contribute with ideas and research topics that can be used for innovation of self-driving technologies.

---

[2] https://www.bbc.com/news/business-48965315.

This paper has six sections: Related Research, Methodology, Result and Analysis, Discussions, and Conclusions and Future Research.

2 Related Research

Scholarly literature is increasingly growing yet idea generation becomes even harder [13]. Generating innovative ideas is possible through trend analysis, and trends are used for decision making in the industry and academia [14]. Moreover, the elicitation of topics about emerging trends is vital for decisions making [15]. Scholarly literature datasets could be analyzed using scientometric and text mining techniques for generating insights and foresight. For example, Marçal et al. applied scientometric techniques on a dataset obtained from Web of Science to identify research gaps in research about self-driving vehicles [21].

On the other hand, a review of trend identification techniques using text mining revealed that trend detection is applied in diverse research domains [22]. Technology foresight primarily relies on non-transparent and inefficient qualitative approaches despite the availability of increasing external data sources and machine learning techniques [23]. Examples of external data sources include online news data and social media data. Online news data can be used for market prediction through the identification of trends using text mining [24]. Similarly, social media data can be exploited to identify trends and predictions using social network analysis on brands and other topics [25].

Finally, there are a few recent research activities in academia that elicits insights and foresight using predictions about self-driving cars. For example, Ayele and Akram identified emerging trending and temporal patterns about self-driving cars using scientific literature by applying the scientometric and visual analytics tool, Citespace, in combination with topic modeling [20]. Moreover, topic modeling to identify hidden topics about self-driving cars was performed by [6]. Insights were identified in [20] using trends and it could be exploited to identify potential foresight, also themes in self-driving cars were identified using a non-time based topic modeling in [6].

3 Methodology

The process of extracting trends through DTM is done following a data mining process model shown in Fig. 1 below. The data mining process, illustrated in Fig. 1 below, is informed by CRISP-DM (CRoss Industry Standard Process for Data Mining). The CRISP-DM is a standard and general process model which can be used independently of the technology used and the industry sector selected [26].

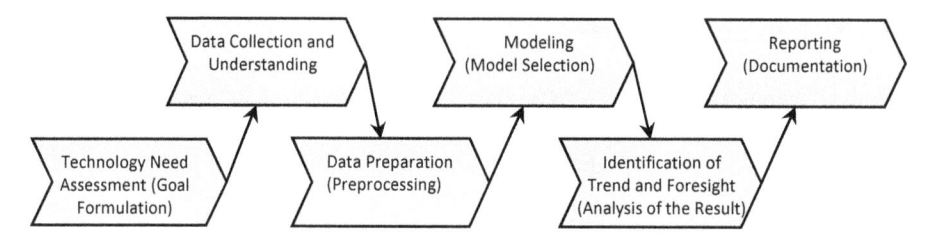

Fig. 1. The data mining process following CRISP-DM.

3.1 Technology Need Assessment: Goal Formulation

The goal of this research is to identify the emerging trends of self-driving cars in academia. The identification of emerging trends could help technology incubators and accelerators, innovation organizers, research and development of tech companies to evaluate ideas and to choose the best ideas to commercialize.

We used Python, Excel, and R as computational tools and a PC with 8 GB RAM, a 64 bit Windows 10 Operating System, and an Intel i7 CPU with 2.7 GHz.

3.2 Data Collection and Understanding

The literature search was conducted using keyword search in Scopus. Scopus is arguably the most comprehensive database listing the majority of journals and conference proceedings for our area of interest. For example, Scopus has more recent and larger datasets of scientific literature than Web of Science [27]. Also, compared to Web of Science, Scopus has better coverage of academic journals [28]. The query used to collect data is illustrated below.

> (TITLE-ABS-KEY (*"self driv*"*) OR TITLE-ABS-KEY (*autonomous*) AND TITLE-ABS-KEY (*car* OR *automobile*)) AND (LIMIT-TO (DOCTYPE, *"cp"*) OR LIMIT-TO (DOCTYPE, *"ar"*)) AND (LIMIT-TO (SUBJAREA, *"ENGI"*) OR LIMIT-TO (SUBJAREA, *"COMP"*) OR LIMIT-TO (SUBJAREA, *"MATH"*) OR LIMIT-TO (SUBJAREA, *"SOCI"*)) AND (LIMIT-TO (LANGUAGE, *"English"*)) AND (LIMIT-TO (PUBYEAR, *2019*) OR LIMIT-TO (PUBYEAR, *2018*) OR LIMIT-TO (PUBYEAR, *2017*) OR LIMIT-TO (PUBYEAR, *2016*) OR LIMIT-TO (PUBYEAR, *2015*) OR LIMIT-TO (PUBYEAR, *2014*) OR LIMIT-TO (PUBYEAR, *2013*) OR LIMIT-TO (PUBYEAR, *2012*) OR LIMIT-TO (PUBYEAR, *2011*) OR LIMIT-TO (PUBYEAR, *2010*))

A total of 5425 documents were retrieved. Since Scopus limits the maximum downloadable documents to 2000, the data was extracted in batches and stored with file names reflecting years of publication from 2010 to 2019 in separate CSV files.

3.3 Data Preparation: Preprocessing

Preprocessing of the data was done using Python and included the removal of stopwords, punctuations, and numbers. The stopword list was updated by identifying irrelevant and trivial terms after running term frequency analysis, and using word clouds. Also, lemmatization of terms was done to convert terms to their root forms.

3.4 Modeling: Model Selection

Dynamic Topic Modeling (DTM): The growing collections of scholarly literature [13] can be explored using machine learning techniques [17]. DTM is a topic trend identification model based on the Latent Dirichlet Allocation (LDA) model [18]. LDA is a topic modeling technique proposed to elicit hidden topics regardless of time [29]. In contrast, the DTM proposed by Blei and Lafferty is a probabilistic time-series model for analyzing the evolution of topics over time. Also, it uses a state space-model to represent topics using multinomial distributions, and to infer latent topics approximations it uses Kalman filters vibrational approximation and non-parametric wavelet regression [18]. In this paper, DTM was used to elicit the evolution of topics about self-driving cars. The implementation of the DTM was written in Python[3].

Evaluation of the Model by Assessing the Result: First, an overall assessment of the coherence of terms under each topic was done to select the best model in terms of topic size. The evaluation of the model was done by comparing the coherence score of several numbers of topics and choosing the maximum coherence score, as illustrated in Fig. 2 below. The calculation of the coherence score and the visualization to identify the maximum coherence score value was done based on the algorithm by [30], and the implementation was based on the Gensim[4] Python library.

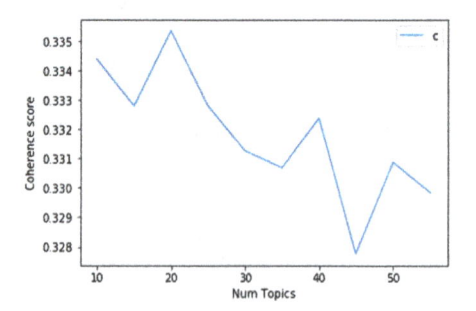

Fig. 2. Topic coherence score for different number of topics.

3.5 Identification of Trends and Foresight: Analysis of the Result

Labeling of Topics: Labeling was done by attaching meaningful names to topics identified as suggested by [29, 31]. However, labeling of topics is difficult when abbreviations and acronyms are present [32]. Therefore, abbreviations and acronyms were identified and interpreted accordingly.

---

[3] https://radimrehurek.com/gensim/models/ldaseqmodel.html.
[4] https://radimrehurek.com/gensim/models/coherencemodel.html.

Identifying Trends and Foresight: The identification of trends throughout the timeline was carried out using graphical visualization of the probability of terms generated by DTM similar to [18, 33]. The selection of time-series model for generating predictions depends on the number of observations. For example, it is recommended to have at least 50 observations and advisability more than 100 observations to use the common autoregressive integrated moving average (ARIMA) model [34]. In the contrary, it is arguably sufficient to use ordinary least square linear regression model to observation of at least [35]). Therefore, we used linear regression on LIDAR, to generate forecasts since we have 10 observations for 10 years. To identify the correlation of LIDAR with radar and cloud we generated correlation graphs.

Analysis of Trends (Idea Elicitation and Evaluation): AI, analytics, and machine learning, in our case unsupervised topic modeling, can be used to elicit ideas by predicting relevance and providing valuable insights about ideas [19]. Similarly, it is possible to identify temporal patterns and predict trends using scientometric, visual analytics, and machine learning [20].

Visualization and thematic analysis were used to elicit trends, analyze elicited trends, extrapolate ideas by implications, and generate reports. Also, extracted ideas and elicited trends can be used for evaluating the potential of ideas collected in an idea bank by analyzing the timelines of the ideas. The thematic analysis of evolving topics generated was done to identify, analyze, and elicit trends, and report themes and patterns of generated data [36]. Decision making in science and technology could be supported by elicitation of topics about emerging trends [15]. Forecasting trends in technology can be done through data analysis [16], and doing so in combination with idea generation improves idea evaluation [12].

3.6 Reporting: Documentation

In this section, the result of the DTM is presented. Also, the analysis of the topic evolution is presented using visualizations of term evolution in a meaningful, interpretable, and usable way. Also, lessons learnt, implementation codes, use cases followed and other best practices are documented for future analysis.

4 Results and Analysis

In this chapter, a summary of the result is presented in Table 1. Also, the analysis of the result is presented in Sect. 4.1. The full list of the visualization of the topics is available in Appendix 1. Some of the terms in the visualizations have conspicuously high probabilities which make it harder for the trends of the other terms to be noticeable. Therefore, some of the topics have two graphs such as Topic 1 (Topic 1A and Topic 1B), Topic 2 (Topic 2A and Topic 2B), and so on with one of the graphs containing all terms. The major topics identified are shown in Table 1 below.

Table 1. Topics identified for details refer to Appendix 1

Topic #	Topics labeled following [29–31]
Topic 1	**Software system architecture and design** Systems, in general, is the highest in this topic see Appendix 1 – Topic 1B. Safety has a positive trend. Similarly, issues about requirement emerged to be discussed and gain more attention after 2013. Architecture and complexity are essential until 2019, but they are steadily decreasing
Topic 2	**Brake system and safety** In this topic, the most dominant trending terms are: break, collision, and pedestrian. Terms such as pedestrian, collision, safety, crash, accident, and risk are gradually increasing in recent years
Topic 3	**Mobile robot cars** In this topic, the term robot has the highest probability. Additionally, learn, and robotic-car are gradually increasing in importance along with the term algorithm
Topic 4	**Transportation traffic communication network** Network, road, connect, and smart gradually increased from 2010 to 2019. Security rose dramatically in 2016. However, safety and control are less discussed compared to other terms. In this topic, car, vehicle, and system have a higher probability than all other terms
Topic 5	**Maneuvering in self-driving cars** Steering or maneuvering is the most discussed term in this topic. Also, track, control, and angle are gradually gaining attention
Topic 6	**Self-driving path planning algorithms and methods** In this topic, problem and method are gradually increasing in relevance from 2010 to 2019. On the other hand, most of the terms available in the topic are more or less continually appearing with more or less consistent rate throughout the years. The terms path and obstacle decrease gently
Topic 7	**Self-driving technology energy utilization** Electric, charge, electric-vehicle, and efficiency gently shows a pickup or increase. Similarly, terms such as low, consumption and reduce also increased in values showing that reducing electric consumption is gaining more attention. However, battery decreased. This probably indicates that battery technology has matured or is gaining lesser attention
Topic 8	**Navigation in self-driving** Terms listed in the order of significance: lane, road, method, increased in significance with a positive change of trend. Estimation, algorithm, and propose gradually growing in significance
Topic 9	**Control and controller model design in self-driving cars** Control is discussed in this topic more than any other terms with at least a 10% probability difference. The term propose gently increases since 2010
Topic 10	**Driver experience in autopilot driving** This topic deals with driver experience in semi-self-driving cars. Trust, level, and automate gently increased in the timeline. Other terms do not show a significant trend which could lead to a meaningful interpretation that other terms in this topic are more or less consistent in the timeline
Topic 11	**Modeling traffic simulation** Model and traffic are the most discussed terms in topic 11 with model dominating in throughout the timeline. Next, simulation is also discussed throughout the timeline, with a higher probability than other terms next to model and traffic. This increasing trend indicates that modeling traffic system is gaining more attention

(*continued*)

Table 1. (*continued*)

Topic #	Topics labeled following [29–31]
Topic 12	**Neural network for object detection and image processing (computer vision and machine learning)** In this topic, the significance of terms such as object and image decreased in the timeline. On the other hand, terms such as network, learn, and deep increased throughout the timeline. Indicating neural network, machine learning, and deep learning are gaining attention in academia
Topic 13	**Unmanned underwater vehicle** Topic 13, uses terms related to vehicle, design, testing, underwater, and so on, indicating that it more about unmanned underwater vehicles. In this topic, platform, use, and system increases
Topic 14	**Research and development in the industry** There is a smooth and slowly increasing concern regarding the future of self-driving cars. Similarly, industry, artificial also gently and gradually increased. Also, the term issue gained some attention, after 2016 and artificial intelligence after 2018
Topic 15	**Road, traffic sign, and features detection (computer vision)** The term detection has a positive trend with higher probability value than other terms. Traffic and classification has more attention, with terms such as feature, sign, and light increasing gently. Road decreased unlike others
Topic 16	**Decision making by human_driver in self-driving cars** Topic 16 is about decision making by a human driver using self-driving. Decision making increases gently while other terms remain relatively constant
Topic 17	**Driver assistance system** Driver and system are the most discussed terms, and it is illustrated in Topic 17A. Terms assistance_system and ADA (Advanced Driver Assistance) systems are discussed from 2010 to 2019 and gained progressively increasing attention in academia
Topic 18	**Sensor data processing** Sensor and data are the most dominant terms in this topic with sensor gently decreasing from 2013 and data keeping it at a constant pace, refer to Topic 18A and 18B. However, sensor and data are ranked top throughout the years in terms of the probability of appearing in the topic. Radar is increasingly discussed from 2010 to 2019. LIDAR senor – increased gently from 2010 to 2019. Information, wireless and network decreased while other terms such as application, system, base, and measurement appear to be constantly appearing with similar probability
Topic 19	**Real-time 3D Data Processing** 3D, algorithm, application, and process are continuously and more or less constantly discussed. Data, point, cloud, and object increased gently. LIDAR gained more attention from 2014 to 2019
Topic 20	**Parking in urban transportation system** Research activity about parking sharply decreased from 2010 to 2019. On the contrary, AV (autonomous vehicle), sharing, and the public sector are increased in significance after 2017. Service, mobility, transport, urban, and transportation are more or less constantly studied from 2010 to 2019

4.1 Analysis of the Result

In this section, the analysis of the result, which enabled us to identify the subsequent suggested ideas for innovation and research agendas, is presented. LIDAR sensor is the most dominant term in intelligent transportation, sensors, and computer vision. We run regression analysis, see Fig. 3 below, on *LIDAR*, *Radar and Cloud* with time in Topics 18 and 19. Also, LIDAR correlates positively with *Radar and Cloud* as illustrated in Fig. 4.

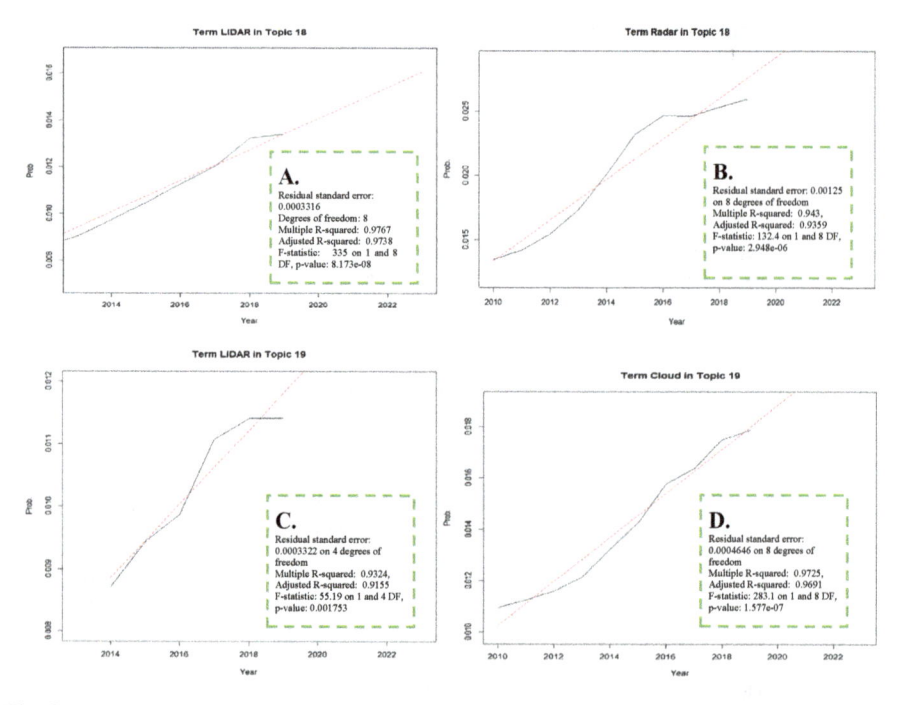

Fig. 3. Linear regression and prediction of terms in Topic 18 (A, LIDAR and B, Radar) and Topic 19 (C, LIDAR and D, Cloud) with a p-value less than 0.05 (99%) indicating a meaningful prediction.

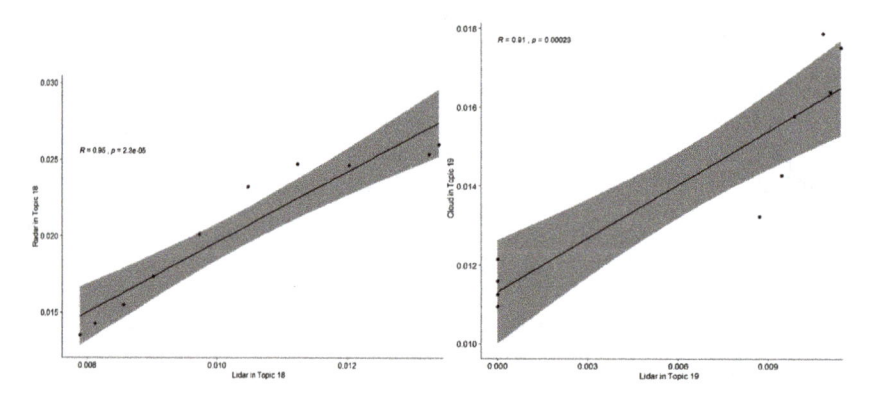

Fig. 4. On the left, correlation of *LIDAR* vs *Radar* in Topic 19, and on the right, correlation of *LIDAR* vs *Cloud* with statistical significances of at least 99%.

Intelligent Transportation: In Topic 15, traffic sign, feature selection increases with light while road slightly decreases. Classification, however, increases indicating the increasing importance of machine learning in feature selection. Smart transportation system (Intelligent Transportation) is discussed in Topic 4. Also, the terms network, road, and connection show a slight increase. The evolution of Topic 11 shows that modeling traffic simulation is important. Topics 18 and 19 illustrate that LIDAR increases in its significance. Similarly, in Topic 20, urban transportation, mobility, self-driving cars sharing, and sharing other resources, such as parking, increase in significance. Topic 7 shows that efficient energy utilization and reduction of electric power consumption are important.

Possible ideas for innovation and commercialization of ideas
- Innovation in road design, traffic sign detection, identification of features for smart cars with different light intensities.
- LIDAR sensor has gained importance over the years, and forecasts and correlations illustrated in Fig. 3 and 4 indicates that studies taking into consideration environmental conditions and the adequacy of lighting intensity in the ambience of self-driving technology is important. This is because adequate roadside lights entail easier feature detection such as the detection of road lanes and other traffic features.

Research agendas
- Parking, mobility, sharing of resources such as vehicles (shared vehicle), shared parking are also relevant topics.

Computer Vision: In Topic 18 and 19, LIDAR increased. Besides, in Topic 19, realtime 3D data processing and light also increased. Similarly, road, traffic sign, feature selection increases with light and classification (classification is part of supervised machine learning). Topic 12 shows that deep learning and neural networks as part of machine learning are increasing in significance for image processing.

Possible ideas for innovation and commercialization of ideas
- Machine learning and AI for object detection (feature classification).
- Deep learning and neural network for object detection.
- Identification of features with different road light intensity using machine learning and evaluating its efficiency.

Research agendas
- A comparative study of deep learning with legacy machine learning for feature detection with different ambience.

Control and Safety: In Topic 2, break, collision, pedestrian, safety, crash, risk increases, and in Topic 5, maneuvering, track, control, and angle increase in trends. In Topic 1, safety has a positive trend, and the concern regarding trust increases in Topic 10. On the other hand, track, control, lane, and angle increase in trend. Similarly, the trend of Topic 10 indicates that control and controller design has also a positive trend.

Possible ideas for innovation and commercialization of ideas
- Identification and implementation of safety standards for self-driving cars.
- Design and development of state of the art maneuvering and control algorithms

Research agendas
- Assessment of accident, trust and safety concerns about self-driving cars.
- How to address safety in the software design of self-driving cars.

Machine Learning and Algorithms: In Topic 3, AI increases, in Topic 8 algorithms and method for navigation increases. In Topic 2, deep learning, neural networks, and machine learning increases for object and image detection. Similarly, feature detection using machine learning increases. In Topic 14, AI increased with industry.
Possible ideas for innovation and commercialization of ideas
- Organizing training data for machine learning to improve
Research agendas
- Identifying efficient classifier for object detection.
- Algorithm design for navigation

Sensors: In Topic 18, sensor and data are the most discussed terms, and in Topic 19, cloud, and data increased. LIDAR has a positive trend in both 18 and 19, i.e. sensor data processing and realtime 3D data processing. Detection increases with road, light, sign, image, feature, traffic, and classification in Topic 15.
Possible ideas for innovation and commercialization of ideas
- Innovation in sensor and realtime 3D data processing.
Research agendas
- Research agenda listed under **Machine learning and algorithms** above.

Navigation: In Topic 4, road, communication, and security have positive trends currently. Path planning, algorithms, have persistently more or less non-negative or positive trends. In Topic 8, lane, road, method, estimation, and algorithms have positive trends. Finally, the evolution of Topic 15 shows that traffic sign, feature light, and classification have positive trends.
Possible ideas for innovation and commercialization of ideas
- Innovation in navigation system design.
Research agendas
- Analysis of the impact of light intensity and environmental weather condition of navigation.

Human Driver Interaction: In Topic 17, the assistance system, and ADA – Advanced Driver Assistance system increases in trend. Also, decision making by drivers in Topic 16, trust in semi-self-driving technologies in Topic 10, security in Topic 4, safety and requirement in Topic 1, AI and industry in Topic 14 increased in trend.
Possible ideas for innovation and commercialization of ideas
- Design and design requirements dealing with human driver interaction.
Research agendas
- Human car interaction issues.
- Design of self-driving cars to deal with driver aspects such as disability.

5 Discussion

The purpose of this research was to find innovative ideas about self-driving cars. The result of this paper contributes to the future development and innovation of self-driving technologies by unveiling valid research agendas to academia and creative ideas to innovation intermediaries such as contest organizers, incubators, and R&D centers.

A recent prediction of self-driving vehicle implementations indicates that self-driving technologies will be affordable and available for more independent mobility in the future [37]. Similarly, Topic 20 shows there is a consistent trend that indicates that there are research activities about mobility in academia. Furthermore, increased safety as in Topic 1 (Software system architecture and design) and Topic 2 (Brake system and safety), reduced traffic as in Topic 11, 6 (path planning) and 15, energy conservation as in Topic 7, issues regarding planning decisions similarly as Topic 16, optimal road as in Topic 4, 8 and 15, optimal parking as in Topic 20 are presented by [37].

Robot-car is a self-driving or driverless car as discussed by [37], Robot-car and algorithm increase gradually in importance over the years in Topic 3. Algorithms are the decisive parts of self-driving cars [38], and self-driving cars have more complex algorithms compared to self-driving planes and self-driving air force jets. For example, self-driving cars have about 100 million lines of code while Boeing 787 Dreamliner has about 6.5 million lines of code an F-22 Air Force jet fighter has 1.7 million lines of codes [37].

Topic 17 indicates that driver assistance systems are gaining academic attention during 2010 to 2019. This coincides with the emergence of a law introduced to the amendment in 2015 to the Americans with Disabilities Act[5]. Moreover, disabled drivers are able to purchase semi-self-driving vehicles with parallel parking options since 2010. Also, projects were disclosed in the same year to produce self-driving vehicles, yet these technologies need to be perfected for persons who have disabilities [39]. Current research indicates that self-driving cars will be affordable during 2040 to 2050, and people with disabilities are expected to travel using these technologies [37].

6 Conclusions and Future Research

The following conclusions were derived from the analysis of the result. Through the trending terms, we can identify areas that need innovation. These areas are intelligent transportation, computer vision, control and safety, machine learning and algorithms, sensors, navigation, and finally, human driver interaction, and legal issues. The results indicate that sensors are the most important parts of self-driving technology which makes us conclude that it is essential to continue innovation activities related to sensors. Especially LIDAR and radar have positive correlations (Topic 18) as well as LIDAR and cloud environment (Topic 19). Also, the forecast indicates that these terms will increase in importance the next three years.

---

[5] https://www.ada.gov/pubs/adastatute08.pdf.

Based on the analysis of the results we suggest the following research agendas:

- Parking, mobility, shared vehicles and shared parking.
- Evaluation of feature detection methods in a variety of ambient conditions
- Legacy machine learning vis-à-vis deep learning for feature detection, given different weather conditions and light intensities (ambiances). Also, designing algorithms for navigation.
- Identification and implementation of safety standards
- Assessment of accident, trust, and safety concerns in the use of self-driving cars; also, how to address safety in the software design
- Human car interaction issues, also considering design for disabled people.
- Identification legal issues as requirements for the design of self-driving cars

For future research, it is suggested to formalize the approach into a method for idea evaluation adapted to practice needs. In addition, it is suggested to analyze additional data sources, such as social media and patent data.

Acknowledgment. The authors wish to thank Alexey Voronov (Ph.D.) and Mahedre DW Amanuel (M.Sc.) at RISE for their support in interpreting the results of the analysis. We would also like to thank Panagiotis Papapetrou (Professor) for offering constructive feedback. This study was conducted as part of the research project IQUAL (2018-04331) funded by Sweden's Innovation Agency (Vinnova).

Appendix 1

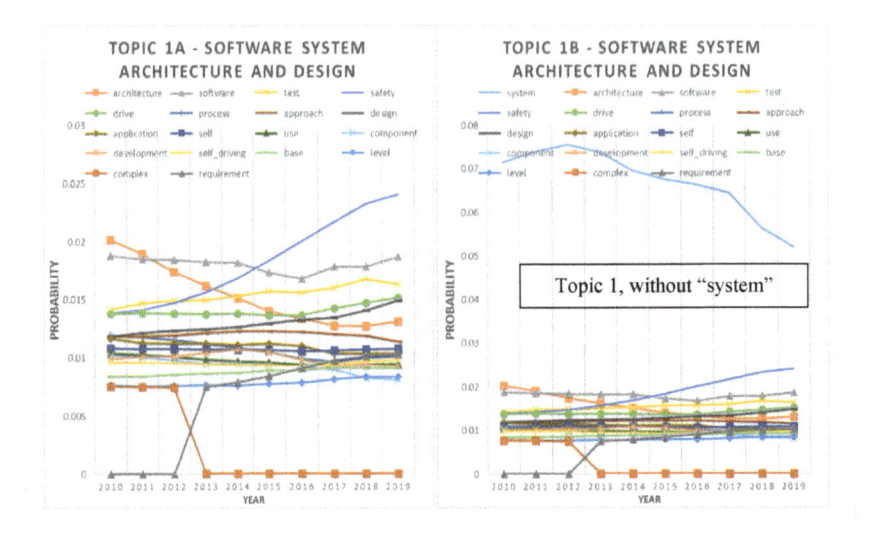

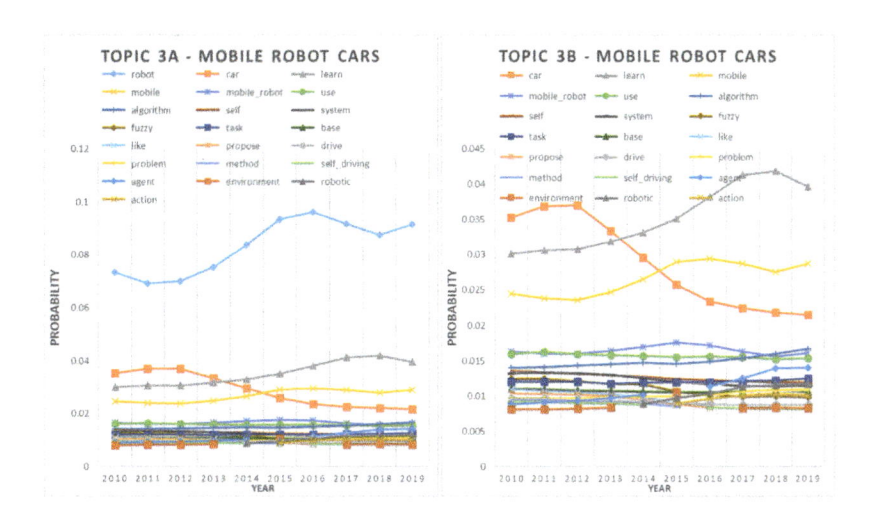

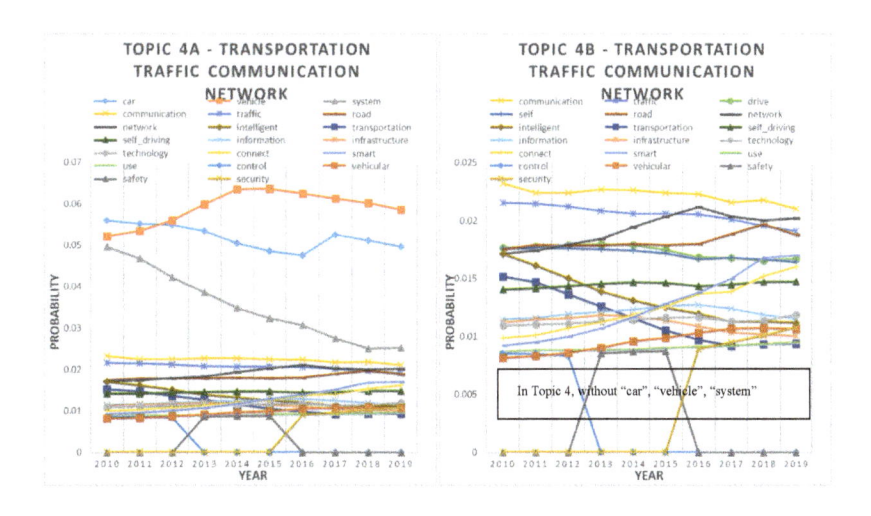

Eliciting Evolving Topics, Trends and Foresight 501

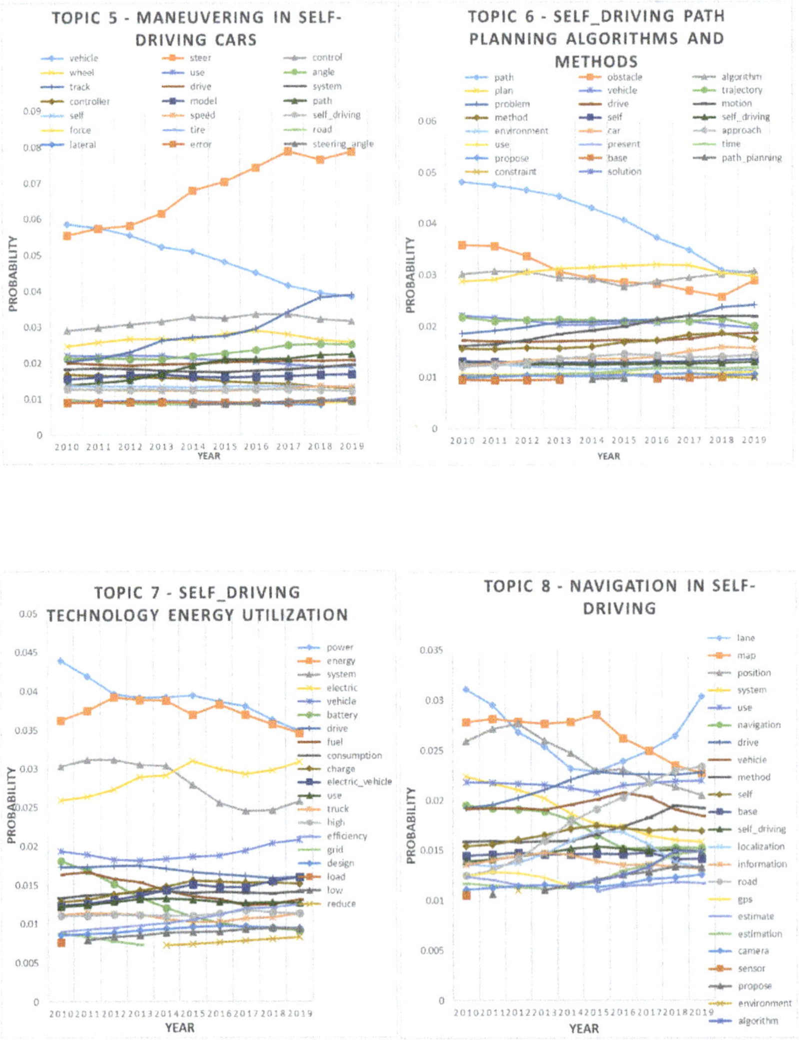
TOPIC 5 - MANEUVERING IN SELF-DRIVING CARS
vehicle
steer
control
wheel
use
angle
track
drive
system
controller
model
path
self
speed
self_driving
force
tire
road
lateral
error
steering_angle
PROBABILITY
YEAR
2010 2011 2012 2013 2014 2015 2016 2017 2018 2019
TOPIC 6 - SELF_DRIVING PATH PLANNING ALGORITHMS AND METHODS
path
obstacle
algorithm
plan
vehicle
trajectory
problem
drive
motion
method
self
self_driving
environment
car
approach
use
present
time
propose
base
path_planning
constraint
solution
PROBABILITY
YEAR
TOPIC 7 - SELF_DRIVING TECHNOLOGY ENERGY UTILIZATION
power
energy
system
electric
vehicle
battery
drive
fuel
consumption
charge
electric_vehicle
use
truck
high
efficiency
grid
design
load
low
reduce
PROBABILITY
YEAR
TOPIC 8 - NAVIGATION IN SELF-DRIVING
lane
map
position
system
use
navigation
drive
vehicle
method
self
base
self_driving
localization
information
road
gps
estimate
estimation
camera
sensor
propose
environment
algorithm
PROBABILITY
YEAR

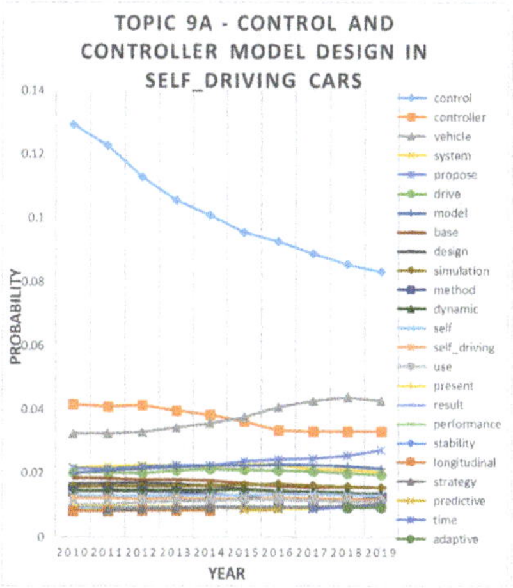
TOPIC 9A - CONTROL AND CONTROLLER MODEL DESIGN IN SELF_DRIVING CARS
PROBABILITY
YEAR
control
controller
vehicle
system
propose
drive
model
base
design
simulation
method
dynamic
self
self_driving
use
present
result
performance
stability
longitudinal
strategy
predictive
time
adaptive

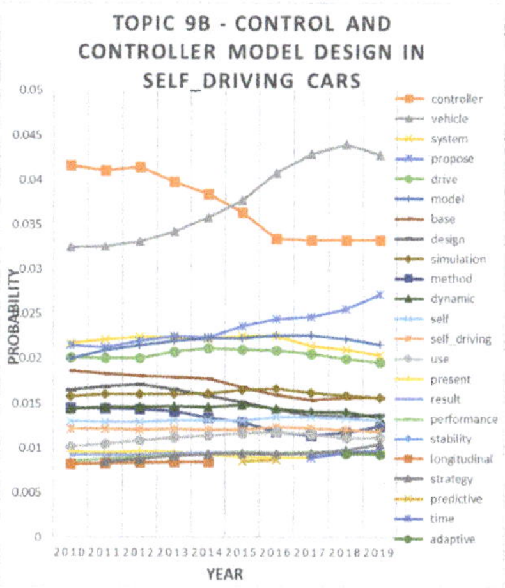
TOPIC 9B - CONTROL AND CONTROLLER MODEL DESIGN IN SELF_DRIVING CARS
PROBABILITY
YEAR
controller
vehicle
system
propose
drive
model
base
design
simulation
method
dynamic
self
self_driving
use
present
result
performance
stability
longitudinal
strategy
predictive
time
adaptive

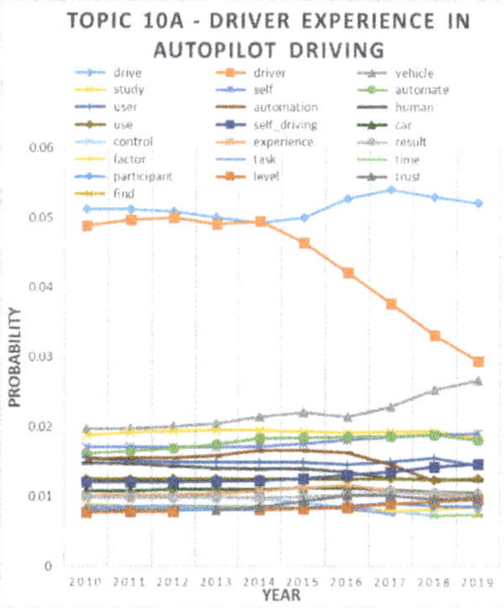
TOPIC 10A - DRIVER EXPERIENCE IN AUTOPILOT DRIVING
PROBABILITY
YEAR
drive
driver
vehicle
study
self
automate
user
automation
human
use
self_driving
car
control
experience
result
factor
task
time
participant
level
trust
find

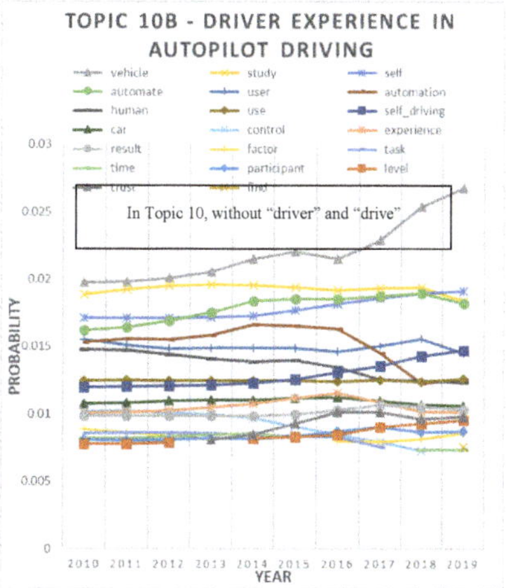
TOPIC 10B - DRIVER EXPERIENCE IN AUTOPILOT DRIVING
PROBABILITY
YEAR
vehicle
study
self
automate
user
automation
human
use
self_driving
car
control
experience
result
factor
task
time
participant
level
trust
find
In Topic 10, without "driver" and "drive"

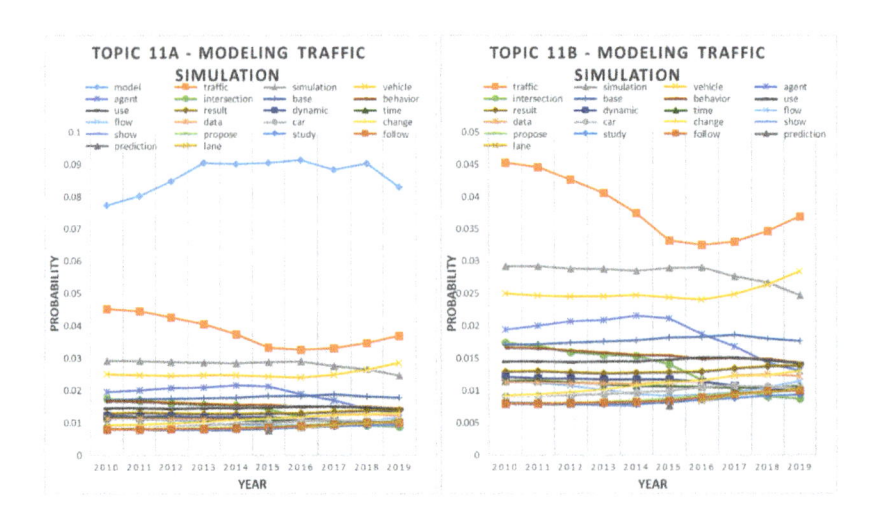

TOPIC 11A - MODELING TRAFFIC SIMULATION

TOPIC 11B - MODELING TRAFFIC SIMULATION

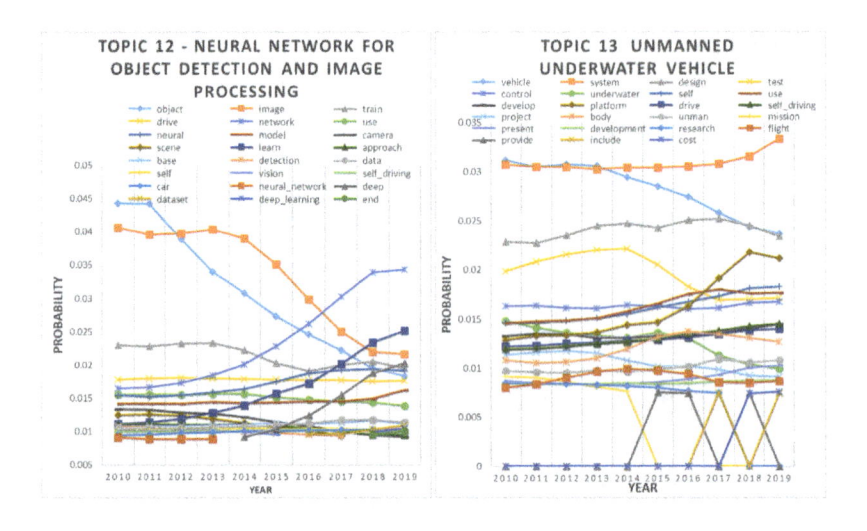

TOPIC 12 - NEURAL NETWORK FOR OBJECT DETECTION AND IMAGE PROCESSING

TOPIC 13 UNMANNED UNDERWATER VEHICLE

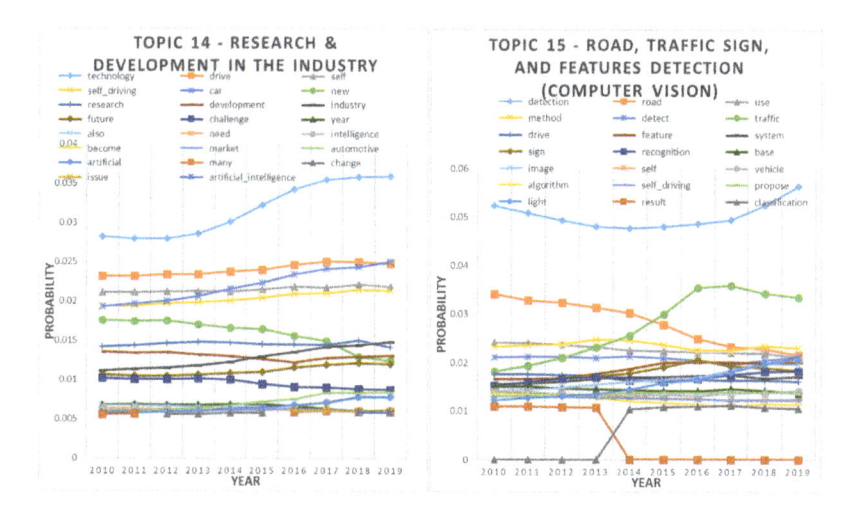

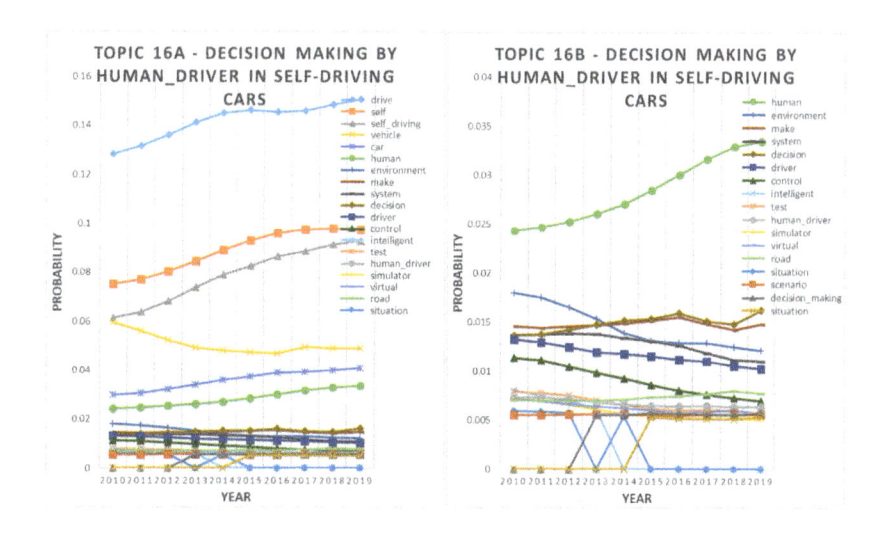

Eliciting Evolving Topics, Trends and Foresight 505

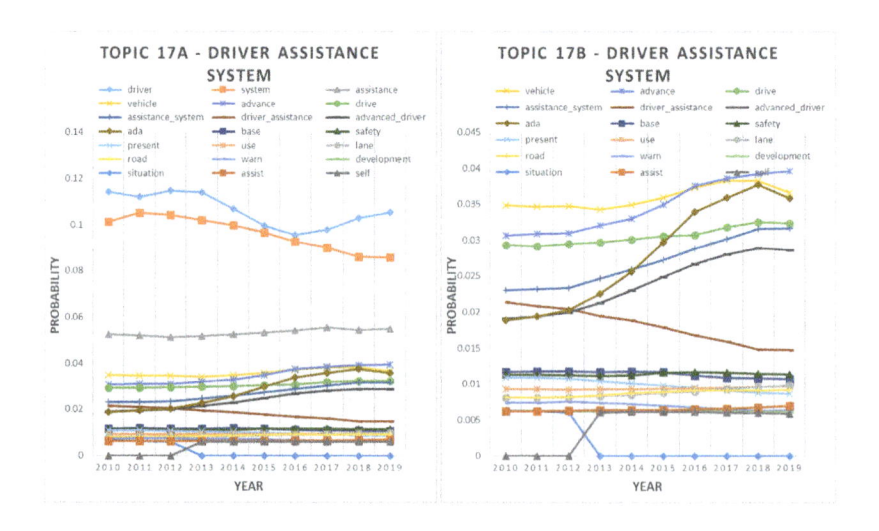

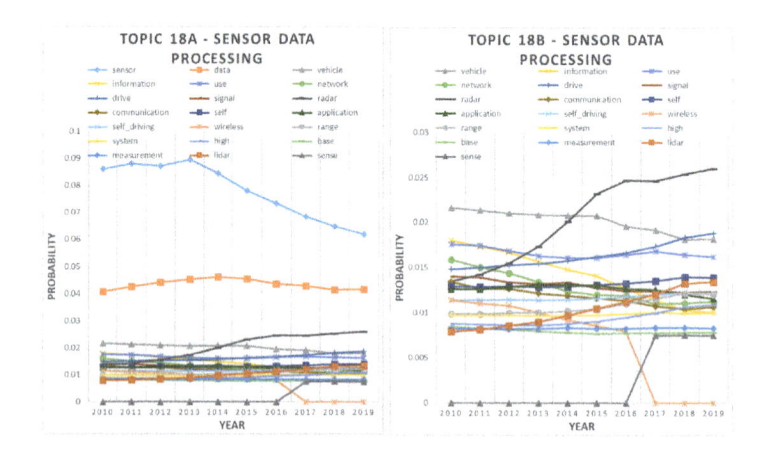

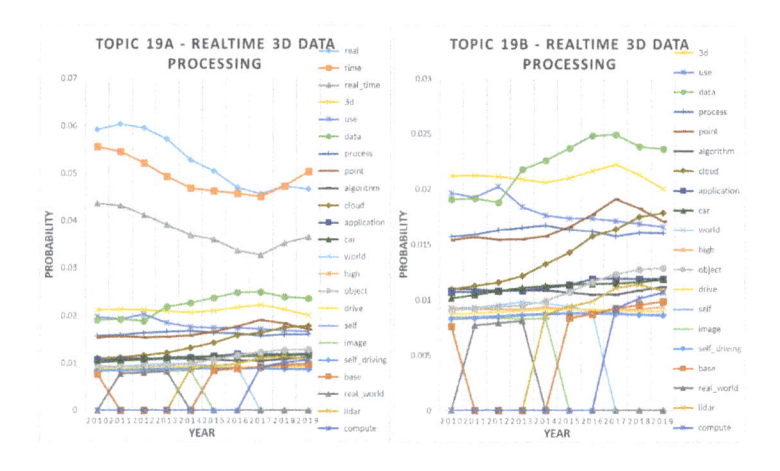

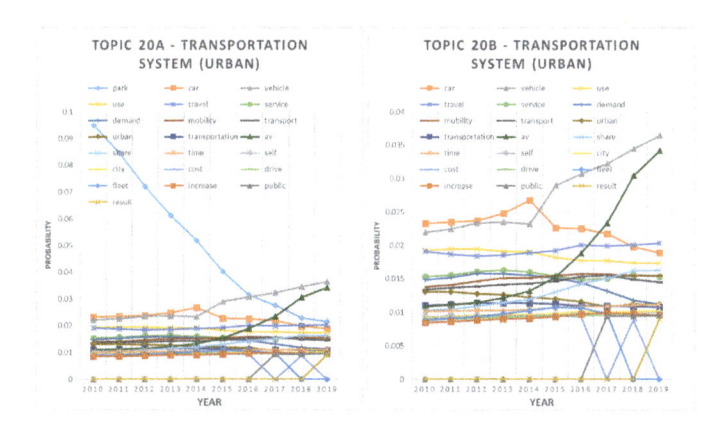

References

1. Howard, D., Dai, D.: Public perceptions of self-driving cars: the case of Berkeley, California. In: Transportation Research Board 93rd Annual Meeting, vol. 14, no. 4502 (2014)
2. Swan, M.: Connected car: quantified self becomes quantified car. J. Sens. Actuator Netw. **4**(1), 2–29 (2015)
3. Anthopoulos, L., Janssen, M., Weerakkody, V.: A unified smart city model (USCM) for smart city conceptualization and benchmarking. In: Smart Cities and Smart Spaces: Concepts, Methodologies, Tools, and Applications, pp. 247–264. IGI Global (2019)
4. Gretzel, U., Werthner, H., Koo, C., Lamsfus, C.: Conceptual foundations for understanding smart tourism ecosystems. Comput. Hum. Behav. **50**, 558–563 (2015)
5. Lin, P.: Why ethics matters for autonomous cars. In: Autonomous Driving, pp. 69–85. Springer, Heidelberg (2016)

6. Ayele, W.Y., Juell-Skielse, G.: Unveiling topics from scientific literature on the subject of self-driving cars using latent Dirichlet allocation. In: 2018 IEEE 9th Annual Information Technology, Electronics and Mobile Communication Conference (IEMCON), pp. 1113–1119. IEEE, November 2018
7. Gassmann, O., Zeschky, M., Wolff, T., Stahl, M.: Crossing the industry-line: breakthrough innovation through cross-industry alliances with 'non-suppliers'. Long Range Plan. **43**(5–6), 639–654 (2010)
8. Ayele, W.Y., Juell-Skielse, G., Hjalmarsson, A., Johannesson, P.: Unveiling DRD: A method for designing and refining digital innovation contest measurement models. Systems, Signs Actions **11**(1), 25–53 (2018)
9. Ayele, W.Y., Juell-Skielse, G., Hjalmarsson, A., Johannesson, P., Rudmark, D.: Evaluating open data innovation: a measurement model for digital innovation contests. In: PACIS, p. 204, July 2015
10. Juell-Skielse, G., Hjalmarsson, A., Juell-Skielse, E., Johannesson, P., Rudmark, D.: Contests as innovation intermediaries in open data markets. Inf. Polity **19**(3+4), 247–262 (2014)
11. Villani, E., Rasmussen, E., Grimaldi, R.: How intermediary organizations facilitate university–industry technology transfer: a proximity approach. Technol. Forecast. Soc. Chang. **114**, 86–102 (2017)
12. McIntosh, T., Mulhearn, T.J., Mumford, M.D.: Taking the good with the bad: The impact of forecasting timing and valence on idea evaluation and creativity. Psychology of Aesthetics, Creativity, and the Arts (2019)
13. Bloom, N., Jones, C.I., Van Reenen, J., Webb, M.: Are ideas getting harder to find? (No. w23782). National Bureau of Economic Research (2017)
14. Salatino, A.A., Osborne, F., Motta, E.: AUGUR: forecasting the emergence of new research topics. In: Proceedings of the 18th ACM/IEEE on Joint Conference on Digital Libraries, pp. 303–312. ACM, May 2018
15. Small, H., Boyack, K.W., Klavans, R.: Identifying emerging topics in science and technology. Res. Policy **43**(8), 1450–1467 (2014)
16. You, H., Li, M., Hipel, K.W., Jiang, J., Ge, B., Duan, H.: Development trend forecasting for coherent light generator technology based on patent citation network analysis. Scientometrics **111**(1), 297–315 (2017)
17. Blei, D.M., Lafferty, J.D.: Topic models. In: Text Mining, pp. 101–124. Chapman and Hall/CRC, Boca Raton (2009)
18. Blei, D.M., Lafferty, J.D.: Dynamic topic models. In: Proceedings of the 23rd International Conference on Machine Learning, pp. 113–120. ACM, June 2006
19. Steingrimsson, B., Yi, S., Jones, R., Kisialiou, M., Yi, K., Rose, Z.: Big data analytics for improving fidelity of engineering design decisions (No. 2018-01-1200). SAE Technical Paper (2018)
20. Ayele, W.Y., Akram, I.: Identifying emerging trends and temporal patterns about self-driving cars in scientific literature. In: Arai, K., Kapoor, S. (eds.) Advances in Computer Vision. CVC 2019. Advances in Intelligent Systems and Computing, vol. 944. Springer, Cham (2020)
21. Marçal, R., Antonialli, F., Habib, B., Neto, A.D.M., de Lima, D.A., Yutaka, J., Luiz, A., Nicolaï, I.: Autonomous Vehicles: scientometric and bibliometric studies. In: 25th International Colloquium of Gerpisa-R/Evolutions. New technologies and Services in the Automotive Industry (2017)
22. Kontostathis, A., Galitsky, L.M., Pottenger, W.M., Roy, S., Phelps, D.J.: A survey of emerging trend detection in textual data mining. In: Survey of Text Mining, pp. 185–224. Springer, New York (2004)

23. Stöckl, S.Q.J.: The next big thing: the use of text mining analysis of crowdfunding data for technology foresight. Master's thesis, University of Twente (2018)
24. Nassirtoussi, A.K., Aghabozorgi, S., Wah, T.Y., Ngo, D.C.L.: Text mining for market prediction: a systematic review. Expert Syst. Appl. **41**(16), 7653–7670 (2014)
25. Gloor, P.A., Krauss, J., Nann, S., Fischbach, K., Schoder, D.: Web science 2.0: identifying trends through semantic social network analysis. In: 2009 International Conference on Computational Science and Engineering, August 2009, vol. 4, pp. 215–222. IEEE (2009)
26. Wirth, R., Hipp, J.: CRISP-DM: towards a standard process model for data mining. In: Proceedings of the 4th International Conference on the Practical Applications of Knowledge Discovery and Data Mining, pp. 29–39. Citeseer, April 2000
27. Aghaei, C.A., Salehi, H., Yunus, M., Farhadi, H., Fooladi, M., Farhadi, M., Ale, E.N.: A comparison between two main academic literature collections: web of science and scopus databases (2013)
28. Mongeon, P., Paul-Hus, A.: The journal coverage of web of science and Scopus: a comparative analysis. Scientometrics **106**(1), 213–228 (2016)
29. Blei, D.M., Ng, A.Y., Jordan, M.I.: Latent dirichlet allocation. J. Mach. Learn. Res. **3**(Jan), 993–1022 (2003)
30. Röder, M., Both, A., Hinneburg, A. Exploring the space of topic coherence measures. In: Proceedings of the Eighth ACM International Conference on Web Search and Data Mining, pp. 399–408. ACM, February 2015
31. Hindle, A., Ernst, N.A., Godfrey, M.W., Mylopoulos, J.: Automated topic naming to support cross-project analysis of software maintenance activities. In: Proceedings of the 8th Working Conference on Mining Software Repositories, pp. 163–172. ACM, May 2011
32. Maskeri, G., Sarkar, S., Heafield, K.: Mining business topics in source code using latent dirichlet allocation. In: Proceedings of the 1st India Software Engineering Conference, pp. 113–120. ACM, February 2008
33. Ha, T., Beijnon, B., Kim, S., Lee, S., Kim, J.H.: Examining user perceptions of smartwatch through dynamic topic modeling. Telemat. Inf. **34**(7), 1262–1273 (2017)
34. Box, G.E., Tiao, G.C.: Intervention analysis with applications to economic and environmental problems. J. Am. Stat. Assoc. **70**(349), 70–79 (1975)
35. Simonton, D.K.: Cross-sectional time-series experiments: some suggested statistical analyses. Psychol. Bull. **84**(3), 489 (1977)
36. Braun, V., Clarke, V.: Using thematic analysis in psychology. Qual. Res. Psychol. **3**(2), 77–101 (2006)
37. Litman, T.: Autonomous Vehicle Implementation Predictions, p. 28. Victoria Transport Policy Institute, Victoria (2019)
38. Stilgoe, J.: Machine learning, social learning and the governance of self-driving cars. Soc. Stud. Sci. **48**(1), 25–56 (2018)
39. Van Roosmalen, L., Paquin, G.J., Steinfeld, A.M.: Quality of life technology: the state of personal transportation. Phys. Med. Rehabil. Clin. **21**(1), 111–125 (2010)

Quality Assurance/Quality Control Engine for Power Outage Mitigation

The Challenge of Event Correlation for a Smart Grid Architecture Amidst Data Quality Issues

S. Chan[✉]

Vit Tall Energy Unit & IE2SPOMTF, San Diego, CA 92192, USA
schan@vittall.org

Abstract. Energy is a key driver of economic growth. A key thematic for distribution utilities has centered upon how to better capitalize upon existing available amounts of energy within an electrical grid. Pragmatic opportunities reside within the distribution system of the grid. By means of examining and validating assumed/purported key nodes as well as properly contextualizing actual key nodes for when expeditious fault location is needed, it is possible to enhance power outage mitigation as well as facilitate more expeditious recovery and restoration of energy. We introduce a Data Visualization Analytics System (DVAS), which leverages a Quality Assurance/Quality Control (QA/QC) Engine to assist with correct identification of key nodes that have high impact on distribution utility performance indices, thereby directly affecting resiliency compliance decision-making within a Smart Grid and reporting to regulatory agencies.

Keywords: Energy efficiency · Power loss optimization · Cause code analysis · Fault location · Outage mitigation · Data visualization · Resiliency compliance · Event correlation · Smart grid architecture · Data quality

1 Introduction

Current business practices at various distribution utilities are challenged with properly conducting forensic investigations (that could yield critical information to identify root causes and, ultimately, inform mitigating courses of action) for a robust analysis of power outages. Without a greater emphasis on some of the core tenets for synchronizing and correlating electrical grid system data, there will persist a gap in knowledge, thereby perpetuating the status quo for not robustly addressing the actual causes of power outages; for example, an examination of the operational data for the studied distribution utility shows that the predominant cause of power outages was actually not the posited cause code "005" (downed trees, vegetation overgrowth, etc. affecting the power lines). Rather, upon analysis, the cause codes affecting the greatest number of customers were actually cause codes "006" (overload) and "004" (scheduled outages). This knowledge gap remains even as new policies and programs are implemented at a

Quality Assurance/Quality Control Engine for Power Outage Mitigation   511

country, regional, and international level, particularly as pertains to "Resiliency Compliance Plans" (RCPs).

This paper posits a quality assurance/quality control (QA/QC) engine schema that endeavors to mitigate against various data quality single points of failures. Section 2 describes global critical infrastructural areas at an international, regional, and country level. Section 3 presents the inefficiencies at the generation, transmission, and distribution levels within the electrical grid. Section 4 posits the urgent need to enhance electrical grid resiliency (and efficiency), via outage mitigation improvements as well as adaptability for timely recovery and restoration. Section 5 provides a lexicon primer to understand electrical grid resiliency opportunities at the distribution level and presents electrical grid resiliency opportunities: prevention or anticipating issues (by way of indicators and warnings, such as by Grid Monitoring Capabilities or GMC), recovery or maintaining power stability (by pinpointing affected areas, such as by Fault Location Capabilities or FLC), and survivability or prioritizing energy (such as by Energy Prioritization Capabilities or EPC). Section 6 progresses through key concepts for an exemplar distribution utility so as to grasp the significance of properly correlating distribution transformers with corresponding primary poles and customer impact numbers. Section 7 discusses the issue of data quality and the validation of data for proper event correlation, via a QA/QC engine schema. Section 8 presents some experimental findings, and Sect. 9 puts forth an interim conclusion and future work.

2 Global Critical Infrastructural Areas

Certain common interest critical infrastructural areas (e.g. international airports), such as in Southeast Asia and Oceania, often operate with insufficient and brittle commercial electrical grids that are vulnerable to various natural and man-made hazards, thereby potentially placing normal as well as exigency operations at a high level of risk. Accordingly, this issue is addressed, in certain ways, internationally, regionally, and locally (e.g. country-level).

2.1 International Level

At the international level, there are the internationally agreed upon themes embodied by the Sustainable Development Goals (SDGs) (a.k.a. Global Goals for Sustainable Development), which are a collection of 17 global goals set by the United Nations (UN). The SDG resolution (an intergovernmental agreement among the 193 Member States of the UN) is articulated in paragraph 54 of United Nations General Assembly Resolution Seventieth Session (A/RES/70/1) of 25 September 2015 (these goals supplanted the Millennium Development Goals [MDGs], which concluded in 2015), and pertinent SDGs include: SDG#7: Renewable Energy or Affordable and Clean Energy; SDG#9: Industry, Innovation and Infrastructure; SDG#11: Sustainable Cities and Communities; and SDG#17: Partnerships for the Goals. The significance of these particular SDGs is that this particular amalgam has resulted in energy being classified by the Association of Southeast Asian Nation (ASEAN) member nations under the rubric of economics, as energy is deemed to be a key engine powering a country's overall goal of economic resilience; indeed, this notion of "energy as an economic driver" seems to resound across the ASEAN.

2.2 Regional Level

At the regional level, there are several venues, such as the Asia Pacific Economic Cooperation (APEC), which have focused upon "energy as an economic driver". For instance, APEC Philippines 2015 was a year-long hosting of the Asia-Pacific Economic Cooperation Summit, which concluded with the APEC Economic Leaders' Meeting held between 18–19 November 2015 in Pasay City, Metro Manila, Philippines. One outcome of this APEC summit was that the Philippines was selected as co-chair to lead the Task Force on Energy Resiliency, and "energy resiliency" has indeed become a core thematic both regionally within the Asia Pacific and at the country-level within the Philippines.

2.3 Country Level

At the country level, by way of example, the Philippines initiated an aggressive plan to substantially increase the country's power supply so as to spur economic development. Under normal circumstances, key actors from the Philippine Government involved within the energy sector include the Philippine Department of Energy (PDOE) and the Energy Regulatory Commission (ERC), which approves energy supply deals as well as the Philippine Distribution Code (PDC) pertaining to energy (the Electric Power Industry Reform Act or EPIRA had created the ERC as an independent, quasi-judicial regulatory body tasked with promoting competition, encouraging market development, ensuring customer choice, and penalizing abuses of power within the electricity industry). However, given the hitherto protracted approvals process for the energy sector within the Philippines, the President issued Executive Order (EO) #30 (a.k.a. "EO 30") on 28 June 2017 to assist with accelerating matters within the energy sector. EO 30 established the Energy Investment Coordinating Council (EICC), which leads the Philippine Government's effort to streamline the regulatory process for energy investments. Chaired by the PDOE, the other members of the EICC stem from various government agencies involved with the permitting process for energy projects. Also, EO 30 mandates relevant government agencies to act upon applications for permits involving Energy Projects of National Significance (EPNS) (i.e. major PDOE-endorsed energy projects that support the specific goals of the Philippine Energy Plan) within thirty days; in summary, time is of the essence when it comes to operationalizing projects related to energy. As additional affirmation to this fact, Nationally Appropriate Mitigation Actions (NAMAs) are formulated to address the constraints hindering economic growth and development within the Philippines. At this point in time, the Philippines is projected to sustain economic growth ranging from 6.5% to 7.5% within the next six years [1], and the principal foreseeable constraint simply centers upon the available amount of energy to fuel this growth. Accordingly, an analysis has been performed on how to better capitalize upon existing available amounts of energy in the Philippines, and efficiency has emerged as a key thematic as pertains to the generation, transmission, and distribution system levels of the electrical grid.

3 Inefficiencies of the Generation, Transmission, and Distribution Levels Within the Electrical Grid

3.1 Generation Level

The PDOE has asserted that, "the Philippines intends to undertake Greenhouse Gases (GHG) and Carbon Dioxide equivalent (CO2e) emissions reduction of about 70% by 2030 relative to its "Business-As-Usual (BAU) Scenario of 2000–2030" [2]. The PDOE also asserts that Philippines' current energy supply-mix will be diversified to minimize the import dependencies upon fossil fuels, which currently meet the country's energy needs to a substantive degree. The posited approach is to address 57–60% of the country's energy needs with renewable energy by 2040 [3]. However, renewable energy introduces heightened instability into the electrical grid, and the Philippines electrical grid is already beset with an ongoing trend of rolling blackouts (an indicator of a potentially unstable system) [4]; thus, the introduction of further renewables into an already unstable system can be seen as paradoxical by some. However, if this indeed is the plan, there is a need to increase the stability of the system, such as at the transmission and distribution (T&D) level and, particularly, at the distribution level.

3.2 Transmission Level

Historically, since the Electric Power Industry Reform Act (EPIRA) of 8 June 2001 (Republic Act or RA No. 9136) took effect, millions of electricity consumers in the Philippines have been charged on their monthly electric bills for system losses (which includes natural calamities and transmission inefficiencies) under the "Recoverable Systems Loss Act". The estimated inefficiency of the Philippines transmission system is quite high, and just recently, the Energy Efficiency Act (RA No. 11285) was signed by the Philippines President given the belief that "additional power supply will not come in the form of new power plants, but in savings from energy efficiency and conservation" [5].

3.3 Distribution Level

As specified in Article 3.4 of the PDC, three of the main classifications as pertains Distribution System Losses include: Technical Loss (e.g. the inherent electricity losses due to electrical equipment as well as the devices and conductors used for the physical delivery of electricity), Non-Technical Loss (e.g. the electricity lost due to pilferage, tampering of meters, erroneous meter readings, etc.), and Administrative Loss (e.g. the electricity attributed to the normal, proper operations of the distribution system—by the distribution utility—as well as any unbilled electricity that is used for community-related activities).

Overall, there exist numerous inefficiencies within the generation, transmission, and distribution systems (leading to the Philippines experiencing numerous outages while having one of the most expensive power rates in the world), but the greatest opportunities—as identified by the Asian Development Bank—might reside with improving the hitherto relatively high inefficiency distribution systems [6].

4 Urgent Need to Enhance Electrical Grid Resiliency

4.1 Recurring Natural Disasters: Earthquakes

Historically, the Philippines has experienced numerous seismic events, such as the 16 July 1990 magnitude 7.9 Luzon Earthquake, which resulted in over 1,600 deaths and P10 billion (USD 191,545,000) worth of damage to property. Moving along the timeline to 15 October 2013, a magnitude 7.2 earthquake struck Bohol. On 2 December 2014, a 6.6 struck near the Moro Gulf. On 3 July 2015, a 6.1 struck Siargao. On 24 September 2016, a 6.3 struck near Maisan. On 29 April 2017, a 6.9 struck Davao and Sarangani. On 29 December 2018, a 7.1 struck near Davao Oriental. On 22 April 2019, a 6.1 struck Castillejos. On 23 April 2019, a 6.4 struck San Julian. Given the recent string of earthquakes afflicting the Philippines, the notion of the "Big One" impacting Metro Manila and other parts of the Philippines has arisen again [7]. The Philippine Institute of Volcanology and Seismology (PHIVOLCS) and the Metro Manila Earthquake Impact Reduction Study (MMEIRS) affirm this notion. Given this prospective disaster, the pressing issue of power outage mitigation comes into play.

4.2 Electrical Grid Resiliency to Mitigate Impacts of Disaster

It should, therefore, be of no surprise that the PDOE has prudently promulgated the adoption of energy resiliency in the planning and programming of the energy sector to mitigate potential impacts of disasters, via Department Circular No. DC2018-01-0001. Indeed, the Philippine Disaster Risk Reduction and Management Act of 2010 (RA No. 10121) conveys the intent to build "a disaster-resilient nation and communities." Even without considering disasters, the DOE Act of 1992 (RA No. 7638) stipulates a "continuous, adequate, and economic supply of energy with the end in view of ultimately achieving self-reliance in the country's energy requirements." The denotes a high degree of energy efficiency, and currently there is high degree of inefficiency, particularly on the distribution side, as noted by several University of the Philippines studies. EPIRA (Section 37 of RA No. 9136) calls for a "restructuring of the electricity sector" so as to segue to an "efficient supply … of energy." Also, EPIRA has promulgated "resilient energy infrastructure," which is defined as "the ability to restore and sustain availability and accessibility of energy in the most timely and efficient manner," and private utilities have been encouraged to adopt a Resiliency Compliance Plan (RCP) to address "response capabilities" including the following aspects:

(1) Incorporate *outage mitigation improvements*.
(2) Improve operational and maintenance standards and practices to ensure *expeditious restoration of energy supply*.
(3) Effectuate a paradigm of *timely recovery and restoration*.
(4) [Incorporate a paradigm of] *Adapt[ability]* to withstand adverse conditions and disruptive events.

The aforementioned policy guidance can be translated into various operationalizable technological capabilities. For example, "outage mitigation improvements" can equate to grid monitoring capabilities (GMC), fault location capabilities (FLC), and energy prioritization capabilities (EPC). "Expeditious restoration of energy supply" as well as "timely recovery and restoration" can both equate to FLC and EPC. "Adapt[ability]" can equate to GMC, FLC, EPC and more. Collectively, GMC, FLC, EPC comprise a layered defense-in-depth approach for power outage mitigation design. However, EPC necessitates FLC, and FLC is not possible if there is prevalent key node mis-identification; for example, what are the key poles supplying power to the airport, hospitals, and supermarkets? If this identification is incorrect, how is FLC and EPC possible? Along this vein, if there is mis-identification of these key pole nodes, this begets the questions as to whether GMC were properly deployed and of value-added proposition. **Also, would the reporting to the regulatory agencies be correct?**

5 A Lexicon Primer to Understand Electrical Grid Resiliency Opportunities at the Distribution Level

5.1 Electrical Grid

An electrical grid is defined as an electrical power system network comprised of, among other components, generating station(s) (a.k.a. power plant), transmission lines, substations, transformers, distribution lines, and consumer(s). Between the ends (i.e. generating station, consumer), electrical power may flow through substations at various voltage levels. Ideally, this is architected so as to minimize the power loss along the generation-transmission-distribution pathway by maintaining a higher voltage whenever possible. From generation to distribution, the voltage is progressively stepped down for safe usage.

5.2 Distribution Utility

An electric utility is a company within the electric power industry (often a public utility) that engages in any of electricity generation, transmission, and/or distribution as pertains to an electric grid. A distribution utility constructs and maintains the distribution wires connecting the transmission system to the final electricity consumer.

5.3 Substation

Within an electrical grid, substations are a key component of the constitutive generation, transmission, and distribution systems comprising the involved grid. A substation may include transformers to change voltage levels at the interconnection of two different transmission voltages or between higher transmission voltages and lower distribution voltages.

5.4 Single Line Diagram

In accordance with international convention, a Single Line Diagram (SLD) (a.k.a. one-line diagram) is the most basic of the set of diagrams that are used to document a distribution substation's electrical paradigm (i.e. power flow); a robust SLD is fundamental for maintenance, upgrades, and emergency response activities. In particular, an SLD should be readily updated as new components are introduced to the system.

5.5 Electrical Grid Resiliency Opportunities

Efforts toward electrical grid "resiliency" tend to focus upon three elements: prevention, recovery, and survivability. Along this vein, the distribution utility should: (1) anticipate issues (by way of indicators and warnings, such as by GMC) to avoid outages, (2) maintain power stability (by pinpointing affected areas, such as by FLC), and (3) prioritize energy (i.e. EPC) for critical elements amidst disaster (e.g. maintain the provision of power to prioritized recipients of electricity, such as traffic signals, hospitals, grocery stores, and faith-based organizations). One opportunity for enhancing grid resiliency comes by way of well understanding the involved distribution grid and its key nodes. After all, EPC necessitates FLC, and FLC is not possible if there is prevalent key node mis-identification; if there is misidentification of these key poles, this begets the questions as to whether the related analysis is correct.

6 An Exemplar Distribution Utility

6.1 Distribution Substation

The purpose of a distribution substation is to transfer power from the transmission system to the distribution system of an area. In addition to transforming voltage, distribution substations also regulate voltage (although for long distribution lines [i.e. circuits], voltage regulation equipment may also be installed along the circuits) and isolate faults. Several distribution substations may comprise a distribution utility. For this case study, the three distribution substations (DS) shall be referred to as DS#1, DS#2, and DS#3.

6.2 Distribution Feeder

Feeders represent the power lines through which electricity is transmitted within power systems. A distribution feeder represents one of the circuits emanating from a DS, and it transmits power from a DS to the designated distribution points serving electricity to the consumer. Typically, there are several distribution feeders per DS (e.g. Feeder A or FA, Feeder B or FB, etc.).

6.3 Primary Distribution Line

Each of the aforementioned distribution feeders segue into multiple Primary Distribution Lines (PDLs), which carry power to distribution transformers located near the electricity consumer. For this case, there are 2287 PDL Segment IDs (SIDs) emanating from 6 Distribution Feeders. Each PDL SID is derived by conjoining the "From Pole ID" with a "-" and the "To Pole ID".

6.4 Distribution Transformer

A transformer is an electrical device consisting of two or more coils of wire that transfer electrical energy between two or more circuits by means of a varying magnetic field (a varying current in one coil of the transformer produces a varying magnetic flux, which, in turn, induces a varying electromotive force across a second coil wound around the same core).

A distribution transformer provides the final voltage transformation within an electric power distribution system. Prior to the distribution point, the rationale for stepping up the voltage to a much higher level is that higher distribution voltages segue to lower currents for the same power and therefore lower power (i.e. $I^2 \cdot R$) losses along the transmission lines. At the distribution point, the basis for stepping down the voltage is to reduce the shock hazard of high voltage (and provide a suitable voltage level for local distribution) as well as to increase the current capacity for use by the electricity consumer. If a distribution transformer is mounted on a utility pole, it is known as a pole-mount transformer; if the distribution transformer is mounted on concrete, it is known as a [distribution tap] pad-mount transformer. Both pole-mount and pad-mount transformers convert the higher [main or primary] voltage of overhead or underground distribution lines to the lower [secondary] voltage of the distribution wires inside the associated fed structures. The larger [primary] distribution lines have three wires (which translates to three phases), while smaller [lateral] distribution lines may include one or two phases, but eventually, the electricity consumers are served with single-phase power; this takes the form of Secondary Distribution Lines (SDLs), which carry "low" voltage power to the electricity consumer. For this case, there are 2287 PDL SIDs entering 767 distribution transformers, and many more SDL SIDs (actually 3264) emanating from these distribution transformers, and the associated numbers for the top distribution transformers with the largest number of SDL SIDs are noted.

6.5 Customer

The types of electricity consumers can be expressed by, among others, that of: (1) residential (e.g. homeowners), (2) commercial (e.g. storefronts), (3) industrial (e.g. factories), and (4) municipal (e.g. water plants, sewer plants, etc.). Despite the differences, the commonality is that they constitute customers. Accordingly, they all play a role in determining the performance indices of their associated electricity supply services. System Average Interruption Frequency Index (SAIFI) (the average number of interruptions that a customer would experience during the measurement period), System Average Interruption Duration Index (SAIDI) (the average duration of interruption for each customer served during the measurement period), and Momentary Average Interruption Frequency Index (MAIFI) (the average number of times a customer experiences a momentary interruption during the measurement period) are some of the indices used to measure distribution system reliability; the Customer Average Interruption Duration Index (CAIDI) (the average length of a sustained customer interruption during the measurement period) is simply SAIDI divided by SAIFI.

Among the discussed indices used to measure distribution system reliability, the System Average Interruption Frequency Index (SAIFI) can be calculated as shown in Eqs. (1a) and (1b):

$$\text{SAIFI} = \frac{\text{Total number of customers Interrupted}}{\text{Total number of customers Served}} \quad (1a)$$

$$\text{SAIFI} = \frac{\sum(N_i)}{N_T} \quad (1b)$$

where N_i = Total number of customers interrupted for each sustained interruption event and N_T = Total number of customers served. Another index is that of System Average Interruption Duration Index (SAIDI), which can be calculated as shown in Eqs. (2a) and (2b):

$$\text{SAIDI} = \frac{\text{Sum of all customer interruption durations}}{\text{Total number of customers served}} \quad (2a)$$

$$\text{SAIDI} = \frac{\sum(r_i \times N_i)}{N_T} \quad (2b)$$

where r_i = Restoration time (in minutes), N_i = Total number of customers interrupted for each sustained interruption event, and N_T = Total number of customers served. Yet another index is that of Customer Average Interruption Duration Index (CAIDI), which can be calculated as shown in Eqs. (3a) and (3b):

$$\text{CAIDI} = \frac{\text{Sum of all customer interruption durations}}{\text{Total number of customer interruptions}} \quad (3a)$$

$$\text{CAIDI} = \frac{\sum(r_i \times N_i)}{\sum(N_i)} \quad (3b)$$

where r_i = Restoration time (in minutes) and N_i = Total number of customers interrupted.

As N_i plays such an instrumental role with regards to the SAIFI, SAIDI, and CAIDI calculations, it becomes axiomatic to examine the key distribution utility components affecting N_i. Accordingly, we examine which distribution feeders affect the greatest number of customers, such as presented in Fig. 1, which also provides us with the value of N_T = Total number of customers served = 18,934.

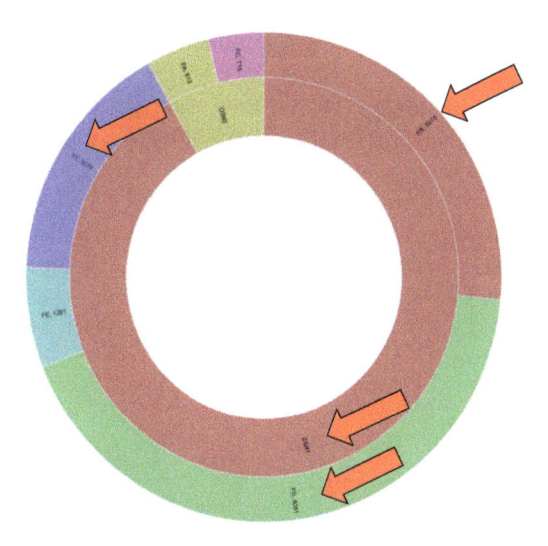

Quality Assurance/Quality Control Engine for Power Outage Mitigation 519

Fig. 1. DVAS vantage point: customers served by various distribution substations and various Distribution Feeders. Distribution Feeder D (FD) services 8051 customers and is, therefore, a key node. FD services 5015 customers and is also a key node. FF services 30779 customers and is also a key node. As DS#1 supplies FD, FB, and FF, it is also a key node.

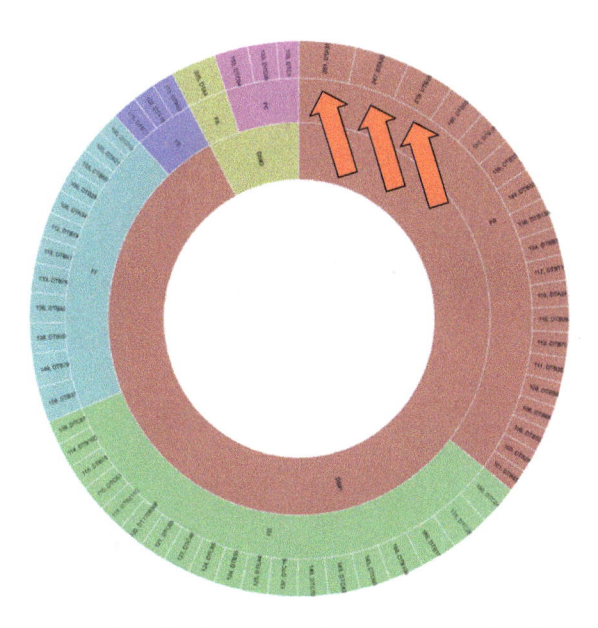

Fig. 2. DVAS vantage point: distribution transformers associated with various distribution feeders and various distribution substations. Distribution transformer DTA31 services 261 customers and is, therefore, a key node.

An analysis of Fig. 1 indicates that distribution feeders FD, FB, and FF are key nodes, which have the potential to impact 8051, 5015, and 3079 customers, respectively. The distribution transformers associated with these distribution feeders are presented in Fig. 2. An analysis also indicates that distribution transformers DTA31, DTA30, and DTB35 are Key Nodes, which have the potential to impact 261, 247, and 219 customers, respectively. Accordingly, it is noted that Distribution Feeder B (FB), which supplies DTA31, DTA30, and DTB35 is a potential Achilles heel and, therefore, a Key Node as well. DTA31 is presented in Fig. 3, DTA30 is presented in Fig. 4, and DTB35 is presented in Fig. 5. These distribution transformers also correspond to the Primary Pole IDs of A123, A096, and B343, respectively, which are also represented in the aforementioned figures.

Fig. 3. Distribution transformer ID: DTA31 and corresponding pole ID: A123.

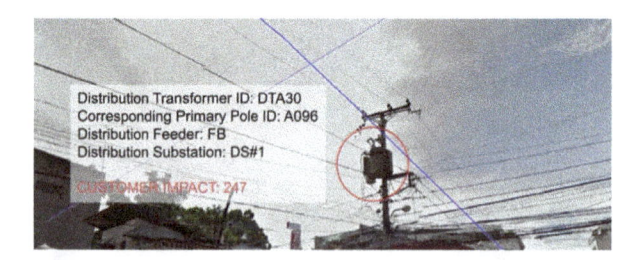

Fig. 4. Distribution transformer ID: DTA30 and corresponding pole ID: A096.

Fig. 5. Distribution transformer ID: DTB35 and corresponding pole ID: B343.

The impact of the aforementioned components can best be evidenced by way of Eq. (1a), which was previously introduced above. Hence, if FD (with its 8051 customers) were to become inoperable, the SAIFI calculation would become as shown in (4a) and (4b):

$$\text{SAIFI} = \frac{\sum(N_i)}{N_T} \tag{4a}$$

$$\text{SAIFI} = \frac{8051}{18934} = .4252 \tag{4b}$$

This would represent a percentage difference (PD) of 54.00%, as shown in (5a) and (5b):

$$\text{PD} = \frac{|V1 - V2|}{|(V1 + V2)/2|} \times 100 \tag{5a}$$

$$\text{PD} = \frac{|18934 - 10883|}{\frac{18934 + 10883}{2}} \times 100 = \frac{|8051|}{14908.5} = 54.00\% \tag{5b}$$

If, for example, DTA31 (with its 261 customers) became inoperable, the SAIFI calculation would be similarly affected. Just as the total number of customers interrupted affects SAIFI, the corollary with regards to total interruption duration also affects the SAIDI and CAIDI calculations. An overview of the various distribution transformers, their corresponding primary poles, and their customer impact numbers are presented in Fig. 6 below.

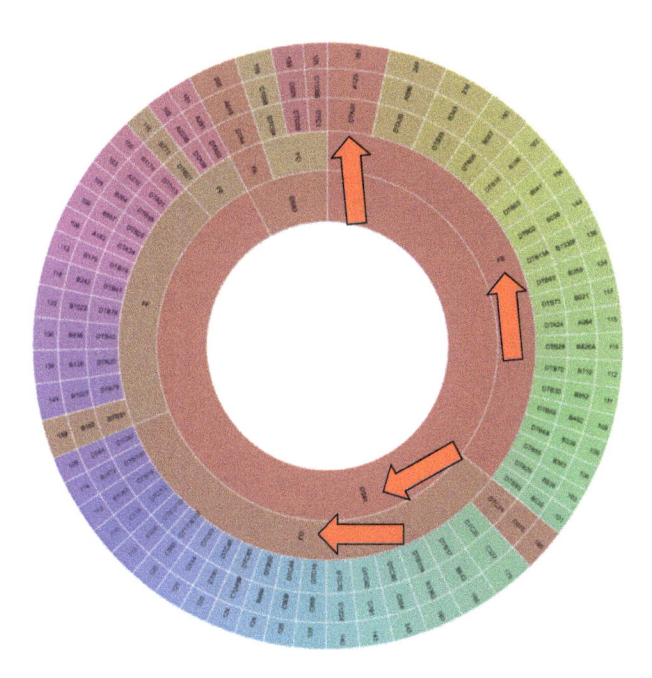

Fig. 6. DVAS vantage point: distribution transformer IDs, corresponding primary pole IDs, and the customer impact numbers.

In essence, Distribution Substation #1 (DS#1) services 17,406 customers and is, therefore, a Key Node. Distribution Feeder D (FD) services 8051 customers and is, therefore, also a Key Node. Distribution Transformer DTA31 services 261 customers and is, therefore, also Key Node. **The significance of DTA31 also underscores the point that Distribution Feeder B (FB) should not be overlooked, whereas FD was more obvious.**

7 Task of Feature Correlation at a Distribution Utility to Better Operationalize EPC, via Validating FLC and Enhancing GMC at Key Nodes

The shift toward a "Smarter" Grid commenced ontologically with providing designator names in the form of "Also Known As" (a.k.a.) for each Document Name or File Name. In this way, the data assembled by distribution utility substation staff into various Documents or Files could be referenced, via a relatively persistent identifier, as shown in Table 1. Extrapolating upon these identifiers, a designated Quality Assurance (QA)/Quality Control (QC) Engine could proceed to validate the various Data Sources. For example, the number of unique distribution transformers listed on Data Source #1c (a.k.a. System Short Code, Tab 3 "TRANSFORMER") should match that of Data Source #1d (a.k.a. System Short Code, Tab 4 "SECONDARY"). Other key feature correlation examples are presented as follows.

Table 1. Document name/file name with corresponding "Also Known As" (a.k.a.).

Document name	File name	Also Known As
System Short Code List		
• System Short Code List, Tab 1 "TREE DIAGRAM"	System_Short_code.xlsx	Data Source #1a
• System Short Code List, Tab 2 "PRIMARY"	System_Short_code.xlsx	Data Source #1b
• System Short Code List, Tab 3 "TRANSFORMER"	System_Short_code.xlsx	Data Source #1c
• System Short Code List, Tab 4 "SECONDARY"	System_Short_code.xlsx	Data Source #1d
• System Short Code List, Tab 5 "CUSTOMER"	System_Short_code.xlsx	Data Source #1e
• Sectional Code List	Sectional_code.xlsx	Data Source #2
• Outage Event Record	Outage_Event.xlsx	Data Source #3
Distribution Network Diagram		
• Distribution Network Diagram FA	FDR A.pdf	Data Source #4a
• Distribution Network Diagram FB	FDR B.pdf	Data Source #4b
• Distribution Network Diagram FC	FDR C.pdf	Data Source #4c
• Distribution Network Diagram FD	FDR D.pdf	Data Source #4d
• Distribution Network Diagram FE	FDR E.pdf	Data Source #4e
• Distribution Network Diagram FF	FDR F.pdf	Data Source #4f

(*continued*)

Table 1. (*continued*)

Document name	File name	Also Known As
Distribution Feeder List		
• Distribution Feeder A	FA.xlsx	Data Source #5a
• Distribution Feeder B	FB.xlsx	Data Source #5b
• Distribution Feeder C	FC.xlsx	Data Source #5c
• Distribution Feeder D	FD.xlsx	Data Source #5d
• Distribution Feeder E	FE.xlsx	Data Source #5e
• Distribution Feeder F	FF.xlsx	Data Source #5f
Keyhole Markup Zip (KMZ) File		
• Communications Network	CommunicationsNetwork.kmz	Data Source #6a
• Feeder Lines	FeederLines.kmz	Data Source #6b
• Pole IDs	PoleIDs.kmz	Data Source #6c
• Substations	SubstationLocations.kmz	Data Source #6d

7.1 Feature Correlation Example

Our software implementation of a Quality Assurance (QA)/Quality Control (QC) Engine endeavored to extract the Distribution Transformer ID from the Circuit Number column within Data Source #3 (a.k.a. Outage Event Record) (please refer to Module 7, block 7a of Fig. 12 below), which comes as set of letters and numbers that should correspond with a specific "Distribution Feeder" and "Sectional Code" (for this case, the "Old [Sectional] Code" as contrasted to the "New [Sectional] Code") (e.g. Circuit No. FC-Street#1-Street#2-A is—theoretically—a designator for Distribution Feeder C [a.k.a. FC] and an ensuing Sectional Code).

From Data Source #3 (a.k.a. Outage Event Record), we take, as an example, Old [Sectional] Code FC-Street#1-Street#2-A. We then endeavor to locate this Old [Sectional] Code FC-Street#1-Street#2-A, within Data Source #2 (a.k.a. Sectional Code List) (please see Module 1, block 1a of Fig. 12 below), so as to ascertain its corresponding New [Sectional] Code. However, the QA/QC Engine was unable to locate it, as Data Source #2 is rife with anthropogenically-introduced errors. The closest match in Data Source #2 (a.k.a. Sectional Code List) is Old [Sectional] Code FB- Street#1-Street#2 (whose corresponding "New [Sectional] Code" is CAL-115).

Presuming that this Data Source #2 case of Old [Sectional] Code FB-Street#1-Street#2 actually corresponds with the Data Source #3 (a.k.a. Outage Event Record) case of Circuit No. FC-Street#1-Street#2-A, the QA/QC Engine makes the supposition that the New [Sectional] Code is CAL-115. If this is indeed true, the corresponding location is that of Street#1 COR. Street#2, which is the distribution utility substation staff notation for "corner." The QA/QC Engine then endeavors to match this New [Sectional] Code CAL-115 against the various Distribution Network Diagrams, which includes FA through FF (a.k.a. Distribution Network Diagram for Distribution Feeder A [FA] through Distribution Network Diagram for Distribution Feeder F [FF]) (please refer to Module 2, block 2f of Fig. 12 below).

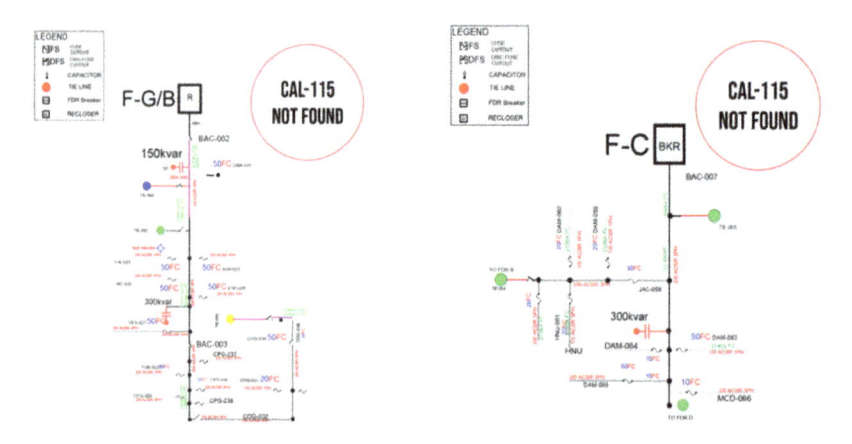

(7a) New [Sectional] Code CAL-115 not found on Distribution Network Diagram for Distribution Feeder B (FB), which is inconsistent with the Distribution Utility Substation Data.

(7b) New [Sectional] Code CAL-115 not found on Distribution Network Diagram for Distribution Feeder C (FC), which is inconsistent with the Distribution Utility Substation Data.

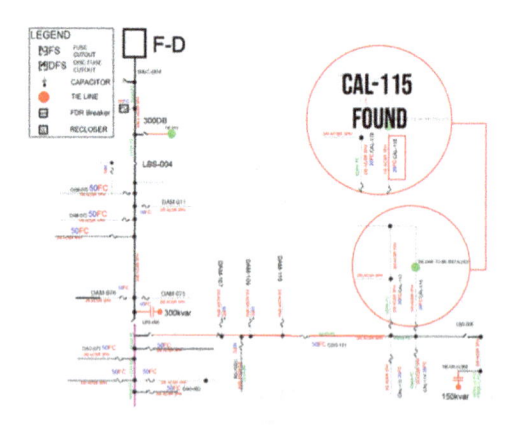

(7c) New [Sectional] Code CAL-115 found on Distribution Network Diagram for Distribution Feeder D (FD), which is inconsistent with the Distribution Utility Substation Data

Fig. 7. QA/QC engine vantage point: New [Sectional] Code CAL-115 is not found on the Distribution Network Diagrams (either FB or FC), pursuant to the inconsistent distribution utility substation data; rather, it is found were it is not supposed to be—on the Distribution Network Diagram for Distribution Feeder D (FD).

For this case, the QA/QC Engine is able to determine that CAL-115 is not located on Distribution Network Diagram for Distribution Feeder C (FC) (as alluded to in Data Source #3 [a.k.a. Outage Event Record], via the FC prefix), and it is also not located on Distribution Network Diagram FB (as alluded to in Data Source #2 [a.k.a. Sectional Code List], via the FB prefix). Instead, the QA/QC Engine determined that CAL-115 is actually located on Feeder D (a.k.a. FD). **It should be noted that the QA/QC Engine utilizes natural language processing (NLP) and computer vision machine learning (CVML) to be able to make this determination, as shown in** Fig. 7 (7a through 7c).

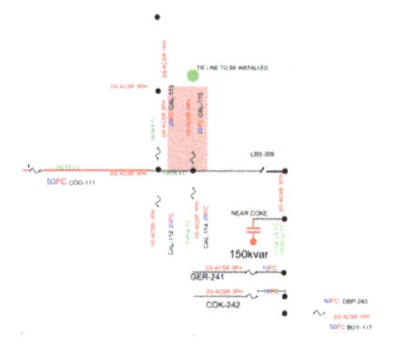

(8a) Left "Small" Section Constraining Boundary for New [Sectional] Code CAL-115.

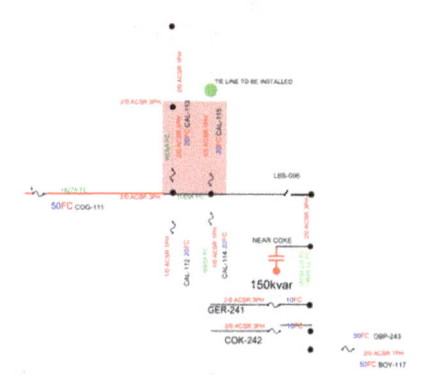

(8b) Left "Medium" Section Constraining Boundary for New [Sectional] Code CAL-115.

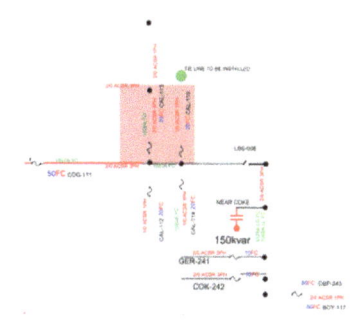

(8c) Left "Large" Section Constraining Boundary for New [Sectional] Code CAL-115.

Fig. 8. QA/QC engine vantage point: exemplar potential "Small," "Medium," and "Large" section constraining boundaries for New [Sectional] Code CAL-115 given unspecified boundary for CAL-115.

The QA/QC Engine also endeavored to match the referenced New [Sectional] Code CAL-115 against the various Data Source #5 (a.k.a. Distribution Feeder List for Distribution Feeder A (FA) through Distribution Feeder F (FF)) (please refer to Module 2, block 2 h of Fig. 12 below). For this case, it was able to determine that CAL-115 is indeed not located within Data Source #5c (a.k.a. Distribution Feeder C) (as alluded to in Data Source #3 [a.k.a. Outage Event Record] via the FC prefix) and is not located within Data Source #5b (a.k.a. Distribution Feeder B) (as alluded to in Data Source #2 [a.k.a. Sectional Code List] via the FB prefix). However New [Sectional] Code CAL-115 was located within Data Source #5f (a.k.a. Distribution Feeder D). **This confirmed the findings of the QA/QC Engine's NLP and CVML.**

The QA/QC Engine further endeavored to identify the coverage of the New [Sectional] Code CAL-115 within the Distribution Network Diagrams (e.g. FB, FC, FD), but it could not explicitly identify the constraining boundaries, such as roughly exampled by the below delineated, in Fig. 8, "small" section, "medium section," "large section," etc. in any of various directions. An exemplar left "small" section is shown in Fig. 8(a), an exemplar left "medium" section is shown in Fig. 8(b), and an exemplar left "large" section is shown in Fig. 8(c).

The ambiguity of the unspecified boundary of New [Sectional] Code CAL-115 segues to potential overlap with other new [sectional] codes. Indeed, the varied exemplars of Fig. 8(a) through 8c reflect the fact that New [Sectional] Code CAL-115 does overlap with other New [Sectional] Codes, such as New Sectional Codes CAL-113, CAL-112, CAL-114, LBS-006, GER-241, and COK-242, as can be seen in Fig. 9 below.

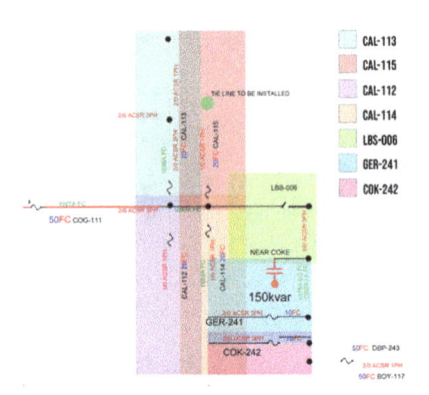

Fig. 9. QA/QC engine vantage point: unspecified boundaries of New [Sectional] Code CAL-115 overlaps with other New [Sectional] Codes.

Quality Assurance/Quality Control Engine for Power Outage Mitigation   527

The QA/QC Engine then utilizes the actual street name of "Street#1" and the cross street of "Street#2" from the "Location" column of Data Source #2 (a.k.a. Sectional Code List) to estimate the location of the target New [Sectional] Code CAL-115 within the pertinent Keyhole Markup Zip (KMZ) file (please refer to Module 1, block 1e of Fig. 12 below), as shown in Fig. 10 below.

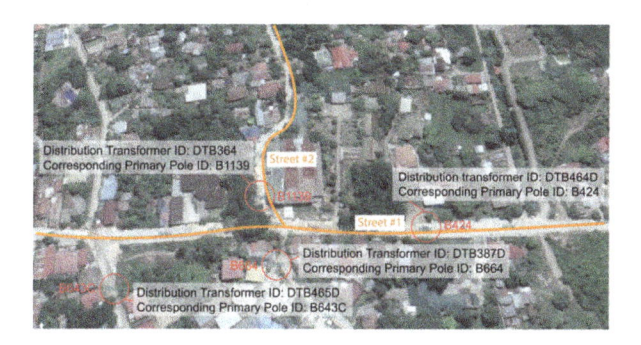

Fig. 10. QA/QC engine vantage point: estimating the target New [Sectional] Code CAL-115 within the KMZ file, via the actual street name for "Street#1" and the actual cross street name for "Street#2."

Upon completing the location estimation task, the QA/QC System determines that there are four Distribution Transformers (DTB37, DTB387D, DTB364, and DTB142D) that might be associated with the target New [Sectional] Code CAL-115. These Distribution Transformers, which are contained within Data Source #1c (a.k.a. System Short Code, Tab #3 "TRANSFORMER"), are depicted in Fig. 11 below.

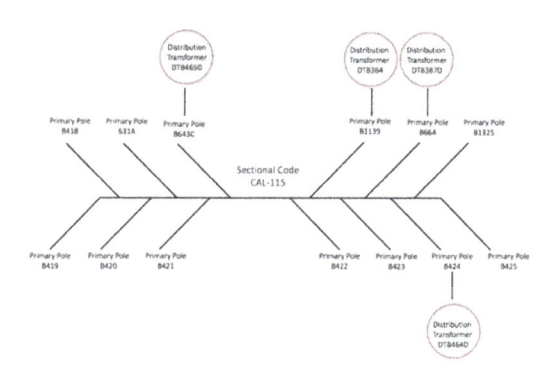

Fig. 11. QA/QC engine vantage point: four distribution transformers (DTB37, DTB387D, DTB364, and DTB142D) that might be associated with New [Sectional] Code CAL-115.

As can be surmised, the Distribution Network Diagrams currently utilized (provided by the distribution utility substation staff) do not provide the requisite resolution of a more robust SLD. **Given the high degree of ambiguity and anthropogenically-introduced errors, relevant Key Nodes are often mis-classified, thereby making robust FLC quite difficult. Accordingly, we utilize the discussed QA/QC Engine to process all the data supplied by the distribution utility substation staff.** The QA/QC architectural schema, divided into various modules, is presented in Fig. 12 below.

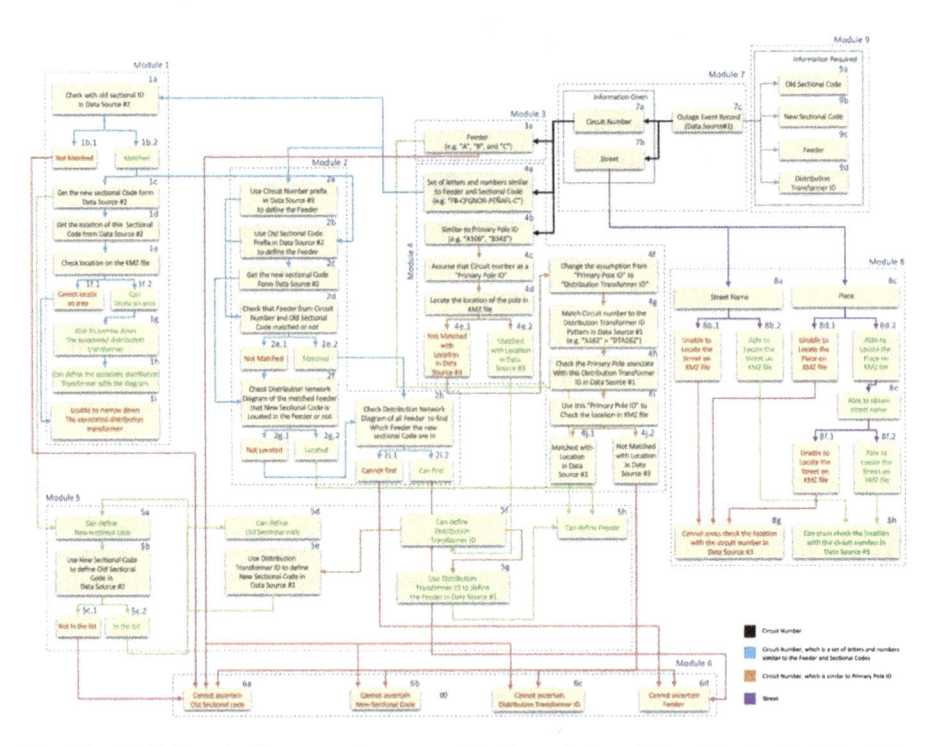

Fig. 12. QA/QC engine vantage point: unspecified boundaries of New [Sectional] Code CAL-115 overlaps with other New [Sectional] Codes.

It should be noted that, in addition to NLP and CVML, the QA/QC Engine utilizes a gazeteer (i.e. a geographic index or dictionary) to match placenames as well as a synonym application programming interface (API) to handle placenames referenced by jargon or slang. For example, some of the data from Data Source #3 (a.k.a. Outage Event Record), under the "Street" column (please refer to Module 7, Block 7b in Fig. 12) are actually establishments (vice a street) (please refer to Module 8, Block 8c in Fig. 12). In many cases, there are typographical errors, so a "fuzzy search" function is utilized. Oftentimes, the benefit of specifying the power outage event location by establishment is that it helps to narrow down beyond simply a street name; however, some of the establishments that have been provided cannot be found in Data Source #6a through Data Source #6d (Keyhole Markup Zip [KMZ] file). Accordingly, despite the efficacy from the combined use of the DVAS and the QA/QC Engine, a 100% "Data Quality" number will not be reached. This will be further discussed in Sect. 8.

8 Experimental Findings

The limitation of the study centers upon a bounded problem set that exists within some member countries of the Association of Southeast Asian Nations (ASEAN); in particular, there exists a common challenge to event correlation (and the ensuing data visualization for decision-making) due to data quality issues. Data cleansing is a key first step for a data science or machine learning workflow. It is estimated that most data scientists spend approximately 20% of their time on actual data analysis and 80% of their time on data cleansing [8]; after all, incorrect or inconsistent data leads to false conclusions. However, an automated QA/QC Engine can be quite useful in freeing up a data scientist's time and reducing human error. There are various parameters for data quality, such as accuracy (the degree to which the data is close to the true values), completeness (the degree to which all required data is known), consistency (the degree to which the data is consistent both within the same dataset and across multiple datasets), and uniformity (the degree to which the data is specified using the same unit of measure). This paper presented a Quality Assurance/Quality Control (QA/QC) Engine to assist with the iterative process of data cleansing—inspection (the detection of unexpected, incorrect, and inconsistent data), refinement (the removal or correction of anomalies discovered), verification (the inspection of the results to verify correctness), and memorialization (the documentation of the changes made and the notation of

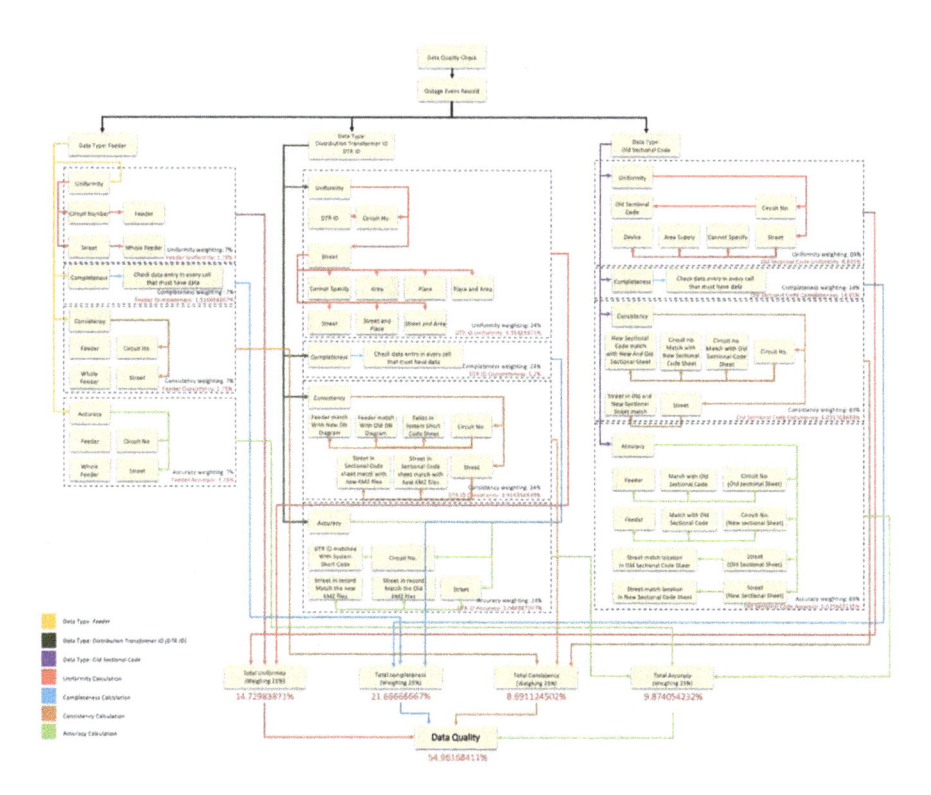

Fig. 13. Data quality parameters and "Data Quality" number

the enhanced quality of the currently stored data)—so as to facilitate correct reportage by a distribution utility to the involved regulatory agencies as well as to facilitate better decision-making for the utility. Figure 13 provides some insight as to the "Data Quality" of the original, uncleansed data when viewed against the aforementioned parameters. As can be seen, the "Data Quality" number for the original dataset provided by the distribution utility substation staff is approximately 54.96%. It should be of no surprise that after the automated data cleansing, which includes NLP and CVML, by the QA/QC Engine, the "Data Quality" number changed to 98.9%.

9 Conclusion

An operating expense (OPEX) is an expense required for day-to-day functioning. In contrast, a capital expense (CAPEX) is an expense incurred to create a benefit in the future. Current Resiliency Compliance Plan (RCP) requests by regulatory agencies for "outage mitigation improvements", "expeditious outage response capabilities", and "resilient, adaptable architectures" necessitate CAPEX expenditures amidst the current paradigm of OPEX expenditures being mis-prioritized. An examination of the operational data for various distribution utilities shows that the predominant cause of outages was actually not the posited cause code; rather, upon analysis, the cause codes affecting the greatest number of customers were actually different cause codes. **Without addressing the varied OPEX mis-prioritizations, it is unlikely that CAPEX outlays will result in the RCP mandated improvements to grid resiliency.** The multi-threaded, CUDA-enabled Graphics Processing Unit (GPU) comprising the utilized Data Visualization Analytics System (DVAS) was quite useful in correcting mis-characterizations, via Figs. 1, 2 and 6. The findings of the QA/QC Engine's NLP and CVML were also quite useful in flagging mis-characterizations (i.e. mis-correlated data), such as delineated in (7a) and (7b) of Fig. 7. Overall, the combined use of DVAS and the QA/QC Engine transformed the "Data Quality" number of 54.96% to 98.9%. Questions that remain center upon whether distribution utilities will be averse to properly re-characterizing their numbers based upon improved "Data Quality" numbers. If this is able to be facilitated, efficiencies and resiliency at the distribution level can be greatly enhanced.

Oftentimes, resiliency plans are being formulated based upon specious logic (i.e. flawed assumptions). For example, prioritization is given to certain nodes (e.g. poles), as there is a belief that they have a profound impact on SAIFI, SAIDI, and CAIDI calculations. Accordingly, these nodes are also given priority for grid monitoring capabilities (GMC), when in reality, other nodes that truly have a profound impact on the SAIFI, SAIDI, and CAIDI calculations, may not be receiving GMC. This leaves the entire involved distribution network exposed and perpetuates a misunderstanding of how to best react during a power outage. In many cases, the root cause of the misunderstanding is the poor "Data Quality" of the distribution utility's substation data, which is oftentimes not subject to scrutiny. Consequently, the reporting to regulatory agencies may also be flawed. The quantitative findings facilitated by the DVAS and QA/QC Engine directly contribute to improving this paradigm as well as better power outage mitigation planning and more expeditious recovery during a blackout, via a more robust understanding of the correlations within the involved distribution grid. The DVAS and QA/QC Engine combination serves as an effective planning and prioritization tool and could prove useful for various resilience compliance planning initiatives. Future work will involve a review of updated techniques for benchmarking purposes as well as the potential involvement of other useful algorithmic modifications for improving the CVML within the posited architectural schema.

References

1. Tubayan, E.: WB Cites PHL Growth Promise, Risks. BusinessWorld (2018)
2. Akanle, T.: Argentina, Belize, Botswana, India, Lao People's Democratic Republic, Mozambique, the Philippines, Samoa, San Marino, Sierra Leone, Thailand Submit INDCs. SDG Knowledge Hub, International Institute for Sustainable Development (IISD) (2015)
3. Mondal, A., Rosegrant, M., Ringler, C., Pradesha, A., Valmonte-Santos, R.: The Philippines energy future and low-carbon development strategies. Energy **147**, 142–154 (2018)
4. Yap, C.: Persistent Power Outage in Philippines Triggers Senate Probe. Bloomberg (2019)
5. Casayuran, M.: Signing of Energy Efficiency Act 'A Win for Ordinary Customers' – Gatchalian. Manila Bulletin (2019)
6. Rivera, D.: Philippine Electricity Rates Still Highest in Southeast Asia. The Philippines Star (2017)
7. Inocando, A.: The Big One: Philippines Project a Death Toll of 34,000 from Possible Quake. The Science Times (2019)
8. Ruiz, A.: The 80/20 Data Science Dilemma. InfoWorld (2017)

Smart Policing for a Smart World Opportunities, Challenges and Way Forward

Muhammad Mudassar Yamin[(✉)], Andrii Shalaginov, and Basel Katt

Department of Information Security and Communication Technology,
Norwegian University of Science and Technology,
Teknologivegen 22, 2815 Gjøvik, Norway
{muhammad.m.yamin,andrii.shalaginov,basel.katt}@ntnu.no

Abstract. Our world is getting evolved to smart world day by day. This smart world is being developed to make people life easier through the data generated by the smart devices. Data is the fuel that powers the smart world evolution, however, making things smart have its consequences. Smart devices are inherently vulnerable to cyber attacks, that's why we are observing an increase in crimes related to cyber space comparing to physical space. To address these crimes, police of the future need to evolve as well and data will be at the center stage of this evolution. In this contribution we are proposing a data centric policing proposal for smart cities. We analyzed current and developing technologies and the opportunities they offered for smart policing for a smart world.

Keywords: Smart cities · Police · IOT · Crime · Machine Learning

1 Introduction

Due to the density of data acquisition devices such as trivial environmental sensors or more complex video cameras, the future *Smart City* serves as a Panopticon of potential surveillance. This acts as a deterrent to discourage criminal activity since data may easily be acquired to facilitate an investigation against the individual in the future. With an assumption that a number of heterogeneous devices and sensors are owned by or have agreements with law enforcement, future Software Defined Networks (SDN) and Virtual Network Functions (VNF) may be able to overcome such data heterogeneity. The data that have been acquired may be stored in a cloud to facilitate spikes in user demand. Machine Learning as a Service can be used to deliver processing for prediction of future actions based on the historical data, anomaly detection, activity classification, human behaviour recognition, or object recognition [8].

Generally, the function of the police is the responsibility for the well-being of its citizens. Technology is expected to become more reliable when it comes to providing automated reasoning and security, but we cannot rule out the fact that humans are the weakest links in modern Information and Communication Technologies (ICT) societies. With the evolution of a digitally-dense future cities, the importance of the systems used and the reliance of humans on those systems will increase manifold times. To curb the number of incidents, the new-age police will have a challenge to keep the people informed, have the computational capacity to process all collected data from sensors/autonomous systems and adapt to the new opportunity model [25] provided by smart cities (directly competing with the adversaries). Worryingly, critical infrastructure can be influenced by either party. If these challenges are met, it will enable future police in investigations (current and retrospectively) and frame smart responses to the forthcoming crime.

Police organizations worldwide are looking to leverage emerging technologies to fight against physical and cyber crimes. Such crimes are more prevalent than ever before due to the Internet of Things (IoT) and the development of smart cities with corresponding digitally-dense infrastructure. Mark Goodman, a man who has had a career in law enforcement, a futurist for the FBI, and senior advisor to Interpol stated that *"More connections to more devices mean more vulnerabilities"* [20]. And the increasing dangers are clear: in 2017 IoT attacks increased by 600%, and in a study 70% of businesses believe their security risk increased significantly. As it stands today, more than 4,000 ransomware attacks occur each day, and statistically grow more than 350% every year. These vulnerabilities affecting smart cities are not new, but now, their consequences can be far more severe due to the cyber-physical nature of smart cities. Now, hackers do not only pose a threat to data, but to human lives as well. So, there can be anticipated new technological challenges that people and police in Smart Cities of future will face in addition to societal challenges already solved through "Big Data for Social Good" initiatives [42].

Even with the change in the *modus operandi* of the Police in smart cities resulting in Smart Police; the functions of the Law Enforcement Agency (LEA) remain somewhat the same i.e. *To ensure Public order, Public safety, Crime prevention, Detection and Investigation.* The infrastructure, that the new age technology will bring-forth, will potentially change the dynamics of Policing. A whole new Opportunity structure will come to surface. Both the LEAs as well as the adversaries will be presented with new means to safeguard or to attack the systems in use. All the challenges and opportunities will revolve around a very central aspect – real-world non-synthetic *Data* and its corresponding utilization in Intelligent Decision-Support Systems (IDSS) [21]. While there is a high likelihood of discovering new vulnerabilities in smart cities hardware ans software, this creates new opportunities to both the criminals and the law enforcement agencies. It is our belief, that LEA must and can take a more proactive approach to preventing, detecting, and investigating crime in both the cyber and physical realms in future smart cities. Anticipated challenges and corresponding opportunities are given below:

Challenges

1. Due to huge amount of data that needs to be collected through a large number of autonomous systems and sensors, the obscurity and complexity may create a novel attack vector.
2. Who will own this data? - who is the data custodian, processor, access control models on information (processed day). The "chain of custody" during criminal investigation needs to be maintained.
3. The integrity of the data can be compromised during capture storage attributed to Public functions.
4. With a large volume of data to be processed, huge computational capabilities will be required resulting in latency with, often critical, decisions.
5. Data processing and police notification of reasoning now becomes a challenge.
6. Historically, when crime in a city reduces, it increases on the outskirts. Since the technology is deployed in the main city with dense population, law enforcement will have to be done in conventional ways to ensure public order outside center.
7. With a wide range of ad-hoc systems and a need to respect privacy of the citizens, the information sharing can pose a significant obstacle on regional, national and international levels.

Opportunities

1. Over the course of history, the role of the Police has predominantly been *reactive*. However, if the data can be provided to the LEAs in time, they can be *proactive* in preventing the crime.
2. Autonomous and semi-autonomous vehicles, unmanned aerial vehicles (UAV)s and robots will make law enforcement not only more efficient, but safe for the police officers, delegating life-threatening tasks to machines.
3. Application of Computational Intelligence provides ways to outsource time consuming manual tasks to automated models, allowing police officers to prioritize other decisive tasks.
4. The data acquisition sources can be shared by various agencies, based on their body function, to use it to their advantage. (*For example:* if an adversary is planning an attack on an asset of national relevance and the contextual data is captured; it can be sent to the national agencies to prevent the act from happening in the first place like big weapon consignment captured on surveillance, a drug deal etc.)
5. The role of the community and so-called "crowd-sourcing" cannot be ruled out in crime prevention and investigation [11]. Platforms like social media, apps for smart phones and citizen awareness campaigns can be a key to an aware and well informed society.
6. With strong control measures in place, the criminals will be discouraged to do a crime in the first place due to high price/high efforts.

In this contribution the researchers present the winning idea[1] of INTERPOL Thinkathon 2018 challenge in which we proposed a semi-autonomous crime

---
[1] The idea has been originally prepared and presented by the team from the Norwegian University of Science Technology during the competition at the Interpol in October, 2018: https://twitter.com/INTERPOL_GCI/status/1039803723071905793.

detection, prevention, and investigation system [39]. That can assist law enforcement agents by processing the data acquired from heterogeneous sensors, and disseminate and render relevant information to the law enforcement agencies or even make autonomous decisions in the appropriate circumstances. To be concise, our solution is essentially to apply a city-wide Intelligent Intrusion Detection-like system as an integral component to our modern age Smart Cities. To tie the ideas together, we illustrated some example scenarios depicting some potential outcomes of our proposed solution. The *first* scenario exhibits a physical crime that leaves digital traces. The *second* scenario stages a cyber-physical crime, that demonstrates how smart appliances and internet enabled technology may be abused. The *third* scenario is an example of a crime occurring purely in cyberspace, to round off the cyber-physical relationships.

This paper is organized as follows: the researchers start with discussing Smart City technologies and their applications in the Sect. 2. Afterwards, we focus on data, wherein we consider issues pertaining to the acquisition of data, its handling, and processing. This also includes potential methods of analysis to be performed on the processed data presented in the Sect. 3. The synthesis of these ideas drives our attention to the newly spawned opportunity model in the context of smart city policing, where we discuss relevant challenges and opportunities with the help of three crime scenarios in the Sect. 4. After that we share future technologies that can make this fantasy a reality in the Sect. 5 and conclude the paper with the Sect. 6.

2 State of the Art: Data as a Key Component of Smart Cities' Infrastructure

Future Smart Cities will ensemble large network of ubiquitous heterogeneous devices scattered across physically-distributed locations. This will lead to many technological opportunities such as increase mobility, fine-tuned geographical location detection and improved situational awareness aiding police investigators to tackle crimes with aid of data analytics.

2.1 Data Sources

Smart City infrastructure intrinsically implies distributed network of sensors and resource-constrained end-point devices that can be used to acquire data and model cyber situational awareness [2,6]. There can be mentioned several general categories of data sources found in the literature with respect to their usage motivation [7,26]:

1. Specialized public services such as access Radio-frequency identification (RFID) tags, stationary Closed-circuit television (CCTV) cameras [40,41], cameras on movable objects (like cards, bikes, police officers);
2. Personal information flow such as bank statements, shopping transactions, call data, Global Positioning System (GPS) data (car, on foot, bicycle, public transportation), health sensors;

Fig. 1. Data sources in smart cities

3. Environmental sensors such as waste and garbage, air pollution level, traffic congestion, parking, noise level, humidity, temperature, etc.

Large fraction of such data can be considered as publicly-available, however, some contain sensitive and Personal Data as defined by recent General Data Protection Regulation (GDPR) implementation [18]. The publicly accessible data sources which can be used in smart city policing is represented in Fig. 1.

2.2 Data Handling/Storage

Cheap deployment of technologies and high bandwidth networks will boost presence of versatile devices in every aspect of every day's life. As result, small pieces of data samples collected at regular intervals across all end-points, once put together, will form a considerable flow of information that cannot be handled by small-scale data storage solutions [14,33]. This will enable every aspect of Big Data paradigm: *Variety, Variability, Volume, Velocity and Value*. *Volume* and *Velocity* are successfully handled by hardware and software solutions for large data storage and high-speed network connections. These are so-called "Data Lakes" capable of handling enormous loads with guaranteed performance, integrity and availability. At the same time, *Variability* and *Variety* of data become a real challenge when it comes to extracting the *Value* from the collected data for the Smart Police of future. Therefore, it is important to look for an advanced method capable of handling such challenges and bringing down the amount of manual labor required by police officers.

2.3 Data Acquisition

The data from aforementioned low-level components travel through middleware, including corresponding IoT hubs and services across Public Internet and Smart Grid into the data storage centers, either locally in the Smart City or globally [17]. Energy-efficient multi-agent systems ensure reliable communications in resource-constrained environment with nearly real-time performance as well as guaranteed freshness of collected data. The main challenges related to acquisition of the data is related to (i) communication protocols, (ii) data formats, levels of granularity and metadata, (iii) interaction model (subscription, regular, etc.) and intervals, (iv) abstraction and representation level for the management and decision makers (e.g., video stream of pool of identified objects) [38]. However, once tackled, the data acquisition will provide unlimited opportunities for versatile data analytic needed in context of safety and security in Smart Cities of future. Finally, there will be a possibility to harmonize previous historical data from different storage engines, including paper based reports.

2.4 Data Processing

Since the amount of data generated by various components of Smart Cities is growing dramatically every year, mankind faces problems related to the fact that data are so complex, sparse and different. Existing data processing methods will not be able to handle it in near future. Therefore, one of the focus areas should be on data fusion, cleansing, correlation and processing to advance threat intelligence and information processing by Law Enforcement Agencies [16,24]. In addition to this, such focus will enable development of new methods and simulation of possible criminal scenarios based on the available historical information. Further on, information rendering is important aspect of future policing focusing on information sharing between LEAs and future reporting not only internally, yet also externally to public.

2.5 Decision Support/Privacy Aware Analytics

State-of-the-Art analytics tools used in modern Crime Investigation are based on the keyword search and manual content exploration, while recent developments also added approximate pattern matching and so-called e-Discovery that helps and simplifies manual work of forensics investigations. Undoubtedly, this offers flexibility and speed in crime investigations, yet will not meet growing demand in advanced data analytic and decision support systems. Therefore, Big Data problems need a different approach to tackle large-scale data seized or correlated for a criminal case. As result, there is a strong need to apply Computational Intelligence methods capable of fast training and timely data processing [32]. This will enable future police in Smart Cities to effectively detect of new attacks, discover new adversarial trends and predict any future malicious patterns from the available historical data. Finally, intelligent data handling will result in building and providing timely response to ongoing crimes and best possible assistance in investigation of committed crimes.

Another aspect that became increasingly demanded in modern cyber-based society is protection of personal and sensitive personal information. Intelligent applications used for decision support will be able to handle Personal Data through application of homographic encryption methods that allows usage of computational methods over encrypted data and text [29]. However, this is an important issue that future police will face due to the fact that personal data protection regulations differs regionally and culturally. Notwithstanding, police main function will be to maintain public order, enforce the law, prevent any illegal crime activities before they happen and ensure reliable and timely crime investigation. Therefore, it is important to understand that colossal and agile data collection in Smart Cities will contribute to safer and secure environment despite multiple social and technical challenges affiliated with automated data processing and decision support in legal framework as it exist now.

3 Proposed Approach to Facilitate Smart Policing

Access to information before and after the crime plays a pivotal role in crime prevention, response and investigation. The researchers analyzed major crime incidents and identified that information sharing between multiple police levels is not ideal [22]. This is due to the fact that multiple entities with in the police may have access to information which according to them may not be relevant or confidential to other entities within the police. *Our proposed solution works by collecting information from multiple sources and creating a big picture of the environment.* In such way, it will be possible to identify relevant information and share it with relevant police entities for appropriate response It is expected that traditional criminals with increased surveillance in the smart city will be hesitant to commit a crime within city vicinity [27]. Traditional criminals will most likely focus on city suburbs where surveillance capabilities of police will be limited to city centers [12]. However, due to the transformation of city to a smart city by digital devices, provides an opportunity to cyber criminals to commit crime within the city center, Due to the inherent anonymity provided by cyberspace and crypto currencies it would be very difficult for the police to catch cyber criminals. Our proposed solution considers the smart city a big Intrusion Detection and Prevention System. The researchers used the NIST cyber security functions when considering the applications of proposed system on smart cities [30]. NIST cyber security framework is presented in Fig. 2.

In the system, first activity profiling of citizens will be performed. The activity profiling will be based upon human out of the loop autonomous technologies to preserve citizens privacy. Data will be collected from multiple sensors present within the smart city. Through the activity profiling a baseline of normal citizen behavior will be created. The baseline will be used to Identify and Protect citizens based upon historical criminal activity. Warnings will be issued for citizens performing activities in crime infested areas and police presences will be marked

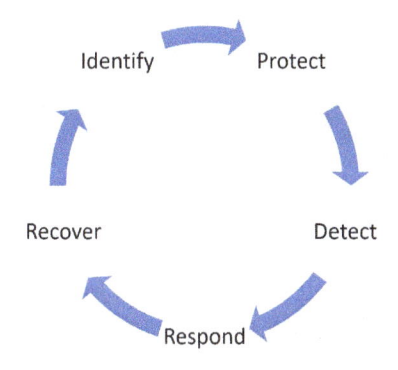

Fig. 2. NIST cyber security functions

in crime infested areas to protect citizens by deterring criminals in performing malicious activity. Outliers from the baseline will be detected, and alerts will be generated based upon suspicious activity without human in the loop. Outlier and alert information will be passed on to human investigators for decision making and setting next line of actions. The system will suggest the next line of actions with possible outcomes in order to facilitate human decision maker to respond to a threat in an effective manner. The System is designed to deploy swarm of autonomous vehicles on land, air, sea and cyber space to perform intelligence surveillance and recon sense operations to respond and recover crime evidence and when the need arises to protect human life can also engage criminal with human assisted decision making. The system will provide unprecedented situational awareness to the police officers of the future. Which will enable the police officers to make smart decision based upon evolving threats. This will reduce the number of man hours required by police officer to do their jobs. A schematic diagram presenting working of proposed system can be seen in Fig. 3.

3.1 Smart Policing: Anticipated Methods

Following intelligent methods can be used to facilitate automated reasoning and proactive approach in Smart Police of future.

Human Out of the Loop Activity Profiling. A lot of data will be generated by the sensors and devices present within the smart cities, and making sense of that data for crime prevention, responses and investigation will be challenge. Moreover, Data ownership, Data Sovereignty and citizen privacy related issues need to be addressed. Analyzing the smart city generated data by humans is difficult and privacy alarming, therefor research need to be carried out in privacy preserving technologies on citizen data. Human out of the loop technologies, can assist in analyzing citizen data while preserving citizen privacy. These technologies are autonomous systems that operate independently without human interference. Their decision making is based upon predefined

Fig. 3. Concept for future smart city policing

rule set up by humans [35]. The rule set can be applied on a general set off data for identification of benign and malicious activities. For this identification of behavior benign and malicious citizen activity profiling is required. Activity profiling is the process of establishing individual citizen profile based upon a set of behavior metrics of citizen activities within the smart city. The behavior metrics includes multiple parameters like travel history, transaction details, internet activity and other parameters data which can be collected from the sensors and devices present in the smart city. The activity profiling will then be utilized to generalize the citizens activity and establishing a baseline for normal citizen activity. The baseline of citizen activity profile will then be used to identify anomalies in citizens behavior for a proactive approach of crime prevention and response. In case the was not detected by the system the activity profile can be utilized for the investigation of crime and for fine tuning the system for similar future crimes.

Information and Computational Intelligence Model Fusion. Data fusion is the process of aggregating data from multiple data sources in order to produce more comprehensive and accurate information compare to an individual data source [10]. As multiple sensors are available in the smart city their functionality is increased by fusing data from multiple sensors it in commonly preferred as sensor fusion. For crime prevention, response and investigation data fusion will play a pivotal role in smart cities, as it will provide high level situational awareness to the autonomous system for analyzing information and supporting decision making in an evolving threat landscape.

Baseline Comparison. Behavior metrices of each and every citizen in the city can be compared with baseline of normal citizen activity developed during the activity profiling. The baseline comparison method will quantify the differences of normal and abnormal citizen activity created over a specific period of time which will then help to identify person of interest from a large population data set of citizens activity profiles. This can be achieved through state-of-the-art outlier detection algorithms. Other mean of information gathering like autonomous vehicles then can be deployed for citizen specific information gathering which can be used in proactive investigation of not normal behavioral activity.

Outlier Detection. Outlier value are extreme deviations from the data under observation which are considered as anomalies in the data. There are two types of outlier univariate and multivariate. Univariate outliers are identified by single feature in data distribution values while multivariate outliers are identified by multiple features in data distribution values. In a smart city data will be generated from multiple data sources so outliers can be detected with multivariate outlier detection algorithms. However, it is very hard to identify outlier on large amount of multi-dimensional data set therefore new and efficient methods will be required to for the detection of anomalies. Advances in machine learning algorithms can assist in solving this problem.

Anomaly Detection with Machine Learning. Machine Learning (ML) is a data analysis method which automates the process of analytic model development. Machine learning is a part of artificial intelligence in which systems are developed that can learn from data to identify patterns in data for making decision with minimal and no human interaction. Two main approaches of machine learning are developed to learn from data these are supervised and unsupervised learning algorithms. In supervised learning algorithms the input data is labeled and is trained over a set of training data with input and desired output. In unsupervised machine learning algorithms set of data which only contains inputs is given to the algorithm. The algorithms learn from data that has not been labeled or categorized which then identifies a structure and pattern in the data, like grouping or clustering of data points. Unsupervised learning algorithms identifies similarities in the data points and classify data by presence and absence of those similarities. In case of anomaly detection in citizen behavior both type machine learning algorithm can play their role. Particular threshold of specified several behavioral matrices of a citizen can be set for the identification of anomalies in a supervised manner and data classification. Unsupervised learning algorithms can identify anomalies without specifying a behavioral matrices however it consider the data which classification is in majority percentage as normal. A semi supervised machine learning approach can assist in analyzing large amount of data with minimum human interferences. Despite the advantage the machine learning algorithms offers in data classification and decision-making process a black box, explain ability of that decision-making process will going to be a big challenges criminal proceeding [36]. Theory of argumentation can assist to solve this problem.

Explanation of Decision Based upon Arguments. Theory of argumentation [4] focused on reaching conclusion based upon sound arguments. As the machine learning algorithms will provide a lot of data classification methods their effective usage in crime investigation prevention and response will require sound reasoning and it very difficult to achieve because law of the land is different at different places to argue on those reasons. Specific rule set need to define which are according to the law of the land for making argument based upon data classification for criminal jurisdiction in an autonomous manner. AI that can identify admissible evidence from set of given data will assist in reducing the time requirements for persecution. The technologies that are need to developed for this purpose are need to be transparent [9]. The inherent black box processing nature of machine learning algorithms is therefore not suitable for this purpose [28]. Hence it is expected that theory of argumentation can play key role in developing such technologies.

Cyber Threat Intelligence and Threat Level Assessment. After identification of anomalies in citizen behavior the system needs to assign a threat level for a better response to the anomaly. The threat level assignment needs to consider multiple factors like the time period in which the anomaly becomes dangerous to other citizens, risks which the anomaly pose, the resources required to investigate the anomaly etc. This will help in better resource utilization on by assigning resources on immediate threats of crime. Every city has different threshold value of threat for different crimes, so the threat level assignment system is need to be developed according to specific city requirements. Sharing of that intelligence information in an autonomous manner will also play a key role [45] as there will be massive amount of information generated by the devices present in the smart city.

3.2 Modern Experimental Technologies and Collaborations

Vision of the future LEA needs in Smart Cities has been addressed by corresponding development of novel technologies and human interaction approaches.

Police Drones. Police drones can play a vital role in future smart cities crime prevention, investigation and response. They offer quite a lot of benefits compare to human police officers. First, they remove human officers out of the danger zone, second the drones are not affected by emotional break down and racial stereotypes and finally they are dispensable. More ever police drones are arguably more efficient in executing similar tasks assigned to a human police officer. This makes them viable in law enforcement and crime investigation in future smart cities, however, these drones only operate in physical world for cyber world radical new technologies will be required for law enforcement and crime investigation.

Cyber Police Bots. As digital infrastructure is being rapidly expanding in the smart cities, security of this digital infrastructure will be a big challenge. Traditional police force seems ill equipped considering recent major cyber incidents. Technologies that enable police to petrol cyber world for the identification, detection and prevention of cyber incident need to be developed. Due to the massive digital footprint of smart cities it is humanly not possible for securing each and every device from the new vulnerabilities which are discovered every day. Autonomous technologies that can petrol the cyber space and identify and prevent potential vulnerabilities exploitation will enhance police capabilities in tackling cybercrimes.

Public Private Partnership. Bruce Schneier once said *If you think technology can solve your security problems, then you don't understand the problems and you don't understand the technology.* We can design and build sophisticated technologies but if it is not human centric then the technology will create more problems then solutions. Methods to incentives information sharing between public and LEA are under development [3]. This will help to improve the performance of predictive police [37] mechanism which are powered by advance machine learning algorithms [43].

Smart Cities Standards. Multitude of work is going on developing smart cities standards, the govern technical, process/management and strategic/leadership aspects of smart cities. For technical aspects PAS 212 is developed for automatic resource discovery of IOT Devices. In term of process/management ISO/TS 37151, ISO 37100, ISO/IEC 30182 and ISO 37156 are developed. They provide key performance indicators, vocabulary,data concept and exchange models respectively. In term of strategic/leadership ISO 37101 is developed for the overall management of the smart city and its infrastructure. in order to measure he maturity level of smart cities infrastructure 37153:2017 [5] is developed. ISO 37153:2017 govern the infrastructure standards and used to measure the maturity level of smart cities infrastructure.

3.3 Challenges and Drawbacks

While implementation of such system is within the technical grasp of modern technologies some challenges still needs to be tackled with. Citizen privacy will be the biggest challenge in the implementation of such system as the system is working on collecting and processing citizen data its security need to be ensured with privacy preserving technologies like multi-party computation [19]. Data tampering for the data sources will create false anomaly condition with in the system which will cause false alert, so integrity of the data produced at the sensors must need to be ensured methods like template protection [23] can be useful to tackle this problem. More ever with increased data fusion from multiple data sensors data dimensions will increase which will make it very difficult in data processing for anomaly detection with current existing computational methods.

Machine learning bias will also be a challenge for training the algorithms in a more neutral manner.

4 Use-Case Scenarios in Smart Cities: Modus Operandi and Police Response

To support the ideas of future policing, there have been suggested the following scenarios that comprehend general categories of crimes found in both digital and physical realms.

4.1 Physical Crime Scenario for 'Detection' Aspect

Before omnipresence of the Internet, all of the crimes were committed in a physical domain with all the traces left by perpetrators and victims.

Storyline. The city centers will contains most of the audio/visual data collection sensors so criminals will be hesitant in conducting crime in city centers. In a hypothetical scenario a individual is doing hiking in city suburb and provide an easy opportunity for a criminals. The criminal snatched the smartphone and additional items from the robber and ran away.

Infrastructure. Unknown to criminal the hiking trail was equipped with International Mobile Subscriber Identity (IMSI) catchers and based upon the previous activities of the hiker on the hiking trail an alert is generated that something is not right and an individual was identified moving with a different two different IMSI. The police officer can dispatch a drone for further investigation and tackling the criminal which is autonomously identified by his own subscriber identity number.

4.2 Cyber-Physical Crime Scenario for 'Prevention' Aspect

With growing inclusiveness of ICT technologies, crimes a re moving towards involvement of new technologies and previously unseen attack vectors.

Storyline. The crime opportunity structure in smart cities will be changed digital devices will be prone to vulnerabilities and cyber criminals will take the advantage of Those vulnerabilities. In a hypothetical scenario cyber criminals plans to launch a cyber physical ransomware attack on smart locks, to lock the citizens and demand ransomware.

Infrastructure. Autonomous cyber policing bots detected communication on dark net about a particular smart lock vulnerability the information is then correlated with scans from public exploit scanning utilities which indicated working and work in progress of the vulnerability weaponization of smart locks. The information is then pass on to smart lock OEM (Original Equipment Manufacturer) to quickly launch an update to allow smart lock users unlock the lock with another backup key.

4.3 Cyber Crime Scenario for 'Investigation' Aspect

Future perpetrators will find more elaborate and sophisticated ways of attacking system and breaching privacy of individuals as well as harming security on a state level.

Storyline. In a hypothetical scenario cyber criminal performs a bank heist this time it was not detected or prevented.

Infrastructure. Data collection from both physical and cyber sensors is utilized for the identification of cyber criminals. Forensic investigation of attack chain will be done that how attackers penetrated the bank firewalls, Intrusion Detection Systems (IDS)/Intrusion Prevention Systems (IPS) through security logs then the data is correlated with ISP traffic history for the identification of command and controls channel of attackers.

5 Way Forward for Smart Cities Research

Considering aforementioned challenges and demands of the policing in future Smart Cities, there can be seen following measures to increase both awareness and readiness of general public and specialized response forces.

5.1 Cyber Ranges

Cyber Ranges are platforms that are designed to emulate the network infrastructure of an organization. The emulated network infrastructure can be then used for testing, training, education and experimentation purposes. Cyber ranges provide the capability to develop a digital twin of smart city in which the digital infrastructure of smart city can be tested and experimented with. As stated earlier in the crime opportunity structure the focus of the crimes is shifting towards cyber domain, so this capability can be utilized to proactively develop and test new crime scenarios ranging from cyber to cyber physical domain. This will help the future smart city police to develop strategies to counter crimes before they happen.

Cyber ranges can be good for cyber security exercises and experimentation [44,46] however to be utilized in an effective manner for the betterment of society, they also need to incorporate the societal view [47]. Integration of societal and technical prospective in cyber range will provide the opportunity to deal with the human aspect associated with the cyber domain in a time of crisis [31]. This will enable the future police to better train in complex scenarios and tackle crime by incorporating the human prospective as well. This results in better and well prepared police force of the future that can handle crime in a proactive manner not in a reactive manner.

5.2 Digital Forensics Readiness and Incident Response

Another important approach that needs to be integrated in policies and future technology deployment organization- and cities-wise is forensics readiness. This is a concept that represents a model where devices are properly configured to retain evidences and data for possible cyber crime investigation on the basis of "need-to-know" principle. Rowlingson in 2014 [34] defined ten measures and activities that are required to ensure so-called Network Forensics Readiness. Those include aspects such as data handling, legal compliance, policy development and employees training. Future Smart Cities will enable huge data traffic that will not be possible to keep in "Data Lakes" for a long time. Therefore, proper Digital Forensics-friendly data collection and processing nodes need to designed to ensure timely investigation of the incidents without any impact on the human-oriented services of Smart Cities. Moreover, measures that will be implemented should aim not only at so-called "extremistan" events and crimes (high-impact low-likelihood), yet also at more frequent events with lower impact to be able to provide adequate response to crimes [1,15].

The next important aspect is Incident Response, which will require corresponding knowledge of the previous crime historical information with corresponding action plans, involving individuals, organizations and cities. Current State of the Art include integration of such regular and ad-hoc teams as Special Weapons and Tactics (SWAT), Community Emergency Response Team (CERT) and Hazardous Materials Management (HAZMAT). While, those have been introduced since 1960 and proved efficiency, Smart Cities put new demands towards cross-departments and cross-response teams collaboration [13]. This requires rethinking of the traditional physical crimes model towards more advanced cyber-oriented components of the city that might be affected. Omnipresent Situational Awareness and Threats (Physical and Cyber) Intelligence will give new insights in addition to conventional digital forensics readiness in place.

6 Conclusion

Preventative measures and reactive measures for fighting crime are blending. The cyber and physical world are increasingly blending as well. Development of such system will present opportunity to enable future police officers to use data to prevent/investigate crime. This will keep LEA ahead compare to the adversaries in new crime opportunity structure. As indicated in opportunity structure model and enabling factors such as the availability of relevant information and technology that will enable future smart police to work **proactivaly**.

As Bejamin Franklin once stated *"Security without liberty is called prison"*, security by consensus [25] should be followed when such technology is developed. Else it will give immense power to people controlling such technologies. These technologies can solve allot of problems that we are facing today in term of physical and digital crime, however developing such technologies may have unintended consequences in wrong hands.

Acknowledgment. The authors are grateful for the support by the Department of Information Security and Communication Technology, Faculty of Information Technology and Electrical Engineering, Norwegian University of Science and Technology (NTNU). The researchers would like to acknowledge the NTNU team participation in the Interpol Thinkathon 2018 competition: Danny Lopez Murillo, Kyle Porter and Ivan Talwar, who equally contributed in development of the winning idea. The researchers would also like to thank the valuable feedback from our coaches Basel Katt, Jens-Petter Sandvik, Katrin Frank and Stewart James Kolwaski, without which this work would not be possible.

References

1. Ab Rahman, N.H., Glisson, W.B., Yang, Y., Choo, K.-K.R.: Forensic-by-design framework for cyber-physical cloud systems. IEEE Cloud Comput. **3**(1), 50–59 (2016)
2. Abu-Matar, M., Davies, J.: Data driven reference architecture for smart city ecosystems. In: 2017 IEEE SmartWorld, Ubiquitous Intelligence & Computing, Advanced & Trusted Computed, Scalable Computing & Communications, Cloud & Big Data Computing, Internet of People and Smart City Innovation (SmartWorld/SCALCOM/UIC/ATC/CBDCom/IOP/SCI), pp. 1–7. IEEE (2017)
3. Agrawal, V.: Information security risk management practices: community-based knowledge sharing (2018)
4. Alexy, R.: A theory of legal argumentation: the theory of rational discourse as theory of legal justification (2009)
5. Anthopoulos, L., Giannakidis, G.: Policy making in smart cities: standardizing city's energy efficiency with task-based modelling. J. ICT Stand. **4**(2), 111–146 (2016)
6. Bačić, Ž., Jogun, T., Majić, I.: Integrated sensor systems for smart cities. Tehnički Vjesn. **25**(1), 277–284 (2018)
7. Bartoli, A., Hernández-Serrano, J., Soriano, M., Dohler, M., Kountouris, A., Barthel, D.: Security and privacy in your smart city. In: Proceedings of the Barcelona Smart Cities Congress, vol. 292 (2011)
8. Boucher, P.: How artificial intelligence works. http://www.europarl.europa.eu/at-your-service/files/be-heard/religious-and-non-confessional-dialogue/events/en-20190319-how-artificial-intelligence-works.pdf
9. Bryson, J., Winfield, A.: Standardizing ethical design for artificial intelligence and autonomous systems. Computer **50**(5), 116–119 (2017)
10. Castanedo, F.: A review of data fusion techniques. Sci. World J. **2013** (2013)
11. Cvijikj, I.P., Kadar, C., Ivan, B., Te, Y.-F.: Towards a crowdsourcing approach for crime prevention. In: Adjunct Proceedings of the 2015 ACM International Joint Conference on Pervasive and Ubiquitous Computing and Proceedings of the 2015 ACM International Symposium on Wearable Computers, pp. 1367–1372. ACM (2015)
12. Dawid, I.: As cities become safer crime decamps for the suburbs. Planetizen. https://www.planetizen.com/node/59987
13. Efthymiopoulos, M.-P.: Cyber-security in smart cities: the case of Dubai. J. Innov. Entrep. **5**(1), 11 (2016)
14. Eggers, W.D., Bill, R.S.: Deloitte, Center for Government Insights, and Deloitte Consulting LLP. Building the smart city with data, digital, and design—deloitte us, January 2018

15. Elyas, M., Ahmad, A., Maynard, S.B., Lonie, A.: Digital forensic readiness: expert perspectives on a theoretical framework. Comput. Secur. **52**, 70–89 (2015)
16. EY. Smart policing for smart cities (2015). http://www.governancenow.com/files/FICCIReport-SMARTPolicingforSmartCities.pdf
17. Gaur, A., Scotney, B., Parr, G., McClean, S.: Smart city architecture and its applications based on IoT. Procedia Comput. Sci. **52**, 1089–1094 (2015)
18. GDPREU. GDPR - personal data. https://gdpr-info.eu/issues/personal-data/
19. Goldreich, O.: Secure multi-party computation. Manuscript. Preliminary version, vol. 78 (1998)
20. Goodman, M.: Future crimes: everything is connected, everyone is vulnerable and what we can do about it. Anchor (2015)
21. Gottinger, H.W., Weimann, H.-P.: Intelligent decision support systems. In: Methodology, Implementation and Applications of Decision Support Systems, pp. 1–27. Springer (1991)
22. Hollywood, J.S., Winkelman, Z.: Improving information-sharing across law enforcement: why can't we know? National Criminal Justice Reference Service (2015)
23. Kelkboom, E.J.C., Zhou, X., Breebaart, J., Veldhuis, R.N.J., Busch, C.: Multi-algorithm fusion with template protection. In: 2009 IEEE 3rd International Conference on Biometrics: Theory, Applications, and Systems, pp. 1–8. IEEE (2009)
24. Kovacevic, A.: Police are using big data to predict future crime rates, November 2018
25. Kowalski, S.: The SBC model as a conceptual framework for reporting it crimes. In: Proceedings of the IFIP TC9/WG9. 6 Working Conference on Security and Control of Information Technology in Society on Board M/S Illich and Ashore, pp. 207–226. North-Holland Publishing Co. (1993)
26. Lim, C., Kim, K.-J., Maglio, P.P.: Smart cities with big data: reference models, challenges, and considerations. Cities **82**, 86–99 (2018)
27. McWhirter, C., Fields, G.: Crime migrates to suburbs, December 2012
28. Michie, D., Spiegelhalter, D.J., Taylor, C.C., et al.: Machine learning. Neural and Statistical Classification, vol. 13 (1994)
29. Minelli, M.: Fully homomorphic encryption for machine learning. Ph.D. thesis, PSL University (2018)
30. NIST. Cybersecurity framework. https://www.nist.gov/cyberframework
31. Østby, G., Yamin, M.M., Al Sabbagh, B.: sieMS in crisis management: detection, escalation and presentation–a work in progress. In: 5th InterdisciPlinary Cyber Research Conference 2019, p. 38 (2019)
32. PredPol. Machine learning and policing. http://blog.predpol.com/machine-learning-and-policing
33. Rotuna, C., Cîrnu, C.E., Smada, D., Gheorghiţă, A.: Smart city applications built on big data technologies and secure IoT. Ecoforum J. **6**(3) (2017)
34. Rowlingson, R.: A ten step process for forensic readiness. Int. J. Digit. Evid. **2**(3), 1–28 (2004)
35. Shahriari, B., Swersky, K., Wang, Z., Adams, R.P., De Freitas, N.: Taking the human out of the loop: a review of Bayesian optimization. Proc. IEEE **104**(1), 148–175 (2016)
36. Shalaginov, A.: Advancing neuro-fuzzy algorithm for automated classification in largescale forensic and cybercrime investigations: adaptive machine learning for big data forensic. Ph.D. thesis, Norwegian University of Science and Technology (2018)
37. Shapiro, A.: Reform predictive policing. Nat. News **541**(7638), 458 (2017)

38. Sivarajah, U., Kamal, M.M., Irani, Z., Weerakkody, V.: Critical analysis of big data challenges and analytical methods. J. Bus. Res. **70**, 263–286 (2017)
39. Trædal, T.J.: Norsk forskning på framtidens politi til topps i interpol, October 2018
40. Ullah, M., Alaya Cheikh, F.: A directed sparse graphical model for multi-target tracking. In: Proceedings of the IEEE Conference on Computer Vision and Pattern Recognition Workshops, pp. 1816–1823 (2018)
41. Ullah, M., Cheikh, F.A., Imran, A.S.: Hog based real-time multi-target tracking in Bayesian framework. In: 2016 13th IEEE International Conference on Advanced Video and Signal Based Surveillance (AVSS), pp. 416–422. IEEE (2016)
42. UNICEF. Big data for social good. http://unicefstories.org/tag/big-data-for-social-good/
43. Wang, T., Rudin, C., Wagner, D., Sevieri, R.: Learning to detect patterns of crime. In: Joint European Conference on Machine Learning and Knowledge Discovery in Databases, pp. 515–530. Springer (2013)
44. Yamin, M.M., Katt, B: Inefficiencies in cyber-security exercises life-cycle: a position paper. In: AAAI Fall Symposium: ALEC, pp. 41–43 (2018)
45. Yamin, M.M., Katt, B.: A survey of automated information exchange mechanisms among certs. In: CERC, pp. 311–322 (2019)
46. Yamin, M.M., Katt, B., Torseth, E., Gkioulos, V., Kowalski, S.J.: Make it and break it: an IoT smart home testbed case study. In: Proceedings of the 2nd International Symposium on Computer Science and Intelligent Control, p. 26. ACM (2018)
47. Yamin, M.M., Katt, B., Kianpour, M.: Cyber weapons storage mechanisms. In: International Conference on Security, Privacy and Anonymity in Computation, Communication and Storage, pp. 354–367. Springer (2019)

Midnight in Tokyo: Mobility Service for Bar-Hopping

Chihiro Sato[✉] and Naohito Okude

Keio University Graduate School of Media Design,
4-1-1 Hiyoshi, Kohoku, Yokohama, Japan
{chihiro,okude}@kmd.keio.ac.jp

Abstract. This research indicates a service application providing citizens unique values when autonomous driving technologies are available as daily mobility and woven into the city fabric. The navigation service *Tokyo 27:00* supports young-beings to hang around, drink out, and chill by bar-hopping in night-time metropolitan areas. Users will enjoy choosing the next bar while inside the mobility which generates routes integrating plural possible destinations and displays it as if an arcade on the mobility front-window interface. This paper shows initial user studies with our initial prototype, and reveals findings for potentiality of driverless mobility embedded in a metropolitan city yet to come.

Keywords: Mobility service · Service design · Route navigation · Design thinking · City life

1 Introduction

Life with driverless cars is just around the corner. Autonomous driving technology has evolved rapidly in the past years; most car manufacturers around the globe have released automated driving features on their new automobiles. Although there still remains safety and regulatory issues to overcome as shown in the recent Uber tragedy accident in Arizona, driverless mobility will become a new value to the citizens and shall become new market within the following decade. When those days come, what kind of values can citizens benefit as the mobility service?

Our research in collaboration with a mobility-oriented corporation seeks for service applications that the city fabric can provide when autonomous driving technologies are available as daily mobility; in particular, bar-hopping in the night-time metropolitan areas. Our driverless navigation service *Tokyo 27:00* supports those metropolitans to hang around, drinking out and barhop till 3 am. It consists of an interface that displays the possible destinations (in this case the nearest bars around) and a route generator that integrates several destinations to go and find out on yourselves with the driverless car. This service was designed upon the integration of design thinking approach and service dominant logic perspectives: (1) ethnography research of how locals hang around in towns, series of iterations on how cities should be experienced, (2) prototypes with interfaces and databases for decision making on where to head next, and (3) user studies with our navigation system prototype for value co-creation (though with a "driver-with" car and not a driverless one). Our findings reveal the potential power of driverless mobility service that can propose value towards the metropolitan young-beings. For those who have submitted a session proposal, they shall be categorized accordingly.

Not only will driverless technologies provide the convenience and quickness, but shall also provide the enjoyment as a value to those that desire mobility. Through this service design, we highlighted requirement and criteria for a simple autonomous vehicle shall consist for daily use as "low speed" but "fun", in comparison with the current trend of autonomous driving for long-distance highways. In this paper, we reveal findings which driverless automobiles may provide as value proposing towards those living in metropolitan cities as an enjoyable experience. We also present the values the various actors will give and perceive through the bar-hopping (or *Yoidore* in Japanese) experience. Since the development of mobility has always been a main component for the urban infrastructure development [1], this research contributes not only to the service design academia but also to the mobility industry seeking for driverless technology applications.

2 Related Studies

2.1 City Nightlife

Nightlife in cities is often overlooked in the academic field of urban studies, even having a term for night blindness [2]. Some state the lack of daylight giving traction to emotions and practices such as criminal acts, lover rendezvous, or nonconventional rebellions [3]. Others note the nighttime ease of flow giving a relaxed and permissive social atmosphere [4]. However, nightlife is much more than the relaxed emotional experiences; the intenseness, pleasure, excitement, and adventure [5]. Nightlife areas with bars and clubs offer many emotionally charging activities [6]. What happens in the city at night is more than the economy, but meeting others, creating identities, and having fun.

A dinner with friends is not just about getting some food into your body, but is to engage feeling close with others. Wine and cheese provides pleasant chances for friends to enjoy their relatedness [7]. Such nightlife spaces are important to cities, subcultures, music preferences, especially for young-beings [8].

2.2 City Mobility and Transportation

What provides individuals to participate in social and community life to engage with others and the world is mobility [9]. While objective indicators such as safety, accessibility, time-saving, and frequency have been mainly discussed, there still lacks focus on social factors in the field of transportation research [10]. Social factors should be respected: social capital, community connectedness, and most of all well-being [11]. Transport mobility predicted subjective well-being through the mediating variables of environmental mastery, positive relations with others and self-acceptance [12].

This is where automotive interaction design (AID) can take substantial role in delivering well-being, from advanced driver assistance systems to social media [7]. Some in-car interface design seeking for driver-passenger collaboration maps the spatial arrangement of passengers to the form factor of a tablet [13].

3 Concept

3.1 Tokyo 27:00 Service

The navigation service Tokyo 27:00 supports young-beings to hang around, drink out, and chill in metropolitan areas, particularly in Tokyo. Since trains are the most popular transportation in Tokyo, citizens usually decide to meet up at a particular train station and hang out within the walking distance radius from the station. When at night, Tokyoites must beware of the final trains which usually depart at around midnight to go back home. If they want to keep enjoying, they must take taxis which are usually extremely expensive. Tokyo 27:00 stimulates such young Tokyoites to expand their hanging-around activity area and time limits with driverless mobility technologies and route generations.

Users will enjoy choosing the next bar to head for within the mobility which generates routes integrating plural possible destinations and displays it as if an arcade on the mobility front-window interface. When the group hops into the mobility after their first gathering (or *ichijikai* in Japanese), they can see several bars around the area to head next (called *nijikai* in Japanese) and select upon their feelings at the moment. The mobility will then generate a route that can go around top 6 candidates that the group may want to head as a unicursal route. The information will be displayed on the front window as if an arcade *yokocho* which can leverage the excitement and expectation. The mobility will drive them to the front of the bars one by one, and the group can check out if their expectation met the reality or not. If yes, they can go in the bar, but if not, they can continue the journey for the next candidate with the mobility. This can continue for the next bar (called *sanjikai* in Japanese) and the following as well, until they feel content enough to go back home, at around 27:00.

This service was designed upon the design thinking approach of (1) ethnography research of how locals hang around in towns, series of iterations on how cities should be experienced, (2) prototypes with interfaces and databases for decision making on where to head next, and (3) user studies with our navigation system prototype for value co-creation (though with a "driver-with" car and not a driverless one).

3.2 Concept Components

This experience can be delivered with the integration of the 5 elements: a database including bars and restaurants around the area, a classifier enabling users to decide how they feel at the moment, an extractor giving sequential steps on which bar to head next, a route generator combining some potential candidates they want to head unicursally, and a front window interface that displays the potential candidates into an *Yokocho* (arcades in Japanese) style – being able to peek inside the *noren* (store curtains in Japanese) and see what's going on at the store.

Bars-Around-the-Area Database. The database for the bars and restaurants in the area were generated from the researcher's three perspectives: (1) what ambience it possesses, (2) balance of food and drink, and (3) the budget. The initial data were collected based on *Tabelog*[1] website which is a commonly used for restaurant-hunting in Japan, but the researchers transformed this with the additional perspectives.

Ambience represents either lively or chill. The data were divided by the name of the bar, signboard at the entrance, the façade, the interior, the brightness of the lights, how the food is set out and dished up, how the guests are spending their time, and word-of-mouth information on *Tabelog*. The balance of food and drink were divided upon store-genre into three categories of drink-heavy (bar, wine store, standing bar), well-balanced (*izakaya*, bistro, Spanish/Italian bars), or food-heavy (restaurant, sushi, ramen). Budget was also divided into three categories of either under 3000 yen, up till 6000 yen, and over 6000 yen. All the parameters were integrated into one database.

Our initial database focused only on Nakameguro to Sangenjaya area of nearly a square kilometer wide, which is just southwest of Shibuya area in Tokyo. See Fig. 1 for detailed image of the database itself.

Fig. 1. Database image

How-We-Feel-Now Classifier. The system selects a bar from the database, and displays four photo images as a montage representing the ambience. The users can either like or dislike the montage depending on how they feel at the moment. This interface is inspired from *Tinder*[2], the social date searching app which lets users to swipe right when like and swipe left when dislike. While the users select their likes/dislikes, the

---

[1] Tabelog, a restaurant searching website https://tabelog.com/.

[2] Tinder, an online dating matching app https://tinder.com/.

system calculates how they feel, for example they want food-heavy, chill ambience, and so on. The interface is shown in Fig. 2.

Fig. 2. Destination classifying interface

Where-to-go-Next Extractor. Researchers wanted to emphasize the bar-hopping experience, which means hopping plural bars in a night (called *hashigozake* in Japanese) and not just one bar. This meant that sequence was a key to the hopping experience. Researchers had 78 Tokyoites in their 20 s–30 s to select the sequential order of what bars they would prefer going from our database. This data is embedded into the classifier so the users would not have to input their preference every single time they re-ride the mobility, but rather have something to start with based on other people's data that are similar to your preference. This classifier can pick up bars with highly ranked possibility of the users wanting to head for, just by inputting what restaurant they were at previously.

Fig. 3. Selection of bar-hopping sequential order (in Japanese for development reasons)

***Yoidore* Route Generator.** With the classifier picking up the possible bar candidates, the route generator integrates the six most highly ranked bars into one route. This route generation itself is dependent on Google API algorithm. Researchers insisted the users to hop inside the mobility and enjoy heading to several candidates in order to provide the *yoidore* bar-hopping experience. Figure 4 is an example of a generated route with a unicursal way integrating the six candidates.

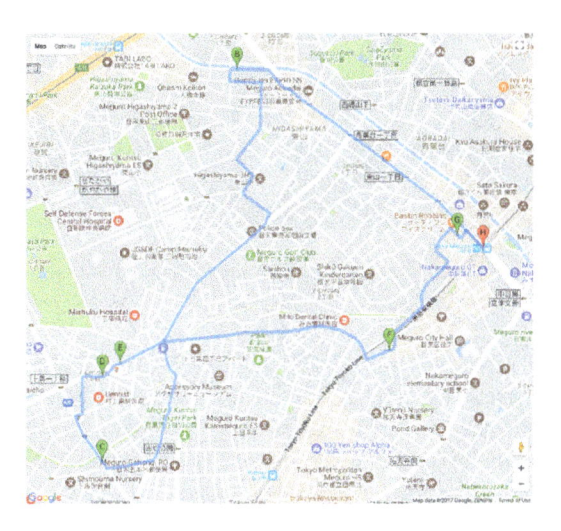

Fig. 4. Generated router

***Yokocho*-Style Window Display Interface.** Since we are assuming a driverless mobility vehicle, the front window of vehicles shall turn into a display interface at times when scenery is not necessarily important. Through this front window display, an ordinary road will transform into a *yokocho*-style arcade by displaying the six highly ranked bars on both sides of the road. As the mobility runs, the bars on the display will come close one by one. Users inside the mobility will be able to sense ambience of the bars with visual displays of signboards, the menus, and moving images of each bar, where Fig. 5 shows the drawn image. With these elements, users can experience bar-hopping in a real *yokocho* as if peeking inside real physical bars one by one. The order of the bars correspond the generated route.

Fig. 5. Mobility interface showing destinations

3.3 Design Process

Ethnography Research. Researchers conducted a series of ethnography in locals of the Tokyo area. We started with researching how young-beings especially in the early twenties are hanging out in reality. We followed a group of threesome males in Shibuya area on a Friday night. What has been decided beforehand was the fact that would meet up at Shibuya station and that they will drink out, but a particular bar has not been decided. After meeting up, all of them took out their smartphones, and searched *Tabelog* (a common Japanese restaurant searching app) inputting the area as "within 500 m from the current GPS location". But then it eventually became only one person within the group who did all the research, while the other two started chatting different topics. After all, it was one person that made the decision on where to head to while the other two just followed him. The group mentioned that this decision process was a languorous moment; all moments during the gathering should be more exciting and interactive among the members. This was something that researchers wanted to overcome by providing a different experience of destination selecting.

Another research focused on a certain town in Tokyo. We spotlighted Nishi-Ogikubo area of western Tokyo, which still possesses the post WWII era atmosphere. The condensed area consists 20 narrow traditional arcades (*yokocho* in Japanese) with micro mom-and-pop bars all unique and original, each having the traditional red lanterns on the storefronts. Locals in the area are very close to one another, each store-owner knowing each other well which is not the case in most Tokyo urbanized areas. This Nishi-Ogikubo town was a model that the researchers wanted to abstract; traditional arcade-style, relationship among the stores, and people being able to barhop the entire day from morning to way past midnight.

During the research, researchers encountered a gentleman in the forties that has been in Nishi-Ogikubo ever since he was born. He owned a café and did real estate business specializing in housing apartments. We decided to conduct ethnography with a master-apprenticeship relationship [14] of how he introduces apartments to any new-comers of the town. What was astonishing was that he did not start with introducing the rooms, but took a walk in the town together with the new-comer instead. Before starting the walk, he asked new-comers what their preferences were; what kind of stores they like, what lifestyles they admire, or how they want to live their lives. He pinned several stores that might suit their favors, and quickly generated a walking route that can drop by all of them. While actually taking the walk, he shared backstories of stores or streets and carefully looked at how the new-comers were reacting to anything they encountered during destinations, and changed routes accordingly to adapt to their behaviors. When reaching destinations, he quickly peeked into the stores and checked the current situation. If stores are busy, he quickly proposed an alternative destination and continued the walk. New-comers were impressed at the smoothness of the walk even though it was not exactly the initial plan but took situated actions [15], along with the excitement he provided with stories during the walk before arriving at destinations.

Researchers analysed this ethnography by writing thick-description [16], looking from various perspectives on contextual design [17], and abstracted mental models [18, 19] of this gentleman as a local town guide. Generating routes adaptively depending on dynamic preferences, changing destinations quickly according to current situations, and having stories along the way to destinations were key points to the experience researchers wanted to provide young-beings to enjoy towns.

Sketching the Experience. Researchers then sketched out the entire experience as a journey, sequentially from the moment they finish their dinner at the first restaurant [20]. From the ethnography research of young-beings selecting where to head, the experience of deciding was the key point of design. The process of being able to chat during the decision of neither this nor that was the core experience this service wanted to provide; how to enjoy this time was the way of communicating within the group. Figure 6 shows the entire customer journey that Tokyo 27:00 service aimed to provide as a general overview.

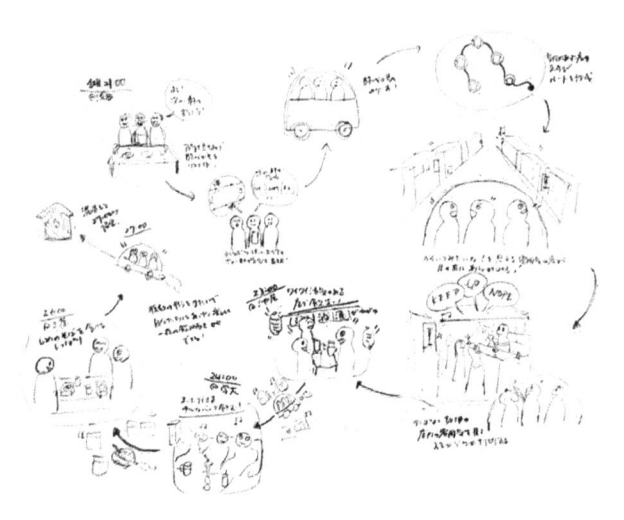

Fig. 6. Drawing the entire customer journey experience (in Japanese for development reasons)

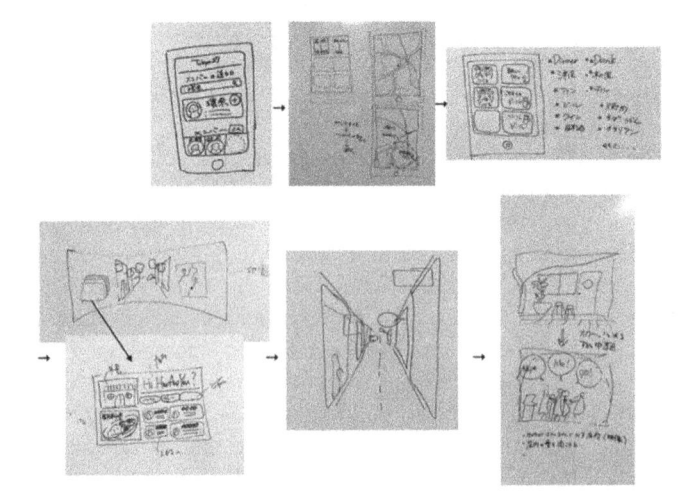

Fig. 7. Mobility interface keypath scenario

How to trigger communication within the user group was the next challenge. Various interface of the mobility had to impress the group and spark conversations. The swiping interface for the how-we-feel-now classifier was an essential touchpoint we wanted to provide the young-beings. The yokocho-style interface with narrow streets and red lanterns on the storefront for the window display was a fundamental concept that researchers crystalized during the Nishi-Ogikubo area fieldwork research. Figure 7 shows the keypath scenarios that the service composed of.

3.4 Design Process

Of the five components which Tokyo 27:00 consists, *yokocho*-style window display interface could not be embedded in real life, due to current regulations of driverless vehicles. Instead, we created a mixed reality style prototype using head-mount display Microsoft Hololens[3]. The virtual bar objects were created with Unity, and plotted as a cyberspace arcade. Figure 8 shows how the user can see through the Hololens glasses display. Moving images start playing when the Hololens sensor recognizes the eye gaze suiting perfectly with the object.

The remaining components were prototyped enough to validate the Tokyo 27:00 concept and bar-hopping experience. Figures 1, 2, 3 and 4 are actual images of the actual prototype. The classifier interface was embedded on a laptop display for system implementation issues.

---

[3] Microsoft Hololens https://www.microsoft.com/en-us/hololens.

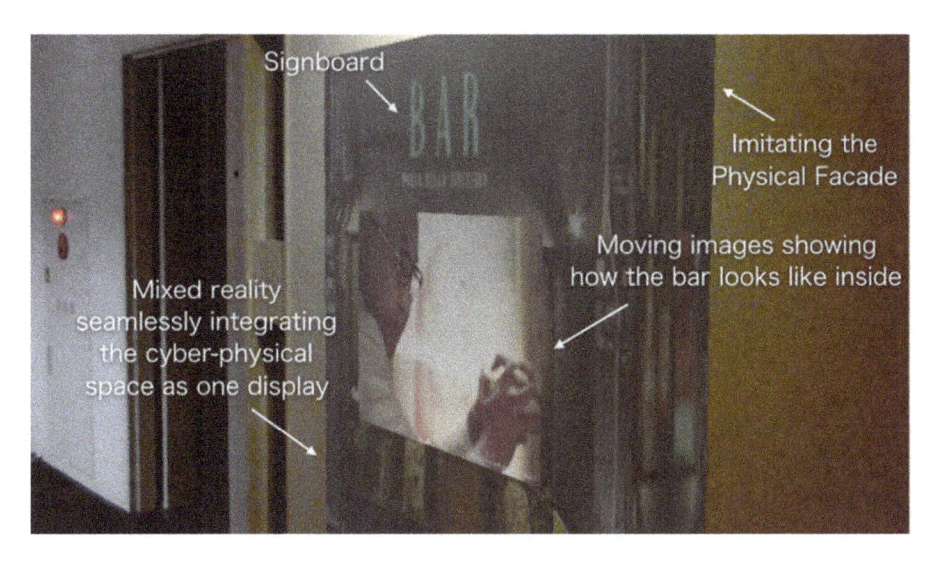

Fig. 8. Window display prototype using Hololens

4 User Study and Findings

4.1 Procedure

Researchers recruited one group of Tokyoite youngsters enjoy bar-hopping with the Tokyo 27:00 on a Saturday night. Since driverless autonomous cars are yet to be legal on the street, instead we had one of our researchers drive a normal car, pretending as if it was driverless. Our system was implemented in a laptop for the proof-of-concept user study.

The Tokyoite group consisted of 2 gentlemen and 1 lady, met up at a random restaurant in the Nakameguro-Sangejaya area called Ikejiri. The user study started after they finished their first gathering (*ichijikai* in Japanese) and then hopped onto the car. We had them input how they feel about the several candidates of the next bar, showing one by one each time selecting if the like it or not. As a result, the system listed six recommendations and integrated a route which goes through all of them.

Researcher drove the group exactly as the recommended route. When reaching the first candidate, they found it not as good as they expected. The next candidate was rather nice, but they thought there would be more to come in the following route, so they did not make their decision at the time. As a result, they saw all six candidates, decided on the bar in Nagameguro area that looked kind of like a secret base, which was something that they would not have selected if they searched on their own. Figure 9 shows a picture of the three Tokyoites deciding on where to head to using the service inside the "driver-with" car for the user study. They enjoyed the night drinking out, and went back to home at 2am; the researcher drove them all back to each of their homes.

Fig. 9. User study of Tokyo 27:00 with a "driver-with" car

4.2 Feedbacks

The following day, researchers conducted an interview following up on what they experienced and how they all felt about the night. Here we highlight five main aspects that were mentioned; being able to encounter good bars, wanting to keep drinking out, enjoying the selection process, conversations not stopping at all, change in the way of hanging out in town. In addition, there were some comments to make the service better.

Encountering Good Bars. One gentleman X mentioned that he hardly knew any bars or restaurants in the area, but this time he could find 7–8 new places that he would want to go, and he even wanted to keep finding more. Another gentleman Y noted the fact that "encountering new bars that I would have never had on my own was wonderful". He continued "I usually drink out near my own neighborhood, Koenji area, which I've gone to about 50 different bars so far. But recently, I've been only going to about 10 of my favorites among them. I haven't had enough courage to go find new ones on my own. You know, I want to have fun when drinking out, I don't want too big of a risk for that."

Wanting to Drink Out More. Gentleman X said the fact of knowing more bars will make people want to go for more, perhaps till 3 or 4 am. The lady Z agreed and added that this shall encourage me to go out more with more people. Y stated that "this service lets me get drunk; you don't have to think about so many extra things when getting drunk using this, you can just trust and rely everything". This perspective of was something new and interesting to the researchers; building the trust relationship between mobility service and the users, in other words, credence goods.

Enjoying the Selection Process. Lady Z mentioned that this service allows us to really decide where to go intuitively. "Usually in unfamiliar neighborhoods, someone in the group must search online with their smartphones where to go next, which is usually me, and that is a lot of effort surfing the internet with smartphone and calling the bars one by one. With this service, we can all decide together, and it automatically takes the group to the location." Y said that "even if the first choice did not work, you still have the remaining 5 to try". Having a wide enough selection population give users less stress than just one candidate. X agreed and continued that "listing 6 selected bars leaves the experience of deciding where to go to but cuts down the cost of searching from the big internet ocean". This perspective of search goods was also new to the researchers.

Continuing the Conversations. X said that the knowing that the service will provide the next destination candidates can reassure them and they can continue their own discussion without thinking about where to head next. "It even condenses the time because the conversation was rich and dense." Y agreed and mentioned "this time we felt that we were talking 30 times more than usual drinking out because we can concentrate on our discussion and not think about where to go next".

Changing the Way of Hanging Out. Y clearly stated that this will change the way he sees towns. "I only had basically 3 towns to hang out: Shibuya, Shinjuku, or Ikebukuro. But with this service, I can feel that I will have like 10 towns to hang out.". X agreed and said that the only reason those 3 towns are selected is because there are many bars condensed within the walking distance. However, with this service it can enhance the drinking out area with its mobility, thus "will definitely change the selection of towns to drink out". Y also mentioned that "this can lead to rediscovering values of towns and neighborhoods".

5 Conclusion

This research aimed to identify what values driverless mobility may provide through a bar-hopping experience. With our initial prototype, results showed just enough that there are potential markets in metropolitan cities at night when hanging around, where mobility can act as a key role of enjoyable experience. Such mobility can provide the next research question of how credit and relationship can be created in such service. This is an important perspective that driverless mobility service must keep thinking, and this is definitely where designing services can contribute.

References

1. Ratti, C., Claudel, M.: The City of Tomorrow: Sensors, Networks, Hackers, and the Future of Urban Life. Yale University Press, New Haven (2016)
2. van Liempt, I., van Aalst, I., Schwanen, T.: Introduction: geographies of the urban night. Urban Stud. **52**(3), 407–421 (2015)
3. Williams, R.: Night spaces: darkness, deterriotorialization, and social control. Space Cult. **11**(4), 514–532 (2008)
4. Melbin, M.: Night as frontier. Am. Sociol. Rev. **43**(1), 3–22 (1978)
5. Hubbard, P.: The geographies of 'going out': emotion and embodiment in the evening economy. In: Emotional Geographies, pp. 117–137. Ashgate, Aldershot (2005)
6. Lovatt, A., O'Connor, J.: Cities and the night-time economy. Plan. Pract. Res. **10**(2), 127–134 (1995)
7. Hassenzahl, M., Laschke, M., Eckoldt, K., Lenz, E., Schumann, J.: "It's more fun to commute"—an example of using automotive interaction design to promote well-being in cars. In: Meixner, G., Müller, C. (eds.) Automotive User Interfaces. Human-Computer Interaction Series, pp. 95–120. Springer, Cham (2017)

8. Gallan, B.: Night lives: heterotopia, youth transitions and cultural infrastructure in the urban night. Urban Stud. **52**(3), 555–570 (2015)
9. Schaie, K.: Mobility for what? In: Aging in the Community: Living Arrangements and Mobility, pp. 18–27. Springer (2003)
10. Lucas, K., Markovich, J.: International perspectives. In: New Perspectives and Methods in Transport and Social Exclusion Research, pp. 223–239. Emerald (2011)
11. Stanley, J.K., Hensher, D., Stanley, J.R., Vella-Brodrick, D.: Mobility, social exclusion and well-being: exploring the links. Transp. Res. Part A **45**, 789–801 (2011)
12. Vella-Brodrick, D., Stanley, J.: The significance of transport mobility in predicting well-being. Transp. Policy **29**, 236–242 (2013)
13. Meschtscherjakov, A., Krischkowsky, A., Neureiter, K., Mirnig, A., Baumgartner, A., Fuchsberger, V., Tscheligi, M.: Active corners: collaborative in-car interaction design. In: Designing Interactive Systems (DIS 2016), pp. 1136–1147. ACM, New York (2016)
14. Lave, J., Wenger, E.: Situated Learning: Legitimate Peripheral Participation. Cambridge University Press, Cambridge (1991)
15. Suchman, L.: Human-Machine Reconfigurations: Plans and Situated Actions. Cambridge University Press, Cambridge (2007)
16. Geertz, C.: The Interpretation of Cultures. Basic Books, New York (1977)
17. Beyer, H., Holtzblatt, K.: Contextual Design: Defining Customer-Centered Systems (Interactive Technologies). Morgan Kaufmann, San Francisco (1997)
18. Craik, K.: The Nature of Explanation. Cambridge University Press, Cambridge (1967)
19. Goodwin, K.: Designing for the Digital Age: How to Create Human-Centered Products and Services. Wiley, New York (2009)
20. Buxton, B.: Sketching User Experiences: Getting the Design Right and the Right Design (Interactive Technologies). Morgan Kaufmann, San Francisco (2007)

Urban Sensibilities, Sharing, and Interactive Public Spaces: In Search of a Good Correlation for Information and Communication in Smart Cities

H. Patricia McKenna(✉)

#541, 185-911 Yates Street, Victoria, BC V8V 4Y9, Canada

Abstract. The purpose of this paper is to explore the relationship between information and communication in the context of smart cities. A review of the research literature for information and communication in relation to sensibilities, sharing, and interactive public spaces in smart cities is provided in formulation of a theoretical perspective for this work. Using a case study approach in combination with an explanatory correlational design, this paper examines the relationship between communication, in the form of sharing and interactive public spaces, and information in the form of urban sensibilities in technology-rich smart cities. Findings reveal a glimpse of the complexities associated with the communication and information relationship in urban environments. This work is significant in that it contributes to the research literature for smart cities; points to the value of identifying patterns and relationships through involving people more directly and meaningfully in understandings of smart cities; and highlights the importance of identifying correlations to inform the development of smart cities and future cities going forward.

Keywords: Affect · Affective computing · Correlation · Human sensing · Interactions · Patterns · Relationships · Sensibilities · Sharing · Smart cities

1 Introduction

The main purpose of this paper is to explore the relationship between information and communication in the context of smart cities and future cities. Gil-Garcia, Pardo, and Nam [1] suggest that a smart city "be seen as a continuum in which local government officials, citizens, and other stakeholders" are encouraged to "think about the initiatives that attempt to make the city a better place to live." A review of the research literature for information and communication in relation to sensibilities, sharing, and interactive public spaces in smart cities is provided in formulation of a theoretical perspective for this work. Noting that "cities vary considerably in size", Borsekova et al. [2] explored whether "size influences the level of selected variables in smart cities", finding "unexpectedly" that "medium-sized cities are more open-minded than larger cities." Using a case study approach in combination with an explanatory correlational design, this paper examines the relationship between communication, in the form of sharing and interactive public spaces, and information in the form of urban sensibilities in

© Springer Nature Switzerland AG 2020
K. Arai et al. (Eds.): FICC 2020, AISC 1129, pp. 563–575, 2020.
https://doi.org/10.1007/978-3-030-39445-5_41

technology-rich smart cities. Findings reveal a glimpse of the complexities associated with the communication and information relationship in urban environments. This work is significant in that it contributes to the research literature for smart cities; points to the value of identifying patterns and relationships through involving people more directly and meaningfully in understandings of smart cities; and highlights the importance of identifying correlations to inform the development of smart cities and future cities going forward.

In terms of background and context, Fidel [3] describes information, from a human information interaction (HII) perspective, as "a concept with various types, where each type caters to a certain context." For Fidel [3], information "is a string of symbols that has meaning, is communicated, has an effect, and is used for decision making" such that "interacting with information is a way to overcome an obstacle in solving a problem." Rickert [4] observes that, "digital technologies are increasingly enmeshed with our everyday environment" such that "information vitalizes our built environs and the objects therein, making them 'smart' and 'capable of action'." Rickert [4] claims that, "we are entering an age of ambience" where "boundaries between subject and object, human and nonhuman, and information and matter dissolve." In short, digital technologies, according to Rickert [4], "not only impact our environment and how we interact with and within it but transform our knowledge about self and world." Guthier et al. [5] identify the need for finding "new methods of sensing affect in a smart city" such as "the repurposing of existing sensors in a creative way to derive affective states." As such, Guthier et al. [5] claim "affect-aware cities" as "a field of research" deserving of attention with "applications in smart city initiatives" having the potential to be "used as a tool in emotion research." Glaser et al., [6] claim that "as the city evolves, so does the field of urban planning" arguing that "we are no longer planning cities – we are reinventing, reusing, and living within them." In conceptualizing smartness in government, Gil-Garcia, Zhang, and Puron-Cid [7] note that, "objects of interaction include data, information, and knowledge" and, citing the work of Nam and Pardo [8], "interaction activities can be sharing, communication or integration." Batty [9] describes the importance of information and the digital in the context of "inventing the future of cities." Abdel-Basset and Mohamed [10] claim that, "using information effectively is going to be a main factor for success in smart cities."

The purpose of this work, together with the background and context provided, gives rise to the main research question under exploration in this work: *How and why are information and communication related in the context of contemporary and future technology-pervasive urban environments and regions?*

2 Theoretical Perspective

In developing a theoretical perspective for this work, a review of the research literature is provided for information in relation to smart cities in terms of sensibilities and then, for communication in relation to sharing and interactive public spaces.

2.1 Sensibilities and Smart Cities

Glaser et al. [6] address the "human scale" by focusing on "the city at eye level" as in, "ground floors that negotiate between the inside and the outside, between the public and the private" as well as "physical components" such as "the façade, building, sidewalk, street, bikeways, trees" and "also the emotional and social aspects." Abaalkhail et al. [11] highlight the value of ontologies for "affective states and their influences" in "allowing a flow of communication" between people and systems "by unifying terms and meanings." Abdel-Basset and Mohamed [10] note that "generated information from independent and distributed sources can be imprecise, uncertain, and/or incomplete in real life" arguing that "models, experts, and sensors" as "sources of information" need to be "reason, perfect and complete." To this end Abdel-Basset and Mohamed [10] propose a role for "single valued neutrosophic sets and rough sets in smart cities" as a model for handling "imperfect and incomplete information systems."

Capturing the people component of smart cities, Batty [9] observes that "the city is many times more complex than a single organism in that it is a collective of many pulses all firing at different rates but that is ultimately coordinated by our own human life cycles and rhythms." Where other researchers [10] highlight reason, perfect, complete, this work focuses on affect and what people sense in the city and the information associated with human sensibilities. In a recent Pew Research study on artificial intelligence (AI) and what it means for the future of humanity, Anderson, Rainie, and Luchsinger [12] identify human agency, data abuse, job loss, to name a few, as emerging areas of concern by experts while suggested solutions include "human collaboration across borders and stakeholder groups" and the need to "develop policies to assure AI will be directed at 'humanness' and common good." In the context of "interactions in virtual and augmented reality", Stephanidis et al. [13] refer to "related technologies that are part of the current and emerging information landscape" as having "the potential to alter the perception of reality, form new digital communities and allegiances, mobilize people, and create reality dissonance" contributing to "the evolving ways that information is consumed, managed, and distributed." In the context of privacy, Stephanidis et al. [13] discuss concerns associated with personal information in online social networks (OSN), citing work by Beye et al. [14], in terms of "who can view one's private information, the ability to hide information from a specific individual or group, the degree to which other users can post and share information about an individual, data retention issues, the ability of the employees of the OSN to browse private information, selling of data, and targeted marketing." Concepts include information sharing, information exchange, transparent information, incomplete information, sensitive information (as in, "user habits or health data"), automated information analysis, information distortion and pollution, information as input data (as in, "users' location, posture, emotions, habits, and intentions") and filtered information.

2.2 Communication, Sharing, Interactive Public Spaces and Smart Cities

Fidel [2] claims that, "*sharing information* is another mode of human information interaction, and one that is common in people's lives." Guthier et al. [5] advance the notion of an "affect-aware city that is able to understand, interpret and adapt" to people and their emotions such as "fear and anger" as indicators of "danger", claiming that "affect-awareness enables new forms of communication." It is worth noting that Glaser et al. [6] push the edges of interactive public spaces with the notion of *hybrid zones,* described by van der Ham [15] as "the space between the private and public realm" as well as "the place where these two meet". Shutters [16] highlights the importance of decision-making, informed by "recognizing relationships" between data and information and "explaining patterns" accompanied and guided by theory. Stephanidis et al. [13] argue that "new technologies bring increased complexity and escalate the need for interaction and communication in numerous ways", citing work by Ng [17] where "digital technologies contribute to the development of twenty-first century skills", one of which is said to be communication. Further, Stephanidis et al. [13] identify communication as one of the elements required for the formation and continuous development of trust in relation to ethics, privacy, and security, the third of seven grand challenges for HCI (Human-Computer Interaction). Regarding privacy in online social networks, Stephanidis et al. [13] refer to the General Data Protection Regulation (GDPR) pertaining to "processing of the data, and the rights of the data subject, including transparent information and communication for the exercise of rights."

2.3 Information and Communication in Smart Cities

A summary of key elements associated with information and communication emerging from the literature review are highlighted in Table 1, organized by year and covering the time period of 2011 to 2019.

Table 1. Summary of key elements associated with information and communication.

Author(s)	Year	Communication	Information
Nam & Pardo	2011	Interaction activities	
Beye et al.	2012		Personal
Fidel	2012		Sharing, interaction
Rickert	2013		Digital technologies
Guthier et al.	2015		Sensing affect
Ng	2015	21st Century skills	
Gil-Garcia et al.	2016		Objects of interaction
Glaser et al.	2016	Common goal of urbanity	
Abaalkhail et al.	2018	Ontologies for flow	
Abdel-Basset and Mohamed	2018		Smart city success
Anderson et al.	2018		AI
Batty	2018		Digital
Shutters	2018		Relationships
Stephanidis et al.	2019	Privacy, transparency, trust	Evolving approaches

For communication, a glimpse of key elements presented in Table 1 includes: interaction activities; 21st century skills; the common goal of urbanity; ontologies for flow; and privacy, transparency and trust. For information, a glimpse of key elements presented in Table 1 includes: personal; sharing; interaction; digital technologies; sensing affect; objects of interaction; smart city success; AI (artificial intelligence); digital information; information relationships; and evolving approaches to how information is consumed, managed, and distributed. Common to both communication and information is the element of interaction while the digital seems to figure strongly for information.

2.4 Framework for Information and Communication in Smart Cities

Based on the review of the research literature provided in this work, a conceptual framework for information and communication in smart cities is formulated. As depicted in Fig. 1, in the context of the interactive dynamic of people – technologies – cities, activities from such interactions focus on information, encompassing data and knowledge while the objects associated with such interactions focus on communication, encompassing integration and sharing. Emergent patterns and relationships inform affective, social, and many other aspects of smart cities with implications for society more generally.

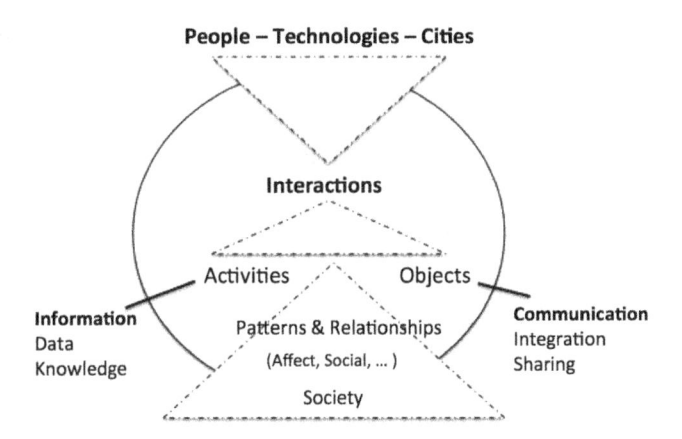

Fig. 1. Conceptual framework for information and communication in smart cities.

The research question articulated in the Introduction of this paper is restated here as a proposition under investigation in this work, as follows:

P1: Information and communication are intricately interwoven within a people – technologies – cities dynamic, based on interactions and associated activities (information) and objects (communication), generating patterns and relationships, extending to affective, social and other dimensions, informing understandings of smart cities with implications for society more broadly, now and going forward.

3 Methodology

The methodology for this work combines the use of an exploratory case study approach with an explanatory correlation design. Yin [18] identifies the "complementarity of case study and statistical research" claiming that "case studies have been needed to examine the underlying processes that might explain a correlation." Data collection involved the use of a pre-tested survey instrument and a pre-tested in-depth interview protocol, designed for people to assess their experience and understanding of the city where they live, as smart.

A website was used to describe the study, invite participation, and enable sign up. Demographic data, including location, age range, and gender, were gathered during registration for the study and people were able to self-identify in one or more categories (e.g., educator, learner, community member, city official, business, etc.). Registrants were invited to complete a survey containing 20 questions as an opportunity to think about smart cities in relation to proxies for information (e.g., heightening urban sensibilities) and communication (e.g., sharing, interactive public spaces), among many other elements. The study attracted a wide range of individuals from small to medium to large-sized cities, mostly in Canada and extending also to cities in other countries (e.g., Europe, Israel, and the United States).

In parallel with this study, evidence was also gathered through group and individual discussions with a broad spectrum of people from diverse sectors across multiple cities (e.g., Toronto, Greater Vancouver, and Greater Victoria). Perspectives emerged from those in business (architectural design, ecology, energy, information technology (IT), tourism), government (city councilors, policy makers, IT staff), education (secondary and post-secondary, researchers, IT staff), and from students (post-secondary – engineering, design, computing, education, media), and community members (IT professionals, urban engagement leaders, urban designers, and policy influencers).

Using content analysis for qualitative data, inductive analysis enabled the identifying of emerging terms from the data collected while deductive analysis enabled the identification of terms emerging from the review of the research literature. Data were then analyzed for patterns and emergent insights using a set of data analytic tools. The analysis of quantitative data involved the use of descriptive statistics for the correlation of variables using the Spearman's rank correlation coefficient. Qualitative evidence from in-depth interviews and open-ended survey questions, along with data systematically gathered from discussions in parallel with this study, supported the analysis, comparison, and triangulation of data, contributing to insight and rigor.

Overall, data were analyzed for an $n = 76$ spanning the age range of people in their 20s to their 70s, consisting of 41% females and 59% males.

4 Findings and Discussion

Findings are presented and discussed based on the correlating of survey results for information and communication proxies and then for a word analysis of interviews, open-ended survey data, and group and individual discussions pertaining to information and communication.

4.1 Correlations for Information and Communication in Smart Cities

For the purposes of this work, as depicted in Table 2, the notion of "heightening urban sensibilities" is used as a proxy for information. Similarly, "sharing" is used in this work as a proxy for communication.

Table 2. Information and communication proxy correlations in smart cities.

Proxies	Information	Communication
Heightening Urban Sensibilities	25% (5); 25% (6); 50% (7)	
Sharing		25% (6); 75% (7)
Correlation		*.81*
Interactive Public Spaces		25% (5); 75%(7)
Correlation		*.81*

When asked to assess the extent to which "city-focused social media and other aware technologies give rise to possibilities for *heightening urban sensibilities*", on a scale of 1 (not at all) to 7 (absolutely), 25% responded at 5 on the scale, 25% at 6 on the scale, and 50% at 7 on the scale. Regarding possibilities for *sharing*, participants tended to respond toward the upper end of the scale with 25% at 6 on the scale and 75% at 7. Using the Real Statistics Software [19] Resource Pack add-in for Microsoft Excel that contains a range of supplemental statistical functions and data analysis tools, the results for information (heightening urban sensibilities) and communication (sharing), when correlated, show a positive Spearman correlation at .81. Creswell [20] indicates that correlations within the .66 to .85 range, are "very good" and "good prediction can result from one variable to the other." With this in mind, a second correlation was conducted using *interactive public spaces* as a proxy for communication and heightening urban sensibilities as a proxy for information with the same result, showing a positive correlation at .81. In view of these findings, it may be worth considering the notion by Rickert [4] of the dissolving of boundaries and whether and if this is affecting the information communication relationship in digital and aware environments.

Exploring further, the people involvement dimension of smart cities is used focusing on emotion/affect where an early stage adaptation of Anderson's [21] body insight scale (BIS) is employed. Emotion/affect as a type of body communication pertaining to feelings of safety was correlated with heightening urban sensibilities as information. For example, people were asked to respond on a 7 point scale (1 = not at all to 7 = absolutely) to the question: *Regarding your body awareness in your city, would you agree that your body lets you know when your environment is safe?* As depicted in Table 3, responses for safety (affect) emerge with 25% at position 2 on the scale, 25% at position 5 and 50% at the upper end of the scale at position 7. The direction of the correlation is shown to be positive with the strength of the relationship considered to be "useful for limited prediction", falling as it does within the .35 to .65 range (Creswell 2018), based on a Spearman correlation coefficient of .50.

Table 3. Information as emotion/affect and communication correlation in smart cities.

Affect	Information	Communication
Safety (affect)		25% (2); 25% (5); 50%(7)
Heightening Urban Sensibilities	25% (5); 25% (6); 50% (7)	
Correlation		.50

Further highlighting the interweaving of information and communication in urban environments, it is worth noting that through open-ended survey questions, when asked, "what contributes to the making of a smart city?" one respondent stated, "participation more than anything, and usefulness of what might be provided through technology and access to information." When asked, "How does technology help you in the city?" 50% of respondents chose the option "for communication" while 75% selected the option "to access city resources" which could be said to contain the inherent notion of information. It is worth noting that through the "other" option, one individual pointed to the value of information through the addition of "mobile apps for orientation and information, not just for navigation, e.g. travel apps, etc." Thinking about city size, one respondent speculated that "one's senses are more technologically aware in large cities than small" adding "but that may be incorrect because in small cities you may have to be technologically smart to get even with the big city in terms of accessing information."

4.2 Word Term Frequencies, Trends, and Correlations for Information and Communication in Smart Cities

In-depth interviews, open-ended survey questions, and group and individual discussions provide qualitative insights through text rendered visually in Fig. 2.

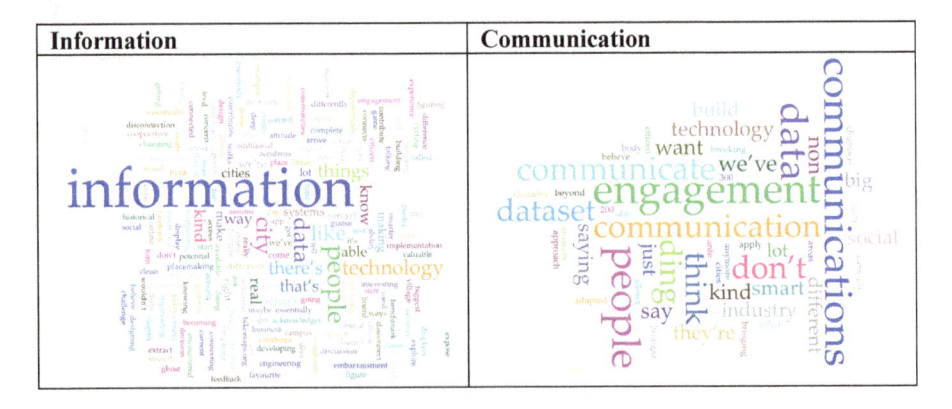

Fig. 2. Smart cities: word clouds for information and communication and related terms.

For example, using the web-based digital text reading and analysis environment of Voyant Tools [22] developed by Sinclair and Rockwell, 40 qualitative segments of text pertaining to information and communication are analyzed, of which 8 are communication text segments and 32 are information text segments. Using Voyant tools to provide an analysis of text, findings based on the *word clouds* tool featured in Fig. 2 show results on the left for information and on the right for communication, along with associated terms for each. The word clouds provide a quick visualization of the top frequency words in each of the documents. According to Voyant developers, "the word cloud positions the words such that the terms that occur the most frequently are positioned centrally and are sized the largest." An underlying algorithm "goes through the list and continues to attempt to draw words as close as possible to the center of the visualization" and "also include small words within spaces left by larger words that do not fit together snugly." For both information and communication, at a glance, terms such as data, people, and technology figure strongly. In the case of information, the term city also figures strongly while for communication, the term engagement figures strongly, and smart and social are also noteworthy. It is worth noting that in one small to medium-sized context, city information technology (IT) staff commented that "I don't think we've really taken full advantage of the data that we currently have" adding that "we're starting to look at the tools to help us mine the data that we already have an interest in" so "we're very much immature in that overall data sense."

Combining the sets of data for communication and information, the Voyant *summary* tool provides an overview of the text as containing 1,846 total words and 600 unique word forms with an average of 41 words per sentence and the most frequent words highlighted as information (39), people (14), data (13), city (10), and like (10). The *trends* tool in Fig. 3 "shows a line graph depicting the distribution of a word's occurrence" across the information document on the left and the communication document on the right. It is also worth noting that the *trends* tool provides a visual view of the *summary* tool data.

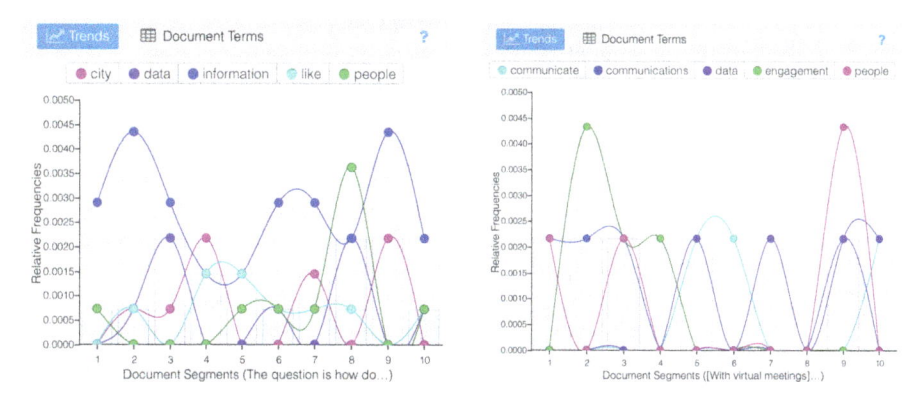

Fig. 3. Smart cities: trends for information and communications - distribution of word occurrences.

Regarding the *trends* view of the information text on the left of Fig. 3, the most frequent words are highlighted as information (38), city (10), people (10), data (9), and like (9). For the communication text on the right of Fig. 3, the most frequent words are highlighted as communication (4), data (4), people (4), and communicate (3).

The *correlation* tool in Voyant "enables an exploration of the extent to which term frequencies vary in sync" as in "terms whose frequencies rise and fall together or inversely." Using the *correlations* tool to determine the correlation, as depicted in Fig. 4, a positive relationship emerges between *information* and *making* at .62 with a significance of .05 where ".05 or less" is said to be indicative of "a strong correlation" [22]. Also of note is the positive relationship between *communications* and *dataset* at .64 with a significance of .04. Other examples of correlations showing significance include *app* and *data*; *social* and *technologies*; and *social* and *use*.

Term 1	←	→	Term 2	Correlation	Significance (p)
app			data	0.6296296	0.051082764
create			data	0.6296296	0.051082764
city			information	0.44600385	0.19637318
communications			dataset	0.6475183	0.042952873
information			making	0.62812	0.0518123
social			technologies	0.61237246	0.059837874
social			use	0.61237246	0.059837874

Fig. 4. Smart cities: correlations for term frequencies associated with information and communication.

The Voyant developers [22] caution that "these values should be used with care" while, as noted earlier in this paper, Creswell [20] claims that correlations between .35 and .65 "are useful for limited prediction." Indeed, as cities develop their skills, budgets, and perspectives on the importance of urban data, the *city* and *information* correlation depicted here at .44 with a significance of .19 may show greater improvement going forward.

While a good correlation for information and communication proxies in smart cities was achieved in this work, when emotion/affect was introduced through the involving of people directly in the exploration of feelings of safety in the city, a correlation "useful for limited prediction" emerged.

In summary, key elements emerging in this exploration of information and communication patterns and relationship in smart cities are outlined in Table 4. For information, associated elements include city, data, heightening urban sensibilities, making (involving participation and engagement), people, technology, and urban apps. Elements associated with communication include data, feelings of safety (emotion/affect), people, social, technology, and urban apps. Elements associated with both information and communication, include, but are not limited to, data, people, technology, and urban apps.

Table 4. Emergent elements associated with information and communication in smart cities.

Elements (associated)	Information	Communication
City	✓	
Data	✓	✓
Feelings of safety (people – emotion/affect)		✓
Heightening urban sensibilities	✓	
Making (participation, engagement)	✓	
People	✓	✓
Social		✓
Technology	✓	✓
Urban apps	✓	✓

5 Challenges, Limitations, Mitigations and Future Directions

A key limitation of this work is the small sample size and this is mitigated, in part, by in-depth and rich detail from a wide range of individuals across small to medium to large urban centers and regions in several countries. Another limitation of this work is the focus on only two variables (e.g., information and communication) and this is mitigated by opportunities to enlarge the exploration to include additional variables going forward along with more participants, more cities, and more range in city size. The use of proxies for the information variable (e.g., heightening urban sensibilities) and the communication variable (e.g., sharing, interactive public spaces) resulted in good correlations (e.g., .81). However, involving people more directly through assessments of emotion/affect in urban spaces, focusing on feelings of safety as a communication proxy in relation to heightening urban sensibilities as an information proxy, resulted in a correlation of .50 with "limited prediction" potential. Going forward, potential exists to explore the information and communication variables directly rather than through the use of proxies for each. Through this work, a space is also opened for further exploration and validation of the importance of emotion/affect in smart cities research and practice in relation to information and communication.

6 Conclusion

This work undertakes an exploration of the relationship between information and communication in contemporary urban environments as a starting point for learning more about elements contributing to the success of smart cities. A hybrid case study and correlational research design was used, informed by a review of the research literature for information and communication in relation to smart cities. In support of the proposition under exploration in this work, the information and communication variables were found to be intricately and highly interwoven. While findings show a good correlation at .81 for the proxies used in this work for information and communication, the introduction of emotion/affect shows a correlation of .50 with only limited predictive potential. Going forward, more work is needed to revisit the information and communication correlations and explore further whether there is a need to consider these two variables more directly in data analysis in the context of smart environments. Also, this paper sets the stage for further work focusing more directly on information and communication in urban spaces and the importance of emotion/affect. This paper is significant in that it contributes to the research literature for information and communication in the context of smart cities; demonstrates the use of correlation as an approach to exploring patterns and relationships in smart cities; and extends urban theory and methodological approaches through taking people-related correlations into consideration.

Acknowledgments. This work was supported in part by an Independent Scholars Grant (2019) from the Canadian Academy of Independent Scholars (CAIS) through Simon Fraser University (SFU), Vancouver, British Columbia, Canada.

References

1. Gil-Garcia, J.R., Pardo, T.A., Nam, T.: A comprehensive view of the 21st century city: smartness as technologies and innovation in urban contexts. In: Gil-Garcia, R., Pardo, T.A., Nam, T. (eds.) Smarter as the New Urban Agenda: A Comprehensive View of the 21st Century City. Springer, San Antonio (2017)
2. Borsekova, K., Koróny, S., Vaňová, A., Vitálišová, K.: Functionality between the size and indicators of smart cities: a research challenge with policy implications. Cities **78**, 17–26 (2018)
3. Fidel, R.: Human Information Interaction: An Ecological Approach to Information Behavior. MIT Press, Cambridge (2012)
4. Rickert, T.J.: Ambient Rhetoric: The Attunements of Rhetorical Being. University of Pittsburgh Press, Pittsburgh (2013)
5. Guthier, B., Abaalkhail, R., Alharthi, R., El Saddik, A.: The affect-aware city. In: Proceedings of the International Conference on Computing, Networking and Communications, pp. 630–636 (2015)
6. Glaser, M., van 't Hoff, M., Karssenberg, H., Laven, J., van Teeffelen, J., (eds.): The City at Eye Level: Lessons and Street Plinths, 2nd and Extended edn. Eburon, Delft (2016). https://thecityateyelevel.files.wordpress.com/2013/01/the-city-at-eye-level.pdf. Accessed 30 June 2019
7. Gil-Garcia, J.R., Puron-Cid, G., Zhang, J.: Conceptualizing smartness in government: an integrative and multi-dimensional view. Gov. Inf. Q. **33**(3), 524–534 (2016)
8. Nam, T., Pardo, T.A.: Smart city as urban innovation: focusing on management, policy, and context. In: Proceedings of the 5th International Conference on Theory and Practice of Electronic Governance, pp. 185–194, Tallinn, Estonia, 26–28 September 2011
9. Batty, M.: Inventing Future Cities. MIT Press, Cambridge (2018)
10. Abdel-Basset, M., Mohamed, M.: The role of single valued neutrosophic sets and rough sets in smart city: imperfect and incomplete information systems. Measurement **124**, 47–55 (2018)
11. Abaalkhail, R., Guthier, B., Alharthi, R., El Saddik, A.: Survey on ontologies for affective states and their influences. Seman. Web **9**(4), 441–458 (2018)

12. Anderson, J., Rainie, L., Luchsinger, A.: Artificial intelligence and the future of humans. Pew Research Center, Internet & Technology, Washington, DC (2018). https://www.pewinternet.org/2018/12/10/artificial-intelligence-and-the-future-of-humans/. Accessed 07 June 2019
13. Stephanidis, C., Salvendy, G. (Chairs), Antona, M., Chen, J.Y.C., Dong, J., Duffy, V.G., Fang, X., Fidopiastis, C., Fragomeni, G., Fu, L.P., Guo, Y., Harris, D., Ioannou, A., Jeong, K., Konomi, S., Krömker, H., Kurosu, M., Lewis, J.R., Marcus, A., Meiselwitz, G., Moallem, A., Mori, H., Nah, F.F.-H., Ntoa, S., Rau, P.-L.P., Schmorrow, D., Siau, K., Streitz, N., Wang, W., Yamamoto, S., Zaphiris P., Zhou, J. (Members of the Group): Seven HCI Grand Challenges. Int. J. Hum. Comput. Inter. **35**(14), 1229–1269 (2019). https://doi.org/10.1080/10447318.2019.1619259
14. Beye, M., Jeckmans, A. J., Erkin, Z., Hartel, P., Lagendijk, R. L., Tang, Q.: Privacy in online social networks. In: A. Abraham (ed.) Computational Social Networks: Security and Privacy, pp. 87–113. Springer-Verlag, London (2012). https://doi.org/10.1007/978-1-4471-4051-1_4
15. Van der Ham, S.: Hybrid zones. In: M. Glaser, M. van 't Hoff, H. Karssenberg, J. Laven, J. van Teeffelen (eds.) The city at eye level: Lessons and street plinths, 2nd and extended edn. Eburon, Delft, Netherlands (2016)
16. Shutters, S.T.: Urban science: putting the "smart" in smart cities. Urban Sci. **2**, 94 (2018)
17. Ng, W.: Digital Technology in Education. Springer, Cham (2015). https://doi.org/10.1007/978-3-319-05822-1_1
18. Yin, R.K.: Case Study Research and Applications: Design and Methods. Sage, Thousand Oaks (2018)
19. Zaiontz, C.: Real Statistics using Excel (2019). www.real-statistics.com
20. Creswell, J.W.: Educational Research: Planning, Conducting, and Evaluating Quantitative and Qualitative Research, 6th edn. Pearson, Boston (2018)
21. Anderson, R.: Body intelligence scale: defining and measuring the intelligence of the body. Humanist Psychol. **34**, 357–367 (2006)
22. Sinclair, S., Rockwell, G.: Voyant tools (2016). https://voyant-tools.org. Accessed 8 June 2019

Visual Odometry from Omnidirectional Images for Intelligent Transportation

Marco Marcon[✉], Marco Brando Mario Paracchini, and Stefano Tubaro

Dipartimento di Elettronica, Informazione e Bioingegneria,
Politecnico di Milano P.zza Leonardo da Vinci, 32, 20133 Milan, Italy
marco.marcon@polimi.it
http://marcon.net

Abstract. In this article we use omnidirectional images obtained from equirectangular panoramas of Google Maps^{TM} to estimate camera egomotion. The systems was also tested using a 360 camera. The goal is to provide an effective and accurate positioning system for indoor environments or in urban canyons where GPS signal could be absent. We reformulated classical Computer Vision geometrical constraints for pinhole cameras, like epipolar and trifocal tensor, to omnidirectional cameras obtaining new and effective equations to accurately reconstruct the camera path using couples or triplets of omnidirectional images. Tests have been performed on straight and curved paths to validate the presented approaches.

Keywords: Visual odometry · Ominidirectional images · Trifocal tensor

1 Introduction

Omnidirectional images are images where the field of view exceeds 180° and thus they cannot be acquired by a camera described by the traditional pinhole model. Omnidirectional images are produced by a traditional camera facing an especially crafted mirror, such as in the case of catadioptric cameras, or by stitching a number of images synchronously acquired by traditional cameras in a panorama [9]. Nowadays 360° cameras, capable to provide high resolution omnidirectional images, are available off-the-shelf from many producers, e.g. VuzeCam [2–5], however just a few algorithms focus on these cameras for SLAM or Visual Odometry [7,18]. In this paper, we use omnidirectional images to perform visual odometry along a path. Omnidirectional images offer several advantages over planar images when estimating camera motion or the 3D model of the scene. Firstly, the large field of view (in our case, the full 360° panorama) allows us to find a larger number of correspondences between pair of images along the path with respect to planar cameras. Moreover, as we will show in the following sections, once the omnidirectional image is mapped on the surface of a sphere, no further camera calibration is needed.

The estimation of the essential matrix between two omnidirectional images and the trifocal tensor between three omnidirectional ones can be obtained from the extension of the work of [17]. We tested two different estimation strategies, one based on the essential matrix combined with an heuristic for scale normalization, and one based on the trifocal tensor.

We tested the two strategies on two different datasets: a collection of images collected from the Google Street View API [1] and a dataset of 1878 images, provided by a private company that used a 360 professional camera, spanning several city streets in Milan. The Google Street API allows downloading 360° images (henceforth called "spherical images") as a set of tiles at increasing resolution. The tiles for a set of 12 spherical images in two streets in Italy were downloaded and stitched together to form images of 833 × 1666 pixels.

The outline of the paper is as follows. In Sect. 2, we review relevant work. In Sect. 3, we show briefly how to adapt the geometry estimation algorithm to the omnidirectional case. In Sect. 4, we describe two methods for estimating the motion path of an omnidirectional camera. Finally, in Sects. 5 and 6 we present the experimental results and the conclusions.

2 Related Work

To our knowledge, the literature on trifocal tensor estimation from omnidirectional images is quite limited. The most complete discussion is in the work of Torii *et al.* [17] and [18] showing how the intuitions in the planar case translate in the omnidirectional case.

More examples rely on adaptations of well-known algorithms to the omnidirectional case, and especially the use of epipolar constraints for geometrical estimation. A visual odometry pipeline based on omnidirectional images is presented by Torii, Havlena and Pajdla in [16], where the 5-point algorithm for estimating the essential matrix is coupled with bundle adjustment to improve the accuracy of the estimate. Pretto *et al.* [14] developed a dense 3D reconstruction algorithm using a catadioptric camera while Tardif *et al.* [15] stitched multiple images acquired by a set of car-mounted cameras in a sequence of panoramas.

Dense reconstruction is also explored by Mei *et al.* [12] were a variant of the Lucas-Kanade dense tracking algorithm is developed for spherical projections, accounting for the non-linearities in the transformation.

3 Trifocal Tensor on Omnidirectional Cameras

The traditional pinhole camera model projects the points in the 3D space on a image plane and it cannot represent cameras with a field of view larger than 180°. Omnidirectional images are instead represented in *spherical coordinates* spanning the whole 360° angle in *azimuth* (angle θ) and 180° in *elevation* (angle ψ) for the full-sphere description.

Differently from the pinhole case, no intrinsics need to be defined because the images are first converted from the specific representation in the acquisition

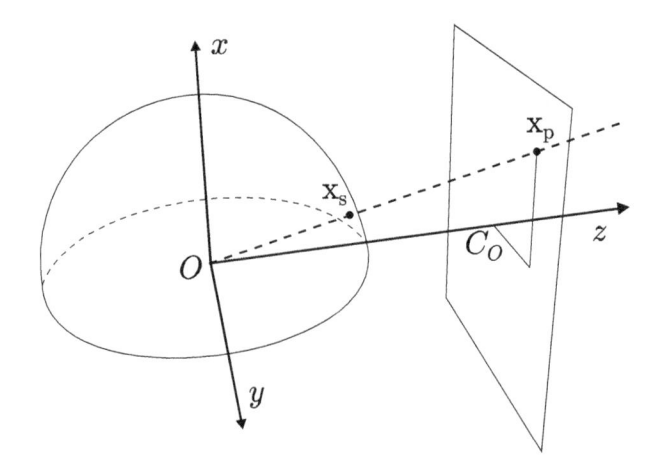

Fig. 1. Conversion from spherical image to planar image: the point on the sphere $\mathbf{x_s}$ can be thought as a 2D homogeneous point with $z_{\mathbf{x_p}}$ as the homogeneous coordinate; dividing by the homogeneous coordinate we recover the point on the image plane $\mathbf{x_p}$.

system to normalized spherical coordinates. In the case of panorama images, the intrinsics are automatically resolved in the stitching process; in the case of catadioptric images the camera or mirror calibration defines a mapping between pixel space and spherical coordinates [13].

An omnidirectional camera is thus fully described by its extrinsic parameters, the translation $\mathbf{t}$ and the rotation R of the world system with respect to camera coordinates. As a sphere spans all directions, R does not affect the content of the image acquired by the sensor, as in the planar case; however, R still affects the *parametrization* of the image in spherical coordinates.

As in the planar case, the 3D coordinates $\mathbf{X}$ of a point in the space need to be converted in the local coordinate system $T = [R|\mathbf{t}]$ of the camera. The projection function is simply the normalization of the Cartesian coordinates of $\mathbf{X}$ with respect to the reference frame of the camera:

$$\mathbf{x_s} = \frac{R\mathbf{X} - \mathbf{t}}{\|R\mathbf{X} - \mathbf{t}\|} \qquad (1)$$

The normalized coordinates $\mathbf{x_s}$ are interpreted as image point on a omnidirectional image. Having unitary norm, they can be converted to spherical coordinates (θ, ψ) where θ is the azimuth and ψ is the elevation.

Image points from pinhole cameras can also be converted to the omnidirectional representation by normalizing the calibrated image coordinates in homogeneous notation:

$$\mathbf{x_s} = \frac{K^{-1}\mathbf{x_p}}{\|K^{-1}\mathbf{x_p}\|} \qquad (2)$$

where K is the intrinsic camera matrix according to [10] and with the resulting spherical image having the z coordinate normal to the image plane.

Conversely, the spherical image point $\mathbf{x_s}$ with $z_{\mathbf{x_s}} \geq 0$ can be interpreted as the image point in homogeneous coordinates of a calibrated pinhole camera with optical center in the sphere center and with image plane perpendicular to the z axis; the points with $z_{\mathbf{x_s}} = 0$ are points at infinity on the image plane (see Fig. 1).

Henceforth for simplicity we will assume that all the images are in the normalized spherical representation.

3.1 Line Representation

In spherical coordinates, the image of a straight line in the space is a great circle on the sphere. A duality exists between points and great circles on the sphere, analogous to the duality between points and lines on the image plane.

A great circle C identifies a plane π, whose normal $\mathbf{a}$ corresponds to a point on the spherical image; conversely, the point $\mathbf{a}$ defines a plane whose intersection with the sphere is a great circle. The image of a line can then be considered as the intersection with the surface of the sphere of the plane defined by the line and the center of the sphere $\mathbf{O}$.

This result mirrors an analogous result for planar cameras, where the homogeneous coordinates $\mathbf{p}$ of an image point can also represent a straight line (dual line) defined by $\mathbf{p}^\top \mathbf{x} = 0$, where $\mathbf{x}$ are the homogeneous coordinates of a generic point on the dual line. By replacing dual lines with dual great circles, the results from the pinhole case are easily translated in the omnidirectional case.

The dual to a point $\mathbf{a} = [a_1, a_2, a_3]^\top$ has a nice analytical form in spherical coordinates (θ, ψ). By imposing the constraint $\mathbf{a}^\top \mathbf{x} = 0$ on the point $\mathbf{x}$ represented in spherical coordinates, we obtain in implicit formulation:

$$a_1 \cos\psi \cos\theta + a_2 \cos\psi \sin\theta + a_3 \sin\psi = 0 \qquad (3)$$

The previous equation can obviously be interpreted as the intersection between a sphere with unit radius and a plane passing through the origin with normal unit vector $\mathbf{a}$.

3.2 Epipolar Constraints

In this section we show how the essential matrix and the trifocal tensor are derived by exploiting the duality between great circles and points.

The epipolar constraint in the omnidirectional case is derived following the same procedure as the planar case [17]. The extrinsics of the first camera can be defined as $[I|\mathbf{0}]$ and the second as $[R|\mathbf{t}]$; where $\mathbf{x}$ and $\mathbf{x}'$ are the projections of the 3D point $\mathbf{X}$ on the two spherical cameras (see Fig. 2). The two image centers O and O' and X define a plane parallel to $\mathbf{t}$, $\mathbf{x}$, $\mathbf{x}'$ and $R\mathbf{x}$; its normal in the reference system of the second camera is:

$$\mathbf{n} = \mathbf{t} \times R\mathbf{x} = [\mathbf{t}]_\times R\mathbf{x} \qquad (4)$$

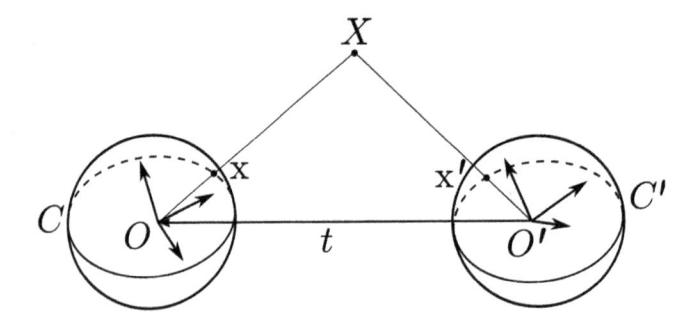

Fig. 2. Derivation of the essential matrix from the projections **x** and **x**′ of a 3D point **X**. The epipolar lines are great circles **C** and **C**′ on the spherical images.

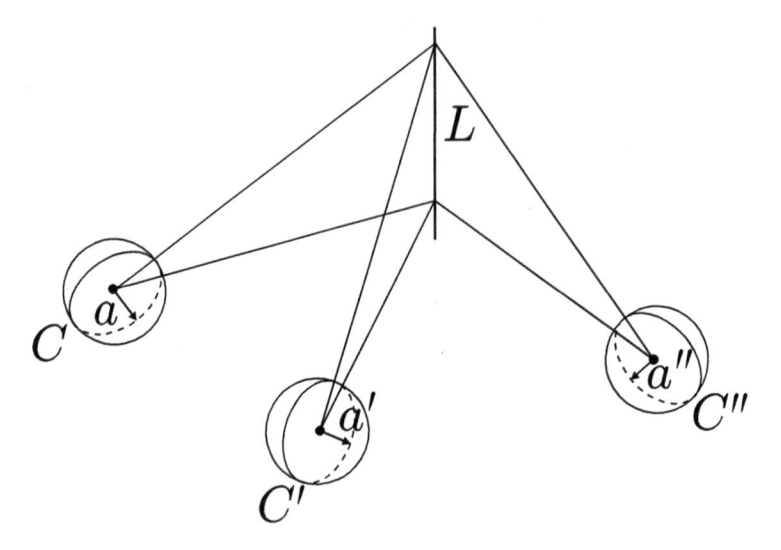

Fig. 3. Derivation of the trifocal tensor from the images of a line, in the case of omnidirectional images. The lines are great circles on the spherical images.

The cross product is formulated using the skew-symmetric matrix associated to **t**. As **x**′ is also parallel to the plane, we have:

$$\mathbf{x'}^\top \mathbf{n} = \mathbf{x'}^\top [\mathbf{t}]_\times R\mathbf{x} = \mathbf{x'}^\top E\mathbf{x} = 0 \qquad (5)$$

We want to stress that the derivation is based only on points and their duals and thus it is independent from the type of camera and its intrinsics. In the planar case, the set of points representing all the possible images of **x** on the second camera would be an epipolar *line*; in the spherical case, the set would be an epipolar *great circle*. See Fig. 4 for a representation of the epipolar circles in equirectangular coordinates, using the formula 3.

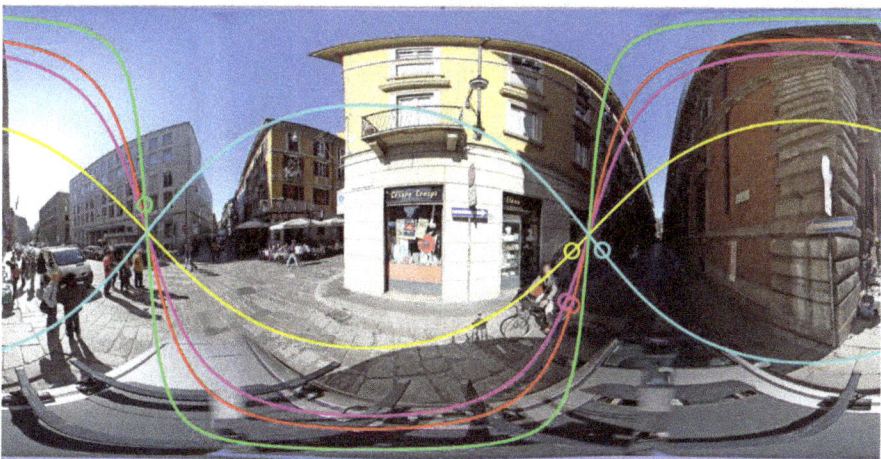

Fig. 4. Epipolar lines superimposed on the first (top) and the second (bottom) frame of a sequence of spherical images. Corresponding points and their respective epipolar lines are identified by the same color; the image of the points in the two images is represented by a small circle. The rotation between the images is reflected in the horizontal shift of the epipoles.

3.3 Trifocal Tensor

The derivation of the trifocal tensor in the case of omnidirectional images follows closely the derivation for planar images. The extrinsics for the first, second and third camera can be defined as $T = [I|\mathbf{0}]$, $T' = [R'|\mathbf{t}']$ and $T'' = [R''|\mathbf{t}'']$ (see Fig. 3). The analysis of the mutual relations between the images of a line on the three cameras follows the trifocal principle: [10], Chap. 15 and [17].

The image of a line is a great circle that lies on the plane passing through the line and the optical center of the camera. In the camera's coordinates, the plane is defined by the set of points normal to the great circle's dual $\mathbf{a}$. If the camera has extrinsics $[R|\mathbf{t}]$, the equation of the plane in the reference system of the first camera can be derived:

$$\mathbf{X'}^\top \mathbf{a} = \mathbf{X}^\top R^\top \mathbf{a} + \mathbf{t}^\top \mathbf{a} = \pi^\top \begin{bmatrix} X \\ 1 \end{bmatrix} \quad (6)$$

where in the last step the homogeneous notation is adopted. The images $\mathbf{a}$, $\mathbf{a}'$ and $\mathbf{a}''$ of a line in the three cameras thus define three planes, represented by the following homogeneous vectors:

$$\pi = \begin{bmatrix} \mathbf{a} \\ 0 \end{bmatrix} \quad \pi' = \begin{bmatrix} R'^\top \mathbf{a}' \\ \mathbf{t}^\top \mathbf{a}' \end{bmatrix} \quad \pi'' = \begin{bmatrix} R''^\top \mathbf{a}'' \\ \mathbf{t}^\top \mathbf{a}'' \end{bmatrix} \quad (7)$$

As the planes all intersect on a line, the matrix having π, π' and π'' as columns has rank 2. By imposing $\pi = \alpha\pi' + \beta\pi''$, deriving the factors α and β from the fourth coordinate of π and substituting back in the first three coordinates, we finally obtain the formula:

$$a_i = \mathbf{a}''^\top T_i \mathbf{a}' \quad \text{with} \quad T_i = r'_i t''^\top - t' r''^\top_i \quad (8)$$

where r_i are the columns of the rotation matrices. The equation (8) is the *line-line-line* relation of the trifocal tensor in omnidirectional camera. The result has two important differences compared to the planar case:

- The lines **l** on the image plane are replaced by the duals **a** of great circles on the image sphere.
- The projection matrices are replaced by Euclidean transformation matrices (i.e. where only translations and rotations are present); this will allow to use the trifocal tensor for a metric reconstruction, and not only for a projective one.

The second point will be especially useful when we will have to reconstruct the path of the camera from the trifocal tensor.

4 Visual Odometry from Omnidirectional Images

We used the algorithms for essential matrix and trifocal tensor estimation in order to reconstruct a path of a camera in the 3D space. As input of the system we collected a number of omnidirectional images from the Google Street View[TM] service.

The point correspondences were extracted by the (Scale-Invariant Feature Transform) SIFT algorithm [11] on the image mapped in equirectangular coordinates; matches were computed only between pairs of consecutive images; tests were performed also using FREAK algorithm [6] obtaining very similar results. As the regions near the poles have significant distortion, 35% of the lower part and 15% of the higher part of the images were excluded from interest point detection. The values were experimentally found as the ones non affecting significantly the number of correct matches.

We tested two different approaches for estimating the translation and rotation of the camera:

- Use the 8-point algorithm on pairs of consecutive images.
- Use the trifocal tensor on consecutive triplets of images (two images shared between triplets).

We compared the results with the ground truth given by differential GPS data. In both cases, we apply RANSAC (RANdom SAmple Consensus) [8] to remove outliers.

In the first approach the essential matrix between a pair of images is obtained using the 8-point algorithm and the coordinate transformation (R, t) between the second and the first camera can be computed from the essential matrix.

In the second approach the trifocal tensor between triplets of images is obtained using the point-point-point correspondences, following for the most part the "gold standard" algorithm for trifocal tensor estimation (algorithm 16.2 in [10]). We use the standard Levenberg-Marquardt minimizer to solve the constrained minimization problem.

4.1 Scale Fixing and Scale Transfer

The essential matrix or trifocal tensor estimation can only recover the transformations between camera only up to a similitude transformation; it is thus necessary to find the scale factor between the estimate and the ground truth by a process of *scale fixing*. In both approaches, we set the scale by comparing the estimated distance between the first two cameras with its ground truth value.

Moreover, as we estimate the transformations between pairs and triplets of images, we need method of *scale transfer*, which propagates the scale across different estimates. We use two different method for the approaches based on essential matrix estimation and on trifocal tensor estimation: in the first case, there is no overlap between pairs and thus we use an heuristic based on the distribution of the triangulated 3D points; in the second case, a scale ratio is implicitly defined between the first and second pair of images in the trifocal triplet and we use this ratio to propagate the scale between different triplets along the path.

The heuristic in the essential matrix estimation is based on the barycenter of the 3D points triangulated on two pairs of images. We select the set of points X^i belonging to the transitive closure of the matching point of the pairs (I_{k-2}, I_{k-1}) and (I_{k-1}, I_k). We compute the centers of gravity for the two sets of estimates; the scale factor is the ratio of the barycenters computed in the two reference systems:

$$s_k = \frac{\bar{X}^i_{k-1,k-2} - \mathbf{t_{k-1}}}{\bar{X}^i_{k,k-1}} \quad (9)$$

where $\bar{X}^i_{k,k-1}$ represents the barycenter of points triangulated from the images I_k and I_{k-1}, the superscript 'i' indicates that the barycenter is evaluated only among points belonging to both 3D reconstructions and $\mathbf{t_k}$ is the camera metric displacement from image I_{k-1} to image I_k

In the case of the trifocal tensor, the essential matrices are recovered from the epipoles (computed during estimation) and the tensor (accordingly to [10]):

$$E_{k-1,k-2} = [e'_k]_\times [T_1, T_2, T_3] e''_k$$
$$E_{k,k-2} = [e''_k]_\times \left[T_1^\top, T_2^\top, T_3^\top\right] e'_k \qquad (10)$$

The non-null singular values $\{\sigma_i\}$ of the essential matrices implicitly define the ratio between the translations in the pairs (I_{k-2}, I_{k-1}) and (I_{k-2}, I_k). We take the average of the two non-null singular values $\bar{\sigma}$, because even if the non-null singular values should be the same, noise can lead to slightly different values.

$$s_{k,k-2} = \frac{\bar{\sigma}_{k,k-2}}{\bar{\sigma}_{k-1,k-2}} \qquad (11)$$

4.2 Scale Propagation

Once computed, the scale factor allows to iteratively compute the camera position with respect to the first camera for all the subsequent images, by setting conventionally $s_1 = 1$.

For the two-views approach, accordingly to Eq. 9, we can write:

$$\hat{\mathbf{t}}_\mathbf{k} = \hat{\mathbf{t}}_\mathbf{k-1} + s_k \hat{R}_{k-1}^{-1} \tilde{\mathbf{t}}_\mathbf{k} \qquad (12)$$

where $\tilde{\mathbf{t}}_\mathbf{k}$ is the estimated translation unit vector while $\hat{t}_k$ represents the overall translation from the starting point to frame k and similarly the cumulative rotation is computed:

$$\hat{R}_k = R_k \hat{R}_{k-1} \qquad (13)$$

While for the three-views approach, accordingly to Eq. 11, we can write:

$$\mathbf{t_{k,k-1}} = \mathbf{t_{k,k,-2}} - \mathbf{t_{k-1,k-2}} = \mathsf{s_{k,k-2}} \tilde{\mathbf{t}}_{\mathbf{k-2,k}} - \mathbf{t_{k-1,k-2}} \qquad (14)$$

where $t_{b,a}$ indicates the translation vector from frame a to frame b and $\tilde{t}$ indicates a unit vector. Accordingly to Eq. 15 the global translation will be:

$$\hat{\mathbf{t}}_\mathbf{k} = \hat{\mathbf{t}}_\mathbf{k-1} + \hat{R}_{k-1}^{-1} \mathbf{t_k} \qquad (15)$$

and Eq. 13 will generate the cumulative rotation.

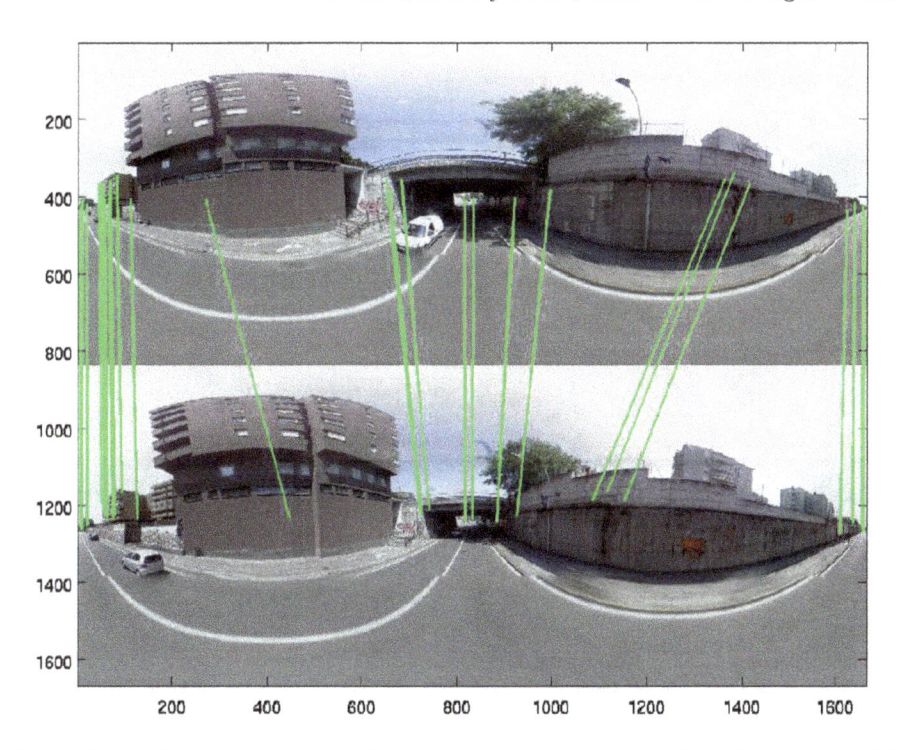

Fig. 5. Inliers between two pairs of images in the estimation of the trifocal tensor. All the outliers are removed.

5 Results

We discuss here the results for two notable cases: a straight road in "Via Sforza" near Lodi, Italy a road turn near "Piazza Castello" in Turin, Italy. The first configuration is usually challenging for Structure from Motion algorithms given the degeneracy of 3D points arranged on the direction of motion; the second configuration tests the robustness of the method, as the images in Street View are separated by long distances (up to 15 m), limiting the overlap and the number of correct matches.

We used RANSAC with an error threshold of 0.002, corresponding to roughly 2000 iterations.

Even if the SIFT or FREAK descriptors of the points are extracted from the equirectangular map of the sphere, which is a non-linear projection, the number of matches is still large enough for a good geometrical estimation (see Table 1), as the equirectangular projection is locally affine and the regions with high distortions are discarded from matching.

In the case of the trifocal tensor, the estimation results are good. By selecting only the correspondences found in three consecutive views a large number

Table 1. Number of interest points, matches and inliers for some frames of the "Via Sforza" and "Piazza Castello" sequences.

Frame	Via Sforza			P.za Castello		
	Pts.	Match	Inl.	Pts.	Match	Inl.
1	1800	-	-	1510	-	-
2	1721	-	-	1480	-	-
3	1707	85	21	1612	71	32
4	1681	96	31	1691	102	32
5	1740	112	35	1683	55	32
6	1597	100	33	1560	63	35

Fig. 6. "Via Sforza" path reconstrunction, a straight path is typically an ill-posed problem for classical cameras egomotion estimation, here the blue pins represent the ground truth from Google Street view GPS positioning system while the yellow path represents the two-views cumulative path reconstruction and the magenta line with red dots the three-views path reconstruction

of outliers is already discarded (see Fig. 5) and the robust estimation of the tensor is virtually outlier-free. In Fig. 6 we show the camera path reconstruction using both methods and the ground truth positions of the Google Street View acquisitions, as can be seen the three-view approach performs a little better with less the 2 m error in more that 180 m path with 17 reconstruction steps. In this case there are two high walls close to each street side dense with posters and graffiti that allow a good reconstruction accuracy. In Fig. 7 we face a much more wide environment where most of the matched points are far from cameras. In this case the three-view approach results in just 1 m error after 8 reconstruction steps while the two-views approach gives 2.5 m error. In particular, analyzing multiple Google Street View paths in different conditions we got an average error standard deviation of 4.5% for the translation estimation at every step using the two-views approach while, using the three-view approach the average error is 2.9%: These values have been obtained from 16 sets of Google Street View images where the distance between acquisitions is below 15 m.

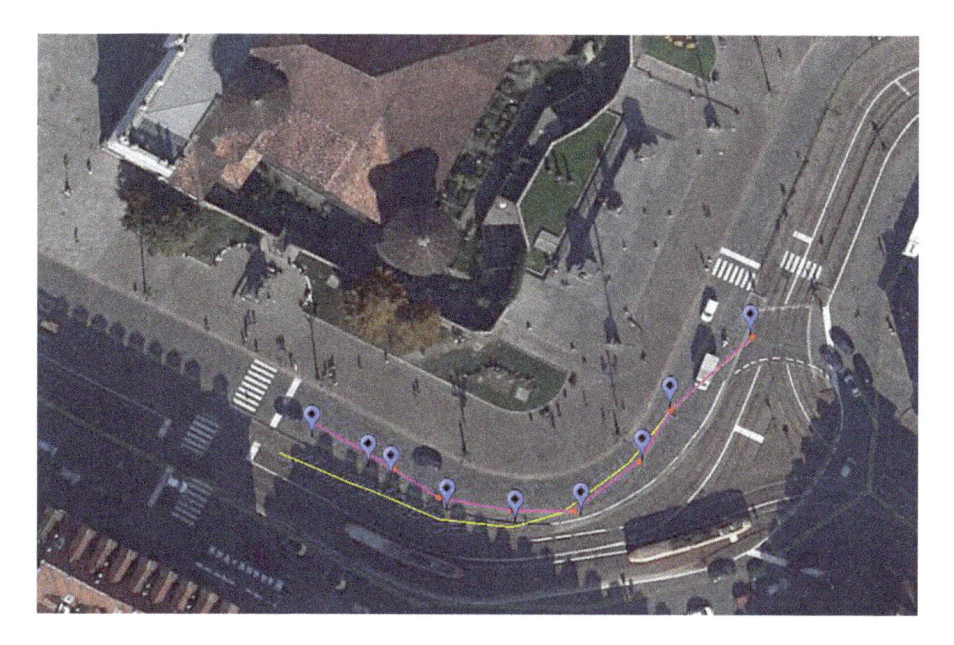

Fig. 7. "Piazza Castello" path reconstrunction, the Google car goes through the ground truth path represented by blue pins while the yellow path represents the two-views cumulative path reconstruction and the magenta line with red dots the three-views path reconstruction

6 Conclusions

In this paper we present a reformulation of epipolar and trifocal tensor constraint for Omnidirectional images. Some significant simplifications, e.g. the disappearance of the intrinsics matrix, reduce the degrees of freedom for such images and, at the same time, visual odometry can take advantage from this kind of images. In fact the opportunity to find points correspondences in all directions around the camera can significantly improve a typically ill-posed problem like visual odometry along straight paths. Tests have been performed using Google Maps^{TM} georeferenced spherical images and accurate results have been obtained both on straight and curved paths in restricted or wide environments.

References

1. Google Street View Image API
2. Garmin virb360 (2018). https://explore.garmin.com
3. GoPro fusion (2018). https://gopro.com
4. Insta360 evo (2018). https://www.insta360.com
5. Vuze 3D 360 video camera from humaneyes (2018). https://vuze.camera
6. Alahi, A., Ortiz, R., Vandergheynst, P.: Freak: fast retina keypoint. In: 2012 IEEE Conference on Computer Vision and Pattern Recognition (CVPR), pp. 510–517, June 2012

7. Aqel, M., Marhaban, M.H., Saripan, M.I., Ismail, N.: Review of visual odometry: types, approaches, challenges, and applications. SpringerPlus **5**, 12 (2016)
8. Fischler, M.A., Bolles, R.C.: Random sample consensus: a paradigm for model fitting with applications to image analysis and automated cartography. Commun. ACM **24**(6), 381–395 (1981)
9. Gledhill, D., Tian, G.Y., Taylor, D., Clarke, D.: Panoramic imaging–a review. Comput. Graph. **27**, 435–445 (2003)
10. Hartley, R., Zisserman, A.: Multiple View Geometry in Computer Vision, 2nd edn. Cambridge University Press, Cambridge (2004)
11. Lowe, D.G.: Distinctive image features from scale-invariant keypoints. Int. J. Comput. Vision **60**(2), 91–110 (2004)
12. Mei, C., Benhimane, S., Malis, E., Rives, P.: Efficient homography-based tracking and 3-D reconstruction for single-viewpoint sensors. IEEE Trans. Rob. **24**(6), 1352–1364 (2008)
13. Mei, C., Rives, P.: Single view point omnidirectional camera calibration from planar grids. In: Proceedings 2007 IEEE International Conference on Robotics and Automation, pp. 3945–3950, April 2007
14. Pretto, A., Menegatti, E., Pagello, E.: Omnidirectional dense large-scale mapping and navigation based on meaningful triangulation. In: 2011 IEEE International Conference on Robotics and Automation, pp. 3289–3296. IEEE, May 2011
15. Tardif, J.-P., Pavlidis, Y., Daniilidis, K.: Monocular visual odometry in urban environments using an omnidirectional camera. In: 2008 IEEE/RSJ International Conference on Intelligent Robots and Systems, pp. 2531–2538. IEEE, September 2008
16. Torii, A., Havlena, M., Pajdla, T.: From Google street view to 3D city models. In: 2009 IEEE 12th International Conference on Computer Vision Workshops ICCV Workshops, pp. 2188–2195 (2009)
17. Torii, A., Imiya, A., Ohnishi, N.: Two- and three- view geometry for spherical cameras. In: OMNIVIS (2005)
18. Valiente, D., Gil, A., Reinoso, S., Juliá, M., Holloway, M.: Improved omnidirectional odometry for a view-based mapping approach. Sensors **17**(2), 325 (2017)

Drivers and Barriers for Open Government Data Adoption: An Isomorphic Neo-Institutional Perspective

Henry N. Roa[1(✉)], Edison Loza-Aguirre[2], and Pamela Flores[2]

[1] Facultad de Ingeniería, Pontificia Universidad Católica del Ecuador,
Av. 12 de Octubre 1076 y Roca, P.O. Box 17-01-2184, Quito, Ecuador
`hnroa@puce.edu.ec`
[2] Facultad de Ingeniería en Sistemas, Escuela Politécnica Nacional,
Ladrón de Guevara, P.O. Box 17-01-2759, E11-253 Quito, Ecuador
`{edison.loza,pamela.flores}@epn.edu.ec`

Abstract. By making government data available to all, Open Government Data (OGD) initiatives promote transparency, accountability and value creation. However, these initiatives face several problems affecting their implementation throughout its adoption process. This study focuses on the forces driving or hindering the adoption of OGD in a developing country at its early stages. In depth, through the analysis of qualitative data, which was collected from seven governmental institutions, we highlight that OGD adoption is mainly hindered by the lack of a comprehensive legal framework. On the other hand, the adoption of OGD is mainly driven by the participation of the governmental practitioners on professional networks and by the transmission of success stories from other countries with similar characteristics to the country studied in this research.

Keywords: Open government data · Neo-Institutional theory · Isomorphic pressures · Adoption · Information systems

1 Motivations

Open data refers to data made freely available by any organization to be used by anyone, without copyright restrictions [1]. To promote transparency, accountability and value creation, governments release their data to be used by citizens [2] with the hope that the latter could create some value from it. The government data released in this fashion is then referred as Open Government Data (OGD).

Moreover, OGD refers not only to Information Technologies but also to a political philosophy, which promulgates several values as follows. First, the access to government information as a pillar for democracy [3]. Second, the engagement of citizens to participate in projects that involve government data [4]. Finally, the belief that data correctly released would increase transparency and improve decision making [5].

Although OGD offers, in principle, several advantages to society, in practice, OGD adoption faces many problems that hinder the implementation of these initiatives. Previous research shows that data quality and legal issues are, by far, the main barriers for initiatives of OGD [6–8]. Nonetheless, the influence and intensity of these barriers are dependent upon the stage of the adoption process of OGD (i.e. early stages versus appropriation stages), and it is also linked to a social setting where OGD is implemented [9]. Furthermore, the identification of the guiding barriers or drivers refers to identify the forces influencing the occurrence of such drivers or barriers. However, studies focused on the influencing forces leading or hindering the OGD adoption are scarce.

This study explores what the forces driving or hindering the adoption of OGD are. We focused at the early stages of the adoption process because the problems happening during the implementation and usage (adoption and post-adoption) of OGD has been already studied in literature (e.g. [10–13]). Contrarily, studies concerning pre-adoption of OGD (the construction of a favourable attitude toward the implementation), are very rare or even non-existent. Since pre-adoption is subject to an institutional influence of external and internal stakeholders [9, 14], this study is based on the neo-institutional theory as the theoretical lenses for studying the forces that lead or block the adoption of OGD. We studied this phenomenon through an in-depth analysis of the discourse of several governmental agencies in a developing country, which is driving its efforts for implementing OGD.

This paper is organized as follows. First, we introduce concepts of drivers and barriers for Information Systems (IS) adoption and the influence on the adoption process of isomorphic pressures (through the lenses of the neo-institutional theory). Then, we present our methods for collecting and analysing data. In Sect. 4, we offer our results. Afterward, we discuss our results and present our conclusions and future research opportunities. We must mention that this research is part of a large research process focused on studying the OGD adoption.

2 Motivations

In this section, we briefly introduce the concepts used to understand the pressure factors driving or hindering the adoption of OGD. We start presenting the adoption process of an IS in order to limit the scope of the research. Then, we present related contributions dealing with drivers and barriers for the adoption of such systems. Finally, we introduce the neo-institutional isomorphic pressures as the theoretical lenses for studying the forces that lead or block the OGD adoption.

2.1 The Adoption Process of Information Systems

The adoption process refers to the stages tailed when a new information technology or system is introduced into a social system in order to support operations, management, and decision-making [15]. Even when this adoption can be studied from different angles, IS adoption research is mostly based on the contributions from the Diffusion of Innovations Theory [16]. According to this theory, the IS adoption process is divided in three stages [17]:

a. The *pre-adoption stage* that refers to the realization and recognition of a need, the information gathering of possible systems or technologies for responding to that need, and the development of a favourable or unfavourable attitude toward these solutions;
b. The *adoption stage* that covers decision making of individuals concerning the adoption of a system or technology, and its implementation;
c. The *post-adoption stage* that is a phase of routinization and assimilation of a system by the members of the organization.

Since contributions dealing with OGD adoption have focused mainly on the adoption and post-adoption stages, in this study, we are interested in the pre-adoption stage of the IS adoption process. Even when authors have not found a consensus concerning the activities involved in this stage (Table 1), it is commonly acknowledged that pre-adoption stage covers the awareness of the introduction of a new IS, the efforts from individuals to learn about this IS, and the development of a favourable or unfavourable attitude toward it [9].

Table 1. Pre-adoption activities as reported in literature

Author	Pre-adoption		
Klonglan and Coward [18]	Awareness. An individual might become aware of the introduction of an innovation	Information. Individual may actively seek out information about the innovation	Evaluation. Individual may decide that the innovation is suited to his/her needs
Rogers [16]	Knowledge. Exposure to the innovation and an understanding of how it functions		Persuasion. An attitude is formed toward the innovation
Kwon and Zmud [19]; Cooper and Zmud [20]	Initiation. Companies justify the need for adopting Information Technology (IT). They perform an active and/or passive scanning of organizational problems/opportunities and IT solutions are undertaken. Finally, a match is found between an IT solution and its application in the organization		
Rai, Brown, and Tang [21]	Awareness. Key decision makers are aware of a new IT	Interest. The firm is committed to actively learn more about the IT	
Swanson and Ramiller [22]	Comprehension. Through the efforts of its members, the firm learns more about an IT innovation and develops an attitude or stance toward it and positions itself, in a basic way, as a prospective adopter or non-adopter		
Zhu, Kraemer and Xu [23]	Initiation. Evaluating the potential benefits of IT to improve a firm's performance in value chain activities such as cost reduction, market expansion, and supply chain coordination		
Hameed et al. [17]	Initiation. It consists of activities related to recognizing a need, acquiring knowledge or awareness, forming an attitude towards the innovation and proposing innovation for adoption		

The construction of a favourable attitude or an unfavourable one, responds to the influence of several pressures acting as drivers or barriers as it will be discussed in the next section.

2.2 Drivers and Barriers for Pre-adopting an Information System

To define a driver and a barrier, in this article, we adopted the definitions of Lesca et al. [9]:

a. A driver is an internal or external pressure which influences the evaluation of a new IS as a solution to the needs of the organization and promotes its pre-adoption.
b. A barrier is an internal or external pressure which influences the evaluation of a new IS and slows or hinders its pre-adoption.

In literature, there exists several contributions studying the drivers and the barriers that an IS faces while moving on each of the adoption stages. However, we were unable to find articles focusing on the drivers and barriers for OGD adoption.

Concerning drivers for adopting IS, most of the ones found in the literature correspond to the implementation stages, leaving aside or less visited, the pre-adoption stage and its influence on the final success of the systems. The results also differ when we focus on a specific stage of the adoption process. For instance, while the pursuit of efficiency or the need to improve inter-organizational relationships have been reported as drivers for the adoption phase of an IS [24], the search for competitiveness or even legitimacy of an organization in a field are examples of specific drivers reported for the stage of pre-adoption of an IS [9].

Regarding barriers, the results also differ when we look at the stage of the adoption process that is the subject of this study. For instance, Ngwenyama and Nielsen [24] cited the following barriers for the IS adoption phase: the lack of formal influence over the organization implementing the system, lack of an organizational memory where past errors linked with IS implementation could be stored, or even the lack of high management support. Also, the misalignment with strategic priorities, difficulties for defining expectation and objectives, lack of resources and skills and the abscess of regulatory incentives are examples of reported barriers hindering IS specifically in the pre-adoption stage [9].

Besides the differences between drivers and barriers according to the phase of the adoption process, and as shown in the examples above, the pre-adoption stage is subject to the influence of external and internal stakeholders [14]. Thus, we base on the neo-institutional theory as the theoretical lenses for studying the forces that lead or block the pre-adoption of OGD as presented in the next section.

2.3 A Neo-Institutional Approach for Studying Drivers and Barriers to Pre-adopting IS

The pre-adoption stage in IS has been studied from the view of the Symbolic adoption model [14, 25] or even the Diffusion of Innovations theory [26, 27]. However, both theories relate more to the development of an attitude toward a technology than to pressures driving this attitude. Alternatively, the neo-institutional theoretical framework [28], and more specifically the theory related to the institutional isomorphism [29], show that the IS pre-adoption results does not merely respond to rational needs, but it can also be explained by the presence of institutional pressures [9]. This situation can explain why organizations that share the same activities tend toward homogenization or standardization in their operations, processes, behaviours or even culture [30].

Through the lenses of the neo-institutional theory, the homogenization of organizations on the same domain of activity is due to the presence of isomorphic pressures. Indeed, there are three types of isomorphic pressures [30]:

a. Mimetic pressures. It results from the fact that, when faced with uncertainty and bounded rationality, organizations tend to imitate one another. The imitation can be unintentional, resulting from the transfer of employees or the intervention of consultants or intentional as a source of legitimization [31].
b. Normative pressures. While in a single organization the jobs are different from each other, they are very similar to the jobs of counterparts in other organizations. This is due to the standardization of education programs and the development of professional networks, which leads to the homogenization of professions [32].
c. Coercive pressures. They are due to the formal or informal political influence exercised by the government, the society or any influential player. Organizational structures exposed to these pressures tend to reflect the rules and the dominant standards of such influencers [33].

The three types of isomorphic pressures, when they are present, influence positively (drivers) or negatively (barriers) in the evaluation of an IS as a solution and, consequently, in the construction of a favourable or unfavourable attitude toward its adoption. Thus, we will use this theoretical framework to study the drivers and barriers for the pre-adoption of OGD in a developing country. To do this, we will follow a qualitative approach to analyse the discourse of representatives of several governmental agencies.

3 Research Method

The aim of this study is to explore which are the pressures driving or hindering the adoption of OGD. To meet our research objectives, we conducted a survey research following the approach described below:

a. Data collection. We conducted semi-structured interviews to eight Chief Information Officers (CIO) or Information Technology managers within seven governmental agencies in a developing country. One of the agencies acts as the promoter of the OGD adoption in the country. All the interviews were fully recorded and then transcribed. These interviews were conducted until a saturation point was reached [34].

b. Coding. To identify the pressures driving or hindering OGD adoption, we coded interviews based on the three isomorphic pressures presented in Sect. 2.3 (mimetic, normative and coercive). Flexible meaning units (e.g. phrases and paragraphs) were used because we were interested in statements reflecting a complete idea [35].
c. Data analysis. After coding, we performed a thematic analysis of coded items to allow the emergence of topics inside each category [36]. Afterwards, the topics emerged were classified according to its influence toward the pre-adoption of OGD as follows: positive influencers were classified as drivers and negative influencers were classified as barriers.

4 Results

First, Table 2 presents the results of the coding process by organization and by the type of isomorphic pressure. According to Table 2, it is possible to notice that the promoter organization was the one that mentioned more drivers and barriers than the others. The totals of coding by isomorphic pressure are presented in Fig. 1. According to Fig. 1, coercive pressures were, by far (35 out 62), the dominant pressure identified from interviews. Second, Table 3 shows the results of the thematic analysis. We identified seven topics that grouped our coded items. After the analysis of each topic and its influence for pre-adopting OGD, we classified each one as driver or barrier. Fortunately, no ambiguity was identified during the analysis of the influence of each topic, which ensured a right classification. Finally, Fig. 2 presents the totals of drivers and barrier items identified. According to Fig. 2, the total number of barriers is higher than the total number of drivers (39 out 62 coded items).

Table 2. Coding results by governmental institution

Organization	Code name	Coercive	Mimetic	Normative	Total
Promoter institution	PROM	11	4	3	18
Governmental institution 1	INST1	2		1	3
Governmental institution 2	INST2	3	5	5	13
Governmental institution 3	INST3	7	1	2	10
Governmental institution 4	INST4	4	1		5
Governmental institution 5	INST5	5		1	6
Governmental institution 6	INST6	3	2	2	7

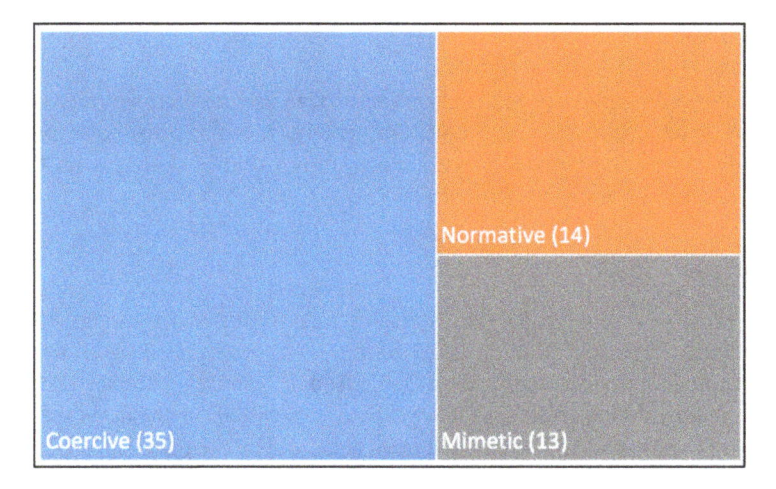

Fig. 1. Coding totals by isomorphic pressure

Table 3. Resulting topics from thematic analysis and barrier/driver classification

Category		Topic		Barrier/Driver
Coercive	35	Lack of a legal framework	23	Barrier
		Lack of a government mandate	6	Barrier
		Failure for identifying society data needs	6	Barrier
Normative	14	Participation in professional networks	9	Driver
		Common professionalization carriers	5	Driver
Mimetic	13	Experiences from other countries	9	Driver
		Experiences from other agencies	4	Barrier

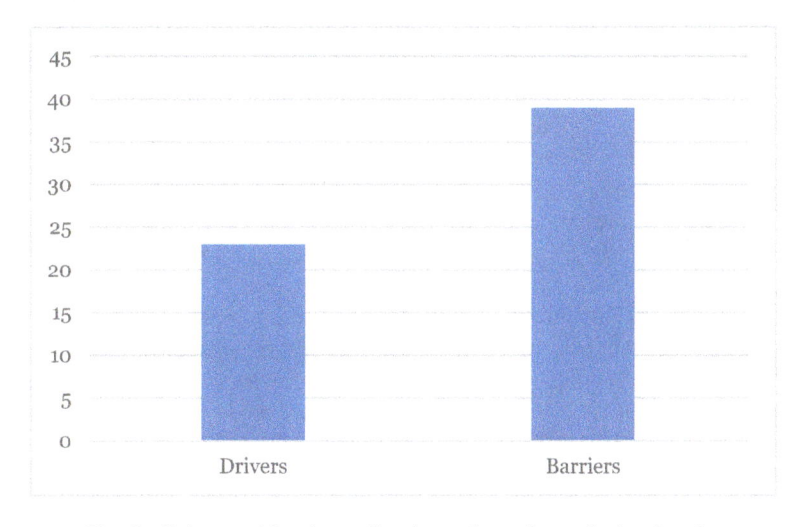

Fig. 2. Driver and barrier coding items from the analysis of topics

5 Discussion

Based on the coding process and analyzed results, in this section, we offer an in-deep discussion of our main observations. Unfortunately, it is not possible to discuss our results with other contributions because there are not studies dealing with OGD pre-adoption as it was mentioned before.

5.1 Barriers

Coercive pressures were, by far, the pressures most mentioned by interviewees. The lack of a legal framework guiding OGD adoption was recognized by all the interviewees as the main barrier for implanting OGD. Both the promoter agency and the other governmental institutions wait for the legal base that will allow them, on the one side, to legitimize their initiatives concerning open data, and by the other side, to ensure that their initiatives are aligned with both the governmental strategy and parameters of implementation (i.e. data formats and privacy considerations). Accordingly, the government promoter of OGD implementation declared: "*We analysed the problematic of open data. We concluded that we need a strong instrument to force its implementation, but at the same time, to allow value creation*" [PROM]. An agency CIO said: "*Talking about the government, the agenda and the guidelines are still in work. As a governmental institution, we need these guidelines because we want to avoid wasting our efforts*" [INST3].

Also, the lack of a governmental mandate (a written policy), about OGD hinder its development. "*Much will therefore depend on the government expectancy [...] We need that the government gives us a strong impetus. We cannot forget that in the public sector, we need that everything should be written*" [INST3]. Since the policy is currently being developed, the governmental institutions are still blocked concerning OGD implementation.

Failure for identifying society data needs was also cited as a relevant barrier. The governmental institutions declared that they aim respond to the information needs rose from the society. However, these needs are not clear for the institutions nowadays. Nonetheless, agencies have plans to involve the public on the definition of their needs in a regular basis. As their interest lies on value creation, governmental institutions do not want to waste time and resources on opening all their data, but on making available the data that could create a value. The initiative of involving the public looks like an interesting cost-effective approach for implementing OGD. It should be followed to capture learning about it.

The other barrier identified refers to the lack of success stories from other agencies in the same country. This mimetic isomorphic pressure is present as a negative influence. Indeed, it is the lack of success stories from other agencies or the lack of promotion of their OGD initiatives that hinder new projects in the field. However, it is not surprising that particular initiatives are isolated and not broadcast since there is not a government policy driving OGD adoption.

5.2 Drivers

Participation in professional networks and success experiences for other countries are the most representative drivers cited by interviewees. Concerning networks, several groups have been conformed in recent years to motivate the use of open data: *"We found several groups already working on open data topics without being a formal effort. In the last two years we established a multidisciplinary group with members coming from different venues"* [INST5]. It is inside the networks were most of the knowledge concerning OGD is shared. The government adopted an approach seeking to involve these groups in a co-creation process for defining OGD policies and efforts. This co-creation process is another interesting process to be followed.

The success stories of other countries are other drivers inspiring OGD adoption. Most of the institutions have followed the reports and articles highlighting the contributions, but also the difficulties on OGD implementations. An interesting point is that the chosen references are developing countries such as the one we are studying. Hence, all the actors cited success stories on countries of the region instead of those of developed countries.

Finally, professionalization was identified as both, a source of institutional pressure and a way for diffuse OGD use in the society. As a normative pressure, local universities include already in their curriculum several courses were open data is promoted. The novelty lies on the intentions of the government concerning the involvement of secondary schools for promoting the use of open data: *"We hope that middle education becomes an actor for the use of open data. Now the tendency is that students go to the internet, search information, but they do not analyze it, they are not researching. So, we want to change actual research paradigms at middle level. We want that the approach for information searches should be based in official public sources"* [PROM].

6 Conclusions

Even when OGD offers several advantages to society, in practice, its adoption faces many problems that hinder these initiatives. In this research, we explored which are the pressures driving or hindering the adoption of OGD. We studied such pressures under the light of the isomorphic forces declared on the discourse of several governmental agencies in a developing country.

On the one hand, our results highlight that OGD adoption is mainly hindered by the lack of a comprehensive legal framework. Other coercive pressures such as the absence of a clear government mandate or the difficulties to understand the information needs of populations have been cited during our study. Thus, government agencies declared that the existence of a legal framework is critical for start OGD initiatives.

On the other hand, OGD is driven by the participation in professional networks and by the broadcast of success stories from other countries with similar characteristics that the country studied in this research. A mimetic approach then has been identified. This result suggests that faced to lack of official guidelines, agencies turn to professional networks or the example of others, in the quest of legitimate their initiatives. Due to the presence of these mimetic behavior, a question about the development of the organizing vision about OGD emerges as a recommendation for future surveys.

Another two opportunities for future research have been identified in the context of OGD introduction in developing countries. The first one relates to the initiative of identifying the population information needs before starting to make available any data. The second one concerns the co-creation process for establishing guidelines for OGD initiatives. Both topics would contribute to understand the contributions of participatory policy development on the OGD arena.

Acknowledgment. This research was supported by the Pontificia Universidad Católica del Ecuador through the grant research project O13050. In addition, we acknowledge and thank the Anonymous Reviewers for their valuable recommendations, which contributed to improving the quality of this paper.

References

1. Sadiq, S., Indulska, M.: Open data: quality over quantity. Int. J. Inf. Manage. **37**(3), 150–154 (2017)
2. Bertot, J., Gorham, U., Jaeger, P., Sarin, L., Choi, H.: Big data, open government and e-government: Issues, policies and recommendations. Inf. Polity **19**(1–2), 5–16 (2014)
3. Allen, K.B.: Access to government information. Gov. Inf. Q. **9**(1), 67–80 (1992)
4. Kassen, M.: A promising phenomenon of open data: a case study of the Chicago open data project. Gov. Inf. Q. **30**(4), 508–513 (2013)
5. Dawes, S.S.: Stewardship and usefulness: policy principles for information-based transparency. Gov. Inf. Q. **27**(4), 377–383 (2010)
6. Roa, H.N., Loza-Aguirre, E., Flores, P.: A survey on the problems affecting the development of open government data initiatives. In: Proceedings of the Sixth International Conference on eDemocracy & eGovernment (ICEDEG), Quito, Ecuador (2019)
7. Conradie, P., Choenni, S.: On the barriers for local government releasing open data. Gov. Inf. Q. **31**(S. 1), S10–S17 (2014)
8. Janssen, M., Charalabidis, Y., Zuiderwijk, A.: Benefits, adoption barriers and myths of open data and open government. Inf. Syst. Manage. **29**(4), 258–268 (2012)
9. Lesca, N., Caron-Fasan, M.-L., Loza-Aguire, E., Chalus-Sauvannet, M.C.: Drivers and barriers to pre-adoption of strategic scanning information systems in the context of sustainable supply chain. Systèmes d'Information et Manage. **20**(3), 9–46 (2015)
10. Krishnamurthy, R., Awazu, Y.: Liberating data for public value: the case of data. Gov. Int. J. Inf. Manage. **36**(4), 668–672 (2016)
11. Styrin, E., Luna-Reyes, L.F., Harrison, T.M.: Open data ecosystems: an international comparison. Transform. Gov. People, Process Policy **11**(1), 132–156 (2017)
12. Nam, T.: Challenges and concerns of open government: a case of government 3.0 in Korea. Soc. Sci. Comput. Rev. **33**(5), 556–570 (2015)
13. Yang, T.-M., Lo, J., Shiang, J.: To open or not to open? Determinants of open government data. J. Inf. Sci. **41**(5), 596–612 (2015)
14. Verra, L.G., Karoui, M., Dudezert, A.: Adoption symbolique d'un Réseau Social pour entreprise: Le cas de Bouygues Construction. In: 17ème colloque de l'AIM, Bordeaux, France (2012)

15. Baskerville, R., Pries-Heje, J.: A multiple theory analysis of a diffusion of information technology case. Inf. Syst. J. **11**(3), 181–212 (2001)
16. Rogers, E.M.: Diffusion of Innovations. The Free Press, New York (1983)
17. Hameed, M.A., Counsell, S., Swift, S.: A conceptual model for the process of IT innovation adoption in organisations. J. Eng. Technol. Manage. **29**(3), 358–390 (2012)
18. Klonglan, G.E., Coward Jr., E.W.: The concept of symbolic adoption: a suggested interpretation. Rural Sociol. **35**(1), 77–83 (1970)
19. Kwon, T.H., Zmud, R.W.: Unifying the Fragmented models of information systems implementation. In: Boland, R.J., Hirschheim, R.A. (eds.) Critical Issues in Information Systems Research, pp. 227–251. Wiley, Chichester (1987)
20. Cooper, R.B., Zmud, R.W.: Information technology implementation research: a technological diffusion approach. Manage. Sci. **36**(2), 123–139 (1990)
21. Rai, A., Brown, P., Tang, X.: Organisational assimilation of electronic procurement innovations. J. Manage. Inf. Syst. **26**(1), 257–296 (2009)
22. Swanson, E., Ramiller, N.C.: Innovating mindfully with information technology. MIS Q. **28**(4), 553–583 (2004)
23. Zhu, K., Kraemer, K.L., Xu, S.: The process of innovation assimilation by firms in different countries: a technology diffusion perspective on e-business. Manage. Sci. **52**(10), 1557–1576 (2006)
24. Ngwenyama, O., Nielsen, P.A.: Using organizational influence processes to overcome IS implementation barriers: lessons from a longitudinal case study of SPI implementation. Eur. J. Inf. Syst. **23**(2), 205–222 (2014)
25. Fui-Hoon Nah, F., Tan, X., The, S.H.: An empirical investigation of end-users' acceptance of enterprise systems. Inf. Resour. Manage, J. **17**(3), 32–53 (2004)
26. Grover, V., Goslar, M.D.: The initiation, adoption, and implementation of telecommunications technologies in U.S. organisations. J. Manage. Inf. Syst. **10**(1), 141–163 (1983)
27. Karahanna, E., Straub, D.W., Chervany, N.L.: Information technology adoption across time: a cross-sectional comparison of pre-adoption and post-adoption beliefs. MIS Q. **23**(2), 183–213 (1999)
28. Hofer, A.R., Hofer, C., Eroglu, C., Waller, M.A.: An institutional theoretic perspective on forces driving adoption of lean production globally: China vis-à-vis the USA. Int. J. Logist. Manage. **22**(2), 148–178 (2011)
29. Di Maggio, P.J., Powell, W.W.: The iron cage revisited: institutional isomorphism and collective rationality in organizational fields. Am. Sociol. Rev. **48**(2), 147–160 (1983)
30. Abdennadher, S., Cheffi, W.: L'adoption du vote par internet aux assemblées générales des actionnaires de sociétés cotées en France: une perspective institutionnaliste. Systèmes d'Information et Manage. **16**(2), 35–71 (2011)
31. Mizruchi, M.S., Fein, L.C.: The social construction of organizational knowledge: a study of the uses of coercive, mimetic, and normative isomorphism. Adm. Sci. Q. **44**(4), 653–683 (1999)
32. Scott, W.R.: Institutions and Organizations: Foundations for Organizational Science. Sage Publications, California (1995)
33. Slack, T., Hinings, B.: Institutional Pressures and Isomorphic Change: An Empirical test. Organ. Stud. **15**(6), 803–827 (1994)
34. Fusch, P.I., Ness, L.R.: Are we there yet? data saturation in qualitative research. Qual. Rep. **20**(9), 1408–1416 (2015)
35. Saldaña, J.: The Coding Manual for Qualitative Researchers. Sage Publications, London (2009)
36. Bardin, L.: L'analyse de contenu. Presses Universitaires de France, Paris (2007)

"Seeking Privacy Makes Me Feel Bad?": An Exploratory Study Examining Emotional Impact on Use of Privacy-Enhancing Features

Hsiao-Ying Huang[✉] and Masooda Bashir

University of Illinois at Urbana-Champaign, Champaign, IL 61802, USA
hhsiaoying@gmail.com

Abstract. With the increasing prevalence of privacy invasions and data breaches, more and more users have been seeking protection for their online privacy, which makes privacy-enhancing features and technologies more important than ever. However, these features and technologies are still not widely adopted by users. In the privacy literature, there seems to be an assumption that online users' privacy behaviors are based on rational decision-making. However, previous research has shown that users' decision-making involves not only rational thoughts but also emotions, which play an important role. To explore that unknown territory, this empirical study focuses on human emotions and examines their impact on users' adoption of privacy-enhancing features. Our study design is based on two theoretical frameworks: feelings-as-information theory and the Technology Acceptance Model. We used private browsing as a case study and conducted an online survey experiment to investigate what types of emotions are elicited in users by private browsing mode and how these emotions affect their acceptance of private browsing. Interestingly, we found that the interface design of private browsing mode provokes both positive and negative emotions in users. Also, these elicited emotions influence users' behavioral intentions. Based on these results, we propose design recommendations for privacy-enhancing features.

Keywords: Usable privacy · Privacy-preserving technology · Interface design · Human-information interaction

1 Introduction

While advances in Internet technologies have positively influenced how people communicate, connect, and share their daily lives with others around the world, ubiquitous and constant data collection and analyses also pose potential threats to users' online privacy. With rapid tracking technologies, privacy violations are no longer just involved with targeted advertising but can even imperil our political democracy. For instance, Cambridge Analytics harvested 50 million Facebook users' psychological profiles and used them for advertising for political campaigns [11, 26]. The increasing reports regarding this type of data breach and hacking have raised public awareness about online privacy and have motivate users to seek protection for personal privacy.

Both academia and industry have responded with a variety of efforts to develop technologies and strategies to preserve users' privacy [12, 33, 38]. While many technologies and policies have been proposed and established, we still have limited understanding of the inconsistencies that the online users' exhibit in their attitudes and behaviors when it comes to enhancing their privacy online. Privacy literature has termed this phenomenon the "privacy paradox" [7].

The privacy paradox is the phenomenon in which an inconsistency lies between people's attitudes and behaviors regarding online privacy. People often express concerns about their online privacy yet show little regard to privacy in their daily online behaviors [1]. That is to say, people are aware of the potential privacy risks but still make 'irrational' choices. Previous research has examined people's irrational privacy behaviors from a rational calculus perspective [49]. However, as indicated by Acquisti et al. [1], privacy-related decision-making is not only affected by people's rational thoughts but also their bounded rationality, social norms, heuristics, and emotions. Studies have also evidenced that emotion plays a fundamental role in our decision-making and behaviors [4, 47]. Therefore, we postulate that an important, missing piece in the online privacy literature is the role of emotion.

Human emotion is a challenging concept that is difficult to quantify and often debated within the disciplines of psychology, philosophy, and neuropsychiatry [40]. In this study, we view emotion as a subjective and conscious experience with various degrees of mental intensity [10, 19]. In addition, previous research has indicated that emotional responses have informational value to people's decisions [37]. Furthermore, from a Human-Computer Interaction (HCI) perspective, the design of technology can affect users' emotions and further influence their adoption of technology [34]. Following this rationale, we argue that emotion plays a mediator role between users' attitudes and behaviors when it comes to the use of privacy-enhancing features.

Several privacy-enhancing features and technologies have been designed and developed in the last decade. These privacy-enhancing features range from interface design [e.g., 13, 28] to system development (e.g., The Onion Router by [17]; Privacy in Online Social Network by [21]). However, these features and technologies have not achieved mainstream use [25]. If we were to examine users' adoption of privacy-enhancing features solely from a rational point of view, we may conclude that users do not adopt or constantly use privacy-enhancing features because they perceive these features to be useless or too difficult to use [14]. Nevertheless, as prior studies have suggested [6, 36], user behavior not only involves cognition but also our emotional state. Perhaps users do not use privacy-enhancing features simply because these features make them feel uneasy and strange. Although the role of users' emotion as it relates to online privacy has begun to be noticed in academia [40], it has not been the main focus of any empirical study in the adoption of privacy-enhancing features. Hence, the aim of this study is to investigate how users' emotional responses influence their acceptance of privacy-enhancing features.

To study this topic, we propose hypotheses motivated by the theoretical framework of feelings-as-information theory and the Technology Acceptance Model. We used private browsing as a case study and designed a survey experiment to investigate how private browsing elicits users' emotions and how these emotions affect their acceptance of private browsing.

2 Related Work

Usable privacy is a field of study that focuses on improving privacy technologies from a Human-Computer Interaction (HCI) perspective. Thus, it is important to examine users' emotions from an HCI point of view. Users' emotions have become an important consideration in the field of HCI because an increasing number of studies demonstrate its impact on users' experiences with technologies [9, 34]. Previous studies in HCI have focused on several emotion-related topics, including measuring emotional dimensions in user experience [2, 29], emotional effects on the use of technologies [8], designing affective user interface [27], positive emotion-driven design [23] and the emotional deduction from the artifact (e.g., mouse cursor) [24]. Although the importance of emotion has been recognized in the HCI literature, its role in the context of privacy has been under-studied [40]. In the following sections, we will review the most relevant work on how emotion is associated with interface design, users' technology acceptance, and privacy from an HCI perspective.

2.1 Emotion and Interface Design

An interface, as the external appearance of a system, can influence how we feel, perceive, and interact with a system [15]. Previous studies have shown that the visual design elements of an interface, such as the shape, color, and texture, can affect users' emotions [9, 15, 29]. For instance, a study by Kim et al. [29] found that the mixed use of shapes and colors in a menu bar can create a mystic feeling; yet the mixed use of shapes in a background can make users feel confused. In addition, users' emotional responses to design are usually immediate and consistent across time [42]. While the emotional effect on users' perceptions and behaviors depends on the task and the context [46], emotion elicited by the design of a system remains a fundamental and influential factor on users' decision-making in technology acceptance.

2.2 Emotion and Technology Acceptance

Users' emotional states can directly or indirectly influence their attitudes, decision-making, and behaviors [6, 36]. When it comes to the use of technology, research has shown that positive emotions elicited by system design have positive effects on users' acceptance and use of technology [45]. Conversely, negative emotions elicited by system design result in negative effects on users' perceived ease-of-use [45] and technology adoption [8]. Furthermore, users who experience negative emotions will be more likely to spread negative word of mouth [24], which may further decrease users' acceptance of technology.

2.3 Emotion in Privacy Context

The role of users' emotions in the context of privacy can be discussed from two perspectives: (1) privacy as an emotional state [51], and (2) users' emotions as an antecedent to privacy attitudes and behaviors [30].

Privacy as Emotion

Psychologists have argued that privacy is an emotional state that provides psychological functions for humans [35, 51], since a person can 'feel' when his or her privacy is invaded without a rational reason. In privacy literature, creepiness is one particular emotional reaction that is often expressed in response to tracking technologies. Creepiness has been described as a combination of fear, uneasiness, strangeness, and disturbance [41, 43, 52]. An interview study by Ur et al. [43] found that users report creepy feelings about online behavioral tracking technologies even though they consider these technologies to be intelligent. Furthermore, privacy-enhancing features may elicit feelings of creepiness, which further can trigger users' privacy concerns. For instance, Zhang et al. [52] conducted an online experiment to investigate the mediating effects of creepiness between privacy cues (e.g., privacy nudge) on a privacy permission interface and users' privacy attitudes in the setting of a social mobile application. They found that privacy cues may increase users' alertness and elicit creepiness emotions. The creepy feeling further makes users more concerned about their information privacy, perceive less control, and feel less comfortable disclosing personal information. In other words, privacy-enhancing features can trigger negative emotions that may discourage the use of technology.

Emotion as a Privacy Antecedent

Prior research has found that users' emotions can be a significant predictor for privacy attitudes and behaviors. For example, Li et al. [30] found that users who rate higher in joyful emotion believe that their privacy is more protected and perceive fewer privacy risks. On the other hand, users who feel more fears perceive more privacy risks. Another study by Anderson and Agarwal [3] found that users who have more negative feelings (e.g., sadness, anger, and fear) toward their health condition are more willing to disclose their personal health information. However, our understandings of how users' emotional states influence privacy attitudes and behaviors is still limited.

Based on previous work, we identified three research gaps in HCI and privacy research, and proposed research questions to address each gap:

1. Does privacy-enhancing feature elicit different types of emotions in online users?: Although prior studies [43, 52] have pointed out that privacy-enhancing features can provoke an emotion of creepiness, we still have very limited knowledge about how users feel about privacy-enhancing features. Do privacy-enhancing features only elicit users' negative emotions, or do they also evoke positive emotions?
2. Does the design of privacy-enhancing features affect users' emotions, attitudes, and acceptance?: The literature in HCI has indicated that the design of a system can influence users' emotions [9, 15, 29]. Yet, very few studies focus on how the design of privacy-enhancing features affect users' emotions.
3. Does users' emotional state affect their acceptance of privacy-enhancing features?: Research [52] has pointed out that users' emotions can affect their privacy attitudes and use of technology. However, how users' emotions influence their attitudes and acceptance of privacy-enhancing features remains unclear.

Our research questions are as follows:

- RQ1: What types of users' emotions are elicited by the interface design of a privacy-enhancing feature?
- RQ2: How does the interface design of a privacy-enhancing feature affect users' emotions, attitudes, and acceptance of those features?
- RQ3: How do users' emotions elicited by a privacy-enhancing feature affect their attitudes and adoption of those features?

3 A Case Study of Privacy-Enhancing Feature: Private Browsing Mode

An abundance of privacy-enhancing features and functions have been developed as technologies become smarter and more personalized. These privacy-enhancing features include a wider range of applications from interface design to system development. One of the popular privacy-enhancing features is private browsing mode. As far as we know, only four studies have examined from users' perspective [20, 22, 39, 50]. All previous studies focused on examining cognitive influence on users' use of private browsing mode. No empirical research investigates the impact of interface design on users' emotions and their adoption of private browsing mode, which will be our focus in this study.

Private browsing mode has been offered as a built-in function in most of today's leading browsers, including Google's Chrome, Mozilla's Firefox, Apple's Safari and Microsoft Edge/IE. The goal of private browsing mode is to prevent one's browsing history and personal information from being accessed by either local adversaries or tracking sites [31]. Although most browsers provide similar features in private browsing mode, the design and functionality still vary among browsers. As shown in Fig. 1, the interface designs for private browsing mode in these four browsers are different from each other. We think that private browsing mode is an appropriate case study for this study for two reasons. First, it is easy to access by users; and second, it encompasses similar functions across browsers but varies in interface design, which is useful for comparing how the design of privacy-enhancing features can elicit users' emotions that affect their attitudes and acceptance. We thus select private browsing mode as a research subject in this study. In the next section, we address our hypotheses based on a proposed theoretical framework.

3.1 Theoretical Framework and Hypotheses

Human emotion is a conscious experience that involves mental activity at different levels of intensity and arousal [10, 19]. According to social psychology, human emotion is a type of information affecting people's attitudes and behaviors. For instance, according to feelings-as-information theory [37], people use their emotions as information to guide their evaluations when they determine that their emotional

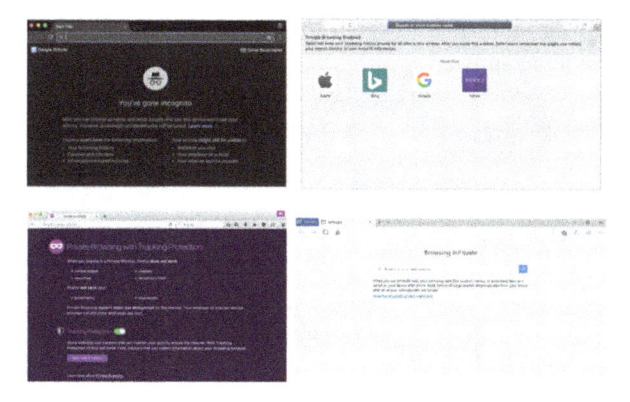

Fig. 1. Interface design of private browsing mode on four leading browsers (Chrome Incognito mode, Safari's private browsing, Firefox's private window, Edge's InPrivate mode)

reactions are valuable to their decision-making [37]. That is, if an individual believes that her/his emotional responses are reasonable bases for judgment, s/he will use them to form attitudes, and vice versa. However, people will depend less on their emotions when they have high expertise in the field of judgement [37]. Applying this concept, human emotion is an affective information that can shape people's attitudes and decisions when it comes to an unfamiliar topic or object. Although psychologists have evidenced emotional impact on people's attitudes and decision, users' emotion seems to be an understudied factor in the premise of Technology Acceptance Model (TAM) [14].

TAM have been a leading model in examining users' acceptance of technology from a psychological perspective. TAM postulates that users' perceived ease of use and usefulness influence their attitudes toward the technology that further affect their behavioral intentions toward using it [14]. Although TAM has been a dominant model predicting users' adoption of a technology, it mainly focuses on the cognitive effects. The role of users' emotions has been understudied in TAM research [32]. As pointed out by prior research [32], it is necessary to have a deeper understanding in examining additional influences such as emotions in the TAM. Therefore, in this study we mainly focus on emotional effects on users' perceptions, attitudes, and behavioral intentions toward the use of private browsing mode. Figure 2 exhibits the extension of the model of technology acceptance proposed in this study. We next address the hypotheses of this study.

Based on aforementioned models and literature, we think that the design of private browsing mode can provoke users' emotions. For instance, the indicator on private browsing mode may remind users of their private state, which may make them feel that they are 'hiding something'. Thus, our first hypothesis is that private browsing mode will provoke users' negative emotional responses.

H1: Compared to the normal browsing mode, private browsing mode will provoke more negative emotions in users.

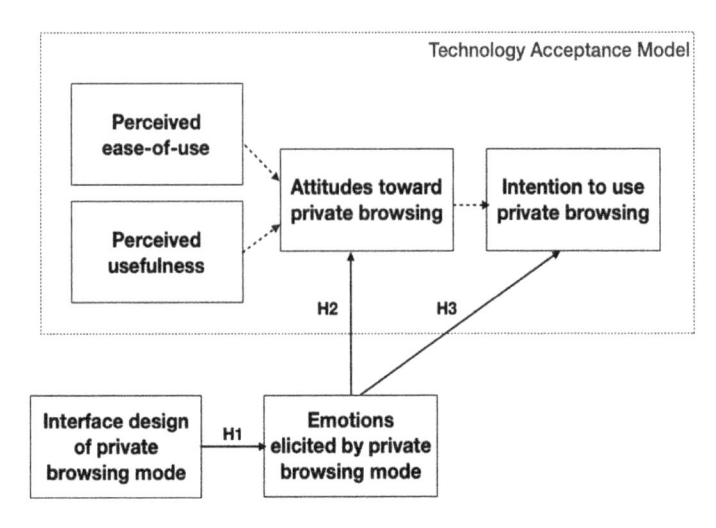

Fig. 2. Theoretical framework and hypotheses

Furthermore, users' negative emotion can further affect their attitudes and behavioral intention to use private browsing. Consequently, our second and third hypotheses are that users' negative emotion provoked by private browsing mode will impact users' attitudes and behavioral intention toward private browsing.

H2: Users' negative emotions provoked by private-browsing mode will adversely influence users' attitudes toward private browsing.

H3: Users' negative emotions provoked by private-browsing mode will adversely influence users' intentions to use private browsing.

4 Study Design

To address the proposed hypotheses, we conducted a mixed-design online survey experiment with one between variable (types of browser) and one within variable (browsing modes: public and private mode). For browser types, we selected four leading browsers as testing subjects: Google's Chrome, Mozilla's Firefox, Apple's Safari and Microsoft Edge/IE [48].

4.1 Survey Procedure

Prior to the beginning of the survey, participants were asked to answer four screening questions about what operating system, browser type, and browsing mode they were familiar with and currently use for their online activities. Considering that familiarity with the browser may bias users' emotional response, participants were randomly assigned to assess one of the browser interfaces that they are not currently using or familiar with. In addition, previous research shows that the majority of online browsing takes place in the default or public mode while private mode is only utilized when there is a specific browsing need [20]. Therefore, based on this inference, we simulated our study design to mimic a real-world scenario where participants were first presented public (default) browsing mode and then private browsing mode. Table 1 exhibits each condition of this survey experiment.

Table 1. Design of survey experiment.

Condition	Public mode	Private mode
Condition 1 (n = 59)	Chrome	Incognito mode
Condition 2 (n = 72)	Firefox	Private window
Condition 3 (n = 86)	Safari	Private browsing
Condition 4 (n = 88)	Edge	InPrivate mode

The survey includes four sections. In Sect. 1, participants were first asked about how they felt at the moment. Then they answered questions about their use of the Internet, previous user experience and knowledge of private browsing, as well as a short form about personality traits. In Sect. 2, participants were first presented with the browser interface in normal mode and were asked to rate their emotions about the interface. Then participants repeated the same procedure for the interface in private browsing mode. In Sect. 3, participants were asked to rate their degree of satisfaction with the interface and answered open-ended questions about the interface design. In Sect. 4, participants' attitudes and behavioral intention toward private browsing mode were measured. At the end of this survey, participants answered demographic questions. The survey conducted during his survey experiment was approved by Internal Review Board (IRB).

4.2 Participants

We recruited 305 participants through Amazon Mechanical Turk, 158 of whom were female. The average age was 49.39 (SD = 17.20). For education, 9.8% of participants had only a high school degree, 35.7% had completed some college or associate's degree, 35.7% had earned bachelor's degrees, and 18.9% had earned advanced degrees. All of our participants were from the United States. The average amount of daily Internet use reported was 6.47 h (SD = 4.34). For private browsing, 68.5% (n = 209) of people reported experience with private browsing mode; 20.7% (n = 63) reported no experience; and 10.8% (n = 33) did not know what private browsing mode is. It is worth noting that although over half of participants had experience in private browsing, it remains unclear how often they used it.

After completing the study, participants were given $0.55 compensation for their participation. To encourage more senior participants, we provided higher compensation ($0.85) for those who were above 55 years old. All senior participants were verified by Amazon Mechanical Turk, which prevent participants from lying about their age.

4.3 Measurement

Emotions About Privacy

To the best of our knowledge, no previous studies have developed measurements to assess people's emotions about privacy. Since human emotion is a multifaceted concept, it is very important for us to clarify which aspect of emotion we measured. In this current study, we focused on measuring participants' subjective 'feelings', which are a fundamental part of emotion [16]. Therefore, readers will see the terms of emotions and feelings used interchangeably in following sections. In addition, users' emotions are subjective feelings, which can be mainly measured by self-report [16]. Thus, we adopted a self-report approach and instruments to measure users' emotions.

To ascertain emotions in users that are related to privacy, we conducted a pilot study with 20 online participants who were asked to list up to 10 types of 'feelings' that they would associate with the concept of privacy. A total of 124 words were generated by these participants to describe their emotions about privacy. We first examined these words and selected the words that described basic emotions [18]. The basic emotions addressed by our participants included sadness, anger, fear, and disgust. Then, the words that did not fit into the basic emotion category were labeled as privacy-related emotions. The privacy-related emotions included feeling private, safe, not being watched, not legitimate, and hiding.

We applied the above two categories of emotions (basic emotions and privacy-related emotions) to measure participants' emotions toward private browsing. We adopted self-reported Basic Emotion Assessment [5] to measure 6 types of basic emotions (happy, sad, angry, fearful, excited, disgusted) using a 6-point unipolar scale. To measure privacy-related emotions, we employed the emotions that were generated by our study participants as mentioned above. A total of 7 items were developed, including: *(1) private-public, (2) safe-vulnerable, (3) not being watched-being watched, (4) not censored-censored, (5) feeling like a legitimate citizen-feeling like a criminal, (6) ordinary-feeling unique, (7) having nothing to hide-having something to hide.* Participants were asked to rate on 5-point bipolar scale. To test the internal consistency of the scale, we conducted a reliability analysis to measure Cronbach's alpha, which is 0.79.

Attitudes Toward Private Browsing

To measure participants' attitudes toward private browsing, we adopted and developed 5 items based on the framework of TAM. These items included: Private browsing is *(1) useful, (2) effective, (3) easy-to-use, (4) an appropriate way, (5) a secure way* to protect my online privacy (Cronbach's alpha = .91). Participants were asked to rate their attitudes on a 7-point Likert scale.

Intention to Use Private Browsing Mode

We developed 5 items to measure participants' intentions to use private browsing mode using a 7-point Likert scale. These items were: I intend to: *(1) use private browsing mode to protect my online privacy, (2) use private browsing mode to secure my personal information online, (3) use private browsing mode more often than before, (4) use the private browsing as the default on my browser, (5) recommend others to use private browsing mode* (Cronbach's alpha = .88).

5 Results

5.1 Emotion Elicited by Private Browsing Mode

Our first hypothesis is that the design of private browsing mode will provoke users' negative emotions (H1). The nonparametric Wilcoxon signed-rank test was conducted to test participants' self-reported emotions in normal and private browsing mode.

For basic emotions, we found that participants rated significantly higher in fearful feeling toward private mode than normal mode (V = 1187, p < .0001). However, there are no significant differences in other basic emotions (Fig. 3). For privacy-related emotions, participants reported feeling less public (V = 22056, p < .0001), vulnerable (V = 11858, p < .0001), and being watched (V = 12790, p < .0001) in private mode than normal mode. In other words, private browsing mode provoked emotions of feeling private, safer, and not being watched. Nevertheless, private browsing mode also significantly elicited more emotions in users such as feelings like a criminal (V = 3518, p = .0002), feeling unique (V = 2468, p < .0001), and feeling hiding something (V = 4423, p < .0001) when compared to normal mode (Fig. 3). We found no significant difference in feeling censored. These results suggest that private browsing mode interfaces did provoke certain negative emotions in participants. Accordingly, the H1 is supported.

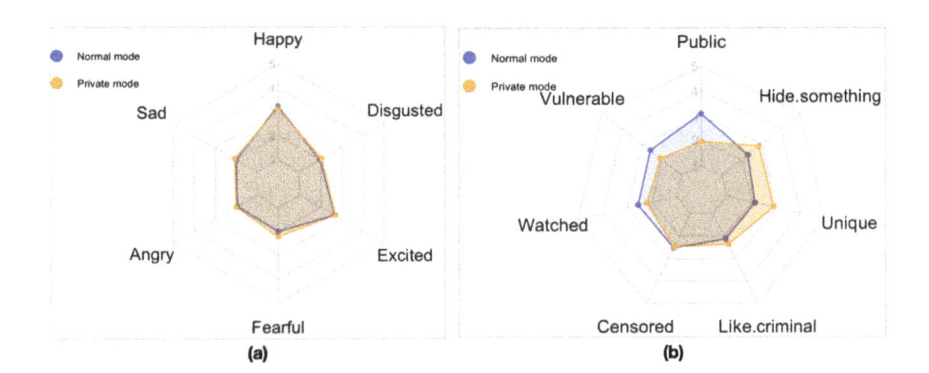

Fig. 3. Participants' emotional responses: (a) Basic emotions and (b) privacy-related emotions) to the interface design of normal and private browsing mode.

5.2 Interface Design of Private Browsing Mode and Emotion

Participants' Emotional Response Toward Interface Design

We were also interested in how users feel distinctively toward different private browsing mode interfaces in the leading browsers. Considering the non-normalized distribution of our data, we used the Kruskal-Wallis H test to determine if the medians of emotion, attitudes, and intention to use toward private browsing mode are equal across these four browsers. When the Kruskal-Wallis H test was significant, we further conducted a post-hoc analysis by using Dunn's test with Bonferroni correction to determine which browser has significant differences from other browsers while comparing users' emotions, attitudes, and behavioral intentions.

In terms of basic emotions toward private browsing, we found that the four browsers showed significant differences in three basic emotions, including happiness ($X^2 = 14.20$, df = 3, $p = .003$), fear ($X^2 = 14.48$, df = 3, $p = .002$), and excitement ($X^2 = 25.31$, df = 3, $p < .0001$). Our post-hoc analyses reveal that Firefox's Private window evoked more feelings of happiness than Edge's InPrivate mode ($p = .001$) (see Fig. 4). Moreover, Chrome's incognito mode provoked more fearful feelings in participants than Edge's InPrivate mode ($p = .020$). On the other hand, Firefox's Private window provokes more excited feelings in participants than all other browsers (Chrome's incognito mode, $p = .017$; Safari's Private browsing, $p = .020$; Edge's InPrivate mode, $p < .0001$).

Our results also indicate significant differences in privacy-related emotions depending on browser types. These emotions include feelings of being censored ($X^2 = 14.30$, df = 3, $p = .003$), like a criminal ($X^2 = 14.30$, df = 3, $p = .003$), unique ($X^2 = 15.59$, df = 3, $p = .001$) and having something to hide ($X^2 = 15.30$, df = 3, $p = .002$).

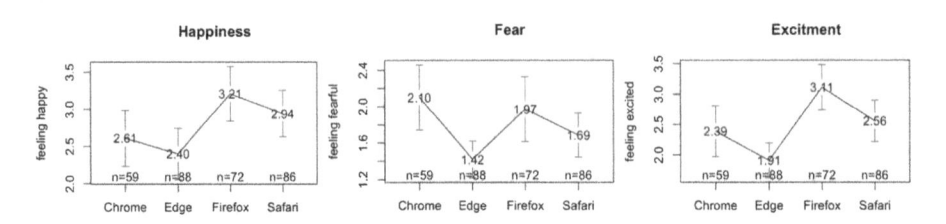

Fig. 4. Mean of basic emotions by browsers.

As shown in Fig. 5, Chrome's incognito mode provoked more emotions of being censored in participants than Edge's InPrivate mode ($p = .003$) and Safari's Private browsing ($p = .007$). In addition, participants reported feeling more like a criminal in Chrome's incognito mode than other private browsing modes (Firefox: $p = .002$; Safari: $p = .004$, Edge: $p = .008$). Furthermore, Chrome's incognito mode elicited more feelings of uniqueness when compared to Safari's Private browsing ($p = .021$) and Edge ($p = .013$). Participants also felt more like having something to hide in incognito mode than in Firefox ($p = .003$) and Safari ($p = .005$). These results seem to suggest that Chrome's incognito mode design may induce more negative emotions than other browsers.

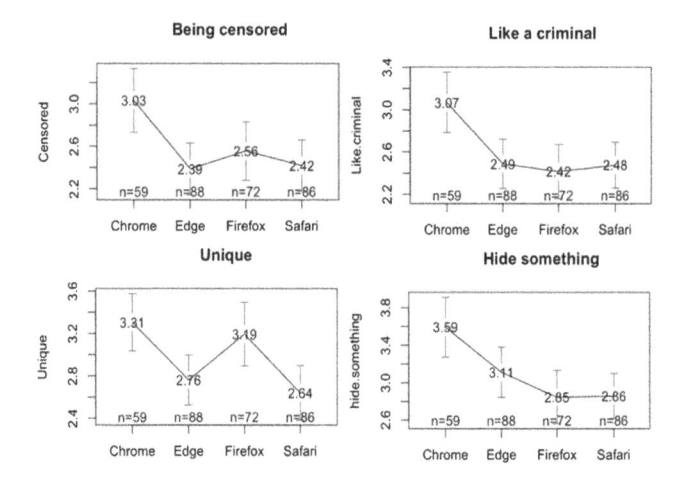

Fig. 5. Mean of privacy-related emotions by browser.

5.3 Emotional Effects on Attitudes Toward Private Browsing Mode

To examine users' emotional effects on their attitudes toward and intentions to use private browsing mode, we conducted a linear regression analysis by employing Generalized Linear Model (GLM) in R. Considering that our variables are not normally distributed, the GLM model is particularly useful because it can transform the response variable by defining the link function. We used the 'identity' link function in the model due to the continuous nature of our variables.

The GLM model for attitudes includes 6 predictors of basic emotions and seven predictors of privacy-related emotions by controlling browser types, age, gender, and users' prior experience with private browsing. The model explained a significant proportion of variance in attitudes toward using private mode ($R^2 = .181$, $F(20,284) = 3.15$, $p < .0001$). According to the analysis, the only significant predictor of users' attitudes toward private browsing mode is the emotion of happiness with positive estimates ($\beta = .131$, $t = 2.20$, $p = .028$). This indicates that users who feel happier in private browsing mode would show more positive attitudes toward it. On the other hand, we found no significant effects of negative emotions on users' attitudes toward private browsing. Hence, our second hypothesis that negative emotions provoked by private browsing mode will adversely influence users' attitudes toward private browsing (H2) is rejected.

5.4 Emotional Effects on Behavioral Intention Toward Private Browsing Mode

The third hypothesis is that negative emotions provoked by private-browsing mode will adversely influence users' intentions to use private browsing (H3). To test this hypothesis, we conducted a second GLM model to examine emotional effects on users' behavioral intention of using private browsing. The second model had the same set of predictors but users' attitudes toward private browsing was controlled. The model also explained a significant proportion of variance in behavioral intentions toward using private mode ($R^2 = .391$, F (21,283) = 8.66, p < .0001).

The results show that the feelings of 'hiding something' ($\beta = -.192$, $t = -2.43$ $p = .016$) and 'vulnerable' ($\beta = -.251$, $t = -2.28$, $p = .023$) are significantly negative predictors of participants' intention of use. That is, participants who felt that they are hiding something or vulnerable while using private browsing mode would be inclined not to use it. Hence, H3 is supported. In addition, participants who felt more excited ($\beta = .164$, $t = 2.44$, $p = .015$) toward the private browsing mode showed significantly higher intentions to use it.

6 Discussion

Emotion is an essential consideration in the use of privacy-enhancing features since it has not only immediate but also consistent influences across time [42]. Accordingly, studying emotional impacts on the acceptance of privacy-enhancing features is important to usable privacy literature. In this study, we used private browsing mode as a case study and examined whether private browsing mode would elicit uses' emotions, which would further affect their use of private browsing mode, by conducting an online survey experiment. Next, we discuss our findings and applications to the design of privacy-enhancing features.

6.1 Users' Emotions Provoked by Private Browsing

Private Browsing Provokes Not only Positive but also Negative Emotions in Users

When compared to public browsing mode, our findings show that users feel more private, safer, more unique, and less watched in private browsing mode. However, users also feel more fearful, like a criminal, and like they are hiding something while using private browsing mode, which supports our hypothesis (H1). These findings further indicate that private browsing mode provokes two conflicting feelings in users. On the one hand, private mode elicits feelings of privacy and safety; on the other hand, it elicits users' emotions of feeling fearful, like a criminal, and having something to hide. We think that these conflicting emotions result from the interface design of private browsing mode, which provokes both negative and positive emotions in users. Another possible explanation is that the use of private browsing mode may already have negative associations, and therefore influence users' emotions. Meaning, the request to use private browsing mode may itself be perceived to be a negative behavior and thus influence their emotions in a negative manner prior to even evaluating the private mode. Therefore, their subsequent emotional response may be elevated after the actual evaluation due to this initial bias. Since this explanation needs further investigation and validation, future studies are needed.

Interface Design Styles of Private Browsing Mode Can Provoke Different Emotions in Users

Interface design is the users' first and most direct interaction with a system, which can determine users' emotions toward the system [15]. In this study, we analyzed users' emotions in response to the interface design of four browsers (Chrome, Firefox, Safar, and Edge), which represent different design styles. One of the main differences between the interface designs of private browsing modes across these browsers lies in their color hues and tones, which range from darker shades to brighter shades.

As displayed in Fig. 6, Chrome's incognito design is the darkest; both Safari and Edge's interface designs use lighter colors. Since the colors used in interface design have critical influences on human emotions [9], an explanation for why Chrome's incognito mode provokes more negative emotions (e.g., feeling fearful, feeling more like a criminal) than other browsers may be because of its use of darker color tone for its interface. On the other hand, when compared to other browsers, Firefox's design elicits more positive emotions (e.g., happiness, excitement) in users although it also uses the darker color tone. This may be due to the use of the color purple, which has been shown to elicit feelings of pleasure (e.g., happiness) and arousal (e.g., excitement) [44].

Fig. 6. Spectrum of color tones for each browser's private browsing mode (dark to bright).

Overall, our results corroborate prior studies' findings that interface design can provoke users' emotions [9, 15, 29]. That is, the use of color embedded in interface design may bear symbolic meanings and may imbue a system with positive or negative associations [15], which may further influence users' acceptance of technologies.

Positive Emotions Enhance Users' Acceptance of Private Browsing
We further investigated how users' emotions provoked by private browsing mode affect their acceptance of it. We found that negative emotions do not have a significant influence on users' attitudes. Conversely, positive emotion (e.g., happiness) provoked by private browsing mode is directly related to users' positive attitudes toward it.

For users' intention to use private browsing mode, we found that both positive and negative emotions have influences. Users who feel more excited, less vulnerable, and less like they are hiding something because of private browsing are more inclined to use it in the future, which also supports our hypothesis (H3). These findings further indicate that positive emotions toward a privacy-enhancing feature may motivate users to adopt it, which corresponds to the idea that a design works better when it can elicit positive emotions [34].

6.2 Privacy, Emotion, and Interface Design: Ethical Considerations and Challenges

Based on our findings, when a privacy-enhancing feature elicits more positive emotions in users, users are more likely to adopt it. However, how best to design for positive or negative emotional response depends on the goals of a system. For instance, if the goal of the design is to promote the use of a privacy-enhancing feature, then designers should consider eliciting positive emotions in users. On the other hand, if the goal of the design is to deter the use of the feature, then designers can utilize the interface to provoke negative emotions and hence users not adopting it. Nevertheless, this approach further raises a critical question: is it ethical for a system to provide the feature while discouraging the use of the feature by manipulating users' emotions in a passive or an implicit way? It is also important for designers to consider eliminating or minimizing emotional harms or negative consequences when it comes to the design of a feature the system offers. Therefore, another ethical consideration in designing interfaces should be respecting users' preferences and consent when it comes to eliciting their emotions. Perhaps designing the system in a neutral way is one appropriate way forward.

Furthermore, the design of privacy-enhancing features may encounter an inherent challenge because privacy-seeking behavior may have prior, negative connotations in a given cultural context [38]. In other words, the decision to seek online privacy itself may have already brought up negative emotions, regardless of the interface design. Therefore, an ethical design for privacy-enhancing features should allow users to be aware of being in a private state while not eliciting more negative emotions. While this remains a new territory for privacy-enhancing design, further research is needed to verify these recommendations.

6.3 Limitations and Future Research

We acknowledge there are some limitations in this work. First, our findings may only apply to the specific sample group and cultural context because we only recruited U.S. participants via an online crowdsourcing platform. For example, people with different cultural backgrounds may perceive interface design elements (e.g., meaning of color tones) in alternative ways. We suggest that future studies can replicate this current study in different cultural contexts and further compare the results. In addition, the generalizability of the results to other privacy-enhancing features may be limited since we only focused on private browsing in this study. We recommend future research to adopt a similar approach to investigate other types of privacy-enhancing technologies, such as anonymous browsers, privacy notifications, or search engines.

7 Conclusion

Emotion has been an understudied factor in usable privacy literature. To fill this gap, this current study investigated the role of emotion in the acceptance of privacy-enhancing features by conducting an online survey experiment on private browsing. Our results suggest that the interface design of private browsing does elicit both positive and negative emotions in users. These elicited emotions further affect users' attitudes and behavioral intentions toward its use. Generally speaking, positive emotions seem to encourage the use of privacy-enhancing features. However, it also raises an ethical challenge whether privacy-enhancing features should be designed to elicit emotions in users. Another design challenge for privacy-enhancing features is that online privacy seeking behavior may provoke negative emotions regardless of interface design, because it may already have certain negative connotations in a given cultural context. While there is still much to be done in order to gain a comprehensive understanding of emotional impacts on the use of privacy-enhancing features, our work reveals an interesting and novel phenomenon. It also provides insights in the design of privacy-enhancing features and it can help inform better approaches for future research in usable privacy.

References

1. Acquisti, A., Brandimarte, L., Loewenstein, G.: Privacy and human behavior in the age of information. Science **347**(6221), 509–514 (2015)
2. Agarwal, A., Meyer, A.: Beyond usability: evaluating emotional response as an integral part of the user experience. In: CHI 2009 Extended Abstracts on Human Factors in Computing Systems, pp. 2919–2930. ACM, April 2009
3. Anderson, C.L., Agarwal, R.: The digitization of healthcare: boundary risks, emotion, and consumer willingness to disclose personal health information. Inf. Syst. Res. **22**(3), 469–490 (2011)
4. Andrade, E.B., Ariely, D.: The enduring impact of transient emotions on decision making. Organ. Behav. Hum. Decis. Process. **109**(1), 1–8 (2009)
5. Basic Emotion Assessment, Therapy aid. https://www.therapistaid.com/therapy-worksheet/basic-emotion-assessment. Accessed 31 May 2019
6. Baumeister, R.F., Vohs, K.D., Nathan DeWall, C., Zhang, L.: How emotion shapes behavior: feedback, anticipation, and reflection, rather than direct causation. Pers. Soc. Psychol. Rev. **11**(2), 167–203 (2007)
7. Barnes, S.B.: A privacy paradox: social networking in the United States. First Monday **11**(9), 11–15 (2006)
8. Beaudry, A., Pinsonneault, A.: The other side of acceptance: studying the direct and indirect effects of emotions on information technology use. MIS Q. **34**, 689–710 (2010)
9. Brave, S., Nass, C.: Emotion in human–computer interaction. Hum. Comput. Interact. **53** (2003)
10. Cabanac, M.: What is emotion? Behav. Proc. **60**(2), 69–83 (2002)
11. Cadwalladr, C., Graham-Harrison, E.: The Cambridge analytica files. The Guardian **21**, 6–7 (2018)
12. Cranor, L., Langheinrich, M., Marchiori, M., Presler-Marshall, M., Reagle, J.: The platform for privacy preferences 1.0 (P3P1. 0) specification. W3C Recommendation 16 (2002)
13. Cranor, L.F., Guduru, P., Arjula, M.: User interfaces for privacy agents. ACM Trans. Comput.-Hum. Interact. (TOCHI) **13**(2), 135–178 (2006)
14. Davis, F.D.: Perceived usefulness, perceived ease of use, and user acceptance of information technology. MIS Q. **13**, 319–340 (1989)
15. Demirbilek, O.: Evolution of emotion driven design. In: Emotions and Affect in Human Factors and Human-Computer Interaction, pp. 341–357 (2017)

16. Desmet, P.: Measuring emotion: development and application of an instrument to measure emotional responses to products. In: Funology 2, pp. 391–404. Springer, Cham (2018)
17. Dingledine, R., Mathewson, N., Syverson, P.: Tor: The second-generation onion router. Naval Research Lab, Washington, DC (2004)
18. Ekman, P.: An argument for basic emotions. Cogn. Emot. **6**(3–4), 169–200 (1992)
19. Ekman, P.E., Davidson, R.J.: The Nature of Emotion: Fundamental Questions. Oxford University Press, Oxford (1994)
20. Gao, X., Yang, Y., Fu, H., Lindqvist, J., Wang, Y.: Private browsing: an inquiry on usability and privacy protection. In: Proceedings of the 13th Workshop on Privacy in the Electronic Society, pp. 97–106. ACM, November 2014
21. Guha, S., Tang, K., Francis, P.: NOYB: privacy in online social networks. In: Proceedings of the First Workshop on Online Social Networks, pp. 49–54. ACM, August 2008
22. Habib, H., Colnago, J., Gopalakrishnan, V., Pearman, S., Thomas, J., Acquisti, A., Christin, N., Cranor, L.F.: Away from prying eyes: analyzing usage and understanding of private browsing. In: Fourteenth Symposium on Usable Privacy and Security, SOUPS 2018, pp. 159–175 (2018)
23. Hassenzahl, M.: The hedonic/pragmatic model of user experience. Towards a UX manifesto, p. 10 (2007)
24. Hibbeln, M., Jenkins, J.L., Schneider, C., Valacich, J.S., Weinmann, M.: How is your user feeling? Inferring emotion through human–computer interaction devices. Group **1000**, 248 (2017)
25. Hourcade, J.P., Cavoukian, A., Deibert, R., Cranor, L.F., Goldberg, I.: Electronic privacy and surveillance. In: CHI 2014 Extended Abstracts on Human Factors in Computing Systems, pp. 1075–1080. ACM, April 2014
26. Isaak, J., Hanna, M.J.: User data privacy: Facebook, Cambridge analytica, and privacy protection. Computer **51**(8), 56–59 (2018)
27. Johnson, D., Wiles, J.: Effective affective user interface design in games. Ergonomics **46**(13–14), 1332–1345 (2003)
28. Kelley, P.G., Bresee, J., Cranor, L.F., Reeder, R.W.: A nutrition label for privacy. In: Proceedings of the 5th Symposium on Usable Privacy and Security, p. 4. ACM, July 2009
29. Kim, J., Lee, J., Choi, D.: Designing emotionally evocative homepages: an empirical study of the quantitative relations between design factors and emotional dimensions. Int. J. Hum. Comput. Stud. **59**(6), 899–940 (2003)
30. Li, H., Sarathy, R., Xu, H.: The role of affect and cognition on online consumers' decision to disclose personal information to unfamiliar online vendors. Decis. Support Syst. **51**(3), 434–445 (2011)
31. Liou, J.C., Logapriyan, M., Lai, T.W., Pareja, D., Sewell, S.: A study of the internet privacy in private browsing mode. In: Proceedings of The 3rd Multidisciplinary International Social Networks Conference on SocialInformatics 2016, Data Science 2016, p. 3. ACM, August 2016
32. Marangunić, N., Granić, A.: Technology acceptance model: a literature review from 1986 to 2013. Univ. Access Inf. Soc. **14**(1), 81–95 (2015)
33. Nissenbaum, H.: Privacy in Context: Technology, Policy, and the Integrity of Social Life. Stanford University Press, Palo Alto (2009)
34. Norman, D.A.: Emotional Design: Why We Love (or Hate) Everyday Things. Basic Civitas Books, New York City (2004)
35. Pedersen, D.M.: Psychological functions of privacy. J. Environ. Psychol. **17**(2), 147–156 (1997)
36. Russell, J.A.: Core affect and the psychological construction of emotion. Psychol. Rev. **110**(1), 145 (2003)

37. Schwarz, N.: Feelings-as-information theory. In: Handbook of Theories of Social Psychology, vol. 1, pp. 289–308 (2011)
38. Solove, D.: Understanding Privacy. Harvard University Press, Cambridge (2008)
39. Soghoian, C.: Why private browsing modes do not deliver real privacy. Center for Applied Cyber security Research, Bloomington (2011)
40. Stark, L.: The emotional context of information privacy. Inf. Soc. **32**(1), 14–27 (2016)
41. Tene, O., Polonetsky, J.: A theory of creepy: technology, privacy and shifting social norms. Yale J. Law Tech. **16**, 59 (2013)
42. Tractinsky, N., Cokhavi, A., Kirschenbaum, M., Sharfi, T.: Evaluating the consistency of immediate aesthetic perceptions of web pages. Int. J. Hum. Comput. Stud. **64**(11), 1071–1083 (2006)
43. Ur, B., Leon, P.G., Cranor, L.F., Shay, R., Wang, Y.: Smart, useful, scary, creepy: perceptions of online behavioral advertising. In: Proceedings of the Eighth Symposium on Usable Privacy and Security, p. 4. ACM, July 2012
44. Valdez, P., Mehrabian, A.: Effects of color on emotions. J. Exp. Psychol. Gen. **123**(4), 394 (1994)
45. Venkatesh, V.: Determinants of perceived ease of use: Integrating control, intrinsic motivation, and emotion into the technology acceptance model. Inf. Syst. Res. **11**(4), 342–365 (2000)
46. Venkatesh, V., Ramesh, V.: Web and wireless site usability: understanding differences and modeling use. MIS Q. **30**, 181–206 (2006)
47. Vohs, K.D., Baumeister, R.F., Loewenstein, G. (eds.): Do Emotions Help or Hurt Decisionmaking?: A Hedgefoxian Perspective. Russell Sage Foundation, New York City (2007)
48. W3schools, Browser statistics. https://www.w3schools.com/browsers/default.asp. Accessed 31 May 2019
49. Wilson, D., Valacich, J.S.: Unpacking the privacy paradox: Irrational decision-making within the privacy calculus (2012)
50. Wu, Y., Gupta, P., Wei, M., Acar, Y., Fahl, S., Ur, B.: Your secrets are safe: how browsers' explanations impact misconceptions about private browsing mode. In: Proceedings of the 2018 World Wide Web Conference on World Wide Web, pp. 217–226, April 2018
51. Young, J.B.: Introduction: a look at privacy. In: Privacy, vol. 1, p. 2 (1978)
52. Zhang, B., Xu, H.: Privacy nudges for mobile applications: effects on the creepiness emotion and privacy attitudes. In: Proceedings of the 19th ACM Conference on Computer-Supported Cooperative Work & Social Computing, pp. 1676–1690. ACM, February 2016

An Adaptive Security Architecture for Detecting Ransomware Attack Using Open Source Software

Prya Booshan Caliaberah[1], Sandhya Armoogum[1(✉)], and Xiaoming Li[2]

[1] School of Innovative Technologies and Engineering, University of Technology Mauritius, La Tour Koenig, Pointes-aux-Sables, Port Louis, Mauritius
asandya@umail.utm.ac.mu
[2] School of Computer Science and Technology, Tianjin University, Tianjin, China

Abstract. Ransomware is a serious security threat faced by organizations and individuals today, and ransomware attacks are on the increase. There is no infallible solution for protecting against ransomware as the malware code uses metamorphic and polymorphic algorithms to generate different versions thus evading signature detection. Ransomware also uses domain generator algorithms (DGA) to generate new domains for the command and control server (C&C), they constantly exploit new vulnerabilities, and they use various infection vectors. Thus, for an organization to protect itself, an adaptive security architecture is required to constantly monitor the network so as to detect new ransomware infection at an early stage such that it can be blocked before encryption of files occur. This approach is a defence in depth approach which supplements the network defences such as patch management, anti-virus software, intrusion detection, firewalls, and content filtering. A framework for the implementation of the adaptive security architecture model using open source software is presented and the proposed framework is tested against the WannaCry and Petya ransomware. The proposed framework was successfully able to alert of the ransomware attack and by the use of the AppLocker feature on Windows, it was even possible to prevent the Petya ransomware from executing on the victim host.

Keywords: Ransomware · Adaptive Security Architecture (ASA) · Monitoring · WannaCry · Petya · Snort · OSSEC · AppLocker

1 Introduction

Ransomware is a sophisticated technique of cyber-extortion which allows cybercriminals to gain profit from unsuspecting victims. It is one of the fastest growing threats in cyber security; it can create havoc in organizations and disrupt businesses. According to Symantec, the rate of ransomware infections has increased by 36% between 2015 and 2016. 470, 000 infections were blocked in 2016, while 319, 000 infections have been blocked only in the first six months of 2017, out of which 28% of infections in

May and 21% of infections in June were largely due to the WannaCry and Petya ransomware, respectively [1]. WannaCry and Petya are a new breed of ransomware as they are self-propagating and can spread to infect computers in the network in a very short period of time. Moreover, it was also reported that in 2017, there was a shift regarding the victims. In 2015 and 2016, around 70% of infections were targeted at individual internet users and around 30% infections were targeted at enterprises. Whereas in 2017, only 58% of infections were targeted at users while 42% infections were targeted at enterprises. Payment of ransom may vary from USD 294 to up USD 1077, and even then, there is no guarantee that the effect of ransomware is reversible, especially for ransomware that encrypt files [2]. Furthermore, in a survey conducted in 2016 [3], it was observed that the attackers were more motivated to hit industries which is mostly dependent on business-critical information and thus were assured of a guaranteed ransom payment. Typical examples of such industries include healthcare services, financial services, manufacturing industries and governments.

Ransomware is thus a serious threat to organizations where IT is a core business component and protecting against ransomware becomes important. Ransomware can lead to momentary or permanent loss of information resulting in interruption of operations, and potential negative impact to an organization's reputation. Ransomware, being commonly sent to victims via emails, or when users visits malicious or compromised websites, several preventive measures are recommended such as user awareness and training programs, enabling strong spam filtering to block phishing emails, scanning incoming and outgoing network packets to block executables from reaching internal computers [4]. Ransomware also infects computers by exploiting security vulnerabilities such as backdoors in a computer. Hence, common security mechanisms such as the use of anti-virus software, firewalls, intrusion detection systems (IDS), intrusion preventions systems (IPS), strict access control, vulnerability scanning to detect and address vulnerabilities, patching of software to fix existing bugs, and disaster management (regular backups) are also good measures for protecting against malware including ransomware. Despite, these preventive measures in place, systems may still be at risk of ransomware, as these measures are not completely effective at protecting against ransomware attacks. This is because ransomware is constantly evolving and new strains of ransomware are created at an alarming rate. Anti-virus software, and intrusion detection systems often relies on signature detection for detecting and protecting against malware. If the attack signature is not yet known, the anti-virus software, IDS and IPS are unable at detecting attacks. Furthermore, ransomware exploits zero-day flaws. A zero-day exploit or flaw refers to new vulnerabilities which are discovered and which has not yet been patched up. According to the Fortinet's Threat Landscape Report for Q1 2018 [5], the zero-day market is maturing; and there was a total of 214 zero-day threats discovered in 2017, while 45 zero-day threats were found only in the first quarter of 2018.

Because of ineffectiveness of existing tools to completely protect against ransomware, new approaches for tackling the issue of ransomware must be investigated. Ransomware attacks are complex and often involve several steps. In this work, a framework is proposed to capture and correlate security logs to detect ransomware attacks in a network. Open source tools were used for implementing the proposed framework. Once, the ransomware attack is detected early on by using the adaptive security architecture proposed, the parameters regarding the attack can be used to set up effective measures to block the ransomware attack. This paper is organized as follows. In Sect. 2, an overview of ransomware and related research works for detecting ransomware attacks is discussed. The proposed framework for capturing security logs from different network security tools namely Snort, OSSEC and Graylog is presented in Sect. 3. The implementation of the proposed framework and experiments conducted are provided in Sect. 4. Section 5 concludes this paper.

2 Background

A ransomware is a type of malware program that infects and takes control of the infected system mainly by using encryption, then demanding a ransom in order for the victim user to regain access [6]. To get back access to the system or for file restoration, the victim is compelled to pay a ransom, more often in the form of digital currency or cryptocurrency such as Bitcoins. Thus, the aim of the attacker is to extort money from the victim. Unlike conventional malwares that are stealthy, ransomware makes it presence known quite rapidly by displaying a message asking for ransom.

Ransomware became popular with the emergence of Cryptolocker in 2013 which defined the business model and proved the financial opportunity for attackers. Between September and December 2013, it was estimated that more than 250,000 systems were infected by Cryptolocker and their perpetuators earned more than USD 3 million [7]. From then on, particularly as from year 2015, there has been a phenomenal rise in ransomware whereby new variant of ransomware were discovered every month [8].

2.1 Ransomware Classification

According to [9], ransomware can be classified based on its severity, platform and target. Different types of ransomware under the platform classification involve the PC, Mobile, Cloud and IoT ransomware. The target category of ransomware distinguishes between ransomware targeted towards consumers or organizations. Ransomware types in the severity category involves scareware and detrimental ransomware whereby detrimental ransomware can be of two types namely locker-ransomware and crypto-ransomware, the latter being the most popular type of ransomware.

Locker-ransomware does not encrypt user files but rather restricts access to the files and it could be removed by using system restore, an offline antivirus scanner or a combination of both. Locker-ransomware thus declined around 2012 as it was less successful at extorting money from victims. Crypto-ransomware on the other hand uses strong encryption to prevent accessibility of files on the victims' hard drive. Decryption becomes difficult without the proper key which leaves the end-user no choice but to pay the ransom in exchange of the decryption key. Crypto-ransomware can encrypt user data on an infected computer and any data storage attached to it i.e. hard disk drives, USB disks, virtual mapped drives or even cloud drives like Dropbox. Contrarily to locker ransomware, the crypto-ransomware doesn't block access to the computer or system files but rather encrypts all user data files such that the data is unusable.

2.2 Crypto-Ransomware

Crypto-ransomware can be further classified as Symmetric Crypto-Ransomware (SCR), Asymmetric Crypto-Ransomware (ACR), and Hybrid Key Crypto-Ransomware (HCR) based on the type of encryption algorithms used. A crypto-ransomware attack involves five stages [10]: (1) the crypto-ransomware infects the computer and installs itself on the machine; (2) the crypto-ransomware establishes connection with its command and control (C&C) server; (3) a communication channel (often secure) is established with the C&C server, which generates and sends encryption key(s) to the ransomware; (4) the ransomware encrypts the victim's files; and (5) the ransomware displays the message with time limit to pay the ransom.

2.3 Hybrid Key Crypto-Ransomware (HCR)

The notorious ransomware Cryptolocker (2013), was designed for Windows computers and it is an HCR as it uses both symmetric (the Advanced Encryption algorithm, AES-256) and asymmetric/public key cryptography (the RSA algorithm with a 2048-bit key pair) for encryption of files, and Bitcoin cryptocurrency for payment. Cryptolocker generates many 256-bit AES keys on the victim's computer. Each of the victims' files is encrypted with a specific and unique AES key. These keys are then further encrypted using a uniquely generated RSA-2048 public key which is sent from the C&C server [11]. The encrypted AES key is then concatenated with the encrypted file. This mechanism ensures that the private key is never transferred over the network. The use of strong keys and proven algorithms, as well as the use of a hybrid cryptosystem makes it difficult to decrypt the encrypted files without the private RSA key. The Cryptolocker's C&C server has been shut down since 2014, but variants of Cryptolocker still persist to this day.

The Locky ransomware (2016) [12] operates in a similar manner using 2048-bit RSA public keys obtained from the C&C server to encrypt the 128-bit AES keys which are used to encrypt the user's files.

The WannaCry (2017) ransomware also uses a hybrid cryptosystem involving the RSA algorithm and the AES algorithm in CBC mode with a key length of 128 bits to encrypt the user files. It differs slightly from Cryptolocker and Locky, as it does not obtain the RSA public key from the C&C server but rather it uses the Windows CrytoAPI to generate the RSA key pair [13]. The private key is then sent to the C&C server and the public key is used to encrypt the AES keys.

The Petya ransomware (2016) uses Elliptic Curve Cryptography (ECC), instead of RSA for securing the AES file encryption key, as ECC is more difficult to break than RSA. Petya also differs from the other ransomware as it does not encrypt all the user files but rather it encrypts the machine's entire master file table (MFT), which prevents access to any files on the system [14]. Later variants of Petya, also encrypts the 1 MB of user files to prevent recovery of files. Petya also has an offline cryptosystem design i.e. it does not communicate with a C&C server. The ransomware has a 192-bit ECC public key hardcoded in its binary. When executing, it uses this public key to generate a shared secret using the Elliptic Curve Diffie-Hellman (ECDH) algorithm. This shared secret is then used to encrypt the AES file encryption keys. However, the attacker requires knowledge of the shared secret to be able to decrypt the files after ransom have been paid. Thus, the ECC public key is combined with the shared secret using a Base58 binary-to-ascii encoding scheme to generate a "decryption code". This code is displayed on the ransom payment message and must be sent to attackers when making the payment. Thereby, this design effectively removes the need to communicate with a C&C server during the infection process. Another recent variant of Petya, called the NoPetya, emerged in June 2017, which encrypts the MFT as well as encrypting a number of other files, generates a random "decryption code" resulting in the fact that data cannot be decrypted even after ransom has been paid.

2.4 Infection Vector

Until recently, the ransomware infection vector was through exploit kits, malvertising, drive-by-download, email-phishing campaigns or via social engineering attacks. For instance, CryptoLocker and Locky were distributed via spam email campaigns and the first Petya ransomware spread through phishing emails containing links to a malicious Dropbox download. However, in 2017, the WannaCry ransomware which exploits the vulnerability in the Windows implementation of the Server Message Block (SMB) protocol, behaved as a worm as it was able to scan other computers in the network for the same vulnerability and propagate itself to the vulnerable computers. It was of an unprecedented scale as it quickly infected tens of thousands of computers across the world, including banks, universities, telco giant Telefonica in Spain and the National Health Service in the United Kingdom. The NoPetya variant also has similar worm propagation behaviour and it uses a variety of techniques to spread to other computers. Table 1 summarizes the features of the four ransomware discussed.

Table 1. CryptoLocker, Locky, WannaCry and Petya/NoPetya features.

Ransomware	Encryption scheme	C&C server	Infection vector
CryptoLocker (2013)	RSA & AES	Yes	Spam email
Locky (2016)	RSA & AES	Yes	Spam email
WannaCry (2017)	RSA & AES	Yes	Worm (SMB exploit)
Petya/NoPetya (2017)	ECC & AES	No	Phishing email/Worm

2.5 Communication Protocol

Ransomware often communicate with their C&C server for getting the encryption keys that they require and for payment details. Previously, malwares were using some specific protocol such as the IRC protocol for communicating with the C&C server. But recently the communication protocols used by the ransomware have evolved. Today, malwares utilize widely used protocols such as HTTP, SSH, DNS and HTTPS. The use of SSL/TLS protected traffic for communication with the C&C server makes it difficult for firewalls to examine and thus block such traffic. Many ransomware nowadays also use the anonymous TOR network for communication whereby the encrypted traffic is designed to look like normal HTTPS which makes TOR traffic difficult to detect. The TOR browser code is often packaged in the ransomware code. Moreover, a static URL of a C&C server can be found and disabled by law enforcement. Thus, ransomware often use the Domain Generation Algorithm (DGA), to generate random domain names to be used for the C&C server.

2.6 Ransomware Prevention and Recovery

Traditional security measures can help prevent against malware. These include the use of anti-malware software, firewalls, intrusion detection and prevention, and vulnerability scanning as well as security mechanisms such as authentication, access control, patch management, disaster planning, content filtering and user awareness and training.

Given that crypto-ransomware uses strong encryption, it is in general very difficult and almost impossible to recover without the decryption key. Author in [15] proposes a mechanism deployed on hosts which dynamically hooks in the Windows CryptoAPI to proactively secure symmetric keys generated in a separate key vault. Ransomware often uses the CryptoAPI to generate symmetric keys to encrypt files, which are themselves then encrypted by a public key from the C&C server. In this case, keys from the vault can be used to decrypt files and the victim would not have to pay the ransom. Palisse et al. proposes a similar approach in [16]. In [17], it has been proposed to rename the vssadmin.exe which handles shadow copies such that ransomware cannot find and delete restore points. After an infection, restore points can be used to recover user files. Moreover, offline backups remain the most guaranteed way of recovering from a ransomware attack.

2.7 Ransomware Detection

By analyzing ransomware or malware in general, anti-malware software companies are able to provide enhanced detection capabilities. Malware analysis consists of performing static analysis and dynamic analysis. Static analysis is when the malware code is analyzed without running the malware. Such techniques include statistical analysis of opcodes using machine learning techniques [18, 19]. This allows to understand the functionality of the malware and also to generate the signature of the malware for detection. Dynamic analysis involves executing the malware or ransomware to observe its behaviour and identify further technical indicators which can be used to build detection signatures for the malware. Such parameters as the CPU usage, memory usage, network usage and system calls statistics are examined during the dynamic analysis. During the dynamic analysis, the communication of the ransomware with the C&C server for command and download of additional code and keys is also observed. Such information can be used to identify and block the C&C server. Law enforcement may also locate and shutdown the server. Thus, information gathered by performing ransomware analysis is useful for enhancing anti-malware software detection capability and for putting in place the appropriate rules for blocking ransomware packets from entering the network. However, these techniques do not protect an organization from unknown and new ransomware which may have different features and characteristics.

Gómez-Hernández et al. [20] proposed the use of honeyfiles (FIFO like file) which can be deployed around the network to catch the ransomware. Ransomware would be blocked once it starts reading the honeyfile. In [21] and [22], the authors argue that when that ransomware is executing, it most certainly makes system level calls which are facilitated through the Windows Application Programming Interface (API) on the windows platform. Thus, monitoring API calls can effectively allow the early detection of ransomware. Similarly, in [23] the authors demonstrate that by using machine learning algorithms such as the Decision Tree (J48) classifier, it is possible to analyze network traffic to detect ransomware. Moreover, analysis of exploit kits payload for features such as redirection count, IP addresses and file type can be used to detect malware including ransomware [24]. RansomWall is a windows defence solution against crypto-ransomware. It integrates different approaches for protecting the machine namely: static and dynamic analysis, honeyfile and traps, machine learning, and file backup [25]. In [26], it has been proposed to move important user files into a safe zone which is implemented as a zip file. This file is kept in an open non-stop write mode such that the operating system prevents the file from being changed by another process, typically from ransomware processes running on the machine. In [27], the author described methods that can be devised to detect and block ransomware using the Microsoft File Server Resource Manager feature AppLocker and EventSentry on a Honeypot folder. While no live result was recorded from a ransomware attack, alerts were generated for potential threat to warn administrators.

Ransomware attack on mobile devices is also a concern. In [28], the authors proposed to detect ransomware by monitoring the power consumption footprint on android devices. The authors [29] propose a hybrid static and dynamic approach for detecting ransomware on mobile devices.

3 Adaptive Security Architecture Model

Early real-time detection of ransomware is necessary before the encryption of files begins. Kharraz et al. in their paper [30], "Cutting the Gordion Knot: A look Under the Hood of Ransomware Attacks" showed that by monitoring irregular file system activity, it may be possible to detect and stop a large number of ransomware attacks. Similarly, the authors in [31] demonstrated how the ransomware (in this case the CryptoWall) can be detected and mitigated through the use of Software Defined Networking to disrupt the connection between the infected machine and the C&C server in a timely manner to prevent encryption of user files. However, such monitoring should be done continuously as the risk of ransomware attack is prevalent due to their polymorphic nature and the fact that they use different kinds of evasion techniques.

According to Gartner [32], the traditional "prevent and detect" approaches are inadequate and the Adaptive Security Architecture (ASA) model should be adopted to continuously monitor and remediate any security incidents that occurs in the system.

As systems become more complex with time, and the nature of security threats evolve, for maintaining security a continuous, contextual and coordinated approach is required. With the ASA model defensive, detective and responsive measures have to be implemented as well as mechanisms to analyze security breaches so as to be able to adjust the security mechanisms in place to better protect the system. Sinno et al. [33], uses the Unisys Stealth security solution and the LogRhythm analytics platform to design an adaptive security architecture. The LogRhythm solution also performs User and Entity Behaviour Analytics (UEBA) which allows to detect attacks by spotting small changes in behaviour that can be helpful in detecting ransomware attack. UEBA tools profile and baseline the activity of users, and processes to correlate user and other entity activities and behaviours, to detect anomalous patterns. In this work, the ASA approach is adopted for ransomware mitigation and a framework is proposed for implementing the architecture using open source tools. The following open source tools have been used: Snort (IDS) to monitor network traffic, OSSEC for monitoring the hosts on the network, GrayLog for log management and analysis, Opnsense firewall for reacting to attacks in a timely manner, and AppLocker on windows to control and set rules for specific types of applications.

3.1 Open Source Tools for Implementing ASA

Snort is a popular open-source network intrusion detection and prevention system (NIDS) which is widely used for Linux and Windows systems [34]. It adopts the signature-based attack detection, protocol analysis together with anomaly-based inspection to monitor network traffic in real-time. It maintains an updated set of rules which allows it to detect emerging threats and it can also be used to scrutinize the packet payload to search for malicious content. Suspicious activities are logged and the network administrator alerted. To complement Snort, the OSSEC host intrusion detection system (HIDS) is used. OSSEC [35] can be used on most operating systems and it allows detecting attacks on computers by performing log analysis, file integrity checking, Windows registry monitoring, and rootkit detection. As Snort, it also provides real-time alerts when threats are detected. Graylog is an integrated open source log capture and analysis solution for operational intelligence [36]. It integrates Elasticsearch [37] search engine. Elastic Stack (ELK) is another open source alternative for log management and analysis. ELK stack constitutes of Elasticsearch, Logstash (data-collection and log-parsing engine), and Kibana (analytics and visualization platform). In our proposed model, Graylog or ELK can be used for capturing all alerts and logs captured from Snort and OSSEC, for analysis to detect ransomware attack early on during the infection. Using GrayLog, the system administrator can analyze logs gathered from the whole network to identify any off-benchmarked event and react to any imminent attacks. GrayLog also has the ability to generate email alerts which can be sent to various network administrators, system administrators and the incident response team. Many ransomwares communicate with a C&C server prior to the encryption of files. Thus, the proposed ASA framework also includes the Opnsense firewall which can be used to take reactive measures by editing the firewall rules to block specific network traffic as soon as intrusion is detected. Opnsense offers a very captive interface where configurations can be easily carried out. It provides the feature of adaptive whitelisting and blacklisting of network traffic. It provides flexibility as security can be enhanced without altering configurations and can be easily carried without interrupting the firewalling process [38]. Opnsense also includes build-in reporting and monitoring tools for monitoring system health based on the round robin data (RRD) collected. Figure 1 depicts the proposed ASA framework. The proposed ASA framework also includes the use of the AppLocker feature of Windows operating system for blocking ransomware attacks. AppLocker allows to create rules to permit or prevent apps from running on a system based on unique identities of files (e.g. executable files (.exe and .com), scripts (.js, .ps1, .vbs, .cmd, and .bat), Windows Installer files (.mst, .msi and .msp), DLL files (.dll and .ocx), and packaged apps and packaged app installers (appx)) [39]. It also allows specifying which users or groups can run those applications. AppLocker has the ability to enforce policies in an audit-only mode where all application access activity can be recorded in event logs. These logs can be collected for further analysis and subsequently updating of rules.

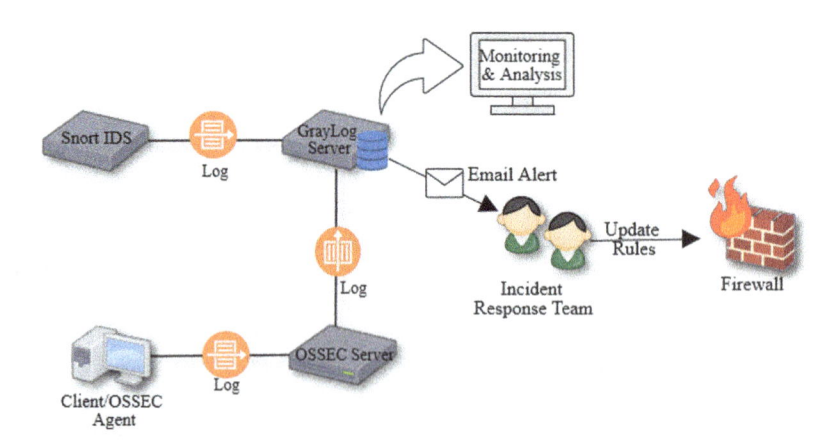

Fig. 1. The proposed ASA framework using open source software

4 Experimental Setup: Implementation of ASA

4.1 Experimental Setup for Implementing the Proposed ASA Framework

A small LAN was designed to implement the ASA framework in a virtualized environment on VMWare. Figure 2 shows the LAN created in VMWare. SNORT was installed in NIDS mode and as a Daemon (as a Service) on a Linux Ubuntu Server. Graylog was also installed on a Linux Ubuntu Server. Since GrayLog uses MongoDB database for storing logs and ElasticSearch for log analysis, MondoDB and

ElasticSearch were installed prior to installing the GrayLog on the server. GrayLog is also set to run at system startup as a Daemon.

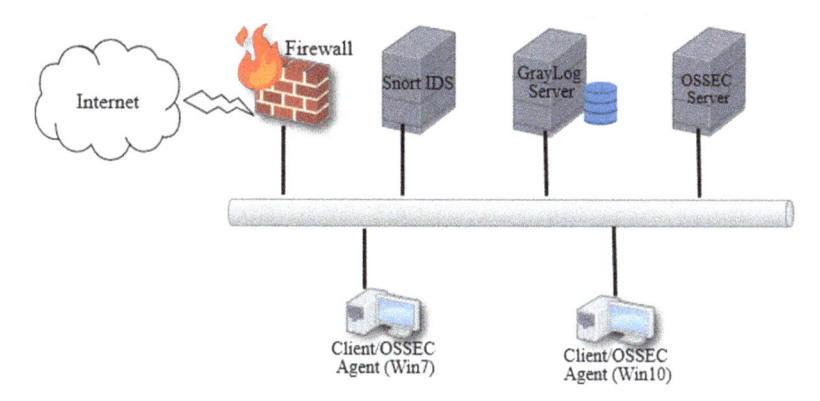

Fig. 2. LAN setup using VMware to implement the ASA framework using SNORT, OSSEC, GrayLog and Opnsense firewall

The Snort logs (alerts) were configured to be sent to the GrayLog server. Snort by default stores log data in files each time the service is started. The directory is /var/log/snort containing a snort.log.timestamp file. Instead of capturing the logs from these files, a syslog service (rsyslog v8-stable) was installed and configured. The RSYSLOG is the rocket-fast system for log processing [40]. The snort.conf file is edited to add the line output alert_syslog: LOG_LOCAL5 LOG_ALERT which redirects the logging to /var/log/syslog. Thus, Snort sends logs to the syslog service which then forwards it to the Graylog server. To allow the GrayLog server to retrieve logs from Snort's syslog, the rsyslog.conf file needs to be configured accordingly.

OSSEC was configured in the client-server model. OSSEC agents were installed and configured on two client nodes (Windows 7 and Windows 10) and the OSSEC server was installed on a Linux Ubuntu server. To enable the OSSEC agents to communicate with the OSSEC server to send logs from the Windows system, the agent needs to be registered with the OSSEC server (IP address of OSSEC server configured on agent) and provided an authentication key. Hence, the agents need to be registered with the server first. This is achieved by using the command /var/ossec/bin/manage-agents, whereby the authentication key for each registered agent is created and made available. The GrayLog server was also configured to retrieve the logs from OSSEC server.

The Windows AppLocker was also configured on the Windows client machines. This can be done via the Local Security Policy Wizard where AppLocker is available under the Application Control Policies. Rules can be created to define security policies. In the context of ransomware attack prevention, rules were created such that only applications and executables found in the Program Files and Windows folders are to be allowed to run. Similarly, a Deny rule was created that prohibits execution of executables or applications that users can run from their profiles. This rule will apply to all exe files located under C:\users*. Once the rules have been created, the Application Identity service needs to be started and must be set to run at startup via the services console. When these configurations are done, the Windows Event Viewer will start populating events and logs when applications are being run. Events are captured for permitted and denied applications. These logs can be captured by the OSSEC agent and sent to the OSSEC server. The GrayLog server will in turn pull these logs to display them on the Web interface of the administrator. Opnsense was installed on another virtual machine and the internal IP address was set to 192.168.142.100.

Some test packets were used to verify the logging mechanism of the GrayLog server. Logs were observed on the Graylog dashboard. Alert emails were also recorded in the configured email accounts. Both inbound and outbound traffic was successfully detected and logged by Snort. The OSSEC agent deployed on the Windows client machines has for purpose to scan and report file changes. Tests were done to verify that logs were sent to the GrayLog server and can be monitored via the GrayLog's interface. Table 2 summarises the tests carried out on the proposed ASA framework.

Table 2. Testing the ASA implementation

Test case	Scenario	Observations
Testing internal/external packet logging with Snort and the GrayLog Server	Pings test from internal nodes outbound & ping tests from external nodes to internal nodes	Snort starts to trigger alert logs, GrayLog server starts acquiring real-time snort logs
Testing OSSEC agent	A text file is created and later edited on the Win7 client	OSSEC Agent report the file changes, GrayLog server display the OSSEC logs
Testing AppLocker	Application and Exe files located in the allowed folders are executed	AppLocker allow execution of these applications, Logs are seen under Windows Event Viewer. No logs sent to OSSEC and therefore no info pulled by GrayLog
Testing AppLocker	Application is run from a prohibited folder	AppLocker block execution of the application, Logs seen under Windows Event Viewer. Logs sent to OSSEC and pulled by GrayLog

4.2 Testing the ASA Implementation for Ransomware Attack

The proposed ASA framework was tested using Wannacry and Petya as shown in Table 3. These two ransomwares were chosen as they are different in the way they operate. WannaCry encrypts users' files and communicate with a C&C server, Petya infects the MBR and encrypts file but does not communicate with a C&C server.

The WannaCry ransomware was sent to the Windows 7 client while AppLocker was disabled on the system. Since AppLocker was not enabled for this test, the ransomware attack was successful on the client. However, alerts logs were successfully generated to notify the administrator of the attack, such that the attack can be quickly isolated and eliminated. Snort running on NIDS mode was configured to detect TOR SSL traffic in its rules. Detections would create logs and these were pulled by GrayLog Server and displayed on the web interface. Figure 3 depicts the TOR SSL communication detected between the C&C server and the client. From the messages, IP addresses were obtained as well as the geo-localization of the command server. GrayLog has the ability to provide GPS information and country codes based on public IP addresses. In this test, TOR SSL traffic has been identified and communication established towards public IP addresses 37.187.112.64 and 78.47.18.110. These public IP addresses were those from command servers located in France and Germany. These IP addresses of the C&C servers identified were blocked from the network by updating the rules of the Opnsense firewall.

Table 3. Ransomware attacks on the experimental network.

Attack scenario	Observations
Wannacry attack on a Windows 7 client	Network intrusion detected by Snort, Logs appear on GrayLog Web interface. Alerts trigger email notification
Petya Attack on a Windows 10 client OS	AppLocker blocks program execution, Logs appear on GrayLog Web interface. Alerts trigger email notification

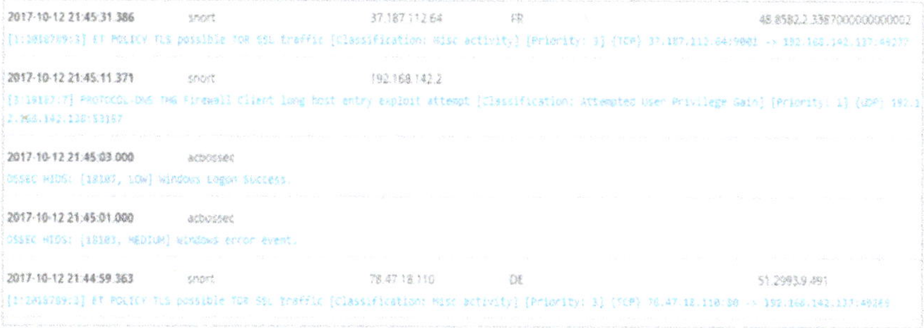

Fig. 3. Excerpt of the Graylog web interface during the Wannacry ransomware attack on the Windows 7 client.

Alerts were also triggered by both OSSEC and Graylog to notify about the ongoing attack. Email notification was sent from the Graylog server is shown in Fig. 4. The email alert notification received contained information from log captured by Snort and pulled by GrayLog. It shows clearly ET POLICY TOR SSL Traffic suggesting that there has been a potential Ransomware communication between the computer and a C&C server. The source IP address identified is 37.187.112.64 and has been located in France (country iso code – FR).

```
source: snort | message: [1:2018789:3] ET POLICY TLS possible TOR SSL traffic [Classification: Misc activity] [Priority: 3] {TCP} 37.187.112.64:9001 ->
192.168.142.137:49277 { snort_protocol: TCP | level: 1 | gl2_remote_ip: 192.168.142.25 | gl2_remote_port: 49320 | streams:
[5967a3869737b6049b6fd61f, 00000000000000000000001] | gl2_source_input: 59679eb39737b6049b6fd0e7 | snort_priority: 3 | dst_ip:
192.168.142.137 | src_ip: 37.187.112.64 | snort_classification: Misc activity | application_name: snort | src_ip_country_code: FR | full_message: <169>0
2017-10-12T21:45:31.386130+04:00 snort snort - - - [1:2018789:3] ET POLICY TLS possible TOR SSL traffic [Classification: Misc activity] [Priority: 3]
{TCP} 37.187.112.64:9001 -> 192.168.142.137:49277 | snort_message: ET POLICY TLS possible TOR SSL traffic | src_ip_city_name: N/A |
gl2_source_node: 467a97d6-faf1-4d7d-b500-a8dac040e680 } | _id: 24598351-af75-11e7-be43-000c292b7dcd | src_ip_geolocation:
48.8582,2.3387000000000002 | facility: local5 | timestamp: 2017-10-12T17:45:31.386Z }
```

Fig. 4. Excerpt of the email Alert received from the Graylog server during the Wannacry ransomware attack on the Windows 7 client

The second ransomware, namely Petya, was executed from the desktop of the Windows 10 machine with AppLocker enabled and configured (assuming that the ransomware was downloaded as email attachment and stored in desktop). The attack was successfully blocked by the AppLocker feature. This execution is captured by Windows and can be observed from the Windows event viewer as shown in Fig. 5. This event was also sent to the OSSEC server via the OSSEC agent on the Windows 10 client. This event was represented by a log on the GrayLog server. Alerts triggered email notifications sent from OSSEC server and GrayLog server are shown in Fig. 6.

Fig. 5. Excerpt of the Windows Event Viewer during the Petya ransomware attack on the Windows 10 client.

Fig. 6. Email Alert Notification during the Petya ransomware attack on the Windows 10 client.

5 Conclusion

According to the ASA model, to protect against security breaches, a continuous and coordinated monitoring of the system is required. This is even more applicable for protection against ransomware attacks which are complex multi steps attacks, characterised by morphing codes and which uses different techniques to infect a computer. In this work, a framework is proposed to implement an adaptive security architecture for detecting and mitigating the risks of ransomware attacks using a combination of opensource tools namely the network intrusion detection and prevention system, Snort, the host intrusion detection system, OSSEC, rsyslog and Graylog for capturing log, the AppLocker feature on Windows client machines and the Opnsense firewall. Based on the combined use of these open source tools, a network can be effectively monitored and email alerts sent to respective administrators in case of suspicious traffic or dangerous events triggered in the system. Such a monitoring system can provide critical information about the whole system and allows detecting and preventing many attacks.

Two different ransomwares were executed in the implemented ASA environment to observe the ability of the system to effectively detect attack. In the case of the WannaCry ransomware, logs on Graylog showing potential dangerous TOR communication on port 9001 was detected. Such early alerts can allow the system administrator to limit the damage caused by the ransomware. IP addresses of the identified C&C server can be blocked by updating the Opnsense firewall. In the case of the Petya ransomware, the AppLocker feature of the Windows client was also able to effectively block the ransomware from executing on the machine. Given that the Petya ransomware does not communicate with a C&C server, logs from Snort NIDS may not be effective in detecting the attack, however the OSSEC HIDS definitely logs changes on files, which can be used to detect the attack. The AppLocker was able to block the attack, as the ransomware was executed from desktop. This would have worked even if the ransomware were to be executed from the Downloads or Documents folder.

The proposed framework for deploying an adaptive security architecture using opensource thus demonstrates the effectiveness of the model for protection against ransomware attacks. The coordinated logging mechanism in place provides visibility on the network which can help to increase the resilience of the network to face malware attacks and also minimizing the reaction time towards ransomware infection. The AppLocker feature is only available on Windows clients and thus our implementation is for the Windows environment only. Future works involve the integration of log analysis and the use of machine learning to evaluate patterns and predict attacks in a timely manner as ransomware remains unpredictable and the characteristics of a new zero-day attack could be easily go undetected.

References

1. O'Brien, D.: Symantec Internet Security Threat Report, Special Report: Ransomware. Symantec (2017)
2. Arsene, L., Gheorghe, A.: Ransomware, A Victim's Perspective: Bitdefender, A Study on US and European Internet Users. BitDefender (2016)

3. Osterman Research: Understanding the Depth of the Global Ransomware Problem, Survey Report. MalwareBytes (2016)
4. Lord, N.: Ransomware Protection & Removal: How Businesses Can Best Defend Against Ransomware Attacks, 06 March 2018. https://digitalguardian.com/blog/ransomware-protection-attacks. Accessed 15 June 2018
5. Fortinet: Threat Landscape Report Q1 2018. Fortinet (2018)
6. Kassner, M.: Ransomware: Extortion via the Internet, 11 January 2010. https://www.techrepublic.com/blog/it-security/ransomware-extortion-via-the-internet/
7. Thada, V.: A primer on ransomware: extortion on the internet. Int. J. Future Revolut. Comput. Sci. Commun. Eng. **3**(9), 63–69 (2017)
8. F-Secure: State of CyberSecurity 2017 (2017). https://www.f-secure.com/documents/996508/1030743/cyber-security-report-2017
9. Al-rimy, B.A.S., Maarof, M.A., Shaid, S.Z.M.: Ransomware threat success factors, taxonomy, and countermeasures: a survey and research directions. J. Comput. Secur. **74**, 144–166 (2018)
10. Zorabedian, J.: Anatomy of a ransomware attack: CryptoLocker, CryptoWall, and how to stay safe, 03 March 2015. https://news.sophos.com/en-us/2015/03/03/anatomy-of-a-ransomware-attack-cryptolocker-cryptowall-and-how-to-stay-safe-infographic/
11. Elise: CryptoLocker – a new ransomware variant, 10 September 2013. https://blog.emsisoft.com/en/1615/cryptolocker-a-new-ransomware-variant/
12. Avast Intelligence: A closer look at the Locky ransomware, March 2016. https://blog.avast.com/a-closer-look-at-the-locky-ransomware
13. CERT-MU: The WannaCry Ransomware White Paper. CERT-MU, Mauritius (2017)
14. Malwarebytes Labs: Petya – Taking Ransomware to the Low Level, June 2017. https://blog.malwarebytes.com/threat-analysis/2016/04/petya-ransomware/
15. Kolodenkerz, E., Koch, W., Stringhiniy, G., Egele, M.: PayBreak: defense against cryptographic ransomware. In: Conference on Computer and Communications Security, Asia (2017)
16. Palisse, A., Le Bouder, H., Lanet, J.-L., Le Guernic, C., Legay, A.: Ransomware and the legacy crypto. In: International Conference on Risks and Security of Internet and Systems (CRiSIS) (2017)
17. Weckstén, M., Frick, J., Sjöström, A., Järpe, E.: A novel method for recovery from crypto-ransomware infections. In: 2nd IEEE International Conference on Computer and Communications (2016)
18. Woo, S.-U., Kim, D.-H., Chung, T.-M.: Method of detecting malware through analysis of opcodes frequency with machine learning technique. In: Advances in Computer Science and Ubiquitous Computing. Lecture Notes in Electrical Engineering, vol. 421. Springer, Singapore (2017)
19. Yewale, A., Singh, M.: Malware detection based on opcode frequency. In: International Conference on Advanced Communication Control and Computing Technologies (ICACCCT) (2016)
20. Gómez-Hernández, J.A., Álvarez-González, L., García-Teodoro, P.: R-Locker: thwarting ransomware action through a honeyfile-based approach. J. Comput. Secur. **73**, 389–398 (2018)
21. Ravi, C., Manoharan, R.: Malware detection using windows API sequence and machine learning. Int. J. Comput. Appl. **43**(17), 12–16 (2012)
22. Hampton, N., Baig, Z., Zeadally, S.: Ransomware behavioural analysis on Windows platform. J. Inf. Secur. Appl. **40**, 44–51 (2018)

23. Alhawi, O.M.K., Baldwin, J., Dehghantanha, A.: Leveraging machine learning techniques for Windows ransomware network traffic detection. In: Cyber Threat Intelligence. Advances in Information Security, vol. 70. Springer, Cham (2018)
24. Gangwar, K., Mohanty, S., Mohapatra, A.K.: Analysis and detection of ransomware through its delivery methods. In: Data Science and Analytics, REDSET 2017. Communications in Computer and Information Science, vol. 799. Springer, Singapore (2018)
25. Shaukat, S.K., Ribeiro, V.J.: RansomWall: a layered defense system against cryptographic ransomware attacks using machine learning. In: 10th International Conference on Communication Systems & Networks (COMSNETS) (2018)
26. Baykara, M., Sekin, B.: A novel approach to ransomware: designing a safe zone system. In: 6th International Symposium on Digital Forensic and Security (ISDFS), Antalya, Turkey (2018)
27. Moore, C.: Detecting ransomware with honeypot techniques. In: Cybersecurity and Cyberforensics Conference (CCC), Amman, Jordan (2016)
28. Azmoodeh, A., Dehghantanha, A., Conti, M., Choo, K.-K.R.: Detecting crypto-ransomware in IoT networks based on energy consumption footprint. J. Ambient Intell. Hum. Comput. **9**, 1141–1152 (2017)
29. Ferrante, A., Malek, M., Martinelli, F., Mercaldo, F., Milosevic, J.: Extinguishing ransomware - a hybrid approach to android ransomware detection. In: Foundations and Practice of Security. Lecture Notes in Computer Science. Springer (2017)
30. Kharraz, A., Robertson, W., Balzarotti, D., Bilge, L., Kirda, E.: Cutting the gordian knot: a look under the hood of ransomware attacks. In: Detection of Intrusions and Malware, and Vulnerability Assessment, DIMVA 2015. Lecture Notes in Computer Science, vol. 9148. Springer, Cham (2015)
31. Cabaj, K., Mazurczyk, W.: Using software-defined networking for ransomware mitigation: the case of cryptowall. IEEE Netw. **30**(6), 14–20 (2016)
32. van der Meulen, R.: Build adaptive security architecture into your organisation, 30 June 2017. https://www.gartner.com/smarterwithgartner/build-adaptive-security-architecture-into-your-organization/
33. Sinno, S., Negri, F., Goldhammer, S.: Designing an adaptive security architecture with unisys stealth and logrhythm. White Paper, Unisys Corporation, USA (2017)
34. Snort (2018). https://www.snort.org/
35. OSSEC: OSSEC (Open Source HIDS SEcurity) (2018). https://www.ossec.net/
36. GrayLog (2018). https://www.graylog.org/
37. Elasticsearch (2018). https://www.elastic.co/
38. OPNsense (2018). https://opnsense.org/
39. Microsoft: What is AppLocker? (2018). https://docs.microsoft.com/en-us/windows/security/threatprotection/windows-defender-application-control/applocker/what-is-applocker
40. Rsyslog: The rocket-fast system for log processing (2018). https://www.rsyslog.com/

Shilling Attack Detection Scheme in Collaborative Filtering Recommendation System Based on Recurrent Neural Network

Jianling Gao[1,2], Lingtao Qi[1,2], Haiping Huang[1,2(✉)], and Chao Sha[1,2]

[1] Nanjing University of Posts and Telecommunications, Nanjing, JiangSu, China
hhp@njupt.edu.cn
[2] Jiangsu High Technology Research Key Laboratory for Wireless Sensor Networks, Nanjing 210003, China

Abstract. With the prosperity of the modern electronic business, merchandise recommendation system has become an important tool for online shopping. However, the threat of shilling attack caused by injecting fake rating records into the system cannot be ignored. To deal with shilling attacks, many methods especial user profile-based detection methods have been proposed. But there are still remainder challenging problems in those methods: (1) detection attributes need to be designed in advance; (2) instability of detection effect when faced with variety of attack models; (3) other aspects such as low accuracy, high computing cost and failure in detecting some special shilling attacks. Therefore, a shilling attack detection scheme based on neural network is proposed in this paper in order to address these challenging problems. In this scheme, the LSTM model is used to learn the historical rating records, and then predict the ratings for the next period when the desired accuracy is achieved. Chi-square test is used to determine whether the item is under attacks by comparing the predicted ratings and the actual ones within the time period. The simulation of experimental results on MovieLens 20M dataset show that our proposal is feasible and effective, and it improves the detection performance.

Keywords: Recommender system · Shilling attack detection · Recurrent neural network · Chi-square test · LSTM

1 Introduction

The collaborative filtering recommender system (CFRS) [1] has been widely used by many e-commerce platforms. CFRS recommends products to customers that they might like according to their history rating records. However, in the real world, malicious users may inject fake rating records into the system to control the results of recommendation to achieve their goals (the scenario is called "shilling attack" [2]). Shilling attack will have a great negative impact on CFRS because CFRS defaults the history rating records to be authentic and credible. To deal with the problem above, many researches are absorbed in detecting shilling attack, but most of them focus on users' profiles, and one of the remainder challenging problems is that user profile-based detection needs to manually determine the features of attacking profiles, which bring about the extra cost. Another challenge caused by user profile detection lies in the diversity of attack models such as random attacks, average attacks and so on. Actually, some researches have proved that focusing on items may be also helpful to detect shilling attack especially to shield the difference among various attack models. Furthermore, other factors such as low accuracy and failure in detecting some special shilling attacks of current detection methods motivate us to design a more effective solution.

Aiming at the challenging problems mentioned above, the main contributions of this paper can be summarized as follows:

1. An item-based shilling attack detection scheme based on LSTM-RNN is proposed to achieve more intelligent detection. Our proposal doesn't need to design detection attributes in advance, and it can realize the automatic detection.
2. Compared with other classic schemes, it achieves a better performance and a higher accuracy and stability for variety of attack models.

The rest of this paper is organized as follows. Section 2 reviews related works. The preliminary knowledge regarding shilling attacks is described in Sect. 3. Section 4 describes the framework and details of attack detection scheme based on LSTM-RNN. Simulation experiments and results analysis are shown in Sect. 5. Section 6 concludes the whole paper.

2 Related Work

Based on the research progress of shilling attack detection, it can be classified into three categories: shilling attack detection based on supervised learning, shilling attack detection based on unsupervised learning, and shilling attack detection based on semi-supervised learning. All of them rely on the pre-set detection attributes.

Shilling attack detection based on supervised learning is based on the following assumptions: there are relatively obvious differences in statistic features between attacking profiles and real profiles. Usually, detectors can distinguish attacking profiles from normal users' profiles by some statistic attributes such as Rating Deviation from Mean Agreement (RDMA), Weighted Degree of Agreement (WDA), and Weighted Deviation from Mean Agreement (WDMA). Zhang and Zhou [3] introduced Hilbert-Huang Transform (HHT) to conduct specific changes to users' rating matrix, which outlines the attacking profiles' features. And then, SVM algorithm is used to detect shilling attack based on the extracted features. However, it will take a lot of time to collect labeled samples and complete training.

Shilling attack detection based on unsupervised learning need some priori knowledge, instead of training samples. Zhou et al. [4] improved RDMA and Degree of Similarity with Top Neighbors (DegSim) and proposed a kind of unsupervised algorithm: RD-TIA algorithm. This algorithm uses the combination features of rating pattern and attacking profiles to detect attacks, which obtains a desirable detection result. Compared with supervised methods, unsupervised methods involve much less computation. However, sometimes it is difficult to obtain priori knowledge.

Shilling attack detection based on semi-supervised learning need not only some labeled samples, but also some prior knowledge. They use both labeled users' and unlabeled users' profiles in the same time to conduct multiple modeling [5]. Cao et al. [8] proposed a shilling attack detection scheme based on semi-supervised learning (called Semi-SAD). This algorithm combines Bayes Classifier with an augmented Expectation Maximization (called EM-λ) to learn labeled and unlabeled users' profiles. Although this method performs well in accuracy, when it comes to mixed samples attack, it has a poor detection performance.

From the motivation of detection scheme, shilling attack detection technology can be divided into two categories: shilling attack detection based on users' profiles and anomaly item detection based on items. At present, most of research is based on users' profiles. However, a small group of researchers take the opinion that the time distribution of an item's ratings can also reveal the existence of various kinds of shilling attack, and some reasonable assumptions about time sequence are made. Gao et al. proposed an anomaly item detection algorithm based on dynamic partition (called D-Window) [7]. They firstly used piecewise linear representation to preprocess the rating time series, then partition them based on the important points, and determine whether it is abnormal by using chi-square test. The main disadvantages of this scheme are that it's complex to partition important points, which results in a high time cost. Besides, experiment results showed that it has a relatively high false detection rate.

3 Shilling Attack Model

Shilling attack refers to a kind of attack that malicious users inject fake rating records into the CFRS to control the results of recommendation.

Generally speaking, a fake profile in a shilling attack may consist of four parts, as is shown in Fig. 1. Where IS refers to selected items, in different shilling attack models they will be selected and given scores in different ways. IF refers to filling items, which is concerned with the filling scale set when the attacking profile is created. IØ refers to void items, which means items that have not been rated. IT refers to target items, which are items that the attacker wants to control.

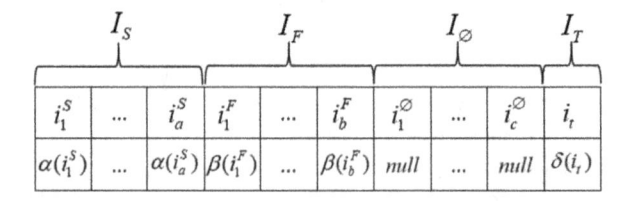

I_S			I_F			$I_\varnothing$			I_T
i_1^S	...	i_a^S	i_1^F	...	i_b^F	$i_1^\varnothing$	...	$i_c^\varnothing$	i_t
$\alpha(i_1^S)$	...	$\alpha(i_a^S)$	$\beta(i_1^F)$	...	$\beta(i_b^F)$	null	...	null	$\delta(i_t)$

Fig. 1. Composition of an attacking profile

From the attacker's motivation, shilling attack can be classified into two groups: push attack and nuke attack. Push attack means the attackers expect to make the target items appear in the recommendation list of more users, and usually gives a high score to the target items in the fake profile. And conversely, nuke attack frequently gives a low score to the target items to make them recommended to less users. Common shilling attack models usually include random attack, average attack, bandwagon attack, mixed attack, etc.

4 Shilling Attack Detection Based on LSTM-RNN

4.1 Scheme Design

In our scheme, we firstly extract the data from the dataset and arrange them in time order. After normalization, part of the data is used to the LSTM-RNN (Long Short-Term Memory-Recurrent Neural Network) training model. After the tracking accuracy of this model can reach a certain degree, we use the model to predict the rating data of next period. And then, chi-square test is used to compare the predicted rating data and the original data. If the data is under shilling attack, the result of chi-square test will exceed the normal value. This designed scheme can be executed as shown in Table 1.

Table 1. The execution of the designed scheme

Input: The rating data of item A

Output: 0 for attack detected, 1 for no attack detected

1. Rank the data in time order, and divide the data into S_1 and S_2;

2. S_1 = normalization (S_1), S_2 = normalization (S_2);

3. LSTM-RNN. train(S_1);

4. S_2' = LSTM-RNN. prediction ();

5. χ^2 = χ, 2-test (S_2, S_2');

6. return if ($\chi^2 < \chi_0^2$).

Since this scheme does not rely on the rating data of other items when carrying out the detection of a single item, it can adopt the method of "parallel processing" for the detection of multiple items. With the rapid development of parallel computing, multi-item detection can still achieve high detection efficiency.

4.2 Rating Tracking and Prediction

Although RNN model has memory function, as the time sequence becomes longer, the memory ability of the later time node will decline compared with the previous time node, just like the human memory for a long time ago will gradually blur or even disappear [6]. Fortunately, this problem has already been solved by LSTM [7]. LSTM introduces an important element, the cell, to achieve the goal of learning long-term dependent information. The cell can save the information of the previous state, thus it can combine previous states, current memories, and current inputs. It has been proved that the network structure is very effective in dealing with long time sequence dependence problems. Its structure can be seen in Fig. 2.

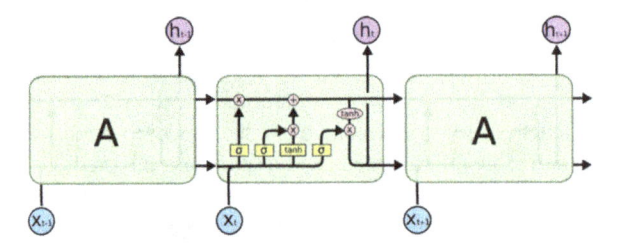

Fig. 2. Structure of LSTM

The meanings of the icons in LSTM are shown in Fig. 3. Wherein a line with arrow means a transfer of a vector from a cell's output to another cell's input, a circle means point-by-point operations such as addition of two vectors, the combined arrowhead lines mean connections of vectors, and separate double arrow lines indicate that the content is copied and distributed to different locations.

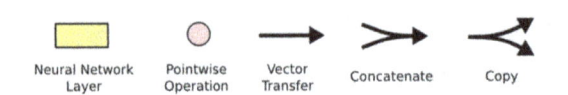

Fig. 3. Meanings of the icons in LSTM

LSTM-RNN model is used to rating tracking and prediction. The first step is to determine which information to be discarded from the cell state through the forget gate. The forget gate uses sigmoid function to normalize the data into the interval [0,1] (0 means "totally forget", whereas 1 means "totally remember"). The input of the gate is h_{t-1} and x_t, and the output is a vector whose values are all between 0 and 1 (the length of the vector is the same as that of the cell state C_{t-1}), which means the passing proportion of each part of C_{t-1}. Wherein, 0 refers to "prevent any information from passing", and 1 refers to "let all of the information pass". Formula (1) shows the function of forget gate, where W_i is the weighting vector, h_{t-1} is the output of the previous cell, $\mathbf{x}_t$ is the output of the current cell, σ means sigmoid function and $\mathbf{b}_i$ refers to the offset vector.

$$\mathbf{f}_t = \sigma(\mathbf{W}_i \cdot [\mathbf{h}_{t-1}, \mathbf{x}_t] + \mathbf{b}_i) \qquad (1)$$

Then it decides which new information to pass to the cell state, which will be implemented by the input gate. The input gate consists of two key layers: the sigmoid layer and the tanh layer. The sigmoid layer decides which information to be updated, as is shown in formula (2), and the tanh layer generates alternative content $\tilde{\mathbf{C}}$ for updating, as is shown in formula (3).

$$\mathbf{i}_t = \sigma(\mathbf{W}_i \cdot [\mathbf{h}_{t-1}, \mathbf{x}_t] + \mathbf{b}_i) \qquad (2)$$

$$\tilde{\mathbf{C}} = \tanh(\mathbf{W}_\mathbf{C}\mathbf{X}_t + \mathbf{W}_\mathbf{C}\mathbf{h}_{t-1} + \mathbf{b}_\mathbf{C}) \qquad (3)$$

The next step is to combine the outputs of forget gate and input gate to update the state of cell. The cell state of $\mathbf{C}_{t-1}$ will be updated to $\mathbf{C}_t$, as is shown in formula (4). $\mathbf{C}_{t-1}$ is multiplied by the output $\mathbf{f}_t$ of the forget gate to discard the information that need to be discarded, and then it is added the output value of input gate to obtain the updated state.

$$\mathbf{C}_t = \mathbf{f}_t * \mathbf{C}_{t-1} + \mathbf{i}_t * \tilde{\mathbf{C}}_t \qquad (4)$$

Finally, the output gate is used to determine the output. First, the sigmoid function is employed to determine what information of $\mathbf{h}_{t-1}$ and $\mathbf{x}_t$ will be outputted as $\mathbf{O}_t$, as is shown in formula (5). Second, the tanh function is adopted to deal with the state of cell $\mathbf{C}_t$ (the result is between -1 and 1), and then it is multiplied by $\mathbf{O}_t$ to obtain the final output $\mathbf{h}_t$, as is shown in formula (6).

$$\mathbf{o}_t = \sigma(\mathbf{W}_\mathbf{o}\mathbf{x}_t + \mathbf{W}_\mathbf{o}\mathbf{h}_{t-1} + \mathbf{b}_\mathbf{o}) \qquad (5)$$

$$\mathbf{h}_t = \mathbf{o}_t * \tanh(\mathbf{C}_t) \qquad (6)$$

4.3 Chi-Square Test

Chi-square test is a kind of hypothesis testing method which can be widely applied. It is usually used to compare two or more components, or to do some analysis of the correlation between two categorical variables. In this paper we use chi-square test to compare the predicted data with the test data in order to determine whether the data is under shilling attack. Its calculation expression can be seen below:

$$\chi^2 = \sum_{i=1}^{k} \frac{(A_i - np_i)^2}{np_i}, i = 1, 2, 3, \cdots, k \qquad (7)$$

Where A_i is the horizontal observation frequency, n is the total frequency, p_i refers to the i_{th} horizontal expected frequency. When n is relatively large, χ^2 will be approximately subject to the chi-square distribution of k-1 degrees of freedom. According to formula (7), when the observed frequency is exactly the same as the expected frequency, $\chi^2 = 0$. The smaller the difference between the observed frequency and the expected frequency is, the smaller the χ^2 value will be. Conversely, the greater the difference between the two, the larger the χ^2 value will be. Therefore, χ^2 is a measure of the distance between the frequency and the expected frequency, and it is also a measure of the hypothesis.

5 Experiments and Analysis

5.1 Dataset

The Movie Lens 20M dataset is selected as the experiment dataset, which is released in April 2015 and contains 138,000 users' 20 million ratings of 27,000 movies. All the rating values are between 1 and 5, where 5 is the maximum score and 1 is the minimum. Since the dataset is very large, taking our hardware conditions into consideration, we only select the items which have over 200 rating records.

5.2 Experimental Setup

After the data normalization operation, we can transfer the data to LSTM model and train the model to reach a certain accuracy to carry out prediction.

First, we establish the LSTM model. The number of hidden neurons in the first layer is 50. At the same time, the number of input neurons is set to 1, the number of output neurons is set to 1, and the number of model iterations is set to 50. We introduce Sequential, Dense, Activation, Dropout and LSTM function library in keras frame and add layers to the model one by one by using the *add()* method. In the first layer of LSTM, the stride of input layer is *input_dim* and the stride of output layer is *output_dim*. We also use Dropout function in the hidden layer, which will select 20 percent of nodes to discard when the weight is updated during each round. Activation function is set as linear, and loss function is set as *mse*, which means *mean_squared_error*, and the optimizer parameter is set to *rmsprop*. The values of the parameters can be seen in the Table 2.

Table 2. Values of parameters

Parameters	Values
Number of neurons in the first layer	50
Number of output neurons	1
Stride of output layer	*output_dim*
Number of input neurons	1
Stride of input layer	*input_dim*
Loss function	*mse* (mean squared error)

The second step is to train the established LSTM model by using training data set. According to the prediction quantity, the training set and the test set are divided with the proportion of 9:1. The training data set is the sample set for learning; the model is established by matching corresponding parameters, while the test data set is used to evaluate the performance of the model. To obtain the accuracy of the model, we can compare the original data with the data predicted by the model. Due to the large amount of data, the model training time is relatively long, but the principle of data presentation is similar.

5.3 Evaluation Metrics

In this paper, the detection accuracy and the detection error rate are used as metrics to measure the performance of our proposal, where the detection accuracy refers to the ratio of the number of attacks detected to the total number of attacks. The calculation expression is shown in formula (8):

$$Accuracy = \frac{DetectedAttacks}{Attacks} \quad (8)$$

Error detection rate refers to the ratio between the number of normal items judged to be attacked items and the total number of normal items. The calculation formula is shown in formula (9). A low error detection rate indicates that less normal items were judged as abnormal items.

$$ErrorRate = \frac{FalseDetection}{NormalItems} \quad (9)$$

5.4 Comparison with Other Schemes

In order to verify the detection performance of the scheme we proposed above, in this section four schemes involved HHT-SVM [3, 8], RD-TIA [4], Semi-SAD [6] and D-Window [9] are selected to make comparisons with our proposal. To ensure the accuracy of the experiments, each scheme will be performed for fifty times independently and the average value will be taken as the result.

In our experiments, the filling scale is set to 3%, the attack size is set to 1%, 3%, 5%, and 10%, respectively. On the premise of push attack, the accuracy rate and detection error rate of each scheme under four attack models including random attack, average attack, popular attack and mixed attack (composed of random attack, mean attack and popular attack with a ratio of 1:1:1) are detected. The experiment results are shown in Figs. 4, 5, 6, 7, 8, 9, 10, and 11.

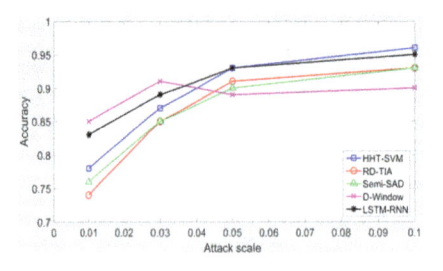

Fig. 4. Detection accuracy in random attack

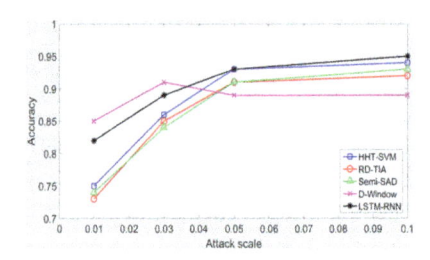

Fig. 5. Detection accuracy in average attack

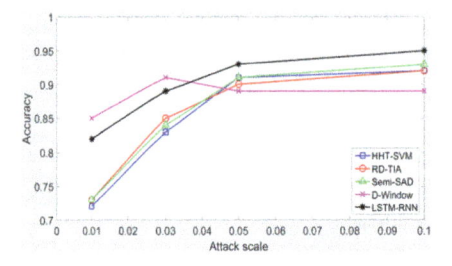

Fig. 6. Detection accuracy in bandwagon attack

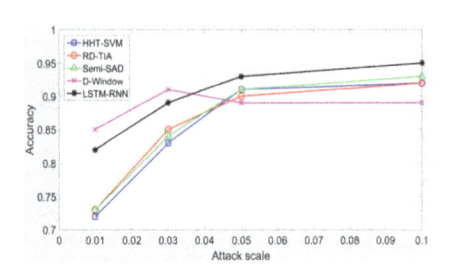

Fig. 7. Detection accuracy in mixed attack

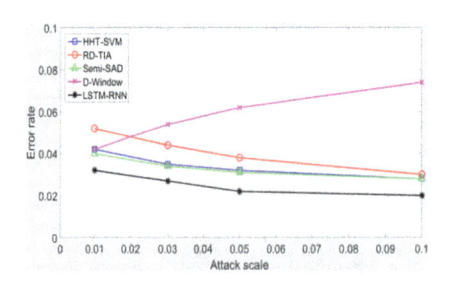

Fig. 8. Error rate in random attack

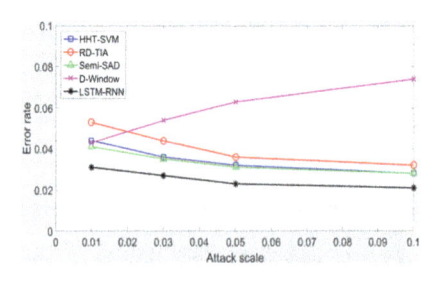

Fig. 9. Error rate in average attack

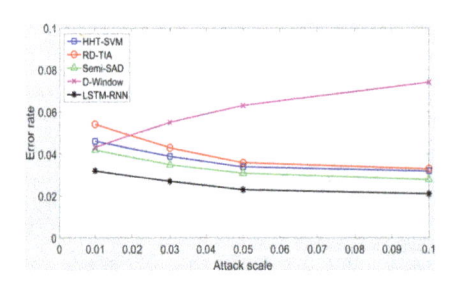

Fig. 10. Error rate in bandwagon attack

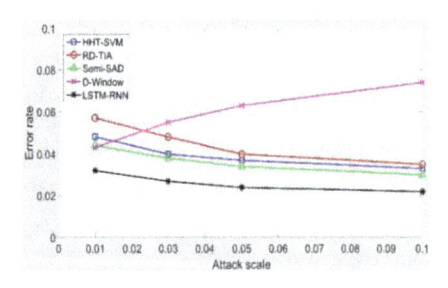

Fig. 11. Error rate in mixed attack

As is shown in the figures, we can see that under the purpose of push attack, LSTM-RNN model's performance is as good as HHT-SVM, but it is more stable. As for mixed attack, there are remarkable declines in the accuracy of HHT-SVM, RD-TIA and Semi-SAD, while there are no significant changes in the performance of D-Window and LSTM-RNN.

It's also important to note that almost no changes to the accuracy of D-Window and LSTM-RNN when detecting different attacking models, which is one of the advantages of anomaly item detection scheme. But the performance of D-Window in a large dataset like MovieLens is not as good as LSTM, the reason may be that there are lots of rating records at the same time, which will cause the detector can't work well. However, LSTM-RNN can improve the detection accuracy in the process of continuous accumulation of rating data, since large amount of data will improve the learning process of recurrent neural network.

6 Conclusions

This research aims at the challenging problems such as traditional shilling attack detection technology requires manual selection of detection attributes, and proposed a kind of shilling attack detection scheme based on recurrent neural network to avoid manual operation and meanwhile can make full use of the vast amount of rating data to achieve a high efficiency. This scheme uses LSTM model to learn history rating data and predict the rating trend for the next period after obtaining the desired accuracy. And then it determines whether the item is under shilling attacks by using chi-square test to compare predicted data with real one. The experimental results in Movie Lens 20M dataset have shown the feasibility, effectiveness and stable performance of our proposal when faced with various attack models. Based on the characteristic of LSTM, our proposal is suitable for big data scale recommendation system, whose large amount of ratings may be beneficial for improving the accuracy of prediction. However, there are still some problems to be solved. For example, in the real world, the CFRS may vary greatly in data volume, rating structure, user structure and other aspects, thus the impact of shilling attack may also be various. How to select the specific detection strategy would be the future work.

Acknowledgment. This work was supported by the National Natural Science Foundation of P. R. China (No. 61672297), the Key Research and Development Program of Jiangsu Province (Social Development Program, No. BE2017742).

References

1. Koren, Y., Bell, R.: Advances in collaborative filtering. In: Recommender Systems Handbook, pp. 145–186 (2015)
2. Mehta, B., Hofmann, T., Nejdl, W.: Lies and propaganda: detecting spam users in collaborative filtering. In: 12th International Conference on Intelligent User Interfaces, pp. 14–21 (2007)

3. Zhang, F., Zhou, Q.: HHT-SVM: an online method for detecting profile injection attacks in collaborative recommender systems. Knowl.-Based Syst. **65**, 96–105 (2014)
4. Zhou, W., Wen, J., Xiong, Q., et al.: Abnormal group user detection in recommender systems using multi-dimension time series. In: Lecture Notes of the Institute for Computer Sciences, Social-Informatics and Telecommunications Engineering, vol. 201, pp. 373–383 (2017)
5. Yang, Z., Cai, Z.: Detecting anomalous ratings in collaborative filtering recommender systems. Int. J. Digit. Crime Forensics **8**(2), 16–26 (2016)
6. Cao, J., Wu, Z., Mao, B., et al.: Shilling attack detection utilizing semi-supervised learning method for collaborative recommender system. World Wide Web **16**(5–6), 729–748 (2013)
7. Gao, M., Tian, R., Wen, J., et al.: Item anomaly detection based on dynamic partition for time series in recommender systems. PLoS ONE **10**(8), 135–155 (2015)
8. Verma, A., Sharma, S., Gupta, P.: RNN-LSTM based indoor scene classification with HoG features. Commun. Comput. Inf. Sci. **955**, 149–159 (2019)
9. Bram, B.: Reinforcement learning with long short-term memory. Adv. Neural Inf. Proces. Syst. **7**(6), 1475–1482 (2001)

Runtime API Signature for Fileless Malware Detection

Radah Tarek[✉], Saadi Chaimae, and Chaoui Habiba

System Engineering Laboratory, ADSI Team,
National School of Applied Sciences, Ibn Tofail University, Kenitra, Morocco
tarekradah@gmail.com, chaimaesaadi900@gmail.com,
habiba.chaoui@uit.ac.ma, mejhed90@gmail.com

Abstract. Nowadays, cybercriminals become sophisticated and conducting advanced malware attacks on critical infrastructures, both, in the private and public sector. Therefore, it's important to detect, respond and mitigate such threat to digital protection the cyber world. They leverage advanced malware techniques to bypass anti-virus software and being stealth while conducting malicious tasks. One of those techniques is called file-less malware in which malware authors abuse legitimate windows binaries to perform malicious tasks. Those binaries are called Living Off The Land Binaries (LOLBINS). That being said, during the execution of the attack it is not used any malicious executable and, consequently, the antivirus is unable to identify and prevent such threats. This paper focuses on defining rules to monitor the binaries used by threat actors in order to identify malicious behaviors.

Keywords: Malware analysis · Fileless malwares · Malware detection

1 Introduction

In [1], authors defined a computer virus as a program that can 'infect' other programs by modifying them to include a possibly evolved copy of itself. They explained how it is theoretically difficult to detect and prevent virus infection. The same constraint is applied to malware in general. Malware is becoming the preferred toolkits for cybercriminals, as it offers them a simple and effective way to infiltrate and damage computers without the user's consent. The term malware is a general term covering different kinds of threats to device safety, such as rootkits, RAT, botnet, etc.

Security researchers perform malware analysis to understand the behavior of malware and reveal their sophisticated techniques. Typically, the method by which malware is usually analyzed falls under two types: Static Malware analysis [2] and, Dynamic Malware analysis [3].

Traditional analysis and detection techniques are not suitable when dealing with advanced malware techniques. This paper aims to introduce a new way to recognize fileless and advanced malware by establishing custom rules for detection. Those rules are what we have called a dynamic signature.

This paper is organized as follows. Section 2 discusses the related works. Section 3 presents common malware detection and evasion techniques. In Sects. 4 and 5, paper describes in detail our proposed detector and the experimental results. Finally, in Sect. 6, we conclude with conclusion and future work.

2 Related Works

In this section, as related works, we present significant researches relevant to malware analysis and detection. As static analysis is not relevant to today's threats, many researchers have attempted to extract the dynamic information by executing files in a controlled environment also known as a sandbox. Authors in [3] proposed an API call sequence similarity to detect malware. Using API Hooking, they were able to trace API call sequences at runtime. Based upon this they were able to classify malware in different categories using API call sequences. This approach is useful to overcome obfuscation techniques used by malware. Another work [4] focused on machine learning detection, and showed how Adversarial Machine Learning used in malware detection can be under an attack. Using a classifier with the input of Windows Application Programming Interface (API), [5] presented a useful evasion attack model by considering different contributions of the features to the classification problem. To overcome this failure, they propose a secure-learning paradigm for malware detection, which present effectiveness against this kind of attacks. Regarding dynamic analysis evasion techniques, [6] proposed a hierarchical, classification of conventional Dynamic Analysis Evasion techniques. In the highest level, a Category of evasion means either the technique is for bypassing Automated or manual dynamic analysis process. For each category, several tactics exist. Tactics are the specific maneuvers or approach for evasion with the specified attitude of its parent category, for manual detection tactics are: Direct detection, Deductive detection, and debugger escaping. For the automated dynamic analysis evasion, tactics are either Detection-Dependent or Detection-Independent. Each tactic can be achieved using various techniques; techniques are the various practical implementations of those tactics. Author in [7] described how we could relate to various malware campaign based on similarities within binaries. They show through similarities that the developers of malware that hit several Middle East infrastructures have reutilized the same code of the old samples and the differences found in the malware binaries indicates that the developers have adapted their tools to evade systems. Authors in [8] were using YARA [9] rules in order to detect the four most relevant ransomware categories [10, 11] and [12]. Authors in [13] presented a dynamic API call sequence-based detector using machine learning with a 3rd order Markov chain that combines iterative learning process with run-time monitoring; Their proposed system has shown a higher rate of accuracy. Finally, in [14], authors presented a malware analysis framework for dynamic and static analysis of malware sample-based behavior, using Cuckoo Sandbox [15] for API call extractions. However, all the previously mentioned research was in an attempt to present a generic detection technique to identify malicious software. That is why we see the need for a technique that will give defenders the ability to decide which behavior is malicious or not, allowing them to prevent fileless malware infection. The main novelty of our proposed method is to allow malware researchers, implement their solutions to mitigate malware without having to write an antivirus, by using a dynamic signature. Malware analyst also can use it as an alternative to YARA rules as it allows them to overcome the use of packers and obfuscation techniques [16].

3 Common Malware Detection and Evasion Techniques

3.1 Detection Techniques

Signature-based detection is the most basic technique to detect malicious software [17], creators of AV software having previously identified and recorded information about viruses; as would a dictionary, the AV can detect and locate the presence of a virus. This dictionary is called the viral definition database, which contains the virus signatures. Before scrutinizing the static detection techniques, we will have a brief introduction to the PE file component [18]. Building upon this understanding, we can better understand and elaborate on the evasion techniques.

PE file: PE file or portable executable is the file format for object code, DLL (Dynamic link library) and others used in multiple versions of Windows operating system. The PE format represents a data structure that contains the necessary information for windows loader to handle the executable code, resolve sections, load the appropriate libraries, manages resources and various PE file component.

The static analysis relies on information extracted from the PE file in order to understand the program behavior and capabilities. Mainly that information is:

- Loaded DLLs: Dynamic link library is a kind of PE file in which several windows API function are stored and regrouped depending on the kind of operation functions can perform. For example, wininet.dll [19] provides functions API for network communication using the HTTP protocol.
- Imported Windows API function: From PE file we can extract windows API functions used by the program. Most Windows API functions are self-explanatory. For example, CreateProcess function under Kernel32.dll is used to create a new process; WriteFile is to write data to a file, etc. Windows API functions are well documented under the MSDN [19], but unfortunately, Microsoft does not provide documentation for some windows API, especially those related to NTDLL library (The lowest library in the user-space of operation system). Malware authors were observed using NTDLL functions in order to perform an obscure operation and to evade detection; however, a non-official and out of date documentation for NTDLL API exist on the internet [20].
- Extracted ASCII strings: ASCII strings can be extracted from a PE file, using various techniques [21], they are mainly located in the ".data" section of a PE file. Examining strings can help understand the program operations.

Resources: In a PE file, the resources section contain various data that will be used by the program at runtime, for example, icons, images, etc. In a malware analysis perspective, the entropy of those sections is taken into consideration [22]. For example, a PE file that contains a data section with higher entropy means that an encrypted, compressed or obfuscated data are embedded within the PE file.

3.2 Evasion Techniques

In this sub-section, we will present some bypass techniques used by malware authors making the traditional signature ineffective:

- String obfuscation: A malware author can hide the presence of ASCII string using simple XOR encryption. This is called string obfuscation. Even though there is some tools which can extract obfuscated string [23], those tools are ineffective with complex encryption schemes.
- API Calls obfuscation: In C and C++, rather than the static way in which a function can be called, functions can also be called dynamically at runtime [24]. This is done by first, loading the appropriate DLL in which the function is stored using LoadLibrary function, and then extracting a function handle using GetProcAddress. This operation is done using two API functions (GetProcAddress and LoadLibrary from kernel32.dll) which are legitimate and used by all windows programs. Another stealthier way is dynamically resolving GetProcAddress and LoadLibrary by parsing the PEB structure [25].
- A wide range of legitimate software uses resources sections; they contain images, and icons used by the program at runtime. Resources may also contain other PE file in case of an installer. That being said, it is not possible to rely on resources sections to detect a potential malware.

4 Contribution

To overcome the limitation of the static signature we will introduce the dynamic signature as a new way to represent a malware behavior. The dynamic signature represents the API call sequence of a given process including the past arguments. Using Microsoft detours [26], we can monitor a given process using API hooking [27].

Once the signature is obtained, we can compare it to other signatures stored in a database to decide either the monitored process is malicious or not.

Compared to the traditional signatures, dynamic signature detection cannot be bypassed using the previous evasion techniques. Dynamic signature is also effective against Fileless attacks [28] due to the use of legitimate software to accomplish malicious tasks.

4.1 Dynamic Behavior Extraction

The behavior of the analyzed software is retrieved, using API hooking. A technique that will allow us to intercept API function calls of a given process as described in Fig. 1. Giving as control over the way the process behaves within the network and the operating system. Once a process is created, this will trigger the monitor component of the solution. The monitor, in turn, will inject a DLL into the newly created process to perform hooking. The DLL is responsible for logging API calls and arguments and sends them back to the Behavior analyzer.

Fig. 1. Hooked function execution

4.2 Matching

The extracted behavior from the previous section will be compared with already stored signatures. If any of the signatures match the extracted behavior, this will trigger the action mentioned in the signature.

API call sequence from a signature is converted into an integer sequence and presented as a Markov chain [29]. The Markov chain will have N possible state, where N is the number of API call sequence within a signature.

4.3 Signature Format

Signatures will be represented using YAML syntax [30], a commonly used human-readable data-serialization language for configuration files, but could be used in many applications where data is being stored or transmitted.

Fig. 2. DLL injection detection using a dynamic signature

As shown in Fig. 2, a given signature is represented with:
- **Signature name:** The name of the signature
- **Action:** The action to be performed if any process behavior match this signature
- **Process name:** (OPTIONAL) the name of the process.
- **Parent process:** (OPTIONAL) the name of the parent process

- **Function_1:** Monitored function
  1. **Arg1:** Function argument
  2. **Arg2:** Function argument
- **Function_2:**
  1. Arg1: Function argument
- **Function_3**
- **MACROS:** such as PID, PROC_NAME …

This allows as looking for particular behaviors rather than merely detecting a specific tool.

4.4 Architecture

Figure 3 represents a brief overview of the architecture of our detector. The monitor component is a Windows driver that operates at a kernel level and is responsible for injecting DLL into user-mode processes using APC [31]. The injected DLL will hook the interesting API calls and send results to the Behavior Analyzer using named pipes. Behavior analyzer will compare the received data with signatures stored in the database, and then perform the appropriate action.

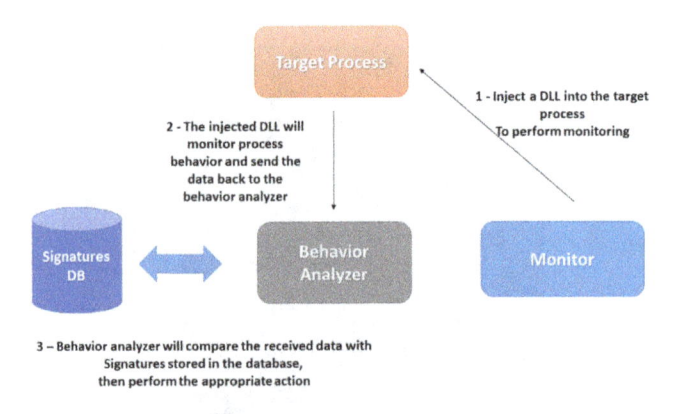

Fig. 3. Dynamic signature based detector architecture

5 Experimental Results

The experiments are conducted using Windows 10 operating system, Intel Core i7, and 8 GB ram. The components of our detector are written in C++, and compiled with Visual Studio 2017. API hooking is performed using EasyHook library [32], and yaml-cpp [33] is used for YAML parsing. Our detector uses PsSetLoadImageNotifyRoutine [34] to register a driver callback that will notify the monitor whenever an image is loaded or mapped into memory, this will allow the monitor to inject a DLL in very early process initialization stage in order to perform API hooking. For experimental purpose, we have written three signatures to identify three different malicious behavior. For the sake of simplicity, we have hooked the highest level API. Consider hooking lowest level API usually found in the ntdll.dll when dealing with malware.

5.1 AMSI Bypass Detection

Antimalware Scan Interface, AMSI [35] is an interface provided by Microsoft that allows applications and services to integrate with antimalware products. AMSI is capable of detecting and blocking of script-based attacks. Malware authors aim to bypass AMSI in order to execute any Powershell script without being detected. Figure 4 shows a signature that we can apply to the AMSI bypass technique available on the internet [36].

```
- name : AMSI Bypass
  description : AMSI bypass behavior
  action: Kill

  API_Calls:

        - API_Call : LoadLibraryA
          args:
             arg_0 : amsi.dll

        - API_Call : GetProcAddress
          args:
             arg_1 : AmsiScanBuffer

        - API_Call : VirtualProtect
          args:
             arg_2 : 0x40

        - API_Call : WriteProcessMemory
```

Fig. 4. AMSI bypass signature

5.2 Lsass.exe Process Dump Detection

Malware dump Lsass.exe process, in order to obtain windows credentials [37]. The signature in Fig. 5 aims to prevent any process from dumping lsass.exe memory.

```
---
- name : LSASS Dump
  description : LSASS Dump behavior
  action: Log

  API_Calls:

        - API_Call : OpenProcess
          args:
             arg_0 : PROCESS_ALL_ACCESS
             arg_2 : PID("lsass.exe")

        - API_Call : NtReadVirtualMemory
```

Fig. 5. LSASS dump behavior

5.3 Process Hollowing Detection

Process hollowing [38] allows the injection of entire executable files into a target process. Malwares use this technique to evade anti-virus detection. We aim to detect and prevent this technique using the signature form Fig. 6.

```
- name : Process Hollowing
  description : Process Hollowing behavior
  action: Kill

API_Calls:

    - API_Call : CreateProcess
    - API_Call : NtUnmapViewOfSection
    - API_Call : VirtualAllocEx
    - API_Call : WriteProcessMemory
    - API_Call : GetThreadContext
    - API_Call : SetThreadContext
    - API_Call : ResumeThread
```

Fig. 6. Process-hollowing signature

The following Table 1 represents a synthesis of the experimental results:

Table 1. Synthesis of results

Behavior	API call count	Result	False positive
AMSI bypass	4	OK	NO
LSASS dump	2	OK	YES
Process hollowing	7	OK	NO

Using the three signatures, we were able to identify the malicious process (Result "OK"); however, "LSASS dump" signature can lead to some false positive (False positive "Yes").

6 Conclusion and Future Works

A dynamic signature model is proposed to represent the behavior of a fileless malware. The signature contains API call sequence that identifies a behavior along with past arguments. Additional information could be mentioned in the signature, like the process name or the parent process name. A detector takes this signature as input then identifies malicious process using API hooking. Our detector showed good results for identifying fileless malware tricks. However, it can lead to a false positive alert. When a new malware technique appears, malware analyst can write a specific signature to identify it so that defenders can update their signatures database to include mitigation for the new techniques. This will reduce malware spreading considerably.

Improvements will be made to the signature structure in order to handle much more behavior criteria and reduce the false positive rate.

References

1. Cohen, F.: Computer viruses. Computers & Security (1987)
2. Moser, A., Kruegel, C., Kirda, E.: Limits of static analysis for malware detection. In: Twenty-Third Annual Computer Security Applications Conference (ACSAC 2007), Miami Beach, FL, pp. 421–430 (2007)
3. Egele, M., Scholte, T., Kirda, E., Kruegel, C.: A survey on automated dynamic malware-analysis techniques and tools. ACM Comput. Surv. **44**(2), 1–42 (2012)
4. Ki, Y., Kim, E., Kim, H.: A novel approach to detect malware based on API call sequence analysis. Int. J. Distrib. Sensor Netw. **11**(6), 659101 (2015)
5. Chen, L., Ye, Y., Bourlai, T.: Adversarial machine learning in malware detection: arms race between evasion attack and defense. In: 2017 European Intelligence and Security Informatics Conference (EISIC) (2017)
6. Afianian, A., Niksefat, S., Sadeghiyan, B., Baptiste, D.: Malware Dynamic Analysis Evasion Techniques: A Survey (2018)
7. Moubarak, J., Chamoun, M., Filiol, E.: Comparative study of recent MEA malware phylogeny. In: 2017 2nd International Conference on Computer and Communication Systems (ICCCS) (2017)
8. Naik, N., Jenkins, P., Savage, N., Yang, L.: Cyberthreat hunting - part 1: triaging ransomware using fuzzy hashing, import hashing and YARA rules. In: 2019 IEEE International Conference on Fuzzy Systems (FUZZ-IEEE), New Orleans, 23–26 June 2019 (2019)
9. Yara.readthedocs.io: Welcome to YARA's documentation! — yara 3.8.1 documentation (2019). https://yara.readthedocs.io
10. Trautman, L., Ormerod, P.: Wannacry, ransomware, and the emerging threat to corporations. SSRN Electron. J. (2018)
11. Homayoun, S., Dehghantanha, A., Ahmadzadeh, M., Hashemi, S., Khayami, R.: Know abnormal, find evil: frequent pattern mining for ransomware threat hunting and intelligence. IEEE Trans. Emerg. Top. Comput. (2017)
12. Cabaj, K., Mazurczyk, W.: Using software-defined networking for ransomware mitigation: the case of cryptowall. In: IEEE Network, vol. 30, no. 6, pp. 14–20, November–December 2016
13. Ravi, C., Manoharan, R.: Malware detection using windows API sequence and machine learning. Int. J. Comput. Appl. **43**(17), 12–16 (2012)
14. Sethi, K., Chaudhary, S., Tripathy, B., Bera, P.: A novel malware analysis framework for malware detection and classification using machine learning approach, pp. 1–4 (2018)
15. Cuckoosandbox.org: Cuckoo Sandbox - Automated Malware Analysis (2019). https://cuckoosandbox.org/
16. Yan, W., Zhang, Z., Ansari, N.: Revealing packed malware. In: IEEE Security & Privacy, vol. 6, no. 5, pp. 65–69, September–October 2008
17. Sai, S.V., Kohli, P., Bezawada, B.: Signature generation and detection of malware families, vol. 5107, pp. 336–349 (2008)
18. Docs.microsoft.com: PE Format - Windows applications (2019). https://docs.microsoft.com/en-us/windows/desktop/debug/pe-format

19. Docs.microsoft.com: About WinINet - Windows applications (2019). https://docs.microsoft.com/en-us/windows/desktop/wininet/about-wininet
20. Undocumented.ntinternals.net: NTAPI Undocumented Functions (2019). https://undocumented.ntinternals.net/. Accessed 10 May 2019
21. Shafiq, M.Z., Tabish, S.M., Mirza, F., Farooq, M.: PE-Miner: mining structural information to detect malicious executables in realtime. In: Kirda, E., Jha, S., Balzarotti, D. (eds.) Recent Advances in Intrusion Detection. RAID 2009. Lecture Notes in Computer Science, vol. 5758. Springer, Heidelberg (2009)
22. Katja, H.: Robust Static Analysis of Portable Executable Malware, Master Thesis in Computer Science, HTWK Leipzig
23. GitHub: fireeye/flare-floss (2019). https://github.com/fireeye/flare-floss/blob/master/doc/theory.md
24. Blackhat.com (2019). https://www.blackhat.com/docs/us-15/materials/us-15-Choi-API-Deobfuscator-Resolving-Obfuscated-API-Functions-In-Modern-Packers.pdf
25. Docs.microsoft.com. (2019). _PEB. https://docs.microsoft.com/en-us/windows/desktop/api/winternl/ns-winternl-_peb
26. Detours: Binary interception of Win32 functions. In: Hunt, G., Brubacher, D. (eds.) Third USENIX Windows NT Symposium. USENIX, July 1999
27. Marhusin, M.F., Larkin, H., Lokan, C., Cornforth, D.: An evaluation of API calls hooking performance. In: 2008 International Conference on Computational Intelligence and Security, Suzhou, pp. 315–319 (2008)
28. Mansfield-Devine, S.: Fileless attacks: compromising targets without malware. Netw. Secur. **2017**(4), 7–11 (2017)
29. Chan, K.T., Lenard, C., Mills, T.: An Introduction to Markov Chains (2012)
30. Yaml.org: The Official YAML Web Site (2019). https://yaml.org/
31. Sikorski, M., Honig, A.: Practical malware analysis. San Francisco (California, EEUU) (2012)
32. Easyhook.github.io (2019). EasyHook. https://easyhook.github.io/
33. GitHub: jbeder/yaml-cpp (2019). https://github.com/jbeder/yaml-cpp
34. Docs.microsoft.com: PsSetLoadImageNotifyRoutine function (ntddk.h) - Windows drivers (2019). https://docs.microsoft.com/en-us/windows-hardware/drivers/ddi/content/ntddk/nf-ntddk-pssetloadimagenotifyroutine
35. Docs.microsoft.com: Antimalware Scan Interface (AMSI) - Windows applications (2019). https://docs.microsoft.com/en-us/windows/desktop/amsi/antimalware-scan-interface-portal
36. Blog, Z.: How to bypass AMSI and execute ANY malicious Powershell code, zc00l blog (2019). https://0x00-0x00.github.io/research/2018/10/28/How-to-bypass-AMSI-and-Execute-ANY-malicious-powershell-code.html
37. Blog.gentilkiwi.com: mimikatz| Blog de Gentil Kiwi (2019). http://blog.gentilkiwi.com/mimikatz. Accessed 10 May 2019
38. Attack.mitre.org: Technique: Process Hollowing - MITRE ATT&CK™ (2019). https://attack.mitre.org/techniques/T1093

Efficient Implementation and Computational Analysis of Privacy-Preserving Auction Protocols

Ramiro Alvarez and Mehrdad Nojoumian[✉]

Department of Computer and Electrical Engineering and Computer Science,
Florida Atlantic University, Boca Raton, FL 33431, USA
{ramiroalvare2015,mnojoumian}@fau.edu

Abstract. Auctions are a key economic mechanism for establishing the value of goods that have an uncertain price. Nowadays, as a consequence of the ubiquitous emergence of technology, auctions can reach consumers, and as a result, drive market prices on a global scale. Collection of private information such as losing bids exposes more information than desired. In fact, the leaked information can be analyzed to provide auctioneers or competitors with advantages on future transactions. Therefore, the need to preserve privacy has become a critical concern to reach an accepted level of fairness and to provide market participants with an environment in which they can bid true valuations. This paper focuses on constructions of sealed-bid auctions based on cryptographic protocols. Instead of solely focusing on theoretical aspects of sealed-bid auctions, this paper dives into implementation details and demonstrates communication and computational analysis and how different settings affect performance.

Keywords: Sealed-bid auctions · Privacy-preserving protocols · Complexity

1 Introduction

For many years auctions have existed as mechanism to trade commodities. At a minimum, an auction involves two groups: bidders and auctioneers. Auctioneers arrange the auctions, set the rules, and declare a winner after the price evaluation. Bidders are parties with an interest to acquire a certain good, or in the case of contractors, provide a service. The literature provides different classification for auctions, some of which are concisely described below.

The most common type of auction is the classical *English auction* involving an act where the bidders speculate and overestimate the price of an item in an open fashion. Usually seen in the sale of antiques and significant artwork, the auction start off at the reserve price and increases in an ascending manner when a bidder bids at that price. It proceeds increasing until no one is willing to pay a greater amount. The winner is the bidder willing to pay the highest price. In contrast, an auction with a price adjustment in a decreasing direction is labeled as *Dutch-style*. Most commonly, decreasing price auctions are observed in perishable markets such as fish and flowers where the interest is to sell an item before it expires. To begin, the auctioneer asks a high price for the item. The price is decreased if no one chooses to buy. The price continues to decrease in intervals until someone is willing to pay. The first willing volunteer will be the winner. Therefore, Dutch-style auctions inherently protect the loosing bids. Last but not least, sealed-bid auctions are the primary focus of this paper. The game changing rule is that bidders have one single opportunity to decide on a bid. This enforces a bidder to drop possible speculation and also reduces the bidding strategy to a true valuation. When the winner pays the standing price, this is referred to as the first-price. On the other hand a *Vickrey auction* states that the winner, being the highest bidder, pays the second highest price.

1.1 Our Motivation and Contribution

As stated in the literature, the main motivation for *sealed-bid actions* is to protect the losing bids as they can be used by auctioneers to maximize their revenues in future auctions. Although many works in the literature discuss theoretical approaches to sealed-bid auctions, the research on implementation is limited. Herein, our aim is to elucidate the intricate details of implementation and provide computational and communication analyses of several protocols. The protocols that were considered are [11,23,24]. For [11], two version were implemented with and without verification. For [23], there are also two implementations based on ElGammal and RSA. Finally, we conclude with an implementation of [24].

Rational for Our Selection: The main reason for our selection is because the first-price auction is still the most usable auction type that is widely used in settings such as e-commerce and ad auctions. Besides, the selected auctions utilize the most popular and strongest cryptographic primitives such as secure multiparty computation (MPC), public-key encryption, and undeniable signature schemes. The newer papers in the literature mainly focus on specific types of auctions, e.g. clock-proxy auctions and multi-unit auctions, or costly cryptographic primitives. These are two main reasons that we selected these protocols.

Our Main Contributions: An efficient implementation of the selected protocols is our primary contribution. We utilized the latest efficient implementations of cryptographic primitives such as SHA1, RSA and ElGamal, primitive root generator, safe prime generator, etc. Moreover, we separated the initialization time from the verification time to have a realistic comparisons among these protocols. Usually, initialization can be done offline without any limitation and that is why it should be evaluated separately. Finally, the way that we compared these protocols, i.e., price range parameters 25%, 50%, 75% and 100%, is our unique evaluation methodology that truly reveals the cons and pros of the selected protocols. In other words, an in-depth assessment of each auction is generated by modifying parameters such the modulus size, the number of bidders, and the price range. The results show that the complexity increases with modulus sizes as expected for all sealed-bid auctions. Interestingly, for [11], the complexity increases more with increases in the price ranges. We also observed that [24] has the highest communication complexity, as discussed later.

2 Brief Literature Review

Due to lack of space, we briefly review the literature. However, a comprehensive survey on privacy-preserving protocols for sealed-bid auctions is provided in [2].

Franking and Reiter introduced one of the early designs for sealed-bid auctions in [8]. The protocol relies on verifiable signature sharing [7] used to prevent bidders from denying their bids. Kikuchi et al. created a first-price sealed-bid auction [12] based on the addition gate of secure MPC. Later, the protocol was modified by [13] to improve privacy, and by [21] to increase anonymity, fairness, and robustness. Nojoumian and Stinson [19] constructed efficient sealed-bid auction protocols based on addition and multiplication operations of verifiable secret sharing (VSS). The proposed solution works for a wide range of sealed-bid auctions. The authors of [6] use a homomorphic public-key encryption scheme. However, this protocol only works with pairwise comparison. With a similar approach, [23] utilizes a public-key cryptosystem to protect the losing bids. Another line of research focuses on sealed-bid Dutch-style auctions such as [18] that utilizes a multicomponent commitment scheme (MCS). The authors of [14] implemented and analyzed different Dutch-style sealed-bid auction protocols in both computationally and unconditionally secure models.

Under the category of Vickrey auctions, [16] proposes a protocol that relies on an oblivious third party. In this construction, the bidders must communicate between two servers. Bids are encrypted and the decryption keys are shared among auctioneers using secure MPC. The efficiency is improved by [15] substituting MPC with a homomorphic scheme. The problem of one of the two servers cheating is addressed in [10] by splitting the bid so that each server doesn't hold the entire information. Another set of protocols is designed to implement $(M+1)$-*price auctions*. This is a type of auction in which M highest bidders take the prize and it is equivalent to the Vickrey auction when $M = 1$. Research works such as [1,5,11,28] contain details on designing $(M+1)$-price auctions. The approach in [11] is to hide the bidding price in the degree of polynomials. The polynomials are shared and the the summation operation is used to construct a polynomial that holds the highest price in the degree. Another auction type, know as *combinatorial auction*, allows the bidders to place evaluations on a bundle of items. Combinatorial auctions can be multi-unit, linear good and general. For references on these auctions, we refer the readers to [9,22,25,26].

3 Sealed-Bid Auctions' Properties

As stated earlier, sealed-bid auctions describe a mechanism in which the bidders are given a chance to decide their final bidding values. Due to the nature of the mechanism, the bidders cannot gain knowledge from other bidders, thus they cannot create a strategy other than to bid a true valuation. Furthermore, the invention of the Internet connects users across nations and removes geological and space constrains, which exist in physical auctions. At the very least, an electronic sealed-bid auction must ensure concealment of losing bids, non-repudiation, verifiability, correctness and fairness, as illustrated in [17].

- **Privacy of Losing Bids:** Determination of the winner should not arrive at the expense of opening or decrypting the losing bids. In fact, a proper protocol should not reveal losing bids at all since this provides information to sellers, which can turn the tables in their favor in subsequent auctions.
- **Non-repudiation:** A bidder should not be able to deny sending a bid which was truly submitted.
- **Verifiability:** The winning bid should be recognized by all the bidders in the auction as being the true winning bid, and every bidder should have a method of verifying that others have followed the auction protocol accordingly.
- **Correctness:** The protocol should not determine an incorrect winner or winning price under any circumstances.
- **Fairness:** A subset of players should not have any advantage over others due to the manner that the protocol is constructed. Also the auctioneer should not have any advantage.

It is worth mentioning that if the security of the underlying cryptographic primitives rely on hardness of well-known mathematical problems such as integer factoring or discrete logarithm, the protocol will be *computationally secure*. However, if we don't have this condition, the protocol will be *unconditionally secure*.

4 Protocol Description

Next, we explain in detail the selected protocols that are the subject of our study.

4.1 Hiroaki Kikuchi's Protocol

The proposed construction, as shown in Fig. 1, is based on the addition operation of secure MPC [3]. The protocol relies on a simple fact that if f and h are polynomials of degree t and s respectively, then $f + h$ has degree $max(t, s)$. First, a prime number p of order q is chosen. Auctioneers publish a price list W. Each bidder chooses a random polynomial from a finite field and the bidding value is equal to the degree of the polynomial. Each auctioneer receives a share from a bidder using secret sharing and computes a total sum, denoted as F over the shares. Because polynomials were chosen so that F's constant term was equal to zero, any party can find the smallest subset that produces a polynomial containing the highest bid as the degree of the polynomial.

The second version of the protocol, as explained in Fig. 2, is made stronger with verifiable secret sharing of [20]. First, a prime p of order q is chosen same way as before, but now also two distinct primitive roots g_1 and g_2 are chosen and

---

Initialization

1. Establish field Z_p^* by choosing primes p and q such that q divides $p - 1$, and operations are done using modular p.
2. The ith bidder chooses $b_i \in \{1, ..., k\}$ and it is concealed by a random polynomial with degree $t_i = b_i + c$ and $a_0 = 0$, where c is the number of faulty auctioneers.

$$f_i(x) = \sum_{j=1}^{t_i} a_j x^j$$

Bid Submission

1. Each bidder evaluates and sends $f_i(\alpha_j)$ to auctioneers A_j for $j = 1, ..., m$.
2. Each auctioneer adds the shares received from the bidders and publishes $F(\alpha_j)$ by a commitment scheme.

$$F(\alpha_j) = \sum_{j=1}^{n} f_i(\alpha_j)$$

Winner Determination

1. Using the public values $F(\alpha_1), ..., F(\alpha_n)$, any entity can use Lagrange interpolation and construct a polynomial of degree $max(t_1, ..., t_n)$.
2. By subtracting parameter c, the highest bid is recovered.

Fig. 1. Hiroaki Kikuchi's protocol.

made publicly available. Bidder b_i chooses two randomly generated polynomials f and h. Each bidder makes a commitment of the polynomial by sending the multiplication of the powers of the primitive roots with the coefficients of the polynomials. For example, for polynomial $f(x) = a_1 x + a_2 x^2 + ... + a_t^t$ and $h(x) = b_1 x + b_2 x^2 + ... + b_s^s$, the bidders send $g_1^a g_2^b 1$, $g_1^a g_2^b 2$, until a commitment is made on all the coefficients. The sums F and H are calculated for the shares received from f and h respectively. Auctioneers calculate $Y = g_1^F$ and $Z = g_2^H$ and publish YZ. The Lagrange Interpolation once again generates a polynomial that contains the highest bid in the degree. The main difference is that, with the extra computation to incorporate the verification protocol, neither auctioneers, nor bidders are permitted to insert fake values. Committing at every step of communication makes cheating detectable.

4.2 Kazue Sako's Protocol

Two practical cryptosystems, ElGamal and RSA, are used in the computationally secured sealed-bid auction of [23], shown in Fig. 3. For each price in the price list, there is an associated public-key. For example, for price list $V = \{v_1, v_2, ..., v_L\}$, there are public-keys $PubK = \{pubk_1, pubk_2, ..., pubk_l\}$, and the auctioneers hold private keys $Pk = \{pk_1, pk_2, ..., pk_l\}$. In order to bid, a bidder uses the key associated with a price and encrypts the bid with that key. During the winner determination phase, the auctioneers pick the private-key associated with the highest price on the list and try to decrypt every

Initialization

1. Establish field Z_p^*, primes p and generators g_1 and g_2.
2. The ith bidder chooses bid $b_i \in \{1, ..., k\}$ and must commit on two polynomials. The polynomials are $f_i(x)$ of degree $t_i = b_i + c$ and $h_i(x)$ of degree $s = k + c$.

$$f_i(x) = \sum_{j=1}^{t_i} a_j x^j \text{ and } h_i(x) = \sum_{j=1}^{s} b_j x^j$$

Bid Submission

1. Each bidder sends shares $f_i(\alpha_j)$ and $h_i(\alpha_j)$ to each participating auctioneer A_j for $j = 1, ..., m$.
2. Each bidder publishes public values that serve as commitments of his own polynomial.

$$E_{i,j} = g_1^{a_1 b_1}, ..., E_{i,t_i}$$
$$E_{i,t_i+1} = g_2^{b_{t_i+1}}, ..., E_{i,s}$$

Verification Step

1. Auctioneer j can verify that the share of bidder i is correct by verifying the polynomial commitment according to the following equation.

$$g_1^{f_i(\alpha_j)} g_2^{h_i(\alpha_j)} = \prod_{l=1}^{s} (E_{i,j})^{\alpha_j^t}$$

2. If verification holds, the auctioneers can proceed to compute and publish the sum of shares on $f(x)$ and $h(x)$.

$$F(\alpha_j) = \sum_{j=1}^{n} f_i(\alpha_j) \text{ and } H(\alpha_j) = \sum_{j=1}^{n} h_i(\alpha_j)$$

3. The computed sum of shares can be verified by any entity using the following equality:

$$Y_j Z_j = g_1^{F(\alpha_j)} g_2^{H(\alpha_j)}$$

Winner Determination

1. The highest bid is the first element in the price list $\{1, ..., k\}$ that satisfies the equality below.

$$g_1^{F^{(t*)}(0)} = 1$$

where $F^{(t*)}(0)$ is obtained using Lagrange Interpolation

$$F^{(s)}(0) = \sum_{j=1}^{s} \prod_{i \neq j \in A_s} \frac{\alpha_i}{\alpha_i - \alpha_j} \tag{1}$$

Fig. 2. Verifiable version of Hiroaki Kikuchi's protocol.

Initialization

1. Establish a price list and generate private and public-key pairs.
2. Publish the set of public-keys and arrange it such that each key matches a price in the price list.

Bid Submission

1. Each bidder chooses a public-key according to the price he intends to bid on.
2. The bidder encrypts the price with the key that is associated to that price according to the established mapping.

Winner Determination

1. Starting with the private-key that is associated with the highest price, the auctioneers decrypt each encryption that they have received. The elements decrypted to the value of the price in the current round are the highest bids.

Fig. 3. Kazue Sako's protocol.

submitted bid. A successful decryption determines the winner. If no winner is found at a specific price, the auctioneers pick the next highest private-key and reiterate the process. In order to make the protocol stronger, the authors suggest using the Shamir's secret sharing scheme [27] to split the keys into n shares and then give a share to each auctioneer to provide a mechanism of resilience against dishonest auctioneers who may decrypt bids.

4.3 Sakurai's and Miyazaki's Protocol

The authors in [24] describe an auction, Fig. 4, that can be built using a convertible undeniable signature scheme [4]. First, the protocol requires a safe prime p, a subgroup generator α and a one-way hash function. The prime p and the primitive root α are used to create secrets that are computationally secure based on the discrete logarithmic problem. The auction proceeds in a Dutch-style fashion. The verifier (auctioneer) and the prover (bidder) engage in several rounds of communications to prove equality or inequality against the standing price. Determination of equality or inequality does not reveal any private information since the auctioneer concludes based on the comparison of two discrete logs. At the time of winner determination, the bidders reveal their private keys, which further confirms correctness of the protocol.

5 Implementation Results

Our simulation was developed under JetBrains CLion environment. Our implementations were written in C++ and compiled under GNU GCC compiler. For an efficient implementation, we utilized the Crypto++ library. For instance,

Initialization

1. Establish p, q and α, where p and q are large prime numbers such that $p = 2q+1$ and α is a generator of field Z_p^*.
2. Each bidder j generates a public-key from private-key S_j as follows $P_j = \alpha^{S_j} (mod\ p)$.
3. Auctioneer(s) publish the different price choices $\{w_1, ...w_m\}$.

Bid Submission

1. Each bidder chooses random numbers $x, k \in Z_q^*$ and computes $h, r, \tilde{r}, c, s$ based on the following equations:

$$h = \alpha^x (mod\ p)$$
$$r = \alpha^k (mod\ p)$$
$$\tilde{r} = r^x (mod\ p)$$
$$c = Hash(w_k, \tilde{r})$$
$$s = k - cS_j (mod\ q)$$

Winner Determination

1. Auctioneer(s) order price list in descending order and select price w_i starting with the highest price w_m.
2. Auctioneer(s) determine, based on the equality or inequality of two discrete logs, if the bidder(s) committed to a price according to the following:

$$u, v, w \in Z_q^*$$
$$\tilde{s} = k - (v+w)x (mod\ q)$$
$$\beta = \alpha^s P_j^{Hash\ (w_m,\ \tilde{r})}$$
$$\beta^{\tilde{s}} \tilde{r}^{v+w} = \beta^k\ (mod\ q)\quad (equality)$$
$$\beta^{\tilde{s}} \tilde{r}^{v+w} \neq \beta^k\ (mod\ q)\quad (inequality)$$

3. The bidder(s) that committed to the price w_i are the winners. If bidders cannot prove the commitment, the next price from the list is selected and the process reiterated.
4. Once a bidder is found to have commitment to a price w_i, the auctioneer(s) verify the validity with the following equation and conclude the auction:

$$\tilde{r} = (\alpha^s P_j^{Hash(w_k, \tilde{r})})^x\ (mod\ p)$$

Fig. 4. Sakurai's and Miyazaki's protocol.

we made extensive use of hashing algorithm SHA1 and utilized public crypto systems such as RSA and ElGamal. In addition, Crypto++ provides necessary blocks for generating random safe primes, prime numbers and primitive roots of a cyclic group. In order to generate random polynomials, we applied random number generators for each coefficient to generate n-elements, and then, we applied modular reduction operation to keep the elements in the finite field Z_p. Polynomials were stored as a vector where the first element mapped to the constant term and the last element mapped to the leading coefficient.

Our experiments were centered on computational complexity and communication overhead around initialization and verification. For each protocol, we tested initialization based on four distinct modulus sizes: 128-bit, 256-bit, 512-bit and 1024-bit. Verification time was measured by allowing the bidders to have different bidding preferences and by manipulating the number of bidders present during the auction. For instance, we measured verification complexity in scenarios containing 25, 50, 75 and 100 bidders. At the same time, the bidders could choose a bid b_i at random from the whole set if the bidding parameter was 100%, otherwise they would have to choose only from a subset. For example, if the set contained 100 elements in numerical order, then bidding parameter 75% essentially meant the bidders would ignore bidding the top 25 elements, and instead, he would choose a bid b_i at random from the remaining 75 lower order elements. Hereafter, we will refer to this bidding preference, which we tested at 25%, 50%, 75% and 100%, as the *price range* parameter.

For simplicity, the experiments were created using a Command Line Interface and a common graphical user interface. Our computation model was synchronous, thus it introduced some delay. However, the delays introduce by this model was relative. Some modules that were affected by the synchronous model were creation of public/private-key pairs, generation of polynomials and publishing the results to our simulated bulletin board. In terms of our computing power, we conducted our executions using a computer with Intel Core $i74810MQ$ CPU @ 2.80 GHz and 16GB RAM. While running the experiment, we ended every process that would steal computing power from the operating system. Besides, we shut down the network and closed down any ports, i.e., we made the operating system solely focus on running the experiment.

5.1 HK's Protocol Based on Secure MPC

The first set of results correspond to the non-verifiable protocol of [11]. Naturally, the initialization increased with an increase in the modulus size, Fig. 5. The first reason for an increase is that generating a random prime number becomes increasingly expensive with a greater bit size, and at best, the algorithms produce only probabilistic prime numbers. The second reason for the observed increase is that, for each polynomial, we chose random numbers based on the size of the field, thus the modular reduction increased and the numbers became larger.

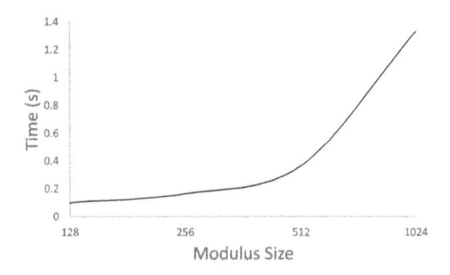

Fig. 5. Initialization of HK's protocol.

Figure 6 demonstrates the required time for different number of bidders and varying price ranges. The verification seemed to be more affected by the price range and the bidder size rather than the bit size of the modulus. Two factors contributed to the increased in time. First, in this protocol, the private value of the bidder is concealed in the degree of a polynomial meaning that, in order to conceal a bidding value of $100, we need to construct a polynomial of degree 100, i.e. more computation when evaluating shares and using the addition operation of MPC. Secondly, with increasing price, we need to increase number of auctioneers if it is required to have the same security threshold. However, increasing number of auctioneers increases communication complexity during the sharing and reconstruction parts of the protocol.

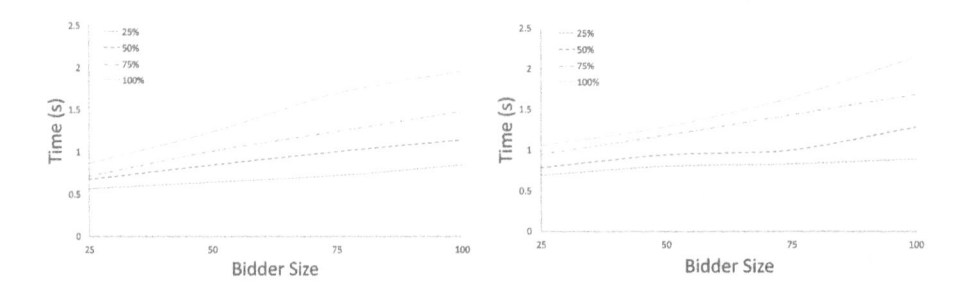

Fig. 6. Verification of HK's protocol: 128b, 512b.

Figure 7 shows the initialization time for the same protocol when verifiability is introduced as a method to prevent the bidders from repudiation and the auctioneers from casting false bids with the purpose of inflating prices. The result of adding extra layers of robustness is an increase in computation. Unlike the simpler approach, we must generate a prime and two distinct generators of the cyclic group. Also, the bidders must generate two distinct polynomials $f(x)$ and $h(x)$. For every coefficient, each bidder must submit commitments. The auctioneers must verify the submitted commitments before they establish a trust in the bidder. The result is an increase in time not only due to a larger modulus size but also the extra numbers of precautions added to the protocol.

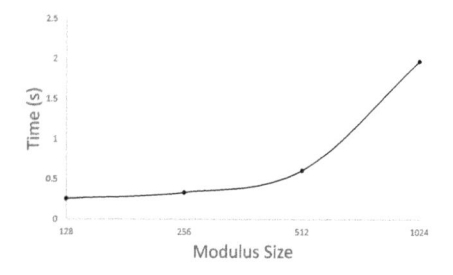

Fig. 7. Initialization of 2nd HK's protocol.

The same pattern as the simple implementation can be observed in the verifiable protocol, Fig. 8. That is, with a larger number of bidders and a greater number of price ranges, we note an increase in time. The extra accumulated time is due to communication complexity and extra verification steps. The auctioneers must publish two constants Y_j and Z_j based on the multiplication computed from $g_1^F g^H$ for each corresponding α_j. As another additional step, the bidders must verify that the auctioneers computed their values correctly.

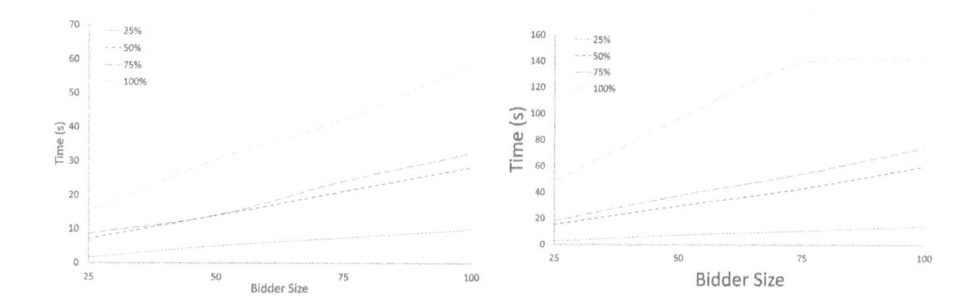

Fig. 8. Verification of 2nd HK's protocol: 128b, 512b.

5.2 KS's Protocol Based on Pub-Key Encryption

Figure 9 demonstrates the initialization time of [23] for two different implementations using ElGamal and RSA schemes respectively. In both cases, the initialization time rises with modulus sizes as it is expensive to generate large primes. The time for ElGamal is higher than RSA since the prime generation algorithm needs to find prime p for which $(p-1)/2$ is also a safe prime, whereas RSA only requires two large unrelated primes. Clearly, the number of price ranges also affects the total initialization time.

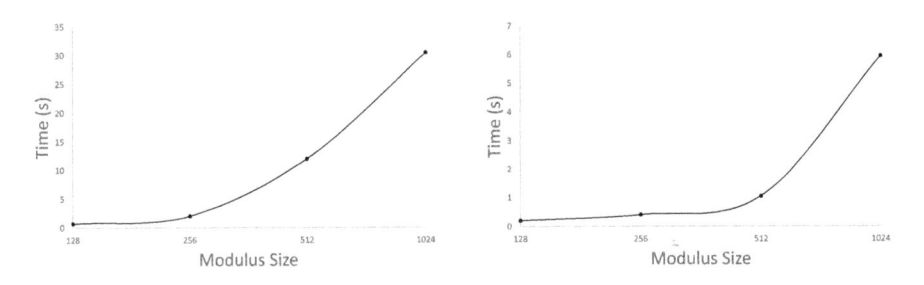

Fig. 9. Initialization of KS's protocol: ElGamal (left) and RSA (right).

One expected outcome is that, increasing the modulus size increases the computational complexity since we have bigger keys. More subtle, however, is the fact that we observed an increase in verification time for lower price ranges, Figs. 10 and 11. We can justify this because of the Dutch-style nature of the auction. When we set the price range to 100% and we have 100 bidders, it is very likely that one of the bidders will bid the highest price. Since decryption occurs from the highest decryption key to lowest, essentially we will greatly reduce the cost if we find a bidder that bids equal to or very close to the highest price. On the other hand, when the price range parameter is set to 25%, the auctioneers lose a significant amount of time decrypting values for which nobody placed a bid. Therefore, for 100 bidders and 100% price range, the result is minimized whereas it is maximized for 100 bidders and 25% price range.

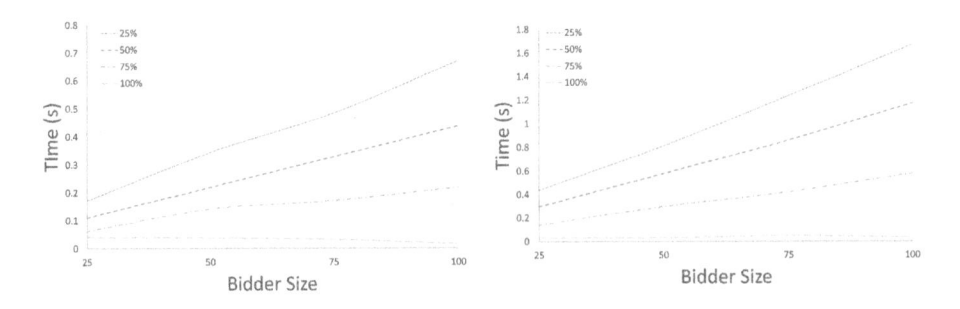

Fig. 10. Verification of KS's protocol-ElGamal: 128b, 512b.

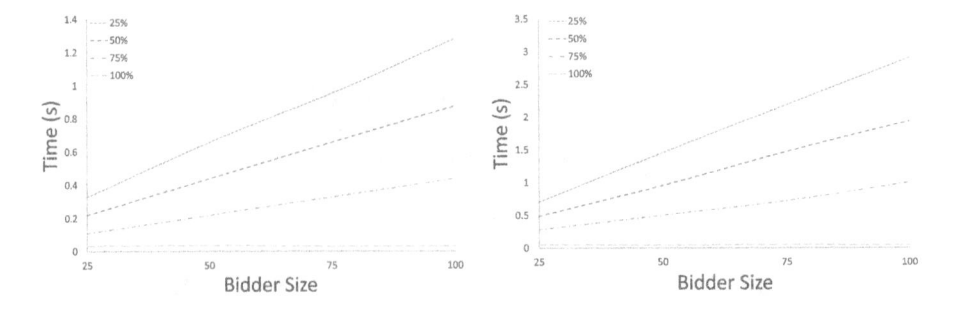

Fig. 11. Verification of KS's protocol-RSA: 128 & 512 bits.

Our choice for ElGamal and RSA is due to the fact that these two are well-known public-key encryption schemes. At the end, the verification times were very similar with RSA being a bit slower than ElGamal. Although encryption is faster for ElGamal, RSA has significant advantage during the initialization steps because of a slow prime generation in ElGamal.

5.3 S-&-M's Protocol Based on Commitment

In this section, we examine the protocol proposed in [24]. The first step in the design of the protocol is to generate two primes p and q where $p = 2q + 1$ and α is a generator. The bidders must compute private and public values. Because no encryption or decryption is required during the initialization phase, this protocol is the fastest to be initialized. Obviously, the complexity increase observed with an increase in the modulus size, shown in Fig. 12.

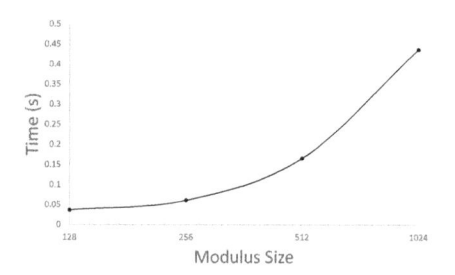

Fig. 12. Initialization of S-&-M's protocol

During the bidding process, the bidders commit to a bid and attach a digital signature. In the opening phases, the auctioneers start at the highest possible price and receive a proof from each bidder showing equality or inequality. Therefore, the verification time required for the protocol is proportional to the number of bidders and inversely proportional to the price range, as shown in Fig. 13.

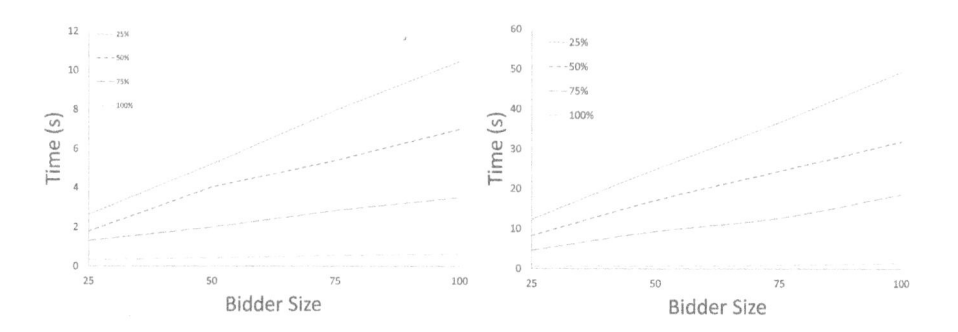

Fig. 13. Verification of S-&-M's protocol: 128b, 512b.

It appears that the modulus size significantly affects the verification time. To better understand the result, we simply analyze one important step of this protocol specifically when the auctioneers compute $\beta = \alpha^s P_j^{H_i(w_m, r)} (mod\ p)$. In fact, we need to perform hashing and then we have exponentiation of large numbers. Afterward, the auctioneers and bidders execute a protocol for proving equality or inequality of two discrete logs. Since the auction is constructed in a Dutch-style manner, the auctioneers and bidders must exchange information over several rounds before defining the winner. Once a bidder successfully proves the equality of his bid with the current standing price, an extra step is required to prove confirmation. In the confirmation step, the bidder must send the auctioneers his private exponent x in the discrete log, and finally, the auctioneers can confirm if the parameter satisfies $r = (\alpha^s P_j^{H_i(w_m, r)})^x (mod\ p)$.

5.4 Comparing All Three Protocols

In Figs. 14 and 15, we provide a unified comparison of all protocols. To fit the entire result in one visible plot, we applied a logarithmic scale for the time axis. For initialization times, we considered 128, 256, 512 and 1024 bits. For verification times, we have plotted 512 bits, 50 bidders, and price range 50%.

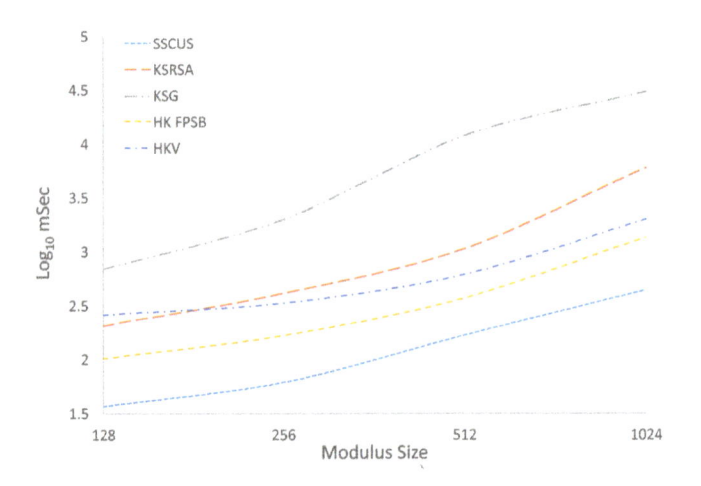

Fig. 14. Initialization time

Within the initialization plot, the KS-ElGamal is the most expensive protocol and the S-&-M is the fastest one. It should be noted that the slow initialization is not that important since preparation of the key-pairs can be performed prior to the auction. Also, if the keys are created in parallel, it would be significantly faster. In the case of verification, the most expensive protocol is the 2nd version of the HK protocol and the fastest is KS-ElGamal. In other words, HK-Verifiable is very expensive since we reconstruct a polynomial using Lagrange interpolation

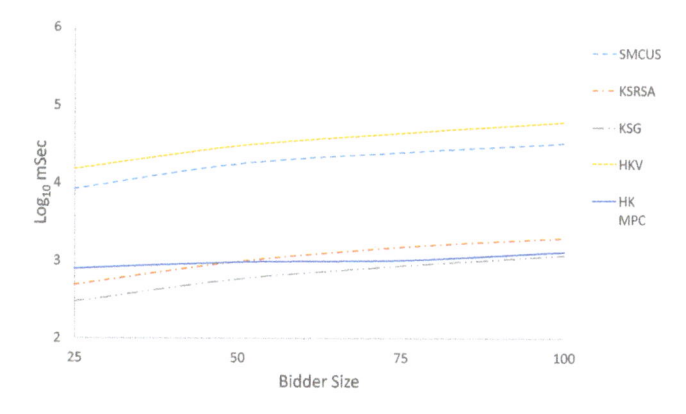

Fig. 15. Verification time

for polynomials with up to 50 terms where each coefficient has around 500 bits. After this protocol is S-&-M since for each round, the determination of equality or inequality requires many calculations and there could be many rounds.

6 Concluding Remarks

We have analyzed five different protocols in the literature of sealed-bid auctions. The protocols consisted of different approaches. Namely, we studied protocols using secure MPC, public-key encryption and commitment scheme. We noticed that protocols using MPC will have a large number of communication rounds. In the case of using public-key encryption schemes, we can encrypt and decrypt bids efficiently, however, we must realize that there is risk of auctioneers opening all bids since they hold all the decrypting keys. Commitment schemes also suffer from communication complexity since, at every round, the bidders must prove that their commitments are not equal to the standing price. In conclusion, a protocol can be fast such as HK and KS, but in order to provide higher security and robustness, such as HK-Verifiable or S-&-M protocols, we must spend extra computation and/or communication rounds. As a part of our future work, we intend to analyze unconditionally secure sealed-bid auction protocols.

Acknowledgments. We gratefully acknowledge our research sponsors, College of Engineering and Computer Science (COECS) and Institute for Sensing and Embedded Network Systems Engineering (I-SENSE) at FAU, for making this research work possible.

References

1. Abe, M., Suzuki, K.: M+ 1-st price auction using homomorphic encryption. In: International Workshop on Public Key Cryptography, pp. 115–124. Springer (2002)
2. Alvarez, R., Nojoumian, M.: Comprehensive survey on privacy-preserving protocols for sealed-bid auctions. Comput. Secur. (C&S) **88**, 101502–101515 (2020)
3. Ben-Or, M., Goldwasser, S., Wigderson, A.: Completeness theorems for non-cryptographic fault-tolerant distributed computation. In: Proceedings of the Twentieth Annual ACM Symposium on Theory of Computing. pp. 1–10. ACM (1988)
4. Boyar, J., Chaum, D., Damgård, I., Pedersen, T.: Convertible undeniable signatures. In: Conference on the Theory and Application of Cryptography, pp. 189–205. Springer (1990)
5. Brandt, F.: A verifiable, bidder-resolved auction protocol. In: Proceedings of the 5th International Workshop on Deception, Fraud and Trust in Agent Societies, SI on Privacy and Protection with Multi-Agent Systems, pp. 18–25 (2002)
6. Cachin, C.: Efficient private bidding and auctions with an oblivious third party. In: Proceedings of the 6th ACM Conference on Computer and Communications Security, pp. 120–127. ACM (1999)
7. Franklin, M.K., Reiter, M.K.: Verifiable signature sharing. In: International Conference on the Theory and Applications of Cryptographic Techniques, pp. 50–63. Springer (1995)
8. Franklin, M.K., Reiter, M.K.: The design and implementation of a secure auction service. IEEE Trans. Softw. Eng. **22**(5), 302–312 (1996)
9. Fujishima, Y., Leyton-Brown, K., Shoham, Y.: Taming the computational complexity of combinatorial auctions: optimal and approximate approaches. In: IJCAI, vol. 99, pp. 548–553. DTIC Document (1999)
10. Juels, A., Szydlo, M.: A two-server, sealed-bid auction protocol. In: International Conference on Financial Cryptography, pp. 72–86. Springer (2002)
11. Kikuchi, H.: (m+1) st-price auction protocol. IEICE Trans. Fundam. Electron. Commun. Comput. **85**(3), 676–683 (2002)
12. Kikuchi, H., Hakavy, M., Tygar, D.: Multi-round anonymous auction protocols. IEICE Trans. Inf. Syst. **82**(4), 769–777 (1999)
13. Kikuchi, H., Hotta, S., Abe, K., Nakanishi, S.: Distributed auction servers resolving winner and winning bid without revealing privacy of bids. In: 7th International Conference on Parallel and Distributed Systems, pp. 307–312. IEEE (2000)
14. Krishnamachari, S., Nojoumian, M., Akkaya, K.: Implementation and analysis of Dutch-style sealed-bid auctions: computational vs unconditional security. In: 1st International Conference on Information Systems Security and Privacy, pp. 106–113 (2015)
15. Lipmaa, H., Asokan, N., Niemi, V.: Secure vickrey auctions without threshold trust. In: International Conference on Financial Cryptography, pp. 87–101. Springer (2002)
16. Naor, M., Pinkas, B., Sumner, R.: Privacy preserving auctions and mechanism design. In: Proceedings of the 1st ACM Conference on Electronic Commerce, pp. 129–139. ACM (1999)
17. Nojoumian, M.: Novel secret sharing and commitment schemes for cryptographic applications. Ph.D. thesis, Department of Computer Science, University of Waterloo, Canada (2012)

18. Nojoumian, M., Stinson, D.R.: Unconditionally secure first-price auction protocols using a multicomponent commitment scheme. In: 12th International Conference on Information and Communications Security. LNCS, vol. 6476, pp. 266–280. Springer (2010)
19. Nojoumian, M., Stinson, D.R.: Efficient sealed-bid auction protocols using verifiable secret sharing. In: 10th International Conference on Information Security Practice and Experience. LNCS, vol. 8434, pp. 302–317. Springer (2014)
20. Pedersen, T.P.: Non-interactive and information-theoretic secure verifiable secret sharing (1998)
21. Peng, K., Boyd, C., Dawson, E., Viswanathan, K.: Robust, privacy protecting and publicly verifiable sealed-bid auction. In: International Conference on Information and Communications Security, pp. 147–159. Springer (2002)
22. Rothkopf, M.H., Pekeč, A., Harstad, R.M.: Computationally manageable combinational auctions. Manag. Sci. **44**(8), 1131–1147 (1998)
23. Sako, K.: An auction protocol which hides bids of losers. In: International Workshop on Public Key Cryptography, pp. 422–432. Springer (2000)
24. Sakurai, K., Miyazaki, S.: A bulletin-board based digital auction scheme with bidding down strategy-towards anonymous electronic bidding without anonymous channels nor trusted centers. In: Proceedings of International Workshop on Cryptographic Techniques and E-Commerce, pp. 180–187 (1999)
25. Sakurai, Y., Yokoo, M., Kamei, K.: An efficient approximate algorithm for winner determination in combinatorial auctions. In: Proceedings of the 2nd ACM Conference on Electronic Commerce, pp. 30–37. ACM (2000)
26. Sandholm, T.: Algorithm for optimal winner determination in combinatorial auctions. Artif. Intell. **135**(1–2), 1–54 (2002)
27. Shamir, A.: How to share a secret. Commun. ACM **22**(11), 612–613 (1979)
28. Suzuki, K., Yokoo, M.: Secure multi-attribute procurement auction. In: International Workshop on Information Security Applications, pp. 306–317. Springer (2005)

Dynamic Programming Approach in Conflict Resolution Algorithm of Access Control Module in Medical Information Systems

Hiva Samadian[1(✉)], Desmond Tuiyot[1], and Juan Valera[2]

[1] Colgate University, Hamilton, NY 13346, USA
hsamadian@colgate.edu
[2] Ana G Mendez University, Gurabo, PR 00777, USA

Abstract. Organization assets and resources are administered to be accessed by some members and not by others. The high sensitivity of assets (e.g. patients' health record and sensitive medical devices) in medical centers, requires the managers to pay special attention to deploy reliable authorization models. A reliable authorization model must be able to resolve the contingent conflicts that can occur due to different authorization assignments to subjects (e.g. technicians). Resolving conflicts is quite a challenge due to the existence of sophisticated inheritance hierarchies that might cause an exponential number of conflicts (in terms of the number of subjects in the organization hierarchy) and the diversity of ways to combine resolution policies. The need to an approach that can handle as much contingent conflicts and resolution policies as possible and work in an appropriate time emerges here. An existing work has presented an exponential algorithm for resolving all conflicts in accordance to all existing policies. This paper develops a dynamic programming (DP) algorithm with a polynomial time complexity for the same conditions. The two approaches were compared by doing three different experiments with both algorithms and comparing the results. The experiments show that the average time decreased to 1/10 on small SDAGs with maximum number of edges. The improvement for large sparse SDAGs is more significant (3/1000). The average time of determining the authorization of a subject over 500 objects is just 52.56 s.

Keywords: Medical information system · Health information system · Conflict resolution · Access control · Authorization · Security policies · Conflict detection

1 Introduction

Ever since health-care information systems have been implemented, their security is being considered an important issue, especially because their data are deemed to comprise extremely sensitive information [1]. The sensitivity of health-care information has been addressed in many resources, including [2–4], and many others. In particular, financial and health information have been considered the most sensitive data attributes [5]. In the medical centers, in addition to data, hardware access is also considered a sensitive issue. Besides the costs that an unauthorized access to sensitive devices can impose on the organization, it eventually may cause a health issue for some patients. In advanced medical centres, accesses to the sensitive data and devices can be controlled automatically using the information provided by access control module of information systems. These modules determine whether access to the hardware assets and sensitive data is authorized or unauthorized. In medical information systems, warehousing applications, management information systems, electronic health record, and other information systems, there is a need for an access control module that deploys authorization modules.

By implementing authorization modules, access by individuals or parties (subjects) to the assets (objects) of the organization can be managed automatically. The output of an authorization module could be seen as some tables (*EfM*) in which each row is a subject, each column is an object, and the intersection of row *s* and column *o* is 1 if, for a certain right, *s* is authorized to access *o*, and 0 otherwise. Such tables are subject to change over and over again especially in big organizations [6]. For example, the policies, employees, hierarchical relations, and organization roles might change, or new roles might be added; thus, providing such tables is not a one-time task. In particular, due to growth of decentralized healthcare systems, many parties such as insurance companies, patients, clinics, and correlated centers might be authorized to access some objects; thus, their changes reflect in the system as well. Moreover, composing the tables explicitly requires the decision makers to decide on the access of each subject to each object, for all access rights, which usually is not practical. Instead, the decision makers decide on the access of some of the subjects to some of the objects for each access right explicitly which results in creation of sparse tables called Explicit access control Matrix (*ExM*). Determining the rest of authorizations is done automatically by authorization modules, making use of the explicit authorizations. By taking advantage of authorization modules, decision makers do not need to compose the whole table. Furthermore, in case any modification is needed, the table can be recalculated easily and efficiently.

In some systems, conflicts are resolved based on the "negative authorization takes precedence" policy [8], while in some other systems such as public information applications, the conflicts are resolved based on the "positive authorization takes precedence" policy. However, authorization modules in access control may include the hybrid models [7] in which both positive (+) and negative (−) authorization modes are supported. This paper introduces an authorization module with a hybrid model that supports both modes. The module receives an *ExM* and provides the fulfilled table (called effective matrix: E*f*M) within the *propagation* process. Table 1 summarizes the terms which are used in this paper.

Table 1. Nomenclature

Subject	Any person or group in the organization hierarchy
Object	Any asset of the organization including information or hardware
SDAG	Subject direct acyclic graph is a DAG which represents the hierarchical relations of the subjects of the organization in regard to their authorization inheritance for all access rights over all objects. The nodes are the subjects, and an edge from a node s_1 to a node s_2 means that s_2 may inherit any explicit authorization of s_1 and ancestors of s_1 in the SDAG
ExM	Explicit access control matrix for a certain right is a sparse matrix with the subjects as rows, objects as columns and the entries (s, o) as the authorization mode of subject s to object o for the respective right. Each entry has a value of 0/1 indicating −/+ explicit modes respectively or null indicating the blanks. (Terms "value" and "mode" are used interchangeably.)
Propagation	The process of filling the blank spaces of an ExM with the authorization values using its filled cells (explicit authorizations) and based on the hierarchical relation of subjects in the SDAG of the organization after finding all contingent conflicts and resolving the conflicts according to a policy
EfM	The result of exerting propagation on ExM

1.1 Conflicts in the Authorization Module and Conflict Resolution Policies

To propagate the values on an *ExM* by determining the modes of authorization (− or +) of blank spaces of the table, for any blank cell at row *s* and column *o*, we first observe the position of *s* in the SDAG hierarchy. Then, we look up explicit authorization modes (the filled entries of *ExM*) of other subjects that are ancestors of *s* in the hierarchy. Such authorizations can potentially be inherited to the respective blank cell. If a negative mode is inherited to the cell, *s*'s authorization mode will be negative; if a positive mode is inherited, *s*'s mode will be positive. However, a negative mode might be inherited from some nodes and, at the same time, a positive mode from some other nodes; this is called a *conflict*. Each path to s from an ancestor of *s* with explicit value assigns a value (− or +) to *s*. Finding all possible paths for *s* with a naive approach could cause an exponential time for identifying the conflicts.

When a conflict occurs, several conflict resolution policies—such as Default, Locality, Majority, and Preferred [9]—are used to resolve the conflict and decide on one single value to enter in the respective *EfM* cell. These policies have been articulated by many researchers and appear in various real-world applications, but they are typically discussed independently and not in combination.

Default policy (D) is the policy of specifying a default mode (either + or −) merely to root subjects of SDAG for which no authorization has been defined. This policy is deterministic which means at any time a root node has one default authorization mode either default + or default −, and no other policies can interfere and change its mode. A closed policy recommends a default negative mode for roots that are not explicitly permitted to access a certain object [10].

Locality policy (L) is a distance-based policy which states that the most specific authorization takes precedence. Thus, for a given subject s, when both + and − authorization modes can be derived from different ancestors, the one that is closer to s wins (we might give preference to the one that is farther; in this case we indicate Locality policy with letter G, which stands for Globality). This policy requires us to know all the distances (length of the paths in SDAG) of explicit valued ancestor nodes of s from s. This policy is not deterministic since no authorization wins when the distances are equal for both modes.

Majority policy (M) states that the conflict can be resolved based on votes, and the authorization that has the majority wins. This policy is also non-deterministic since it can result in a tie. Finally, the Preferred policy (P) determines which authorization wins when both + and − authorization modes can still be derived for a particular subject after performing other three policies. Notice that Preferred policy is deterministic.

Combining the existing conflict resolution policies gives the possibility of taking advantage of all of them. Chinaei and Zhang [9] present all 48 possible meaningful combinations of the major conflict resolution policies (D, L, M and P) in a Unified Conflict Resolution Framework (UCRF). Designers of access control modules typically choose a single approach to conflict resolution and incorporate a hardwired conflict resolution method. Consequently, if an enterprise subsequently decides to choose an alternative conflict resolution strategy, the whole system has to be replaced. Maintaining separate software for multiple strategies is expensive for software providers. This work takes advantage of a unified conflict resolution that overcomes these problems.

1.2 Contribution and Related Works

In this paper, the SDAG and *ExM*s for a typical medical centre (e.g. a hospital) have been simulated as the raw data for the *propagation* process. An efficient dynamic programming algorithm for *propagation* process that lists and solves all contingent conflicts in a reasonable practical time has been developed and implemented. The implementation has been applied on the set of simulated entries in the context of medical centres and the emerged conflicts were resolved by all combined policies specified in UCRF. Results show that although a naive algorithm runs exponentially and is quite useless, this approach works in a reasonable time in practical instances.

An algorithm for *propagation* process that extracts the maximal subgraph of SDAG in which the target subject is the sole sink node has been presented by [11]. In each subgraph, a search is done to list the entire set of possible authorization assignments for the subject. In contrast, in this work we first search the whole SDAG only once regardless of s and the *ExM* in question. This is good because we use its result for propagating any *ExM*. Unlike their work, here the search starts from the roots of SDAG and spans the entire graph. Searching the graph is done with a dynamic programming approach that proves to be an efficient approach with improvement in the results. The algorithm then prunes the results of each node s to obtain its desired list of paths in regard of the *ExM*. We have also redefined the background concepts like SDAG, *propagation*, and conflict resolution.

Some existing solutions for computing effective authorizations assume that the explicit authorizations are propagated on tree structured data [12, 13]. This trivializes conflict resolution since there is only one path between any ancestor and a leaf. Moreover, the number of ancestors for a leaf is bounded by the height of the tree, which is usually a small value in real world data [14]. This paper supports the DAG structure which is more general and is practiced in most of the real-world applications.

Our approach is different from the combining algorithms in XACML [12], in which the resolution model relies on the object hierarchy rather than the subject hierarchy. It also differs from the approaches [15] in which the combined policies–so called rules– are not independent as it is so here in our work. Some other related works [16] have addressed more distributed systems and expand control of access to the users; which is a different perspective than our work.

Author in [17] addresses attribute-based access control (ABAC) systems where attributes are subjects, objects and rights. Our works share this property. ABAC is gaining momentum for adoption due to its flexibility, granularity, context-awareness, externalization of authorizations, and many other benefits [18, 19]. They addressed the challenge of the complexity of resolving anomalies such as inconsistency and the fact that by the time of their work the existing implementations suffer from the absence of tools that reply to this complexity. The authors propose a special framework that is based on a mathematical formalism of the idea of conflict between access control policies. They provide a method of partitioning access requests into classes such that requests in the same class are handled by the same policies. Making use of this representation of access queries, they develop special tools, including a conflict/anomaly solver which outputs a conflict-free policy set.

The authors of [20] use a modification of a data classification algorithm to provide an efficient method that detects conflicts in access control policies and identifies situations where no provided policy can apply. However, the focus of that paper is the detection of such anomalies and not their resolution. Our paper, on the other hand, develops an algorithm that both detects and resolves the policy conflicts.

Author in [21] presents an access control system that is intended to operate in a cloud computing environment. Part of their proposed system is a policy model which includes a conflict resolution component. Each policy is assigned a priority by the data owner and/or the cloud service provider which is used to calculate and assign weights to them. Conflicts are then resolved by choosing the policy with the highest calculated weight. Unlike their work, our paper makes use of a combination of 4 major resolution policies as explained in Sect. 1. Also, our paper differs from their work in at least two other major aspects. First, they consider the relation between access rights while we consider the rights independently. Second, they search among a set of given conflicts while we detect the conflicts.

Author in [22] extends the classic role-based access control framework (RBAC) to solve new problems raised by virtualization, which has become increasingly widespread. The authors modeled the policies of the new virtual RBAC (VRBAC) into concepts of description logic and made use of inference rules to detect policy conflicts. Part of their focus then is conflict detection; our work, on the other hand, attempts to resolve conflicts in addition to detecting them.

1.3 Organization of the Paper

The rest of the paper is organized as follows: Sect. 2 presents the details and an example of SDAG, ExM, and propagation using UCRF. Section 3 explains the dynamic programming algorithm for propagation. Section 4 outlines the results of the experiments. Finally, Sect. 5 presents the conclusion of this paper and expresses possible ideas for a future work.

2 Example of SDAG, *ExM* and Improved *Propagation* (Using UCRF)

Figure 1(a) illustrates an example of a small SDAG that represents the hierarchical relations of 9 subjects regarding their authorizations (for any object and right). Figure 1 (b) illustrates an *ExM* for authorization *read* in which nodes s_2 and s_4 have explicit + authorizations to read a certain object o_2, and s_5 has an explicit - authorization respecting same action and the other nodes do not have explicit values. (Explicit values of subjects to read other objects are not shown). To propagate the values of the *ExM*, we need to determine the authorization value of all its blank cells (e.g. node *user* to read object o_2). For each node such as *user,* it goes through the following steps: (1) it first must create a table including all possible values for *user*'s authorization called *allrights* table. The *allrights* table for right "read" of subject *user* over object o_2 is shown in Fig. 1(c). To determine all possible values, one needs to check all ancestor nodes of user who have explicit values. Here, they are s_2 and s_5. (There is no path from s_4 to user means that it is not user's ancestor and its authorization cannot be inherited to user.) If the policy for conflict resolution includes Default policy, we know that all roots would have a default explicit value. Therefore, we must include all root nodes to be checked. There is more than one path from s_2 to user; each of them imposes same mode but with different distances. Thus, one possible mode/value for user is +/1 through a direct path from s_2 to user with length (distance) 1, another is +/1 through the path s_2-s_3-s_5-user with length (distance) 3. For s_5, however, there is one path to user with mode -, so a possible mode for user is - with distance 1 as it is shown in the allrights table. The modes corresponding to the root nodes that get their explicit value through the default policy are indicated as d (see Fig. 1(c)). It is because, later on, we check both possibilities for d to be either − or +.

Step (2): After determining all possibilities for the value of user's authorization if all the possibilities are the same value, that value is user's value; if not, we have conflict(s) and use a conflict resolution policy. There are 48 combined policies in UCAF which are listed in Fig. 1(d). For example, D+LMP+ is a policy in which blank roots get their default explicit mode as +. If by this movement all possible modes for user are + the ultimate mode of user would be +. If after applying D+ still the conflict exists, which in this example is so, we try L if it resolves the conflict we do not go further if not the next is M. L in this example doesn't resolve the conflict because for the minimum distance which is 1 we have both modes + and − so we need to check M. The same way if P+ is needed we force user to have mode + but it is not needed here because the M resolves the conflict since after applying D+ the number of + modes are

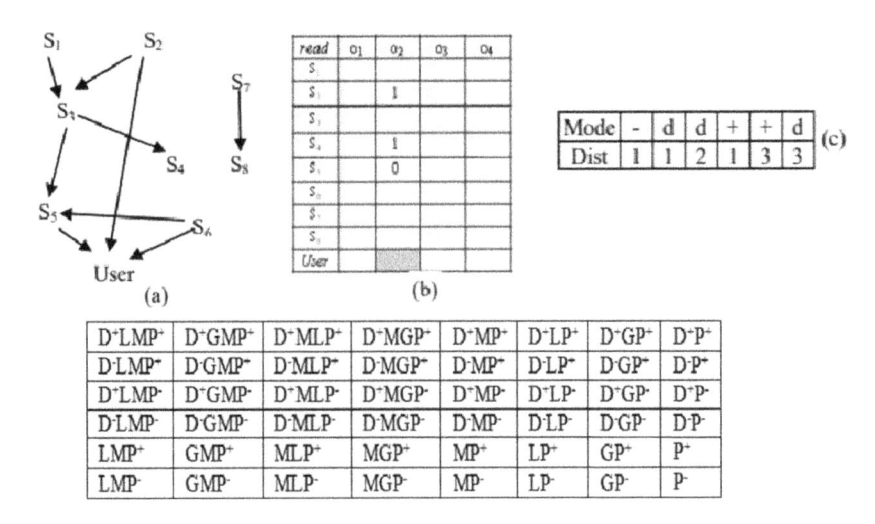

D⁺LMP⁺	D⁺GMP⁺	D⁺MLP⁺	D⁺MGP⁺	D⁺MP⁺	D⁺LP⁺	D⁺GP⁺	D⁺P⁺
D⁻LMP⁺	D⁻GMP⁺	D⁻MLP⁺	D⁻MGP⁺	D⁻MP⁺	D⁻LP⁺	D⁻GP⁺	D⁻P⁺
D⁺LMP⁻	D⁺GMP⁻	D⁺MLP⁻	D⁺MGP⁻	D⁺MP⁻	D⁺LP⁻	D⁺GP⁻	D⁺P⁻
D⁻LMP⁻	D⁻GMP⁻	D⁻MLP⁻	D⁻MGP⁻	D⁻MP⁻	D⁻LP⁻	D⁻GP⁻	D⁻P⁻
LMP⁺	GMP⁺	MLP⁺	MGP⁺	MP⁺	LP⁺	GP⁺	P⁺
LMP⁻	GMP⁻	MLP⁻	MGP⁻	MP⁻	LP⁻	GP⁻	P⁻

Fig. 1. (a) SDAG sample, (b) Respective ExM for right read over the objects o_1–o_4, (c) allrights table for the authorization of user for reading object o_2. (d) List of combined conflict resolution in UCRF

more than − modes. Some of the unified policies in UCAF don't include all basic policies. In particular, some of them such as MGP- do not have D. In these cases, we simply ignore giving explicit values to roots that means for example in Fig. 1(c) we ignore the existence of columns with mode d.

3 Improved *Propagation* Algorithm

The first phase of *propagation* in an *ExM*, as it was explained in Sect. 2, requires a search for all paths in SDAG that starts from an explicit valued node or a root and ends in *s*. We first search for all paths ending in *s*, for any *s* in SDAG by spanning the SDAG only once and store the list of such paths in a variable called *pathslist*. A brute-force search for finding all paths in a graph with *m* node takes O(mm). A naïve recursion search for the paths ending in *s* takes O(m!). In our approach, the search starts from the roots which are the base case of our search with the empty *pathslist*. We put all nodes that all of their parents have their *pathslist* different than null in the frontier. For nodes *s* in frontier for all their parents *p* the algorithm adds edge *ps* to paths of *p*'s *pathslist* and adds the resulting paths to the *pathslist* of s. Dynamic programming suggests storing the values (list of desired paths for each node s of SDAG) to avoid repeating same calculations over and over. The repetition is plausible because s might appear in the desired paths of many subjects and for each subject, we need to calculate the value of s. For propagating an *ExM* we need to prune the *pathslist* of required subjects by removing all those paths that doesn't start from an explicit valued node or a root. Having this pruned list, we have the *allrights* table for the subject. Table 2 presents our *propagation* algorithm.

Table 2. Propagation algorithm

Propagation algorithm

```
1   propagation (ExM, SDAG, rule)                                           // output: EfM of ExM.
2      findpathslist (SDAG) if it is not already found
3      for each (s, obj)
4         allrights (ExM,SDAG, s, obj)
5         resolve (rule, s-table) ----------------------------------------------- End of propagation --------------------------
6   findpathslist (SDAG)                                                    // output: true if done.
7      for each root set its pathslist with a single path with length 0     //pathslist of s is the listof all paths that end in s
8      for each node s if all parents of s has its pathslist different than null   //After running findpathslist each s in SDAG has its pathslist
9         for each parents par of s
10           for each paths pat in par.pathslist
11              create a new path pat-s by adding an edge (par,s) to pat and adding 1 to its length and add pat-s to s.pathlist
12      go back to line 8 until there is no s left that satisfies condition in line 8---------------------End of findpathslist-----------------------------
13  allrights (ExM,SDAG, s, obj)                                            // output: allrights table for (s, obj)
14     prune (s, obj)          // prune takes the pathslist of s, remove those paths which start with a non-root non-explicit valued (for obj)
15     return allRightTable  which is a list of all pairs (m, l)            // m:explicit mode of the first node of each path, l: length of path
End of allrights ---------------------------------------------------------------------------
resolve (rule, s-table) as in [11]          //resolves the conflicts of s-table (output of allrights) in accordance to rule; outputs EfM
```

4 Experiments and Discussion

We compared our work with a similar work [11] in which they solved the same problem with a different algorithm. We did the same experiment as what they performed in their paper and compared the results. They used three KDAGs (SDAGs that has the maximum number of edges) with 10, 15, and 20 nodes, and for each of them they considered a node and calculated the value of that node. They repeated this 10 times, each time with the assumption that a portion of nodes in the KDAG (from 0.5% of all nodes to 9.5% as it is shown in Fig. 2(a)) have the explicit values randomly. We have done the same experiment by a system with CPU Intel® Core 2 Duo 2 GHz with 4G RAM which is relatively ordinary; therefore, the influence of difference between the powers of computers is negligible. Figure 2(a) compares our results which introduces an improvement of 1/10 in CPU time (The average of times for each KDAG then average of the three numbers is one tenth of the same number in their experiment.). As it can be seen in this figure, the increment of the CPU time by the increment of the percentage of explicit valued nodes are much less with our algorithm. The size of the SDAGs in these experiments are small. With bigger sized SDAGs the algorithm of [11] will fail to run in a reasonable time since the algorithm is exponential in asymptotic running time, while this is not the case with the algorithm introduced in our work. Second experiment has been done in which the SDAG doesn't have the maximum number of edges, but it has 8000 node and the ExM is sparse with 20% explicit values. We used the D-LP- policy to be able to compare our results with the result of the same experiment in [11]. Awing the improvement in this paper we could provide EfMs in 7.5 s while this time for their work is about 40 min. Our third experiment simulates more real situations in which not only the number of subjects is big, we resolved ExMs with 500 objects. Figure 2(b) illustrates the results which show the rate of increment of the elapsed time. This also shows the average of 52.56 s for determining the authorization of a subject over 500 objects which is an excellent help in rapidly growing decentralized medical centers with huge number of subjects and objects.

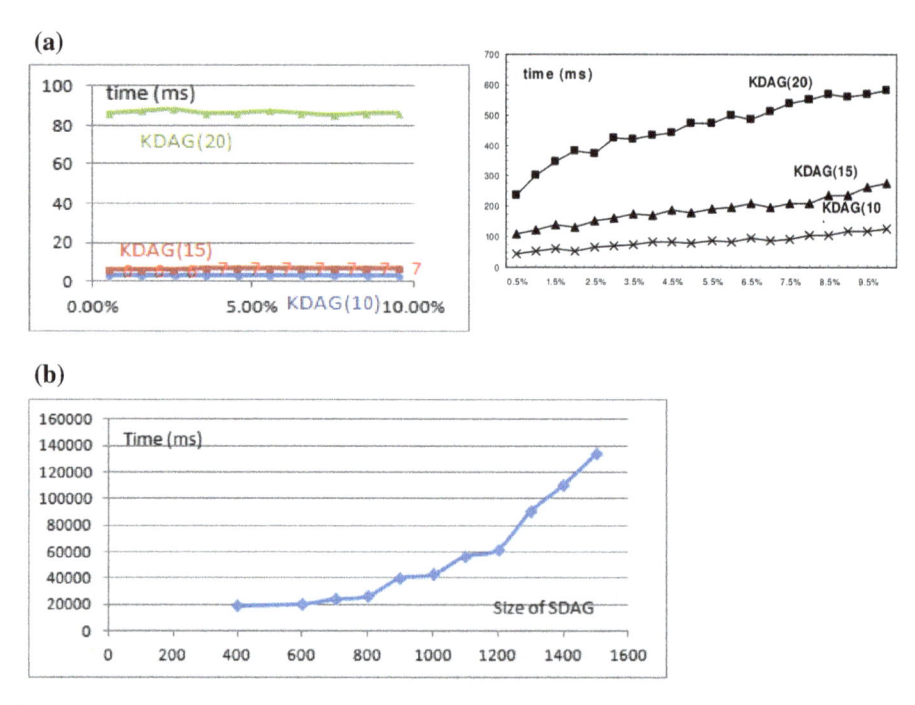

Fig. 2. (a) KDAG calculation comparison. Left: our results, Right: the result of [11] (b) Time for resolving E*x*M with 500 objects in different SDAG sizes

5 Conclusion and Future Work

Since the problem in hand consists of a search among a graph that encounters repetitions in visiting the nodes and has recursive behavior, it bears the features that meet the required conditions of a DP approach. Thus, a DP approach was deployed in our work and paid off with a significant improvement in time of the process. The experiments show that the average time decreased to 1/10 on small SDAGs with maximum number of edges. The improvement for large sparse SDAGs is more significant (3/1000). The average time of determining the authorization of a subject over 500 objects is just 52.56 s. The developed algorithm can be used to obtain some useful statistics such as the dependency of the speed of the process on the rate of explicit values in *ExM*, the rate of subjects with explicit values and their level in the SDAG. In addition, it is possible to find out how restricted or flexible the resulting *EfM* is based on the aforementioned statistics and the policies used. These statistical studies have been left as a topic for a separate paper in the future.

Acknowledgment. We thank Dr. Amirhossein Chinaei for introducing the topic of the research and his valuable contribution on the review of the work in its early stages. We also thank Colgate University Faculty Research Council for financially supporting the work.

References

1. Smith, E., Eloff, J.: Security in health-care information systems—current trends. Int. J. Med. Inf. **54**(1), 39–54 (1999)
2. Xiao, Q., Wang, Z., Tan, K. L.: LORA: link obfuscation by randomization in graphs. In: VLDB Workshop, Seattle (2011)
3. Banerjee, M.K.R., Wu, L., Barker, K.: Quantifying privacy violations. In: VLDB Workshop, Seattle (2011)
4. Deng, M., Nalin, M., Petkovi, M., Baroni, I., Abitabile, M.: Towards trustworthy health platform cloud. In: 9th VLDB Workshop, Istanbul (2012)
5. Westin, A.: Social and political dimensions of privacy. J. Soc. Issues **59**(2), 431–453 (2003)
6. Leitner, M., Rinderle-Ma, S.: A systematic review on security in process-aware information systems – constitution, challenges, and future directions. Inf. Softw. Technol. **56**(3), 273–293 (2014)
7. Jajodia, S., Samarati, P., Sapino, M.L.: Flexible support for multiple access control. ACM Trans. Database Syst. **26**(2), 214–260 (2001)
8. Bertino, E., Jajodia, S., Samarati, P.: A flexible authorization for relational data management systems. ACM Trans. Inf. Syst. **17**(2), 101–140 (1999)
9. Chinaei, A.H., Zhang, H.: Hybrid authorizations and conflict resolution. In: 3rd VLDB Workshop on Secure Data Management (SDM 2006), Seoul (2006)
10. Harrison, M.A., Ruzzo, W.L., Ullman, J.D.: Protection in operating systems. Commun. ACM **19**(8), 461–471 (1976)
11. Chinaei, A.H., Chinaei, H.R., Tompa, F.: A unified conflict resolution algorithm. In: 4th VLDB Workshop, SDM 2007, Vienna (2007)
12. Moses, T.: eXtensible access control markup language version 2.0. OASIS Standard (2005)
13. Zhang, H., Zhang, N., Salem, K., Zhuo, D.: Compact access control labeling for efficient secure XML Query evaluation. In: 2nd International Workshop on XML Schema and Data Management (2005)
14. Mignet, L., Barbosa, D., Veltri, P.: The XML web: a first study. In: WWW 2003 Proceedings of the 12th International Conference on World Wide Web (2003)
15. Koch, M., Mancini, L.V., Parisi-Presicce, F.: Conflict detection and resolution in access control specifications. In: 5th International Conference on Foundations of Software Science and Computation Structures (2002)
16. Calvillo, J., Roman, I., Roa, L.M.: Empowering citizens with access control mechanisms to their personal health resources. Int. J. Med. Inf. **82**(1), 58–72 (2013)
17. Yahiaoui, M., Zinedine, A., Harti, M.: Deconflicting policies in attribute-based access control systems. In: IEEE 5th International Congress on Information Science and Technology (CiSt), Marrakech (2018)
18. Hu, V.C., Chandramouli, R., Ferraiolo, D.F.: Attribute-Based Access Control. Artech House Inc., Norwood (2003)
19. Axiomatics. https://www.axiomatics.com/. Accessed 15 June 2019
20. Shaikh, R.A., Adi, K., Logrippo, L.: A data classification method for inconsistency and incompleteness detection in access control policy sets. Int. J. Inf. Secur. **16**(1), 91–113 (2017)
21. Habiba, M., Islam, R., Ali, A.B.M.S., Islam, Z.: A new approach to access control in cloud. Arab. J. Sci. Eng. **41**(3), 1015–1030 (2016)
22. Luo, Y., Xia, C., Lv, L., Wei, Z., Li, Y.: Modeling, conflict detection, and verification of a new virtualization role-based access control framework. Secur. Commun. Netw. **8**(10), 1904–1925 (2014)

Two-Factor Authentication Using Mobile OTP and Multi-dimensional Infinite Hash Chains

Uttam K. Roy[1(✉)] and Divyans Mahansaria[2]

[1] Department of IT, Jadavpur University, Kolkata, India
royuttam@gmail.com
[2] Tata Consultancy Services, Kolkata, India
divyansmahansaria@gmail.com

Abstract. Hash chains are often used to implement One Time Password based authentication systems. Some use finite hash chains that require frequent system re-initialization. Some use computationally-intensive public-key algorithm to achieve infiniteness. Eldefrawy et al. proposed a hash-based infinite chain but has limited ability to resist pre-play and guessing attack. This paper provides a smartphone-based two-factor authentication system nRICH that uses both knowledge (password) and possession (seed) based information. The OTP is generated perpetually from a multi-dimensional infinite hash chain that eliminates the limitations of other techniques. It is superior to resist pre-play attack. The hard challenge is a random path from origin to a random point inside a multi-dimensional moving hypercube. We have rigorously performed the security analysis and compared with other techniques w.r.t. various metrics and found suitable to be implemented in even low-end devices. The only drawback is the increased length of the challenge to be typed by the user. We propose to use QR code to avoid this problem.

Keywords: One Time Password · Authentication · Hash chain · One-way encryption · Security

1 Introduction

The authentication mechanisms rely on three kinds of data: (i) *knowledge* (what user knows e.g. password) (ii) *possession* (what user has e.g. credit card) and (iii) *inherence* (what user is e.g. fingerprint). Among these, *biometrics* (inherence-based) is often considered as the superior authentication measure. However, the principle, biometric follows isn't any different from others; an identity thief can steal biometric data from a network/computer in the same way passwords are stolen. And if it is compromised; the effect is either *non-recoverable* or takes long time to fix it. We can change our password instantly; get a new credit card from the bank in two weeks; but how to get new fingerprints/iris to replace the stolen ones? Since, biometrics cannot be changed; widespread use of hackable biometrics-based authentication may make the situation critical.

This work is partially supported by the project entitled "QR code-based Authentication Using Mobile OTP and..." under RUSA 2.0 (Ref. No. R-11/668/19)

© Springer Nature Switzerland AG 2020
K. Arai et al. (Eds.): FICC 2020, AISC 1129, pp. 682–700, 2020.
https://doi.org/10.1007/978-3-030-39445-5_50

This reality enforces researchers to concentrate on the *knowledge* and/or *possession* based systems which use 'tokens' that can at least be changed if stolen. The users have better/quicker control on the former one and probably that's why the Internet is still using the simple but elegant *password-based* authentication systems.

These systems allow users to use the same password for multiple authentications. Although, strong systems enforce users to change their passwords after periodic interval; in the meanwhile intruders may still eavesdrop the line connecting the user and the system to get the password and can *impersonate* the user to the system subsequently.

To make eavesdroppers powerless, instead of a single password, researchers suggested to use a new one [1] for every authentication. These passwords are often called *One Time Password* (OTP). However, it is not a good idea to ask the novice users themselves to create a 'difficult-to-guess' new password for each authentication. Instead, the users may use a *device* to do the same on their behalf. The device should be *portable* so that the user can carry it with him/her and use it whenever required. The most obvious portable device the users already have and carry almost all the time is a *smartphone*. Smartphones [2, 3] support ubiquitous authentication without any extra device to be brought and deployed.

Although SMS-based OTP solutions [4–6] are often used today, they have several problems. First, the SMSs are crackable [7, 8] in spite of all security measures taken while designing Global System for Mobile (GSM) communications. Second, turning off the SMS service (say during roaming) leaves the system useless. Third, neither the delivery nor the delivery time for SMS is guaranteed [9]; the messages may be dropped by the underlying networks due to the lack of capacity or if the recipient's phone is out of service area. Fourth, there is a considerable cost involved [9, 10] sending an SMS for every authentication (say for huge online bank transactions).

To overcome these problems, Eldefrawy [11] suggested a system where smartphones are used only as the OTP generator and to offer interface independence. OTPs are generated based on an ever-changing *seed* (synchronized between client and server) and an ever-changing *challenge* thrown by the server.

Although, it overcomes some problems of SMS-based methods and some previous proposals [1, 12, 13], ability to *resist predictable attacks* is not strong enough. Specifically, probability of guessing next challenge is $1/m^2$ where m is the range of *x* and *y* coordinates of a challenge *(x, y)*. For example, if $m = 5$, there are only *25* possibilities of next challenges/OTPs.

In predictable attacks, attackers try to predict challenges or OTPs to break the system. So, the number of possible OTPs and challenges that flow back-and-forth through the system must be as large as possible. In this paper, we've primarily extended the idea of [11] with several modifications to enhance the ability to *resist predictable attacks* significantly keeping all other advantages unchanged. The system is backward compatible and incurs negligible overhead. We have developed an experimental prototype called n-dimensional Running Infinite Chain Hash (nRICH[TM]) authentication system using Java socket technology, but may be implemented using other technologies as well. The performance of this system is compared with other state of the art systems and found to be superior.

2 Related Works

In 1981, Leslie Lamport first proposed to use a new password (One Time Password) for every authentication [1] instead of a single one. This eliminates replay attack—discover the password and use it subsequently.

The idea was implemented by Haller [12], where a finite sequence of OTPs is generated using a finite hash chain from a single seed s as:

$$h^N(s),\ h^{N-1}(s),\ h^{N-2}(s), \ldots h^1(s)$$

In general, at i^{th} authentication, the server throws challenge i and the client replies with $OTP_i = h^{N-i}(s)$. The server applies h one more time on it and checks with the previous OTP [i.e. $OTP_{i-1} = h^{N-(i-1)}(s)$] sent by the client. If result does not match then the authentication fails, else authentication succeeds and OTP_i is stored to be used for $(i + 1)^{th}$ authentication. The major disadvantage is that the system is limited to certain pre-determined number N of authentications. After that it requires a system *re-initialization*. It is also susceptible to *small challenge attack* [14] where an attacker takes the host's role, sends a small sequence number to the user to know the hash chain *initial value(s)* and calculates further OTPs; thus by breaking the system. Although, computation cost for later authentications is less (few hash operations), it is significant (many hash operations) for former ones.

Goyal et al. [13] reduced this computational cost where the server aids client computation by inserting R breakpoints in hash chain. They called it as *N/R system*. Server computation cost remains unchanged. Although it *increases* the number times a client can login before system *re-initialization*, (i) re-initialization is still needed and (ii) N is still limited.

To avoid finiteness of hash chain, Bicakci et al. [15] used *public-key cryptography* to get effectively a *forward infinite non-invertible* function. It used RSA [16] public key algorithm and a pair of public key e and private key d. The ith OTP is $OTP_i = A^i(s, d)$ and verification is done as $A(OTP_i, e) = OTP_{i-1}$. However, public-key encryption and decryption algorithms are computationally intensive and applying them increasing number of times making them infeasible to use in limited computing devices.

In [17], Yeh et al. claimed that they enhanced *N/R system* eliminating server spoofing attacks, pre-play attacks, and off-line dictionary attacks. However, Chefranov [14] showed that it is essentially same as Lamport [1] but transmits some sensitive information across networks using XOR operation. It was criticized and proved to be vulnerable to pre-play attack by Yum et al. [18].

In [11], Eldefrawy et al. proposed a two factor authentication (2FA) scheme based on 2/3 nested hash chains; one is used for seed update and all are used for OTP generation. Although it results an infinite forward hash chain, strength to resist guessing attack is poor. This paper is primarily is an extension to Eldefrawy's work.

3 The nRICH System

We are only concerned with *external* active/passive (i.e. who have access to the networks) and *insider* passive (who have access to the computers and try to elevate privileges to get sensitive information) attackers. However, insiders who overload network or computer storage or processing capacity, leading to system crash are out of the scope of this work.

The nRICH system is an extension to the Eldefrawy's work [11] with several modifications in order to significantly increase the ability to resist predictable attacks. It can be hooked to the present Internet system seamlessly with little modifications.

3.1 Assumptions

Our system makes following assumptions:

Authentication Helper Device: The users to be authenticated have a *smartphone* with them (if they already don't have one). The smartphone is used as the OTP generator and offers interface independence.

No Secret Algorithms: The security algorithms used by an authentication system should not be kept *hidden*. A hidden algorithm, if somehow exposed, will make the system useless. We shall allow every other to explore and evaluate its cryptographic strength in all possible ways.

No Stored Secrets: In a strong authentication system, any secret information (such as password) should be stored in the host as it is. This may draw attacker's attention and may increase the chance of wide-spread security breach. For example, Unix/Linux systems store hashed passwords in the password file *(/etc./shadow)* preventing intruders to directly use them. Our system does not also store any secret information to any host as it is but as encrypted form.

3.2 Operational Principle

There are two parts of the system: (i) a *server* that authenticates clients; (ii) remote *clients* who want to get authenticated.

Apart from the password, the first factor of authentication, they share a common synchronized *seed*. The server throws a *challenge* to the client. The client generates *OTP* based on this challenge and sends back to the server. The server itself calculates OTP and compares with the received one. To make eavesdroppers powerless and to defend pre-play attack, the next challenge and OTP space are kept as large as possible.

The OTP dimension is no longer limited to 2 or 3; it is extended to n. The only requirement is the availability of n hash functions $h_0, h_1, h_2, \ldots, h_{n-1}$, one along each dimension. With the advent of cryptography, finding 30 different excellent hash functions currently is no more difficult. The SHA family itself has more than 15 (SHA-0, SHA-1, SHA-256, SHA-512, SHA-224, SHA-384, SHA512/224, SHA512/256, SHA3-224, SHA3-256, SHA3-384, SHA3-512, SHAKE128, SHAKE256) hash functions. There are many others family of hash functions such as BLAKE, MD,

RIPEMD etc. However, proposed method is *backward compatible* and allows using a lower dimension if adequate number of hash functions is not available.

We consider an *n-dimensional hypercube* (Fig. 1). Suppose $m = \{m_i\}, 0 \leq i < n$, be the bound of the hypercube. The system works as follows: Upon receiving an authentication request from a client, the server generates a *challenge* to be used to generate OTP from a shared *seed*. The *challenge*, we propose, is a *random forward path* from origin to a random point $p[=(c_i), c_i$ be the ith *coordinate*, $0 \leq i < n,]$ within the hypercube boundary. It is basically a sequence of hash functions to be applied to the current *seed* to get the OTP. Since hash functions are *one-way* functions, we choose only forward paths (i.e. have only coordinate increase).

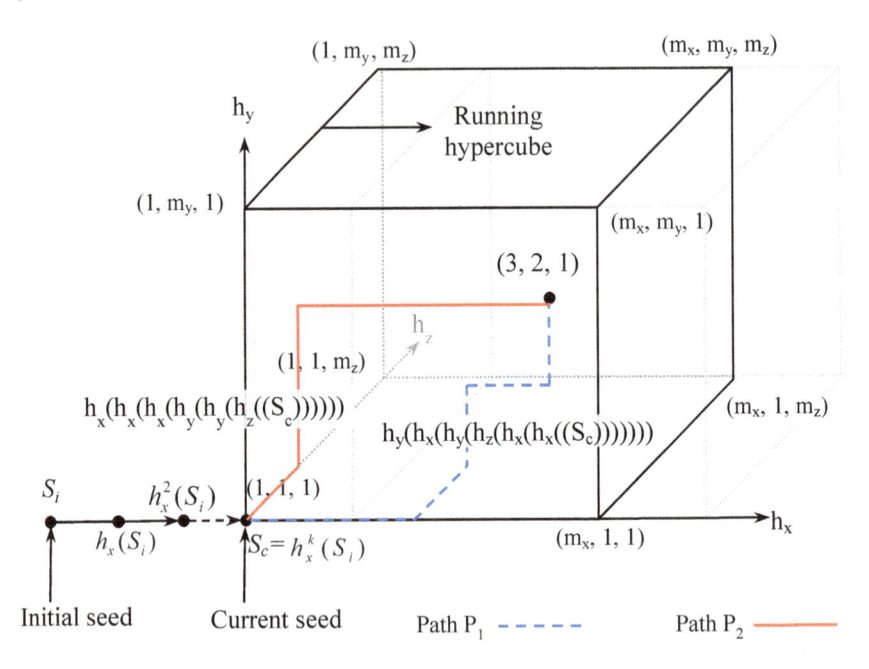

Fig. 1. Three-dimensional hash chain for OTP generation. Seed is updated using an infinite chain of h_x. Two possible paths (challenges) from (0, 0, 0) to (3, 2, 1) are shown.

The *seed* is also updated by applying h_0 after each successful authentication that effectively moves the hypercube to a new position. So, the set of next possible challenges is different at different time. Ideally, the hypercube should be moved by m_0 along 0-th axis.

Since, the number of such paths (challenges/OTPs) in a moving hypercube is substantially larger (see Sect. 4 for details) than the number of points considered in [11]; it has higher ability to resist predictable attacks. The OTPs are of variable length as a combination of hash functions is used; its length depends on the last hash used. For example, If SHA256 and SHA512 are used; the OTP length will be either 256-bit or 512-bit if SHA256 or SHA512 is used as the last hash respectively. This further decreases the chances of predictable attacks. The exact strength calculations have been discussed in detail in Sect. 4.

Moreover, the dimension n may be optimized to increase total number of challenges keeping the maximum number of allowable hash operations unchanged. For example, in 2-D, if range along each dimension is 5, the maximum number of hash operations is $5 + 5 = 10$ and number of possible next challenges (=no. of points) is $5 \times 5 = 25$. However, it may be reconfigured with several other possibilities. We choose the optimal one (that gives maximum number of paths), which has 3 dimensions with limit (3, 3, 4). This effectively increases the challenge set size from 25 to 650 (see Table 1 for details).

Furthermore, instead of changing seed for every authentication, it may be kept fixed for certain times (say 100) and may be changed after that using a 'seed update' procedure. This eliminates the burden of seed synchronization.

3.2.1 Registration Phase

We assume that the messages exchanged during registration phase (happens once per user) go through a secured channel (SSL/TLS may be used). The *client* uploads login id (email-address may be used) li, a strong password pw, first factor of authentication and some information unique to it.

The unique information may be any one or combination of 15-digit International Mobile Equipment Identity (IMEI), 15-digit International Mobile Subscriber Identity (IMSI), 60-bit Electronic Chip ID(for iPhones), 48-bit Wi-Fi MAC address(if available), Bluetooth device address(if available), etc.

The *server* uses unique information of client, its own unique information (e.g. 48-bit Ethernet address) and current date and time to calculate the *initial seed* S_i. The way to make the seed strong is beyond the scope of this paper. In the simple case, concatenation followed by a hash function may be used. The server sends S_i and n *hash functions* (as distinct as possible) $h = \{h_0, h_1, h_2, \ldots, h_{n-1}\}$ to the client. The user configures his mobile app with this initial seed and hash functions and keeps S_i in a safe place so that in can be used in emergency (e.g. mobile is lost).

The server also chooses limits of each dimension $m = \{m_0, m_1, m_2, \ldots, m_{n-1}\}$. The server saves li, hashed password $pw_e = h_0(pw)$ and S_i for further use. Secret password is not stored as it is; instead its hash value is stored disallowing even the server to retrieve it.

For each client registered, the server maintains three variables:

S_i → The initial seed
^sS_c → The current seed at server end; initialized to S_i
$C_s\ (\geq 0)$ → A counter that keeps track of number of times the hash applied on S_i to get ^sS_c; initialized to 0 and synchronized after each *successful* authentication.

User's mobile app also maintains two similar variables:

S_i → The initial seed.
^cS_c → The current seed at client end; initialized to S_i and updated as $^cS_c = h_0(S_i)$ after each *successful* authentication.

$C_c\ (\geq 0) \rightarrow$ A counter that keeps track of number of times the hash applied on S_i to get ^cS_c i.e. $^cS_c = h_0^{C_c}(S_i)$. It is initialized to 0 and gets incremented by 1 after each *successful* authentication.

The pseudo-code at the server for registration phase is shown in Fig. 2.

1. **register(li, pw, e_c)**
   Input: Client's login id li, password pw and Ethernet address e_c
   Output: Client gets registered and gets initial seed and n hash functions
2. Initialize n, $m_0, m_1, m_2, ..., m_{n-1},\ h_0, h_1, h_2, ..., h_{n-1}$;
3. $m \leftarrow \{m_0, m_1, m_2, ..., m_{n-1}\}, h \leftarrow \{h_0, h_1, h_2, ..., h_{n-1}\}$;
4. $S_i = concat(e_c, e_s, currentDate)$;   //$e_s$=server's unique information
5. $C_s=0;\ ^sS_c \leftarrow S_i;\ pw_e=h_0(pw)$;
6. Store li, $pw_e, ^sS_c, C_s, S_i$ for further use
7. Send S_i, and h to client

Fig. 2. Steps followed by the server during registration phase

3.2.2 Authentication Phase
The message flow sequence during a steady state authentication is shown in Fig. 3.

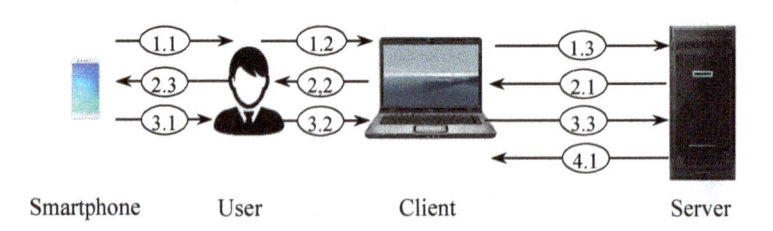

Fig. 3. Message flow during authentication

1.
   1.1. The user gets current seed counter C_c from mobile app
   1.2. User types *login id*, *password*, the first authentication credential, and current seed counter C_c in the authentication page
   1.3. Client remembers *login id* and sends C_c along with the *login id* and *password* to the server.

   Note that as an additional layer of security, counter C_c and OTP are not sent together.

2.
   2.1. The server notes this C_c and chooses a random point $pt = (c_0, c_1, c_2, ... c_{n-1})$ inside the n-dimensional hypercube. The challenge is a random path from origin

to this point with path length $pl = (c_0 + c_1 + c_2 + \ldots + c_{n-1})$. A path is a sequence of dimension number as follows:

$$path = \{d_1, d_2, d_3, \ldots d_{pl}\}, \quad 0 \leq d_i < n \quad (1)$$

The algorithm to generate the challenge is shown in Fig. 4. We used *Fisher–Yates shuffle* algorithm that generates a random permutation of given a sequence. It has the linear time complexity $O(n)$, n being the no. of items being shuffled. In our Case it is $O(mn)$ [We assume, $m = m_0 = m_1 = m_2 = \ldots = m_{n-1}$].

1. **challenge(m)**
   Input: A vector m containing limits of a n-dimensional hypercube
   Output: A random path from origin to a randomly chosen point inside the hypercube
2. path ← [];
3. for i=0 to n-1
4.   generate a random integer $c : 1 \leq r \leq m[i]$
5.   Add i c number of times to path;
6. endfor
7. // Implementing Fisher–Yates shuffle
8. for i = length(path)-1 downto 1
9.   generate a random integer $k : 0 \leq k \leq i$
10.   swap(path[k], path[i])
11. endfor
12. store path;
13. return path;

Fig. 4. Steps to generate a challenge

Server saves this challenge with respect to client's *login id* and sets a timer and throws it to the client. The client displays this challenge and asks for the OTP.

2.2. User gets this challenge and gives it to the mobile app.

3.

   3.1 The mobile app calculates OTP_c (by applying hash functions repeatedly) and returns to the user (Fig. 5).

$$OTP_c = h_{d_{pl}}(\ldots(h_{d_3}(h_{d_2}(h_{d_1}(^cs_c))))) \quad (2)$$

1. **generateOTP(path, S_c, h)**
    Input: path, current seed and k hash functions
    Output: next One Time Password
2. $OTP \leftarrow S_c$;
3. for each i in path
4. $\quad OTP = h_i(OTP)$
5. endfor
6. return OTP;

Fig. 5. Generation of One Time Password

3.2. User provides this OTP_c to the client
3.3. Client sends OTP_c along with *login id* to the server

4.
  4.1 On receiving OTP, the server checks the timer and if it is expired, authentication request fails. Otherwise, it computes a temporary seed to be used to calculate OTP as:

$$S_t = \begin{cases} h_0^{(C_c - C_s)}(^sS_c) & \text{if } (C_c - C_s) \geq 0 \quad (3) \\ h_0^{C_c}(S_i) & \text{otherwise} \quad (4) \end{cases}$$

Ideally, for the first time authentication $(C_c - C_s)$ will be 0; $(S_t = {}^sS_c = S_i)$ and subsequently $(C_c - C_s)$ should be 1 (i.e. C_c is 1 ahead of C_s). So hash function is applied at most once. However, if $C_c < C_s$ (e.g. client loses C_c and starts from 0), server can't use (3) as hash functions are *non-invertible*. In that case, the server can still calculate S_t from the initial seed S_i using (4). Albeit it may require applying hash several times but it is expected to occur infrequently.

If the server crashes, it has to recover only S_i. C_s and ^sS_c will get their initial values 0 and S_i respectively. This *robustness* prevents attackers from taking advantage of *crashing* system and client machines.

Note that ^sS_c itself is not yet updated as the request may come from an imposter and the authentication may fail during OTP verification. It is updated after successful authentication.

In [11], server synchronizes its seed with the client's seed by repeatedly applying hash function on initial seed. If a client gets authenticated frequently, the current seed status becomes very large that results a costly seed update. Our method eliminates the problem by synchronizing the counters at both sides. If server goes out of synchronization/loses its counter, it starts from the scratch. If client loses its counter, it can start from 0. Either way it works.

This S_t is then used to calculate the OTP_s.

$$OTP_s = h_{d_{pl}}(\ldots(h_{d_3}(h_{d_2}(h_{d_1}(S_t)))))\quad (5)$$

1. **authenticate(li, C_c, OTP_c)**
2. retrieve C_s, ^sS_c, pw_e corresponding to li;
3. $S_t = h_0^{(C_c - C_s)}(^sS_c)$
4. OTP_s = **generateOTP(p, S_t, h)**
5. if $OTP_c = OTP_s$
6.   authentication succeeds;
7.   //update seed and counter atomically;
8. $^sS_c = S_t; C_s = C_c$;
9. endif
10. else authentication fails;

Fig. 6. Steps followed by the server to authenticate a client.

If $OTP_c = OTP_s$ the client is authenticated (Fig. 6) and server updates its current seed and counter atomically as:

<atomic>

$$^sS_c = S_t;$$
$$C_s = C_c; \quad (6)$$

</atomic>

Client updates its ^cS_c and C_c after every successful authentication as

<atomic>

$$^cS_c = h_0(^cS_c);$$
$$C_c = C_c + 1; \quad (7)$$

</atomic>

So, the current seed gets updated using an infinite chain of h_0.

3.3 Numerical Illustration

We shall assume, during illustration, that client's MAC, server's MAC and current time is used to generate initial seed. Suppose, client's MAC address C_{MAC} = "40-B8-9A-17-6B-EB", server's MAC address S_{MAC} = "78-24-AF-67-A8-58" and current time T_c = "Thu Aug 30 15:39:57 IST 2018". Suppose MD5 hash algorithm is used, then $S_i = h_{MD5}(C_{MAC} + S_{MAC} + T_c)$ = "d71a45189bf8a150df219316214e245". The server initializes ^sS_c and C_s with S_i and 0 respectively, chooses n (say 2 for simplicity) hash functions h_0 and h_1 and limits $m_0 = m_1 = 10$ (say). The server sends names of two hash functions and S_i to the client. Client initializes ^cS_c and C_c with S_i and 0 respectively.

After a few successful authentications, C_c becomes 5(say). The value of C_s should ideally be 4. Client sends 5 to the server. Server generates a random challenge (0, 1, 1, 0, 0, 1), remembers it and sends to the client. Client computes OTP_c as $h_1(h_0(h_0(h_1(h_1(h_0(^cS_c))))))$ and sends it back to the server. Server computes S_t as $h_0(^sS_c)$

and calculates OTP_s as $h_1(h_0(h_0(h_1(h_1(h_0(S_t))))))$. If OTP_s and OTP_c are same, server sends 'authentication success' message to the client and updates $C_s = 5$ and $^sS_c = S_t$. The client, upon successful authentication, updates $C_c = 5 + 1$ and $^cS_c = h_0(^cS_c)$.

4 Security Analysis

The proposed system can defend *off-line guessing attack* as strong hash functions are used. It can also resist *replay-attack* as a new password is used for every authentication. In the following sections, we shall assess security strength of the proposed system w.r.t various attacks.

4.1 Pre-play Attack: Probability Guess

In pre-play attack, an *intruder* guesses *next challenge*, impersonates the server to the client and throws the challenge. The client responds with an OTP which is memorized and played when the server actually throws the challenge next time. In our system, probability of successful guessing of next challenge is almost zero. Following discussion explains it.

Suppose, $p(c_0, c_1, c_2, \ldots, c_{n-1})$ is a point with an n-dimensional hypercube boundary. Then the number of possible *paths* from origin to this point is given by the multinomial coefficient:

$$n_p = \binom{(c_0 - 1) + (c_1 - 1) + (c_1 - 1) + \ldots + (c_{n-1} - 1)}{(c_0 - 1), \ (c_1 - 1), \ (c_2 - 1), \ \ldots, \ (c_{n-1} - 1)}$$

$$\Rightarrow n_p = \frac{((c_0 - 1) + (c_1 - 1) + (c_2 - 1) + \ldots + (c_{n-1} - 1))!}{(c_0 - 1)!(c_1 - 1)!(c_2 - 1)!\ldots(c_{n-1} - 1)!}$$

$$\Rightarrow n_p = \frac{\left(\sum_{i=0}^{n-1}(c_i - 1)\right)!}{\prod_{i=0}^{n-1}(c_i - 1)!} \tag{8}$$

So, total number of possible paths from origin to any point in the hypercube with dimensions $(m_0, m_1, m_2, \ldots, m_{n-1})$ of length l $(n \leq l \leq \sum_{i=0}^{n-1} m_i)$ is:

$$N(m_0, m_1, m_2, \ldots, m_{n-1}) = \sum n_p \quad \forall p \text{ in hypercube}$$

$$\Rightarrow N(m_0, m_1, m_2, \ldots, m_{n-1}) = \sum_{p_{n-1}=1}^{m_{n-1}} \ldots \sum_{p_2=1}^{m_2} \sum_{p_1=1}^{m_1} \sum_{p_0=1}^{m_0} \frac{\left(\sum_{i=0}^{n-1}(p_i - 1)\right)!}{\prod_{i=0}^{n-1}(p_i - 1)!} \tag{9}$$

For simplicity, let's take $m_0 = m_1 = m_2 = \ldots = m_{n-1} = m$. Then the probability of successful guessing of next challenge for [11] is:

$$P(m) = \frac{1}{m^n} \qquad (10)$$

Probability of successful guess of next challenge for nRICH system is:

$$P'(m) = \frac{1}{N} \qquad (11)$$

Since $N \gg m^n$, (see Table 1) $P'(m) < < P(m)$. So, in terms of pre-reply attack, proposed algorithm is superior to Eldefrawy's work.

Table 1. No. of next challenges for different scenarios.

Max no. of hash operations permissible	No. of possible next challenges/OTPs			
	[11]	Proposed method		
	n = 2		For optimal n	
4	4	5	5	(n = 2)[2, 2]
6	9	19	19	(n = 2)[3, 3]
8	16	69	90	(n = 3)[2, 3, 3]
10	25	251	650	(n = 3)[3, 3, 4]
12	36	923	7365	(n = 4)[3, 3, 3, 3]
14	49	3431	75331	(n = 4)[3, 3, 4, 4]
16	64	12869	1207288	(n = 5)[3, 3, 3, 3, 3]
18	81	48619	21295783	(n = 6)[3, 3, 3, 3, 3, 3]
20	100	184755	492911196	(n = 5)[4, 4, 4, 4, 4]

Specifically, consider $m = 10$ in 2-D. The maximum permissible hash operations $m + m = 20$ (last row of the table). Then for [11], number of possible challenges is $m^2 = 100$. In our case, it is *184755* for 2-D and *492911196* for optimized dimension [n = 5 with boundary (4, 4, 4, 4, 4)]. In fact, almost the same number (90) of challenges could have been generated using only *8* hash operations (3rd row) using 3-D with boundary [2, 3, 3] that requires only 3 hash functions.

We define the *strength* of pre-play attack of our algorithm w.r.t. [11]

$$S = \frac{P'(m)}{P(m)} \qquad (12)$$

Following table (Table 2) compares the strength to resists pre-play attack for 2-D. For m = 10, nRICH system has 1847 times better strength to resist pre-play attack.

Table 2. Comparison between proposed method and [11]

n = 2					
m	m^n	N	$P(m) = \frac{1}{m^n}$	$P'(m) = \frac{1}{N}$	$S = \frac{N}{m^n}$
2	4	5	0.25000	0.20000	1.25
3	9	19	0.11111	0.052632	2.11
4	16	69	0.062500	0.014493	4.31
5	25	251	0.040000	0.0039841	10.04
6	36	923	0.027778	0.0010834	25.64
7	49	3431	0.020408	0.00029146	70.02
8	64	12869	0.015625	7.7706e−05	201.08
9	81	48619	0.012346	2.0568e−05	600.23
10	100	184755	0.010000	5.4126e−06	1847.55

4.2 Small Challenge Attack

In *small challenge attack* [14], an attacker takes the host's role, impersonates the host to the client and tries to know initial hash chain values by throwing a suitable (small) challenge. These values are then used to calculate subsequent OTPs (if they depend on the previous values) and impersonate the client to the host. This way the system is broken.

Our system calculates OTPs using a too complicated multi-dimensional forward hash chain and no OTP depends on any other. Hence, this type of attack is not possible.

4.3 Forgery Attack

A forgery attack may of two types: *selective forgery* and *universal forgery*.

In *selective forgery*, an enemy has the capability to generate OTP for a challenge *selected* by the server. Since, the enemy has no way to know and/or predict challenge path (hash functions, their number and order), it cannot generate an acceptable OTP; Hence, proposed method can defend selective forgery attack.

An *universal forger*, has the capability to generate OTP from a challenge chosen by itself. First, since, the forger does not know what dimension (n) and hash functions (h_0, h_1, h_2, ... h_{n-1}) are used it cannot create a valid challenge. Second, since the forger neither knows the *initial seed* S_i nor the *current seed*, it cannot generate OTP from challenge. So, the nRICH system can defend both forgery attacks.

4.4 Stolen Device

It is important to prevent stolen device from remaining an OTP generator for the same user [19]. Stealing the smartphone will not help the attackers as to generate OTP both current seed and password is required. It is very unlikely that the attacker has both the smartphone and the password stolen. In that case, the victim immediately asks the server to invalidate him/her and re-registers with the host. However, to convince the host the victim will have to provide enough proof to prove his/her identity. Password and initial seed S_i (which was kept secured in a safe place) which are only known to the user may be used to prove his/her identity.

Note that if an attacker can somehow get all information used for the authentication, there will be no way to identify who the actual user is except physical appearance. That's why no authentication system is full-proof; they reduce the probability of compromise.

Also note that if the attacker tries to obtain the initial seed by a dummy registration with mobile's Ethernet address, it will get a new one; not the original S_i as server's current time is used to generate initial seed.

4.5 Insider Threat

Insiders have legitimate access to computer systems and *abuse* it to steal information. To defend them, all secret information is stored in non-invertible hashed form. So, insiders remain powerless to retrieve a single piece of information.

However, they can try to impersonate a user to a different host using OTPs they have seen so far and generation formula $OTP_c = h_{d_{pl}}(\ldots(h_{d_3}(h_{d_2}(h_{d_1}(^c s_c)))))$.

However, OTP's *seed* is created from unique properties (such as Ethernet address, date and time of registration) of both *user* and different *hosts*. The hash set $h = \{h_0, h_1, h_2, \ldots, h_{n-1}\}$ and its cardinality (dimension) n may be different for different hosts. Moreover, a random permutation $\{d_1, d_2, d_2, \ldots, d_{pl}\}$ of different length of hash functions is used at different time to produce OTPs.

The *insiders* neither know the *dimension* nor the *hash set*, nor the combination of hash nor the *seed*. So they can never derive the OTPs except off-line guessing attack whose probability is almost zero.

4.6 Synchrony Attack

The system is robust enough to defend attackers from taking advantage of *system crash*. Note that the current seeds and their counters are updated only upon successful authentication. Even if they lose their counters, they have ability to synchronize them. If the client loses current seed and counter, it initializes them with initial seed and 0 respectively. If server loses current seed and seed counter, it can retrieve them from initial seed. Either way it works.

5 Performance Analysis

We shall now analyze storage and computational cost of our algorithm. We implemented an experimental prototype of nRICH system using Java socket technology, but there is nothing specific to Java. The client and server parts are tested in two similar computers having following configurations:

Intel(R) Core(TM) i3-5010U CPU @ 2.10 GHz
No. of core = 2, Number Of Logical Processors = 4
L2 Cache = 256 KB, L3 Cache = 3072 KB

Storage Requirement

The following is a sample Java method for OTP generation using built-in hash implementations. When tested in a PC with 6 costly hash algorithms: $h = \{$ "SHA-256", "MD2", "MD5", "SHA-1", "SHA-384", "SHA-512"$\}$ with 1000 challenge length, it costs about 134 KB of RAM. If hash algorithms are implemented directly, it is expected to take much less memory.

```java
static byte[] generateOTP(int[] challenge, String sc, String[] h) {
    byte[] bytes = sc.getBytes( StandardCharsets.UTF_8 );
    try {
        MessageDigest md;
        for(int i=0;i<challenge.length;i++) {
            md = MessageDigest.getInstance( h[challenge[i]] );
            md.update( bytes );
            bytes = md.digest();
        }
        return bytes;
    } catch(Exception e) { return null;}
}
```

Time Requirement

Our system requires exactly same number of hash operations as Eldefrawy's system. Specifically, if $x_t = y_t = 10$, both use $x_t + y_t = 20$ hash operations. So, in terms of time complexity it is equally efficient. When tested in a PC with configuration shown earlier, we obtained the execution time as shown in Fig. 7.

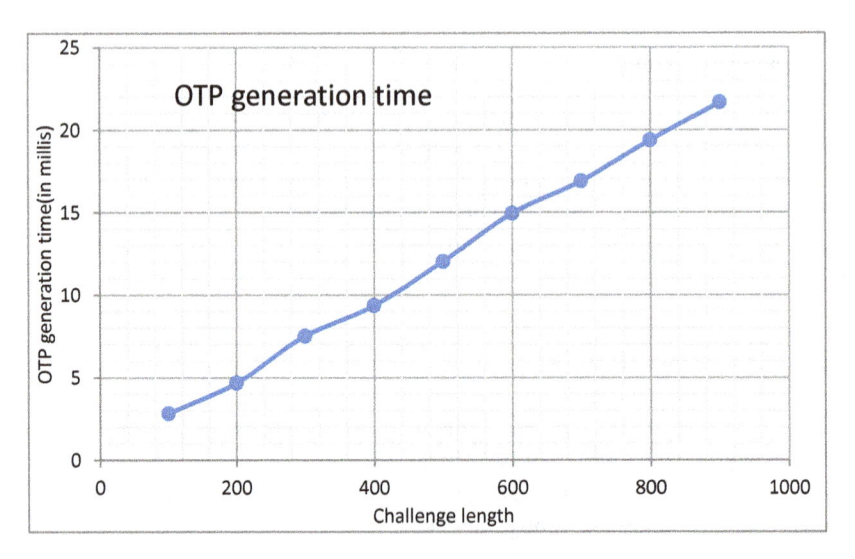

Fig. 7. Time taken for different challenge length (=number of hash function invoked)

The challenge generation incurs a negligible cost (see Table 3 and Fig. 8) to the authentication server. However, an authentication server is expected to be a relatively heavyweight machine to tolerate this negligible cost. When tested on a low end i3-based PC, it incurred ≈197 μs even for fairly large (10,000) challenge.

Table 3. Challenge generation time

m	n = 2	n = 4	n = 6	n = 8	n = 10
100	0.01185	0.01238	0.01606	0.01867	0.02317
200	0.00809	0.01616	0.02493	0.03178	0.04052
300	0.01210	0.02493	0.03519	0.04825	0.05883
400	0.01578	0.03529	0.05272	0.06203	0.07847
500	0.01909	0.03962	0.05882	0.07835	0.09836
600	0.02336	0.04753	0.07019	0.09551	0.12001
700	0.02757	0.05572	0.08256	0.10967	0.13719
800	0.03072	0.06298	0.09462	0.12480	0.15795
900	0.03594	0.07062	0.10632	0.14139	0.17887
1000	0.03940	0.07807	0.11727	0.15853	0.19758

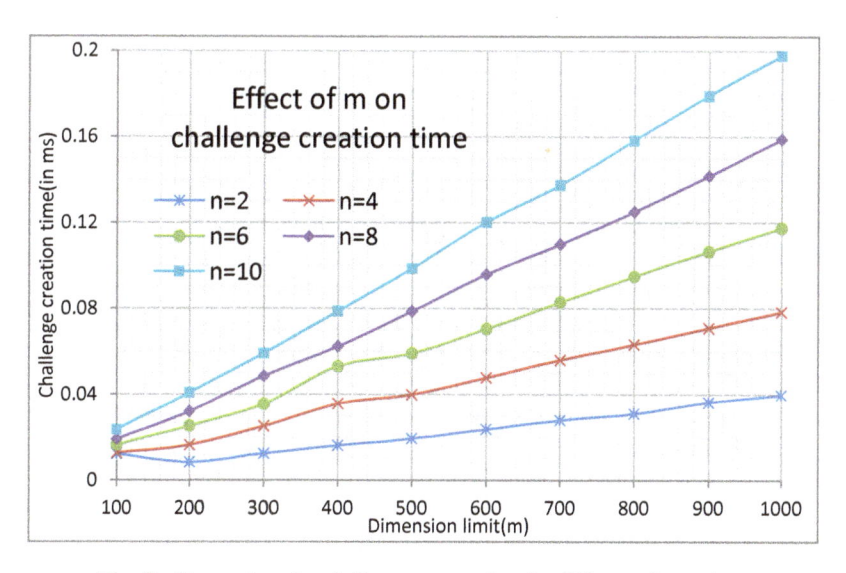

Fig. 8. Time taken for challenge generation for different dimensions

5.1 Discussion

The overall performance comparison of nRICH system with other authentication systems that use hashing is shown in Table 4.

Table 4. Performace comparison of nRICH system with other authentication systems

Metric ↓	System →	Lamport's system [1]	Goyal's system [13]	Yeh's work [17]	Eldefrawy's system [11]	nRICH system
Resistance to eavesdropping		Yes	Yes	No	Yes	Yes
Resistance to server compromise		Yes	Yes	No	Yes	Yes
Number of authentications before re-initialization		Low	Moderate	Low	Unlimited	Unlimited
Messages exchanged		2	3	5	3	3
Chain steps per session		$N - t$	$(N - t) \mod R$	$N - t$	$x_t + y_t$	$x_t + y_t$
Technique		Hash	Hash	Hash	Hash	Hash
Resistance to pre-play attack		Low	Low	Low	Moderate	Very high
Resistance to offline guessing attack		Low	Low	Low	Moderate	Very high
Client computational requirements		High	High	High	High	Low

Lamport's system requires system re-initialization after center number of authentications. The number of hash functions applied per authentication is also high initially; decreases for subsequent authentications. Although Goyal's and Yeh's systems increase the number of authentications before system re-initialization, system restart is still required. Yeh's work cannot also deal with eavesdropping and server compromise as sensitive information is passed back and forth during authentication. Eldefrawy's system eliminates all the problems except the resistance to pre-play attack which is low. Although, Bicakci's method (not shown in the table) provides an infinite hash chain, it uses gradually increasing number of computationally intensive public key algorithm making it impractical to use in limited computing devices. Specifically, when encrypted using RSA public key, we got 210 times slow down. The nRICH system eliminates all the problems and is superior to the existing techniques.

6 Conclusions and Future Work

This paper proposed a smartphone-based two-factor authentication system viz. nRICH that uses both *knowledge* (something the user and only the uses knows) and *possession* (something that the user and only the user has) based information. The hard OTP is generated from a multi-dimensional infinite hash chain that eliminates the limitations of other techniques specifically pre-play attack. The difficult-to-guess challenge is a random path from origin to a random point inside a multi-dimensional moving hypercube. We have rigorously analyzed the security performance and compared with other techniques w.r.t. various metrics and found suitable to be implemented in even low-end devices.

The only drawback is the increased length of the challenge to be typed by the user. We shall propose an enhancement to use Quick Response (QR) code to avoid this problem. Smartphone's camera will be used to read the challenge from client's computer and client's web cam will be used to read the OTP generated by the smartphone.

References

1. Lamport, L.: Password authentication with insecure communication. Commun. ACM **24** (11), 770–772 (1981)
2. Cha, B., Park, S., Kim, J.: Cluster Comput. **19**, 1865 (2016). https://doi.org/10.1007/s10586-016-0666-6
3. Cha, B.R., Kim, Y.I., Kim, J.W.: Telecommun. Syst. **52**, 2221 (2013). https://doi.org/10.1007/s11235-011-9528-y
4. Holtmanns, S., Oliver, I.: SMS and one-time-password interception in LTE networks. In: 2017 IEEE International Conference on Communications (ICC), Paris, pp. 1–6 (2017). https://doi.org/10.1109/icc.2017.7997246
5. Hallsteinsen, S., Jorstad, I., Thanh, D.-V.: Using the mobile phone as a security token for unified authentication: systems and networks communication. In: International Conference on Systems and Networks Communications, pp. 68–74. IEEE Computer Society, Washington, DC (2007)
6. Indu, S., Sathya, T.N., Saravana Kumar, V.: A stand-alone and SMS-based approach for authentication using mobile phone. In: 2013 International Conference on Information Communication and Embedded Systems (ICICES), Chennai, pp. 140–145 (2013)
7. Mulliner, C., Borgaonkar, R., Stewin, P., Seifert, J.P.: SMS-based one-time passwords: attacks and defense. In: Rieck, K., Stewin, P., Seifert, J.P. (eds.) Detection of Intrusions and Malware, and Vulnerability Assessment. DIMVA 2013. Lecture Notes in Computer Science, vol. 7967, pp. 150–159. Springer, Heidelberg (2013)
8. Siddique, S.M., Amir, M.: GSM security issues and challenges. In: Proceedings of the Seventh ACIS International Conference on Software Engineering, Artificial Intelligence, Networking and Parallel/ Distributed Computing, SNPD 2006. IEEE Computer Society, Washington, DC (2006)
9. Wang, H.: Research and design on identity authentication. System in mobile-commerce, pp. 18–50. Beijing Jiaotong University (2007)
10. Laukkanen, T., Sinkkonen, S., Kivijarvi, M., Laukkanen, P.: Segmenting bank customers by resistance to mobile banking. In: International Conference on the Management of Mobile Business, p. 42. IEEE Computer Society, Washington, DC (2007)
11. Eldefrawy, M.H., Khan, M.K., Alghathbar, K., Kim, T., Elkamchouchi, H.: Mobile one-time passwords: two-factor authentication using mobile phones. Secur. Commun. Netw. **5**, 508–516 (2012). https://doi.org/10.1002/sec.340
12. Haller, N.: The S/KEY one-time password system. In: Proceedings of the ISOC Symposium on Network and Distributed System Security, San Diego, CA, pp. 151–157, February 1994

13. Goyal, V., Abraham, A., Sanyal, S., Han, S.: The N/R one time password system. In: Proceedings of International Conference on Information Technology: Coding and Computing, ITCC 2005, vol. 1, pp. 733–738. IEEE Computer Society, Washington, DC (2005)
14. Chefranov, A.: One-time password authentication with infinite hash chains. In: Novel Algorithms and Techniques in Telecommunications, Automation and Industrial Electronics, pp. 283–286. Springer, Heidelberg (2008)
15. Bicakci, K., Baykal, N.: Infinite length hash chains and their applications. In: Proceedings of the 11th IEEE International Workshops on Enabling Technologies: Infrastructure for Collaborating Enterprises, WETICE 2002, pp. 57–61. IEEE Computer Society, Washington, DC (2002)
16. Rivest, R.L., Shamir, A., Adleman, L.M.: A method for obtaining digital signatures and public-key cryptosystems. Commun. ACM **21**(2), 120–126 (1978)
17. Yeh, T., Shen, H., Hwang, J.: A secure one-time password authentication scheme using smart cards. IEICE Trans. Commun. **E85–B**(11), 2515–2518 (2002)
18. Yum, D., Lee, P.: Cryptanalysis of Yeh–Shen–Hwang's one-time password authentication scheme. IEICE Trans. Commun. **E88–B**(4), 1647–1648 (2005)
19. Raddum, H., Nestås, L., Hole, K.: Security analysis of mobile phones used as OTP generators. In: Proceedings of the Fourth IFIP Workshop in Information Security Theory and Practice, WISTP 2010, pp. 324–331. Springer, Heidelberg (2010)

IT Security for Measuring Instruments: Confidential Checking of Software Functionality

Daniel Peters[1(✉)], Artem Yurchenko[1], Wilson Melo[2], Katsuhiro Shirono[3], Takashi Usuda[3], Jean-Pierre Seifert[4], and Florian Thiel[1]

[1] Physikalisch-Technische Bundesanstalt (PTB), Berlin, Germany
{daniel.peters,artem.yurchenko,florian.thiel}@ptb.de
[2] National Institute of Metrology (INMETRO), Quality and Technology, Rio de Janeiro, Brazil
wsjunior@inmetro.gov.br
[3] National Institute of Advanced Industrial Science and Technology (AIST), Tsukuba, Japan
{k.shirono,takashi.usuda}@aist.go.jp
[4] Security in Telecommunications, Technical University Berlin, Berlin, Germany
jpseifert@sect.tu-berlin.de

Abstract. Legally supervised measuring instruments, like supermarket scales or utility meters for the supply of electricity, to name just a few, need to be checked in most countries. In this regard, smart meters are a fitting example for distributed systems that need to fulfill many IT security requirements. It is of utterly importance to make sure that the functionality of these measuring devices is preserved, with the goal to enhance trust in the market, protect the consumer of fraud, and preserve privacy. Normally, legally controlled measuring devices are checked before commissioning by so-called Notified Bodies, and afterwards cyclically by market surveillance officers. The hardware is scrutinized by manually testing the sensors. This paper looks more closely at the software testing aspect and highlights how current methods can be enhanced to check correct software functionality. We describe alternatives that will pave the way to a more secure and trustworthy market, which additionally, grants more flexibility to patch software bugs without the need for recertification, as long as the core functionality of the device remains the same. In our framework the functionality checking can be done automatically, while preserving confidentiality on all ends. Based on this framework, it is no problem to allow remote displays, e.g., smartphones, or, a completely distributed measuring instrument, e.g., with many sensors in different locations connected over the Internet. Our approach is of general nature, but perhaps most interesting for smart meter infrastructures.

Keywords: Smart meters · Metrology cloud · Software integrity checking · Homomorphic encryption · Functional encryption · Probabilistically checkable proofs · Legal metrology · Blockchain technology

1 Introduction

What are legally supervised measuring instruments? First of all, it should be noted that most of these instruments are software controlled and many are in use today, e.g., more than 130 million legally relevant meters in Germany alone [1]. These measuring instruments are employed for commercial or administrative purposes or for measurements that are of public interest. The majority of them are used for business purposes, in particular they are commodity meters for the supply of electricity, gas, water or heat. An important example are smart meters and their security gateways, which already can be seen as distributed measuring devices. More classical measuring instruments, with which the end user comes into contact, are, e.g., counters in petrol pumps, scales in the food sector, or speed and alcohol meters (hopefully the latter two not that much).

Nowadays, the person executing or being affected by an official measurement cannot check the determined result; the parties concerned must rather rely on the correctness of the measurement. Of course there are measures in place to check the instruments beforehand, by so-called Notified Bodies, which are mostly national metrology[1] institutes. In general, Notified Bodies are denominated by their states and are considered to be trusted entities that must be embraced by the manufacturers of the instruments. Additionally, these devices are also tested cyclically by market surveillance officers, to make sure that no misbehavior has occurred in the meantime. For software this is done by looking at the so-called software identifier, which should be calculated out of the important files saved on the device and should always have the same value if nothing has changed, or a different one if software modifications took place. This procedure is error-prone because most devices are very powerful and run a lot of dynamic software parts with full grown operating systems, which could at anytime influence the measurement, without changing the files that are implementing the actual measuring algorithm. Furthermore, as stated at the beginning, the correct evaluation of the individual measurement results cannot be checked.

Misbehavior of these devices can lead to huge loss of money for the state and all the stakeholders influenced by the measurement results. Since software is one of the key components of such devices [1] and developments in the overall IT market like Cloud computing and the Internet of Things are progressing into the legal metrology market, one must expect measuring instruments to become more and more integrated into open networks. Therefore, correct (distributed) software functionality becomes a top priority, nowadays, and will be even more important in the future. With these facts in mind, we will describe a framework for checking measuring instruments software that enables devices and sensors to be securely connected to the Internet and even be partly distributed (see Sect. 4). Additionally, we will make sure that the checking procedure will not leak any information about the algorithm used or the inputs and respective outputs that are generated. The procedure will just yield true, if the algorithm executed correctly without revealing any other information, keeping consumer

---

[1] Metrology is the science of measurement, not to be confused with meteorology.

data and manufacturer intellectual property private. For this purpose, we first describe theoretically, how measuring algorithms can be evaluated for correctness on every measurement by the use of probabilistically checkable proofs, while also protecting the consumer's and manufacturer's privacy through homomorphic and functional encryption. In the second step, we show by a practical smart contract evaluation of a homomorphic encryption algorithm that for tailored scenarios like accumulating data of different smart meters no overhead is produced for keeping data confidential and distributed.

2 Software Testing in Legal Metrology

2.1 Important Documents

Internationally, the most important software guide for measuring instruments is the OIML D 31 document[2]. The International Organization of Legal Metrology (OIML) was set up to assist in harmonizing regulations across national boundaries to ensure that legal requirements do not lead to barriers in trade. And as mentioned, software requirements for this purpose are formulated in the OIML D 31 document. Another document that is equally important, but mostly used in European countries, is the WELMEC 7.2 Software Guide[3]. It is used mainly in Europe, because WELMEC is the European committee which promotes cooperation in the field of legal metrology.

A special case are smart meters. A smart meter is an electronic device that records, depending on what it is measuring the consumption of electric energy, gas or water, respectively. It communicates the information to the provider for monitoring and billing and it enables two-way communication. In some states, e.g., in Germany, the communication has to go through a special central device which is called Smart Meter Gateway. It is the interface between consumers and energy providers and can either be placed near the smart meter or integrated into the smart meter itself. The function of the gateway is to store, process and then send the measurements in anonymized, encrypted format to the consumers and energy providers, respectively. For the whole smart meter infrastructure many IT documents in different countries were developed, to describe how security can be achieved, examples are CEN/CLC/ETSI/TR 50572:2011, Smart Metering Equipment Technical Specifications (British smart metering implementation programme), NISTIR 7628, NISTIR 7823, NTA 8130 (Dutch Smart Meter Requirements), BSI-CC-PP-0073 (Protection Profile for the Gateways of a Smart Metering System), BSI TR-03109 (German requirements) and the PTB-A 50.8 (German IT security requirements from the metrology point of view), to name just a few.

In this paper we will show, how the smart meter security functionality can easily be integrated into the individual smart meters, so that the central device, e.g., the smart meter gateway, theoretically, does not need to fulfill any software requirements, as long as it can proof to the verifier that the correct calculations were executed.

---

[2] https://www.oiml.org/en/files/pdf_d/d031-e08.pdf.
[3] https://www.welmec.org/documents/guides/72/.

2.2 Current Mechanisms to Check Software

To describe current mechanisms to check legally relevant measuring instruments' software, we will look more closely at the WELMEC 7.2 Software Guide. The guide describes acceptable software solutions in more detail than the OIML D31, but both documents formulate similar requirements, hence the general requirements are the same, i.e. software protection against deliberate modifications with sophisticated software tools and ambitious software assessment with examination of the source code has taken place.

All of these testing is done by a Notified Body. The challenge today is that a software-controlled device is subject to many threats because only a small system error in one layer, i.e. from the hardware elements such as the executing chips or the necessary hardware devices (e.g., memory, attached devices, etc.), through the firmware, bootloader to the operating system and overlying programs, leads to an untrustworthy system. To counteract this, one tries to formulate appropriate security requirements for each system layer. Mostly WELMEC 7.2 and also OIML D31 explain how through modularisation and clear software separation, security can be created up to a certain level. Works like [2] enhance these ideas and try to achieve the highest level of software security using separation kernels and virtualization technologies.

The disadvantage of these methods and even the general requirements from above is that they are concentrated on module/file integrity and not functional correctness. For example, most measuring instruments just show, when requested, a short identifier to the verifying authority (e.g. market surveillance) on the screen. The calculation of this identifier just tries to spot changes in the legally relevant files. Sometimes not even securely, because it is often based on insecure checksum algorithms, e.g. CRC-32. This restrictive approach is severely limiting manufacturers in our modern age. For example, Cloud applications and external displays can rarely be evaluated for conformity because the hardware can no longer be controlled. This is considered to be inadequate because users, out of convenience, often want to access data via the Internet and their smartphone. It should also be noted that the legal metrology requirements do not address confidentiality, which in the interconnected world becomes more and more an issue because users want to be able to preserve their privacy. Therefore, in our opinion, another approach of testing software should be taken. We think, it is enough to just verify the core algorithm that is responsible for the measuring procedure and not the whole software stack. If we can make sure that we can confidentially, without revealing any user or manufacturer data, check this core algorithm for correctness on every measurement, it is not necessary anymore to check the whole software stack for protection against modifications, which from today's perspective, seems to be impossible. Software programs are usually much too large to be completely verifiable and the entire hardware structure alone is too complex.

3 Fundamental Building Blocks

We assume throughout the paper that there are temper-proof identifiers, e.g., that one-way functions exist. All cryptographic methods are normally based on this assumption and for the sake of simplicity we also assume that NP $\neq$ P.[4]

To construct our approach for verifying correct algorithm evaluation for distributed measuring devices, without leaking any information of the algorithm used and also hiding the inputs and outputs, we need four building blocks, namely, homomorphic and functional encryption, probabilistically checkable proofs and distributed ledger technology, which are described below.

3.1 Homomorphic Encryption

The purpose of homomorphic encryption (HE) is to allow computation on encrypted data. Computing on encrypted data means that if a user has a function f and wants to obtain $f(m_1, ..., m_n)$ for some inputs $m_1, ..., m_n$, it is possible to instead compute on encryptions of these inputs, $c_1, ..., c_n$, obtaining a result, which decrypts to $f(m_1, ..., m_n)$. A more formal description of the needed functions for an HE scheme is given below.

- $(pk, sk, evk) \leftarrow Gen(1^\lambda, \alpha)$ is the key generation algorithm. It takes two inputs, a security parameter λ in unary and an auxiliary input $\alpha \in A$, and outputs a key triple (pk, sk, evk), where $pk \in K_p$ is the key used for encryption, $sk \in K_s$ is the key used for decryption and $evk \in K_e$ is the key used for evaluation. A is the auxiliary space and the Ks are the key spaces.
- $c \leftarrow Enc(pk, m)$ is the (PPT) encryption algorithm. As input it takes the encryption key $pk \in K_p$ and a plaintext $m \in P$. Its output is a ciphertext $c \in X$. P is the plaintext space and X the ciphertext space for the encryption function.
- $o \leftarrow Eval(evk, f, \mathbf{c})$ is the evaluation algorithm. It takes as inputs the evaluation key $evk \in K_e$, a function $f \in F$ (F is the space of accepted functions) and a tuple of inputs $\mathbf{c} := (c_1, c_2, ..., c_n) \in Z^n$. It produces an evaluation output $o \in Y$, where Y represents the evaluation ciphertext space, and $X \cup Y = Z$ (i.e., Z is the space of all ciphertext).
- $m \leftarrow Dec(sk, c)$ is the decryption algorithm. It takes as input the decryption key $sk \in K_s$ and either a ciphertext from the encryption function ($c \in X$) or from an evaluation output ($c \in Y$) and produces a plaintext $m \in P$.

A HE evaluation scheme (Gen, Enc, Eval, Dec) is said to correctly decrypt, if for all $m \in P$, $\Pr[Dec(sk, Enc(pk, m)) = m] = 1$, where sk and pk are outputs of $Gen(1^\lambda, \alpha)$ and Pr denotes the probability. This means that we must be able to decrypt a ciphertext to the correct plaintext, without error. Additionally, it should correctly evaluate all functions in F if for all $c_i \in X$,

---

[4] More exactly, we assume that the one-way functions used in this paper are at least NP complete, and that NP $\neq$ BQP, i.e., not even quantum computers can find their solutions in polynomial time.

where $m_i = Dec(sk, c_i)$, for every $f \in F$, and some negligible function ϵ, $\Pr[Dec(sk, Eval(evk, f, \mathbf{c})) = C(\mathbf{m})] = 1 - \epsilon$, where sk, pk and evk are outputs of $Gen(1^\lambda, \alpha)$. This means that with overwhelming probability, decryption of the homomorphic evaluation of a permitted circuit yields the correct result. If F is the space of all computational functions, the HE scheme is called fully homomorphic encryption (FHE) scheme (see [3] for a nice overview of different HE schemes).

From now on, we assume that our used HE schemes are correct, i.e., have correct evaluation and correct decryption. Additionally, out of simplicity, we will normally omit the evaluation key evk and will treat it as a part of the public key, so if $evk \neq pk$, it can be easily derived out of pk, else $evk = pk$. The same applies to the spaces X, Y, and Z, where we just assume $X = Y = Z$.

In HE, confidentiality is naturally preserved through input and output encryption. Still, strict algorithm confidentiality/privacy is not available from the definition, i.e. the evaluator knows the function but not the inputs and outputs. From here on out, we will call this the function skeleton. There are possibilities to even hide the function skeleton by for example making it an additional encoded input. If the overall evaluation function is a universal Turing Machine (TM), it can process the algorithm with its input without revealing it. Of course one must take into consideration, that the run-time and the processing of the algorithm itself, can reveal information about the input data. Solutions for these challenges are given partly in [4], where the authors also describe how oblivious TMs can come in handy.

3.2 Functional Encryption

Functional encryption allows a secret key to be issued using a master key, dependent on a function f. Given a ciphertext, the secret key allows the user to learn the value of f applied to the plaintext and nothing else [5]. Computing functions on encrypted data links the two concepts of homomorphic and functional encryption together. A notable difference is the way the functions are applied. FE grants control over what functions can be applied to the data via a master key holder, who issues keys based on a decision of the appropriateness of the function. A key can be used to obtain the plaintext result of the function applied on the encrypted data. Instead, FHE permits functions to be run by anyone possessing the evaluation key, however only the owner of the secret key can decrypt the result into plaintext. A more formal description of FE is given below.

A functional encryption scheme (FE) for a functionality F defined over (K, X) is a tuple of four PPT algorithms (Setup, Gen, Enc, Dec) satisfying the following correctness condition for all $k \in K$ and $x \in X$:

- $(pp, mk) \leftarrow Setup(1^\lambda)$ generates a public and master secret key pair.
- $sk \leftarrow Gen(mk, k)$ generates a secret key for predicate k.
- $c \leftarrow Enc(pp, x)$ encrypts the message x.
- $y \leftarrow Dec(sk, c)$ uses sk to compute $F(k, x)$ from c.

with $\Pr[y = F(k,x)] = 1$. The scheme as defined above must be enhanced for our proposes, namely we would like to allow some kind of multi-inputs or a merge function for n inputs from different sources, e.g. as described in [6,7]. For this purpose we modify our Enc and Dec functions as follows:

- $c \leftarrow Enc(pp, x, i)$ encrypts message $x \in X_i$, where index $i \in \{1, 2, ..., n\}$ and $\bigcup_{i=1}^{n} X_i = X$.
- $y \leftarrow Dec(sk, c_1, ..., c_n)$ uses sk to compute $F(k, x_1, ..., x_n)$ from $c_1, ..., c_n$.

For our purpose all inputs can be encoded under the same pp key. It should be also noted that k can normally be inferred out of sk but mk remains secret. As before $\Pr[y = F(k, x_1, ..., x_n)] = 1$ should be true to achieve correctness. We assume that the pp key is kept secret from the evaluator, who is executing $Dec(sk, c_1, ..., c_n)$.

3.3 Probabilistically Checkable Proofs

Probabilistically Checkable Proofs (PCP) are used to show that a solution to a problem is correct, whereby a verifier can check the proof generated by the prover faster than finding the solution itself, i.e., the proof can be checked by a randomized algorithm reading a bounded number of bits of the proof. A more formal description can be found below.

Given a decision problem L, a probabilistically checkable proof system for L consists of a prover and a verifier. Given a claimed solution x with length n, which might be false, the prover produces a proof π which states x solves L. The verifier is a randomized oracle Turing Machine V that checks the proof π for the statement that x solves L and decides whether to accept the statement. The system has the following properties:

- Completeness: For any $x \in L$, given the proof π produced by the prover, the verifier accepts the statement with probability at least $c(n)$.
- Soundness: For any $x \notin L$, then for any proof π, the verifier mistakenly accepts the statement with probability at most $s(n)$.

For completeness $c(n)$ and soundness $s(n)$, following should hold, $0 \le s(n) \le c(n) \le 1$.[5] Additionally, for the computational complexity of the verifier, the randomness complexity $r(n)$ can be given to measure the maximum number of random bits that V uses, and the query complexity $q(n)$ of the verifier is the maximum number of queries that V makes to π, over all x of length n.

To achieve confidentiality zero knowledge properties can enhance PCPs. Hereby, a prover can prove the correctness of an assertion without leaking any extra information. More formally speaking:

- Zero Knowledge: There exists an expected PPT algorithm which can, without communicating with the real prover, produce transcripts indistinguishable from those resulting from interaction with the real prover.

---
[5] From here on forward we will assume $c(n) = 1$.

Additionally, proof systems (normal and zero knowledge) are categorized into two domains, interactive and non-interactive. Loosely speaking, an interactive system can be seen as an interrogation-style proof system, where the prover and verifier interact, with the verifier asking questions and the prover responding. At the end the verifier decides whether or not to accept the input. In the non-interactive system the prover sends just one overall proof to the verifier. A non-interactive proof system can be more convenient because no further communication on the network is necessary and the proof can be checked offline. For our purpose there is no restriction concerning the network traffic, hence, both (interactive and non-interactive) proof systems can be used.

3.4 Distributed Ledger Technology

Distributed Ledger Technology (DLT) is a concept that has become popular with the growing interest in *Blockchains* [8–10]. Initially associated with cryptocurrency markets due to the Bitcoin popularity [8], Blockchain-based architectures have been proposed for a wide set of application areas, including sensor networks, Internet of Things (IoT), smart cities, among others [9].

A Blockchain can be described as a distributed append-only data structure (called a *ledger*) which is replicated and shared among a set of network peers [9,11]. This structure consists of a sequence of blocks and each new block is linked to the previous one using a cryptographic function. Consequently, the block i cannot be changed without also modifying all subsequent blocks $i+1, ..., n$. The availability of data in a Blockchain does not depend on trusted third parties, since it is a decentralized model, which can greatly save costs [10]. Its integrity is ensured by consensus among the peers, which implies an agreement about any new block that is going to be appended to the ledger [11,12].

A Blockchain can store virtually any digital asset, from data to self-executing scripts, whereby the latter are called *smart contracts*. This enhances Blockchain's ability from being just a reliable data storage solution, to being a complete distributed platform for proper automated workflow [9], in which smart contracts are executed by every assigned network peer that has the permission, in an independent and automatic manner.

In general, Blockchain platforms can be classified as *permissionless*, in which anybody can join and participate in the network consensus; or *permissioned*, in which consensus is achieved by a set of known and identifiable peers [12]. Bitcoin and Ethereum are examples of permissionless Blockchains. Permissioned Blockchains are particularly interesting in business applications in which the parties need to identify each other [12], and in our opinion also for legal duties where some of the stakeholders can be regarded as trustworthy, e.g. Notified Bodies in legal metrology. Also, permissioned Blockchains consensus protocols usually expend less computational resources and can reach better transaction latency and throughput. Out of these reasons we choose a permissioned Blockchain for our practical evaluation (see Sect. 6).

4 Combining Everything

For our approach we want to make sure that the identifier, which is output by the device, is not calculated over its important files, but over its relevant functionality, i.e. the core algorithm. We will describe our approach that combines the schemes explained in Sect. 3. We do not describe the trivial variants in detail, i.e., in which the algorithm with all its inputs and outputs is known to the verifying authority (e.g. if the manufacturer does not care about Intellectual Property and the consumer does not worry about privacy issues). In that case, the algorithm can be just send to the verifying authority. If the algorithm is signed by a trusted entity, it is reexecuted on the inputs (known by the verifying authority), and the generated output can directly be checked with the output shown on the device.

4.1 Initial Setup

We assume that the program to be executed is previously tested by a trusted entity. This is already the case for legal metrology, where conformity assessment bodies (Notified Bodies) are representing the trusted entity. These Notified Bodies review the relevant parts of the software, i.e., the parts that are responsible for the measurement result. Once a manufacturer has completed her measuring programs and divided them into their legally relevant functions (core functionality), the trusted entity can review the program. After the verification has been successfully passed, the trusted entity stores a secure unique identifier of the program (e.g., hash of the binary image of the program or part of a proof) to a secure storage. The secure storage may be a Cloud managed by trusted stakeholders, but can also be a distributed ledger that only trusted parties can write to, e.g., a permissioned Blockchain, or that prevents by other means the unauthorized changing of values, e.g., ensuring that an "honest majority" exists at all times. If the sensors are checked and sealed, a part of the secure storage can be in the sensors themselves, shown as small green boxes labeled as trusted computing bases (TCB) in Fig. 1.

4.2 The Distributed Measuring Instrument

We assume, that the core algorithm receives inputs from some input sensors which are connected to a central device. These sensors are checked by the trusted entity (Notified Body before commissioning and market surveillance afterwards in the market) and also sealed to make tempering as hard as possible. We therefore, consider them to be secure. Additionally, the algorithm used by the central device is also being verified to make sure that it calculates the correct output over the inputs. It is not assumed that this algorithm is actually being executed by the central device, because the central device is considered untrustworthy. Nevertheless, mechanism are being described in this paper that allow for a verification authority at any time to check correct algorithm execution for every measurement, also for measurements lying in the past. A general outline of what we define as our distributed reference network is shown in Fig. 1, whereas all parts depicted in green are assumed not attack-able (as secure as possible by sealing and being checked constantly for intrusions, etc.).

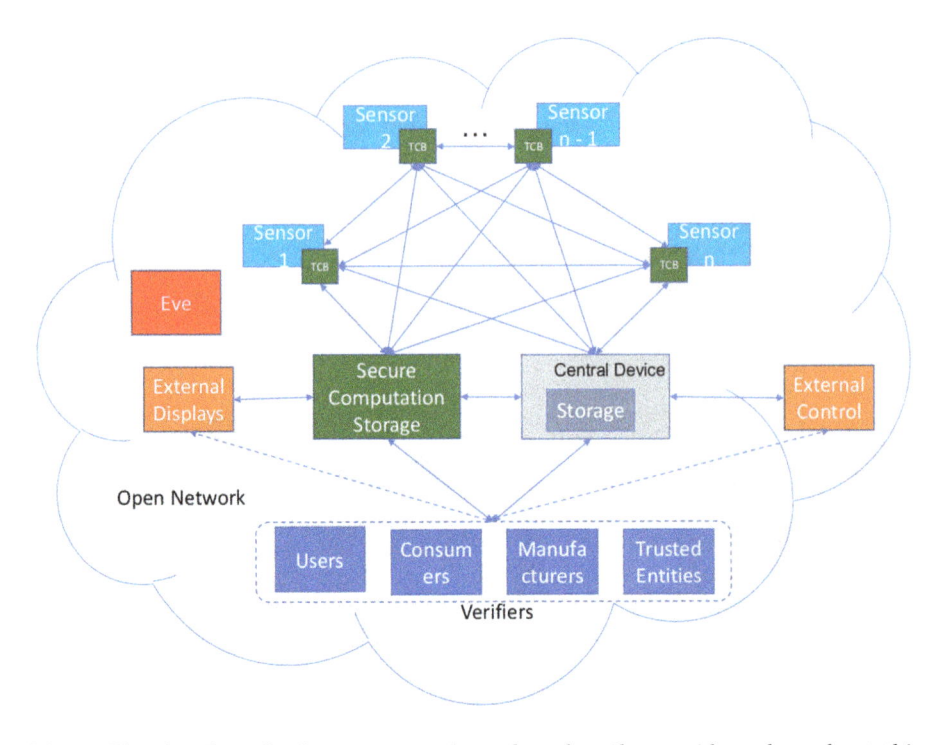

Fig. 1. The distributed reference network used to describe our ideas throughout this paper.

In the figure, all parties are connected through an insecure open network, e.g., the Internet, which also consists of malicious attackers, labeled as Eve. We assume that Eve cannot change or stop transmissions of packages between the sensors and the Secure Computation Storage and between the sensors themselves. The secure computation storage normally just stores data, but can also execute some tasks, e.g., handling a PKI. The Trusted Computing Bases (TCB)s of the sensors and the Secure Computation Storage are, as mentioned depicted in green, and therefore, assumed secure through constant testing. Eve is just able to take over the External Display, the External Control software and the Central Device. Furthermore, Eve can try to impersonate the four entities, User, Consumer, Manufacturer and Trusted Entity, whereas the Secure Computation Storage stores the certificates and public keys of the trusted parties that are necessary to make secure communication possible. The TCBs in the sensors (and also the Secure Computation Storage) need to make sure through these certificates that they are accessed and controlled by entities that bear permission rights, which is also assumed.

There are applications that allow the access to the Central Device, depicted as External Control, and a possibility to read out the Secure Computation Storage through the External Display process. An example of the External Display could be a web front end that can be accessed securely through the Secure Computation Storage, meaning that the device accessing it, i.e. running the External Display client software, could be manipulated, but not the External Display server software, hence, a not manipulated, correctly executing client will always show correct information.

The aforementioned entities are the ones concerned by the measurements and want to verify correct processing. From here on out, they will be referenced as verifiers, if no differentiation is necessary. Legal metrology normally speaks of consumers and users as different entities, e.g.. a supermarket scale is bought from the manufacturer by the supermarket owner who is the "User", but actually mostly used by the "Consumer" which is the client buying the weighed products. Another example are smart meters, where the "User" is actually the provider and the "Consumer" is the person paying the bills. If not otherwise indicated, we will just use the word users and consumers as an equivalent. Only when we describe the smart meter case, we will differentiate and call the "User" from Fig. 1 provider. Trusted Entities, as mentioned above, are normally denominated by states and are, for example, Notified Bodies and market surveillance entities.

4.3 Complete Confidentiality Without Trusting the Central Device

Previous works explained how homomorphic signature (HS) schemes can be used to preserve sensor/consumer confidentiality in smart grids [13,14]. In general, a homomorphic signature scheme consists of the normal signature algorithms for generating public and secret keypairs, signing data and verifying the signature of data. But HS schemes additionally, support some evaluate algorithm that translates functions on messages to functions on signatures, i.e., if o is a valid set of signatures on data m, then $Evaluate(f, o)$, with f being a supported function of $Evaluate$, should be a valid signature for $f(m)$.

In HS, data confidentiality of individual households can be preserved by in-network aggregation, i.e., the measured data is send from smart meter to smart meter, and each smart meter adds its input to the result. Thus, the result reported to the provider only consists of information about the overall measurement but hides the individual metering data. To prevent that the meters can see the intermediate results the measurements can be encrypted using a homomorphic encryption scheme. The HS scheme is used to sign the encrypted metering data, and to aggregate the signatures along with the corresponding ciphertexts at each meter. This allows for the energy provider to verify the correctness of the evaluation by checking the consistency between the aggregated result and the aggregated signatures. The main drawback of this procedure is that the central device is not used at all, and all computations are outsourced to the meters. Additionally, it must be made sure that the provider has no access to the individual meters and their communication, because she owns the secret key that can decrypt any intermediate results.

We will describe a new procedure that uses HE, FE and PCPs to check the legally relevant algorithms for correctness. It keeps data confidential on all ends while leaving the main calculations to the central device and not the meters:

1. Each sensor uses a common FE secret key (known by all sensors) to encrypt its output with a FE scheme and transmits this encrypted value to the Secure Computation Storage, additionally the sensors secretly make the function key known to the provider.
2. Each sensor uses a common HE secret key (known by all sensors and the consumer) to encrypt its output with a HE scheme and transmits this encrypted value to the Central Device.
3. The Central Device manipulates this data through a HE evaluation function, which if legally relevant, was beforehand tested by a Trusted Entity. Through generating a PCP, the Central Device generates a proof for every evaluation function it executes and transmits this proof to the Secure Computation Storage.
4. With the help of the proof a verifier can check if the evaluation function was executed correctly.
5. If all data needed for the overall function is available on the Secure Storage, the provider can use the secret function key to evaluate the functional encrypted values that belong together with the overall function, e.g. the calculation of the overall result that she needs to bill the consumer.

In the initial phase, before these five steps can be started, the Trusted Entity receives the core functionality (algorithm f) of the Central Device from the manufacturer, which fulfills legally relevant duties. This core algorithm is tested for its correctness, and then formed into a proof system (which is complete and sound), by constructing first a decision problem L, whereby $x \in L$, iff $f(i) = m$, with $x = (f, i, m)$ (for all inputs i and outputs m, only for the specific function f). A verifier and a prover algorithm are produced that for any $x \in L$, given the proof π produced by the prover, the verifier accepts the statement. If we want f to be hidden to the verifier, we specify the proof system to be zero knowledge, meaning that no manufacturer IP is leaked by the proof π. If the Central Device is handling encrypted data homomorphically f represents a HE function (circuit), to achieve consumer privacy. To give just one example of an implementation (many other are possible), the decision problem would only give a positive response, i.e. the verifier algorithm, if the pre-calculated hash of the, by the Trusted Entity tested, function f is equal to the function f_x, with $x = (f_x, i, m)$ and $f_x(i) = m$, whereas i and m can be homomorphically encrypted values, if f is HE function.

Furthermore, a functional encryption scheme and a homomorphic encryption scheme are needed. The FE scheme must allow multi-inputs like described in Sect. 3.2. Hereby, the sensors synchronize themselves (through a secure channel, or once when they are setup by a Trusted Entity) and generate a common public and master secret key pair, $(pp, mk) \leftarrow Setup(1^\lambda)$, which both are only known to them. Afterwards they generate a function key $sk \leftarrow Gen(mk, k)$, i.e., a secret key for predicate k, whereby k is the representation of f', the function the provider needs to compute on the data to get the necessary information, e.g. for billing purposes, the overall sum without relying on the consumer to send it. The function f' can be also tested for correctness by the Trusted Entity, if it fulfills legal duties. The function key sk is securely send to the provider. This FE procedure, beginning from regenerating (pp, mk), can be repeated for every new set of measurements that belong together, so the provider can evaluate f' only on these values, by always receiving a new sk. For the HE scheme the consumer generates $(pk, sk) \leftarrow Gen(1^\lambda, \alpha)$.[6] The public key is then given to the Central Device and sensors, e.g. by sending it to the Secure Computation Storage, whereas the secret key is only known to the consumer.

Now, in the first two steps the sensors can individually generate $c_{FE} \leftarrow Enc(pp, x, i)$, which functionally encrypts data x, where index $i \in \{1, 2, ..., n\}$ corresponds to the individual sensor, and encrypt the same data $c_{HE} \leftarrow Enc(pk, x)$ under a HE scheme. In the third step, the Central Device manipulates the data, but while preserving confidentiality, i.e. $o_{HE} \leftarrow Eval(pk, f, \mathbf{c_{HE}})$. Additionally it shows that the result correspond to the decision problem L by generating a proof π. Afterwards, in the forth step, the verifier can check the proof π and calculate $a \leftarrow Dec(sk, o_{HE})$ which yields the overall result. At the end, in step five, the consumer could send the output a, securely to the provider to pay. But to counter malicious consumers that do not send the result or a wrong calculation, the FE scheme can help. The provider can simply calculate $y \leftarrow Dec(sk_{FE}, \mathbf{c_{FE}})$ using sk_{FE} to compute $f'(k, x_1, ..., x_n)$.

5 The Implementation Setup

5.1 The Secure Computation Storage as a Blockchain

A promising approach of an upcoming platform for legal metrology is the "European Metrology Cloud" (EMC), a coordinated European digital quality infrastructure for innovative products and services [15]. In general, the Metrology Cloud can be regarded as a distributed network because every stakeholder that participates, manages her own node/server to participate in the network and keeps her data locally save. This aspect makes Blockchain technology a great candidate to achieve consensus between the individual stakeholders, and implement the secure computation storage as a service for the EMC.

One of the main features of a Blockchain is the possibility of embedding self-executing software into the ledger (called smart contracts). That enables a Blockchain to store not only data but also business rules, and automated workflows [9]. Our framework implementation takes advantage of these Blockchain properties to execute homomorphic evaluation using *chaincodes* (as smart contracts are called in Hyperledger Fabric, the Blockchain platform we use). The main idea is to store encrypted measurements into the Blockchain. That assures

---

[6] As mentioned in Sect. 3.1, we consider the evaluation key to be equal to the public key $evk = pk$, out of convenience.

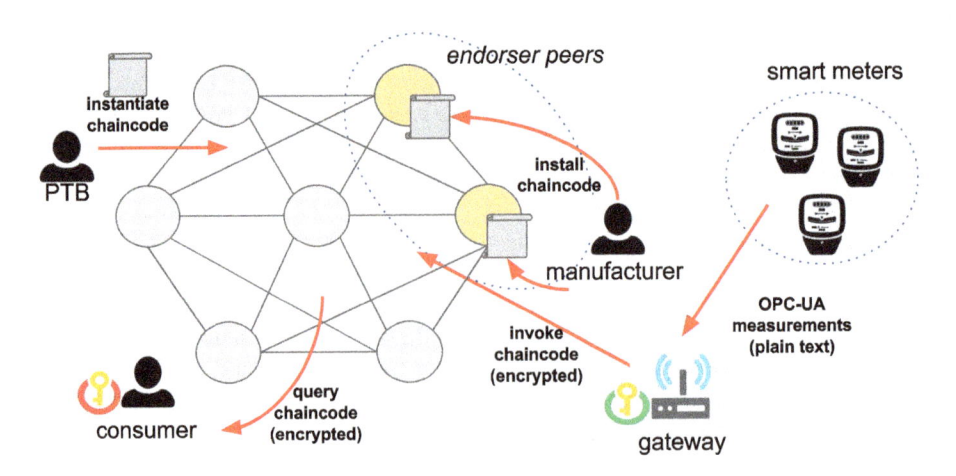

Fig. 2. How we use Hyperledger Fabric features to implement our idea

confidentiality for any entity. Furthermore, we want to work with these measurements in terms of executing operations on them. We develop an experiment based on Fig. 2. Here, smart meters send energy consumption measurements to a Blockchain network. The network peers receive these measurements and keep an updated total consumption record for all meters combined. We chose a permissioned Blockchain system for our implementation, which as already discussed, seems to be the natural selection in legal metrology [16]. Our implementation uses Hyperledger Fabric (short just Fabric), which is a complete platform for implementing permissioned Blockchains able to deal with more than 2,000 transactions per second.

5.2 The Smart Meter Infrastructure

For our practical test we constructed a demonstration smart meter infrastructure which consists of a pool of smart meters connected to a gateway. Figure 2 depicts our Blockchain network setup which maps our theoretical structure from Fig. 1 concretely to smart meters. We focused our tests only on a HE implementation running on a Blockchain, to show in the first step that for tailored problems that do not need fully homomorphic evaluation, operations can be fast even on encrypted data. Based on our findings, we think that it make sense to also implement PCP and FE on top of this framework as described in Sect. 4, but we leave this up to further research. Here, we tested how fast plaintext computations are running compared to HE, and we treated gateways as normal meters for simplicity, i.e. we always run the same client application and send the transactions to the Blockchain directly. It should be noted that normally, the smart meters would encrypt their measurements before they send them to the gateway, so the gateway never sees plaintext measurements, but we omit this here to have an easier way of comparing the times needed for plaintext compared to homomorphic encryption evaluations without always having to switch the client

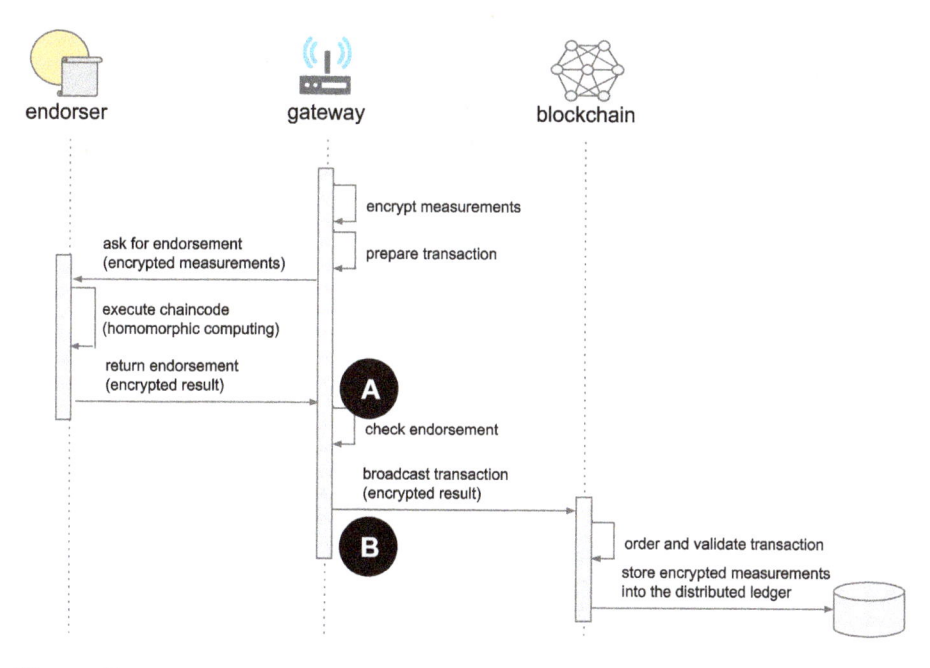

Fig. 3. The transaction workflow in Fabric. A and B are points, where performance evaluations are executed in our experiment.

applications. Nevertheless, like also normally done, every time a client submits a transaction in Fabric, she needs to ask one or more so-called endorsers, which are the only peers that are allowed to execute chaincodes. We outline the process in the sequence diagram in Fig. 3. The gateway sends a transaction proposal to a group of endorsers, and waits for a reply. Each endorser executes the respective chaincode. If the transaction is feasible, the endorser returns its result in a signed endorsement package. The gateway collects all the necessary endorsement packages and finally submits its transaction to the *orderer service*. The orderer peers implement the Blockchain consensus by ordering the transactions, creating the blocks, and propagating them to all the other committer peers in the network. Every committer *validates the transactions* and commits the block to its local copy to the ledger. However, *committer peers do not execute the chaincode again*.

6 Proof-of-Concept

For our proof of concept we implemented the aforementioned smart meter infrastructure virtually on one server using Fabric. Our benchmark hardware consists of an Intel® Server S2600CW with 256 GB of RAM and two Xeon® E5-2650 v4 CPUs running at 2.2 GHz, with twelve cores per processor and two threads per core (resulting in 48 "virtual" cores). This server hosts all the Blockchain network peers and the concurrent client instances. We assigned four CPU cores to two special containers, called *peer0* and *dev*. The *peer0* works as an endorser and a *commiter*. It receives the transactions from clients and endorses these transactions after executing the respective chaincode. Later, it also receives the blocks generated by the Blockchain consensus, validates the transactions, and commits them into its local ledger. The *dev* container encapsulates all the tasks related to the chaincode execution.

Additionally, there are *orderer* peers and client applications. The *orderer* implement the Fabric orderer service (i.e. create consensus). They receive endorsed transactions from the clients, generate ordered transaction blocks, and replicate these blocks to the peers (in our case, only to *peer0*). The *orderer* containers and the client application instances are freely distributed among the remaining CPU cores. Our client implementation uses the Python Multiprocessing package which enables concurrent processes and threads while optimizing the execution on multicore systems.

For our test, we concentrate on achieving confidentiality by just running HE to see if the overhead compared to the overhead from the Blockchain, like the communication among clients, peers, the orderer consensus, and also the block replication among the committer peers, is not too big. Considering that, we evaluate our implementation to see if homomorphic evaluation generates considerable overhead atop the Blockchain overhead by measuring the time for plain text evaluation, and by also using different key sizes for HE.

We implemented a chaincode which checks the energy consumption related to a particular meter in the ledger and increases this consumption with a new measurement (adding them together) provided by the *invoker* client (depicted as gateway in Fig. 3).

As mentioned, the chaincode has two specific modes to do that. The first one works with measurements in plaintext, calculating ordinary sums and storing the updated consumption value also in plaintext. The second mode works on encrypted measurements and performs homomorphic evaluation on them by using the Paillier cryptosystem [17], which is homomorphic for the addition of numbers. The encrypted total outcome value is afterwards stored in the ledger. For our second mode, we executed chaincode that implements homomorphic encryption/evaluation with different key sizes, i.e. of 512, 1024 and 2048 bits.

We evaluate the described cases by taking metrics at two different points A and B (see Fig. 3). In point A, we measure the throughput – given in transactions per second (tps) – and the latency – provided by the average time required to conclude the transaction, in seconds – without submitting the transactions to the Blockchain network. In point B, we measure throughput and latency during the complete transaction workflow. That produces the overhead resulting from the Blockchain network consensus (orderer service), the new block replications among the peers, and the respective validation of each transaction on every block.

In these tests, we setup our Blockchain network with a minimal configuration of only two peers, to which many clients can be connected. Each peer consists of a Docker[7] container from the official Hyperledger Fabric 1.4 LTS[8] distribution.

6.1 Experimental Results for HE

We run our experiment with different transaction workloads produced by concurrent instances of a client application implemented in Python[9]. Each client instance simulates a gateway that generates at least one transaction per second and tries to submit it to the Blockchain. By increasing the number of concurrent clients, we gradually test the capacity of our Blockchain network by measuring how it deals with the high number of transactions. Since our network consists of only one endorser peer, we can easily estimate how much of its computational effort is generated by homomorphic evaluation.

Our benchmark gathers throughput, latency and CPU usage for different benchmark rounds. Each round tests a different workload (from 400 to 1600 simultaneous clients) and different Paillier cryptographic key sizes (512, 1024, 2048 bits, compared to plaintext evaluation). Also, the rounds execute in two different modes, according to the measuring point (A or B) described in the previous subsection. Each round consists of the following steps:

1. We start with a fresh instance of our Blockchain network. That means the ledger is empty, so previous data does not affect the peers' performance.
2. We prepare our ledger by inserting a dataset of 100 unique meter IDs for each client instance.
3. We simultaneously start all the concurrent client instances designated to the round. Each one runs over an individual thread and submits transactions using *peer0* as the endorser. We also run a script in the background that logs the amount of CPU usage for each Fabric container.
4. The endorser peer simulates each concurrent transaction by executing the chaincode. After the execution, the endorser returns an endorsement package to the respective client.
5. If the round is set to take measurements at point A, each client instance finishes its transactions after receiving the respective transaction endorsement package and logs the elapsed time. If the execution mode is set to take measurements at point B, the client concludes the transaction after submitting the endorsement package to the Blockchain network.
6. Each round continues for 120 s. After that, all client instances are stopped, and the elapsed time logs are used to calculate the throughput and latency.

Figure 4a and b summarize the CPU usage for test points A and B, respectively. We can see that the container *peer0* reaches the max CPU usage with a

---

[7] https://www.docker.com.
[8] https://hyperledger-fabric.readthedocs.io/en/release-1.4.
[9] https://www.python.org.

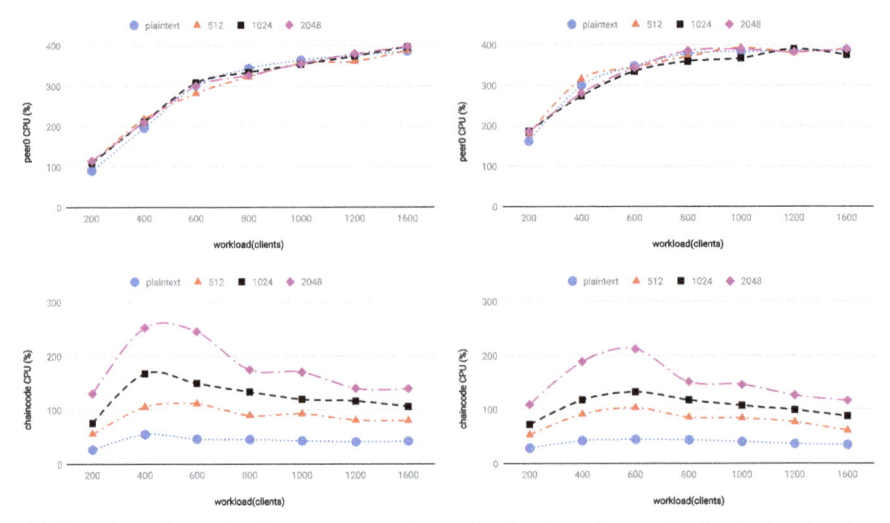

(a) Benchmark results for tests at point A. The graphs shows the CPU usage in both *peer0* (top) and *dev* (bottom) containers.

(b) Benchmark results for tests at point B. The graphs shows the CPU usage in both *peer0* (top) and *dev* (bottom) containers.

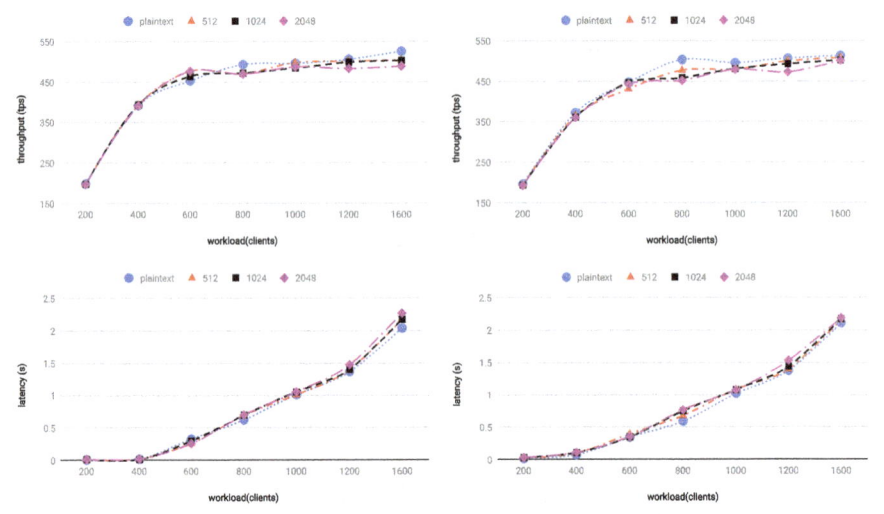

(c) Transactions benchmark results for tests at point A. The graphs shows the throughput rate (top) and latency (bottom).

(d) Transactions benchmark results for tests at point B. The graphs shows the throughput rate (top) and latency (bottom).

Fig. 4. Experimental evaluations. For the CPU usage, 400% means maximum usage, i.e. all 4 cores are used to 100% by the peer

workload of approximately 800 concurrent clients. At this point, *peer0* uses practically all the capacity of its four reserved CPU cores (400%). Since the test cases with different cryptographic key sizes present similar results, we can conclude that our homomorhic chaincode implementation does not affect *peer0* behavior. On the other hand, the *dev* container barely uses half of the capacity of its 4 CPU cores. One can notice that *dev* gets the best CPU usage within a workload of around 600 concurrent client instances. After that, *dev* CPU usage is reducing, indicating that *peer0* degrades its capacity in managing the transactions. Although the CPU usage increases when using longer cryptographic keys, a significant amount of CPU remains idle, even when we have a Paillier key size of 2048 bits. In practice, these results confirm that ordinary Blockchain tasks (e.g., consensus, and transaction validations) still exceed homomorphic evaluation in terms of computational costs. We also evaluate the throughput and latency of the transactions. The graphs in Fig. 4c and d present these again for measurements at points A and B. It can be seen that the throughput increment is steady until a workload of 800 client instances is reached. After that, throughput only increases slowly, reaching a state of around 450 tps. The latency increases in an inverse proportion, by confirming that the performance degrades after *peer0* reaches its max CPU usage. It should be noted that the performance in terms of throughput and latency is the same, independent of if we use homomorphic or plaintext addition.

7 Conclusion

In resume, we can affirm that our benchmark tests show that the use of HE combined with Blockchain technology is a feasible and good approach to achieve confidentiality for distributed measuring instruments. The tests using the Paillier cryptosystem point out that the computational overhead does not exceed the computational effort required by common tasks demanded by the Blockchain. Based on these findings we argue that even the correct evaluation of the execution of code, without re-executing it or even knowing it, is feasible using the described mechanisms in this paper like PCPs and zkProofs. Through FE, it is even possible, to make sure that the provider/seller can bill the client without trusting her to send him any measurements in plaintext, i.e. without knowing the individual consumption of a specific meter, he can still compute the final billing result. Hence, we come to the conclusion that although normally, the practical implementations, of PCPs, FE, and FHE schemes create challenges concerning very large and complex algorithms - e.g. all FHE schemes today have a large computational overhead especially for multiple users - these mechanisms can still be very helpful if tailored to specific use-case scenarios. Especially for legally supervised measuring instruments, this seems to be the case, because often their algorithms are very simple, e.g., just adding different sensor data together.

References

1. Esche, M., Thiel, F.: Software risk assessment for measuring instruments in legal metrology. In: FedCSIS, vol. 5, pp. 1113–1123 (2015)
2. Peters, D., Peter, M., Seifert, J.P., Thiel, F.: A secure system architecture for measuring instruments in legal metrology. Computers **4**(2), 61–86 (2015)
3. Armknecht, F., Boyd, C., Carr, C., Gjøsteen, K., Jäschke, A., Reuter, C.A., Strand, M.: A guide to fully homomorphic encryption. IACR Cryptology (2015)
4. Goldwasser, S., Kalai, Y.T., Popa, R.A., Vaikuntanathan, V., Zeldovich, N.: How to run turing machines on encrypted data. In: Canetti, R., Garay, J.A. (eds.) Advances in Cryptology - CRYPTO 2013, pp. 536–553. Springer, Heidelberg (2013)
5. Boneh, D., Sahai, A., Waters, B.: Functional encryption: definitions and challenges. In: Ishai, Y. (ed.) Theory of Cryptography, pp. 253–273. Springer, Heidelberg (2011)
6. Iovino, V., Żebrowski, K.: Mergeable functional encryption. In: Okamoto, T., Yu, Y., Au, M.H., Li, Y. (eds.) Provable Security, pp. 434–451. Springer, Cham (2017)
7. Goldwasser, S., Gordon, S.D., Goyal, V., Jain, A., Katz, J., Liu, F.H., Sahai, A., Shi, E., Zhou, H.S.: Multi-input functional encryption. In: Nguyen, P.Q., Oswald, E. (eds.) Advances in Cryptology - EUROCRYPT 2014, pp. 578–602. Springer, Heidelberg (2014)
8. Nakamoto, S.: Bitcoin: a peer-to-peer electronic cash system (2008). https://bitcoin.org/bitcoin.pdf
9. Christidis, K., Devetsikiotis, M.: Blockchains and smart contracts for the internet of things. IEEE Access **4**, 2292–2303 (2016)
10. Zheng, Z., Xie, S., Dai, H.N., Wang, H.: Blockchain challenges and opportunities: a survey. Int. J. Web Grid Serv. 1–24 (2017)
11. Sousa, J., Bessani, A., Vukolić, M.: A Byzantine fault-tolerant ordering service for the hyperledger fabric blockchain platform. In: 48th Annual IEEE/IFIP International Conference on Dependable Systems and Networks (DSN) (2018)
12. Vukolić, M.: Rethinking permissioned blockchains. In: Proceedings of the ACM Workshop on Blockchain, Cryptocurrencies and Contracts - BCC 2017 pp. 3–7 (2017)
13. Li, F., Luo, B.: Preserving data integrity for smart grid data aggregation. In: 2012 IEEE Third International Conference on Smart Grid Communications (SmartGridComm), pp. 366–371 (2012)
14. Traverso, G., Demirel, D., Buchmann, J.: Homomorphic Signature Schemes: A Survey, 1st edn. Springer, Heidelberg (2016)
15. Thiel, F., Esche, M., Grasso Toro, F., Oppermann, A., Wetzlich, J., Peters, D.: The European metrology cloud. In: International Congress of Metrology (2017)
16. Melo Jr., W.S., Bessani, A., Neves, N., Santin, A.O., Carmo, L.F.R.C.: Using blockchains to implement distributed measuring systems. IEEE Trans. Instr. Meas. 1–12 (2019)
17. Paillier, P.: Public-key cryptosystems based on composite degree residuosity classes. In: Stern, J. (ed.) Advances in Cryptology - EUROCRYPT 1999, pp. 223–238. Springer, Heidelberg (1999)

Hardware Transactional Memory as Anti-analysis Technique for Software Protectors

Federico Palmaro[1(✉)] and Luisa Franchina[2]

[1] Prisma, Rome, Italy
f.palmaro@prismaprogetti.it
[2] Hermes Bay, Rome, Italy

Abstract. Software protectors aim to shield executable file against reversing and cracking, by implementing sophisticated mechanisms capable of hiding real binary code and by inserting, inside an executable file, pieces of control code created to reconstruct software to its original state only during execution. This process of hiding and restoring protected bytes is composed by complex code fragments which perform all essential operations, keeping in general a constant structure for all protected software. The goal of this paper is to enhance security of these types of software against crackers, using Hardware Transactional Memories (HTM) by exploiting their features in a way for which they were not designed for. The proposed methodology offers a good level of defence of the protector core part by implementing new specific checks which use hardware processor features, ensuring reliability and a good level of performance with respect to different software implementations and which can be inserted inside a just present chain of checks, enhancing whole programs protection.

Keywords: Software protectors · Hardware transactional memories · Reverse Engineering · Software security

1 Introduction

Binary code of Portable Executable file can be easily studied by crackers with Reverse Engineering techniques, which enhance their power during the time. In this way, some important information, like secret data or just Intellectual Property can be stolen, causing serious damage to software authors. These Reverse Engineering techniques starts with disassemblers [1] and decompilers which try to reconstruct in some cases original source code, managing to understand how it works. A special tool which performs these actions and makes analysts able to execute disassembled code step-by-step is called debugger, that is the main weapon for a reverse engineer. Other tools used are software monitoring technologies like Dynamic Binary Instrumentation [2] and so on. A strong defence line [3] against these types of analysis consists in software protectors, which perform compression and encryption of original bytes driven by a decompression routine that manages the reconstruction of original software.

The industry of Software Protectors constantly evolves to ensure better protection for its customer's executables, trying to implement always new methodologies and tricks for better performance, combined to improved compression and encryption mechanisms. The main model of protection is composed by building a single wrapping layer around an executable [4] such as UPX, Enigma and Armadillo which is responsible for creating new section with compressed and encrypted data, reconstructing real code in a second time during execution.

More recently, virtualized packers are deployed such as VMProtect and Themida which consist in routines that embed original logic of program inside a custom Virtual Machine (VM) interpreter [5]. The difficulty of this techniques is to understand the logic behind VM to reconstruct original flow of code. Unfortunately, to enhance protection, protectors add some random elements inside bytecode which make analysis more complicated.

A lot of works try to restore the original state of the executable [6–8], excluding the protectors actions. The problem mentioned in this paper is the need of breakpoints, either software or hardware, in some specific point of the protected program in a way to help the reconstruction routines. These breakpoints are special opcodes inserted by a debugger before a selected instruction in order to stop execution and analyze the current state of the machine. Since this is a well-known mechanism, protectors have implemented different type of shields against it, exploiting anti-debugging techniques [9–11] which use both software and hardware facilities in order to obtain information on the software attached to them. Usually these techniques consist of system queries to check some specific information which make protector software able to identify the presence of hostile tools around it. Other specific techniques are also exploited, more sophisticated compares to the previous ones. They can exploit time measurement, exceptions tricks and other debugger-specific techniques aimed to detect the presence of a running debugger.

At this point, we can introduce a new identification exploit aimed to detect the main problem described before, which is a presence of breakpoints, placed by a user of a debugger tool: if Software Protectors are able to place a strong check around core code fragments of their reconstructor routines, their protection against malicious people will be enhance a lot.

Related to this purpose, we use an hardware feature present in a recent processors, for instance Intel which are used in this paper, called Transactional Synchronization Extensions (TSX-NI) [12], which is an extension of x86 Instruction Set Architecture, adding support to Hardware Transactional Memories (HTM). These type of instructions attempt to enhance concurrent programming facilities by allowing a small set of load and store operation, called transaction, in an atomic way. The mechanism behind execution of transactions has some restriction deriving from its intrinsic implementation, putting some limitations in code that can be inserted inside transactions. In fact, transactions fail in case some particular events happen during execution. For the case of the paper, we are interested in one particular point inside this event list, that is number 4 called _XABORT_DEBUG. How it can be deducted from definition, in case a software or hardware breakpoint was hit during execution, transaction fail its task and return a code of event happened.

Linking this explanation with the just defined algorithm used by Software Protectors, a new methodology for protecting core part of decompression and decryption can be introduced, offering a new approach based on technologies designed for other task and exploitable for our purpose.

2 State of the Art

2.1 Anti-debugging Techniques

War between software reversers and authors of protection code is constantly evolving in the field of analysis identification. If on one side analysis tools become more sophisticated in terms of functionalities, on the other side the environment around executable files suffers small changes derived from the presence of these tools that in a specific manner modify elements inside the system.

We have three types of techniques which make software able to recognize a presence of invasive tools:

- **Query on the system:** This area incorporates all techniques based on queries about the presence of some strange "artifacts" in the environment by which the system is composed. These queries can be performed at low level as direct system calls aimed to check possible flags changed by debugger, extracted directly from the system data structure. Examples can be found in Al-Khaser code which explains how system calls can be easily called to explore the environment and obtain precious information.
- **Exception handling:** Exceptions are errors which may occur during execution of a software [13]. These signals are first intercepted by the program itself which should be able to manage them in order to restore the correct state of the software, otherwise they are captured by the operating system that provides for the correct termination of the program itself. Debuggers in order to execute correctly a program, they must manage exceptions in a correct way if they want to remain transparent respect to the whole execution state. The problem in this case refers to the presence of some type of exceptions which are not correctly managed by debugger, causing a strange behavior that is different respect a execution performed in a clean system. Exceptions can also deceive analyses based on calling context information [14] on the execution stack or fine-grained ones tracking basic clocks inside functions [15].
- **Time:** During execution inside debugger, the whole system time continue to flow in a normal way, making time measurements a possible exploitation: duration of an action like function call or assembly instruction usually take a quite constant value that is different respect to the same action executed under analysis software [16]. There are a lot of ways to perform these types of measurements which make their detection problematic.

2.2 HTM in Software Security

Hardware transactional memories are designed to manage concurrency between threads directing whole set of research articles which mentioned them in a discussion on how to use HTM in concurrent security aspects or a more recent HTM used in Virtual Machine Introspection (VMI) [17]. They are used to ensure data protection, for instance, during cryptographic processes [18] avoiding that other process avoiding that other processes access cached sensitive data. Side effects obtained by key limitations imposed by developers in HTM is a recent idea proposed in software security, exploiting them in a way to guarantee some security checks with low impact in performance. For instance, some works [19,20] exploit the cache constraints in a way to avoid side-channel attacks, protecting systems from possible data theft. In fact, in a transaction, all data used by application in that moment must be in cache fields, making sure that there is a possibility to monitoring all data movement between cache and system memory.

3 Overview

3.1 Hardware Transactional Memories: What They Were Created For?

Parallelism in modern software is a key point to improve performance exploiting high number of cores and threads [21] present in a new x86 processors. Programmers usually manage concurrency in software level, using a lot of facilities just offered by software library and system API, but this modality has a penalty when performance aspects are considered. Hardware implementations were created to overcome this weakness of software support, designing the atomicity of these transactions directly in the microcode. Architectures which support transactional memories are:

- Intel transactional synchronization extensions (TSX)
- AMD advanced synchronization families
- POWER
- Sun's Rock processor

In this paper we focus on the Intel TSX family which offers the support of these instructions from high-end Haswell architecture, and is going to enable it even in the mid-range CPU.

A set of load and store operations supported by atomic transactions will then be executing in isolation respect to the other executions present in other CPU threads, avoiding the programmer's concerns to manage the concurrency.

To illustrate how the flow of transaction is wrapped by HTM, below is present a snippet of code which represents how to use transactional memories:

```c
#include <immintrin.h>
#include <stdio.h>


int main() {
    int n_tries, max_tries = 10;
    unsigned status = _XABORT_EXPLICIT;
    int count = 0;

    for (n_tries = 0; n_tries < max_tries; n_tries++)
    {
        status = _xbegin ();
        if (status == _XBEGIN_STARTED || !(status & _XABORT_RETRY))
            break;
    }
    if (status == _XBEGIN_STARTED)
    {
        /**** ALL CODE OF TRANSACTION HERE ****/
        count++;
        _xend ();
    }
    else
    {
        count--;
    }
    printf("Status: %d", status);
    printf("Count: %d\n", count);
    return 0;
}
```

As we can see from the code snippet, between _xbegin() and _xend() calls all the instructions are performed in an atomic way only if the limitation imposed by the transactional memories developer are respected.

Due to hardware implementation, only some Intel CPUs supports this mechanism: as we write the architecture supported are those in a Table 1.

Table 1. Processor families that support HTM

CPU generation	Year
Haswell	2013
Broadwell	2014
Skylake	2015
Kaby lake	2016
Coffee lake	2017

3.2 Intersection Between HTM and Software Security

Back to the Software Protectors, all mechanisms of data decompression and decryption are composed by a set of load and store operations which manage the system memory in order to modify bytes of the altered protected software and restore the initial state of the program.

At this point the core idea of the paper is to detect all these main code fragments inside restore routines and wrap their instructions with a transaction at hardware level. The importance of this mechanism refers to the core analysis method for reverse engineer which consists in putting some breakpoints at the specific instructions addresses, reaching these points and continuing analysis with the knowledge of the new context. Avoiding that someone can put breakpoint (both hardware and software) in the core routines and can navigate them in complete freedom, the difficulty in manual restoring original software grows in a sparkling way.

After having avoided breakpoint placements, another action of reverser engineers can be avoided inside protected core code fragments: the strongest claim in fact is that also single-stepping inside a transaction will be recognized by microcode, triggering the detection mechanism to deviate path, that is for instance on fake routines. When a debugger perform single-step, it virtually places a software breakpoint before the next address that will be detected by CPU if it is executed inside transactions.

3.3 Practical Disarming

For instance, let's examine the UPX packer, given its free and open-source nature: there is a simple way to disarm its protections. As first step, instruction *pushad* must be found at the beginning of Entry Point, putting a breakpoint in order to stop execution at the specific address. After seeing memory locations where registers have been placed, a hardware breakpoint must be inserted in order to reach that specific memory location after a whole unpacking routine has been completed, reaching instruction called *popad*. Entire process therefore needed two breakpoints, one of which is hardware. After finding core part of the unpacking process, we can think of applying our Hardware Transactional Memories in a way to wrap, for instance, the *popad* instruction (Fig. 1). As soon as the debugger reaches the incriminated address, the transaction fails and the alternative branch will be executed, denoting the presence of a reverse engineer behind the scenes.

3.4 Comparison with Other Detections

The power of this wrap layer, an HTM transaction, stands out respect to the others described in *Sect.* 2.1 thanks to an asynchronous process implemented in microcode that performs all checks needed as side effect.

- The first type of anti-debugger technique of which we have written is based on a simple call to the system which can be made in different ways but they can be skipped by an expert analyst if you leave him free to use his classic work tools. As we can see from the snippet of code below, for an analyst, modify the result of a specific library call or system call is not a problem,

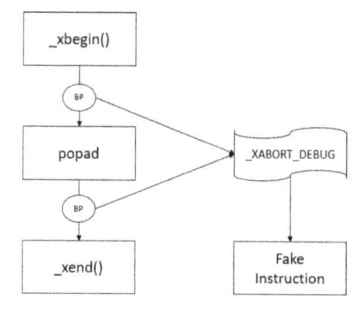

```
          _xbegin()
             |
            BP
             |
           popad  ----->  _XABORT_DEBUG
             |                  |
            BP                  |
             |                  v
          _xend()           Fake Instruction
```

Fig. 1. Redirection of flow when breakpoint was hit

putting a breakpoint after the call have been executed and modify the *flag* value in "FALSE".

```
CheckRemoteDebuggerPresent( GetCurrentProcess(), flag );
if( flag )
    ExitProcess(0);
else
    //DoEvil
```

- The second type of technique manages exceptions in a fashion way, being however relegated to the trigger of signal to analyst which could trigger an alarm in the mind of those trying to unpack protected code and let malicious people still use their tool with full functionalities. In the following snippet of code, there are many possibility to jump this check, for instance jumping directly to the next instruction inside "try" block or forward exceptions to the program directly from debugger.

```
__try {
    __asm int 3
}
__except(1){
    return FALSE;
}
return TRUE;
```

- The last category of detection refers to time measurements which is maybe the most heterogeneous at all, having a lot of different way to perform measurements inside running machines. Problem with it is likely exception mechanisms, that is the possibility for the person behind analysis tool of use all its features to fight back the tricks included in software protectors.

Surely the whole set of techniques described above is however not simple to avoid all times, causing a necessary barrier inside each software protector. The power of using HTM as anti-debugger techniques is in avoiding that functionalities like breakpoint and consequently single-stepping inside a debugger can be used, increasing time and attention that a software reverser should use when he tries to reconstruct original application code.

4 Evaluation

4.1 Aim of the Test

To evaluate the effectiveness of the technique described above, two different types of test are prepared: we have used five different debuggers to test if all of these tools impact in the same way on the protected binaries, while the second part of the test consist in a comparison of performance with and without our protection applied, showing both effectiveness against analysis tool and a low impact on the execution performance.

The platform used in the tests is an Intel machine with Coffee lake 6 physical cores and 12 hardware threads using hyper-threading processor named i9-8950HK, running at 2.90 GHz and 32 GB of RAM. We use Lubuntu 18.04 as operating system on which are installed all debugger tools.

Debuggers Set Test: Debuggers used for our test are the most famous and used by reverse analysts, which are GDB debugger, IDA Pro, Immunity Debugger, WinDBG, Olly 1.0 and x64dbg. The test consists in a simple program, which is structured as the snippet of pseudocode below, executed inside each debugger present in the list and in which two phases are performed: placing a single breakpoint, first software and than hardware, and test if the fake branch was taken while the second part consist in a chain of single-step actions inside a critical section, still to analyze which execution branch is taken.

```
1: function DEBUGGER RESISTANCE TEST
2:     pushad
3:     status <- _xbegin()
4:     if status == _XBEGIN_STARDED then
5:         popad
6:     else
7:         undo pushad
```

Performance Evaluation: This is a critical aspect that we need to analyze, because of tight requirements imposed to good software protectors. A decompression routines needs to have a very low impact on a whole execution, using techniques that are not hungry for time in a first place. Hardware transactional memories, by their nature, are optimized due to their implementation made directly in the microcode of the processor, and should have a very low impact in time consumption. The test will be performed taking a self-made program which simulates the behaviour of a possible critical section of a software protection routine. As we can see in the following snippet of pseudocode, offering a simple array filling, a loop of transaction based on movement of bytes in memory will be implemented, comparing the time with and without the support of HTM.

```
1: function PERFORMANCE TEST
2:     counter <- 100000
3:     while counter != 0 do
4:         status <- _xbegin()
5:         if status == _XBEGIN_STARTED then
```

6:      $v[counter] < -0$
7:      **else**
8:      $v[counter] < -1$
9:      $counter < -counter - 1$

4.2 Results and Discussion

After providing a description of the tests, we divide their results by category, summarizing each result in Table 2:

Breaking Debuggers: This snippet of assembly code is the test software used for our tries. It is composed by transaction _XBEGIN_STARTED, transaction _XEND, normal _XABORT flow and _XABORT_DEBUG case. Experiments are developed interacting with debuggers described before, putting breakpoint both hardware and software inside transaction block (lines [7,13]) and analyzing if control flow redirect on _XABORT_DEBUG after fail. In the test suite, we tried to put breakpoints inside all blocks, single-stepping throw basic blocks and execute only few instructions at time.

```
1  int  main(){
2       ...
3         mov    eax,  0FFFFFFFFh
4         xbegin $+6                              ;begin of transaction
5
6       loc_401433:
7         mov    [esp+18h], eax
8         cmp
   dword ptr [esp+18h], 0FFFFFFFFh;if _XBEGIN_STARTED
9         jnz    short loc_401450
10        mov    eax, ds:_done
11        add    eax, 1
12        mov    ds:_done, eax
13        xend                                    ;end transaction
14        jmp    short .exit
15      loc_401450:
16        mov    eax, [esp+18h]
17        and    eax, 10h
18        test   eax, eax
19        jz     short .exit
20        mov    eax, ds:_done   ;if _XABORT_DEBUG
21        sub    eax, 1
22        mov    ds:_done, eax
23       ...
24  }
```

As supposed in description, all mentioned debuggers trigger HTM signal if breakpoint is present or single-stepping is performed only inside lines [7,13] which redirects execution flow to the abort path. Keeping delicate operations inside the transaction proves itself a success compared to other techniques which let people behind analysis use their favourite features. In fact with this technique, change some values in the debugger or modify control flow of a program is not enough to break protection due to transaction invalidation caused by HTM intrinsic.

Time of Execution: Results of this test enhance the power supposed of this idea, strengthening about a possible insertion inside of a software protection, mixed maybe with other techniques actually in use. Average time difference between executions was almost zero, providing the same results in multiple executions. Hardware transactional memories are well optimized for this load type, making insertion of small security checks which exploit them a low impact patch in a whole protection mechanism.

As future work we plan to explore techniques such as on-stack replacement [22,23] to make the fake alternative path hardly distinguishable in terms of effects on the program state to further deceive the adversary, introducing minor transformations on the state for unpacking.

Table 2. Comparison between HTM checks and others, focusing on time overhead and debugger in which such check successfully executed.

	Systemcall	Exceptions	Time
Time overhead	=	>	>
Number of debugger supported	<	<	=

5 Conclusions

We presented this technique as a novel approach that exploits the power of hardware transactional memories support, providing a strong level of protection against malicious people who want to stole information placed inside of protected software. As strong requirements impose a good level of performance to software protectors, this mechanism is effective against more powerful debuggers with a penalty in performance that approaches to zero.

References

1. Kruegel, C., Robertson, W., Valeur, F., Vigna, G.: Static disassembly of obfuscated binaries. In: Proceedings of the 13th USENIX Security Symposium (2004)
2. D'Elia, D.C., Coppa, E., Nicchi, S., Palmaro, F., Cavallaro, L.: SoK: using dynamic binary instrumentation for security (and how you may get caught red handed). In: Proceedings of the 2019 ACM Asia Conference on Computer and Communications Security, Asia CCS 2019, pp. 15–27. ACM, New York (2019)
3. Kim, M., Lee, J., Chang, H., Cho, S., Park, Y., Park, M., Wilsey, P.A.: Design and performance evaluation of binary code packing for protecting embedded software against reverse engineering. In: 2010 13th IEEE International Symposium on Object/Component/Service-Oriented Real-Time Distributed Computing, pp. 80–86, May 2010

4. Rolles, R.: Unpacking virtualization obfuscators. In: Proceedings of the 3rd USENIX Conference on Offensive Technologies, WOOT 2009, p. 1. USENIX Association, Berkeley (2009)
5. Coogan, K., Lu, G., Debray, S.: Deobfuscation of virtualization-obfuscated software: a semantics-based approach. In: Proceedings of the 18th ACM Conference on Computer and Communications Security, CCS 2011, pp. 275–284. ACM, New York (2011)
6. Lindorfer, M., Kolbitsch, C., Milani Comparetti, P.: Detecting environment-sensitive malware. In: Sommer, R., Balzarotti, D., Maier, G. (eds.) Recent Advances in Intrusion Detection, pp. 338–357. Springer, Heidelberg (2011)
7. Paleari, R., Martignoni Giampaolo, L., Roglia, F., Bruschi, D.: A fistful of red-pills: how to automatically generate procedures to detect CPU emulators (2009)
8. Ugarte-Pedrero, X., Balzarotti, D., Santos, I., Bringas, P.G.: SoK: deep packer inspection: a longitudinal study of the complexity of run-time packers. In: 2015 IEEE Symposium on Security and Privacy, pp. 659–673, May 2015
9. Branco, R.R., Barbosa, G.N., Drimel, P.: Scientific but not academical overview of malware bugging, anti-disassembly and anti-VM technologies (2012)
10. Leitch, J.: Anti-debugging with exceptions (2011)
11. Ferrie, P.: The ultimate anti-debugging reference. Technical report (2011)
12. Herlihy, M., Eliot, J., Moss, B.: Transactional memory: architectural support for lock-free data structures. In: Proceedings of the 20th Annual International Symposium on Computer Architecture, pp. 289–300, May 1993
13. Rin, N.: Virtual machines detection enhanced. https://artemonsecurity.com/vmde.pdf
14. D'Elia, D.C., Demetrescu, C., Finocchi, I.: Mining hot calling contexts in small space. Softw.: Pract. Exp. (2015)
15. D'Elia, D.C., Demetrescu, C.: Ball-Larus path profiling across multiple loop iterations. In: Proceedings of the 2013 ACM SIGPLAN International Conference on Object Oriented Programming Systems Languages and Applications, OOPSLA 2013, pp. 373–390. ACM, New York (2013)
16. Chen, X., Andersen, J., Mao, Z.M., Bailey, M., Nazario, J.: Towards an understanding of anti-virtualization and anti-debugging behavior in modern malware. In: 2008 IEEE International Conference on Dependable Systems and Networks With FTCS and DCC (DSN), pp. 177–186, June 2008
17. Liu, Y., Xia, Y., Guan, H., Zang, B., Chen, H.: Concurrent and consistent virtual machine introspection with hardware transactional memory. In: 2014 IEEE 20th International Symposium on High Performance Computer Architecture (HPCA), pp. 416–427, February 2014
18. Li, C., Guan, L., Lin, J., Luo, B., Cai, Q., Jing, J., Wang, J.: Mimosa: protecting private keys against memory disclosure attacks using hardware transactional memory. IEEE Trans. Dependable Secure Comput. 1 (2019)
19. Gruss, D., Lettner, J., Schuster, F., Ohrimenko, O., Haller, I., Costa, M.: Strong and efficient cache side-channel protection using hardware transactional memory. In: 26th USENIX Security Symposium (USENIX Security 17), pp. 217–233. USENIX Association, Vancouver, August 2017
20. Chen, S., Liu, F., Mi, Z., Zhang, Y., Lee, R.B., Chen, H., Wang, X.: Leveraging hardware transactional memory for cache side-channel defenses. In: Proceedings of the 2018 on Asia Conference on Computer and Communications Security, ASIACCS 2018, pp. 601–608. ACM, New York (2018)

21. Yen, L., Bobba, J., Marty, M.R., Moore, K.E., Volos, H., Hill, M.D., Swift, M.M., Wood, D.A.: LogTM-SE: decoupling hardware transactional memory from caches. In: 2007 IEEE 13th International Symposium on High Performance Computer Architecture, pp. 261–272, February 2007
22. D'Elia, D.C., Demetrescu, C.: Flexible on-stack replacement in LLVM. In: Proceedings of the 2016 International Symposium on Code Generation and Optimization, CGO 2016, pp. 250–260. ACM, New York (2016)
23. D'Elia, D.C., Demetrescu, C.: On-stack replacement, distilled. In: Proceedings of the 39th ACM SIGPLAN Conference on Programming Language Design and Implementation, PLDI 2018, pp. 166–180. ACM, New York (2018)

Development of the Technique for the Identification, Assessment and Neutralization of Risks in Information Systems

Askar Boranbayev[1(✉)], Seilkhan Boranbayev[2], and Askar Nurbekov[2]

[1] Nazarbayev University, Astana, Kazakhstan
`aboranbayev@nu.edu.kz`
[2] L.N. Gumilyov Eurasian National University, Astana, Kazakhstan
`sboranba@yandex.kz, nurbekoff@gmail.com`

Abstract. The article is devoted to the development of methods for identifying, assessing and neutralizing risks in order to ensure the reliability and security of information systems. The regulatory requirements for risk analysis in information systems have been developed. The methodology for analyzing information security risks in the banking sector has been developed and analyzed. Effective risk reduction strategies were used. Studied methods allow the user to receive a quantitative risk assessment of the system. This makes it possible to eliminate the need to use expensive resources to identify risks. Research was conducted on building an Information Security Risk Management System, measures and procedures for identifying, measuring, monitoring, controlling and minimizing information security risks. The purpose of the Information Security Risk Management System is to prevent and reduce the threat of negative consequences associated with the operation of information systems, as well as external factors affecting information systems. It is aimed at minimizing the risks in bank's activities related to the violation of the integrity, confidentiality and availability of information systems.

Keywords: Safety · Risk analysis · Risk assessment · Method · Reliability

1 Introduction

The execution of the order of the Government of the Republic of Kazakhstan, in terms of legal regulation of digital innovations and implementation of the State Program called "Digital Kazakhstan", leads to accelerated application and implementation of digital innovations, including blockchain technologies, open data, Yellow Pages Rules, robotization and artificial intelligence, biometric authentication, crowd-funding, big data, etc. The pace of development of digital technologies necessitates an increase in the level of information security and reliability of those being implemented nologies [1].

Description of the significance of many studies on the creation of tools for effective risk management is given in [2–4]. Given the above need, consider the two developed methods of risk analysis:

1. The method of analyzing the risks of failure of information systems;
2. Information security risk analysis methodology for the banking system.

As part of the implementation of the first methodology, a software system for risk analysis [5] was developed earlier. It was made in the form of a web application and provides for the round-the-clock provision of information risk analysis services for information systems. This web application was developed based on risk analysis methods in the early design stages [6, 7] and neutralizing the effect of a risk event [8, 9].

The works of other authors [10–13] on the use of the above methods in other fields of activity were also studied. The methods help determine which individual components the user must examine to ensure reliability before the information system is commissioned, i.e. at an early stage of product development [14]. As a result of using the methods, the user receives a quantitative risk assessment of the system under study. This method makes it possible to eliminate the need to use expensive resources to identify risks [15].

To successfully mitigate risks, effective risk reduction strategies are used, which are stored in the web application databases.

As part of the implementation of a study to develop a second methodology, research was conducted on building an information security risk management system (SURIB) - a set of measures and procedures for identifying, measuring, monitoring, controlling and minimizing information security risks. The subject of risk analysis in the banking sector is one of the most important topics that many researchers pay attention to these days [16]. The purpose of SURIB is to prevent and/or reduce the threat of negative financial and non-financial consequences associated with the operation/use of information systems, as well as external factors affecting information systems. SURIB corresponds to the external operating environment, strategy, size, nature and level of complexity of operations and is aimed at minimizing the risks in the bank's activities related to the violation of the integrity, confidentiality and availability of information systems resulting from intentional destructive impact from employees and (or) third parties.

2 Method of Research and Discussion of Research Results

In this article we will consider in more detail the implementation of the methodology for identifying and analyzing risks for the banking sector, due to the high criticality of the need to protect the information systems of banks. Also, the relevance of this issue is enhanced by the approval in 2018 of the Resolution of the National Bank of the Republic of Kazakhstan "On Approving Requirements for Ensuring the Information Security of Banks and Organizations Carrying out Certain Types of Banking Operations, the Rules and Deadlines for Providing Information on Information Security Incidents, including information about violations, failures in information systems."

Below are the main stages of identifying and processing information security risks for banks.

2.1 Identify Information Security Risks

Information security risks are identified based on vulnerability analysis for each specific information system and (or) business process. The following methods are used to determine the presence of information asset vulnerabilities:

- Specialized software for the automatic detection of vulnerabilities in information systems and equipment.
- Monitoring of open sources of information for the availability of information on the use of certain methods in cases of realization (or attempts to implement) threats.
- The database of incidents (a database of risk events) that took place in the past.
- Operational information from employees about the lack of functioning of the information system in terms of information security, and (or) the presence of potential vulnerabilities in the business processes of the bank.
- Conducting a self-assessment of risks, through which it identifies and assesses potential risks (vulnerabilities), assesses the effectiveness of control over identified risks (vulnerabilities) and determines the level of residual risk.

2.2 Information Security Risk Assessment

For risk assessment, the developed methodology is used, the construction of which uses both the experience of experts, the requirements of the legislation of the Republic of Kazakhstan, and generally accepted risk assessment models as basic and requirements for such standards and commercial tools for risk assessment, such as:

- Recommendations of the National Institute of Standards and Technology NIST SP 800-30: 2002.
- International Standard for Information Security Risk Management Systems BS 7799-3;
- CRAMM (the UK Government Risk Analysis and Management Method) - a method developed by the UK Security Service, is the UK government standard.
- OCTAVE (Operationally Critical Threat, Asset, and Vulnerability Evaluation) is a risk analysis methodology that includes the implementation of all phases in identifying and evaluating critical assets, threats and vulnerabilities for an organization.

The method for calculating quantitative risk indicators is developed and regularly evaluated based on the expert method.

Method of calculating quantitative risk indicators.

Based on the results of the risk analysis, the expert works on each threat to carry out the following work:

(a) Identification of the type of threat. An example of identifying certain types of threats is given in Table 1. The list of threats is updated as possible in the future or existing threats that create conditions for disrupting the performance of the entire information system on information security attributes (C - confidentiality, I - integrity, A - accessibility).
One example for implementing these types of threats are:

- Theft of information with subsequent use in order to obtain commercial benefits;
- Theft of information in order to cause damage (financial, reputational);
- Theft of money;
- Violation of technological processes in order to cause damage (financial, reputational).

(b) Determination of the level that determines the increased risk of a threat, put on a five-point scale: 1 - "risk is minimized", 2 - "risk level is below average", 3 - "average risk level", 4 - "risk level is above average", 5 - "High level of risk" (Fig, 1).

Table 1. Determination of types of threats.

№	Types of Information Security Threads	Violations C, I, A
Physical Threats:		
1	Theft of information carriers with confidential information for the bank	C
2	Theft of paper documents by insiders with confidential information for the bank	C
3	Theft, modification, destruction of information	C, I, A
Misuse of computer equipment and the Internet by employees of the organization		
4	Unintentional modification (destruction) of information representing C/I/A bank employees	C, I, A
5	Using network tools in an unauthorized manner	I, A
Threats to leak confidential information		
6	Leakage of confidential information from the network via communication channels (e-mail, web, etc.)	C
...	...	...

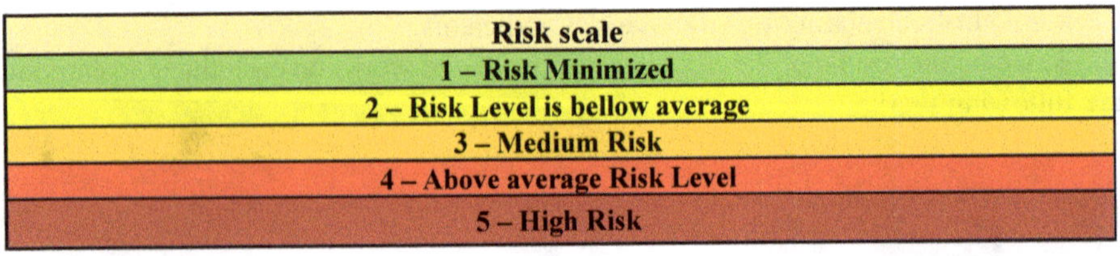

Risk scale
1 – Risk Minimized
2 – Risk Level is bellow average
3 – Medium Risk
4 – Above average Risk Level
5 – High Risk

Fig. 1. The scale of risks to the assessment of threats to protected objects

(c) Carrying out an analytical calculation for each model of offenders. The maximum score depends on the level of possibility to reduce the risk in this area, that is, for example, the quality of the audit trail does not require a calculation of five digits or five points, it is enough to accept the existence of an opportunity to reduce the risk of either 1- "NO" or 2 - "YES" zero (0) - accepted in the event that a particular model of potential violators and types of threats have no protection.

For example, in the event of such a threat as "Deliberately sending information outside the bank, the bank employees accept the possibility of reducing risk "Quality of the audit trail", the value "1" or "2" is taken, where 1 - partial audit trail is present, limited by possibilities, 2 - the audit trail is present in full and allows maximum protection from an adverse event.

If the threat requires a more in-depth analysis for conducting calculations, it is acceptable to calculate based on 5 (five) points, where the number 5 is taken as 100% of the whole value of the point, that is, maximum.

For example, "Infection of personal computers of a bank with viruses by malicious software via e-mail" in this case, the threat requires a more detailed analysis of the possibility of reducing risk such as "Organizational complexity of implementation by the attacker", values are taken from "1" to "5", where 1 - implementation complexity is minimal, 2 - implementation complexity above minimum, 3 - implementation complexity is average, motivation is partially present, 4 - implementation complexity is above average level, motivation is limited, 5 - the complexity of implementation is maximum, there is no motivation. Then, according to the MAX column, all the points are summed and the total amount is determined and so on in all categories, according to the total, when all categories are calculated, the total amount of MAX points is divided by the total summed total amount for all models of potential offenders and in the lowest field we have the level of other risk; The resulting figures, which determine the level of risk, are transferred to the risk map and are marked in colors from "Green" to "Red" (Fig. 1). For this or that color value, certain measures are taken, for example, "Acceptable zone" - no measures are allowed, "Yellow zone" - to increase control, "Orange zone" - Critical situation - needs attention, "Red zone" - Very critical situation - requires urgent intervention.

Risk assessment for each information system and analysis of the adequacy of risk management measures are carried out by experts.

Risk assessment takes into account but is not limited to the following factors:

- The size, nature and complexity of the bank;
- The state of the information system, technology and their capabilities;
- The qualifications and experience of the personnel involved in the risk management process;
- Legislation of the Republic of Kazakhstan;
- Probability of realization of identified risks;
- Severity of the consequences of the identified risks;
- Implemented organizational and technical measures to minimize risks.

The table of maintaining the risk assessment matrix in a bank at a lower/increasing level is determined in accordance with the calculation matrix using the example of Table 2.

Table 2. Calculation matrix.

Type of threat against the protected object	Assigned points to downside risk	Categories of potential offenders
Deliberately sending information outside the bank by bank employees	MAX	A | B | C | D | E | F | G | H | K | L | M
(1) Quality of audit trail		
(2) The presence of control periodic (post fact)/operational (intersection)		
...		

Quantitative indicators of the identified risk for each category of the model of potential violators take into account:

- theoretical possibility of realization of the threat;
- the facts of the threat in the bank (incidents);
- information in open sources about the facts of the threat from third parties;
- technical complexity of the threat implementation;
- the organizational complexity of obtaining benefits by attackers from the realization of the threat;
- mass character (possibility of single incidents or mass cases);
- the presence of direct or indirect damage to the bank;
- availability of organizational protection measures;
- availability of technical protection measures;
- the motivation of intruders;
- the quality of the audit trail;
- availability of control (periodic or operational);
- the cost of risk realization by an attacker.

2.3 Mapping Information Security Risks

In order to realize the awareness and continuity of the risk management process, the bank develops and maintains up to date a risk map, which is the basis for determining response measures aimed at managing the identified risks. The form of the risk map is displayed in Table 3.

Table 3. Risk map form.

№	Types of information security threats	Object at risk	I, II, II, IV quarters (year 20..)										
			Categories of potential offenders									
			A	B	C	D	E	F	G	H	K	L	M
1	...	...	...	...	...	...	...	...	...	...	...	...	...

2.4 Measures and Response Measures

For each identified risk, SURIB ensures that the bank takes response measures depending on the availability of technical, organizational and financial resources: taking risks, avoiding risks, minimizing risks or transferring risks.

The Bank develops risk management action plans for each identified risk indicated in the risk map. In the process of coordinating the action plan, an analysis of the adequacy of risk management measures, the necessary resources and the timing of their implementation is carried out. According to the results of the implementation of the action plan, quarterly reports are generated. In the case of the emergence of new threats, recommendations are developed for adjusting action plans.

2.5 Self-assessment of Information Security Risks

Risk self-assessment is a tool through which a bank identifies and assesses potential risks, assesses the effectiveness of control over identified risks and determines the level of residual risk. Self-assessment of risks is formed from the following stages:

- Familiarization with the questionnaire (Table 4) regarding consideration of group and particular indicators of information security.
- Bringing the computational values of the estimates into an average numerical index.
- Drawing up a plan to minimize the main risks.
- Approval of the final report.

Table 4. The form of the questionnaire on self-assessment of risks.

№	Group indicator (Name of group indicator)						
	Private indicator (Name of private indicator)	Grading of the private indicator					Total average of information security
		1	2	3	4	5	
1							

3 Conclusion

Failure to operate an information system, especially in the banking system, can lead not only to disruption or cessation of its operation, but also to more global consequences in the form of large-scale monetary loss, its irreversible negative destruction, or a significant decrease in the company's security level for a long period of time. The possibility of disruption of such systems creates corresponding risks. In turn, the presence of risks inevitably leads to the need to find effective methods of managing them. More complete and detailed information about the methods and approaches to ensure the reliability and security of information systems using expert assessments is given in [17–31].

The developed methodology is designed to identify, evaluate and neutralize risks in banking systems, in order to increase their reliability. The technique is implemented in the developed web application to identify and mitigate information security risks. Also, this technique can be refined for use in other information and automated systems: economic, industrial, financial and economic, etc.

References

1. Boranbayev, A., Boranbayev, S., Nurusheva, A., Yersakhanov, K.: The modern state and the further development prospects of information security in the republic of Kazakhstan. In: 15th International Conference of Information Technology, Information Technology – New Generations, pp. 33–38 (2018)
2. Boehm, B.W.: Software risk management: principles and practices. Softw. IEEE **8**, 32–41 (1991)
3. Charette, R.N.: Software engineering risk analysis and management. Intertext Publications (1989)
4. Higuera, R.P., Haimes, Y.Y.: Software Risk Management. DTIC Document (1996)
5. Boranbayev, A., Boranbayev, S., Yersakhanov, K., Nurusheva, A., Taberkhan, R.: Methods of ensuring the reliability and fault tolerance of information systems. In: 15th International Conference of Information Technology, Information Technology – New Generations, pp. 729–730 (2018)
6. Lough, K.G., Stone, R., Turner, I.: The risk in early design method. J. Eng. Des. **20**(2), 155–173 (2009)
7. Lough, K.G., Stone, R.B., Tumer, I.Y.: Implementation procedures for the risk in early design (red) method. J. Ind. Syst. Eng. **2**(2), 126–143 (2008)
8. Krus, D., Grantham, K.: Generated risk event effect neutralization: identifying and evaluating risk mitigation strategies during conceptual design. In: INCOSE, Rome, pp. 1225–1237 (2012)
9. AlKazimi, M.A., Altabbakh, H., Murray, S., Grantham, K.: Evaluating generated risk event effect neutralization as a new mitigation strategy tool in the upstream industry. Proc. Manuf. **3**, 1374–1378 (2015)
10. Grantham Lough, K., Stone, R., Tumer, I.: Prescribing and implementing the risk in early design (RED) method. In: Proceedings of DETC 2006, Philadelphia, USA, - Philadelphia, pp. 431–439 (2006). https://doi.org/10.1115/detc2006-99374
11. Krus, D., Grantham, K.: Failure prevention through the cataloging of successful risk mitigation. Strategies **13**, 712–721 (2013). https://doi.org/10.1007/s11668-013-9728-8
12. Krus, D.A.: The risk mitigation strategy taxonomy and generated risk event effect neutralization method. Ph.D. thesis. Missouri, p. 176 (2012)

13. Lough, K.G., Stone, R.B., Tumer, I.Y.: The risk in early design (RED) method: likelihood and consequence formulations. In: Proceedings of DETC'06: ASME 2005 International Design Engineering Technical Conferences and Computers and Information in Engineering Conference, pp. 1–11 (2007). https://doi.org/10.1115/detc2006-99375
14. Vucovich, J.P., et al.: Risk assessment in early software design based on the software function-failure design method. In: Proceedings of the 31st Annual International Computer Software and Applications Conference, Institute of Electrical and Electronics Engineers (IEEE) (2007)
15. Grantham, K., Elrod, C., Flaschbart, B., Kehr, W.: Identifying risk at the conceptual product design phase: a web-based software solution and its evaluation. Mod. Mech. Eng. **2**, 25–34 (2012)
16. Yousefi, H.: Risk assessment and risk analysis in information security. In: 7th Conference on Quality & Productivity in Electronic Industry (2008)
17. Boranbayev, A.S.: Defining methodologies for developing J2EE web-based information systems. J. Nonlinear Anal.: Theory Methods Appl. **71**(12), 1633–1637 (2009)
18. Boranbayev, A., Mazhitov, M., Kakhanov, Z.: Implementation of security systems for prevention of loss of information at organizations of higher education. In: Proceedings of the 12th International Conference on Information Technology: New Generations (ITNG 2015), Las Vegas, Nevada, USA, pp. 802–804, 13–15 April 2015
19. Boranbayev, A., Boranbayev, S., Altayev, S., Seitkulov, Y.: Application of diversity method for reliability of cloud computing. In: Proceedings of the 2014 IEEE 8th International Conference on Application of Information and Communication Technologies-AICT2014, Astana, Kazakhstan, p. 244–248, 15–17 October 2014
20. Turskis, Z., Goranin, N., Nurusheva, A., Boranbayev, S.: A fuzzy WASPAS-based approach to determine critical information infrastructures of EU sustainable development. Sustain. (Switz.) **11**(2), 424 (2019)
21. Turskis, Z., Goranin, N., Nurusheva, A., Boranbayev, S.: Information security risk assessment in critical infrastructure: a hybrid MCDM approach. Inform. (Neth.) **30**(1), 187–211 (2019)
22. Boranbayev, S., Goranin, N., Nurusheva, A.: The methods and technologies of reliability and security of information systems and information and communication infrastructures. J. Theor. Appl. Inf. Technol. **96**(18), 6172–6188 (2018)
23. Boranbayev, A., Boranbayev, S., Nurusheva, A.: Analyzing methods of recognition, classification and development of a software system. Adv. Intell. Syst. Comput. **869**, 690–702 (2018)
24. Boranbayev, A., Boranbayev, S., Nurusheva, A.: Development of a software system to ensure the reliability and fault tolerance in information systems based on expert estimates. Adv. Intell. Syst. Comput. **869**, 924–935 (2018)
25. Boranbayev, A., Shuitenov, G., Boranbayev, S.: The method of data analysis from social networks using apache Hadoop. Adv. Intell. Syst. Comput. **558**, 281–288 (2018)
26. Boranbayev, A., Boranbayev, S., Nurusheva, A., Yersakhanov, K.: Development of a software system to ensure the reliability and fault tolerance in information systems. J. Eng. Appl. Sci. **13**(23), 10080–10085 (2018)
27. Boranbayev, S., Nurkas, A., Tulebayev, Y., Tashtai, B.: Method of processing big data. Adv. Intell. Syst. Comput. **738**, 757–758 (2018)
28. Akhmetova, Z., Boranbayev, S., Zhuzbayev, S.: The visual representation of numerical solution for a non-stationary deformation in a solid body. Adv. Intell. Syst. and Comput. **448**, 473–482 (2016)

29. Boranbayev, S.N., Nurbekov, A.B.: Development of the methods and technologies for the information system designing and implementation. J. Theor. Appl. Inf. Technol. **82**(2), 212–220 (2015)
30. Hritonenko, N., Yatsenko, Y., Boranbayev, S.: Environmentally sustainable industrial modernization and resource consumption: is the Hotelling's rule too steep? Appl. Math. Model. **39**(15), 4365–4377 (2015)
31. Akhmetova, Z., Zhuzbayev, S., Boranbayev, S., Sarsenov, B.: Development of the system with component for the numerical calculation and visualization of non-stationary waves propagation in solids. Front. Artif. Intell. Appl. **293**, 353–359 (2016)

Estimation of the Degree of Reliability and Safety of Software Systems

Askar Boranbayev[1(✉)], Seilkhan Boranbayev[2], and Askar Nurbekov[2]

[1] Nazarbayev University, Astana, Kazakhstan
aboranbayev@nu.edu.kz
[2] L.N. Gumilyov Eurasian National University, Astana, Kazakhstan
sboranba@yandex.kz, nurbekoff@gmail.com

Abstract. The tasks of ensuring the reliability and security of a software system, managing risk and organizing risks are important. The purpose of the developed software system was to identify, evaluate and neutralize the risks of information and automated systems. Ensuring the confidentiality of the information stored in the system, including historical data on errors and risks of the systems, plays an important role. Considering the importance of protecting this system from risks of information security, studies have been conducted using the method of protection from DDoS attacks and management of risks. Approaches are based on the adaptation of certain methods to assess the risks of software systems, help reduce risks by using risk mitigation and assessment strategies based on collected historical data on risk reduction. The study assembled a network of virtual machines. Machines performed various tasks. The first machine was a server analyzing the amount of incoming traffic. The second machine - served as the target server. The third machine performed the task of the attacker and it attacked the second machine. The main purpose of the article is to present the results of reducing information security risks, namely, distributed attacks on the developed software system.

Keywords: Distributed attack · Software system · Reliability · Security · Risk · Method

1 Introduction

Today, the government program called "Digital Kazakhstan" is being actively implemented in our country. It is aimed at the progressive development of digital structures, which in turn entails economic growth, increasing the competitiveness of the economic growth of the country. It can be assumed that it will bring favorable changes in the quality of life of the population of our country. It is also expected that the implementation of this program will improve the investment climate, the level of health care, the quality of education, and will ensure employment of the population. It is also expected that the development of information technologies in the country will entail an increase in the number of information security risks.

In [1, 2], the approaches were considered that were based on risk assessment and their neutralization in order to improve the reliability of information systems. These approaches allow for early risk assessment stage of the software development process and determine the most effective mitigation strategies for them. Approaches are based on the adaptation of the RED [3, 4] and GREEN [5] methods to assess the risks of software systems, help reduce risks by using risk mitigation and assessment strategies based on collected historical data on risk reduction. Based on these approaches - the software system was developed. The developed software system stores data of an organization, including confidential information, the leakage of which can affect the security of the information system of the organization.

In this study, the developed software system was protected against various types of intrusions and attacks in order to preserve the integrity, confidentiality and availability of information stored in this software system.

The main purpose of the article is to present the results of reducing information security risks, namely, distributed attacks on the developed software system.

2 Ensuring the Reliability and Security of the Software System

The system for analyzing the risks of information systems was developed on the basis of the following approaches:

(1) *Risk assessment approach.* The used approach is called RED (Risk in Early Design) [6]. It was previously used in industrial and other industries, including the field of electromechanical design. This approach was described in more detail in [1, 7] as an approach to ensure the reliability of information systems. Also, the RED approach, along with ETA and FTA methods, was described in [8, 9].

The RED method allows analysis and assessment of each risk found. In order to increase the reliability of software systems, it is not enough to analyze and detect failures, it is also necessary to use methods to reduce or eliminate the failures that have occurred. The sooner failures are identified and neutralized, the less organizations will incur costs and it will be much easier to implement risk reduction strategies.

Since each system is unique, the failures that appear in systems are very diverse and accordingly, it is difficult to find unique strategies for all existing systems. Strategies to reduce risks are based on the experience of system experts (analysts, developers, etc.) who are already familiar with the failures in the system and are directly engaged in debugging. However, there is a problem that after the failure has been corrected, the information on the risk neutralization strategy is not saved in the bug report and the new employee who is not yet familiar with the operation of the system has to spend a lot of time to find a suitable strategy when similar failures occur.

In order to facilitate the work in the future, organizations need to have information about previously applied strategies and compare them with new system failures.

Under the reduction of risks, we understand the process of transfer, reduction of the level of danger or complete elimination of risk in the information system.

(2) *The approach to assess the neutralization of the effect* of the event of the generated risk (GREEN). This approach was previously used as a new tool of assistance for engineers and managers in choosing an appropriate approach to reduce identified and estimated risks. The approach was also used in the industrial sector, including the extractive industry [5, 10, 11].

To implement the GREEN approach and to effectively select appropriate risk reduction strategies, the developed software system stores a historical database of failures identified through the RED approach. GREEN, using this database, indicates the most appropriate strategy for minimizing the consequences.

The developed software system plays an important role in preventing possible failures. It uses information about the information system, such as *functions, components, failures* and *the degree of failure*.

These data are formed with the help of experts in the field of the information system being studied. The data generated by experts is stored in the knowledge base and used for risk analysis and assessment. As a result, the software system allows even inexperienced developers to easily perform risk analysis in the early stages of work on the information system that is being checked.

Let's proceed to a more detailed description of the functionality of the developed software system:

(a) *Loading data about the tested information system.* Download can be carried out in two ways: manually (by entering information through the keyboard or selecting from the system's database) and downloading using a pre-prepared CSV file.
(b) *Output of the results.* The software system displays the results of the risk in two forms: a graphical diagram and/or a text file with detailed information on the risks.
(c) *Keeping strategies.* The software system stores a complete database with risk reduction strategies. The software system allows you to select the appropriate strategy from the existing database, or add your strategy to the database.
(d) *Uploading data.* The user can upload the results to a CSV file for later use.
(e) *Output of TOP risk, users and developers.* Based on the data stored in the database, you can derive the most popular ones: TOP-10 repeated failures; Top-10 users; Top-10 developers.
(f) *Description of the program information.* The section "About the program" explains to the user the each step of working with the software system. In addition to text information and images, the user can see the operation of the software system by viewing a special video lesson.

The software system has a requirement for the reliability of the information system. It consists in the following: the probability of occurrence of a potentially dangerous situation, and even more so of an accident, due to the failure of the information system should be minimal. So, for example, if a failure can lead to serious consequences, and even more so to cause an emergency, it is unacceptable. Usually such a failure occurs due to the fault of the system developer. Therefore, during the development of the system and at the stage of its design, it is necessary to conduct appropriate procedures for ensuring reliability.

As can be seen from the functionality of the software system, it stores the information of the organization, including confidential information, the leakage of which can affect the security of the entire information system of the organization.

To ensure the protection of the developed software system from a distributed attack, the following work was carried out:

- The spread and development of DDoS attacks were analyzed.
- The network of virtual machines is organized.
- The Zabbix network monitoring system was studied.
- The network capabilities of various Linux operating systems (CentOS, Ubuntu, Kali) have been studied.
- DDoS attack is organized.
- The network-based security features of the Linux-based operating system are configured to provide protection against DDoS attacks.
- The applicability of the method of protection against distributed attacks for the protection of the software system was investigated.
- Testing the method of protection against distributed attacks for the developed software system.
- Analysis of the results obtained.

3 The DDoS-Attacks, Mechanisms of Their Implementation and Ways to Protect

The popularity of DDoS attacks is explained by several reasons:

(1) The DDoS-attacks are fairly simple to implement. To implement this type of attack it is not necessary to have knowledge in the field of information security, it is enough to install special software and specify the victim's IP-address.
(2) Distributed denial of service attacks are very effective. After some time - the service, on which the attack is being implemented, becomes unavailable. Thus, it is possible to withdraw from the game your direct competitor or harm someone's business [12].

There are several types of DDoS attacks. Often the difference is due to the difference in the mechanisms of their implementation. Accordingly, the types of DDoS-attacks are named after the packets that fill the bandwidth. For example, UDP-spoofing uses UDP connections and so on. For the organization of DDoS-attacks, botnets are most often used.

In addition, DDoS-attacks can be strengthened in some ways. In an attack on Spamhaus in 2013, the amplifier used DNS resolvers. In these cases, an attacker, substituting his IP-address and IP-addresses of machines located in the botnet at the victim's address, sends requests to a large number of DNS servers. The servers send a response to the target server, and probably can overload it.

Not only computers are used as infected devices for the botnet. In recent years, the popularity of IoT (Internet of Things) has increased - these are devices that are used in everyday life and have access to the Internet. Most often, standard passwords are used in these devises, and there is no possibility of installing antivirus software, so they can be conveniently exploited.

Another way to strengthen DDoS attacks is to use *memcached* servers. *Memcached* servers are used by many sites to simplify the work with large databases. As it turned out recently, they have a vulnerability, which makes it possible to use them in DDoS-attacks. With this vulnerability, the largest DDoS attack in history was organized, with a traffic flow of 1.35 TB/s to the GitHub service.

It should be noted that to date there is no solution that would completely protect itself from DDoS attacks. However, there are various measures to protect yourself from accidental and weak attacks. Administrators of hosting companies try to ask their providers for the maximum possible width of the channel. Many resources have already moved to cloud storage, thereby protecting themselves from DDoS attacks. In addition, there are various resources, such as Cloudflare and Incapsula, which provide protection services against DDoS attacks.

Let us consider another way to protect against DDoS attacks - *packet filtering*. To implement this method, first of all it was necessary to organize the network of several computers (in this work - virtual machines). Attack was organized using special software. Then, the system configuration of the machine on which the attack was conducted was carried out in order to change the packet filtering rules. There are changes in incoming traffic using a Zabbix monitoring system that was installed on one of the machines.

In this study, an attack was organized, and packet filtering settings were implemented to protect against the DDoS attacks of the software system.

4 Modeling DDoS-Attacks and Organizing Protection Against It

4.1 Organizing the Work of the Network of Virtual Machines

During our study we organized a network of virtual machines. We called it a "Workspace" ("polygon"). Then we ran DDoS attack on it.

The study was implemented on virtual machines Oracle VM Virtual Box software.

After installing the Oracle VM VirtualBox, based on this software, 3 virtual machines were created. On all machines, various operating systems of the Linux family were installed.

- On the first virtual machine we installed operating system Linux Ubuntu. The attack was applied on this machine (on one of the open ports). In order to open the port to receive packets, the Apache HTTP Server was installed on the virtual machine. Also, in this virtual machine, to protect from the DDoS attack, we have modified the firewall configurations for filtering packets.
- The second virtual machine was based on the CentOS operating system on which the Zabbix monitoring system was installed.
- On the third machine, we installed the operating system Kali Linux to implement the DDoS attack itself. For this, the Torshammer software was used.

Next, we established the network between all the virtual machines. The Oracle VM VirtualBox software allows us to find the physical Ethernet connections on the main machine, and on their basis build the network structure of virtual machines.

Network settings in this program allows us to select several options for distributing computers in the network of virtual machines. If necessary, you can even throw them in different VLANs.

In this case, the settings of the created NAT network were used. It allows virtual machines to be in the same network and at the same time have an Internet connection (via a physically connected Ethernet connection).

Then, the received settings were checked on the machines themselves, in order to make sure that they are on the same subnet, and that there is no IP address conflict. As a result, the server on CentOS received the address 10.0.2.5, Linux Ubuntu with the open port received the IP address 10.0.2.4 (Fig. 1), and Kali Linux 10.0.2.6.

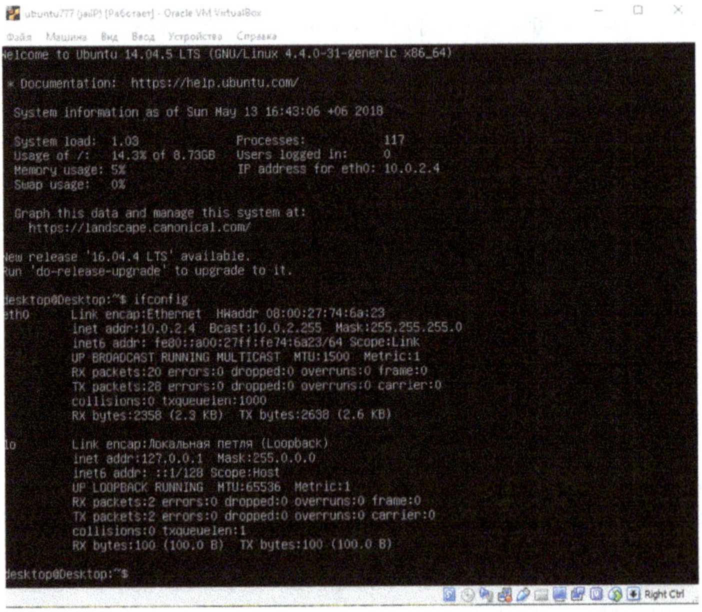

Fig. 1. IP address of the virtual machine

4.2 Installing Zabbix Computer Network Monitoring System

As already described above, the Zabbix monitoring system was installed on the server, allowing us to monitor incoming traffic. The result of the DDoS attack was monitored using it. Zabbix is considered as a universal monitoring system, as it consists of several parts:

(1) The Server of monitoring.
(2) Databases (MySQL, PosthreSQL, SQLite or Oracle).
(3) Web-interface, presented in PHP.
(4) Zabbix agent.

The Server of monitoring periodically receives data, polling the port, performs processing, analysis and has the ability to run alert scripts.

Databases allow to store the data requested from the port. Then, based on the results of this data, graphs are created, which are observed in the Monitoring section.

The web interface allows to easily observe all the events that occur. Find out the status of the server, check the CPU usage, check the amount of incoming and outgoing traffic, etc.

The agent can be run on monitored objects, for data analysis by the server. The agent has several different options, we can monitor it using several protocols, such as SNMP, HTTP, SSH, FTP.

Then Zabbix, installed on the server, communicates with Zabbix agent. Zabbix-agent is pre-installed on the necessary machine (Ubuntu with an open port).

The result of these commands is the Zabbix agent associated with the system. On the Zabbix web page, information about the host appears (Fig. 2).

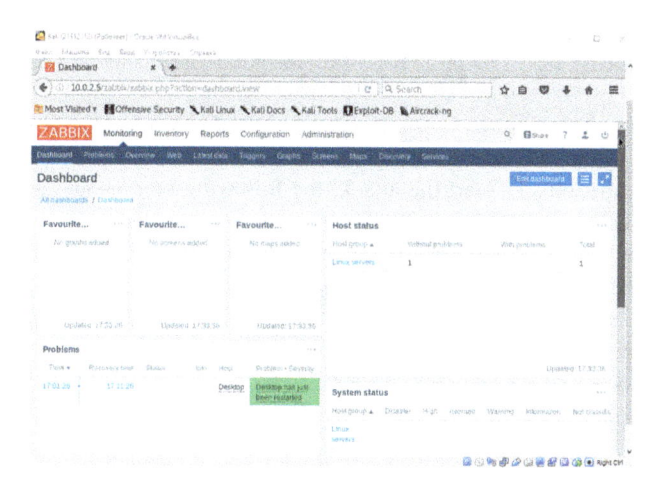

Fig. 2. Zabbix web interface hosts window

4.3 Conducting a DDoS Attack

To monitor the loading of the port, a transition was made to the Monitoring Tab and the Latest Data Tab was selected. The article considers the Incoming Network Traffic graph.

As can be seen in Fig. 3, at the time when the experiment began, the amount of incoming traffic was 1.4 Kb/s.

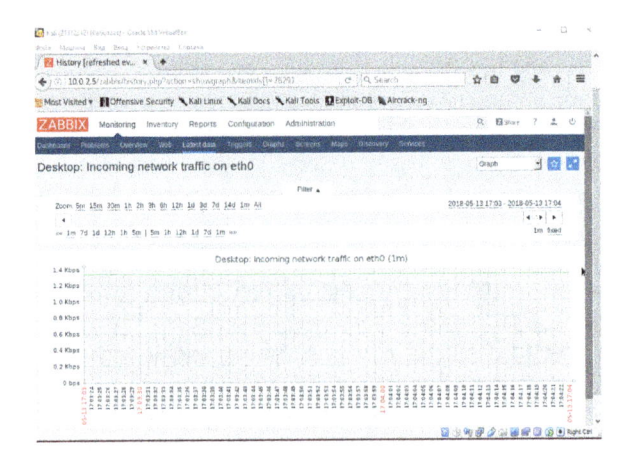

Fig. 3. Inbound traffic prior to the attack

To implement the DDoS-attack we installed the special software called Torshammer.

After running Torshammer, a window with a list of commands appears, with which we could specify the IP address of the server we want to attack, select the port, the number of threads, etc. To start the attack, the following command was entered:

./torshammer.py -t 10.0.2.4 -r 2000

As you can see, the address of the "victim" was specified in this command - 10.0.2.4 and the number of threads - 2000.

As can be seen from Fig. 4, the amount of incoming traffic on the graph increased to 120–140 Kb/s. Thus, we can say that at this point the server is under a DDoS attack.

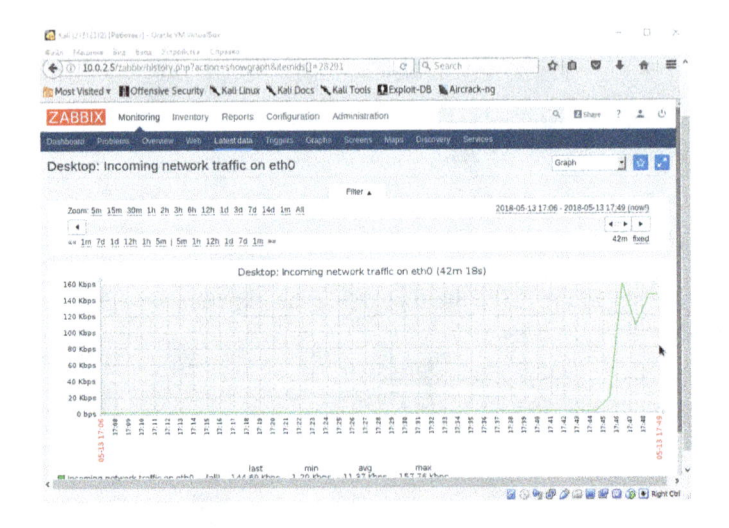

Fig. 4. Inbound traffic after an attack

4.4 The Organization of Protection from DDoS-Attacks

From the side of the virtual machine, on which the DDoS attack was organized, the connection to this port was checked (Fig. 5).

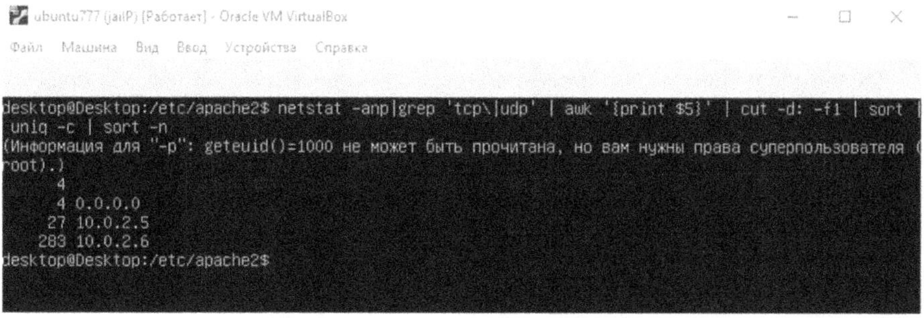

Fig. 5. Checking connections to the port

This command displays the number of attempts to *tcp* and *udp* connections since the last time it was turned on. Thus, you can determine how often we get a "knock" from the IP-address of the machine with which the DDoS-attack is carried out.

At this point, the attacker is recognized, it could be limited to banning traffic exchange coming from this IP-address, the so-called blacklisting, but this does not solve the main problem - the organization of protection against DDoS attacks.

Next, we configure the filtering of packets of incoming traffic. In the Linux operating system, this setting is done using the *Iptables* utility, which can configure the Kernel core to filter packets of incoming traffic.

In *Iptables*, the following tables are used to implement various tasks:

(1) The Filter table is one of the most widely used tables in *Iptables*. The filter table is used to decide whether to send the packet to its destination or to reject its request. In the language of the Firewall, this is called "filtering" the packets.
(2) The *NAT* table is used to implement network address translation rules. When packets enter the network stack, the rules in this table determine whether and how to change the source or destination addresses of the packet to affect how the packet is routed. This is often used to route packets on the network when direct access is not possible.
(3) The *Mangle* table is used to change the IP headers of a packet in various ways. For example, you can set the TTL (Time to Live) value of a packet by increasing or decreasing the number of valid network transfers that the package can support. Similarly, you can change the other headers of the IP protocol.
(4) Table *Raw*. The *iptables* has the status of *stateful*, which means that packets are evaluated relative to their relation to previous packages. The connection tracking functions built on top of the netfilter framework allow iptables to view the packets as part of the current connection or session, rather than the flow of discrete,

disjointed packets. Logic of connection tracking is usually applied very soon after the packet enters the network interface.

Considering this and analyzing the information known about how DDoS attacks work, a list of rules for configuring incoming packets was prepared. In these rules, the allowed size of incoming packets is changed, the waiting time for receiving a TCP connection confirmation, the number of allowed TCP and UDP streams is reduced, etc. Figure 6 shows a screenshot of the Linux operating system configuration file.

To add these rules in the *Iptables* configuration, the *sysctl.conf* file was changed, which is located in the directory etc. To edit this file, the following command was entered into the terminal: *sudo nano /etc./sysctl.conf.*

Fig. 6. Iptables configurations on the Linux operating system.

After the changes were made, the DDoS attack was started again. And this time it was possible to observe that the amount of incoming traffic decreased and fluctuates around 20–25 kb/s. Accordingly, we can conclude that the filter works correctly, rejecting the forbidden packets. Figure 7 shows a screenshot, which shows that incoming traffic has significantly decreased.

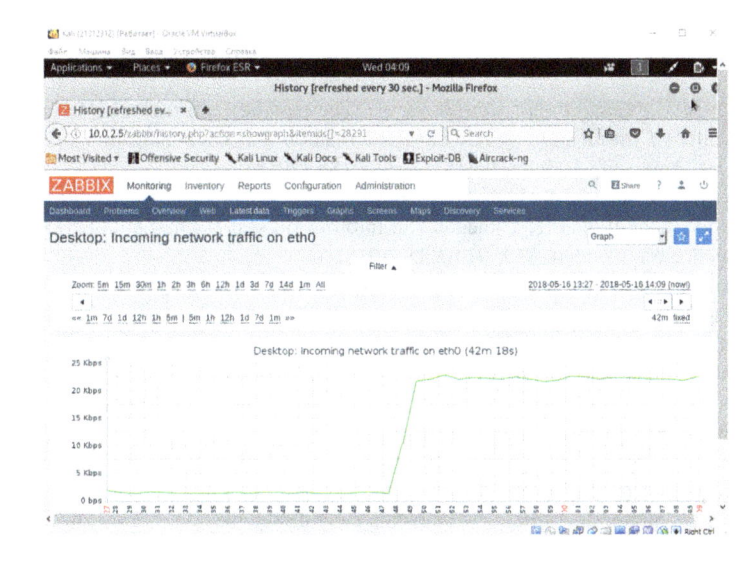

Fig. 7. Inbound traffic after protection.

5 Conclusion

Every year, organizations that provide services on the Internet allot more and more financial resources to protect the information security of their systems. DDoS-attacks are very popular type of attacks on information resources and are fraught with financial losses. Practice shows that most businesses are afraid of this type of attack on their information resources. Today, many enterprises have already experienced DDoS-attacks, and, of course, the protection against such attacks is quite an urgent issue.

Methods described in this article were applied on the developed software system based on the RED and GREEN approaches, as a protection against DDoS attacks. More details about these methods and approaches of providing reliability and security of information systems were described in works [13–26].

This article presents the results of reducing information security risks, namely, distributed attacks on the developed software system. The study gathered the network of virtual machines. These machines performed various tasks. The study has shown that configuration changes have reduced the port's load (the result: the graphs built by the Zabbix monitoring system). Initially, the graphs showed values in the region of 120–140 kb/s, and with the settings changed showed 20–25 kb/s. From here it can be concluded that the load of the port by incoming packets has decreased by almost 6 times.

This paper presented only some ways of protecting against distributed attacks of denial of service, and we are working on improving them. Presented methods showed their effectiveness. The "Workspace" (polygon) was organized to test the methods and approaches. It can also be used to test other ways of protecting against DDoS attacks.

References

1. Boranbayev, S., Goranin, N., Nurusheva, A.: The methods and technologies of reliability and security of information systems and information and communication infrastructures. J. Theoret. Appl. Inf. Technol. **96**(18), 6172–6188 (2018)
2. Boranbayev, A., Boranbayev, S., Nurusheva, A.: Development of a software system to ensure the reliability and fault tolerance in information systems based on expert estimates. Adv. Intell. Syst. Comput. **869**, 924–935 (2018)
3. Grantham Lough, K., Stone, R.B., Tumer, I.: Prescribing and implementing the risk in early design (RED) method. In: Proceedings of DETC 2006, Number DETC2006-99374, Philadelphia, PA, September 2006
4. Grantham Lough, K., Stone, R.B., Tumer, I.: The risk in early design (RED) method: likelihood and consequence formulations. In: Proceedings of DETC 2006, Number DETC2006-99375, Philadelphia, PA, September 2006
5. Krus, D.A.: The risk mitigation strategy taxonomy and generated risk event effect neutralization method, Doctoral dissertations (2012)
6. Lough, K.G., Stone, R., Turner, I.: The risk in early design method. J. Eng. Des. **20**(2), 155–173 (2009)
7. Boranbayev, A., Boranbayev, S., Yersakhanov, K., Nurusheva, A., Taberkhan, R.: Methods of ensuring the reliability and fault tolerance of information systems. In: 15th International Conference of Information Technology, Information Technology – New Generations, pp. 729–730 (2018)
8. Krus, D.A., Grantham Lough, K.: Risk due to function failure propagation. In: International Conference on Engineering Design, ICED 2007, pp. 787–788 (2007)
9. Krus, D.A., Grantham Lough, K.: Function-based failure propagation for conceptual design. In: Artificial Intelligence for Engineering Design, Analysis and Manufacturing, pp. 409–426. Cambridge University Press (2009)
10. Krus, D., Grantham, K.: Failure prevention through the cataloging of successful risk mitigation strategies. J. Fail. Anal. Prev. **13**, 712–721 (2013)
11. Krus, D., Grantham, K.: Generated risk event effect neutralization: identifying and evaluating risk mitigation strategies during conceptual design. In: INCOSE 2012, Rome, pp. 1225–1237 (2012)
12. Raghavan, S., Dawson, E.: An Investigation into the Detection and Mitigation of Denial of Service (DoS) Attacks: Critical Information Infrastructure Protection. Springer, Heidelberg (2011)
13. Boranbayev, A., Boranbayev, S., Nurusheva, A., Yersakhanov, K.: The modern state and the further development prospects of information security in the Republic of Kazakhstan. In: 15th International Conference of Information Technology, Information Technology – New Generations, pp. 33–38 (2018)
14. Boranbayev, A.S.: Defining methodologies for developing J2EE web-based information systems. J. Nonlin. Anal. Theory Methods Appl. **71**(12), 1633–1637 (2009)
15. Boranbayev, A., Mazhitov, M., Kakhanov, Z.: Implementation of security systems for prevention of loss of information at organizations of higher education. In: Proceedings of the 12th International Conference on Information Technology: New Generations (ITNG 2015), 13–15 April 2015, Las Vegas, Nevada, USA, pp. 802–804 (2015)
16. Boranbayev, S., Boranbayev, A., Altayev, S., Seitkulov, Y.: Application of diversity method for reliability of cloud computing. In: Proceedings of the 2014 IEEE 8th International Conference on Application of Information and Communication Technologies-AICT2014, Astana, Kazakhstan, 15–17 October 2014, pp. 244–248 (2014)

17. Turskis, Z., Goranin, N., Nurusheva, A., Boranbayev, S.: A fuzzy WASPAS-based approach to determine critical information infrastructures of EU sustainable development. Sustainability (Switzerland) **11**(2), 424 (2019)
18. Turskis, Z., Goranin, N., Nurusheva, A., Boranbayev, S.: Information security risk assessment in critical infrastructure: a hybrid MCDM approach. Informatica (Netherlands) **30**(1), 187–211 (2019)
19. Boranbayev, A., Boranbayev, S., Nurusheva, A.: Analyzing methods of recognition, classification and development of a software system. Adv. Intell. Syst. Comput. **869**, 690–702 (2018)
20. Boranbayev, A., Shuitenov, G., Boranbayev, S.: The method of data analysis from social networks using apache Hadoop. Adv. Intell. Syst. Comput. **558**, 281–288 (2018)
21. Boranbayev, A., Boranbayev, S., Nurusheva, A., Yersakhanov, K.: Development of a software system to ensure the reliability and fault tolerance in information systems. J. Eng. Appl. Sci. **13**(23), 10080–10085 (2018)
22. Boranbayev, S., Nurkas, A., Tulebayev, Y., Tashtai, B.: Method of processing big data. Adv. Intell. Syst. Comput. **738**, 757–758 (2018)
23. Akhmetova, Z., Boranbayev, S., Zhuzbayev, S.: The visual representation of numerical solution for a non-stationary deformation in a solid body. Adv. Intell. Syst. Comput. **448**, 473–482 (2016)
24. Boranbayev, S.N., Nurbekov, A.B.: Development of the methods and technologies for the information system designing and implementation. J. Theoret. Appl. Inf. Technol. **82**(2), 212–220 (2015)
25. Hritonenko, N., Yatsenko, Y., Boranbayev, S.: Environmentally sustainable industrial modernization and resource consumption: is the Hotelling's rule too steep? Appl. Math. Model. **39**(15), 4365–4377 (2015)
26. Akhmetova, Z., Zhuzbayev, S., Boranbayev, S., Sarsenov, B.: Development of the system with component for the numerical calculation and visualization of non-stationary waves propagation in solids. Front. Artif. Intell. Appl. **293**, 353–359 (2016)

A Novel Image Steganography Using Multiple LSB Substitution and Pixel Randomization Using Stern-Brocot Sequence

Md. Abdullah Al Mamun[1], S. M. Maksudul Alam[1], Md. Shohrab Hossain[1(✉)], and M. Samiruzzaman[2]

[1] Department of CSE, Bangladesh University Of Engineering and Technology, Dhaka, Bangladesh
{1405077.muam,1405087.smma}@ugrad.cse.buet.ac.bd,
mshohrabhossain@cse.buet.ac.bd
[2] Research scientist, BioNanoTech, London, UK
samiruzzaman@gmail.com

Abstract. Steganography is a technology of hiding confidential information within regularly used data files such as image, audio, video files. There have been several research works on the image steganography. Most of the works used only single LSB or multiple LSB of RGB components of each pixel. However, using only one LSB of the color components is an obstacle to convey much information. Again, if the pixels are not selected randomly, then the pixels used in steganography can be predicted very easily, thereby exposing the hidden information. In this paper, we have proposed a novel approach to inject information into an image in such a way that it is very difficult (almost impossible) for others (except sender and receiver) to detect and extract the hidden information within an image. We have encrypted our secret message before embedding it within an image. Pixels of the image have been selected in a randomized way using Stern-Brocot Sequence. Multiple LSBs of colors components (RGB) are used while inserting the encrypted message into the randomly selected pixels. We have implemented our proposed algorithm and showed our experimental results on the basis of PSNR and MSE for the stego images. Our results show that the quality of the images changes with different size of our secret messages and the quality of the stego image does not change significantly after applying our proposed algorithm. We have successfully extracted our secret messages from the stego images at the receiver side. Our proposed algorithm of image steganography can be useful for establishing a secret communication over a public communication using that public communication. As our proposed algorithm provides two step protection layer, it is quite impossible to extract the actual secret message from the image.

Keywords: Steganography · Stern-Brocot sequence · Stego image · LSB method · PSNR · MSE

© Springer Nature Switzerland AG 2020
K. Arai et al. (Eds.): FICC 2020, AISC 1129, pp. 756–773, 2020.
https://doi.org/10.1007/978-3-030-39445-5_55

1 Introduction

Steganography is getting popular day-by-day because of its capacity of providing highest security to data. In this era of communication, information security is one of the biggest challenges. The modern digital technologies are expected to protect highly confidential information. Steganography opens a new window in the field of privacy and security. Steganography is a method of hiding secret data or messages within ordinary data. Steganography can be used to keep data more secure as it provides higher level of privacy to the confidential data. Using steganography, it is difficult to intercept data that is sent to the receiver without knowing the exact algorithm used in it. Moreover, it is almost impossible to extract the data hidden within images.

Normally, media files (image, audio, video files which are large in size) are used in steganographic transmission. The image steganography is the most popular and the easiest method of all. Image steganography refers to hide the information within regular images. The sender embeds the secret messages within an image as a stream of bits. The change (due to embedding of secret messages) in the image file is so subtle that someone, who is not specifically looking for the secret message, is unlikely to notice the change in image file. Thus, It provides the opportunity to send secret information over a regular communication channel.

An Image consists of a number of pixels and each pixel consists of a number of bits within its color components. Replacing the least significant bits of the color components in the pixels within an image does not affect the overall quality of the image significantly. Actually, the effect cannot be easily detected without the help of technology. That is why, if a number of least significant bits are used to convey an information, the information can easily be hided within the image and it cannot be understood. In this paper we have proposed an algorithm using multiple LSBs substitution and randomized pixel selection with the help of Stern-Brocot sequence along with encryption-decryption for embedding a secret message within an image.

There have been several research works on the image steganography. Most of the works have introduced the usage of only a single LSB of RGB components of each pixel [1–5] or only multiple LSB [6] of RGB components of each pixel or single LSB [7] of RGB components of each pixel with pixel randomization. Zhou et al. [8] has used cryptography for performing image steganography. However, those works do not include the method of using multiple LSB substitution along with encryption-decryption and randomized pixel selection altogether. Using only one LSB of the color components of every pixel is an obstacle to convey much information. Again if the pixels are not selected randomly then the pixels used in steganography can be predicted very easily and then the hidden information can be retrieved easily. So the goal of steganography fails.

To the best our knowledge, there has been no earlier work on replacing multiple LSB along with encryption-decryption of the secret message with randomized pixel selection using Stern-Brocot sequence altogether. Laskar et al. [7] is the only previous work that proposed a method to use only one LSB of color components by selecting the pixels randomly within image for embedding hidden message.

The major *limitations* of these existing works are that they do not contains all possible protection layer for a secret message. Therefore, it is essential to propose an appropriate algorithm which ensures the maximum security for the secret message sent from a sender side to a receiver side. In that case we need an algorithm that consists of all possible protection layer to embed the secret message within the image.

We have proposed a secured image steganography algorithm that will provide two step protection layer. First layer is the encryption-decryption of the secret message and the second layer is randomized pixel selection within an image.

The *objective* of this work is to perform steganography in such a way that ensures highest possible security to the embedded secret message within an image so that no intruder can extract the original message within the image. It also increases the efficiency of using space within image as we are using multiple LSBs so that more information can be embedded within the image. So, the extension of steganography with the encryption-decryption along with the use of multiple LSBs and the way of our pixel selection makes the steganography much more reliable.

The main *contributions* of our work are as follows:

- We have proposed a secure algorithm that provides two step protection layer, i.e., encryption-decryption and randomly pixel selection using Stern-Brocot Sequence, to the embedded secret message within image.
- We have implemented the proposed algorithm using Java programming language.
- Observing the PSNR and MSE of the stego image (the image after embedding hidden messages) the findings of our work is that quality of the images does not change significantly by applying all the steps of our algorithm to different images.

Our proposed algorithm ensures high level security of the secret message so that no intruder can extract the secret message from the image despite of knowing that there is a hidden message within the image as the intruder does not know in which sequence the pixel were exactly selected while embedding the messages, it will be much more secured.

Our results show that there is a very little difference between original images and stego images which cannot be detected with eyes. We have implemented the algorithm and experimented on some different images using (i) 2 bit LSB, (ii) 3 bit LSB, (iii) 3 bit LSB with random pixel selection using Stern-Brocot Sequence. We have done the quality estimation with PSNR on the stego images which is mostly used in order to measure the quality of reconstruction of lossy compression codes (e.g., for image compression). The signal in this case is the original data, and the noise is the error introduced by compression. We have also showed mean squared error (MSE) of the stego image. PSNR is most easily defined via the mean squared error. Despite of using multiple LSB and pixel randomization, Peak signal-to-noise ratio (PSNR) of stego images are satisfactory which ensures a little difference between original image and stego image. As a result, we can send secret message efficiently within an image in a more secured way than that of previous. The main advantage is that if any intruder knows about the hidden data within the image but it is about impossible for the intruder to crack the hidden secret message from the image which ensures high level security as the secret message was encrypted from the sender side.

The rest of the paper is organized as follows. Section 2 describes some basic terminologies of image steganography. In Sect. 3, we list previous research works on image steganography, followed by the proposed approach in Sect. 4. Section 5 contains the algorithm of our proposed method. In Sect. 6, we present the results showing the performance of our algorithm. Finally, Sect. 7 has the concluding remarks.

2 Terminologies

There are some important terminologies which will be used throughout this paper for describing our proposed algorithm on image steganography. They are discussed below in details.

2.1 Stern-Brocot Sequence

Stern-Brocot Sequence can be generated from stern-Brocot tree. The very first two elements of Stern-Brocot Sequence is 1, 1. Then adding the first pair of the sequence gives 2 (1 + 1) and then copying the last one of that pair after it, yields a sequence 1, 1, 2, 1. Then the next pair is 1, 2. We get 3 by adding them and copy the last one of the pair, which is a 2. So, now the sequence is 1, 1, 2, 1, 3, 2. Now the pair is 2, 1 and if they are added, we get another 3. Then copying the last one of the pair is a 1. It yields the sequence 1, 1, 2, 1, 3, 2, 3, 1. The numbers of this infinite sequence is called Stern-Brocot Number. Now from the beginning of the Stern-Brocot numbers and we take them in pairs to form a fraction. Let's take the first one of the fraction. That's $\frac{1}{1}$. The next one as a fraction is $\frac{1}{2}$. So, we are taking them, numerator-denominator, all the way down in pairs. The next one's $\frac{2}{1}$ and then $\frac{1}{3}$, $\frac{3}{2}$, $\frac{2}{3}$ and so on. If we carry that out then it will list out every possible fraction. This will list every single ratio of whole numbers. It will always list them in their most simplified form and it will only ever list them once each. So no fraction will appear twice. That means if there is $\frac{1}{2}$; no other fraction will ever appear which equals $\frac{1}{2}$. This sequence will give us every single rational number in existence.

2.2 Stego Image

Stego image is the result of a image steganography process. Information is hidden behind the simple appearance of a stego image. The pixel bits of a stego image are manipulated and the information is injected using different ways of image steganography. The pixel bits are changed in such a way that normal human eye sight can easily ignore the changes. So we can say that a stego image is an image with the hidden message.

2.3 LSB Method

All the pixels of an image are consist of the binary data of the image and each digit of the binary data is called a single bit. According to significance, the left most bit of a bit stream is most important because it contributes most to the value of the bit stream. The significance gradually decreases from the left to the right of the bit stream. Hence, the right most bit of a bit stream is the least significant bit and it contributes the least to the values of the bit stream. So if we want to change the bit stream of pixels of an image, least significant bits (LSBs) are our best option. In LSB method, we change the LSBs of bit streams of pixels to inject our message. That is why, changing LSBs of bit streams of pixels does not affect the appearance of our image and the message is embedded into the stego image.

2.4 PSNR (Peak to Signal Noise Ratio)

PSNR refers to peak signal-to-noise ratio. It is the logarithm of the ratio of squared maximum pixel value and MSE value. The measurement is used to compare the quality of a reconstructed noisy image. We have constructed injecting information into it. So there will be some noise in our stego image. Our goal is to keep the PSNR value high as much as possible. The higher PSNR value means the higher quality of image.

2.5 MSE (Mean Square Error)

MSE means mean squared error. It measures the average of the squares of the errors. It is the average squared difference between the estimated value and the actual value of a pixel. It is also a measure of quality of a constructed noisy image. The lower the value of MSE the higher the quality of the image. So, we have tried to keep the MSE value of our stego image as lower as possible.

3 Previous Works

A few works have been performed in recent years that are closely related to our image steganography. Liao et al. [9] have proposed a new medical JPEG image steganographic scheme based on the dependencies of inter-block coefficients. Rahim et al. [10] discusses about image steganography technique by using encoding method base64 that converts the same binary data to the form of a series of ASCII code. Sheidaee et al. [11] proposed an image steganography algorithm to embed the secret message based on Discrete Cosine Transform

(DCT) compound and LSB method. Arun et al. [12] introduces image Steganography using the method of LSB XOR Substitution. Surse et al. [13] explains the use of data compression technique DWT along with different transforms and algorithms in Image Steganography. Yadav et al. [14] provides a suitable method about 3-Level Security Based Spread Spectrum Image Steganography. Jain et al. [15] provides method of Mask Encryption Based Highly Secure Image Steganography. Zhou et al. [8] have proposed a LSB based color image steganography technique using cryptography. Laskar et al. [7] proposed a method to work on the randomized selection of the pixels in a image but they have used only one LSB of color components. Trivedi et al. [16] the analysis of Several Image Steganography Techniques in spatial Domain. Subhedar et al. [17] explains current status and key issues in image steganography. Swain et al. [18] analyzes variable group of bits substitution. Patel et al. [6] has used multiple LSB of colors of each pixels that are used instead of single LSB of colors of each selected pixels. They have discussed the technique of using multiple LSBs of colors of selected pixels and some issues related to attacks.

Sugathan and Sherin [1] introduces an algorithm for LSB replacement for RGB color images. Arora et al. [2] proposed an idea was about to substitute the LSBs of color components of the pixels of image. Bhardwaj et al. [3] has used three levels of protection that is given to the secret message using steganography. Jiang et al. [4] combines the MSB matching and LSB substitution. Akhtar et al. [5] has actually provided the improved LSB method of image steganography. The technique uses modulus function for hiding the secret message. Aljamea et al. [19] explains how to detect hidden url in all types of images. They have proposed an algorithm about how to detect hidden URL within images using single LSB substitution.

4 Proposed Approach

We have embedded the data byte within an image in a predefined order in sender side and we have extracted the data in the same order in which it was inserted within stego image in the receiver side. So no way of guessing the actual order except sender and receiver.

We have used 24 bit color image. 3 LSB bits for each of the color components of RGB in a pixel have been used. So, we needed only one pixel in order to store 9 bits. Thus one pixel is enough for inserting one character (8 bits) of texts. Suppose, we want to embed a message "em" within an image and our encryption-decryption key is 2. So, now the encrypted message will be "go". For maintaining consistency 1 byte for each character have been used always. The bit pattern of character 'g' is (01100111) and 'o' is (01101111). Total 16 bits message bit pattern is (0110011101101111). So the message of bit pattern is 16 bits. In addition, We have also sent the total of our message using first 32 bits of our message within the image. That means first 32 bits of our message implies how many bit stream we have to read from the image using the predefined sequence from sender side. After reading the whole bit streams it should be kept in mind that our encrypted hidden message starts from after 32 bits of the whole bit stream read from the image. So we need total 48 bits: 32 bits for specifying message bit and 16 bits for hidden message "go". So the bit patterns to be sent is (000000000000000000000000110000011001110101111) where the first left 32 bits specifies the total length of the bit streams sent from the sender to receiver (here bit pattern is 48) and next 16 bits specifies the message bits. To send 48 bits, we need to select minimum 6 pixels since each pixel can store 9 bits. Suppose, We have chosen six pixels randomly using Stern-Brocot tree $P_{2,1}, P_{1,3}, P_{3,2}, P_{2,3}, P_{3,1}, P_{1,4}$, respectively which will be generated from stern brocot tree.

$$P_{2,1} = \begin{cases} 10110\mathbf{000} \ (R) \\ 11101\mathbf{000} \ (G) \\ 10000\mathbf{000} \ (B) \end{cases}$$

$$P_{1,3} = \begin{cases} 11110\mathbf{000} \ (R) \\ 10010\mathbf{000} \ (G) \\ 10010\mathbf{000} \ (B) \end{cases}$$

$$P_{3,2} = \begin{cases} 00110\mathbf{000} \ (R) \\ 11111\mathbf{000} \ (G) \\ 10100\mathbf{100} \ (B) \end{cases}$$

$$P_{2,3} = \begin{cases} 10111\mathbf{101} \ (R) \\ 00101\mathbf{100} \ (G) \\ 10000\mathbf{000} \ (B) \end{cases}$$

$$P_{3,1} = \begin{cases} 10111\mathbf{110} \ (R) \\ 10101\mathbf{001} \ (G) \\ 10101\mathbf{111} \ (B) \end{cases}$$

$$P_{1,4} = \begin{cases} 11110\mathbf{011} \ (R) \\ 00101\mathbf{001} \ (G) \\ 11110\mathbf{111} \ (B) \end{cases}$$

Thus, message bits are embedded within the image from the sender side and the image is sent to the receiver side. The receiver receives the stego image and generates the indexes of the pixel using Stern-Brocot Sequence. According the randomly generated index of the pixel, Receiver determines the length of the total message from first 32 bits (contains the total length of bits to be read) from first 4 pixels as each pixel contains 9 bits of hidden message in its LSB. Then receiver ignores first 32 bits as it specifies the total size of message bit stream. Thus An appropriate receiver receives message very securely.

4.1 Sender Side Flow Chart

The sender side flowchart has been shown in Fig. 1. Sender sends the secret message within an image in a secured way. That is why the sender follows a sequence of procedures for providing highest security to the secret message for sending it to the receiver.

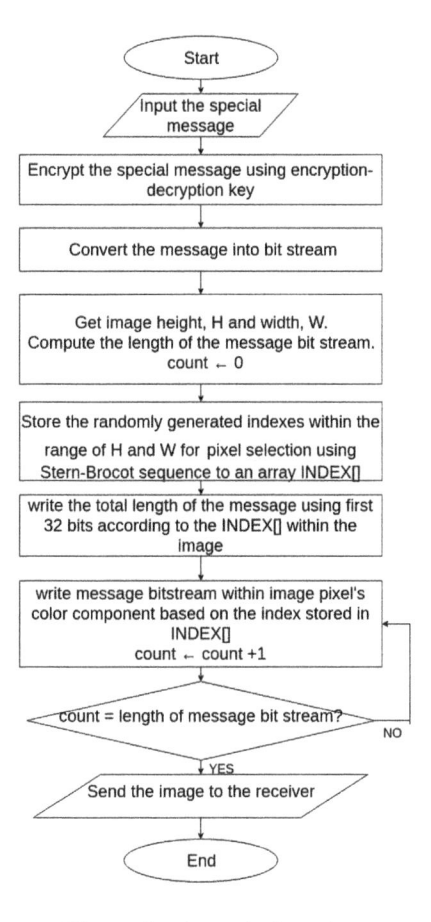

Fig. 1. Sender side flow chart

The steps used by the Sender are as follows:

- **Firstly,** Sender will encrypt the secret message with a predefined specified key which is also known to the receiver.
- **Secondly,** Sender converts the encrypted secret message into bit streams. Then sender finds out the image details and computes the total length of the encrypted message.
- **Then** Sender generates the indexes for randomized pixel selection within the image using Stern-Brocot sequence according the image details and stores all of those indexes into an array INDEX[].
- **After that** Sender uses first 32 bits of the pixels selected from INDEX[] for specifying the total size of the secret message so that receiver can understand how much bit need to read. After 32 bits the bit stream of secret message starts.
- **Finally,** after embedding the secret message along with its size within the image, Sender sends the stego image to the receiver.

4.2 Receiver Side Flow Chart

In order to extract the secret message from the stego image at the receiver side, The receiver also follows a sequence of procedures to retrieve the actual secret message that is shown in Fig. 2 below:

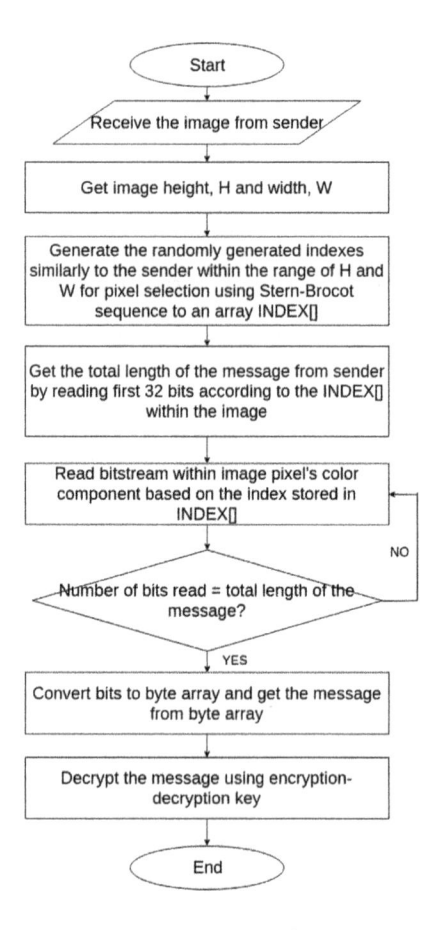

Fig. 2. Receiver side flowchart

The steps used by the Receiver are as follows:

- **Firstly,** Receiver receives the stego image from the sender and reads the basic information such as height, width of the stego image.
- **Secondly,** The receiver generates the indexes for pixel selection just similar to the sender side using Stern-Brocot sequence to extract the secret message from the stego image and stores the sequence of pixel in an array INDEX[].
- **Then** the receiver reads first 32 bits according the pixel sequence stored in INDEX[] for extracting the total length of the secret message sent from the sender.

- **After that** Receiver reads the specific bits from the pixels' bit stream according the INDEX[]. Receiver continues to read the data until the length of the extracted bit stream is equal to the length of the secret message.
 - **Finally,** Receiver decrypts the secret message with the predefined key which is only known to the sender and receiver itself.

After successfully completing all of the procedures described above, Receiver successfully finds out the actual secret message that was hidden in the stego image.

5 Proposed Algorithm

In color imaging systems, a color is typically represented by three component intensities such as red (R), green (G), and blue (B). Message can be embedded within images by choosing pixels in a random manner and inserting the messages as a stream of binary data bits using multiple bits from least significant bit of RGB for each of the randomly selected pixels. We have divided our algorithm into six steps at the Sender Side

Step 1: Encrypt and prepare the secret Message
Step 2: Get image details
Step 3: Create random Index for pixel selection
Step 4: Insert the encrypted message within images
Step 5: Send the stego image to the Receiver
Step 6: Extract the actual message in the receiver side

5.1 Encrypt and Prepare the Secret Message

The sender and receiver selects a encryption-decryption key, denoted by α_K, with their early consent. α_K will not be embedded within secret message. So, it is impossible for others to know the α_K except sender and receiver. Thus the security issue becomes harder. Sender encrypts the secret message using α_K and generates the encrypted message α_M. So the sender does the followings:

 - **Firstly,** Create an array of characters named as *message*[] for storing encrypted secret message, α_M, which will have to be sent to the receiver within image.
 - **Secondly,** Convert the ASCII value of each character of $\alpha_M(message[])$ into 8 bit streams because this will help us to convert the secret message(α_M) in a byte array. Suppose, α_M contains the character 'a'. The ASCII value of 'a' is 97 and the bit pattern of 'a' is "1100001" which is 7 bit. As it has 7 bits, it will be padded with one leading zero to fulfil the criteria. After padding, the bit pattern of 'a' will be "01100001".
 - **Next,** Store the bit stream of α_M into *S1*[].

– **Finally,** Create an array of characters $S2[]$ that contains the length of bit stream of $S1[]$. The length of the bit stream stored in $S2[]$ will be 32 always. Hence, padding with leading zeros may be required in the bit stream of $S2[]$. Construct final message array, $M[]$, by appending $S1[]$ at the end of $S2[]$. So, $M[] = S2[] + S1[]$. Here, $M[]$ is the encrypted message with its total size in the 1st 32 bits.

5.2 Get Image Details

Read the detail of image into which the message will have to be inserted. Get height and width of the image. Using the height and width of the image, define a two dimensional array $P[][]$ which will be used to store the whole image.

5.3 Generate Random Indexes for Pixel Selection

Create an array $Index[]$ which will be used to store the randomly generated index from Stern Brocot Sequence (generates all possible unique fractions each of whose numerator and denominator are co-prime) discussed in Sect. 2.1. As each image is a sequence of pixels which can be represented in a two dimensional array. We get each fraction only once from the Stern-Brocot sequence. We have used numerator of the fraction to select the row and denominator of the fraction to select the column of our desired pixel. Stern-Brocot Sequence is 1, 1, 2, 1, 3, 2, 3, 1, 4, 3, 5, 2, 5, 3, ... and so on. The Stern-Brocot Sequence of length nine is 1, 1, 2, 1, 3, 2, 3, 1, 4. So, the fractions generated from Stern-Brocot tree will be in following order $\frac{1}{1}, \frac{1}{2}, \frac{2}{1}, \frac{1}{3}, \frac{3}{2}, \frac{2}{3}, \frac{3}{1}, \frac{1}{4}$ and so on, respectively. Now the pixels will have to be selected in the following order: $P_{1,1}, P_{1,2}, P_{2,1}, P_{1,3}, P_{3,2}, P_{2,3}, P_{3,1}, P_{1,4}$ where $P_{i,j}$ means the pixel of ith row and j^{th} column of the two dimensional array of image. At $P_{i,j}$, if the value of 'i' exceeds the height of image or if the value of 'j' exceeds the width of the image then the particular $P_{i,j}$ generated from Stern-Brocot sequence will be ignored. Stern-Brocot Sequence generates all possible unique fractions each of whose numerator and denominator are co-prime. For example, Stern-Brocot sequence will not generate fraction like $\frac{2}{2}$, $\frac{4}{4}$ etc. But they are also valid index for pixel. So if the sender or receiver needs more pixels to select then they can use the remaining unused pixel of the image sequentially which are not generated in Stern-Brocot Sequence. Moreover, Instead of selecting those remaining unused pixels sequentially, Another predefined sequence can be used to select those remaining unused pixels for introducing second stage pixel randomization with the consent of both sender and receiver. This is the unique advantage of our proposed method compared to previous works because we are introducing double stage randomization for image pixel selection along with encryption-decryption which ensures highest security of our secret message to receiver. If the attackers can steal the encryption and decryption key then still they will not be able to use the same algorithm as the receiver to crack the hidden information because of this two stage pixel randomization where the predefined sequence of this second stage randomization is only known to by sender and receiver.

5.4 Insert the Encrypted Message Within Images

Index[] contains the elements of Stern-brocot sequence. So take the each consecutive pair of *Index*[] sequentially where the first and second element of the pair denotes the row number and column number of the pixel respectively. Thus the pixel is selected by using Stern-Brocot sequence.

Write the total length of secret message using 32 bits before writing the encrypted secret message within the color component of a pixel of an image according to the specified sequence of pixel that is stored in *Index*[]. The sender has all required index in *Index*[] for pixel selection of that image.

- Get next valid index(i, j) where i, j denotes the row and column of the pixel from *Index*[].
- Get RGB component of the pixel P[i][j].
- 3 least significant bits of each color component can be replaced if needed. At first replace the 1st LSB for each color component of RGB with the bits stored in *M*[] (contains messages bit stream) sequentially.
- Then replace the 2nd LSB for each color component of RGB with the next unread bits stored in *M*[] sequentially also if needed.
- After that Replace the 3rd LSB for each color component of RGB with the next unread bits stored in *M*[] sequentially if needed.

Continue the last five steps repeatedly until all the bits in message bit stream, *M*[] are inserted within the image.

5.5 Send the Stego Image to the Receiver

When all bits of the bit stream in *M*[] gets embedded into the image, the sender sends the stego image to the receiver.

5.6 Extract the Actual Message in the Receiver Side

The following sequence of steps should be followed at receiver side to extract the secret message within the image.

- Receive the Stego image.
- Generate the index randomly from the Stern-Brocot sequence, denoted by *Index*[], just like the sender side to get the sequence of the pixel of image in appropriate order.
- Find the 1st 32 bits using the sequence of pixel stored in *Index*[]. From this 32 bits the receiver gets the total length of secret message. Now the receiver knows the total length of secret message. The receiver has all required index in *Index*[] for pixel selection of that stego image.
  - Get next valid index(i, j) where i, j denotes the row and column of the pixel from *Index*[].
  - Get RGB component of the pixel P[i][j].

- Read the 1st LSB for each color component of RGB of the selected pixel sequentially and store it in *Message*[].
- Then read the 2nd LSB for each color component of RGB of the selected pixel sequentially and store it in *Message*[].
- After that read the 3rd LSB for each color component of RGB of the selected pixel sequentially and store it in *Message*[].

The receiver will continue the last five steps until the number of bits read is not equal to the total length of secret message.

- Now, convert the bit stream *Message*[] into byte array, denoted by $B[]$. Now ignore the first four bytes as it specifies the total length of the message. So, read from the second byte where each byte represents a character of encrypted secret message that was sent from the sender side.
- Finally, Receiver decrypts each character of the byte array $(B[])$ with the decryption key which was decided by the consent of both sender and receiver. Then The receiver generates the actual secret message.

Thus receiver extracts the secret message from sender.

6 Results

We have implemented our proposed algorithm described in Sect. 5 using Java programming language. We have performed our experiment on some random images of different dimensions. Original images and stego images resulted from replacing 3 LSB of the color components in a pixel along with pixel randomization using Stern-Brocot sequence are shown below:

(a) Original Image.png   (b) Stego Image.png

Fig. 3. Original vs. Stego image with fish

The images shown in Figs. 3(a), 4(a), 5(a), 6(a) and 7(a) are the original images. Images shown in Figs. 3(b), 4(b), 5(b), 6(b) and 7(b) are the stego images resulted from replacing 3 LSB of the color components in a pixel along with pixel randomization using Stern-Brocot sequence for sending the secret message from sender to receiver. The length of actual message was 140 bytes in size that means total length of secret message was 144 bytes because first 32 bits(4 bytes) contains the total size of the secret message.

(a) Original Image.png   (b) Image.png

Fig. 4. Original vs. Stego image with a flower

(a) Original Image.png   (b) Stego Image.png

Fig. 5. Original vs. Stego image with a dog

(a) Original Image.png   (b) Stego Image.png

Fig. 6. Original vs. Stego image with a leaf

(a) Original Image.png   (b) Stego Image.png

Fig. 7. Original vs. Stego image with a tree

Fig. 8. Change of PSNR for embedding various length of secret message

In Fig. 8, The PSNR (peak to signal noise ratio) of images for different sizes of messages are shown in a bar chart. Here, We have performed our task for five different sizes of message that has been embedded into three different images in Figs. 3(a), 7(a) and 4(a), respectively for observing the PSNR of those images and finding out the difference of image quality by increasing the message size. Here, we have selected three images of different categories on the basis of colorfulness. We have embedded the message of sizes 140 bytes, 280 bytes, 420 bytes, 560 bytes and 700 bytes respectively. We can say from the result that the PSNR varies at very lower scale because of increasing the size of secret message in the case of most colorful images.

In Fig. 9, The MSE (Mean Square Error) of the images for different sizes of messages are shown in a bar chart. Here, We have performed our task for five different sizes of message that has been embedded into three different images in Figs. 3(a), 7(a) and 4(a), respectively for observing the change of image quality by increasing the message size. We have embedded the message of sizes 140 bytes, 280 bytes, 420 bytes, 560 bytes and 700 bytes, respectively. It can be noticed that the MSE varies at very lower scale because of increasing the size of secret message in the case of most colorful images. The Bar Chart is given below:

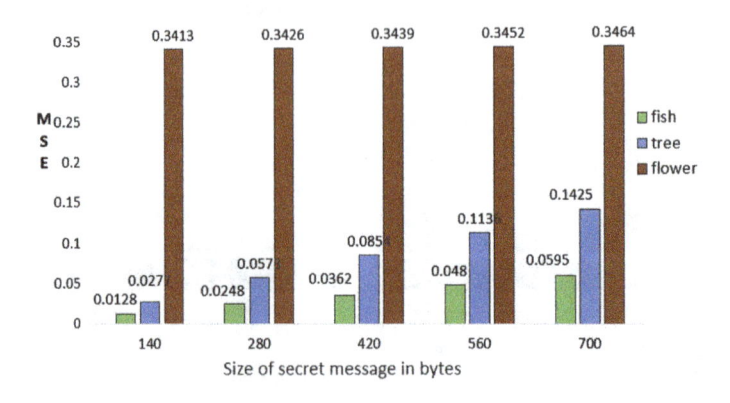

Fig. 9. Change of MSE for embedding various length of secret message

Table 1. PSNR and MSE of the different stego images

	2 bit LSB		3 bit LSB		3 LSB+Random	
	PSNR	MSE	PSNR	MSE	PSNR	MSE
Fish	76.33	0.0044	72.26	0.0116	71.08	0.0123
Flower	52.82	0.3400	52.81	0.3414	52.81	0.3414
Dog	75.67	0.0018	78.33	0.0047	70.67	0.0130
Leaf	72.23	0.0033	80.79	0.0104	69.94	0.0332
Tree	71.81	0.0104	75.33	0.0039	67.44	0.0277

In Table 1, we have showed the difference of PSNR and MSE values by replacing 2 LSB without pixel randomization, 3 LSB without pixel randomization and 3 LSB with pixel randomization. We have compared the results of our proposed image steganography method with the results of 2 LSB and 3 LSB substitution using our algorithm taking five different images. In the case of 2 LSB and 3 LSB substitution, the algorithm is same as our proposed method except the pixel randomization part.

Observing the data in Table 1, it can be stated that Using 3 LSB substitution along with encryption-decryption and randomized pixel selection with the help of Stern-Brocot sequence does not change the original image so much. We can see that the MSE is little bit higher in our proposed method than 2 LSB or 3 LSB substitution method. This is because our proposed method does not choose the pixels sequentially for replacing LSBs, rather it selects the pixels using a predefined algorithm in order to introduce randomization in pixel selection within the image. We are selecting the pixel randomly and the pixel can be different enough than its neighboring pixels. Then if the color component of that pixel is changed a little bit, in that case there is a certain possibility of little uprising the MSE of the stego image. Here, we have to compromise the protection and quantity of secret message with the quality of stego image.

7 Conclusion

In this paper, we have proposed a novel approach for embedding secret messages securely and efficiently in terms of space usage within an image. Since 3 LSB bit of each color component of a pixel has been used, a large message can be sent within a very small image without much changing the quality of image which is not identifiable in open eyes. Most importantly, since message has been sent to receiver by selecting pixel randomly within the image, no one can guess the sequence of selected pixel. The algorithm of pixel selection and encryption-decryption key are only known to the sender and the receiver, thereby providing high level security in message transmission. Thus, no intruder can extract the secret message. In spite of using both 3 LSB bits and selecting the pixels randomly generated from Stern-Brocot sequence, the PSNRs of the corresponding images have not increased significantly than that when only 3 LSB bits were used for message hiding. Even most of the time it has given better performance for selecting pixel index randomly. MSEs of the images was in tolerable limit in our experiment. As we are providing extra security level by selecting the pixel randomly, we have to compromise the picture quality a little bit. However, such degradation is not visible in open eyes.

Acknowledgment. This research was supported by SciTech Consulting and Solutions, Dhaka, Bangladesh.

References

1. Sugathan, S.: An improved LSB embedding technique for image steganography. In: 2nd International Conference on Applied and Theoretical Computing and Communication Technology (iCATccT), pp. 609–612. IEEE (2016)
2. Arora, A., Singh, M.P., Thakral, P., Jarwal, N.: Image steganography using enhanced LSB substitution technique. In: Fourth International Conference on Parallel, Distributed and Grid Computing (PDGC), pp. 386–389. IEEE (2016)
3. Bhardwaj, R., Sharma, V.: Image steganography based on complemented message and inverted bit LSB substitution. Proc. Comput. Sci. **93**, 832–838 (2016)
4. Jiang, W., Guo, Z., Wang, K., Huang, Y.: A self-contained steganography combining LSB substitution with MSB matching. In: 5th International Conference on Computer Science and Network Technology (ICCSNT), pp. 635–640. IEEE (2016)
5. Akhtar, N., Ahamad, V., Javed, H.: A compressed LSB steganography method. In: 3rd International Conference on Computational Intelligence and Communication Technology (CICT), pp. 1–7. IEEE (2017)
6. Patel, R.M., Shah, D.: Multiple LSB data hiding based on pixel value and MSB value. In: Nirma University International Conference on Engineering (NUiCONE), pp. 1–5. IEEE (2013)
7. Laskar, S.A., Hemachandran, K.: Steganography based on random pixel selection for efficient data hiding. Int. J. Comput. Eng. Technol. **4**(2), 31–44 (2013)
8. Zhou, X., Gong, W., Fu, W., Jin, L.: An improved method for LSB based color image steganography combined with cryptography. In: IEEE/ACIS 15th International Conference on Computer and Information Science (ICIS), pp. 1–4. IEEE (2016)
9. Liao, X., Yin, J., Guo, S., Li, X., Sangaiah, A.K.: Medical JPEG image steganography based on preserving inter-block dependencies. Comput. Electr. Eng. **67**, 320–329 (2018)
10. Rahim, R., Nurdiyanto, H., Hidayat, R., Ahmar, A.S., Siregar, D., Siahaan, A.P.U., Faisal, I., Rahman, S., Suita, D., Zamsuri, A., et al.: Combination base64 algorithm and EOF technique for steganography. J. Phys. Conf. Ser. **1007**(1), 012003 (2018)
11. Sheidaee, A., Farzinvash, L.: A novel image steganography method based on DCT and LSB. In: 9th International Conference on Information and Knowledge Technology (IKT), pp. 116–123. IEEE (2017)
12. Arun, C., Murugan, S.: Design of image steganography using LSB XOR substitution method. In: International Conference on Communication and Signal Processing (ICCSP), pp. 0674–0677. IEEE (2017)

13. Surse, N.M., Vinayakray-Jani, P.: A comparative study on recent image steganography techniques based on DWT. In: International Conference on Wireless Communications, Signal Processing and Networking (WiSPNET), pp. 1308–1314. IEEE (2017)
14. Yadav, P., Dutta, M.: 3-Level security based spread spectrum image steganography with enhanced peak signal to noise ratio. In: Fourth International Conference on Image Information Processing (ICIIP), pp. 1–5. IEEE (2017)
15. Jain, T., Shrotriya, A., Verma, V.K., Kumar, H.: Mask encryption based highly secure image steganography. In: International Conference on Intelligent Communication and Computational Techniques (ICCT), pp. 256–261. IEEE (2017)
16. Trivedi, M.C., Sharma, S., Yadav, V.K.: Analysis of several image steganography techniques in spatial domain: a survey. In: Proceedings of the Second International Conference on Information and Communication Technology for Competitive Strategies, p. 84. ACM (2016)
17. Subhedar, M.S., Mankar, V.H.: Current status and key issues in image steganography: a survey. Comput. Sci. Rev. **13**, 95–113 (2014)
18. Swain, G.: Digital image steganography using variable length group of bits substitution. Proc. Comput. Sci. **85**, 31–38 (2016)
19. Aljamea, M.M., Iliopoulos, C.S., Samiruzzaman, M.: Detection Of URL in image steganography. In: Proceedings of the International Conference on Internet of things and Cloud Computing, p. 23. ACM (2016)

Detecting PE-Infection Based Malware

Chia-Mei Chen[1], Gu-Hsin Lai[2(✉)], Tzu-Ching Chang[1], and Boyi Lee[1]

[1] National Sun Yat-Sen University, Kaohsiung 804, Taiwan
`cchen@mail.nsysu.edu.tw`
[2] Taiwan Police College, Taipei 116, Taiwan
`guhsinlai@gmail.com`

Abstract. Advanced Persistent Threat (APT) attacks are notorious in businesses and organizations, performed by highly organized, well-funded hacker groups against specific targets. Attackers employ advanced and customized attack tactics to invade target systems to gain access. DLL injection and PE Infection are common customized tactics to hide the presence of APT in order to retain control and unaware by users for a long duration. The average time for an APT to be discovered is one and a half years and some incidents lasted longer by using DLL injection or PE infection. Signature-based detection fails to discover such malware or unknown malware. This paper presents a hybrid approach combining static and dynamic analysis to discover misbehaviors. The experimental results show that the proposed approach could detect customized PE infected malware efficiently.

Keywords: Malware detection · DLL injection · Advanced Persistent Threat

1 Introduction

As organizations gradually rely more on the internet, they are increasingly vulnerable to cyber attacks. Even though the networks are equipped with layers of defense mechanisms such as intrusion detection system, firewall, anti-virus, and spam filter, the organizations still suffer from the targeted attacks.

Cybercrime has gone global and there has been a growth in the number of tailored attacks towards specific targets. New malware is constantly and rapidly emerging threats. The study [8, 10] indicated that malware increased with a growth rate of 50% each year and 88% of new malware are variants of the known one. Malware writers apply obfuscation skills to evade detection and customize malware based on the target environment. They want to avoid developing malware from scratch every time as well as evade detection, so they reuse attack code by hiding the similarities and obfuscating the code. One common stealth approach [15–17, 19] that malware writers have used these years is to infect executable files and inject injecting malicious code. A piece of malware discovered by the incident response team has infected a client application used throughout the institute and the anti-virus has not been able to identify it. Such infected file has the same file size as the original one, so it is very stealthy and hard for the user to notice it is infected.

As an illustration shown in Fig. 1, the attack applies binary patch tactic to infect an executable file and disguises it as a spam mail. Once a user opens the attachment, the malware injects the malicious code into the memory. Tools such as Ollydbg are available for performing the infection and malware injection. In order to retain the control over the victims, attackers may apply DLL injection technique to inject malicious code into normal processes, such as svchost or iexplore. As such approach crafts new customized malware, the malware would not be identifiable by signature-based anti-virus. An injection-aware malware detection is desire to identify such stealth attacks.

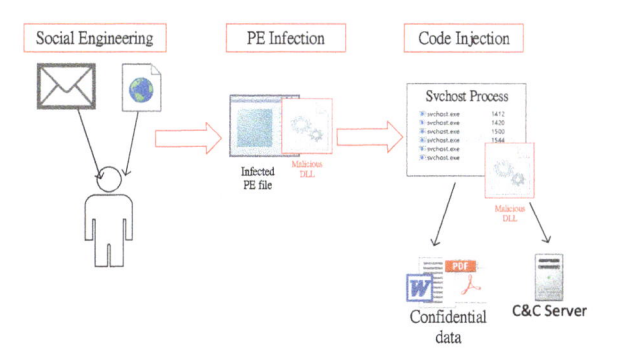

Fig. 1. An illustration of APT Attack.

For PE Infection attack, attackers inject DLL files in target PE files. It's difficult for detection mechanism like anti-virus software to detect either infected PE files or malicious DLL files. Figure 2 shows the content analysis of compromised PE file.

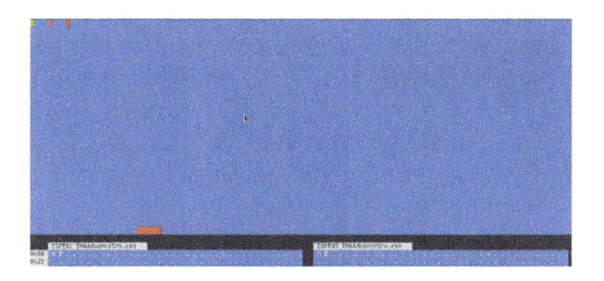

Fig. 2. Content analysis of compromised PE file

In Fig. 2, the red bar part is the section where the attacker modify. The infected file and original file are almost identical. Figure 3 shows the content what attackers inject into a normal PE file.

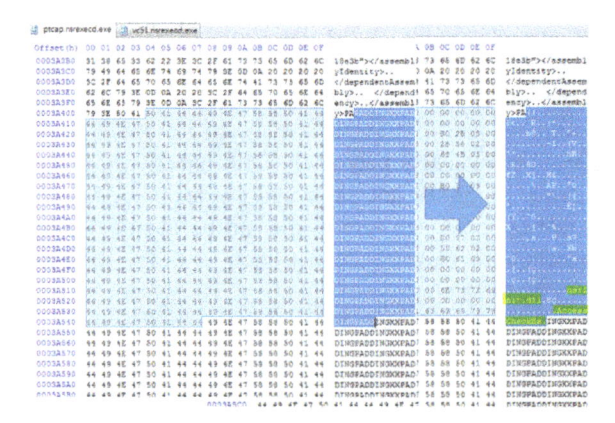

Fig. 3. An example of infected PE file

Figure 3 illustrates that attackers inject the malicious DLL into the PADDING (NOP) area(s). Antivirus could not find out infected PE files, because attackers only inject DLL names and modify pointers of PE files. There is not any specific string or patterns of infected PE files.

According to our preliminary study, a piece of malware [9] had infected an application commonly used in government agencies and anti-virus failed to identify it for over a year. This malware applying PE Infection attack technique has the same file size as the original one, so it is very stealthy and hard to be noticed. Based on the real security incident case studies [22], in order to retain the control over victim machines, attackers may apply DLL injection technique to inject malicious code into normal processes, such as svchost [2, 5] or explorer [6]. A lesson learned from the literature review and incident investigations is that attackers adopted the attack technique to craft the malware customized for specific target organizations and the malware successfully subverted the defense mechanism. Therefore, an efficient detection method for identifying such adversarial attack technique is critical to defense the attacks.

The rest of the paper is structured as follows. Section 2 summarizes the background knowledge of subject-matter executable file structure and the relevant related work to our use cases. Section 3 describes the proposed methodology. Section 4 presents the performance evaluation, followed by the concluding remarks and future work in Sect. 5.

2 Related Work

2.1 Malware Detection

A malware analysis guide [19] suggested features for analyzing malware by extracting useful attributes from PE files, such as antivirus tools, hashes, and strings, functions, and headers of a target file. Antivirus tools rely on the malware signatures, while each PE infection based malware is unique and might not be detected by signature-based detection systems. Hashing provides a fingerprint. Like signature-based solutions, it requires a priori knowledge that the hashes of all the benign software. A string in a program can be a message to be printed out, URL, or file name. Therefore, the strings in a program provide information about program functionality.

Rezaei et al. [21] extracted opcode strings from the code section of the PE files and developed a malware detection method using Misha similarity and edit distance algorithms. The difference between a benign PE file and the infected one has a little discrepancy; the distance might be insignificant to distinguish PE infected malware.

Dynamic analysis could capture the run time behaviors by observing program execution in a control environment, aka sandbox. Polino et al. [20] proposed a dynamic analysis system which reports a descriptive summary for analysts querying the behaviors including invoked API function calls and data-flow dependencies. However, it requires security professions to verify the correctness of the generated semantic tagging.

Given the fact that static and dynamic analysis have their pros and cons. Researchers combine both to improve the performance. Choi et al. [11] proposed a malware detection method based on machine learning classification model. This method suggests applying open source tools to extract features from static and dynamic analysis. However, the study was lack of experiments to demonstrate the proposed idea. Ye et al. [23] summarized the features of static and dynamic analysis: DLLs, APIs, opcode sequences, control flow graph, and strings were extracted from static analysis; dynamic analysis looked for modifications upon system files, registries, and network activities, API call sequence, and system calls. Based on the literature review, the process of detection approaches mainly can be divided into two stages: feature extraction and classification/clustering.

A security study [18] reported that DLL injection technique is commonly used by attackers to hide their malicious behaviors and about half of the malware applied such attack technique. The DLL injection techniques [4, 12–14, 18] can be classified into three types. The first one achieves the purpose of injection by modifying the registry key about DLLs [1], for example by modifying the registry value of AppInit DLL [1]. The DLLs specified in the registry will be loaded by the running process. The second type of DLL injection techniques applies the hooking mechanism originally designed for message-handling, where an application can install a function to monitor the message traffic and perform some tasks [7]. Adversaries may use hooking to load and execute malicious code within the context of another process [3], such as by SetWindowsHookEx(). The third type creates a new thread, such as by CreateRemoteThread(), and injects it into the memory space of a legitimate process.

3 Proposed Detection Method

The proposed detection approach combines static and dynamic analysis as outlined in Fig. 4. The proposed dynamic analysis checks memory to see if a running process invokes an external DLL and inspect the properties of the process, its parent process, and invoked DLLs to see there is an anomaly. The proposed static analysis checks the repetition of RVA in PE header. If repetition RVA is found in PE file, the file is identified as malicious. The detail of each detection method is explained below.

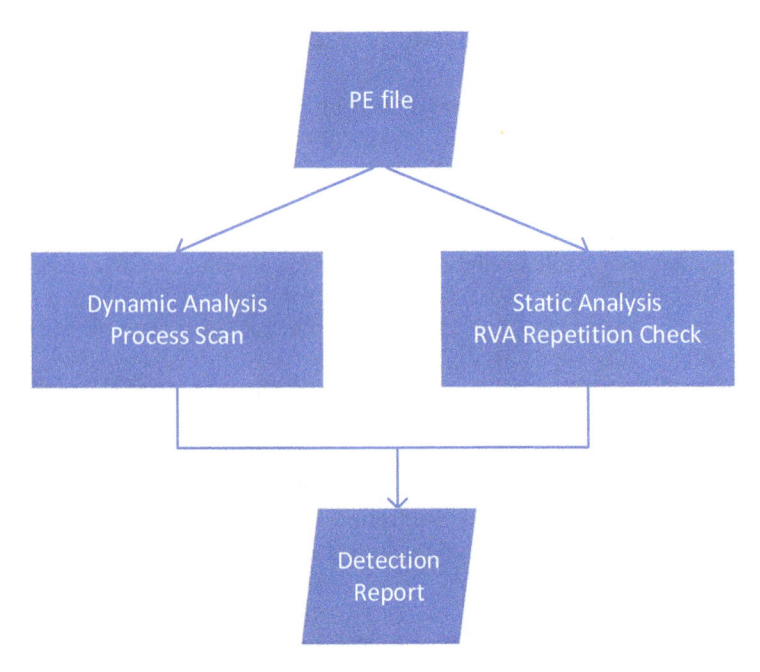

Fig. 4. System architecture of the proposed detection approach.

3.1 Dynamic Analysis

Attackers dynamically inject malicious DLL into the memory space of another process; in most cases, it is a system or commonly used process, such as svchost or explorer. In-memory injection attacks can be detected by dynamic analysis; they need additional information to verify its anomaly, as legitimate software might apply this system feature as well. Based on incident investigation reports [22], the following anomaly behaviors were observed: in-memory DLL injection malware is not located at the regular system folder, even though its process name is the same as a benign process; the program and the associated malicious DLL have no digital signature. To improve the detection performance, the proposed dynamic analysis detection method will further examine the properties of the suspicious processes to flag the anomalies described above. The proposed dynamic analysis consists of two steps: the first step finds the processes inject DLLs dynamically and the second step identifies malicious DLL injection by examining the properties of the process or the DLLs.

3.2 Static Analysis

The static analysis method consists of two steps. Step 1 screens a target binary executable program to see if there is a suspicious API function whose location is apart from the rest of the API functions, where an API function apart from others indicates that it might be a function injected in code cave. Attackers would copy the original RVA Import Table and append new entries for a malicious DLL and its associated functions and redirect the program execution entry point. Therefore, at Step 2, the proposed method searches if the code contains a repetition of the RVA Import Table.

A preliminary study was conducted for determining the threshold of the API distance in the proposed static analysis detection algorithm. It calculated the API distances of all the software in a clean Windows machine and those of PE infected malware samples. The histogram of the API distances is outlined in Fig. 5; the result shows that there is a big gap between the benign and malicious software and hence the threshold for API distance could be determined easily.

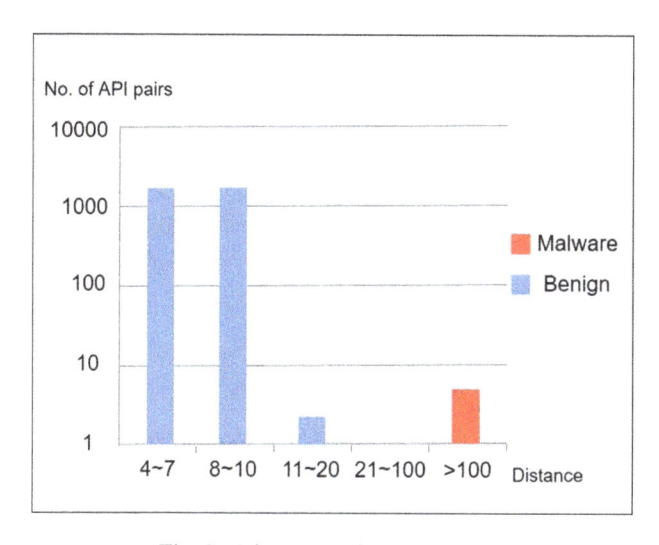

Fig. 5. Histogram of API distances.

4 Performance Evaluation

This study conducted two experiments: Exp 1 is to validate the performance of the proposed dynamic analysis detection method; Exp 2 is to evaluate that of the proposed static analysis detection method.

Exp 1 was conducted under a clean Windows system, a controlled environment with commonly used software installed, to simulate a normal host used by users. Malware samples were created by injecting malicious DLL into benign software to simulate real attack scenarios. The experimental results are depicted in Table 1 and demonstrate that the proposed dynamic analysis detection method could classify the benign and malicious software correctly.

Table 1. Dynamic analysis results.

Software	Status	Detection result
Skype	Clean	Benign
Skype	Injected	Malicious
Calculator	Clean	Benign
Calculator	Injected	Malicious
Poison Ivy	Malware	Malicious

4.1 Evaluation of Static Analysis Detection

PE infected malware samples were generated by injecting malicious code into benign programs through two PE infection attack approaches: manually infecting a PE file and applying PE editor tool such as Stud_PE. The samples were tested by the commercial anti-virus as well and the detection results are summarized in Table 2. The results demonstrate that the proposed static analysis method could flag PE infected malware efficiently, while none of the commercial anti-virus software could flag them.

Table 2. Static analysis results.

Software	Injection method	VirusTotal	Detection result
mspaint	Customized	0/54	Malicious
mspaint	By tool	0/54	Malicious
Calculator	Customized	0/54	Malicious
Calculator	By tool	0/54	Malicious
Notepad	Customized	0/54	Malicious
Notepad	By tool	0/54	Malicious

5 Conclusion

PE infection and Code injection are two popular approaches that APT used. Rather than find out malicious DLL files, in this paper, we propose a system to identify infected PE files. The proposed system checks abnormal distance of DLL and duplications of RVA Import Table to find out infected PE files. Experiment result shows our system is effective and efficient.

The proposed approach is not to replace antivirus software. It is a complementary approach to detect infected PE files. APT attacks also use code injection approach to inject malicious DLL files in existing process. The performance evaluation experiments demonstrate that the proposed detection method outperforms the commercial anti-virus software on unknown malware detection. The proposed system could identify the compromised PE files and injected DLL files. However, the behavior of injected DLL files still unknown. The future work could use machine learning technology to understand the behavior of malicious DLL files.

References

1. AppInit DLLs https://attack.mitre.org/techniques/T1103/. Accessed 7 July 2019
2. Dealing with Svchost.exe Virus' Sneak Attack. https://www.kaspersky.co.uk/resource-center/threats/dealing-with-svchost-exe-virus-sneak-attack. Accessed 7 July 2019
3. Hooking. https://attack.mitre.org/techniques/T1179/. Accessed 7 July 2019
4. Windows DLL Injection Basics. http://blog.opensecurityresearch.com/2013/01/windows-dll-injection-basics.html. Accessed 7 July 2019
5. APT Group Sends Spear Phishing Emails to Indian Government Officials. https://www.fireeye.com/blog/threat-research/2016/06/apt_group_sends_spea.html. Accessed 7 July 2019
6. POISON IVY: Assessing Damage and Extracting Intelligence. https://www.fireeye.com/content/dam/fireeye-www/global/en/current-threats/pdfs/rpt-poison-ivy.pdf. Accessed 7 July 2019
7. Hooks. https://docs.microsoft.com/en-us/windows/win32/winmsg/hooks. Accessed 7 July 2019
8. AsSadhan, B., Moura, J.M., Lapsley, D.: Periodic behavior in botnet command and control channels traffic. In: GLOBECOM 2009–2009 IEEE Global Telecommunications Conference, pp. 1–6. IEEE (2009)
9. Chiu, D., Attack Gains Foothold Against East Asian Government Through "Auto Start". Accessed 7 July 2019
10. Choi, H., Lee, H., Kim, H.: BotGAD: detecting botnets by capturing group activities in network traffic (2009)
11. Choi, Y.H., Han, B.J., Bae, B.C., Oh, H.G., Sohn, K.W.: Toward extracting malware features for classification using static and dynamic analysis (2012)
12. Ki, Y., Kim, E., Kim, H.K.: A novel approach to detect malware based on API call sequence analysis. Int. J. Distrib. Sensor Netw. **11**, 659101 (2015)
13. Kim, S., Park, J., Lee, K., You, I., Yim, K.: A brief survey on rootkit techniques in malicious codes. J. Internet Serv. Inf. Secur. **2**, 134–147 (2012)
14. Kuster, R., Three Ways to Inject Your Code into Another Process. https://www.codeproject.com/Articles/4610/Three-Ways-to-Inject-Your-Code-into-Another-Proces. Accessed 7 July 2019
15. Lakhina, A., Crovella, M., Diot, C.: Mining anomalies using traffic feature distributions. In: ACM SIGCOMM Computer Communication Review, pp. 217–228. ACM (2005)
16. Livadas, C., Walsh, R., Lapsley, D.E., Strayer, W.T.: Using machine learning techniques to identify botnet traffic. In: LCN, pp. 967–974. Citeseer (2006)
17. Lu, W., Rammidi, G., Ghorbani, A.A.: Clustering botnet communication traffic based on n-gram feature selection. Comput. Commun. **34**, 502–514 (2011)
18. Malik, A., DLL Injection and Hooking. https://securityxploded.com/dll-injection-and-hooking.php. Accessed 7 July 2019
19. McGrath, D.K., Kalafut, A., Gupta, M.: Phishing infrastructure fluxes all the way. IEEE Secur. Priv. **7**, 21–28 (2009)
20. Polino, M., Scorti, A., Maggi, F., Zanero, S.: Jackdaw: towards automatic reverse engineering of large datasets of binaries. In: International Conference on Detection of Intrusions and Malware, and Vulnerability Assessment, pp. 121–143. Springer (2015)
21. Rezaei, S., Afraz, A., Rezaei, F., Shamani, M.R.: Malware detection using opcodes statistical features. In: 2016 8th International Symposium on Telecommunications (IST), pp. 151–155. IEEE (2016)
22. TACERT, TACERT Documents. https://tacert.tanet.edu.tw/prog/Document.php. Accessed 7 July 2019
23. Ye, Y., Li, T., Adjeroh, D., Iyengar, S.S.: A survey on malware detection using data mining techniques. ACM Comput. Surv. CSUR **50**, 41 (2017)

Network Security Monitoring in Automotive Domain

Daniel Grimm[✉], Felix Pistorius, and Eric Sax

Institute for Information Processing Technologies, Karlsruhe Institute of Technology,
Engesserstr. 5, 76131 Karlsruhe, Germany
{daniel.grimm,felix.pistorius,eric.sax}@kit.edu
http://www.itiv.kit.edu/

Abstract. With the development of autonomous vehicles, the networking of vehicles with their surroundings continues to increase. On the one hand, wireless interfaces enable vehicle owners to communicate with other vehicles or infrastructure to use new applications such as smart parking services in car parks. On the other hand, external communication interfaces impose vulnerabilities to vehicles that can be exploited by cyber threats. The worst case scenario would be that unauthorized persons remotely take control of driving functions. The development of suitable countermeasures is increasingly coming into the focus of industry and research. In addition to authentication and encryption algorithms for the CAN (Controller Area Network) bus system, methods for monitoring network security in vehicles, for example by means of intrusion detection systems, are a current field of research. At the moment, CAN is the most popular bus system in automotive in-vehicle communication, but new technologies such as Automotive Ethernet arise. Hence, security for modern vehicles has to deal with various bus systems inducing different challenges.

In this work, we introduce a classification of techniques to monitor vehicle communications for security purposes to the automotive domain. Typical security measures in enterprise information technology are systematically compared with the state of the art in vehicle security. Our work serves to identify open fields of research and to classify future work.

Keywords: Automotive · Security · Network · Monitoring

1 Introduction

The vehicles of the future will be intelligent, autonomous and connected. The reason for this is the increasing dynamic interaction with other vehicles as well as the surrounding infrastructure. Another recent technical progress is the introduction of autonomous driving functions. Today, this is leading to more than 100 million lines of software code in each vehicle fulfilling the various functions [1]. The code is running on at least 80 microprocessors [2,3] on Electronic Control Units (ECUs) in the automotive domain. These new concepts will play an ever-increasing role in our society. According to forecasts, even by 2020, more than 90% of new vehicles will be connected to the Internet [4] summing up to about 470 million connected vehicles till 2025 [5]. The vehicle of the future possesses wired connections to the outside world like On-Board Diagnostics (OBD) as well as wireless such as Wi-Fi, Bluetooth and V2X (Vehicle-to-Vehicle, Vehicle-to-Infrastructure) communication. Connection to the internet is established either directly using cellular networks, Vehicle-to-Infrastructure communication or via the integration of smartphones into the vehicle [6]. On the one hand, this increased functionality will lead to a better integration of the car into daily life, for example smart parking services [7], or remote maintenance. On the other hand, this also means that the attack surface for cyber threats will increase due to the variety of external interfaces in the vehicle.

Intrusions via ECUs which have connections to the outside world are the most apparent possibility to compromise the vehicle [8]. Hence, external communication has to be monitored and analyzed. Compromised ECUs with external interfaces not only communicate externally, but can also compromise other internal ECUs by communication within the vehicle. Therefore, internal communication must also be secured and monitored. There exist multiple internal communication bus systems that are suited for different demands such as entertainment or safety functions. For entertainment a higher bandwidth is required, whereas safety functions, e.g. the brake pedal, have the need of real-time communication. With the introduction of sensors such as cameras and radar that generate massive amounts of data, the processing architecture changes. ECUs with higher performance, new communication systems and different ECU network architectures get into the focus of industry and research [9]. Consequently, vehicles of the future will have to be equipped with different measures to monitor the ECUs themselves as well as the various communication systems inside the vehicle and to external communication partners. On that account, the National Highway Traffic Safety Administration (NHTSA) published a guidance to improve the cyber security of vehicles in 2016 [10]. However, this guideline does not recommend which specific measure and protection method is needed and where it should be used.

1.1 Motivation

Security measures from enterprise IT enjoy growing interest in automotive research and development e.g. intrusion detection systems [11,12]. But with the changing network architectures and emerging new communication systems the question arises, if the current research is appropriate to secure future electric/electronic architectures of in-vehicle communication. In order to identify open fields for research it is necessary to analyse where with state-of-the-art techniques a holistic view of vehicle security is not possible in the future. Therefore, a general security monitoring approach for vehicles would be beneficial.

Security monitoring is defined as "the automated process of collecting and analyzing indicators for potential security threats, then triaging these threats for appropriate action" [13]. In our context, this includes techniques to actively gather and process information on the state of the vehicles security rather than authentication methods and encryption. In automotive domain, the generation of Message Authentication Codes (MAC) using the Advanced Encryption Standard 128 (AES128) algorithm is given as a prominent example, which is included in the Secure on-board Communication (SecOC) [14] Authentication methods and encryption are necessary to fulfil security goals but are not able to monitor the general vehicle security state or give further insight on security breach causes. Thus, they cannot leverage the process of choosing appropriate reactions. As outlined in the recent report "Encrypted Traffic Analytics" by Cisco [15], with the introduction of encryption new threats emerge that have an enormous economic impact, because already in 2019 half of all attacks use a form of encryption to mask their malicious behavior. Hence, we focus on using techniques that collect, analyze and monitor the ECUs and communication inside of the vehicle instead of encryption and authentication techniques. Consequently, this work introduces security monitoring to vehicle industry and aims to classify the current related security methods in automotive domain regarding this new concept.

This paper is structured as follows: In the next section, we give a short overview on the current in-vehicle communication and networking architecture. Afterwards, the state of the art in automotive security as well as in the enterprise information technology is summarized. Section 5 introduces a new categorization of the common automotive security measures. In the last chapter, we briefly summarize our work and identify the most necessary fields for future research.

2 Electronic and Communication Architecture in Automotive Domain

2.1 Communication Networking

As mentioned in Sect. 1 ECUs are not working isolated but communicate via bus systems and hence form a distributed system. In the course of time, various bus systems and network technologies have developed in the automotive industry as a result of rising demands on bandwidth and different requirements. The most important ones are outlined in the following.

Controller Area Network (CAN): Due to the high number of electronic components in vehicles that are increasingly safety-relevant, Bosch developed in the 1980s the Controller Area Network (CAN) bus system that fulfills a reliable data exchange between ECUs. CAN is an asynchronous multi-master serial bus and

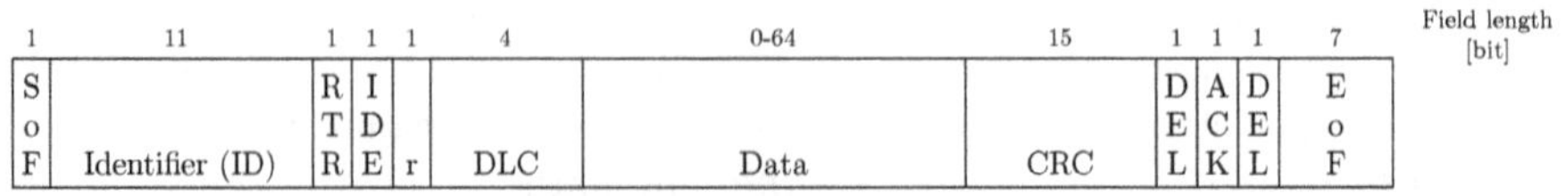

1	11	1 1 1	4	0-64	15	1 1 1	7	Field length [bit]
SoF	Identifier (ID)	RTR IDE r	DLC	Data	CRC	DEL ACK DEL	EoF	

Fig. 1. CAN frame

the transmitted messages contain up to 8 Byte of payload data. Each message contains a unique identifier that defines the priority of a message. The payload is structured in different signals of arbitrary length, but the structure of signals of a message with a specific identifier is fixed during the vehicle development. The structure of a CAN message is shown in Fig. 1. The CAN standard differentiates two different data rates: High Speed CAN with up to 1 MBit s^{-1} and Low Speed CAN providing up to 125 kBit s^{-1} [16]. In addition, Bosch specified in the year 2012 the extension CAN with Flexible Data Rate (CAN-FD) that allows for the transmission of up to 64 Byte per CAN message.

Local Interconnect Network (LIN): The single master serial bus system Local Interconnect Network (LIN) is used to connect simple sensors and actuators, such as a window regulator. Low costs for wiring and bus drivers are the main requirement for this bus system, and the bandwidth, which is 20 kBit s^{-1}, is correspondingly low [16].

FlexRay: The redundantly dimensioned (two physical channels) and time-controlled FlexRay bus system was developed for particularly safety-critical applications [16]. It allows a maximum data rate of 10 MBit s^{-1} per channel but in comparison to CAN and LIN for higher costs. It is mainly used where a mechanical coupling of sensors and actuators with strict latency demands is to be replaced. For example, the mechanical connection between steering wheel and steering gear can be replaced by a serial electrical connection with FlexRay. This serial bus systems uses a combination of Time Division Multiple Access (TDMA) and Flexible TDMA (FTDMA) to enable both deterministic as well as dynamic communication. As a multi-master architecture, global time synchronization for TDMA and FTDMA is performed cooperatively on local time bases.

Media Oriented Systems Transport (MOST): For multimedia applications such as the transmission of audio or video data within the vehicle, all the bus systems mentioned above do not offer sufficient bandwidth. Therefore, the serial bus system Media Oriented Systems Transport (MOST) was developed with a maximum bandwidth of 150 MBit s^{-1} using optical transmission technology [17]. The control units are arranged in a ring topology. The necessity of optical cabling makes this bus system very expensive to implement compared to the alternatives. To ensure synchronous data transmission one ECU serves as the timing master.

Ethernet: The automotive industry uses Ethernet- and IP-based communication since several years as external interface for diagnostic access and flashing of ECUs. In the meantime, this technology is also increasingly being used for in-vehicle communication. The main reason for its use is the increasing demand for transmission bandwidth for infotainment applications and driver assistance systems. For example, CAN or FlexRay are not suitable for the transmission of camera or radar sensor data because the bandwidth is insufficient. However, the MOST standard is expensive and not designed for safety-critical applications. Modern Ethernet-based communication is a switched collision-free point-to-point connection, which means that each participant is only communicating with its respective switch. It is known as a flexible and low-cost transmission standard from the IT sector in Local Area Networks (LANs), can provide up to 1000 MBit s^{-1} and is therefore an interesting alternative for applications in the vehicle. The 100BASE-T1 standard [18] with 100 MBit s or alternatively 1000BASE-T1, which provides 1 GBit s^{-1} [19], is used on the physical level using a single unshielded twisted pair of wires. Data rates of more than 1 GBit s^{-1} are conceivable in the future [20]. If the costs of the 10 MBit s^{-1} variant are low enough, this standard is a viable alternative to CAN [21].

While the physical layer of Ethernet-based communication is different in LANs and automotive domain, both rely mostly on the Internet Protocol (IP), Transmission Control Protocol (TCP) and User Datagram Protocol (UDP). On top of this standard TCP/IP stack, automotive-specific protocols such as Diagnostic over IP (DoIP) and Scalable service-Oriented MiddlewarE over IP (SOME/IP) are used. For real-time capable communication via Ethernet Time Sensitive Networking (TSN), which is a set of protocols and standards, was introduced as an extension to the standard IEEE 802.1Q [22]. Regarding external communication of the vehicle (V2X, Internet) the wireless protocols of the IEEE 802.11 define the physical layer [6].

2.2 Communication Paradigms

The communication via the aforementioned bus systems is mostly a so-called signal oriented data transmission. Lately, with Ethernet and IP-based communication the service-oriented transmission of information has found its way into the vehicle. CAN, LIN and FlexRay operate signal-oriented only, Ethernet is used in both paradigms.

Signal-Oriented Communication: A signal depicts a specific information such as the speed or ignition state and is identified by a signal name or number. A frame in signal-oriented communication can contain multiple signals and is identified by a frame identifier accordingly. The transmission of data is initiated by the sender of the information, e.g. if a new measurement value of a sensor is available. Irrespective of whether this data is currently required by a receiver in the network, the frame is sent. As the bus systems CAN, LIN and FlexRay conduct a broadcast-based data transmission, every bus subscriber is able to read all data. Hence, a certain network recipient has to decide on its own if received data is relevant for itself.

Service-Oriented Communication: In a service-oriented software architecture, the communication is typically organized in a publish-subscribe pattern. This means, individual ECUs offer services and use or subscribe to the services of other ECUs. The transmission of a specific information can be considered as a service. A more practical approach to define services is function-oriented, for example for the function "open right side window". Here, the transmission of all information necessary for the specific vehicle function "open right side window" is regarded as a service. In addition to solely transmitting information, remote functions can be called using service-oriented communication. The two main implementation alternatives in automotive domain are SOME/IP and recently, Data Distribution Service (DDS). SOME/IP [23] was specifically developed for the automotive industry whereas DDS is standardized by the Object Management Group (OMG) and used across many industries [24]. Both support transmission via TCP or UDP, which is why these standards rely in a vehicle on the use of Ethernet and IP-based communication. Using DDS, data is published to a specific topic, and topics are logically grouped in domains.

2.3 Operating Systems

Depending on the scope of the task performed by a particular ECU, the requirements placed on the deployed software vary. Intelligent sensors and actuators often only have a dedicated task, therefore these ECUs often do not use a complete operating system but the software runs directly (bare-metal) on the microcontroller. For typical automotive ECUs with 16 or 32 bit microcontrollers, a real-time operating system is usually used. This allows the execution of several tasks and abstracts parts of the underlying hardware. Historically, the first open and standardized operating system was OSEK/VDX. Based on that, AUTOSAR (AUTomotive Open System ARchitecture) evolved [25], which specifies an open and standard software architecture for embedded real-time systems. The application is developed independent of the hardware and executed on a middleware, the run-time environment (RTE). Via the RTE an application can use basic software that provides services and abstracts the ECUs microcontroller hardware and network topology. Two different versions exist: AUTOSAR *classic* which is based on OSEK/VDK, and AUTOSAR *adaptive* that uses a restricted POSIX operating system [25].

For ECUs with infotainment purpose, the GENIVI alliance develops an open-source software architecture based on embedded Linux [26]. Existing open source software is supplemented with automotive-specific components to support the typical platforms in vehicles. GENIVI systems can operate and integrate with AUTOSAR ECUs since the same definition language for software components is used.

2.4 Network Architecture

The ECUs in a vehicle are networked in different, manufacturer- and model-specific topologies. In general, past network architectures are typically a domain architecture in which mechanical systems as well as ECUs and the respective bus systems are grouped according to their functions. Examples are the powertrain domain, consisting of a CAN bus for motor and gearbox, or the chassis & safety domain using a FlexRay bus. Besides powertrain and chassis, typically the three further domains body electronics, human-machine-interaction (HMI) and telematics can be distinguished [27]. Powertrain and chassis & safety are domains including functions necessary for the vehicles movement and stability, whereas body, HMI and telematics contain the systems that are related to the passengers comfort, multimedia access and the vehicles head unit. In Fig. 2 an example for the domain-based architecture is shown, which includes a central gateway ECU that is responsible for the connection and routing between different domains. With the introduction of advanced driver assistance systems (ADAS) the safety domain is increasingly important in the vehicles network architecture. As ADAS systems rely on sensors that produce large amounts of data, we included an Ethernet-based ADAS domain in our example architecture. Navale et al. [9] propose the use of Ethernet for all functional domains, which are powertrain, body, infotainment, vehicle motion & safety, camera systems and diagnostics in their work. In addition, they emphasize the use of Ethernet for an architecture that is based on domain controllers connected via an Ethernet backbone network. Domain controllers act as gateways between the central backbone network and their respective domain sub-network. The backbone architecture opens the development towards a hierarchical concept instead of the current heterogeneous approach. As also outlined by Navale et al. further centralization of functions in a vehicle computer that fuses the domain controllers functions is possible. Figure 3 shows a centralized approach with two vehicle computers, that are responsible for resource-intensive tasks of the five past vehicle domains, categorized in vehicle- and passenger-related functions.

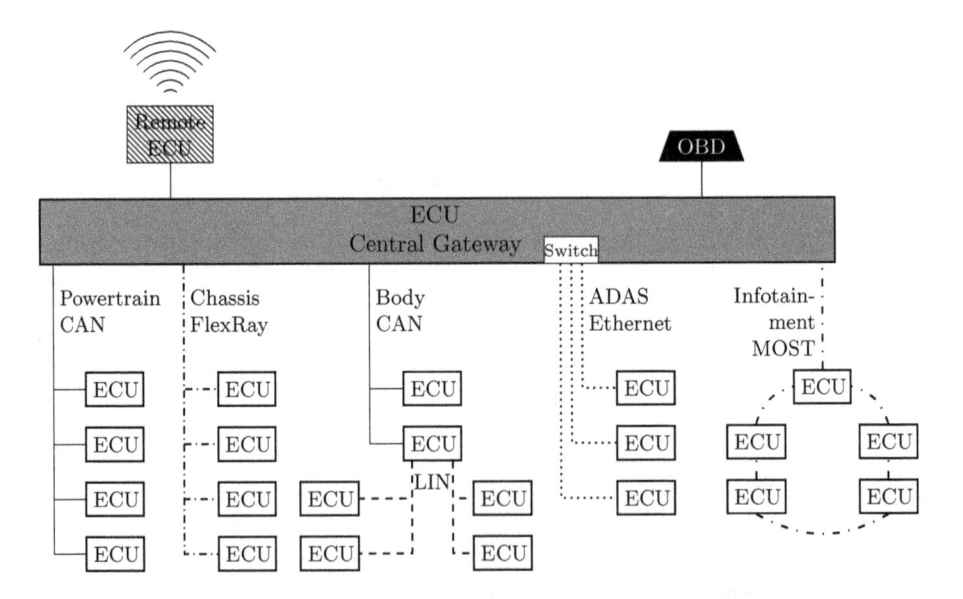

Fig. 2. Architecture with central gateway.

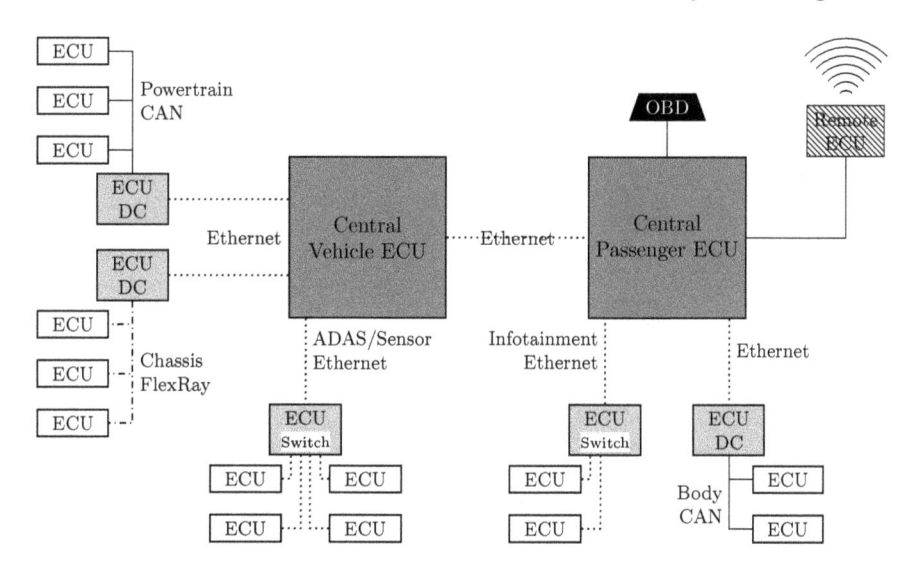

Fig. 3. Centralized architecture with two main control units.

3 Communication Architecture in Enterprise IT

As a reference of structured enterprise IT networks, we outline the fundamentals of the Cisco hierarchical internetworking model, which is an industry wide adopted architecture [28]. Local Area Networks (LAN) in enterprises evolved from a flat or meshed network to a structured design with hierarchies and modularity that allow for resiliency and flexible adoption. Flat networks do not allow for scaling and adding of new participants and any change affects many systems, whereas a hierarchical design constrains changes to a subset of the whole network. The scale of the LAN depends on the size of the enterprise, hence a general network architecture includes different choices for both small and large companies. A detailed guide [29] describes how to design a modern enterprise network following the hierarchical approach. The design consists of the three layers *access*, *distribution* and *core*, but for smaller networks not all layers are necessary. The access layer includes end-devices such as computers, laptops, or wireless access points and grants the devices access to the network. Access to the network is granted by Layer 2 switches. In the distribution layer, the access layers switches are aggregated, routing via Layer 3 switches is performed and connection to services such as internet or intranet applications is provided. The core layer is used to reduce the network complexity (number of wired connections between distribution switches) in large-scale networks. It acts as backbone responsible for high-speed switching and availability. In Fig. 4, an example that uses all three layers is shown. As the size or physical distribution of the network may require an extension of the core, networks can still have three logical layers but more than three physical layers of switches. As shown in the figure, on each layer networking devices with different capabilities such as routing or firewalls are used.

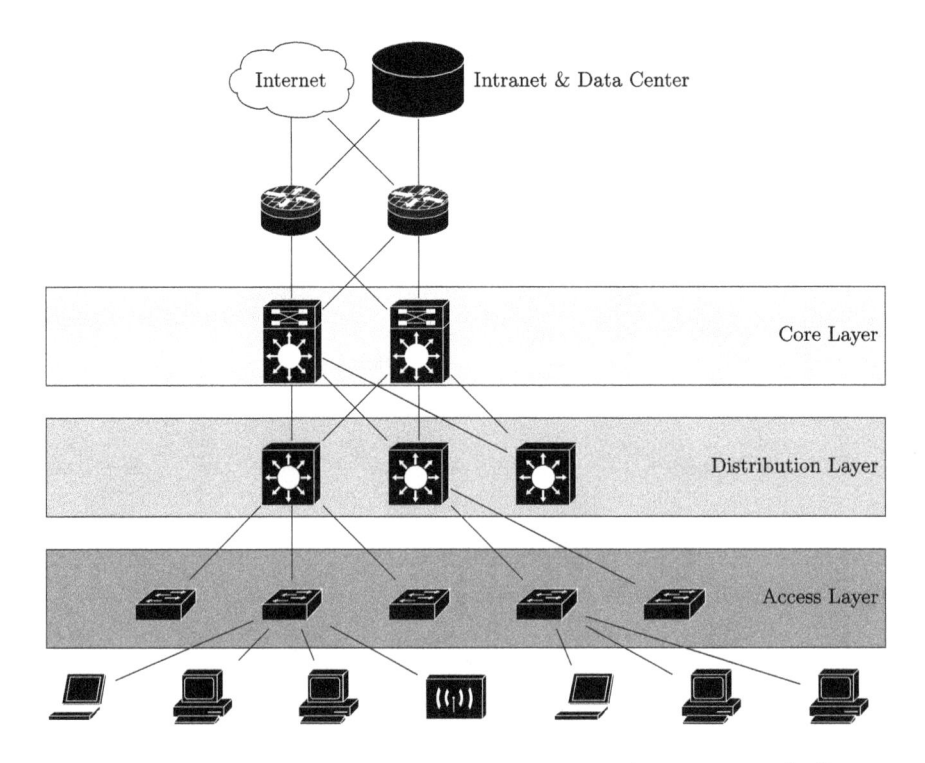

Fig. 4. Three-layer hierarchical architecture in IT (adapted from [28]).

4 State of the Art in Network Security Monitoring

4.1 Security Monitoring in Enterprise IT

A typical instrument of network security monitoring is a sniffer. These tools directly monitor the payload of the network traffic. An example is tcpdump which captures TCP segments that are switched over IP. In addition, there are numerous other software solutions that perform a similar function, such as Wireshark [30] and others. A more sophisticated option is network flow monitoring. A flow is a unidirectional sequence of packets that share certain attributes like the IP addresses and TCP or UDP ports. At least, a network flow record aggregates the data of a flow by calculating the duration, number of packets and sum of payload bytes. However, most flow collectors do not incorporate the payload data. One big advantage is that flows include additional statistical information on the captured data. NetFlow is one of the most popular flow monitoring techniques and protocols in the world today [31]. A similar alternative is the IPFIX standard, which is a manufacturer independent extension of NetFlow [32]. Moreover in enterprise IT there exist Intrusion Detection Systems (IDS) like Snort [33], Zeek (formerly called Bro, [34]) or OSSEC [35]. These systems can be differentiated in network-based and host-based Intrusion Detection Systems (NIDS/HIDS) by the data they operate on. Furthermore, IDS can be classified into signature-based and anomaly-based systems. A network-based IDS monitors and analyzes the network traffic, whereas a host-based IDS observes the behavior of a specific computer on the level of operating system and running applications. In signature-based IDS, the detection is based on a pre-defined misbehavior that is searched for. In contrast, anomaly-based IDS are modeling the normal behavior of a host or the network traffic and detect deviations of this model of normality. All possible combinations of monitoring focus (network/host) and functionality (signature/anomaly) are widely used in IT systems. In addition, Firewalls are a common measure that is, for example, included in every modern operating system. Additionally, they can be deployed as network firewalls residing on network switches or routers.

4.2 Security Monitoring in Automotive Domain

With the use of Ethernet, the open interface to the outside world and the use of higher operating systems such as Linux, modern automotive network and communication architectures are approaching the classic IT systems. On this account, common methods from traditional IT, such as intrusion detection systems or firewalls, are currently the focus of research in the automotive industry. Müter et al. [36] discuss specific challenges in the realization of an automotive IDS and present a categorization of network-based IDS. This categorization is based on the definition of eight different anomaly detection sensors and focuses on the CAN bus. In addition, the appropriate placement of an IDS in the vehicle and the data to be analyzed are briefly addressed. Most work is done in the field of anomaly-based Intrusion Detection Systems (IDS), where various publications exist [11,12,36–39]. The focus is the identification of anomalies on the CAN bus. New IDS technologies using physical properties such as voltage or clock skew instead of network traffic are arising [40,41]. Besides the CAN bus, Ethernet is becoming a viable alternative in automotive networks. In contrast to CAN, only little research is done on IDS systems for Ethernet in the automotive domain [42,43]. Various companies offer firewall and IDS/IPS solutions that are specifically tailored to the application [44,45]. However, Firewalls and IDS/IPS are currently still the exception in vehicle networks and ECUs and the publications in automotive domain concentrate on specific threats or application domains.

Some research groups are working on the structured detection of security breaches on a whole vehicle fleet. One example is given by Security Management of Services in Connected Cars (SeMaCoCa) which is a Security Information and Event Management System (SIEM) located in the backend of the connected car [46]. Another publication that proposes a SIEM in the backend is using a machine-learning based technique called Sand-Sprinkled Isolation Forest to analyze in-vehicle communication and log data [42].

5 Categorizing Network Security Monitoring in Automotive In-vehicle Networks

The ever-increasing number of electronic devices inside of vehicles that are able to communicate with each other and thus form networks has the consequence that rising demands are placed on the network security monitoring. This is clearly shown by the amount of ECUs in a vehicle which is more than 80 [2,3]. However, network security monitoring is a core area of information technology outlined in a vast amount of literature. In the following chapter we introduce a classification of network security monitoring systems to the special characteristics of automotive in-vehicle information technology, inspired by the work of Collins [47]. The different methods for network security monitoring are referred to as network security sensors. An overview of this classification is given in Table 1 and further details are outlined in the subsections below.

5.1 Classification Regarding the Vantage

The term *vantage* of a sensor means its positioning in the considered network architecture. A common network includes not only terminals, but also many intermediate systems, e.g. switches or routers that are also involved in the data transfer. Each communication participant, as well as the individual interfaces in between, can serve as a sensor, if network security monitoring measures are deployed. It is important to mention; however, that the different sensors can pick up different information depending on their placement. Thus, e.g. a sensor whose vantage point is a computer or other end-device can only monitor data related to that device. Such sensor has no access to the network traffic that is available to a sensor on a layer 3 switch, e.g. in the core layer. On the other hand, a sensor in the core layer is not able to monitor the end-devices behavior. In order to protect the network against attacks as efficiently as possible, it makes sense to optimally select the sensor vantage points.

In automotive in-vehicle networks, the main vantage points we can distinguish are the central gateway, which is able to observe all the traffic that has to be routed between different domains. Furthermore, the telematics and/or infotainment ECUs have to be mentioned as they are usually forming the connection point to the internet. In terms of the bus systems CAN and FlexRay one may distinguish between main domain controllers and terminal ECUs on a specific bus. The domain controller can monitor the communication between a backbone network and the in-domain communication. On the other hand, a sensor on a terminal ECU may only observe the traffic on the in-domain network. Furthermore, a sensor on a terminal ECU may observe the applications and functions running on itself whereas any other ECU would not be able to monitor the applications on the terminal ECU.

With the introduction of Ethernet to the vehicle the communication paradigm changes towards stricter hierarchies with switched point-to-point communication instead of broadcast-based bus signaling. This enables us to use the same differentiation of possible vantage points as in enterprise IT. According to

Table 1. Classification of network security monitoring in automotive industry

Classification aspect	Category	Main representatives in automotive networks
Vantage	External Communication	Telematics unit, OBD, Ethernet diagnostic port
	Root of communication network	Central gateway, backbone network, central ECUs
	Sub-network root	CAN bus domain controller ECU, FlexRay domain controller ECU, ECUs that serve as LIN master
		Network switch ECU on Ethernet
	Terminals	Terminal point on Ethernet
		ECU on CAN, LIN, FlexRay bus
Operational Area	Network	CAN, LIN, Ethernet, FlexRay, MOST
	Host	AUTOSAR Classic, AUTOSAR Adaptive, other standards, proprietary/no OS
	Application Data	Single signals in CAN/FlexRay messages, a vehicle function, a specific SOME/IP service, a DDS topic, Port-to-Port connection via Ethernet
Action	Report	CAN logs and traces from other bus systems, OS logs
	Event	IDS systems
	Control	Anti-Virus system, Firewalls, Intrusion Prevention Systems (IPS)

that, we may distinguish between sensors located on switching ECUs, capable of observing all the connections they switch, and sensors on terminal ECUs.

5.2 Classification Regarding the Operational Area

In addition to the positioning of the sensors, security sensors can be classified according to the *Operational Area* on which they operate. In contrast to Collins, who refers to this categorization aspect as *domain* [47], we refer to the *Operational Area* since in automotive industry this term is related to the functional groups such as powertrain.

The first category of this aspect are the network sensors, which monitor and operate on the traffic in a network. Examples are sniffers, which were mentioned before, as well as network intrusion detection systems (NIDSs). Host sensors provide data describing the operation of a host. Therefore, they differ from the network sensors. The last group in the domain categorization are application data sensors (service sensors in [47]). They work with structured data (such as log files) generated by individual applications. An example in this case would be a sensor that monitors e-mail traffic in a system. In matters of automotive domain we can use the same categories, but the characteristic is different. Network sensors are operating on the specialized automotive networks. The typical representatives in this category are the Controller Area Network (CAN), Local Interconnect Network (LIN), FlexRay, Media Oriented Systems Transport (MOST) and Ethernet. Regarding the host we can distinguish ECUs running on

AUTOSAR classic, AUTOSAR adaptive, other standardized operating systems like GENIVI [26] and ECUs that do not use an operating system. ECUs running on AUTOSAR adaptive may be further subdivided according to the underlying POSIX system they use. A service occurring in the in-vehicle information processing is, for example, a single signal transmitted via a bus system as signals depict the most fine-granular unit of information. It must be noted, that we do not consider frames on bus systems (collection of multiple signals) as a specific service since they do not serve an individual task. Instead, they serve multiple duties – the transmission of multiple signals. Moreover, a frame is always related to the network it is sent on. Other services may include an actual service of SOME/IP or the data of a port-to-port connection via TCP or UDP on Ethernet.

5.3 Classification Regarding the Action

The third classification aspect concerns the interaction of a sensor with the data it has collected, known as a sensors *Action*. It describes three categories of sensors. Report sensors provide direct information about the phenomena observed by each sensor. They have a relatively simple structure, but can greatly facilitate the detection of irregularities or anomalies in network traffic because they include information at the payload level. Some examples from this category are libpcap-based tools mentioned above as well as flow collectors, which aggregate packets with similar metadata (e.g. IP addresses and TCP or UDP ports) into new data units (network flows). In addition, the data of report sensors may be of great importance for development of event or control sensors, which are described in the following.

Event sensors differ to report sensors since the data they observe is aggregated in the form of an event. Manually defined rules or automated algorithms determine when an event has to be generated. Intrusion Detection Systems (IDS) are an example of this category: They search for irregularities in network traffic or host behavior and issue an event message if such deviations occur.

Control sensors are similar to event sensors in that they also use a variety of data to perform their function. On the other hand, they can also trigger certain countermeasures, such as blocking of packets. Examples of such sensors are antivirus systems, firewalls or intrusion prevention systems (IPSs).

In automotive domain the actions of security sensors can be classified in the same way. However, the report sensors are specialized to the networks and operating systems that are present in vehicles. For event and control sensors in vehicles very restrictive demands on the quality of the sensors outputs must be placed. A control action like blocking traffic from the brake pedal ECU may be only applied in case of extremely high certainty. Thus, control actions are merely suited for non-safety critical data. In addition, event sensors that are producing unnecessary events or false alarms on a frequent basis are not useful as they will probably be ignored.

6 Conclusion

Nowadays, due to outsourcing to suppliers and purchasing sub-systems vehicle manufacturers do not know the software running on an ECU exactly, hence security of such systems can not be judged. Because of the user a car manufacturer will know even less in the future through apps, devices like smartphones and watches and OBD dongles. In addition, the encryption of data makes analyzing payload data difficult. If we compare the given categories of Network Security Monitoring with the published state of the art security methods in automotive domain, we can see that the most work is published in a very restricted subset. The CAN bus was identified in different publications as the main vulnerability in automotive security. Hence, network based CAN anomaly detection, as the most popular research area in automotive security, spans only the Operational Areas of *Network* or *Application Data* with one Action (*Event*). However, the network operational area of security sensors in automotive industry is far away of being totally dealt with. The most communication bus systems are not in the focus of research today, as well as the *Host* category. The *Vantage* is not considered in detail in the referenced publications. Regarding the sensors action, the *Report* sensors are not under investigation to date. The classification also shows that the operational area *Application Data* is focused on one application group: the signal-based transmission. The focus is currently on the development of vulnerability centric solutions for specific domains or vehicles which do not provide a holistic view of vehicle security. We have to keep in mind that no prevention method can guarantee one hundred percent security. Therefore, the focus must evolve from vulnerability based special solutions for few domains and disregarded vantage to threat-centric development. This, with a network monitoring concept for the whole vehicle which puts more emphasis on data collection and detection enables us to provide a holistic and generic monitoring architecture [48].

7 Summary and Future Work

In this work, we introduced the concept of security monitoring to automotive domain. Security monitoring is a methodology of traditional IT to get a holistic view on the security of a communication network. Typical measures of security monitoring were outlined and compared to the current state of the art in automotive security. A classification scheme of traditional IT which considers the operational area, the vantage and the action of a security measure was used to structure the topic of automotive network security monitoring. With regard to that classification, we identified new possible fields of research. One very interesting point is the investigation of logging sensors in automotive domain, since to date logging of vehicle information was not for the purpose of analysis during the vehicles daily use, but only for vehicle development. Another topic is the development of security measures in the operational area of application data, but with focus on service-oriented communication or vehicle function-related analysis, instead of monitoring only signal-based communication. To sum up, a holistic view of the vehicle security is currently not possible due to missing solutions in various areas.

Regarding possible next steps, we have to keep in mind that automotive security is not limited to a single vehicle. To enable a vehicle manufacturer to gain further insights on threats and to develop viable updates for vulnerabilities, security monitoring must consider analysis and correlation of data of many vehicles. Especially the monitoring of a vehicle fleet poses new challenges to industry and research. Analysis of vehicle fleet data can only be performed on a backend server since the amount of data we have to analyze is enormous due to thousands of vehicles as well as historical data we have to include. An effective and efficient data collection of security related information inside of vehicles is needed to enable analysis in backend servers. Furthermore, security solutions for all operational areas have to be developed. Especially for Ethernet-based and service-oriented communication, which will dominate future network architectures, further steps have to be taken.

References

1. McCandless, D., Doughty-White, P., Quick, M.: Code bases: millions of lines of code (2015). http://www.informationisbeautiful.net/visualizations/million-lines-of-code/. 13 Sept 2019
2. Charette, R.N.: This car runs on code. IEEE Spectr. (2009). https://spectrum.ieee.org/transportation/systems/this-car-runs-on-code. 13 Sept 2019
3. Wyglinski, A.M., Huang, X., Padir, T., Lai, L., Eisenbarth, T.R., Venkatasubramanian, K.: Security of autonomous systems employing embedded computing and sensors. IEEE Micro **33**(1), 80–86 (2013)
4. Hyundai Media Newsroom: How will the Internet of Things transform the car industry? Hyundai Media Newsroom (2017). https://www.hyundai.news/eu/technology/how-will-the-internet-of-things-transform-the-car-industry/. 13 Sept 2019
5. PwC: The 2017 PwC's strategy & digital auto report (2017). https://www.strategyand.pwc.com/media/file/2017-Strategyand-Digital-Auto-Report.pdf. 13 Sept 2019
6. Coppola, R., Morisio, M.: Connected car. ACM Comput. Surv. **49**(3), 1–36 (2016)
7. Hartmann, F., Pistorius, F., Lauber, A., Hildenbrand, K., Becker, J., Stork, W.: Design of an embedded UWB hardware platform for navigation in GPS denied environments. In: 2015 IEEE Symposium on Communications and Vehicular Technology in the Benelux (SCVT), pp. 1–6. IEEE, Piscataway, NJ (2015)
8. Miller, C., Valasek, C.: A survey of remote automotive attack surfaces (2014). http://illmatics.com/remote%20attack%20surfaces.pdf. 13 Sept 2019
9. Navale, V.M., Williams, K., Lagospiris, A., Schaffert, M., Schweiker, M.A.: (R)evolution of E/E architectures. SAE Int. J. Passeng. Cars Electron. Electr. Syst. **8**(2), 282–288 (2015)
10. National Highway Traffic Safety Administration: Cybersecurity Best Practices for Modern Vehicles (2016). https://www.nhtsa.gov/sites/nhtsa.dot.gov/files/documents/812333_cybersecurityformodernvehicles.pdf. 13 Sept 2019

11. Weber, M., Pistorius, F., Sax, E., Maas, J., Zimmer, B.: A hybrid anomaly detection system for electronic control units featuring replicator neural networks. In: Arai, K., Kapoor, S., Bhatia, R. (eds.) Advances in Information and Communication Networks, Advances in Intelligent Systems and Computing, vol. 887, pp. 43–62. Springer, Cham (2019)
12. Hoppe, T., Kiltz, S., Dittmann, J.: Applying intrusion detection to automotive it-early insights and remaining challenges. J. Inf. Assur. Secur. (JIAS) **4**(6), 226–235 (2009)
13. Hewlett Packard Enterprise: What is security monitoring - HPE definition glossary (2019). https://www.hpe.com/emea_europe/en/what-is/security-monitoring.html. 13 Sept 2019
14. AUTOSAR Foundation: Specification of secure onboard communication. Document Identification No. 654 (2017)
15. Cisco Systems: Encrypted traffic analytics (2018). https://www.cisco.com/c/dam/en/us/solutions/collateral/enterprise-networks/enterprise-network-security/nb-09-encrytd-traf-anlytcs-wp-cte-en.pdf. 13 Sept 2019
16. Zimmermann, W., Schmidgall, R.: Bus systeme in der Fahrzeugtechnik: Protokolle, Standards und Softwarearchitektur. ATZ/MTZ-Fachbuch, Springer Vieweg, Wiesbaden, 5., aktual. und erw. aufl. edn. (2014)
17. MOST Cooperation: MOST specification (2010). https://www.mostcooperation.com/publications/specifications-organizational-procedures/request-download/mostspecification-3v0e2pdf/. 13 Sept 2019
18. IEEE Standards Association: ISO/IEC/IEEE International Standard - Part 3: Standard for Ethernet - Amendment 1: Physical Layer Specifications and Management Parameters for 100 Mb/s Operation over a Single Balanced Twisted Pair Cable (100BASE-T1), 8802-3:2017/Amd 1-2017 (2018)
19. ISO/IEC/IEEE: International Standard - Information technology – Telecommunications and information exchange between systems – Local and metropolitan area networks – Specific requirements – Part 3: Standard for Ethernet Amendment 4: Physical Layer Specifications and Management Parameters for 1 Gb/s Operation over a Single Twisted-Pair Copper Cable, 8802-3:2017/Amd 4-2017 (2017)
20. IEEE Standards Association: Standard for Ethernet Physical Layer Specifications and Management Parameters for Greater Than 1 Gb/s Automotive Ethernet (scheduled for fall 2019), p802.3ch (2019). https://standards.ieee.org/project/802_3ch.html
21. IEEE Standards Association: IEEE Draft Standard for Ethernet Amendment 5: Physical Layer Specifications and Management Parameters for 10 Mb/s Operation and Associated Power Delivery over a Single Balanced Pair of Conductors (scheduled for fall 2019), p802.3cg (2019). https://standards.ieee.org/project/802_3cg.html
22. IEEE Standards Association: Timing and Synchronization for Time-Sensitive Applications in Bridged Local Area Networks, 802.1AS-2011 (2011). http://www.ieee802.org/1/pages/tsn.html
23. AUTOSAR Foundation: SOME/IP protocol specification: release 1.1.0. document ID 696 (2017)
24. Object Management Group: Data Distribution Service (DDS): Version 1.4 (2015). http://www.omg.org/spec/DDS/1.4. 13 Sept 2019
25. AUTOSAR Foundation: AUTOSAR - AUTomotive Open System ARchitecture (2003). https://www.autosar.org/. 13. Sept 2019
26. GENIVI Alliance: GENIVI open source platform (2009). https://www.genivi.org/. 13 Sept 2019

27. Navet, N., Simonot-Lion, F.: Automotive Embedded Systems Handbook, 1st edn. CRC Press Inc., Boca Raton (2008)
28. Cisco Networking Academy: Connecting Networks Companion Guide, 1st edn. Cisco Press, Indianapolis (2014)
29. Cisco Systems: Campus LAN and Wireless LAN Design Guide (2018). https://www.cisco.com/c/dam/en/us/td/docs/solutions/CVD/Campus/CVD-Campus-LAN-WLAN-Design-Guide-2018JAN.pdf. 13 Sept 2019
30. Combs, G.: Wireshark (1998). https://www.wireshark.org/. 13 Sept 2019
31. Claise, B.: Cisco system NetFlow services export Version 9 (2004). https://rfc-editor.org/rfc/rfc3954.txt. 13 Sept 2019
32. Quittek, J.: Requirements for IP flow information export (IPFIX) (2004). https://www.rfc-editor.org/rfc/rfc3917.txt. 13 Sept 2019
33. Roesch, M.: Snort - lightweight intrusion detection for networks. In: LISA 1999: Proceedings of the 13th USENIX Conference on System Administration, vol. 132, p. 411. USENIX Association, Berkeley, CA, USA (1999)
34. Paxson, V.: Bro: a system for detecting network intruders in real-time. Comput. Netw. **31**(23–24), 2435–2463 (1999)
35. Daniel, B.: Cid: OSSEC - open source HIDS SECurity. https://www.ossec.net/docs/manual/ossec-architecture.html. 13 Sept 2019
36. Muter, M., Asaj, N.: Entropy-based anomaly detection for in-vehicle networks. In: IEEE Intelligent Vehicles Symposium (IV), 5–9 June 2011, Baden-Baden, Germany, pp. 1110–1115. IEEE, Piscataway, NJ (2011)
37. Stabili, D., Marchetti, M., Colajanni, M.: Detecting attacks to internal vehicle networks through Hamming distance. In: Infrastructures for Energy and ICT: Opportunities for Fostering Innovation, pp. 1–6. IEEE, Piscataway, NJ (2017)
38. Taylor, A., Japkowicz, N., Leblanc, S.: Frequency-based anomaly detection for the automotive CAN bus. In: 2015 World Congress on Industrial Control Systems Security (WCICSS), pp. 45–49. IEEE, Piscataway, NJ (2015)
39. Weber, M., Klug, S., Sax, E., Zimmer, B.: Embedded hybrid anomaly detection for automotive CAN communication. In: Proceedings of the 9th European Congress on Embedded Real Time Software and Systems, ERTS^2 2018, Toulouse, France, 31st January–2nd February 2018, pp. 1–10 (2018)
40. Cho, K.T., Shin, K.G.: Fingerprinting electronic control units for vehicle intrusion detection. In: 25th 5USENIX6 Security Symposium (5USENIX6 Security 16), pp. 911–927 (2016)
41. Cho, K.T., Shin, K.G.: Viden: attacker identification on in-vehicle networks. In: Thuraisingham, B. (ed.) CCS 2017, pp. 1109–1123. Association for Computing Machinery, New York, NY (2017)
42. Haga, T., Takahashi, R., Sasaki, T., Kishikawa, T., Tsurumi, J., Matsushima, H.: Automotive SIEM and anomaly detection using sand-sprinkled isolation forest. escar Europe (2017)
43. Grimm, D., Weber, M., Sax, E.: An extended hybrid anomaly detection system for automotive electronic control units communicating via ethernet efficient and effective analysis using a specification- and machine learning-based approach. In: Helfert, M., Gusikhin, O. (eds.) VEHITS 2018, vol. 2018-March. SCITEPRESS - Science and Technology Publications Lda, Setúbal, Portugal (2018)
44. Argus Cyber Security: Argus Solution Suites (2019). https://argus-sec.com/de/argus-solution-suites/. 13 Sept 2019
45. Arilou Technologies Ltd.: Solutions - Arilou. https://ariloutech.com/solutions/. 13 Sept 2019

46. Berlin, O., Held, A., Matousek, M., Kargl, F.: POSTER: anomaly-based misbehaviour detection in connected car backends. In: IEEE Vehicular Networking Conference, VNC, pp. 1–2 (2017)
47. Collins, M.: Network Security Through Data Analysis: Building Situational Awareness, 1 Million Log Records at a Time. O'Reilly, Beijing (2014)
48. Sanders, C.: Applied Network Security Monitoring: Collection, Detection, and Analysis. Syngress an Imprint of Elsevier, Waltham (2014)

Android Malware Detection in Large Dataset: Smart Approach

Qudrat E. Alahy, Md. Naseef-Ur-Rahman Chowdhury[(✉)], Hamdy Soliman, Moshrefa Sultana Chaity, and Ahshanul Haque

New Mexico Institute of Mining and Technology, Socorro, NM 87801, USA
{alahy.ratul,naseef.chowdhury,moshrefasultana.chaity, ahshanul.haque}@student.nmt.edu, hss@nmt.edu

Abstract. As the most widely used operating system for smartphones, Android is still growing, with many applications deployed in the mobile space, as well as other Android-based Internet-of-Things devices. A major side effect of the unprotected usage of such apps is the security loophole allowing app developers to access users' critical data on their devices. Hence, the lack of modern, precise validation of Android apps necessitates a new technique for malware detection. Proposed is a new smart mechanism that utilizes several machine learning models to analyze Android app behavior. More than 100 thousand Android application packages (APKs) containing more than 80,000 malware variants from 179 different families (in addition to benign Android apps) were collected. For added robustness, the model was trained with various malware found between 2006 and 2018. In consideration of the utilized app-dataset size, our smart model is poised as a very fast processing method for vast amounts of apps, unimplemented by other works in the field. The proposed smart Android malware detector obtained a very encouraging accuracy, ranging between 95% and 97%, on average for around 100 thousand analyzed APKs.

Keywords: Malware detection · Machine learning · Android · Security

1 Introduction

Recent smartphone usage has significantly increased, led by the most popular mobile operating system (OS) Android, with a market share of 84.7% [1]. On these Android devices, users can install a wide variety of applications. There are several ways to install applications on an Android device. An application can be installed via an official place such as the Google Play Store, or unknown sources. Unlike Apple iOS devices, Android does not restrict app installation to a specific source. Hence, malware authors are more tempted to target recipient devices with malware for running malicious code via Android apps. They might inject malicious code into a popular application via a technique called repackaging [26], with the intention to damage, steal or disrupt data at the user-side.

Much active research is carried out and still on-going for malware prevention on personal computers [4]. Yet, in the case of smartphones, research is comparatively lagging. Reference [3] surveyed more than 3.5 million Android applications available in the marketplace. Unfortunately, the spread of Android malware is proportional to the popularity of the Android system, and there is no guarantee that even Google Play store apps are malware-free.

There are many types of attacks via malware such as: phishing, trojans, spyware, SMS-fraud, botting, etc. Moreover, malware recently has been seen mining for cryptocurrency on mobile devices. According to Peiravian et al. [19], the extraneous, malicious code might be smuggled inside many apps by intruders, and go undetected by the Google Play store.

There are three main types of infiltration mechanisms that malware use to infect Android systems:

- Repackaging
- Updating
- Downloading

Repackaging: Intruders lure users by creating similar apps akin to some of the most popular ones, tempting them with a variety of useful features. These apps are repackaged, cloned versions of the original apps, containing malicious code [26].

Updating: Many authors include an update component in their app which can later download malicious code, while the app is in use [26].

Downloading: Malware developers lure users into downloading attractive apps, falsely promising them a variety of amazing features, with malicious code inserted inside such apps [26].

There are many malware detection mechanisms [36] mostly based on content signatures, which compare an app's signature to a database of known malware signature definitions. This mechanism can only detect known malware, not new ones. Research has shown that signature-based approaches never keep up with the speed of new malware development [5]. Hence, there is an urgent need to research and develop solutions to alleviate non-detection of malware on the Android platform. Examples of effective solutions can include characteristic and behavioral-based (static or dynamic) methods [37]. One of the most popular static behavioral-based methods of malware detection is based on analyzing its requested permission list and resource usages, e.g. Location Services, Contact Information, WiFi, etc. In our approach, we extracted data from more than one thousand user-defined permissions. A more powerful approach is dynamic behavioral-based, which dynamically observes the behavior of applications while running, e.g. dynamic API calls, capturing the run-time activities of the application. But, such an approach is complex and time-consuming [27].

Thus, we are proposing a new framework for classifying Android applications, utilizing machine learning (ML) techniques for large and diverse datasets. Our framework compiles a set of more than one thousand different app-requested permissions. Permissions are encoded to train ML classifiers to detect malicious applications. Experimentation using different quality/quantity ML input-datasets for training and testing achieved detection accuracy as high as 97.46%. The remainder of the paper is structured as follows. Section 2 depicts related works. Section 3 briefly describes the Android application structure and its security approach. Our methodology is introduced in Sect. 4. Experimental results are reported in Sect. 5, and our conclusion is stated in Sect. 6.

2 Related Works

A large amount of ongoing research has been done in the field of Android malware detection utilizing different ML models [9–18]. Primarily, there are two types of malware detection approaches, namely Static & Dynamic Analysis Technique (SAT & DAT).

SAT analyzes the Android APK without executing it. The most common two methods of SAT are namely signature & permission based. The signature-based malware detection method was introduced in the mid-90s [20]. The authors method utilized ML modeling on 525 malicious and 122 benign applications, achieving an accuracy of 86.56% in malware detection, when overlapping testing and training samples. Yet, in [20] it was noticed that this method performs poorly when separating the training and testing samples. In permission-based methods, the requested set of permissions from the user when installing/operating the app is examined. Talha et al. [21] proposed a technique called "APK Auditor" that has three components: a signature database, which stores extracted information and analyzes results; the Android client which is used by the end user to grant application analysis requests; and a central server to connect the signature-based database with the Android client. The APK Auditor obtained an overall 88% accuracy when experimenting with 8,762 applications, of which 6,909 were malicious. Li et al. [20] developed a lightweight tool to run in the Android environment, extracting many features from applications at runtime. After feature extraction, another feature selection process, based on PCA (Principal Component Analysis), reduces the feature dimensions. Then the Support Vector Machine (SVM) model is utilized for malware detection. Recent research by Zhou et al. [2] aims to detect malware by analyzing Android malware installation methods, the nature of malicious payloads, malware activation mechanisms, and so on. Their analysis is based on the evaluation of four representative mobile security softwares over more than 1,200 collected samples. Experimentally, they obtained a 79.6% best-case detection accuracy, and in worst-case 20.2%. Based on such results, they criticized the weakness of current malware detection solutions and called for the need to develop next-generation anti-mobile-malware solutions.

DAT is done by executing an APK and analyzing its behavior. Once the application is installed and executed on the device, the app's log or trace is generated by Android OS and then recorded by the malware detection model. A dataset is then generated through the analysis of recorded data by extracting information from the log file. It requires much more time and resources to analyze an application's dynamic behavior (vs. static techniques).

3 Android Security Approach

Android's security model highly relies on its static, permission-based mechanism. An Android application requires several permissions to operate. There are about 130 permissions that govern access to different resources [22]. While installing (on older OS versions) or running (in newer OS versions) onto a mobile device, an Android application's requested permissions must be first approved. Prior to installation, the system prompts a list of permissions requested by the application, seeking user confirmation in the older OS versions. However, in the newer OSs all the requested permissions are granted whenever a user wants to use the application. Although permission requests are useful for users to prevent possible misuse of resources by apps, users often rarely have the knowledge to determine if permissions might be harmful or not. For example, requesting network access, including WiFi and short message service (SMS) are normal for generic apps, whereas some malware misuses these services to steal bandwidth or message information. Therefore, analyzing only permission requests does not guarantee detection of malware.

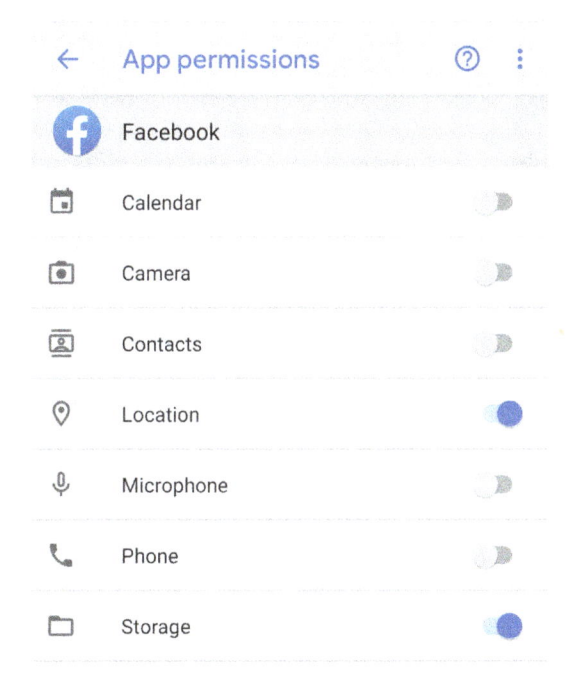

Fig. 1. The Android permission screen [8]

At the system level, Google has announced that an app security checking mechanism is applied to each application uploaded to their Android market [27]. The open design of Android OS allows a user to install downloaded applications from any source, even untrusted. Nevertheless, the permission list is still the basic guard for some users to detect harmful applications when they notice strange requests for their personal information (e.g. phonebook). Figure 1 shows user interface to convey user permissions.

Google categorizes Android permissions into four protection levels [28]:

Normal Permissions: Normal permissions are those permissions an app needs to access data or resources outside of the application's sandbox. These permissions are not risky. Whenever an app declares these permissions in the "Android-Manifest.xml" the system automatically grants the permissions for the app at installation time. The system neither asks user permission nor can they be revoked. For example, *SET_ALARM*, *SET_WALLPAPER*, *SET_WALLPAPER _HINTS*, etc. are considered normal permissions.

Signature Permissions: Protect access to the most dangerous privileges. The system grants the permission only if the requesting application is signed with the same certificate as the application that declared the permission.

Dangerous Permissions: Regulate access to potentially harmful API calls that could give access to private user data. For example, permissions to read the location of a user (ACCESS FINE LOCATION) or WRITE CONTACTS are classified as dangerous.

System Permission: Permissions that the system grants only to applications that are part of the Android OS.

A straightforward idea to classify a harmful app is to check whether the app requests permissions at the dangerous or higher level. Although Android adopts an authorized permission model to control access to its components, there is no clear evidence demonstrating how good or bad it is at detecting a malicious app based on permissions or combinations of permissions. It should be clarified that the permissions shown to a user during an installation process are requested permissions rather than required permissions. The requested permissions are declared by an app developer manually. However, not all declared permissions are required by the app. In addition to Google's methods for protecting Android from malicious applications, many security software companies have launched their own security apps.

3.1 Android Permission Settings

The manifest file can declare permissions that the application must have in order to access protected parts of the API and interact with other applications. It also declares the permissions that others are required to have in order to interact with the application's components. These permissions are the raw data for any static experiments, so to obtain the AndroidMainfest.xml file, Apktool [6] was used, which is a tool for reverse engineering APK files, and can generate the Android-Mainfest.xml from an APK file. Upon obtaining the AndroidMainfest.xml, it is possible to parse the file and extract requested permissions from it for each APK.

To demonstrate that permission settings are indeed relevant to the behaviors of benign apps and malware, we compared top permissions requested by these malicious apps in the dataset with top permissions requested by benign apps.

By analyzing our dataset, we have found that *INTERNET, ACCESS_NET -WORK_STATE, READ PHONE STATE, SEND_SMS*, and *ACCESS_WIFI_ STATE* permissions are frequently requested in both malicious and benign apps. The first two are typically needed to allow for embedded code and libraries to function properly. But malicious apps clearly tend to more frequently request SMS-related permissions, such as *READ_SMS, SEND_SMS, RECEIVE_SMS*, and *WRITE_SMS*. Figures 2, 3 and 4 represents top permissions used by the all apps, benign apps and malware respectively from our dataset.

Also, we noticed that malicious apps tend to request more permissions than benign apps overall. In our dataset, the average number of permissions requested by malicious apps is 13.7 while the average number requested by benign apps is 9.1. These results are consistent with the fact that the pattern of used permissions in malware applications is noticeably different from benign applications.

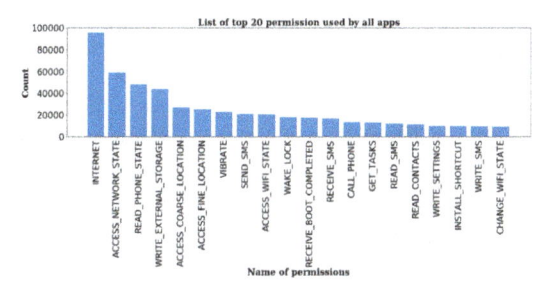

Fig. 2. List of top 20 permission used by all apps in training dataset.

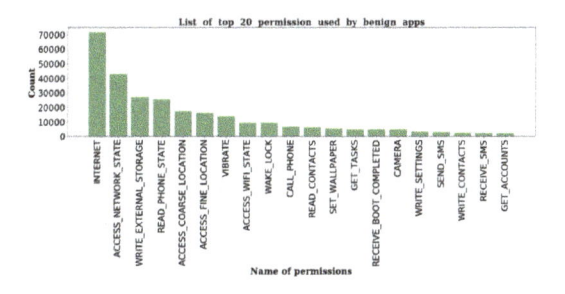

Fig. 3. List of top 20 permission used by benign apps in training dataset.

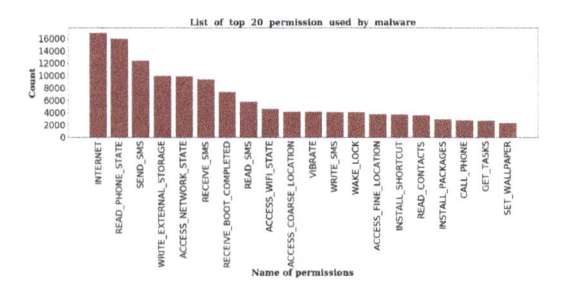

Fig. 4. List of top 20 permission used by malware in training dataset.

4 Our Approach

An overall workflow of the experiment is explained next. Figure 6 illustrates the experiment workflow structure. At the beginning, Android APK files from different malware families over the last decade are collected. Then we extract the "AndroidManifest.xml" file for both benign and malware apps, and utilize them along with the extracted permission list to build the ML training vector dataset. These training vectors are then labeled malware or benign, based on our prior knowledge, to train different machine learning classifiers. The following subsections depict the overall process detail phase-by-phase. Our workflow is represented in Fig. 5.

4.1 Data Collection

This phase covers the APK data collection process which is used to train our machine learning model. Such data will have variation regarding time span and malware families. We focused on four sources of APK data based on varying criteria. Our emphasis is on large dataset size for our ML models to be amenable for efficient, future application. Moreover, we considered popular datasets in order to compare our results to other models using similar data. Our targeted large and popular datasets are:

Marvin Dataset. The Marvin dataset [7] contains over 135,000 total mobile applications of which only 15,000 are malware, which were collected in 2012 to train classification models with older versions of malware.

Virus Total Dataset. Virustotal [23] is a website launched in June 2004. It aggregates various antivirus products and scans app files online, reporting the voting results of various antiviruses on each submission (malware or not). The website stores the tested apps along with their voting results, of which we collected 8,966 files (including goodware and malware) from 2018 and 3,212 files from 2017. We used the collected data to train and test our model with modern (2017–18) application behavior as well.

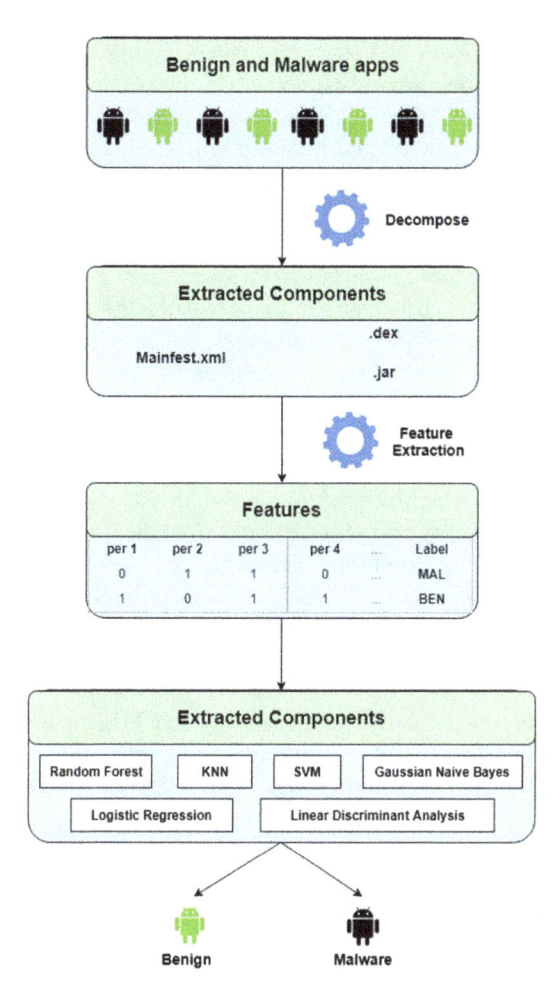

Fig. 5. Vector generation from decomposing Android APKs

Drebin Dataset. The Drebin [24,25] dataset contains 5,560 malware applications from 179 different families which were collected between 2010–2012. Although for our experiment we didn't detect family for the malware, we can analyze behavior between different malware families.

Malgenome Dataset. This dataset [1,2] contains more than 1200 malware from different families and was collected between Aug, 2010–Oct, 2011.

4.2 Formation of Training Dataset Vectors

In the first phase, we collect data to build our ML model training vector sets for classifying *malware* and *benign* Android permissions. Formation of training vectors is carried as follows. We start by extracting the manifest file from each existing APK file in the previously collected data set. Then we prepare the permission lists (in binary) from extracted "Android Manifest" files for each APK. At the end of the process, each vector will have three fields, namely APK number, the binary permission list (as explained in Subsect. 3-B above), and associated label (mal/goodware). The resulting vectors will be used to train our different classification ML models.

Given a set of permissions obtained for each app, a simple way to characterize the app is to use each permission as a feature. As a result, every app can be represented as a binary vector, namely P, where $p_i = 1$ if and only if the "AndroidManifest.xml" file has the ith permission and $p_i = 0$ if corresponding "AndroidManifest.xml" does not include the permission. As will be shown in the experiments section, these features can be used to differentiate benign and malware with moderate detection accuracy. However, we intend to advance the detection accuracy by combining permission analysis and Android APIs calls.

To achieve this efficiently we store the permissions as key-value pairs in *json* format. We then convert it to one-hot encoding [29] and finally develop a sparse matrix.

4.3 Machine Learning Classifier Phase

For our experiment we used different classification ML models including: Support Vector Machine (SVM) [30], Random Forest (RF) [31], K-Nearest Neighbors(KNN) [32], Linear Discriminant Analysis (LDA) [33], Logistic Regression (LR) [34] and Decision Tree (DT) [35]. The performance/accuracy of different ML classification models is a function of the dataset size. Hence, we utilized ML models that perform relatively better over large datasets.

Random forest is an ensemble algorithm, which works with a set of decision trees from a randomly selected subset of training data. It aggregates the votes from different decision trees to decide a final class. Random forest was chosen over decision trees because voting for the result performs better for larger datasets.

KNN is a non-parametric, lazy algorithm; it does not make any assumptions about the underlying data distribution. It classifies by majority vote of its neighbors. KNN assigns the class most common among its K nearest neighbors by a distance function. For KNN we varied value settings to analyze overall behavior in respect to the number of neighbors. The graph is shown in Fig. 6.

We noticed that a large k does not guarantee a better accuracy of malware detection in the KNN model.

Linear Discriminant Analysis assumes that data is Gaussian. Another assumption is that each attribute has the same variance.

Logistic Regression transforms its outputs using a logistic sigmoid function to return a probability value which can be matched to discrete classes. We utilized logistic regression to classify malware and goodware because of the simplicity of the model [34].

A Support Vector Machine (SVM) is a discriminative classifier, defined by a separating hyperplane. For training data if the input is labeled, the algo-

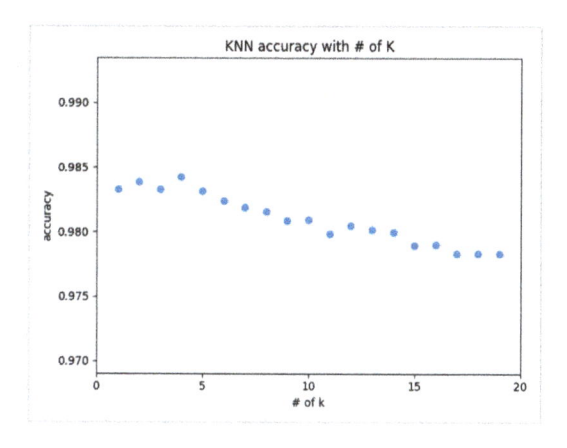

Fig. 6. Accuracy vs. number of neighbors in KNN

rithm outputs an optimal hyperplane which categorizes new examples. In two-dimensional space this hyperplane is a line dividing a plane in two parts wherein each class lays on either side.

Decision Tree (DT) is a ML technique where the data is continuously split in the node based on parameters. In DT, the leaves denotes the actual outcome [35].

We trained our dataset with various sizes of input. We varied the families and origin of applications for training and testing purposes. Some machine learning models outperform for larger datasets, others outperform on small sets. If the generation and family is known for the testing dataset, some machine learning algorithms perform better than the others.

5 Experiment Results

For the experiment we used various combinations of dataset based on size and generation. Datasets of different sizes are chosen randomly from different generations and families of malware and goodware. Our intention is to train the network for every possible combination, so that we can get a better result if the training and testing datasets are totally different. For the primary analysis we collected 109,480 samples from Drebin, Malgenome and Virus Total datasets. We trained the network with 75% of the sample, with the 25% remainder of the sample used as a testing dataset. From the APKs we process one-hot encoding for generating the vectors. For efficiency, we convert the original matrices to dense matrices. Table 1 represents the accuracy of our experiment by analyzing 109,480 apps.

To analyze the proficiency of our model, we conducted another set of experiments where the generation of testing data is completely different from the training dataset. This experiment has two parts.

Table 1. Accuracy using 109,480 apps for analysis

Classifier	Train acc.	Test acc.	Precision	Recall	F-score
SVM	96.57	96.53	97	97	96
LR	96.38	96.38	98	97	98
RF	98.70	98.19	98	98	98
DT	98.75	97.85	98	98	98
LDA	95.81	95.62	96	96	96
KNN	98.02	97.79	98	98	98

The first part of the experiment used the VirusTotal 2018 dataset as a testing dataset containing 8,966 applications. For training our model we used 99,254 applications collected from the Drebin and Marvin datasets. The result set is shown in Table 2:

Table 2. Accuracy using VirusTotal 2018 as a testing dataset

Classifier	Train acc.	Test acc.	Precision	Recall	F-score
SVM	97.35	77.45	100	77	87
LR	97.22	75.06	97	78	80
RF	99.18	61.02	100	61	76
DT	99.22	64.27	100	64	78
LDA	96.59	73.47	100	73	85
KNN	72.29	74.62	72	73	72

Another experiment was done using the Malgenome dataset containing 1,260 applications (collected between 2010–2011) as a testing dataset; for training the model we used 99,254 applications collected from the Drebin and Marvin datasets. The result of the experiment is shown in Table 3:

Table 3. Accuracy using Malgenome as a testing dataset

Classifier	Train acc.	Test acc.	Precision	Recall	F-score
SVM	97.35	76.53	100	77	87
LR	97.22	75.18	97	78	81
RF	99.18	96.27	100	96	98
DT	99.22	96.27	100	96	98
LDA	96.59	65.34	100	65	79
KNN	72.29	96.99	98	82	86

By analyzing the results we can observe that most of the ML models perform poorly with unknown application behaviors. Notably, KNN, Random Forest, and Decision Tree performs much better than the others.

To analyze how the Random Forest Tree is making decisions through calculation we examine the gini index (first 10 indices are listed below in Table 4, which have most impact). As we know in random forest every time a split of a node is made on variable m, the gini impurity criterion for the two descendent nodes is less than the parent node. Adding up the gini decreases for each individual variable over all trees in the forest gives a quick variable-importance that is often very consistent with the permutation importance measure.

Table 4. Permission name vs. their gini index

Permission name	gini index (descending order)
SEND SMS	0.25
INTERNET	0.15
RECEIVE SMS	0.07
INTERNET	0.06
UNNAMED: 0	0.06
READ PHONE STATE	0.03
INSTALL PACKAGE	0.02
READ SMS	0.02
UNINSTALL SHORTCUT	0.018

5.1 Evaluation Measures

The performance of a ML model cannot be determined only by accuracy. So, we analyze appropriate performance metrics below. We measure the below performance from our confusion matrix:

$$Precision = \frac{TP}{TP + FP}$$

$$Recall = \frac{TP}{TP + FN}$$

$$F - Score = \frac{2 \times Recall \times Precision}{Recall + Precision}$$

Here True Positive (TP) represents number of goodware correctly classified and True Negative (TN) is the number of malware correctly classified. On the contrary, False Positive (FP) is the number of goodware classified as malware and False Negative (FN) is the number of malware classified as goodware.

6 Conclusions

In this paper, we proposed a technique for detecting Android malware in large scale datasets using machine learning. Our model can detect malware from different families. The framework is also successful in distinguishing between both malware and benign apps that were collected over a period of 12 years (between 2006–2018), regardless of being from the Android marketplace or not. In the design process of our detection model we have trained various machine learning models by varying the size of dataset and generation and family of Android malware. We found out that some machine learning models outperformed other models in detecting malware for different configurations. Through the design process, we have extracted information and develop datasets through the analysis of the static behavior of more than 100 thousands Android applications from different Android malware families and diverse time frames. In our future work we will further improve our detection model's efficiency/accuracy via incorporating/adding the powerful dataset dynamic-analysis capability.

References

1. Zhou, Y., Wang, Z., Zhou, W., Jiang, X.: Hey, you, get off of my market: detecting malicious apps in official and alternative Android markets. In: Proceedings of the 19th Annual Network & Distributed System Security Symposium, February 2012
2. Zhou, Y., Jiang, X.: Dissecting android Malware: characterization and evolution security and privacy (SP). In: 2012 IEEE Symposium on Security and Privacy (2012)
3. Cheng, J., Wong, S.H., Yang, H., Lu, S.: SmartSiren: virus detection and alert for smartphones. In: International Conference on Mobile Systems, Applications, and Services (MobiSys) (2007)
4. Sanz, B., Santos, I., Laorden, C., Ugarte-Pedrero, X., Bringas, P.G., Alvarez, G.: PUMA: permission usage to detect Malware in Android. In: Advances in Intelligent Systems and Computing (AISC) (2012)
5. Wang, J., Deng, P., Fan, Y., Jaw, L., Liu, Y.: Virus detection using data mining techniques. In: Proceedings of IEEE International Conference on Data Mining (2003)
6. Android-Apktool, a tool for reverse engineering Android APK files. https://code.google.com/p/android-apktool/
7. MARVIN: efficient and comprehensive mobile app classification through static and dynamic analysis
8. Android Platform Architecture. https://developer.android.com/guide/platform/
9. Chen, X., Andersen, J., Mao, Z., Bailey, M., Nazario, J.: Towards an understanding of anti-virtualization and anti-debugging behavior in modern malware. In: DSN (2008)
10. Jidigam, R.K., Austin, T.H., Stamp, M.: Singular value decomposition and metamorphic detection. J. Comput. Virol. Hacking Tech. **11**(4), 203–216 (2014)
11. Fredrikson, M., Jha, S., Christodorescu, M., Sailer, R., Yan, X.: Synthesizing near-optimal malware specifications from suspicious behaviors. In: SP 2010 Proceedings of the 2010 IEEE Symposium on Security and Privacy, pp. 45–60 (2010)

12. Kolbitsch, C., Comparetti, P.M., Kruegel, C., Kirda, E., Zhou, X., Wang, X.: Effective and efficient malware detection at the end host. In: USENIX Security (2009)
13. Lanzi, A., Balzarotti, D., Kruegel, C., Christodorescu, M., Kirda, E.: AccessMiner: using system-centric models for malware protection. In: CCS (2010)
14. Palahan, S., Babic, D., Chaudhuri, S., Kifer, D.: Extraction of statistically significant malware behaviors. In: ACSAC (2013)
15. Paleari, R., Martignoni, L., Roglia, G.F., Bruschi, D.: A fistful of red-pills: How to automatically generate procedures to detect CPU emulators. In: USENIX WOOT (2009)
16. Rieck, K., Trinius, P., Willems, C., Holz, T.: Automatic analysis of malware behavior using machine learning. J. Comput. Secur. **19**(4), 639–668 (2011)
17. Singh, T., Di Troia, F., Corrado, V.A., et al.: J. Comput. Virol. Hack. Tech. **12**, 203 (2016). https://doi.org/10.1007/s11416-015-0252-0
18. Zhu, D.Y., Jung, J., Song, D., Kohno, T., Wetherall, D.: TaintEraser: protecting sensitive data leaks using application-level taint tracking. SIGOPS Oper. Syst. Rev. **45**(1), 142–154 (2011)
19. Peiravian, N., Zhu, X.: Machine learning for Android malware detection using permission and API calls. In: 2013 IEEE 25th International Conference on Tools with Artificial Intelligence (2013)
20. Li, X., Liu, J., Huo, Y., Zhang, R., Yao, Y.: An Android malware detection method based on Android Manifest file. In: International Conference on Cloud Computing and Intelligence Systems (CCIS), pp. 239–243 (2016)
21. Talha, K.A., Alper, D.I., Aydin, C.: APK Auditor: permission-based Android malware detection system. Digit. Invest. **13**, 1–14 (2015)
22. Arasavalli, S., Sravya, Y., Venuturumilli, S., Tottempudi, P., Ramakoteswarrao, G.: Securing Android devices from snooping apps. ICOEI 2018 (2018)
23. Virus Total. https://www.virustotal.com/gui/graph-overview
24. Arp, D., Spreitzenbarth, M., Hubner, M., Gascon, H., Rieck, K: Drebin: efficient and explainable detection of Android Malware in your pocket. In: 21th Annual Network and Distributed System Security Symposium (NDSS), February 2014
25. Spreitzenbarth, M., Echtler, F., Schreck, T., Freling, F.C., Hoffmann, J.: MobileSandbox: looking deeper into Android applications. In: 28th International ACM Symposium on Applied Computing (SAC), March 2013
26. Zheng, M., Sun, M., Lui, J.C.S.: Droid analytics: a signature based analytic system to collect, extract, analyze and associate Android malware. In: 2013 12th IEEE International Conference on Trust, Security and Privacy in Computing and Communications (2013)
27. Choudhary, M., Kishore, B.: HAAMD: hybrid analysis for Android malware detection. In: 2018 International Conference on Computer Communication and Informatics (ICCCI) (2018)
28. https://developer.android.com/guide/topics/permissions/overview
29. Cassel, M., Lima, F.: Evaluating one-hot encoding finite state machines for SEU reliability in SRAM-based FPGAs. In: 12th IEEE International On-Line Testing Symposium (IOLTS 2006) (2006)
30. Hearst, M.A., Dumais, S.T., Osuna, E., Platt, J., Scholkopf, B.: Support vector machines. IEEE Intell. Syst. Appl. **13**(4), 18–28 (1998)
31. Breiman, L.: Mach. Learn. **45**, 5 (2001). https://doi.org/10.1023/A:1010933404324
32. Liao, Y., Vemuri, V.R.: Use of K-nearest neighbor classifier for intrusion detection. Comput. Secur. **21**(5), 439–448 (2002)
33. Li, M., Yuan, B.: 2D-LDA: a statistical linear discriminant analysis for image matrix. Pattern Recogn. Lett. **26**(5), 527–532 (2005)

34. Haifley, T.: Linear logistic regression: an introduction. In: IEEE International Integrated Reliability Workshop Final Report (2002)
35. Navada, A., Ansari, A.N., Patil, S., Sonkamble, B.A.: Overview of use of decision tree algorithms in machine learning. In: 2011 IEEE Control and System Graduate Research Colloquium (2011)
36. Wu, D.-J., Mao, C.-H., Wei, T.-E., Lee, H.-M., Wu, K.-P.: DroidMat: Android malware detection through manifest and API Calls tracing. In: 2012 Seventh Asia Joint Conference on Information Security (2012)
37. Seshardi, V., Ramzan, Z., Satish, S., Kalle, C.: Using machine infection characteristics for behavior-based detection of malware. https://patents.google.com/patent/US8266698B1/en

Design and Implementation of an e-Voting System Based on Paillier Encryption

Miaomiao Zhang[✉] and Steven Romero

Manhattan College, Riverdale, NY 10471, USA
miaomiao.zhang@manhattan.edu

Abstract. With the rapid development of cloud computing technology, cloud services are gaining wider application space. It gives users the advantage of computing power and storage space that were beyond the reach of the past. However, user privacy and data security are the main problems in the application and promotion of cloud system. How to ensure the privacy of data and ensure its availability in the process of calculating data is a major problem. Ensuring both the privacy and the usability of the data in the process of calculation remains a major challenge. As a promising tool for solving such problem, homomorphic encryption is a hot topic in both academia and industry in recent years. The purpose of this research is to demonstrate the effectiveness of the Paillier encryption and its homomorphic properties implemented in an electronic voting system. We describe an e-voting system based on Paillier homomorphic encryption along with other cryptographic tools such as blind signatures and zero-knowledge proof. The proposed scheme guarantees the general voting system requirements such as eligibility, accuracy, simplicity, privacy, robustness and verifiability. The scheme is implemented in C++ using GMP, an arithmetic multiprecision library, and the "Paillier" library. This implementation uses the CPU to make the calculations necessary during encryption, decryption, vote validation, and tallying. Some portions of the proposed e-voting scheme such as signing the blinded ballots, checking for valid votes and counting up the votes could be made to run in parallel in order to improve the e-voting system performance. We also implement a "small" version of the voting system using CUDA and demonstrate that it is possible to use the GPU's processing power to accelerate the speed of this e-voting system.

Keywords: e-Voting · Homomorphic tallying · Paillier cryptosystem · Blind signature · Parallelization

1 Introduction

Voting is a method for a group to make a collective decision on a matter that is under consideration. It is fundamental to any consensus-based society. With the increase in population size and coverage of larger constituencies, the traditional paper-based voting becomes cumbersome for large-scale voting. In addition, there are many people who do not vote. In the 2016 United States presidential election, there were about 231,556,622 eligible voters in the United States. Out of all of these people only about 60%, voted. This means that a significant 92,671,979 or 40%, did not vote [3]. There are many factors that might hold people back from voting: people do not have the time to go to one of the designated voting location; the percentage of the youth that vote out of the total is significantly lower than the percentage of people from older age groups [1]; people with illness or disabilities are often neglected during elections not intentionally but due to the fact that some of the voting locations are inaccessible [2]; the security and privacy concerns by the voters [2].

Electronic voting (aka. e-voting) is an online process in which the electronic ballots allow registered users to cast their votes from an electronic device and to transmit them via Internet towards the electronic electoral urn where tallying is also done by using electronic means. Compared to paper-based voting, e-voting has advantages such as accuracy, convenience, efficiency and low cost. In addition, the essential property required for a e-voting system is the security.

E-voting is an active area of research due to the fact that it is expected to reduce the cost of elections and increase the voter turnout. Over the past two decades, the computer security field has studied the possibility of electronic voting systems, with the goal of fulfilling the security requirements and minimizing the cost. Various cryptographic [8,10,21] and networking technologies [6,16] that exist and some which are still developing (e.g. distributed ledger technologies [15,17]) come handy in resolving many of the security issues in an e-voting system. In particular the cryptographic tools such as [20] offer a number of benefits to electronic voting and counting solutions. In most of the cases, one or more trusted third party (TTP) is involved to make e-voting systems more easily implemented and controlled. Generally speaking, the majority cryptographic e-voting schemes are aiming to achieve a list of security properties which can be classified into the following four aspects, based on the assumption that all parties are limited to polynomial-bounded computational resources:

- Democratic: Only eligible voters can vote and each eligible voter can cast at most one vote that counts.
- Voter privacy: No one can tell how a voter actually voted.
- Robustness: The voting system is able to tolerate some faulty behavior by the voters, i.e., the voter cannot be coerced/bribed to vote a particular way.
- Verifiability: This includes individual verifiability and public verifiability. Individual verifiability ensures each voter may verify his/her vote, and public verifiability ensures that any party can verify that the election is fair and the published final result is correctly mapped from the received votes.

1.1 Our Contribution

In this paper, we consider the usage of bulletin board. We present a secure cryptographic e-voting system designed to minimize cost and complexity. Our scheme is based on Paillier homomorphic encryption, blind signatures and zero-knowledge proof. Tally requires only one decryption per candidate, thanks to the homomorphic property. The security of the system is based on the hardness of integer factorization. Along the way, we present a concrete construction of Σ-protocol (OR-proof) using the Paillier homomorphic encryption. We discuss the security parameters requires for the proposed e-voting system and potential of parallelization. Some portions of the proposed e-voting scheme such as signing the blinded ballots, checking for valid votes and counting up the votes could be made to run in parallel for speed up.

We estimated the performance of our scheme and implemented it in C++ using the GNU Multiple Precision (GMP) Arithmetic Library and the Paillier library [5]. The implementation for large security parameters (i.e. $|n^2| = 4096$) is done using CPU to make the calculations necessary during encryption, decryption, vote validation, and tallying. Due to lack of support for CUDA, none of the existing multiple precision arithmetic libraries used on CUDA can meet the Paillier security parameter requirements of $|n| = 2048$ bits. We implemented "small" versions of our voting system (that support n up to 32 bits) on CPU and GPU, and thus to compare the speedup gain. The result demonstrate that the improvement factor increases with the size of input data (votes). In the tally phase, with a voter number of 10K, the GPU implementation gains approximately 400 speed up factor over the sequential CPU implementation.

1.2 Outline of This Paper

The paper is organized as follows. In Sect. 2, we introduce the model of our e-voting system. In Sect. 3, we cover some cryptographic components of the proposed e-voting system. An OR-proof based on Paillier encryption is presented in Subsect. 3.3. Section 4 present the e-voting scheme, and it is analyzed in Sect. 5. The implementation and performance analysis for the e-voting scheme is presented Sect. 6. Finally, we conclude in Sect. 7.

2 The Election Model

Now we describe our bulletin board model for elections. The e-voting system involve the following several entities. For the sake of simplicity each entity consist only one or two individuals. We do not consider distributing authorities, though they can be thresholdized.

Candidates. In the simplest case, the election aims to select a winner from several candidates. The candidate can be a physical person, yes/no decision, or any set of propositions among which a choice has to be made. They can be further designated by the number $\{1, 2, \ldots, k\}$.

Voter. A voter is a registered person who is allowed to express one vote for one candidate. Each ballot has the same weight in the final result.

Registration Authority. To cast their ballots, voters present them to a centralized registration authority with some identification to authenticate themselves. The registration authority checks that the voter has valid identify and has not cast a vote yet, but the registration authority does not get to see the vote itself. The Registration Authority who holds a signing key will sign the ballots of authorized voters.

Counting Center. The counting center include a bulletin board B and a tally authority T. The bulletin board is assumed to function as follows: voters can make post requests for items of their choice, potentially subject to some access or verification policy, for example verify the signatures on the votes from the registration authority and validate that each voter only voted for one candidate. The validated item will be posted on B where any party can read. All voters can access all messages written on B, but can only identify their own messages. This can all be implemented in a secure way for instance using an already existing public key infrastructure and server replication to prevent denial of service attacks. When the voting system is closed, the tally authority T count all valid votes and publish the result.

Security Requirements. Our e-voting scheme requires the following properties:

- Correctness (democratic and robustness):
    - only eligible voters are able to cast votes, i.e. registered voters.
    - No voter can vote more than once.
    - No voter can replace, change or bias votes, i.e. non-malleability.
    - A vote cannot be changed once published on the bulletin board.
- Privacy: The privacy of the voters ensures that a vote will be kept secret from the registration authority. Also a particular vote is not linkable to any voter.
- Verifiability: A voter is able to verify if his/her vote is counted in the final tally; and any party including observers can convince himself that the election is fair and the published the tally is correctly computed as the sum of the ballots that were correctly casted.

3 Cryptographic Building Blocks

3.1 Paillier Cryptosystem

Our e-voting scheme relies on the Paillier public-key cryptosystem which provides semantically secure encryption, efficient decryption and additive homomorphism by ciphertext multiplication. The details of Paillier cryptosystem can be found in [20]. In this subsection, we provide a brief summary below.
The Paillier cryptosystem contains three algorithms.

- $KeyGen(1^k)$: generate two safe primes p and q. The secret key sk is a tuple $(p, q, \lambda), \mu$, where $\lambda = lcm(p-1, q-1)$, $\mu = (L(g^\lambda \mod n^2))^{-1} \mod n$, where $L(x) = \frac{x-1}{n}$. The public key pk includes $n = pq$ and $g \in \mathbb{Z}_{n^2}$ such that $g \equiv 1 \mod n^2$. Oftentimes, if p and q are chosen of the same length, then $g = n+1$.

- $Enc_{pk}(m, r)$: encrypt a message $m \in \mathbb{Z}_n$ with randomness $r \in \mathbb{Z}_{n^2}$ and public key pk as $c = g^m r^n \mod n^2$. It is often denoted as $c = Enc_{pk}(m)$ when the randomness r is not important to the explanation.
- $Dec_{sk}(c)$ decrypt a ciphertext $c \in \mathbb{Z}_{n^2}$. Compute

$$(L(c^\lambda \mod n^2) \cdot \mu) \mod n = \frac{L(c^\lambda \mod n^2)}{L(g^\lambda \mod n^2)}$$

Homomorphic Property. Paillier cryptosystem has additive homomorphic properties.

$$\begin{aligned} Enc_{pk}(m_1) \cdot Enc_{pk}(m_2) &= (g^{m_1} r_1^n) \cdot (g^{m_2} r_2^n) \mod n^2 \\ &= g^{m_1+m_2} \cdot (r_1 r_2)^n \mod n^2 \\ &= Enc_{pk}(m_1 + m_2) \end{aligned}$$

The additive homomorphism enables

$$Dec_{sk}(Enc_{pk}(m_1) \cdot (Enc_{pk}(m_2)) = m_1 + m_2$$

and

$$Dec_{sK}(Enc_{pk}(m)^k) = k \cdot m$$

Basically, the result of the multiplication of two ciphertexts equals to the ciphertext of the result of the addition of two messages. These "algebraic properties" enable to compute with encrypted values without knowing the content of ciphertexts. They are useful when the anonymity of users is required. For example, in electronic voting schemes, each voter will encrypt their votes and write the encrypted result on the bulletin board. The counting center can compute the tally without decrypting each vote and therefore can guarantee the privacy of voters and improve the efficiency.

3.2 Blind Signature

Blind signatures are a robust solution to attain voter authorization while maintaining anonymity. In the vote processes, the list of active voters must authenticate with an authority who holds a signing key and signs the ballots of authorized voters. Unfortunately, a straightforward use of an ordinary digital signatures here would reveal everyone's votes to the authority. In order to guarantee the anonymity of the votes and prevent double voting, one can use blind signature [7]. Each voter fills in his ballot and blinds it – we will define this formally in a moment but think of blinding for now as placing the ballot in an envelope – then authenticates himself to the authority. The registration authority signs the blinded ballot without knowing its contents. The voter then turns the signature on the blinded ballot into a signature on the real ballot and casts this ballot along with the signature.

Blind signatures requires two security properties.

- *Unforgeability*: no one can forge the signatures on messages that the signer has not blind-signed, even though the signer doesn't know which messages she has signed.
- *Blindness*: the signer cannot learn which messages she has signed.

We consider a blind signature scheme that is based on RSA [22] which includes five algorithms:

- $KeyGen(1^k)$: Generate two large primes p and q; Compute $n = pq$ and $\lambda(n) = lcm(p-1, q-1)$; Choose an integer e such that $1 < e < \lambda(n)$ and $gcd(e, \lambda(n)) = 1$, i.e., e and $\lambda(n)$ are co-prime; Determine d as $d \equiv e^{-1} \mod n$, i.e., d is the modular multiplicative inverse of e modulo $\lambda(n)$. The verification key (public key) vk consists of the modulus n and the encryption exponent e. The signing key (private key) $sk = (d, p, q, \lambda(n))$.
- $Blind(m, vk)$: The blind algorithm picks a random $r \in \mathbb{Z}_n^*$, computer $b = H(m) \cdot r^e \mod n$.
- $Sign_{sk}(b)$: as for the basic RSA signature, the signer computes the blind signature $s = b^d \mod n$.
- $Unblind(s, r, vk)$: computes $\sigma = b/r \mod n$.
- $Verify_{vk}(m, r)$: as for the basic RSA signature, checks if $\sigma^e \equiv 1 \mod n$. The interactive process between the user and signer is shown in the diagram below.

Figure 1 shows the steps of using an RSA-based blind signature in e-voting. The blind signatures can be applied to a voter's ballot so that it is impossible for anyone to trace the ballot back to voter.

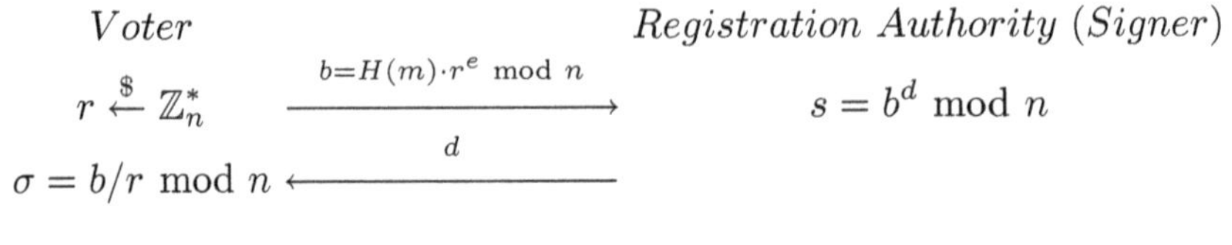

$$
\begin{array}{lcl}
Voter & & Registration\ Authority\ (Signer)\\
& b = H(m) \cdot r^e \mod n & \\
r \xleftarrow{\$} \mathbb{Z}_n^* & \longrightarrow & s = b^d \mod n \\
& d & \\
\sigma = b/r \mod n & \longleftarrow &
\end{array}
$$

Fig. 1. A blind signature based on RSA

3.3 Zero-Knowledge (ZK) Proof

Zero-knowledge proofs [13] are a technique that allow you to prove that you have done a certain operation correctly, without revealing more than that fact. In a voting system, we want to prevent voter's misbehavior such as submitting multiple votes for one candidates, negating a vote or bias the voting result. The solution is that whenever a participant submits some information (say, a ballot) they must submit two things: the ballot and a proof that they have made a correct ballot. "Zero-knowledge" means that the proofs reveal nothing beyond that the ballot is correct.

Zero-knowledge is a relatively broad field in cryptography with many special instances of zero knowledge defined. In an e-voting scheme, we will essentially use the honest-verifier zero-knowledge, which is also known as Σ protocols. A Σ-protocol is required to be zero-knowledge against an honest verifier only. Σ-protocols are versatile building blocks. One basic construction with Σ-protocols is called the OR-proof. It allows a prover to show that given two inputs x_0, x_1, he knows the secret w, such that either $(x_0, w) \in R$ or $(x_1, w) \in R$ without revealing which is the case, where R is a binary relation. Theoretically speaking, such OR-proof construction exist for any R [9]. Given encryption of Paillier, we present an concrete OR-proof construction of Σ-protocol proving that the encrypted message lies in a given pair of message (m_0, m_1). We present an interactive version of the protocols below. Using the Fiat-Sharmir paradigm [12], the verifier can be replaced by a hash function to form a non-interactive proof. Our construction can be viewed as special case of the proof that an encrypted message lies in a given set of messages [4].

Let n be the RSA modulus and $c = g^{m_i} r^n$ mod n^2 and encryption of m_i where i is secret. In this protocol the prover convinces the verifier that c is an encryption of either m_0 or m_1.

Running the protocol in Fig. 2 t times, the chance of forgery (completeness error) can be reduced to $1/A^t$. To obtaining completeness error to be less than 2^{-80}, we requires that $t \geq \frac{80}{\log_2 A}$. In our implementation in Subsect. 6.1, we can adjust the value of A to reduce the number of repetitions of the interactive proof, e.g. when $A = 4096$, only $t = 7$ rounds are needed.

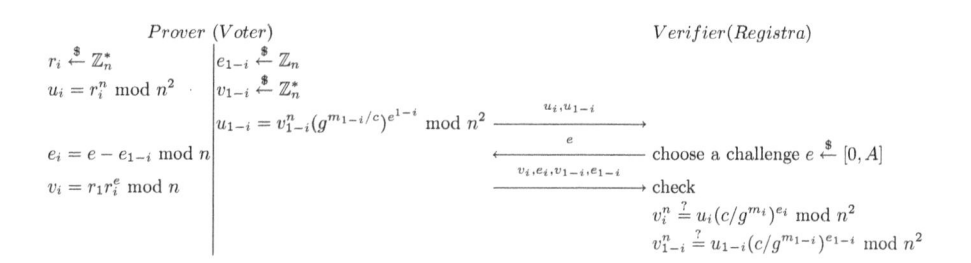

$$
\begin{array}{lll}
\textit{Prover (Voter)} & & \textit{Verifier(Registra)} \\
r_i \xleftarrow{\$} \mathbb{Z}_n^* & e_{1-i} \xleftarrow{\$} \mathbb{Z}_n & \\
u_i = r_i^n \bmod n^2 & v_{1-i} \xleftarrow{\$} \mathbb{Z}_n^* & \\
& u_{1-i} = v_{1-i}^n (g^{m_{1-i}}/c)^{e_{1-i}} \bmod n^2 & \xrightarrow{u_i, u_{1-i}} \\
e_i = e - e_{1-i} \bmod n & & \xleftarrow{e} \text{choose a challenge } e \xleftarrow{\$} [0, A] \\
v_i = r_1 r_i^e \bmod n & & \xrightarrow{v_i, e_i, v_{1-i}, e_{1-i}} \text{check} \\
& & v_i^n \stackrel{?}{=} u_i (c/g^{m_i})^{e_i} \bmod n^2 \\
& & v_{1-i}^n \stackrel{?}{=} u_{1-i} (c/g^{m_{1-i}})^{e_{1-i}} \bmod n^2
\end{array}
$$

Fig. 2. ZK-proof that an encrypted message is either m_0 or m_1

4 The Proposed e-Voting Scheme

Given the primitives of the previous section we are now ready to assemble a simple and secure election scheme. The participants in the e-voting protocol are n voters $V_1, V_2, \ldots, V_n$ and k candidates $C_1, C_2, \ldots, C_k$. The registration authority A, bulletin board B and the tally authority T.

Setup. We consider 1-out-of-k e-voting, where one winner is chosen among k candidates. In the initialization phase, the registration authority generates his RSA key pair and certifies his public key with an independent certification authority. We use $pk = (e, n)$ for the public key and $sk = (p, q, d)$ for the private key, where p, q are large primes of approximate the same length. The private key sk is only know by the registration authority A which will be used for blindly sign a valid ballot, and the public key pk can be used to verify the signature signed by A, which is published on the bulletin board B and known by any party in the system.

The tally authority T in the counting center also generated his public/private key pair (PK, SK) which is used for Paillier encryption/decryption. Based on Subsect. 3.1, $PK = (N, g \in \mathbb{Z}_{N^2})$, where $g \equiv 1 \mod N^2$; and $SK = (P, Q, \lambda, \mu)$, where P, Q are large primes that are different from p and q. The public key PK will be certified with an independent certification authority (which can be the same as the registration authority) and published on the bulletin board B, i.e., any party in the system will know PK and the voters can use PK to encrypted his/her vote. The private key SK is only know by T, which will be used in the final phase to decrypt the tallies.

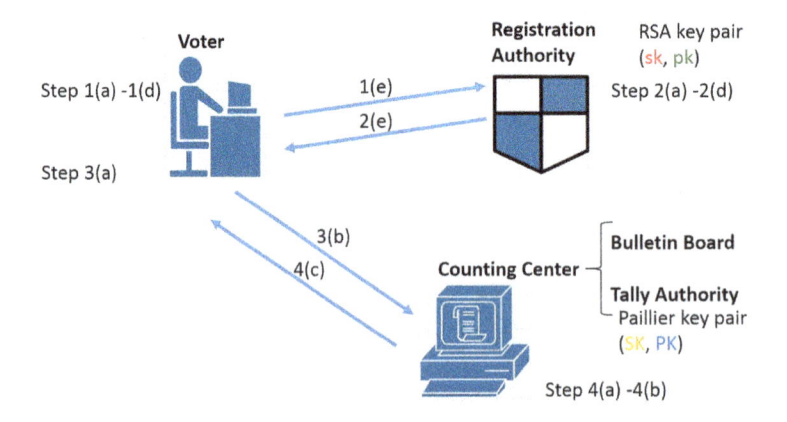

Fig. 3. The voting phase

The Voting Phase. Consider a voter V_l who wishes to vote for the i-th candidate. He issues a ballot and interacts with the registration authority A in the following way:

1. The voter
   (a) Downloads from the bulletin board the two public keys pk and PK
   (b) Decide on his vote. Let the unit vector $e_i = (0, 0, \ldots, 0, 1, 0, \ldots, 0, 0)$ of length k denote the vote, where only the i-th entry is one denoting that the voter vote for the i-th candidate. Using the tally authority's

public key PK, V_l encrypts each entry in e_i with Paillier scheme and generates the encrypted version of the vote E_l which contains a sequence of k ciphertexts $c_1, c_2, \ldots, c_k$, where c_i is the encryption of 1 and $c_j, j \neq i$ is an encryption of 0.
   - (c) Generates k proofs to convince any verifier that $(c_1, c_2, \ldots, c_k)$ is a valid vote, i.e. each ciphertext c_i is an encryption of either 0 or 1. This is done using the OR-proof presented in Fig. 2. We use $ZKProof_{V_l}(E_l)$ to denote the zero-knowledge on E_l created by voter V_l.
   - (d) To attain the signature from the registration authority on his vote e_i, the voter V_l generate a random number to mask e_i as describe in Fig. 1.
   - (e) Send $(V_l, E_l, ZKProof_{V_l}(E_l), m'_{e_i}, aux)$ to the registration authority A for authentication and verification, when aux contains some auxiliary information for checking the eligibility, m'_{e_i} denote the masked version of the vote.
2. Upon receiving the tuple $(V_l, E_l, ZKProof_{V_l}(E_l), m'_{e_i}, aux)$ from the voter V_l, the registration authority A does the following.
   - (a) Check V_l's eligibility by examine aux.
   - (b) Check if the proof $ZKProof_{V_l}(E_l)$ is valid.
   - (c) Generate a voting id for V_l.
   - (d) Execute his part of blind signature scheme as shown in Fig. 1 to generate the blind signature s using the private key sk.
   - (e) Send s together with id back to V_l. Notice that A signs exactly one ballot for each voter.
3. Once receive the blind signature s on e_i, V_l does the following.
   - (a) Unblind s and get the RSA signature σ on the plaintext version of vote e_i. σ will be used to verify the ballot by the bulletin board.[1]
   - (b) Cast the ballot anonymously on the bulletin board B by sending the tuple (id, E_l, e_i, σ) to the counting center. The communication is secured by SSL that has been agreed when the connection is set up (SSL handshake).
4. Upon receiving a vote from V_l, the Bulletin board B does the following:
   - (a) Checks the ballot is from a registered voter by verifying the signature σ on e_i.
   - (b) The encrypted version of a valid ballot E_l is posted together with id on the bulletin board.
   - (c) The bulletin board acknowledges voter by sending the id, date and time of voting and these information is signed on under his own private key.

The voting phase can be summarize in Fig. 3.

4.1 The Counting Phase

In this subsection, we present the computation of tally. First, we describe the bulletin board of the counting center which can be accessed by all parties for example from the website of the counting center.

---

[1] σ is a valid signature on e_i by the registration authority A, but e_i is never revealed to A.

Voter ID \ Candidates	C_1	C_2	C_3
id_1	$Enc(0)$	$Enc(0)$	$Enc(1)$
id_2	$Enc(1)$	$Enc(0)$	$Enc(0)$
~~id_3~~	~~$Enc(1)$~~	~~$Enc(1)$~~	~~$Enc(0)$~~
id_4	$Enc(0)$	$Enc(1)$	$Enc(0)$
id_5	$Enc(0)$	$Enc(0)$	$Enc(1)$
id_6	$Enc(0)$	$Enc(0)$	$Enc(1)$
Sum	$Enc(\Sigma_1) = Enc(1)$	$Enc(\Sigma_2) = Enc(1)$	$Enc(\Sigma_2) = Enc(3)$

Fig. 4. Bulletin board

Figure 4 shows such a bulletin board with the votes from $n = 6$ eligible voters for $k = 3$ candidates. Each row records an voter's ID number and his encrypted ballot. The encryption algorithm Enc denotes the Paillier homomorphic encryption which is randomized. The notation $Enc(0)$ is short for $Enc(0,r) = g^0 r^n$ mod n^2 where $r \in \mathbb{Z}_{n^2}$ is freshly chosen for each run of the encryption algorithm. Therefore, all encryptions of 0s look different, so that an attacker will not be able to tell whether two voters made the same vote by simply checking whether the ciphertexts are the same. Similarly, all encryptions of 1s look different in the table.

Recall that the voter ID, e.g., id_l is assigned to an eligible voter V_l by the registration authority A. The connection between voter id_l and V_l is only known by the voter V_i and A. No one else knows which ballot (row) comes from V_i. This guarantees the voter's anonymity.

When the voting is closed, each voter can check if his ballot is correct publish on the bulletin board. Tallier T will compute the voting result as follows.

1. For each row, check if the encrypted ballot is a valid vote for one candidate. Based on the additive homomorphic property of Paillier ($Enc(m_1) \cdot Enc(m_2) = Enc(m_1+m_2)$), T multiply all ciphertexts and decrypt the result using PK. If the decryption is 1, this is a valid vote. If it is greater than 1, discard this vote. For example, in Fig. 4, the vote from id_3 is invalid as it voted for both candidates C_1 and C_2, which resulting a decryption of 2. Therefore, the row for id_3 is crossed out on the bulletin board.
2. For each candidate C_i, compute the product of the correct votes and write it into the corresponding entry in the last row of B. The product is the encrypted tally for C_i.
3. Decrypt the ciphertexts in the last row to get the tallies for all candidates. Checks if the sum of the tallies equals to the number of valid ballots in B. Compare the sum Σ_j, for $1 \leq j \leq k$ and announce the winner.

5 Analysis

5.1 Complexity

Let us now estimate the computation complexity on a voter V, the registration authority A and tally authority T in the proposed e-voting scheme. Assume there are k candidates and all n voters are honestly follow the protocol and cast one correct vote.

On the voter side, for performing the a ballot encryption, a voter needs to do k encryption using Paillier. The voter also need to generate an OR-proof for each of the k encryptions. By using Fiat-Sharmit heuristic [12], we can make the OR-proof non-interactive. The voter needs to perform 3 random number generation, 4 exponentiation and 3 multiplication, one addition and one hashing. To bring the completeness error down to a negligible amount, such step need to repeat l times. To blind the original ballot, the vote need to generate one random number and perform one hashing, one exponentiation and one multiplication. Upon receiving the blind signature from A, the voter need to unblind, which takes one multiplication.

On the registration authority side, for each voter, A need to validate the voter's eligibility and check the OR-proof. The checking process takes one random number generation, 4 exponentiation, 4 multiplication. To blindly sign the ballot, A performs one multiplication.

The tally authority T first need to perform row check before counting. To check a row (ballot) posted on B, T takes $k-1$ multiplication and one Paillier decryption. To get the tally for each candidate, T performs $n-1$ multiplication[2] and one Paillier decryption. So in total, $(k-1)+k\cdot(n-1) = k\cdot n-1$ multiplications and $k+n$ decryptions. Table 1 summarizes the computation complexity of different operations in the e-voting systems we discussed above. Note that our calculations only consider heavy weight operations such modular exponentiation and multiplication in the voting and counting phases, where operations such as key generation, sampling random number, hashing, addition, validating the eligibility of voters and acknowledging voters are not included.

Table 1. Computation complexity of proposed e-voting scheme ($\mathbb{E}_P$: Paillier encryption, $\mathbb{D}_P$: Paillier decryption, Exp: exponentiation, Mul: multiplication)

Operation	Party		
	V	A	T
$\mathbb{E}_P$	k		
$\mathbb{D}_P$			$k+n$
Exp	5	$4n$	$k\cdot n-1$
Mul	4	$5n$	

---

[2] This step can be asymptotically improved from $O(n)$ to $O(\log n)$ by a parallel algorithm that apply multiplication operation in a pairwise fashion using a binary tree.

5.2 Security Analysis

Now we show that the proposed e-voting scheme is secure, i.e., it satisfies correctness (including No vote duplication, eligibility, fairness), privacy and verifiability.

1. Correctness of the proposed e-voting scheme:
   - No fake vote and the sum of valid ballots are accurately counted.
   - No vote duplication: a voter can vote twice.
   - Eligibility: only eligible voters can vote.

   Proof: All encrypted ballots posted on the bulletin board are verified by the bulletin board first. Invalid ballots, i.e., ballot that contains encryption of neither 1 or 0 cannot be added. Before tallying, the tally authority will further check if a ballot only voted for one candidates. Invalid vote will be crossed out on the bulletin board. Therefore, only valid ballot will be tallied. Due to the additive homomorphic properties of the Paillier encryption, the final tally is the sum of all valid ballots.

   Once a voter registered with the registration authority A, he has a record in the database; any attempt to vote the second time, his additional encrypted ballot will be rejected by A. Without the signature from A, the duplicated ballot cannot be casted on the bulletin board. Therefore, the proposed e-voting scheme satisfies no vote duplication.

   The voter's identity must be registered and checked by A before signed blindly. Everyone can verify that the voter who has cast a ballot on the bulletin board is included in the list of all eligible voters.

   Hence, the proposed e-voting scheme satisfies completeness.

2. Privacy: All ballots are secret (not linked to voters).

   Proof: Since the ballot is encrypted with the public key of the tally authority T, the encrypted ballot cannot be individually decrypted from its corresponding ballot by any party except T. The encrypted ballot is sent anonymous to the counting center accompanied by the voter registration id. Since id is issued by registration authority A, no one else can link a ballot on the bulletin board to the voter, except the voter himself or A. In addition, Paillier encryption is a randomized encryption scheme, which produces different ciphertexts for a single vote (based on a freshly chosen random number). For efficiency and security purpose, tally authority T does not decrypt single vote. Instead, T only decrypted the tallies for each candidates. Therefore, the proposed e-voting scheme ensures privacy.

3. Verifiability: the public can verify the voting system and a voter is able to verify if his/her vote is counted in the final tally.

   Proof: The final ballot and the proof of validity are posted on the bulletin board therefore, anyone can verify the validity of the final vote, the correctness of the ballot collection and the final result. Thus, valid votes are counted correctly.

Due to the page limit, we only presented an informal argument about the security of our e-voting scheme above. Formal security can be proved based on the security results of the OR-proof used, the semantic security of Paillier encryption scheme and the unforgeability and blindness of the RSA blind signatures.

6 Implementation

In this section, we first provide a description on how we implemented the e-voting system on CPU in Subsect. 6.1. After we explain how we have setup the programs to simulate different steps in the e-voting system, we will provide the results and analysis of our program in terms of efficiency, and discuss how certain aspects of this implementation may be improved by parallelization. Then in Subsect. 6.2, we present the speed-up result on GPU using CUDA.

6.1 Implementation on CPU

Experiment Setup. We performed all testing on a system with an Intel Core i7-8700 CPU clocked at 3.20 GHz, and the system uses Debian Linux 9 (Stretch) Operating System. Since our e-voting system is based on the hardness of factoring large integer problem, for security purpose, the module n^2 in Paillier should be of 4096 bits long. In order to handle the parameter n of 2048 bits, the "GNU Multi Precision Arithmetic Library" (GMP) is necessary. We created our own RSA library written in C using GMP. It only contains methods that are necessary for this e-voting system, specifically for the blind signature and zero-knowledge proof steps. We used a Paillier library implemented using GMP created by Bethencourt [5], and we make some minor modifications so that it can be used together with our RSA library.

Some Details. In our e-voting scheme as described in Sect. 4, two pair of public/private keys must be generated written in put separate files: the RSA keys used for blind signatures and the keys for Paillier for voting and counting. The SHA-1 hashing method created with GMP is used in our implementation In the blind signatures step, the random value r needed to be generated in slightly different way since Paillier has the random value generated to be co-prime with n^2. The important thing to note is that the random number cannot be less than half of n. This ensures that the random number chosen will properly be encrypted and decrypted.

Testing and Performance. We run our test in the case of three candidates for 10, 100, and 1000 voters. The public parameter n used in both Paillier and RSA are of length 2048 bits. That is the secret parameters p and q are primes of length 1024 bits. The output was recorded for each of the following steps: Paillier key generation, RSA key generation, Paillier encryption for one vote, OR-proof, blind signature, row counting, tallying votes. These steps were repeated multiple to get an average running time.

The testing result shows that the average time for generating the Paillier keys for n^2 size of 4096 bits is 17 s. The average for generating the RSA keys for n of size 2048 bits is 13 s. The key generation steps can be down before the voting in an off-line manner. The average time for encryption using Paillier of 1 vote is 296 ms. The average time for interactive OR-proof between one voter and the registrar authority is 8.4 s when setting $A = 5000$ and the repetition number $t = 7$. Table 2 below shows the results for the row checking, tallying, and blind signatures steps for 10, 100, and 1000 voters.

Table 2. Elapsed time (in seconds) as no. of voters increase

Operations	#Voters		
	10	100	1000
Row Check	0.979	10	98
Tallying	0.295	0.306	0.379
Blind Signatures	0.02	0.2	2

Discussion. As suggested by the data, the time required to perform Row Checking and Blind Signature grows at the same rate as the number of voters. This means that when the number of voter is one million, there is an estimated 27 h for Row Checking and 33 h for Blind Signature. These steps will take too long and do not meet the standards for speed. The remote e-voting schema promises more instantaneous results. Therefore, it is necessary to look for ways to speed up the process. One possible solution which will provide a significant speed up could be done using Nvidia's graphical processing units along side with CUDA. However, there is currently no multi-precision arithmetic library implementation from CUDA which can work similarly to the GMP library. Since a good implementation of this e-voting system heavily relies on multi-precision arithmetic, these steps with the same parameter length cannot be yet performed in the same way with CUDA. In next subsection, we present our implementation result for a "small" version of the e-voting system on GPU using CUDA.

6.2 Speed-Up Using CUDA

In the Sect. 5, we discussed the complexity of the proposed e-voting scheme and showed that computation cost grows linearly as the number of voters grows. This is also demonstrated in Subsect. 6.1. Since signing the blinded ballots, checking for valid votes and counting up the votes are independent processes, they could be made to run in parallel in order to improve the efficiency for large-scale elections. The GPU is designed to do multiple tasks simultaneously. If the tasks can be made massively parallel, then the GPU computation will be several orders of magnitude faster.

We have done some research on possible ways to implement this e-voting system using CUDA. Technically speaking, it is possible to translate most of the GMP library to be used for CUDA. However, this task will prove to be a difficult and time consuming effort. There are a few libraries currently available which might prove to be useful if properly documented or further studied.

The CUMP library [23] and XMP (accelerated(X) Multi-Precision) [19] both claimed to enable batched multi-precision operations in CUDA. CUMP is designed to be faster on NVIDIA GF100/GF110 GPUs. The advantage in performance is reached by using 64-bit fullwords as a basic arithmetic type. It also uses a register blocking technique, and it uses "little endian" order format. Unfortunately, both libraries lack of documentation which made them hard to be used. Further investigation should be made if time allowed.

Cuda-fixnum [18] is an extended-precision modular arithmetic library that is designed to be used on CUDA. The developers claim that the library can create efficient functions to operate on vectors on n-bit integers. The library currently supports integers of size 32, 64, 128, 256, 512, 1024, and 2048. Therefore, it will not meet out Paillier requirement of size 4098 bits for the module n^2. This library also does not contain proper documentation, and there are not many instructions to how to use some of the functions featured in the library. This library is fairly new compared to the other two libraries mentioned, and the authors claim that this library will continue to receive updates. We hope that we may be able to use this library in the near future to be able to apply for our e-voting system.

Designing our own library will most likely prove to be the best and most efficient method. Next step is to dedicate time to developing some library to provide multi precision arithmetic. At the least, the functions that are necessary to run the Blind Signature and Row Counting step should be implemented as part of this library. There has been some research done recently by Emmart [11] on high-performing multi-precision arithmetic using graphical processing units. The author presents many algorithms that can be used to implement fast arithmetic with larger numbers. It also discusses the use of some parallel algorithms. We will do some additional research such as this paper to help us develop an efficient multi-precision arithmetic library for CUDA.

Implementation of a "Small" Version on GPU. In order to show the speed-up of our e-voting scheme in parallelism, we implemented our scheme in C++ on CUDA, which enables GPU for general purpose processing. We choose to use the datatype *unsigned long long* that allows us to store integer value up to $2^{64} - 1$ which is the biggest integer value it can support in standard C++ (on 64-bit platform). In this "Small" version, the primes p and q used in Paillier must be of length 16 bitsm and hereby the module n^2 is of 64 bits.

We compare the performance of our e-voting system on both the CPU and the GPU using an Intel(R) CoreTM PC (i7-8550U, 8 GB RAM with Windows 10 64-bit Operating System) and a NVIDIA GPU (NVIDIA GeForce GTX 1080, 8 GB on device memory). We measure and compare the execution time for the steps of blind signatures and the tallying phase. As expected, the usage of GPU significantly improves the speed of execution of Signing, Paillier decryption and module multiplication. The improvement factor increases with the number of voters increase. Our experiment shows that when test in the case of three candidates with 10,000 voters, the GPU implementation on blind signatures and tallying are 410 times faster than the CPU implementation.

As a remark, the simulation for concurrent setting of zero-knowledge proof is quite tricky. The OR-proof we presented in our scheme is similar to the The Guillou-Quisquater (GQ) identification scheme [14] that is based on RSA. The poofs of security concurrent attacks is unknown. In our implementation, we did not take into account of the parallelization for this step. Further investigation is needed.

7 Conclusion

In this paper we proposed a secure and simple e-voting scheme that is based on Pailler homomorphic encryption. A classical CPU-based implementation of our voting system is presented. The implementation is based on GMP, with the Paillier module n^2 of length 4096 bits. We discovered that the counting of the votes is a fast process compared to the Row Checking and Blind Signature steps. In order to improve efficiency of these steps, we suggest to implement these steps using Nvidia GPUs. Due to the lack of user friendly multi-precision arithmetic library for CUDA, we are unable to proceed with such an implementation. However, a "small" version is implemented to demonstrate the acceleration.

Our next step is to develop or discover a multi-precision arithmetic library similar to GMP to use for CUDA so that we may be able to implement the Blind Signature and tallying step in parallel. This step is important because the current time needed to perform these steps is high as the number of voters increases. In addition, we want to looking into the concurrent security of the concrete construction of the OR-proof in Subsect. 3.3. Furthermore, we plan to simulate the network aspect and transportation of data over networks, e.g.,transfer of votes from the voter to the registration authority as well as the counting center and estimate the communication complexity of our e-voting scheme.

References

1. Youth vote in the united states. https://en.wikipedia.org/wiki/Youth_vote_in_the_United_States
2. Voters with disabilities: Observations on polling place accessibility and related federal guidance (2017). https://www.gao.gov/products/GAO-18-4
3. 2016 November general election turnout rates (2018). http://www.electproject.org/2016g
4. Baudron, O., Fouque, P.A., Pointcheval, D., Stern, J., Poupard, G.: Practical multi-candidate election system. In: Proceedings of the Twentieth Annual ACM Symposium on Principles of Distributed Computing, PODC 2001, pp. 274–283. ACM (2001)

5. Bethencourt, J.: Paillier library. http://hms.isi.jhu.edu/acsc/libpaillier/
6. Boneh, D., Golle, P.: Almost entirely correct mixing with applications to voting. In: Proceedings of the 9th ACM Conference on Computer and Communications Security, CCS 2002, pp. 68–77. ACM (2002)
7. Chaum, D.: Blind signatures for untraceable payments. In: Advances in Cryptology: Proceedings of CRYPTO 1982, pp. 199–203. Plenum (1982)
8. Cramer, R., Damgård, I., Schoenmakers, B.: Proofs of partial knowledge and simplified design of witness hiding protocols. pp. 174–187, January 1994
9. Damgard, I.: On Σ-protocols (2010). http://www.daimi.au.dk/~ivan/Sigma.ps. Version 2
10. Dossogne, J., Lafitte, F.: Blinded additively homomorphic encryption schemes for self-tallying voting. In: Proceedings of the 6th International Conference on Security of Information and Networks, SIN 2013, pp. 173–180. ACM (2013)
11. Emmart, N.: A study of high performance multiple precision arithmetic. Ph.D. thesis, University of Massachusetts Amherst, March 2018
12. Fiat, A., Shamir, A.: How to prove yourself: practical solutions to identification and signature problems. In: Proceedings on Advances in Cryptology—CRYPTO 1986, pp. 186–194. Springer, London (1987)
13. Goldwasser, S., Micali, S., Rackoff, C.: The knowledge complexity of interactive proof-systems. In: Proceedings of the Seventeenth Annual ACM Symposium on Theory of Computing, STOC 1985, pp. 291–304. ACM, New York (1985)
14. Guillou, L.C., Quisquater, J.J.: A "paradoxical" identity-based signature scheme resulting from zero-knowledge. In: Proceedings on Advances in Cryptology, pp. 216–231. CRYPTO 1988. Springer, Berlin (1990)
15. Heiberg, S., Kubjas, I., Siim, J., Willemson, J.: On trade-offs of applying block chains for electronic voting bulletin boards. IACR Cryptol. ePrint Archive **2018**, 685 (2018)
16. Juels, A., Catalano, D., Jakobsson, M.: Coercion-resistant electronic elections. In: Proceedings of the 2005 ACM Workshop on Privacy in the Electronic Society, WPES 2005, pp. 61–70. ACM (2005)
17. Moura, T., Gomes, A.: Blockchain voting and its effects on election transparency and voter confidence. In: Proceedings of the 18th Annual International Conference on Digital Government Research, DG.O 2017, pp. 574–575. ACM (2017)
18. n1analytics: n1analytics/cuda-fixnum, November 2018. https://github.com/n1analytics/cuda-fixnum
19. NVlabs: Nvlabs/xmp, September 2016. https://github.com/NVlabs/xmp
20. Paillier, P.: Public-key cryptosystems based on composite degree residuosity classes. In: Proceedings of the 17th International Conference on Theory and Application of Cryptographic Techniques, EUROCRYPT 1999, pp. 223–238. Springer (1999)
21. Parsovs, A.: Homomorphic tallying for the estonian internet voting system. IACR Cryptol. ePrint Archive **2016**, 776 (2016)
22. Rivest, R.L., Shamir, A., Adleman, L.: A method for obtaining digital signatures and public-key cryptosystems. Commun. ACM **21**(2), 120–126 (1978)
23. skystar0227: skystar0227/CUMP, October 2012. https://github.com/skystar0227/CUMP

AI-Enabled Digital Forensic Evidence Examination

Jim Q. Chen[✉]

U.S. National Defense University, Fort McNair, Washington DC, USA

Abstract. Digital forensics is crucial for the prosecution of offenders in cyberspace, including nation-state actors and non-nation-state actors. The evidence discovered, verified, and associated during the evidence examination phase serves as the basis for digital forensic analysis and eventually the basis for the verdict of a judge and a jury. However, the current digital forensic evidence examination procedure usually takes a relatively long time, demands a great amount of resources, and requires great efforts from human experts. The biggest challenges in this procedure are accuracy and speed. Nevertheless, in some cases, delay is not allowed, as it may have significant impact upon a critical mission, especially a time-sensitive one. To address this challenge, this paper recommends an AI-based digital forensic evidence examination architecture that is empowered by contextual binding, machine learning, and human-machine teaming. In this approach, human experts are teamed up with artificial intelligence (AI) systems in conducting evidence examination. This approach certainly improves the efficiency and effectiveness of an investigation, thus successfully supporting missions.

Keywords: Digital forensics · Evidence examination · Artificial intelligence · Contextual binding · Machine learning · Human-Machine teaming

1 Introduction

Digital forensics plays a significant role in supporting justice because it provides untampered evidence to the court of law, making evidence speak for facts without any bias. Hence, it is a crucial method in prosecuting offenders in cyberspace, no matter whether they are nation-state actors or non-nation-state actors. Digital forensics is centered around evidence that is admissible in a court of law. It consists of the follows phases: evidence identification, evidence collection, evidence preservation, evidence acquisition, evidence examination, evidence analysis, evidence documentation, and evidence presentation. Chain of custody is used to maintain the integrity of evidence throughout the life-cycle of evidence. This whole process, manned by knowledgeable human digital forensic experts, usually takes long time and requires a lot of resources. In a time-sensitive case such as a cyber conflict, any delay in digital forensic investigation may yield significant consequences. In general, the challenges in digital forensics are accuracy and speed.

Instead of addressing all the phases of digital forensic investigation, this paper is only focused on operations within digital forensic evidence examination, as this operation heavily relies on computer systems and it can be automated with the help of AI.

AI intends to implement human intelligence into computer systems. It supports two modes: the automation mode and the autonomy mode. The automation mode allows computer systems to automate its process under human supervision, while the autonomy mode allows computer systems to make decisions and act on their own accordingly without human intervention. Machine learning is a subset of artificial intelligence. Machine learning mechanisms, as pointed out by Schmidhuber [1], provides "chains of possibly learnable, causal links between actions and effects".

In order to address the challenges of speed and accuracy in digital forensic evidence examination, this paper proposes an AI-enabled approach that is empowered by contextual binding, machine learning, and human-machine teaming. This approach is certainly capable of guaranteeing accuracy while providing fast speed in digital forensic evidence examination. It is of practical significance in cyber conflict and cyber defense since it enables fast and accurate targeting by quickly identifying attackers. Besides, it enables AI-systems to learn quickly and dynamically with the help of various checking and validation processes built into the architecture proposed.

The paper is organized in the following way: In this section, i.e. Section 1, a brief overview of digital forensics is provided, the scope of this study is drawn, and the methodology used is briefly mentioned. In Sect. 2, the issues and limitations of the current approach in digital forensic evidence examination are discussed. In Sect. 3, an AI-enabled digital forensic evidence examination architecture is recommended. In Sect. 4, the implementation and the benefits of this architecture are discussed. Future research is suggested. In Sect. 5, a conclusion is drawn.

2 Related Works

In digital forensic evidence examination, when massive amount of complex data needs to be processed while not enough resources are available to support the effort, delay inevitably occurs.

Irons and Lallie [2] discusses the "large data problem in digital forensics", which, they hold, is caused by "a continual growth in cybercrime, increasing magnitudes of storage, a multitude of data evidence sources and continual increases in computational power". They argue that the large data problem leads to the increase in the backlog in digital investigation.

Lakade [3] shares the same view. He thinks that "a major challenge to digital forensic analysis is the ongoing growth in the volume of data seized and presented for analysis".

Simou et al. [4] examine the challenges in digital forensic investigation from another perspective, i.e. cloud forensics. In this new and complex environment, the challenges are not only from the volume, velocity, and variety of data but also from the identification of the physical locations of data and the acquisition of evidence in the cloud. "Investigators have to conduct digital forensic investigation on cloud computers to identify, preserve, collect, and analyze all the evidentiary so as to acquire accurate results and properly present them in a court of law". Likewise, Lopez et al. [5] address the digital forensic challenges in cloud computing. They mention the huge amounts of data as well as the platform that may be used "to distribute malware, conduct scams and perform other criminal activity". In addition, they mention other technical challenges such as encryption, steganography and anti-forensic tools. Among all these challenges, they point out that "nonetheless, according to most forensic practitioners, the biggest issue they need to deal with is the enormous amount of data they need to examine. Additionally, when dealing with digital evidence, almost every action can modify the evidence or leave digital traces that may have legal significance. Hence, forensic examinations need to be undertaken by highly qualified staff." Also, in a cloud environment, they hold that "forensic investigators might not have absolute control of the evidence".

Alqahtany et al. [6] maintain that the main challenges in cloud forensics are dependence upon cloud providers, time analysis and evidence correlation of data sets from multiple sources, cross border issues, lack of control on environments, and jury's technical comprehension.

Other than the challenges from data such as the volume, velocity, and variety, there are challenges from the current methods and tools utilized in digital forensic evidence examination.

As pointed out by Beebe [7], the "cost of human analytical time spent sifting through non-relevant search" is a big issue. To deal with these issues, Beebe demands smart analytical algorithms that are able to successfully ignore noise and quickly get to relevant data with the help of contextual information. Beyond doubt, this is a good idea. However, Beebe does not further explore ways of implementing this idea.

Pollitt [8] argues that a huge amount of data does not equal to a huge amount of knowledge, or relevant pieces of evidence in this particular case. The exact term Pollitt uses is "data glut, knowledge famine".

All these issues and challenges hinder prompt and accurate digital forensic evidence examination. To address these issues and challenges, an AI-enabled digital forensic evidence examination architecture, which is empowered by contextual binding, machine learning, and human-machine teaming, should be set up and resorted to as a solution. This effort can help to materialize what Marti and Reinelt [9] dream for, i.e. obtaining a solution that is optimal and accurate with reasonable computational effort.

This recommended solution is discussed in the next section.

3 An AI-Enabled Digital Forensic Evidence Examination Architecture

As shown in the previous section, the major challenges in digital forensic evidence examination are speed and accuracy in identifying targets or suspects with the support of relevant and reliable evidence discovered from a huge amount of input data sets. To reach this goal, a huge amount of data sets, especially those in cloud environment, has to be processed with AI-enabled systems. Mere manual examination is too slow to be practically useful at all, even though it may still be needed for checking and validation. Undoubtedly, the volume, velocity, and variety of data have to be handled; different aspects have to be scrutinized; and relevant evidence has to be quickly identified.

Among these tasks within this particular phase of investigation, the most challenging ones are to quickly identify relevant pieces of evidence, use them to recreate crime scenes, and accurately identify suspects.

To accomplish these tasks, the unique capabilities of humans and the unique capabilities of machines (i.e. computer systems) have to be recognized. Machines are much better than humans in dealing with the challenges of the volume, velocity, and variety of data, and in quickly identifying patterns among the available pieces of data. Humans are much better than machines in figuring out heuristic approaches with the help of intuition, guts feeling, and the sixth sense. The unique capabilities that humans possess can help generate intelligent hints that effectively shorten the process of evidence examination conducted by machines in discovering relevant evidence and associating evidence to suspects.

It needs to be mentioned that there are at least two types of evidence examination. One type is target-aware evidence examination, and the other is target-unaware evidence examination. The former refers to the discovery of specific artifacts clearly listed in a subpoena or a warrant. So to speak, it is a guided search. The latter refers to the discovery of suspicious artifacts that may eventually lead to the discovery of a suspect. However, at the very moment, there is no specific artifacts listed in a subpoena or a warrant. So to speak, it is an unguided search.

In a guided and target-aware evidence search, supervised learning may be effectively utilized. Likewise, in an unguided and target-unaware evidence search, unsupervised learning can be successfully employed.

To satisfy the unique requirements for operations within this phase of investigation, a unique architecture is called for. Within this architecture, a competent of human-machine teaming is needed. The formation of this team makes it possible to fully utilize the strengths of humans as well as the strengths of machines.

In normal situations when fast-pace in data processing should be achieved, machines take the lead in looking for evidence based on instructions or requirements from a subpoena or a warrant. Machine learning algorithms are used to speed up this process. Meanwhile, occasions are reserved for humans to offer guidance or intelligent hints should there be any need. Occasions are also reserved for humans to verify the correctness of the results generated by machines whenever necessary. This way the accuracy and speed are guaranteed simultaneously. To this end, humans and AI-enabled machine systems are provided with opportunities to inject intelligent hints that help to reduce the search time in identifying relevant evidence and then discovering targets.

In order to gain a holistic view in digital forensic evidence examination, different types of input data sets from various sources are collected and processed. They consist of textual data sets, image data sets, voice data sets, and video data sets. The AI-enabled systems are capable of processing all these types of data sets. Besides, the systems are capable of associating relevant pieces of evidence discovered from different types of sources. These capabilities are very useful in identifying suspects accurately and promptly.

These capabilities are empowered by the following key components, which together serve as the engine of the AI-enabled digital forensic investigation system. The key components consist of the Revised Restrictive Contextual Binding Condition proposed in Chen [10], a priority list for examination, and relevant machine learning mechanisms.

The Revised Restrictive Contextual Binding Condition requires each variable should be properly bound by its corresponding contextual operator. This binding condition is listed below:

Assume X is an entity, and CO is a contextual operator.

In a specialized time, location, environment, and background, if X is directly related to CO with respect to all the attributes such as action-initiator (who), action (what), action-recipient (who/what_recipient), time (when), location (where), method (how), and purpose (why) in such a setting:

$CO_i[WHO_1, WHAT_2, WHAT\_RECIPIENT_3, WHEN_4, WHERE_5, HOW_6, WHY_7]$
$\{\ldots\ldots X_i[WHO_1, WHAT_2, WHAT\_RECIPIENT_3, WHEN_4, WHERE_5, HOW_6, WHY_7]\ldots\ldots\}$

then X_i is contextually bound by CO_i in a restrictive way.

As pointed out in Chen [10], this is a typical representation of Type 1 Binding as all the attributes in the variable are contextually bound by the attributes in the contextual operator. "If one contextual attribute in the variable is not directly related to the corresponding attribute in the contextual operator, the variable is not contextually bound by the contextual operator in the restrictive sense."

In the digital forensic evidence examination, once a variable entity is contextually bound by a contextual operator, it is immediately identified with certainty as a target because all the conditions for being a target are satisfied.

Besides, a priority list is required to accelerate evidence examination. This list can be revised based on the change of contexts.

Chen [11] proposes a context-based heuristic approach, which makes use of the value of weight assigned to the key attributes in determining the priority list in evidence search. The key attributes are comprised of "who", "why", "how", "when", "where", and "what". By default, the weight attribution is as follows: the attribute "who" has 30% weight, the attribute "why" has 25% weight, the attribute "how" has 15% weight, the attribute "when" has 10% weight, the attribute "where" has 10% weight, and the attribute "what" has 10% weight.

The formula for this default environment is as follows:

$$\begin{aligned} P(X) &= \sum_{i=1}^{6}(Xi * Wi) \\ &= (X_1 * W_1) + (X_2 * W_2) + (X_3 * W_3) + (X_4 * W_4) \\ &\quad + (X_5 * W_5) + (X_6 * W_6) \\ &= (1 * 0.3) + (1 * 0.25) + (1 * 0.15) + (1 * 0.1) \\ &\quad + (1 * 0.1) + (1 * 0.1) \\ &= 0.3 + 0.25 + 0.15 + 0.1 + 0.1 + 0.1 \\ &= 1 \end{aligned} \quad (1)$$

Given this weight attribution, the attribute "who" will be the first one to be searched; the attribute "why" will be the second one to be searched; and the rest of the attributes will be searched in the order based on the weights that they bear. A search based on the combination of attributes is also allowed. The order is based on the total weight assigned to the relevant attributes. For instance, the attribute combination of "who" and "why" has a priority higher than the attribute combination "how" and "what" as the former possesses 55% weight while the latter occupies 25% weight.

The value of weight is adjustable based on requirements within the customized contextual environment. For instance, if a subpoena or a warrant just asks for an individual, the automated forensic investigation system uses the default priority list shown above since it is focused on the evidence about that individual. If a warrant asks for what that individual did in the virtual world, the attribute combination of "who" and "what" thus has the highest priority. The corresponding weight will be reassigned. In this particular case, the attribute "who" has 30% weight, the attribute "what" has 20% weight, the attribute "when" has 15% weight, the attribute "where" has 15% weight, the attribute "why" has 10% weight, and the attribute "how" has 10% weight. This weight attribution guarantees that the attribute combination of "who" and "what" should have the highest priority so that it will be searched first.

The search process is led by AI-enabled systems so that speed can be guaranteed. During this process, the object that bears the file name listed in the subpoena or the warrant is the target hunted for. This hunt examines all the locations that the suspect can get access to and can save files to, including local space, network space, cloud space, unallocated/unpartitioning space, and movable space.

To assist this hunt, the timestamps of artifacts, such as files, are examined. These timestamps include at least the file creation time, the file last access time, the file modification time, and the file last saved time. The different patterns of the combinations of these timestamps are analyzed to find out whether the files in question were created on this device or they were copied from another device to this device. These timestamp patterns are also analyzed to find out whether these files were modified after they had been created or copied from another device.

In addition, should there be no direct match between the file names listed in a subpoena or a warrant and the file names in a system under investigation, the factors that may reveal the inconsistency of a file are then hunted for in the next round of evidence examination. These factors include but are not limited to the size of a file, the location of a file, the frequency of the revision of a file, the type of a file, the way in which a file was revised or manipulated, the consistency of a file in its environment, and the behavior of a file. These factors may lead to the discovery of a potential target file. Depending upon the order of the factors used in a hunt, the priority of the attributes can be adjusted accordingly. In a sense, the value of weight can be changed to support a new type of search every time when the contextual environment gets changed.

In an unguided and target-unaware search, specific target artifacts are not provided in a subpoena or a warrant. In this environment, the default weight attribution listed in (1) is utilized. During the phase of digital forensic evidence examination, some unexpected abnormal objects may be discovered. Unsupervised learning is thus employed to assist the categorization of the new discovery and to initiate new rounds of examination.

To support all these operations in a systematic way, the AI-enabled digital forensic evidence examination architecture is proposed. It is displayed below in Fig. 1.

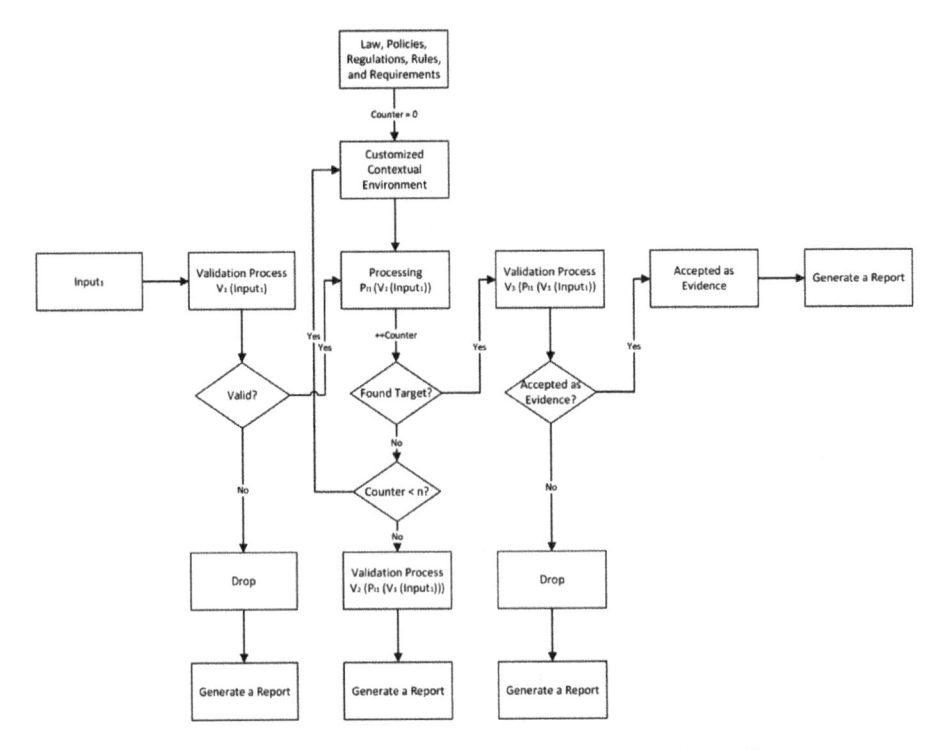

Fig. 1. The AI-enabled digital forensic evidence examination architecture.

There are three points that should be made with respect to this architecture. First, accuracy is evidently emphasized. For each possible result, validation mechanism is used before a forensic examination report is generated. This validation mechanism is led by humans while being assisted by AI-enabled systems. Human investigators are held responsible and accountable for the accuracy of the examination. Second, this AI-enabled automated digital forensic evidence examination process is fully supported by human-machine teaming, which guarantees speed and accuracy. Machines are responsible for speed while humans are responsible for accuracy. This unique combination can successfully support the mission of digital forensic examination by quickly and accurately identifying relevant forensic evidence. Third, intelligent hints provided by humans or learned by machines can be injected into the customized contextual environment. This makes it possible to shorten search paths or to reduce the number of rounds in searches in the process of looking for relevant artifacts. Thus, heuristics is successfully implemented in searches with the help of human-machine teaming.

In the next section, the implementation of this AI-enabled digital forensic evidence examination architecture is discussed. So are its benefits.

4 Discussion

This AI-enabled digital forensic evidence examination architecture can be used for both target-aware searches and target-unaware searches. In a target-aware search, the priority list within the customized contextual environment needs to be modified to satisfy the requirements in a subpoena or a warrant. In a target-unaware search, the default priority list is used because no specific target is listed in a subpoena or a warrant. While in its search, unsupervised learning helps to identify inconsistency of particular objects with respect to file timestamps, size, location, type, environment, behavior, revision frequency, way of manipulation, and other relevant factors. The inconsistency is then categorized and further investigated to find out different types of suspicious and hidden artifacts, such as files hidden in unallocated partition space, files hidden within other files, files with file type code changed, and files with file extension changed. At the time when a new category is created, the priority list within the customized context environment is adjusted accordingly. As a result, a new focused evidence examination is initiated.

Some specific features in the AI-enabled digital forensic evidence examination architecture are worth mentioning.

One is the validation of the input data sets. Before they are processed, the input data sets are validated to make sure they have not been tampered and findings from the data sets meet the qualification of forensic evidence. Should the data sets fail in this validation, they are immediately dropped from the digital forensic evidence examination process. Only the data sets that are validated can be further processed. Symbolically, an input data set that is in the validation process is represented as "$V_1(Input_1)$".

Another feature is the direct association of the investigator who conducts the evidence examination to the validated data set being examined. This holds the investigator accountable even if the automated process is being utilized. This makes sure that the investigator should check and validate the automated procedure before it is put into use and the investigator should check and validate the results returned by the automated procedure. This checking and validation mechanism guarantees the validity of the evidence, which can eventually be accepted by a court of law. Symbolically, the validated target found is represented as "$V_3(P_{II}(V_1(Input_1)))$".

Still another feature is the validation of the results from digital forensic evidence examination. Before the results are accepted as evidence, another round of checking and validation should be conducted. Again, humans are taking the lead in this validation process. Only those that pass this round of checking and validation are accepted as evidence.

Obviously, this architecture has integrated multiple rounds and different levels of checking and validation. This certainly guarantees the quality of the evidence, making sure that it be accepted by a court of law. Meanwhile, the architecture promotes human-machine teaming in the use of artificial intelligence. In digital forensic evidence examination, humans are assigned with the leading role while machines are assigned with the assistant role. This means that humans are held liable for the result of evidence examination. This guarantees accuracy in the examination. Meanwhile, automated procedures enhanced by intelligent hints from humans and machines are utilized to accelerate the evidence examination procedure. This guarantees fast speed in digital forensic evidence examination.

Beyond doubt, this architecture has the following advantages.

First, the AI-enabled digital forensic evidence examination architecture, empowered by contextual binding, machine learning, and human-machine teaming, can successfully address the challenges of speed and accuracy in evidence examination. It not only speeds up the process in locating the evidence but also guarantees the quality of the evidence.

Second, this architecture is of practical importance, especially in cyber conflict and cyber defense. Quickly locating the relevant evidence in identifying attackers helps fast and accurate targeting, which forcefully supports the defense forward strategy that "moves our cyber capabilities out of their virtual garrisons, adopting a posture that matches the cyber operational environment", as stated by Nakasone [12]. In other words, the defense forward strategy employs deterrence and varied cyber operations that disrupt or halt malicious cyber operations at its source, or at least degrade them before they hit their intended victims.

Third, this architecture employs various checking and validation processes. This not only guarantees the quality of evidence, but also quickly finds out the errors that AI systems make should there be any. This helps the AI systems to learn from the errors that they make and then figure out ways of correcting these errors. As a result, this enhances the learning capability of these systems. From this perspective, this architecture serves as a good venue for the improvement of machine learning.

As shown previously, the AI-enabled digital forensic evidence examination architecture is of great significance in digital forensic investigation. In future research, this architecture needs to be tested with various types of forensic input data sets so that its practical value can be evidently evaluated. Besides, this architecture may be further fine-tuned and adjusted to satisfy other unique needs in cyber operations. In addition, this architecture can be further extended to cover other phases in digital forensic investigation, such as digital forensic evidence analysis.

5 Conclusion

Digital forensics is crucial for the prosecution of offenders in cyberspace. The evidence discovered, verified, and associated during the evidence examination phase serves as the basis for digital forensic analysis and eventually serves as the basis for the verdict of a judge and jury. In cyber conflict, it makes quick and accurate targeting possible. However, the current digital forensic evidence examination procedure usually takes a relatively long time, demands a great amount of resources, and requires great efforts from human experts. The biggest challenges in this procedure are accuracy and speed. Accuracy is so critical because the evidence will not be accepted by a court of law should there be any doubt about it. Likewise, speed is so critical because an opportunity may be missed in forward defense should there be a delay in targeting.

To address these two major challenges, recommended is an AI-enabled digital forensic evidence examination architecture empowered by contextual binding, machine learning, and human-machine teaming. With contextual binding, multiple rounds of focused and structural evidence examination are made possible. With the use of machine learning methods, such as supervised learning methods, unsupervised learning methods, and reinforcement learning methods, the AI-enabled system can have various ways of handling various types of data sets and quickly making sense out of them. With human-machine teaming, various built-in checking and validation processes can be enabled; various heuristic and intelligent hints can be promptly inserted into the AI-enabled system. All these mechanisms make it possible for the AI-enabled system to learn quickly and dynamically so that it can be more intelligent and can rapidly identify relevant targets in various environments.

As shown in this paper, this approach is certainly capable of guaranteeing accuracy while providing fast speed in digital forensic evidence examination. It is of practical significance in cyber conflict and cyber defense because of the unique capability of fast and accurate targeting that it provides. Besides, it can make AI systems more intelligent and quickly adapt to new environments. As a result, an efficient and effective investigative procedure is ensured, and missions in cyberspace can successfully be accomplished.

References

1. Schmidhuber, J.: Deep learning in neural networks: an overview. In: Technical Report IDSIA-03-14. Switzerland: The Swiss AI Lab IDSIA, University of Lugano & SUPSI (2014)
2. Irons, A., Lallie, H.: Digital forensics to intelligent forensics. Future Internet **6**, 583–596 (2014). https://doi.org/10.3390/fi6030584
3. Lakade, S.: Digital forensics: current scenario and future challenges. Int. J. Inf. Secur. Cybercrime. **4**(2) (2015). https://doi.org/10.19107/ijisc.2015.02.02
4. Simou, S., Kalloniatis, C., Gritzalis, S., Mouratidis, H.: A survey on cloud forensics challenges and solutions. Secur. Commun. Netw. **9**, 6285–6314 (2016). https://doi.org/10.1002/sec.1688
5. Lopez, E., Moon, S., Park, J.: Scenario-based digital forensics challenges in cloud computing. Symmetry **8**(107), 1–20 (2016). https://doi.org/10.3390/sym8100107
6. Alqahtany, S., Clarke, N., Furnell, S., Reich, C.: Cloud forensics: a review of challenges, solutions and open problems. In: Proceedings of the 2015 International Conference on Cloud Computing (ICCC), Riyadh, Saudi Arabia, pp. 1–9, 27–28 April 2015 (2015)
7. Beebe, N.: Digital forensic research: the good, the bad and the unaddressed. In: Peterson, G., Shenoi, S. (eds.) Advances in Digital Forensics V, pp. 17–36. Springer (2009)
8. Pollitt, M.: The good, the bad, the unaddressed. J. Digit. Forensic Pract. **2**(4), 172–174 (2009). https://doi.org/10.1080/15567280902882852
9. Marti, R., Reinelt, G.: Heuristic methods. In: The Linear Ordering Problem, Exact and Heuristic Methods in Combinatorial Optimization, vol. 175, pp. 17–40. Springer, Berlin (2011). https://doi.org/10.1007/978-3-642-16729-4_2
10. Chen, J.: Contextual binding and intelligent targeting. In: Proceedings of the 2016 IEEE/WIC/ACM International Conference on Web Intelligence, pp. 701–704. IEEE (2016)
11. Chen, J.: An intelligent path towards fast and accurate attribution. In: Arai, K., Kapoor, S., Bhatia, R. (eds.) Intelligent Computing: Proceedings of the 2018 Computing Conference, vol. 2, pp. 1072–1082. Springer (2018)
12. Nakasone, P.: A cyber force for persistent operations. Joint Force Q. **92**(1), 10–14 (2019)

Stained Visible Watermarking: A Securely Tunable Way of Joint Image Copyright and Privacy Protection

Xiaoming Yao[1(✉)] and Hao Wang[2]

[1] Hainan University, Haikou 570228, HN, China
xiaomingyao@163.com
[2] Nsfocus Information Technology Co., Ltd, Haikou 570228, HN, China

Abstract. Tunability of the visual degradation of the publicized image may impact the security of the protected content when selective encryption and visible watermarking are separately used to protect the image copyright and privacy. Either the consistency inside the visible watermark or the relevance of the public data to the encrypted one can be exploited for the unauthorized recovery without keys. In this paper, we propose a novel framework, referred to as stained visible watermarking, that integrates the perceptual cryptography and visible watermarking into a secure and tunable system to protect both the copyright and the privacy. Key-generated noises as tunable stains work together with the visible watermark to disguise the protected data with the required level of intelligibility and break the consistency of neighboring pixels inside the visible watermark to enhance the robustness of the watermark. The tunability of the visual degradation is fully controlled by the strength parameters of the stains and watermark. Data normalization is introduced for reversibility to avoid potential under/over-flow. Experimental results and security analysis have demonstrated that our proposed scheme can jointly protect the privacy and copyright of images that are posted online, and the publicized image can be securely fine-tuned to adapt to the requirement of the applications.

Keywords: Copyright protection · Visible watermarking · Privacy-preserving · Security · Selective encryption · Tunability

1 Introduction

With the rapid development of deep learning techniques, images and videos that are posted online are getting more and more easily pirated, analyzed and even seamlessly tampered for various private benefits, making compromisation of ownership, fake photos, and privacy leakage as the dominant security and privacy challenges of the publicized multimedia content [1, 3, 6].

In order to keep the private content in secret while sharing securely the other content online, selective encryption needs to separate the content into two parts [8]. The first part is the public part, it is left intact. The second part is the protected part, it is encrypted. For the purpose of showing something with the images/videos to the public, the tunability of the visual degradation is desirable for viewers without the keys to understand partially the content.

In conventional selective encryption, it is usually achieved by choosing some subset of the protected data to encrypt while keeping the other data intact. But the encrypted part must not be predictable with the publicized data for security unless it is decrypted with the keys. A good scheme of selective encryption should have a good trade-off between the security and the required tunability.

With the rapid development of deep learning networks, previously secure schemes are getting more vulnerable.

It is the redundancy of the protected data instead of the visual degradation that determines the security level. Kuo et al. pointed out [5], intelligibility tends to be scattered among all frequency components in a transformed domain. In this case, even one remaining coefficient can reveal "the outline of Lena and her hat". For this reason, they proposed to integrate selective encryption into the entropy coding [10].

Lian et al. proposed a scheme with high security for JPEG2000 compressed images or videos to adjust the visual degradation under the control of quality factor by encrypting coefficients' signs, permuting data blocks in different color planes, and permuting wavelet coefficients [7]. However, since neighboring relations among the blocks to be permuted and the 4-bit columns within the $X \times Y$-sized bit-plane remain unchanged, It is likely to recover the protected data using deep learning techniques.

Mosaic (or pixelation), a technique mostly used for blurring the region while preserving its perceptual meaning by replacing it with its low resolution version to hide the secrets [3], is recently reported vulnerable to unauthorized recovery since some patterns of the neighboring pixels before and after pixelation can be used as clues for the super-resolution recovery [1].

Li et al. proposed a scheme [6] that attains recoverable privacy protection for video content distribution, but its security still depends on that of the low-resolution image, which is similar to Mosaic.

Visible watermarking is a more active way to protect the copyright, especially when the embedded watermark is recognizable but unobtrusive, as shown in Fig. 1(b). However, the consistency among neighboring pixels inside the watermark makes it vulnerable to removal attacks [2, 4]. Tsai and Chang proposed a scheme to enhance its security by adding uniformly distributed noise to the nonzero bits of the binary watermark to break the consistency [9].

When selective encryption and visible watermarking are separately used to protect the image copyright and privacy except that the watermark and the encrypted data are in the same region, it is more likely that the tunability would impact the security of the protected content and the robustness of the copyright. The encrypted data can act as noise to break the consistency within the watermark.

In this paper, we propose a novel framework of selective encryption by considering all the protected data but with randomly distributed weights, referred to as stained visible watermarking, that integrates selective encryption and visible watermarking into a secure and tunable system to protect both the copyright and the privacy. A spatial scheme to reduce the redundancy and consistency among neighboring pixels is thus presented. Key-generated noises, acting as tunable stains with random weights, work together with the visible watermark to disguise the protected data with the required level of intelligibility and enhance the robustness of the watermark against removal attack, as shown in Fig. 1. The tunability of the visual degradation is fully controlled by the strength parameters of the stains and watermark, i.e. alpha and beta. In Fig. 1(c) and (d), when alpha = 0.8 and beta = 380, the watermark and the protected image looks much clearer. They can be precisely recovered back to the original image with the authorized keys since data normalization is introduced to avoid potential under/over-flow.

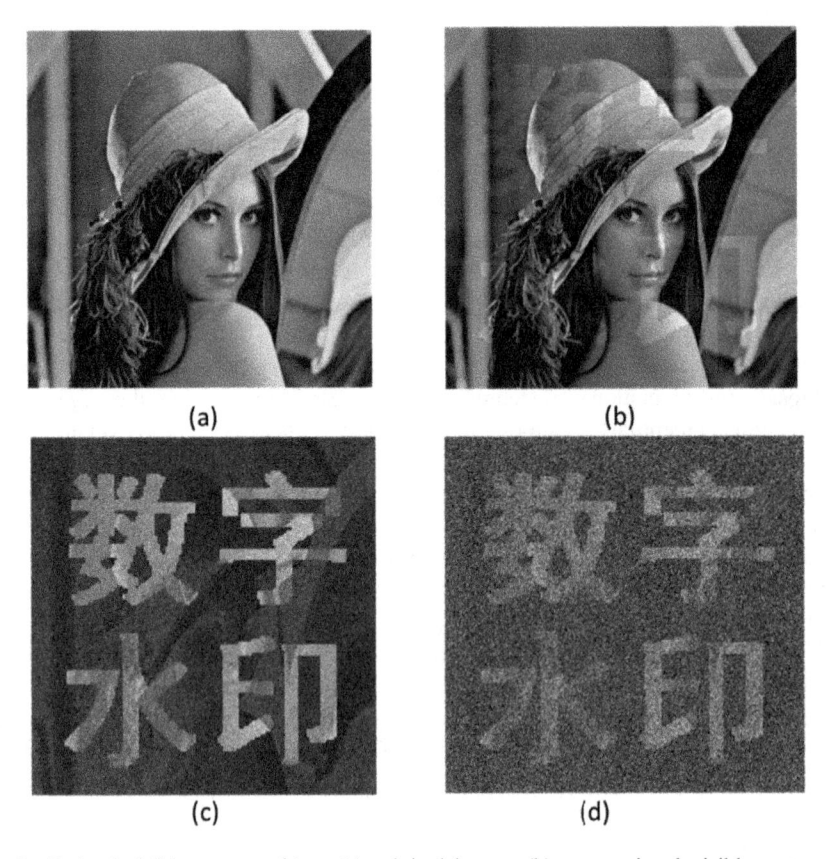

Fig. 1. Stained visible watermarking: (a) original image; (b) conventional visible watermark; (c) alpha = 0.8 and beta = 380; (d) alpha = 0.4 and beta = 750.

We study and evaluate the visual degradation of the publicized images and the redundancy among neighboring pixels to assess security and tunability with multiple metrics such as stained degree, histogram, and frequency spectrum. Experimental results have demonstrated the effectiveness of our proposed scheme.

Our main contribution is in the proposal of a novel framework that can address the issue of balancing the tunability and security within the sphere of selective encryption while enhancing the robustness of visible watermarking by using only a single mechanism. It is simple and low cost, suitable for most applications of multimedia content distribution with some private secrets to protect.

We proceed by presenting the spatial scheme, followed by experimental results. Then we analyze and discuss the watermark robustness and security under attacks of recovery without keys. We conclude in Sect. 5.

2 Stained Visible Watermarking: A Spatial Scheme

The aim of stained visible watermarking is to break the consistency of the neighboring pixels within the visible watermark by adding key-generated noises so that the visible watermark is unable to remove without keys and the perceptual details are stained.

The system is illustrated in Fig. 2.

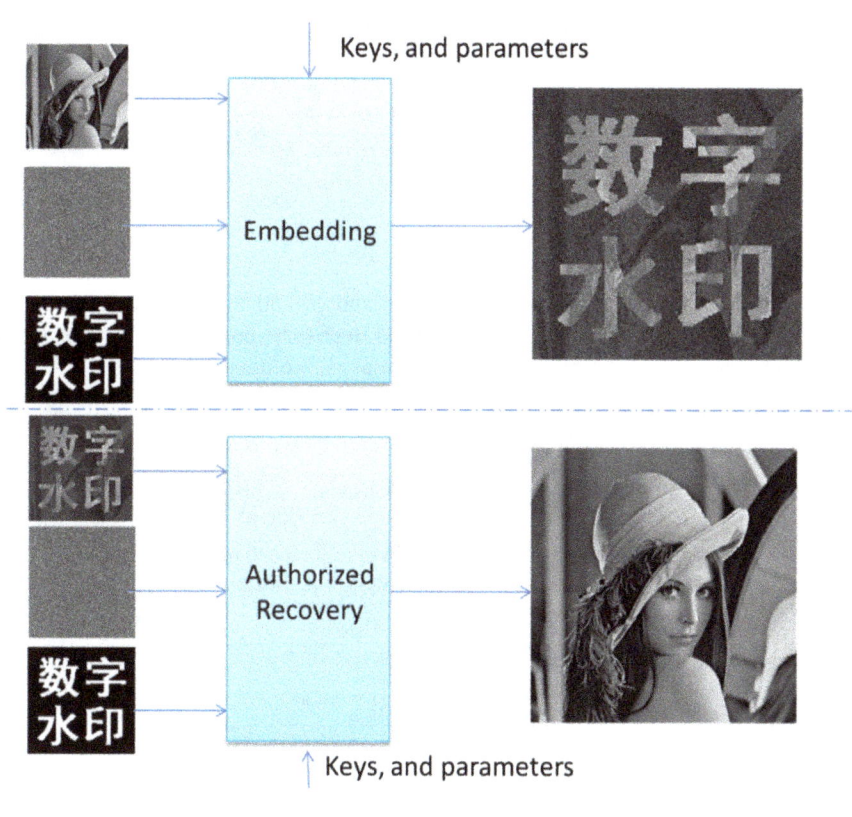

Fig. 2. Illustration of our proposed scheme: the embedding phase, and the authorized recovery phase

2.1 Embedding Phase

For the protected host image A, given the watermark W, the multiplicative noise N_1, and the additive noise N_2 are uniformly distributed pseudo-random noises that are generated using different seeds, where $N_1 \in (3, 4)$ and $N_2 \in (-1, 1)$.

The embedding procedure is formulated as follows, i.e.

$$A'(p) = A(p)(1 + \alpha * W(p) * N_1(p)) + \beta * N_2(p), \qquad (1)$$

where $p = (x, y)$ is the pixel location, α and β are respectively the strength parameter of the watermark and the noise. The noise N_1 acts as the random weights of the watermark W.

The binary watermark W is represented with a bi-polar expression, i.e. its zero-bit is expressed as -1. In this case, noting that when α is close to 0.333 and the watermark is -1, the term $1 + \alpha * W(p) * N_1(p)$ is approaching 0.

Then, data normalization is required to restrict $A'(p)$ within the range of $[A_{min}, A_{max}]$, and we have the final watermarked image:

$$A''(p) = (A_{max} - A_{min}) * \frac{A(p) - A'_{min}}{A'_{max} - A'_{min}} + A_{min} \qquad (2)$$

2.2 Authorized Recovery Phase

If the watermark W, the strength parameters α and β, and the seeds to generate the noises N_1 and N_2 are known, the authorized recovery process is reverse to the embedding phase. Since A'' has the same range as A, we have,

$$A_{max} = maximum(A''), A_{min} = minimum(A'') \qquad (3)$$

A'_{min}, and A'_{max} are computed using (1) as follows,

$$\begin{aligned} A'_{min} &= A_{min}(1 - \alpha * minimum(N_1)) + \beta * minimum(N_2) \\ A'_{max} &= A_{max}(1 - \alpha * maximum(N_1)) + \beta * maximum(N_2) \end{aligned} \qquad (4)$$

Then the watermarked image A'' can be transformed back to A', i.e.

$$A'(p) = (A''(p) - A_{min}) * \frac{A'_{max} - A'_{min}}{A_{max} - A_{min}} + A'_{min} \qquad (5)$$

The original image A is precisely recovered since we have,

$$A(p) = \frac{A'(p) - \beta * N_2(p)}{1 + \alpha * W(p) * N_1(p)} \qquad (6)$$

Please note that this scheme requires the amplitude range (i.e. A_{min}, A_{max}, A'_{min}, and A'_{max}), strength parameters α and β, the watermark W, and the seeds to generate the noises for the recovery, which can be transmitted with the publicized image using information hiding techniques as reported in [6].

3 Experimental Results

We use MATLAB 7.1 as the simulation tool to implement the scheme, and all test images are directly from the USC-SIPI image databases [11] without changing the filenames.

In this section, we study and evaluate the tunability of visual degradation of the publicized image, the histogram, variation of the consistency among the neighboring pixels after stained watermarking, correlations of the low-resolution images, and the variation of the frequency spectrum amplitude. Then, we present a simple security analysis against attacks of unauthorized recovery based on the experimental results.

3.1 Tunability of Visual Degradation

To evaluate if the visual degradation of the watermarked image is tunable by changing the strength parameters, image quality metrics are required. In our experiments, we use the peak signal-to-noise ratio (PSNR) and subjective quality for this purpose.

Peak Signal-to-Noise Ratio (PSNR)

As a measure that can evaluate the visual degradation between the publicized image A'' and the original one A, PSNR is defined as follows:

$$PSNR = 20 log \frac{255}{\sqrt{MSE}} \quad (7)$$

where MSE (mean square error) is a measure that evaluates the difference between A'' and A and is defined by

$$MSE = \sum (A''(p) - A(p))^2 / M \times N \quad (8)$$

where $p = (x, y)$ is the pixel location, M and N are respectively the height and width of A.

Subjective Quality

This result is obtained directly from the viewers when the score ranging from 1 to 5 expresses the quality from the worst to the best.

Tunability of Visual Degradation

Generating the watermarked and the recovered images. Let $\alpha = 0.5 : 0.5 : 3.5$ and $\beta = 50 : 200$, 1057 watermarked images are obtained by following the guideline of embedding the stains and the watermark without changing the seeds and the watermark. Then 1057 recovered images are obtained by following the authorized recovery procedures using the seeds, the watermark, and the amplitude range.

Computing the PSNR and plotting the curves. It is found that the PSNR is unable to show the visual degradation when it is less than 15 dB, as illustrated in Fig. 3.

In Fig. 3, for specific α, the PSNR is increasing with the increase of β, which is false. We turn to examine the 500 watermarked images obtained herein one by one for subjective assessment. We found that the subjective quality of the watermarked images varies from 1 to 3 when strength parameters α and β are properly chosen. Therefore, it is secured that the visual degradation of the watermarked image is perceptually tunable.

Most PSNR of the recovered images is more than 40 dB, which is acceptable for most applications of multimedia distribution. However, when β is too large for a specific α or vice versa, the recovering precision decreases. For a better visual quality of the recovered image, the beta-to-alpha ratio should be restricted to some certain value up to the applications, e.g. 1000.

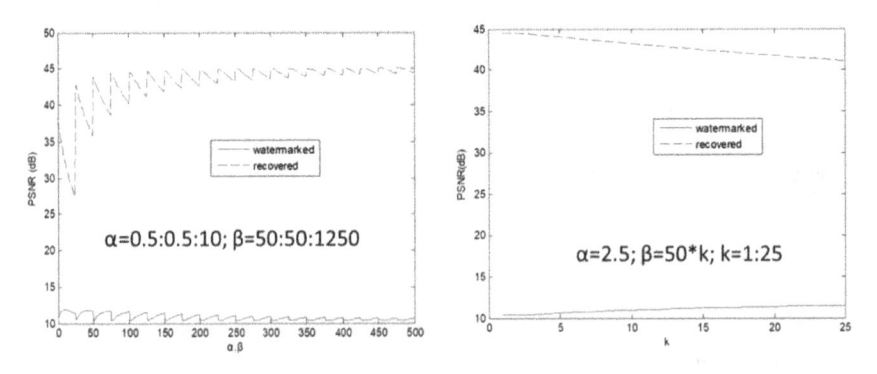

Fig. 3. Variation of PSNR with strength parameters: on the left, for specific α, β is starting from 50 and gradually increased to 1250; on the right, a partial curve on the left is extracted where α is fixed to 2.5 and β is changing from 50 to 1250 with a gradual increase of 50.

3.2 Security Against Unauthorized Recovery

The largest security risk of selective encryption is unauthorized recovery. In this case, the unencrypted data can be exploited to predict the encrypted part. For this purpose, we evaluate our proposed scheme from three aspects: the histogram, the frequency amplitude spectrum, and the low-resolution PSNR.

Histogram

The histogram of the image can show the statistics of pixels of the image with the number of occurrences. Scrambling operations may change the intelligibility of the image while preserving its histogram, which makes it vulnerable to known-plaintext attack. Hence, the histogram is helpful to evaluate security.

Frequency Amplitude Spectrum

Kuo et al. have pointed out [10] that all frequency coefficients are connected to the image content. In other words, if frequency coefficients are partially modified, e.g. the high-frequency coefficients, the mid-frequency coefficients, or the low-frequency coefficients, the unchanged parts can be perfectly extracted by respective band-filtering.

Cumulative absolute frequency amplitude (CAFA) is thus defined as a measure to evaluate the variations of the frequency spectrum in a polar coordinate system, i.e.,

$$C_a = \int\int_{\rho \geq 0, \theta=0}^{\rho=f_{max}, \theta=2\pi} |A(\rho, \theta)| d\rho d\theta \tag{9}$$

where ρ is the frequency spectrum radius to the center, and θ is the phase angle of the coefficients on the circle.

Low-Resolution PSNR

The content generated from the Fourier low-frequency spectrum of the image represents the low-resolution version of the image. It has been recently reported that it is likely to be recovered by using deep learning techniques [1] once the low-resolution version is at hand.

Low-resolution PSNR referred to as LrPSNR, is defined as the PSNR of the low-resolution versions. The low-resolution versions of the images, the watermarked and the original image, can be computed using the following formula:

$$f(x,y) = \sum\sum_{|u| \leq r, |v| \leq r} F(u,v) e^{-j2\pi(xu+yv)/r^2} \tag{10}$$

where $F(u, v)$ is their Fourier form in the frequency domain, r is the half-width of the selected window, and u, v are the horizontal and vertical frequency, respectively.

In [1], a facial image of size 8×8 can be recovered to a plausible high-resolution version of size 32×32. It is justifiable to infer that for a test image of size 512×512, its low-resolution version of size 128×128 with 100% similarity can be compromised. We set the maximum window size of the spectrum for our experiments be 64×64, and the minimum window size 8×8.

We compare our results of the histogram with that of the original image and that of the conventional visible watermarking, as shown in Fig. 4.

It is seen in Fig. 4 that for a specific alpha, our proposed scheme entails a significant change of the histogram when the beta is greater than 350 while conventional visible watermarking almost preserves its statistics.

To evaluate how strength parameters impact the CAFA, we first fix the alpha, and then the beta. The results are illustrated in Fig. 5.

In Fig. 5, it is noticed that all frequency coefficients change with the strength parameters, either increasing or decreasing to some extent.

The LrPSNR is used to measure the similarity between the lower-resolution versions of specific window size. To compute them, the original and the watermarked image should be transformed to the frequency domain using Fourier Transform, and only the coefficients within the given window centered in the spectrum are inversely transformed using (10) to obtain the respective low-resolution versions. The window size can be 8×8, 16×16, 24×24, 32×32, 40×40, 48×48, 56×56 and 64×64.

In Fig. 6, it is observed that a larger difference is between the original and the watermarked images of smaller window size. With the increase of the window size, the range of the difference is getting wider with the increase of the strength parameters. It means that if the strength parameters are chosen properly, then the LrPSNR can be kept to a smaller value. An increase in the window size as a whole improves the LrPSNR. However, when the window size is of larger than 40×40, this improvement is negligible. We can thus conclude that the strength parameters have significant impacts on the LrPSNR over the low- and mid-frequency coefficients, and it is unlikely to recover the super-resolution version of the original image by using this stained low-resolution one.

3.3 Robustness of Visible Watermarking

Visible watermarking is vulnerable to removal attack since the watermark is visible and there is a consistency among the neighboring pixels inside the watermark [2, 4].

As well known, neighboring pixels of natural images bear strong similarity, i.e. their differences are statistically very small. Hence, we evaluate the statistics after stained watermarking.

The stained degree is defined based on the standard deviation to evaluate how the neighboring pixels are dispersed by the stains added to them. To compute it, the image should be segmented into several non-overlapped sub-blocks, and for each sub-block, we have a standard deviation, σ_i, then the stained degree, S_d, is the average, namely,

$$S_d = \frac{1}{m} \sum_{i=1}^{m} \sigma_i \tag{11}$$

where m is the number of sub-blocks inside the image.

For watermarked images that we obtain by setting the strength parameters, the stained degrees of those images are computed and shown in Fig. 7, where the sub-block is of size 2 × 2.

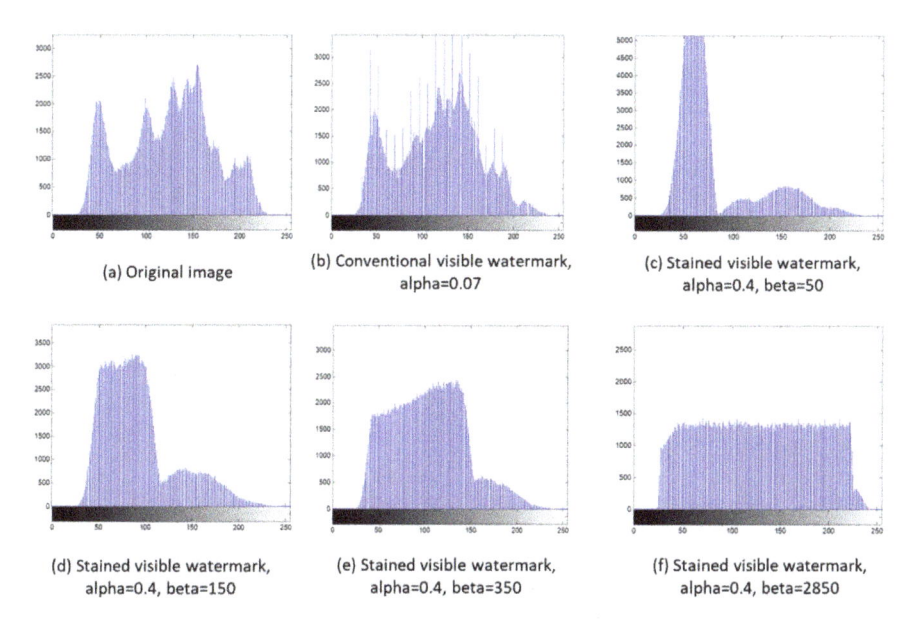

Fig. 4. Comparison of the histograms: (a) the original; (b) conventional visible watermarking, alpha = 0.07; (c) stained watermarking, alpha = 0.4, beta = 50; (d) stained watermarking, alpha = 0.4, beta = 150; (e) stained watermarking, alpha = 0.4, beta = 350; (f) stained watermarking, alpha = 0.4, beta = 2850.

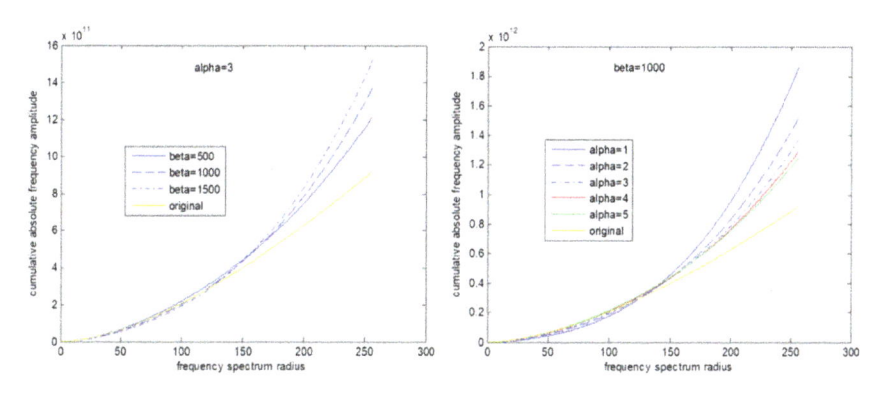

Fig. 5. Variation of CAFA with the strength parameters: On the left, alpha is 3, the increase of the beta entails the decrease at the part of the low-frequency coefficients but then increases abruptly when the radius is more than 100; on the right, beta is 1000, the decrease of the alpha entails the decrease at the low-frequency but increases after that.

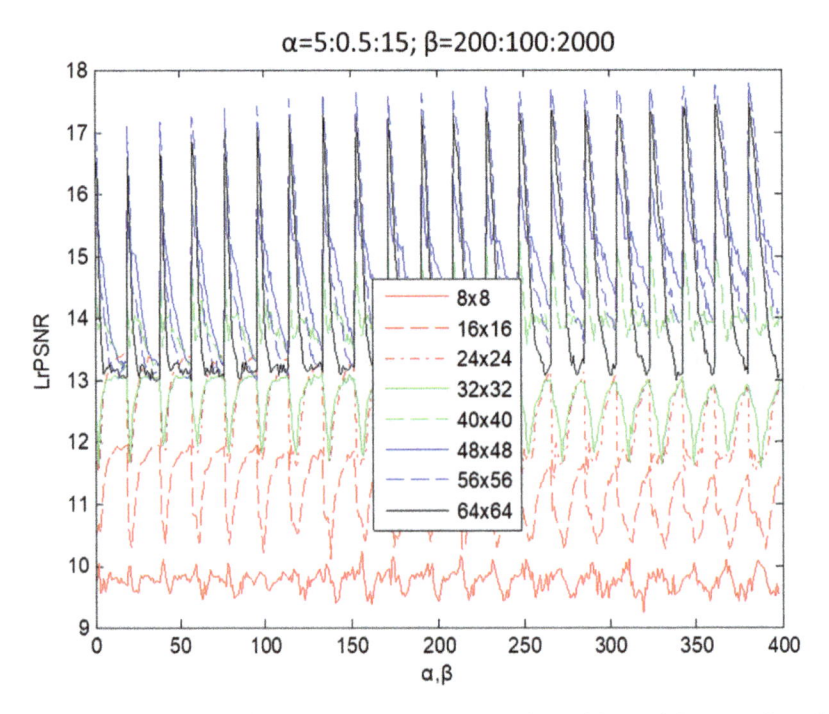

Fig. 6. Variations of LrPSNR with strength parameters alpha and beta: alpha starts from 5 and increases to 15 with a step of 0.5; beta starts from 50 and increases to 200 with a step of 50. The window size changes from 8 × 8 to 64 × 64.

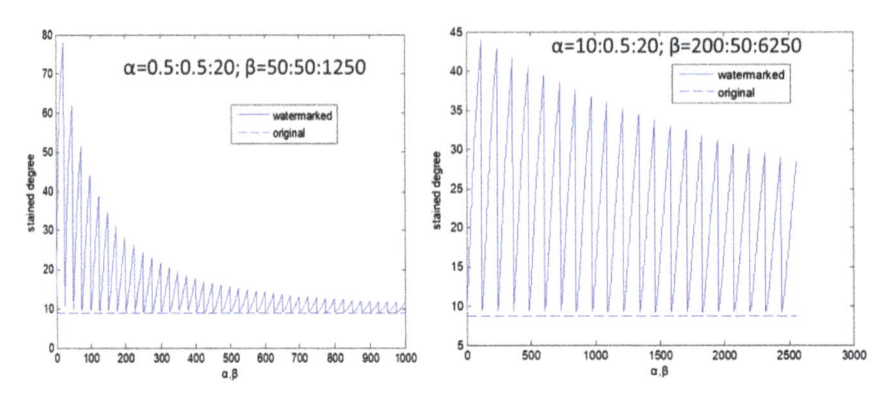

Fig. 7. Variation of stained degree with the strength parameters: for specific α, an increase of β leads to an increase of the stained degree; for specific β, an increase of α leads to a decrease of the stained degree. The stained degree of the original image is nearly 8.9.

In Fig. 7, the stained degree of the watermarked image depends on the strength parameters, and it can be far more than that of the original one. This means the stained watermarking increases the stained degree and breaks the consistency among the neighboring pixels inside the watermark. We can conclude that the robustness of the visible watermark depends on the strength parameters α and β.

Since the strength parameters act upon the metrics in a very complicated way, we choose specific values for them and evaluate how those metrics vary with only one parameter when another one is fixed. By setting α to be 2.5, we compute the LrPSNR and the stained degree when β varies from 500 to 6500 with a step increase of 500.

The main results can be summarized in Table 1. While the visual degradation is tunable via subjective assessment, the LrPSNR is basically below 14 dB and the stained degree is double more than that of the original 8.9!

Table 1. Brief summary of experimental results

β	LrPSNR								Stained degree
	8 × 8	16 × 16	24 × 24	32 × 32	40 × 40	48 × 48	56 × 56	64 × 64	
500	9.8	11.7	13.1	13	13.7	13.8	13.3	13.1	21
1000	9.7	11.9	13.4	13	13	13.2	13.3	13	33.5
1500	9.7	11.9	13.5	13	13	13.2	13.5	12.9	43
2000	9.7	11.9	13.5	13	13	13.2	13.7	12.8	50.3
2500	9.7	11.9	13.5	12.9	13	13.2	13.7	12.8	56.1
3000	9.6	11.9	13.5	12.9	13	13.2	13.7	12.8	60.9
3500	9.6	11.9	13.5	12.9	13.1	13.2	13.7	12.8	64.8
4000	9.6	11.9	13.5	12.9	13.1	13.1	13.7	12.8	68
4500	9.6	11.9	13.5	12.9	13.1	13.1	13.7	12.8	70.9
5000	9.5	11.9	13.5	12.9	13.1	13.1	13.8	12.8	73.3
5500	9.5	11.9	13.5	12.9	13.1	13.1	13.8	12.7	75.4
6000	9.5	11.9	13.6	12.9	13.1	13.1	13.8	12.7	77.2
6500	9.5	11.9	13.6	12.9	13.1	13.1	13.8	12.8	78.9

4 Security Analysis

We evaluate and analyze the following typical attacks, i.e. the unauthorized recovery by denoising, the known key attack, and the brute force attack.

4.1 Unauthorized Recovery by Denoising

Unauthorized Watermark Removal

As reported in [9, 10], watermark removal attack is based on the fact that the neighboring pixels inside the watermark are correlated. Indeed, the embedding of the visible watermark has no impact on the consistency of the neighboring pixels of the host image. In this case, the watermark area can be extracted using the technique of regional growth, so that the visible binary watermark can be recovered. Then the watermark can be removed using independent component analysis technique.

However, in our proposed scheme, the stains change the consistency a lot and make such watermark region extraction very difficult. As illustrated in Table 1, the stained degree of the watermarked image is more than double that of the original one, the consistency among the neighboring pixels inside the watermark is badly broken.

It is thus justifiable to infer that conventional visible watermark removal attack would not work in this case.

Unauthorized Recovery by Denoising

Since sensitive details are primarily obscured by the stains as a hybrid additive and multiplicative noise, the most effective attack is to remove the stains using denoise tools. The attack goes like this. First, a low-resolution version is obtained using low-pass filtering. Then the high-resolution version is obtained with the deep learning super-resolution tools in [5], which is assumed the recovered image without the stains.

Unfortunately, this low-resolution version of the watermarked image is quite different from that low-resolution version of the original one since the best LrPSNR is no more than 14 dB. Therefore, the result of denoising is unable to recover the original image even if the super-resolution tools are very effective.

4.2 Known Key Attack

The known key attack occurs whenever the same key is used to encrypt all images and captured by an attacker. In our proposed scheme, the key refers mainly to the seeds for generating the pseudo-random noises. As for the strength parameters, it is recommended that the first parameter α can be set manually while the second one β should be automatically determined with the help of the viewer to decide the visual degradation of the watermarked images.

In this case, even if the key and the first parameter α are disclosed because of using the same key, the attacker is unable to find β. Please note that β can be quite different for different images to reach the same visual degradation.

Moreover, since the watermark is visible of the same size as the stained region, we assume that the watermark is likely to be duplicated by an attacker with current image processing tools such as Photoshop.

According to the authorized recovery process, it is unable to recover the A'_{min} and A'_{max} in (4). Therefore, it is unable to continue the next recovery process.

We conclude that our proposed scheme is secure against the known key attack.

4.3 Brute Force Attack

In our proposed scheme, the strength parameters can be of any real numbers, there are restrictions on relations between them to avoid potential information loss though. This means the key-space is infinite. Actually, for any specific α, some β can be found that the same visual degradation can be achieved as other different pairs of α and β.

On the other hand, for the same visual degradation of the watermarked images, possible pairs of α and β are infinite as well.

Suppose that the attacker can acquire a large collection of the watermarked images via the internet, the statistics of the stains can be found out by sampling the stained regions. This information is useless when generating pseudo-random noises without the seeds because they are significantly different if their seeds are different even if the statistics are the same. The key-space of the seeds is infinite.

Therefore, we can conclude that the brute force attack fails without the strength parameters and seeds.

5 Conclusions

When images shared online contain privacy content, it is legally required that the secret details of this content should be blurred while the rest should remain visible for better usability. Current tools such as mosaic, scrambling, and other selective encryption approaches may provide adjustability to some extent based on the scramble level as reported in [3], but suffer from the vulnerability to removal attacks. In this paper, we propose a novel framework of secure and tunable selective encryption, referred to as stained visible watermarking, that deliberately adds stains with the visible watermark in the sensitive region to hide the secret details while preserving some of the tunable perceptual meaning. In this case, the watermark and the stains work together to contaminate the private content and break the consistency among the neighboring pixels inside the watermark. Since it is very difficult to recover the stained region without the related keys and parameters, it can jointly protect the copyright and private secrets of the image at the same time.

Extensive experimental results and security analysis have demonstrated that the visual degradation of the watermarked images can be fine-tuned with the strength parameters and typical attacks with only partial or without keys can be resisted effectively.

In the future, we plan to study the potential vulnerability under attacks using techniques of generative adversarial networks assuming that a large collection of similar stained watermarked images online are available.

Acknowledgment. This project is supported by the National Natural Science Foundation of China under Grant 61462023.

References

1. Dahl, R., Norouzi, M., Shlens, J.: Pixel recursive super-resolution. In: Proceedings of IEEE ICCV 2017, Venice, Italy, pp. 5449–5458 (2018)
2. Dekel, T., Rubinstein, M., Liu, C., Freeman, W.T.: On the effectiveness of visible watermarks. In: IEEE Conference on Computer Vision and Pattern Recognition (CVPR 2017), Honolulu, Hawaii, USA, pp. 6864–6872 (2017)
3. Honda, T., Murakami, Y., et al.: Hierarchical image scrambling method with scramble-level controllability for privacy protection. In: Proceedings of IEEE International Midwest Symposium on Circuits and Systems, OH, USA, pp. 1371–1374 (2013)
4. Huang, C., Wu, J.: Attacking visible watermarking schemes. IEEE Trans. Multimedia 6(1), 16–30 (2004)

5. Kuo, C., Chen, M.: A new signal encryption technique and its attack study. In: Proceedings of IEEE International Conference on Security Technology, Taipei, Taiwan, pp. 149–153 (1991)
6. Li, G., Ito, Y., et al.: Recoverable privacy protection for video content distribution. EURASIP J. Inf. Secur. **2009**, 1–11 (2009)
7. Lian, S., Sun, J., Wang, Z.: Perceptual cryptography on JPEG2000 compressed images or videos. In: Proceedings of IEEE CIT 2004, Wuhan, China, pp. 78–82 (2004)
8. Massoudi, A., Lefebvre, F., De Vleeschouwer, C., et al.: Overview on selective encryption of image and video: challenges and perspectives. EURASIP J. Inf. Secur. **2008** (2008). Article no. 179290
9. Tsai, H., Chang, L.: A high secure reversible visible watermarking scheme. In: Proceedings of IEEE ICME 2007, Beijing, China, pp. 2106–2109 (2007)
10. Wu, C., Kuo, C.: Fast encryption methods for audiovisual data confidentiality. In: Proceedings of SPIE 4209, Multimedia Systems and Applications III, 22 March 2001. https://doi.org/10.1117/12.420829
11. The USC-SIPI image databases. sipi.usc.edu/database/database.php

sD&D: Design and Implementation of Cybersecurity Educational Game with Highly Extensible Functionality

Yoshiyuki Kido[1,2(✉)], Nelson Pinto Tou[2], Naoto Yanai[2], and Shinji Shimojo[1,2]

[1] Cybermedia Center, Osaka University,
5-1, Mihogaoka, Ibaraki-shi, Osaka 5670047, Japan
{kido,shimojo}@cmc.osaka-u.ac.jp
[2] Graduate School of Information Science and Technology, Osaka University,
1-5, Yamadaoka, Suita-shi, Osaka 5650871, Japan
{nelson,yanai}@ist.osaka-u.ac.jp

Abstract. Cybersecurity issues have gained more attention due to the rapid development of new technology. In many cases, cyber theft happened due to a lack of understanding of how to secure a single piece of private information. In the present paper, we attempt to provide an inexpensive and easy tool for learning cybersecurity through a board game simulation called Security Defense and Dungeons (sD&D). The basic concept is not only to introduce the latest cybersecurity equipment and solutions, but also to provide cybersecurity education for players to raise awareness regarding cybersecurity threats. In addition, we attempt to train users in good practices concerning team communication and human resources. The simple game interface includes cybersecurity task scenarios, featured intuitive game design for users, and easy exploration of the provided scenario. In addition, the proposed game, in which players can share game resources and bits of knowledge, communicate via a chat room. In the present study, in order to evaluate the proposed method and prototype implementation, we performed a user study with student users. The questionnaire results were mostly positive comments and opinions.

Keywords: Gaming education · Cybersecurity education · Security awareness

1 Introduction

With the development of information and communications technology in recent years, it is now possible to manage and share information across the globe. However, information systems and the Internet not only bring convenience to our lives but also introduce various cybersecurity issues, such as attacks by malicious users, as well as leakage of personal information and cyberbullying, which can have serious repercussions. Therefore, information literacy in users is a key component for individuals or companies and organizations to improve protection against attacks on cybersecurity measures. However, there are few educational game tools aimed at nurturing cybersecurity and improving the information literacy of individuals. Studies have suggested that students effectively learn what they have practiced [11]. Fundamentally game-based learning allows a learner to better engage the real world. Such games involve an artificial mockup or simulation of the environment of the real world. In the present study, we proposed the design of network multiplayer games and solve a scenario related to security issues. In comparison with analog and other digital game scalability and flexibility, achieving this remains challenging. For example, space for the player to play in analog games is a scalability issue that prevents the addition of a new player or setting up a new game environment. Furthermore, the flexibility to accommodate new security issues without significant modification of the game has not yet been adequately explored. Therefore, we attempt to build a cybersecurity game with scalability and flexibility, which will provide context-changeable security games for education.

There are many cybersecurity analog games that attempt to train users to improve their cybersecurity literacy. These games allow people to learn about general knowledge of cybersecurity while having fun. Although most of these games provide useful information on cybersecurity the games do not really give the user a sense of fully comprehending the tools or how to identify a security breach as early as possible. For example, methods by which to use these tools to perform a cybersecurity attack or to protect network resources are not shown in cybersecurity analog games. On the other hand, there are several programming contests and competition for cybersecurity learning called "Capture the Flag Unplugged" [6], which involves a very difficult competition regarding how to find security holes in an information system environment prepared by the competition organizer. This competition attempts to train cybersecurity engineers, rather than young people, with more professional skills. This is one motivation for our team to design a cybersecurity board game.

Another motivation for building this game application is the occurrence of cyberbullying targeting young people in social media communication [8]. In addition, according to a study on the definition and concept of cyberbullying, raising the awareness of how users share their information on the internet is one solution for cyberbullying prevention. Recently, a serious game of enhancing privacy awareness in social network tools has been developed [2]. In this game, users can learn how to behave correctly as well as cyber safety in social networks. However, their approach does not treat several technical issues in cybersecurity. Young people and technology beginners need knowledge of information and network technology. In order to understand the behavior of cybercrime more deeply, users should understand the mechanism of information technologies, such as authentication, access controls, and other cybersecurity mechanisms. The reason for this is that, if social network users do not know the behavior of the information and network system, each user cannot determine who obtained information regarding his/her personal identify.

Therefore, we herein consider the design of Security Defense and Dungeons (sD&D), which is a cybersecurity board game. The main contribution of the present paper is to raise the awareness of cybersecurity education and to consider game scalability and flexibility in context-changeable cybersecurity game education. As such, we have evaluated the game effectiveness regarding our intended goals through a user study.

In the present paper, we briefly introduce cybersecurity issues and describe the basic idea and games approach to expose people to thinking about security in Sect. 2. Sections 3 and 4 show in detail the goals of the proposed games, architecture, and prototype design. In Sect. 5, we describe the evaluation of user impressions. Finally, we discuss task for future research and conclude the paper in Sect. 6.

2 Related Research

Cybersecurity educational games have shown significant efforts to provide end users with education. Their concepts include comprehending cybersecurity, which impacts privacy against malicious attacks, in particular to increase cybersecurity awareness for a user that is not an IT expert. Although we experienced raised cybersecurity awareness while playing the games, the games did have some weak points. For example, playing the game takes a long time, the game is complicated, and there is difficulty applying the game tools in real cybersecurity scenarios.

Several studies have been carried out to incorporate an analog board game as part of cybersecurity education to increase user understanding of cybersecurity attacks. This includes the Control-Alt-Hack [4] tabletop board game, which was designed to address a variety of current technologies and actual threats regarding cybersecurity issues. An interesting part of this game is the role of characters in the cybersecurity field (social engineering) and addressing the core of cybersecurity issues. The design of the cards enforces exposure risks of various cybersecurity threats as well as mitigation or avoidance methods. For example, a cookie-blocked card describes the topic of writing a web browser extension to circumvent tracking cookies. However, the game may last longer if a player does not fully comprehend the game's rules or fails to complete a mission, which leads to loss of credibility. in addition, the game mechanism of rolling dice to determine whether a task was successfully carried out by the player, seems not to be a very intuitive approach.

Android: Netrunner [1] is a one-on-one commercial science fiction security card game. The game provides the concept of a megacorporation and secures their premises, while a runner or hacker tries to break into the premises. Unlike Android: Netrunner, the proposed game design focuses on simulating current security issues and presents simple rules to users. For example, each task in the proposed game attempts to engage a player to understand and use the property described in the task to attack or defend specific security faults.

D0x3d! [7] introduce network cybersecurity terminology, attack and defense mechanic and basic computer concepts. The game is a collaborative game with up to four players. Throughout the game, the user gains cybersecurity knowledge from the pictogram shown on a card. For example, a pictogram featuring a broken border provides the notion of a reinforcement rule that a hacker can use to enter or exit a node.

Previous studies on digital games explain the cybersecurity theme to teach computer cybersecurity principles. For example, computer game simulation was used to enumerate computer cybersecurity lessons, as described in studies of "teaching objectives of a simulation game for the computer system" [3,10]. Such research suggested that digital game design can be used for positive training to identify cybersecurity threats easily [11]. Furthermore, as discussed in "Design and preliminary evaluation of a cybersecurity Requirements Education Game (SREG)" [12] revealed the need to integrate organizational tools into the game. Game-based learning not only accelerates the learning process, but also emulates a user's engagement to more actively identify a vulnerability attack, through social engineering, for example.

3 Research Goal and Concepts

3.1 Goal of the Present Research

The principal goals of the cybersecurity game are to raise awareness in security education. As described in the Introduction, there is a need to raise user awareness regarding security education in the era of information and technology, as this is essential to protect the user from cyber-thief and cyberbullying attacks. The awareness goals we hope to achieve are as follows.

First of all, we believe the game will be fun if it is easy to play and requires a low skill level related to the game theme. Our goal is to design a game that is easy to play and omit complex rules to provide a wide range of audiences of different skills and ages the opportunity to play the developed game. Furthermore, in order to increase the players' understanding of security tools, the design incorporates an easy scenario with adequate information displayed in the card window.

Next, the security game development quest of bringing the security scenario up to date was attempted including setting of the game configuration and behavior to provide security. We proposed simple rooms and cards, where each component can be added by the file configuration. In the future, we intend to provide a flexibility feature for adding a security scenario.

Finally, in many developing nations, for example, Timor-Lest the information and technology infrastructure are still not advanced compared to in developed countries. Impact tools, such as servers and network components, which provide an experimental environment for the students, are not affordable for some educational institutes. With this network game simulation, we hope students in such difficult situations can experience a simulation environment to better prepare for security considerations.

3.2 Why a Game?

Similar to Control-Alt-Hack [4], we assume that a game can provide excellent tools for the user to learn security education in a fun environment, rather than under pressure, and to engage in communication in a multiplayer game. With the digital game, network tools and security simulation are achievable. In particular, in delivering the freedom to experience threats without repercussions, we enable identification of cybersecurity threats, understanding, and interpretation of cybersecurity issues in terms of risk ramifications, and how to take proper avoidance actions.

In addition, game-based education is proving to be effective in altering user behavior [9]. Furthermore, Eagle [5] designed "Wu's Castle", a role-playing game in which students can use the C++ code to solve in-game problems. This game allows a student, to independently solve logic-related programming issues.

4 Design and Implementation

The basic idea of this game reutilizes an idea from conceptual card board games, which consist of a card deck that contains security information as a logical element for a player to explore. We combine the properties of a card game with current security attack threats in order to create scenarios for a player to play in our game. Our design implements room and card conceptually. A room is space for players to perform their tasks and include resource, trap and task rooms. The card components include hacker tools or network properties for a player to accomplish their mission.

4.1 Game Design

Figure 1 shows the proposed user interface of sD&D. The interface is composed of a player information window as well as a window showing the positions of the other players, rooms, cards and a chat room for interaction between players. Initially, the author explored a number of games to perceive the rules and mechanic for the sD&D design objectives. In making sD&D, the focus was on designing simple cybersecurity game rules as well as a game mechanism to enforce learning processes throughout the scenario provided in the game. Therefore, the sD&D design approach uses an $N \times N$ room to compose different game scenarios and properties for multiple players to explore the cybersecurity game as a team.

Element: Room. The concept of room design is as space for the player to perform their mission, to move around, and to pick up available cards. There are three different types of rooms scattered according to the number of rooms defined in the game. Each room provides a distinct learning goal for the player while the player is accomplishing his/her task or learning a new security concept. A trap room is a room with no access door. The design of the trap room is such that once a player enters the room, a correct answer must be provided in order to

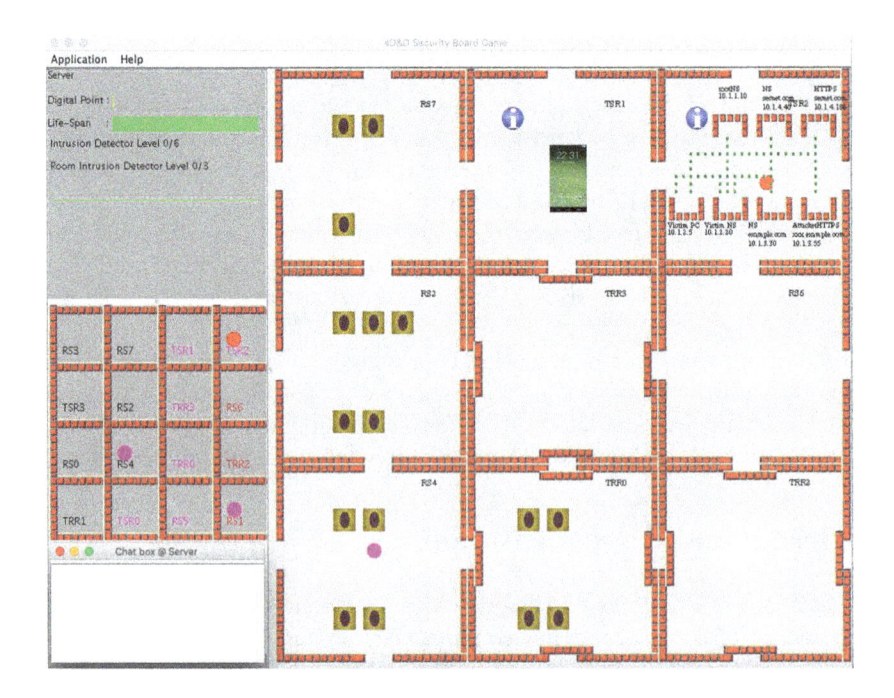

Fig. 1. Main window of sD&D

unlock the door. The trap room may be an empty room or may contain essential cards required by a player to solve a cybersecurity task. As a prerequisite to unlocking a door, the player must drop a card. Therefore, when entering this room, the player has to have cards as resources and knowledge as the key to further proceed in the game, otherwise the game is ended.

A cybersecurity-related quiz is loaded dynamically via file configuration. As shown in Fig. 2(a), "What is the main purpose of a DNS server?" is an example of a cybersecurity question in the quiz. The game's administrator can quickly add new quiz entries by modifying the file configuration.

The second type of room is a resource room. A resource room can be assumed to be a deck of cards in card games. In sD&D design, the cards are scattered in the room face down rather than being placed in a pile. This method allows for multiple players to pick up cards simultaneous without waiting for the turn to advance. Figure 2(b) shows the resource room design used in this game.

The third type of room is a task room, as shown in Fig. 2(c). A task room is an attempt to map the cybersecurity simulation as close as possible to a real cybersecurity attack. The cybersecurity scenario composed in the task room is designed to raise the awareness of cybersecurity issues and broaden the understanding of software faults, which can potentially lead to a cybersecurity breach. Furthermore, a simulation with a specific example of a cybersecurity topic triggers the user to think deeply about the issues and inspires the user to come up with a solution to the problem. Figure 3 shows the simulation window image of the cybersecurity scenario in DNS cache poisoning. In this window, a user can perform DNS spoofing on target DNS servers along the scenario script (DNS cache poisoning scenario is as shown below). Throughout the scenario, a player is expected to grasp basic network design and to comprehend the DNS information flow and ultimately understand the danger of DNS cache poisoning combined with other cybersecurity attacks, such as phishing.

– DNS cache poisoning scenario script

```
Your mission: Eavesdropping user information utilizing DNS
cache poisoning Goal:

Inject cache poisoning to the cache server, by changing the
legitimate IP address to attacker IP address. First, your
task is to build network configuration according to the
following information:

1. Cache Server: 10.1.2.20 pointed to the root DNS

2. Root DNS Server with all legitimate NS record

3. DNS secnet.com: 10.1.4.40 with NS record secnet.com ->
10.1.4.180

4. DNS attacker: 10.1.3.30 with ns record secnet.com ->
10.1.3.55 and xxx.example.com -> 10.1.3.55. Your second task
is injecting DNS cache poisoning to the cache server using
''dnsspoof'' to change the legitimate IP address 10.1.4.180
to attacker web IP address 10.1.3.55. A successful mission
shown ping to secnet.com will get a reply from the IP
address of attacker IP address: 10.1.3.55

Necessary tools (cards):

1. PC for testing

2. Server for root DNS

3. Server for victim DNS

4. Server for attacker DNS

5. Server for DNS secnet.com

6. Server for HTTPS secnet.com

7. Server for attacker HTTPS

8. dnsspoof tools

Instruction:

Following are list of task player have to do:

1. Server setup.

2. Run the dnsspoof tool.
```

```
3. Ping the attacker legitimate web address xxx.example.com.
the attacker DNS will reply legitimate NS record for xxx.
example.com but also make a query request and reply for
secnet.com domain

4. After some time the DNS cache poisoning eventually
succeed.

5. From PC victim ping secnet.com and verify reply is IP
address of attacker PC.
```

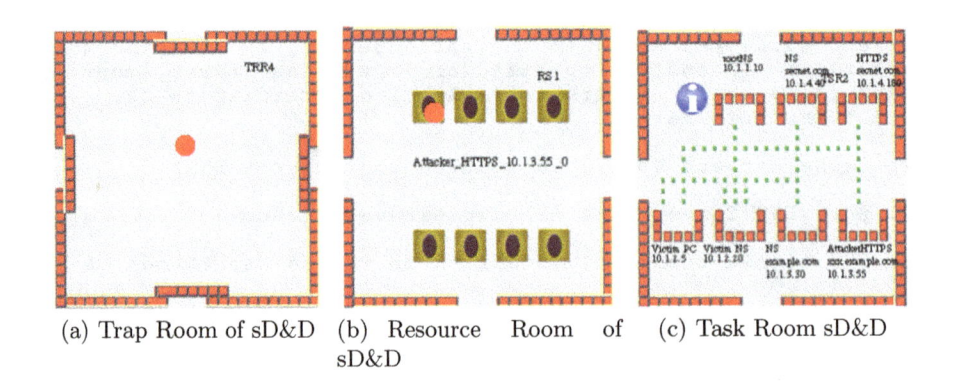

(a) Trap Room of sD&D   (b) Resource Room of sD&D   (c) Task Room sD&D

Fig. 2. Overview of room design of sD&D

Communication is a tool for successful collaboration in any aspect of a collaboration task. It is essential to interact with other players in order to win the game as a team, especially in network multiplayer games. In order to facilitate collaboration, such as requesting an answer to a question or exchanging cards between players a chat room is available.

Element: Card. The card design in sD&D is primarily inspired by dox3d! [7] and the Control-Alt-Hack [4]. However, in the present study, the author designates three types of cards, resource cards, intrusion detection cards, and magic patch cards, with detailed descriptions of the information of security tools. Each card provides specific cybersecurity tools for attack and defense. For example, a resource card can consist of a hacking tool, such as AirPlay-NG which is used to attack a wireless access point. The details of the card are as shown in Fig. 4(d). Also, resource cards such as DNS Server in Fig. 4(c), are used to set up the environment for the security incident simulation. The intrusion detection card is a type of card for the system administrator to provide defense of an information system, and an example of this card is shown in Fig. 4(a). Finally, the magic patch card, the counter-attack for the administrator defense card, is shown in

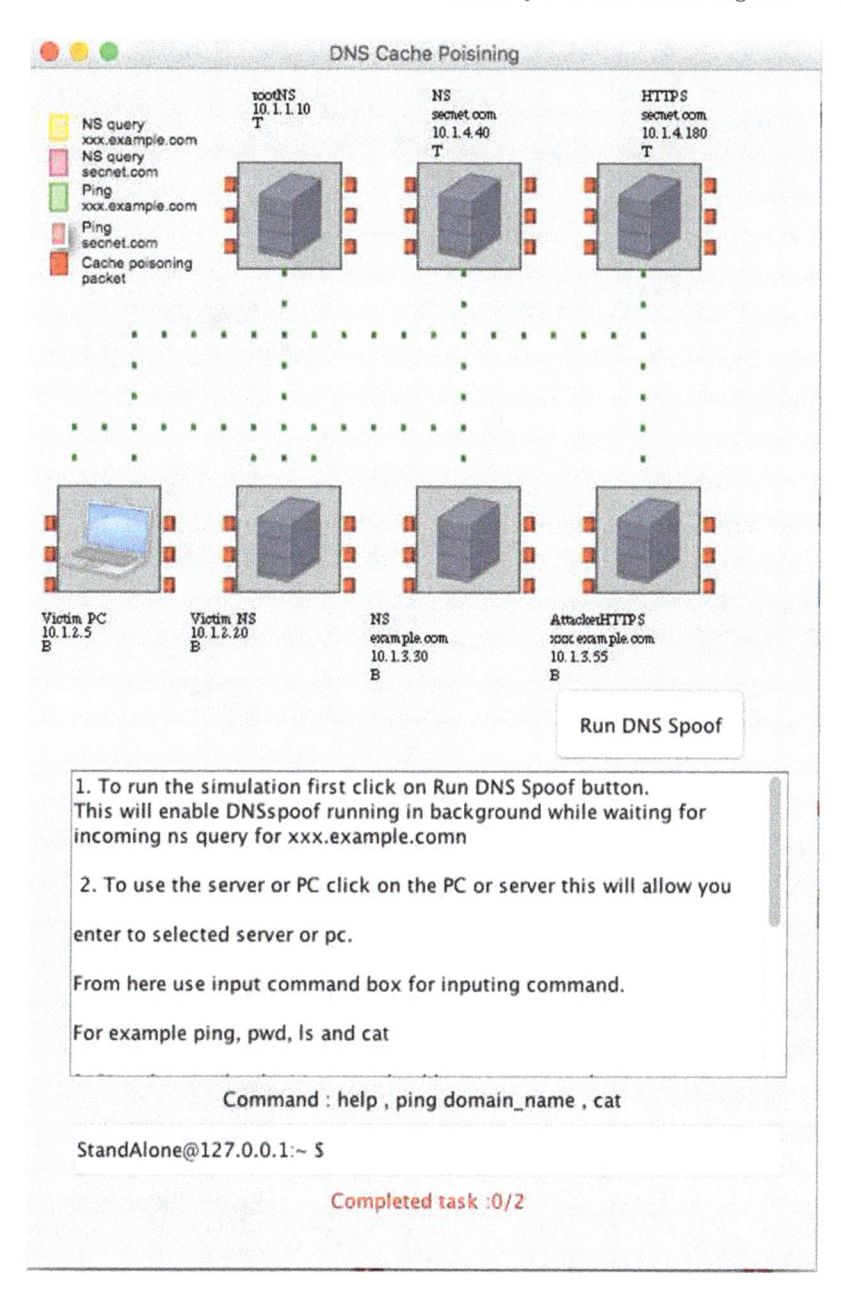

Fig. 3. DNS cache poisoning simulation

detail in Fig. 4(b). For example, during the game, if a player has collected an intrusion type of card, the player is detected by the system administrator, and, as consequence of the intrusion, the detection level is raised. Therefore, the player's lifespan is decreased, and the level of difficulty of the game also increases. However, the player can nullify the intrusion detected by using a magic patch card as a counter-attack.

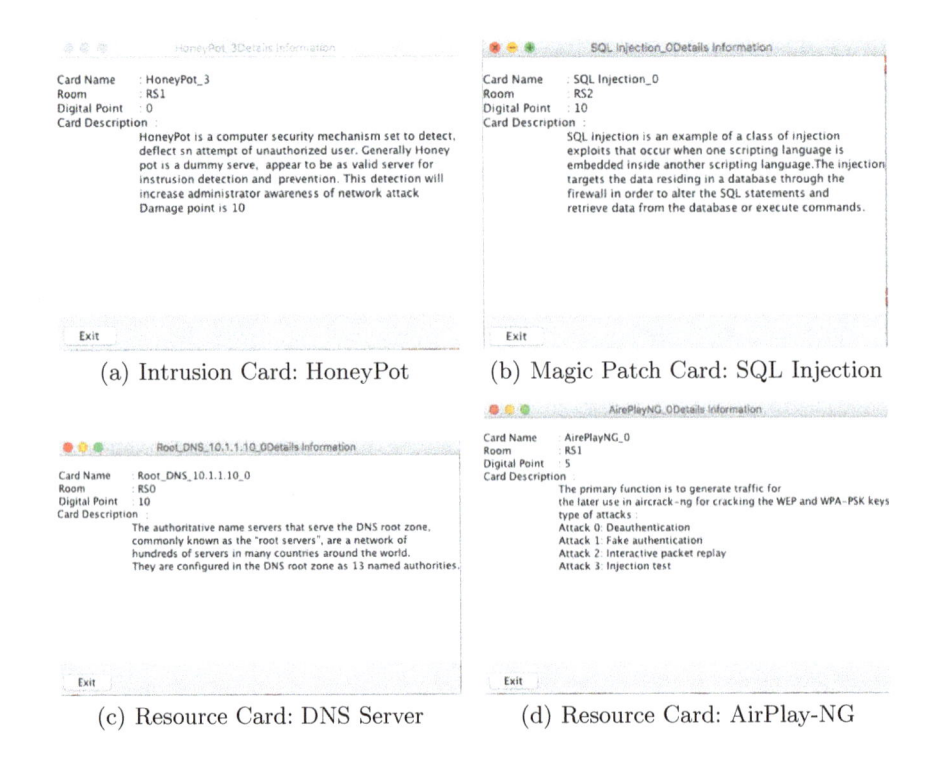

(a) Intrusion Card: HoneyPot

(b) Magic Patch Card: SQL Injection

(c) Resource Card: DNS Server

(d) Resource Card: AirPlay-NG

Fig. 4. Overview of sD&D card design

4.2 Scalability and Flexibility Concept

In this research, we define the scalability of the sD&D design as the suitable for the implementation of additional components (card, room, and task) in games via XML file configuration without significant modification of the core of the game framework. Therefore the structure of the XML file should provide ease of user in adding new rooms, cards, and scenarios at the same time, subject to the constraints and rules of sD&D.

One of the most challenging objectives of sD&D is to provide cybersecurity awareness regarding current cybersecurity contextual issues through game simulation. Since current cybersecurity issues evolve rapidly simulating a scenario for a player to experience a cybersecurity threat, such as how an attack method can damage an information system as well as possible early detection to avoid further risk, is not an easy task. Of course, adding such new simulation often requires the additional of a feature to the game framework, which results in a significant modification of the game framework. In order to overcome significant changes in the game framework, the sD&D design provides flexibility to the user with administrator privilege to design, modify and implements new scenarios in the game's file configuration.

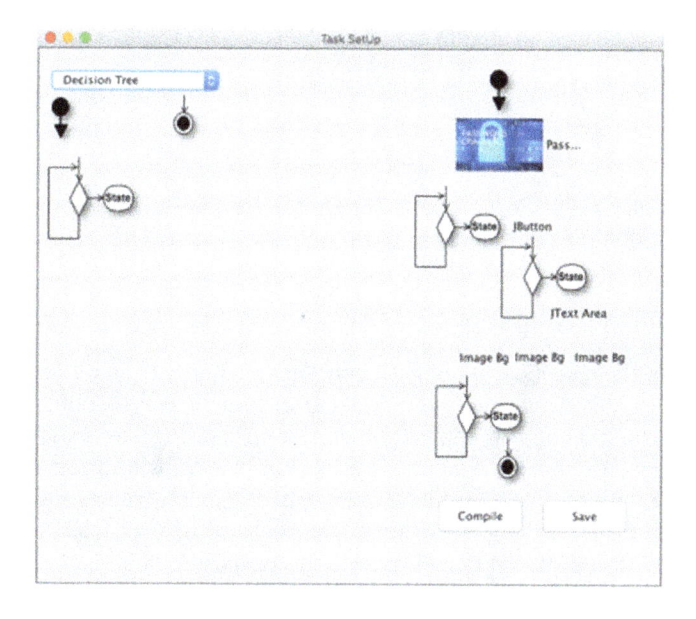

Fig. 5. Cybersecurity scenario design sD&D

Figure 5 shows the sD&D scenario design and configuration window. A user can define a cybersecurity scenario and simulation diagrams using this window before starting the game. In the present research, the author assumes the flexibility concept as a capability of sD&D in order to clarify the various inputs from the users, subject to modification and the use of simulation components in the sD&D platform, to generate new or modified cybersecurity simulation scenario.

5 Evaluation

The author measured the performance of the proposed sD&D game by comparing the CPU and memory loads. Furthermore, the estimated time consumption of multicast packet interchange between nodes with respect to the number of players and rooms determine the scalability of the game. A user study was conducted with a number of students to evaluate the accessibility and the ability to achieve the design goals of the sD&D.

In this evaluation, the author assumes that the game is scalable if it can support a minimum of 36 different rooms with reasonable computation and network resources. The 36 rooms are capable of supporting up to approximately

Table 1. Computing node specifications

	Spec
CPU	Intel Core i7, 2.2 GHz
Memory	8 GB, 1600 MHz DDR3
Network	AirPort Extrem (0x14E, 0x117), 54 Mbps
OS	macOS High Sierra version 10.13.3
Java	Version 1.8.0_77

208 different cards, assuming that 26 of the rooms are resources and trap room with maximum eight cards in each room and 10 of the rooms are task rooms.

5.1 Scalability Evaluation

Table 1 shows the specification of each computing node on the performance evaluation test. In this evaluation, the game size limit of 6×6 rooms (36 rooms) is based on the author's assumed criteria for measuring the scalability of the proposed game. In the performance evaluation, the consumption of CPU and memory by each computing node is shown in Fig. 6. As the number of rooms increases, the CPU load increases to a maximum of 52% of CPU consumption upon initialization of the game. However, the proposed game is capable of minimizing the CPU load while the game is played.

In order to reduce the usage of computational resources during the game, sD&D only draws a default matrix of 3×3 with the respective room properties to the screen. Each player screen is applied to map the whole room dimension in a different panel, respectively to room name attribute and another player location.

Figure 7 illustrates the evaluation of the multicast packet round trip average with respect to the numbers of players. The highest round trip time (RTT) occurred when request authentication was performed. This is because, upon an authentication request, a number of procedures need to be accomplished to build the game environment.

As the number of players increases, the RTT of the packet also increases. However, in this game, the packet size is constant, ranging from 32 to 107 bytes. A packet is sent throughout the network only if it is triggered by player action. As a result, an increase in the number of RTT packets with respect to the number of players is not expected to exceed 50% of the average RTT.

5.2 User Study Case

A survey was conducted for six students after playing the sD&D game. The participants were five male students and one female student. Of these, two were international students and four were local students. Mean age was 25. Most of

(a) Game Size: 3 × 3 rooms

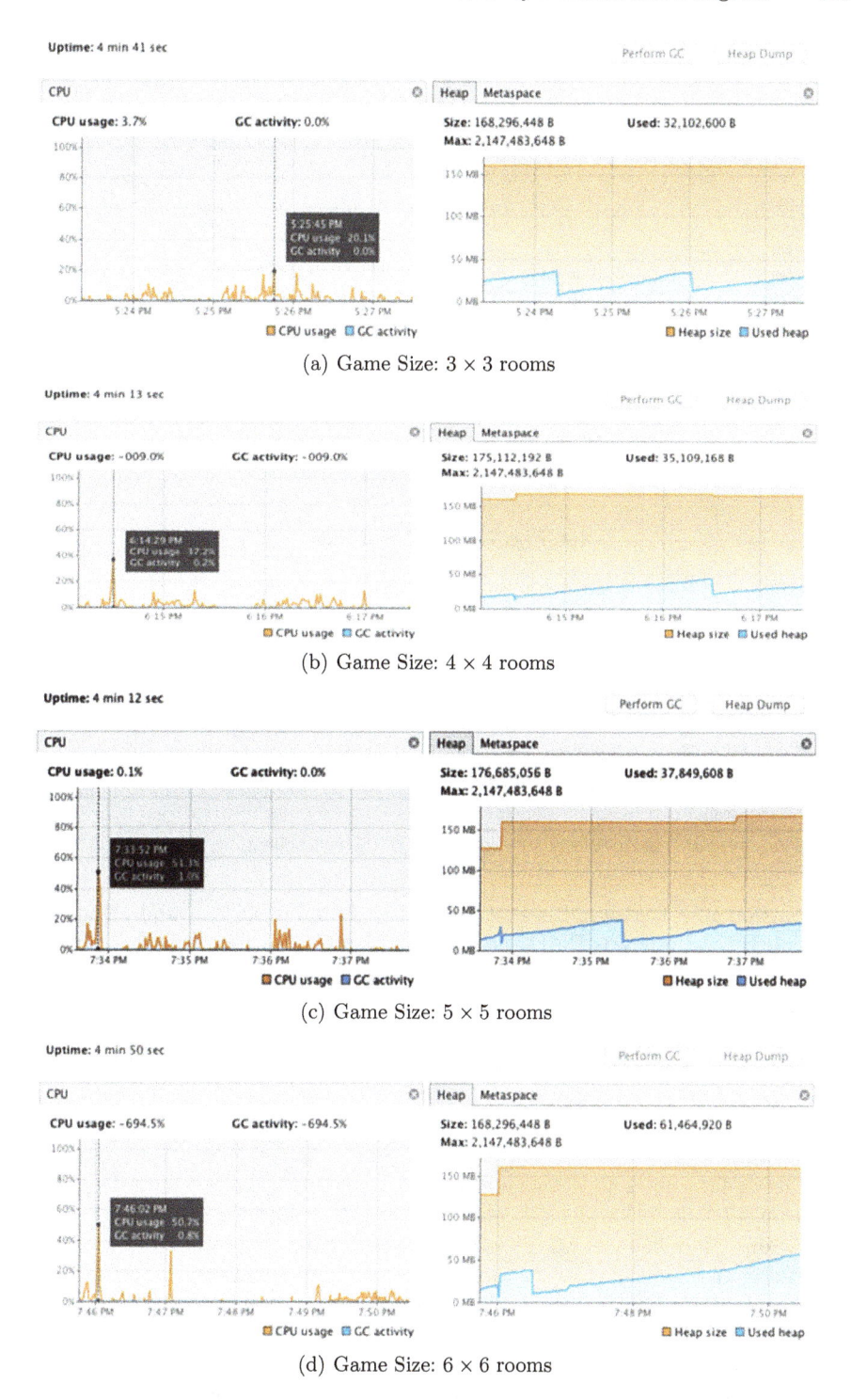

(b) Game Size: 4 × 4 rooms

(c) Game Size: 5 × 5 rooms

(d) Game Size: 6 × 6 rooms

Fig. 6. CPU and memory consumption

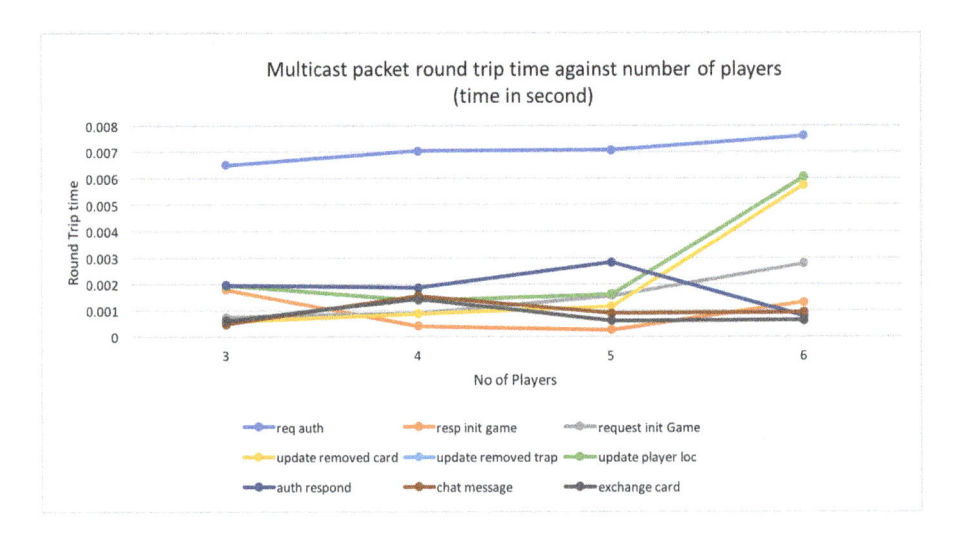

Fig. 7. Round trip time

Table 2. Survey question and result

Survey question	Rating
Game design interface according to ease of task completion	4/6
Understanding the rule of the game	3/3
Ease of following the game instruction	5/6
Appropriateness of using the key input in the game	5/6
The card design provides sufficient information on a particular security topic	5/6
The ease of room design allows for exploration of the components in the room	6/6

the participants were graduate students in computer science and information technology, and one was Ph.D. candidate student studying in applied physics.

Table 2 shows the level of satisfaction (participant rating) after playing the sD&D game. The scale ranges from 0 to 10 points, and a satisfactory rating is considered to be a participant score of over 5 points. Overall, most of the participants have a positive response after playing the game, as shown in Table 2. However, question 1 (Game design interface according to ease of task completion) and question 2 (Understanding the rules of the game) still need significant improvement.

Table 3 lists open questions and the participant reason for the rating. Open questions include: "How does the game increase the players' awareness of security education?" and "Do you want to play the game again or recommend the game to a friend or critic of the game?". Most of the participants agree that the proposed game raises their knowledge of cybersecurity education and is accessible to play. However, one of the participants found it quite challenging to start the game due to lack of information on the rules and the startup procedure. This small problem occurred because the participant did not go through the help menu.

Table 3. Survey open question and result

Open question	Answer P4	P5	P6
Briefly explain the reason for your ranking evaluation	Because of the lack of an initial explanation as to how to win this game and use resource cards in missions, players have difficulty in understanding what to do first	I think it is a fun game that also has valuable cybersecurity knowledge elements in it	I like this game, it easy to understand and fun to play. Also, the game requires you to use your mind to solve the problems
What topic (security scenario) are most influenced you in understanding security issues?	Social media attacking	Hacking other people's account	
How did the tools (room and card) in the game help you to understand the security domain?	They were helpful in understanding that there are many tools for attacking IT devices. However, I could not understand the meanings of the commands and option parameters	I was able to spend some more time playing in order to gain a deeper understanding of security as a whole	The game is really good. The security domain is explained in game fashion, and you have to complete tasks to understand the domain
How would you describe your knowledge of the security domain after playing this game?	I came to understand two things by playing this game. First, WPA2 keys are not always safe (WPA2 keys might be weak to dictionary attack). Second, leaving a personal electronic device (e.g., iPhone) somewhere people freely come and go (e.g., cafe or office) can increase the risk of a social media attack	Moderate	My knowledge about the security domain has increased. Before playing game, I wasn't much aware of the security protocol, but now I know what can happen if I don't follow some instructions
Do you want play the game again?	Yes	Yes	Yes
Will you recommend this game to a friend?	Maybe	Yes	yes
Critics of the game	I think this game can be helpful in learning about cybersecurity. In order for it to be more friendly to users, an initial explanation about this game is necessary	The game is a great game to play but simplified instructions are needed for non-tech-savvy users	The game was good

6 Conclusion

In the present study, we proposed a cybersecurity game prototype that will be made available on our web site[1]. The game attempts to raise education awareness of cybersecurity issues, and primarily strengthens user knowledge of essential cybersecurity tools and software flaws that can be used in a cybercrime. By understanding the method of cybersecurity breaches, the prevention of information theft while using internet services can be avoided. Moreover, students who have studied networks and cybersecurity can fully comprehended the issues. As a result, a new approach to solving cybersecurity breaches can emerge.

The proposed mechanism characterized the simulation of a current cybersecurity network, implemented using the multicast socket developed in the Java programing language, combined with the tabletop card game concept. The simple game interface consists of rooms, cards, and a multiplayer display of a matrix of three by three rooms to be explored by the player during the game. Furthermore, evaluation experiments indicated that the scalability and performance of the game are quite reasonable.

Acknowledgments. A part of this research work was supported by the JSPS KAKENHI Grant Number JP18K11355.

References

1. Andriod: Netrunner. https://www.fantasyflightgames.com/en/products/android-netrunner-the-card-game/
2. Cetto, A., Netter, M., Pernul, G., Richthammer, C., Riesner, M., Roth, C., Sänger, J.: Friend inspector: a serious game to enhance privacy awareness in social networks. CoRR abs/1402.5878 (2014). http://arxiv.org/abs/1402.5878
3. Cone, B., Thompson, M., Irvine, C., Nguyen, T.: Cyber security training and awareness through game play. In: Security and Privacy in Dynamic Environments (SEC 2006). IFIP International Federation for Information Processing, vol. 201, pp. 431–436 (2006)
4. Denning, T., Lerner, A., Shostack, A., Kohno, T.: Control-alt-hack: the design and evaluation of a card game for computer security awareness and education. In: Proceedings of the 2013 ACM SIGSAC Conference on Computer and Communications Security, pp. 915–928 (2013)
5. Eagle, M.: Level up: a frame work for the design and evaluation of educational games. In: Proceedings of the 4th International Conference on Foundations of Digital Games, pp. 339–341 (2009)
6. Ford, V., Siraj, A., Haynes, A., Brown, E.: Capture the flag unplugged: an offline cyber competition. In: Proceedings of the 2017 ACM SIGCSE Technical Symposium on Computer Science Education, pp. 225–230 (2017)
7. Gondree, M., Peterson, Z.N.: Valuing security by getting [d0x3d!]: experiences with a network security board game. Presented as Part of the 6th Workshop on Cyber Security Experimentation and Test (2013). https://www.usenix.org/conference/cset13/workshop-program/presentation/Gondree

---

[1] https://viscloud.ais.cmc.osaka-u.ac.jp/sdandd/.

8. Grigg, D.W.: Cyber-aggression: definition and concept of cyberbullying. Aust. J. Guid. Couns. **20**(2), 143–156 (2010)
9. Gustafsson, A., Katzeff, C., Bang, M.: Evaluation of a pervasive game for domestic energy engagement among teenagers. Comput. Entertain. (CIE) **7**(4) (2009)
10. Irvine, C.E., Thompson, M.: Teaching objectives of a simulation game for computer security. In: Proceedings of the Informing Science and Information Technology Joint Conference (2003)
11. Jin, G., Tu, M., Kim, T.H., Heffron, J., White, J.: Game based cybersecurity training for high school students. In: Proceedings of the 49th ACM Technical Symposium on Computer Science Education, pp. 68–73 (2018)
12. Yasin, A., Liu, L., Li, T., Wang, J., Zowghi, D.: Design and preliminary evaluation of a cyber security requirements education game (SREG). Inf. Softw. Technol. **95**, 179–200 (2018)

Mobile System for Determining Geographical Coordinates for Needs of Air Support in Cases of GPS Signals Loss

Karol Jędrasiak[1(✉)], Aleksander Nawrat[2], Przemysław Recha[2], and Dawid Sobel[2]

[1] The University of Dabrowa Gornicza, Dąbrowa Górnicza, Poland
jedrasiak.karol@gmail.com
[2] Silesian University of Technology, Institute of Automatic Control, Gliwice, Poland

Abstract. The article presents the concept of implementing a mobile system for determining the geographical location of the observed point using a network of unmanned flying objects tethered, called virtual masts. The presented concept was developed taking into account the possible loss of GPS signal and the related need to efficiently determine the geographical location of the observed points, e.g. for emergency purposes. As part of the publication, the current status of work on the system was presented: currently a unique unmanned tethered flying platform called a virtual mast has been successfully implemented, which is capable of controlled hovering of unlimited length at a height of 100 m and observing the environment using high quality stabilized cameras. The second part of this publication presents the concept of determining the location of objects of interest using algorithms in the field of computer vision. The concept was tested at the time in a simulation environment. Promising results were obtained that will form the basis for further work on the mobile system of virtual masts for the needs of the Border Guard.

Keywords: Virtual mast · Unmanned aerial vehicle · Positioning

1 Introduction

Unmanned aerial vehicles are increasingly used to obtain information in both military and civilian applications. There is currently a trend aimed at simplifying the way of managing groups of unmanned aerial vehicles by gradually increasing the level of their autonomy. Image recognition acquired by unmanned aerial objects is a reliable source of intelligence information, in particular when it is integrated with the system for the interpretation and analysis of geospatial data. In Poland, both hardware and algorithmic solutions are being developed to create complex management systems for large groups of unmanned aerial vehicles. From the user's point of view, the main advantages of using a large number of unmanned objects flying over a single object are: fast task execution time, increased fault and interference resistance, and the ability to implement cooperation between unmanned objects. Unmanned aerial vehicle management research is being conducted using a Polish solution, the diagram of which is shown in Fig. 1.

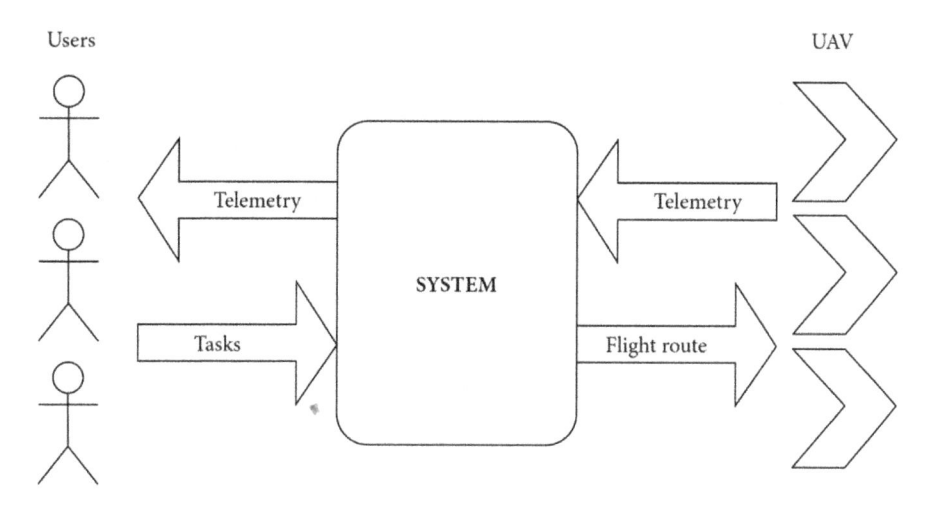

Fig. 1. A simplified operational scheme of group management systems for Unmanned Facilities - Virtual Masts.

Due to the specific construction of the unmanned aerial vehicle system, which using its engines maintains hovering at a given height and is permanently connected to a ground vehicle providing it with the necessary power, in the rest of this publication such objects will be called virtual masts.

With the development of space technologies, the need to defend satellite installations, including a set of satellites used for terrestrial navigation, such as GPS or GLONASS, is increasingly being talked about. In the event of their failure, whether as a result of cosmic radiation, cosmic skirmish, power supply failure, space debris, or the use of a ground jamming system, the effects would be extremely noticeable for ground installations. For this reason, as part of these scientific investigations, this article presents a positively verified concept of using a mobile network of virtual masts, thanks to which it is possible to determine the geographical location of operator's points of interest, which may be of key importance when it is necessary to provide first aid in a given area or to take advantage of air support.

2 Solutions Overview

The problem of determining the geographical location for the needs of GPS signal decays using virtual masts has not been directly discussed in literature so far. Probably mainly due to the small amount of construction of unmanned aerial vehicles that can be hanged at a given height and capable of successfully completing the task (Fig. 2). Descriptions of tethered UAV systems can be found in US patents.

First from 2013, US008590829B2 [1] describes a method of controlling a system of flying unit that is connected to ground unit. Flying unit include propeller and a moveable steering unit that affects orientation of the flying unit. Flying unit is equipped with interface module that connects flying unit to ground unit. Flying unit is powered though connection with ground unit. Connecting element is made of a flexible cable and connecting element manipulator on ground unit can wind up and wind out flexible cable.

In US008695919B2 [2], from 2014, flying unit beside main propeller is equipped with additional propellers that control flying unit attitude. Flying unit can move in relation to ground unit, while flexible cable is maintained in tensed status by connected element manipulator.

Patent US009056687B2 [3] describes similar tethered drone technology like two patents mentioned earlier, but adds more information about controlling flying unit. Flying unit and ground unit are equipped with GPS sensors. Flying unit can use ground unit's GPS sensor as reference position to enhance own position sensor accuracy.

Real example of such technology is HoverMast [4] produced by Sky Sapience company. It consists of drone equipped with visible light camera and infrared camera. It is tethered to ground unit with flexible cable. Flexible cable provides power supply to flying unit and wideband data link for transmitting camera views down to human operator in ground unit. HoverMast is adapted to defense applications. It can be operated from a variety of small vehicles or marine vessels. It can hover up to 150 meters above the host vehicle. While landed it flying unit is kept in compact housing unit. Human operator can send command to flying unit to take off. After that command flying unit automatically leaves its housing unit, then extends rotor arms and finally take off and rises to designated height. Flying unit is completely autonomous. If some problems occur during flight, human operator can take over control of flying unit or send command to immediately land on ground unit.

Fig. 2. Examples of tethered drone (a) (b) Embention TS150, (c), (d) Sky Sapience HoverMast.

3 Description of the Solution

As part of the research work carried out, a system has been developed to determine the geographical location of points of interest for the system operator. The system consists of two main elements: a network of mobile virtual masts and software responsible for image stabilization and determining the geographical location of operator's points of interest.

It is worth noting that no system of this type is known so far, which confirms the lack of known publications and only three patents in the area close to the thematic scope, but not encroaching on the scope discussed in the publication.

3.1 Multiple Instances of Mobile Virtual Masts

Hovering watchtower system was developed (Fig. 3). It was called mobile virtual mast. It consists of ground vehicle, capable of driving in rough terrain equipped with housing that covers lift and hexacopter. Hexacopter is connected to ground unit with cable that is kept tense while hexacopter is flying. That cable transmits power to drone and provides data transmission link with ground control. Hexacopter is equipped with laser rangefinder and two cameras: one operating in visible light and second operating in infrared spectrum. On human operator command drone is lifted above vehicle roof, then extends rotor arms and then is ready to fly. Hexacopter can fly 100 m above ground vehicle. Both ground vehicle and drone are equipped with navigational sensors. Using Real Time Kinematic technology provides very high precision of determining drone position (on avg $\pm$ 1 cm).

Drone operates in fully autonomous mode, keeps position and altitude. In case of emergency human operator can take over control of drone and land it safely or commence autonomous landing procedure. Ground vehicle is equipped with screens, where views from drone's cameras and telemetry are presented. Operator can remotely rotate cameras.

Many instances of such system can be use together to estimate geographic position of a point of interest, such as buildings, vehicles, people. Every unit can determine distance to common target using laser rangefinder. Those readings can be sent between units using radio link. After gathering all this readings and using mulitlateration a position of the target can be determined and its geographical coordinates can be estimated.

Fig. 3. Tethered drone system – virtual mast - for border guards.

3.2 Determination of the Geographical Location for the Purpose of Air Support

For the purpose of determining geographical location, a system has been developed that consists of a stabilized video camera, an operator's position for indicating points of interest, and location-determining software. At the beginning of the task of estimating the geographical location of the point of interest, the System Operator begins the procedure by triggering the message to all virtual masts observing the point of interest in order to manually direct their observation heads to the object of interest. Then the operator using wireless communication receives a set of images on which he indicates objects of interest. Then begins the process of estimating the location of the objects of interest for the needs of; e.g. rescue operations using air support.

Communication
Communication between virtual masts is being done via the wireless network. Network connects drones and startup platform with operator panel. On operator computer works server application which exchange messages synchronly with client application works on drone board. From platform it is possible to control drone position and camera rotation. Messages were obtained from drone with camera image. It is also possible to order suspension drone in the air, in one position and mechanical optical stabilizer provide not shaky video stream.

Estimate Virtual Masts Real-world Position
It is possible estimate the position of camera from single view if the dimensions and real word coordinates of observed object characteristic points are known [D1]. It is required to know some number of key points of the object in its coordinate system. Because each points lay on the ray from camera coordinate system center, through the

camera imager to the feature real-word position and there are more features it is possible to reconstruct the pose of the object relative to the camera. It is assumed that the minimal number of features required for attempting distance estimation is three.

Three-dimension Real-world Point Estimation

It is possible to estimate the pose of three-dimensional object with multiple cameras. It is required to know the location of cameras and on each camera it is required to find characteristic points, which are possible to find on views from other cameras. When the position and rotation of drone cameras are known therefore it is possible to estimate the position and distance to an object. An object should be visible in at least two cameras. It is very similar to the way to how eyes works and estimate distance to an objects. To increase accuracy of position estimation it is required to calibrate every camera.

Camera calibration relies on finding the internal and external camera parameters [5, 6]. Internal parameters are related to the optical and electronic properties of the camera and they include: the focal length, radial distortion and tangent coefficients and the actual coordinates of the camera image center. External parameter are those that describe the transformation, that is the rotation and transformation of the system associated with the observed 3D scene to camera system. When camera model is known it is possible to undistort camera images. The next step is adjustment for angles and distances between cameras named rectification because cameras are not aligned, image planes are no coplanar with each other and optical axes are not exactly parallel as in ideal stereo rig so it is required to rectify the left and right images onto parallel arrangement. After performing this step, it is possible to point the correspondence features on images (Fig. 4).

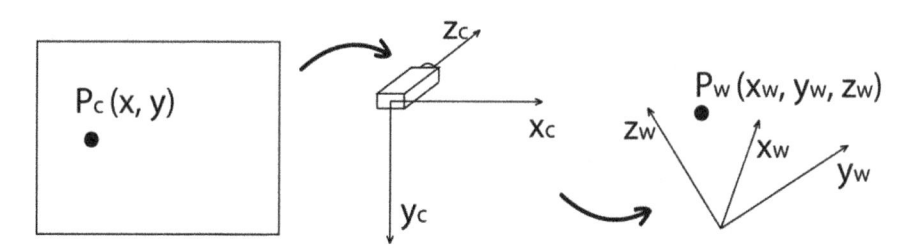

Fig. 4. Transformation of multiple 2D coordinates into 3D coordinates.

In the test case points were marked by drones operator. The output of this step is disparity map, pixel difference a pair of stereo image. The last step is triangulation for calculate point distances from disparity map.

4 Tests Under Simulation Conditions

The drone's camera pose estimation in known object's coordinate system and an object pose estimation in drones coordinate system was simulated in virtual environment. The Unity3D provide camera parameterization represents the real world camera model.

Proper shader to add radial and tangential camera image distortions was also implemented and included. Because virtual camera and object positions are known it is possible to compare estimated pose result with "real" positions and rotations (Table 1). The operator interface was implemented to select correspondences in camera and make estimation and calibration actions. Pose estimation and stereo calibration algorithm was also implemented. The screens from application are shown in Fig. 5.

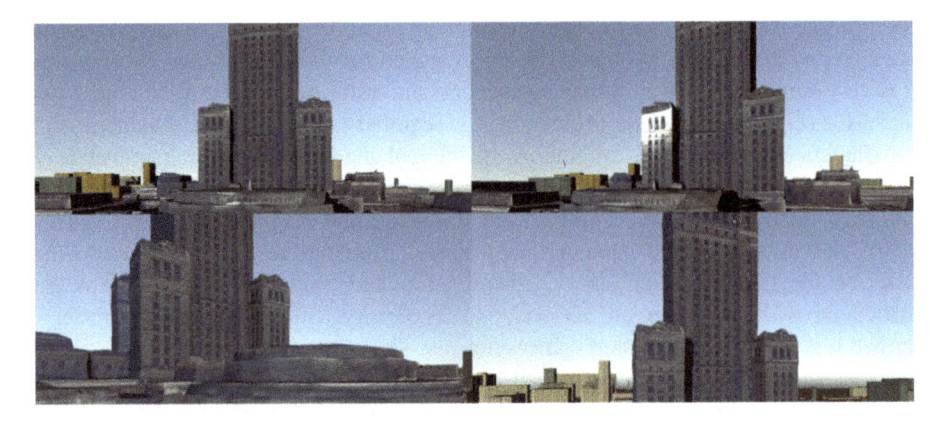

Fig. 5. A set of simulated images from cameras of four virtual masts used for the estimation of the geographical location of the system operator's point of interest.

The top view illustrating the situation is presented in Fig. 6. It is worth noting that as part of the implementation of the works, a simulation environment was created containing mapped floor dimensions of all buildings in Poland and selected buildings that are specific landmarks of cities in detail.

Fig. 6. A situational view from above showing a fragment of the virtual world map used for the simulation.

Table 1 shows the result of the comparison for a selected case of distance estimation from individual virtual masts numbered VM1 to VM4 together with real values calculated using the known true value in a simulation environment.

Table 1. Table presenting the average error of estimated distances with true values for individual virtual masts (VM).

Distance estimation [m]				Ground truth distance [m]				Avg. Error	Error Std. Dev.
VM1	VM2	VM3	VM4	VM1	VM2	VM3	VM4		
41,36	57,34	33,12	88,17	40	55	32,50	82,50	2,50	2,23

5 Summary and Conclusions

Summing up the research process, it should be noted that the Polish defense industry is able to offer some solutions in both training and image recognition using digital technologies. Of course, these are tactical solutions. Of course, the presented solutions are purely hypothetical and are dual-use technologies.

As part of the work carried out so far, the concept of a virtual mast has been developed, its main components: unmanned aerial vehicle, tensioning rope, vehicle, power supply system, stabilized head. As part of the publication, the concept of using a group of virtual masts for the purpose of determining the location of the operator's object of interest was presented. The presented considerations were supported by simulation verification, the result of which the authors of the publication consider as promising and constituting the basis for further research using real virtual masts, because of course simulation tests do not replace real tests; however, they allow repeatable verifications in conditions similar to real ones, while reducing the costs of conducting tests.

As part of further research, it is expected to conduct research in real conditions using a group of four virtual masts and further development of algorithms to enable automatic tracking of mobile objects of interest using the currently developed algorithms for tracking objects in video streams [7–10].

Acknowledgment. This work has been supported by National Centre for Research and Development as a project ID: DOBBIO9/24/02/2018 "The Virtual Interactive Center for Improving the Professional Competences of Border Guard Officers".

References

1. Keidar, R., Cohen, S.: U.S. Patent No. 8,590,829. U.S. Patent and Trademark Office, Washington, DC (2013)
2. Shachor, G., Cohen, S., Keidar, R.: U.S. Patent No. 8,695,919. U.S. Patent and Trademark Office, Washington, DC (2014)
3. Shachor, G., Cohen, S., Keidar, R., Yaniv, Z.: U.S. Patent No. 9,056,687. U.S. Patent and Trademark Office, Washington, DC (2015)

4. HoverMast. https://www.skysapience.com/defense-products
5. Sobel, D., Jedrasiak, K., Daniec, K., Wrona, J., Nawrat, A.: Camera calibration for tracked vehicles augmented reality applications. In: Nawrat, M.A. (ed.) Innovative Control Systems for Tracked Vehicle Platforms. Springer, Cham (2014)
6. Official OpenCv. http://opencv.org/. Accessed July 2019
7. Jedrasiak, K., Andrzejczak, M., Nawrat, A.: SETh: the method for long-term object tracking. In: International Conference on Computer Vision and Graphics, pp. 302–315. Springer, Cham (2014)
8. Nawrat, A., Jędrasiak, K.: Fast colour recognition algorithm for robotics. Problemy Eksploatacji **3**, 69–76 (2008)
9. Bieda, R., Jaskot, K., Jędrasiak, K., Nawrat, A.: Recognition and location of objects in the visual field of a UAV vision system. In: Nawrat, A., Kuś, Z. (eds.) Vision Based Systems for UAV Applications, pp. 27–45. Springer, Heidelberg (2013)
10. Jędrasiak, K., Nawrat, A.: Image recognition technique for unmanned aerial vehicles. In: International Conference on Computer Vision and Graphics, pp. 391–399. Springer, Heidelberg (2008)

Cyber Manhunt

Evaluation of Technologies and Practices for Effective Community Development and Maintenance

Sonali Chandel[✉], Yanjun Chen, Jiale Dai, and Jianyan Huang

College of Engineering and Computing Sciences,
New York Institute of Technology, Nanjing, China
{schandel,ychen105,jdai05,jhuang33}@nyit.edu

Abstract. Internet and its related technologies have become one of the most exciting innovations in the history of social development in recent times. However, along with many positive aspects of this virtual world, it has many side effects as well, which poses many risks and threats to its users and society. When it comes to the privacy of people, the damage the Internet can make should never be underestimated. Various governments and private organizations all over the world are spending a lot of time and money on cybersecurity options to make sure that their netizens are safe, and their data is protected in the cyberspace. However, despite all these efforts, some significant cyber issues are still not getting enough attention. "Cyber Manhunt" is one of these issues. Therefore, we plan to do our research to find the cause of cyber manhunt from various technology perspectives such as dark web, IP location, information filter, and some hacker's technique. We will also investigate the positive aspects of cyber manhunt, along with the damages caused by cyber violence and information leakage during the process. We will study the current technologies that are misused to do cyber manhunt and suggest some solutions and improvements so that it can be used efficiently to minimize the damage caused to the victims. We hope that the results of our research can help individuals deepen the understanding of cyber manhunt. The results of this research can be used in studying the ethical issues of cyber manhunt and improve the technology behind information search and filtering in the future.

Keywords: Cyber manhunt · Cyber violence · Cybersecurity · Information leak · Privacy · Dark web · Information filter · IP location · Hacker

1 Introduction

Cyber manhunt, also called as 'Human Flesh Search' or 'Internet Mass Hunting,' is a unique search engine that uses the participation of Internet users to provide, filter, and find the requested information. In recent times, cyber manhunt has been used in many hot issues of social concern, which makes this process a very controversial topic.

The development of the Internet and its related technologies has given birth to many cyber issues that damage social stability and personal safety on a massive scale. A cyber manhunt is one of the cyber issues that is either usually ignored or not given enough attention in society. There are many essential reasons for the phenomenon of cyber manhunt. However, the most significant driving force of cyber chase is the demand of the netizens. In the meantime, social contradictions are the catalyst for this phenomenon. Some technologies also dominate extensively when it comes to cyber manhunt, assaulting people's privacy, and affecting social stability. Also, to some extent, the loopholes in the existing laws have accelerated the development of cyber manhunt.

The first recognized human flesh search dates back to March 2006, when netizens from Tianya Club (an online community) in China, collaborated to identify an online celebrity named Poison [24]. Poison's true identity as a high-level government official was dug out by the efforts of the netizens. It has been more than a decade, and the development and content of the cyber manhunt had extended largely beyond what anyone had imagined when it was used for the first time. Initially, a cyber manhunt was mostly used for entertainment to search for information about a particular person on the Internet.

However, nowadays, the cyber manhunt has developed from its original purpose to the role of public opinion's supervision. Cyber manhunt is one of the online security vulnerabilities that need to be evaluated for maintaining the safety of the community. However, in some cases, cyber manhunt can be considered advantageous when it is used to provide a new way for people to search for information about someone who is at large after committing a crime. Thus, helping society and the police to solve a criminal case or a social issue.

While researching the most popular sites for academic papers and journals, we did not find many papers focusing on the topic of cyber manhunt. This clearly shows that even in the research community, how neglected this important cyber issue is. Hence, we are motivated to study this topic comprehensively. Our research aims to find out the main reasons behind cyber manhunt, evaluate the damages it can cause, and offer some solutions to deal with this issue.

For instance, China has become a significant Internet market in recent times. According to the Chinese State Information Center, China will have 1.1 billion Internet users by 2020 [22]. Such a vast number of the population, depending on social media, makes cyber manhunt a powerful search system with so many resources readily available to everyone who can connect to the Internet. Therefore, our study on the cyber manhunt becomes very critical to raise its awareness in the society to be able to make sure that in the future, it is used for the right reasons in the most appropriate way, causing the least damage.

Our paper is divided into ten sections. Section 1 introduces the necessary information about the cyber manhunt. Section 2 presents related work. Section 3 analyzes the characteristics of cyber manhunt. Section 4 offers the elements leading to a cyber manhunt. Section 5 discusses the reasons causing a cyber manhunt using a technological perspective. Section 6 presents the negative aspects of cyber manhunt, which includes the damages caused online and offline both. Section 7 talks about the positive aspects of cyber manhunt. Section 8 offers some feasible solutions for individuals and the government to minimize the harmful effects of cyber manhunt. Section 9 presents some ways for ordinary people to protect themselves from becoming a victim of cyber manhunt. The last part presents the conclusion and put forward the future work for this topic.

2 Related Work

M. Marchiori introduces the connectivity of a search engine used by the hackers to excavate the privacy of users [4] and presents a lot of hacker's technology used for a cyber manhunt. Thomas [13] provides a strong pertinence in cyber manhunt compared with other articles based on Cheswick's work [12]. At the same time, it also introduces more accurate methods and countermeasures of hacking into a firewall than Pipkin's paper [14]. Also, it profoundly excavates the reason why hackers snoop on other people's privacy based on Connor's writing [15]. Based on Chi's work, we found more reasons that contribute to cyber manhunt [41]. Shen's paper talks about the laws and regulations in public aspects, along with government regulations [21]. Zhang's work shows some useful suggestions to reduce the incidence of cyber manhunt [17]. Capone's paper presents the cases and examples that are old and have less meaning now since the techniques used today are so advanced and different [6]. New examples are included in our paper, which is contemporary. We also found some positive and negative aspects of cyber manhunt based on Doody's work [7]. We also continued on Zhuo's work [9] by adding anonymity, globalization, and psychoanalysis in the features section. Most of the authors mentioned above have not presented enough options as a solution to solve this social and cybersecurity issue. All these shortcomings in the previous works give us more possibilities to research this topic. By analyzing and comparing the solutions from different papers, specific recommendations have been made to regulate cyber manhunt in our article.

3 The Characteristics of Cyber Manhunt

In the process of cyber manhunt, the personal data of the victim is always made up of small and related fragments of information, which is mostly provided by cyber citizens [31]. What the searcher can track is the history and the trace of the net browsing history of the target people. It means all the pictures that people post on their social media or the location they allow to be shown on some applications and websites visited can be their information fragments. Moreover, if someone relays those pictures or locations, the information can be continuously used until the victim is found, creating a diffused node [34]. Therefore, the more people forward someone's information, the easier it becomes to victimize someone for the cyber manhunt. The process of cyber manhunt is mostly performed as follows [31]:

(1) Select the target (Get ID or photo)
(2) Post it online to get attention (Search pieces of information)
(3) Find related information about the target
(4) Collect information to find the true identity
(5) Release the identity online.

3.1 The Mixed and Typical Process

Not all the cyber manhunt stories can successfully draw the attention of society. In most cases, the interest fades very fast because very few people care about it. By analyzing the successful cases, we find that a major event or some specific person in some particular cases usually triggers a complete cyber manhunt process. Based on the event and the person, the search process of the cyber manhunt is started, and many netizens follow up, providing clues. In the end, the target person is exposed. Figure 1 shows how a typical cyber manhunt works [13].

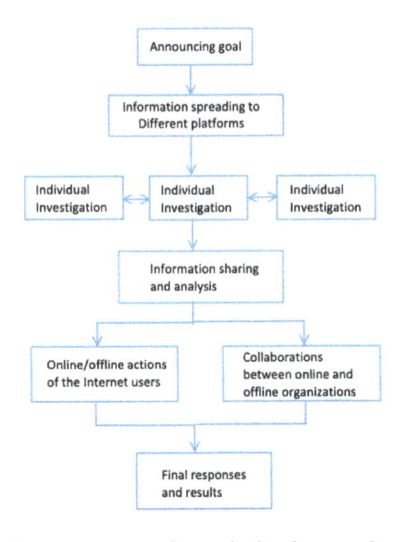

Fig. 1. Process of a typical cyber manhunt

3.2 Anonymity

The possibility of having the anonymity of the network results in the generation of anonymous groups. In an unidentified group, it is effortless to be influenced by one's opinion [24]. Human beings tend to follow other's thoughts and views as an easy way out instead of taking their thoughtful consideration. During the cyber manhunt, most of the netizens participating in the search use a virtual identity in a virtual environment. Therefore, it can be quickly concluded that a cyber manhunt is an event that is driven by very little emotional connection and more on societal influence.

In most of the cyber manhunt cases, most people blindly support an opinion. Most commonly, the same or very similar thoughts dominate in most of the comments related to a cyber manhunt without even considering its authenticity. In many cases, such a biased and wrong attitude results in cyber violence, which includes making aggressive, provocative, or insulting remarks on the Internet and causing emotional, mental, and social damage to the victims.

3.3 Psychological Analysis

The following points present the psychological factors involved in a cyber manhunt:

- **Customary law.** Customary law means the established pattern of behavior that can be objectively verified within a particular social setting. People like using their traditional judgment to decide whether one's behavior is right or wrong. In real life, if people confront someone who arouses their dissatisfaction and scorn, they tend to "teach them a lesson". The same human habit can also be seen on the Internet. Degrading others, and even sending a death threat to others is a widespread phenomenon that people show on the Internet. When someone's behavior causes rage among the public, more people will join the online movement to make personal attacks on the target without even knowing them. This process is also called as Cyber Cannibalism [10].
- **Group awareness.** In some cases, when people feel that their views are a minority in public, then they usually are not that willing to spread their opinions as quickly as when they think that their opinions are as per the majority of the people. In the latter case, most of them will bravely say whatever they feel like saying, without much hesitation. This theory is called the "Spiral of Silence" theory. It is reflected very naturally in the cyber manhunt. This kind of group awareness influences many people to change their minds in many cases, even when they have a different view compared to the majority of people in the beginning. This may explain why there are always many influential voices appearing during a cyber manhunt [11].

4 Elements Leading to Cyber Manhunt

The emergence and development of cyber manhunt are influenced by various social factors. Crowd psychology and the power of public social opinions are indispensable influences and driving factors for changes and growth. People who participate in cyber manhunt are generally divided into casual participants and regular contributors. Figure 2 indicates the main motivations of a cyber manhunt, which was studied involving 680 participants in total [5].

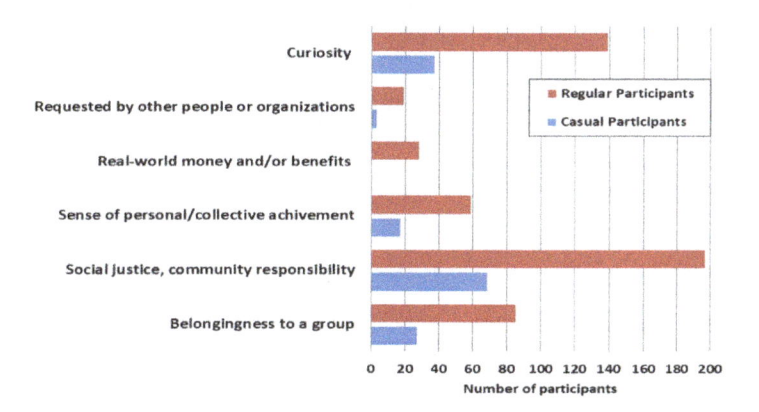

Fig. 2. Main motivations behind a cyber manhunt from 2 types of participants

Divided into casual and regular types, these two types of participants did the cyber manhunt for their pursuits in three main aspects: a sense of social justice, curiosity, and belongingness to a group. When it comes to the difference between two types, some of the regular participants will earn money or benefits by making purposeful remarks to change or control the development of particular events. Having 22.5% casual participants and 77.5% regular contributors, the percentage indicates that most contributors participated in more than one event. Combined with comparative data of different motivations, it can be seen that cyber manhunt has a high correlation with the social events that the public cares about. This also contributes to having participants that are more regular. Especially in China, which has a normalized cyber manhunt in social media, where the public will take the initiative to join events that they think are beneficial to society.

China is also one of the leading countries in the world where cyber manhunt events take place on a massive scale. In the first decade after the emergence of cyber manhunt (2001–2012), the leading cyber manhunt events in China were classified by types presented in Fig. 3 [30]. Since the Internet was not very well developed in that decade as it is now, the lack of recognized network moral constraints made cyber manhunt incidents take place in almost all parts of society.

As per the data presented in Fig. 3, we can see that among the seven aspects, people care the most about moral, public power, and legal issues. These three types of problems are most likely to arouse people's attention, and these issues are usually considered very beneficial to social justice and development. Entertainment issues account for 15.5% of all issues. This kind of cyber manhunt issues are generally intended to satisfy people's curiosity, so it is difficult to predict the results. If some individuals or organizations guide public opinion maliciously, it is likely to have a terrible influence on people involved, or even incur cyber violence to a very great extent among the netizens.

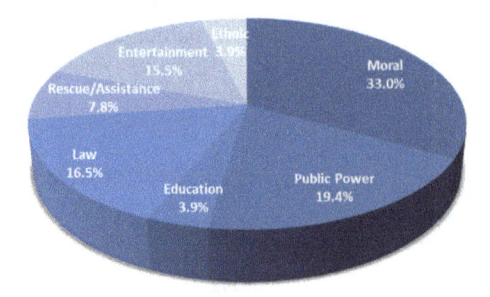

Fig. 3. Type distribution of China's first decade's cyber manhunt events

4.1 People

When it comes to individuals, awareness about self-protection is very crucial when it comes to protection from becoming the victim of a cyber manhunt. To a considerable extent, neglecting the importance of personal data is inevitably likely to exert an incredibly negative influence on the users. The problems could be because of inadequate protection of personal information and redundant information harassment. Following points can explain the reasons behind the relatively weak security of information by the people:

- **Relevant agency causes leakage.** Sometimes, cyber manhunt utilizes the service of a proper agency to get access to the personal data of the target. In most cases, the staff working in these agencies lacks occupational moral principles [38]. In case someone wants the report about a specific person from them, the team may not strictly enforce the process of identifying and preserving customer identities. Besides, under such circumstances, it is impossible to determine which agency has leaked the customer's information.
- **Lack of belief in legal recourse.** Some people are unwilling to easily count on legal recourse [39]. This happens mainly because sometimes, even when the perpetrators are caught, they do not get any punishment. In most cases, the person stealing others' information will just be fined. Sometimes people do not even want to complain as the cost of a lawsuit is relatively high, and the lawsuit procedure is quite cumbersome.
- **Lack of cybersecurity knowledge.** Many netizens are not aware of standard methods of protecting their privacy online. Many applications have the information push service that can share the information that the users are most interested in by analyzing user's personal log information such as type of resources, access mode, and frequency of access to the sites that users visit regularly [17].
- **Lack of awareness of cyber laws.** Lack of knowledge about the cybersecurity laws in case of a cybercrime ends up creating many cyber manhunt victims [39]. As a result, many netizens do not know where to go to ask for help when they get involved in cyber violence like cyber manhunt. This causes many cybercriminals to do whatever they want without worrying about being caught or punished.
- **Carelessness in using the browser.** The information stored by the cookies on a user's computer can easily cause the leakage of user's private information. In many cases, the user's data can be manipulated by experienced hackers and used for cyber manhunt. Besides, people tend to leak their information unintentionally. For example, email id, phone number, use of online shopping information, filling questionnaires and taking online surveys, etc. All this can be taken advantage of during a cyber manhunt [25].

4.2 Social Media

A very significant reason why cyber manhunt takes place also comes from social media [18]. Following are the reasons from the social media perspective that leads to a cyber manhunt:

- **Information leakage due to social networking:** One of the most common ways of a user's information leakage online is while using social networking apps or sites. Some social media sources may sell their user's data on purpose for monetary benefits. Alternatively, sometimes the third-party apps or websites connected to these social media apps or sites can use some information tracking technology to

steal people's private data and sell it to other companies for commercial use. Many times a user itself is not careful about their privacy settings on these sites and apps and ends up becoming a victim as a result.
- **Theft and Internet fraud:** Some companies offer information selling services when someone pays them the money they ask for. This kind of company can help find anyone's real name and real-life information. This process is straightforward to implement, as all that needs to be done is to search the information about the target and release it for a certain amount of fee that is not very high compared to the loss and misery it brings to the victims. The lower price directly leads to more cases of severe cyber manhunt.
- **Posting personal life information online:** Due to the user's lack of awareness and the rapid development of the Internet, many people tend to post information about their life online as a public post. The readily available detailed personal information that is just a mouse click away can easily be manipulated during a cyber manhunt.
- **The dark web sells information:** The dark web is an online space that is mostly used for secret and illegal online deals. The resources there cannot be accessed through normal hyperlinks and browsers. It needs to be obtained through dynamic web technologies [25]. It is a pervasive source of accessing user's stolen personal information for just a meager price. Many times, if the information needed for a cyber manhunt is hard to find, people turn to the dark web to get the information they need.

4.3 Influence of Cultural Differences

After studying the fact that cyber manhunt started in China, we wanted to find out if there exists the same situation in other countries as well. What we found very interesting was the fact that most of the cyber manhunt cases happen only in China. This can be attributed to the cultural difference that exists between China and other parts of the world.

In China, cyber manhunt is more about social morality. In most cases, cyber manhunt starts with someone's inappropriate actions or misbehavior that creates social anger. This causes the netizens to hunt for the person behind the screen and make that person public. For example, in July 2007, a post titled "the most vicious stepmother in history beat her daughter to the point of spitting blood" caused outrage among Chinese netizens. Netizens posted notices on the Internet, and it turned out that the woman who was being victimized was not even a stepmother. It was an artificially insinuated case that had no commercial value involved, but it still became viral [20].

In other parts of the world, cyber manhunt is mostly used for criminal investigation. In May 2017, a cyber manhunt was launched to find hackers behind WannaCry ransomware attacks that crippled the systems the world over. Many European cybercrime specialists from affected countries supported this manhunt and used all the tools they could to bring the hackers to justice [29]. In the end, the hackers were found and punished severely.

In many countries, actions are taken to restrain cyber manhunt. For example, the federal Electronic Communications Privacy Act (ECPA) or Children's Online Privacy Protection Act (COPPA) in the US protects against the violation of a person's privacy by intercepting, accessing, or leaking stored communications. The US also has some related laws to protect citizens' online privacy. The European Union (EU) also pays much attention to the legislation to protect the user's data. In 1999, the EU formulated the general principles of privacy protection for individuals on the Internet and other relevant laws and regulations to establish a unified legal and regulatory system related to online privacy protection in member countries. In 2018, the EU released a new law called GDPR (General Data Protection Regulation). It not only applies to the data that resides on any server inside EU countries, but it also applies to any data that passes through the EU [2]. The new law changed a lot in the whole world when it comes to the data protection provided by the companies, mainly because the penalty for not taking this law seriously is very high.

Cyber manhunt can also serve as the means for the supervision tool in many countries. Any netizen can participate in a cyber manhunt to find the person who breaks the national law or cause some social unrest in the country. Cyber manhunt is undoubtedly becoming a prevailing trend all over the world because of the tremendous growth of the Internet in recent years [6].

5 Analysis of Related Technologies Leading to Cyber Manhunt

The appearance of a cyber manhunt in our society has a deep relationship with the development of technology. Social media is just the product of human's living habits and technology. Internet technologies give a chance to social media to become the right place for people to store their information online without much hesitation. In this section, we mainly talk about some standard techniques and technologies used for a cyber manhunt.

5.1 Methods to Track User's Information Online

- **IP Searching:** Cyber manhunt usually begins from IP searching. It is the first approach to retrieve the target's information. The most simple authentication procedures use the IP address as an index. The IP address is the universal identification index on the Internet [48]. The process is that the hackers will track the IP address of the user by making use of PHP code and will store that address and information retrieved from an IP address like location, last visit, session time, website visited and all sorts of online activities in a database [3]. This means that when people spend their time scanning the homepage of their social media, hackers can get almost all of their private information at the same time. The uniqueness of IP address provides excellent convenience to hackers. If the hacker gets specific administrative authority in a network, then they can even find out the target's physical address in the real world.
- **E-mail Attack:** E-mail attacks are mainly divided into two parts. One is e-mail bombing, which refers to sending thousands of spams to one mailbox with a fake IP address and email address. The other one is email spoofing, where the attacker

usually pretends to be a system administrator and sends an e-mail to the user, asking the user to click on a link for account security or reward collection. The link in the e-mail usually hides the virus. Once the user clicks the fake link created by the attacker, the computer will be hacked. Hacker always uses SMTP (Simple Mail Transfer Protocol) to achieve their goal. This result in user's information is being exposed to the hackers; thus, having a high risk of being leaked to the public [47].

- **Place a Trojan Horse Program:** Trojan horse programs can hack directly into a user's computer. It is often disguised as a program or a game. Once the user opens these programs, it hides an application in the computer system that can be executed quietly when the computer starts. When people connect to the Internet, the program will notify the hacker about the user's IP address and preset port. Hackers will use the available information to hack into the computer [15].
- **WWW Spoofing Technology:** The hacker will create a fake URL of the web page that the user visits, leading to their server. When the user sends a request to the original page, their request is redirected to the hacked server. Therefore, the hacker can hack the computer, network, or the database by using pharming and phishing. When the users logs in or register an account, all the personal information they input, such as their name and password, etc. is readily available for the hackers to manipulate and misuse [13].
- **Search engines:** On the Internet, there exists a code called 'Ghost Embedding' written in HTML. A ghost component in a web object is a part of the code where text can be inserted that, in all likelihood, will never be observed by a user when viewing the web object using a web browser. For instance, HTML comments are ghost components, as well as Meta description tags [4]. Therefore, hackers will make a new website and upload it to the Internet. When users surf the web, the code runs as usual, and users' personal information can be easily acquired. Hackers can use the web search engine database to hack into it and get the information they need. Also, every social media has quite a robust connectivity with each other, which means that the more websites people scan, the more information they will release unintentionally. Nowadays, with the rapid development of image recognition, combining the search engine with image recognition also causes a significant breakthrough in cyber manhunt [32].
- **Exploiting Security Holes:** There are two kinds of system vulnerabilities that can be directly manipulated by hackers to obtain certain privileges. These are remote vulnerabilities and local vulnerabilities [12]. Most of the hacker attacks are made by exploiting a remote vulnerability. However, they need to cooperate with local vulnerabilities as well to extend the opportunities obtained, often to the system administrator rights. Most of the time, it is either the network terminals or the modem servers or both that are a beneficial commodity to the hacker [13]. With the highest administrator privileges, a hacker can create attacks like sniffing and session hijacking and exploit local vulnerabilities on the system. They can also use some spoofing programs such as Trojans to get administrator's rights and passwords. Many systems have security vulnerabilities of one kind or another, some of which are owned by the operating system or the application itself and are generally challenging to defend against hackers before patches are developed. Usually, more service a system provides, the more vulnerabilities exist in it [13].

After researching these related technologies, it turns out that IP searching, Trojan horse, and WWW spoofing technology are the methods for skillful hackers, which can cause significant damage and lead directly to people's information leakage.

6 Aftereffects of Cyber Manhunt

This section discusses the aftereffects caused by the process of the cyber manhunt on the victims and the society in general.

6.1 Online Damage

- **Privacy Leak:** Privacy leak is one of the most common hazards of human flesh search. Under most circumstances, a cyber manhunt is based on social aspects other than the technology, which more likely may cause the privacy leak. There are many forms of privacy violations. The most common are usually unauthorized publication of photos, use of images, and dissemination of personal contact information [6]. For example, an American soldier named David Motari became a victim of cyber manhunt because he dropped a puppy off a cliff in 2008 [37]. To hunt him down, his complete private information was revealed on the Internet by the netizens. Netizens did not think twice that their online behavior while trying to bring justice would have an impact on not only the soldier's individual life but also, it could also bring the risk of information leak to the U.S. military. As a result, the U.S. military had to step in to investigate the case to protect the state secrets. This shows that the scope of influence created by cyber manhunt is much broader than the netizens can ever think of.
- **Trolling:** Trolling refers to the online insult where people post violent and nasty comments about someone or makes fun of someone to hurt them on the Internet. The Internet allows everyone to hide their real information making them free and less accountable to post their thoughts and opinions. Therefore, people are more likely to insult others on the Internet than in reality. Trolling is widespread that happens on the Internet every day on a vast scale. It is quite common to see many posts full of anger, hate, and abusive words. Many trolls on the Internet are not focused on the facts at all but only on the personality of the trollers. For example, China Global Television Network (CGTN) host Xin Liu became the victim of a cyber manhunt and was heavily trolled by the Chinese netizens in May 2019 for her comments on U.S-China trade war and Intellectual property theft with the Fox Business Network host Trish Regan [44].
- **Target Exaggeration:** Target Exaggeration means that once netizens use cyber manhunt to expose one person, more people around the victim will also get involved in the case. For example, in 2018, a BMW driver was seen arguing with a worker at a petrol station in Singapore. The driver only wanted $1.42 of gasoline as he had already sold that car, but the worker filled the gas worth $19.12. As a result, the driver refused to pay. However, when a witness posted the video of this incident, the whole thing suddenly turned into an issue of rich people bullying

workers. Viewers who did not know the truth made the innocent driver a victim of the cyber manhunt, and in just one day, the driver's real name, occupation, workplace, contact information, and even personal photos were released on the Internet. Although the truth was finally revealed, the incident still caused irreparable damage to the car driver and the worker, the gas station, and the used-car dealer involved, were affected as well [27].

- **Slander:** Slander refers to the use of violence or other means that damages the reputation of others. On August 20, 2018, a female doctor named Ann had a conflict with a 13-year-old boy and his parents at a swimming pool in Sichuan province of China. A cyber manhunt followed on Ann as a result, and she, along with her husband, had to face verbal insults and all sorts of personal revelation. The 35-year-old female doctor could not stand this kind of Internet violence and chose to commit suicide [26]. The rapid spread of rumors depends on many netizens. Some anonymous users like to fabricate facts on the Internet to raise the public's attention. Also, many observers are likely to make rash comments and spread the posts without verifying the authenticity of the information, which will inevitably enlarge the event's impact.

6.2 Offline Damage

Cyber manhunt may decline the quality of information on various media sources as well. This happens because of the negligence of the media network and the mainstream media's unwillingness to report the truth. The radical opinion of others may significantly affect people's judgment. As a result, the netizens on social media may care less about the fact [8]. Some network media regard cyber manhunt as a kind of publicity gimmick to attract more eyeballs for the absence of ethics. Under such circumstances, events that have nothing to do with cyber manhunt are also labeled as a cyber manhunt for drawing people's attention.

In some countries, once the information related to the politically charged content is published or is likely to be used for fostering dissent and organizing anti-government protests, the governments will control the information distribution [43]. Because of the restriction from the government, even if the reporters reach the scene, people will not see the news reported in the media in any form.

7 The Positive Effects of Cyber Manhunt

7.1 An Online Weapon to Fight Against Corruption

The cyber manhunt can prove to be advantageous as far as it is used as a weapon to fight against corruption [7]. For example, when netizens notice that the property owned by a government official looks inconsistent with his or her actual income level, then they can assume that the corruption might exist in that case. The netizens can point out the reasons for suspicion and cause heated discussion on the Internet. During the discussion, more evidence and details can be explored by the efforts of the participating netizens leading to the cyber. Netizens can dig into the background and history of corrupt officials to force the government to take action against the corrupt official.

7.2 Clues for Searching for a Particular Information

Cyber manhunt can play a crucial role when it comes to helping people find some specific information. Compared to using a traditional search engine, cyber manhunt is more advantageous in looking for someone specific. For the conventional search engines, it is very efficient when searching for information that already exists on the Internet. On the contrary, if the required information does not exist on the Internet, the traditional search engine will not work well. This is because the conventional search engine's working principle is based on word filtering, machine learning, and deep learning to make it more suitable to the search keywords.

Cyber manhunt can operate both technically and manually while taking proper advantage of all the people involved in the process when encountering problems in looking for certain information. For example, there was a piece of news reported by South China Morning Post that, due to a cyber manhunt, a Chinese-American child called Kylee Bowers, who was lost in China in 2005, finally found her parents in 2018 [19]. So, the value of cyber manhunt can be reflected in searching for something that cannot be found easily in its usual way using conventional search engines.

7.3 A Platform for Expressing Oneself

One of the positive aspects of the human flesh search phenomenon is that it helps to balance one's emotions [24]. The feudal subject ethics emphasizes the obedience of the public, which still exists today in some countries. This can make some people depressed in the actual society. The virtual network society provides a relatively free platform for individuals to express themselves and release the accumulated dissatisfaction in the actual community, which is conducive to the development of an individual and society.

8 Effective Community Development and Maintenance by Public and Government

By analyzing the common phenomenon of online behavior and social media's immoral actions, we discuss the active community development in terms of personal and social aspects.

8.1 Positive Utilization of Searching Technology

The use of a search engine should be regulated. Most of the ordinary people are not able to find out other's personal information without technical help. Therefore, it is of great importance to restrict the use of search engines and other useful technologies to reduce the possibility of cyber manhunt. Also, the government and related organizations should encourage ordinary people to utilize the searching technology positively. With the help and regulations from the administration, correct online searching rules can be created to give the community a better social media environment.

People can also get information and help they want from other people by posting questions online. This kind of search method is also widely used in a cyber manhunt that can be a breakthrough in information transfer. With the help of the crowd on social media, a target can be found quickly. According to this advantage, a cyber manhunt can also help the police arrest some criminals. For example, in 2018, a 21-year-old flight attendant was killed by her taxi driver on her way home from the airport in Henan province of China [42]. The taxi company immediately offered a reward of 14 thousand dollars for people providing clues on the suspect's whereabouts and posted suspect's photos on China's biggest social media platform called Weibo. With the help of cyber manhunt, the suspect was found within a few days. The murderer had already committed suicide by the time he was found, but cyber manhunt still helped the police to track the identity of the criminal.

8.2 Enhancing the Awareness of Self-protection

For cyber manhunt, the network becomes a useful way to dig anyone's information online. As a result, the security of people's personal information is not guaranteed. It is nearly impossible to have online privacy, even if a user is alert and actively using different ways of protecting their data. Apps and websites can also predict what a user wants to buy in the future based on their online behavior tracking records. Therefore, the current problem is not how to prevent privacy leaks, but how to avoid human flesh search, how to live in an environment without privacy. Here, we provide some suggestions about online self-protection methods:

- **Be discreet with providing the information online:** People should be cautious about the publicity of their mail id and other personal information when they register on a website [16]. Nowadays, many sites choose to use email address verification to let a user use their services. In that case, the vast majority of the email address has to be real and valid. Therefore, if the researcher can get to the people's registration information, they can find the registered mail id as well, which can help them easily find the information about their target. Even a blog post can lead a researcher to the target's personal information. In case the target is not active online, the researcher can still find the online information of the target's friends to find the target's real information. All these things can be taken advantage of during a cyber manhunt.
- **Separate Accounts for Work and Personal Life:** For most people, they like to post their recent life events and private photos on social media [33]. Meanwhile, many people may deliberately show their life happenings to not only close friends but also their acquaintances. To prevent professional connections and strangers from observing people's personal lives or get their real identity, it is suggested to create two social media and mail accounts, one for work and one for daily life.
- **Set Strong passwords:** Having a strong password is essential as it is directly related to a user's privacy. Nowadays, there exists a kind of website called "Social Engineering Database" [46]. In hacker's circles, it is a means to obtain information

for hacking. Cyber manhunt on the Internet is an application of social engineering. Figure 4 [23] shows the statistics of the global data leakage in different fields of society in 2018. We can see that the leakage covers an extensive range of areas that includes every aspect of people's daily lives. Most of the leaked data is entered into the Social Engineering Database and shared by hackers all around the world. From the chart, we can see that the Internet industry has the highest number of leaks (28%).

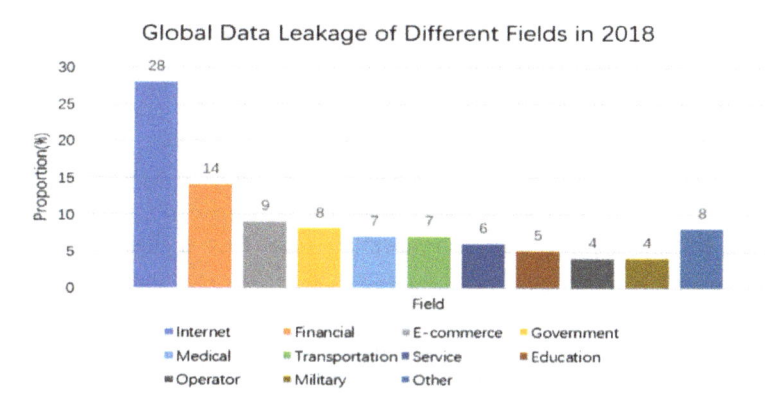

Fig. 4. Global data leakage in different fields in 2018

8.3 Supervision from Governments

Cyber manhunt is a very hot topic for the governments, as already mentioned in earlier sections of the paper. In 2017, China started enforcing the People's Republic of China Network Security Law. It explicitly strengthens the protection of personal information. No individual or organization may steal or otherwise illegally obtain personal information, or unlawfully sell or provide personal information to others [28]. In May 2018, GDPR by the European Union came into force that sets strict personal data management standards for all companies that trade with EU countries or hold EU citizens' data [2].

Strengthening the protection of network information can effectively reduce cyber manhunt issues that obtain a victim's personal information through network information. Under the influence of GDPR, the US Government Accountability Office (GAO) also released a report in January 2019, suggesting Congress to enact a comprehensive federal interoperable Internet privacy law. At the same time, the report highlighted the importance of striking a balance between the benefits of collecting data and protecting user's privacy [40].

On the one hand, cyber manhunt is the manifestation of the citizen's supervision right. On the other hand, the behavior of some people and the website already breaks the moral and legal boundaries. This kind of behavior causes the abuse of cyber manhunt gradually evolving into a tool infringing on others' rights and interests. Therefore, the key for the government is how to face the negative aspects of cyber manhunt and to find ways to reduce or even avoid cyber manhunt. To deal with the cyber manhunt, the government can take the following steps:

- **The lack and difficulty of establishing the law:** Making strict laws are an excellent way to limit cyber manhunt. However, as a form of expression, the anonymity of cyber manhunt is the essential point to help protect the freedom of speech [45]. Since the freedom of speech is of great importance in the constitutional system in the world, the limitations of the modern state on the expression of Internet freedom can only be limited. Even if some countries have sound and sophisticated legal systems for the protection of related information and personal privacy, they still do not have the relevant law to regulate cyber manhunt. To establish the rules, the government is supposed to have a clear understanding of how cyber manhunt works. It is necessary for the government to keep a balance between freedom of speech and invasion of privacy, personal information, or even slander.
- **The drawbacks of making up afterward:** The current discussion on cyber manhunt is focused on the identification and investigation of responsibilities. According to the real-life cases, we have seen that even if the victim wins, the judgment is far from making up the damage and loss the victim endures during the events. Therefore, if the government can help prevent the cost and the loss from the beginning instead of fixing the problem after it happens, it will maintain the network's regular order and save many avoidable mishaps.
- **The innovation of establishing the law about cyber manhunt:** China's "Kitten Killer" case in 2006 was the first cyber manhunt case in the world. It is also the first example of a Chinese court addressing the issue of cyber violence. This warned people who preferred to use the human flesh search engine for hunting down individuals [1]. This shows the great necessity of the judicial system to handle cyber manhunt. The following are the points about how to make innovation in establishing laws to protect personal information on the Internet:
  - **Monitor major human flesh search sites and portals:** Websites should be under watchful eyes of the government to control the human flesh search. However, it is impossible to monitor so many sites, so the government should have a warning mechanism. They can use artificial intelligence to make the filtering mechanism, judging if the personal information is involved in the keyword and the critical information. Once an event that may cause cyber manhunt happens, the government can start the warning mechanism. Related websites could be informed that cyber manhunt may occur, and the required actions to defend can be taken.
  - **Establish protection consciousness and strengthen the industry's self-discipline:** The people in charge of the forums or websites can take care of the content and decide whether the material can be posted on the Internet or not. In that case, the government should make sure the leaders are familiar with the laws of the cyberspace.
  - **Guide websites to foster opinion leaders in virtual communities:** In the transmission of the information on the Internet, especially for some news, the opinion leaders in the virtual communities are essential. They can comment on some issues and affect others' attitudes. Generally, the opinion leaders are

mostly celebrities or some political leaders. The government should attach great importance to the influence of the opinion leaders and have more connections with them.
- **Technical assistance and mature supervision mechanism for the website:** The government should use technology to filter and block harmful information on the Internet. In addition, the online report and complaint mechanism should be used to make sure everyone has the chance to take part in fighting against cyber manhunt.

8.4 The Regulation of the ISP

Imposing liability on Internet Service Provider (ISP) can control cyber manhunt to a great extent as well [1]. Since most of the online posts are anonymous, so whenever there is a case of a data breach, the victims cannot find who leaked their information. In this case, the ISPs need to bear responsibility. ISPs can be held responsible for the legality of each online post, or they need to build a real-name system for every netizen to be able to track down the origin of leak or attack to cause the leak.

9 Ways to Escape Cyber Manhunt

9.1 Case Study of Satoshi Nakamoto

Although the process of cyber manhunt is compelling, there is still some approach to evade from the eyes of the Internet. One person who has successfully escaped cyber manhunt until today is Satoshi Nakamoto. This person has contributed a lot to the field of bitcoin and blockchain, but he or she has not been found in person yet. People have been trying to find out his or her identity for many years. Despite so many records left everywhere on the Internet; Satoshi Nakamoto still managed to survive the cyber manhunt successfully. This person is an excellent example of protecting himself or herself from cyber manhunt.

We analyzed his or her behavior pattern and found out that the most effective way to hide one's identity on the Internet is to never reveal the real information about oneself. When people were trying to search for Satoshi Nakamoto, the information they found was so confusing. That person claimed that he was 43 years old male living in Japan. However, he never used Japanese on any site. On the contrary, the language he used online was British style. As a result, people guessed that the real identity of Nakamoto might be of a British person.

At the same time, there also exists some evidence indicating that Nakamoto might be an American. According to the statistics of Nakamoto's active hours online, his or her online behavior falls between 9 am and 11 pm in the US time zone [36], which perfectly matches the work and rest time of a typical American person. Just because there are so many pieces of evidence with conflicting information, people even believe that Satoshi Nakamoto might not be a single person but a team of people [35].

What the case tells us is that using misleading information to cover the users' information is an effective way to save oneself from becoming the victim of a cyber manhunt. For some websites, the use of real information for registration is not compulsory. In that case, users should use false information to register. Even if some websites controlled by the government request the actual data, it is better to avoid binding that current account to the social media account. Doing this may lower the possibility of people undertaking cyber manhunt and find more information about someone.

9.2 Delete the Online Records

For ordinary people to get rid of the cyber manhunt, it is essential to clear their history records on the Internet. Firstly, if the account and the password are remembered online, one should directly sign in and delete all the information posted before. Secondly, if the id of the account is easy to identify, it is better to cancel the account. Thirdly, nowadays, many search engines have a function which supports image search. If one's photo is found and uploaded to the Internet, then their detailed information can be collected by the search engine. However, according to the cyber laws of many countries, a user can apply to the website for deleting the photo or ask them to remove the search results about that particular information or picture. Finally, sensitive words are certain words or phrase that is set by the website administrators, which cannot be searched online. So, when people are creating the username, they can choose certain words that might be the sensitive word to avoid appearing in the search results. By doing this, users can mightily diminish the chance of becoming the victim of a cyber manhunt.

10 Conclusion and Future Work

As a form of information exchange and sharing, cyber manhunt causes considerable damage to society and individuals. The appearance of the significant data era provides various kinds of technologies for people to search for information no matter where and why they plan to use it. People not having a real and comprehensive understanding of the Internet, and managers not establishing an effective regulation breeds danger, panic, and crime on the Internet. Like a double-edged sword, the positive influence on cyber manhunt can guide people to use the Internet in the right way while the adverse effects of cyber manhunt can help people increase their conscience about online privacy and help companies to improve their technology and administration to decrease the frequency of cybercrime.

The application of information filters can help build a faster and safer search engine. The evaluation of cyber manhunt shows both positive and negative effects. The most common cause for casual participants and regular contributors is the same for justice. This shows the possibility of effective regulations and improvements in cybersecurity management. Besides, cyber manhunt is used for different reasons in different countries. For China and some Asian countries, the public uses cyber manhunt primarily for social justice. For other countries, organizations or governments use it to find criminals.

By analyzing some commonly related technologies, we find that the use of the search engine is the primary cause of cyber manhunt since people do not need to learn technical skills to master it. Apart from regular contributors, most people participate in a cyber manhunt by using the search engine. This leads to the victim's information becoming widespread. Meanwhile, the misuse of the search engine from social media companies shows the importance of having a government regulation by utilizing the positive use of cyber manhunt and innovation of establishing the laws against cyber manhunt.

For the future study on cyber manhunt, the analysis of the typical cases offers good examples for focusing on the moral impacts of the cyber manhunt. The analysis of damage will help people find the reason for cyber manhunt and help provide useful strategies for an institution and social media companies. The use of some AI techniques can also provide better control over the whole process from going out of hand.

References

1. Cheung, A.S.Y.: A study of cyber-violence and internet service providers' liability: lessons from China. Pac. Rim Law Policy J. Assoc. **1835**(2), 336 (2009)
2. https://eur-lex.europa.eu/eli/reg/2016/679/oj/
3. Aggarwal, V.: Data security approach for right to information of developer. Int. J. Adv. Res. Comput. Sci. 4, 175–246, 292–293 (2010)
4. Marcher, M.: Security of World Wide Web Search Engines, p. 14 (1997)
5. Cheswick, W.R., Bellovin, S.M., Rubin, A.D.: Firewalls and Internet Security, 2nd edn, p. 143. Addison-Wesley Longman Publishing Co., USA (2003)
6. Capone, V.: The Human Flesh Search Engine: Democracy, Censorship, and Political Participation in Twenty-First Century China, p. 70 (2012)
7. Doody, J.: China's Expanding Cyberspace. ECFR Asia Centre (2014)
8. Wei, J.: Why is 'Human Flesh Search' Only Popular in China (2011)
9. Zhuo, F.: A Behavioral Study of Chinese Online Human Flesh Communities: Modeling and Analysis with Social Networks (2012)
10. Yan, D., Cao, Y.: Cyber Cannibalism, China Daily, 2 January 2012. http://usa.chinadaily.com.cn/china/2012-01/02/content_14370175.html
11. Elisabeth, N.: The spiral of silence a theory of public opinion. J. Commun. **24**(2), 43–51 (1974)
12. Cheswick, W.R., Bellovin, S.M., Rubin, A.D.: Firewalls and Internet Security, 2nd edn, p. 143. Addison-Wesley Longman Publishing Co., Boston (2003)
13. Thomas, D.: Hacker Culture, p. 48. University of Minnesota Press, Minneapolis (2002)
14. Pipkin, D.L.: Halting the Hacker: A Practical Guide to Computer Security, vol. 2, p. 143. Prentice Hall Professional Technical Reference, Upper Saddle River (2002)
15. Connor, J.: Hacking Become the Ultimate Hacker Computer Virus, Cracking, Malware, IT Security, p. 13 (2015)
16. https://www.zhihu.com/question/48691691
17. Zhang, L.: On strategies of personal information protection in the personalized information service in big data times. ITM Web Conf. **7**, 03002 (2016)
18. https://www.cbsnews.com/news/the-data-brokers-selling-your-personal-information/
19. http://www.yidianzixun.com/article/0JRvihvn
20. http://news.163.com/10/0601/14/683OE4T200014AED.html
21. Shen, W.: Online Privacy and Online Speech: The Problem of the Human Flesh Search Engine, vol. 12, p. 44 (2017)
22. http://tech.ifeng.com/a/20150420/41061128_0.shtml

23. http://blog.sina.com.cn/s/blog_14143d9640102zhun.html
24. https://ozziessay.com.au/essay-on-the-cyber-manhunt/
25. https://bokonads.com/data-is-fallout-not-oil/
26. https://baijiahao.baidu.com/s?id=1612195281113107306&wfr=spider&for=pc
27. http://www.zaobao.com/zopinions/editorial/story20180420-852323
28. http://www.npc.gov.cn/npc/xinwen/2016-11/07/content_2001605.htm
29. https://www.thepaper.cn/newsDetail_forward_1685751
30. He, Y.H., Zhou, F.: The first decade of human flesh search – an empirical study based on the theory of collective behavior. Mod. Commun. **3**, 129–134 (2013)
31. Nian, F.Z., et al.: A Human Flesh Search Algorithm Based on Information Puzzle, pp. 1243–1248. School of Computer & Communication Lanzhou University of Technology (2018)
32. Lu, L., Huang, H.: A hierarchical scheme for vehicle make and model recognition from frontal images of vehicle. Trans. Intell. Transp. Syst. **20**(5), 1774–1786 (2019)
33. https://www.cert.org.cn/publish/main/9/2018/20180605080248533599764/20180605080248533599764_.html
34. Wang, F.Y., et al.: A Study of the Human Flesh Search Engine: Crowd-Powered Expansion of Online Knowledge. The IEEE Computer Society, August 2010
35. https://medium.com/cryptomuse/how-the-nsa-caught-satoshi-nakamoto-868affcef595
36. https://www.sohu.com/a/251655420_100112719
37. https://www.infowars.com/david-motari-alleged-puppy-killer-tracked-down/#inline-comment
38. http://news.ifeng.com/a/20150522/43814423_0.shtml
39. http://www.doc88.com/p-7798661090872.html
40. United States Government Accountability Office, "Internet Privacy," June 2019. https://www.gao.gov/assets/700/696437.pdf
41. Chi, E.H.: Information seeking can be social. Computer **42**(3), 42–46 (2009)
42. https://www.whatsonweibo.com/suspect-in-chinese-stewardess-didi-chuxing-murder-case-jumped-into-river/
43. https://www.zhihu.com/question/31060463
44. http://www.soundofhope.org/gb/2019/06/04/n2933086.html
45. Yu, J.: Study on Legal Issues of Human Flesh Search, p. 23. Shandong University, April 2011
46. https://m.jiemian.com/article/1526703.html
47. https://max.book118.com/html/2017/1001/135363839.shtm
48. Anonymous: Maximum Security: A Hacker's Guide to Protecting Your Internet Site and Network, Kybernetes, vol. 29, no. 1, p. 500 (2000)

Smart Agriculture with Advanced IoT Communication and Sensing Unit

David Krcmarik[1(✉)], Reza Moezzi[1,2], Michal Petru[1], and Jan Koci[1]

[1] Institute for Nanomaterials, Advanced Technologies and Innovation, Technical University of Liberec, Studentská 2, Liberec, Czech Republic
David.Krcmarik@tul.cz
[2] Faculty of Mechatronics, Informatics and Interdisciplinary Studies, Technical University of Liberec, Studentská 2, Liberec, Czech Republic

Abstract. In this practical project, an advanced reconfigurable communication unit is studied which is well-matched to be used in harsh agriculture environment. The unit transfers all data through SQL database and back-end with a UF (user friendly) web interface. It collects many data online from different sources like accelerometer, tensometers, CAN, temp. sensors and analog or digital inputs. The communication is based on Global System for Mobile Communications (GSM) and It uses GPS data to obtain accurate real-time localization for knowing where the machine is. The unit is powered with three different power sources. An external motor generator, a solar battery or a standard 12 V tractor battery. To enable the device working even without any of the above-mentioned power sources, an internal battery is also provided as a secondary source. The heart and novelty of the unit is a multifunctional Linux based Printed Circuit Boards (PCB) which would be described in detail. Several simultaneously working units, can be simply addressed using private standard network. The back-end lets the operators to organize chosen activities. One can select desired data to be detected. It is conceivable to arrange the time span between two different packages and the data acquisition rate which would be sent to back-end. Information are accessible via web interface and can be copied in CSV format for post processing. The whole system is also designed to capture user-defined data constraints and if such a constraint is touched, a warning message would be sent to a pre-defined contact.

Keywords: Internet of Things (IoT) · Smart and precision agriculture · Sensors · GSM · Big data

1 Introduction

New trends in smart or precision agriculture (PA) have been introduced in the past years. It involves different approaches and implications in robotics, artificial intelligence (AI) to IoT (Internet of Things) actuators and sensors in various type of network configurations. Precision agriculture empowers human to exploit the production of agricultural goods with respect to identical time preserving the resources and the same amount of inputs. This article would emphasis on the IoT part in PA. Contemporary precision agriculture is massively reliant on massive quantity of gathered data.

IoT based popular infrastructure networks, makes easy for data acquisition in this domain. IoT devices require to involve sensors and the technique of sending or receiving the data to back-end. IoT based strategies are regularly situated in distant areas where connection based on cable is not achievable. GSM, LoRa, WiLD, SigFox and Zigbee [1] are common network strategies which can be handled for communication. For instance, the WiFi in Long Distance (WiLD) network is implemented to connect IoT devices for communication [2–4] in the countryside areas for low-cost PA. But the WiFi-based system has its weaknesses in term of limitation of its coverage. To prevent such difficulties, GSM data (3G, 4G or GPRS) is considered as a valuable way for communication, because such networks are accessible almost everywhere and they are common. One must also consider in term of energy performance concepts. The finest strategy is once the IoT instruments are independent from external energy sources, and they are able for self-charge by residual energy generated or solar power. The data acquired from several devices endure some problems due to variation of protocols or time delay. Therefore, management system is a vital part to collect data. This management systems can be a Node-Red (for small-scale & user-friendly purposes) or a dedicated system like SmartFarmNet [5]. The tendency of implementation of IoT grounded resolutions in smart agriculture, is undoubtedly apparent and increasing in years. Amongst different solutions, primarily nursing of various issues like irrigation, moisture, concentration and air flow are the most interested situations [6]. To handle those issues, in many cases monitoring and capturing actuators with IoT based devices are involving. One of the most common situations is to control smart greenhouses [7]. It is roughly due to the situation, that unlike monitoring, to control the case is required typically further resources for higher energy consumption demanded. On the other hand, it also must take into account processing of massive numbers of data which brings the big data concept to deal with it.

A brief review of the prior and recent research in this field can be given as follow. Popovic et al. [8] have offered an experimental strategy to be used in agricultural irrigation, monitoring and ecological nursing. They have made their explanation based on low-priced and commonly offered Raspberry Pi (RPI) and Arduino with various Libelium sensors. They urged a forthcoming implementation of protocols like CoAP or MQTT and they have motivated to use programming languages like statistical R or Python. Another study [9] focused on various options used in real circumstances. In the paper, the PA is explained as a grid-network which standardized protocols can be used for different purposes in such a way that the nodes are assigned to a degree intelligent for self-configurable based on the environmental conditions. Presently, most of the practical situations are from such ranges: water quality supervision, control of irrigation, pest prevention, cattle measure, monitoring or control of greenhouses and using UAV (unmanned aerial vehicle) for smart field assessment. In [10] a useful list of IoT ready control nodes and related sensors are presented by authors. The data is directed to a chief hub through wireless-based technologies, then the back-end can be whichever constructed from scratch or one can use available infrastructure like Plotly, Xively or ThingSpeak for certain operations. The attained data are necessary to be processed and being evaluated. Thanks to Big Data infrastructure for handling such problems. Regarding Big Data and data analytics, a detailed research is done in [10]. In the management part, a few agriculture management info systems are accessible like

OnFarm, Easyfarm, Farmx, Cropx, etc. Data analytics is currently one of the important concepts that can assist with estimation from vast amount of data and decision making. From attained data, useful investigations can be concluded to perform the desired design. these data can be complemented with vision-based camera photos [8]. In all above-mentioned studies, a unit which can handle all operations in a compact way, has not been achieved, thus presented applied IoT unit is introduced in current study.

In this practical and novel paper, the authors would offer an advanced unit to be used in harsh agriculture in the domain of IoT. The PCB introduced in the paper is designed to carry out many tasks for sensing by connected sensors and communication with various farming and agricultural devices monitoring and transferring data by GSM information service.

This article is managed within six parts. First an introduction is given. In Sect. 2, the IoT structure and monitoring unit charging are discoursed. In Sect. 3, the practical communication unit and tested sensors are presented. Section 4 dives into the back-end which implements Microsoft SQL Server (MSSQL) DB architecture. Performing the big data idea is discussed in Sect. 5 and lastly, final section concludes and completes the paper.

2 Charging of Monitoring Unit

The unit presented in this article, is used with agricultural vehicles like tractors to supervise and monitor CAN interface, tensometers, accelerometers, analog or digital inputs and GPS data. It is valuable to monitor the tractor maneuver (whether it is working properly according to optimal state). The units involve an energy management structure and three different bases of inner battery charging. The unit should work independently from the vehicle therefore battery management system is an important fact in unit design. In the sunny condition, the unit can be charged by a solar panel above the vehicle. In another case, a generator mounted on a rotating piece of the machine can charge the unit battery. Third option would source the unit directly from the battery of the vehicle. For the situation when none of the above-mentioned energy sources are available, the unit has been provided with an extra internal battery source unit which permits all-in-one functioning for critical situation (Fig. 1). In Fig. 1, it is observable a flexible solar panel mounted directly on the top of the box or in another proper position. It has been evaluated the probable operational solar energy through the year. In Europe central region (Czechia) the chart showing operated photovoltaic gains is in Fig. 2. There is a noteworthy growth in energy when considering MPPT (max. power point tracking) controller. The minimum amount of energy is touched in January – about 20 to 25 kWh/m^2. If we choose mentioned minimum values and reduce it a bit (due to dust on the panel or optimal orientation angle) to rate 15 kWh/m^2 the selected (SO28-v.1) as solar panel with surface area 0.07 m^2 should deliver approximately 1 kWh in January. The total power consumption of the communication unit and its sub-electronics devices are less than the dedicated power by solar panel.

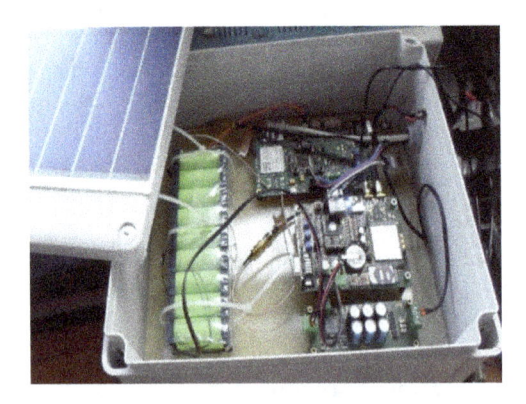

Fig. 1. The case box, containing charge management unit (right bottom), battery (left), & communication and sensor unit (right top)

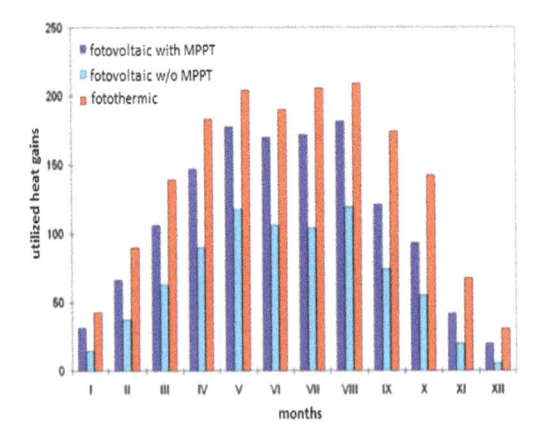

Fig. 2. Achievable heat gain during one year (MPPT)

The unit is attached to the suspension part of the agriculture machinery running in operations like seeding, tillage, fertilizing, etc. The consistency of the energy charging can be considerably improved when a generator showing in Fig. 3 is mounted within the rotational part of suspension system. In our unit, a low-cost compacted generator called NEMA17, is used for mentioned reason. A suitable gearbox connected to generator, is selected to provide proper voltage about 24 V when the vehicle is moving with 15 km/h speed. The charging current for the sweep of vehicle speeds is measured. For that reason, we have also used an especially designed charging management part (Fig. 4). The generator gives minimum power at 5 km/h speed and grasps its extreme current at speed of 8 km/h. It preserves the utmost amperage current at 40 km/h. when the stated speed is topped, the power management blocks the unnecessary power to avoid damage of battery cells in the block. From tested- field, we have come up with standard speed of agriculture mechanization during 11 to 20.5 km/h.

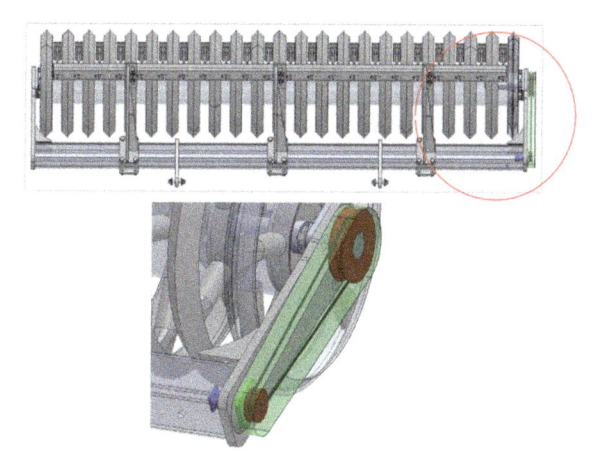

Fig. 3. Rotational piece of machine suspension and mounting detail

Fig. 4. Unit for power management

3 Sensor and Communication Unit

The core and the most important part of this design is a considered 4-layer PCB. It consists of a GSM-GPS module (in our case Quectel EC 25-E) and a Atmel SMART SAMA5D3 (ARMCortex-A5@536 MHz) processor which is running a Myrmica 2.0 Linux (Fig. 5). Connectors in the PCB are necessary for agriculture operations: embedded accelerometer, 3 tensometers, temp. sensors, analog & digital pins, tamper pins, USB connection and CAN.

Fig. 5. 4-layer PCB communication and sensor unit

In order to obtain data from tensometers and accelerometer it was necessary to sample the data more regularly than other sensors. So, we have captured characteristic tensometer data profile shown in Fig. 6 and it has been revealed at least 50 Hz as sampling frequency is required in order to detect the peaks.

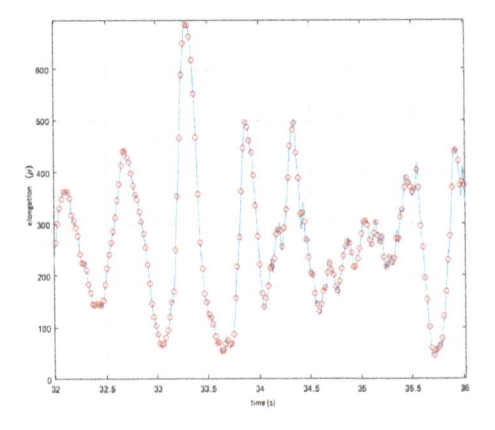

Fig. 6. Measurement of tensometer to catch the proper sampling points and peak

For calibration of the unit, a test tensometer training is designed according to its schematic shown in Fig. 7.

(a)   Circuit schematic of Tensometer

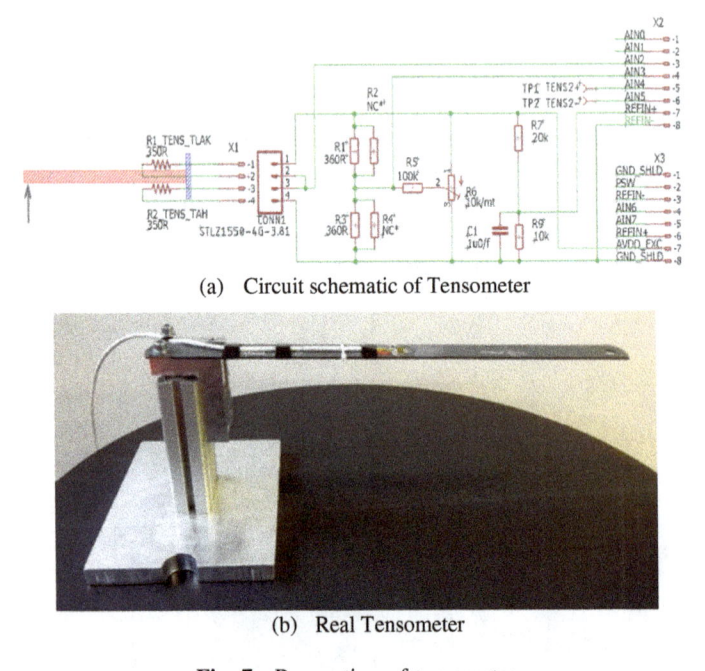

(b)   Real Tensometer

Fig. 7. Preparation of tensometer

4 Back-End

Data are transferred to a back-end through standard TCP/IP via GSM connection. Back-end involves a MSSQL database to store raw data. This interface is accessible over a web service. A user interface comprises the front-end (web pages) where all required data and possible configuration of the units is performed. Two user groups are defined: administrator and a regular user. Administrator is able to create a new configuration or making new units. he can give a definite configuration to an especial unit. Administrator is also able to allocate certain regular users to be able to observe and take data from a particular unit. The front-end runs on a predefined web address. In the setting, an administrator can pick which data, should be transferred from unit to back-end. He can select either the interval for data collections (IDC) or interval for data sending (ISE). As an example, when ISE is 540 and IDC is 4, it means the data are scanned every 4 s and after 9 min data are collected into a JSON POST and redirect to a web end-point service. It is possible also to set the Access Point Name (APN), phone number for SMS sending and setting wakening from sleep mode (AWK).

In general, its valuable to mention that sleeping modes are important in IoT units. For example, in our case study, when the suspension agriculture mechanism is not in operational state the unit would switch to the sleeping or hibernating mode to save energy consumption. Therefore, we have organized the unit in 3 scenarios. programming scenarios on board are possible on small STM32 processor with very low energy consumption (<1 μW). The unit is continuously monitoring generator voltage and acceleration. It consists of its own real-time clock consequently the GSM module is constantly accessible for SMS reception. One can select whether the unit jumps to operation upon a certain generator voltage/acceleration is grasped, or after some predefined time of idleness or by reception of wakening alert like a SMS. This is constructed as a string into the parameter AWK. For occasion, the string "accel 3G 10 min" indicates that the unit awakens where at least 3G acceleration is detected on the axis and goes to sleep whereas no acceleration is measured at least 10 min.

All the measured and transferred data using the web interface can be visualized by user (Fig. 8). He can pick a time-span where it is interested. Moreover, He is able to send out the data into an Excel format like .cvs for post processing reasons.

Fig. 8. Web interface illustrating the unit movement in the map (top pic) and selected data variations in the graphs.

5 Big Data

The core goal of the data collection over this IoT unit is to accumulate adequate data for further AI (artificial intelligence) processing. Behind the back-end, a big data engine is running to evaluate the measured data. Principally, this is essential because the amount of data is rising, and It must be considered a back-end with suitable processing power and large amount of RAM to relate various measured quantities and associate them. The test has been done with Apache Spark and Elasticsearch. We have found out that Elasticsearch has shown better performance to find certain incidences of the events within our data, however Spark can be possibly more appreciated to find some unexpected correlations. In order to get some valued results, it is required to gather the data over larger time-spans. Once some data averts expressively and same time the agriculture mechanism suspension is stated to be failed, such an unexpected incident get come in into the system. If such incidents happen more frequently, then the whole system can be particularly self-trained for predicting specific scenarios by machine learning model-based and henceforth increase the trustworthiness of the devices. A feasible library called MLlib, has been implemented to inform beforehand situations which were trained previously.

6 Conclusions

Smart agriculture is defined where the plants get enough nursing they require, determined with appropriate precision. thanks to the modern technology specifically IoT concept. Current study presented an advanced IoT unit used in a harsh agricultural environment. It is developed to apply for monitoring the suspension mechanisms of agricultural vehicle (tractor). It is well described how energy sources are used to run the mechanism and several methods how to re-charge the battery source. Every unit is accessible via a private web address and the settings can be organized independently. The PCB 4-layer board used in this practice works as core of the device. whole data from all units in the farm are composed into a MS-SQL DB where the recorded data can be recovered for export, post processing or visualization. The authors strongly believe the Apache Spark joint with library MLlib can be a convenient tool in order to diagnose unexpected fault behavior and predict error in the system. Future study can be conducted into the direction of energy [11] and economy [13] consideration or applying augmented/virtual reality [12] innovations to expand the IoT domain in agriculture.

Acknowledgment. The authors would like to thank Ministry of Education, Youth and Sports in Czechia and the European Union for financing this research study in the framework of the project "Modular platform for autonomous chassis of specialized electric vehicles for freight and equipment transportation", Reg. No. CZ.02.1.01/0.0/0.0/16_025/0007293.

References

1. Jawad, H.M., Nordin, R., Gharghan, S.K., et al.: Energy-efficient wireless sensor networks for precision agriculture: a review. Sensors **17**(8), 1781 (2017)
2. Bhagwat, P., Raman, B., Sanghi, D.: Turning 802.11 inside-out. ACM SIGCOMM Comput. Commun. Rev. **34**(1), 33–38 (2004)
3. Chebrolu, K., Raman, B.: FRACTEL: a fresh perspective on (rural) mesh networks. In: Workshop on Networked systems for developing regions, pp. 8:1–8:6. ACM (2007)
4. Hussain, M.I., Ahmed, Z.I., Sarma, N., Saikia, D.: An efficient TDMA MAC protocol for multi-hop wifi-based long distance networks. Wirel. Pers. Commun. **86**(4), 1971–1994 (2016)
5. Jayaraman, P.P., Yavari, A., Georgakopoulos, D., et al.: Internet of things platform for smart farming: experiences and lessons learnt. Sensors **16**(11), 1884 (2016)
6. Talavera, J.M., Tobon, L.E., Gomez, J.A., et al.: Review of IoT applications in agro-industrial and environmental fields. Comput. Electron. Agric. **142**, 283–297 (2017)
7. Tzounis, A., Katsoulas, N., Bartzanas, T., et al.: Internet of things in agriculture, recent advances and future challenges. Biosys. Eng. **164**, 31–48 (2017)
8. Popovic, T., Latinovic, N., Pesic, A., et al.: Architecting an IoT-enabled platform for precision agriculture and ecological monitoring: a case study. Comput. Electron. Agric. **140**, 255–265 (2017)
9. Ray, P.P.: Internet of things for smart agriculture: technologies, practices and future direction. J. Ambient Intell. Smart Environ. **9**, 395–420 (2017)
10. Elijah, O., Rahman, T.A., Orikumhi, I., et al.: An overview of internet of things (IoT) and data analytics in agriculture: benefits and challenges. IEEE Internet Things J. **5**(5), 3758–3773 (2018)
11. Rahimian Koloor, S.S., Karimzadeh, A., Yidris, N., Petrů, M., Ayatollahi, M.R., Tamin, M.N.: An energy-based concept for yielding of multidirectional FRP composite structures using a mesoscale lamina damage model. Polymers **12**(1), 157 (2020)
12. Cyrus, J., Krcmarik, D., Moezzi, R., Koci, J., Petru, M.: Hololens used for precise position tracking of the third party devices - autonomous vehicles. Commun. - Sci. Lett. Univ. Zilina **21**(2), 18–23 (2019)
13. Minh, V.T., Moezzi, R., Owe, I.: Fuel economy regression analyses for hybrid electric vehicle. Eur. J. Electr. Eng. **20**(3), 363–377 (2018). https://doi.org/10.3166/EJEE.20.363-377

Author Index

A
Abhari, Kaveh, 235
Abhari, Maryam, 235
Abomhara, Mohamed, 113
Abreu, David, 431
Ahmed, Javed, 113
Aishwarya, N., 392
Akhat, Bakirov, 420
Al Mamun, Md. Abdullah, 756
Al Raweshidy, Hamed, 23
Al Wahaibi, Fawziya, 23
Alahy, Qudrat E., 800
Alam, S. M. Maksudul, 93, 756
Aldawsari, Bader A., 300
Alogna, Yuri, 353
Al-Shoshan, Abdullah I., 79
Altahat, Zaid, 313
Alvarez, Ramiro, 655
Amakata, Masazumi, 139
Anastasiya, Grishina, 420
Apaza-Condori, Jeferson, 218
Armoogum, Sandhya, 618
Ayele, Workneh Y., 488

B
Bashir, Masooda, 600
Bhakta, Malaykumar Shitalkumar, 60
Bianchi, Luigi, 40
Boranbayev, Askar, 733, 743
Boranbayev, Seilkhan, 733, 743
Bussooa, Aadarsh, 372

C
Caliaberah, Prya Booshan, 618
Castro-Gutierrez, Eveling, 218

Cedillo, Priscila, 443
Chaimae, Saadi, 645
Chaity, Moshrefa Sultana, 800
Chan, S., 510
Chandel, Sonali, 159, 883
Chang, Tzu-Ching, 774
Cheggou, Rabea, 257
Chen, Chia-Mei, 774
Chen, Jim Q., 832
Chen, Yanjun, 883
Chowdhury, Md. Naseef-Ur-Rahman, 800
Chun, Shu Hong, 1
Córdova, Andrés, 443

D
Dai, Jiale, 883
Di Paola, Francesco, 353
Dinara, Matrassulova, 420
Dingli, Luo, 400

E
Edeko, Frederick O., 272
Effah, Emmanuel, 320
Ehley, David, 313
Eisa, Saleh M., 31
El-aal, Shereen A., 184
Elezaj, Ogerta, 113
Estrada, Miguel, 313

F
Ferhah, Kamila, 257
Flores, Pamela, 589
Franchina, Luisa, 721
Fujii, Junichiro, 139

G
Gad-Elrab, Ahmed A. A., 184
Gao, Jianling, 634
Garcia, Jordi, 281
Ghali, Neveen I., 184
Giang, Le Ngoc, 1
Gonçalves, Ramiro, 431
Grigoriy, Mun, 420
Grimm, Daniel, 782

H
Habiba, Chaoui, 645
Han, Meng, 172
Haque, Ahshanul, 800
Hegazy, Ola, 205
Hossain, Md. Shohrab, 93, 756
Hosseini, Hossein, 313
Huang, Haiping, 634
Huang, Hsiao-Ying, 600
Huang, Jianyan, 883
Hung, Vu Thai, 1

I
Ibhaze, Augustus E., 15, 272
Ibragim, Suleimenov, 420
Inzerillo, Laura, 353
Ishii, Akira, 139
Issa, Hanady H., 31

J
Jędrasiak, Karol, 874
Joshi, Shital, 60
Juell-Skielse, Gustaf, 488
Jukan, Admela, 281

K
Katt, Basel, 532
Khoumeri, El-Hadi, 257
Kido, Yoshiyuki, 857
Koci, Jan, 903
Krcmarik, David, 903

L
Lai, Gu-Hsin, 774
Laufenberg, Daniel, 172
Lee, Boyi, 774
Lee, Christopher, 51
Li, Lei, 172
Li, Xiaoming, 618
Loza-Aguirre, Edison, 589

M
Mahansaria, Divyans, 682
Mamun, Md Abdullah Al, 93

Marcon, Marco, 576
Marín-Tordera, Eva, 281
Mariya, Kostsova, 420
Martins, Constantino, 431
Masip-Bruin, Xavi, 281
Matos, Paulo, 431
McKenna, H. Patricia, 563
Melo, Wilson, 701
Moezzi, Reza, 903
Mungur, Avinash, 372

N
Nawrat, Aleksander, 874
Nojoumian, Mehrdad, 655
Nowostawski, Mariusz, 113
Nurbekov, Askar, 733, 743
Nyarko, Kofi, 456

O
Obba, Paschal, 15
Okakwu, Ignatius K., 15
Okoyeigbo, Obinna, 15
Okude, Naohito, 550
Orukpe, Patience E., 272

P
Palmaro, Federico, 721
Paracchini, Marco Brando Mario, 576
Peters, Daniel, 701
Petru, Michal, 903
Ping, Kuang, 400
Pistorius, Felix, 782

Q
Qi, Lingtao, 634

R
Ramachandra, Raghavendra, 113
Recha, Przemysław, 874
Ren, Guang-Jie, 281
Roa, Henry N., 589
Roberts, Ronald Anthony, 353
Rocha, João, 431
Romero, Obeth, 251
Romero, Steven, 815
Roy, Uttam K., 682

S
Samadian, Hiva, 672
Samara, Kamil, 313
Samet, Saeed, 130
Samiruzzaman, M., 93, 756
Samson, Uyi A., 15
Santos, Filipe, 431

Sato, Chihiro, 550
Sax, Eric, 782
Schott, Gareth, 477
Seifert, Jean-Pierre, 701
Semwal, Sudhanshu Kumar, 51
Semwal, Sudhsnahu Kumar, 467
Serrano, Javier Benedicto, 467
Sha, Chao, 634
Shahriar, Hossain, 172
Shalaginov, Andrii, 532
Shebl, Noha K., 31
Shedu, Emmanuel, 456
Shehata, Khaled A., 31
Shimojo, Shinji, 857
Shirono, Katsuhiro, 701
Shukla, Parth Anand, 130
Soares, Hugo, 431
Sobel, Dawid, 874
Soliman, Hamdy, 800
Souza, Vitor Barbosa, 281
Srinivasan, Saikrishna, 477
Stewart, Joseph, 313

T
Tarek, Radah, 645
Thiare, Ousmane, 320
Thiel, Florian, 701
Tou, Nelson Pinto, 857

Tubaro, Stefano, 576
Tuiyot, Desmond, 672

U
Usuda, Takashi, 701

V
Valdez, Wilson, 443
Valera, Juan, 672

W
Wang, Hao, 842
Wu, Hanwen, 159

Y
Yamaki, Hayato, 338
Yamin, Muhammad Mudassar, 532
Yanai, Naoto, 857
Yao, Xiaoming, 842
Yasuno, Takato, 139
Yildirim, Sule, 113
Yurchenko, Artem, 701

Z
Zaghrout, Afaf A. S., 184
Zhang, Miaomiao, 815
Zhang, Song, 159

CPSIA information can be obtained
at www.ICGtesting.com
Printed in the USA
BVHW060905090420
577268BV00003B/54

9 783030 394448